D1128371

BIOCHEMISTRY
A FUNCTIONAL APPROACH

Third Edition

ROBERT W. McGILVERY, Ph.D.

Professor and Chairman, Department of Biochemistry,
University of Virginia School of Medicine

in collaboration with

GERALD W. GOLDSTEIN, M.D.

Professor of Internal Medicine
University of Virginia School of Medicine

W. B. SAUNDERS COMPANY

Philadelphia / London / Toronto / Mexico City / Rio de Janeiro / Sydney / Tokyo

W. B. Saunders Company: West Washington Square
Philadelphia PA 19105

Listed here is the latest translated edition of this book together with the language of the translation and the publisher. 7/9/86; 3rd ed., Spanish, Nueva Editorial Interamericana, Mexico D.F., Mexico

Library of Congress Cataloging in Publication Data

McGilvery, Robert W.
Biochemistry, a functional approach.

Includes bibliographies and index.
1. Biological chemistry. I. Goldstein, Gerald, 1922– . II. Title. [DNLM: 1. Biochemistry. QU 4 M145b]
QP514.2.M43 1983 574.19′2 82-42557
ISBN 0-7216-5913-6

BIOCHEMISTRY A Functional Approach ISBN 0-7216-5913-6

© 1983 by W. B. Saunders Company. Copyright 1970 and 1979 by W. B. Saunders Company. Copyright under the Uniform Copyright Convention. Simultaneously published in Canada. All rights reserved. This book is protected by copyright. No part of it may be reproduced, stored in a retrieval system, or transmitted in any form or by any means, electronic, mechanical, photocopying, recording, or otherwise, without written permission from the publisher. Made in the United States of America. Press of W. B. Saunders Company. Library of Congress catalog card number 82-42557.

Last digit is the print number: 9 8 7 6 5 4

To Alice

PREFACE TO THE THIRD EDITION

This book is intended for those who wish to understand living organisms, especially those who strive to define and ameliorate human ills. Biochemistry has become an essential tool for these purposes. It would be almost impossible for a student to survey on his own the massive body of existing knowledge, constantly augmented by a remarkable torrent of brilliant discoveries that shows no signs of diminishing. The purpose of the book, then, is to organize our knowledge into something that can be comprehended in a relatively short time and still convey a reasonably complete picture of the chemical structure and function of man. Readability without sacrifice of coverage has been a prime goal; I still agree with Ernest Hooton that you ought to treat science seriously, but you don't have to act as if you are in church. An important device in gaining that goal is to keep attention constantly focused on function, showing how individual chemical structures and reactions contribute to maintenance of the whole being, and how deviations in these chemical entities produce disease.

Keeping the whole subject within bounds has involved some sacrifices. The subject is paramount, and the author's interests are secondary. Much of our present understanding of biochemistry is sound; it will still be valid for students a century hence, whereas much of the experimentation used to develop that understanding will shortly be obsolete. It is that basic knowledge of body function, rather than the experiments by which it was developed, that is covered in this book. These experiments have been exciting stuff, and it is painful to acknowledge that they rapidly acquire the musty stigma of historical detail to most students. However, some methodologies are so widely used or have such important implications that they warrant general familiarity; these are covered.

I hope the book serves its purpose well, but it clearly would have had much less chance of doing so without the intensive collaboration of Gerald Goldstein. Skilled internist by profession, perpetual student at heart, he has the most important qualification for a great teacher: the desire to share his joy upon learning new things. He contributed to the draft of every chapter, and every page has had the benefit of his repeated critical review. However, he is not to be blamed for any defects; I always had the last word. I had additional support from Joyce Hamlin, who reviewed Chapters 4 through 7, and Gary Balian, who reviewed Chapter 9.

I am grateful to those who gave me permission to use published materials in this edition; these are individually acknowledged where they appear. Peter Agre, Robert J. Fletterick, Jane S. Richardson, and Michael J. Stock went out of their way to supply drawings or prints for inclusion in the book. Struther Arnott clarified the vagaries of DNA structure, and Michael Thorn offered his helpful analysis of renal acid-base compensation. Bernard L. Trumpower attempted my conversion to the new faith on mitochondrial electron transport. In addition, Tom Maniatis and Roscoe Brady clarified particular points. Many colleagues and students throughout the world have continued my own education through comments on the previous editions and copies of manuscripts and publications.

Roberta Kangilaski again acted as editor of the book. Once more, her enthusiasm, skill, and conscientious monitoring made her a true colleague. The book also had the benefit of the copy editing of Edna Dick, design by Lorraine Kilmer, and the skilled competence of the production departments at the W.B. Saunders Company and York Graphic Services under the patient direction of Frank Polizzano.

Finally, my biggest debt continues to be to my wife, Alice, diplomatic counselor and smoother of the path.

R.W. McGILVERY
Charlottesville, Virginia

CONTENTS

CHAPTER 1
INTRODUCTION

Organisms are marvelously complex assemblages of chemical compounds constantly involved in intertwining arrays of reactions, and one of the great delights in the study of biochemistry is the realization that it is possible to organize this complexity into comprehensible patterns. Much is to be learned, but one sees many large truths about all living things along with patches of finely detailed knowledge.

Biochemistry is essential to an understanding of living beings. Knowledge of biochemistry, like knowledge of mathematics, has become a routine tool for students of many disciplines, who in turn often make significant additions to that knowledge.

However, as with other branches of learning, there are ill-defined boundaries to biochemistry beyond which its value as a device for description and explanation diminishes, even though its principles are determinative. We do not discuss the vagaries of the London gold market in terms of the biochemistry of the participants any more than we would discuss that biochemistry in terms of the physics of the constituent atoms upon which the chemistry rests. As we go from molecule to cell to tissue to organ to individual to society, we must introduce at each stage a more simplified system of statistics to handle the larger domain, statistics that we label as anatomy, physiology, psychology, economics, history, music, and so on. Even so, the basic biochemistry sometimes shows through in previously unsuspected ways. "Chemistry" has long been blamed for the human's chronic rut, but newer psychochemistry and its byproduct drug culture have exposed the extent to which distinctively human activity can depend upon the concentration of single chemical compounds.

To illustrate the boundaries of application of biochemistry, consider the type behavior of the surgeon and the internist. The surgeon manipulates what he can see, and he thinks of his procedures in terms of anatomy and physiology, even in cases in which the ultimate effectiveness of his efforts is gauged in biochemical terms. On the other hand, the stock in trade of the internist largely consists of chemicals. He injects them or asks the patient to swallow them in order to alter the biochemistry of the patient or of some foreign organism, and biochemical language is often the natural medium for discussing what is happening.

Of course, we are making a very simplistic distinction between the specialties. The modern surgeon includes sophisticated biochemical thinking among his tools, and the internist always has probed and poked, and in modern times called in the radiologist, in order to visualize structure.

What kind of prospectus can we offer about biochemistry as a useful medium for describing the living? To begin, biochemistry describes the origin of form. The chemical constituents and the forces developed between them determine in describable ways the microscopic anatomy—the nature of cells and their constituent organelles. Biochemical principles are the tools of choice for that purpose. Biochemistry also describes the forces involved in the association of cells, its language merging into the language of anatomy as more complex levels of organization are considered.

Biochemistry describes heredity. The information on the nature and behavior of an organism that is passed from generation to generation and the mode of its transmission are molecular in nature. The complex summation of that information constitutes genetics, but the details are biochemistry.

Biochemistry describes much more. It is a social nicety to greet an acquaintance after several years of separation with the words, "You haven't changed a bit." We secretly note the coarsened skin, the deepened lines, the graying hair, but in a quantitative physical sense this conventional courtesy is frequently not far from the truth. Given a primitive environment during his absence, this one man, using the simplest of tools, could have diverted a stream, cut down hectares of forest, and in other ways changed his environment so drastically that even the most unobservant would know of his existence. Yet at the end of all this, we might well be hard put to measure more than a small change in the physical dimensions or chemical composition of the man himself.

All of this is well known, even trite, but many of the phenomena that biochemistry can address are contained in the small tale. Motion is a molecular phenomenon resulting from the cyclical formation and cleavage of chemical bonds. The energy for that motion is derived from reactions of compounds that enter the organism in a constant stream, with the products excreted into the environment. The entire process of energy generation and utilization is described by biochemistry.

The major point of the tale is the ability of the organism to maintain its character over long periods of time while acting as a chemical machine to change the environment, and this is also an important part of our story. We shall see how chemical reactions are used to constantly rebuild nearly all parts of the body; the molecular structure and the reactions proceeding within it are subject to continual review of need—a kind of zero-base budgeting, which enables adaptation to changing circumstances.

Finally, we see that the machinery does wear out. The species has built within it limiting devices that say, "Enough. Let the next generation take over." Most attempts at explaining aging are made in biochemical terms, but we know too little to say how effective this approach ultimately will be.

CHAPTER 2

AMINO ACIDS AND PEPTIDES

Much of the chemistry of living organisms concerns five major classes of compounds: carbohydrates, lipids, minerals, nucleic acids, and proteins. We shall begin our study with an exploration of the nature and formation of proteins because they are the compounds that define most of the properties we ascribe to life. They determine our metabolism, form our tissues, give us motion, transport compounds, and protect us from deleterious invasion. Even the heredity of an organism is nothing more than an expression of its ability to make various kinds of proteins at different rates.

What are proteins? In operational terms they are structures for placing reactive chemical groups in particular three-dimensional patterns and for controlling access to these groups. Looking at proteins in this way, we have three tasks. We must understand how proteins are made; that is, how their various three-dimensional patterns or conformations are created. We must understand the constitution of these patterns—the chemical groups that are available and how they can be distributed in space. Finally, we must describe how these patterns perform biological functions. Only then can we properly appreciate these functions and their response to disease.

In chemical terms, a protein is a polymer of α-amino acids, that is, 2-amino carboxylic acids. We are familiar with man-made polymers; the ingenuity of the organic chemist has made available a selection with a wide range of properties, but this selection is trivial compared to the diversity possible with polymers of the natural amino acids. An adult human contains thousands upon thousands of different proteins. Consider for the moment only three of these myriads: collagen, myosin, and hemoglobin, each of which is discussed in more detail later.

Collagen is a long rod that associates into stiff fibers used to strengthen bone, cartilage, skin, and so on by resisting deformation. Myosin is also a long rod, but it ends in a two-headed bundle. It is deliberately constructed to change shape because a bending of the molecule is responsible for the contraction of muscles. Hemoglobin, on the other hand, is a compact globular protein built to wrap around iron complexes that will bind oxygen and to be very soluble in water so that it can be packed into circulating cells.

Here we see examples of proteins used for structure, mechanical motion, and transport. How can polymers constructed on the same basic pattern—a linear combination of α-amino acids—perform such diverse functions? The answer lies in the diversity of the amino acids themselves. Let us first examine the nature of these building blocks.

monoamino, monocarboxylic

unsubstituted

glycine **L-alanine** **L-valine** **L-leucine** **L-isoleucine**

heterocyclic

L-proline

aromatic

L-phenylalanine **L-tyrosine** **L-tryptophan**

thioether

L-methionine

hydroxy

L-serine **L-threonine**

mercapto

L-cysteine

carboxamide

L-asparagine **L-glutamine**

monoamino, dicarboxylic

L-aspartate **L-glutamate**

diamino, monocarboxylic

L-lysine **L-arginine** **L-histidine**

THE NATURE OF AMINO ACIDS

Proteins are constructed from 20 different amino acids. Tissues contain substantial amounts of each of these, sometimes as much as several millimoles per kilogram. This pool of free amino acids must be present if proteins are to be made, but amino acids also serve other important purposes. Some are used as chemical messengers to transmit impulses between nerves; others are actively metabolized to form products with important physiological functions. Derangement of amino acid metabolism frequently has severe consequences.

The amino acids have this general structure:

$$H-\overset{\oplus}{N}H_2-\underset{H}{\overset{R}{C}}-C(\cdots O)_2^{\ominus} \quad \text{or} \quad {}^{\oplus}H_3N-\underset{H}{\overset{R}{C}}-COO^{\ominus}$$

The molecule is shown in two ways, the first indicating every bond, and the second using a common kind of shorthand notation in which substituent hydrogen atoms are lumped together without indicating bonds, and individual C=O bonds also are not drawn.

Amino acids draw their properties from a hydrogen atom and three substituent groups on C-2, the α-carbon atom. It is the nature of the R group that gives character to an individual amino acid, and these groups, usually called the **side chains,** are all-important in determining the properties of proteins.

Properties Conveyed by Side Chains

The amino acids are commonly classified according to the character of their side chains, as is shown on the opposite page. Let us survey the nature and function of these side chains in general terms now and consider them in more detail as we encounter their specific effects in subsequent chapters. The chemical functions in the structures can be recognized without the necessity of rote memorization.

Glycine has no side chain:

$${}^{\oplus}H_3N-\underset{H}{\overset{H}{C}}-COO^{\ominus}$$

glycine

Therefore it occupies the least space of all of the amino acids. This is an important property in itself; the polymer chain can be packed tightly where glycine occurs.

Hydrophobic Bulk. Many of the amino acids are built to take up space without interacting with water. They are especially useful in shaping the interior of protein molecules. Side chains that serve in this way include alkyl hydrocarbon groups:

$${}^{\oplus}H_3N-\underset{H}{\overset{CH_3}{C}}-COO^{\ominus}$$

L-alanine

$${}^{\oplus}H_3N-\underset{H}{\overset{CH(CH_3)_2}{C}}-COO^{\ominus}$$

L-valine

$${}^{\oplus}H_3N-\underset{H}{\overset{CH_2-CH(CH_3)_2}{C}}-COO^{\ominus}$$

L-leucine

$${}^{\oplus}H_3N-\underset{H}{\overset{H-C(CH_3)-CH_2-CH_3}{C}}-COO^{\ominus}$$

L-isoleucine

aromatic rings:

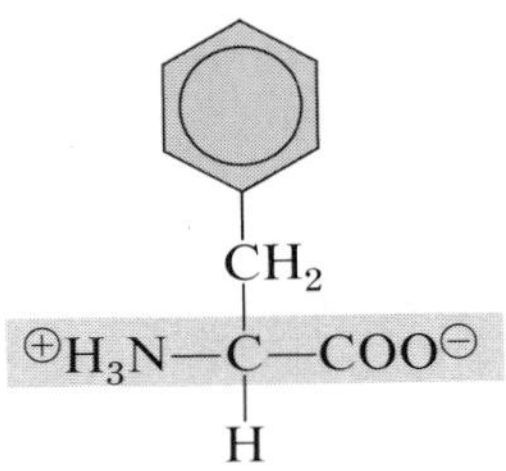

indole ring

L-phenylalanine 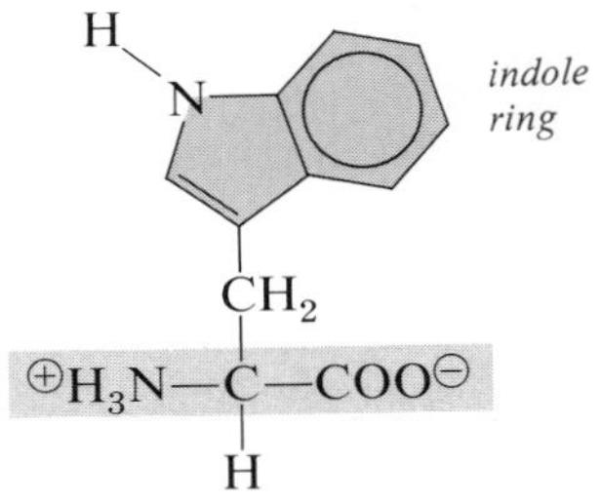 **L-tryptophan**

a heterocyclic ring, in which the side chain is also attached to the ammonium group on C-2:

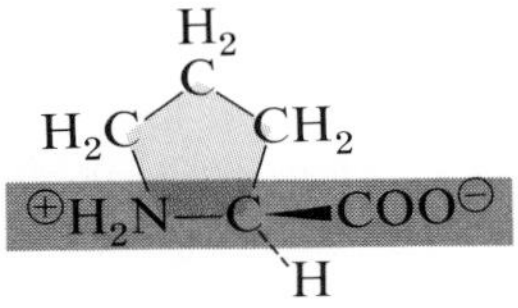

L-proline

and a thioether:

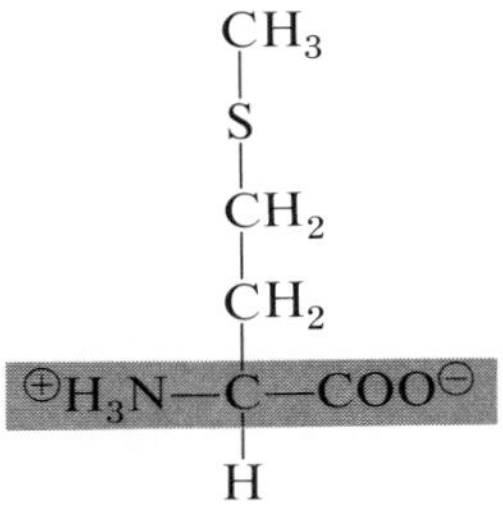

L-methionine

π-Bond Interaction. When aromatic rings are stacked side by side, the π-electrons of the rings interact to form weak bonds. Some amino acids have aromatic rings that bond in this way with each other or with other flat resonant structures.

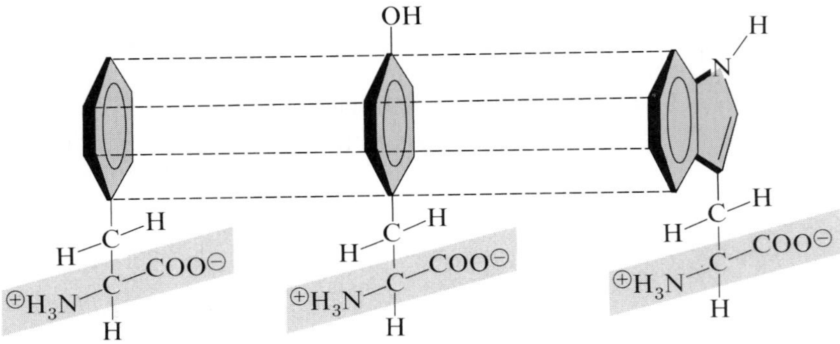

L-phenylalanine **L-tyrosine** **L-tryptophan**

(The dashed lines are intended to convey the existence of interaction between the rings.)

Ionized Side Chains. Some amino acids have ionized groups in their side chains that cause strong affinity for water where they occur in proteins. These include amino acids with carboxylate groups:

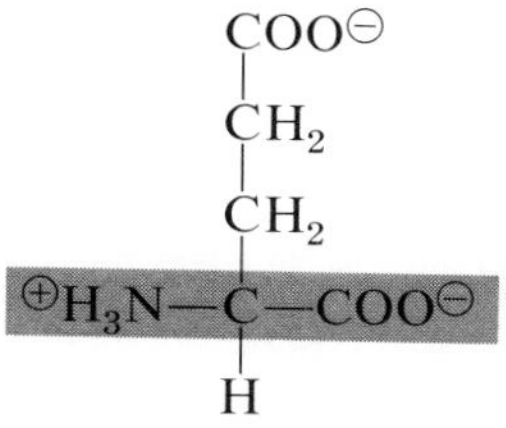

L-glutamate

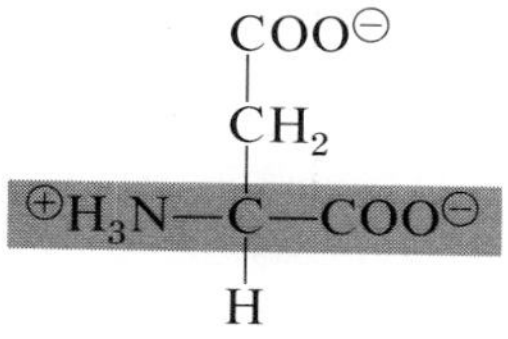

L-aspartate

and amino acids with positively charged groups containing nitrogen atoms:

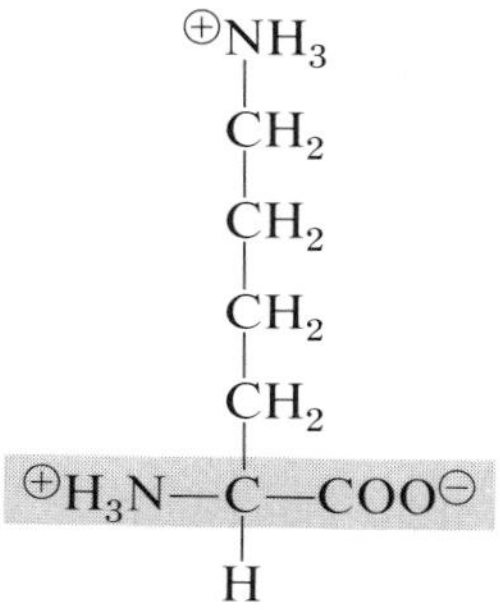

guanidinium group

L-lysine

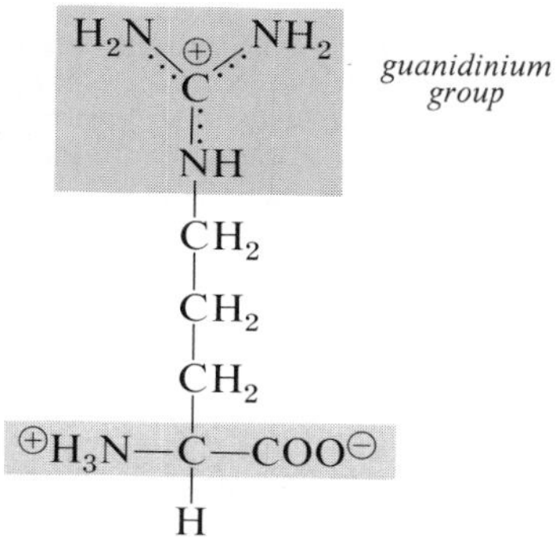

L-arginine

The positively and negatively charged side chains can form bonds through electrostatic interaction.

Binding of Metallic Cations. Atoms with unshared electrons are sometimes used to bind metals, or groups containing metals. For example, histidine occurs in hemoglobin at positions at which its side chain can bind an iron atom, and the thioether group of methionine has a similar purpose in other proteins:

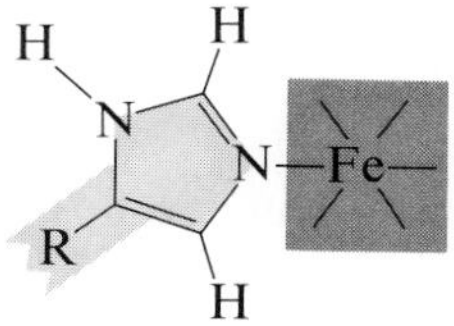

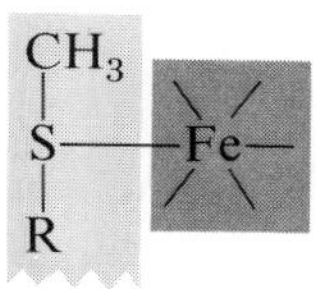

The carboxylate group of side chains in aspartate or glutamate is sometimes used to bind zinc or calcium:

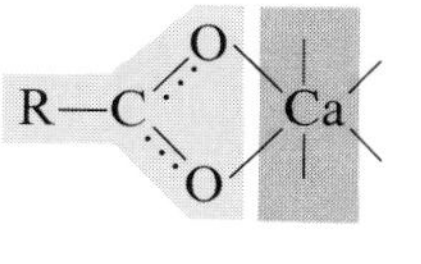

or

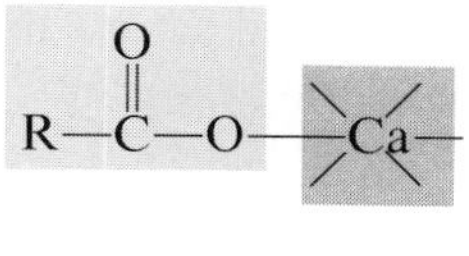

Synthetic amino acids have been developed to bond metallic ions with the carboxylate and uncharged amino groups. One that is widely used both as a reagent and as a drug is ethylenedinitrilotetraacetate, or EDTA (also known as ethylenediaminetetraacetate):

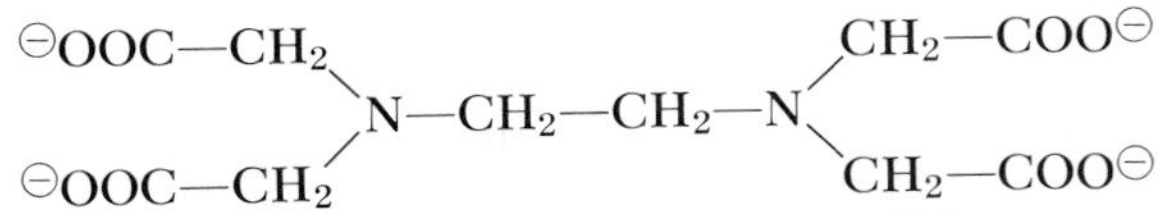

$$\begin{array}{l} {}^{\ominus}OOC{-}CH_2 \qquad\qquad\qquad\qquad\qquad CH_2{-}COO^{\ominus} \\ \qquad\qquad\quad \diagdown \qquad\qquad\qquad\qquad\quad \diagup \\ \qquad\qquad\qquad N{-}CH_2{-}CH_2{-}N \\ \qquad\qquad\quad \diagup \qquad\qquad\qquad\qquad\quad \diagdown \\ {}^{\ominus}OOC{-}CH_2 \qquad\qquad\qquad\qquad\qquad CH_2{-}COO^{\ominus} \end{array}$$

ethylenedinitrilotetraacetate (EDTA)

EDTA has a high affinity for metal cations with two or more positive charges. It is used to treat lead poisoning because the lead chelate is soluble and can be excreted. It is necessary to give EDTA in excess in order for it to compete with the many reactive groups in the body that also have a high affinity for lead. If the tetrasodium salt were given, any excess remaining after lead was bound would then bind calcium and remove it. To avoid this undesirable event, EDTA is administered as the disodium calcium salt. This will still remove lead, since EDTA has a higher affinity for lead than it does for calcium, which will be displaced:

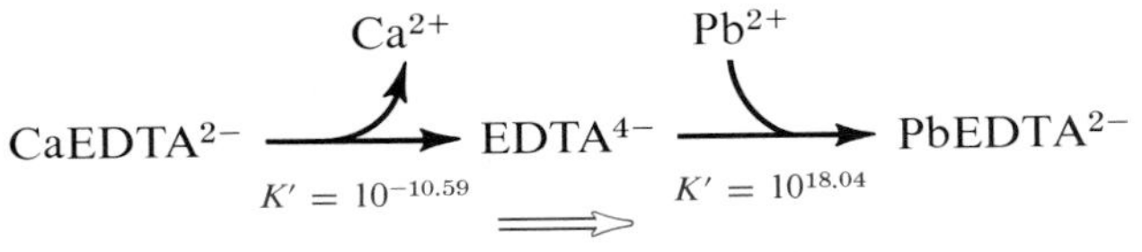

$$CaEDTA^{2-} \xrightarrow[K' = 10^{-10.59}]{\nearrow Ca^{2+}} EDTA^{4-} \xrightarrow[K' = 10^{18.04}]{Pb^{2+} \searrow} PbEDTA^{2-}$$

$$\Longrightarrow$$

Hydrogen Bonding. The hydrogen bond is one of the most important in forming the structure of proteins. It is created when a proton is shared between two atoms containing unpaired electrons, such as O, N, or S:

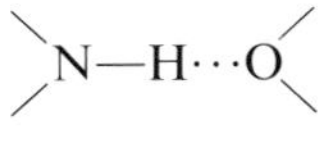

$$>N{-}H\cdots O<$$

Not only do many of the constituent amino acids have groups in their side chains containing such atoms, but they are also present in the α-amino and carboxylate groups through which the amino acids are polymerized, and all of these groups, in either side chain or backbone, may form hydrogen bonds.

Amino acid side chains containing N and O are frequently exposed at the surface of proteins where they can interact with water, not only because of their greater polarity but also because they form hydrogen bonds with water. This tendency is often counteracted by participation in internal hydrogen bonding. For example, burial of the alcoholic hydroxyl group of serine or threonine and the phenolic hydroxyl group of tyrosine is often facilitated by hydrogen bonding, sometimes to an adjacent carbonyl group:

$$\mathrm{R{-}O{-}H \cdots O{=}C\!\!<}$$

Even the strongly polar side chains, nearly always in contact with water, are sometimes buried through formation of hydrogen bonds. These include not only those with charged groups mentioned earlier, but also those of asparagine and glutamine, which are amide derivatives of aspartate and glutamate:

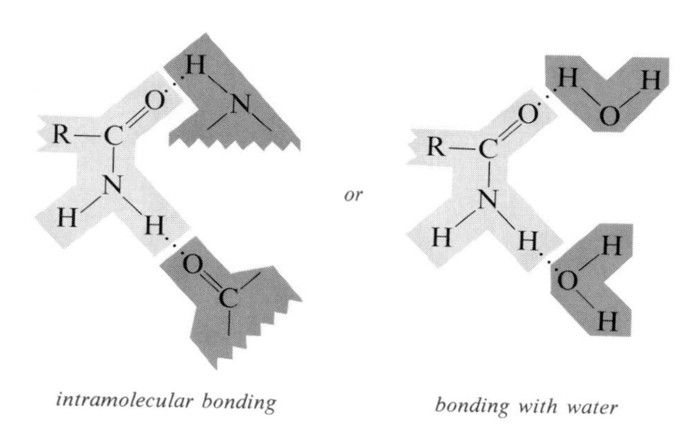

intramolecular bonding *bonding with water*

Amino Acids as Acids and Bases

Amino acids can both donate and accept protons; they are therefore said to be **amphoteric.** Every amino acid in neutral solution can behave as an acid because it contains at least one charged ammonium group from which a proton can dissociate:

$$\mathrm{R{-}NH_3^{\oplus} \rightleftharpoons R{-}NH_2 + H^{\oplus}}$$

Similarly, it behaves as a base because it contains at least one charged carboxylate group that can accept a proton:

$$\mathrm{H^{\oplus} + R{-}COO^{\ominus} \rightleftharpoons R{-}COOH}$$

Consider the simplest amino acid, glycine. It can equilibrate with H^+ in two ways:

$$\underset{\substack{\textbf{\textit{A} form}\\ \textbf{\textit{(acidic cation)}}}}{\mathrm{{}^{\oplus}H_3N{-}CH_2{-}COOH}} \xrightleftharpoons[K_1 = 10^{-2.35}]{\mathrm{H^{\oplus}}} \underset{\substack{\textbf{\textit{Z} form}\\ \textbf{\textit{(zwitterion)}}}}{\mathrm{{}^{\oplus}H_3N{-}CH_2{-}COO^{\ominus}}} \xrightleftharpoons[K_2 = 10^{-9.78}]{\mathrm{H^{\oplus}}} \underset{\substack{\textbf{\textit{B} form}\\ \textbf{\textit{(basic anion)}}}}{\mathrm{H_2N{-}CH_2{-}COO^{\ominus}}}$$

The form shown in the middle is a **zwitterion,** meaning hermaphrodite ion, because it has equal numbers of positive ammonium groups and negative carboxylate groups, although its net charge is zero. It behaves as a base because the carboxylate groups will combine with increasing concentrations of H^+ to form uncharged COOH groups. The remaining ammonium group then gives the molecule a net positive charge (cationic form). On the other hand, the zwitterion can behave as an acid because the ammonium group will lose H^+ when the concentration of H^+ is lowered, leaving an uncharged amino group. The molecule then has a net negative charge from the remaining carboxylate group (anionic form).

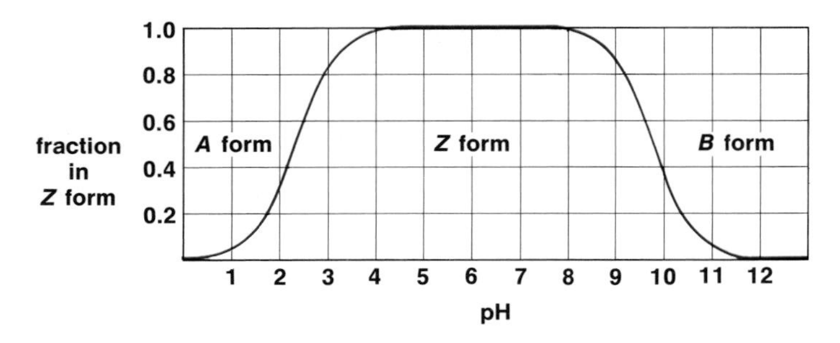

FIGURE 2–1

A plot of the fraction of glycine present as the zwitterion at various pH values. The fraction not present as a zwitterion exists as a cation in acidic solutions *(left)*, and as an anion in basic solutions *(right)*. Most monoamino, monocarboxylic acids behave in a similar manner upon titration.

What is the physiological form of glycine? The fraction of zwitterion present at a given H^+ concentration is shown in Figure 2–1. We see that glycine exists mostly as a zwitterion over a broad range centered near 10^{-6} M H^+ (pH 6). In general, amino acids with one ammonium group and one carboxylate group exist mainly as the zwitterion in physiological fluids. For example, here is the distribution of the various forms of glycine at pH 7.4 ($[H^+] = 10^{-7.4}$ M), which is the normal pH of blood plasma:

$^+H_3N—CH_2—COO^-$	zwitterion	99.58%
$H_2N—CH_2—COO^-$	anion	0.41%
$^+H_3N—CH_2—COOH$	cation	0.00089%
$H_2N—CH_2—COOH$	uncharged	0.0000037%

Other amino acids with one ammonium group and one carboxylate group behave as acids and bases in much the same way as does glycine, with the zwitterion being the physiological form of each.

Why are the molecules called amino acids? Well, the original investigators thought that the uncharged form predominated. This form has an authentic amino group and a carboxyl group, which makes it both an amine and a carboxylic acid. We have known for many decades that it is almost nonexistent, but many still draw amino acids that way.

Acidic and Basic Side Chains. The carboxylate and substituted ammonium groups on the side chains of some amino acids also behave as acids and bases. This is still true when the amino acids are combined to form proteins, so these are the groups mainly responsible for giving amphoteric properties to proteins. The dissociation constants of the groups change, sometimes 10-fold or more, when they are incorporated into proteins, but the general principles can be grasped by examining the behavior of the free amino acids containing such side chains.

CARBOXYLATE SIDE CHAINS. Consider aspartic acid:

$$\underset{A\ \text{form}}{\overset{\oplus}{H_3N}—C(COOH)(H)—CH_2—COOH} \underset{K_1 = 10^{-1.99}}{\overset{H^+}{\rightleftharpoons}} \underset{Z\ \text{form}}{\overset{\oplus}{H_3N}—C(COO^{\ominus})(H)—CH_2—COOH} \underset{K_2 = 10^{-3.90}}{\overset{H^+}{\rightleftharpoons}} \underset{B_1\ \text{form}}{\overset{\oplus}{H_3N}—C(COO^{\ominus})(H)—CH_2—COO^{\ominus}} \underset{K_3 = 10^{-9.90}}{\overset{H^+}{\rightleftharpoons}} \underset{B_2\ \text{form}}{H_2N—C(COO^{\ominus})(H)—CH_2—COO^{\ominus}}$$

Here we have three distinct equilibria. The 1-carboxylic group is a stronger acid than the carboxylic group on glycine, but the side chain carboxylic group is a weaker acid. The ammonium group doesn't behave much differently from the ammonium group of glycine. The shift in the proportions of ionic forms with changes in H^+ concentration is shown in Figure 2–2. The zwitterions of aspartate and glutamate occur in acidic solution, but the fully ionized forms with one negative charge occur in tissues.

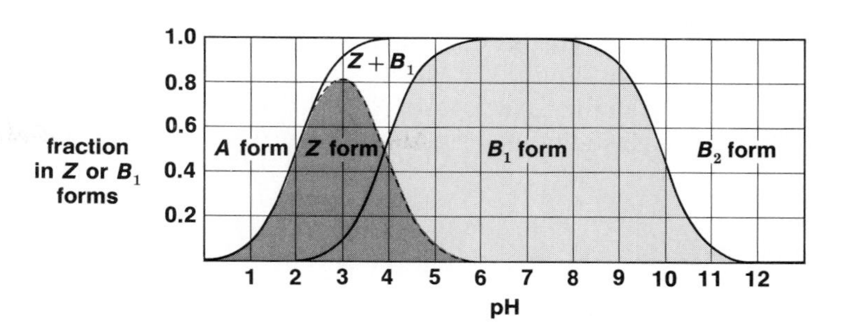

FIGURE 2–2

Monoamino, dicarboxylic acids such as aspartate can exist in four ionic forms. The fraction of the fully ionized form (B_1), with two negative carboxylate groups and one positive ammonium group, is shown by the central solid curve. The solid curve to the left plots the extent of conversion of the cationic form, with neither carboxylic acid group ionized, to more ionized species. The dashed bell-shaped curve is the fraction present as the zwitterion.

Monosodium glutamate gives a nearly neutral solution, and its use as a food seasoning, especially in oriental dishes, illustrates the importance of the biochemistry of the free amino acids. The "Chinese restaurant syndrome" afflicts some sensitive people with severe headaches, numbness, palpitation, and other symptoms of neurological disturbance owing to the effects of abnormally elevated glutamate concentrations. This was discussed with some levity until it was discovered that high glutamate can cause permanent damage to neurons of the embryonic hypothalamus in experimental animals. The use of monosodium glutamate as a flavor additive in baby foods is therefore being curtailed. Excess aspartate causes similar effects, but there has been no reason to add this amino acid in free form to foods.

CATIONIC SIDE CHAINS. Three amino acids have side chains that may be positively charged under physiological conditions. **Lysine** has an ammonium group at the end of a hydrocarbon tail. (We shall see that twitching the tail to sweep the ammonium tuft over the surrounding surface is an important function of lysyl residues in some proteins.) This group is an even weaker acid than the ammonium group on C-2, which means that it will retain its positive charge at even lower H^+ concentrations (higher pH values). The result is that both of the ammonium groups of lysine, in addition to the carboxylate group, are charged in physiological fluids, and the physiological form is therefore a cation with one net positive charge:

$$
\begin{array}{c}
{}^{\oplus}NH_3 \longleftarrow K = 10^{-10.79} \\
| \\
(CH_2)_4 \\
| \\
K = 10^{-9.18} \longrightarrow {}^{\oplus}H_3N—C—COO^{\ominus} \\
| \\
H
\end{array}
$$

The guanidinium group on the side chain of **arginine** is even weaker. Put another way, free guanidine groups are very strong bases, almost as strong as the hydroxide ion itself, and they bind protons avidly. Therefore, the side chain of arginine retains its positive charge in all but strongly alkaline solutions, and the physiological form also is the cation with one net positive charge:

$$
\begin{array}{c}
H_2N \overset{\oplus}{\cdots} C \cdots NH_2 \\
\vdots \\
NH \\
| \\
(CH_2)_3 \\
| \\
{}^{\oplus}H_3N—C—COO^{\ominus} \\
| \\
H
\end{array}
$$

Histidine is Different. The imidazole group in its side chain is approximately half-ionized at pH 6.1 ($[H^+] = 10^{-6.1}$ M)—sometimes as high as pH 7 when the amino acid is used to form proteins. This means that the physiological form of histidine is a mixture of the zwitterion and the cationic forms:

$$\text{imidazole-CH}_2\text{-CH(}^{\oplus}\text{H}_3\text{N)-COO}^{\ominus} \xrightleftharpoons{H^{\oplus}} \text{imidazolium}^{\oplus}\text{-CH}_2\text{-CH(}^{\oplus}\text{H}_3\text{N)-COO}^{\ominus}$$

The facility with which the side chain of histidine can switch from being an acid to being a base is an important feature for many biological functions, including the catalytic properties of enzymes.

pH and Bond Formation. Since changes in hydrogen ion concentration affect the ionic forms in which side chains exist, they also affect the formation of bonds by proteins. In order for an electrostatic bond to exist between a lysine side chain and a glutamate side chain, for example, both positive and negative charges must be present:

too acidic ←

$$R_{Glu}\text{—COOH} \xrightleftharpoons{H^{\oplus}} R_{Glu}\text{—COO}^{\ominus} \quad {}^{\oplus}H_3N\text{—}R_{Lys} \xrightleftharpoons{H^{\oplus}} H_2N\text{—}R_{Lys}$$

→ too alkaline

If the solution is too acidic, the carboxylate group will pick up H^+ and lose its charge. If it is too alkaline, H^+ will leave the ammonium group, which will lose its charge. Such bonds are therefore frequently most stable near neutrality.

Alkalinity favors bonding of metallic ions. The bonds are formed with carboxylate groups or the uncharged form of imidazole or amino groups, and the lower the concentration of H^+, the greater the proportion of these ionic forms available (see opposite page, top).

Isoelectric Point. Raising the concentration of H^+ decreases the number of charged carboxylate ions and increases the number of charged ammonium groups or other similar cationic groups. Lowering the concentration of H^+ has the opposite effect.

It follows that for each compound carrying both carboxylate and ammonium groups there is some value of the hydrogen ion concentration at which the number of negatively charged carboxylate groups will exactly equal the number of positively charged groups. This is true no matter how many of the respective groups there may be on a molecule. The H^+ concentration at which this occurs, usually expressed as a pH value, is known as the isoelectric point for the compound. It is the pH at which the molecule will fail to migrate in an electric field because it has no net charge. Some of the molecules may bear a net negative charge at a given moment, but they will be counterbalanced by an equal number of molecules bearing a net positive

$$R{-}COOH \underset{}{\overset{H^{\oplus}}{\rightleftharpoons}} R{-}COO^{\ominus} \overset{M^{2+}}{\rightleftharpoons} R{-}C(O)_2{\cdots}M^{\oplus}$$

$$R{-}NH_3^{\oplus} \overset{H^{\oplus}}{\rightleftharpoons} R{-}NH_2 \overset{M^{2+}}{\rightleftharpoons} R{-}NH_2{-}M^{\oplus}$$

$$\text{R-imidazolium}{-}N{-}H \overset{H^{\oplus}}{\rightleftharpoons} \text{R-imidazole}{-}N \overset{M^{2+}}{\rightleftharpoons} \text{R-imidazole}{-}N{-}M^{\oplus}$$

charge; the number of molecules that are zwitterions is greatest at the isoelectric point.

At a pH above the isoelectric point of a protein, cationic groups lose H^+; there are fewer positive charges. The number of negative charges is the same or slightly greater, so the compound has a net negative charge and migrates to the positive pole in an electric field. At a pH below the isoelectric point (increasing acidity), previously charged carboxylate groups gain H^+ and lose negative charge. The number of positively charged groups is now the same or slightly greater, so the compound has a net positive charge and migrates to the negative pole in an electric field.

Calculation of the Isoelectric Point (pI). The ionizations of simple monoamino, monocarboxylic acids are described by two acidic dissociation constants:

$$K_1 = \frac{[R{-}COO^-][H^+]}{[R{-}COOH]} \qquad K_2 = \frac{[R{-}NH_2][H^+]}{[R{-}NH_3^+]}$$

The isoelectric point occurs when $[R{-}NH_2] = [R{-}COOH]$. A little algebraic manipulation shows that this happens when $[H^+] = \sqrt{K_1 K_2}$. Put in logarithmic form,

$$\text{isoelectric pH} = \text{pI} = \tfrac{1}{2}(pK_1 + pK_2)$$

in which pK_1 and pK_2 are the negative logarithms of the respective dissociation constants. Consider leucine as an example:

$$K_1 = 10^{-2.36}; \quad K_2 = 10^{-9.60}$$
$$\text{pI} = \tfrac{1}{2}(2.36 + 9.60) = 5.98$$

Most of the monoamino, monocarboxylic acids have isoelectric points near pH 6.

What is the isoelectric point of a dicarboxylic, monoamino acid such as aspartate? It is a pH halfway between the pK values for the two carboxylic acid groups: $\text{pI} = \tfrac{1}{2}(1.99 + 3.90) = 2.95$. At this acidic pH, sufficient protons add to the two carboxylate groups to leave only one remaining negative charge, while the ammonium group (pK = 9.90) is almost totally charged.

NATURE OF SOLUTION

STRONGLY ACIDIC $> 10^{-2}$M H^+	*ACIDIC* $10^{-2} - 10^{-4}$M H^+	*WEAKLY ACIDIC* $10^{-4} - 10^{-7}$M H^+	*WEAKLY BASIC* $10^{-7} - 10^{-9}$M H^+	*BASIC* $10^{-9} - 10^{-11}$M H^+	*STRONGLY BASIC* $< 10^{-11}$M H^+
	monoamino, monocarboxylic				
$^{\oplus}H_3N—R—COOH$ *cation*$^{+1}$	$^{\oplus}H_3N—R—COO^{\ominus}$ *zwitterion*				$H_2N—R—COO^{\ominus}$ *anion*$^{-1}$
		monoamino, dicarboxylic			
COOH / $^{\oplus}H_3N—R—COOH$ *cation*$^{+1}$	COOH / $^{\oplus}H_3N—R—COO^{\ominus}$ *zwitterion*	$COO^{\ominus}$ / $^{\oplus}H_3N—R—COO^{\ominus}$ *anion*$^{-1}$			$COO^{\ominus}$ / $H_2N—R—COO^{\ominus}$ *anion*$^{-2}$
	diamino, monocarboxylic (lysine)				
$^{\oplus}NH_3$ / $^{\oplus}H_3N—R—COOH$ *cation*$^{+2}$	$^{\oplus}NH_3$ / $^{\oplus}H_3N—R—COO^{\ominus}$ *cation*$^{+1}$			$^{\oplus}NH_3$ / $H_2N—R—COO^{\ominus}$ *zwitterion*	NH_2 / $H_2N—R—COO^{\ominus}$ *anion*$^{-1}$
	guanidino (arginine)				
$H_2N\cdots\overset{\oplus}{C}\cdots NH_2$ / NH / $^{\oplus}H_3N—R—COOH$ *cation*$^{+2}$	$H_2N\cdots\overset{\oplus}{C}\cdots NH_2$ / NH / $^{\oplus}H_3N—R—COO^{\ominus}$ *cation*$^{+1}$			$H_2N\cdots\overset{\oplus}{C}\cdots NH_2$ / NH / $H_2N—R—COO^{\ominus}$ *zwitterion*	$HN{=}C—NH_2$ / NH / $H_2N—R—COO^{\ominus}$ *anion*$^{-1}$
	imidazolyl (histidine)				
imidazolium ring (H—N, ⊕, N—H) / $^{\oplus}H_3N—R—COOH$ *cation*$^{+2}$	imidazolium ring (H—N, ⊕, N—H) / $^{\oplus}H_3N—R—COO^{\ominus}$ *cation*$^{+1}$		imidazole ring (N=, N—H) / $^{\oplus}H_3N—R—COO^{\ominus}$ *zwitterion*	imidazole ring (N=, N—H) / $H_2N—R—COO^{\ominus}$ *anion*$^{-1}$	

FIGURE 2–3 Comparative behavior of various classes of amino acids with variation in H^+ concentration. The boundaries between the various ionic forms roughly indicate the (H^+) at which the concentrations of two forms will be equal. The physiological forms are in the central column.

Similarly, the isoelectric point of a monocarboxylic, diamino acid such as lysine is a pH halfway between the pK values for the two ammonium groups: pI = $\frac{1}{2}(9.18 + 10.79) = 9.99$. At this alkaline pH, only one H^+ remains attached to the two amino groups, while the lone carboxylate group (pK = 2.16) is totally ionized for all practical purposes.

Ionization and Solubility. The isoelectric points provide useful information for reasoning about the behavior of amino acids in solution because of the relationship they have to the various ionic forms, as is summarized in Figure 2–3. For example, the presence of charged groups on amino acids and proteins has important effects on their solubility.

Amino acids and proteins are least soluble at the isoelectric point, other things being equal. This is so because the zwitterion has no net charge, and it can therefore crystallize as such. The anionic or cationic forms can crystallize only as salts, such as sodium glycinate or glycine hydrochloride:

$$H_2N{-}CH_2{-}COO^{\ominus}\ Na^{\oplus} \qquad {}^{\ominus}Cl\ {}^{\oplus}H_3N{-}CH_2{-}COOH$$

sodium glycinate **glycine hydrochloride**

Since these salts can freely dissociate in water, they are much more soluble.

Does this mean that amino acids with isoelectric points near the pH of physiological fluid are likely to crystallize in the tissues? No, it does not because the zwitterions, while having no net charge, do have an off-axis distribution of positive and negative charges that create strong dipoles in the molecule, making nearly all quite soluble in water, even though they are less soluble than the ionic forms. This is generally true even when abnormally high concentrations of amino acids result from genetic defects (aminoacidopathies).

There is one conspicuous exception. Cystine is an amino acid that contains two ammonium groups and two carboxylate groups because it is formed by linking two molecules of another amino acid, cysteine, through side chain sulfur atoms (p. 148):

$$(2)\quad {}^{\oplus}H_3N{-}\underset{\displaystyle CH_2{-}SH}{\overset{\displaystyle COO^{\ominus}}{C}}{-}H \xrightarrow[\text{oxidation}]{2\,H\cdot} {}^{\oplus}H_3N{-}\underset{\displaystyle CH_2{-}S}{\overset{\displaystyle COO^{\ominus}}{C}}{-}H \quad H{-}\underset{\displaystyle {-}S{-}CH_2}{\overset{\displaystyle {}^{\oplus}NH_3}{C}}{-}COO^{\ominus}$$

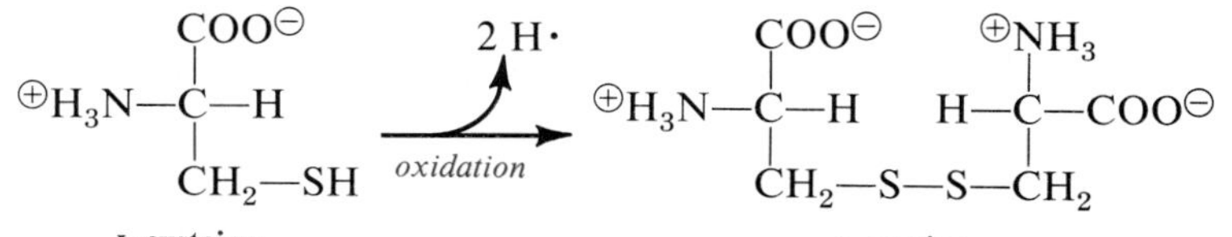

L-cysteine **L-cystine**

The crystal lattice of this molecule is so stable that the zwitterion is only soluble to the extent of 160 mg per liter of water at 37° C.

This low solubility causes trouble in people born with **cystinuria**, a genetic defect that causes them to excrete large quantities of the amino acid in the urine. The H^+ concentration of urine frequently is greater than 10^{-6} M, approaching the isoelectric point for cystine (10^{-5} M H^+, pH 5.0) at which the amino acid is least soluble. The presence of other compounds in the urine increases the solubility of cystine to approximately 300 mg per liter through "salting-in" and complex-formation, but cystinuric urines frequently exceed even this concentration, with the excess crystallizing into stones in the kidneys, ureters, and bladder. The problem is not uncommon; over 1 per cent of the stones found in urinary tracts contain cystine as a major component.

Could something be done to increase the solubility of cystine in the urine? One way of doing this would be to shift the pH from the isoelectric point. If we study Figure 2–1 we see that a shift of nearly 3 pH units above the isoelectric point is required to convert 10 per cent of a typical zwitterion to the anionic form. However, the solubility increases rapidly beyond that point, and it is indeed possible to keep substantially greater amounts of cystine in solution by prescribing repeated doses of sodium bicarbonate so as to raise the urinary pH to 7.5 or more, along with high fluid intake. Unfortunately, the promise of this therapy proved somewhat illusory; diminished acidity also increases the concentration of fully ionized phosphate in the urine, causing an increased precipitation of calcium phosphate as stones. The result is the replacement of one rocky insult to the urinary tract by another.

Stereoisomerism of the Amino Acids

Let us briefly review what organic chemistry tells us about stereoisomers as it applies to the amino acids. All of the amino acids that are introduced into peptide chains by protein synthesis have one carbon (C-2) that is bonded to four different groups, except glycine, which has two H atoms. There are two possible arrangements of groups around such an asymmetric center:

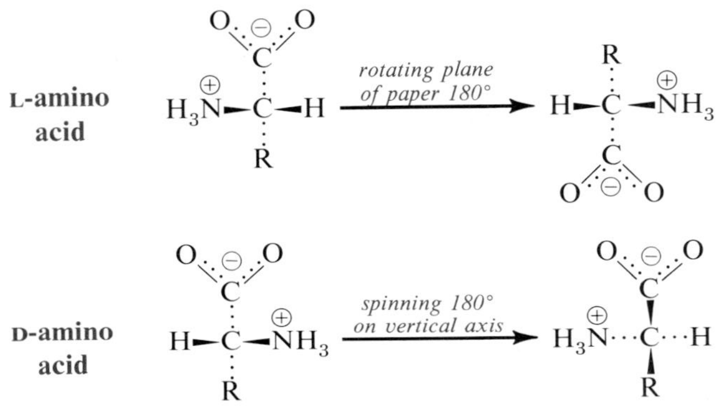

L-amino acid $H_3\overset{\oplus}{N}$—C—H (with $COO^{\ominus}$ above, R below) —*rotating plane of paper 180°*→ H—C—$\overset{\oplus}{N}H_3$ (with R above, $COO^{\ominus}$ below)

D-amino acid H—C—$\overset{\oplus}{N}H_3$ (with $COO^{\ominus}$ above, R below) —*spinning 180° on vertical axis*→ $H_3\overset{\oplus}{N}$···C···H (with $COO^{\ominus}$ above, R below)

Amino acids having one of the arrangements are said to have the L-configuration; those with the other have the D-configuration. The two configurations are mirror images of each other and like all mirror images of asymmetric objects cannot be superimposed no matter how they are turned. They are said to be enantiomorphic isomers, or **enantiomers.**

All of the asymmetric amino acids occurring in proteins belong to the L-configurational family. Even though the D- and L-isomers have many identical chemical and physical properties, they cannot approach a fixed arrangement of groups (such as occurs in another asymmetric compound) in the same way, and most biochemical reactions hinge upon mating arrangements between asymmetric groups. In the case of the amino acids, the distinction is so critical that many microorganisms deliberately use D-amino acids to create peptide antibiotics that are highly toxic to other organisms. The animal kidney has the ability to destroy D-amino acids, apparently to eliminate any possibility of forming toxic peptides.

Dating by Racemization. The L-amino acids slowly isomerize to form an equal quantity of the D-isomers, even when they are polymerized into proteins. This provides a means of dating the age of protein-containing materials such as bones, provided that a history of their environmental temperature is inferable. Since the D/L ratio reaches 0.5 in 10,000 years at 17° C, the method is mainly applicable to specimens that are recent by paleontological standards. One promising use is determination of the true age of allegedly long-lived members of various primitive cultures through examination of extracted teeth.

The Fischer Convention. It is common to designate a stereoisomer in structural formulas by a convention in which all vertical bonds are directed behind the plane of the paper and all horizontal bonds are directed in front of the plane of the paper. We have followed the convention in this book, but a word of caution is necessary. Many people do this so long as they want to draw a carbon skeleton in a vertical position, but they simply turn the drawing sideways if they want the carbon skeleton to be horizontal; of course, this directs the vertical bonds in front of the paper and the horizontal bonds behind. The practice can cause great confusion with branched-chain molecules and has been avoided here.

R and S Nomenclature. Because of the difficulty sometimes created in designating configurational family with many types of compounds, a new nomenclature has been invented. Briefly, each of the four constituent groups about an asymmetric carbon is arranged in order of increasing atomic number of the nearest constituent atom or in order of increasing valence electron density. (N ranks higher than C, and O higher than N; ethylene carbons rank higher than saturated carbons, a $—CH_2—COO^-$ group ranks higher than a $—CH_2—CH_3$ group, and so on.) One looks at the asymmetric center in such a way as to peer directly down on the substituent of lowest rank order, which is frequently —H with the amino acids. When this is done, the remaining three substituents will be arranged as spokes on a wheel, and one goes around the wheel from the lowest rank order to the highest.* If this is a clockwise direction, the configuration is **rectus** or (R); if it is counterclockwise, the configuration is **sinistrus** or (S). The process is repeated for each asymmetric center. Designating the isomers does take some practice in visualization of the structures, but the nomenclature has the advantage of creating an unambiguous designation of the absolute configuration, no matter how many asymmetric centers there are. Under this system, L-threonine is (2S:3R)-threonine, or more systematically, (2S:3R)-2-amino-3-hydroxybutyrate (Fig. 2-4).

* One can also arrive at the configurational designation by having the substituent of lowest rank away from the viewer, like the shaft of a steering wheel, and then going from highest to lowest in rank with the other substituents.

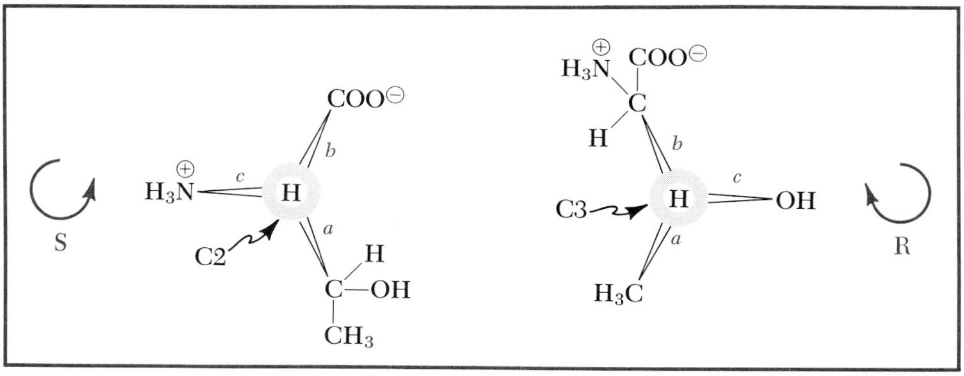

S
H_3N^+ c H
COO^- b
C2 a
H
C—OH
CH_3

H_3N^+ COO^-
C
H b
C3 H c OH
a
H_3C
R

FIGURE 2–4 **Configuration in the (R) and (S) system can be determined by looking down on an asymmetric atom arranged so that the substituent of lowest atomic number is facing the viewer. In the example on the left, one is looking past the substituent H atom at C-2 of L-threonine. In order to go from the lowest to the highest substituent among the remaining three, one must go in a left-hand circle (COO outranks CH—OH[CH_3] because it has two O atoms attached to the carbon, and these higher atomic number atoms outrank the single O and C in the other group; the directly attached N of the amino group has a higher atomic number than C and therefore outranks both of the other groups). The left-hand circle is designated as (S) configuration.**

The example on the right shows the configuration about C-3 of L-threonine. One is again looking down on H as the substituent of lowest rank order in terms of atomic number. In this case, the methyl group is the lowest ranking of the remaining three substituents, and the hydroxyl group is the highest. Traveling in a right-hand circle carries one from lowest to highest ranking substituent, so this carbon has an (R) configuration.

PEPTIDES

A peptide is a compound formed by linking amino acids with amide bonds, using the ammonium group of one molecule and the carboxylate group in another. The combination can be represented in a formal way:

$$^{\oplus}H_3N{-}R{-}COO^{\ominus} + {}^{\oplus}H_3N{-}R'{-}COO^{\ominus} \longrightarrow {}^{\oplus}H_3N{-}R{-}\overset{\overset{\displaystyle O}{\|}}{C}{-}\underset{\underset{\displaystyle H}{|}}{N}{-}R'{-}COO^{\ominus} + H_2O$$

The relationship of peptides to the constituent residues of amino acids is also shown in Figure 2–5. (We shall see that the mechanism of biological synthesis is considerably more complicated.) A compound containing two amino acid residues is said to be a **dipeptide,** one containing three is a **tripeptide,** and so on. Relatively short chains of several amino acid residues are **oligopeptides** (**oligo** = few, little), whereas longer polymers are **polypeptides.** A polypeptide with more than 100 or so amino acid residues is said to be a **protein.** (The distinction is not sharp. Polypeptides with fewer amino acid residues are likely to be called proteins if they ordinarily have a well-defined conformation of the sort described in the next chapter.)

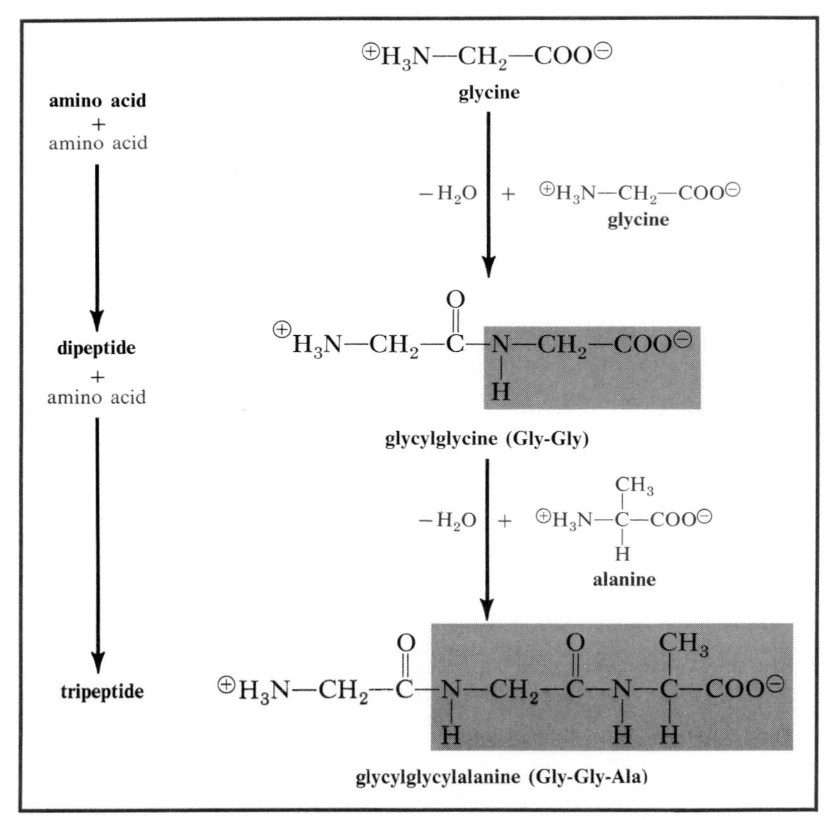

FIGURE 2–5 **Peptides are amides formed between the carboxylate group of one amino acid and the ammonium group of another. They are named from the ammonium end (the N-terminal) in terms of the constituent amino acid residues.**

Nomenclature. Peptides are named from the terminal residue bearing a free ammonium group (the amino or N-terminal residue). The successive groups are designated by the name of the amino acid from which they are derived except for replacement of the suffix with -yl. (Glycyl in place of glycine, aspartyl in place of aspartate, but tryptophanyl, asparaginyl, cysteinyl, and glutaminyl in place of tryptophan, asparagine, cysteine, and glutamine.)

The names are clumsy for all but the shortest peptides and are replaced by standard three-letter abbreviations for most purposes. An example:

methionylvalylaspartyllysyltyrosine

becomes:

Met-Val-Asp-Lys-Tyr.

The free ammonium group is on a methionine residue at one end, and the free carboxylate group is on a tyrosine residue at the other end in this example. All of the abbreviations are the first three letters of the name of the amino acid, except for

Asn—asparaginyl Trp—tryptophanyl
Gln—glutaminyl Ile—isoleucyl

(Note that the abbreviations signify a group, not the free amino acid, in the same way that Ac means acetyl, not acetic acid.)

There is an additional one-letter code designed for archival storage and computer processing that is occasionally encountered. It is tabulated in an appendix (p. 867).

Functions of Peptides. Most of the peptides with biological function are proteins, which typically have about 1000 amino acid residues. The average residue formula weight is near 110 in most proteins, so a molecular weight of the order of 100,000 is common, although the range is from 10,000 for small proteins to several million for large complexes made of several polypeptide chains.

However, the smaller peptides ought not be overlooked. They also can have important functions even though their total content in tissues is small compared to that of the proteins. Some are very potent substances. Most of the toxins in animal venoms and in plant sources are polypeptides, many of them too small to be considered proteins. Minute amounts of some oligopeptides with as few as three modified amino acid residues are effective as hormones. A derivative of a dipeptide, aspartylphenylalanyl methyl ester ("Aspartame") is 160 times as sweet as sucrose and is now being considered as a sugar replacement. (There is some discussion of possible deleterious effects from the extra aspartate and phenylalanine released by breakdown of the compound.)

The Chemistry of Peptides

The character of a protein hinges upon the nature of its constituent chemical groups and their arrangement in space. How does this arrangement occur? One factor is the nature of the peptide **backbone;** there are favored orientations for a bare peptide skeleton. The other factor is the primary structure; varying sequences of amino acids create varying forces because of the different side chains. In other words, the nature of the constituent groups in itself helps determine the way they are arranged.

The Peptide Bond. The characteristic feature of peptides is the recurring amide bond linking C-2 atoms of successive amino acid residues, which is called a peptide bond. The backbone of a polypeptide chain is made of repeating C-2 and amide groups. The critical fact about this peptide bond is that it has some double bond character. We draw it as:

$$-\overset{\overset{\displaystyle O}{\|}}{C}-\overset{\overset{\displaystyle H}{|}}{N}-$$

when it would be more realistic to draw it as:

$$-\overset{\overset{\displaystyle O}{\vdots}}{C}\cdots\underset{\underset{\displaystyle H}{|}}{N}-$$

The important consequence of this double bond character is that it freezes the entire amide group in a single plane. Small deviations can occur, but for our purposes the planar configuration may be regarded as nearly fixed. The only possibilities for free rotation in the peptide backbone occur at the two bonds on the α-carbon atoms.

In other words, the arrangement in space of the backbone of a peptide chain is nearly completely defined by two angles of rotation on each α-carbon atom (Fig. 2–6), which are named Ramachandran* angles. These angles provide a handy way of defining the spatial orientation of amino acid residues in a peptide backbone. The actual numbers need not concern us, but we ought to note that they are not randomly scattered among all possible values. Some angles do not occur because steric hindrance prevents the particular orientation; adjacent carbonyl oxygen atoms collide before the angle can be reached. In addition, there are some orientations that

*G.N.A. Ramachandran (1922–), Indian biophysicist.

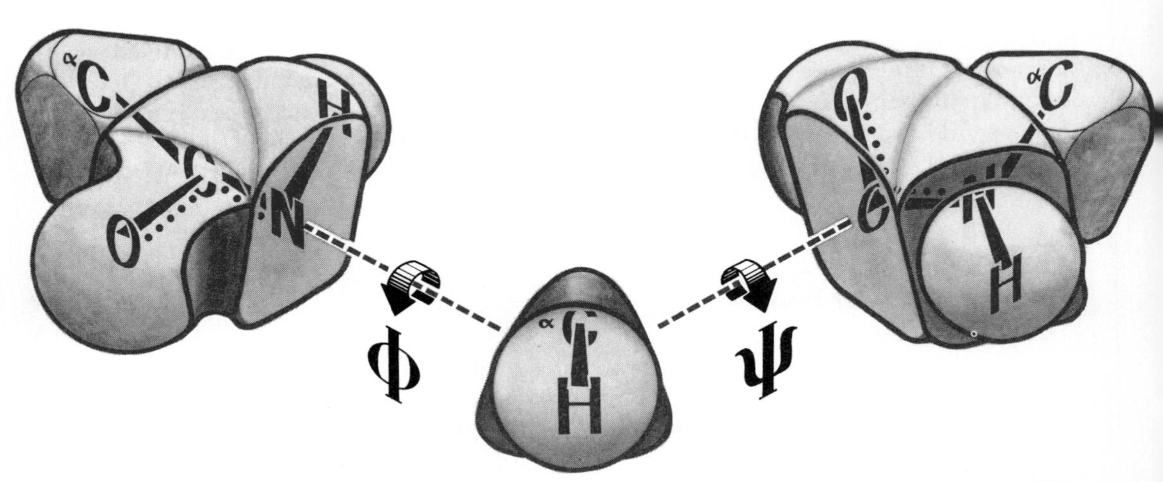

FIGURE 2–6 Free rotation of a peptide chain occurs only around the bonds joining the nearly planar amide groups to the α-carbons. The angles of rotation, known as Ramachandran angles, therefore determine the spatial orientation of the peptide chain. By convention, the Φ and Ψ angles are assigned values of 180° in the illustrated conformation.

occur much more frequently than others equally free of steric hindrance, and we shall see in the next chapter that the favored configurations permit the formation of additional forces between parts of the peptide chain.

Electrophoresis. We have seen that polypeptides have both positive and negative charges, a positive ammonium group on one end and a negative carboxylate group on the other end, but also a variable number of charged groups in the amino acid side chains, depending upon the amino acid composition of the polypeptide. Polypeptides are amphoteric; they titrate as acids and as bases, and they have isoelectric points at which they are frequently least soluble and have the greatest tendency to aggregate. However, the isoelectric points themselves, and the way in which the side chains titrate, differ from one protein to another in a single solution. Therefore, proteins in a sample can often be separated by electrophoresis: migration in an electric field (Fig. 2–7).

BLOOD PLASMA PROTEINS. The technique of electrophoresis is routinely used in clinical laboratories to separate proteins in the blood **plasma,** the extracellular part of blood, or in the blood **serum,** which is the clear fluid exuded after blood clots. These proteins are readily sampled and are commonly measured because they have important functions, and changes in their concentrations reflect disturbances in other tis-

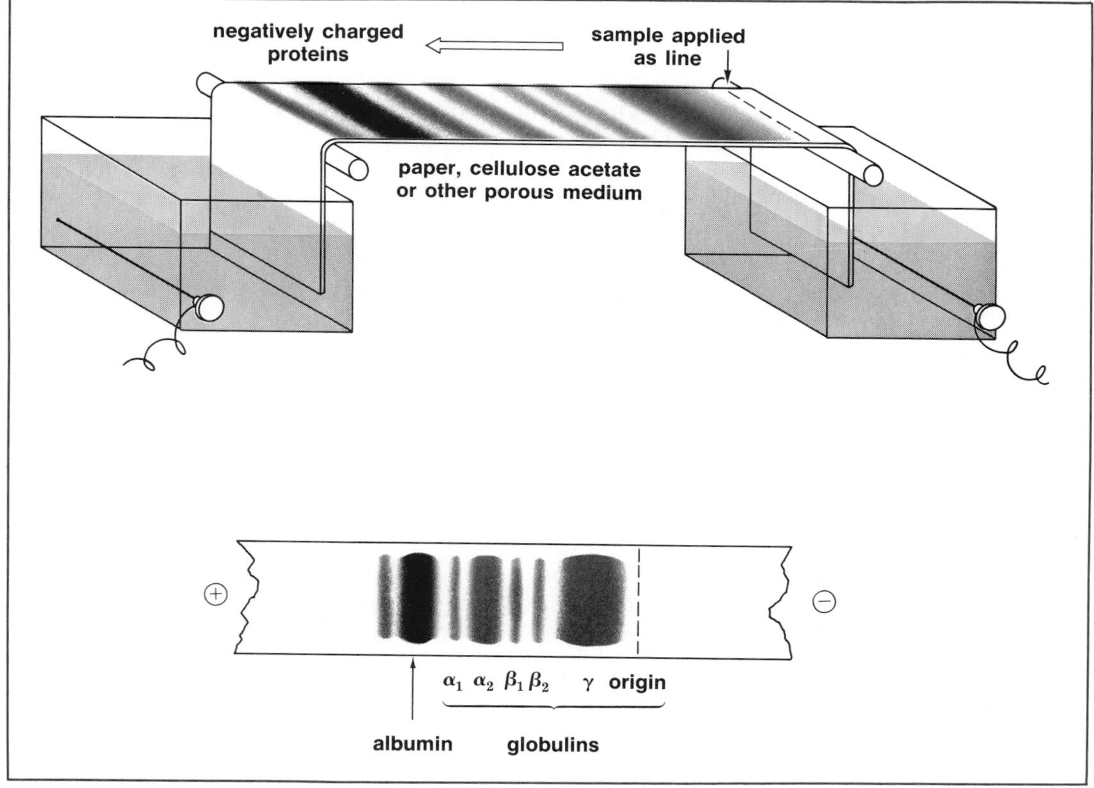

FIGURE 2–7 *Top.* **Zone electrophoresis. A strip of porous material (e.g., paper, cellulose acetate, or polyacrylamide gel) is saturated with a buffer solution and suspended between two containers of buffer that are fitted with inert electrodes. The sample containing a mixture of proteins is applied as a thin stripe; those proteins having different net charges separate when a voltage is applied. The buffer is usually made sufficiently alkaline to create a net negative charge on most proteins. Those with the most charges migrate most rapidly.** *Bottom.* **Results of electrophoretic separation of the proteins in human serum.**

sues. Most are soluble globular proteins; they have compact shapes akin to spheres rather than more rod-like shapes. An important exception in the plasma is fibrinogen, a protein used to form clots and therefore missing from serum. (Blood clotting is discussed in Chapter 18.)

The serum contains most of the major plasma proteins. The most abundant of these serum proteins are named as **albumins** or **globulins.** This is an old terminology stemming from early technics for separating proteins; those soluble proteins that remained soluble in pure water were designated albumins, those that required the presence of dilute salt for solution were designated globulins.

Early application of electrophoresis in free solution to blood plasma revealed only one major kind of albumin, but several major groups of globulins with differing mobilities, which were termed α, β, and γ. Today, we know that there is indeed only one major kind of albumin in serum, but there are hundreds of different proteins among the globulins.

What do these proteins do? For the moment, note that many of the proteins are designed to bind specific small molecules for transport through the blood. Others bind foreign substances as a protective device. Still others catalyze particular reactions.

Characterization of Size by Centrifugation. Peptides range in size from glycylglycine, with a molecular weight of 132, to polypeptides containing as many as 1000 amino acid residues in a single chain. Describing the size of a given peptide, or sorting mixtures of peptides according to size, is a common and useful practice, particularly in the case of the larger polypeptides, or **macromolecules.** (Macromolecule is a term of even more vague definition than protein. Many would say that a macromolecule has a mass greater than a few thousand daltons,* corresponding roughly to the range at which the techniques we are now discussing begin to be commonly applied.)

Similar molecules can be separated according to size by spinning them in a centrifuge. Since net motion of molecules can be observed readily by optical methods, and it isn't easy to determine the molecular weight of a macromolecule with precision, it has become common to compare the size of large molecules in terms of the velocity with which they move through a solution in a centrifugal field. This is a pragmatic practice. The measure that is used is the sedimentation constant given in **Svedberg† units (S).** Svedberg units describe in multiples of 10^{-13} sec the velocity attained per unit of applied force by a particle moving through a liquid medium.

Sufficient force must be applied to the solution so that the molecules are moved detectably faster than they are scattered by brownian motion. Commercially available ultracentrifuges develop forces of the order of $300{,}000 \times g$ (300,000 times the force of gravity) in which it is possible to observe the movement of peptides with molecular weights as low as 10,000.

Several factors other than particle mass determine the sedimentation velocity. Archimedes tells us that centrifugal field will not move a particle unless its mass differs from the mass of the suspending fluid displaced by it. That is, the *partial specific volume* of the particle (the effective volume occupied by unit mass) must differ from that of the solvent. Most simple peptides are heavier than dilute salt solutions and therefore move toward the periphery of a centrifuge rotor ("sink"). However, some proteins have large amounts of fats and similar material attached,

---

*The dalton is a measure of the mass of individual molecules expressed in atomic units based on ^{12}C. That is, one atom of ^{12}C has a mass of 12 daltons. (But the atomic weight of ^{12}C is 12, not 12 daltons. Atomic and molecular weights are dimensionless ratios.)

†T. Svedberg (1884–1971). Swedish Nobel Laureate responsible for much of the theoretical and practical development of centrifugation as a tool for studying large molecules.

and these less dense materials cause such lipoproteins to move toward the axis of the rotor ("float") with a velocity expressed as an S_F.

Given the development of a force on a particle in a centrifugal field, it will accelerate, but there will be increasing frictional resistance to motion through the medium as the velocity increases. Acceleration will continue only until a velocity is reached at which the frictional counterforce just balances the applied centrifugal force. It is this limiting velocity that is described by the sedimentation constant. The sedimentation constant therefore depends upon size of particle and shape of particle as well as partial specific volume. We know from experience that a large chunk of material falls through water faster than an equal mass of fine powder, even though the total accelerating gravitational force is the same. The powder settles more slowly because its larger surface area is in frictional contact with the medium. Since the partial specific volume is nearly the same for the same material, the sedimentation constant is an index of size, provided that the particles have similar shapes. Shape is important because long, thin molecules have greater surface area to generate frictional force than do spherical molecules of the same mass. The asymmetric molecules of high axial ratio (ratio of long dimension to short dimension) therefore have smaller sedimentation constants for a given molecular weight.

In the succeeding chapters, we refer to some cellular components by their S values for purposes of identification. To make sense out of the discussion, it is important to have in mind that S values, unlike molecular weights, are not additive when molecules combine. The union of two 5S particles does not create a 10S particle. (Two cannonballs glued together don't sink twice as fast as one alone.)

Although we can't directly compare sedimentation constants and molecular weights, it is helpful to have a rough idea of the corresponding ranges. A value of 2S will be obtained for nearly spherical proteins with molecular weights somewhat over 10,000; long, thin molecules having sedimentation constants of 2S may have molecular weights over 50,000. Similarly, 4S crudely corresponds to a molecular weight of 50,000 for spherical proteins, 8S to 160,000, and 16S to 400,000; the molecular weights of long, thin molecules with the same sedimentation constants are severalfold greater.

FURTHER READING

Readable Reviews

R.E. Dickerson and I. Geis (1969). The Structure and Action of Proteins. Harper and Row. Excellent introduction to the field.

R. Barker (1971). Organic Chemistry of Biological Compounds. Prentice-Hall. Has useful survey of properties of amino acids.

Detailed References

J.P. Greenstein and M. Winitz (1961). Chemistry of the Amino Acids, vol. 3. Wiley.

A.E. Broadus and S.O. Thier (1979). Metabolic Basis of Renal Stone Disease. N. Engl. J. Med. 300:829.

Ultracentrifugation is treated in the following:

K.E. Van Holde (1971). Physical Biochemistry. Prentice-Hall. p. 98ff.

C.K. Cantor and P.R. Schwimmel (1980). Biophysical Chemistry, part II. Freeman. p. 591ff.

CHAPTER 3

THE CONFORMATION OF POLYPEPTIDE CHAINS IN PROTEINS

PRIMARY STRUCTURE

Proteins are made of one or more polypeptide chains. Some proteins contain other kinds of molecules conjugated with the polypeptides, such as carbohydrates or fats, but a substance isn't a protein unless it contains a polypeptide.

The fundamental principle of protein chemistry is that one protein differs from another because each of its polypeptide chains consists of amino acids polymerized in a particular sequence. The amino acid sequences of the constituent polypeptides are the primary structure of the protein. It is this defined arrangement of amino acid side chains that ultimately gives a protein its properties. Any change in primary structure creates a different protein. Proteins that differ in only a single amino acid residue at some locations may have nearly identical properties, but there are other locations at which substitution of only one amino acid for another can cause changes in the nature of the protein that are drastic enough to be life-threatening if the protein has a critical function.

Number of Different Polypeptides. Although there are only 20 different kinds of amino acids from which proteins are made, there is no danger of exhausting the possible primary structures for the creation of new proteins. A protein made by combining only 100 amino acid molecules is quite small as proteins go, and yet 20 different amino acids can be combined 100 at a time in 20^{100} different ways. This number is so large that every single protein molecule conceivable in our universe, existing now or at any time in the past, could be unique without using more than an infinitesimal portion of the possibilities.

In fact, however, each protein molecule is not unique. Informed guesses at the number of different kinds of proteins that a newborn infant makes range between 10,000 and 100,000. (These figures neglect differences caused by partial hydrolysis or other modifications of a polypeptide chain after it is formed.) Assuming there are 40,000, the infant will have on the average about 1×10^{17} identical molecules of each kind of protein. In addition, every one of the $\sim 1.5 \times 10^{8}$ infants born each year will contain identical copies of many of these kinds of protein molecules. Their common humanity is a result of this fact, and it is a marvelous thing that something like 20,000 atoms, to use a typical number, can repeatedly be assembled with so few errors in exactly the same way in each and every human conceived.

However, some protein molecules will be different in two infants taken at random (but not in identical twins). That is, there are many proteins of which one infant will have 10^{17} molecules and the other infant will have none. The second infant is likely to have molecules of a similar, but not identical, protein in its place. The individuality of the two infants comes from these differences in protein composition.

Similarly, the fly that buzzes around the infant will contain many copies of protein molecules that are identical to molecules found in another fly and some that differ in detail. However, it is almost certain that the human and the fly have no protein molecules in common; their functions are too different to be served in exactly the same way. Proteins are made to satisfy these diverse needs by varying the way in which the 20 kinds of amino acids are put together, that is, by altering the primary structure, and specification of the variations is a major part of the heredity of the organism.

A SEMANTIC POINT

It is common in biochemical literature to use the name of a protein as a class word when talking about properties shared by similar proteins from all biological sources. Thus, the statement, "hemoglobin is red," is automatically understood to apply to all related proteins from all species. Similarly, the term "human hemoglobin" without qualification usually means the collection of proteins found in most humans, even though there are two and sometimes more different compounds in each individual that would be named as hemoglobins. Proteins that differ in only a few amino acid residues are nearly always given the same general name.

SECONDARY STRUCTURE

The spatial arrangement of polypeptide chains is now known in many proteins through the use of x-ray crystallography. Even casual inspection shows that the chains are not bunched up in random masses; most proteins contain many segments with recognizable patterns. These segments of regular geometry are called secondary structures.

Three types of secondary structures are common in proteins: **helixes, reverse turns,** and **pleated sheets.** It is the sequence of amino acid side chains that determines whether a particular portion of the polymer will fold into one of these patterns, but there are some general considerations that apply to all.

Structures will be favored in which there is a minimum of unfilled space. **Dis-**

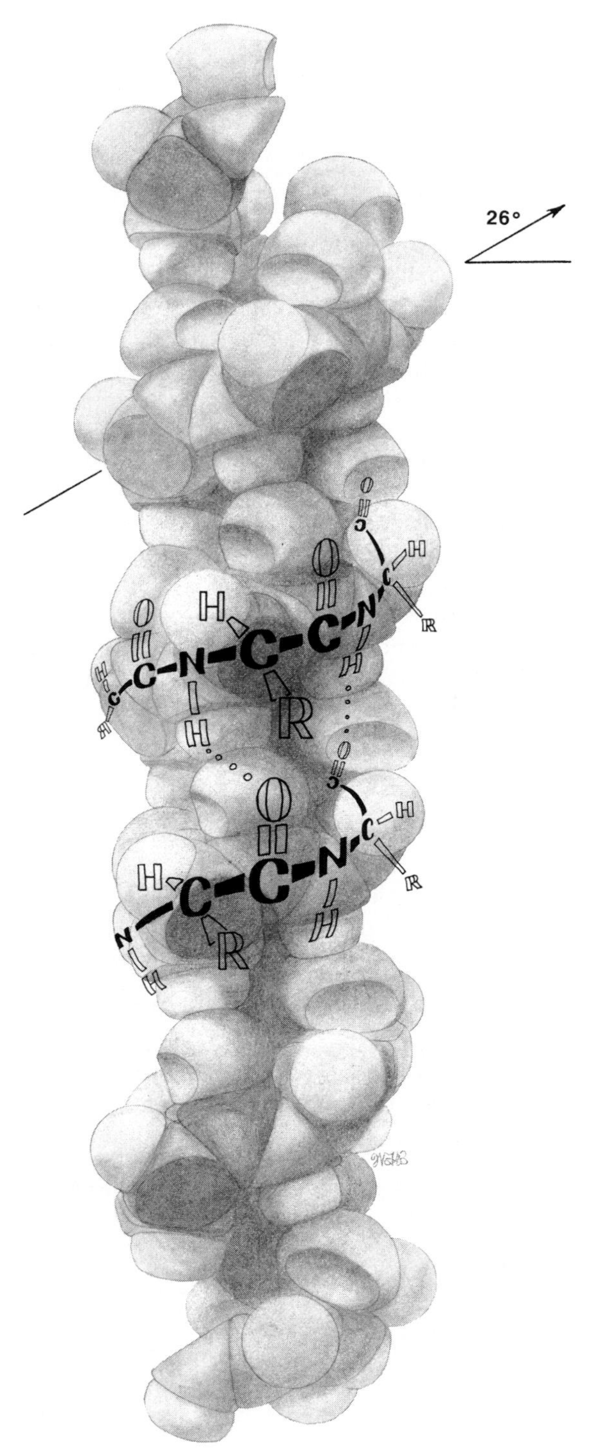

FIGURE 3–1 A model of an α-helix. The constituent atoms are labeled on two turns. The side chains of the amino acids have been omitted, but their positions are indicated by R. The helix has 3.6 residues per turn, bringing the fourth residue into a position where it can form a hydrogen bond to link the turns. Each —NH and C=O group is involved in forming such bonds in the middle of the helix.

persion forces, also called **London forces,** stabilize structures in which atoms are tightly packed. Either the structure itself must be snugly nestled together or it must have crevices large enough to be crammed with solvent molecules.

A structure will be favored if it brings all of the C=O and N—H groups of the polypeptide backbone into positions where they can form **hydrogen bonds** with each other or with side chain groups. Otherwise, the polypeptide would tend to open up so that these groups could form hydrogen bonds with water. Put another way, the protein tends to fold so that the energy released by internal hydrogen bonding of the backbone groups is as great as the energy that could be released by bonding with water.

Helixes

Polypeptides, like any long repeating polymer, tend to form coils. Why is this so? Because any repetitious displacement in space travels a spiral path. To eliminate the effects of differing side chains, imagine a polypeptide made from one kind of amino acid. Except for those residues near the ends of the chain, each will be exposed to much the same environment as its neighbors. Therefore, they will tend to form the same Ramachandran angles relative to each other; the polypeptide backbone will regularly bend in the same way. It so happens that there are particular spiral arrangements with sterically favorable Ramachandran angles that are reasonably compact and also enable all possible hydrogen bonds to be formed with the polypeptide backbone.

The α-Helix. The most common helix in proteins occurs when the polypeptide chain twists into a right-handed screw with the N—H group of each amino acid residue hydrogen-bonded to the C=O group of the fourth following residue, which is in the adjacent turn of the helix (Fig. 3–1). This screw is the α-helix. It is a rod-like structure* coated with amino acid side chains extending outward from its axis. While it is obviously suited for creating long fiber-like molecules, we shall see that it also commonly occurs in globular protein molecules.

Other helixes can be formed by single polypeptide chains so as to hydrogen bond all possible backbone groups. All that must be done in principle is to tighten or loosen the coils of the helix to align the bonding groups at other residues in the chain (Fig. 3–2). For example, the chain can be tightly twisted so that each residue forms hydrogen bonds with the third successive residue, rather than the fourth. Such a helix has exactly three residues per turn, rather than the 3.6 found in the α-helix. It is called a 3_{10} helix because each hydrogen bond creates a 10-membered ring and there are three residues per turn. (If the twisting is somewhat less tight, an N—H group can be bonded to the C=O groups of both the third and the fourth following residues, forming a turn of α_{II}-helix.)

Similarly, untwisting the α-helix so as to align groups on the fifth following residues will create a π-helix.

Of all of these possibilities, the α-helix is the most common. Why? Because it is the one in which the twist is just tight enough to pack the backbone snugly without straining the bonds or leaving a hollow core. Even so, forces occasionally favor localized formation of these other helixes.

---

*Although the polypeptide backbone in an α-helix appears roughly cylindrical, it is nearly square in an end-on view; the helix is a twisted parallelepiped.

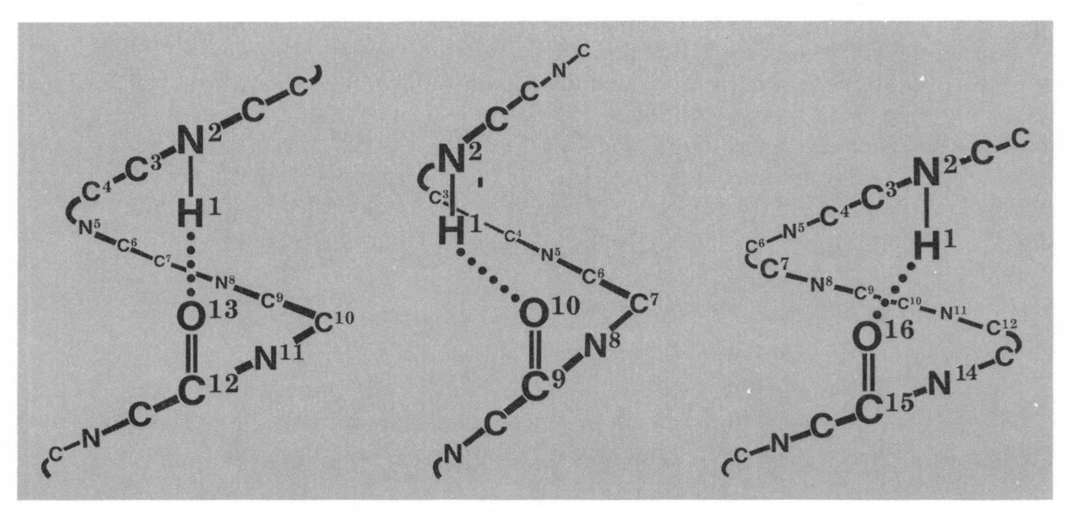

FIGURE 3–2 A schematic illustration of the position of the hydrogen bonds in three types of helixes. Each helix may be designated by the number of residues per turn with a subscript indicating the number of atoms in each ring created by formation of a hydrogen bond. The α-helix is the most stable, but turns of the tighter 3_{10} helix or the looser π-helix are sometimes found at the ends of segments of α-helix.

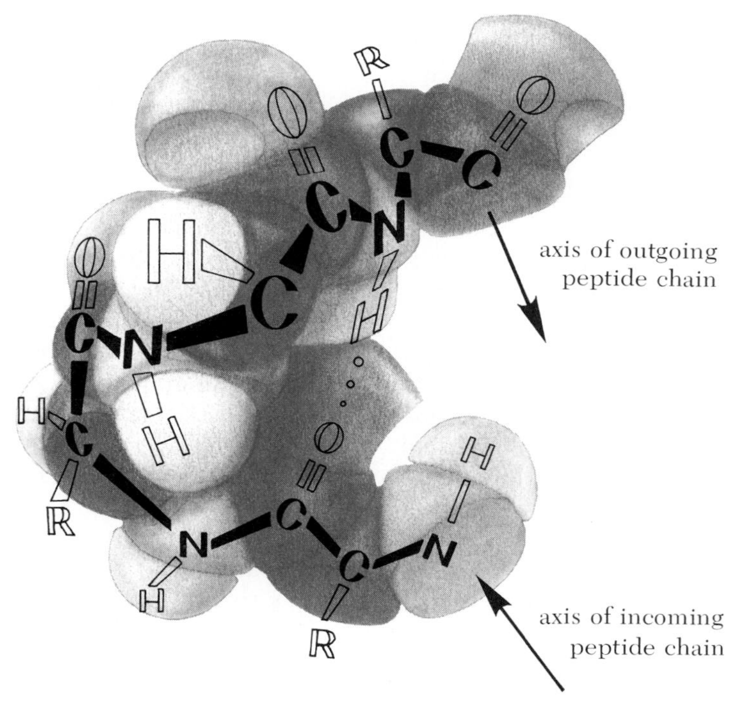

FIGURE 3–3 Some sharp bends in peptide chains are created by forming a hydrogen bond between the carbonyl oxygen of one residue and the amino hydrogen of the third following residue. There are six different ways in which the chain can twist to do this, of which one is illustrated here.

Reverse Turns

Polypeptide chains often fold back upon themselves so as to change, or even reverse, direction. Important forms of these bends involve hydrogen bonding from one residue to the third following residue. One of six ways in which this can be done is shown in Figure 3–3. There are other specific geometries that do not involve hydrogen bonding within the bend, but in either case these "reverse turns" or "chain reversals" occur in specific favored configurations.

Pleated Sheets

Compact sheet-like structures form when segments of polypeptide chains lie side by side at nearly maximum extension (Fig. 3–4). These β-structures, like the α-helix, are made with all possible hydrogen bonds between the segments of backbone. They are also known as pleated sheets because of the zigzag appearance they have when viewed on edge. The sheets may be **antiparallel,** with the chain running in opposite direction in adjacent segments. This is the form taken when a single chain is successively folded back upon itself in the most compact possible way. The sheets also may be **parallel;** this requires that the polypeptide chain be looped back upon itself in some way so that adjacent segments of the sheet run in the same direction, creating a β_P structure.

Sheets in proteins are often formed with both parallel and antiparallel segments, and frequently are somewhat twisted. These configurations are represented by a convention in which the general direction of the backbone is shown by a flat arrow, pointing from the amino terminal of the chain toward the carboxyl terminal (Fig. 3–5). The figure also illustrates the resemblance of a commonly occurring pleated sheet, when depicted in this way, to a design motif often found on ancient pottery—the Greek key.

The striking characteristic of the pleated sheet is that side chains of successive amino acid residues appear on opposite sides of the sheet; the sheets therefore tend to form when every other side chain can interact in some effective way. The intervening side chains are segregated on the opposite side of the sheet.*

A small irregularity known as the β-bulge is sometimes introduced into pleated sheets by insertion of an extra amino acid residue into one segment. The result is a greater distortion of the sheet enabling the formation of some structures to be discussed shortly.

Nonordered Arrangement

Part of a polypeptide chain may lack any recognizable geometrical order of the kinds just described and is sometimes said to be in a random coil. This is a misleading term because the arrangement of a polypeptide chain is seldom random, even if it appears as tangled as a crushed coathanger. Lack of the repetitive elements of design that we call secondary structure simply reflects response to nonrepeti-

---

*We owe much of the credit for recognition of the helical and sheet structures to Linus Pauling (1901–), American physical chemist and Nobel Laureate. He earlier developed the concept of the hydrogen bond. His fame has been tarnished in his later years by his strong advocacy of vitamin C as a wonder drug, but this should not obscure the magnitude of his earlier scientific contributions.

ANTI-PARALLEL (β)

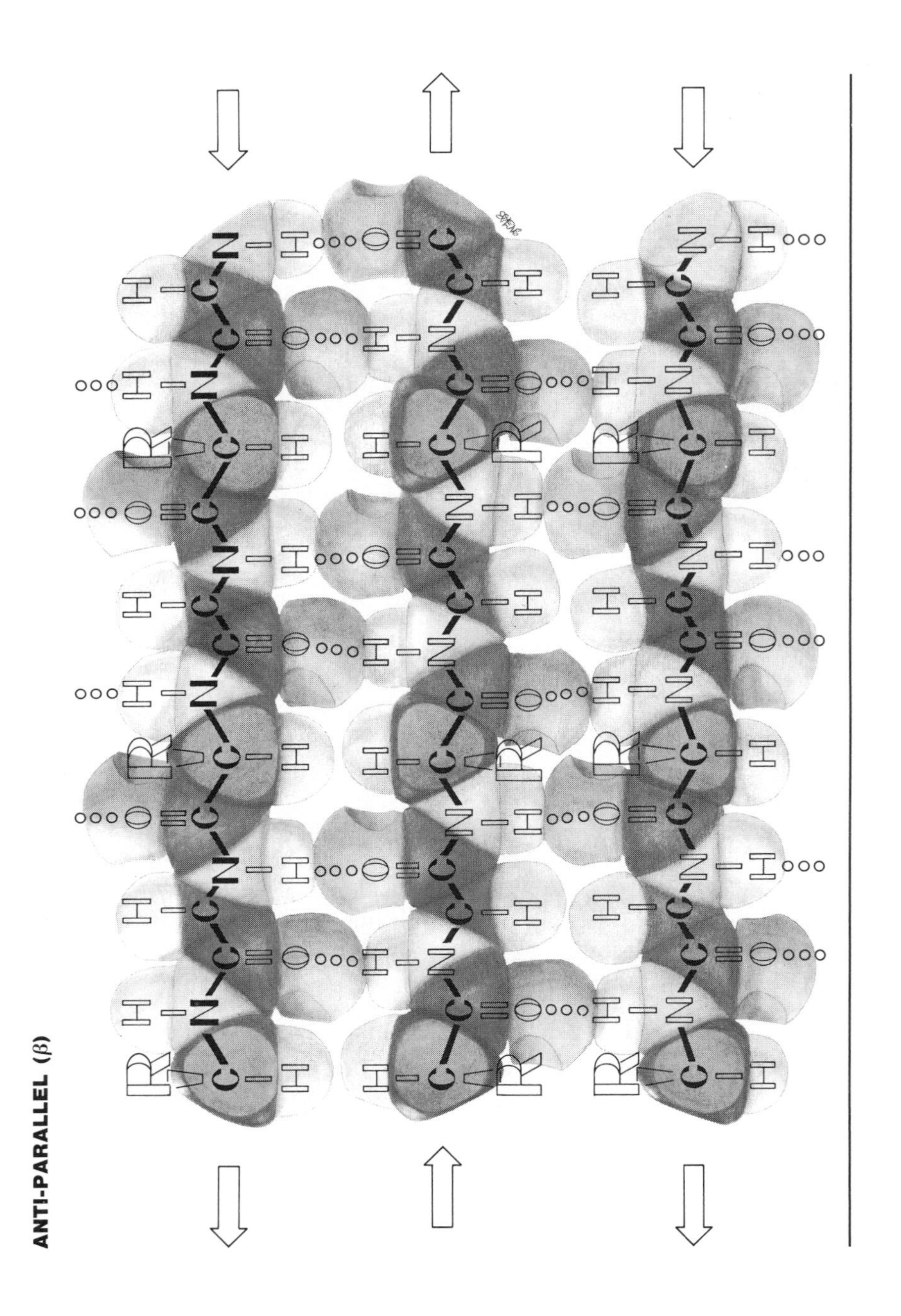

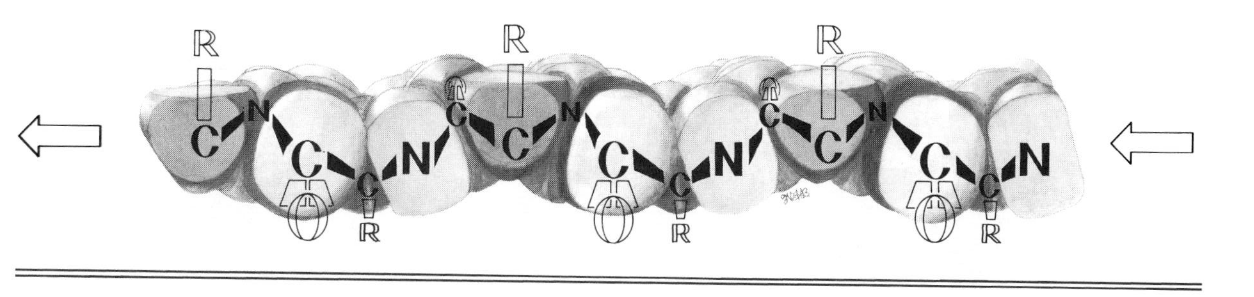

PARALLEL (β_p)

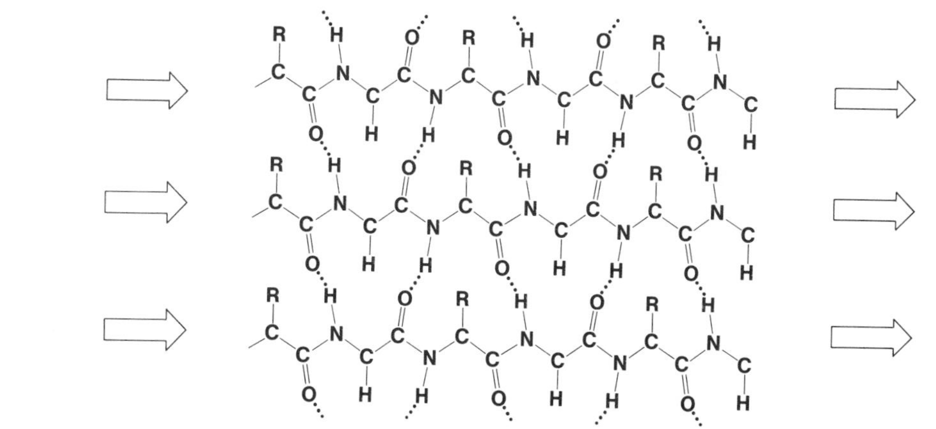

FIGURE 3–4 *Top.* A view of one face of a model of an antiparallel pleated sheet (β-structure). Alternate side chains are directed toward the viewer, and the intervening side chains are out of sight on the other face of the sheet. Every NH and C=O group in the middle of a pleated sheet is involved in a hydrogen bond. *Center.* A nearly edge-on view of the sheet, illustrating the small pleats running the width of the sheet, and the alternating appearance of the side chains on the two sides. The hydrogen bonds are directed toward and away from the viewer. *Bottom.* Diagram of linkages in a parallel pleated sheet (β_p structure). Only the R and H substituents above the plane of the paper are indicated on the α-carbon atoms.

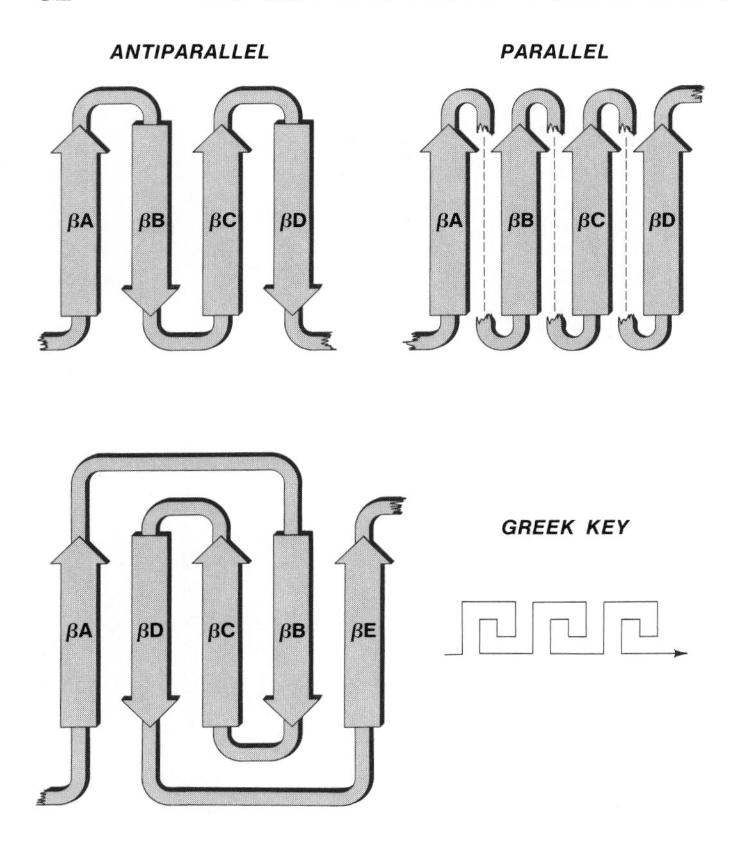

FIGURE 3–5

The course of the polypeptide chain in pleated sheets is represented diagrammatically by broad flat arrows pointing toward the carboxyl terminal of the chain. *Top.* Purely antiparallel (*left*) and parallel (*right*) examples are shown. *Bottom.* An example of an antiparallel sheet joined in a way resembling the Greek key motif on ancient pottery. See J.S. Richardson (1977). β-Sheet Topology and the Relatedness of Proteins, Nature 268:495 for more examples.

tive forces—nonrepetitive because they involve the side chains of the amino acid residues.

TERTIARY STRUCTURE

Two Examples

The tertiary structure of a polypeptide chain is the complete form assumed by the chain—the combination of various secondary structures and nonordered segments into an ordered whole. We shall concentrate for now on globular proteins and consider the more specialized fibrous proteins as we come to them. Figure 3–6 outlines the polypeptide backbone in each of two proteins. The one on top is part of liver alcohol dehydrogenase, a protein that is responsible for the oxidation of ethanol to acetaldehyde. The one on the bottom is the polypeptide of myoglobin (muscle hemoglobin), a protein responsible for binding and transporting oxygen in muscle fibers.

Although both of these proteins are relatively globular, their overall shapes give no clue to the quite different arrangements of the polypeptide chains creating the proteins. The liver alcohol dehydrogenase contains a central core of a bent parallel pleated sheet made from six segments of chain enclosed by four recognizable segments of α-helix. (The segments are lettered according to the order in which they appear from the amino terminal of the chain. βA is the first segment used to form a pleated sheet, αA is the first segment appearing in an α-helix, and so on.) Between these segments are various twists and bends.

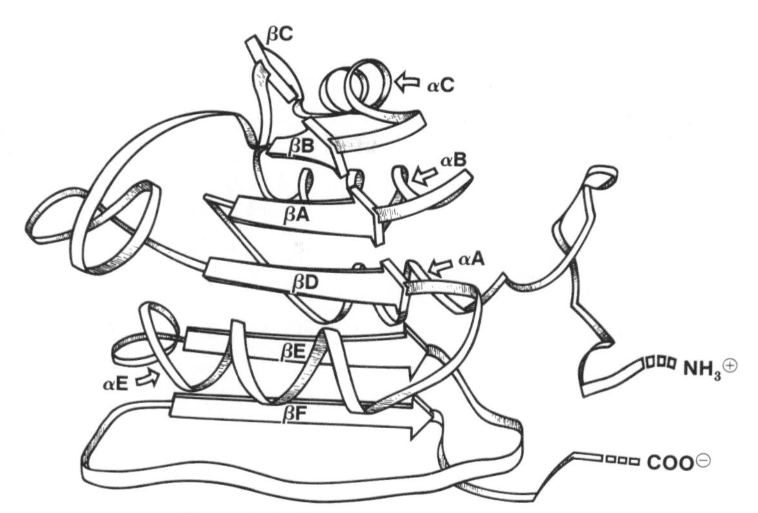

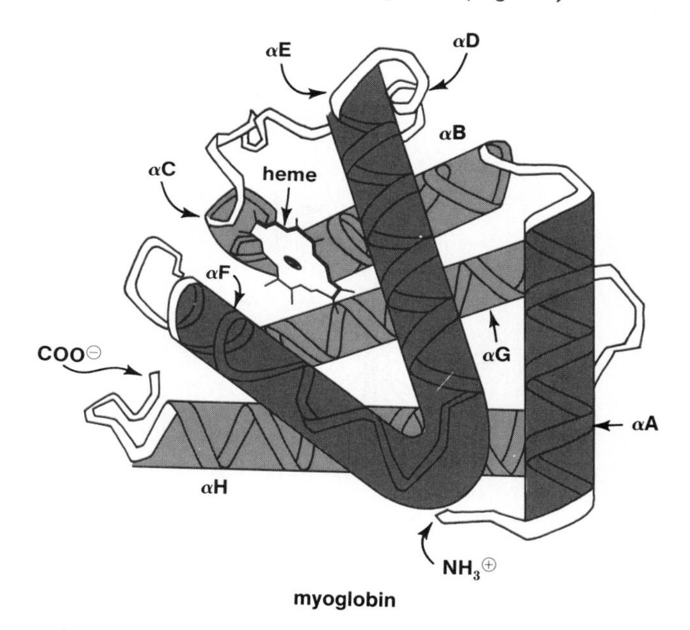

FIGURE 3–6

The tertiary structure of two polypeptide chains. *Top.* Part of the alcohol dehydrogenase found in the cytosol of liver. A central core of twisted parallel pleated sheet is surrounded by segments of α-helix and reverse turns, along with short non-ordered segments. *Bottom.* The complete myoglobin molecule, which binds oxygen in muscle fibers, is made entirely from eight segments of α-helix with short connecting pieces of reverse turns and non-ordered segments. It is designed to fold around a molecule of heme so as to exclude water from it, as if the protein contained a pocket into which the heme slides. The segments in both structures are labeled in the order in which they occur in the primary structure, with αA and βA being the segments of helix and pleated sheet occurring closest to the amino terminal. Drawing of liver alcohol dehydrogenase by B. Furugren from C.-I. Brändén et al. (1973) Proc. Natl. Acad. Sci. U.S.A., 70:2441. Myoglobin chain drawn from view given by R.E. Dickerson in H. Neurath, ed. (1964) The Proteins, Vol. 2. Academic Press, p. 634.

The myoglobin polypeptide, on the other hand, contains no pleated sheet and is made up almost entirely of segments of α-helix, with what appears to be a minimal amount of connecting chain between some segments.

Patterns of Tertiary Structure

The two examples of tertiary structure we used are not at all unusual. The general arrangement shown for liver alcohol dehydrogenase is similar to one found in several other proteins, some of quite different function. The same arrangement of stacked helixes seen in myoglobin is also present in other oxygen-binding proteins, ranging from a hemoglobin found in the nitrogen-fixing nodules of legumes to circulating human hemoglobins.

a. **antiparallel α**

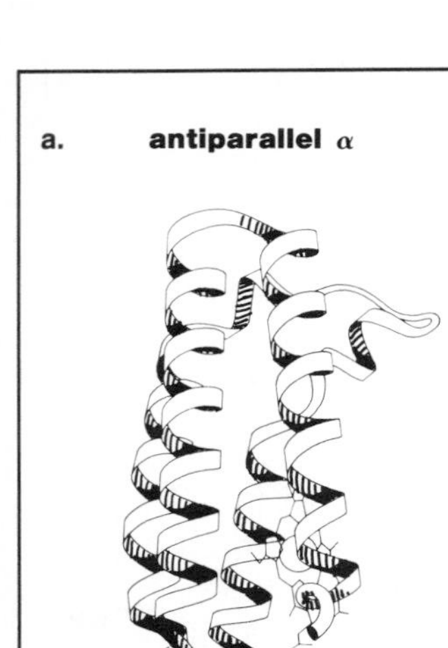

b. **parallel α/β**

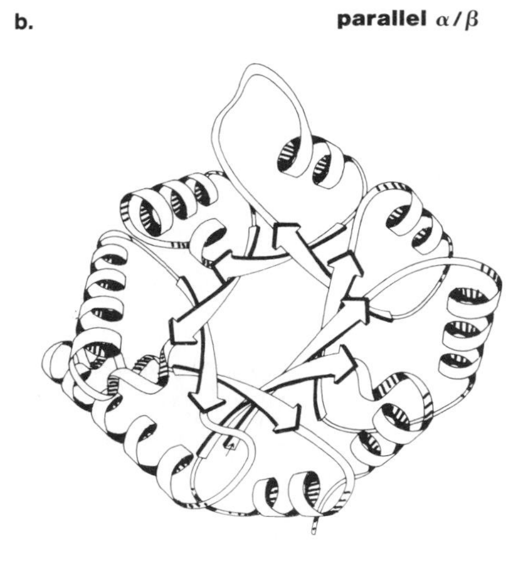

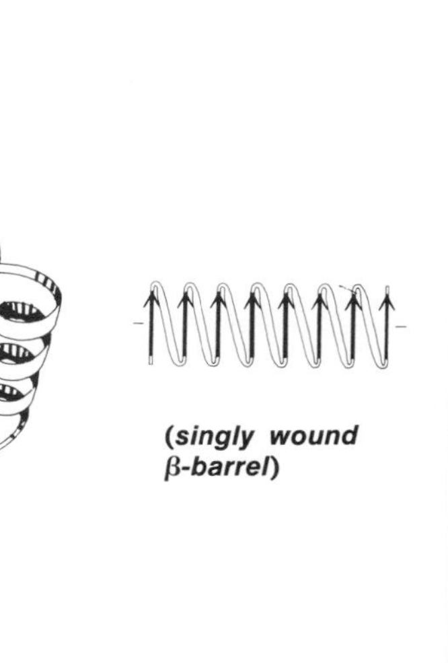

c. **parallel α/β** (*doubly wound sheet*)

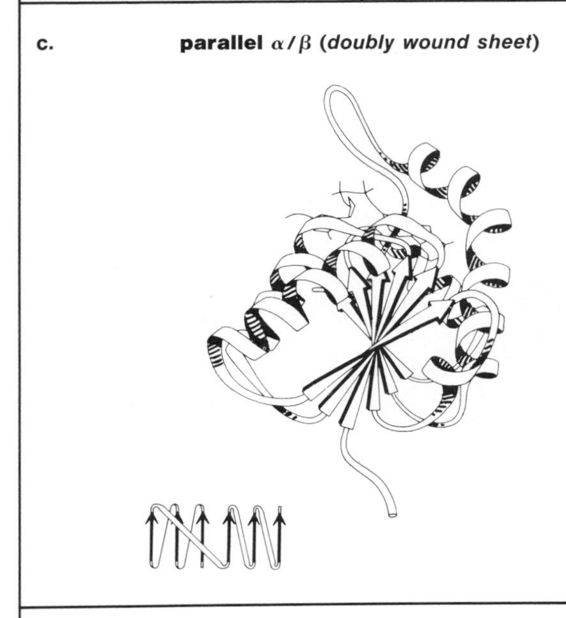

d. **antiparallel β** (*Greek key barrel*)

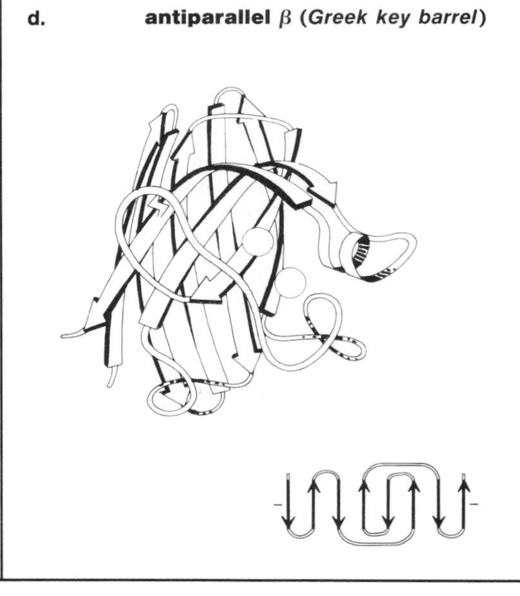

e. **open-face β-sandwich**

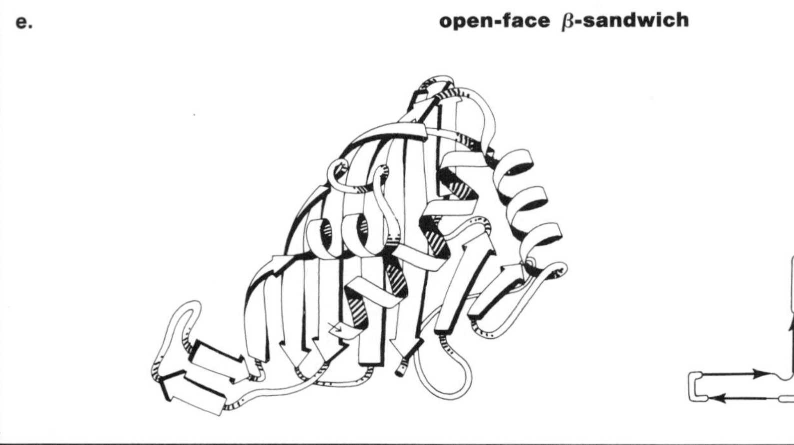

See legend on opposite page

Indeed, it is now possible to recognize several ways in which groups of pleated sheet and helix segments are organized to make globular proteins (Fig. 3–7). Some polypeptide chains fold into stacks of pleated sheets surrounded by bends and nonordered segments. In others, a large pleated sheet is folded into a cylinder (the β-barrel). Many others have pleated sheet cores surrounded by helixes, much like alcohol dehydrogenase, but this can be accomplished in many ways. (The term **super-secondary structure** has been applied to these recurring patterns, but they do not have the fixed geometry of the fundamental secondary structures.)

Domains. One polypeptide chain need not be folded into one compact unit. For example, in alcohol dehydrogenase part of the chain is folded to make the structure shown in Figure 3–6. However, another part of the chain is folded into a similar but not identical structure with a separate function, even though it is covalently linked to the first structure by the connecting region of the polypeptide chain. Domains, then, are recognizable units of tertiary structure resembling separate proteins but formed from one polypeptide chain. They often fold independently of each other and retain their character if the chain linking them is broken.

Origin of the Tertiary Structure

Why do liver alcohol dehydrogenase and myoglobin differ so much in the conformation of lowest energy content? We know the difference in amino acid composition of the two proteins is the root cause, but exactly how do the forces developed by the varying side chains generate the structures? This is a question we cannot answer in detail; it is not yet possible to deduce exactly how a polypeptide chain of known primary structure will fold, but great progress has been made toward the development of the general principles.

Secondary Structures Are Preferred. Since the pleated sheets and helixes saturate the backbone with hydrogen bonds, permitting it to be buried away from water, it is not surprising that many polypeptides preferentially fold into a combination of these shapes, often connected by specific kinds of reverse turns. Even those globular proteins without recognizable sheets or helixes, and they are not common, are mainly made from specific turns.

The Structures Are Compact. The same forces that cause secondary structures to be compact also cause them to combine into snug tertiary structures. Openings are

FIGURE 3–7 **Representative structures of globular proteins. α-Helixes are represented by coils, pleated sheets by flat arrows, and nonrepetitive structures and chain reversals by small ropes. The formal topology of examples containing pleated sheets is also given, with the direction of each sheet segment shown by an arrow. The order in which the segments are connected by helixes and nonrepetitive structures is shown by open bars when in front of the sheet and by lines when behind the sheet. The name of the protein in which the example occurs is also given, with the page on which its function is described in parenthesis.**

a. Packed array of antiparallel α-helixes. ***Bacterial cytochrome c′ (p. 403).***

b. Parallel arrays of α-helixes and pleated sheet forming a barrel. The helixes are wound around the barrel in a single direction. ***Triose phosphate isomerase from muscle (p. 464).***

c. Parallel arrays of α-helixes and a pleated sheet in a twisted plane. The helixes are wound in both directions around the sheet. ***Lactate dehydrogenase from muscle (p. 290).***

d. Antiparallel pleated sheet, connected in the Greek key arrangement to form a barrel. ***Cu-Zn superoxide dismutase, in which the metallic ions are shown by circles (p. 418).***

e. Mixed parallel and antiparallel pleated sheet as the bread in an open-faced sandwich, with segments of α-helix and nonrepetitive structures providing the hydrophobic margarine and polar jelly. (Some flattened β-barrel-like structures have two slices of pleated sheet bread, with the facing amino acid residues representing the hydrophobic margarine; no polar jelly here.) ***Glyceraldehyde-3-phosphate dehydrogenase from muscle (p. 466).***

These are taken from a superb article on protein taxonomy by Jane S. Richardson. ***Adv. Protein Chem.*** **34:168 (1981), except for the diagram of the topology in part (e) and the shading, which have been added. Reproduced by permission.**

usually more than one atom wide and are intended for access of solvent or for combination with other molecules. The protein may have bizarre forms, but it has little unoccupied space. It is not porous.

The Surface Is Polar and the Interior Is Nonpolar. One of the major forces affecting the shape of a protein comes from the tendency of polar side chains to associate with water and of nonpolar side chains to agglomerate in the interior of the molecule. These effects stabilize contortions of the polypeptide that bring the hydrophilic residues to the surface and bury the hydrophobic residues.

Hydrophilic Side Chains. The most hydrophilic amino acid residues are those with net charges, such as aspartate and glutamate with their negatively charged carboxylate groups, and lysine and arginine with their positively charged ammonium and guanidinium groups. The chain must nearly always fold so as to keep them in contact with water. It is only possible to bury these groups when the charge is neutralized by forming the uncharged forms, combining positive and negative groups, or by dissipating the charge through multiple hydrogen bonds. Even when this is done, the folding must be strongly favored elsewhere to remove these polar groups from contact with water. The carboxamide groups of glutamine and asparagine are also polar enough to strongly favor conformations that keep them in contact with water.

Hydrophobic Bonds. The force causing nonpolar side chains to huddle together out of contact with water comes not from any specific attraction of the groups for each other but from the propensity of surrounding water to resist being forced into ordered structures by the presence of the nonpolar groups. The so-called hydrophobic bond is really an association created by spurned water, and it is the same kind of force that causes insoluble oils to form compact spheres upon suspension in water. How does it arise?

Molecules of water have a strong tendency to form hydrogen bonds with each other as is shown by the release of heat when water freezes into the fixed lattice of ice. There is an opposing tendency of all molecules to stay dispersed, to be random and unordered. This tendency, measured as entropy, is increased by a rise in temperature. The two effects just balance at the melting point. At higher temperature, hydrogen bonds persist for shorter and shorter times; each molecule becomes more and more promiscuous.

A nonpolar molecule introduced into liquid water represents a eunuch at an orgy. Since it cannot form hydrogen bonds or other bonds created by polarity, neighboring water molecules lose some of their possibilities for random union, and associate with each other longer than they would in the absence of the nonpolar intruder. However, this longer association represents greater order, a step toward freezing into ice, and we have already said that creation of such order at temperatures above the melting point can be accomplished only by the introduction of energy. It follows that dispersing nonpolar molecules in water requires energy; if they are left to themselves, they diminish their exposed surfaces by clustering, thus impeding fewer random liaisons among water molecules. They become a less effective moral minority, as it were.

SOME NONPOLAR CHAINS ARE ON THE SURFACE. If the degree of polarity of the side chains were the only determinant of structure, a polypeptide would fold so that every group not interacting with water was buried in the interior. The tendency to make this separation is very strong and can aid in shaping complex surfaces, but the separation is not complete. There are usually more nonpolar side chains than can be accommodated in the interior of a protein, and some are exposed on the surface of most proteins. This is an important part of the development of specific function, as well as shape. Having more reactive groups separated by the relatively unreactive nonpolar side chains gives individual character to the surface geometry—the stars are bright because the sky is black.

Generation of Specific Structures

The general nature of the forces acting on a polypeptide chain seem clear, but it is more difficult to understand exactly what influence a particular amino acid sequence may have on protein conformation. However, we have some clear-cut examples with which to make a start.

Proline is a Helix-Breaker. This is the one apparently infallible general rule for prediction of structure. A residue of proline cannot occur in the middle of a straight α-helix:

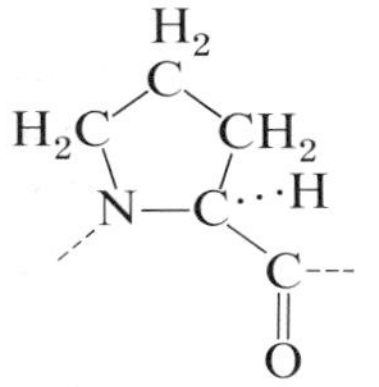

prolyl group

The nitrogen atom of a prolyl group has no attached hydrogen atom with which to bond a preceding carbonyl group, and the fixed orientation of the bulky side chain gets in the way of any preceding turn. Therefore, a prolyl group can only be in the first turn of an α-helix; the preceding amino acid residue must occur in some other configuration, at least bending, and usually breaking, the helix.

Reverse Turns Are Often Exposed. What causes chain reversals? The chain can be bent upon itself at any residue if the bend enables strong bonding elsewhere, but there are some amino acids whose side chains favor reverse turns. Proline, with its odd geometry, is an example. Glycine is another; its lack of any side chain permits some tight folds that are not possible with other amino acids, and it is more likely to occur in a reverse turn than any other:

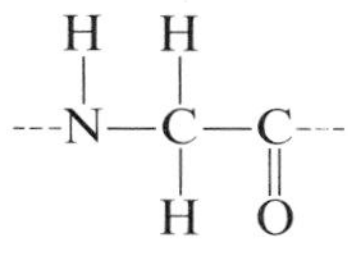

glycyl group

Other amino acids favor the formation of reverse turns because they have hydrogen-bonding groups only a short distance removed from the peptide backbone. These include asparagine, serine, and aspartate:

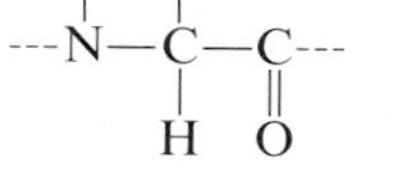

seryl group 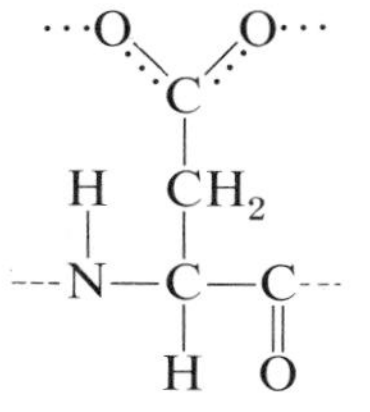 **aspartyl group** 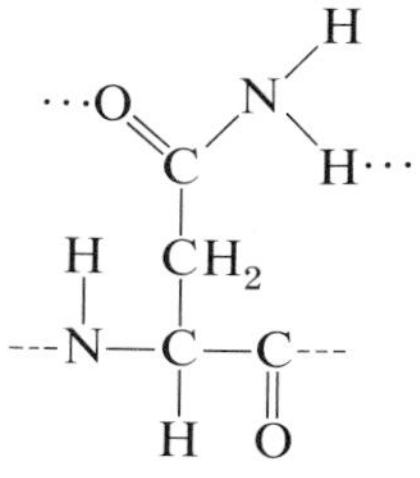 **asparaginyl group**

Here we have it: reverse turns are likely to contain polar side chains or to have a nearly naked backbone where glycine occurs. Either event will favor exposure to water at the surface of the molecule. The elbows of protein molecules are often bared.

Pleated Sheet or α-Helix? Since the pleated sheets and helixes have complete hydrogen bonding of the backbones, the forces favoring one over the other must be

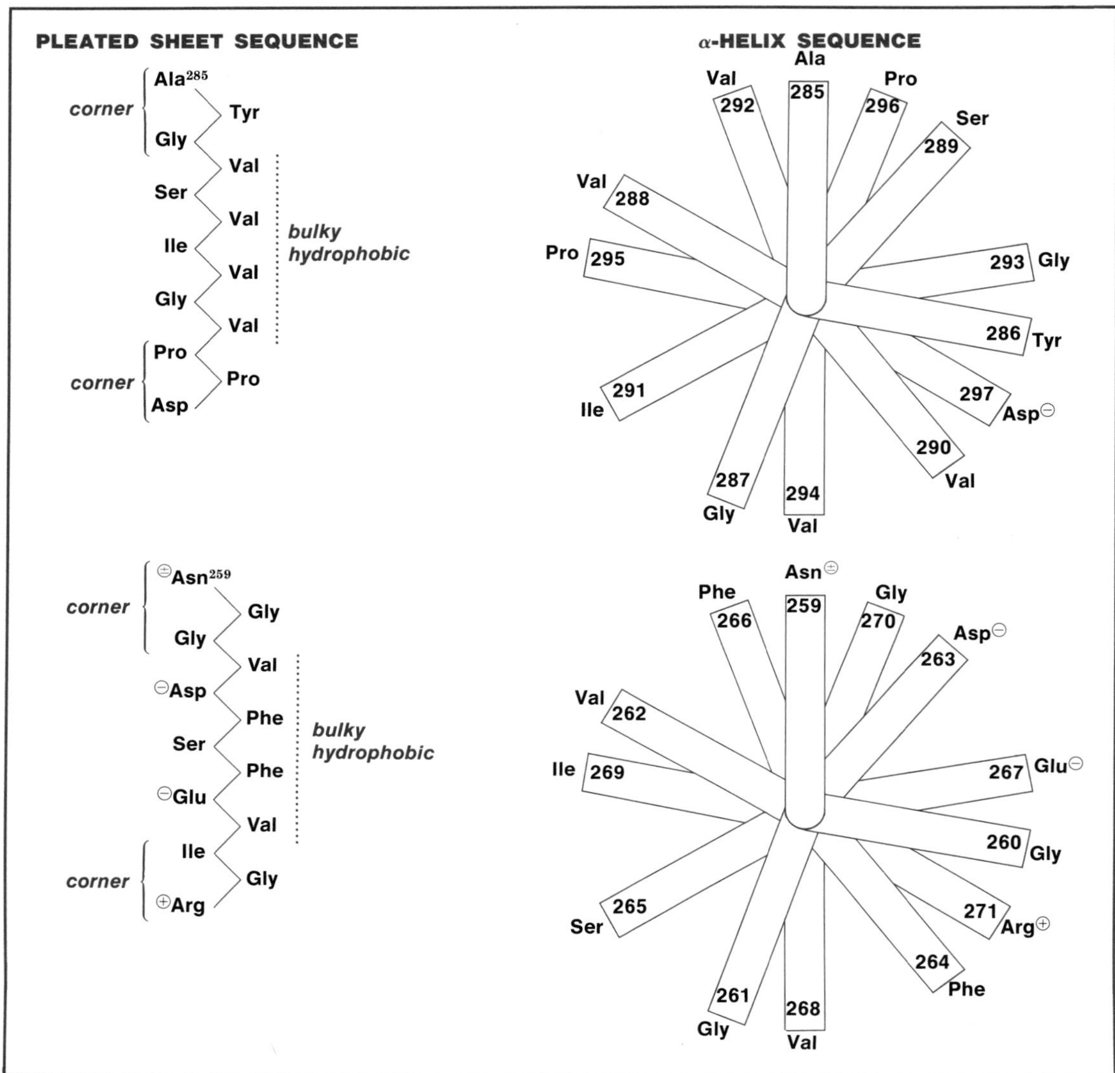

FIGURE 3–8 The forces creating segments of secondary structure can sometimes be visualized by plotting the amino acid sequence in the order it would assume in the structure. *Left.* Two segments of peptide chain from liver alcohol dehydrogenase are plotted as segments of pleated sheet. Such an arrangement creates a massive array of nonpolar side chains on one side of the sheet (*to the right as shown*), which would tend to stay together in contact with another nonpolar surface. The ends of the segments contain prolyl and glycyl groups that facilitate the formation of bends. *Right.* The same segments are plotted as if they were in α-helixes. Successive tabs are spaced 100° apart, and numbered according to the position of the corresponding amino acid residue in the polypeptide chains. The distribution of side chains around the surface of the helix can then be seen as if one were looking down its axis from the N-terminal end. Compact clusters of hydrophobic side chains may indicate an interior contact that stabilizes the structure. No such clusters are evident here.

developed by the differing dispositions of the side chains. For example, generation of a secondary structure with a hydrophobic surface that can be faced toward the interior may facilitate construction of a compact molecule. We can sometimes predict if formation of a pleated sheet or α-helix is more likely to occur by picturing what happens in the two cases.

Alternate amino acid side chains are exposed on the same face of a pleated sheet. Therefore, if we plot the amino acid sequence as a zigzag, those residues to one side will be on one face of a sheet, those on the other side on the other face. The left part of Figure 3–8 shows what happens when this is done with the sequences of the two longest segments of pleated sheet from liver alcohol dehydrogenase, together with the preceding and following residues. Folding these sequences as a sheet creates a massive aggregation of hydrophobic groups from the side chains of valine and phenylalanine:

```
 H3C   CH3
    \ /
 H   CH                    H   CH2
 |   |                     |   |
···N—C—C···             ···N—C—C···
     |  ||                        ||
     H  O                         O
```

valyl group **phenylalanyl group**

The other face of the sheet has smaller or charged side chains. Also notice the nature of the residues at the two ends of the sequences. The bends at these positions are loaded with glycyl, prolyl, aspartyl, and asparaginyl groups. Here we have unusually clear examples of segments of primary structure that will make sharply defined hydrophobic regions through formation of segments of pleated sheet, terminated on each end by an abundance of chain-reversing residues.

Generation of an α-Helix. Since an α-helix has 3.6 residues per turn, the residues are spaced at 100° intervals viewed from the end, as one goes around the helix. We therefore can get an idea of the forces that will be generated by drawing a "helical wheel," in which the positions of successive residues in a chain are set down at 100° intervals around a circle. We are in effect looking down the cylindrical axis of the helix and observing the distribution of side chains around the surface. Figure 3–8 gives a comparison of plotting the same primary structure as a pleated sheet and as a helical wheel. The thing to look for is some indication of the development of specific bonding forces along particular sides of the helix. Could, for example, one side of the helix be facing the hydrophobic interior and another side be facing the aqueous phase? If so, a helix becomes more likely.

When we look at the segments in those terms, they appear rather nondescript. The segments from 12 o'clock to 4 o'clock are relatively polar; the remaining portion is nonpolar, but the groups are too widely scattered to be definitive. The first segment couldn't be a continuous α-helix in any case because of the prolyl residues that end it. The second segment possibly might be, but the apparent bonding forces are nothing like those seen with a pleated sheet configuration.

Now let us look at a sequence from myoglobin (Fig. 3–9). Here the pleated sheet configuration results in a rather random distribution of groups, while the wheel shows that an α-helix will create a well-defined region of bulky hydrophobic groups. In fact, this segment of the chain is a portion of an α-helix, with the hydrophobic

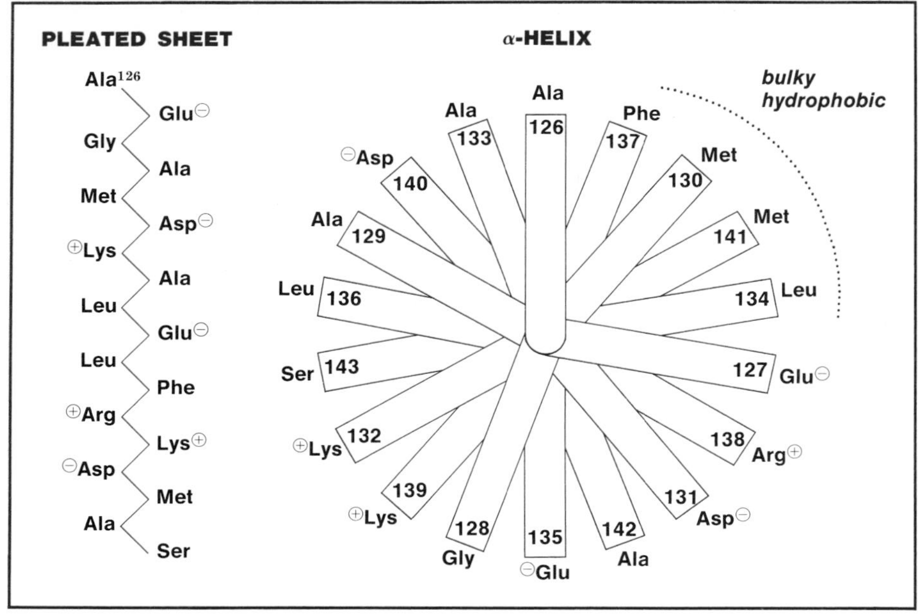

FIGURE 3–9 Plots of an amino acid sequence from human myoglobin as a pleated sheet and an α-helix. There is no indication of separate hydrophobic and polar areas if folded as a pleated sheet, whereas folding as a helix produces a compact area of hydrophobic surface.

segment in contact with other parts of the chain in the interior, and the remainder of the surface being exposed to surrounding water. It is the potential for creating this distribution of groups that makes myoglobin fold in the way it does.

The Heme Pocket

There is another important aspect to the structure of myoglobin. It is a heme protein, meaning that it belongs to a class of conjugated proteins that contain heme groups in addition to polypeptide chains, and the primary structure must be arranged so as to hold the heme in position.

Heme is a combination of iron(II) and protoporphyrin IX (Fig. 3–10). We shall say a great deal more about the porphyrins later. The point of importance now is that the porphyrins contain a highly resonant planar ring system in which four nitrogen atoms are fixed in the center at a spacing that is ideal for bonding of their unshared electrons with a metal ion. The porphyrins therefore act as tetradentate (four-toothed) ligands, and protoporphyrin IX has an especial affinity for iron.

Iron(II) ions have a coordination number of six, which means that they can associate with six electron pairs per atom. When a ferrous ion forms a chelate with a porphyrin, two of its coordination positions are still unfilled, so free heme in aqueous solution will be hydrated as shown in the figure. One of the functions of the polypeptide portion of myoglobin (the "globin") is to provide a histidine residue at a position where one of its nitrogen atoms will link to the fifth coordination position of the iron (Fig. 3–11). The sixth position of the iron is left open and now has the ability to bind oxygen.

The remainder of the peptide chain wraps around the porphyrin ring to hold it in position. The walls of this heme pocket are created by hydrophobic side chains

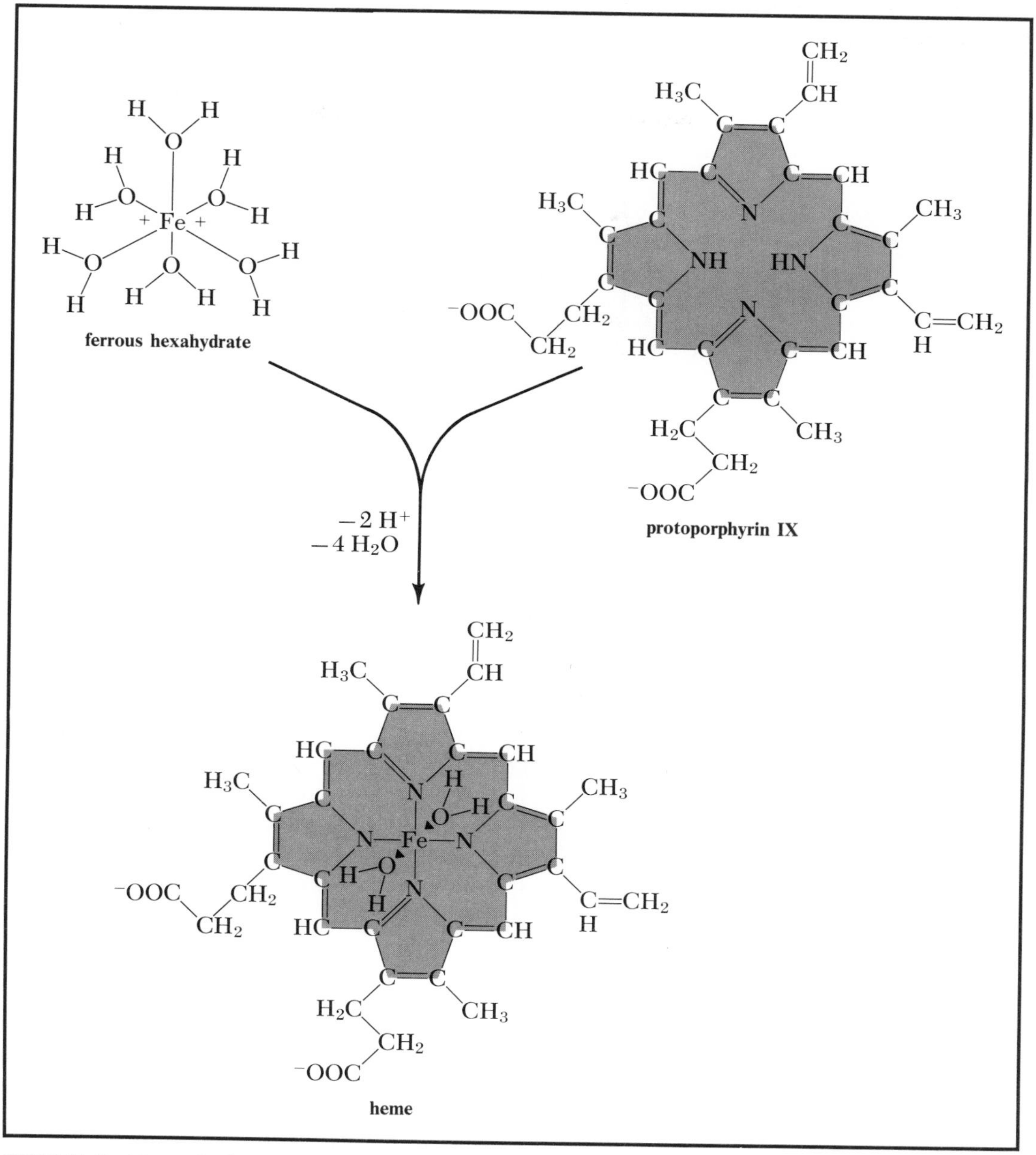

ferrous hexahydrate

protoporphyrin IX

$-2\ H^+$
$-4\ H_2O$

heme

FIGURE 3–10 The formal stoichiometry of the combination of iron (II) and protoporphyrin IX to form heme. The metallic ion has a coordination number of six, and any otherwise unfilled positions will be occupied by water or hydroxide ions.

from several of the surrounding helical segments, closely fitting the hydrophobic porphyrin ring. In addition, the aromatic side chain of a phenylalanyl group lies parallel to the porphyrin for interaction of the π-electrons. The porphyrin has two charged side chains, but these protrude out of the heme pocket to contact the aqueous phase.

In sum, the polypeptide chain of myoglobin has an amino acid sequence that favors the formation of a particular set of α-helices that will snugly nestle around a

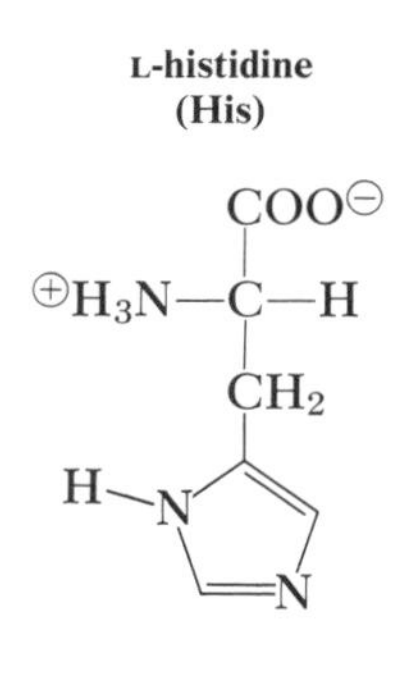

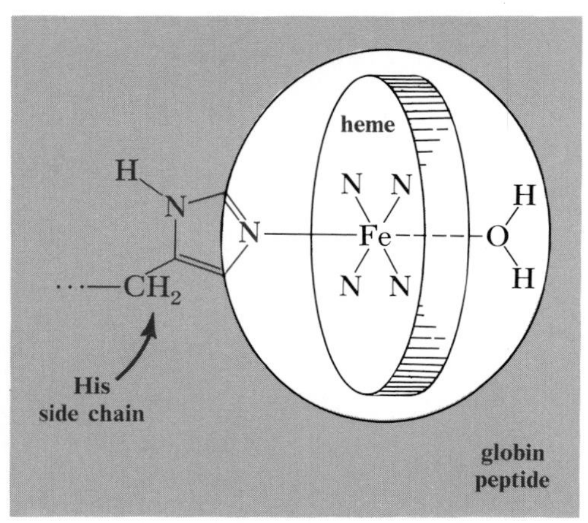

FIGURE 3–11 Schematic illustration of heme in a pocket formed by the globin polypeptide. A histidyl side chain acts as an additional ligand for the iron atom in heme. The free amino acid, histidine, is shown at the left.

heme group and hold it in four ways: chelation of the iron, hydrophobic exclusion of water, interaction of aromatic rings, and close packing of the atoms. Indeed, the polypeptide is built to fold into a particular tertiary structure in the presence of heme. (Otherwise, the hydrophobic heme pocket would be exposed to water.)

QUATERNARY STRUCTURE

Some protein molecules are complexes of more than one polypeptide chain; each chain with its tertiary structure then constitutes a subunit of a larger molecule. The geometry of the combination of subunits into the complete molecule is said to be its quaternary structure.

The forces available for the formation of quaternary structure are the same as those that create secondary and tertiary structure. One of the most important is the tendency for hydrophobic groups to combine so as to exclude water. Suppose that a polypeptide chain folds in such a way as to create a face with exposed hydrophobic side chains. That face would tend to stick to any other hydrophobic surface. If there are two polypeptide chains with areas of hydrophobic surface that can fit closely together, then the two chains will bond into a larger molecule, and this is the way in which many proteins containing multiple subunits are constructed. (Of course, tertiary structure with an exposed hydrophobic face wouldn't be very stable as an isolated entity, but when two or more chains are present simultaneously that can form matching faces, otherwise transient areas of hydrophobic surface will be stabilized by combining with each other.)

Liver alcohol dehydrogenase, for example, is a dimer that contains two identical subunits. The domains with pleated sheet cores that we discussed earlier comprise the tertiary structure of each subunit. However, these subunits have one segment of pleated sheet (segment βC in Figure 3–6) exposed at the surface in such a way that if two subunits abut, a continuous pleated sheet will be formed from one subunit to the other. The subunits are then joined both by association of hydrophobic side chains and by hydrogen bonding.

Hemoglobin

Perhaps the most thoroughly studied example of quaternary structure is that of hemoglobin. The molecules of hemoglobin that carry oxygen in human blood are tetramers made from a pair of each of two kinds of polypeptide chains (Fig. 3–12). These chains are designated α and β in the most abundant hemoglobin in adults, and this hemoglobin A therefore has the chain formula $\alpha_2\beta_2$. Each of these chains has a tertiary structure much like that of myoglobin, which is made from only one poly-

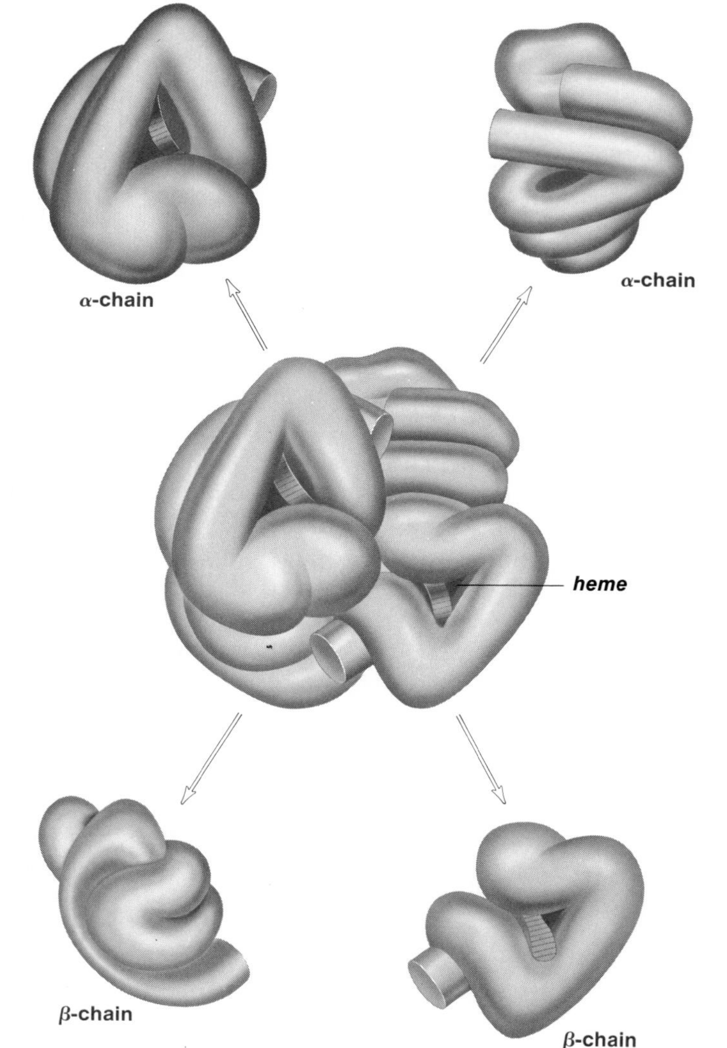

FIGURE 3–12

A schematic drawing of the most abundant adult hemoglobin. The molecule contains pairs of each of two kinds of peptide chains, termed α and β, which are also shown in exploded view at the approximate orientation they have in the complete molecule. Two of the four heme units are visible in pockets formed by folds of the peptide chains.

The drawing is of an idealized molecule—a sort of emaciated hemoglobin in which some of the external side chains have been plucked off to expose the underlying skeleton of the chains. In fact, it is difficult to distinguish chains at the surface of a complete model because the space between them is filled with protruding atoms.

The drawing is based on reports by Cullis, Muirhead, Perutz, and Rossman, Proc. Roy. Soc. (London), ser. A, 265:161 (1962); and by Perutz, J., Molec. Biol., 13:646 (1965); but the angle at which the molecule is viewed differs from those presented in the references.

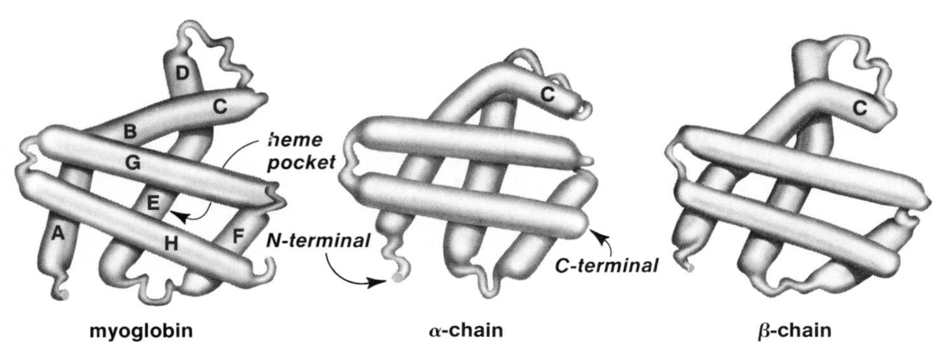

FIGURE 3–13 The arrangement of the peptide backbone is quite similar in myoglobin and in the α and β subunits of hemoglobins. The α-helixes are indicated as cylinders. These highly schematic renditions view the molecule from the opposite side of that given for myoglobin in Figure 3–6.

peptide chain (Fig. 3–13). However, they differ from myoglobin in that they are built to associate in particular ways. That is, α chains tend to bind β chains, and vice versa. The result is a tetrahedral molecule with the four subunits in close contact, except for a hole in the center that is freely accessible to water.

The hemoglobin chains associate and the myoglobin chains do not because some of the amino acid residues on the surfaces are different. The hemoglobin subunits have many additional hydrophobic residues exposed; they can only be buried by bringing the subunits together. They also have different polar side chains at locations where they form hydrogen bonds when the subunits abut. It is the additional surface charge and the lack of hydrogen bonding that keep the myoglobin molecules separated. (The hemoglobin subunit surfaces are also contoured for closer contact.)

On the other hand, the surfaces remaining exposed after formation of the tetramer are designed to minimize any tendency for further association. Red blood cells are stuffed with hemoglobin almost to the limit of its solubility, with adjacent molecules not much more distant than they are in a crystal. Even a moderate tendency to aggregate would not permit such a high concentration, and therefore would diminish the capacity to carry oxygen. We shall see in Chapter 13 that changes in single amino acid residues sometimes cause problems in this way.

What is the point of building hemoglobin, or any other protein, out of multiple polypeptide chains when many respectable proteins have only one? Sometimes proteins are built this way simply to become large enough so that they won't leak through membranes, but we shall see when we turn to the biological functions of proteins that those composed of more than one subunit can respond to changes in environmental concentrations in a more complex way than can those made from only one polypeptide chain. But before we examine these functions we should consider how proteins are made, and this process involves another class of macromolecular polymers, the nucleic acids, discussed in the next chapters.

AMYLOIDOSIS—A PLEATED SHEET DISEASE STATE

Amyloidosis was first recognized as a concomitant of tuberculosis.* Most of the deaths from tuberculosis resulted from damage caused by the invading organism, but a quarter or so of cases ran a different and irreversible course, characterized by gross enlargement of the spleen and liver, damage to the heart, and especially failure of the kidneys causing debilitating loss of protein into the urine and edema (accumulation of fluid). The affected organs had a rubbery consistency and microscopic examination showed striking changes: "Histologically, the picture of amyloid degeneration is one of structural obliteration in a glassy sea, in which islets of unsubmerged tissue survive, to later perish from starvation."†

This rubbery, transparent material being laid down throughout the body was called amyloid by the pioneering pathologist Rudolf Virchow because some samples, like starch (amylo = starch), gave a blue-violet color with iodine. Chemist Friedrich Kekule, of benzene fame, correctly said it was proteinaceous, but only in recent years has the character of the amyloid deposits been clarified.

Amyloid is now known to consist of fibers, each made of paired stacks of antiparallel pleated sheets (Fig. 3–14) and singularly resistant to degradation by normal physiological processes. Relatively inert fibers made from stacked pleated sheets are known in invertebrates (silk is an example) but have not been found to be normal components in vertebrates. Where, then, does it come from? Amino acid sequence shows that there is more than one kind of amyloid, and at least some, probably all, are products of the partial degradation of normal proteins. One kind of protein is converted to amyloid in some patients with chronic inflammatory conditions, such as chronic tuberculosis or rheumatoid arthritis. Another kind yields amyloid in some of those who have an abnormal proliferation of cells forming a particular antibody. Still other amyloids accumulate locally with some malignant proliferations of particular organs. A different amyloid accumulates in the brain of patients made senile by Alzheimer's disease. We have different causes producing somewhat different products with the common structure of stacked pleated sheets. For this reason, the generic term β-fibrilloses has been proposed to replace the single designation of amyloidosis for the diverse conditions.

It now appears that amyloid is deposited because some proteins, although normally soluble, yield during their degradation by hydrolysis pieces that are more stable as part of a stack of pleated sheets. Once the stack is formed, it is impervious to further hydrolytic attack. The appearance of the component with a propensity to stack is an accident of nature. Intermediates with that property do not form during the degradation of most proteins; in most cases too many polar groups are present in the wrong places. That is why only a few people develop amyloidosis. Too much formation of one protein containing a segment of polypeptide chain that by itself will stack as pleated sheets and in a position where it will be released intact by normal degradative processes—that is the story of amyloidosis.

---

*It is now difficult to appreciate why the romantic literature of the 19th century was so often built around the young adult dying of consumption. During that period, roughly one-fourth of all adult deaths in Europe were caused by tuberculosis. Even in 1910, one in 500 of those 20 to 40 years old died each year from this disease in the United States; one in 25 would not survive beyond these prime years.

†P. 37 in H.D. Power and W.W. Hala (1929). Principles of Pathology for Practitioners and Students. Appleton.

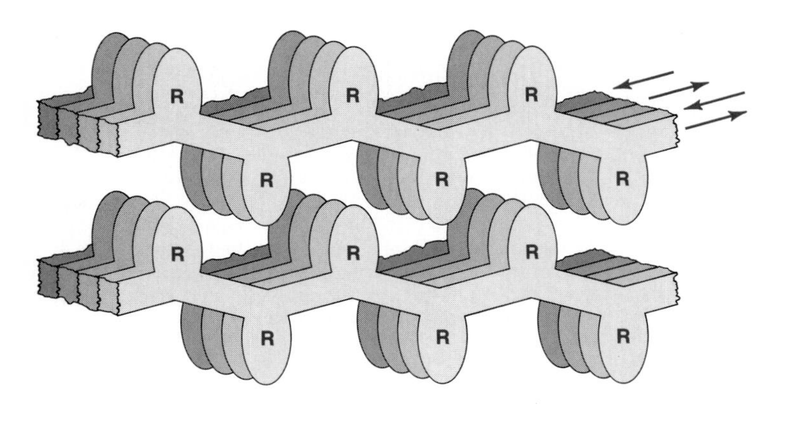

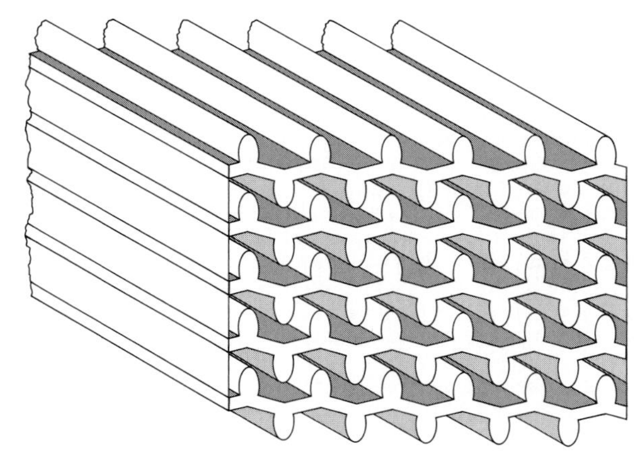

FIGURE 3–14

Schematic diagram of amyloid fiber. *Top.* Detail of stacking of antiparallel pleated sheet viewed down the fiber axis. Two sheets are shown viewed on edge. *Left.* One of the two pairs of stacks in an amyloid fiber viewed from an end. The number of sheets and their average width fits the physical data, but the actual course of the polypeptide chains is not known.

FURTHER READING

Readable Reviews

C.R. Cantor and P.R. Schwimmel (1980). Biophysical Chemistry I. The Conformation of Biological Macromolecules. Freeman. The first part of this volume covers protein structure.

J.S. Richardson (1981). The Anatomy and Taxonomy of Protein Structure. Adv. Prot. Chem. 34:168.

G.G. Glenner (1980). Amyloid Deposits and Amyloidosis: The β-Fibrilloses. N. Engl. J. Med. 302:1283, 1333.

Technical Reviews

M.G. Rossman and P. Argos (1981). Protein Folding. Annu. Rev. Biochem. 50:497.

P.Y. Chou and G.D. Fasman (1978). Empirical Prediction of Protein Conformation. Annu. Rev. Biochem. 47:217.

P.D. Gorevic and E.C. Franklin (1981). Amyloidosis. Annu. Rev. Med. 32:261.

P.S. Kim and R.L. Baldwin (1982). Specific Intermediates in the Folding Reactions of Small Proteins and the Mechanism of Protein Folding. Annu. Rev. Biochem. 51:459.

Technical Articles

J.S. Richardson, E.D. Getzoff, and D.C. Richardson (1978). The β-Bulge: A Common Small Unit of Nonrepetitive Protein Structure. Proc. Natl. Acad. Sci. (U.S.) 75:2574.

M. Levitt (1978). Conformational Preferences of Amino Acids in Globular Proteins. Biochemistry 17:4277.

CHAPTER 4

DNA AND GENETIC INFORMATION

The potential of a fertilized ovum to grow into a human lies in the information carried within it; this single cell contains complete specifications for its own transformation into a human body with 10^{13} nucleated cells of many different kinds molded into specific tissues. (This approximation neglects the red blood cells. An adult human has roughly 2×10^{13} of these small modified cells, which have no nuclei or mitochondria, and therefore no DNA.) The information specifies not only the **form,** but also the **patterns of behavior** of tissues; cells in the adult respond to chemical and physical changes in their surroundings as the result of instructions in the primal cell. Only a mere six picograms—6×10^{-12} g—of deoxyribonucleic acid (DNA) need be present to contain this information, both for structure and for function.

Nothing else in human experience is remotely comparable to the compact packaging of large quantities of information in DNA. Our most sophisticated integrated circuits may have 100,000 electrical elements on silicon chips as small as 5 mm square and 0.2 mm thick, but the average volume occupied by 200 of these elements, microscopic though it is, is sufficient to contain the different molecules of DNA necessary to direct the growth of replacements for all of the medical students, house staff, and medical faculty members presently in the United States—some 180,000 of them.

DNA Is a Polynucleotide. Nucleic acids are polymers of nucleotides, and nucleotides are made by combining nitrogenous heterocyclic rings, sugars, and phosphate groups (ionized derivatives of orthophosphoric acid, H_3PO_4). The heterocyclic constituents are referred to as the **"bases"** of nucleotides.

Nucleic acids are arranged in this way:

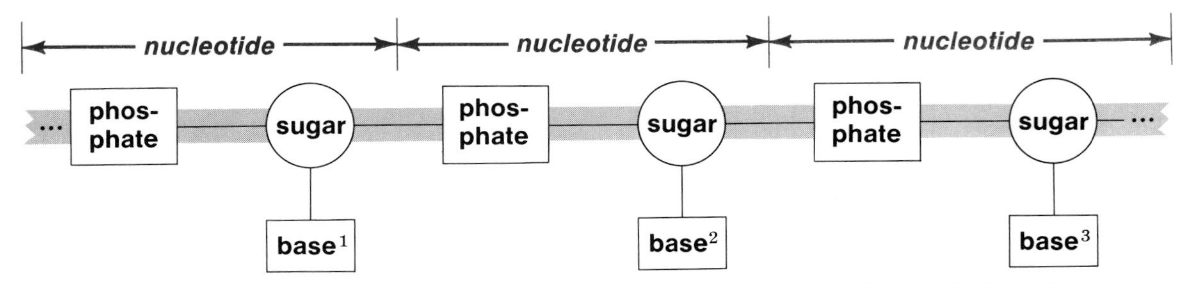

Each sugar residue in a repeating backbone of sugar-phosphate units has a heterocyclic base attached as a side chain.

Two structural classes of nucleic acids occur in cells, differing in the nature of the sugar they contain. Some contain ribose, a five-carbon sugar,* and are known as ribonucleic acids (RNA), while others contain the 2-deoxy derivative of ribose and are known as deoxyribonucleic acids (DNA):

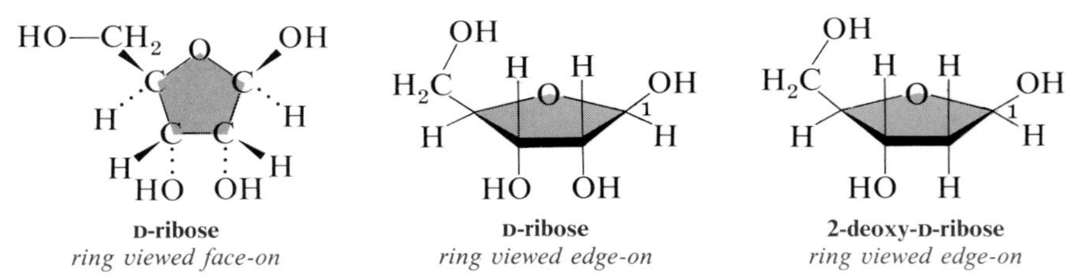

D-ribose
ring viewed face-on

D-ribose
ring viewed edge-on

2-deoxy-D-ribose
ring viewed edge-on

DNA Occurs as a Double Helix. The DNA of cells is made of two different, but related, polynucleotide chains, twisted together into a double helix (Fig. 4–1). In this stable structure, the heterocyclic bases from each chain are stacked in pairs, like plates of atoms, with the stack held together by two ropes of sugar-phosphate backbone running along the outside of the stack.

The information in DNA consists of the sequence in which different base pairs occur in the center of the stack, and the length of a molecule in different organisms is roughly proportional to the amount of information it contains, although there are great variations. Nearly all of the DNA in a fertilized human ovum is in 46 large molecules, one for each chromosome, in the nucleus. (Much smaller molecules occur in the mitochondria and the centrioles.) If these large DNA molecules from one cell were stretched end to end, they would match the height of a tall man—nearly two meters—but they are folded into the small volume necessary to fit within the cell nucleus. They are only a thin two nanometers in diameter.

Since each base pair constitutes a plate only one atom thick—roughly 0.34 nanometers—the total length of two meters in a human cell must be created by stacking some 6×10^9 pairs (1 nanometer $= 10^{-9}$ meter).

Base Sequence Determines Amino Acid Sequence. We have seen that it is the variation in sequence of amino acid side chains along the repeating backbone that conveys individuality to polypeptides. Similarly, it is the sequence of four different bases along the repeating polynucleotide backbone that constitutes information in DNA. Here we have the fundamental postulate of molecular biology:

> **Sequences of bases in DNA eventually determine the sequences of amino acids incorporated into polypeptides. Since the form and function of proteins depend upon the amino acid sequences in their constituent polypeptide chains and these functions include the construction and regulation of all other cellular constituents, DNA in this way specifies the character of the organism.**

---

*The structure of sugars is discussed in Chapter 10.

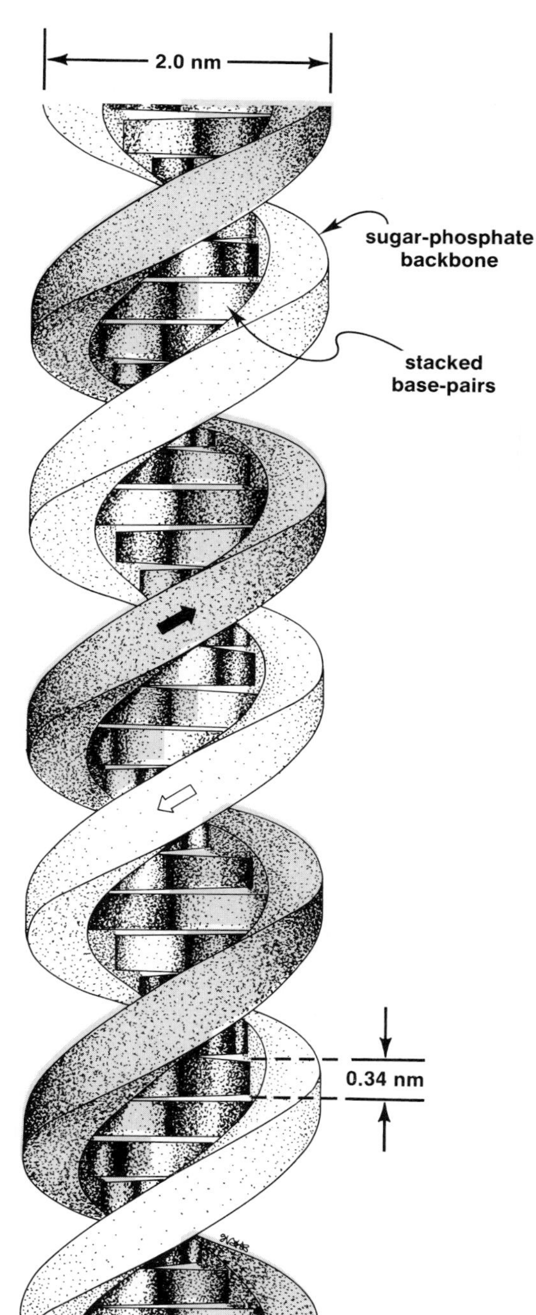

FIGURE 4–1

DNA is composed of two polynucleotide strands running in opposite directions that twist into a double helix so as to bring the bases into contact in the center. The sugar phosphate ester backbones of the strands are exposed on the outside of the helix. Ten pairs of nucleotide residues, with their associated paired bases in the center, form one complete turn of the helix.

The Bases Are Purines and Pyrimidines. Two of the principal bases in DNA contain pyrimidine rings, and the other two contain purine rings (Fig. 4–2). The pyrimidines are **cytosine** and **thymine;** the purines are **adenine** and **guanine.** Note them well because the shuffling of the sequences of these bases in the parent cells is the source of all of the differences between the smallest bacterium and the most sapient human. Other bases—derivatives of the major compounds—are also present. Although they are a very small fraction of the total, they may represent important regulatory signals.

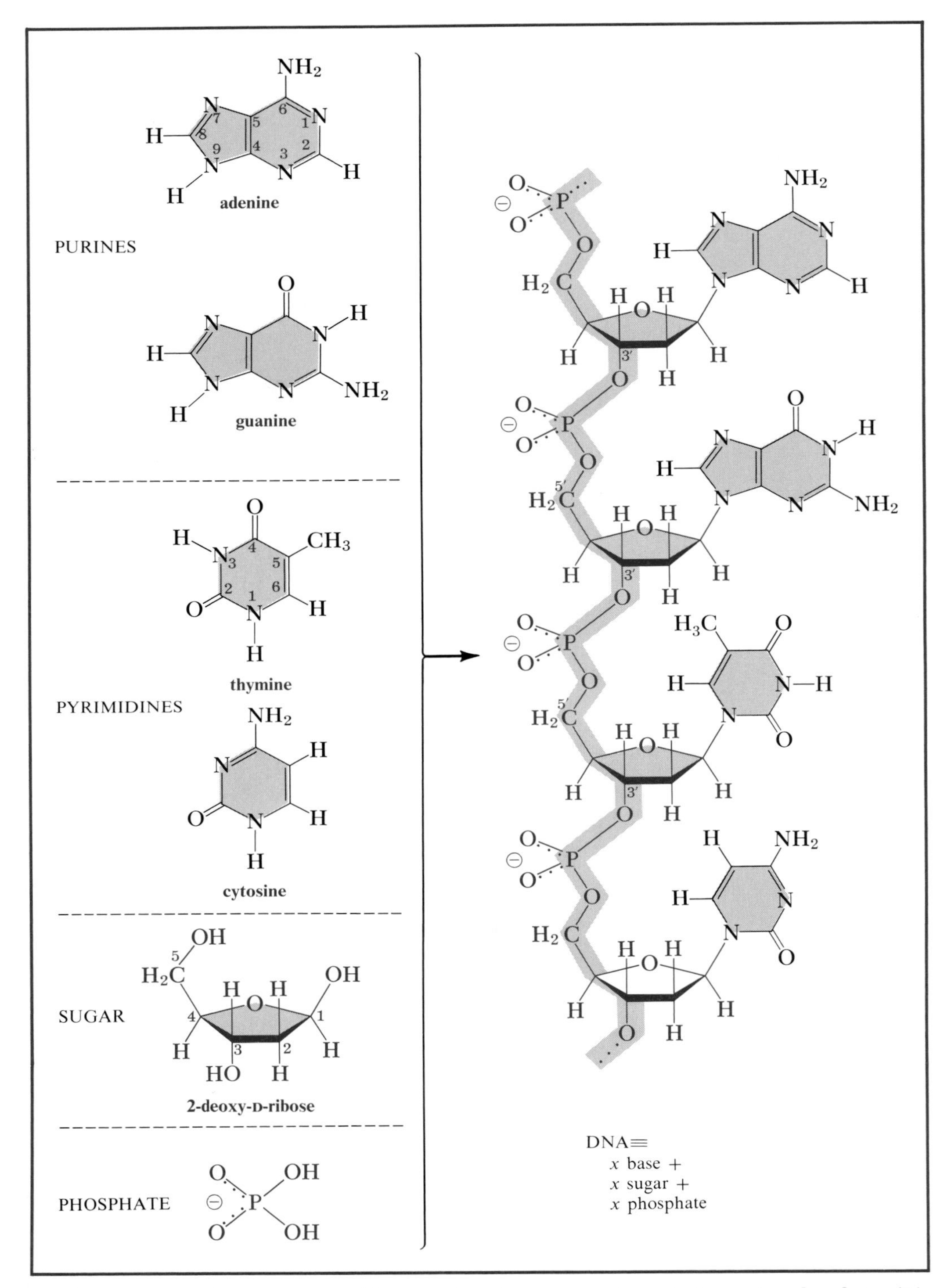

FIGURE 4–2 The formal stoichiometry for the combination of inorganic phosphate (P_i), 2-deoxy-D-ribose, and purines or pyrimidines into deoxyribonucleic acid (DNA). The conventional abbreviation for the polynucleotide fragment shown is d(A-G-T-C-). Adenine and guanine are substituted purines; thymine and cytosine are substituted pyrimidines.

The numbering of atoms in ribose, purines, and pyrimidines is shown. In the nucleotides, the atoms of the sugar are designated 1′, 2′, etc., to distinguish these positions from the numbered positions on the purine or pyrimidine rings.

The critical feature of these heterocyclic compounds is that they form hydrogen bonds when fitted in a particular way (Fig. 4–3). In this configuration, adenine pairs with thymine by two hydrogen bonds, and guanine pairs with cytosine by three hydrogen bonds. Adenine and thymine, guanine and cytosine are said, therefore, to be **complementary bases;** the stacked bases in the center of the double helix in DNA consist of these pairs:

one polynucleotide chain		*other polynucleotide chain*
	adenine = thymine	
	thymine = adenine	
	guanine ≡ cytosine	
	cytosine ≡ guanine	

DNA chains do not associate to form a double helix unless the base sequence in one chain is the complement of the sequence in the other chain, and this association of complements is the basis for the transfer of information.

DNA Contains Equal Amounts of Purines and Pyrimidines. Since double-stranded DNA contains a matching complement for each constituent base, it follows that the content of adenine equals the content of thymine, and the content of guanine equals the content of cytosine, with the total number of purine residues equal to the total number of pyrimidine residues. This is true for the total DNA in a cell and

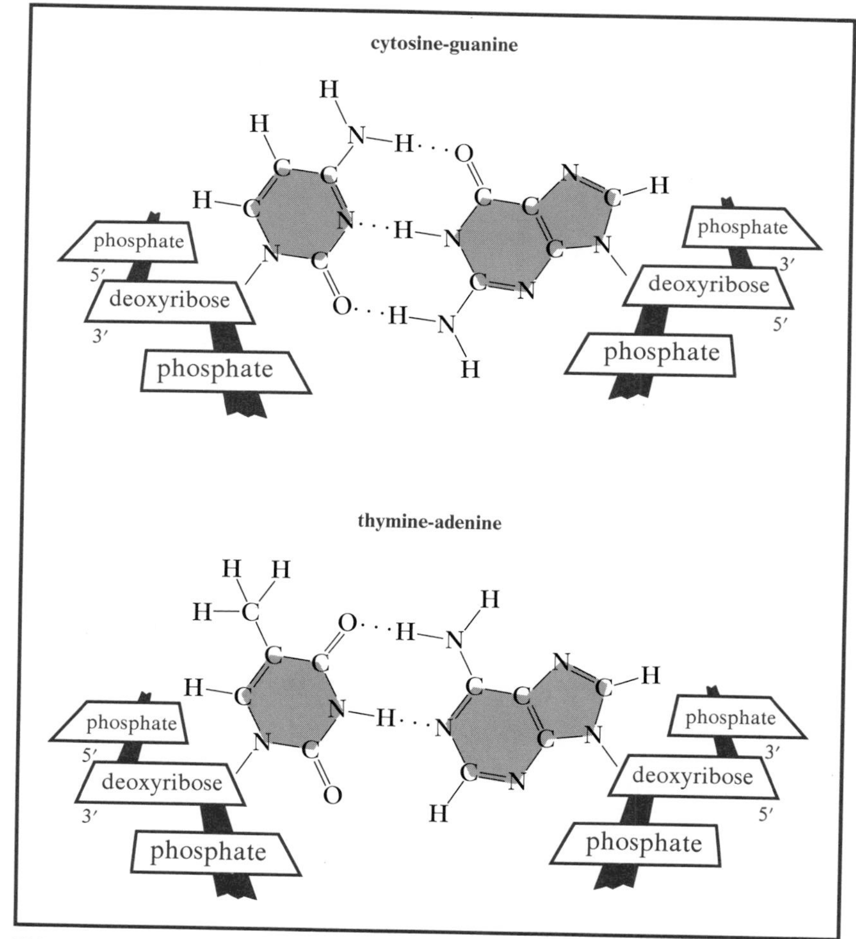

FIGURE 4–3 Hydrogen bonding of complementary bases between adjacent nucleic acid strands. The symbols are spaced to scale the actual positions of the atoms according to Pauling and Corey (1956), Arch. Biochem. Biophys., 65:164.

for any segment of the double helix within it; it is an identifying characteristic of complementary nucleic acid chains.

The Chains in DNA Are Antiparallel. The sugar-phosphate backbone of polynucleotides is not symmetrical; Figure 4–2 showed that the nucleotide units are connected by forming phosphate ester bonds between C-5′ of the deoxyribose residue in one nucleotide and C-3′ of the deoxyribose residue in the next nucleotide.* That is, a polynucleotide backbone has 5′ and 3′ ends that are not mirror images of each other. The chain has a polarity. It so happens that the most stable association of two chains occurs when their bases are complementary in antiparallel sequences, with the sequence of 5′ and 3′ linkages going in opposite directions on the two chains. The two chains then have opposite polarity.

This is an important point, so let us pursue it in more detail. The layout of a particular segment of polynucleotide may be sketched like this:

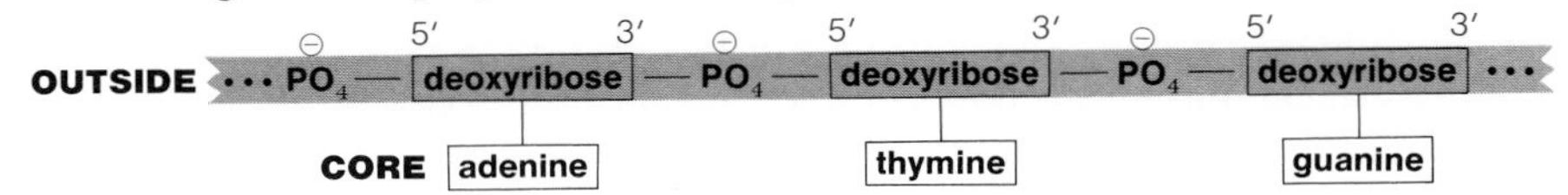

with the chain viewed so that C-5′ is on the left of each deoxyribosyl group, and C-3′ on the right. This is the conventional orientation for viewing polynucleotides. The second polynucleotide chain in DNA is constructed so that each successive base is a complement of the corresponding bases in the first chain when it is oriented with C-5′ to the right:

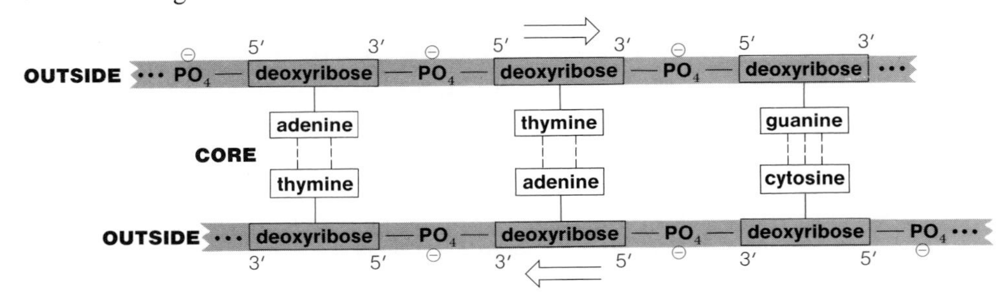

Nomenclature of Nucleotides

Nucleosides are base-sugar compounds.

Nucleotides are nucleoside phosphates (base-sugar-phosphate compounds).

Nucleotides, either free or polymerized into nucleic acids, have names derived from the names of their constituent bases.

---

*The prime mark is used in numbering the sugar components of nucleotides to distinguish positions on them from positions having the same numbers on the base components.

Constituent Base	Ribonucleoside*	Symbol*	Corresponding Nucleotide*
adenine	adenosine	A	adenosine 5′-phosphate
guanine	guanosine	G	guanosine 5′-phosphate
cytosine	cytidine	C	cytidine 5′-phosphate
thymine	thymidine	T	thymidine 5′-phosphate

*Adding the prefix *deoxy-* to the names of the nucleosides or nucleotides designates the corresponding compounds containing deoxyribose. The symbol is then prefixed by d. For example, dA is 2′-deoxyadenosine.

Nucleic Acids. The sequence of polynucleotides is usually given by single letter abbreviations, which formally designate the constituent nucleosides. The connecting phosphate groups are indicated by intervening dashes, and nucleosides containing deoxyribose rather than ribose are indicated by a preceding "d." Thus, the segment used as an example on p. 52 is read from left to right as given (C-5′ of each sugar to the left):

. . . -dA-dT-dG . . .

or

. . . -d(A-T-G) . . .

In practice, the intervening hyphens and the prefix d are often omitted, relying upon context to indicate if a DNA sequence is being given.

To show if the ends of polynucleotide chains have free hydroxyl groups or are phosphorylated, an "HO-" or a "p" may be added to the sequence. pdA . . . indicates a chain beginning with phosphate on C-5′ of a deoxyadenosyl group. HO-dA . . . would indicate the 5′-hydroxyl group is not phosphorylated.

Group names for nucleotide residues are derived from trivial names for nucleotides formerly in more common use:

Present Name	Older Name	Present Group Name
adenosine 5′-phosphate*	adenylic acid*†	adenylyl*
guanosine 5′-phosphate	guanylic acid	guanylyl
cytidine 5′-phosphate	cytidylic acid	cytidylyl
thymidine 5′-phosphate	thymidylic acid	thymidylyl

*The prefix deoxy- is added when appropriate.

†Cognoscenti of structure modify these names so as to indicate the anionic form that actually exists under physiological conditions (adenylate, guanylate, etc.).

Unless otherwise specified, trivial names are understood to designate the commonly occurring nucleoside 5′-phosphates, although 2′- and 3′-phosphates are also known.

Characteristics of the Double Helix

Several kinds of double helixes can be constructed from models of DNA. Two right-handed helical structures, known as **A** and **B helixes,** are consistent with crystallographic and other physical measurements obtained with mixed samples of DNA. A left-handed helix, the **Z helix,** is also feasible with certain base sequences. It now appears that DNA mostly exists in the B helix, the form in which it has been usually depicted, and that is the structure we now describe. However, it ought to be kept in mind that the molecule is not frozen; the extent to which the conformation shifts from one form to another is not clear. The helix can certainly bend.

Polar Exterior and Nonpolar Core. Figure 4–4 shows different views of a model of DNA. All of the phosphate groups are on the outside. Diesters of phosphoric acid are very strong acids and DNA is fully ionized, so it has continuous skeins of negative charge running down each side that must be neutralized with associated cations. The bases, on the other hand, are compactly stacked in the middle so as to exclude water from contact with their hydrophobic faces. The hydrogen bonds connecting the base pairs are therefore in a nonpolar environment where they can approach maximum strength. Isolated nucleosides will not form strong bonds with their complements in aqueous solution, because the faces of the rings are not shielded from disrupting contacts with water molecules; however, they do form tight pairs in nonaqueous solvents, and similar strong bonding is seen in the nonpolar interior of the double helix, relative to the force developed between mismatched bases.

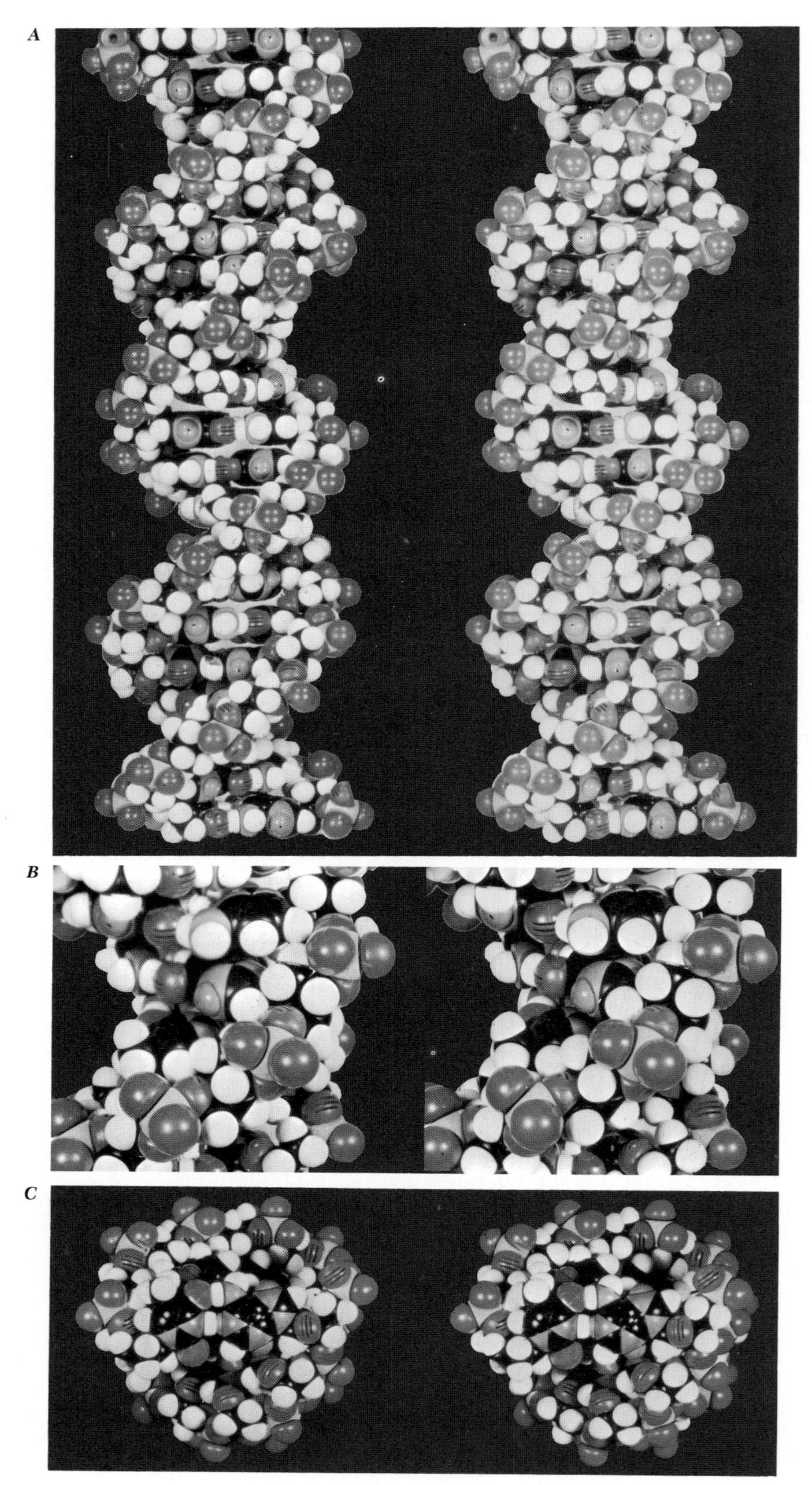

See legend on opposite page

A

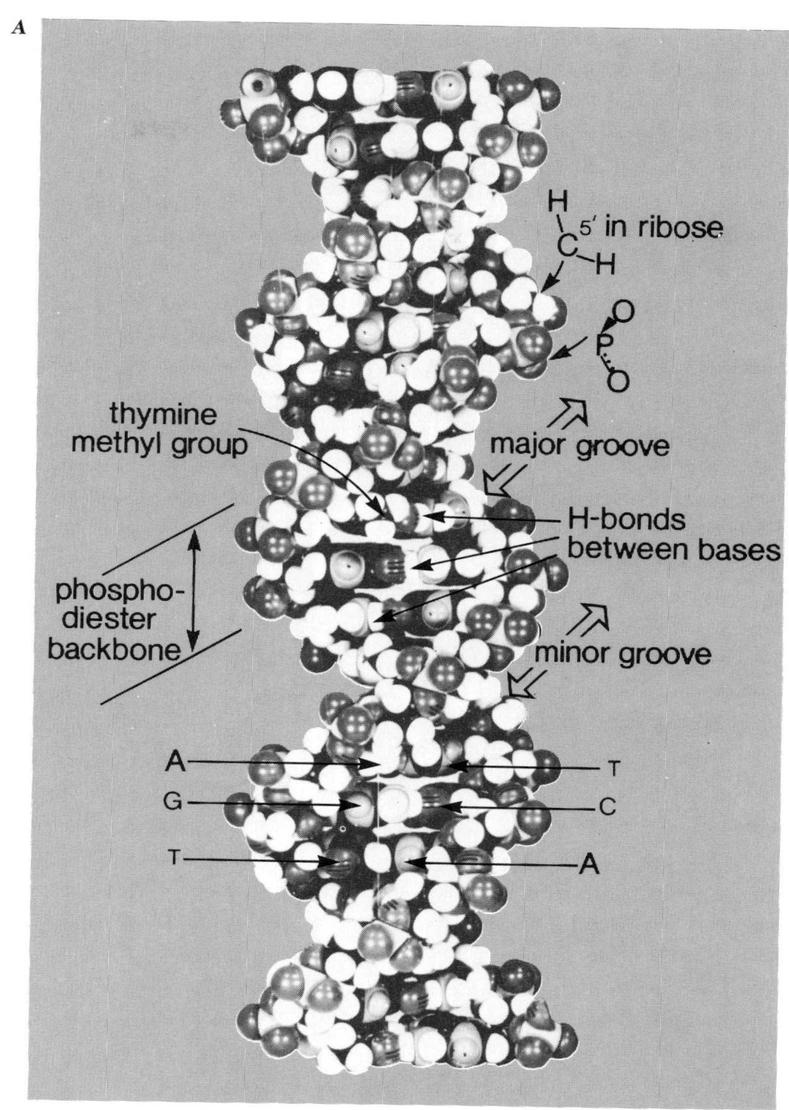

B

C

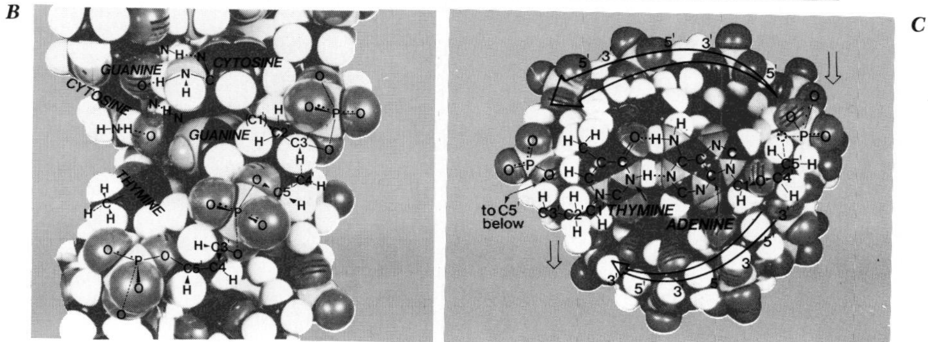

FIGURE 4–4

Stereo photographs of a space-filling model of DNA. Some significant features are indicated in the keys (*right*). *A.* Two full turns of double helix. *B.* A closer view into the major groove showing details of the sugar-phosphate backbone and the stacked bases. *C.* End view, showing base-pairing. Author's model based on S. Arnott (1970), Progr. Biophys. Mol. Biol., 21:267.

A note on viewing: Stereo illustrations are widely used in modern biochemistry. It is helpful to learn to see them without optical aids—the illusion of depth is greater. One ought not expect to do this automatically; persistence is required, but once the skill is developed, it is easily applied. I find it helpful to bring the pair very close to the eyes, staring into the distance through the blurred images until they fuse, then withdrawing them slowly until focus is reached.

The stacking of the bases also develops additional forces through aromatic interactions and the close fit of adjacent rings that stabilize the particular helical configuration.

Wide and Narrow Grooves. A striking feature of the general outline of the double helix is the deep grooves formed between the phosphate-sugar backbones. One groove is substantially wider than the other in DNA, and they are frequently designated as the major and minor grooves. This is not to be taken as an indication of relative functional importance, and the minor groove is no small crack in the surface.

The grooves are the only windows through which the bases of a double helix can be identified by interaction with an external molecule. Each of the four bases has a particular orientation to the axis of the helix, so that the same atoms are always present in each groove. For example, the base of the wide groove is always paved with C-6, N-7, and C-8 of purine rings, and C-4, C-5, and C-6 of pyrimidine rings. The narrow groove is paved with C-2 and N-3 of the purine rings, and C-2 of the pyrimidine rings. It is at least possible that some proteins may be constructed to identify particular locations on the long DNA molecules through interaction with specific combinations of these exposed atoms even though the double helix remains intact, and in this way control the information to be transferred.

Physical Characteristics. The double strands of DNA may be separated **(denatured)** by increasing the temperature of a solution, and the temperature at which the transition is half-completed is the **melting point** of the DNA structure. The term also has some physiological connotations; for example, the transfer of information from DNA involves a separation of the two polynucleotide chains, and the factors necessary for this disruption of the double helix are said to lower the melting point below the temperature of the cell. Values of the melting point of DNA alone under typical experimental conditions range around 85° C; the actual value depends upon base composition, since it requires more energy to break the three hydrogen bonds in G-C pairs than it does the two in A-T pairs.

Among the many properties of DNA, the one most frequently measured as an index of the loss or gain of double helical structure is the ultraviolet absorption spectrum. The nucleosides and nucleotides have characteristic ultraviolet spectra owing to the conjugated bonds in the purine and pyrimidine groups. The absorbancy

TABLE 4–1 SUMMARY OF THE NATURE OF DUPLEX DNA

Two antiparallel polynucleotide chains twisted into a double helix
Negatively charged phosphate-deoxyribose backbones on outside, complementary base pairs on inside
Adenine and thymine, guanine and cytosine paired by hydrogen bonds that are stabilized in non-polar interior
Length of molecules: 14 to 73 mm in human cells, 1.6 mm in the bacterium *E. coli*; 0.34 nm per base pair; 3.4 nm per turn
Diameter: 2.0 nm
Base pairs per complete turn: 10 (36° turn per pair)
Two grooves, 2.2 nm and 1.2 nm across, measured parallel to axis of helix
Buoyant density: 1.660 + 0.098 (fraction of G + C)
Melting point: ranges around 85° C for native molecules, increases with greater G + C content
Mass: 618 daltons per pair of nucleotides (anion form)

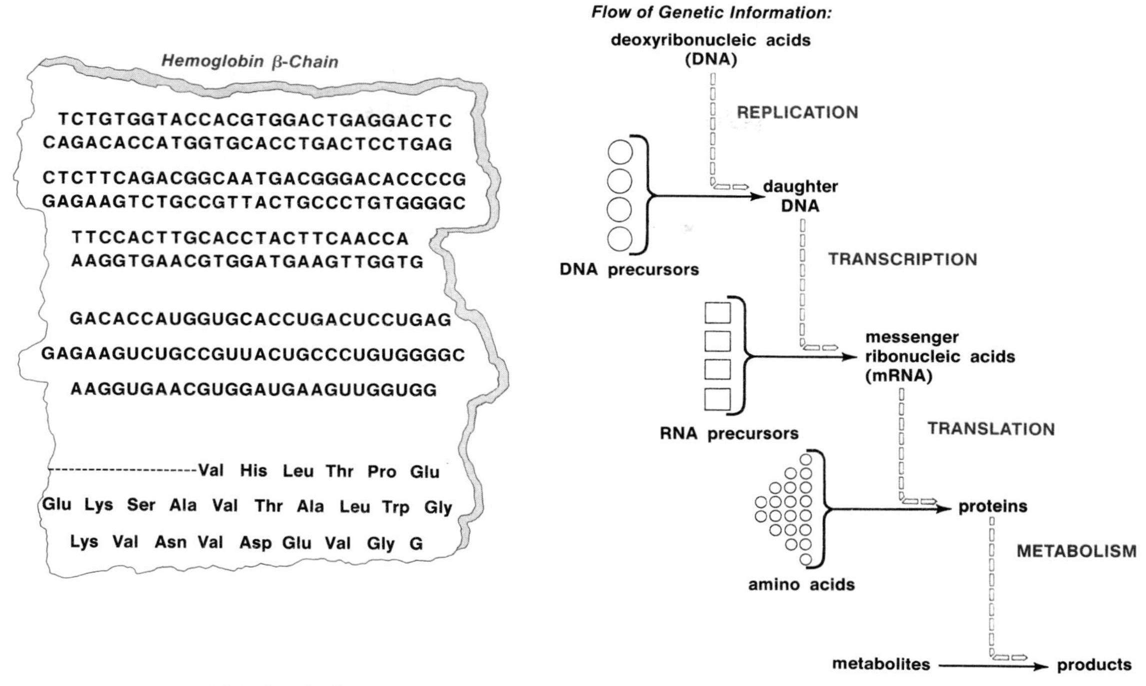

FIGURE 4–5 A biochemical Rosetta stone. Fragments of the genetic information for constructing the β chain of human hemoglobins are shown at the left as matching double-stranded DNA sequence, RNA sequence, and amino acid sequence. They are aligned with corresponding stages in the flow of genetic information shown at the right.

is diminished by the hydrogen bonding occurring in the double helix, and this loss, or hypochromicity, can be used as a quantitative index of the fraction of the molecules existing in helical form.

If solutions of denatured but complementary DNA are held at temperatures just below the melting point, the double helix will again form. This process is referred to as **annealing** the DNA; the rate at which it occurs depends upon the fraction of the DNA molecules that are exact complementary matches—the higher the fraction, the greater the proportion of collisions between molecules that will bring complementary sequences near each other.

The buoyant density, ρ (rho), of DNA in concentrated salt solutions increases with the proportion of guanosine-cytidine pairs:

$$\rho = 1.660 + 0.098\,(\mathrm{G} + \mathrm{C})/(\text{total nucleosides})$$

The change is sufficient to permit the separation of DNA molecules into groups of similar base composition by centrifuging them through a gradient of increasingly concentrated salt solutions. (Cesium chloride, CsCl, is frequently used.) Those molecules with a greater fraction of G-C, and a correspondingly lower fraction of A-T, sediment farther before reaching a solution of equivalent density.

USE OF GENETIC INFORMATION

The genetic information is the sequence of nucleotides in DNA. How can the order of occurrence of only four compounds construct the form and direct the function of even a simple bacterium, let alone a human being? The entire process hinges upon the binding of complementary bases in a way that can be interpreted as amino acid sequences. Figure 4–5 outlines the principal features of the flow of information during this process along with a biological Rosetta stone—a comparison of the languages in which the information appears at each stage during the construction of one polypeptide chain—the β-chain of human hemoglobin A_1.* Two kinds of languages are used: nucleotide sequence represented by double-stranded DNA and single-stranded RNAs and amino acid sequence in polypeptides.

Replication. When cells divide, each daughter cell must have faithful copies of the entire genetic instructions in the original cell. This is accomplished by replication of all of the DNA, in which new complementary polynucleotide chains are created for each of the original chains in the cell. In preparation for mitosis, the strands of each double helix are separated, and a new DNA strand is constructed on each previous strand. Each old strand then has a new complementary partner exactly like the other old strand, if replication proceeds without error.

Transcription. The use of the information in DNA to direct the life of a cell involves the synthesis of ribonucleic acid (RNA) molecules as working copies of segments of DNA.

Some of the RNA molecules are structural components of the apparatus of protein synthesis, for example, the **ribosomes** on which polypeptide chains are built. Other RNA molecules act as **messengers** from DNA in the nucleus, carrying instructions for assembly of cellular proteins; these instructions are used by enzymes catalyzing protein synthesis in the cytosol.

Translation. The translation of the nucleotide sequence on messenger RNA molecules into an amino acid sequence in a polypeptide chain involves attachment of each amino acid to an identifying small RNA molecule, **transfer RNA.** This molecule serves to transfer the amino acids from free solution to the point of assembly of polypeptide chains, but the transfer only occurs when directed by the instructions in messenger RNA molecules.

Metabolism and Other Functions. The ultimate purpose of the proteins is to make structures and carry out functions. Among their active roles is catalysis of thousands of chemical reactions, which constitute the metabolism of the organism.

Arrangement of DNA Sequences

Repetitive and Nonrepetitive DNA. In the total of 6×10^9 pairs of nucleotides representing the human genome, some lengthy segments have a unique sequence. Others may occur as a few identical copies. Still others occur in families as many quite similar but not quite identical copies, while others occur as hundreds of thousands of very similar if not identical copies.

The different degrees of repetition can be recognized in the following way: The total DNA of cells is sheared into relatively small lengths by rapid stirring of a solution. (Even gentle stirring can create sufficient moment on these long molecules to break some covalent bonds.) The fragmented DNA is dissociated into single strands by raising the temperature above the melting point. The strands are then exposed to annealing conditions that favor reassociation into double helixes. The double helixes and remaining single strands can be separated by adsorbents such as nitrocellulose filters or hydroxyapatite columns, which retain single- or double-

---

*The languages were indeed deciphered by the study of the synthesis of hemoglobin—rabbit rather than human, however.

stranded segments, respectively. This enables measurement of the extent of reassociation with increasing time of annealing. The chance that a fragment will encounter another fragment containing long sequences of complementary bases with which it can form a stable double helix is proportional to the number of these similar sequences in the original DNA. The more abundant the sequence, the faster the recombination. A long annealing time is required for recombination of nearly unique sequences.

The **nonrepetitive fraction** of eukaryotic DNA, with at most only a few copies of each sequence, includes the information for constructing most of the polypeptide chains. Many of the eukaryotic genes for polypeptide chains occur in pieces; that is, the DNA sequence with information for the beginning of a polypeptide chain may be separated from the information for the next segment, with a further interruption before additional information occurs. There may be as many as 50 such interruptions in the information for building a single polypeptide chain. The intervening segments of DNA, known as **introns,** may be as long or longer than the **exons** bearing the information later translated into amino acid sequence.

There is considerable similarity between many of the spacer, or intron, segments within genes, and they therefore appear in the **moderately repetitive** portion of the fragmented DNA, along with DNA carrying certain genes that are indeed repeated many times within the genome.

The **highly repetitive sequences,** with short segments occurring in tens of thousands of copies or more, often can be isolated as satellite DNA—DNA that has a density markedly different from the remainder owing to unusual base composition—and therefore separates in a separate **"satellite"** band upon centrifugation. (There is great difference among species in the numbers and relative density of these satellite bands.) Much of the satellite DNA represents the **constitutive heterochromatin,** a DNA fraction associated with the nucleolus, and probably having a structural function.

FURTHER READING

Readable Reviews

C.R. Cantor and P.R. Schwimmel (1980). Biophysical Chemistry I. The Conformation of Biological Macromolecules. Freeman. The latter part of this volume covers nucleic acids.

J.D. Watson (1976). Molecular Biology of the Gene, 3rd ed. Benjamin.

Technical Reviews

H.R. Mahler and E.H. Cordes (1971). Biological Chemistry, 2nd ed. Harper and Row. Includes detailed discussions.

R.M.S. Smellie et al. (1981). Davidson's Biochemistry of the Nucleic Acids, 9th ed. Methuen.

S. Arnott (1970). The Geometry of Nucleic Acids. Prog. Biophy. Mol. Biol. 21:267.

R.E. Dickerson et al. (1982). The Anatomy of A-, B-, and Z-DNA. Science 216:475. The stereo diagrams are in themselves worth the price of admission for those able to use them.

D. Freifelder (1978). The DNA Molecule. Structure and Properties. Freeman. Compendium of important past papers.

D.J. Patel, A. Pardi, and K. Itakura (1982). DNA Conformation, Dynamics, and Interactions in Solution. Science 216:581. Studies showing the effects of additional bases, GT pairs, and other aberrations.

S.B. Zimmerman (1982). The Three-Dimensional Structure of DNA. Annu. Rev. Biochem. 51:395.

CHAPTER 5

RIBONUCLEIC ACIDS: FORMATION AND FUNCTION

THE NATURE OF RIBONUCLEIC ACIDS (RNA)

RNA and DNA are both polynucleotides, but they serve different functions, which must be regulated separately. We have seen that molecules of DNA constitute the master file of genetic information carefully reproduced in successive generations of cells, whereas molecules of RNA are used as working copies and as tools for making proteins. These differences in function are reflected in the life span of the molecules; the cell does not intentionally destroy DNA, whereas it constantly degrades and rebuilds RNA as a necessary part of its adjustment to changing circumstances.

The differing functions of RNA and DNA mean that they participate in different reactions; they have groups that form different bonds even though both kinds of nucleic acids have the fundamental polynucleotide structure with a sugar-phosphate backbone on which bases are attached as side chains. RNA differs from DNA in several ways:

The sugar in RNA is ribose rather than deoxyribose, and the extra hydroxyl group is sufficient in itself to make an RNA chain recognizably different from a DNA chain:

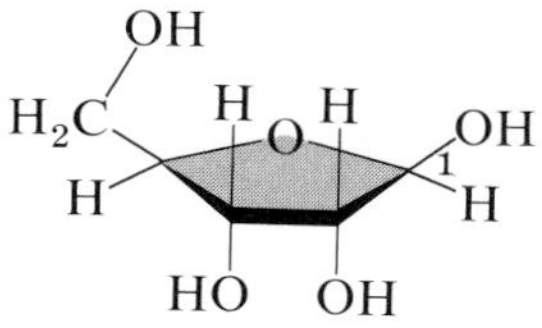

D-ribose 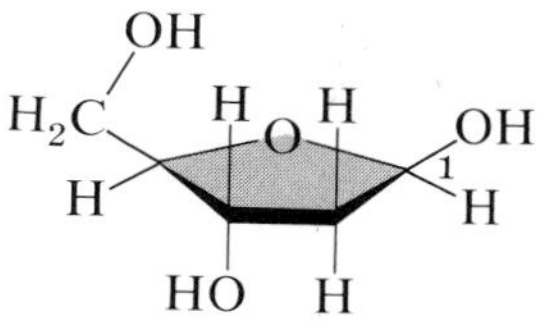 **2-deoxy-D-ribose**

RNA molecules are made with uracil in place of thymine as the base complementary to adenine. Uracil differs from thymine only in the absence of the 5-methyl group, and it forms the same hydrogen bonds:

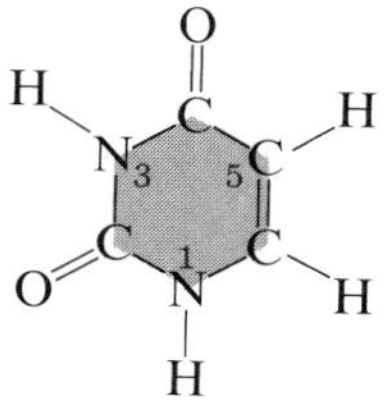

uracil

The other three bases laid down in RNA (adenine, cytosine, and guanine) are the same as those in DNA, although we shall see that these bases are sometimes modified after the polynucleotide is formed.

Molecules of RNA are single-stranded; they do not contain the antiparallel complementary strand found in the duplex structure of DNA. (Viral nucleic acids are often exceptions (p. 117); their parasitic life permits bending the rules.)

Molecules of RNA are smaller than molecules of DNA. Each molecule of RNA is made as an antiparallel complement to only part of a DNA strand. The smallest molecule of DNA in the nucleus of a human cell contains 46,000,000 base pairs. The largest piece of RNA isolated from cultures of human cells has approximately 50,000 nucleotide residues—only a little more than 0.001 the length of the shortest DNA strand—and even it may be an artifactual complex rather than a native single molecule. Some RNA molecules are transcribed with as few as 102 nucleotide residues.

RNA AND PROTEIN SYNTHESIS

The mechanism by which RNA carries genetic information for the synthesis of proteins is sketched in a general way in Figure 5–1. A segment of DNA containing specific instructions for the amino acid sequence of a polypeptide is transcribed to form an oversized precursor of **messenger RNA (mRNA).** The precursor is part of a fraction called **heterogeneous nuclear RNA (hnRNA)** because of the wide range of sizes of the newly made transcripts. The precursor is processed within the nucleus to form a finished molecule of messenger RNA, which is then transferred to the cytoplasm.

Ribosomes in the cytoplasm contain the machinery for making peptide bonds, and they bind to the messenger RNA, which has in its single polynucleotide sequence the information for making one polypeptide chain. One molecule of messenger RNA frequently carries several ribosomes, which appear as a cluster,

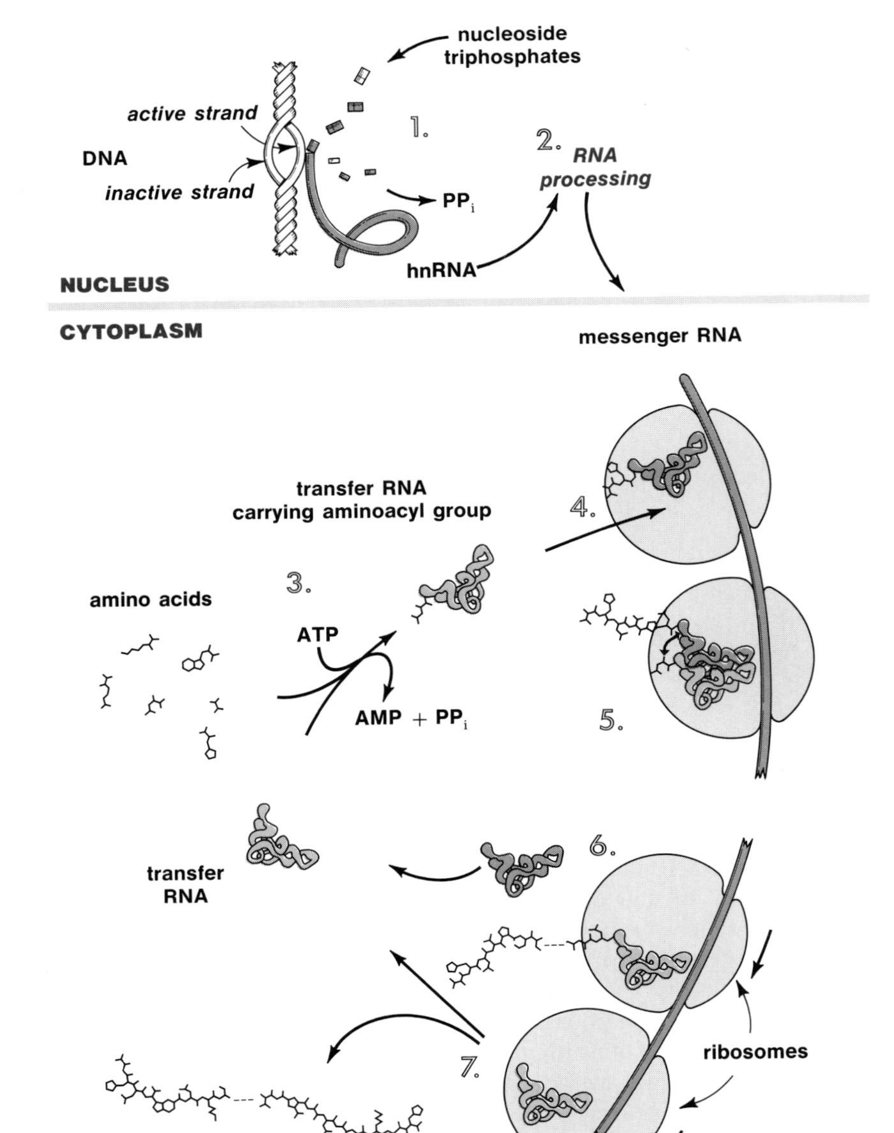

FIGURE 5–1

Schematic summary of protein synthesis. *Step 1.* A molecule of DNA in the nucleus is unwound so that one of its strands can be used as a template to direct the formation of a molecule of heterogeneous nuclear RNA (hnRNA). The RNA is made from nucleoside triphosphates, which lose inorganic pyrophosphate (PP$_i$) as they attach to the growing RNA chain.

Step 2. Messenger RNA molecules are formed in the nucleoplasm by modifying precursors among the hnRNA molecules. The finished messenger RNA moves through the nuclear membrane into the soluble cytoplasm.

Step 3. Meanwhile, amino acids combine with specific molecules of transfer RNA (tRNA) in the cytoplasm by a reaction that also involves the cleavage of adenosine triphosphate (ATP) into adenosine monophosphate (AMP) and PP$_i$.

The following steps are shown on separate ribosomes for clarity, but in fact they are repeated in sequence on each ribosome. The successive ribosomes create longer and longer polypeptide chains as they move down a molecule of messenger RNA.

Step 4. The tRNA molecules, carrying the amino acids as aminoacyl groups, diffuse to the ribosomes, on which the growing polypeptide is attached to another molecule of tRNA. The incoming tRNA, which bears the next group required for the growing polypeptide (in this case a leucyl residue), has a sequence of nucleotides that causes it to complex with mRNA on the ribosome.

FIGURE 5–1 *Continued*

or **polysome,** in electron micrographs. Successive ribosomes in a polysome carry polypeptide chains increasingly near completion.

Amino acids are brought to the site of peptide synthesis by transfer RNA (tRNA), to which they are covalently attached. Each kind of amino acid has at least one specific kind of transfer RNA to carry it, and this transfer RNA has a particular sequence of bases to form bonds with complementary base sequences within messenger RNA attached to a ribosome. The ribosome provides a moving platform upon which the messenger RNA is "read" by binding appropriate transfer RNA molecules to successive base sequences. The ribosome removes the amino acid residues from the transfer RNAs as they appear and adds them one at a time to a growing polypeptide chain. When the polypeptide is finished, it is released.

THE GENETIC CODE

Triplet Coding

Since only four different nucleotides are present to convey information for the placement of 20 different amino acids into peptide chains, each amino acid must be represented by combinations of at least three nucleotides. This is true because there are only 16 different doublets of four nucleotides (4^2), but there are 64 triplets (4^3). In the polynucleotide chain of messenger RNA molecules directing the synthesis of polypeptides and in the segments of DNA chains from which these molecules are transcribed, each successive group of three nucleotides, beginning at a precisely defined location, designates either a specific amino acid or a chain-terminating punctuation mark. This is the molecular basis for the inheritance of protein composition.

Information is transferred by antiparallel binding of complements. The code is read from the 5′ end of messenger RNA to its 3′ end—a happy coincidence of nomenclature and function—but the RNA is built by reading one strand of DNA in the opposite direction, from 3′ to 5′, so as to match bases that are antiparallel complements. The coding triplets in transfer RNA are likewise antiparallel complements of the corresponding triplets in messenger RNA. The triplet sequence in messenger RNA is said to be a **codon** for a particular amino acid, and the matching sequence in transfer RNA is said to be an **anticodon.**

Table 5–1 lists the meaning of the 64 possible combinations of three successive bases in both mRNA and the DNA strand from which the RNA is transcribed. The relationship of the antiparallel complements during transcription in the nucleus and during translation in the cytosol is shown in Figure 5–2.

FIGURE 5–1 *Continued*

Step 5. **When the proper tRNA is in place, the polypeptide is transferred onto the amino group of the new residue brought in by tRNA, so that the chain is now one residue longer.**

Step 6. **When the transfer of the previous step is completed, the previously bound tRNA no longer carries a polypeptide and is free to dissociate from the ribosome, returning to the mixed pool of tRNA in the soluble cytoplasm, where it is available for transport of another molecule of its specific amino acid. The ribosome now moves along the mRNA molecule to the position where the placement of the next amino acid will be directed.**

Step 7. **Steps 4, 5, and 6 are repeated. As each amino acid residue adds to the polypeptide, the ribosome moves down the mRNA molecule. When a ribosome reaches a specific terminator sequence in mRNA, the now complete polypeptide is detached into the soluble cytoplasm. The ribosome itself can then move free of the mRNA and be available for attachment to the beginning of the message in yet another molecule of mRNA (not shown).**

TABLE 5–1 THE GENETIC CODE

The triplets in messenger RNA (codons) are listed in black ink. Beneath them in brown ink are the corresponding triplets in the DNA strand from which messenger RNA is transcribed. The messenger RNA and DNA triplets are antiparallel complements.

Listed by Triplet			
A-A-A –Lys d(T-T-T)	C-A-A –Gln d(T-T-G)	G-A-A –Glu d(T-T-C)	U-A-A –Term* d(T-T-A)
A-A-G –Lys d(C-T-T)	C-A-G –Gln d(C-T-G)	G-A-G –Glu d(C-T-C)	U-A-G –Term* d(C-T-A)
A-A-C –Asn d(G-T-T)	C-A-C –His d(G-T-G)	G-A-C –Asp d(G-T-C)	U-A-C –Tyr d(G-T-A)
A-A-U –Asn d(A-T-T)	C-A-U –His d(A-T-G)	G-A-U –Asp d(A-T-C)	U-A-U –Tyr d(A-T-A)
A-C-A –Thr d(T-G-T)	C-C-A –Pro d(T-G-G)	G-C-A –Ala d(T-G-C)	U-C-A –Ser d(T-G-A)
A-C-G –Thr d(C-G-T)	C-C-G –Pro d(C-G-G)	G-C-G –Ala d(C-G-C)	U-C-G –Ser d(C-G-A)
A-C-C –Thr d(G-G-T)	C-C-C –Pro d(G-G-G)	G-C-C –Ala d(G-G-C)	U-C-C –Ser d(G-G-A)
A-C-U –Thr d(A-G-T)	C-C-U –Pro d(A-G-G)	G-C-U –Ala d(A-G-C)	U-C-U –Ser d(A-G-A)
A-G-A –Arg d(T-C-T)	C-G-A –Arg d(T-C-G)	G-G-A –Gly d(T-C-C)	U-G-A –Term* d(T-C-A)
A-G-G –Arg d(C-C-T)	C-G-G –Arg d(C-C-G)	G-G-G –Gly d(C-C-C)	U-G-G –Trp d(C-C-A)
A-G-C –Ser d(G-C-T)	C-G-C –Arg d(G-C-G)	G-G-C –Gly d(G-C-C)	U-G-C –Cys d(G-C-A)
A-G-U –Ser d(A-C-T)	C-G-U –Arg d(A-C-G)	G-G-U –Gly d(A-C-C)	U-G-U –Cys d(A-C-A)
A-U-A –Ile d(T-A-T)	C-U-A –Leu d(T-A-G)	G-U-A –Val d(T-A-C)	U-U-A –Leu d(T-A-A)
A-U-G –Met d(C-A-T)	C-U-G –Leu d(C-A-G)	G-U-G –Val d(C-A-C)	U-U-G –Leu d(C-A-A)
A-U-C –Ile d(G-A-T)	C-U-C –Leu d(G-A-G)	G-U-C –Val d(G-A-C)	U-U-C –Phe d(G-A-A)
A-U-U –Ile d(A-A-T)	C-U-U –Leu d(A-A-G)	G-U-U –Val d(A-A-C)	U-U-U –Phe d(A-A-A)

**Term* are terminator codons.

Note that amino acids within a block of codons, or in the corresponding position in the same row or column of blocks, are interchangeable by substitution of one nucleotide.

Table continued on opposite page

TABLE 5–1

THE GENETIC CODE *Continued*

Listed by Amino Acid

Amino Acid	Codon	DNA
Ala–	G-C-A	d(T-G-C)
	G-C-G	d(C-G-C)
	G-C-C	d(G-G-C)
	G-C-U	d(A-G-C)
Arg–	A-G-A	d(T-C-T)
	A-G-G	d(C-C-T)
	C-G-A	d(T-C-G)
	C-G-G	d(C-C-G)
	C-G-C	d(G-C-G)
	C-G-U	d(A-C-G)
Asn–	A-A-C	d(G-T-T)
	A-A-U	d(A-T-T)
Asp–	G-A-C	d(G-T-C)
	G-A-U	d(A-T-C)
Cys–	U-G-C	d(G-C-A)
	U-G-U	d(A-C-A)
Gly–	G-G-A	d(T-C-C)
	G-G-G	d(C-C-C)
	G-G-C	d(G-C-C)
	G-G-U	d(A-C-C)
Gln–	C-A-A	d(T-T-G)
	C-A-G	d(C-T-G)
Glu–	G-A-A	d(T-T-C)
	G-A-G	d(C-T-C)
His–	C-A-C	d(G-T-G)
	C-A-U	d(A-T-G)
Ile –	A-U-C	d(G-A-T)
	A-U-U	d(A-A-T)
	A-U-A	d(T-A-T)
Leu–	C-U-A	d(T-A-G)
	C-U-G	d(C-A-G)
	C-U-C	d(G-A-G)
	C-U-U	d(A-A-G)
	U-U-A	d(T-A-A)
	U-U-G	d(C-A-A)
Lys–	A-A-A	d(T-T-T)
	A-A-G	d(C-T-T)
Met–	A-U-G	d(C-A-T)
Phe–	U-U-C	d(G-A-A)
	U-U-U	d(A-A-A)
Pro –	C-C-A	d(T-G-G)
	C-C-G	d(C-G-G)
	C-C-C	d(G-G-G)
	C-C-U	d(A-G-G)
Ser –	A-G-C	d(G-C-T)
	A-G-U	d(A-C-T)
	U-C-A	d(T-G-A)
	U-C-G	d(C-G-A)
	U-C-C	d(G-G-A)
	U-C-U	d(A-G-A)
Thr –	A-C-A	d(T-G-T)
	A-C-G	d(C-G-T)
	A-C-C	d(G-G-T)
	A-C-U	d(A-G-T)
Trp –	U-G-G	d(C-C-A)
Tyr –	U-A-C	d(G-T-A)
	U-A-U	d(A-T-A)
Val –	G-U-A	d(T-A-C)
	G-U-G	d(C-A-C)
	G-U-C	d(G-A-C)
	G-U-U	d(A-A-C)
Term*–	U-A-A	d(T-T-A)
	U-A-G	d(C-T-A)
	U-G-A	d(T-C-A)

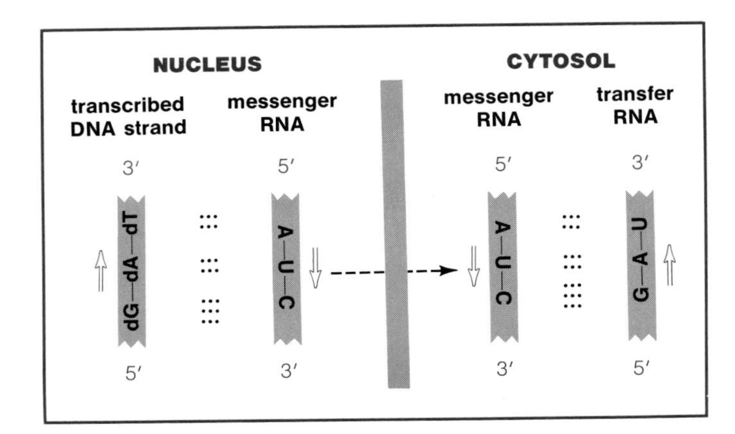

FIGURE 5–2

Linkage of complementary bases in antiparallel sequences during information transfers. Open arrows indicate direction of reading in formal nomenclature.

The Genetic Code is Nearly Universal. The same sequences of three nucleotides in messenger RNA are translated into the same amino acid sequences throughout all organisms. The known exceptions occur in the mitochondria of eukaryotes. These organelles have unusual kinds of transfer RNA, unusual in that they match some amino acids with codons that are different from those used everywhere else. The known exceptions are:

AGA—Term	AUA—Met_f
AGG—Term	UGA—Trp

The other codons are the same as those in the "universal" list. The apparatus of protein synthesis in mitochondria is unusual in other respects to be discussed later. It is still remarkable that simple bacteria and the nuclei of the most complex plants and animals speak the same genetic language.

The Genetic Code and Amino Acid Structure

Nucleotide triplets and amino acids are not randomly matched. Close study of the table of triplets will show that the code is constructed so that substitution of one base for another frequently creates a triplet for the same, or a structurally related, amino acid. This is not only esthetically pleasing, it is also a coldly practical arrangement. Accidental mismatching or chemical alteration of a nucleic acid can cause substitutions of one base for another in the nucleic acids, as we shall discuss in detail in Chapter 7. In some cases, the substitution causes no change in amino acid sequence. For example, introduction of an alanyl group into a protein is coded by d(T-G-C) in DNA. The triplets made by changing the first nucleotide to dC, dG, or dA still code for an alanyl group.

Even in those cases in which a base substitution creates a triplet designating another amino acid, the effects are often minimized by the similarity of the new amino acid. To continue our example, the table shows that a single base substitution in any Ala triplet will result in the incorporation of Ala itself, or Thr, Pro, Ser, Gly, Asp, Glu, or Val. Many of these groups are like the alanyl group in function; they are likely to occur in chain reversals, and they may match the function of an alanyl group closely enough to permit adequate use of the accidentally modified protein.

This conservation of function during accidental alteration of base sequence is especially clear in the case of the amino acids bearing large hydrophobic side chains, which are so important in developing interior geometries in a protein molecule. The triplets for Ile, Leu, Met, Phe, and Val differ by only one base, and at least two thirds of the accidental base substitutions that may occur in coding for these groups will still result in incorporation of one of these amino acid residues at the same location.

Similar relationships can be seen for the polar amino acid residues. The pairs that can be interchanged by single base substitutions include Asn/Asp, Asp/Glu, Gln/Glu, Arg/Lys, Arg/His. In short, the genetic code evolved so as to diminish any adverse consequences of accidental base substitution.

FORMATION OF RNA BY TRANSCRIPTION OF DNA

The Chemical Reactions

RNA is constructed from nucleoside 5′-triphosphates, in which three residues of phosphoric acid are linked by anhydride bonds (Fig. 5–3). These linkages between phosphate groups are pyrophosphate bonds; they are related to inorganic pyrophosphates (PP_i).

Polynucleotides are built by addition of one nucleotide residue at a time, not by a concerted reaction. The 3′-hydroxyl group at the end of the growing polynucleotide chain cleaves through nucleophilic attack the α,β-pyrophosphate bond of the next nucleotide to be added (Fig. 5–4). This extends the polynucleotide chain and releases inorganic pyrophosphate. The 3′-hydroxyl group of the added nucleotide is free and ready for attack on the next precursor nucleoside triphosphate. The growing RNA chain is therefore built from the 5′ terminal toward the 3′ terminal, in the same direction that it is named.

The Information Transfer

The straightforward polymerization of nucleoside triphosphates does not explain the critical feature of RNA synthesis—the perfect synthesis of a complement to part of a DNA chain, beginning and ending at precisely defined points. Much is yet

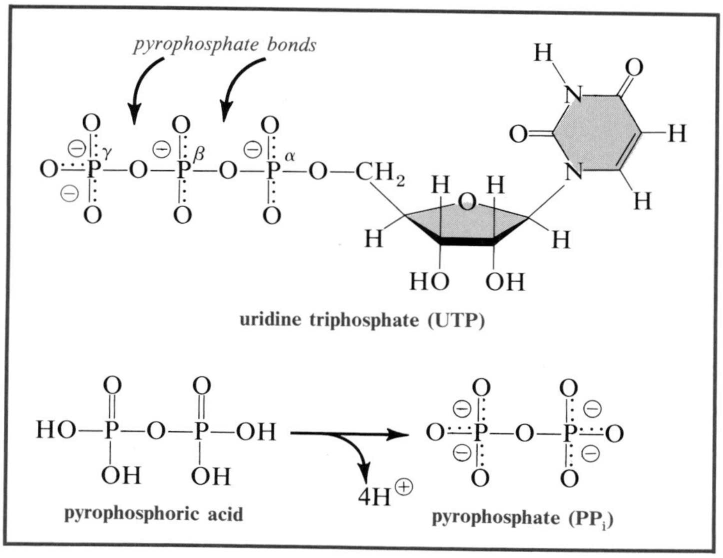

FIGURE 5–3

Nucleoside triphosphates contain three molecules of phosphoric acid linked by anhydride bonds, in addition to ribose and a base, which is uracil in the example given. The phosphoric anhydride bonds are referred to as pyrophosphate bonds because the same structure occurs in inorganic pyrophosphate, the ionized form of pyrophosphoric acid.

pyrophosphate bonds

uridine triphosphate (UTP)

pyrophosphoric acid → $4H^{\oplus}$ → **pyrophosphate (PP_i)**

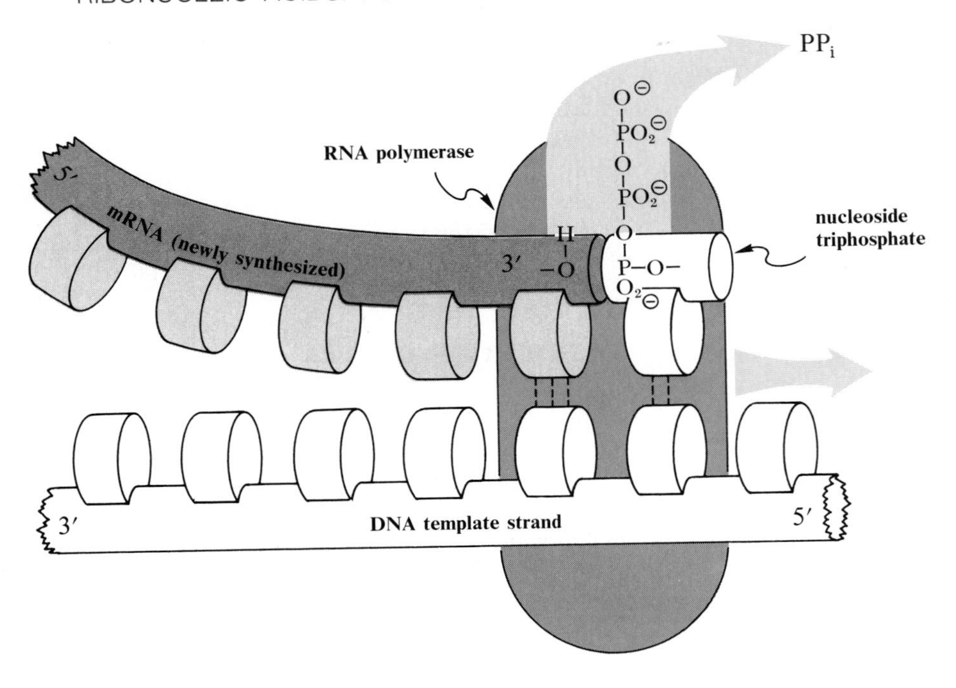

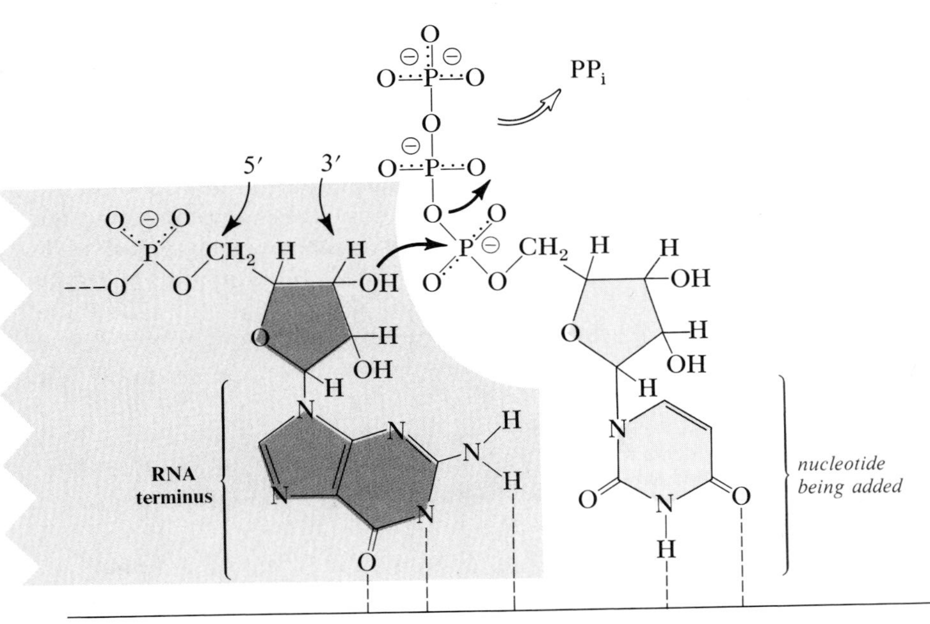

FIGURE 5–4 *Top.* Ribonucleic acids are created by successive addition of nucleoside phosphate groups to the 3′ end of a polynucleotide chain. The particular group to be added is determined by a DNA template in the presence of RNA polymerase, which strengthens hydrogen bonding so that only a ribonucleotide with a base that is the antiparallel complement of the corresponding base on DNA will be bound. The RNA is built from 5′ end to 3′ end, while the DNA template is read from 3′ end to 5′ end. *Bottom.* The precursors used to lengthen the RNA chain are ribonucleoside triphosphates, which lose a pyrophosphate group upon addition to the terminal 3′-hydroxyl group of the growing RNA chain.

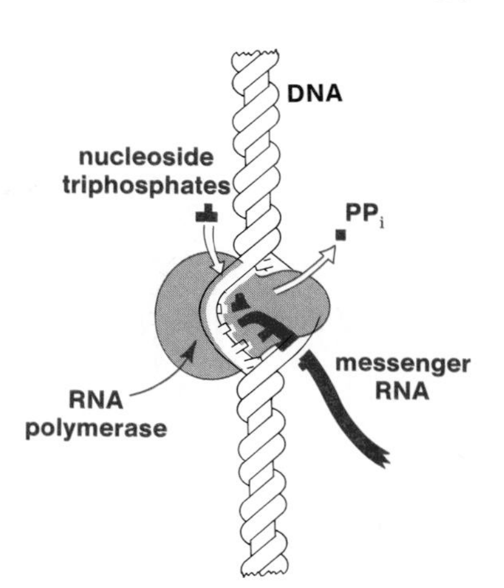

FIGURE 5-5

In order to expose the bases in DNA for use as a template in making RNA, it is necessary to uncoil the DNA double helix as RNA polymerase moves along it. Approximately six bases are exposed at a time, and only one of the DNA strands is used as a template.

to be learned about the mechanism for this process in which mistakes are rare, but we have some sound ideas. There are essentially two separate questions: How is an error-free complement synthesized? How is the transcription begun and ended at precise points?

Perfect transcription hinges upon the presence of specific proteins to catalyze the process, **DNA-directed RNA polymerases.** We won't discuss catalysis by proteins until later, but we can develop some properties of the polymerase at this time.

High accuracy requires "correct" precursors to be bound much more strongly to the DNA template that is being transcribed than are "incorrect" precursors—so strongly that the probability of the incorrect compound displacing the correct compound is negligible. The nucleotide added to RNA at each step is "correct" when its base is the complement of the corresponding base in DNA. In other words, accurate transcription involves very strong bonding of nucleotides carrying complementary bases, and yet we know that the hydrogen bonding between free nucleotides in aqueous solution is relatively weak. The presence of the polymerase must increase the affinity in some way.

How can a protein promote bonding between complementary nucleotides? A rational hypothesis is that it has some nonpolar surfaces that snugly fit complementary base pairs, protecting the hydrogen bonds from competing interactions with water and providing additional forces to stabilize the pairing. Mismatched bases would not fit the protein and would be kept apart by water.

Before any of this can happen, the double helix of DNA must be partially unwound so as to separate the two strands and expose the base sequence that is to be transcribed. This is accomplished by an association of protein with separated DNA strands, which must be stronger than the association of the strands with each other. Present evidence is that approximately 6 base pairs in the DNA double helix are separated in this way, with the polymerase advancing one base at a time as the RNA chain is built (Fig. 5-5). Since the binding of RNA precursor and DNA template is antiparallel, it follows that the RNA polymerase "reads" the template from the 3′ to the 5′ end while building the RNA from 5′ to 3′ end.

Nomenclature of Gene Sequences

Since a DNA strand is read "backward" from 3′ to 5′ during transcription, there is a problem in describing the nucleotide sequences of a gene in DNA by the formal

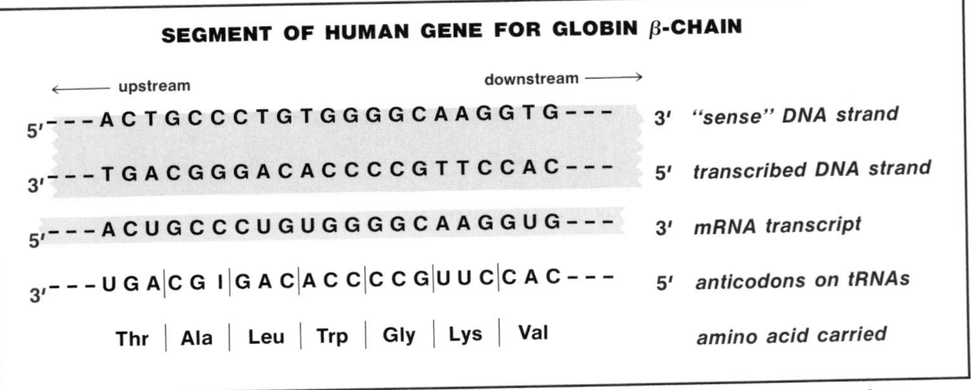

FIGURE 5–6 The nomenclature of gene sequences is based upon the sequence in the transcribed messenger RNA. The strand of DNA having the same sequence except for thymine in place of uracil is called the "sense" strand, and this is the sequence customarily specified even though it is not transcribed to form RNA. The base sequences of the tRNA anticodons corresponding to the mRNA triplets are similar to the corresponding sequences in the transcribed DNA strand. The sequence listed is a segment from the human gene for the hemoglobin β-chain. Sequences farther toward the 5′ end of the sense strand are said to be upstream, those farther toward the 3′ end of that strand are downstream.

rules of nucleotide nomenclature; it is solved by giving the sequence for the other DNA strand, which matches the sequence for messenger RNA except for replacement of U by T (Fig. 5–6). This DNA strand is variously referred to as the **"sense"** or **"mRNA"** strand of a gene, recognizing its greater readability by humans. (Note that triplets for an amino acid in the sense strand are identical to those given for mRNA in Table 5–1 except for the presence of T in place of U.) Similarly, portions of DNA farther toward the 5′ end on the sense strand (toward the 3′ end of the transcribed strand) are said to be **upstream** from a gene, whereas those in the opposite direction are **downstream.**

Multiplicity of RNA Polymerases. Mammalian cells, like other eukaryotic cells, are known to have at least three distinct RNA polymerases in their nuclei. Type I RNA polymerase is associated with the nucleolus and generates large precursors of ribosomal RNA from specific segments of DNA. The types II and III RNA polymerases occur in the nucleoplasm and make precursors of messenger RNA and transfer RNA, respectively. (The type III enzyme also makes a small ribosomal RNA segment and some small nuclear RNAs of unknown function.)

The RNA polymerases can be differentiated experimentally by the action of α-amanitin, a compound found in the mushroom *Amanita phalloides:*

RNA Polymerases	Transcribed Precursor	Effect of Amanitin
I	ribosomal RNA	small
II	messenger RNA, some small RNAs	inhibited by 10^{-9} to 10^{-8} M
III	transfer RNA, other small RNAs	inhibited by 10^{-5} to 10^{-4} M

We see that amanitin has little effect on the RNA polymerase in the nucleolus, which synthesizes the large precursor of ribosomal RNA, but only small concentrations are necessary to stop the action of the polymerase that synthesizes precursors of messenger RNA. Intermediate concentrations block the formation of transfer RNA and other small RNAs.

Amanitin is composed of eight amino acids joined in a ring. (Some of the amino acids are modified from the usual forms.) Its effect has practical consequences because the unwary who eat the Amanita mushroom are preventing the formation of messenger RNA, and thus the formation of proteins, in their cells. Unfortunately, they don't know all is not well until the depletion of the supply of proteins is great enough to cause acute gastrointestinal distress and other indications of damage; by that time the amanitin is thoroughly absorbed. A temporary improvement after the first day sometimes conceals irreversible and lethal loss of tissue function.

AMANITIN POISONING

Produced by *Amanita phalloides*.

Cyclic octapeptide. (8 amino acids joined in ring.)

One 50-gram mushroom has about 7 mg of α-amanitin.

Human oral LD_{50} (50 per cent die after ingestion): about 0.1 mg/kg body weight.

Mechanism of action: one molecule bound by ternary complex of RNA polymerase, messenger RNA, and transfer RNA, preventing further elongation of messenger RNA. Keq $= 10^{-9}$ M.

Characteristic feature is 6- to 15-hour period between ingestion and symptoms; consider α-amanitin poisoning probable when latent period is six or more hours after eating mushrooms.

Acute gastrointestinal symptoms (pain, nausea, vomiting, watery diarrhea) improve rapidly.

Truly serious consequences, deterioration of liver and kidney function, appear three to five days later.

Signals in DNA

The only information in DNA is the sequence of nucleotides. Correct usage of this information requires many signals, all known to the RNA polymerases but only partially known to us.

In this seemingly unpunctuated sequence of bases, any three bases in a row can represent an amino acid, or a polypeptide chain termination signal, or it may be part of the information for a structural RNA molecule. The individual RNA polymerases must recognize sequences that put them to work reading specific parts of a DNA molecule. A one-key shift in the position of the fingers on a typewriter (Jstf eptl, su ½su/) is unreadable, and transcription must begin and end at exactly correct positions for the same reason.

The polymerases only transcribe one strand of DNA in any region. We cannot designate a particular strand as the "transcribed" strand and its complementary strand as the "sense" strand throughout their lengths. Some messages for transcription are on one strand of the duplex molecule and some on the other, which must be read in the opposite direction. Obviously, there is something recognizably peculiar about the base sequence that may cause an RNA polymerase to be oriented in opposite directions when transcribing different parts of the same DNA duplex.

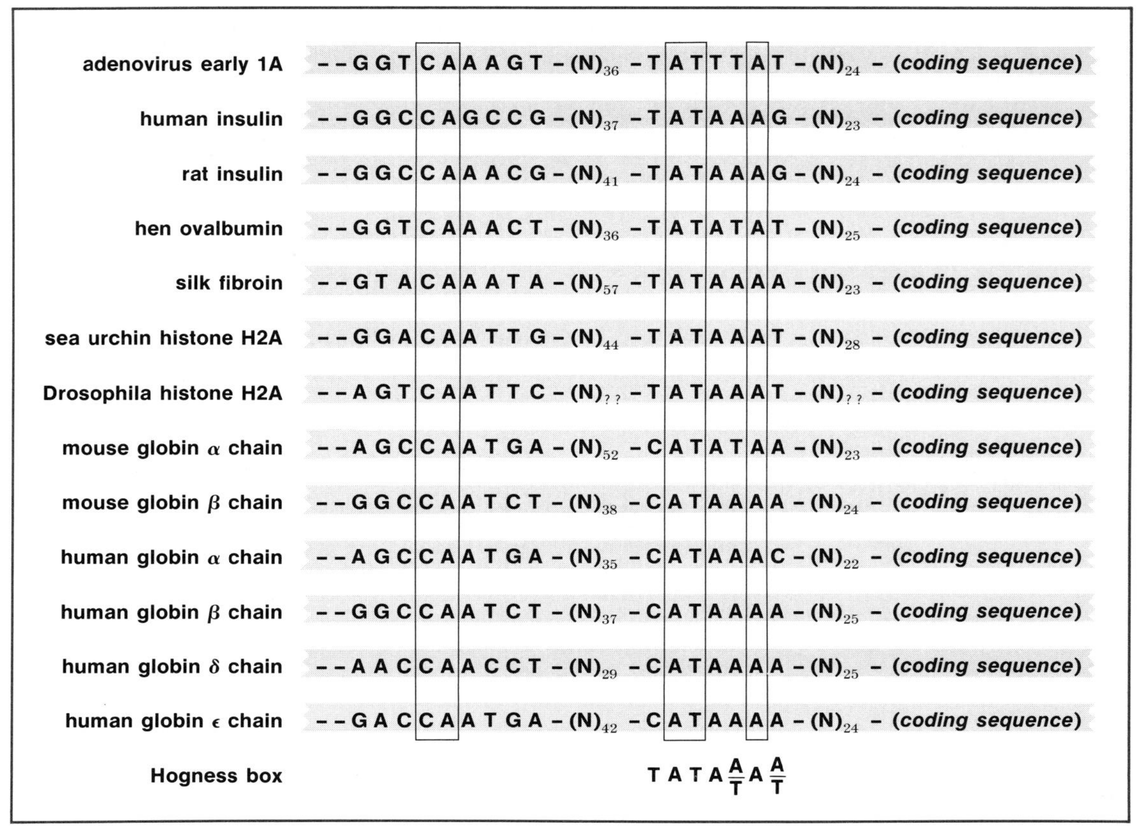

FIGURE 5–7 Upstream base sequences in various eukaryotic genes. The ATAA sequence occurring in many, but incomplete in others, is sometimes called the Hogness box.

Some signals may be specific base sequences. For example, a wide variety of eukaryotic genes are now known to have a pair of signal sequences spaced some distance apart upstream from the point at which transcription begins (Fig. 5–7). Experiments show that these regions facilitate accurate transcription but are not absolutely required for it. Other regions of undefined character still farther upstream appear to be necessary. Whatever the signal sequences may be, they preclude the use of the same coding sequences to represent amino acids in proteins at the same intervals. Otherwise, the polymerase would sometimes bind in the middle of a structural gene.

Some signals may be palindromes. Palindromes are segments of double helix that look identical when viewed from either direction.* They are also known as **inverse repeat segments.** Each strand of a palindromic sequence is complementary to itself as well as to the other strand of the double helix (Fig. 5–8). This makes it theoretically possible for each strand to double back upon itself where duplex DNA has been locally separated. This would form protruding whiskers on the monotonous

---

*Palindrome is borrowed from the name for words or groups of words that read the same in either direction, for example, "level," or "Able was I ere I saw Elba." A spectacular, but less pure, example that neglects spaces and punctuation has been devised by Alastair Reid: "T. Eliot, top bard, notes putrid tang emanating, is sad. I'd assign it a name: gnat dirt upset on drab pot toilet." (Quoted by Brendan Gill in *Here at the New Yorker,* Random House [1975].)

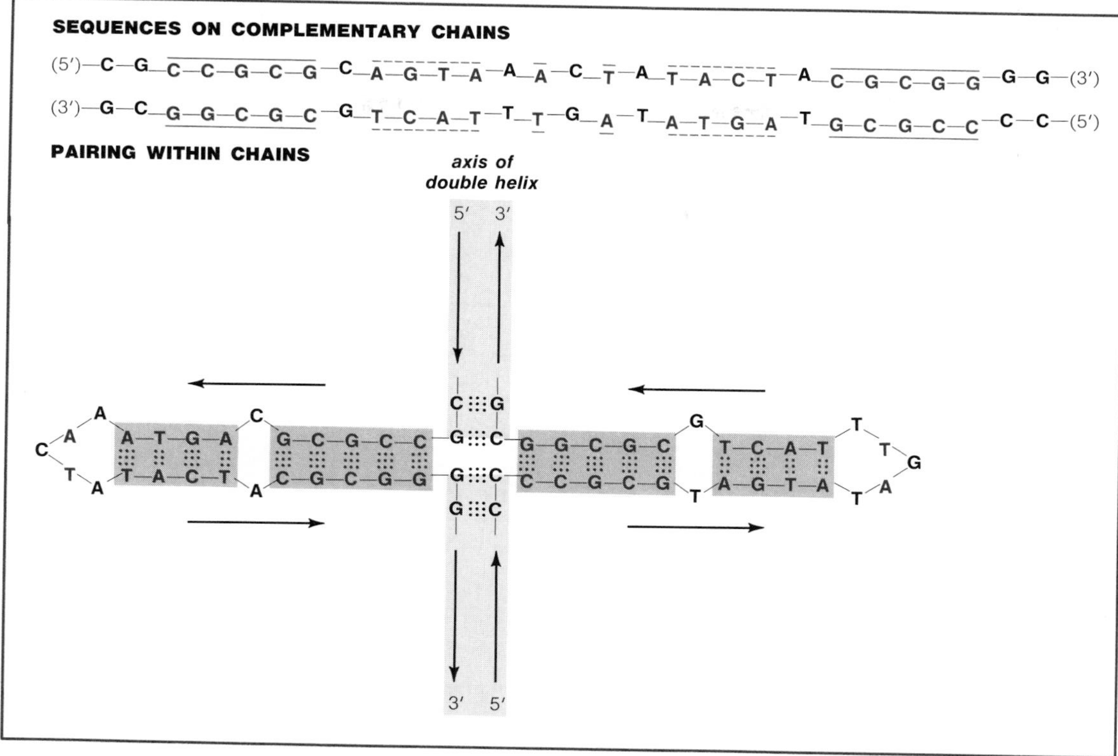

FIGURE 5–8 *Top.* Palindromes in duplex nucleic acid chains are segments with the same sequence of nucleotides when read from opposite directions. Two sets of palindromes are illustrated, under dashed and solid lines, respectively. For example, the sequence C-C-G-C-G, read from the 5′ end, occurs to the left in the top chain, and the same sequence occurs to the right in the bottom chain, read from its 5′ end. In each case, the other chain has the antiparallel complement. Therefore, these two pieces of double helix appear identical from the corresponding 5′ ends. *Bottom.* Palindromes may theoretically form a double helix in two ways. Each strand may bind to the other strand as is usually seen in DNA, or the antiparallel complements within each strand may combine with themselves. Should this happen, the result would be two segments of double helix, each formed by one strand, that protrudes from the long two-strand double helix. Some matching bases would not be able to bond effectively owing to steric considerations, for example, the A-T pairs in the end loops. It is not clear if many of these whiskers protrude from undisturbed DNA, but they may form when the chains unfold during replication, and they can also form in the single chain RNA molecules transcribed from DNA. (See Figure 5–10 for an example.)

long double helix. The whiskers would in themselves contain short pieces of double helix, each made from a single strand folded on itself.

The Operon Concept. The nature of the identifying signals for the action of RNA polymerase is better defined in bacteria. The bacterial RNA polymerase behaves as if the continuous strand of DNA being transcribed were divided into units known as operons. An operon frequently includes coding for several polypeptides; that is, it contains more than one gene. The initiation site for transcription—the **promoter**—is defined by a particular sequence of bases (TAT(A,G)ATA), or a variant, occurring upstream, and known as a **Pribnow box.** In addition, operons often have sites for binding other proteins to regulate the rate of initiation of transcription of new protein chains (Fig. 5-9). In at least some cases, these **initiator** and **operator** sites contain palindromes as signals for binding.

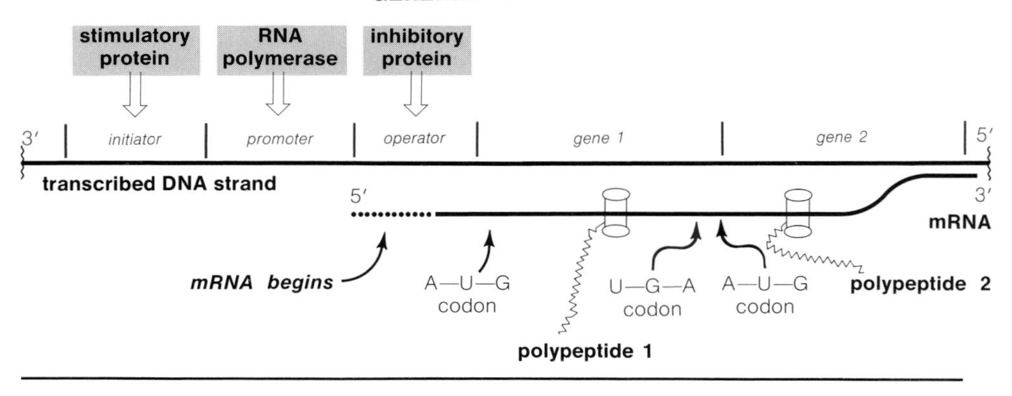

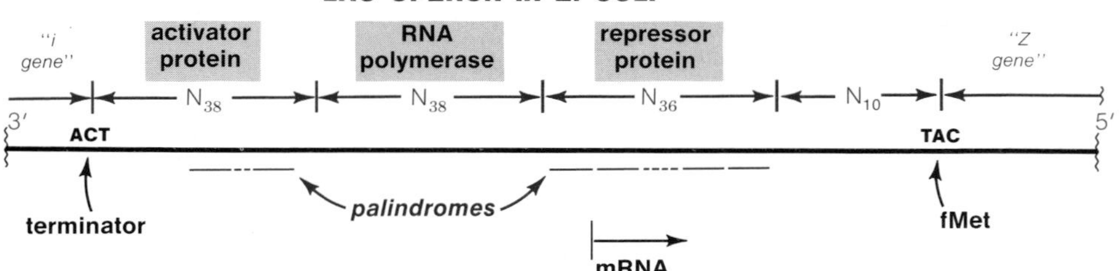

FIGURE 5–9 *Top.* Generalized structure of an operon in a bacterial DNA strand. All operons include a site for attachment of RNA polymerase (the promoter site) on the DNA strand followed toward the 5′ end by sequences that are transcribed into the form of RNA. In the example shown, these sequences include two successive genes. Both genes are transcribed as a single segment of messenger RNA, which has two initiation sites and two termination sites for making two different polypeptide chains, one for each gene.

Operons may also include initiator sites, usually toward the 5′ end from the promoter, at which signaling proteins are bound to stimulate the beginning of transcription by RNA polymerase. Operons frequently have operator sites, usually after the promoter, at which signalling proteins are bound to prevent transcription by RNA polymerase. These various binding sites sometimes overlap. Part of the operator or promoter sites may be transcribed in the initial segment of RNA that precedes polypeptide coding in messenger RNA or the structural sequences in transfer RNA and ribosomal RNA.

Bottom. An actual operon sequence that controls lactose metabolism by the bacterium *Escherichia coli.* (Lactose is the sugar of milk.) The terminator triplet of the preceding gene is followed by a sequence of 38 nucleotide residues at which an activator protein is bound. Another 38 residues included in the RNA polymerase binding site are followed by 36 residues in an operator at which a repressor protein is bound. The repressor protein is removed when the bacterium encounters lactose or related sugars, thereby permitting the polymerase to begin transcription of a messenger RNA that includes instructions for making enzymes necessary for lactose metabolism. Both the activator and repressor regions include palindromic sequences that are likely signals for binding of the specific proteins. Messenger RNA transcription begins well within one of these sets of sequences that precede the initiating A-U-G triplet.

STRUCTURE AND FUNCTION IN RNA

Internal Folding of RNA

RNA molecules can have a secondary structure determined by nucleotide sequence, analogous to the secondary structures of polypeptides determined by amino acid sequence. The secondary structure in this case is a double helix made by antiparallel bonding of complementary base pairs, just as it is in DNA.

However, RNA is not made as separate antiparallel complementary chains, except in certain viruses. Single strands of RNA can form a double helix only when different parts of the chain have antiparallel complementary segments that can bend back upon themselves (Fig. 5–10). Because of the additional hydroxyl groups in the RNA backbone, these are segments of a more open-cored helix, with 11 or 12 base pairs per turn instead of the 10 seen in DNA.*

Furthermore, it is possible to generate tertiary structures in RNA. Bases can be held in proximity by forming still different hydrogen bonds than those occurring in the double helix. The forces creating the tertiary structure of nucleic acids are much like those creating the tertiary structure of proteins.

Here we have an important point: **Part of the base sequence in DNA molecules is present to create specific structures in RNA molecules by direct transcription.** Some of the short RNA molecules twist into compact shapes that are recognized as carriers for amino acids—these are the transfer RNA molecules. Some of the larger RNA molecules twist upon themselves to generate surfaces that associate with particular proteins to form the ribosomes upon which proteins are made.

*The double helixes of polynucleotides with 10, 11, and 12 pairs per turn are known respectively as B, A, and A′ helixes. In addition to the longer pitch, the surface of the A helixes seen with RNA is distinguished by the shallowness of the narrow groove, which is almost absent, and the conversion of the wide groove into a relatively deep cleft.

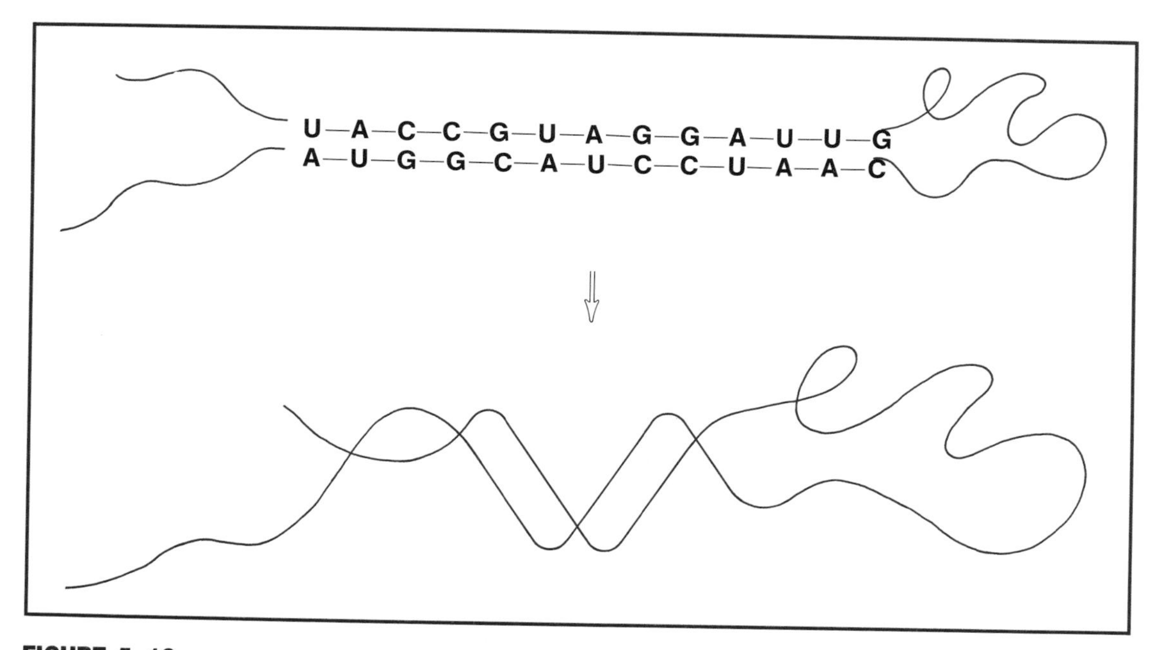

FIGURE 5–10 Origin of secondary structures in ribonucleic acids. Antiparallel complementary sequences within a single chain can cause it to fold back upon itself to create a segment of double helix.

However, part of the nucleotide sequence in DNA contains the information for amino acid sequence in proteins, and the resultant sequence in messenger RNA molecules does not always have bases in the right places for creating compact secondary structures. Molecules of messenger RNA therefore are usually in a less ordered configuration, although they contain some structured segments.

Processing of Transcripts. Transcription rarely makes RNA molecules in their finished forms. All except some very small molecules undergo further processing to create physiologically functional products. Four types of reactions are involved:

1. **Partial hydrolysis** to remove portions of the head or tail of a molecule, or to cleave it into pieces;

2. **Splicing** of parts of one molecule into a new molecule;

3. **Covalent modification** of some bases or sugar residues so as to change the spatial geometry or bonding characteristics;

4. **Extension** of either end of the chain by addition of more nucleotide residues.

Messenger RNA (mRNA)

The precursors of messenger RNA first appear as **heterogeneous nuclear RNA (hnRNA).** Since the genes for many different proteins are being transcribed at any one time, it is reasonable to expect the products of transcription to be a heterogeneous mixture. However, only 10 per cent of the mass of hnRNA survives to be transported into the cytosol as mRNA. Some of the missing material can be accounted for as spacers within genes, and some may be mRNA precursors that are discarded as part of the regulation of gene expression, but we really don't know the purpose of most of it.

Genes in Pieces. Typical precursors of messenger RNA will have the coding sequence broken into pieces. Let us use the mRNA for the beta chain of human hemoglobin as an example:

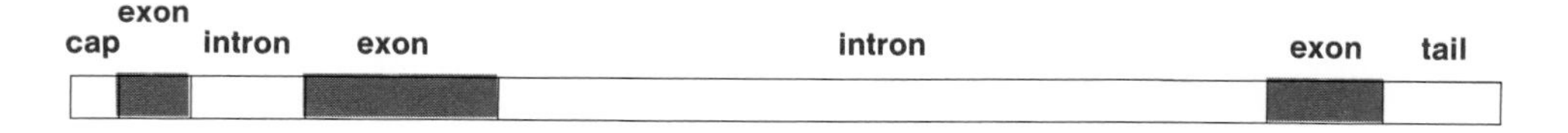

The amino acid sequence of the beta chain is coded in three separate exons, preceded by a leader and followed by a tail. Since the beta chain has 146 amino acid residues, the exons contain $3 \times 146 = 438$ nucleotide residues. However, the transcribed precursor has 1606 nucleotide residues; 53 in the leader, 135 in the 3′ tail, and 980 in the two introns. (The totals for leader and tail include three residues each for initiator and terminator codons discussed later in the chapter.)

Processing of the transcript includes these events:

1. A poly-A terminal with as many as 200 adenosine phosphate groups is added to the 3′ end of the transcript. A poly(A) polymerase in the nucleus adds one group at a time without requiring DNA as a template. It somehow recognizes the candidate mRNAs in the large pool of hnRNA, perhaps through the sequence -AAUAAA- usually present in the 3′ end of mRNA precursors. The purpose of the poly-A tail is not clear; it may promote transport to the cytoplasm or protect the mRNA from destruction.

2. The 5′ end is capped by a guanosine phosphate group through formation of an anhydride phosphate bond. The 5′ end of the added guanosine is connected to the

5′ end of the RNA chain by a triphosphate bridge. This structure is indicated as $G^{5'}$-ppp . . .

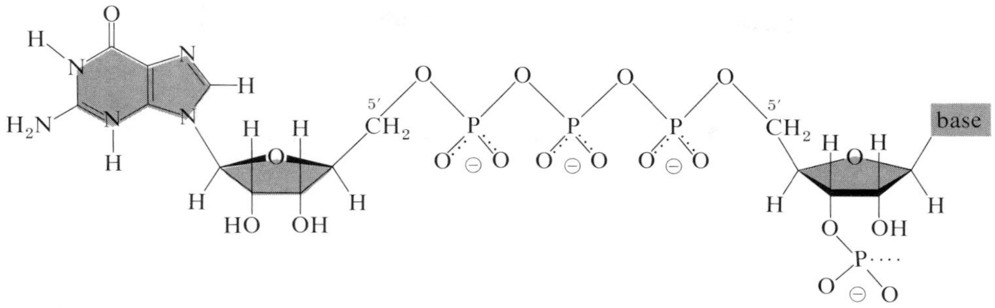

3. The guanosine cap and its adjacent nucleoside group are covalently modified by methylation of both the bases and one of the ribose groups. The completed cap facilitates binding of mRNA by ribosomes.

4. The exons are now spliced into a continuous sequence:

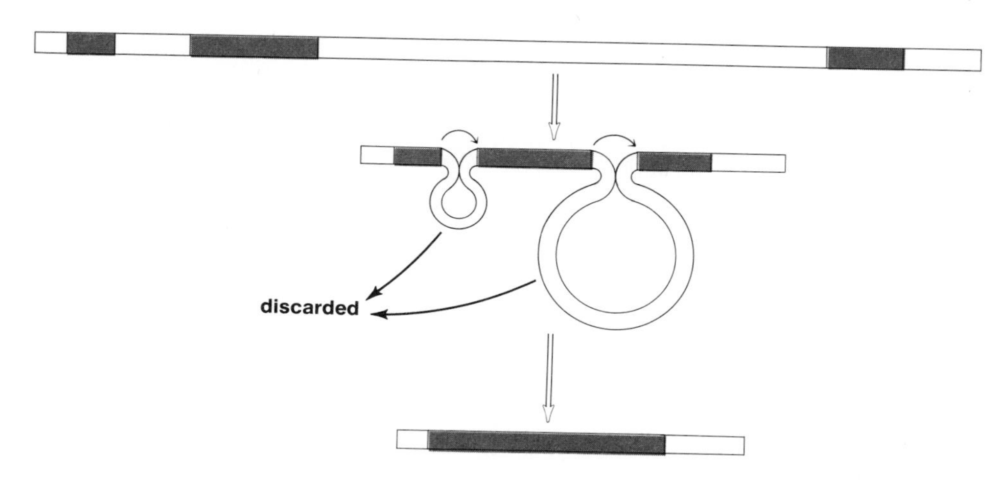

The details of the splicing operation are not known. The introns—the segments to be removed—insofar as is now known always begin with -GU- and end with -AG-. There is obviously more to recognition than that because there are many occurrences of these particular successive bases within the exons themselves. Furthermore, it has been shown with other genes that large parts of the introns can be removed without interfering with proper splicing, but if a complete intron is deleted, a finished mRNA is not formed.

Some molecules of **small nuclear RNA (snRNA)** may have a function in the splicing process, perhaps bridging the gaps created by cleavage of intron-exon junctions through formation of complementary base-pairing bonds. These small molecules (100 to 200 bases) have caps similar to those on messenger RNA and have extensive covalent modifications similar to those in transfer RNA described later. They occur as ribonucleoprotein particles.

The discarded introns are 0.6 of the transcript of the beta chain gene. What is the purpose of this seemingly wasteful procedure? No one knows. Some speculate that genes have been broken into pieces to facilitate evolutionary adaptation; genes for new proteins might be made by reshuffling genes for several old proteins. Others think the discarded introns have information of their own, perhaps to regulate gene expression.

Finally, finished molecules of messenger RNA are transported to the cytoplasm, probably as ribonucleoprotein particles, but the mechanism and regulation of transport is unknown.

Ribosomes

The ribosomes are particles composed of two different subunits, which associate during polypeptide synthesis and dissociate into the separate subunits upon completion of the polypeptide chain. The subunits are made from many different molecules of proteins and four molecules of RNA.

Ribosomal RNA (rRNA)

Formation of Precursors. The four molecules of rRNA are made by transcription of two separate genes. One is quite small (120 bases, 5S) and its transcription begins with the first nucleotide of the finished product. How do we know? Because the 5S piece in completed ribosomes still has a 5′-triphosphate group. The 5S pieces are transcribed in the nucleus by RNA polymerase III, the same enzyme that makes precursors of snRNA and tRNA. The action of this polymerase has an unusual feature: its promoters lie within the genes, not upstream from them.

The other three pieces of rRNA are made as a single 45S transcript (approximately 12,900 bases) by the action of RNA polymerase I within the nucleolus of the cell.

Chromosomes contain many copies of the ribosomal genes so as to enable simultaneous transcription of many identical rRNA precursors. This multiplicity of

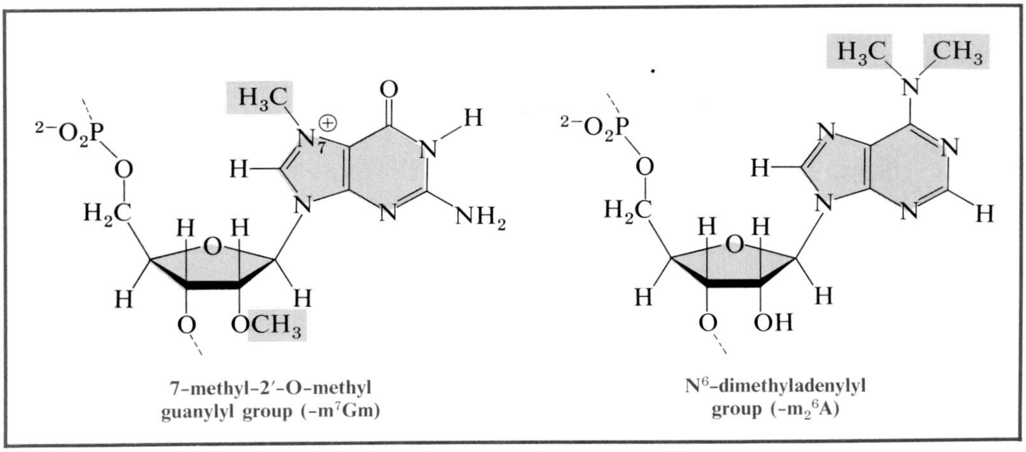

7-methyl-2′-O-methyl guanylyl group (-m^{7}Gm)

N^6-dimethyladenylyl group (-m$_2{}^6$A)

FIGURE 5–11 **Types of methyl derivatives found in ribosomal RNA. Methylation occurs mainly on the ribosyl moiety of nucleotide residues to form 2′-O-methyl derivatives, and to a much lesser extent on the bases, especially the purines. The left formula illustrates a guanine-containing nucleotide modified in both ways, while the right illustrates an adenine-containing nucleotide modified only on the purine by a double methylation. In standard abbreviations, an m before the nucleoside letter indicates a methylated base, with a superscript number to indicate position, and an m after the nucleoside letter indicates methylation of the 2′ oxygen on the ribosyl group. Thus, m$^6_2$A designates an adenosine moiety with two methyl groups on N-6 of the adenine; m^{7}Gm designates a guanosine moiety with a methyl group on position 7 of the guanine ring, and on the 2′ oxygen of its ribosyl group.**

genes is necessary because a typical mammalian cell contains about one million ribosomes and their lifetime is short (approximately 130-hour half-life in rat liver). In order to meet the demand, a cell synthesizes over 10,000 molecules of each of the two rRNA precursors per hour. To facilitate this rapid transcription, DNA in a liver cell contains an estimated 1660 copies of the 5S rRNA gene and 330 copies of the

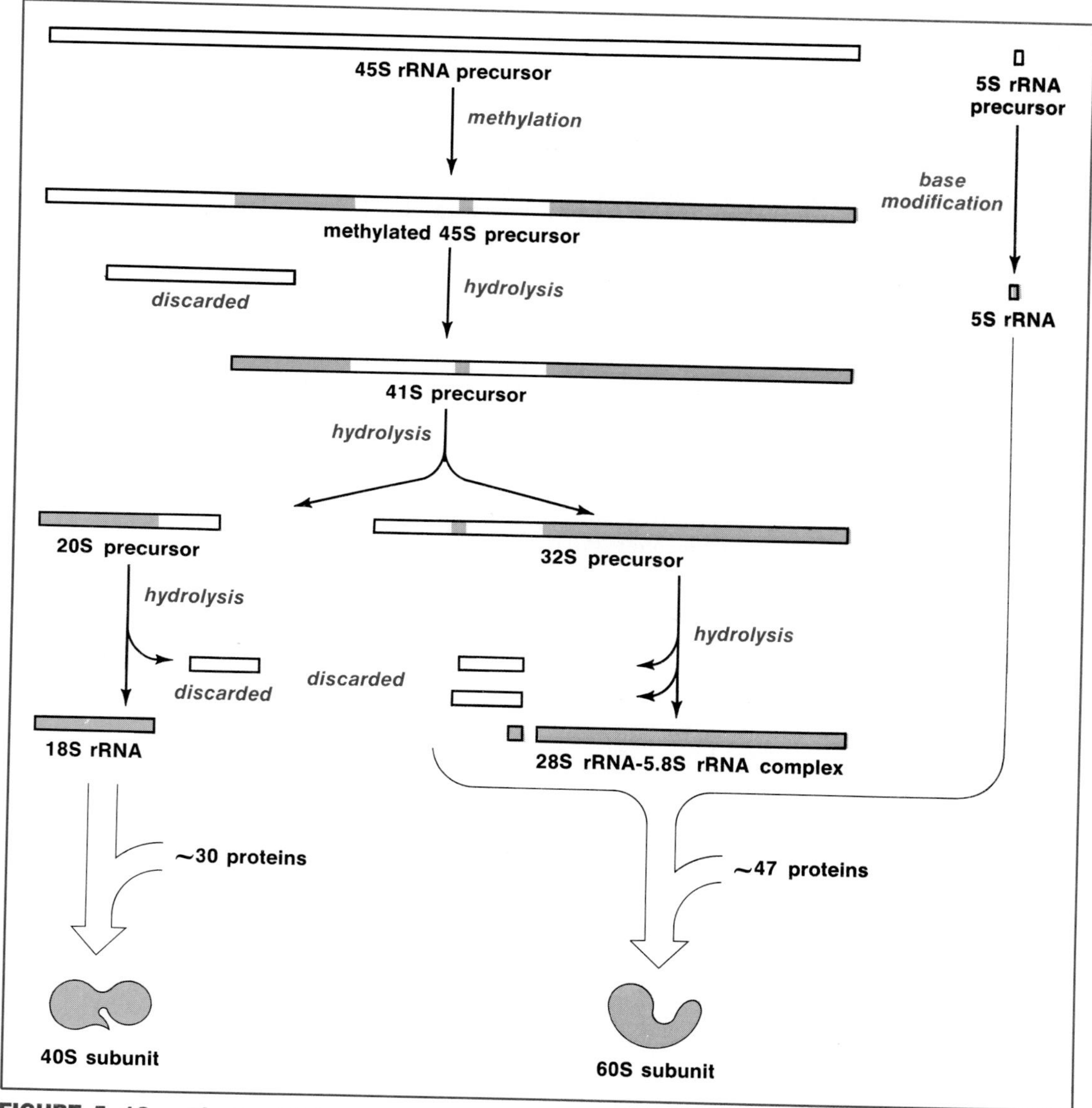

FIGURE 5–12 The four RNA components of ribosomes are specified by two genes. *Right.* A small precursor of 5S rRNA is transcribed from one of the genes by RNA polymerase III in the nucleoplasm and is covalently modified. *Left and center.* The other gene is transcribed in the nucleolus by RNA polymerase I to form a much larger precursor (45S); it includes the sequences of 5.8S, 18S, and 28S rRNAs. These sequences are then methylated. The unmethylated spacer segments are removed by a series of hydrolyses to liberate the finished pieces of ribosomal RNA. Protein molecules combine with the pieces of RNA in an ordered sequence during the various stages of synthesis and processing, finally completing the 40S and 60S ribosomal subunits.

gene for the 45S transcript. These genes occur as **tandem repeats:** identical copies separated by spacers on several chromosomes.

Processing of the 45S rRNA precursor involves covalent modifications, mainly methylations. Of approximately 110 methyl groups added, over 100 appear on the 2′ hydroxyl groups of ribose residues, with the balance mainly going on guanine and adenine groups (Fig. 5–11).

The next steps in processing the 45S ribosomal RNA precursor are a series of hydrolytic cleavages (Fig. 5–12), forming as the final products 5.8S, 18S, and 28S rRNA molecules. All of the methylated bases are in these molecules; the methylations evidently protect them from hydrolysis during processing.

The final rRNA molecules contain only a little over half of the 12,900 nucleotides present in the original transcript: 4850 in the 28S piece, 1740 in the 18S piece, and 160 in the 5.8S piece, as estimated in human HeLa cells.

HeLa cells were originally obtained by culture of a carcinoma of the cervix in a patient named Henrietta Lack; various strains of the progeny are now widely used for experimental purposes.

Each of the finished segments of ribosomal RNA folds back upon itself to form double helixes because it contains many lengths of nucleotides that are complementary in antiparallel order. The result is an elaborate secondary and tertiary structure that is yet to be deciphered for any of the units, no doubt dependent to large extent upon protein-RNA interactions as well as internal bonding. Even the arrangement of double helixes is not established, but one of the proposals for the 5.8S rRNA is shown in Figure 5–13 to illustrate the principle.

The Completed Ribosomal Particles. Some ribosomal proteins or their precursors stick to the 45S RNA chain before it is completed. (Some proteins, like some RNAs, are processed after synthesis, as is discussed in Chapter 8.) As the individual

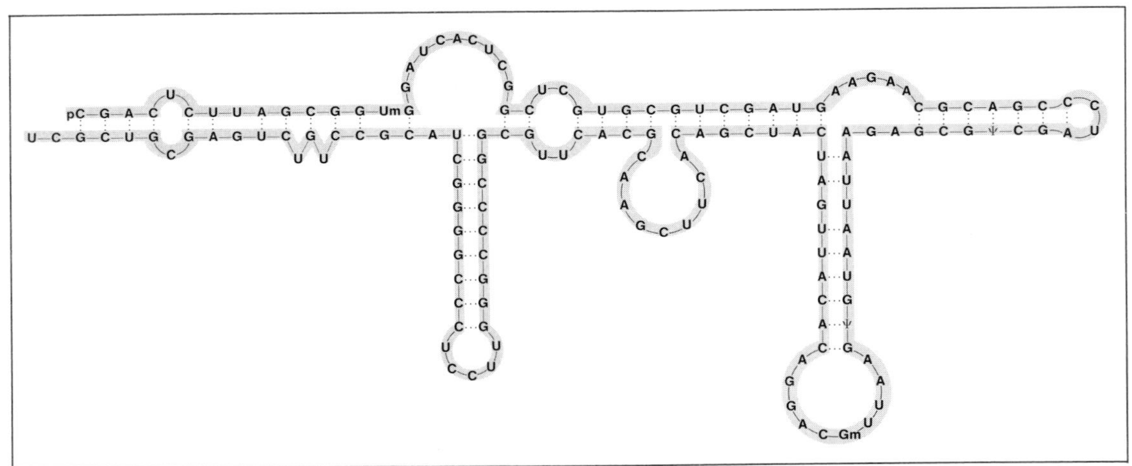

FIGURE 5–13 A proposed secondary structure for 5.8S rRNA. The drawing is modified from Nazar, Sitz, and Busch (1975), J. Biol. Chem., 250:8591. The proposal is based on the primary structure of RNA from Novikoff ascites hepatoma, a malignancy of rat liver that is widely used for experimental purposes. Note that no break in the helical segments is indicated where U-G pairs occur. This is in accord with experimental findings with other kinds of RNA even though U and G are not strongly bonded.

TABLE 5–2

COMPOSITION OF EUKARYOTIC RIBOSOMES

Ribosomal Particle	40S	60S	80S
Mol. Wt. of Particle	1.4×10^6	2.9×10^6	4.3×10^6
Number of RNA Constituents	1 (18S)	3 (5S, 5.8S, 28S)	4
Fraction as RNA	44%	59%	50%
Number of Proteins (approximate)	30	47	77

RNA components appear in their final forms, more of the component protein molecules adhere to the RNA and to each other, making the separate aggregates that constitute the ribosomal subunits.

A complete active ribosome has a sedimentation coefficient of 80S. (The ribosomes of prokaryotes are somewhat smaller, sedimenting at 70S.) The subunits have sedimentation coefficients of 60S and 40S; they spontaneously combine into inactive 74S particles when they are not participating in polypeptide synthesis. (They are lighter than the active 80S particles owing to absence of several protein factors that participate in polypeptide synthesis.) The composition of the active subunits and the complete ribosomes is summarized in Table 5–2.

Ribosomes Contain Pockets. The larger 60S ribosomal particle appears in electron micrographs to have a dished top, with an asymmetrical protuberance holding up the 40S particle when the two are combined to form a complete ribosome (Fig. 5–14). The 40S particle evidently has a pinched waist, which leaves an opening where it is held away from the surface of the 60S particle. This opening is believed to be lined by the sites for binding messenger RNA and the transfer RNA molecules bearing the amino acids that will be combined during translation of the message. The complete ribosome is in effect a sleeve that passes over messenger RNA, transfer RNAs, and the growing peptide chain to effect their contact with the protein factors in the ribosome necessary for the translational process.

Functions of the Ribosomes. What are the significant functional features of ribosomes? Ribosomes provide points of attachment for messenger RNA, and for the transfer RNA molecules bearing amino acids. They provide an environment that strengthens the hydrogen bonding between messenger RNA and the complementary anticodons on various transfer RNAs, probably a hydrophobic environment that stabilizes hydrogen bonds. They provide proteins that catalyze the formation of the peptide bonds and the provision of energy for the entire process. They undergo a

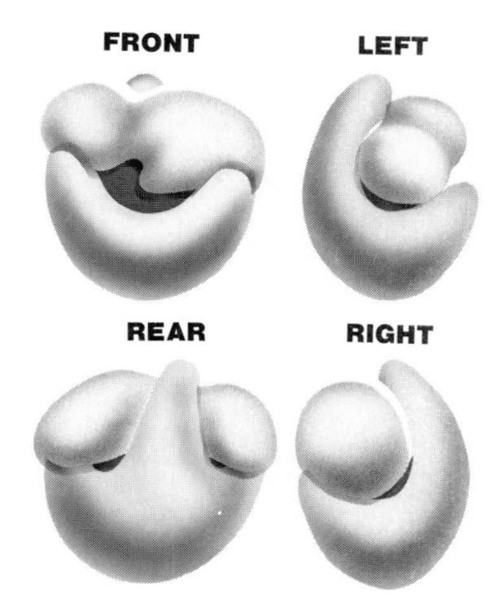

FIGURE 5–14

Ribosomal subunits in *E. coli.* (Eukaryotic ribosomes appear similar, but have not been studied in such detail.) The large subunit (gray) has a trough with a toothed margin on which the small subunit (brown) appears to rest. The small subunit has a constriction. Much study and a little imagination is necessary to generate detailed representations such as these from rather vague electron micrographs. There does appear to be an opening between the subunits that permits the ribosome to surround messenger RNA and its attached transfer RNA, while still being able to move over these structures. (Based on drawings by G. Stöffler and H.G. Wittman *in* H. Weissbach and S. Petska, eds. [1977], Molecular Mechanisms of Protein Biosynthesis. Academic Press, p. 117.)

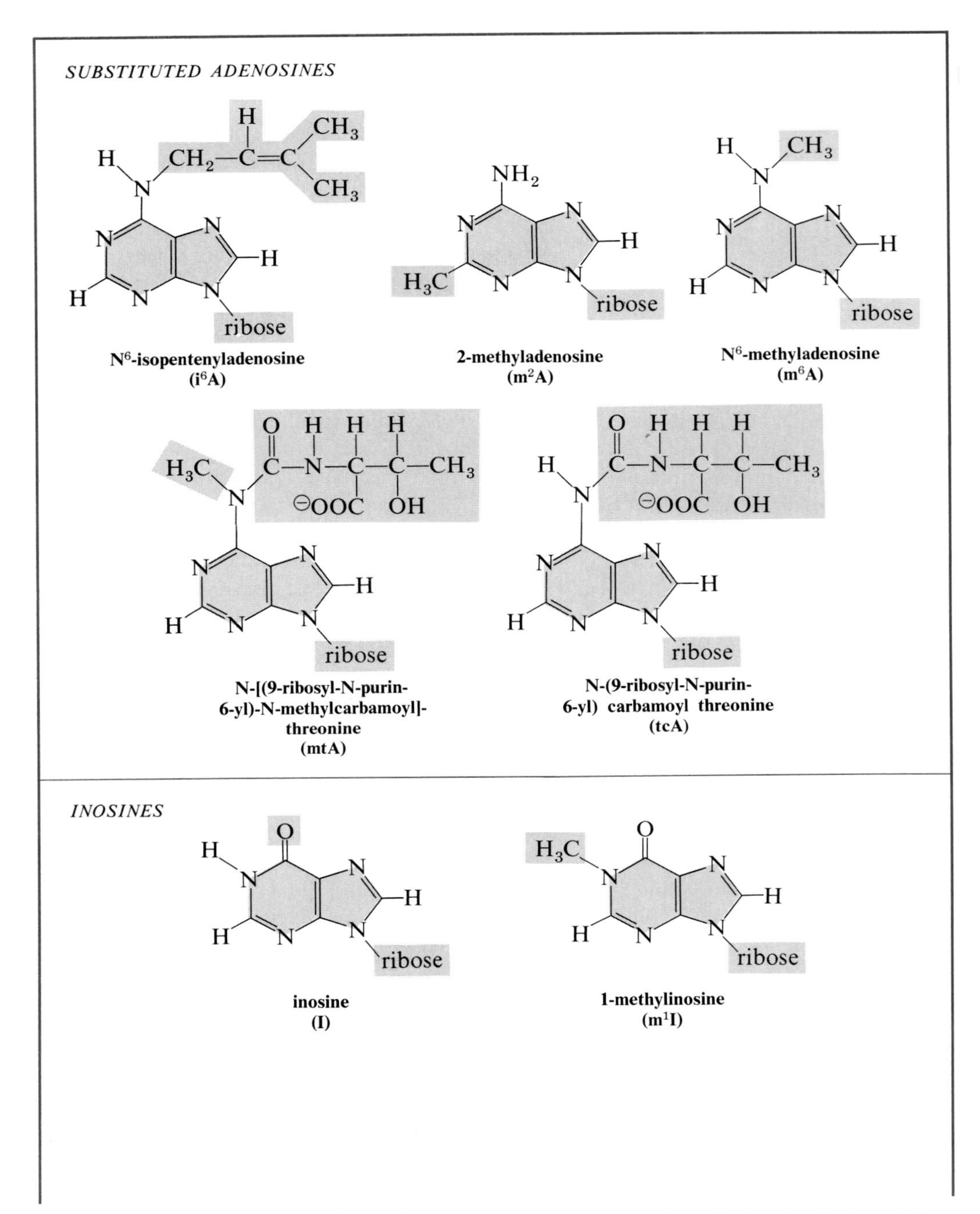

FIGURE 5–15 Some modified bases known to occur in various eukaryotic transfer RNAs and their accepted abbreviations. The examples include most of those found in anticodons and their flanking bases. 7-Methylguanosine (m^7G) and N^6,N^6-dimethyladenosine (m^6_2A) were shown in Figure 5–11. Inosine is a nucleoside containing hypoxanthine formed by hydrolysis of the 6-amino group of adenosine.

Illustration continued on opposite page

SUBSTITUTED GUANOSINES

1-methylguanosine
(m^1G)

N^2,N^2-dimethylguanosine
(m^2_2G)

7-methylguanosine
(m^7G)

2′-O-methylguanosine
(Gm)

N^2-methylguanosine
(m^2G)

α-carboxyamino-β-hydroperoxy-γ-(3-ribosyl-4,9-dihydro-4,6-dimethyl-9-oxo-*1H*-imidazo [1,2-α]-purin-7-yl) butyric acid dimethyl ester
(oyW = peroxywybutosine)

7-{[(*cis*-4,5-dihydroxy-2-cyclopenten-1-yl)amino]methyl}-7-deazaguanosine
(Q = queuosine)

SUBSTITUTED CYTIDINES

3-methylcytidine
(m^3C)

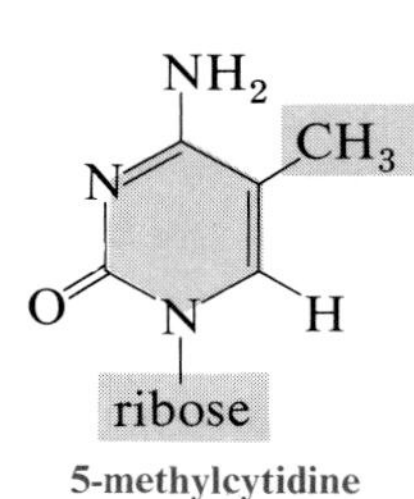

5-methylcytidine
(m^5C)

2′-O-methylcytidine
(Cm)

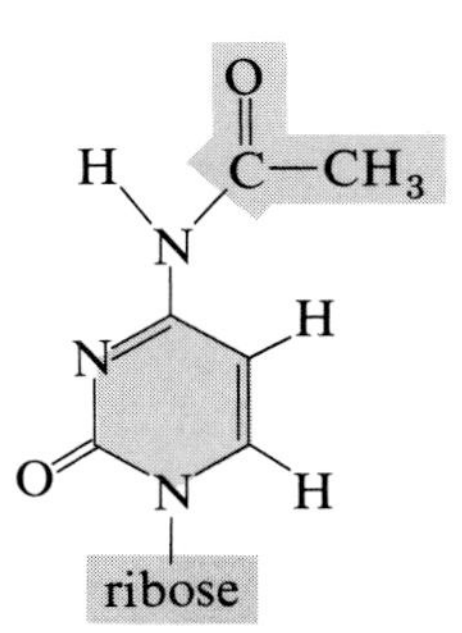

N^4-acetylcytidine
(acC)

FIGURE 5–15 *Continued*

Illustration continued on following page

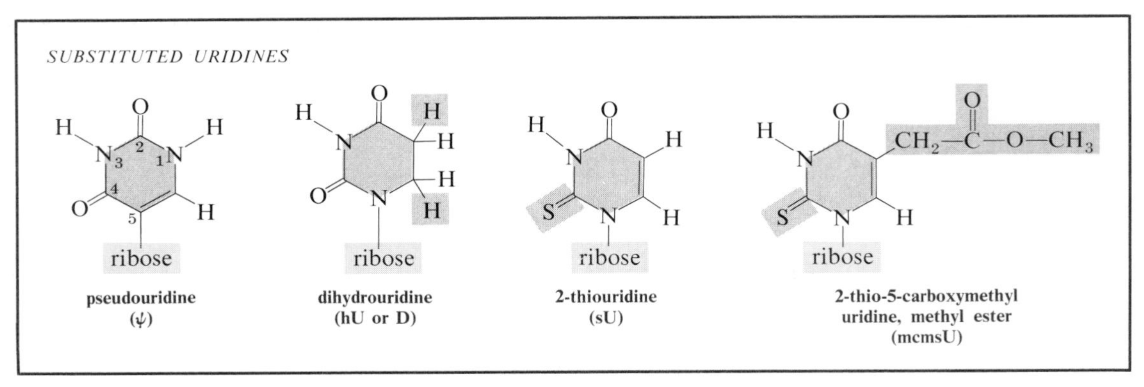

FIGURE 5-15 *Continued*

structural change that moves them along the messenger RNA from triplet to triplet, and finally, they provide devices for recognizing the end as well as the beginning of the coding for a polypeptide chain, and for releasing the finished product.

Transfer RNA (tRNA)

Transfer ribonucleic acids have the small size and high solubility required for transporting amino acid residues; the known examples have from 68 to 93 nucleotide residues. A liver cell contains a swarm of about 10^9 molecules of transfer RNA distributed among some 60 different kinds. There is at least one kind of tRNA for each amino acid, but some amino acids are carried by more than one kind.

Although they are relatively small molecules, the transfer RNAs are critical links in the translation of nucleotide sequences into amino acid sequences. They form highly specific patterns of bonds on three separate occasions:

1. Attachment of the correct amino acid to the appropriate tRNA;

2. Positioning of the loaded tRNA onto a ribosomal subunit so that it may bond to messenger RNA and also donate its amino acid for polypeptide synthesis; and

3. Attachment to the base triplet in messenger RNA corresponding to the amino acid being transported.

These demands for specific recognition are satisfied through the formation of an elaborate tertiary structure containing extensive sequences of secondary structure and including a florid collection of covalently modified bases (Fig. 5-15).

Secondary Structure—the Cloverleaf. When the primary structure of some transfer RNAs was first determined, it was noticed that matching the greatest number of bases as antiparallel complements created an outline something like a cloverleaf (Fig. 5-16), and it now appears that this is a fundamental arrangement of helical segments in all kinds of transfer RNA, except possibly for mitochondrial tRNA (see later discussion).

The three arms of the cloverleaf are always present, again excepting mitochondrial tRNA; there is also a variable arm that is short in some kinds of transfer RNA and long in others. The same bases are always present at certain locations, whereas those at other positions often vary.

The 3′ and 5′ ends of the molecule are paired in a double helix to form the **acceptor stem.** The 3′ tail of the stem always has the sequence -C-C-A-OH, and this is the site at which amino acids are attached for transport.

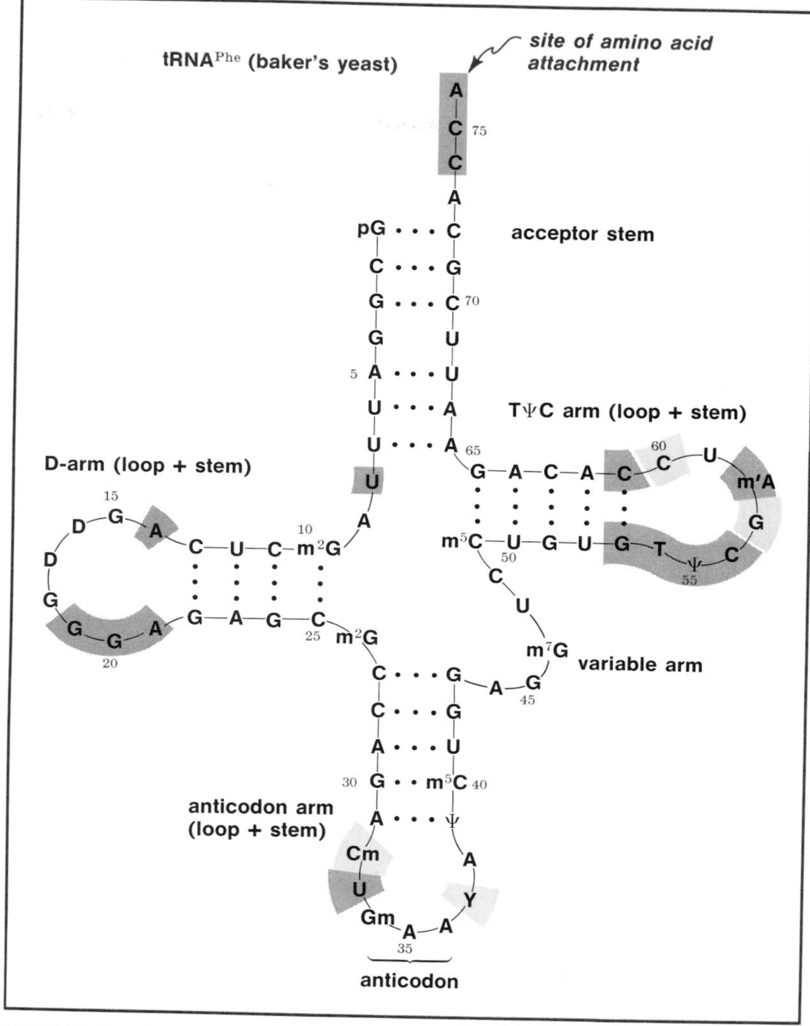

FIGURE 5–16 The secondary structure of a transfer RNA. This example was the first to be determined in detail, and illustrates many of the general features in transfer RNAs from all organisms. *Colored shading:* base always present. Extra Pu before A-4 and extra Py-Pu between G-20 and A-21 in some tRNAs. (Pu, purine nucleoside; Py, pyrimidine nucleoside.) *Gray shading:* base of same type, purine or pyrimidine, always present.

The loop opposite to the acceptor stem in the cloverleaf contains the anticodon that is bound to messenger RNA. This loop, and its supporting stem, is therefore known as the **anticodon arm.**

Tertiary Structure. The cloverleaf is a formal device for depicting paired bases; in the real molecules the segments of double helix are bent back upon themselves and fastened into place by a multiplicity of hydrogen bonds between the bases (Fig. 5–17). The result is a triangular molecule in which the short segments of double helix are stacked upon themselves to create longer sections of helix. The compact structure of this molecule is achieved by more extensive hydrogen bonding than is seen in the usual helical arrangement. Not only do hydrogen bonds form in the customary com-

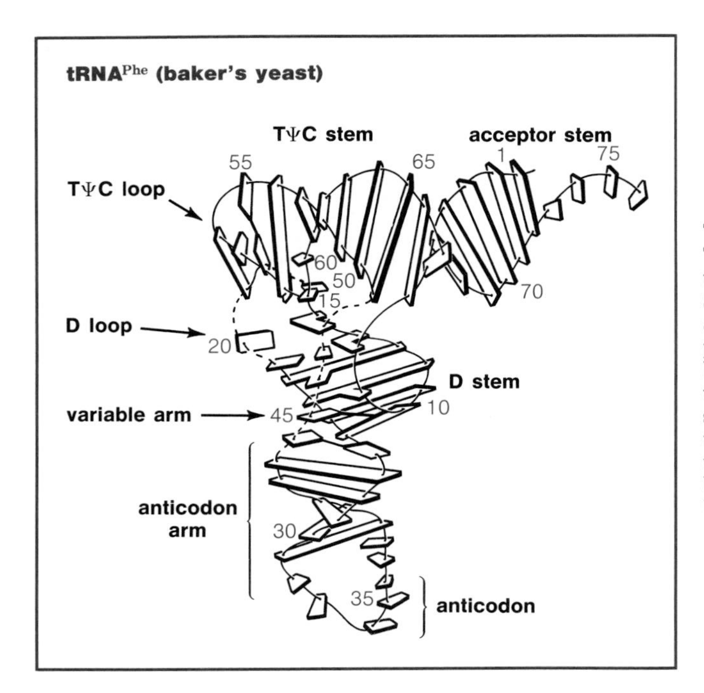

FIGURE 5–17

The tertiary structure of a transfer RNA. Transfer RNAs are bent upon themselves so as to stack the helical segments shown in Fig. 5–16 into two long helical segments at right angles to each other. This figure indicates pairs of bonded bases by bars, with the ribose-phosphate backbone threaded through them. Segments of double helix (secondary structure) are held in position by the formation of additional hydrogen bonds, involving the phosphate-sugar backbone as well as the bases. Modified slightly from an illustration by S.H. Kim (1975), Nature, 256:680.

plementary positions, but additional hydrogen bonds are formed that involve atoms at other positions in the bases, as well as atoms in the sugar and phosphate backbone. Every base in the molecule is involved in hydrogen bonding except for the C-C-A at the 3′ terminal and the three bases in the anticodon. Those that are not a part of antiparallel complements form other types of bonds. We have a preview in these small molecules of the complexity we ought to expect in the structures of the much larger ribosomal RNA molecules, which are yet to be determined in detail.

Loading with Amino Acid. Amino acids are carried on the terminal adenosine residue of the C-C-A sequence as aminoacyl esters of the hydroxyl groups of the ribose moiety (Fig. 5–18). Loading involves an initial cleavage of ATP by the amino acid, forming an intermediate aminoacyl anhydride of AMP and releasing inorganic pyrophosphate (PP_i). The aminoacyl group can then be transferred to a molecule of tRNA. The cleavage of ATP is a necessary part of the process because it makes the formation of aminoacyl tRNA feasible by shifting the overall position of equilibrium. (We shall discuss the rationale of these coupled reactions in detail in Chapter 19.)

Both the 2′ and 3′ hydroxyl groups of the terminal adenosine residue are available for attachment of an aminoacyl group; sometimes one is used, sometimes the other, sometimes either—the choice varies with organism and amino acid, but it

FIGURE 5–18 The formation of an aminoacyl tRNA involves a reaction between the proper amino acid and adenosine triphosphate (*top*), releasing PP_i and forming the aminoacyl adenosine monophosphate, which is a mixed anhydride of carboxylic and phosphoric acids. The aminoacyl group is then transferred to the terminal ribosyl moiety of the corresponding tRNA (*center right*). The group is shown on the 3′ oxygen (*bottom center*), but there is in fact an equilibration between the 2′ and 3′ positions. The AMP that is released (*center left*) is phosphorylated by the processes of oxidative metabolism, generating the original ATP.

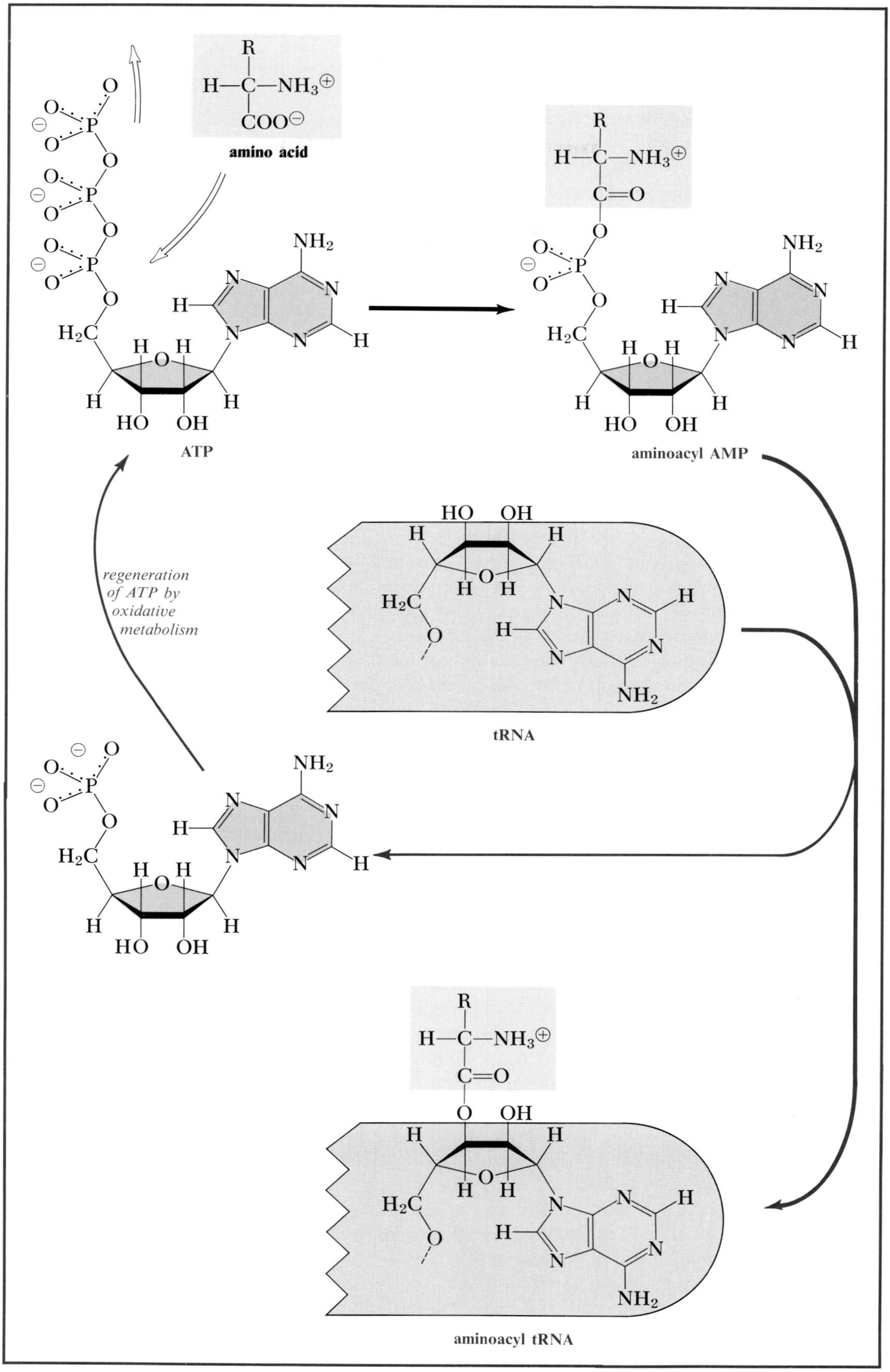

amino acid

ATP

aminoacyl AMP

regeneration of ATP by oxidative metabolism

tRNA

aminoacyl tRNA

FIGURE 5–18 *See legend on opposite page*

makes no difference because there is a rapid spontaneous equilibration to create a mixture of 2′- and 3′-aminoacyl tRNA molecules.

Recognition by Aminoacyl tRNA Synthetases. The ATP-driven attachment of an amino acid to a corresponding tRNA is catalyzed by an enzyme known as an aminoacyl tRNA synthetase. There is one synthetase for each amino acid, and it recognizes all of the kinds of tRNA that are built to carry that amino acid. When we think on it, we see that each of these enzymes must bind molecules of transfer RNA in two ways. It must recognize the C-C-A terminal to which amino acid groups are always attached; since all transfer RNAs are alike in this respect, it is probable that all of the synthetases form the same sort of bonds with the terminal region. On the other hand, the enzyme must recognize some specific regions in the transfer RNAs that are built to carry an individual amino acid; all of the $tRNA^{Val}$, no matter how many there may be, must have some specific surface features for recognition by the synthetase that attaches valine to the tRNA, but that will not be recognized by the other 19 synthetases. These specific features must arise from the variable regions of the molecule, that is, those base sequences that are not alike in all transfer RNAs, which are indicated in Figures 5–16 and 5–17.

Similarly, the aminoacyl synthetase proteins must also bind one kind of amino acid. It is crucial to the accuracy of translation for each synthetase to contain a region forming highly specific bonds with one amino acid and another region forming equally specific bonds with the transfer RNAs that carry the amino acid. Otherwise, the correct correspondence between amino acid and the anticodon also present on the transfer RNA will not exist. (More is said about the origin of specificity of enzymes in later chapters.)

Binding to Ribosomes. We shall see that the processes of translation—the formation of the peptide bonds—involve several protein components of ribosomes. It would be wasteful to have a separate machinery of peptide synthesis for each of the amino acids—the nature of the amide bond is the same in all. The specific binding of anticodon in transfer RNA with triplet on messenger RNA is sufficient identification, and the ribosome is so constructed that once this initial password is given, everything else proceeds in the same way. One amino acid looks like another to the proteins catalyzing the formation of peptide bonds. All transfer RNAs except the initiator tRNA (see later discussion) have a similar sequence of bases in the final loop, which is called the T-ψ-C loop because this particular combination is usually present (Fig. 5–16). This loop appears to bind an antiparallel complementary sequence in the 5S RNA of prokaryotic ribosomes, and perhaps in the 5.8S RNA of eukaryotic ribosomes.

Anticodons. Molecules of transfer RNA become attached to messenger RNA through a triplet of bases, or the anticodon, that is freely available at one apex (Fig. 5–17). The linkage of the antiparallel complements—anticodon and codon—is facilitated by the favorable environment provided in the ribosomes, but the positioning of the anticodon for easy consummation of the union is evidently aided in another way. The anticodons in tRNA, again save one, always occur in this sequence: -Py-U-N-N-N-Pu, in which Py and Pu are pyrimidine and purine nucleosides, respectively, and the Ns are the anticodon nucleosides. The flanking -Py-U- and -Pu- positions may well attach to specific positions on the ribosomes so as to hold the anticodon between them in an approachable posture.

The anticodons themselves have some common characteristics. The base in the initial nucleoside in the sequence—the one that binds to the 3′ end of the codon in messenger RNA—is often modified, and the second nucleoside sometimes contains pseudouridine instead of uridine, but no other alterations are known; the third nucleoside is never altered.

Wobble and Two-Base Coding. It is not necessary to have 64 different molecules of transfer RNA to match the 64 possible triplet codons in messenger RNA. The anticodon in some tRNA molecules will bind to more than one mRNA codon, even though anticodon and codon are not exact antiparallel complements. One way is for the third base of the codon to be ignored during binding. No error will be introduced if the amino acid carried by the tRNA is specified by a family of codons: a block of four codons with identical bases in the first two positions, but any base in the third position. For example, serine is coded by the family U-C-A, U-C-G, U-C-C, and U-C-U, all of which could be recognized by a transfer RNA bearing the anticodon U*-G-A, in which U* contains a modified uracil group. The extent to which this occurs is currently being explored.

Another way in which different codons can be bound is through a deviation ("wobble") of the third base in an mRNA triplet from its customary orientation, a deviation sufficient to accommodate the formation of different hydrogen bonds. The result is that initial nucleosides in the anticodon will bind with final nucleosides in the codon as follows:

Anticodon Position #1	***binds with***	**Codon Position #3**
C		G
G		C, or U
I		A, C, or U
U		A, or G

(I = inosine, Fig. 5–15)

For example, a $tRNA^{Ala}$ carrying the anticodon I-G-C can be bound to three different mRNA triplets, G-C-A, G-C-C, and G-C-U, designating alanyl groups. Only one additional kind of tRNA is required to match the remaining Ala codon (G-C-G).

The use of I-G-C illustrates the way in which modification of a base in the anticodon (A changed to I as a substitute for G) has extended the possibilities for binding. On the other hand, the freedom of binding of the first anticodon base can be limited by some modifications. Changing uridine to 2-thiouridine or methylthiouridines makes it incapable of the wobble binding of guanosine, so it will only match adenosine, its normal complement. Changing guanosine to the highly modified queuosine strengthens the wobble binding of uridine and cripples the binding of cytidine, the normal complement. The known anticodon sequences in eukaryotes are listed in Table 5–3.

Synthesis of transfer RNA involves the full repertory of techniques available to the cell for post-transcriptional processing of RNA. The initial transcript has extra leader and terminal sequences that must be removed by hydrolysis and sometimes has an intron to be bridged by splicing of the two exons. The C-C-A sequence on the 3′ end of every tRNA as a site for attaching amino acids is not in the original transcript; it is added by specific enzymes in the nucleus after a redundant tail originally present on the 3′ end is removed. In addition, the nucleus contains a splendid array of enzymes ready to covalently alter the bases into new forms. Each of these enzymes is very choosy in what it does. Some work on the original full-length transcript, others will only decorate intermediate forms, while others wait until the nucleotide chain is at its final length. All of the enzymes pick out bases in specific defined sequences. One that modifies uracil to 2-thiouracil, for example, will not operate on every uracil group or even at a uracil group in the same position on every tRNA. These highly specific changes not only define the codons to be recognized, as we discussed earlier, they are no doubt involved in recognition by aminoacyl synthetases and by specific parts of the ribosomes.

TABLE 5–3 ANTICODON SEQUENCES IN EUKARYOTES

The first nucleoside of the anticodon binds to the third nucleoside of the codon. When more than one codon will fit according to the wobble rules, the alternatives are separated by diagonals. Mitochondrial anticodons are omitted.

Residue	Sequence	Codons	Residue	Sequence	Codons
Ala	I-G-C	G-C-A/C/U	Leu	m^5C-A-A	U-U-G
Arg	I-C-G	C-G-A/C/U		U-A-G	C-U-A/G
	mcmU-C-U	A-G-A		Cm-A-A	U-U-G
Asn	Q-U-U	A-A-U	Lys	C-U-U	A-A-G
Asp	Q-U-C	G-A-U		mcmsU-U-U	A-A-A
	G-U-C	G-A-C/U	Metf	C-A-U	A-U-G
Cys	G-C-A	U-G-C/U	Met	C-A-U	A-U-G
Glu	mcmsU-U-C	G-A-A		Cm-A-U	A-U-G
	sU-U-C	G-A-A	Phe	Gm-A-A	U-U-C/U
Gly	G-C-C	G-G-C/U	Ser	I-G-A	U-C-A/C/U
	C-C-C	G-G-G		G-C-U	A-G-C/U
His	G-U-G	C-A-C/U	Thr	I-G-U	A-C-A/C/U
Ile	I-A-U	A-U-A/C/U	Trp	Cm-C-A	U-G-G
			Tyr	G-ψ-A	U-A-C/U
			Val	C-A-C	G-U-G
				I-A-C	G-U-A/C/U

Abbreviations:
Cm = 2′-O-methylcytidine
Gm = 2′-O-methylguanosine
I = inosine
m^5C = 5-methylcytidine
mcmsU = 5-(methoxycarbonylmethyl)-2-thiouridine
mcmU = 5-(methoxycarbonylmethyl)-uridine
Q = queuosine or derivative
sU = 2-thiouridine
ψ = pseudouridine

One of the most unusual, and enigmatic, modifications involves the formation of queuosine (see Fig. 5–15). This change involves the replacement of a complete guanine residue by queuine. The strange part is that animals are apparently unable to synthesize queuine, which is therefore obtained from the diet or from gut bacteria. Queuine would appear to qualify as a vitamin. However, no impairment of function has yet been seen in animals kept germ-free and reared on a diet lacking in queuine, even though their tRNA molecules have demonstrably retained G where Q ought to be.

Mitochondrial RNAs

Mitochondria, like bacteria, contain DNA that is closed into a circular chromosome. The complete sequence of the 16,569 base pairs comprising human mitochondrial DNA was announced in 1981, and it was startling.*

---

*An editorial comment in the 1981 issue of *Nature* carrying the announcement (290:443) described it as "a landmark in biology." Even if proximity to the event is later seen to have unduly magnified its importance in the eyes of the commentators (and mine, too), it is still an elegant piece of work, performed by a group from Australia, Germany, and the United States. However, the nexus for this diverse assemblage was the Englishman Frederick Sanger, twice Nobel Laureate. Sanger first showed how to determine amino acid sequences, and then developed some of the more powerful tools by which nucleotide sequences in RNA and DNA are determined (see p. 138).

SUMMARY OF CYTOSOLIC RIBONUCLEIC ACIDS

TABLE 5–4

Relative sizes (shown as if compacted spheres)*

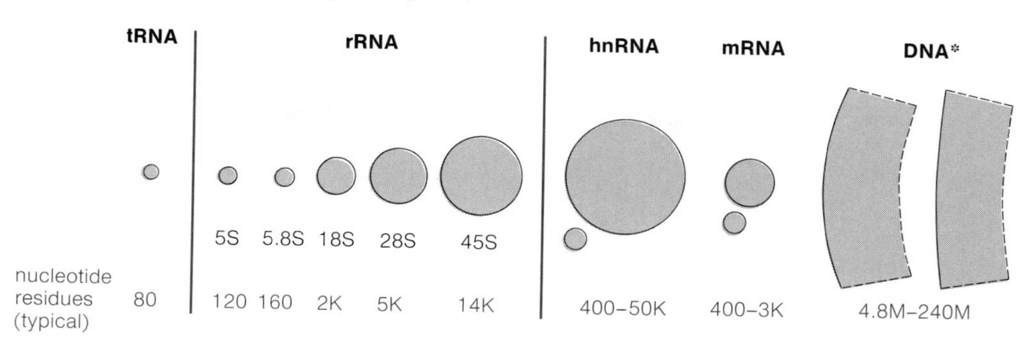

Messenger RNA (mRNA)

Precursors transcribed in nucleoplasm by polymerase highly sensitive to amanitin
Precursors included in heterogeneous nuclear RNA (hnRNA)
Precursors shortened by hydrolysis and methylated without affecting central message for amino acid sequence
Intervening sequences removed by splicing
Methylguanosine cap added to 5′ ends of most
As many as 200 adenylyl groups added to 3′ end
Terminal portions contain signals for initiation and termination of translation as polypeptide chain
Structure not known in detail; 5′ end may have ordered conformation, but central message believed to have less secondary structure than rRNA or tRNA in most cases
Size of molecule varies with length of polypeptide chain it represents; some with 100 nucleotide residues, some with 3,000, and more

Small Nuclear RNA (snRNA)

At least 10 types, some on nuclear skeleton, some in nucleolus, some in nucleoplasm
10^4–10^6 molecules of each per cell; 5–8S, with 90–220 nucleotide residues
Molecules in nucleoplasm may be important in splicing of messenger and transfer RNAs
Occur as ribonucleoprotein particles

Ribosomal RNA (rRNA)

Formed as 45S component in nucleolus by amanitin-insensitive polymerase I, except 5S component formed by polymerase III in nucleoplasm
Functional sequences in 45S component extensively methylated
Intervening spacers in 45S component removed by hydrolysis to create 5.8S, 18S, and 28S components
Internally complementary base sequences form A or A′ helixes (11 or 12 base pairs per turn) within themselves
Ribosomes contain one each of the RNA components and many protein molecules
Ribosomes contain two subunits—a 40S particle and a 60S particle (30S and 50S in bacteria), which combine to form 80S intact ribosomes in cytoplasm
40S ribosomal subunit contains 18S RNA
60S ribosomal subunit contains 28S, 5.8S, and 5S RNAs
Ribosomes have a pocket through which mRNA and tRNA pass during synthesis of polypeptide chain

Transfer RNA (tRNA)

Precursors formed in nucleoplasm by polymerase moderately sensitive to amanitin
Bases and sugars extensively modified, and excess nucleotides on 5′ end removed by hydrolysis
Mature molecule contains 73–85 nucleotide residues (mitochondrial tRNA has as few as 62; prokaryotic tRNA up to 93)
Molecule folds into compact tertiary structure resembling right triangle; anticodon and attachment site for amino acids at opposite ends of dished hypotenuse
Always has -C-C-A at 3′ end
Amino acids attached by ester bonds to terminal ribosyl group on 2′ or 3′ hydroxyl groups
Some areas of constant structure for attachment to ribosomes and maintenance of conformation
Some areas with variable composition to generate specific binding by one aminoacyl tRNA synthetase
One amino acid may be carried by different kinds of tRNA, all of which react with one aminoacyl tRNA synthetase

*Fragments of the largest and smallest human DNA molecules are depicted to show radii of curvature. The entire smaller sphere would be wider than the page at the scale used. None of the nucleic acids are present as compact spheres in real cells.

Human mitochondrial DNA is coded to form 13 mRNAs, 22 tRNAs and 2 rRNAs. The remainder of the proteins found in mitochondria are synthesized in the soluble cytoplasm and transported into the organelles. The mitochondrial genome and its RNA transcripts are organized for strict economy of space. The ribosomal RNAs contain only 954 and 1559 nucleotides; there is no analogue to the 5S or 5.8S RNA of cytosolic ribosomes.

Minimal processing is the rule for mitochondrial transcripts. The messenger RNAs are made with no, or very short, leaders before the initiator A-U-G codon, and have no terminal sequences. Indeed, some do not have a complete terminal codon, and end in U or U-A partial codons, which become U-A-A only upon addition of a poly-A tail.

Similarly, the transfer RNAs are economy models, lacking some of the features found in the cytosol molecules, and binding complete families of codons when possible.

The tRNA genes themselves have dual functions; they are templates for the tRNA transcripts, but they also act as spacers between the other genes in the chromosome. **Both strands of DNA are transcribed completely in mitochondria.** Each of the transcripts contains only part of the "correct" RNA. That is, some of the genes must be read one way in the DNA to get the proper sequence of RNA and some the other way, but willy-nilly, a complete reading of the entire DNA is made in both directions producing two RNA transcripts, each of which has useful pieces of RNA, interspersed with nonsense segments, that are different from RNA pieces of the other transcript. The mitochondria contain enzymes that can recognize the exact beginning and end of the tRNA sequences contained within these two transcripts and catalyze hydrolytic cleavage of the RNA at these points. One of the transcripts (of the H chain—the chain richer in heavy G-C pairs) is apparently broken as the tRNA spacers emerge, but the other appears as a long molecule of RNA that is later cleaved to yield its fewer "sense" segments.

THE TRANSLATION MECHANISM

General Description

Polypeptide chains are built one amino acid residue at a time on ribosomes bound to messenger RNA. The machinery of peptide synthesis contained in the ribosomes cranks through the same sequence of events, over and over again, with the ribosome advancing three nucleotide residues along the messenger RNA as each amino acid is added to the growing polypeptide. Molecules of transfer RNA carry amino acids to the site of peptide synthesis, where they are bound to messenger RNA, anticodon to codon. Each emptied molecule of transfer RNA is discharged from the ribosome before another loaded molecule can be bound.

Initiation of Translation

The building of a polypeptide chain goes along more or less automatically once it is begun, but the placement of the first amino acid residue is something special. This seminal event is complex and requires several participants: the separated subunits of a ribosome, messenger RNA, specific proteins, and a special methionine-bearing initiator transfer RNA that differs in some characteristics from all other transfer RNAs. The ribosomal subunits must be separated in advance and the initiator tRNA charged with a methionyl group before initiation can begin. A quite large protein complex known as **eukaryotic initiation factor 3 (eIF-3)** bonds tightly to the interacting face of the small subunit to make a 43S particle that can no longer bind to the now-free 60S subunit.

Role of Methionine. The initial codon to be translated in messenger RNA is always A-U-G, designating methionine. Two kinds of transfer RNA carry methionine. One, designated $tRNA_m^{Met}$, is used to incorporate all methionyl groups but the initial one. The other, designated $tRNA_f^{Met}$, is used to initiate translation; it lacks the T-ψ-C sequence characteristic of other tRNAs. Prokaryotes and the mitochondria and chloroplasts of eukaryotic cells begin synthesis of their polypeptides with the N-formyl derivative of methionine:

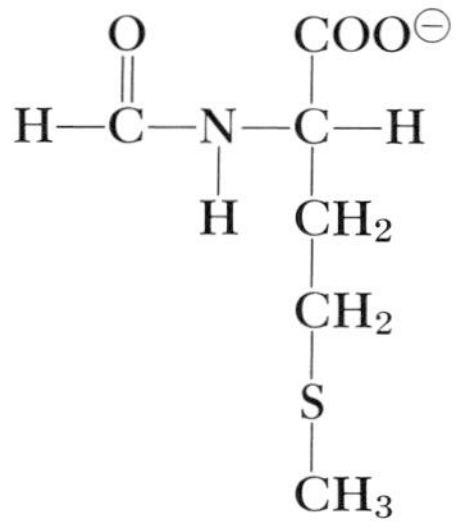

```
      O     COO⊖
      ‖     |
   H—C—N—C—H
         |   |
         H  CH2
             |
            CH2
             |
             S
             |
            CH3
```

N-formyl-L-methionine

This is the reason for the subscript "f" in $tRNA_f^{Met}$, even though the initiating methionine is not formylated in eukaryotes. Once synthesis of the polypeptide is underway, the eukaryotes usually remove the initial methionine residue, whereas the formylmethionyl group is often present at the first position of prokaryotic polypeptides.

Neutrophils are leukocytes (white blood cells) that function as an effective immigration service; they engulf and terminate most aliens. Bacteria can give away their location by making proteins, because the neutrophils have receptors on their outer membrane to detect formylmethionyl polypeptides. The result is a movement of the neutrophil toward the bacterium and the triggering of a number of responses aimed at destruction of the invader that are discussed in later chapters.

Let us now examine the subsequent steps in detail, using the synthesis of a hemoglobin beta chain as an example:

Step One

A molecule of initiator tRNA charged with a methionyl group ($Met\text{-}tRNA_f^{Met}$) combines with an initiation factor (eIF-2) and a molecule of guanosine triphosphate (GTP). The complex that is formed fits a particular location on a light ribosomal subunit:

$$43S + (Met\text{-}tRNA_f^{Met}\text{-}(eIF\text{-}2)\text{-}GTP) \longrightarrow 43S \text{ preinitiation complex}$$

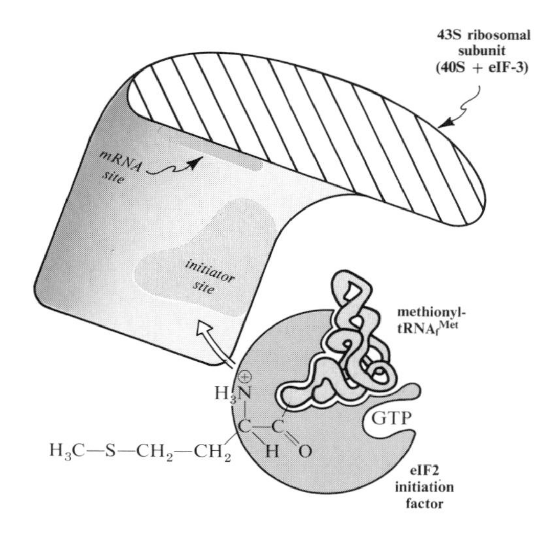

Step Two

A molecule of beta globin messenger RNA is bound to the 43S preinitiation complex. It is not known how the proper end of messenger RNA is recognized. The cap helps, but is not obligatory. In any event, the eukaryotic mRNAs appear to slide from their capped ends on the ribosomal subunit until the first A-U-G sequence is brought into proximity with the C-A-T anticodon on the initiator tRNA already present. At least three additional initiation factors not shown (eIF-4A, 4B, 4E) may participate, and energy is supplied through concomitant hydrolysis of a molecule of adenosine triphosphate (ATP) to adenosine diphosphate (ADP) and inorganic phosphate (P_i).

$$43S + eIFs + mRNA + ATP \longrightarrow ADP + P_i + 48S \text{ complex}$$

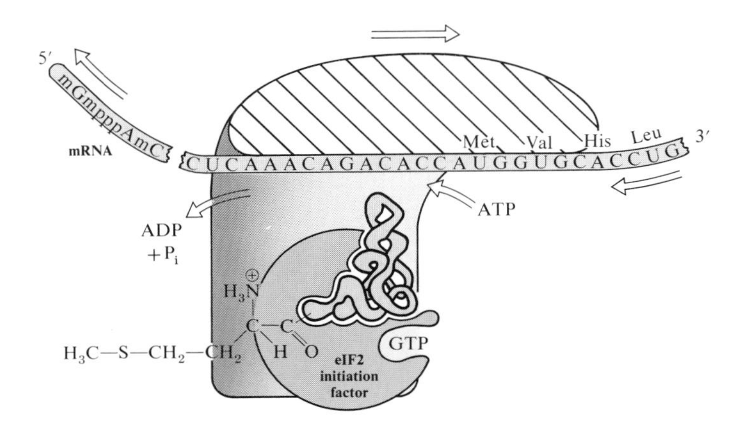

Step Three

Once the mRNA is properly bound in place, formation of the complete ribosome can proceed. The GTP still present on the eIF-2 bound to the small subunit is hydrolyzed to GDP + P_i, releasing many of the initiation factors. Still another initiation factor not shown (eIF-5) is added. This promotes the binding of the heavy (60S) ribosomal subunit to form the complete mRNA-ribosome complex.

$$48S + eIF\text{-}5 + 60S \longrightarrow mRNA\text{-}ribosome + GDP + P_i + eIF\text{-}2, 3, 4$$

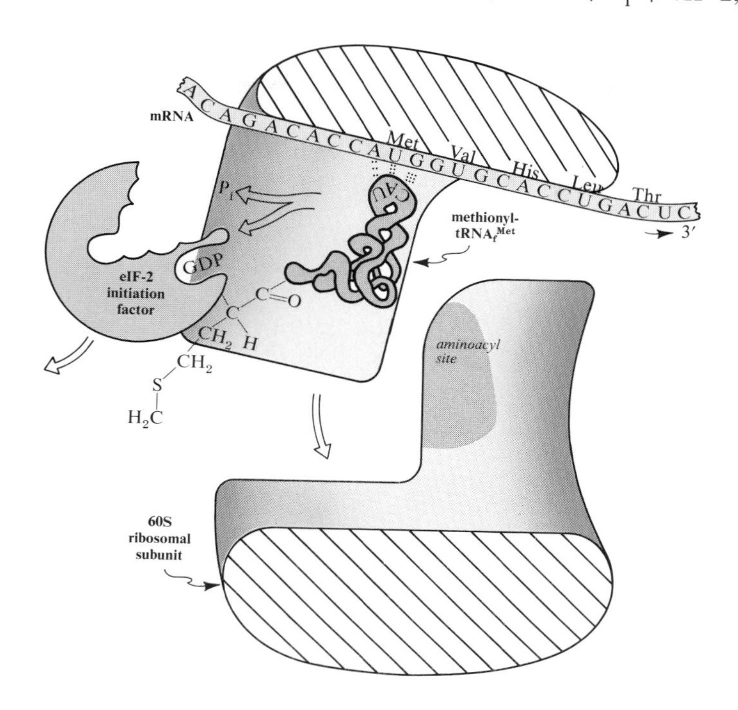

Step Four

Initiation is now complete, and chain elongation can begin. The combination of the two ribosomal subunits around mRNA generates two adjacent sites with an affinity for tRNA, and spaced so that the adjacent tRNAs bond to adjacent triplets in mRNA. The methionyl initiator tRNA already present occupies one site—the **peptidyl,** or **P,** site. The other site, known as the **aminoacyl,** or **A,** site is transiently vacant. A molecule of valyl-tRNA^{Val}, combined with an **elongation factor protein (EF-1)** and GTP, encounters the site. (Valine is the first amino acid in the β chain after the initiating methionine is removed.)

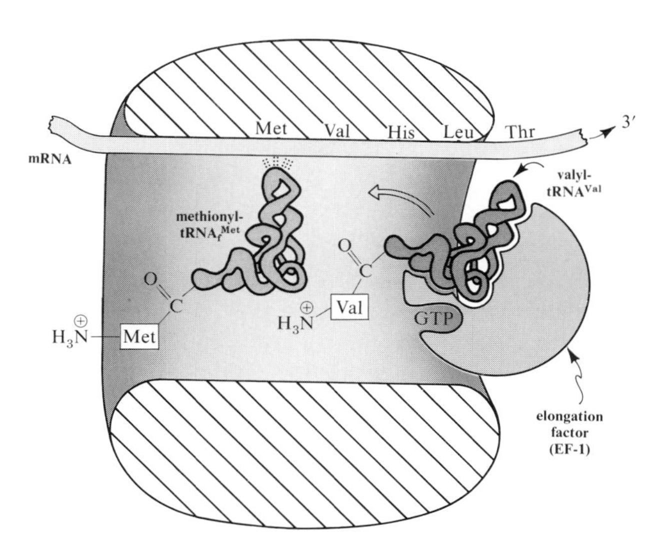

Step Five

Any of the transfer RNAs on elongation factors may come to the aminoacyl site, but only the one containing an anticodon that will fit the corresponding codon on messenger RNA at the site becomes fixed in place. This binding causes hydrolysis of GTP and release of P_i. The elongation factor is shown as leaving with GDP attached; this occurs in prokaryotes, but is not established in eukaryotes.

$$(\text{Val-tRNA}^{\text{Val}})\text{-(EF-1)-GTP} + \text{mRNA-ribosome} \longrightarrow (\text{Val-tRNA}^{\text{Val}})\text{-mRNA-ribosome} + \text{(EF-1)-GDP} + P_i$$

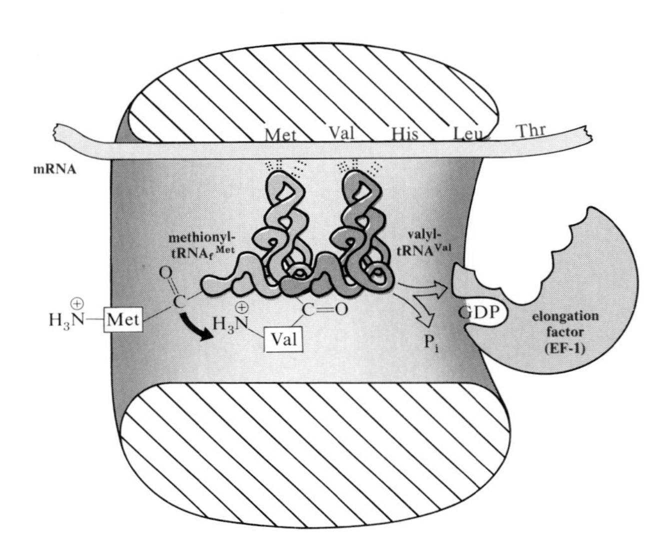

Step Six

The first two amino acid residues are now in place on their respective tRNAs, ready for synthesis of the first peptide bond. Synthesis occurs by moving the methionyl group from its transfer RNA onto the ammonium group of the adjacent valine residue, as shown in the preceding diagram. The reaction is catalyzed by peptide synthetase, a protein in the heavy ribosomal subunit:

$$\text{Met-tRNA}_f^{\text{Met}} + \text{Val-tRNA}^{\text{Val}} \longrightarrow \text{tRNA}_f^{\text{Met}} + \text{Met-Val-tRNA}^{\text{Val}}$$

(All components are bound to the mRNA-ribosome complex.)

Step Seven

At this point, the peptidyl site is occupied by initiator tRNA stripped of its methionyl group; the aminoacyl site contains a tRNA^{Val} loaded with the dipeptidyl group Met-Val. Naked transfer RNAs have a lower affinity for the P-site than those carrying peptidyl groups, and this may be the precipitating factor for subsequent events. The initiating tRNA is released into the solution, ready to be charged with another methionyl group. Another elongation factor (EF-2, not shown) and GTP are bound, while the entire ribosome moves along the messenger RNA for three residues. This translocation brings the tRNA^{Val} with its attached Met-Val peptide onto the peptidyl site, where it is still bound to the Val codon on mRNA. During the translocation, GTP is hydrolyzed, and P_i and the GDP-carrying EF-2 are released.

The codon for the next amino acid, histidine in this case, is now adjacent to the vacant aminoacyl site, which is then filled by a histidyl-tRNA^{His} on an EF-1 molecule with its attached GTP.

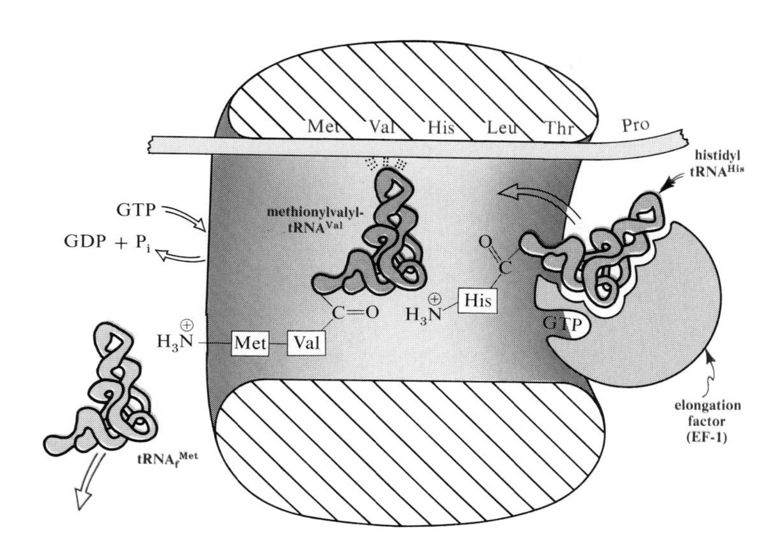

Step Eight

The building of the polypeptide chain proceeds by a repetition of steps 4 to 7 just outlined. The His-tRNA^{His} is fixed to the histidine codon on mRNA. At this point, the P site is occupied by Met-Val-tRNA^{Val}, the A site by His-tRNA^{His}. The Met-Val peptidyl group is then transferred onto the ammonium group of His, creating the second peptide bond.

$$\text{Met-Val-tRNA}^{Val} + \text{His-tRNA}^{His} \longrightarrow \text{tRNA}^{Val} + \text{Met-Val-His-tRNA}^{His}$$

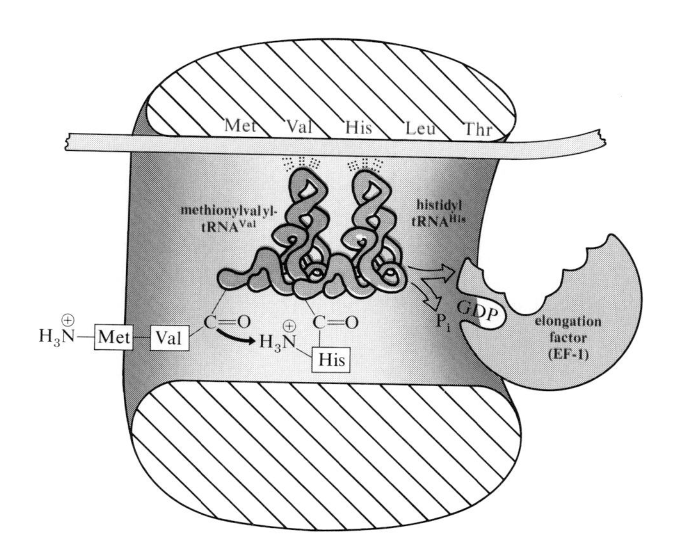

Step Nine

Once more, the now naked $tRNA^{Val}$ on the P site dissociates from the ribosome, and the ribosome moves three nucleotide residues down the messenger RNA chain with a concomitant hydrolysis of GTP. This translocation brings the transfer RNA carrying the newly formed peptidyl group, Met-Val-His-$tRNA^{His}$, onto the P site of the ribosome and exposes the A site to entry by a loaded transfer RNA with an anticodon specifying the next amino acid, leucine in this case, complexed with elongation factor-1 and GTP.

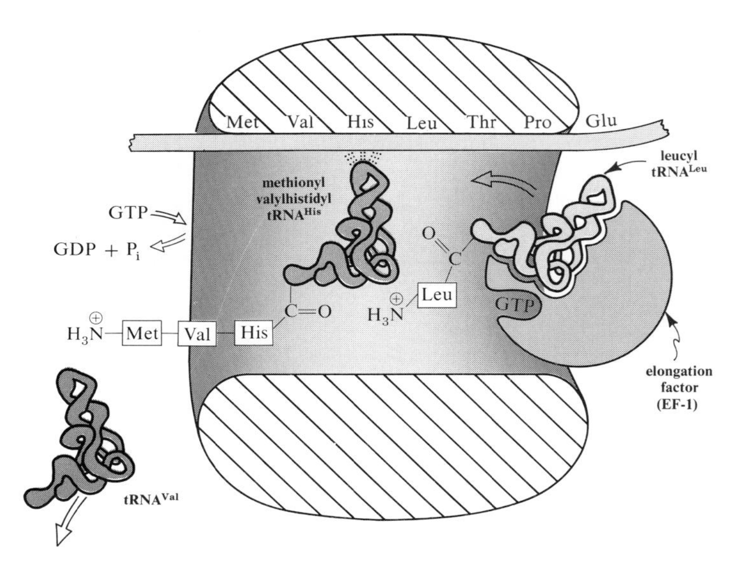

Subsequent Elongation Steps

Peptide bond synthesis, translocation, and charging the ribosome with the next aminoacyl tRNA are repeated until the polypeptide chain is completed; a total of 146 elongation sequences are necessary to create the hemoglobin β chain, ending with a histidine residue. At some point during this process, the methionyl group is hydrolyzed from the amino terminal of the chain, leaving a valyl group in the first position with a free ammonium group.

Chain Termination

The translocation completing the final elongation sequence brings a $tRNA^{His}$ bearing the completed β chain onto the peptidyl site. It is still attached to a His codon in messenger RNA. The next triplet in messenger RNA, adjacent to the aminoacyl site, is a termination codon, U-A-A. The presence of this codon on the A site causes the binding of yet another GTP-bearing protein—a **termination factor.** This factor catalyzes the transfer of the entire polypeptide from tRNA to water in the surrounding cytosol. At the same time GTP is hydrolyzed to GDP and P_i.

Four separate steps—initiation, binding of elongating amino acid, translocation, and chain termination—appear to proceed by related mechanisms involving hydrolysis of GTP to GDP + P_i in the presence of distinct protein factors.

Chain termination causes a concomitant release of the remaining bound protein factors, so the ribosome dissociates into its separate light and heavy subunits, which will recombine into an inactive 74S particle unless prevented by combination with another molecule of eIF-3.

Chain Folding

Generation of the primary structure completes the steps in protein synthesis requiring the ribosome apparatus, but the primary structure must fold into the final conformation. This occurs spontaneously, but not instantaneously. The rate-limiting slow phase for the folding of some polypeptides involves a change in the configuration of proline residues. Earlier, we noted that the peptide bond is most stable in the trans configuration (p. 20). This is not true in all cases when the bond is formed with the nitrogen atom of proline, because it is present in a ring. The ring sometimes fits more readily into a stable protein structure when the amide bond is in its cis conformation:

trans-peptidyl-Pro- ⇌ *cis*-peptidyl-Pro-

Proline peptides are first synthesized in the trans conformation (the most stable form in free solution), so the formation of the final protein configuration must wait for the relatively slow isomerization of trans to cis conformations of some proline peptide bonds.

FURTHER READING

Readable Reviews

T. Hunt (1980). The Initiation of Protein Synthesis. Trends Bioch. Sci. 5:178.

Technical Reviews

H. Weissbach and S. Petska, eds. (1977). Molecular Mechanisms of Protein Synthesis. Academic Press.
R. Breathnach and P. Chambon (1981). Organization and Expression of Eucaryotic Split Genes Coding for Proteins. Annu. Rev. Biochem. 50:349.
P.A. Sharp (1981). Speculations on RNA Splicing. Cell 23:643.
H. Busch et al. (1982). SnRNAs, SnRNPs, and RNA Processing. Annu. Rev. Biochem. 51:617.
L.J. Korn (1982). Transcription of Xenopus 5S Ribosomal RNA Genes. Nature 295:101.
J.W.B. Hershey (1980). The Translational Machinery: Components and Mechanism. p. 1 in D.M. Prescott and L. Goldstein, eds.: Cell Biology, A Comprehensive Treatise, vol. 4. Academic Press.
R. Jagus, W.F. Anderson, and B. Safer (1981). The Regulation and Initiation of Mammalian Protein Synthesis. Prog. Nucl. Acid Res. 25:127.
H.F. Noller and C.R. Woese (1981). Secondary Structure of 16S Ribosomal RNA. Science 212:403.
U. Lagerkvist (1981). Unorthodox Codon Reading and the Evolution of the Genetic Code. Cell 23:305.
S. Altman (1981). Transfer RNA Processing Enzymes. Cell 23:3.
P.R. Schimmel and D. Soll (1979). Aminoacyl-tRNA Synthetases: General Features and Recognition of Transfer RNAs. Annu. Rev. Biochem. 48:601.
M. Sprinzl and D.H. Gauss (1982). Compilation of tRNA Sequences. Nucl. Acids Res. 8:r1.
U. Maitra, E.A. Stringer, and A. Chaudhiri (1982). Initiation Factors in Protein Biosynthesis. Annu. Rev. Biochem. 51:869.
R.H. Pain (1981). Proline and Protein Procrastination. Nature 290:187.

CHAPTER 6

REPLICATION OF DNA AND THE CELL CYCLE

LOCALIZATION OF DNA: CHROMOSOMES

The fundamental event in the reproduction of cells is the formation of new molecules of deoxyribonucleic acid. Replication of DNA ensures that people beget people and that people remain people. Although partial replication occurs in the interval between cell divisions in order to repair damaged DNA molecules, complete duplication of the information occurs only in preparation for cell division. The 1.74 meters of DNA in human cells are folded into compact packages so as to fit in nuclei with typical diameters of 4 to 6 micrometers. These packages are chromosomes, and replication of DNA also involves replication of the complete chromosome package.

Most human somatic cells contain 46 chromosomes. Each chromosome in a resting cell is monotene, that is, it contains one double-stranded fiber of DNA (duplex DNA), a fact that was not appreciated until recent years owing to the difficulty of handling such very long molecules without breaking them by mechanical shear. The largest chromosome contains some 255 femtograms (255×10^{-15} g) of DNA in nearly 80 mm of duplex, while the smallest has 46 fg in 15 mm of duplex. The 31,000 micrometers of duplex in a typical example is compacted to make a chromosome between 4 and 6 μm long. The somatic cells are diploid, meaning that they carry a double set of chromosomes, one derived from each parent, with most polypeptides being independently specified by DNA from each parent.

The Chromatin Fiber

The name chromosome was first applied to the densely stained structures visible in dividing eukaryotic cells with the light microscope. It was later shown that these compact mitotic chromosomes are composed of fibers looped into bundles (Fig. 6–1). A chromosome fiber is made of chromatin, which is a complex of DNA and proteins,

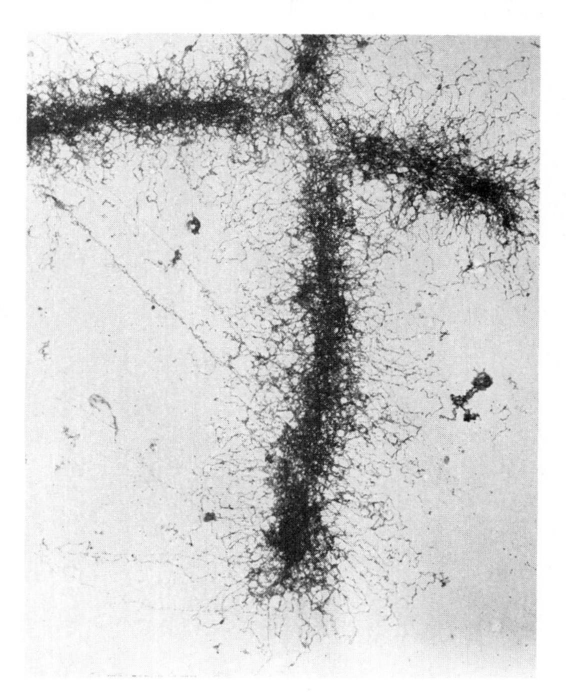

FIGURE 6–1

An electron micrograph of a human chromosome during mitosis (10,000×). This preparation was gently spread out on water to show the individual chromatin fibers that constitute the highly condensed chromosomes. The junction of the two chromatids at the centromere is clearly shown. (From D.E. Comings and T.E. Okada, (1970) Cytogenetics 9:440. Reproduced by permission.)

along with a small amount of RNA. HeLa cell chromatin, for example, contains 1730 daltons of protein and 90 daltons of RNA for each 1000 daltons of DNA.

Histones are the most abundant proteins in chromatin; much of the compaction of DNA is achieved through its association with a nearly equal mass of histones. They are relatively small proteins, 11,000 to 21,000 daltons, with a high proportion of lysine and arginine residues creating a net positive charge. Histones are therefore isoelectric in basic solutions, whereas the other proteins of chromosomes, including many enzymes and regulatory proteins, are usually isoelectric in more neutral or acidic solutions.

Every nucleated cell that has been examined—plant, animal, or protist—contains five general classes of histones that may be distinguished by size and amino acid composition (Table 6–1). The amino acid sequences within each class, except for histone H1, are quite similar in all kinds of organisms. Indeed, the arginine-rich histones H3 and H4 show the least change of all known proteins over the course of evolution; H4 of the common garden pea (*Pisum sativum*) is identical in all except two of its 135 residues to H4 of the common cow (*Bos taurus*), and H3 differs at only four positions in the two proteins. There is evidently little room in these small proteins for deviation from the long-tested functional arrangement.

TYPICAL COMPOSITION OF CLASSES OF HISTONES **TABLE 6–1**

Class	H1	H2A	H2B	H3	H4
Total amino acid residues	210	129	125	135	102
% Lys groups	28	11	16	10	10
% Arg groups	2	9	6	13	14
% Ala groups	25	13	10	13	7
% Gly groups	7	11	6	5	17
Number Cys groups	0	0	0	2	0

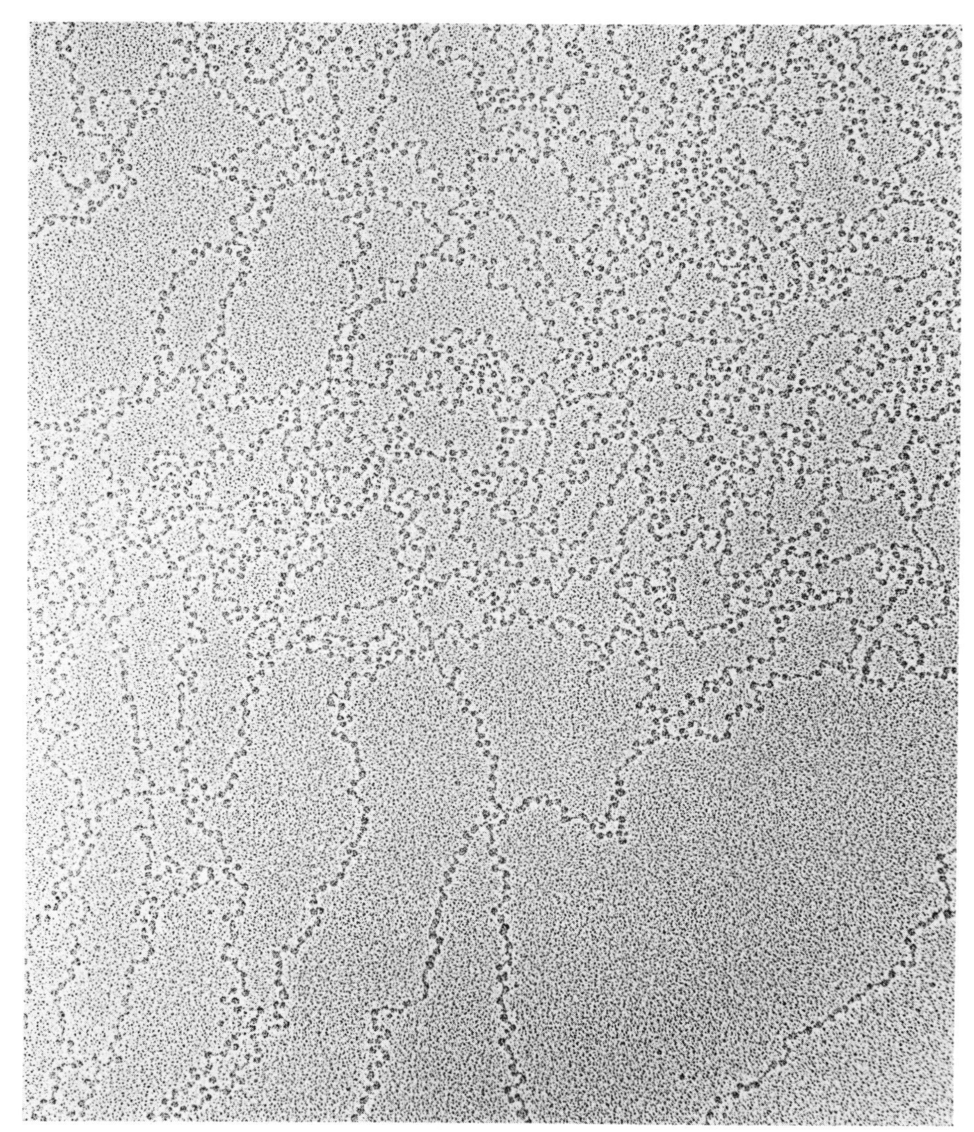

FIGURE 6–2 Dispersed chromatin showing bead-like nucleosomes on a thread of DNA (73,200×). Prepared from growing embryo (blastoderm stage) of *Drosophila melanogaster*. Unpublished print courtesy of S.L. McKnight. Details of the methodology and other superb electron micrographs showing replication forks and nascent ribonucleo-proteins growing along the DNA fibers may be found in S.L. McKnight and O.L. Miller, Jr., (1977) Electron Microscopic Analysis of Chromatin Replication. Cell 12:795.

Nucleosomes: The Repeating Units of Chromosomes. The fundamental structure of chromatin is a ball of histones wrapped with nearly two turns of DNA, with successive balls connected by the DNA fiber much like beads on a string (Fig. 6–2). This repetitious arrangement is not apparent in intact chromatin, which is formed by association with other proteins that cause the "shish kebab" to coil upon itself at least twice to compact the DNA further. It is important to recognize that the repetition of structure has no relation to base sequences in the associated DNA; the determining forces are generated by the histones and the duplex polynucleotide backbone, not by the internal base pairs in DNA.

The nucleosome core is an octamer composed of a pair each of histones H2A, H2B, H3, and H4. Each of these histones has a region rich in positive charges with the remainder of the sequence more typical of globular proteins and forming alpha helixes. It is not known in detail how histones associate, but the final structure is formed into a spiral ramp—a spool—about which DNA winds. The interaction of the positive charges of the histones with the negative charges on the phosphate groups of the DNA backbone is only part of the forces involved in binding DNA to the core. The winding of DNA around and between nucleosome cores causes changes in the helical structure that we previously discussed (p. 53).

A nucleosome core is wrapped with 1.75 turns of DNA. The DNA is further bound to a full two turns where it meets and leaves the core by one molecule of H1 histone. The amino acid sequence of this histone is more variable between species, and it is less tightly bound.

Euchromatin and Heterochromatin

Most of the chromatin fibers constitute the euchromatin and disperse in the nucleus between cell division, but some fibers remain in a condensed form that can be intensely stained by basic dyes and are known as heterochromatin. Some of the condensed chromatin localized to the inner layer of the nuclear envelope in female cells is the remnant of one X chromosome known as the Barr body. (The other X chromosome is dispersed, and its DNA is actively used for transcription of RNA.) We are more concerned with the heterochromatin that occurs in both males and females. Much of this DNA is highly repetitive, containing as many as 10^6 identical sequences of nucleotides.

The function of heterochromatin is still not clear. Most is evidently not transcribed as RNA. It is closely associated with the centromeres of the mitotic apparatus (see later discussion), and it also forms bands in the condensed chromosomes appearing at mitosis; it is therefore likely to have a structural function of some kind.

We already noted (p. 59) that highly repetitive DNA is likely to include satellite DNA, differing in gross base composition, and therefore in buoyant density, from much of the other DNA.

Other Chromosomes

Bacteria contain single chromosomes, in which the double helix is **supercoiled,** with its ends then joined to form a duplex circle. The total length of DNA in *Escherichia coli* is only 0.1 that in the smallest human chromosome, but it still must be tightly folded to fit within the bacterial cell.

Eukaryotic cells also contain **circular chromosomes,** which are found in those cytoplasmic organelles that have their own apparatus for protein synthesis, such as the **mitochondria** in plants and animals, and the **chloroplasts** in plants. These organelles have a separate heredity carried by their circular DNA; they may be descendants of symbiotic prokaryotic organisms that once resided in primitive eukaryotic cells. Their ribosomes, as well as their DNA, are more like corresponding constituents of bacteria than those found elsewhere in the same cell.

Even though human mitochondrial DNA has a molecular weight of only 1.02×10^7, and codes for 13 proteins at most, it represents a significant heredity that can be passed only through the maternal line. (Spermatozoa do not contribute significant numbers of mitochondria to the fertilized ovum.) The importance of this cytoplasmic inheritance, at least in the plant kingdom, was forcefully emphasized when a major fraction of the U.S. corn crop was lost to a blight, owing to altered mitochondrial heredity in some widely used hybrids. Comparable defects in human cytoplasmic genes are yet to be demonstrated.

THE REACTIONS OF REPLICATION

The primary reaction for synthesizing DNA is much like the reaction for synthesizing RNA: a strand of existing DNA is used as a template to guide the condensation of deoxyribonucleoside triphosphates with the growing end of a new DNA

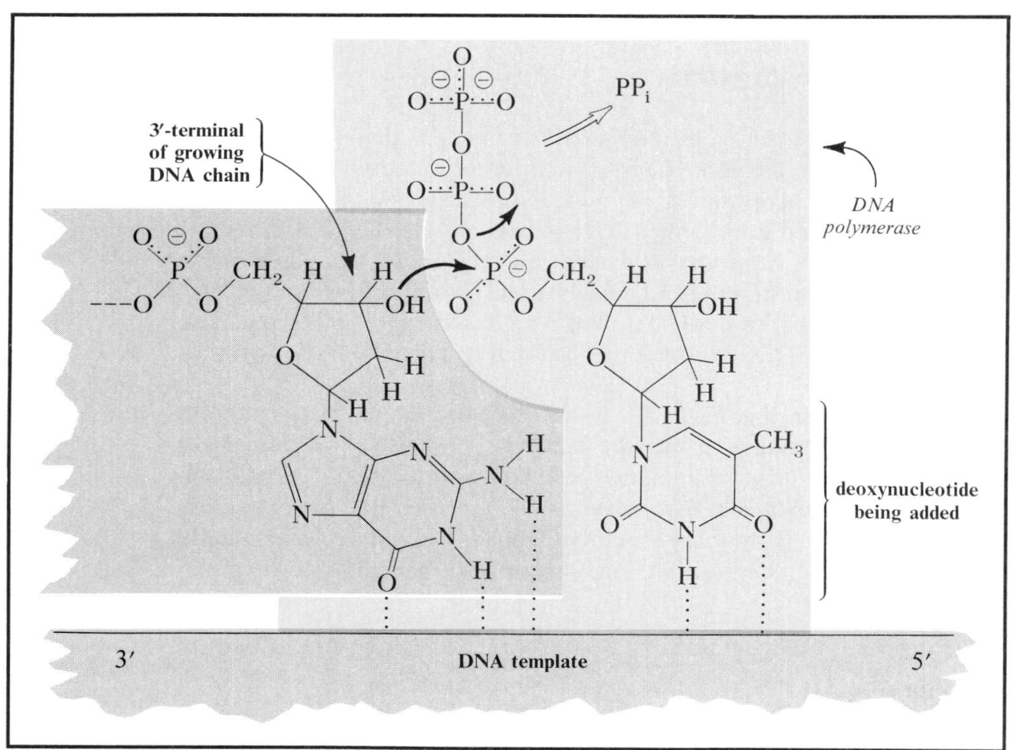

FIGURE 6–3 DNA is replicated by the extension of a new strand in proximity to an existing strand, which is used as a template. Deoxynucleotide residues are brought to the site in the form of deoxyribonucleoside 5′-triphosphates (*right*), and are attached to the growing chain (*left*) at its free 3′-hydroxyl group. The attachment involves the loss of inorganic pyrophosphate (PP$_i$). The selection of the correct complementary nucleotide residue involves the formation of hydrogen bonds between it and the DNA template strand; the bonds are strengthened by the presence of DNA polymerase.

strand, with an accompanying loss of inorganic pyrophosphate (Fig. 6–3). However, the replication of DNA differs from the synthesis of RNA by transcription in that antiparallel complements are made for both of the DNA strands.

Replication Is Semiconservative. When new DNA is synthesized, the old DNA that serves as a pattern is unwound, and each of the old strands acts as a template for laying down a new antiparallel complement:

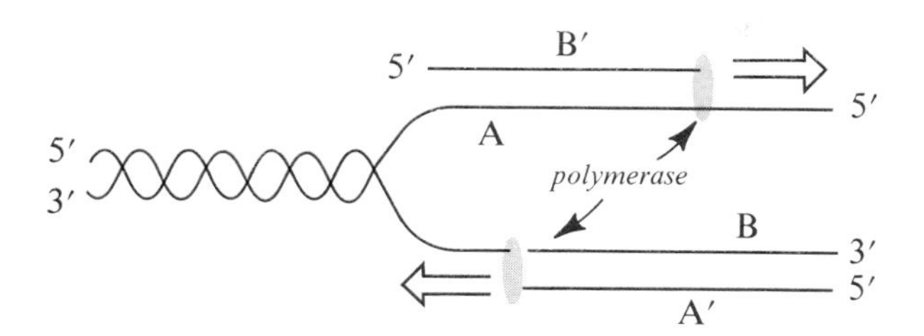

The DNA being replicated contains complementary strands, here labeled **A** (*brown*) and **B** (*black*), which have their 5′ terminals at opposite ends of the double helix. When these are unwound, molecules of DNA polymerase act independently on them to lay down new complements in opposite directions. A new complement, **B′**, is made for old strand **A**, and a new complement, **A′**, is made for old strand **B**. At the completion of replication, the full length of each old strand will be matched by a new complementary strand, and these pairs will twist into two molecules of double helix, each half old and half new (Fig. 6–4). Since replication usually proceeds perfectly, each of the two molecules is identical to the original double helix containing two old strands.

Function of the Two Strands. Since replication, like transcription and translation, makes use of antiparallel complementary binding of bases as the device for information transfer, we can see why DNA, unlike mRNA, must be double-stranded. Within each gene along a DNA molecule, only one strand acts as the active template for transcription into RNA. The other strand, the sense strand for that gene, is usually ignored during the life of the cell. (Of course, some genes in the entire DNA molecule are read from one strand and some from the other so that each strand is in part active and in part inert as a source of information transcribed into RNA.) However, this seeming inertness disappears whenever the information must be replicated; what is to be the active strand of a gene in a new cell must be copied as the antiparallel complement of the sense strand of the parent DNA. The sense strand therefore acts as an information store between cell divisions. We shall see in the next chapter that it also has the same function between cell divisions by providing a template for repair of any damage occurring to the transcribed strand during the life of the cell.

The Steps in Replication

Replication involves more than the simple synthesis of complementary polynucleotide chains. Chromatin must be disassembled so as to expose DNA to the polymerase, and sufficient new histones and other chromatin proteins must be synthesized and reassembled with the two emergent DNA molecules to create new chromatin. It is a complex process of many steps; our knowledge is often sketchy, but let us consider in sequence what is known.

Multiple Initiation Sites—Replicons. A DNA molecule is too large to be made in one piece. Mammalian DNA polymerases create approximately one micrometer

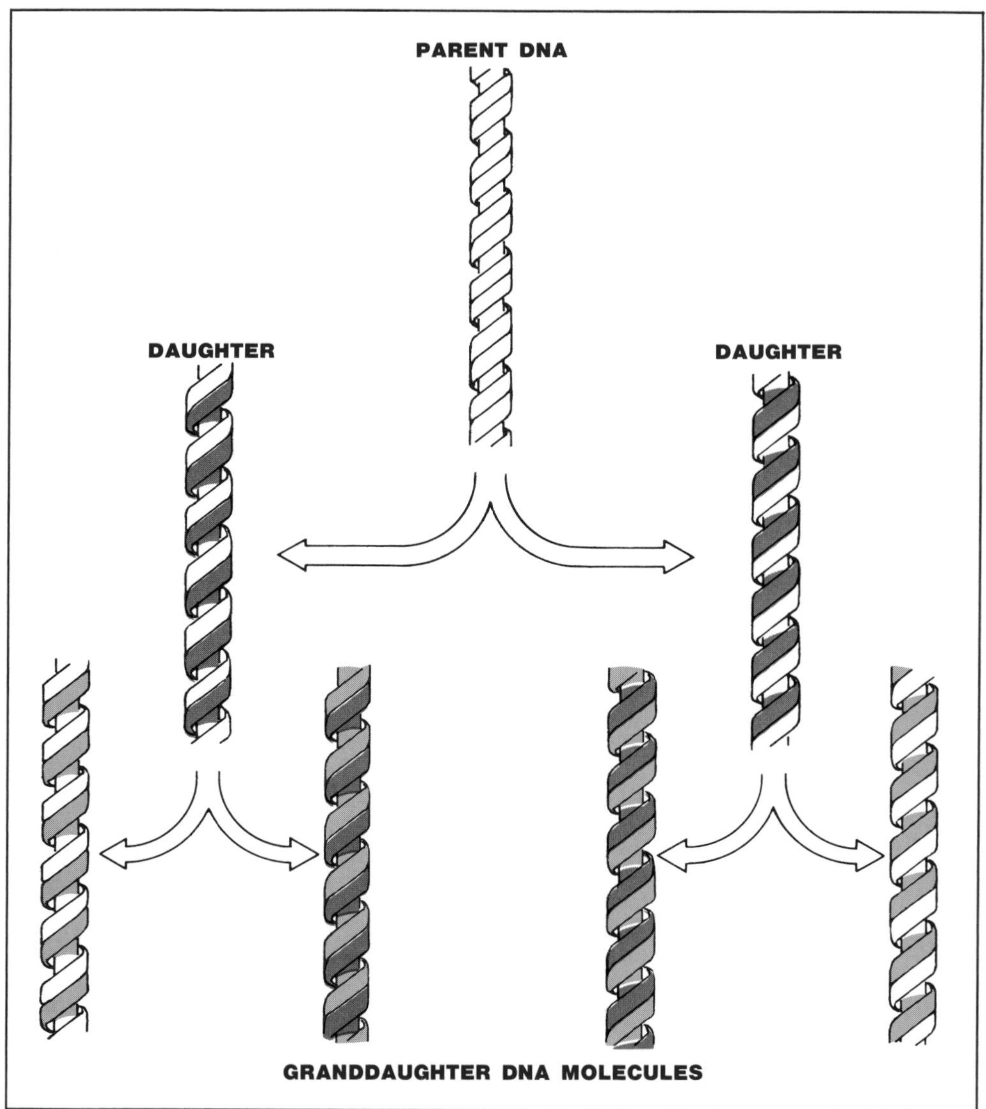

FIGURE 6–4 Replication of DNA proceeds by a semi-conservative mechanism in which a new complementary strand is synthesized to match each parent strand. Each daughter molecule in the first generation contains one parent strand (*white*) and one newly synthesized strand (*brown*). In the next generation, these strands will be in turn separated and distributed among the granddaughter molecules, each of which will contain one newly-synthesized strand (gray).

of DNA per minute. A polymerase molecule would therefore require about 55 days to make its way down the full length of the largest DNA molecule in human cells, whereas the observed time for total replication is frequently between four and eight hours, sometimes less. Replication is completed in a fraction of a day instead of many weeks by making new strands of DNA in short pieces, with polymerization proceeding in several places at once. Each molecule of DNA has several initiation points at which its replication begins. The segments that are replicated from one initiation point are termed replicons; there are perhaps 1000 replicons, each some 15 to 60 μm in length, per chromosome.

Mechanism of Initiation. We don't know how replication begins. Even the nature of the initiation points is unknown. The long chromatin molecules appear to be organized in loops attached to the structural skeleton of the nucleus, and it may be that each loop constitutes a replicon.* One thing is certain: The DNA must be unwound from a nucleosome preceding the replication site, and the duplex DNA itself must be unwound to make both strands available as templates for DNA polymerase molecules. A **helix-destabilizing protein** with a high affinity for single-stranded DNA effectively lowers the melting point of duplex DNA below the ambient temperature.

Unwinding DNA creates a topological problem. When the strands in the middle of a rope are untwisted, the entire rope must rotate to prevent formation of a kink elsewhere. The molecules of DNA are so large that its loose ends can't be whirled around. Bacteria nick one strand during replication so as to avoid strain in their circular chromosomes, which also have a supertwist forming a coiled coil. The mechanism of strain relief during eukaryotic replication may be similar.

Replication Forks. When the central portion of a duplex DNA molecule is unwound, two topologically identical forks are created at the points where the strands diverge:

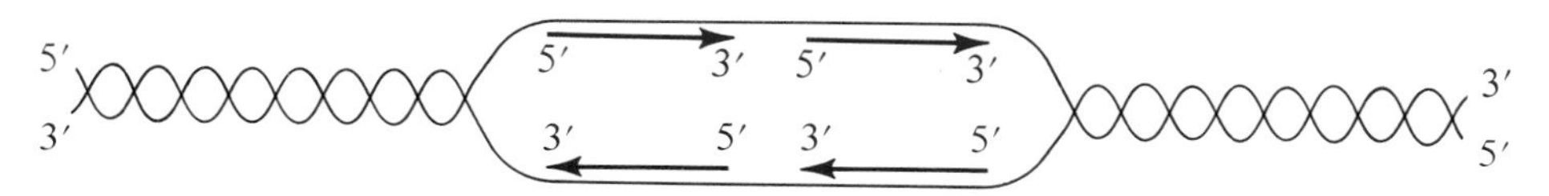

The machinery of replication works on both strands at each of these forks. One strand is replicated toward the fork at one end and away from the fork at the other end, while its antiparallel complement is being replicated in correspondingly opposite directions at the two ends. **Replication is therefore bidirectional,** proceeding toward the ends of the chromosome from each initiation point. Four molecules of DNA polymerase could be functioning at once for each replicon—two per fork. If the number of replicons per DNA molecule is of the order of 1000, replication could be proceeding simultaneously at about 2000 forks, with both strands being replicated at each fork.

RNA Primers and Okazaki Pieces. As a DNA duplex untwists, one strand is replicated in long segments, perhaps a continuous piece, at each fork, with the strand growing toward the fork as the parent molecule is untwisted farther. The other strand is replicated in short discontinuous pieces, known as Okazaki pieces after the man who first discovered them.†

Since the original DNA strands are antiparallel, one of the strands must be read by DNA polymerase traveling away from the replication fork. As further untwisting occurs, the fork in effect moves in the opposite direction from the polymerase. The growing gap can only be filled by the polymerase jumping to a new position closer to the fork and beginning replication of another short piece, which again will grow away from the fork.

However, the DNA polymerase can only add deoxyribonucleotides to an existing strand; it cannot begin a new one. A short RNA primer is first laid down, and new DNA is made as an extension of the primer. Cells that are preparing to divide activate an RNA polymerase to make short pieces (approximately 9 nucleotides) of

---

*One hypothesis is that DNA polymerase is bound at the point of attachment of the loops and that the loops feed through the polymerase site during replication.

†Dr. Reiji Okazaki at age 14 was in Hiroshima when the atomic bomb exploded. He later developed myelogenous leukemia and died August 1, 1975, at age 44.

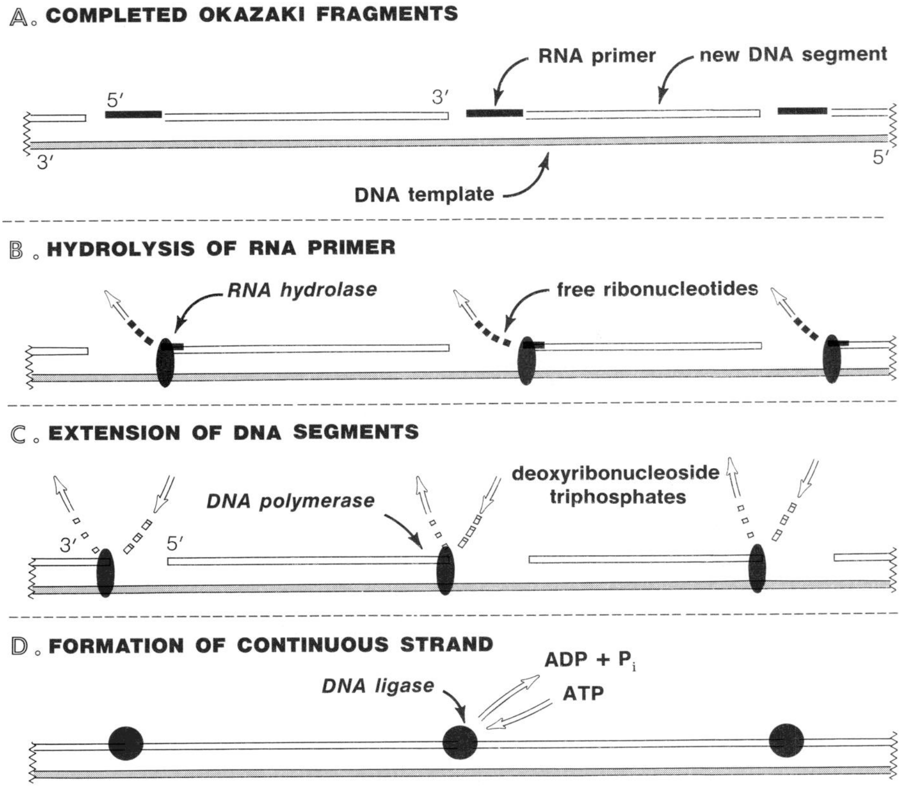

FIGURE 6–5 *A.* DNA is replicated in discontinuous segments, known as Okazaki fragments, each of which is headed by a length of RNA primer. *B.* The RNA is removed by hydrolysis, catalyzed by a hydrolase specific for RNA-DNA hybrids. *C.* The gaps are then filled by increasing the length of the DNA segments through the action of another kind of DNA polymerase. *D.* Finally, the segments are joined through the action of a DNA ligase, which drives the reaction by simultaneously hydrolyzing a molecule of adenosine triphosphate (ATP).

RNA complementary to the exposed single strands of DNA. The RNA may be long enough to make some specific RNA-DNA hybrid structure. In any event, this RNA primer enables DNA polymerase to begin adding deoxynucleotides. The result is a short piece of DNA (40 to 290 nucleotides, average 135) with an 8- to 10-nucleotide RNA head. The tail of the DNA abuts the RNA head of the previous Okazaki piece (Fig. 6–5).

Removal of RNA and Combination of Okazaki Pieces. The ultimate products of replication are continuous strands of DNA equal in length to the original templates. The conversion of Okazaki pieces to continuous strands involves removal of the priming segments of RNA, filling the gaps between the segments by elongating them with more deoxynucleotide residues, and then joining them.

The short RNA primer is removed by an enzyme that specifically catalyzes the hydrolysis of RNA attached to complementary DNA without affecting other RNA molecules in the cell. Once the priming RNA groups are removed, the space they occupied is filled by the action of another DNA polymerase, which extends the DNA chain to cover the now exposed template. After the gaps between DNA segments are filled, the abutting 3′ and 5′ ends are joined by still another enzyme, a DNA ligase, thereby creating a continuous strand. (A molecule of adenosine triphosphate is hydrolyzed to provide the energy for the ligase reaction: the utilization of ATP for such purposes is discussed in detail later.)

Modification of Bases and Restriction Enzymes. After replication of the base sequence is completed, a few of the bases in DNA are modified; methylation of some cytosine residues is especially common. The function of these changes in eukaryotes is not known; in prokaryotes they serve in part as a device for identifying the DNA of the organism. Cells produce restriction enzymes to catalyze the hydrolytic destruction of any DNA molecule that does not have the specific pattern of modified bases peculiar to that cell. If the DNA cannot be identified as a friend, it is taken to be a foe and is destroyed. This not only removes some viral invaders, but it also can remove parts of the cell's own DNA strands that have been damaged. (The segments destroyed by restriction enzymes are usually in or near palindromes.)

Construction of New Chromatin. As the daughter and parent DNA strands emerge from the replication sites, they wrap around histone cores to form precursor nucleosomes. In some cases, an intact histone octamer from parent nucleosomes is used, in other cases a completely new set of histones associates to make the core. The winding of DNA around the core creates a countertwist elsewhere that is believed to be relieved by a **topoisomerase.** This enzyme breaks one strand of DNA without releasing the ends and reseals them on the other side of the intact strand (Fig. 6–6). The effect is to generate a supercoil in the DNA. Further adjustments, probably involving other proteins, occur before the final nucleosome conformation is attained some 5,000 to 40,000 base pairs distant from the replication fork. Details of these steps and of further coiling and compaction of the chromatin are lacking.

THE CELL LIFE CYCLE

A cell activated to undergo cell division performs a series of orderly, integrated biochemical events culminating in the formation of two daughter cells. It has been customary to view these events as occurring in discrete phases. The microscopically observable **mitotic phase (M)** and the biochemically detectable phase of **DNA synthesis (S)** were separated by two phases called **gaps.** The phase between mitosis and DNA replication was called $\mathbf{G_1}$, and the phase between DNA replication and mitosis was labeled $\mathbf{G_2}$.

It is almost self-evident that rigorous controls must exist to limit the size of organs and tissues. Hence, a normal organ or tissue only permits its cells to proceed in the cell cycle as required for its functions at the time. A major purpose of the physical examination of a patient is to determine if the individual organs have in fact exceeded their normally limited size. Enlargement may represent excessive growth, as well as inflammatory swelling.

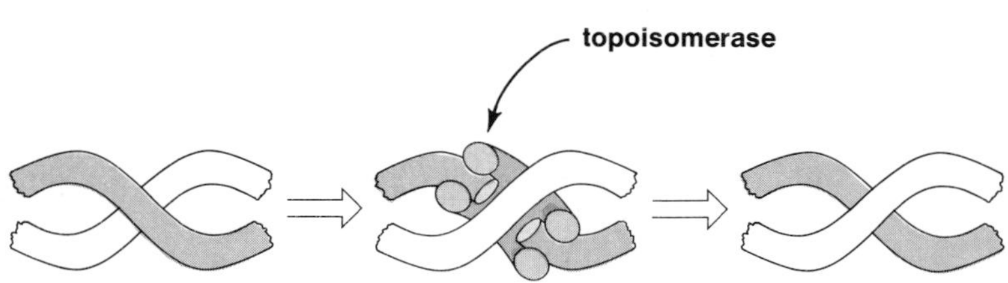

FIGURE 6–6 Diagram of eukaryotic topoisomerase action. The enzyme cleaves one strand of duplex DNA and reseals the ends on the opposite side of the untouched strand. The effect is to introduce a supertwist into the double helix.

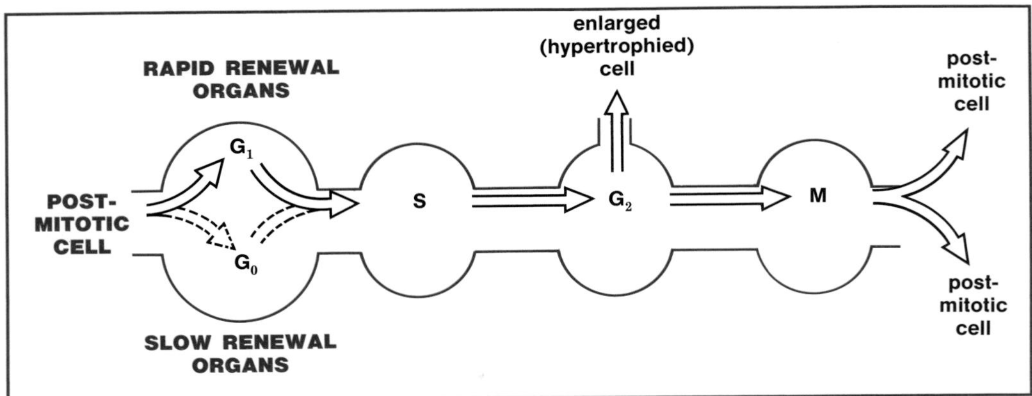

FIGURE 6–7 Outline of the cell cycle. Postmitotic cells in some organs enter a brief gap phase (G_1) as a preliminary to renewed division. In other organs, the cells carry out their function over a long period (G_0) without division. Premitotic events involve DNA synthesis (S phase) and a second gap phase (G_2) from which enlarged cells are sometimes diverted. Otherwise, mitosis (M) proceeds, with the two daughter cells ready to begin the cycle again.

Organs that regularly sustain cell losses, such as the hematopoietic system and the gastrointestinal tract, have a large fraction of their cells in almost continuous cycles of cell replication. These organs appear to have the program leading to cell division continually active, and are considered to enter a G_1 period after mitosis (Fig. 6–7). Other organs such as kidney and liver have a very small fraction of cells in division while the majority of cells have long quiescent periods between cell divisions. Such cells appear to inactivate the program for cell division at the completion of mitosis. A separate phase, G_0, has been proposed for such cells.

Variations in the proportion of cycling cells occur in many organs. The number of cells forming leukocytes increases in response to bacterial infection. Partial hepatectomy rapidly signals the onset of cell division in the remaining liver. Changes in the proliferative activity of the uterus regularly accompany changes in the estrogen and progesterone levels in the body, and are the basis for the menstrual cycle.

Requirements for Macromolecular Synthesis. Interference with ribonucleic acid synthesis or protein synthesis at any time in the cycle up to the onset of mitosis will either destroy the cell or delay the progression through the cycle. The program of cell cycle is arranged to provide for the transcription of various RNAs and the translation into proteins at specific times so that the appropriate components are available when required. In contrast, deoxyribonucleic acid synthesis is scheduled to occur only in the S phase. During mitosis the synthesis of all macromolecules decreases markedly.

Duration of Phases. There is considerable variation in the time spent in G_0, G_1, S, and G_2. Neurons and muscle cells may spend a lifetime in G_0. G_1 may be as short as two hours or last for days. S in human cells is known to vary from 11 to 34 hours; the G_2 phase occupies from 3 to at least 16 hours. Mitosis continues for about 30 minutes to 1 hour.

G_0–G_1 Phase. The fulfillment of metabolic and structural functions by mature cells does not require mitosis, and a majority of the cells in the body are in G_0 or G_1 phase. The decisive early events in the transition from these phases have been difficult to determine. They include activation of genes leading to synthesis of regulatory nonhistone proteins in the cytoplasm. These proteins move to the nucleus, are modified (see Chapter 8), and activate new areas of the genome resulting in the transcription of new RNA. By the sequential unfolding of this pattern, the cell has increased the synthesis of ribosomal RNA as well as new messenger RNA. The enzymes for DNA and histone synthesis appear, and the cell is ready for the S phase.

Putrescine and spermidine, which are aliphatic polyamines, are also produced in increased quantities:

$$\overset{\oplus}{H_3N}-CH_2-CH_2-CH_2-CH_2-\overset{\oplus}{NH_3}$$

putrescine

$$\overset{\oplus}{H_3N}-CH_2-CH_2-CH_2-\overset{\oplus}{NH_2}-CH_2-CH_2-CH_2-CH_2-\overset{\oplus}{NH_3}$$

spermidine

Since these compounds are multivalent cations at physiological pH, they are presumed to react with the negatively charged backbone of DNA or RNA for some important purpose yet to be discovered. (The formation of putrescine and related compounds is discussed on page 656.)

Synthesis of DNA, S phase. When the necessary proteins are at hand, replication of DNA begins. New DNA is not synthesized helter-skelter; there is a defined sequence in the replication of different segments of the molecules, with heterochromatin being made late in S phase when it may be necessary for structural organization of the chromosomes.

Histones are also synthesized during the S phase. In order to achieve the necessary speed of transcription and translation, the histone genes occur in multiple sets. Each set contains one gene for each of the five histones and the sets are repeated in tandem. The messenger RNAs of the histones are unusual in not having polyadenylate tails.

When replication is completed, the cell is tetraploid, with four of each kind of autosomal chromatid. (Autosomes are the chromosomes other than sex chromosomes; a chromatid is the fraction of a chromosome containing one duplex DNA molecule.) The number of sex chromatids also has been doubled.

Second Gap, G_2 phase. Complete replication of DNA is followed by another period in which the necessary RNA and proteins are synthesized. Less is known about the character of these proteins, which may be required to accelerate the condensation of DNA into compact chromosomes. In addition, there is a decrease in the synthesis of the H2A, H2B, H3, and H4 histones.

Some cells may develop differently in G_2, growing larger without division. This may be a common event in muscle, where there are many tetraploid cells—cells with twice the normal content of DNA. In the adult heart, 95 per cent of the cells are tetraploid; only 15 per cent are tetraploid in children under 3 years of age. (An enlargement of tissues without increase in cell number is called hypertrophy; an increase in cell number is a hyperplasia.)

Mitosis, M phase. The terminal event of the cell cycle is the segregation of the constituents of one cell into two daughter cells. It is this phase that is most dramatic under the light microscope, but it is usually complete within one half hour, whereas some 8 to 12 hours are required to prepare for mitosis.

It is not necessary for us to review in detail the complicated sequence of events by which the constituents of one cell are segregated into two cells—the functionally important event from the standpoint of biochemical genetics is the division of the chromosomes. Replicated chromatin condenses into thick diploid chromosomes, in which the chromatin fibers are looped into cylinders some 400 to 800 nm in diameter. The chromosomes are assembled into a planar array and then are split into haploid pairs, with one of each pair drawn into opposite sides of the cell. This assembly and separation is accomplished by the spindle apparatus, made from microtubules (Chapter 9), and including the proteins actin and myosin, which contribute motility (Chapter 20).

Genetic Recombination. Similar DNA molecules are in close proximity during mitosis until chromatid separation, and are subject to exchange of segments between molecules. This may happen, for example, by breaking polynucleotide chains at precisely comparable places on the molecules, and resealing the breaks between the molecules rather than in a single molecule:

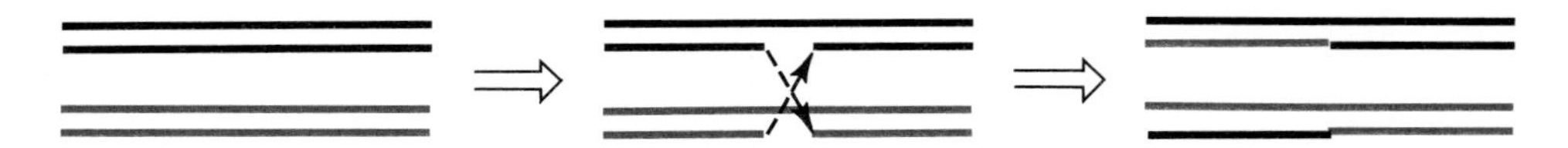

Crossovers rearranging genes from different parental strands are rare in dividing somatic cells. (It would be difficult to detect crossovers between two daughter DNAs; if the crossover and replication are perfect, the products would still be identical.) However, genetic recombination is the all-important device for reshuffling the properties of parents among their children, and the molecular mechanism hinges upon the peculiar life history of germ cells.

Germ cells go through a different sequence of divisions by which the tetraploid chromatin created in the S phase is parceled out among four haploid daughter cells, each of which contains only one copy of one gene. Included in this sequence is a **meiosis**—a division in which replicated sister chromatids from both parents are fused into a single chromosome. The related, but not identical, DNA molecules derived from the parents are then attached to each other, so that exchange of whole gene segments by crossover creates new combinations of genes (Fig. 6–8). We infer that our survival is evidence for precisely aligned crossover regions between genes—our proteins are not a random hodgepodge of pieces. Some parts of the genome are especially prone to reshuffling; the genes that move together are said to be a **transposon.**

To recapitulate, each eukaryotic somatic cell contains two similar, but not identical, molecules of DNA specifying the same genes. Upon replication for **mitosis,** each of these molecules is converted to two identical copies. The cell then contains two different and separate pairs of DNA molecules; each pair is made of identical molecules that are attached to each other until separation of the haploids during mitosis.

During meiosis, two identical copies of each of the two different DNA molecules are also made, but the four resultant molecules are now fused into one grand chromosome, in which exchange of segments occurs among all four. After exchange, the four molecules are all different, with each having a unique combination of genes from the two parent molecules. When these four molecules are distributed among four germ cells, each will contain a different heredity. Since the exchange process itself may occur in many different ways, the odds are high that no two germ cells will be exactly alike.

Therapeutic Disruption of the Cell Life Cycle

Compounds that kill cells include powerful drugs used in the treatment of human diseases as well as useful agents for the analysis of biochemical events during the life of the cell. Many of the effective cytotoxic drugs interact with enzymes involved in replication, transcription, or translation, or with the nucleic acids them-

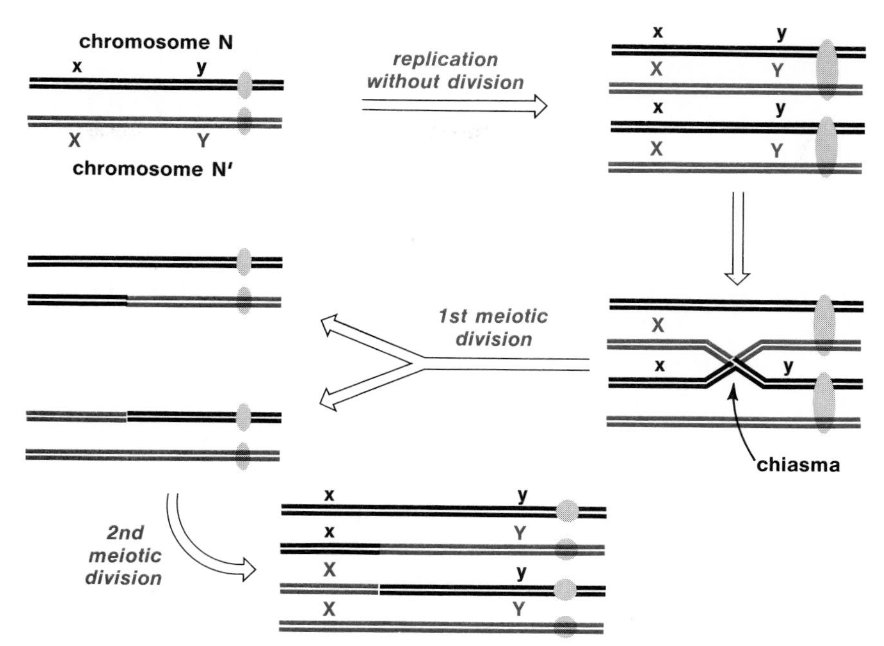

FIGURE 6–8

The crossovers occurring during meiosis recombine segments of DNA so that one molecule contains some genes derived from one parent, and some from the other. When a germ cell precursor replicates without division (*top*), the resultant bivalent contains two identical pairs of duplex DNA molecules. After crossover and the first meiotic division (*center*), the pairs of DNA molecules are no longer identical. Separation of these pairs to create haploid germ cells (*bottom*) results in four different chromatids. Two of these are like the parents on the segment shown, but other segments on these molecules would probably be unlike the parents in their combination of genes.

selves. They are referred to as **chemotherapeutic agents,** although those produced by microorganisms are more specifically called **antibiotics.**

Several categories of clinical situations justify the use of such potentially toxic drugs. These include:

1. Changes of cells into cancerous growths.

2. Circumstances in which the normal immune response has undesirable consequences. In order to permit grafting of one person's kidney or heart into another, it is necessary to treat the recipient with chemotherapeutic drugs, under the alias of **immunosuppressive drugs,** so as to prevent the proliferation of those cells that would destroy the graft (except in an identical twin). Autoimmune diseases, a poorly understood and heterogeneous group in which immune reactions damage apparently normal tissues, are another example.

3. Infections by bacteria, fungi, or other parasites. Drugs used in bacterial infections are selected to be less dangerously toxic to the host by exploiting the differences between eukaryotic and prokaryotic cells, such as different ribosome structure.

Selectivity and Specificity

The use of chemotherapy for the treatment of cancer is still empirical and relatively crude. Even so, it is now possible to cure several kinds of disseminated human cancers. What are the problems preventing more general success? A major one is our ignorance of the biochemical characteristics of malignancy, despite the obvious differences in appearance and behavior between normal and malignant cells. If there is a biochemical event unique to the malignant state, it is yet to be found. Consequently, there is no basis for rational development of a drug specifically affecting malignant cells, and no such drug has appeared adventitiously. Attacks on malignant cells with currently available agents also affect normal cells; the toxicity of these

TABLE 6–2 DRUGS INTERFERING WITH NUCLEIC ACID FUNCTIONS

Agents Used in Cancer Chemotherapy

Breaking DNA strands
bleomycin
procarbazine
gamma radiation
neocarzinostatin (in clinical trials)
mitomycin

Crosslinking DNA strands
alkylating agents
mitomycin
cis-platinum(II)diammine chloride

Removes DNA supercoiling
mAMSA* (in clinical trials)
intercalating agents
cis-platinum(II)diammine chloride
ellipticine (in clinical trials)

Inhibits DNA methylation
5-azacytidine (in clinical trials)

Inhibits DNA polymerase
aphidicolin (preclinical testing)
arabinosyl cytosine

Inhibits DNA synthesis at unknown site
arabinosyl cytosine
5-azacytidine (in clinical trials)
daunorubicin
doxorubicin
vinblastine
vincristine

Inhibits synthesis of 45S rRNA
actinomycin D
daunorubicin
adriamycin
mithramycin
nitrogen mustard
chloro-ethyl cyclohexylnitrosourea (CCNU)
ellipticine (in clinical trials)

Inhibits processing of 45S rRNA
CCNU
bis-chloro-nitrosourea (BCNU)
dihydro-5-azacytidine (in clinical trials)
5-fluorouracil
5-fluorouracil deoxyriboside (5 FUdR)

Inhibits synthesis of hnRNA
actinomycin D
nitrogen mustard

Inhibits mRNA polyadenylylation
3′-deoxyadenosine (experimental)

Inhibits transport or cleavage of mRNA
BCNU

Inhibits tRNA synthesis
formycin (not used in patients)
5-azacytidine (in clinical trials)

Inhibits mitochondrial tRNA synthesis
ethidium bromide (not used in patients)

Disaggregates polyribosomes
BCNU
D-galactosamine (experimental)

Prevents ribosome formation
actinomycin D

Inhibits peptide chain synthesis
emetine (not used in cancer)
cycloheximide (not used in patients)
anguidine (in clinical trials)

Terminates peptide chain synthesis prematurely
puromycin (not used in patients)

Induces synthesis of fraudulent RNA
5-azacytidine (in clinical trials)

Interferes with mitotic spindle
maytansine (in clinical trials)
vincristine
vinblastine

Cross-links DNA with protein
adriamycin

Agents Used in Infections (bind bacterial ribosomes)

chloramphenicol	blocks peptidyl transferase
streptomycin	affects initiation complex, prevents normal triplet recognition
tetracycline	blocks binding of aminoacyl tRNAs

*Amsacrine = 4′-(9-acridinylamino)-methanesulfono-m-anisidine

drugs is therefore a major limitation to their use. Quantitative, rather than qualitative, differences must be exploited, and new insights into biochemical events during the cell cycle have been of critical importance in the design and use of new drugs. (There is also a reverse flow; the effects of the drugs sometimes extend our appreciation of the biochemistry involved.)

The sites of action of many drugs influencing nucleic acid structure or function, including experimental as well as clinically useful examples, are shown in Table 6–2. The same drug may be listed several times because it has several effects of undetermined relative importance. Some drugs are active in quiescent cells (G_0 to G_1) as well as in cycling cells. Some may act on a particular phase of the cycle.

Nitrogen mustards, which are analogues of the sulfur-containing mustard gases used during World War I (1914 to 1918) react with guanine residues (Fig. 6–9). These bivalent compounds form cross-links between adjacent segments of DNA, preventing both replication and transcription.

Intercalating agents, compounds containing a large planar ring system that will slip between base pairs in DNA (Fig. 6–10), also prevent transcription. Actinomycin D is an effective example that fits best between G-C pairs, where both of its two peptide chains can form hydrogen bonds with the guanine amino groups. (Actinomycin D is one of many antibiotics produced by Streptomyces species, which are soil bacteria resembling fungi in their growth habit.)

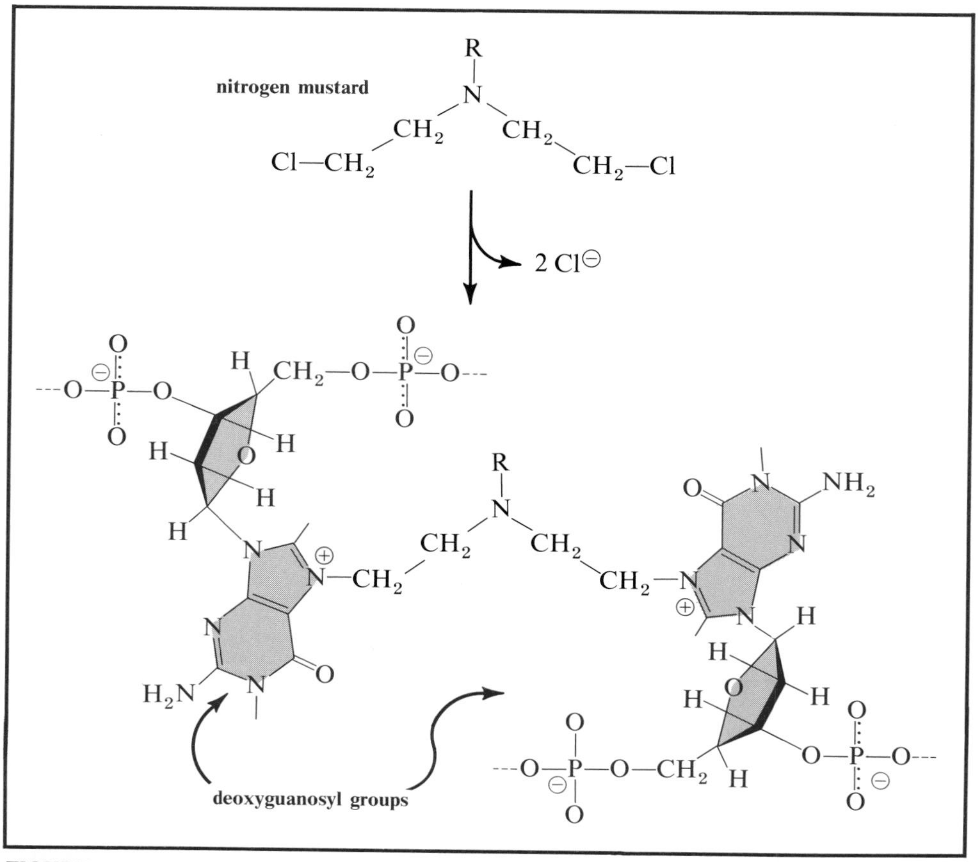

FIGURE 6–9 Nitrogen mustards are alkylating agents that link two deoxyguanosyl residues in DNA, thereby preventing the untwisting necessary for replication or transcription. A principal mode of action is shown here.

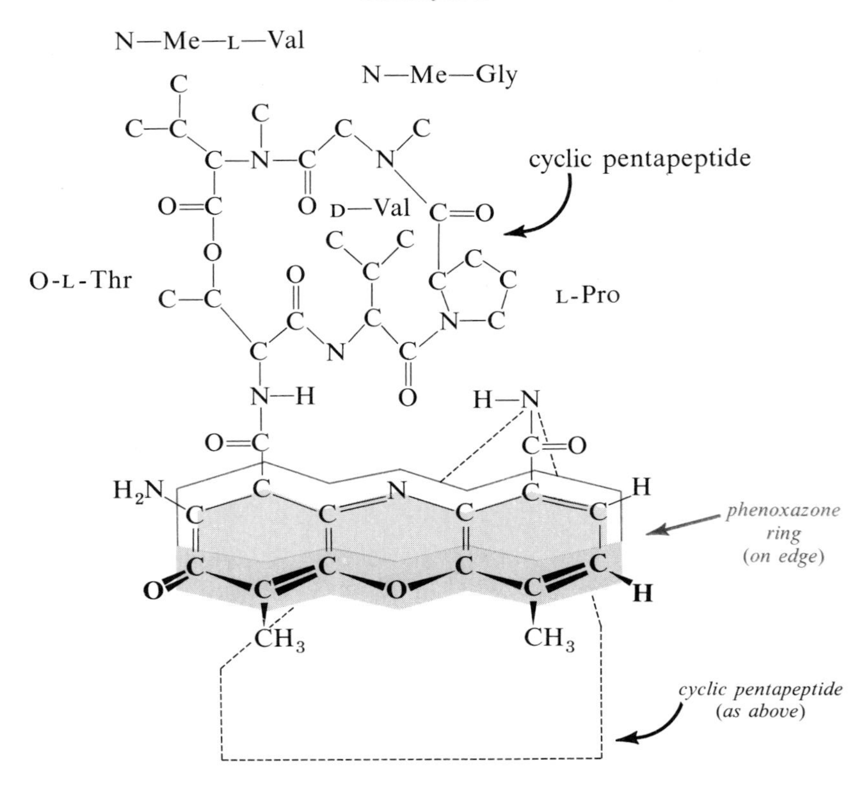

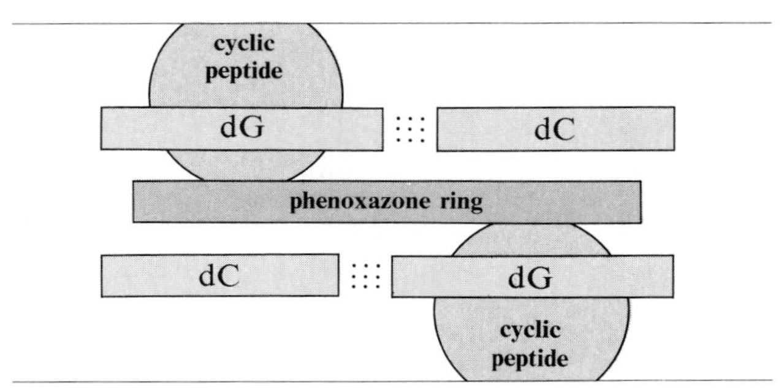

FIGURE 6–10

Actinomycin D is an example of an intercalating agent, containing a planar ring system that will slip between base pairs in DNA, causing it to partially untwist and become rigid. The top formula shows the phenoxazone ring of actinomycin D on edge. Attached to the ring are two identical cyclic pentapeptide units containing modified amino acid residues. (Substituent H atoms are omitted for clarity.) These pentapeptide groups, only one of which is shown in detail, are roughly at right angles to the intercalating ring, and are so disposed that they fit in the grooves of the DNA double helix.

The structure of actinomycin D is such that it forms stable bonds when inserted between two dG-dC pairs, as shown at the bottom.

Chemotherapy for Infections. Many agents are available to block the cell cycle of bacteria without substantial effect on the cell cycle of the human host. Most of these agents act by binding one or more components of the bacterial ribosome. Superficially, this appears to account for the selectivity of the agents, since the bacterial and eukaryotic ribosomes differ in composition and structure. **Chloramphenicol** and **streptomycin,** for example, have little effect on ribosomes from the cytosol of mammalian cells at concentrations that totally prevent protein synthesis by bacterial ribosomes. However, this is not enough to account for the differences in toxicity because the mitochondria of the mammalian cells also have ribosomes that resemble the bacterial ribosome, and isolated mitochondrial ribosomes are susceptible to inhibition by many of the same agents, such as chloramphenicol, streptomycin, and tetracycline. Antibiotics such as these may be safe to use only because they enter mammalian cells less readily than they enter bacterial cells, or because they do not cross mitochondrial membranes. Even so, some of the toxic side effects of these antibiotics may be due to their entrance into host mitochondria.

VIRUSES AND THEIR REPLICATION

Viruses are simple organisms composed of one or more molecules of nucleic acid, usually associated with at least a few protein molecules. The proteins may have only a structural role, such as the coat of proteins that polymerize as a capsid around

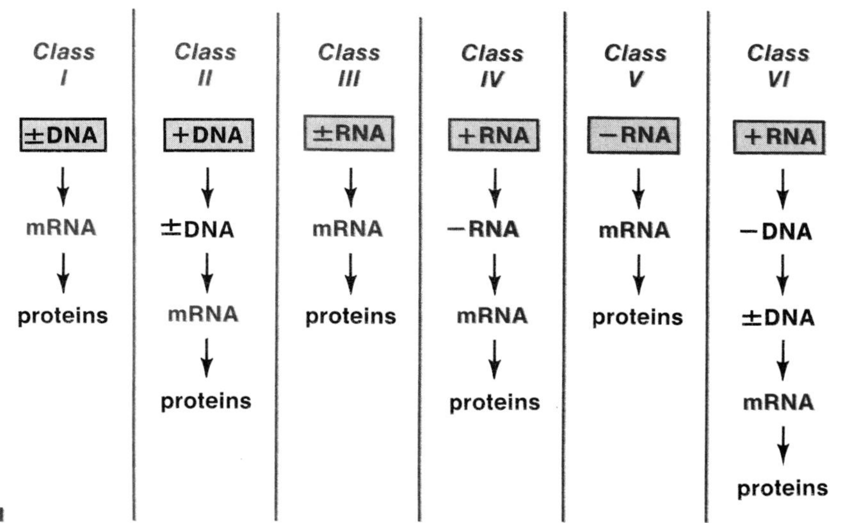

FIGURE 6–11

Information flow from various classes of viruses. The nucleic acid in the viral particle is shown in boxes at the top. DNAs are in black and RNAs in brown throughout. Plus strands have bases in the same sequence found in mRNA, minus strands are the corresponding antiparallel complement. For example, a Class VI virus has an RNA strand containing genes with the same sequence later found in the mRNA for those genes; the virus first causes a DNA strand antiparallel to the viral RNA to be synthesized by an RNA-directed DNA polymerase. This minus DNA is then replicated to form the corresponding plus strand; the duplex DNA (±) can then be incorporated into the host genome, with the minus strand later transcribed to form viral mRNA.

the nucleic acids in some viruses, or they may have a catalytic function during the life cycle of the virus. The viral nucleic acids specify the formation of these proteins, and sometimes others that are necessary for the replication of the virus, but are not included in it. The difference between viruses and other organisms is that the virus must invade other cells and use part of the machinery of the host in order to reproduce itself. Any cell—plant, animal, or protist—is subject to attack by viruses.

Viruses appear to carry their genetic information in every conceivable way. The information may be in DNA or RNA; the nucleic acid may be double-stranded or single-stranded, and it may be circular or linear, in one piece or several.

The DNA viruses may supply their own enzymes of replication, enabling them to multiply in the cytoplasm of the host cell, or they may rely upon the host cell's nucleus for a supply of DNA and RNA polymerases. The RNA viruses must perforce supply novel enzymes to use RNA as a template, either **RNA-directed RNA polymerase** or **RNA-directed DNA polymerase.** The flow of genetic information from the various classes of viruses is shown in Figure 6–11.

The RNA viruses that cause the formation of RNA-directed DNA polymerase have attracted special attention in recent years. They include **oncogenic** (cancer-producing) viruses, and the DNA that is produced can be incorporated into the host cell's genome for an indefinite latent period, replicated in many generations before the viral genome is expressed to create more viral particles. In addition, the RNA-directed DNA polymerase, also known as **reverse transcriptase,** is a powerful laboratory tool, as discussed in the next chapter.

The viruses, simple as they may be, are like other forms of life in exerting some control over the expression of their genetic information. Particular functions may appear early or late in their life cycles.

FURTHER READING

General Reviews

A. Kornberg (1980). DNA Replication. Freeman. Readable account by Nobel Laureate in field.

B. Lewin (1980). Eucaryotic Chromosomes, vol. 2 of Gene Expression, 2nd ed. Wiley. Excellent, thorough survey.

R. Dulbecco and H.S. Ginsberg (1980). Virology. p. 857 in B.D. Davis et al., eds. Microbiology, 3rd edition. Harper and Row. Authoritative.

Technical Reviews

H. Busch, ed. (1974–). The Cell Nucleus. Academic Press. Multivolume authoritative source.

T. Igo-Kemenes, W. Horz, and H.G. Zachau (1982). Chromatin. Annu. Rev. Biochem. 51:89.

A.B. Pardee et al. (1978). Animal Cell Cycle. Annu. Rev. Biochem. 47:715.

L.A. Loeb and T.A. Kunkel (1982). Fidelity of DNA Synthesis. Annu. Rev. Biochem. 51:429.

S. Sperling and E.J. Wachtel (1981). The Histones. Adv. Prot. Chem. 34:1.

K. Geider and H. Hoffmann (1981). Proteins Controlling the Helical Structure of DNA. Annu. Rev. Biochem. 50:233.

R. Yuan (1981). Structure and Mechanism of Multi-Functional Restriction Endonucleases. Annu. Rev. Biochem. 50:285.

L.A. Klobutcher and F.H. Ruddle (1981). Chromosome Mediated Gene Transfer. Annu. Rev. Biochem. 50:533.

D. Dressler and H. Potter (1982). Molecular Mechanisms in Genetic Recombination. Annu. Rev. Biochem. 51:727.

M. Gellert (1981). DNA Topoisomerases. Annu. Rev. Biochem. 50:879.

T. Lindahl (1981). DNA Methylation and Control of Gene Expression. Nature 290:363.

S. Anderson et al. (1981). Sequence and Organization of the Human Mitochondrial Genome. Nature 290:457. Also see the two following papers by Montoya, Ojala, and Attardi.

Also see the reading list for Chapter 4.

CHAPTER 7

ALTERATIONS OF DNA—ACCIDENTAL AND INTENTIONAL

The character of cells and the organisms constructed from them depend upon the DNA that they contain; alterations in the arrangement of chromatin or in the sequence of bases will frequently cause obvious changes in the nature of the creature, be it one of the blessed or a mere bacterium.

The DNA of an organism contains much more information than is used at a given moment. Signals in the form of changing concentrations of particular chemical compounds govern the choice of particular segments of DNA to be transcribed, the number of molecules of each messenger RNA that will be made, and the rate of translation of each mRNA molecule. The character of an organism therefore hinges not only on the peptides that it is able to make but also on the way in which it responds to changes in its environmental circumstances. Environment controls the expression of heredity, but the response to environment is in itself a hereditary characteristic that is subject to mutation.

CHANGES IN THE GENETIC MESSAGE

Damage to DNA

It is reasonable to expect that something might go wrong with the giant DNA molecules. Chemical reactions may change some of the 6×10^9 base pairs present in each somatic cell. Some of the long strands may rupture. An occasional mistake in replication may cause the wrong base to be inserted in a new strand. The wrong strands may be connected during recombination. Such errors do occur. Many of them, maybe most, result in the death of the individual cell, and there are so many cells that the loss is trivial. Some disrupt the genetic program of the cell so as to promote unrestrained division of the cell into a malignant growth and the causes of such errors are therefore said to be carcinogenic or oncogenic (onchos = tumor). Others result in the development of more benign anomalies that are perpetuated as the cell divides. (A similar spectrum of changes may follow insertion of sequences by a virus.)

Errors in the DNA of germ cells are more serious because they may be replicated in all of the cells of the offspring and continued in the descendants. Such errors are said to be **mutations,** and these constitute a major social problem. In a survey of hospitalized children in Baltimore, 6.4 per cent were in the hospital because of mutations in a single gene, 0.7 per cent were there because of a more gross aberration in chromosome structure, and 31.5 per cent were judged to have some gene-influenced condition. Only 53.5 per cent were believed to have a nongenetic condition or disease, with the remaining 8.2 per cent not assigned to any category. Approximately one in ten infants is born with a malformation or other genetic disorder; five per thousand have some disorder that can be transmitted genetically to their offspring.

Data on the low frequency of some genetic conditions can mislead the unwary on the prevalence of the altered genes that cause them. Consider phenylketonuria, to be discussed in detail later (p. 611), which results from one of several mutations, and produces mental retardation if treatment is not begun in early infancy. From 1,000,000 live births in the U.S., only 71 infants will have phenylketonuria. Each of the 71 is unable to make the normal protein. Because the condition is recessive, both chromosomes carrying specifications for the protein must be defective, and both parents must have contributed defective versions of the gene. In such cases the probability that a child will have two defective genes is the product of the probabilities for the presence of one defective gene in each chromosome. Since the gene is autosomal—that is, it does not occur on a sex-determining chromosome—the probabilities are the same for males as they are for females in large populations. We can therefore estimate the gene frequency for phenylketonuria to be $(71 \times 10^{-6})^{1/2}$, which is 8.4×10^{-3}. Approximately one out of every 60 humans has a gene for phenylketonuria in one of his paired chromosomes, an impressive incidence, and there are many other deleterious conditions of similar prevalence.

Origin of Errors

Accidents in Replication. Something occasionally goes wrong with the polymerization of new DNA strands. A nucleotide may be omitted, or an extra one inserted, but the most common error is the incorporation of a nucleotide that is not the complement of the corresponding base on the template strand.

When one purine is substituted for the other, or one pyrimidine for the other, the error is known as a **transition.** The substitution of a purine for a pyrimidine, or vice versa, is known as a **transversion.**

Since the accuracy of replication hinges upon the strength of bonding between complementary nucleotides in a ternary complex with the DNA polymerase and not between themselves alone, accuracy is also affected by the nature of the polymerase. Some polymerases are more likely to make mistakes than others (Table 7–1). The error rate for the α-polymerase responsible for replication in mammals has been variously estimated at about one error per 30,000 base pairs. The RNA-directed DNA polymerase associated with a virus causing myeloblastosis (a form of leukemia) in birds is an example of an enzyme that makes many more errors. In experi-

ERROR RATE OF MAMMALIAN DNA POLYMERASES **TABLE 7–1**

Reciprocal rate given: number of base pairs added per error. The higher the number, the greater the accuracy of replication.

Polymerase	α	β	γ
Rate	23,800 to 30,500	2,930 to 6,660	6,600 to 8,070

(Estimates from Kunkel and Loeb, Science 213:765 (1981).)

ments using synthetic templates, this enzyme made one error in as few as 600 nucleotide pairings. Some argue that a certain rate of error is necessary for adequate evolutionary response to new situations—random modifications of proteins permit the selection of changes enabling the organism to compete better, and rapid changes may be especially appropriate for viruses that must generate new forms in order to survive as their hosts become immune to the original versions. Even if so, flexibility may be gained at a price, for most of the errors in coding are believed to be deleterious rather than advantageous.

Some errors may result from the existence of the bases in more than one form. For example, we draw the ring of adenine in a fully aromatic form but draw the other bases as keto tautomers because these are the most abundant forms. However, a small amount of each exists as the other tautomer at a given moment (Fig. 7–1), and the tautomers may form hydrogen bonds with the "wrong" complementary base. Similarly, if the imide nitrogen of the rings of either guanine or thymine is ionized, these bases may bond with each other.

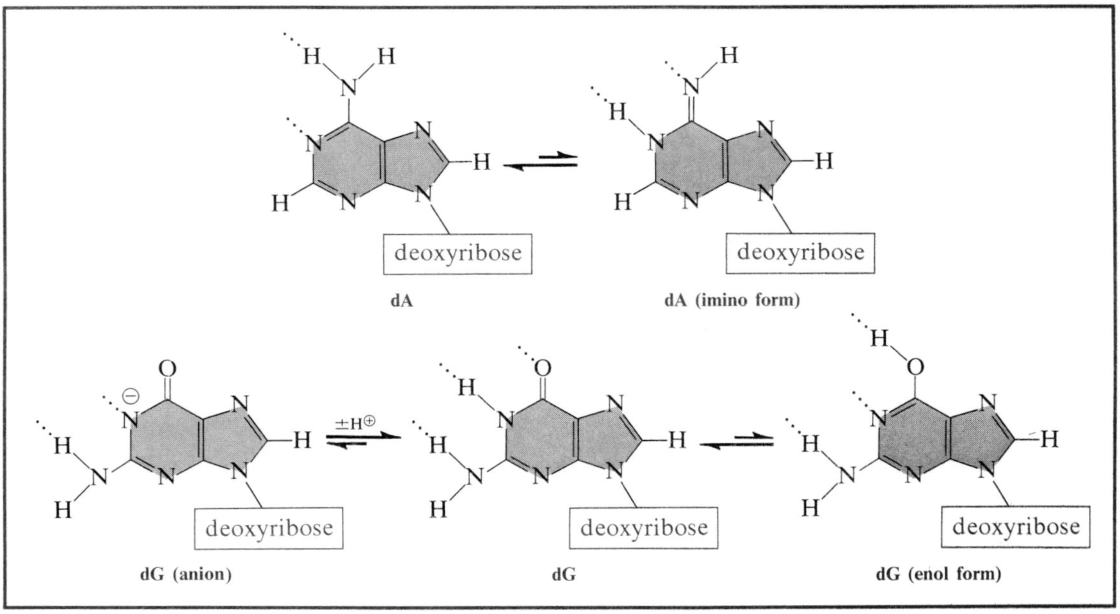

dA ⇌ dA (imino form)

dG (anion) ⇌ ($\pm H^{\oplus}$) dG ⇌ dG (enol form)

FIGURE 7–1 **The bases in nucleic acids have a slight tendency to exist in tautomeric forms, with a corresponding shift of donor and acceptor atoms for making hydrogen bonds. (The paired bases are not shown.) For example, the tautomer of adenine (*above right*) has atoms in position to form two hydrogen bonds with cytosine, whereas the tautomer of guanine (*lower right*) has atoms in position to form three hydrogen bonds involving both carbonyl oxygens and the imide hydrogen of thymine. The bases can also ionize to a small extent with the same result; the anionic form of guanine (*lower left*) can form hydrogen bonds with thymine.**

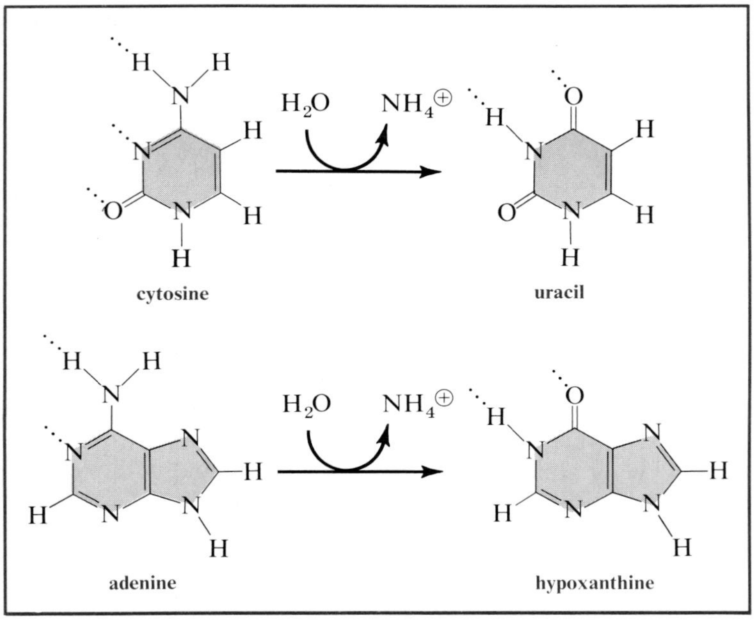

cytosine

uracil

adenine

hypoxanthine

H_2O $NH_4^{\oplus}$

FIGURE 7–2

Top. **Hydrolytic deamination of cytosine creates uracil.** ***Bottom.*** **The less common hydrolytic deamination of adenine creates hypoxanthine, which can form two hydrogen bonds with cytosine (not shown).**

Chemical Modification. Errors in DNA also appear at any time after it is formed. Spontaneous chemical reactions cause some. For example, cytosine residues occasionally undergo a **hydrolytic deamination** that converts them to uracil residues (Fig. 7–2). A similar, but slower, deamination of adenine residues forms hypoxanthine residues, which behave as if they were guanine residues in the formation of complements.

The base-deoxyribose bond in DNA can also be broken by hydrolysis. Purine nucleosides are most susceptible to this spontaneous reaction.

Radiation Damage. Electrons are expelled from DNA by energetic radiation, such as x-rays, gamma rays, or ionized particles. Reactions of the resultant intermediates include breakage of the polynucleotide chain through disruption of phosphodiester bonds and opening of the rings in the bases. When oxygen is present, it is also activated by radiation, and damage is significantly increased through the formation of additional oxidized products.

The destruction of DNA is the basis for the use of x-rays and other radiation against cancers. Approximately one hit out of 300 causes lethal damage to mammalian cells. The absorption of radiation also increases the risk of malignancy; although the risk is not well-defined for either high or low doses, a reasonable estimate is that approximately one additional person in each million people will develop a cancer for each rad absorbed per year. When a radiotherapist is dealing with the immediate and certain threat of cancer, the dosages given are limited more by the potential damage to the vital organs than by the prospect of later causing another cancer. The radiologist is much more cautious in using even low doses of radiation for diagnostic purposes, because there is no definitely safe lower limit. Improvements in technique are constantly diminishing the required exposures.

DNA is also damaged by absorption of photons of ultraviolet, or shorter, wavelengths. Absorption in the ultraviolet range mainly affects the pyrimidines, especially thymine, by activating the ethylene bond in the ring. If the activated pyrimidine is next to another pyrimidine, a bond may form between them, creating a dimer; if not,

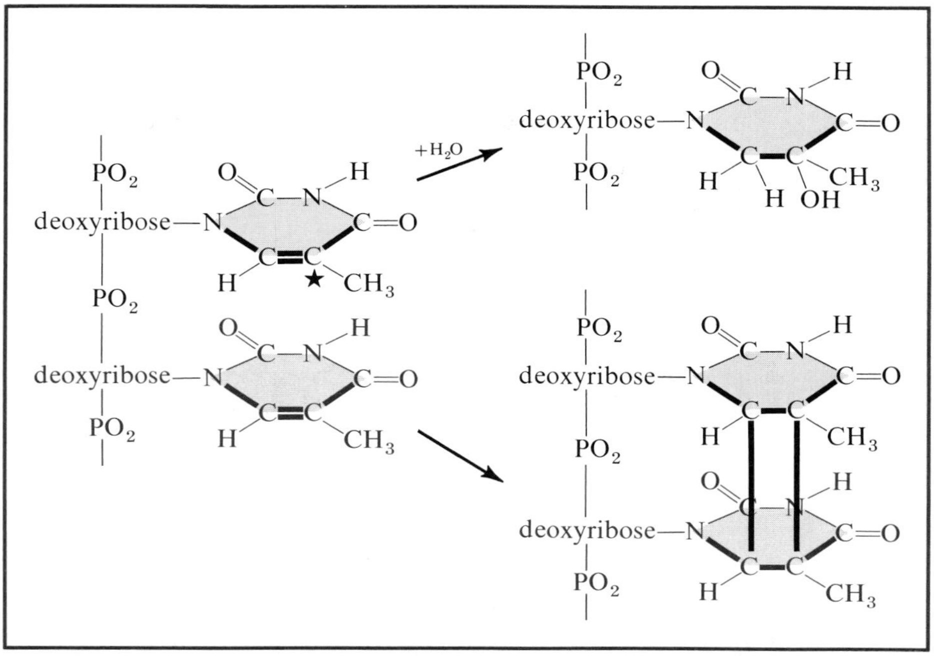

FIGURE 7–3 A residue of thymine (*black*) that is activated by absorption of a quantum of ultraviolet light may react with water (*top*) or with a neighboring second residue of thymine (*brown*). The hydration reaction is relatively innocuous, but the formation of a thymine dimer clamps DNA into a biologically inert configuration.

water may add to the molecule (Fig. 7–3). The formation of a dimer is the much more disruptive change; it may occur within a strand or between strands.

Chemical Mutagenesis and the Ames Test. Compounds in the environment capable of reacting with DNA and causing mutations of genes or of altering gene expression to transform normal cells into malignant ones have always been with us. The affinity of people for flame increased the exposure, particularly through pyrolysis of food, still occurring in backyard barbecues. A major source of concern now is the exposure to many thousands of new compounds that have helped in easing some aspects of the human lot while perhaps aggravating others. It is difficult to eliminate those that are truly dangerous without discarding all and throwing us back to the Stone Age. Many compounds in wide use for long periods without any evidence of adverse effects are described as "generally regarded as safe" (GRAS). Others are examined as facilities and time permit.

The task of screening potentially mutagenic, and therefore perhaps carcinogenic, compounds has been greatly eased by the development of the simple and rapid Ames test.* The test uses a mutant strain of bacteria (*Salmonella typhimurium*) that has lost the ability to synthesize histidine but is prone to reverse mutations through which the ability is restored. A medium deficient in histidine but containing the test compound is inoculated with the mutant. After a suitable interval the number of colonies of bacteria that have recovered the ability to make their own histidine is counted and compared with controls. Mutagenic substances will produce hundreds to thousands of colonies per plate.

*Bruce Ames (1928–). American biochemist.

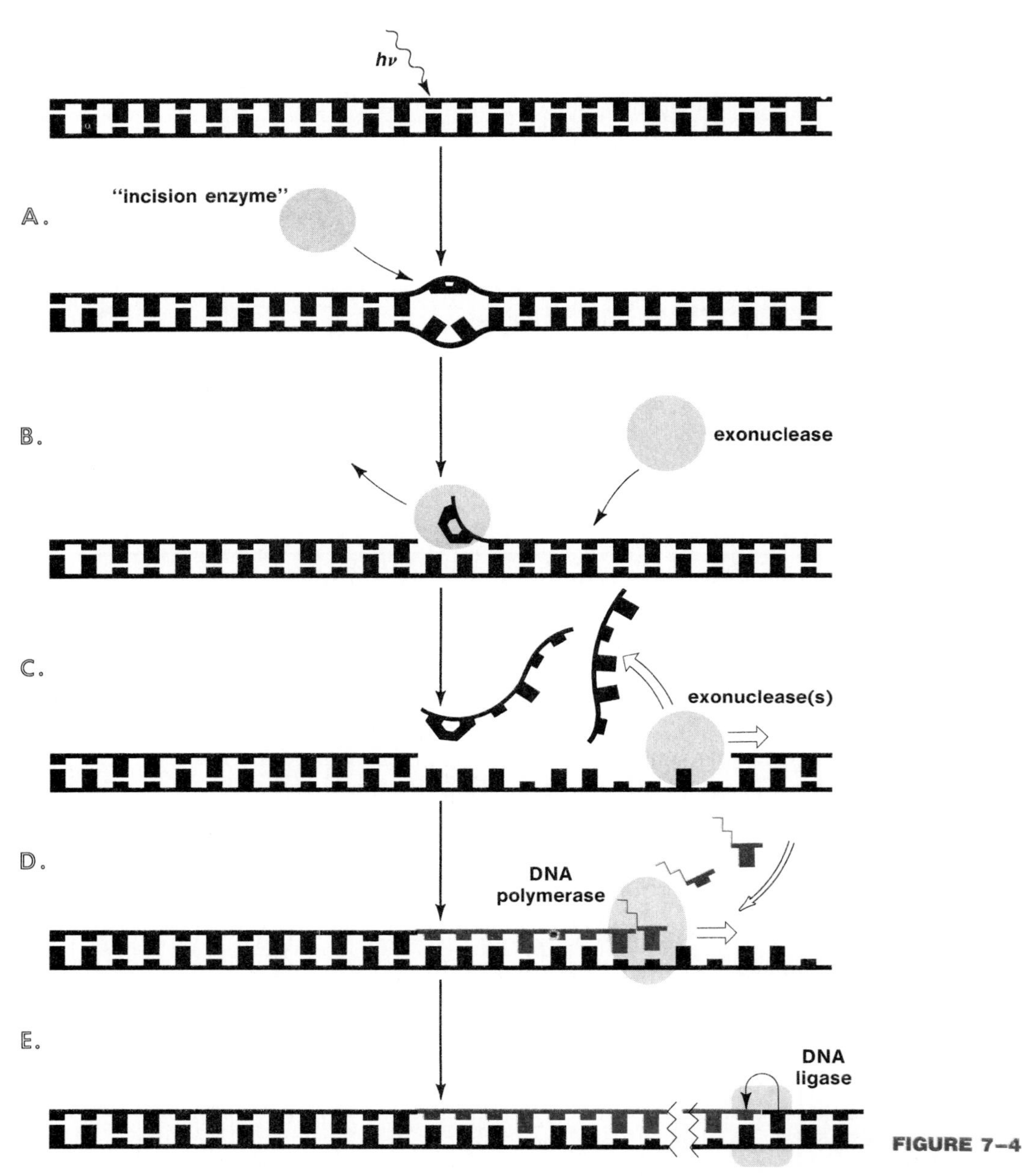

FIGURE 7–4

Thymine dimers and perhaps other modified bases can be removed from DNA by the combined action of several enzymes. The critical first step involves recognition of the defect by an uncharacterized incision enzyme, which hydrolyzes one or more adjacent phosphodiester bonds. After removal of the damaged segment, long-patch repair proceeds by removal of many more nucleotides from the damaged strand. A DNA polymerase fills the gap by synthesis of a new complementary segment, using the intact strand as template. Finally, a DNA ligase covalently links the new piece to the original segment at the end of the gap.

If the suspect compound were tested as such, the test would miss those cases in which it is not the compound itself that is the mutagenic culprit but one or more of its metabolic products. For this reason, a homogenate of rat liver is included in the incubation so as to cause the formation of at least some of those products. About 60 to 90 per cent of known carcinogens give a positive result when tested in this way.

Repair of DNA

The error rate of DNA polymerase is relatively low but still intolerably high for perpetuation of a species. The enzyme in prokaryotes corrects most of its own mistakes with a built-in exonuclease activity that automatically hydrolyzes any mismatched nucleotides, but the eukaryotic enzymes apparently lack this feature. However, both eukaryotes and prokaryotes have other devices for repair of accidents. The result is diminution of the overall mutation rate to an extremely low value; estimates range around one persistent change per 10^9 base pairs per cell division, or one per 10^6 divisions for a 1000-nucleotide gene.

Excision Repair. A general mechanism for repairing damaged DNA involves removal of a segment of the damaged strand, with the gap then filled by the action of a DNA polymerase, followed by the action of DNA ligase (Fig. 7–4). At least two versions of this mechanism exist—**long-patch** or **short-patch** repair—and some details are not known.

The critical step is recognition of the error. One or more enzymes will combine with a DNA strand containing pyrimidine dimers or other abnormal base constituents and hydrolyze the polynucleotide backbone adjacent to the defect. After the DNA strand is broken, exonucleases remove the defective residues and then enlarge the gap by cleaving as many as 100 more residues. The gap is then filled by a DNA polymerase (probably a β-polymerase, not the α enzyme involved in replication), and finally closed by DNA ligase. The rebuilding of DNA strands in this way represents "unscheduled DNA synthesis"—that is, it is occurring without regard to the stage of the cell cycle.

When polynucleotide chains are broken by ionizing radiation, only as few as one or two nucleotide residues are hydrolyzed by an exonuclease before the gap is filled and the break eliminated (short-patch repair). Paradoxically, these seemingly small defects are the least likely to be repaired, although the job is sometimes completed in less than an hour once it is undertaken.

Long-patch repair as described here is more reliable than short-patch repair, even though it requires as much as 20 hours for completion. Both long- and short-patch repair are used to repair accidental changes in one of the DNA strands. Breaks in both strands are likely to be lethal unless they happen to occur between genes in a way permitting continued transcription.

Light-Activated Repair. The formation of pyrimidine dimers by ultraviolet radiation is sometimes reversed by using a thief to catch a thief. At least some cells contain a protein that associates with the dimers so as to change their light absorption spectrum toward longer wavelengths. Absorption of visible light will then activate cleavage of the pyrimidine dimers back into their functional monomeric forms. Since the cells in which the dimers form must be cells that are exposed to light, it follows that light is available in the same cells for a reversal of the process. Not only that, but visible light penetrates deeper into the skin than does ultraviolet light.

Removal of Uracil or Hypoxanthine. We saw that uracil or hypoxanthine residues sometimes appear in DNA by hydrolysis of cytosine or adenine residues. Two

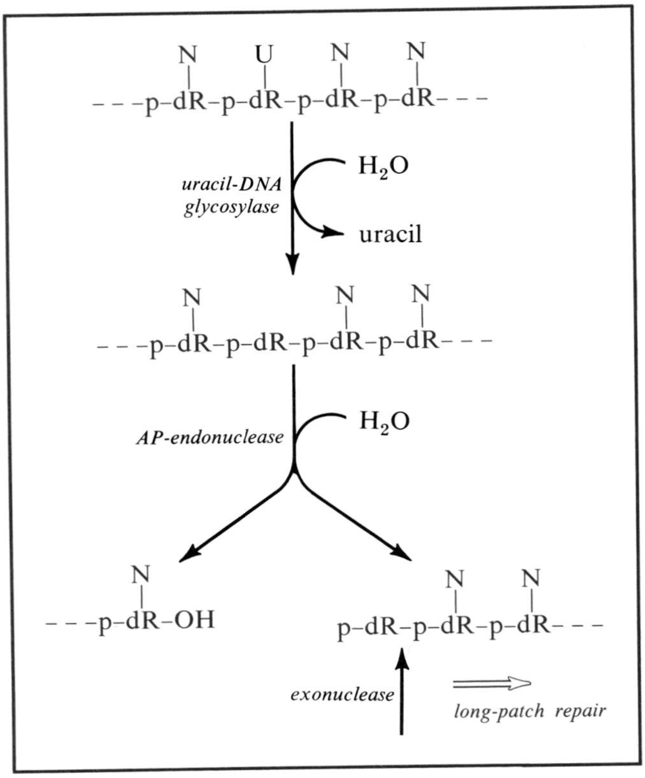

FIGURE 7–5

The replacement of uracil residues appearing in DNA. *Top.* A glycosylase specifically hydrolyzes the uracil-deoxyribose bond, liberating the uracil. *Bottom.* An AP-endonuclease specifically hydrolyzes the backbone phosphodiester bond next to naked deoxyribose residues (apurinic/apyrimidinic residues). After removal of the naked deoxyribose-5′-phosphate group by an exonuclease, long-patch repair can proceed to rebuild the damaged strand.

kinds of enzymes are present to remove these defective bases and trigger the long-patch repair process for their proper replacement (Fig. 7–5). **DNA glycosylases** remove uracil or hypoxanthine by hydrolysis of the base-deoxyribose bond. An **AP-endonuclease** recognizes the naked deoxyribose and cleaves an adjacent phosphodiester group in the DNA backbone (AP = apurinic/apyrimidinic). The exposed baseless sugar is then removed by an exonuclease, and long-patch repair continues with removal of more nucleotides, followed by resynthesis of the correct DNA sequence.

It would be more difficult to correct the relatively common accidental deamination of cytosine if the product, uracil, were a normal constituent of DNA. This is probably the reason for using thymine in place of uracil. Why doesn't the same sort of error occur in RNA? It, and the other mechanisms for error, probably do, but the lifetime of the RNAs is relatively short and there are many copies formed of each kind of molecule with which to perform their proper functions.

Defective Repair. Xeroderma pigmentosum provides the best evidence for the importance of DNA repair to humans. This rare disease is characterized by sensitivity to ultraviolet light (290 to 320 nm), with exposed skin developing a spectrum of disturbances ranging from excessive freckling through horny growths (keratoses), dilated capillary networks (telangiectasia), and ulceration to the appearance of carcinomas in the first decade of life.

Patients with xeroderma pigmentosum have one of at least seven different biochemical defects in the repair mechanism. This can be demonstrated by combining cells from various patients; repair is normal in some combinations. This occurs when a different step is defective in the mixed cells; each then supplies the other with the missing activity.

> **XERODERMA PIGMENTOSUM**
>
> **Incidence of 1 in 250,000 people**
>
> **Autosomal recessive inheritance**
>
> **Widespread separate involvement includes malignancy in skin, with from 2 to 100 separate cancers appearing in a patient; freckles; damage to eye**
>
> **Prevention of damage in afflicted patients by minimizing ultraviolet exposure through screening lotions, clothing, and special glasses and by frequent skin examination**
>
> **Occasionally linked to central nervous system disease with small head, mental deficiency, hearing loss, and faulty gait**

Other conditions are believed to arise from defective repair mechanisms. Patients with **ataxia telangiectasia** are very sensitive to x-radiation, developing severe erythema (reddening of the skin), ulcerative dermatitis, and necrosis (death and dissolution) of other tissues. The condition is characterized by loss of muscular coordination (ataxia), as well as by telangiectasias, and those afflicted have 1200 times the normal incidence of malignancies of lymphoid tissue. Indeed, those carrying one of the various defective genes causing the condition, estimated to be 1 per cent of the population, have been calculated to include 5 per cent of all those developing a malignancy prior to age 45. (No methodology is available to confirm this striking prediction.)

Longevity and the relative effectiveness of DNA repair mechanisms may be linked; at least there is some evidence for an association in mice.

MUTATIONS

Any change in the DNA from which additional cells are constructed represents a mutation. If the term is not qualified, it refers to alterations in the germ cells that affect subsequent generations, but there are also somatic mutations that result in changes in the tissues of one individual. Let us examine the molecular basis for these changes, and some of the results that are manifested by differences in one kind of protein, human hemoglobin.

Errors in Crossover

The process of crossing-over during meiosis introduces the possibility for several types of misadventure, including the loss of segments of DNA from a strand, unequal exchanges from one strand to the other, and exchanges involving segments that are not homologous.

Hemoglobins Lepore. Some humans have hemoglobins with peptide chains containing unusual sequences that were created by faulty crossover. Normal human adults contain two principal kinds of hemoglobin, HbA_1 and HbA_2, in their erythrocytes. Each has the same pairs of α chains, but they differ in the second pair, which is designated β in HbA_1 and δ (delta) in HbA_2. The chain formulas are therefore $\alpha_2\beta_2$ for HbA_1 and $\alpha_2\delta_2$ for HbA_2. The β and δ chains differ in only 10 out of the 146 amino acid residues, but these differences are sufficient to alter function.

The subtleties of chain composition are discussed in Chapter 13; suffice it now

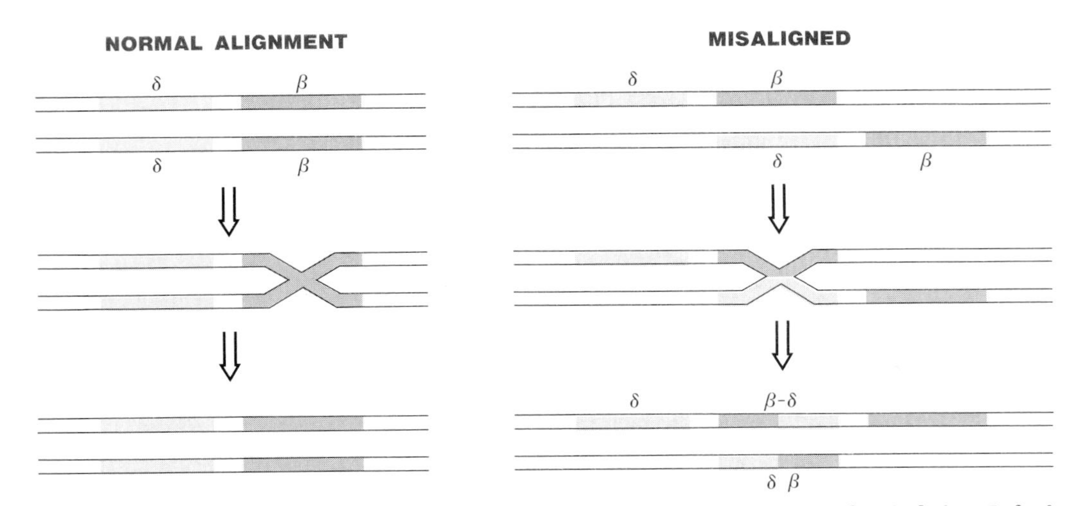

FIGURE 7–6 **Genes for δ chains in hemoglobin are adjacent to the genes for β chains. *Left.* A crossover within one of these genes between sister chromosomes during meiosis ordinarily has no effect on the nature of hemoglobin in the progeny. *Right.* Misalignment of chromosomes before crossover occurs creates one chromosome that contains both β and δ genes as well as a gene for a hybrid hemoglobin chain that begins with the β sequence and ends with the δ sequence. The other chromosome contains only a gene for the other type of hybrid hemoglobin, known as Hb Lepore, which begins with the δ sequence and ends with the β sequence. The mechanism of crossover is more complicated than indicated here.**

to note that the genes for β and δ chains are neighbors on one chromosome, where it is possible for errors in alignment and crossover to occur between them (Fig. 7–6). Such errors sometimes create chains in which one end has the sequence of a β chain and the other end has the sequence of a δ chain. Hemoglobins containing the δ-β hybrid are known as Lepore hemoglobins, those containing the β-δ hybrids as anti-Lepore hemoglobins. (The name is that of the family in which the first examples were discovered.) Several different variants are known of each type, differing in the position of splicing of the genes. Formation of a Lepore gene eliminates the normal genes for both the β and δ chains, so those homozygous to the gene cannot make either HbA_1 or HbA_2. Those with the anti-Lepore gene still have the normal β and δ chain genes, so that the abnormality has little effect.

Frame-Shift Mutations

An error in replication or crossover may cause the insertion or deletion of nucleotide residues in DNA. If the error occurs in a segment from which messenger RNA is transcribed, the result will depend upon the number of residues affected. Suppose that three residues, or some multiple of three, are added or subtracted. The remaining residues will still be in the correct triplet sequence for coding the intended amino acids, so the result will be the formation of a peptide chain that has some residues missing, or some additional residues inserted, in an otherwise correct polypeptide (Fig. 7–7).

On the other hand, error in replication that changes the number of nucleotide residues by one or two, or some multiple not divisible by three, causes all of the remaining length of messenger RNA to enter a ribosome at the wrong place in the

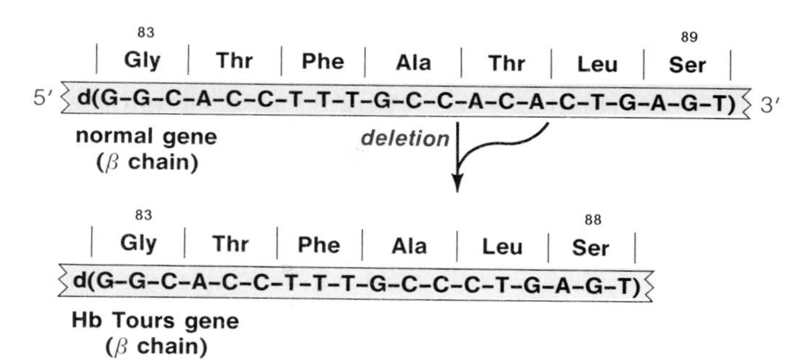

FIGURE 7–7

When faulty crossover or replication creates a gene differing from the parent by blocks of three nucleotides in the exon, the gene product will be a polypeptide differing from the normal only in the absence or presence of some amino acid residues. In the example shown, the deletion of three residues designating a threonyl group in the hemoglobin β chain gene causes the formation of a chain that is normal except for the lack of the one amino acid. Combination of a normal α chain with this shortened β chain creates Hb Tours.

triplet sequence, so that the wrong amino acid will be incorporated into the resultant polypeptide chain at nearly every position. These errors are said to be frame-shift mutations; the product of such mutations will have little of the character of the intended protein unless the alteration occurs near the carboxyl end of the amino acid sequence, leaving the bulk of the structure intact.

Hemoglobins created by both types of nucleotide deletion are known. Some variants lack one or more amino acids with the remainder of the sequence correct, indicating that one or more multiples of three nucleotides have been deleted from the DNA directing hemoglobin formation. Other variants have extensive changes in amino acid composition caused by frame-shift mutations. A well-characterized example is Hb Wayne, which contains five additional residues at the end of an α chain because the normal terminator codon lost its meaning.

Here are the nucleotide sequences in the sense strand of DNA carrying information for the end of α chains and the corresponding amino acids in normal hemoglobins and in Hb Wayne:

NORMAL α-CHAIN DNA

| Lys | Tyr | Arg | Term | *transcribed, but not translated*

5′ d(– – –A–A–A–T–A–C–C–G–T–T–A–A–G–C–T–G–G–A–G–C–C–T–C–G–G–T–A–G–C–A–G– – –)

5′ d(– – –A–A–T–A–C–C–G–T–T–A–A–G–C–T–G–G–A–G–C–C–T–C–G–G–T–A–G–C–A–G– – –)

| Asn | Thr | Val | Lys | Leu | Glu | Pro | Arg | Term |

Hb WAYNE DNA

The deletion of one nucleotide from the Lys triplet in DNA creates a messenger RNA in which all subsequent triplets will be read one residue late. The sequence U-A-A normally read as a terminator codon in RNA will now be read as belonging to the end of one triplet and the beginning of another, each of which designates an amino acid rather than termination. The reading continues on the abnormal mRNA, extending the polypeptide chain until a chance combination of nucleotides falls into a terminator sequence. This chance combination occurs in what is a post-terminator spacer sequence in normal messenger RNA.

Hemoglobin Cranston is an example of a frame-shift mutant created by insertion of nucleotides. In this case, an A-G sequence is repeated, a kind of an error that could be caused by a daughter strand slipping two residues during replication:

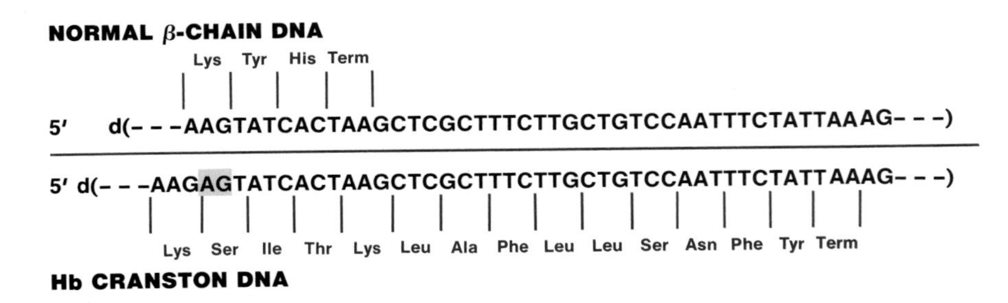

Here the gene for a β chain is altered so that the resultant mRNA will be translated beyond the normal termination point until another terminator sequence is encountered. (Mutant hemoglobins are frequently named after the locality at which they were discovered.)

Point Mutations: Base Changes

A point mutation is the substitution of one base for another in DNA. Since there are many genes coding for each transfer and ribosomal RNA, it is likely that a point mutation in one of them will pass unnoticed. Even though many of the genes coding for messenger RNA are unique, some point mutations in these sequences also are not detectable. This is so because of the redundancy of the genetic code; many base substitutions create triplets coding for the same amino acid. For example, changing d(A-A-A) to d(G-A-A) in DNA merely converts one triplet for phenylalanine to another, and the composition of the resultant polypeptide will not be altered.

The point mutations that do cause alterations in amino acid composition have variable effects; some may never be seen because they so drastically affect the function of critical proteins that they cause early death of the germ cells in which they appear. Others are known only in the heterozygous state, in which the production of an abnormal protein can be tolerated because the corresponding unaltered chromosome continues to direct the formation of messenger RNA from which the normal protein is translated. There is sufficient margin of safety for many functions to permit survival under most circumstances with half the normal activity of a protein.

Still other mutations have little or no discernible effects because the substituted amino acid residue does not alter the function of the protein in any substantial way, and we discussed in Chapter 5 (p. 66) how the code is constructed so as to minimize damage from point mutations. Germ-cell mutations that do not affect function are said to be silent. Only a small fraction of the estimated 7×10^{-9} mutations occurring per nucleotide per year persist in future generations, and most of these are silent mutations.

Point mutations in hemoglobin exemplify the range of effects that are seen. Dozens of different hemoglobins are known in humans that differ in only one amino acid residue from the more common hemoglobins. Some have lost so much function that they are seen only in heterozygous individuals, who still have one normal gene to produce a functional peptide chain. Others are so nearly like the common hemoglobins that their existence is only detected by accident or by massive screening of an apparently normal population. Most variant hemoglobins occur in relatively few people and tend to disappear rapidly from the population; some have conveyed sufficient advantage in special environments for them to spread in the population, as is discussed in Chapter 13.

Hemoglobins also illustrate the persistence of silent mutations in that the β and δ chain genes undoubtedly arose by duplication of some common ancestor, and even though they now differ in *amino acid sequence* at only 10 positions, the exons differ in *nucleotide sequence* at 31 positions. Most of the preserved mutations are silent. Contrariwise, the introns and the regions flanking the ends of the β and δ genes differ widely in sequence; just as the craters on the lunar surface faithfully record the past flux of meteorites, the nucleotide differences in those regions little eroded by evolutionary pressure more accurately reflect the rate of damage the genome has sustained throughout the millenia.

The important point for now is that all of the known single amino acid substitutions in human hemoglobins can be caused by a single change in a coding triplet. For example, variants are frequently created by replacing a histidyl group with a tyrosyl group (His→Tyr). A histidyl group is designated in the sense strand of DNA by d(C-A-T) or d(C-A-C), whereas a tyrosyl group is designated by d(T-A-T) or d(T-A-C), and the His→Tyr mutation would result from an error causing the replacement of dC by dT.

On the other hand, we never see a phenylalanyl group inserted in place of a histidyl group, because the phenylalanyl group is designated by d(T-T-T) or d(T-T-C), and His→Phe would therefore require both a substitution of dT for dA and of dT for dC. Such double mutations are very improbable. Suppose that one in every 100,000 infants is born with a new single base substitution at some position in the cistrons for α and β chains in hemoglobin. (This is a high incidence, requiring a 50-fold faster mutation rate than is usually estimated.) In order for a double mutation to occur, one of these infants also must have a second base substitution in the same triplet altered by the first substitution. This would occur only in fewer than one out of 10^{12} infants, even with the high mutation rate assumed; the odds are that this infant is still to be born.

Point mutations in the terminator triplet frequently result in the continued incorporation of amino acids beyond the normal end of a polypeptide chain. The extension of the chain continues until the chance occurrence of a terminator codon in a region ordinarily not transcribed. The effect is much like that caused by a frameshift mutation, except that the proper amino acids appear in the polypeptide before the terminator position, and the number and kind of extra amino acids incorporated is the same for all point mutations in the terminator. (Why? Because the point mutation does not alter the reading of subsequent triplets, and the next terminator will still occupy the same position.) Examples of such mutations are known in the genes for the α chain of human hemoglobin (Fig. 7–8).

RECOMBINANT DNA

Techniques are now available for changing the genetic composition of cells through the incorporation of precisely defined segments of foreign DNA. Rarely has mere scientific knowledge caused more uproar among laymen in advance of specific applications affecting their lives. Fear of irrevocable and dangerous changes in the inheritance of bacteria or people led to a rare and vehement public debate over questions of scientific fact. This was followed by a wild scramble to purchase stock in companies formed by some of the leading investigators in the field for practical application of their skill.

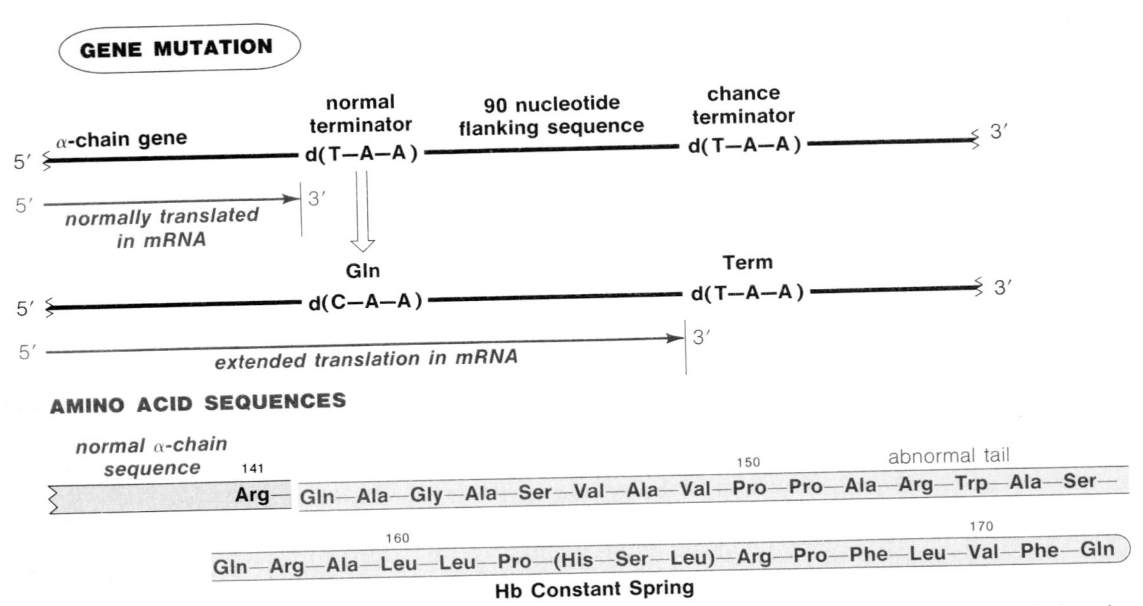

FIGURE 7–8 Point mutations may change the template for a terminator codon to one for a codon designating an amino acid residue. The mutant messenger RNA will continue to be translated as a polypeptide until a chance sequence of nucleotides gives a termination signal. In the gene for the α chain of hemoglobin, 90 nucleotides occur in the spacer segment between the normal terminator and the next chance occurrence of a terminator. Consequently, mutation of the normal terminator to the sequence designating a glutaminyl group results in the incorporation of an additional 31 amino acid residues beyond the normal end of the α chain. (One for the modified terminator, and 30 for the intervening 90 nucleotides.) The lengthened chain, occurring in Hb Constant Spring, is shown with the normal terminus in black, and the added residues in brown.

A major consequence of recombinant DNA technology has been a dramatic increase in the rate of growth of our knowledge. It is now possible to determine the sequences of particular genes, even in the complex human genome, and to study their homology and frequency with relative ease. It is simple in many cases to determine the amino acid sequence of a protein by determining the nucleotide sequence in the nucleic acids directing its formation.

The use of recombinant DNA technology to enable bulk synthesis of specific proteins, such as insulin (p. 150) or growth hormone (p. 735), by incorporating the gene into readily grown bacteria is well underway. The permanent incorporation of new genes into eukaryotic organisms is projected for the future, a prospect that may be as seemingly beneficial as correcting genetic damage in a human, or as risky as altering the capability of a plant to grow in particular environments.

The Tools

The possibilities for application of recombinant DNA technology are so many that it behooves us to understand the principles involved. Let us examine four techniques of special importance.

Restriction endonucleases are used to cleave DNA reproducibly at specific points. Many of these microbial enzymes are specific for palindromes; dozens are now known, and a few of the more commonly used are listed in Table 7–2.

SOME RESTRICTION ENDONUCLEASES

TABLE 7–2

The first letter of the specific enzyme name indicates the genus and the next two the species from which it is obtained. Any additional letter designates a particular strain of organism, and Roman numerals are added if more than one enzyme has been isolated from the same species. Vertical arrows indicate the point of cleavage in particular base sequences.

Enzyme	Sequence Cleaved	Source
Enzymes yielding blunt-ended fragments		
*Alu*I	↓ AGCT TCGA ↑	*Arthrobacter luteus*
*Hae*I	↓ (A,T)GGCC(T,A) (T,A)CCGG(A,T) ↑	*Haemophilus aegyptius*
*Hae*III	↓ GGCC CCGG ↑	
Enzymes yielding sticky-ended fragments		
*Bam*HI	↓ GGATCC CCTAGG ↑	*Bacillus amyloliquefaciens* F
*Mbo*I	↓ GATC CTAG ↑	*Moraxella bovis*
*Eco*RI	↓ GAATTC CTTAAG ↑	*Escherichia coli* RY13
*Hind*III	↓ AAGCTT TTCGAA ↑	*Haemophilus influenzae* Rd
*Mbo*II	↓ GAAGAN_8 CTTCTN_7 ↑	*Moraxella bovis*

Why would bacteria want to hydrolyze palindromes? The restriction endonuclease is a device for attacking foreign DNA. If a cell makes an endonuclease to cleave a particular palindrome, it also makes a methylase of identical specificity to protect its own DNA by methylating at least one base in the palindrome wherever it occurs in that DNA. Only one strand need be methylated, so it might seem that any sequence of bases could serve as a signal. That is true until the DNA is replicated in preparation for cell division. If the signal sequence is a palindrome, both strands will have been methylated because they will have identical signal sequences, and both molecules of daughter DNA will be protected by the presence of one methylated strand until the newly synthesized complementary strand can also be methylated. However, if the signal sequence is not a palindrome, then one parent strand will be bare of methyl groups and unable to protect its newly synthesized complement, which will contain the signal, from cleavage.

Some endonucleases cleave palindromes in the center to make pieces of duplex DNA with blunt ends (Fig. 7–9), but most act off-center to make pieces with sticky ends. That is, the two strands are still held together by a few complementary base pairs between the cleavage sites. The force is weak but sufficient to cause recombination of separated pieces with sticky ends under annealing conditions. In the laboratory, segments of DNA derived from separate sources are held together in this way and then linked by a DNA ligase, as also illustrated in Figure 7–9.

Plasmids or viruses are used as vectors to carry DNA into host bacteria that will make many copies of the DNA. The extra DNA generated by this gene amplification can cause the host cell to make large quantities of a particular protein, such as a human hormone. The DNA also can be isolated in workable quantities for sequence determination. To illustrate the principle, consider the use of bacteriophage λ; a virus infecting *Escherichia coli.*

One third of the DNA of phage λ can be traded in for new genes without affecting the ability of the phage to enter *E. coli* and be reproduced. This has made it possible to create useful vectors by modifying the phage extensively. One series of vectors is the **Charon phages.** (Charon is the mythical boatman who transports the

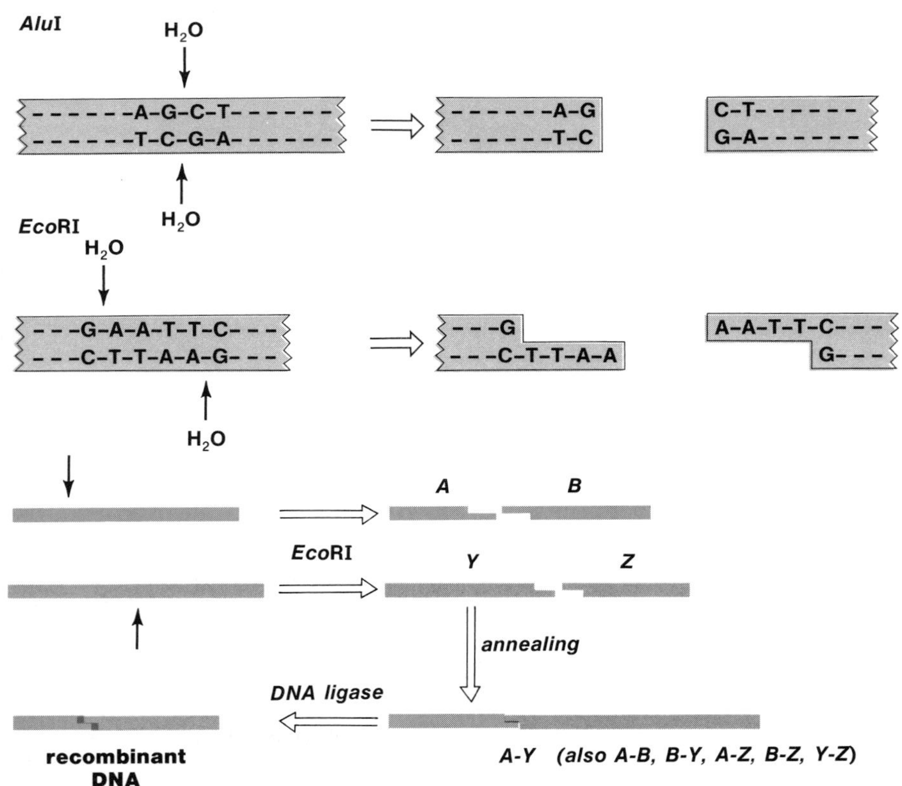

FIGURE 7–9 **Use of restriction endonucleases. *Top.* *Alu*I is an example of a restriction endonuclease that attacks a specific palindrome in the center, creating products with blunt ends. *Center.* *Eco*RI, like most restriction endonucleases, attacks palindromes off-center. The new overlapping ends are said to be sticky because they base-pair perfectly. *Bottom.* When restriction endonucleases act on different DNA molecules, sticky-ended products can recombine in many permutations upon annealing. The junctions can be covalently sealed by DNA ligase. Recombinant DNA made in this way can contain sequences derived from different sources.**

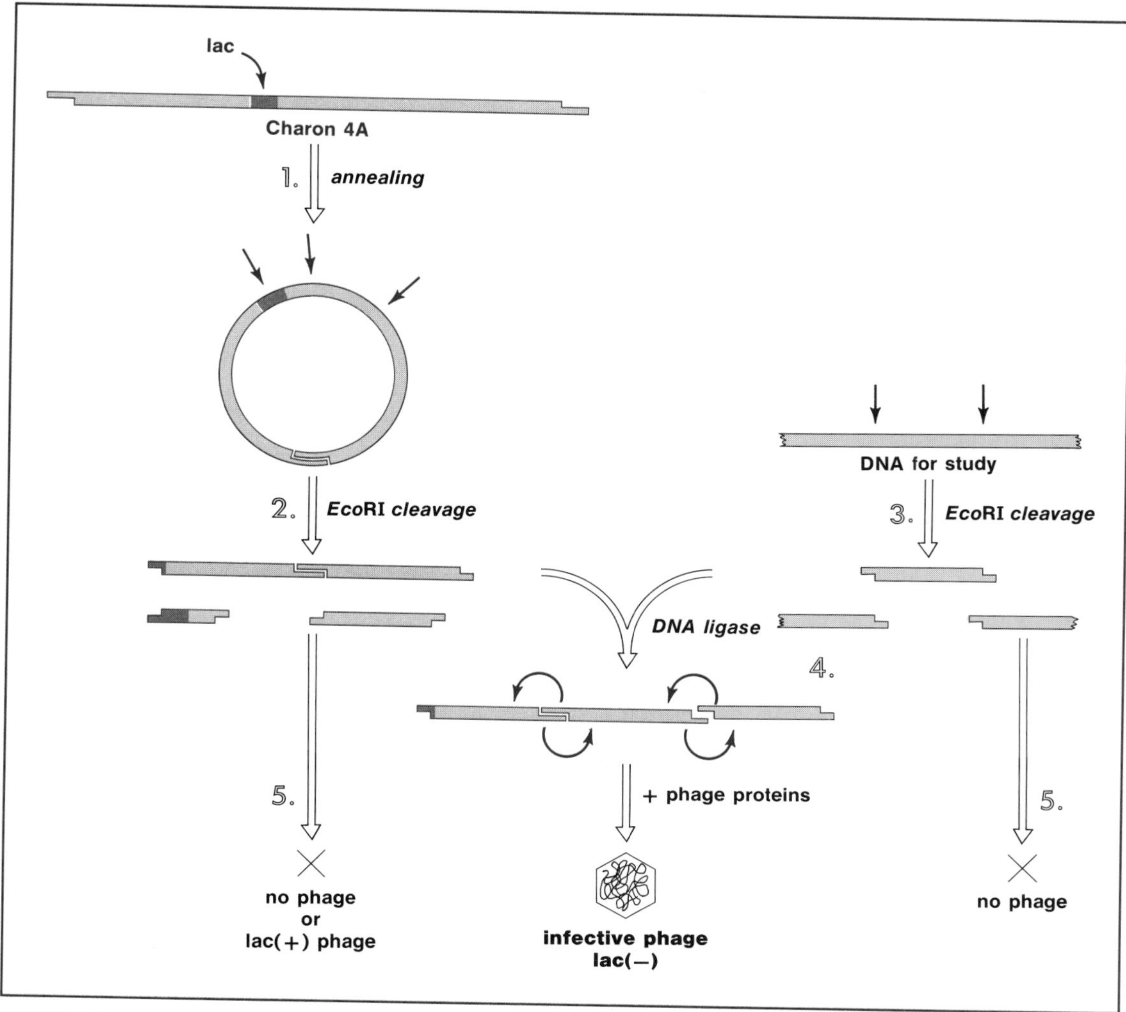

FIGURE 7–10

Use of a phage vector for amplifying DNA in bacteria. (*1*) Lambda phage is linear with 12-base sticky ends that can be annealed to close into circular molecules. Charon 4A is a modified λ phage that includes part of the *E. coli lac* gene.

(*2*) *Eco*RI endonuclease attacks Charon 4A at three sites, removing a segment of DNA in two pieces. This destroys the lac gene, but not the potential infectivity.

(*3*) The DNA containing the sequence of interest is also cleaved with *Eco*RI endonuclease to form a variable number of sticky-ended fragments. (The conditions of exposure are regulated so as to have the bulk of the products be of appropriate size.)

(*4*) The Charon 4A arms and the subject DNA fragments, both with sticky *Eco*RI ends, are mixed, annealed to combine the sticky ends, and covalently joined by a DNA ligase.

(*5*) Phage proteins are added to the mixture of recombinant DNA molecules to form infective phage particles. The pieces of the vector phage are too small to form phage particles; any of the original vector that escaped cleavage will still contain the *lac* gene, which can be identified by use of a color reagent. The DNA to be amplified does not itself carry the instructions of infectivity; only vector-subject DNA recombinants of appropriate size are incorporated into infective phage units. These can be multiplied in cultures of an appropriate strain of *E. coli*.

dead over the River Styx.) A Charon phage includes the viable portions of the phage DNA enclosing a segment of *E. coli* DNA carrying an indicator, the *lac* gene.

The indicator *lac* gene identifies Charon phages surviving later treatment with *Eco*RI endonuclease. The *lac* gene has three sites at which it is attacked by *Eco*RI, but the viable phage DNA in the vector has none. The gene causes formation of an enzyme that hydrolyzes an indole derivative* of galactose, a sugar (p. 178). When the indole is released by hydrolysis, it is auto-oxidized to a bright blue product. The blue color identifies the location of intact Charon vectors.

Use of Charon Phage Vectors (Fig. 7–10). The Charon phage is replicated as a linear sticky-ended DNA duplex, which can be annealed into a circular molecule. *Eco*RI cleaves the circle at different points when it destroys the *lac* gene, forming new sticky ends, but with the phage DNA still held together.

If the sample of DNA for which amplification is desired is also treated with *Eco*RI, it too will be cleaved into pieces with sticky ends. Mixing these with the sticky-ended segments made from Charon phage and annealing will create many united pieces. Repackaging the DNA into intact phage particles is then attempted by incubating the DNA with protein preparations derived from wild-type phages. Some of the DNA will be too small and some too large to make a phage particle. (The particular vector with which we shall be concerned, Charon 4A, is built to carry between 8,200 and 22,200 inserted base pairs.) These will include uncombined pieces, and pieces formed by joining two vector segments. The DNA that is incorporated into new phage particles includes some intact vector DNA and some in which pieces of the DNA of interest are joined with the viable vector.

The phage particles are then multiplied in *E. coli* and grown as separate colonies on lawns of *E. coli* spread on agar plates. Clear lysed patches in the *E. coli* lawns show where phage particles have grown. The color reagent for the *lac* gene is also included in the agar, so those phage colonies containing segments of the DNA of interest can be identified by the absence of blue color. (The other colonies contain workable segments of the *lac* gene, not replaced by new DNA.)

Identification of genes, either free or in vectors, is another important procedure, and it is done by using DNA complementary to messenger RNA (Fig. 7–11). (Note that **complementary DNA (cDNA)** is not synonymous with the gene DNA, because it will lack any introns or flanking signal sequences.) The method hinges upon the availability of relatively pure messenger RNA, and it has therefore been applied to those genes that are the principal ones expressed in specialized cells. (Hemoglobin and immunoglobulin genes in mammals and ovalbumin gene in chickens are examples.)

Messenger RNA is isolated by using cellulose with synthetically attached oligo-dT tails. When a cell extract is passed through it, the poly-A tails of mRNA stick to the oligo-dT on the cellulose; the other components of the extract can be washed through. The mRNAs are then eluted and used to make a complementary DNA strand by the action of a viral RNA-directed DNA polymerase (reverse transcriptase). Use of labeled deoxynucleoside triphosphate precursors makes the cDNA radioactive for use as a probe.

Since the cDNA has a sequence complementary to the mRNA template on which it was made, its sequence is also complementary to parts of the sense strand of the DNA bearing the gene transcribed to make the mRNA. The sense strand in phage colonies carrying segments of this particular gene will therefore bind the radioactive probe, and this is how the gene segments are located among all of the DNA fragments incorporated into Charon vectors.

---

*3-(D-galactosyl)-4-chloro-5-bromoindole

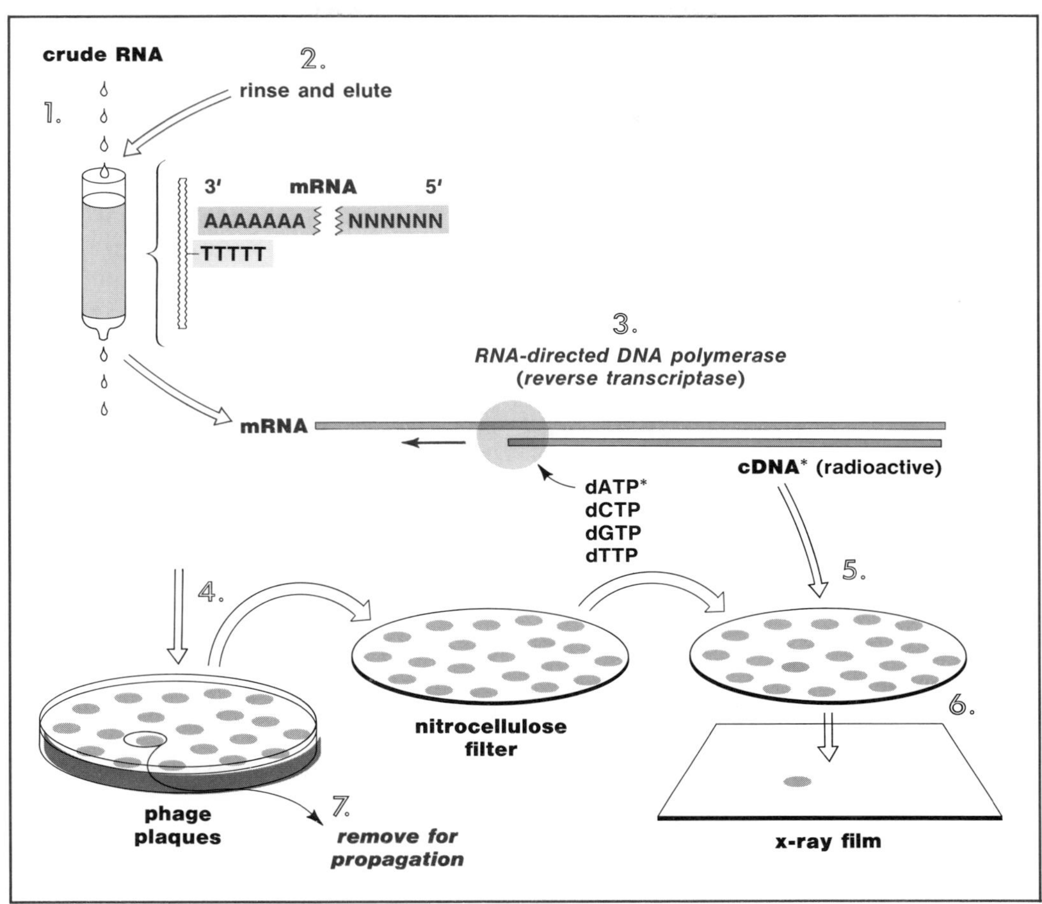

FIGURE 7–11 Use of complementary DNA (cDNA) in detection of gene sequences. (*1*) Crude RNA prepared from cells is passed over an affinity column containing cellulose with covalently attached oligo-dT chains. Messenger RNA bearing poly-A tails will bind to the column through A-T pairing.

(*2*) The column is rinsed free of unbound material and the messenger RNA is eluted.

(*3*) A viral RNA-directed DNA polymerase (reverse transcriptase) synthesizes radioactive DNA complementary to the messenger RNA template.

(*4*) Phage colonies containing segments of the subject DNA are blotted with a nitrocellulose filter. Fragments of DNA and phage are fixed to the filter, forming a mirrored replica of the colonies.

(*5*) The filter is dried and soaked with a solution of radioactive cDNA, which adheres to any phage colonies containing sequences like those in the original mRNA sequences.

(*6*) X-ray film in contact with the rinsed nitrocellulose filter will be exposed wherever radioactive cDNA has been bound to the phage recombinants; this marks the colonies containing DNA of interest.

(*7*) The colonies so designated are removed for further propagation in bacteria.

The mechanics of using the cDNA probe are as follows: Sheets of nitrocellulose are pressed onto the agar plates carrying phage colonies. Some of the phage particles and their constituent DNA stick to the nitrocellulose from each colony, making a mirrored replica. The DNA strands are separated and fixed on the nitrocellulose by transiently making it alkaline. The dried sheet is exposed to the radioactive cDNA probe. After unbound probe has been washed away, x-ray film is used to locate the remaining radioactivity. The film will be darkened at the location of the phage colonies containing gene segments. Phage from the colonies can be propagated separately to harvest larger amounts of each piece of DNA containing a gene fragment.

Determination of the nucleotide sequence in the DNA isolated from a particular vector colony may be done in two ways. One involves the synthesis of varying lengths of a new complementary strand under conditions in which the strand is forced to end with the same modified nucleotide. This is accomplished by including a small fraction of a 2′,3′-dideoxynucleotide in the reaction mixture. Whenever this is built into the growing chain in place of the regular nucleotide, further synthesis is no longer possible. For example, suppose a little dideoxyadenosine triphosphate is included with the four normal deoxynucleoside triphosphate precursors, DNA polymerase, and separated strands of the DNA segment of unknown sequence. (A bacterial polymerase is used that does not require an RNA primer.) Replication will begin, except that occasionally the polymerase will add a dideoxyadenosine residue as a complement to a deoxythymidine residue in the template. When it does so, replication stops, and the polymerase must start synthesis afresh. The lengths of the complementary DNA synthesized vary according to the chance position at which a dideoxy, rather than a monodeoxy, adenosine was inserted. These varying lengths identify the distance along the template at which deoxythymidine occurs. The lengths can be determined by electrophoresis in a gel.

Similarly, the other three bases can be located by separate incubations involving their complementary dideoxynucleoside triphosphates. The four incubated samples, all using radioactive precursors, are run on neighboring lanes in gel electrophoresis, and the base sequence can be read directly from an autoradiogram (Fig. 7–12).

The second method involves selective destruction of particular kinds of bases in the unknown DNA, followed by cleavage of the polynucleotide backbone at the naked sugar residues. Destruction by one procedure cleaves at guanine, by another at guanine or adenine, by another at cytosine, and by the fourth at cytosine or thymine. In this method, either the 3′ or 5′ ends of the unknown DNA is first labeled with one or more radioactive nucleotide residues, so all of the varying-length fragments produced from it that are radioactive have sequences beginning at one end of the original molecule. This method also produces varying lengths of DNA, with the length indicating the location of a particular base in the DNA being degraded. Samples degraded by the four procedures are separated in neighboring lanes by gel electrophoresis. The base sequence can also be read directly from an autoradiogram of the separated mixtures, as is shown in Figure 7–12.

Good autoradiograms enable one to read the sequences directly for more than 200 nucleotide residues. If the original DNA unknown segment is longer than this, as most are, it is first cut into defined pieces by restriction endonucleases, and these pieces are separated and sequenced individually.

Sequencing a Human Hemoglobin Gene

As an aid in understanding how recombinant DNA techniques can be applied, let us examine how the sequence of the human structural gene for the β chain of hemoglobin was determined.

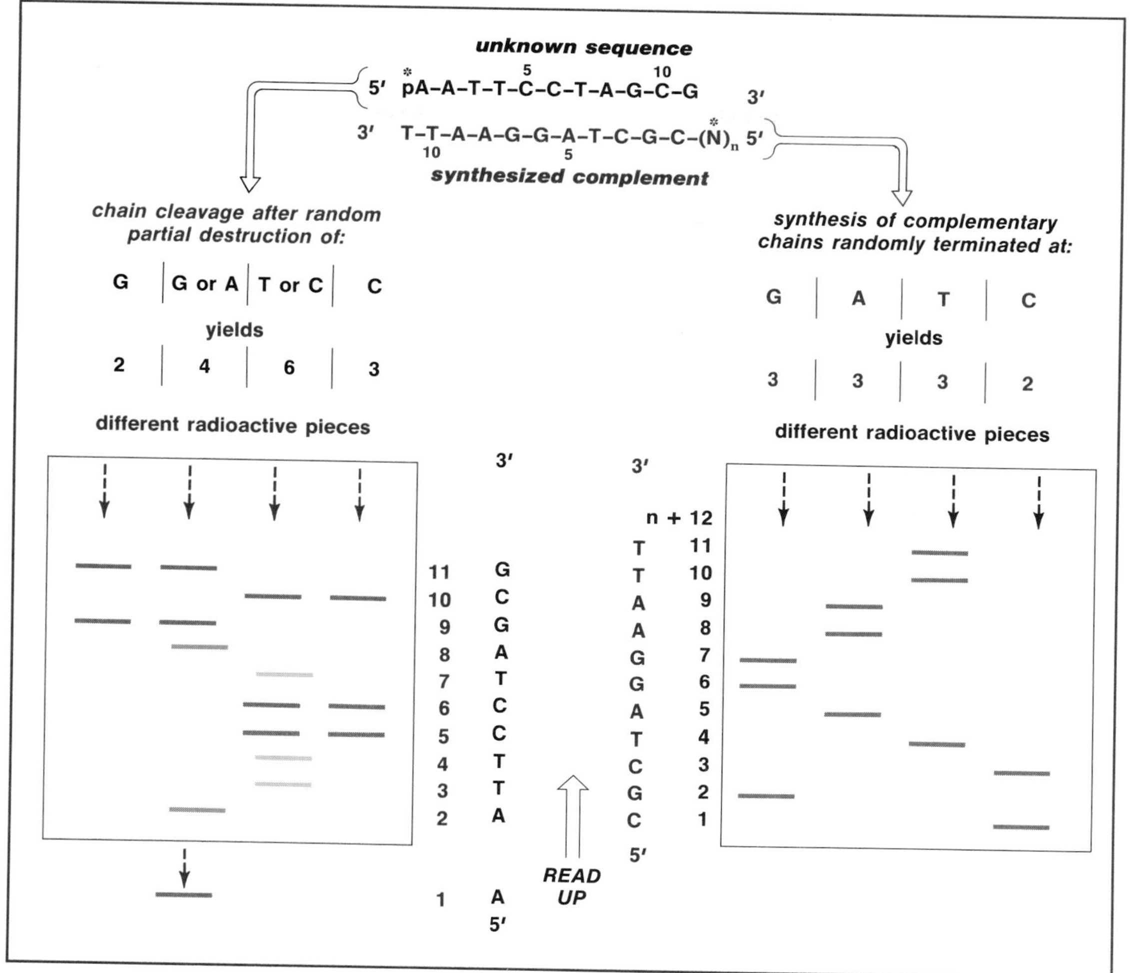

FIGURE 7–12

Sequencing of DNA. *Left.* The unknown sequence is labeled at one end with radioactive phosphate. The sample is divided into four portions. Each portion is treated with reagents causing selective loss of one or two kinds of bases, with chain cleavage at the points of loss. Any piece that is radioactive must include the entire sequence from the 5′ end to the point of cleavage; other pieces are ignored. Since the cleavage occurs at random, all possible segments will be formed. The products of the four treatments are applied to neighboring lanes of a gel and separated by electrophoresis. The smallest piece (one nucleotide in this case) travels fastest, the next smallest (two nucleotides) next fastest. In the example, the fastest-moving radioactivity is in the sample treated so as to destroy dG or dA. There is no corresponding radioactivity in the sample treated to destroy dG alone, therefore the smallest fragment was formed by destroying dA, which must have been one nucleoside removed from the radioactive end. (Destruction of the very first nucleotide results in the liberation of inorganic phosphate (P_i), which migrates so fast that it tends to run off the gel and is detected separately. Here it was found to be moving only in the sample treated to destroy both dG and dA; therefore the first nucleoside in sequence is also dA.) The second fastest band still on the gel is in the sample treated to destroy both dT and dC, but there is no corresponding band in the sample in which only dC was destroyed. Therefore two nucleotides remain after a dT was destroyed, and that dT was the third nucleoside in the chain. By reading on up toward the slowest radioactive band, the entire sequence is determined.

Right. Complements to the subject DNA are synthesized in four batches, each made in the presence of a different 2′-3′-dideoxynucleotide, which stops further synthesis whenever it randomly is inserted into the new chain. The precursors are made radioactive to label each new chain. As in the chemical degradation method, the various lengths of new chain are separated by electrophoresis, with samples from mixtures terminated at each kind of nucleotide run in neighboring lanes. Here the fastest radioactivity appears when synthesis occurred in the presence of dideoxyCTP. Therefore the first nucleotide contains cytosine. The next fastest appears in the dG column, and so on, with the sequence of the complement of the subject DNA being read up. Since it is an antiparallel complement, the original sequence is read down as complementary bases.

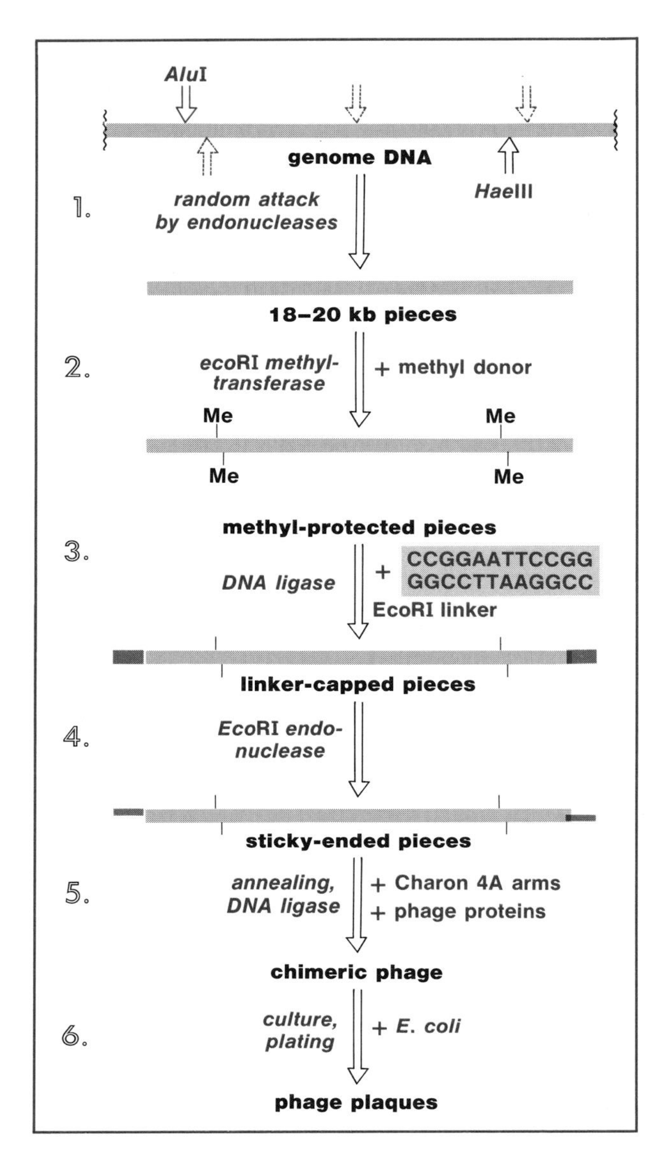

FIGURE 7–13

Preparation of a human genome library. (1) Attack by two restriction endonucleases is interrupted when segments 18,000–20,000 base pairs long are formed in maximum yield.

(2) The mixture of fragments is treated with *Eco*RI methyltransferase and a methyl donor so as to protect all *Eco*RI sites in the genome.

(3) A synthetic 12-base pair DNA linker is attached to each end of the subject DNA fragments by a DNA ligase.

(4) Each linker cap contains an *Eco*RI site (GAATTC), so cleavage with *Eco*RI endonuclease makes each end sticky.

(5) The DNA segments are combined with Charon 4A arms and formed into phage particles to carry the recombinant DNA into *E. coli* for amplification and plating as separate colonies derived from individual phage particles.

Preparation of a genome library, that is, the separation of the complete human genome into manageable pieces of DNA, was a first step (Fig. 7–13). Once such a library is prepared, it is possible to use it repeatedly for the isolation of different genes.

1. DNA isolated from the liver of an aborted fetus was exposed to two restriction endonucleases, *Alu*I and *Hae*III. The products of both of these enzymes are DNA with blunt ends, since they attack their respective palindromes in the center. Because the objective was to make relatively large pieces (15 to 20 kilobase pairs), short exposures to low concentrations of the enzymes were used so that the genome would be cleaved randomly at only a few of the many possible sites where the four-base palindromes occur. Pieces of appropriate size were isolated by centrifugation of the mixture (not shown).

2. The DNA fragments were treated with *Eco*RI methylase in the presence of

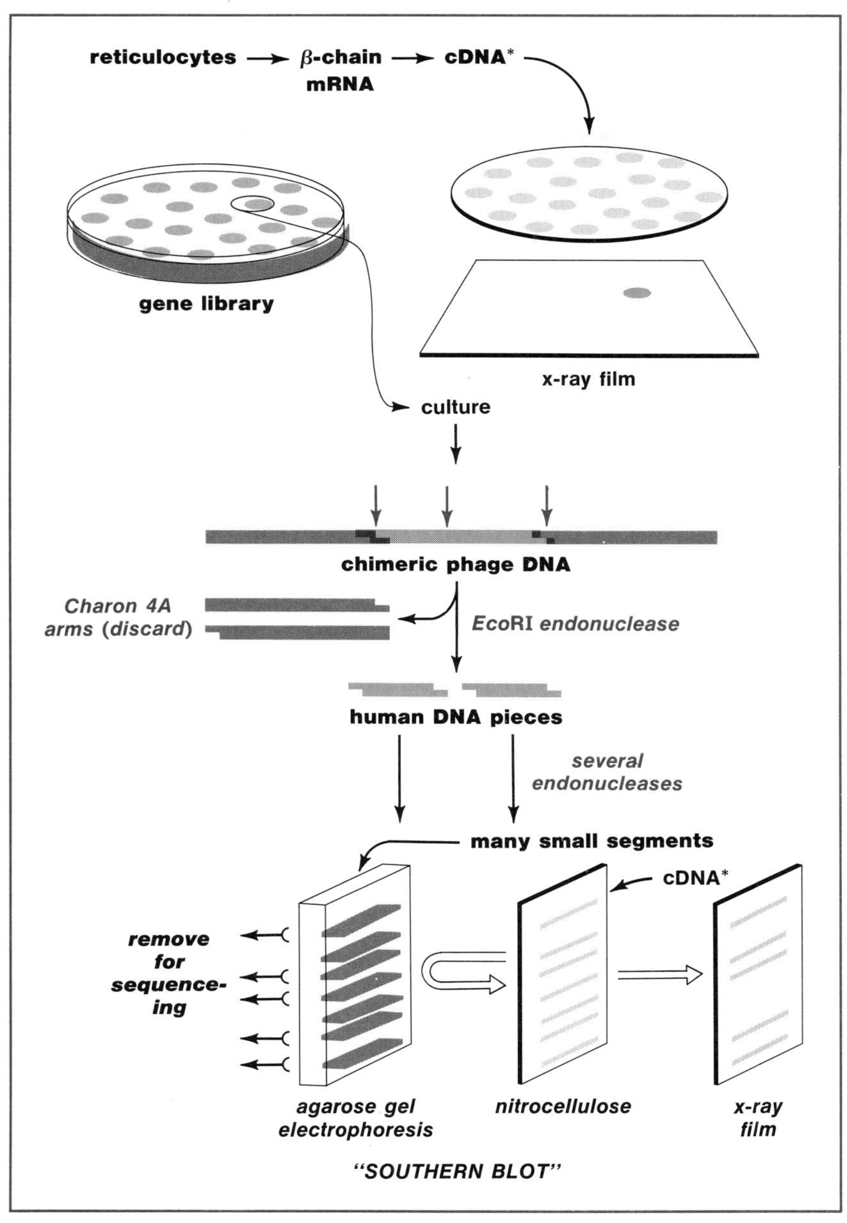

FIGURE 7–14

Isolation of gene for hemoglobin beta chain. *Top.* A radioactive probe is prepared by isolating β chain messenger RNA and synthesizing DNA complementary to it. The probe is used to search replicas of the human genome library for phage plaques carrying the β chain gene. *Center.* The phage containing the gene is amplified by culture in *E. coli* and its chimeric DNA is treated with *Eco*RI endonuclease. Since the amplified DNA is not protected by methylation, the endonuclease not only cleaves the built-in linker sites, but also any other sites that happen to occur in the DNA of interest. The β chain gene has one such site, so the DNA of interest appears in two pieces. *Bottom.* Each of the pieces is treated separately with other endonucleases to break it into small fragments suitable for sequencing. These fragments are separated by electrophoresis on an agar gel. Those containing parts of the β chain exons are identified by the Southern blot technique: blotting with nitrocellulose and testing the transferred DNA with radioactive DNA complementary to β chain mRNA. The bands detected as positive upon autoradiography of the blotted replicas are removed from the gel for sequencing of the DNA.

a methyl group donor (p. 628) to prevent their internal cleavage by *Eco*RI endonuclease in later steps.

3. Small (12-base pair) synthetic segments of double-stranded DNA containing an *Eco*RI endonuclease site were linked to the ends of the DNA fragments through the action of a DNA ligase.

4. Treatment of the DNA with *Eco*RI endonuclease now created sticky single-stranded ends.

5. The DNA was annealed with sticky-ended Charon 4A vectors prepared as outlined previously, and the chimeric* DNA was packaged into phage particles.

6. The phage particles were amplified by growing in host bacteria, and then plated out on lawns of bacteria. (Roughly 10,000 phage particles were added per 31 million cells.) The approximately one million independent phage colonies obtained in this way constitute the genome library.

Isolation of the globin β chain gene (Fig. 7–14). A probe to test for the gene was made by isolating globin messenger RNAs from immature red blood cells (reticulocytes) and making radioactive DNA complementary to the mRNA. The probe was incorporated into a plasmid for convenience in handling, and used to screen all of the genome library for fragments of the β chain gene. The DNA from two colonies was found to hybridize with the probe, indicating that they contained the β chain gene. After growing more phage from one of the colonies, the inserted DNA containing the gene was broken into smaller pieces by endonuclease treatment. After separation of the pieces by electrophoresis and transfer to nitrocellulose, pieces containing the gene were identified by hybridization with the radioactive cDNA probe.

Sequencing of the gene was carried out by the chemical degradation method on each piece after further cleavage of the gene fragments with endonuclease. All that remained was to read the autoradiograms and identify the overlaps between the partial sequences on each set so as to determine the order in which the pieces were originally joined.

PROCESSED GENES

It may be that living cells themselves use some of the techniques that have been described for DNA recombination. Recent surveys have detected some eukaryotic genes in which all introns have been removed from the DNA, and the 3′ end of the gene in the sense strand has a polydeoxyadenylylated tail. It is assumed that a reverse transcriptase has functioned within the eukaryotic cell to make cDNA from processed messenger RNA, in which the introns have been deleted and the poly-A tail has been added. The significance and frequency of this event are unknown.

---

*Interspecific hybrids are termed **chimera** from the Greek name for a fire-belching monster usually appearing with a serpent's tail and a lion's head on a goat's body. None have been seen in recent years, and it may be extinct.

FURTHER READING

Mutations and Repair

T. Lindahl (1982). DNA Repair Enzymes. Annu. Rev. Biochem. 51:61.

J.V. Neel (1979). Mutation and Disease in Humans. p. 7 in V.K. McElheny and S. Abrahamson, eds. Banbury Report 1. Assessing Chemical Mutagens: The Risk to Humans. Cold Spring Harbor Laboratory.

B.N. Ames (1979). Identifying Environmental Chemicals Causing Mutations and Cancer. Science 204:587.
B. Singer and J.T. Kusmierek (1982). Chemical Mutagenesis. Annu. Rev. Biochem. 51:655.
J.D. Hall and D.W. Mount (1981). Mechanisms of DNA Replication and Mutagenesis in Ultraviolet-Irradiated Bacteria and Mammalian Cells. Prog. Nucl. Acid Res. 25:53.
B.M. Sutherland (1978). Photo Reactivation in Mammalian Cells. Int. Rev. Cytol. Suppl. 8:301.
T. Lindahl (1979). DNA Glycosylases, Endonucleases for Apurinic/Apyrimidinic Sites, and Base Excision-Repair. Prog. Nucl. Acid Res. 25:53.
P.C. Hanawalt et al. (1979). DNA Repair in Bacteria and Mammalian Cells. Annu. Rev. Biochem. 48:783.
M.C. Peterson and P.J. Smith (1979). Ataxia Telangiectasia. Annu. Rev. Genet. 13:291.
J.E. Cleaver (1978). Xeroderma Pigmentosum. p. 1072 in J.B. Stanbury, J.B. Wyngaarden, and D.S. Frederickson. The Metabolic Basis of Inherited Disease, 4th ed. McGraw-Hill. A valuable source for reviews of most genetic defects of human metabolism.

Recombinant DNA

J. Abelson et al. (1977). Science 196:159ff. This issue contains many articles describing the techniques of recombinant DNA research, including most of those mentioned in this chapter.
J. Abelson et al. (1980). Science 209:1319ff. Another issue with many articles describing the structure and expression of genes as determined by use of recombinant DNA techniques. Discussions of globin, immunoglobulin, and interferon genes are included.
T. Maniatis et al. (1978). The isolation of structural genes from libraries of eucaryotic DNA. Cell 15:687. Describes formation of the gene library.
R.M. Lawn et al. (1978). The isolation and characterization of linked β- and δ-globin genes from a cloned library of human DNA. Cell 15:1157.
F.E. Baralle et al. (1980). Cell 21:621ff. This and four following papers from other laboratories describe the sequence of the human genes for the β, γ, δ, and ϵ chains of hemoglobin.
T. Maniatis (1980). Recombinant DNA Procedures in the Study of Eukaryotic Genes. p. 563 in L. Goldstein and D.M. Prescott. Cell Biology, vol. 3. Academic Press.
R.J. Roberts (1980). Directory of Restriction Endonucleases. p. 1 in L. Grossman and K. Moldave, eds. Methods in Enzymology, vol. 65. Academic Press.
A.M. Maxam and W. Gilbert (1980). Sequencing End-Labeled DNA with Base-Specific Chemical Cleavages. p. 499; A.J.H. Smith (1980). DNA Sequence Analysis by Primed Synthesis. p. 560; both in L. Grossman and K. Moldave, eds. Methods in Enzymology, vol. 65. Academic Press. These reviews describe the two major methods for determining DNA sequence.
B. Hohn and K. Murray (1977). Packaging Recombinant DNA into Bacteriophage Particles in Vitro. Proc. Natl. Acad. Sci. U.S. 74:3259.

CHAPTER 8

POST-TRANSLATIONAL MODIFICATION OF PROTEINS

A major theme in our discussion to this point has been the expression of genetic information through creation of polypeptide chains with specific arrangements of 20 amino acid residues. However, many polypeptides undergo further covalent reactions not involving the nucleic acids before they become finished mature proteins. Some of the reactions are hydrolyses of portions of the polypeptide chain; others modify the nature of the side chains so that there are more than 20 different kinds of residues. Are these extragenetic determinations of structure? No, they are not. In the modification of proteins, as in the modification of nucleic acids, the modifying agents themselves are created according to genetic instructions, and the sites at which they act on the polypeptide chains are fixed by the original amino acid sequences of the chains. Environment can influence the rate and the time at which these genetically predisposed changes occur but not their nature.

In this chapter, we shall examine typical examples to limn the principles without making a detailed catalog; other examples will be mentioned when we later treat the functions involved.

MODIFICATION BY PARTIAL HYDROLYSIS

Many newly synthesized polypeptide chains are promptly cleaved by hydrolysis, with one or more fragments being discarded. Proteins that are destined for secretion from the cell, and perhaps some that are included in lysosomes or other specialized cellular structures, are particularly likely to be constructed as oversize polymers to be refined later by hydrolysis into active products. Examples are many of the blood plasma proteins, which are synthesized in the liver; several polypeptides that act as hormones; and enzymes that are secreted by the pancreas or the liver. The purposes of this seemingly wasteful procedure are not the same for all such proteins.

Identification for Compartmentation. Some proteins are packaged for export in secretory granules; the proteins cross the endoplasmic reticulum membrane into the cisternal spaces and are transferred to the Golgi apparatus for enclosure. Insulin, the parathyroid hormone, and the digestive enzymes of the pancreas are handled in this way, along with other proteins. Those examples are specifically cited because they

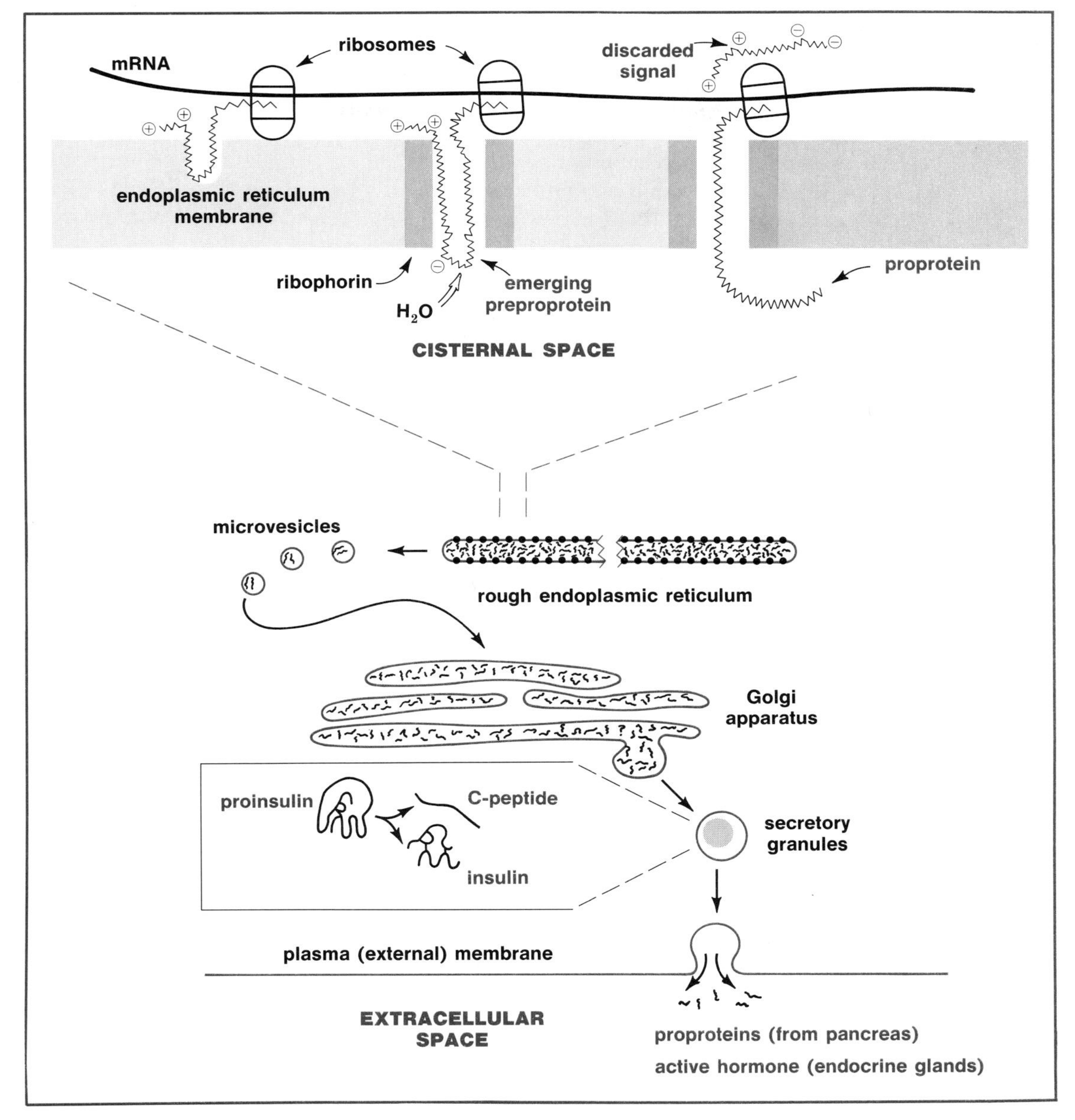

FIGURE 8–1

The synthesis of proteins for export from the cell. *Top.* Many secreted proteins first appear as a preprotein or preproprotein with an initially translated signal sequence having a charged end and a hydrophobic middle. The charged end remains outside the endoplasmic reticulum while the middle loops into the nonpolar endoplasmic reticulum membrane. The loop may align as a pleated sheet with ribophorin proteins in the membrane, effectively opening a pore for passage of the remainder of the polypeptide chain. The signal sequence is removed by hydrolysis when its end emerges into the cisternal space, and the remainder of the proprotein continues to pass through the membrane into the space. *Center, smaller scale.* The proprotein is packaged in microvesicles for transport to the Golgi apparatus. *Bottom.* Secretory granules containing the proprotein are released from the Golgi apparatus and fuse with the plasma membrane for discharge of the contents from the cell. In the case of proinsulin and proparathyroid hormone, the proprotein is partially hydrolyzed to create the active form within the secretory granules and the Golgi apparatus, as illustrated in a magnified view for insulin. The hydrolytic enzymes from the pancreas are not activated until they have been discharged into the lumen of the intestine.

are known to undergo two stages of hydrolysis in transit (Fig. 8–1). The first hydrolysis occurs immediately after formation of the peptide chain; it may even begin before translation is completed. The discarded segment contains a **signal sequence** that initiates movement of the balance of the polypeptide through the membrane, and then has no further function. Consider the formation of **human placental lactogen** as an example. The placenta secretes this hormone to stimulate development of the secretory apparatus in the mammary glands in anticipation of impending demand for their product. It is made as a **preprotein**—a precursor protein—beginning with 25 amino acid residues not found in the circulating hormone:

This is a typical signal sequence with these characteristic features: The first residue is the initiating methionine. The next six residues include a charged amino acid side chain—that of arginine in this case—but sometimes aspartate or lysine in other examples. The following 13 residues are nonpolar, with many bulky hydrocarbon groups. The final five residues preceding the sequence of the mature hormone again include polar side chains, and also amino acids facilitating the formation of reverse turns.

It is postulated that the signal sequence facilitates membrane penetration by forming a loop (Fig. 8–1) with the hydrophobic sequence aligned with membrane proteins, perhaps through pleated sheet formation. Two proteins not shown in Figure 8–1 are necessary for initial attachment of the signal sequence to the membrane. One, a **signal recognition protein,** combines with the sequence when 60 to 70 residues emerge from the ribosome and stops further translation. This prevents release into the cytosol of a protein destined for secretion. However, the signal recognition protein will combine with a **docking protein** on the endoplasmic reticulum, which permits translation to continue, forcing the growing polypeptide chain through the membrane. In any event, when the end of the signal sequence is exposed inside the endoplasmic reticulum, it encounters an enzyme that cleaves it from the following hormone sequence, which is now safely on its way through the membrane.

This picture provides an explanation for the synthesis of many exported proteins by ribosomes attached to the endoplasmic reticulum, that is, by the ribosome-studded rough endoplasmic reticulum. This membrane has proteins, **ribophorins,** not present in the smooth endoplasmic reticulum, and the signal sequences of growing polypeptide chains may combine with the ribophorins, linking ribosomes and membrane.

Masking of Biological Activity. Some proteins are threats to the tissue in which they are made, and enzymes that hydrolyze proteins, such as those involved in digestion or blood clotting (see Chapters 17 and 18), are particularly dangerous. Such enzymes are made with an extra segment of polypeptide that prevents them from attacking other proteins. This inactive form of the enzyme is said to be a **proenzyme,** which is converted to the active form by itself being partially hydrolyzed to remove the inhibiting segment of polypeptide chain.

The first stage of hydrolysis of these enzymes converts a **preproprotein**—the form initially translated and transported into the endoplasmic reticulum—into a **proprotein,** which is secreted from the cell. The second stage of hydrolysis, the conversion of proprotein to the active form, occurs outside the cell under circumstances making it desirable for the protein to carry out its function. For example, digestive

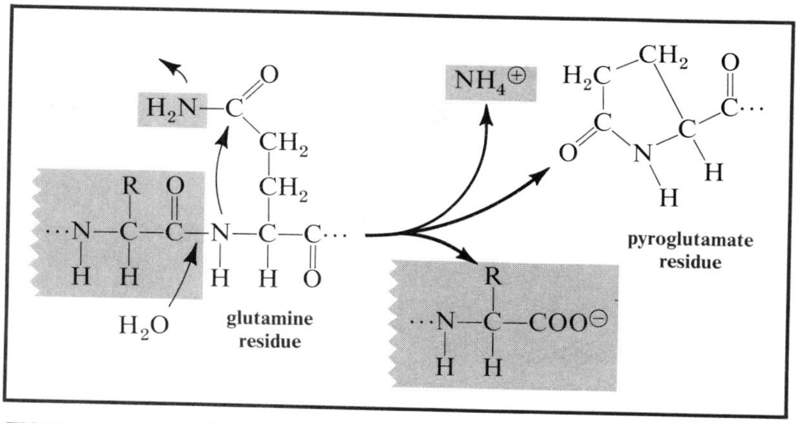

FIGURE 8–2 Hydrolysis of polypeptide chains at the peptide bond preceding a glutamine residue causes conversion of the residue to the uncharged pyroglutamate residue. The effect is an exchange of the peptide nitrogen (which would appear as an ammonium group in hydrolysis of other peptide bonds) for the side chain amide nitrogen of the glutamine. The amide nitrogen is released as a free ammonium ion. The other product of the hydrolysis is the initial segment of the chain, with a free carboxylate group.

enzymes are activated after entry into the intestines; tissue blood clotting enzymes are activated when tissues are damaged.

Pyroglutamate Residues. Some polypeptide chains begin with a residue of the heterocyclic compound variously known as pyroglutamate, oxoproline, or pyrrolidone carboxylate (Fig. 8–2). The presence of this residue is an indicator that the polypeptide is made from a longer precursor because the pyroglutamate residue is formed when a peptide bond preceding a glutamine residue is hydrolyzed. Hydrolyses at other locations do not cause distinctive changes in the terminal groups.

For example, pyroglutamate occurs in some **gastrins,** which are polypeptide hormones secreted by cells from the distal part of the stomach (antrum) and the initial part of the small bowel (duodenum). Gastrins provoke increased secretion of acid by the stomach. The activity of the hormones resides in the C-terminal end (Fig. 8–3), so hydrolytic removal of the initial portions of the chain makes little difference. The first gastrins isolated are now known as little gastrin and minigastrin, with minigastrin formed by hydrolysis of little gastrin. It is obvious that little gastrin is itself the product of hydrolysis of a still larger polypeptide, because the N-terminal residue is pyroglutamate. Such a polypeptide was indeed discovered (big gastrin), with a glutaminyl group in the right place to correspond with the pyroglutamyl group of the little gastrin, but it also began with a pyroglutamate residue, indicating the existence of still another precursor, which has also been demonstrated (big big gastrin), although its amino acid sequence has not been determined.

DISULFIDE BOND FORMATION

Fixation of Tertiary Structure. Most proteins attain their physiological form by spontaneously folding into a conformation of minimal energy content. However, some desirable configurations evidently cannot be attained in this way; the active form of some proteins is not the one with minimal energy content, and it exists only

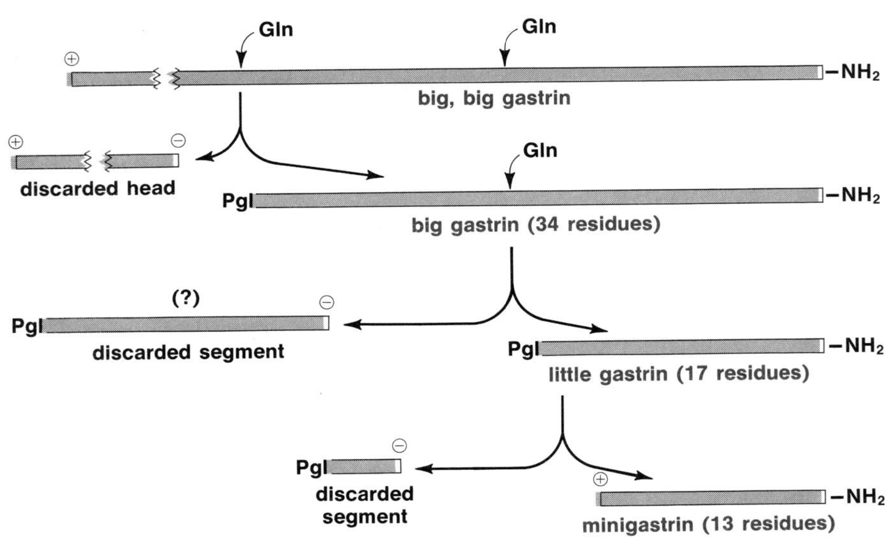

FIGURE 8–3

Various species of the hormone gastrin are created by successive hydrolyses. The first two hydrolyses occur before glutaminyl groups, creating products that begin with pyroglutamate (Pgl) residues. The final hydrolysis occurs at a leucine residue, with the product containing the charged ammonium group typical of most polypeptides. The terminal carboxyl group is present as an amide (see p. 152).

because it is cemented in place with additional covalent bonds between segments of polypeptide chain. Insulin (next section) and collagen (next chapter) are examples.

The problem is to hold the desired conformation long enough for it to be pinned together by some further reaction; it is solved by synthesizing the protein as a proprotein, in which the initial segment contributes forces necessary for proper folding of the entire molecule. The to-be-active segments of the chain are then fixed by forming covalent bridges. When the initial segment has served its purpose and is discarded by hydrolysis, the activity of the remaining molecule is revealed.

One of the ways in which peptide chains can be held in a particular configuration is by oxidizing the sulfhydryl groups on a pair of cysteine residues to form a disulfide bond (Fig. 8–4*A*). Disulfide bonds are created in proinsulin, the precursor of insulin, by a reaction with the oxidized form of a tripeptide, glutathione. The overall effect is a transfer of hydrogen atoms from the thiol groups of cysteine residues in proinsulin to the disulfide bridge of oxidized glutathione.

When disulfide bonds are being formed in proteins that contain several cysteine residues, the "wrong" sulfhydryl groups transiently may be near each other and become covalently linked. However, a mechanism of disulfide exchange is known by which the bonds migrate from one position to another (Fig. 8–4*B*). They will remain in the position that fixes the configuration of least energy content once it is achieved, so every molecule of a given protein will have the disulfide bridges in the same position, assuming that there is some favored conformation for the polypeptide chain.

With this in mind, we can examine the formation of insulin in more detail (Fig. 8–5). Synthesis begins with translation of messenger RNA into a single polypeptide chain, preproinsulin. The gene for preproinsulin is normally expressed only in particular cells, the beta cells of islets of Langerhans in the pancreas. Preproinsulin is 24 residues longer than proinsulin, and these residues are removed by hydrolysis before the peptide passes into the Golgi apparatus.

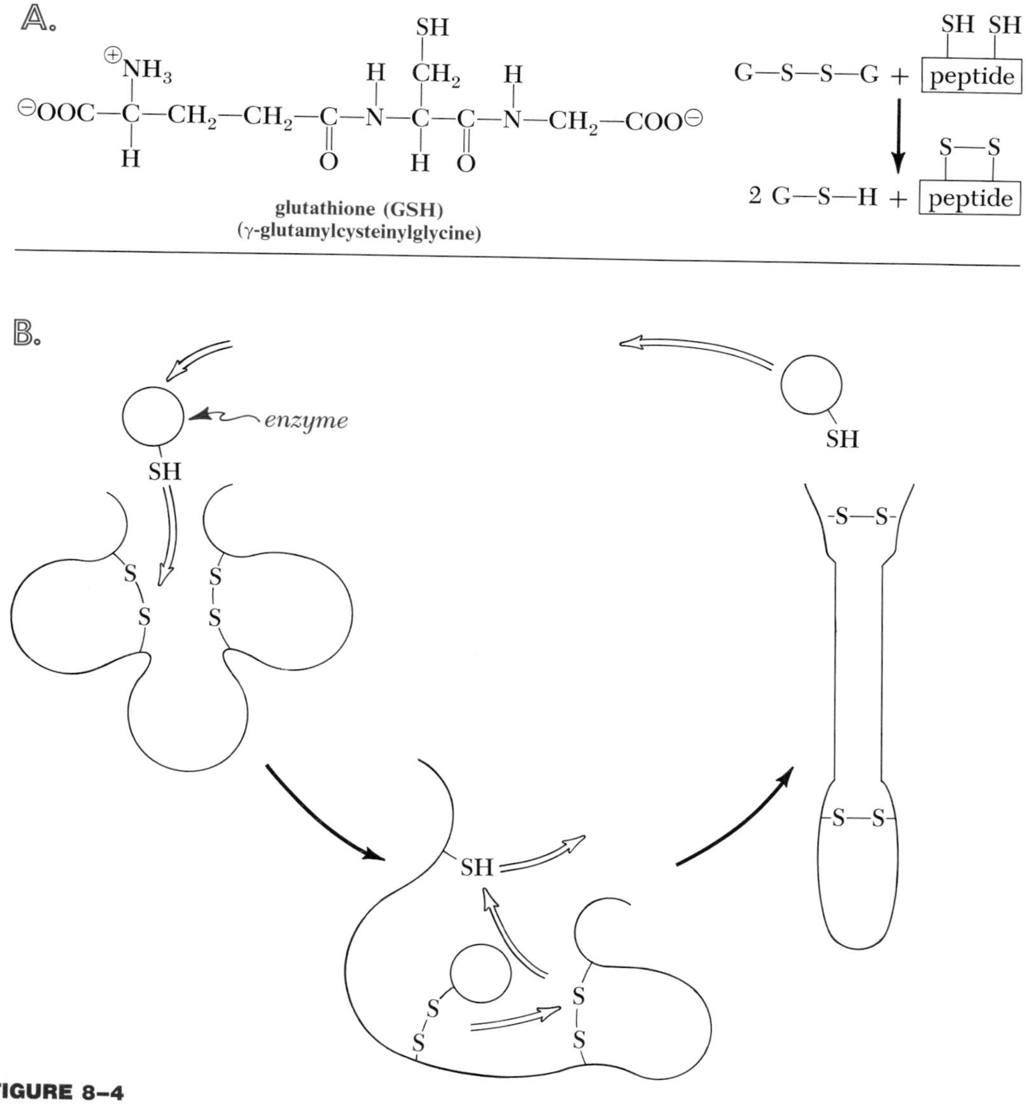

A.

glutathione (GSH)
(γ-glutamylcysteinylglycine)

$$G{-}S{-}S{-}G + \text{peptide(SH)(SH)} \rightarrow 2\ G{-}S{-}H + \text{peptide(S{-}S)}$$

B.

enzyme

FIGURE 8–4

A. Disulfide bridges are formed in proinsulin by the reaction of two sulfhydryl groups with oxidized glutathione, which is a disulfide. The glutathione is reduced to its sulfhydryl form by oxidizing the cysteinyl groups to their disulfide form. B. Disulfide bridges are rearranged by reaction with a sulfhydryl-containing enzyme that splits the various disulfide bonds and reforms them. Successive recombinations occur until the arrangement of the peptide chain with the lowest energy content (greatest internal bonding) is achieved.

At some stage, probably before transit through the membrane is completed but after hydrolysis of the signal sequence, the proinsulin polypeptide begins to fold into its conformation of minimal energy content. The correct pairs of cysteine residues are near each other in this conformation for formation of disulfide bonds.

Hydrolysis begins at two sites in proinsulin after it is packed into secretory granules by the Golgi apparatus. The segment of polypeptide linking the sites of hydrolysis (the C-peptide), along with some additional residues of arginine and lysine, are released. The remainder of the original chain is now in two pieces, which are held together by the disulfide bridges, and this molecule is active insulin,

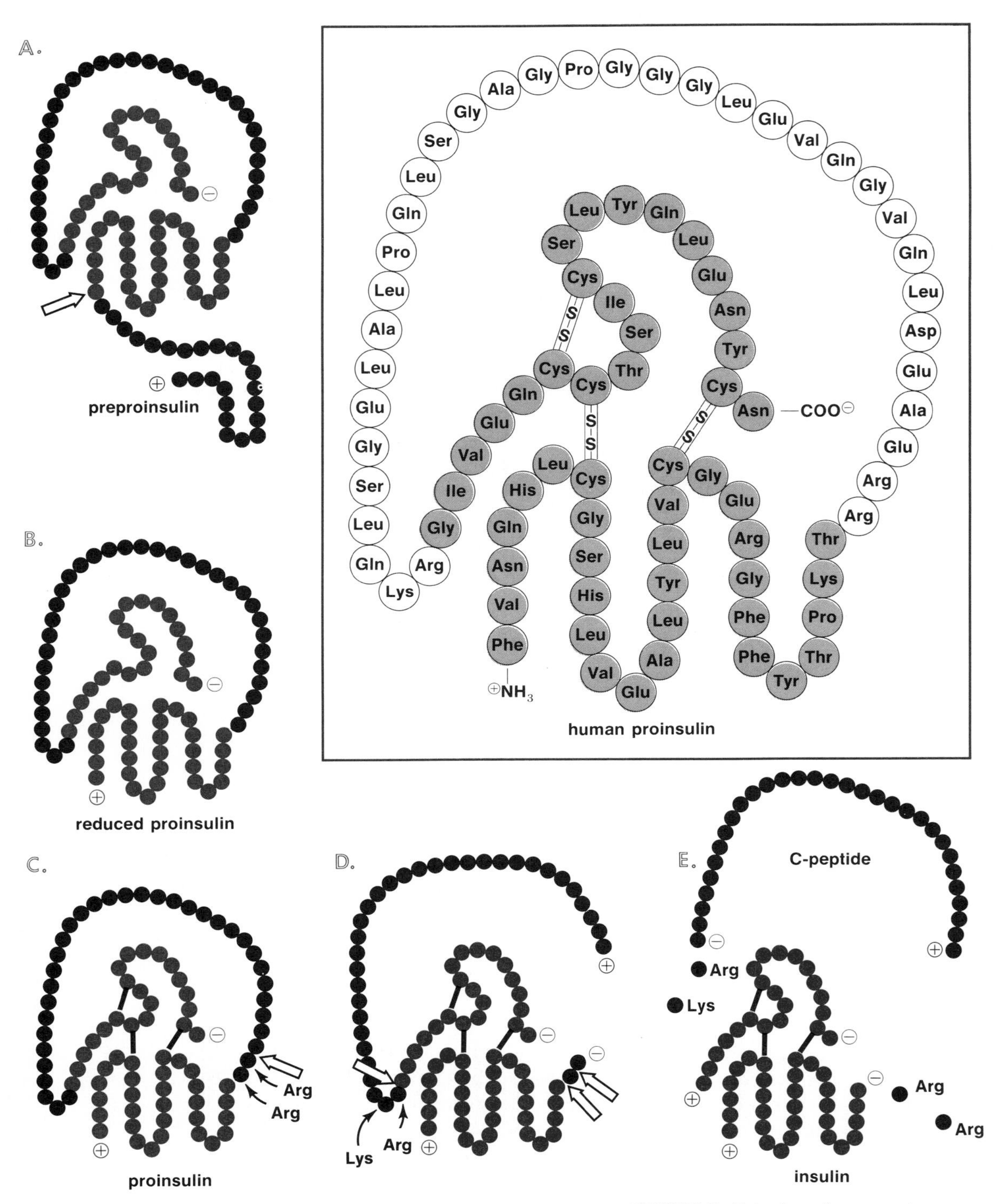

FIGURE 8–5 *See legend on opposite page.*

FIGURE 8–6

The N-terminal portion of ribonuclease A, an enzyme from calf pancreas, occurs in an α-helix. The remainder of the structure is not indicated. When ribonuclease is attacked by subtilisin (a hydrolytic enzyme from bacteria), a single cleavage occurs between residues 20 and 21 (*arrow*). Although the cut ends swing away, the small N-terminal segment remains attached, and its helical structure is retained because the combined structure is still in the conformation of lowest energy content; the cleaved molecule (S for subtilisin-modified) retains its biological activity.

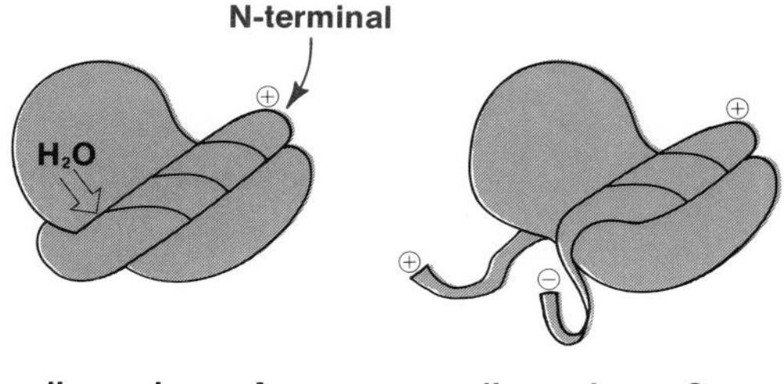

which remains packed in granules until a signal is given for release into the blood (p. 520). Insulin forms a zinc chelate, which crystallizes as a dense hexamer within the granules.

If the disulfide bridges in insulin are broken by reduction, the A and B chains separate. Only a small fraction of the original activity of the insulin can be recovered by oxidizing the mixture of the two chains so as to regenerate disulfide bonds. One might argue that the poor yield is due to the bimolecular reaction necessary for association of the chains; it is energetically less favorable to combine two molecules than it is to rearrange one into a similar conformation.

This notion is not sustained by experiments with other proteins (Fig. 8–6). In a classic example, the **ribonuclease** from calf pancreas (an enzyme hydrolyzing polyribonucleotides) was hydrolyzed at a specific peptide bond by attack with a bacterial enzyme. When the split ribonuclease was separated into two fragments, neither had any remaining activity. However, combination of the two immediately restored full activity, indicating that the pieces could combine in the specific conformation of intact ribonuclease, even though the polypeptide chain was still broken. We can conclude that ribonuclease exists in a natural conformation, whereas the mature insulin molecule is being held in a forced geometry by its disulfide bridges.

COVALENT MODIFICATIONS

Why should the complement of 20 amino acids available for protein synthesis be augmented by further changes after the polypeptide chain is made? Sometimes it

---

FIGURE 8–5 The formation of insulin. *A.* Preproinsulin is formed as a single polypeptide chain by the beta cells in the islets of Langerhans of the pancreas. The first 24 residues in the chain are rich in hydrophobic groups, which aid passage into the endoplasmic reticulum, where the residues are discarded by hydrolysis. *B.* The resultant proinsulin, the amino acid composition of which is shown in the box at upper right for humans, assumes a conformation of minimal energy content. (The drawing is arranged so as to display the covalent alterations to advantage; it does not give the actual arrangement of the peptide chain.) *C.* Disulfide bridges are formed between the three pairs of cysteinyl groups that are adjacent to each other in the favored conformation. The proinsulin is transferred to the Golgi apparatus and packaged in secretory granules. Attack by hydrolytic enzymes begins. An endopeptidase cleaves the single chain near a pair of arginyl groups into two pieces, which are held together by the disulfide bonds. *D.* Another enzyme, an exopeptidase, removes the two arginine residues from the end exposed by the endopeptidase. Meanwhile, the chain is attacked by the endopeptidase at another site, thereby exposing adjoining lysyl and arginyl groups to attack by the exopeptidase. *E.* The result is the removal from proinsulin of a large central segment of polypeptide chain, the C-peptide, along with four basic amino acid residues. The remaining ends of the chain are held together as a single insulin molecule by the previously formed disulfide bonds, and it is this molecule that appears in the blood upon discharge of the secretory granules.

is a device for distinctive recognition, much like some changes in nucleic acids; there is no confusing the site of an altered amino acid with other locations of similar sequence. Other times it is indeed a means of providing more structures with which to fulfill a function. **Perhaps the most common and most important reason for covalent modification is to provide a means of regulation, often reversible.** Groups are added or removed to change the biological function of a protein at specified times, and more examples of this major role of post-translational modification are constantly being discovered.

Removing Charge

We have already seen one way of diminishing charge on a polypeptide chain—hydrolysis to create the uncharged pyroglutamyl group. Charges can be eliminated in a more general way by converting ammonium or carboxylate groups to neutral amides. For example, the terminal ammonium groups or the lysyl side chains may be acetylated:

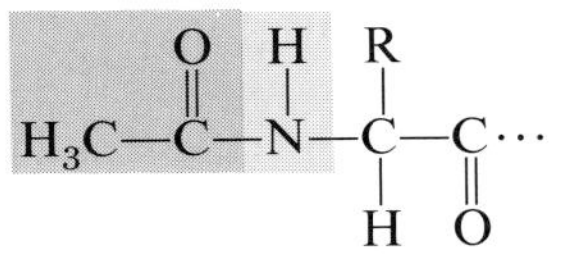

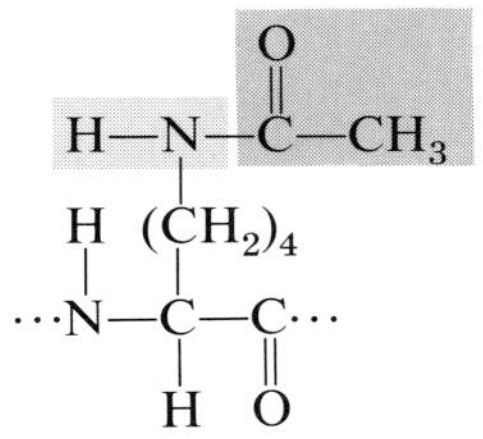

$H_3C{-}C(=O){-}N(H){-}C(H)(R){-}C(=O)\cdots$ $\qquad$ $\cdots N(H){-}C(H)((CH_2)_4{-}N(H){-}C(=O){-}CH_3){-}C(=O)\cdots$

Similarly, a terminal carboxylate group can be converted to a simple amide:

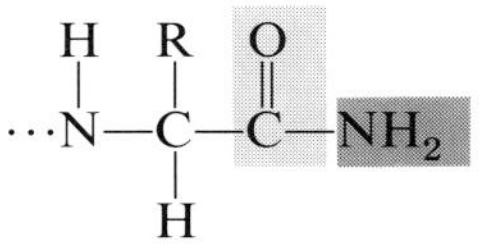

$\cdots N(H){-}C(H)(R){-}C(=O){-}NH_2$

C-terminal carboxamide

Gastrins are examples of peptides modified in this way (Fig. 8–3).

These changes, and those discussed in the following, are not exceedingly rare. Several proteins are known to occur as acetyl derivatives. Terminal amidation occurs in several small peptide hormones. An extreme example of loss of charge is seen in **thyrotropin releasing factor,** which contains only three amino acid residues. The first is a pyroglutamyl group derived from a glutamine residue by hydrolysis of a precursor, and the last is a proline residue with the carboxylate group amidated:

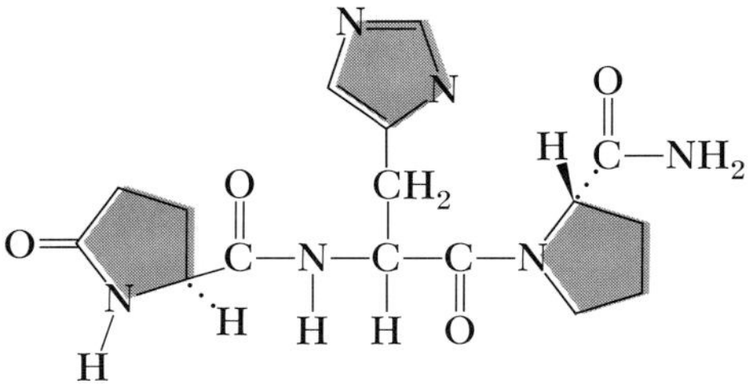

thyrotropin releasing factor
(pyroglutamyl histidyl proline amide)

The releasing factor causes the anterior pituitary gland to release thyrotropin, a hormone that stimulates the thyroid gland to release still another hormone, thyroxine (p. 719).

Adding Charge

Some proteins are modified by adding anionic groups, such as phosphate or sulfate groups, in the form of amides or esters. For example, the hydroxyl groups of serine or threonine residues are common acceptors for phosphate, creating phosphate esters:

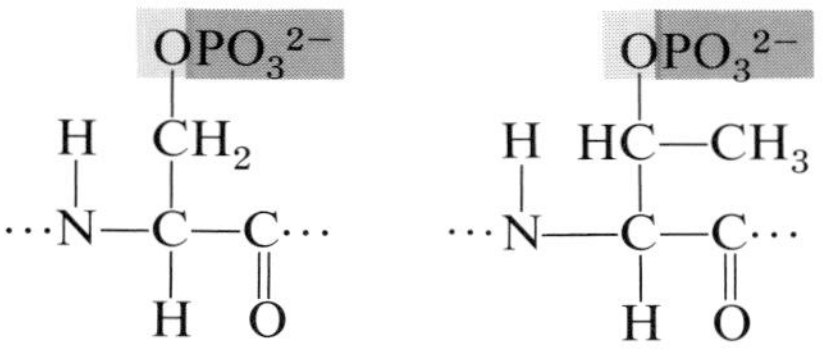

phosphoseryl group **phosphothreonyl group**

The phenolic group of tyrosine is also phosphorylated in some proteins:

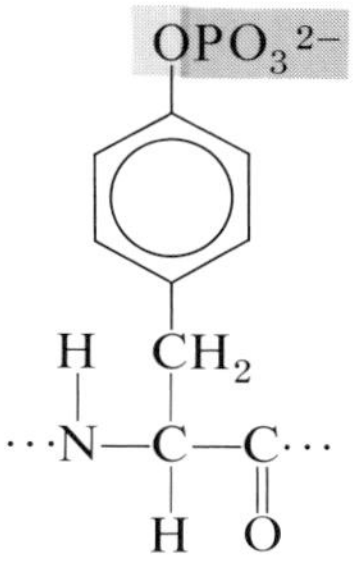

phosphotyrosyl group

Less common but not rare is formation of amides by phosphorylation of lysyl, arginyl or histidyl groups:

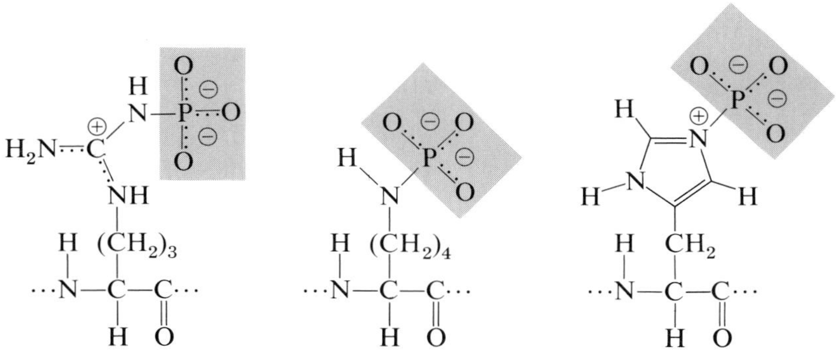

PArg **PLys** **N^{τ}-PHis**

There are several schemes for specifying atoms in the imidazole ring of histidine. Organic nomenclature minimizes the number assigned substituents; therefore the N most distant from the alanine group is #1, the other N is #3. Biochemists reversed these numbers. Still another system extends the α, β, γ, nomenclature into the ring. Another effort is being made to end chaos: Now the N nearest the alanine is N^{π} (π for *pros,* near) and the more distant N is N^{τ} (τ for *tele,* far). What will tomorrow bring? Who knows how extensive the ignorance of the past will be!

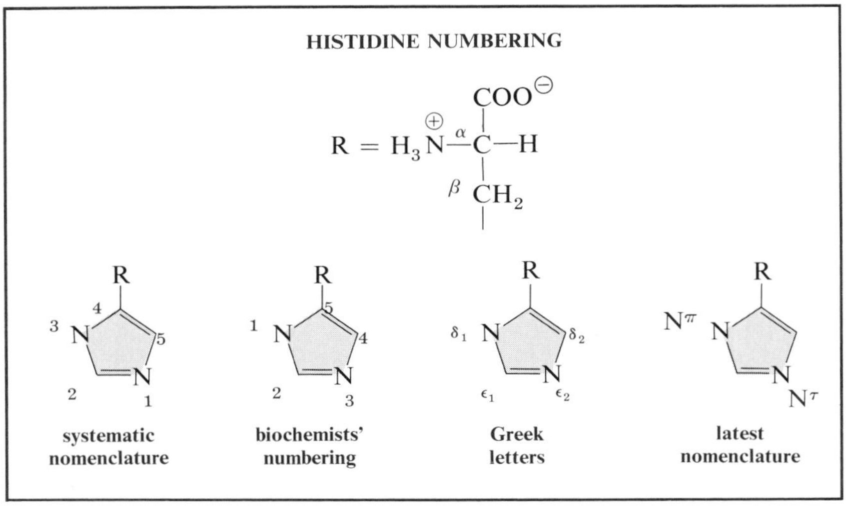

HISTIDINE NUMBERING

$$R = H_3\overset{\oplus}{N}-\overset{\alpha}{C}(COO^{\ominus})(H)-\overset{\beta}{CH_2}-$$

systematic nomenclature: R at 4; N 3, 5, N 1, 2

biochemists' numbering: R at 5; N 1, 4, N 3, 2

Greek letters: δ_1 N, δ_2, N, ϵ_1, ϵ_2

latest nomenclature: N^{π}, N^{τ}

Similarly, proteins are sometimes converted to sulfate esters or amides. Some gastrin molecules, for example, are modified by esterification of their tyrosyl groups with sulfate:

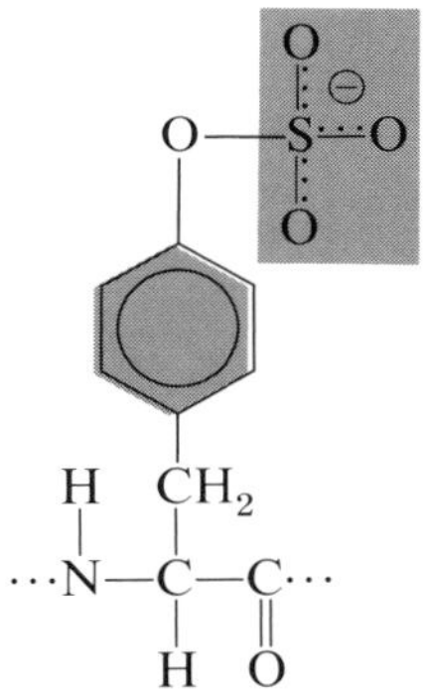

$$\cdots NH-CH(CH_2-C_6H_4-O-SO_3^{\ominus})-C(=O)\cdots$$

Functions of Phosphoproteins

Phosphorylation and dephosphorylation is a major mechanism for controlling the biological activity of proteins. With some proteins, phosphorylation favors a shift into a more active conformation; with other proteins, it is removal of the phosphate by hydrolysis that increases the activity. Control of the cell cycle, the action of some hormones, a shift from fuel storage to fuel mobilization—these are only examples of critical processes regulated through specific phosphorylations of proteins. We shall explore the general process and individual examples of regulation in more detail after examining the nature of enzymes.

Nutrition for the Young. Some proteins are heavily phosphorylated as a device for carrying phosphorus and associated calcium from the mother to the offspring. Animals that reproduce via a closed egg include heavily phosphorylated proteins in the yolk for the nourishment of the embryo, and calcium ion is also provided by chelation with these negative groups. For example, **phosvitin** is a protein with 100 to 120 phosphate groups per molecule of 35,000 daltons. (Ten per cent of the mass of phosvitin is contributed by phosphorus atoms.) The polypeptide chain contains many clumps of seryl residues, which are the sites of phosphorylation. Chicken eggs are important sources of both phosphorus and calcium in the human diet owing to the presence of phosvitin and similar proteins.

Mammalian infants have another source of phosphorus and calcium: milk, which contains casein. Casein is not a single entity; it is a micelle of several different polypeptides in combination with calcium and some small anions. Casein from rats and mice contains three major gene products. The composition varies with species, and the polypeptides from cows are the most thoroughly described. The bovine caseins occur in a micelle that has a core of two classes of phosphorylated polypeptides (α and β), containing from 0.5 to 1.0 per cent P, and they are associated with Ca^{2+}. The core is covered with a disulfide-linked coating of another class of polypeptide (X) containing much less phosphorus. The coating prevents micelles from aggregating. Human casein is believed to have a similar structure, although the components differ from those in bovine milk. (There is also considerable polymorphism of casein within species, indicating that many genes are involved in its formation.)

When casein is delivered to the infant stomach, exposure to the increased $[H^+]$, together with partial enzymatic hydrolysis of the coating X peptides, causes a disruption of the micelles. The constituent polypeptides have little secondary structure, and the length of the polypeptide chain is readily exposed for digestion. The accompanying phosphate and calcium ions are then available for absorption.

ADP-Ribosylation

Nicotinamide adenine dinucleotide is a compound with a central role in metabolism because it acts as an oxidizing agent in a large number of metabolic reactions. We shall consider its structure and function in that connection (p. 289). For now, we are concerned with an auxiliary function of the compound: It is a donor of **adenosine diphosphate ribose (ADP-ribose)** groups for modification of proteins. NAD contains two ribonucleotides linked through their phosphate groups by a pyrophosphate bond:

$$\boxed{\text{nicotinamide}}-\underbrace{\boxed{\text{ribose}}-\overset{\ominus}{PO_2}-\overset{\ominus}{PO_2}-\boxed{\text{ribose}}-\boxed{\text{adenine}}}_{\text{adenosine diphosphate ribose}}$$

Some enzymes can cleave the nicotinamide-ribose bond and transfer the remaining ADP-ribose group onto specific carboxyl groups in proteins. In some cases, additional ADP-ribose groups are added onto the first so as to make a polyADP-ribose side chain on the protein (Fig. 8–7).

In most cases, ADP-ribosylation is used as a regulatory device, but some patho-

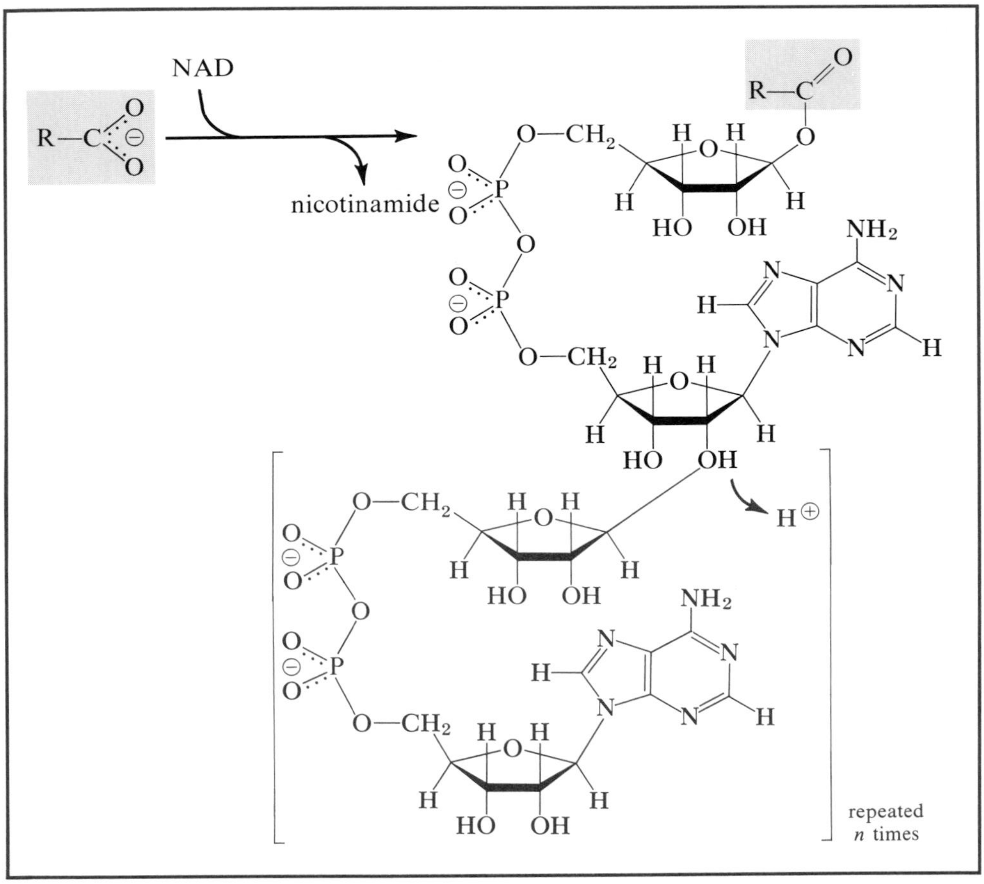

FIGURE 8–7 **ADP-ribose groups are transferred from NAD to carboxyl groups in proteins. After the initial group is attached (*top right in black*), additional groups may be attached to it, making a polymer (*lower right in color*).**

genic microorganisms employ it to bring down their host. For example, *Corynebacterium diphtheriae* has adapted to its own ends a phage that infests it and carries a gene for the production of a particular protein, the **diphtheria toxin.** The toxin protein is a single polypeptide chain forming two domains linked by a disulfide bond. The bacteria hydrolyze the chain between the two domains, and reduce the disulfide bridge. These changes enable the toxin to become active and cause illness in an infected human. One of the domains combines with receptors on the human host's cells normally used for other purposes and causes transport of the second domain into the cell. This second domain acts as an enzyme; it is the true toxin because it uses NAD in the host cell to transfer ADP-ribose onto the host cell's elongation factor 2. The result? No more protein synthesis in the host cell, cell death, and dissolution of the tissue (necrosis), which releases more nutrient for the invading bacteria.

Study of the ADP-ribosylation of elongation factor 2 revealed an unsuspected example of another kind of post-translational modification, this one normal. The affected amino acid residue in EF-2 is a modified histidine residue:

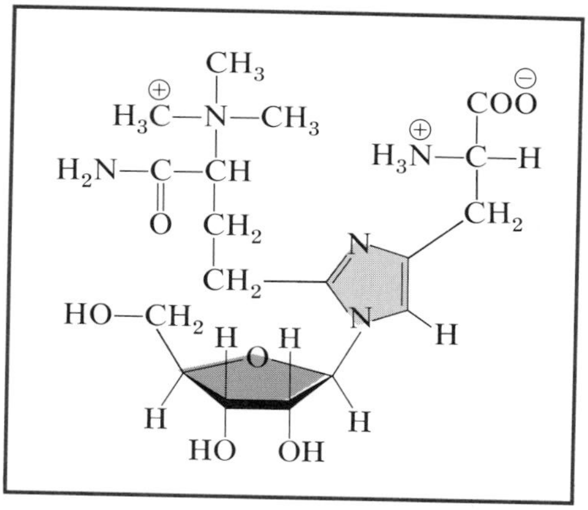

assigned the trivial name of **diphthamide,** although its formation as a normal part of EF-2 has nothing to do with the diphtheria organism.

Methylations

Several proteins are modified by methylation; the affected groups are the charged side chains (Fig. 8–8). Methylation of glutamate carboxyl groups to form the methyl esters is known to be the device by which some bacteria are stimulated to move up a chemical concentration gradient **(chemotaxis),** and it appears to be involved in similar responses by human leukocytes. **Sperm immotility,** a common cause of human infertility, is often caused by defective methylation of sperm proteins.

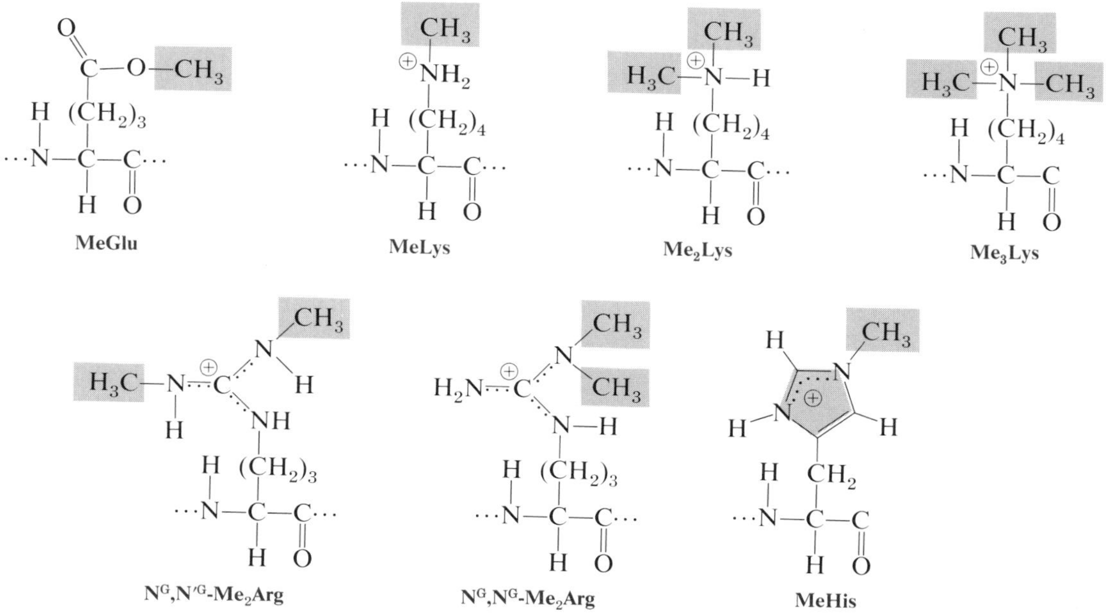

MeGlu MeLys Me_2Lys Me_3Lys

N^G,N'^G-Me_2Arg N^G,N^G-Me_2Arg MeHis

FIGURE 8–8

Carboxylate and nitrogenous side chains in polypeptides are sometimes methylated. More than one methyl group may be substituted on the same side chain, sometimes on the same nitrogen atom.

Methylation of aspartate residues occurs in membrane proteins. Methylation of either glutamate or aspartate residues destroys the negative charge. The methyl esters are readily hydrolyzed to regenerate the original carboxylate groups, so methylations of this type can be used as reversible control devices.

Methylation of the cationic amino acid side chains does not remove the positive charge, and unlike the methyl esters, the N-methyl amino groups are not readily cleaved. Methylation of these groups probably serves recognition and structural functions. Lysine residues of some nuclear proteins may have one to three methyl groups added. Histidine residues in muscle proteins are methylated, especially at N^{τ}, and urinary excretion of N^{τ}-methylhistidine is used as an index of the breakdown of muscle proteins in the study of some disease states. Arginine residues may be methylated on any one of three nitrogen atoms in the side chain. Dimethylarginines occur in various places (cell nuclei, nerve sheaths, sperm tails), and are normally excreted in the urine.

Modification of Histones

The histones are especially subject to modification, and the changes they undergo appear to be important devices for regulating the cell cycle. Histones contain methylated amino acid residues, they undergo charge modification through acetylation and phosphorylation, and they are ADP-ribosylated. Modification of charges would affect the electrostatic interaction between the negative DNA backbone and the positive histones, but there are more subtle effects on conformation yet to be analyzed. Histone H-4 may have two to four acetyl groups added; this causes dissociation of the histone from DNA and increases the rate of transcription of the exposed DNA. Phosphorylation of H1 histone also should diminish charge interaction; however, it occurs during the prophase and evidently promotes condensation of the chromatin. Histones are also made less positive by addition of ADP-ribose groups. One is attached to a glutamate residue on H2B through an ester bond; three may be attached to H1 through ester bonds with two glutamate residues or with the terminal carboxylate group. The esterification removes the negative charge on the carboxylate group, but two negative charges are brought in with the ADP-ribose.

FURTHER READING

General Reviews

F. Wold (1981). In Vivo Chemical Modification of Proteins (Post-Translational Modification). Annu. Rev. Biochem. 50:783.

D.F. Steiner et al. (1980). Proteolytic Cleavage in the Post-translational Processing of Proteins. p. 175 in D.M. Prescott and L. Goldstein, eds. Cell Biology, A Comprehensive Treatise, vol. 4. Academic Press.

J.H. Walsh and S.K. Lam (1980). Physiology and Pathology of Gastrin. Clin. Gastroenterol. 9:567.

Technical Reviews and Articles

R.B. Freedman and H.C. Hawkins (1980–). The Enzymology of Post-translational Modification of Proteins. Multivolume treatise.

M. Zimmerman, R.A. Mumford, and D.F. Steiner, eds. (1980). Precursor Processing in the Biosynthesis of Proteins. Ann. N.Y. Acad. Sci. vol. 343. A symposium with several review articles.

D.L. Meyer, E. Krause, and B. Dobberstein (1982). Secretory Protein Translocation Across Membranes—The Role of the "Docking Protein." Nature 297:647.

N. Ogata, K. Ueda, and O. Hayaishi (1980). ADP-Ribosylation of Histone H2B. J. Biol. Chem. 255:7610ff.

B.G. Van Ness, J.B. Howard, and J.W. Bodley (1980). ADP-Ribosylation of Elongation Factor 2 by Diphtheria Toxin. J. Biol. Chem. 255:10710.

M.S. Springer, M.F. Gay, and J. Adler (1979). Protein Methylation in Behavioral Control Mechanisms and in Signal Transduction. Nature 280:279.

C. Gagnon et al. (1982). Deficiency of Protein-Carboxyl Methylase in Immotile Spermatozoa of Infertile Men. N. Engl. J. Med. 306:821.

M. Neuhauser et al. (1980). Urinary Excretion of 3-Methyl-histidine as an Index of Muscle Protein Catabolism in Postoperative Trauma. Metabolism 29:1206.

CHAPTER 9

FIBROUS STRUCTURAL PROTEINS

In our initial discussion of proteins, we emphasized those of globular form. The ready solubility in water of most globular proteins is not always a desirable property. Some proteins are used to provide structural elements, and maintenance of structural integrity requires low solubility. This is as true for the cytoskeleton of a single cell as it is for the integument of an elephant. Many structural elements also must have mechanical strength. It is desirable for an organism to resist the slings and arrows of outrageous fortune, and it is imperative that it not thaw and resolve itself into a dew.

Molecules meeting these requirements often have a high ratio of length to diameter and are constructed to form fibers by associating side to side. Vertebrate skin contains striking examples. Two populations of cells in the skin are separated by a thin layer made of a network of fibers—the **basement membrane.** Cells in the **dermis** beneath the membrane produce fibers made of the protein **collagen,** and the fibers are crisscrossed in a felt-like mat.

Most of the cells in the **epidermis** above the membrane contain an insoluble protein aggregate, **keratin.**

The division between dermis and epidermis is as sharply marked by the predominance of collagen and keratin as it is by the visible differences between the cells of the two layers under the microscope. Both of these proteins are very insoluble. Both have high tensile strength. Both form fibers. Here the resemblance ends, however, and it is interesting to see how their structures vary to satisfy the separate requirements of the tissues in which they occur.

Fibers and Matrix

Both the dermis and the epidermis contain examples of a general kind of supporting structure made by embedding fibers in an amorphous matrix. Such structures appeared early in the evolution of multicellular organisms. In their now-diverse applications the fibers may range from flexible shafts to rigid rods, and the matrix may be a hard casting or a resilient bed. (Man has consciously used the principle for many millenia; we think of fiber glass and reinforced concrete and tend to forget why the followers of Moses were indignant when required to make bricks without straw.)

The wide range of physical properties in biological combinations of fiber and matrix is achieved by varying the materials in the two phases. The most common constituents, those with which we are concerned, are polymers of amino acids or carbohydrates: proteins or polysaccharides. However, other organisms have different combinations. For example, the connective tissue of woody plants contains fibers of a carbohydrate polymer, **cellulose,** buried in a matrix of condensed polyphenols, **lignin.** The exoskeleton (shell or cuticle) of arthropods contains fibers of **chitin,** another polysaccharide, buried in a matrix of protein drastically modified by tanning. Table 9–1 summarizes some of the fiber-matrix combinations.

We shall discuss collagen, elastin, and keratin in this chapter as examples of fibrous proteins, and examine the carbohydrate matrix in the next chapter.

COLLAGENS

The collagens are proteins that occur as insoluble rigid rods in all organs. They may be dispersed in a matrix when stiffening of a gel is all that is required, as in the ground substance around cells and the vitreous humor of the eye, or they may be neatly bundled in tight parallel arrays when great strength is required, as in tendons.

Indeed, collagens are the predominant type of protein in the human body. They are an important constituent of bone in which the fibers are arranged at an angle to each other so as to resist mechanical shear from any direction. About half the dry weight of cartilage is collagen, from which the cartilage acquires its toughness. Even the cornea of the eye contains a high proportion of collagen arranged in neatly stacked arrays (Fig. 9–1), so as to transmit directly impinging light with minimal

TABLE 9–1

	Fiber	Matrix	Typical Locations
keratin	protein (α-keratin)	protein (keratohyalin)	epidermis, nails, hair, horn, hoof
collagenous tissue	protein (collagen)	polysaccharide (chondroitin sulfate)	intercellular space, cartilage, bone, basement membranes
wood	polysaccharide (cellulose)	polymerized polyphenols (lignin)	higher plants
chitinous shell or cuticle	polysaccharide (chitin)	tanned protein (sclerotin)	exoskeleton of crabs, insects, spiders

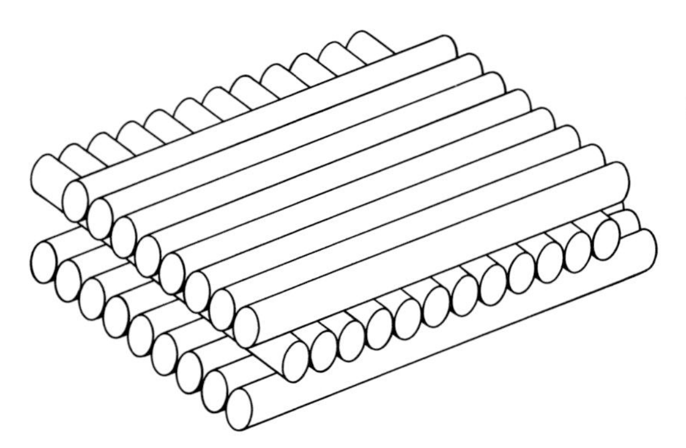

FIGURE 9–1

Collagen fibers are stacked crosswise to give rigid transparency to the cornea of the eye. Normally incident light is little scattered by this arrangement.

scatter, whereas the neighboring opaque sclera has a more disorganized arrangement.

The polypeptides composing collagen are formed in fibroblasts or in the related osteoblasts and chondroblasts of bone and cartilage as soluble precursors, which are secreted into the extracellular space before they are modified to form the final rigid structure. (Solidification within the cell ought to be at least discomforting, if not lethal.) Fibroblasts can lay down collagen required for local strengthening, for example, at the margin of a wound.

Structure of Collagen

The amino acid composition of collagen is distinctive with nearly a third of the residues being glycine. For example, one kind of collagen polypeptide chain has a total of 1,052 residues in which the entire sequence from residue 17 through residue 1,027 can be represented as $(Gly\text{-}X\text{-}Y)_{337}$.

Furthermore, over one in five of the total residues is proline, and nearly half of these are modified by hydroxylation at the 4 position and to a much lesser extent at the 3 position of the ring:

proline residue

3-hydroxyproline residue

4-hydroxyproline residue

These hydroxyprolines, abbreviated Hyp, occur mainly in collagen. A few are found in another connective tissue protein, **elastin,** and still fewer in the C1q component of the complement system of blood (p. 354) and the enzyme acetylcholinesterase (p. 309), but in general the content of hydroxyproline in a tissue is an index of its content of collagen. To summarize, of the total amino acid residues in the known kinds of collagen,

33 to 35 per cent are **glycine**
20 to 24 per cent are **proline + hydroxyproline,** distributed as
6 to 13 per cent proline, and
9 to 17 per cent hydroxyproline

The 18 other amino acids occur in fewer than half of the residues.

The Polypeptides of Collagen Form a Triple Helix. Collagen molecules are made from three polypeptide chains twisted into a rope-like structure (Fig. 9–2). The major factor in creating this structure is the occurrence of glycine as every third residue, because it enables twisting each chain one turn for each three residues, so as to bring all of the glycine residues to the inside of the rope. Since glycine has only a hydrogen atom for a side chain, polypeptides can make a close fit. The twist within chains is left-handed, but the bundle of three chains is strengthened by one right-handed twist for each ten turns of a single chain.

The prolyl and hydroxyprolyl groups are accommodated in a triple helix without distortion and stabilize the structure, whereas we earlier emphasized that they will not fit in the less-tightly twisted α-helix of a single polypeptide chain. In addition, full stability of the triple helix depends upon the formation of hydrogen bonds with the hydroxyl group on residues of hydroxyproline. (Much of the stabilizing force comes from the formation of hydrogen bonds between the peptide backbones; there is one direct bond per turn of a chain and an additional bond involving an inserted water molecule, but these are not quite sufficient in themselves to give a full stability.)

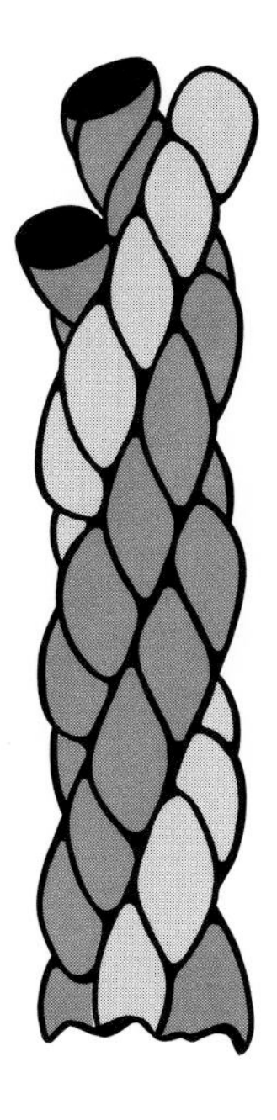

FIGURE 9–2

Collagen is made of three polypeptide chains. Each chain is twisted to the left one turn in three residues, and the three chains are twisted together in a right-handed helix, with ten turns of each chain per turn of the triple helix. Type I collagen contains a pair of one kind of chain, and a different third chain, as shown; other collagens contain three identical chains.

Hydroxylysine and Sugars. Hydroxyl groups are also formed on some of the lysine residues in collagen:

$$\begin{array}{c} COO^{\ominus} \\ | \\ {}^{\oplus}H_3N-C-H \\ | \\ CH_2 \\ | \\ CH_2 \\ | \\ HO-C-H \\ | \\ CH_2 \\ | \\ {}^{\oplus}NH_3 \end{array}$$

5-hydroxy-L-lysine

Hydroxylysine also occurs in the C1q component of complement but not in elastin, so it is even more distinctive than hydroxyproline, although it is not as abundant. The hydroxylysine side chains apparently have two main functions: to participate in the formation of cross-links (next section) and to act as sites for the attachment of sugar groups. (We shall consider glycoproteins—proteins containing attached residues of sugar—at some length in the next chapter.) Collagen contains two sugars, galactose and glucose, linked with the hydroxyl group of hydroxylysine residues. The number of added sugar molecules varies from one type of collagen to another.

Heterogeneity. A given molecule of collagen is not always hydroxylated at all potential sites on proline or lysine residues, and all of the possible carbohydrate residues are not always attached. The result is some variation of composition within one type of collagen, as well as between types. This variation has a functional role.

Different Kinds of Collagen. There are at least seven, perhaps eight, different genetic types of polypeptide chains in collagens, and the collagens of different tissues vary according to the types of chains being made. An unfortunate and confusing nomenclature has developed employing Greek letters entirely differently from the way they are used with other proteins. To avoid further conflict, let us simply designate these chains as "a" through "g."

Our designation	a	b	c	d	e	f	g
Literature designation*	α1 (I)	α2	α1 (II)	α1 (III)	α1 (IV)	α1 (V)	α2 (V)

*The literature uses β for a dimer of triple helical collagen molecules.

The principal collagen of the body, type I, contains two "a" chains and one "b" chain (a_2b). The others, with the possible exception of type V, are made from three identical chains—c_3, d_3, or e_3. We don't know the functional significance of these differences in chain composition, but their probable importance is evident from their different distributions:

Type I collagen is about 90 per cent of the adult total, occurring in bone, tendon, cornea, soft tissues, and scars; it contains the least carbohydrate.

Type II collagen, c_3, occurs in cartilage and is made by chondrocytes.

Type III collagen, d_3, predominates in blood vessel walls and fetal skin; it contains disulfide bridges between chains, whereas types I and II do not. It is present in scars and all adult soft tissues examined, but not in unscarred bone or tendon.

Type IV collagen, e_3, is the collagen of the basement membrane. It also contains disulfide bridges and it has the highest carbohydrate content. The characteristic

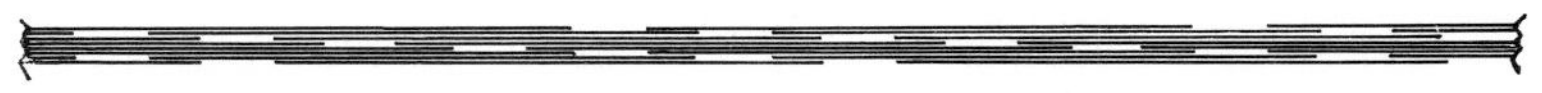

FIGURE 9–3 Individual molecules in a collagen fibril are staggered by ~23 per cent of their length. Molecules in line are spaced by gaps equal to ~14 per cent of their length. This diagram exaggerates the cross-fibril dimension by ~75 per cent, compared to the length, but representing the molecules by lines does not show that each is a triple helix, as shown in Fig. 9–2.

collagen helix is interrupted by several short nonhelical sequences along the length of the type IV collagen molecule.

Type V collagens are not well characterized. They have been found in basement membranes, placental tissues, and the epidermis, where they may form connections between some cells and other fibrous proteins.

Fiber Formation and Cross-Linking

Collagen molecules are stacked together into rigid, strong fibers after secretion from the cell. The molecules are built to be stacked in a very regular parallel way (Fig. 9–3), with the ends of each molecule contacting definite regions on the sides of adjacent molecules. The exact arrangement in the third dimension is not known, but it may be a spiral, as if a sheet of rods had been rolled up like a carpet. The fibers are strengthened through the formation of covalent bonds between adjacent molecules.

These cross-links are formed by oxidatively deaminating the side chains of selected lysine or hydroxylysine residues (Fig. 9–4), creating aldehyde groups. (The modified residues are referred to as allysine or hydroxyallysine. The older literature used lysinal, but this term is now reserved for the synthetic analog of lysine in which the α-carboxyl group is reduced to an aldehyde.) The aldehydes react in a variety of ways with groups on other side chains of adjacent molecules, including other aldehyde groups, amino groups in lysine or hydroxylysine residues, and the imidazole groups of histidine residues. The reactive aldehydes are created mainly near the ends of the polypeptide chains, outside of the triple helical region, and combine with groups that appear to be specifically located for this purpose in the more central portions of adjacent molecules.

Two hydroxyallysine groups can condense with a hydroxylysine group to form a stable pyridinium cross-link, perhaps between three molecules:

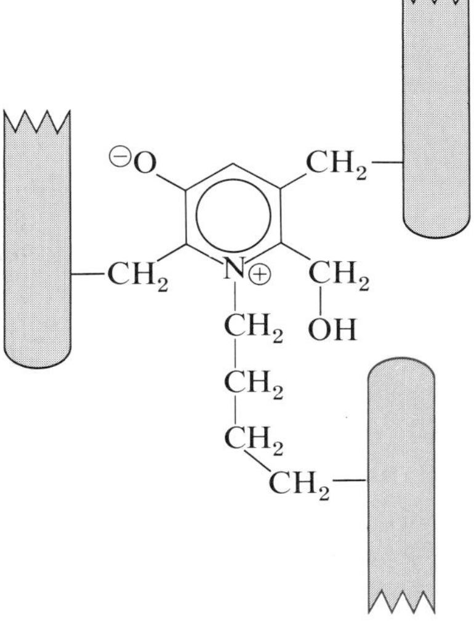

$^{\ominus}O$
CH_2
CH_2
$N^{\oplus}$
CH_2
CH_2 OH
CH_2
CH_2
CH_2

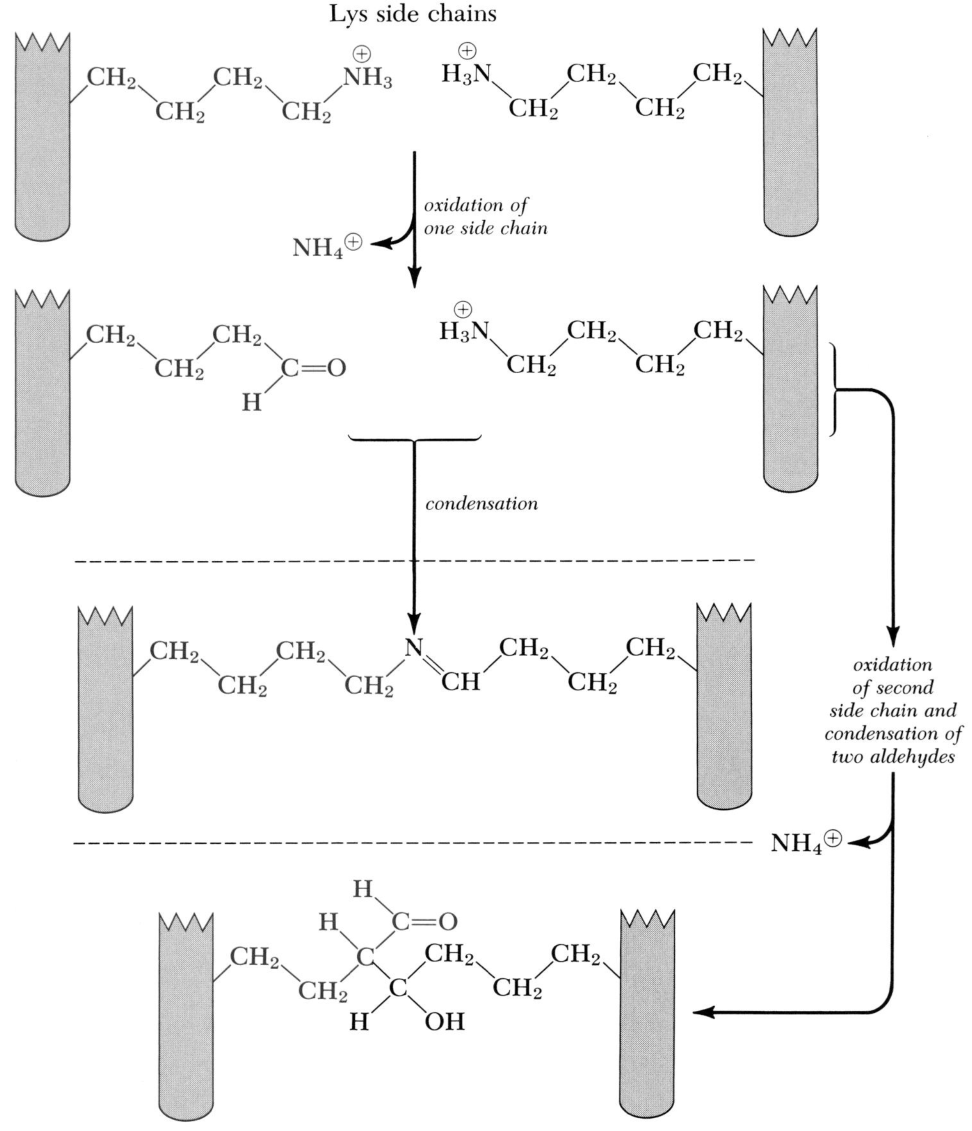

FIGURE 9–4

Cross-links are formed between collagen molecules by oxidizing lysyl side chains to aldehydes. The aldehyde groups may condense with additional unmodified lysyl side chains (*middle*), or with each other (*bottom*), or with histidyl side chains (not shown). Several variants are known.

The number of these cross-links appears to increase with age. Their exact structure and location in a collagen molecule is uncertain.

Synthesis of Collagen

The synthesis of collagen is almost a catalog of various kinds of post-translational modification of polypeptide chains. Some of the modifications begin before the messenger RNAs are completely translated; others do not occur until the molecules have been secreted from the cells, thus protecting against the formation of insoluble fibers within the cell. The temporal sequence of the changes is summarized in Figure 9–5. Let us examine some of them in more detail.

The procollagen polypeptides are first synthesized as preprocollagen chains with unusually long signal sequences. After passage through the rough endoplasmic reticulum, the signal sequences are removed to form procollagen chains, which have long segments at both the amino and carboxyl terminals that are later discarded by hydrolysis (Fig. 9–6). (Some collagen monomers begin with pyroglutamate residues as a result of cleavage of procollagen at a glutamine residue.) These extra segments are

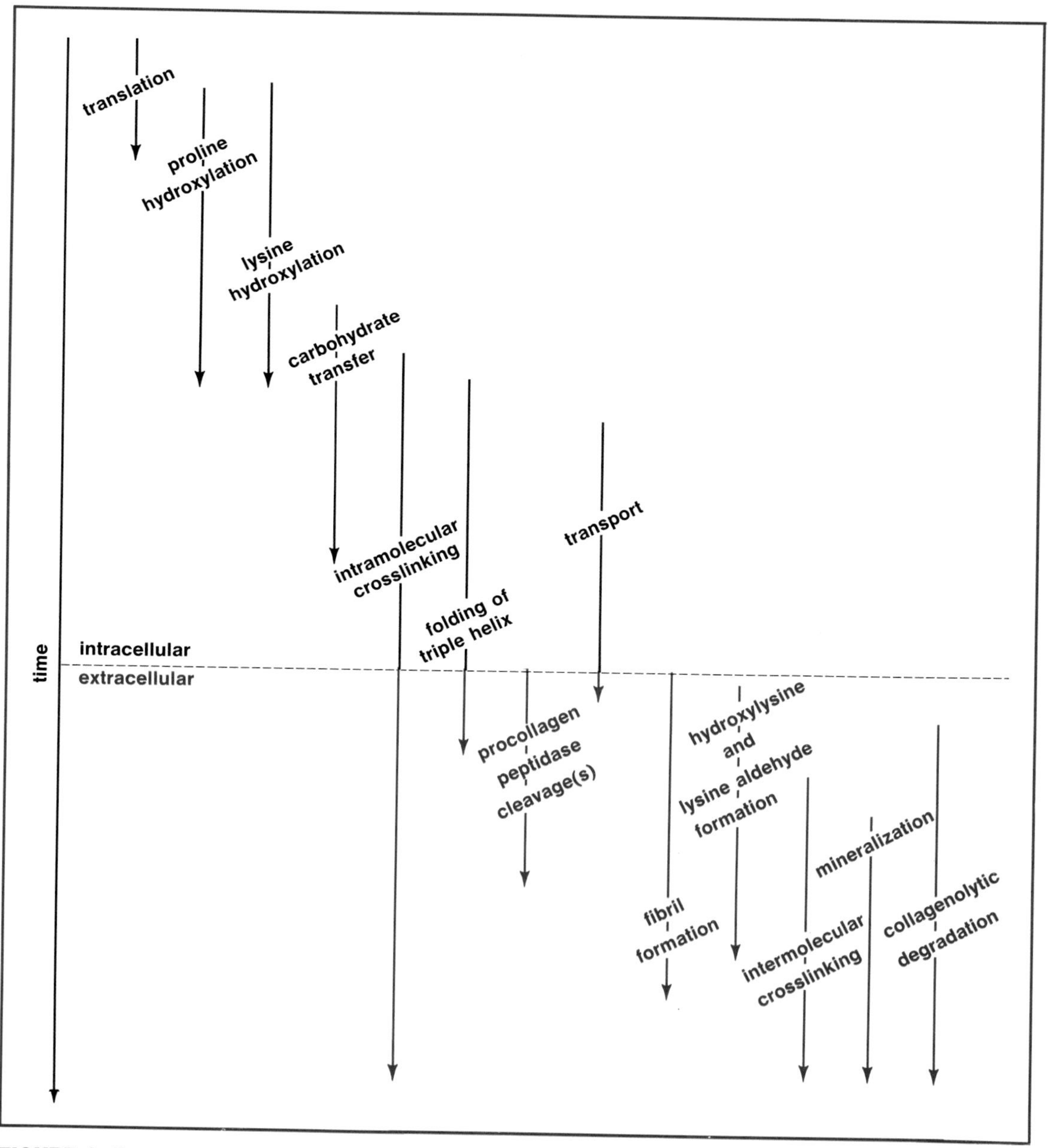

FIGURE 9–5 The time course of events in collagen synthesis. Redrawn from P.M. Gallop and M.A. Paz, (1975) Physiol. Rev., 55:473. © American Physiological Society. Used by permission.

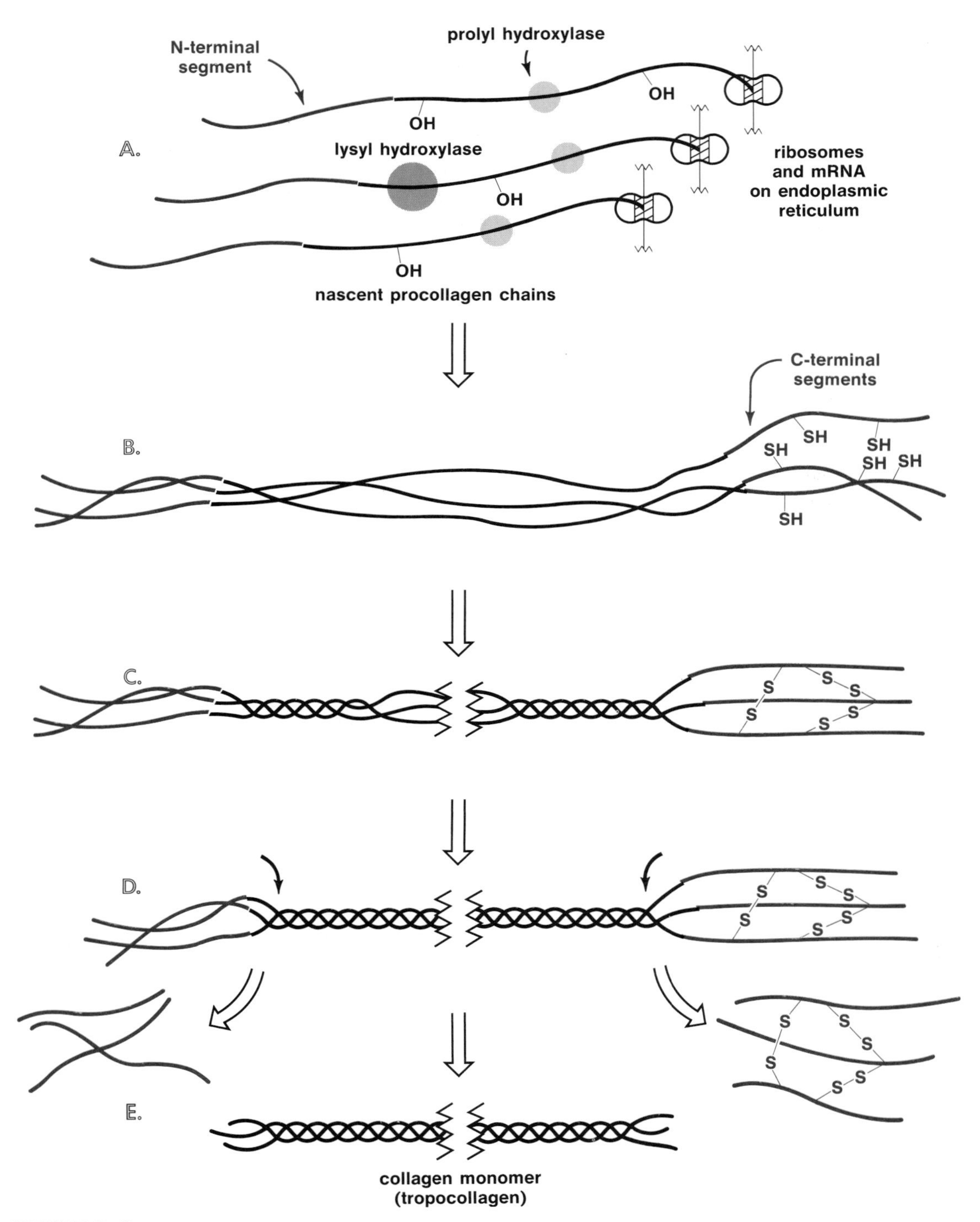

FIGURE 9–6

Collagen molecules are first synthesized as long procollagen polypeptide chains. *A.* Hydroxylation of proline and lysine residues begins before completion of translation. *B.* Disulfide bridges link the chains and promote (*C.*) the formation of triple helix. *D.* During and after secretion of the molecule, specific enzymes catalyze the cleavage of now superfluous amino and carboxyl terminal polypeptide segments, leaving the mostly helical central portion of the molecule available for association into fibers (not shown).

not rich in glycine or proline, and only portions of the extensions are triple helical. The initial N-terminal segment acts as a signal after cleavage to slow synthesis of further collagen. (Mechanisms of regulation of enzymes are discussed in Chapter 16.) The final C-terminal portion contains cysteinyl groups to form disulfide bridges; it is these bridges that initially hold the three polypeptide chains together and facilitate the formation of a triple helix to create a procollagen molecule.

Hydroxylation. The hydroxylation of proline and lysine residues begins before synthesis of the polypeptide chains is completed. The hydroxylation of proline residues is necessary for initiation of triple helix formation in the central portion of the procollagen chains. The hydroxylation reactions (which are discussed further on p. 658) are complex processes involving the simultaneous oxidation of another compound, α-ketoglutarate, to succinate.

The Hydroxylations Do Not Occur At Random. The hydroxylations are under genetic control in that the enzymes catalyzing them affect groups occurring only in specific amino acid sequences. Prolyl groups are converted to **4-hydroxyprolyl** groups when they occur immediately before glycyl groups, (-Gly-X-Pro-Gly-, but not -Gly-Pro-X-Gly-). A few, but not all, prolyl groups are converted to **3-hydroxyprolyl** groups, particularly in type IV collagens, when they occur between glycyl and 4-hydroxyprolyl groups, as in -Gly-Pro-4Hyp- → -Gly-3Hyp-4Hyp-. The determining factors are not known, but there is only one 3-hydroxyproline in an "a" chain, whereas there are as many as 16 in an "e" chain. Similarly, lysyl groups are hydroxylated only at specific sites, which vary in different types of collagen.

Secretion and Partial Hydrolysis. The secretion of procollagen involves transport through the Golgi apparatus where sugar residues are added to selected hydroxylysyl groups. Procollagen, like other secreted proteins, is packaged in vesicles, and partial hydrolysis begins before secretion.

In any event, specific enzymes are secreted along with most procollagens to catalyze the hydrolysis of the N-terminal and C-terminal segments. The loss of the C-terminal portion removes the disulfide bridges between the three polypeptide chains in most collagens, but stabilization is still provided by the prolyl groups that have already been hydroxylated.

Some of the type III collagens in soft tissues apparently retain their C-terminal segments because the finished protein still has disulfide bridges.

The other procollagens are converted by hydrolysis to a collagen monomer, previously known as **tropocollagen,** in which all but a small portion of the residues at the two ends of the chains are in a triple helix.

Fibril Formation and Cross-linking. Once the ends are lost, the collagen monomers can associate side by side to create larger fibrous structures.

Cells forming collagen secrete the enzyme lysyl oxidase to catalyze oxidation of some lysyl and hydroxylysyl side chains, creating the aldehyde groups from which cross-links are formed.

ELASTIN

Elastic fibers in a tissue enable it to stretch without tearing. The vocal cords, large arteries, and some ligaments of the vertebral column are especially rich in these fibers. They are of major importance for movement of the lungs and skin. Mature elastic fibers are mainly composed of elastin. Elastin, like collagen, is nearly one third glycine residues. It also contains a high proportion of alanine, proline, and valine residues; together with glycine these account for nearly 80 per cent of the total. Elastin is unlike collagen in that the glycine residues are not regularly spaced at every third residue, so it does not form a triple helix. Instead, the glycine residues frequently flank residues of valine in more lengthy repetitive sequences. For example, one portion of the polypeptide chain contains 11 repeats of the sequence Val-Pro-Gly-Val-Gly, and another portion has 6 repeats of Pro-Gly-Val-Gly-Val-Ala. The result is a frequent occurrence of reverse turns (p. 37) in the chain, so that most of the molecule is in a random amorphous arrangement, but with some turns occurring in tandem to make a section of 3_{10} helix. The overall structure is similar to the deformable amorphous structure of rubber.

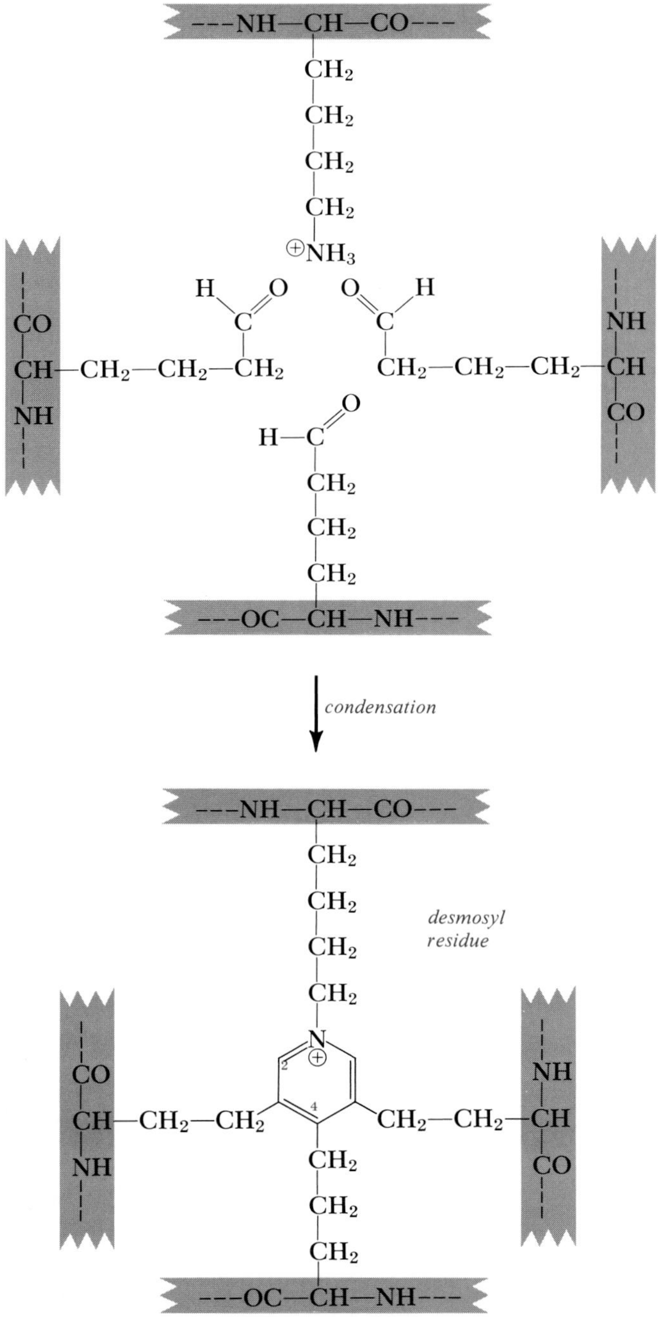

condensation

desmosyl residue

FIGURE 9–7

The formation of desmosine or isodesmosine residues in elastin involves the combination of three allysine residues (the aldehyde derivative of lysine) with one unmodified lysine residue. The result could be covalent linkage of four different polypeptide chains, but it is currently believed that two polypeptide chains are joined by using a pair of lysyl groups from each.

Another important feature of elastin is the occurrence of some residues of lysine in pairs separated by one to three alanine residues and preceded by up to eight Ala in continous sequence, as in:

. . . Ala-Ala-Ala-Ala-Lys-Ala-Ala-Lys-Ala . . .

These paired lysine residues enable formation of cross-links peculiar to elastin. Cells secreting tropoelastin, like those secreting procollagen, also secrete lysyl oxidase to catalyze the oxidation of some lysine side chains to the aldehyde. When the allysine residues are closely paired, as they sometimes are in elastin, they readily condense with each other and an unmodified lysine group to form a heterocyclic ring linking two of the pairs (Fig. 9–7). The result is a residue of **desmosine** or **isodesmosine** made from what were four lysine residues, two from each linked chain. One tropoelastin molecule will be attached to others by four to five of these cross-links, enough to keep the random structure compacted in a deformable ball but not enough to make it rigid.

KERATINS

Keratins are hard and tough proteins in the epidermis and related structures (nail, hair, horn, and hoof). The entire fiber-matrix complex is frequently called keratin, but present usage leans toward designating the fibers as α-**keratins** and the matrix as **keratohyalins.**

The general strategy for creating the keratin complex is first to make the fibers and then to polymerize the matrix around them. Young cells toward the base of the epidermis form internal tonofilaments of α-keratin. These tonofilaments may be free or run through the cytoplasm from one point of contact with neighboring cells to another. Cells become more densely packed with filaments as they mature and move toward the surface of the skin, and synthesis of keratohyalin begins to surround the filaments with matrix, which in turn is enclosed by a quite hard and impermeable modified cell membrane.

Keratin Fibrils

The keratin fibers are composed of bundles of fibrils, each fibril made from three polypeptide chains. As many as seven different precursor chains may be pres-

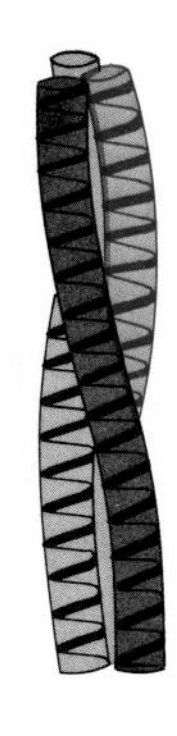

FIGURE 9–8

α-Keratin is made from coiled coils. Each polypeptide chain is in a right-handed α-helix, and a bundle of three α-helixes is given a left-handed twist to create a strong rope-like structure. The three chains are usually different.

ent in a single tissue, so the fibrils are a mixture rather than a single entity, but all appear to have similar properties.

The precursor chains contain regions in which the constituent amino acid side chains favor the formation of α-helix, separated by other regions rich in residues that prevent α-helix formation—proline, glycine, and serine (the N-terminal residue is always N-acetylserine for unknown reasons). The composition of the helical regions is such that three chains associate to form a coiled coil (Fig. 9–8), with the right-handed coils of the individual chains countertwisted into left-handed triple helixes, making a strong, rope-like bundle.

The nonhelical regions presumably favor association of the fibrils with each other and with the matrix. In any event, approximately 10 to 12 fibrils do aggregate into bundles, which usually have an ill-defined core and a surrounding ring.

Keratin fibers are sometimes stabilized by formation of a kind of cross-link we have not previously encountered. It involves a lysine side chain on one fibril and a glutamine side chain on another (Fig. 9–9). An enzyme catalyzes the replacement of the —NH_2 group in the glutamine side chain with the amino group of the lysine side chain. The result is the formation of an amide bond, sometimes called an **isopeptide** bond, between the two polypeptide chains. Such cross-links are relatively rare in skin but are more common in hair, especially in the cells constituting the core (medulla).

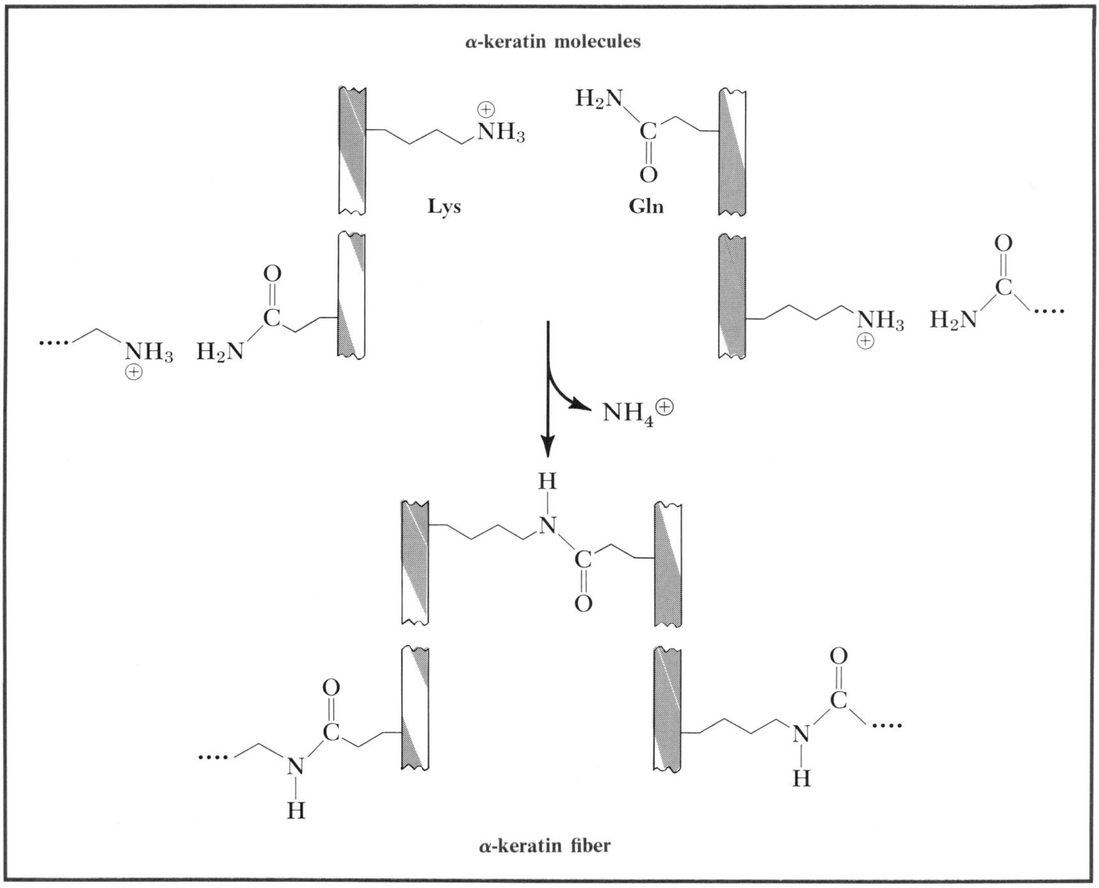

FIGURE 9–9 **The side-by-side association of α-keratin molecules into fibrils is frequently strengthened by the formation of covalent bonds between them. These bonds are created by transferring the glutamyl group of a glutamine residue from its amide nitrogen to the nitrogen atom of a lysine residue. The amide nitrogen of the original glutamine is released as a free ammonium ion.**

Keratohyalin Matrix

At least two proteins are precursors of the matrix. One is rich in basic amino acids and the other is rich in cysteine residues. The character of keratin depends upon the proportion of the precursors. For example, hair and nail keratins are made with more of the sulfur-rich protein. The sulfhydryl groups are oxidized to disulfide bridges as the cells mature, creating a tightly linked hard and horny matrix. The skin keratohyalin, on the other hand, has relatively little sulfur, and the high proportion of the basic precursor protein gives rise to a more flexible keratin.

TUBULIN

Microtubules are important components of many intracellular structures: the mitotic spindle, the cytoskeleton of most, if not all, eukaryotic cells, and more specialized structures such as flagella, spermatozoa, and neurotubules.

Single microtubules are composed of 13 filaments associated side by side to

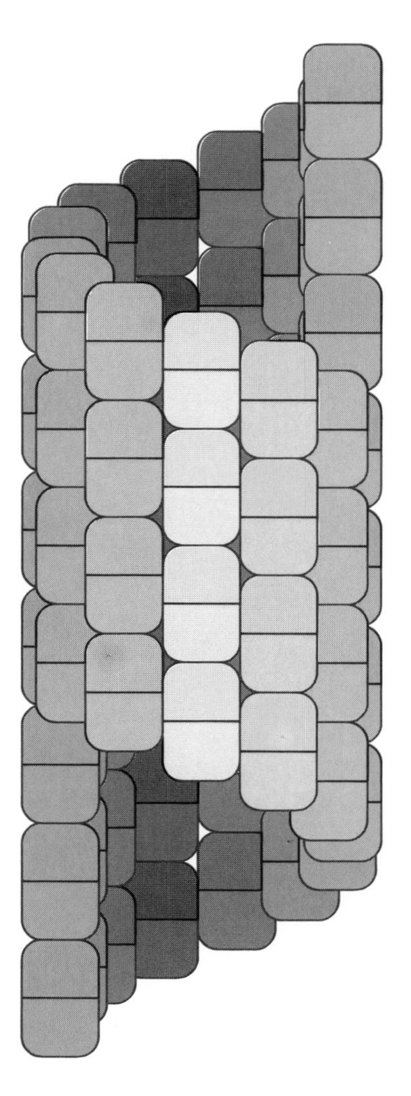

FIGURE 9–10

Schematic arrangement of microtubules. A microtubule is a hollow bundle of 13 filaments, each made from stacked molecules of tubulin. Tubulin is a protein containing two different subunits (*gray and brown shading*). One subunit lies slightly behind the other in the real molecule (not shown here). Adjacent filaments are displaced from each other so as to create the effect of a cylinder made from three stacked helical fibers, but the fundamental unit is a straight filament, not a helix.

form a cylinder (Fig. 9–10). Each filament is an end-to-end polymer of a protein, tubulin, which is a dimer of two polypeptide chains. The filaments are staggered relative to each other, creating a surface appearance of a three-fiber helix, but the microtubule is not a stack of helical ropes; it is a bundle of relatively straight rods. Microtubules appear and disappear through addition of more tubulin units at one end of the cylinder, or their removal at the other end, but the regulation of the process is not well understood.

Some important drugs act by combining tightly with tubulin. The alkaloids **vinblastine** and **vincristine** isolated from a species of *Vinca* (a genus including common ground cover plants) are important in cancer chemotherapy. **Colchicine,** isolated from *Colchicum autumnale,* the autumn crocus, has long been used in treating gout (p. 685), and is also used by plant breeders to induce polyploidy. All of these compounds presumably act by interfering with mitosis in appropriate cells.

DEFECTS IN STRUCTURAL PROTEINS

We have seen that the collagens and elastin are major components of many tissues (Fig. 9–11), and it follows that any disruption in the formation of these proteins may have a variety of consequences, depending upon the relative degree of impairment of particular structures. Specialized disturbances in collagen and elastin synthesis, as opposed to general impairment of protein synthesis, may occur at any stage, from transcription of the genes through the hydroxylation reactions, partial hydrolysis of the polypeptide chains, and oxidation of lysyl side chains to create cross-links. Many clinical entities involving defects in synthesis of structural proteins have been recognized.

Disturbances in Collagen Synthesis

Scurvy is the result of a deficiency of **ascorbate (vitamin C)** in the diet; it is the price we sometimes pay for abandoning the free-living fruit-picking life of our pri-

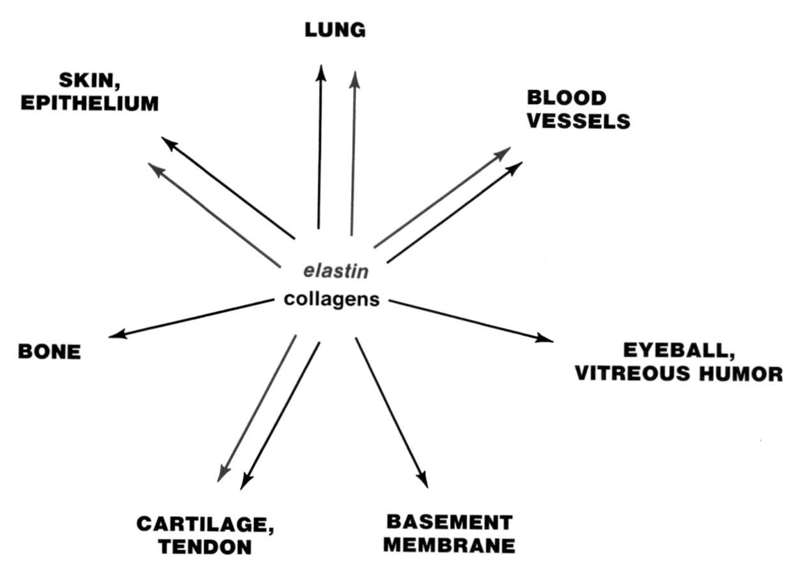

FIGURE 9–11

The key role of collagen as a structural element makes a variety of supportive tissues vulnerable to any defect in the synthesis of the protein. Elastin is less widely used, but it has a vital function in the arteries, in which it is a major fraction of the total protein.

mate ancestors. Most of the effects are those of defective collagen synthesis. Infants are afflicted with painful tenderness in the limbs so that they draw up their legs and lie quietly. Bone formation is defective, and the blood vessels are weakened, with hemorrhages common. Adults also tend to bleed readily, especially from the gums, around hair follicles in the skin, and into the joints. Wounds remain open. The teeth loosen.

Scurvy has been known for millennia as an affliction of organized traveling groups of men: soldiers on the march, sailors at sea, and exploring parties who have depended upon easily shipped supplies of nonperishable foods. Western civilization didn't make the connection between the disease and diet until the eighteenth century, when a series of studies by the British Navy brought scurvy under control on its vessels.

The exact mechanism of ascorbate action is not known. Ascorbate is a ready donor of electrons; that is, it is a good reducing agent, and this property is probably the basis for its physiological function, but definitive proof is lacking. It in some way facilitates the smooth hydroxylation of prolyl and lysyl groups in collagen and elastin. Normal connective tissue cannot be laid down in its absence.

Ehler-Danlos syndrome is a name applied to a group of rare hereditary disorders in which collagen formation is impaired, although each is a distinct condition of different origin affecting different steps in collagen synthesis. For example, one patient may have a defect in the synthesis of type III collagen, another cannot hydrolyze procollagen so as to permit normal association into fibers, another is unable to hydroxylate lysyl groups for normal attachment of carbohydrate residues or normal cross-link synthesis, another is deficient in the lysyl oxidase necessary for cross-link synthesis, and still others have uncharacterized lesions. However, the clinical results are similar for several types of Ehler-Danlos syndrome.

The lack in these people of the rigidity conveyed by collagen without loss of the elasticity conveyed by elastin results in abnormally free motion of the joints (hyperextensibility), excessive stretching of the skin, coupled with easy tearing and ready bruising. The India-rubber man and Etta Lake, the Elastic Lady, both of circus fame, had Ehlers-Danlos syndrome.

Defects in collagen synthesis may arise in a less obvious way. People with a group of conditions known as **osteogenesis imperfecta** have improper development of more rigid collagenous structures. The total incidence is about one in 10,000 births. The tendons are thin and subject to rupture. The opaque sclera of the eye becomes thin enough to allow the choroidal pigments to be seen through it, giving a Wedgewood blue appearance. Multiple fractures of the bones are likely, both before birth and later. The condition is not necessarily totally incapacitating as exemplified by Ivar the Boneless, who could not walk. Nevertheless, he was one of the leaders of the great assault of the Vikings on England in the middle of the 9th century. (He was carried into battle on a shield.)

In some versions of the condition, fibroblasts produce less of type I collagen, and more of type III; in others, the production appears normal, and what is going wrong is not clear in any case.

Not all defects in collagen synthesis have such general effects. One family with **dominant epidermolysis bullosa simplex,** a condition in which skin blisters form after even minor trauma, proved to have a deficiency in the enzyme that completes formation of the carbohydrate side chains on collagen by adding residues of glucose.

Diabetes mellitus is a common and serious condition we shall consider in more depth later; it is a failure to utilize the sugar glucose normally (p. 740). One of the dangerous consequences of diabetes is a thickening of the basement membrane in arterioles generally and in the glomerulus of the kidney specifically. The membrane serves as a filtration barrier and as a scaffold separating parenchymal cells and connective tissue in many organs. Seamless repair of injuries by replacement of damaged cells requires a normal basement membrane; scars form when it is defective. The alteration of this collagenous tissue is the source of some of the irreversible consequences of diabetes that may result despite good management.

Disturbances in Both Collagen and Elastin Synthesis

No diseases peculiarly affecting the formation of elastin have been described. However, a variety of conditions affect the activity of lysyl oxidase and block the formation of normal cross-links in both elastin and collagen to varying degrees. Gene expression and nutrition differ from one kind of cell to another; different events may well affect the same enzyme to different extents in different cells. We already noted that one type of Ehler-Danlos syndrome is caused by a deficiency of lysyl oxidase; why this particular defect primarily impairs collagen synthesis, whereas others we shall now discuss also impair elastin synthesis, is not known.

Since lysyl oxidase requires **copper** for activity, any condition affecting the availability of copper to the cells will influence its action. Dietary deficiencies of copper are rare, but failure to absorb copper can be caused by intestinal disease or by genetic defects such as the kinky- or **steely-hair syndrome,** also known as **Menkes' disease.** Impaired elastin formation causes defects in the arterial intima—the internal layer of the arterial wall. Impaired collagen synthesis is manifested by bone defects similar to those seen in scurvy.

Cutis laxa is a name applied to another genetically heterogeneous group of conditions including defects in lysyl oxidase activity. As the name implies, a common feature is loose, pendulous skin. This is the principal feature in the dominantly inherited type of cutis laxa. The recessive forms have generalized weakness in connective tissues, with hernias, pulmonary emphysema (a breakdown of the small air sacs—the alveoli—of the lungs into larger thick-walled chambers), and diverticula (protuberant pouches) of both the intestines and the urinary tract.

FURTHER READING

Collagen

D.J. Prockop et al. (1979). The Biosynthesis of Collagen and Its Disorders. N. Engl. J. Med. 301:13,77. This review also discusses the basic science and the clinical conditions.

P. Bornstein and P.H. Byers (1980). Disorders of Collagen Metabolism. p. 1089 in P.K. Bondy and L.E. Rosenberg, eds. Metabolic Control of Disease. Saunders. This includes an excellent summary of the nature of collagen as well as genetic defects in its synthesis.

R.B. Freedman and H.C. Hawkins (1980). The Enzymology of Post-Translational Modification of Proteins. Academic Press. See p. 53: K.I. Kivirikko and R. Myllyla. The Hydroxylation of Prolyl and Lysyl Residues; p. 457: J.G. Heathcote and M.E. Grant. Extracellular Modification of Connective Tissue Protein.

E.-R. Savolainen et al. (1981). Deficiency of Galactosylhydroxylysyl Glucosyltransferase, An Enzyme of Collagen Synthesis in a Family with Dominant Epidermolysis Bullosa Simplex. N. Engl. J. Med. 304:197.

D.W. Hollister, P.H. Byers, and K.A. Holbrook (1982). Genetic Disorders of Collagen Metabolism. Adv. Hum. Genet. 12:1.

Elastin

L.B. Sandberg et al. (1981). Elastin Structure, Biosynthesis, and Relation to Disease States. N. Engl. J. Med. 304:56.

P.H. Byers et al. (1980). X-Linked Cutis Laxa. N. Engl. J. Med. 303:61.

J. Uitto (1979). Biochemistry of the Elastic Fibers in Normal Connective Tissues and Its Alterations in Diseases. J. Invest. Dermatol. 72:1.

E. Fuchs and H. Green (1979). Multiple Keratins of Cultured Human Epidermal Cells Are Translated from Different mRNA Molecules. Cell 17:573.

J.D. Lonsdale-Eccles, J.A. Haugen, and B.A. Dale (1980). A Phosphorylated Keratohyalin-Derived Precursor of Epidermal Stratum Corneum Basic Protein. J. Biol. Chem. 255:2235.

Tubulin

B. Lewin (1980). Gene Expression 2, 2nd ed. Wiley. Chap. 2 covers microtubules and tubulin in more detail.

CHAPTER 10

CARBOHYDRATES AS STRUCTURAL ELEMENTS

Carbohydrates include the simple sugars, modified derivatives, and polymers of one or more of these compounds. One of the sugars, D-glucose, is the premier fuel for most organisms, and it is also the most important single precursor of other body constituents. These sweeter functions of glucose will occupy much of our later attention, but we are now concerned with a more sinewy role, the use of carbohydrates as structural elements.

THE NATURE OF SUGARS

The sugars are formally defined as polyhydric aldehydes or ketones—compounds with a hydroxyl group and a carbonyl function on separate carbon atoms. We shall see that even the simple sugars, the **monosaccharides,** exist as a mixture of tautomeric forms in equilibrium with each other, and the free hydroxy aldehyde or ketone is frequently only a minor component. Even so, it is handy to classify sugars in terms of this form. For example, many of the structural carbohydrates are derivatives of three sugars that are considered to be **aldohexoses**—aldehyde sugars with six carbon atoms:

```
        O            O            O
       //           //           //
   H—C          H—C          H—C
     |            |            |
   H—C—OH       H—C—OH      HO—C—H
     |            |            |
  HO—C—H       HO—C—H       HO—C—H
     |            |            |
  HO—C—H        H—C—OH       H—C—OH
     |            |            |
   H—C—OH       H—C—OH       H—C—OH
     |            |            |
    CH2—OH       CH2—OH       CH2—OH
```

D-galactose   D-glucose   D-mannose

These sugars are identical in empirical formula and in the kind of substituents on each carbon atom, which means that they are by definition **stereoisomers,** compounds that differ only in the spatial arrangement of the substituent groups. However, they are not mirror images, which is the special class of stereoisomers known as **enantiomers.** For example, the mirror image of D-glucose is L-glucose, which is very rare in nature, although use of the synthetic compound as a nonmetabolizable sweetener has been proposed:

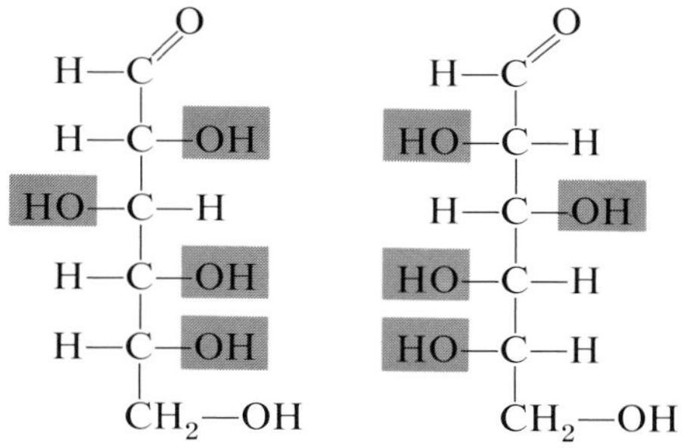

D-glucose	L-glucose
H—C=O	H—C=O
H—C—OH	HO—C—H
HO—C—H	H—C—OH
H—C—OH	HO—C—H
H—C—OH	HO—C—H
CH_2—OH	CH_2—OH

Despite their similar structures, glucose, galactose, and mannose are quite different compounds as is shown by the following physical properties.*

	D-Galactose	D-Glucose	D-Mannose
solubility in water (*g/100 ml*)	10.3 (0°)	32.3 (0°)	
		83 (17.5°)	248 (17°)
melting point (°C)	167°	146°	132°

D-Glucose and D-galactose and D-glucose and D-mannose are related to each other as **epimers**—stereoisomers differing in configuration on only one carbon atom. (D-Mannose and D-galactose are not epimers.)

Furanose and Pyranose. Little of the three aldohexoses exists in the free aldehyde form drawn above because a six-carbon chain readily folds upon itself so as to bring the aldehyde group into proximity with a hydroxyl group on either C-4 or C-5, and the two groups readily react to form a hemiacetal ring (see top figure, next page).

The sugar is then in a furanose or pyranose form, with either a five-membered reduced furan ring or a six-membered reduced pyran ring. It is difficult to represent the steric arrangement of these cyclic forms by a Fischer convention (horizontal bonds in front of the plane of paper, vertical bonds in back), and a **Haworth convention** is frequently used, in which the ring is depicted nearly on edge (see middle figure, next page).

The Haworth convention is useful for many purposes, but even it does not accurately represent the true conformation of the molecule because the rings are not planar. It is becoming more common to indicate the pucker of the ring by a conformational formula; D-glucose, for example, is predominantly in a chair form (see

*We are cheating a little here because the data given apply to the tautomeric form in which these compounds readily crystallize and not to the free aldehyde forms. These tautomers are discussed in the next pages.

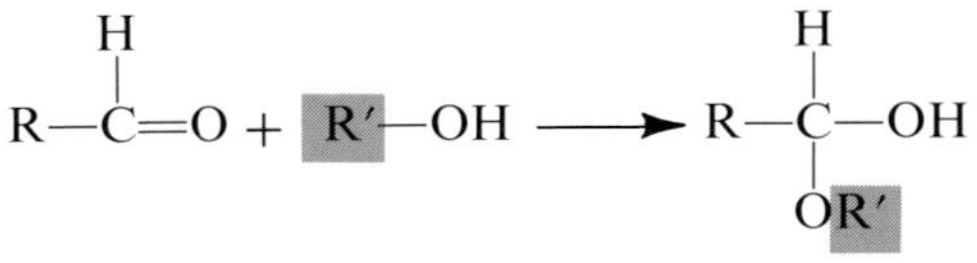

$$R{-}\overset{H}{\overset{|}{C}}{=}O + R'{-}OH \longrightarrow R{-}\overset{H}{\overset{|}{C}}{-}OH \text{ with } OR' \text{ on C}$$

aldehyde alcohol hemiacetal

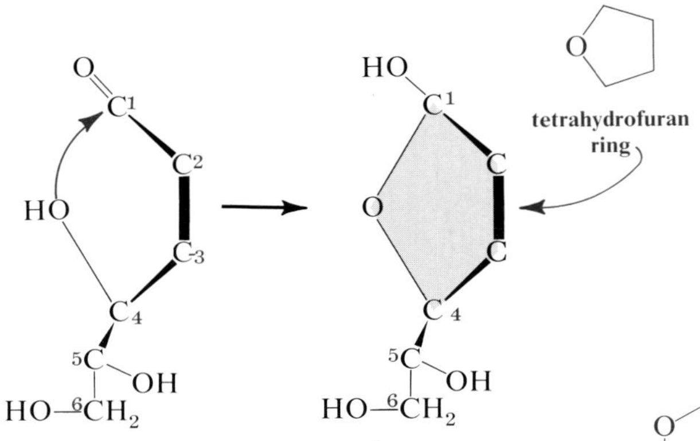

O=C1, C2, C3, C4, HO, 5C, OH, HO—6CH2 → HO, C1, O, C, C, C 4, 5C, OH, $HO{-}^6CH_2$

O (tetrahydrofuran ring)

tetrahydrofuran ring

furanose

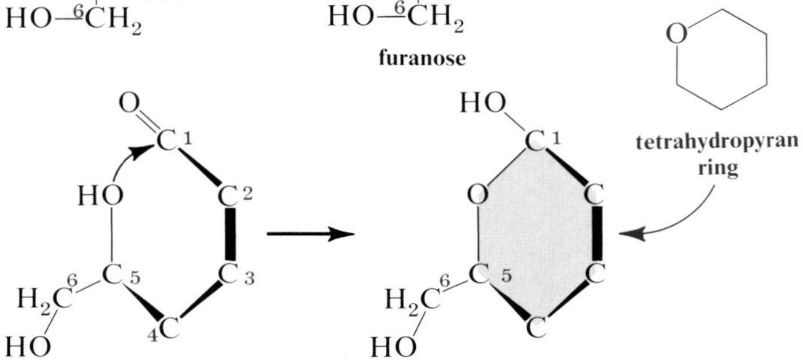

O=C1, C2, C3, 4C, C5, HO, H_2C^6, HO → HO, C1, O, C, C, C, C 5, H_2C^6, HO

O (tetrahydropyran ring)

tetrahydropyran ring

pyranose

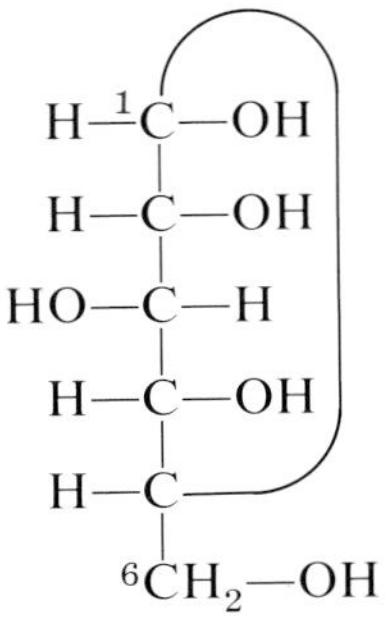

$H{-}^1C{-}OH$

$H{-}C{-}OH$

$HO{-}C{-}H$

$H{-}C{-}OH$

$H{-}C$

$^6CH_2{-}OH$

Fischer convention

H OH

^1C

$H{-}C{-}OH$

$HO{-}C{-}H$ O

$H{-}C{-}OH$

$H{-}C$

$^6CH_2{-}OH$

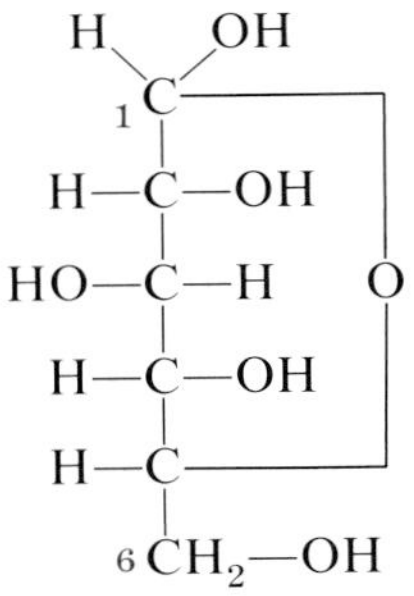

modified Fischer convention

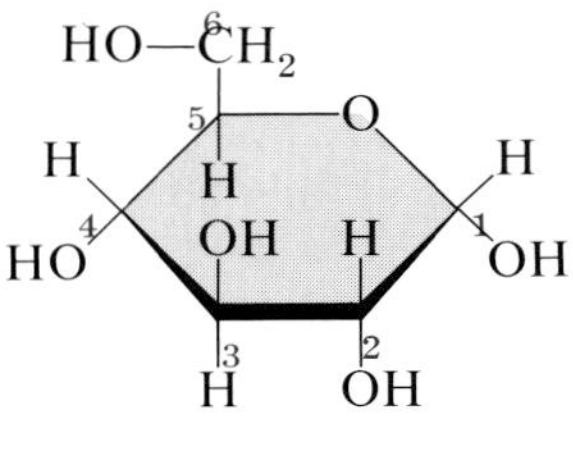

$HO{-}^6CH_2$

5 O

H H H

HO 4 OH H 1 OH

3 2

H OH

Haworth formula

OH

CH_2

O

HO OH

OH

HO

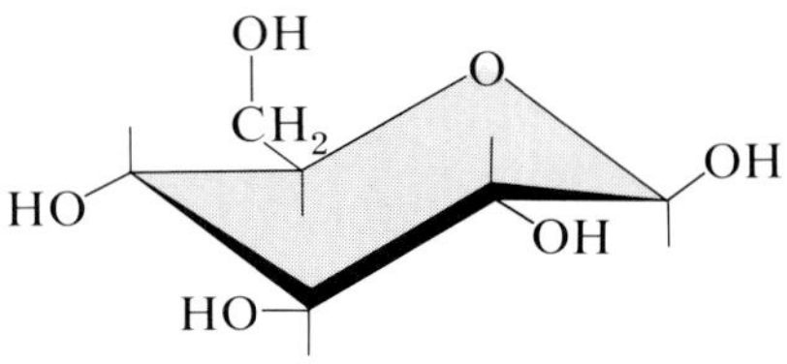

bottom of facing page). One reason for the stability of this conformation is that the hydroxyl groups on C-2, 3, and 4 are **equatorial** (in the general plane of the ring) rather than **axial.** (The equatorial position is not always the most stable at C-1, owing to proximity to the ring oxygen atom.)

Because of its configuration glucose can achieve the most stable conformation (lowest energy content) of all of the aldohexoses. If all of the aldohexoses equilibrated in a primordial soup, there would be more glucose than any other isomer. Perhaps this is why it is the predominant biological sugar.

Configurational Family. The designation of a sugar as D or L hinges on the configuration of the asymmetric carbon most distant from the carbonyl function, which is C-5 in the case of the hexoses. The reason is that the sugars may be regarded in a formal sense as derivatives of either D- or **L-glyceraldehyde,** the simplest aldoses. For example, adding one formaldehyde unit would create the tetroses:

H—C(=O)—H
\+
H—C=O
HO—C—H
CH_2OH
L-glyceraldehyde

H—C(=O)—H
\+
H—C=O
H—C—OH
CH_2OH
D-glyceraldehyde

H—C=O
HO—C—H
HO—C—H
CH_2OH
L-erythrose

H—C=O
H—C—OH
HO—C—H
CH_2OH
L-threose

H—C=O
HO—C—H
H—C—OH
CH_2OH
D-threose

H—C=O
H—C—OH
H—C—OH
CH_2OH
D-erythrose

Further additions would create the pentoses, then the hexoses, and so on. Each addition doubles the number of possible aldose isomers, so there are 16 possible aldohexoses in eight DL pairs.

The notion of glyceraldehyde as the parent compound turned out to be a happy concept; most of the sugars in nature are indeed derived from D-glyceraldehyde by reactions that retain the D configuration on the next to last carbon atom.

Anomers. Closing the ring to form a furanose or pyranose generates a hydroxyl group on C-1 that may be above or below the plane of the ring, yielding isomers that are designated as α or β (Fig. 10–1). (When drawn in the Fischer convention, the α-isomer has the hydroxyl group on the same side as the C-5 oxygen atom.)

The crystalline glucose of commerce is α-D-glucopyranose, but when it is dissolved in water, the ring is free to open into the aldehyde form. The aldehyde form in turn is free to condense into the furanose or the pyranose forms, and either of these may be the α or β isomers, as shown in Figure 10–1. What we regard as a single compound in solution is in fact a mixture of at least five different compounds. In the case of glucose, circumstances are somewhat simplified because less than 1 per cent

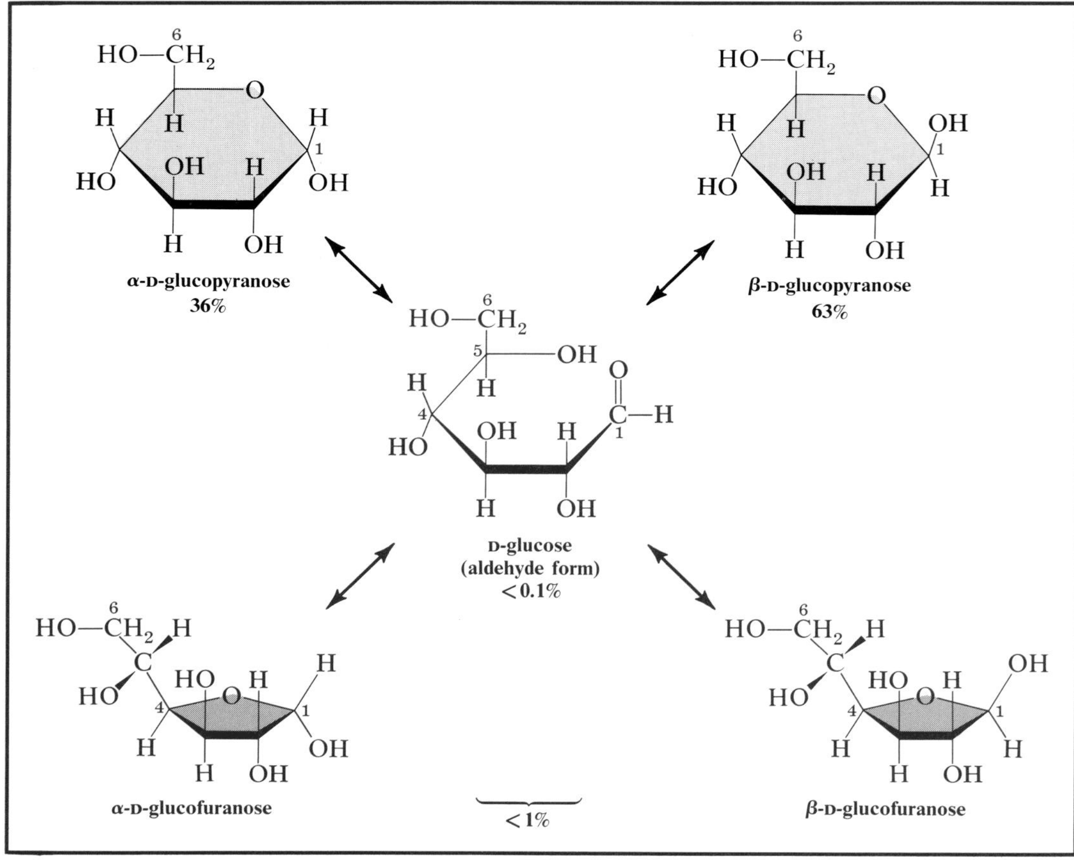

FIGURE 10–1 The aldehyde form of glucose (*center*) reacts to form a pyranose ring (*top*), or a furanose ring (*bottom*). Closure of the ring creates a mixture of two configurations on C-1, designated α and β. All of these forms equilibrate in solution, but the two pyranose forms (*black*) predominate by far.

of the total is in the furanose form and less than 0.1 per cent is in the aldehyde form. A solution of glucose may be regarded as a mixture of α-D-glucopyranose, and β-D-glucopyranose, constantly equilibrating through the formation of trace amounts of the aldehyde. Glucose without qualification is understood to mean primarily the D-pyranose forms. This equilibration can be observed by measuring the optical activity of the solution (its ability to rotate the plane of polarization of light); the rotation decreases as a fresh solution of α-D-glucose comes to equilibrium, and the equilibration of the sugar is called **mutarotation.**

The furanose isomers form more rapidly than do the pyranose isomers, but they are such a small fraction of the total in the case of glucose that the actual concentration is not known. This is not true of all sugars; 5 per cent of galactose is present as furanoses at equilibrium.

MODIFIED SUGARS

Sugars are modified for structural purposes to provide other types of reactive groups. The principal modified compounds are **hexosamines,** in which the oxygen atom on C-2 is replaced by nitrogen; the **hexuronates,** in which C-6 is oxidized to a carboxylate group; and the **deoxy sugars,** in which one or more carbon atoms, usually C-6, has an H in place of the usual OH group.

Hexosamines

The common hexosamines of animal cells are D-glucosamine and D-galactosamine:

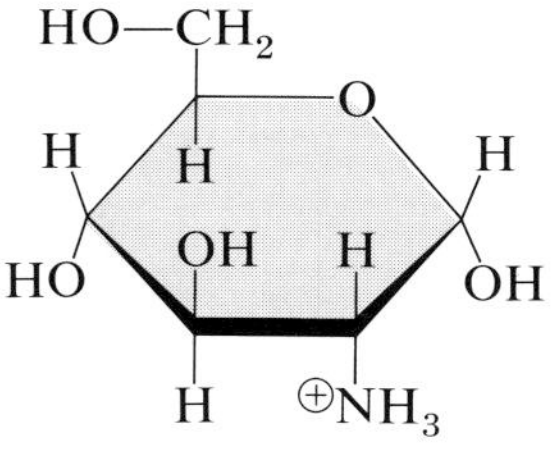

α-D-glucosamine (2-amino-2-deoxy-α-D-glucose)

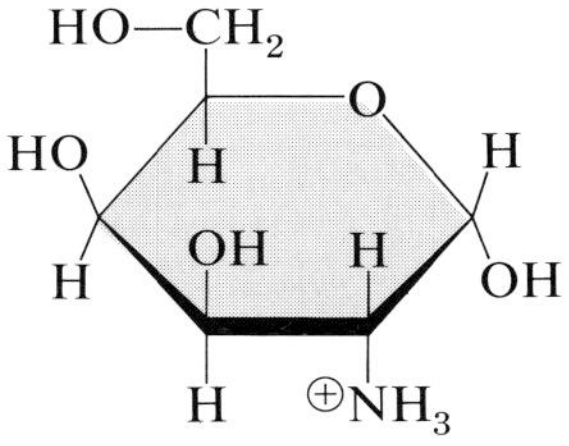

α-D-galactosamine (2-amino-2-deoxy-α-D-galactose)

They sometimes occur as such in structural carbohydrate, but more often as the N-acetyl derivatives:

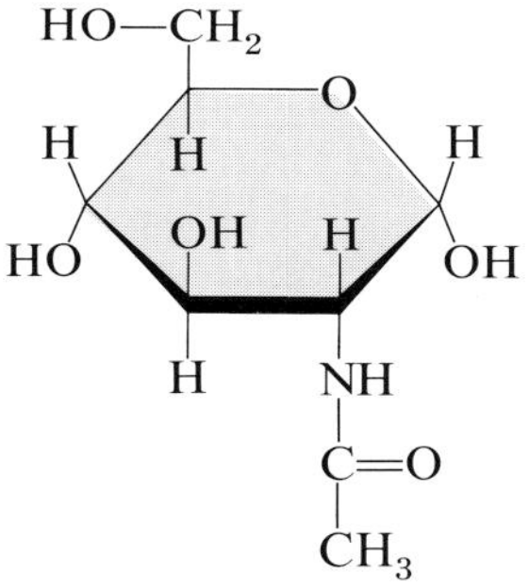

N-acetyl-D-glucosamine

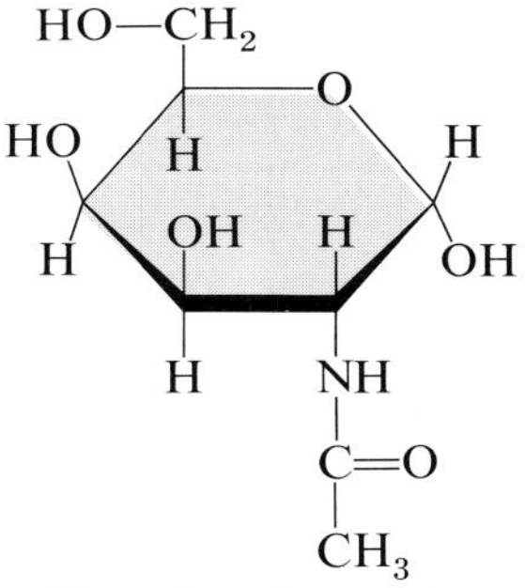

N-acetyl-D-galactosamine

Only one isomer is shown in the preceding illustrations, but these compounds undergo the same equilibration of free aldehyde, pyranose, and furanose with α and β forms as do the simple sugars; that is, they can mutarotate.

Hexuronates

Two hexuronates, in which C-6 is oxidized to the anionic form of a carboxylic acid, are of importance in animals. These are the anions of D-glucuronic acid, derived from D-glucose, and its 5-epimer, L-iduronic acid:

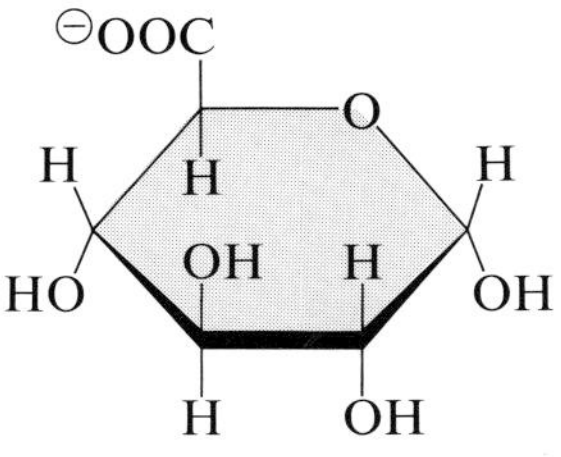

α-D-glucuronate

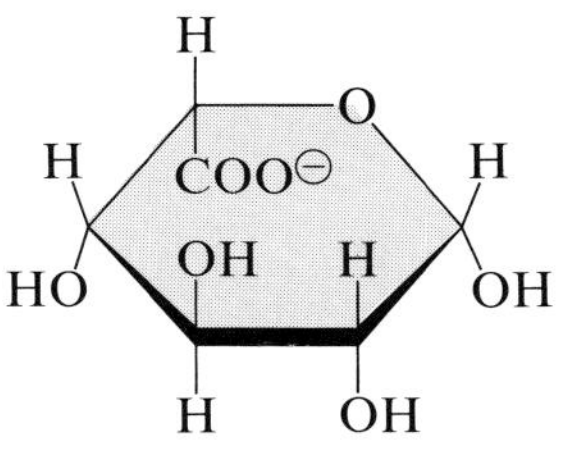

β-L-iduronate

L-Iduronic acid is named as a derivative of the hexose, L-idose, but the L-iduronate residues in structural carbohydrates are formed by epimerization of residues of D-glucuronate (p. 700). The uronates also occur as anomers and can mutarotate.

Deoxy Sugars

Two deoxy hexoses of importance are L-fucose and L-rhamnose:

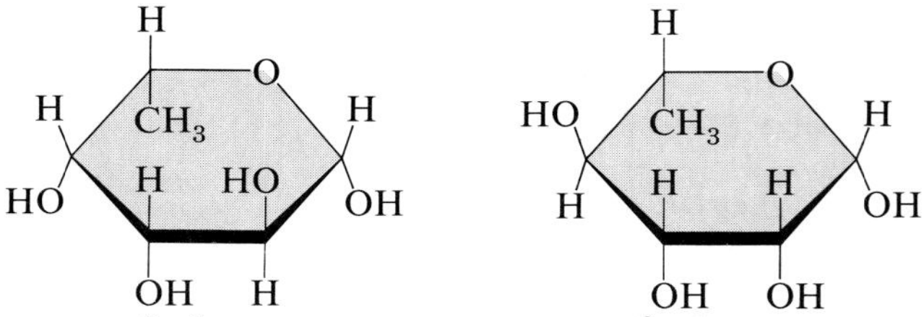

β-L-fucose
(6-deoxy-β-L-galactose)

β-L-rhamnose
(6-deoxy-β-L-mannose)

Although these compounds have a formal relationship to L-galactose and L-mannose, they are formed from D-hexoses (p. 701).

Sialic Acids (Neuraminates)

The final component in our list of major structural carbohydrates, N-acetylneuraminate, or sialate, is the most complex (Fig. 10–2). The terminal six carbons in the nine-carbon chain are derived from N-acetyl-D-mannosamine. (This amino analogue of mannose is not in itself an important structural component.) The initial three carbons are derived from the 2-ketocarboxylic acid, pyruvic acid, in its anionic form.

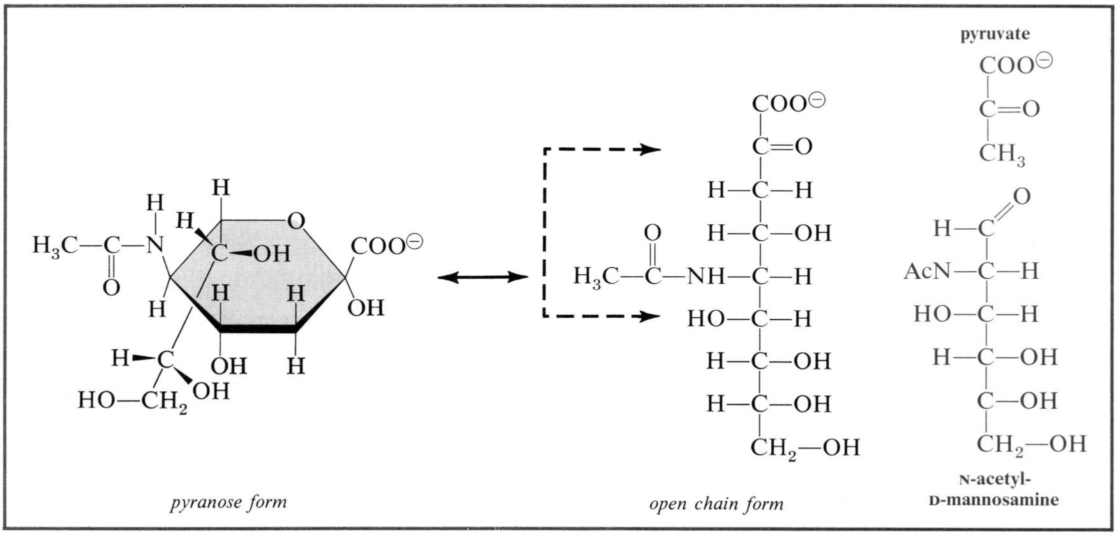

pyruvate

pyranose form

open chain form

N-acetyl-
D-mannosamine

FIGURE 10–2 N-Acetylneuraminate, an important constituent of glycoproteins, exists in a pyranose form. The nine-carbon neuraminate chain is made by condensing derivatives of pyruvate and mannosamine (*right*).

SUGARS

An aldose has an aldehyde group; a ketose has a ketone group.
A hemiacetal (or hemiketal) is the condensation product of a carbonyl group with one alcohol group.
A furanose is the hemiacetal (or hemiketal) form of a sugar with a five-membered furan ring (four carbon atoms and one oxygen atom in ring).
A pyranose is the hemiacetal (or hemiketal) form of a sugar with a six-membered pyran ring (five carbon atoms and one oxygen atom in ring).
An epimer has a different steric arrangement on one asymmetric carbon atom.
Enantiomers are mirror images; they have a different steric arrangement on all asymmetric carbon atoms.
An anomer of a furanose or pyranose is the stereoisomer with a different steric arrangement on the potential carbonyl carbon (C-1 with aldoses, C-2 with ketoses).
Mutarotation is the equilibration of one anomer of a sugar with the other possible forms through the intermediate formation of the free aldehyde form.
Hexosamines are sugars in which the oxygen atom on C-2 is replaced by a nitrogen atom; that is, they are 2-amino-2-deoxy-aldohexoses.
Hexuronates are sugars in which the terminal carbon atom is oxidized to an ionized carboxylic acid group.

POLYMERS OF SUGARS

Glycosides

Sugars polymerize by forming acetals. We know from organic chemistry that the hemiacetals of aldehydes can react with another molecule of alcohol to yield an acetal:

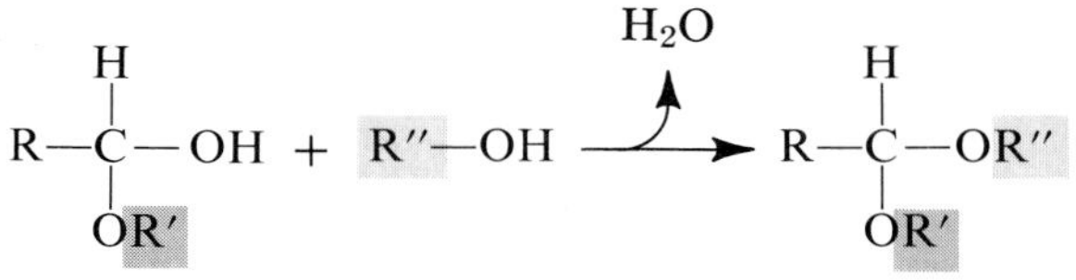

$$\underset{\text{hemiacetal}}{R\text{—}\overset{\displaystyle H}{\underset{\displaystyle OR'}{C}}\text{—}OH} + R''\text{—}OH \xrightarrow{\;H_2O\;} \underset{\text{acetal}}{R\text{—}\overset{\displaystyle H}{\underset{\displaystyle OR'}{C}}\text{—}OR''}$$

The furanose or pyranose forms of sugars are hemiacetals, and they can condense this way in the test tube to form acetals, which are termed **glycosides:**

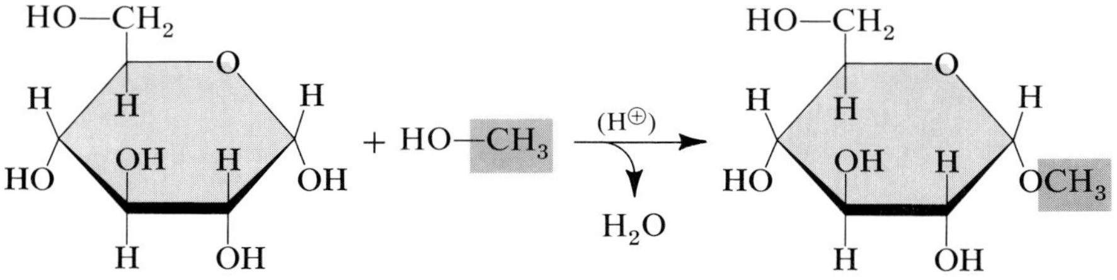

$$\text{sugar (glucose)} + HO\text{—}CH_3 \text{ alcohol (methanol)} \xrightarrow[H_2O]{(H^{\oplus})} \text{glycoside (methyl glucoside)}$$

Many glycosides with this type of structure occur in nature, although they are formed in a less direct way (Chapter 27). Most are much more complex than the simple example shown above. The plant kingdom includes an especially varied assortment, with a great variety of compounds bearing hydroxyl groups combined with many kinds of sugars. Some of these plant glycosides have profound physiological effects at low concentrations in animals, such as the cardiac glycosides derived from *Digitalis,* the common foxglove.

When sugars combine, one molecule contributes the hemiacetal and the other molecule contributes the hydroxyl group. For example, two molecules of glucose can combine to form maltose, a glucose glucoside:

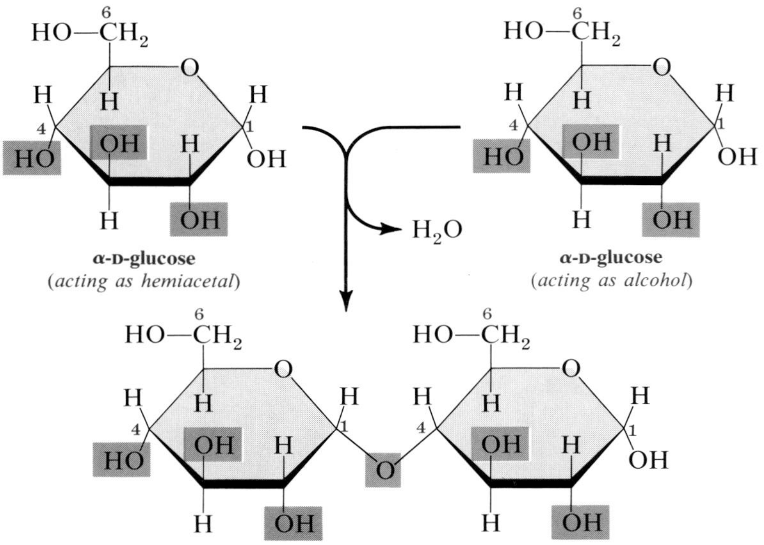

α-D-glucose
(acting as hemiacetal)

α-D-glucose
(acting as alcohol)

H_2O

α-maltose
α-D-glucosyl-(1→4)-α-D-glucose
(O-α-D-glucopyranosyl-(1→4)-α-D-glucopyranose)

Again, we are showing a formal reaction; biological combination occurs by a more indirect route.

Compounds such as maltose that are composed of two simple sugar residues are said to be **disaccharides.** Disaccharides can mutarotate if one of the residues can equilibrate through the free aldehyde form:

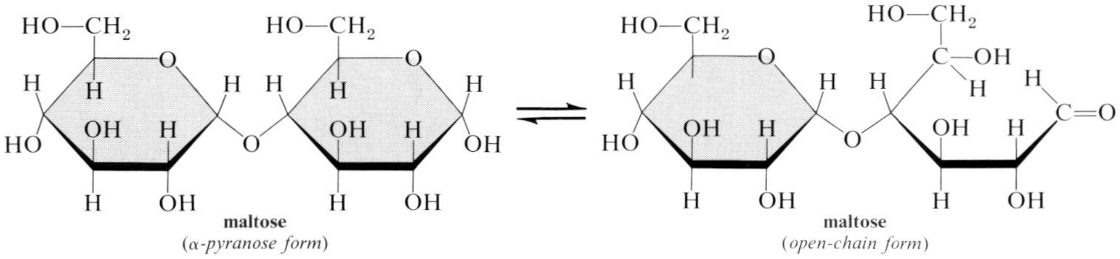

maltose
(α-pyranose form)

maltose
(open-chain form)

Maltose therefore shows mutarotation and exists in α and β pyranose forms.

Maltose is an example of a homooligosaccharide—*homo* because it is a combination of identical sugar residues, *oligo* because it is a small polymer. More specifically, it is a homodisaccharide. (Other oligosaccharides can be trisaccharides, tetrasaccharides, and so on. The simplest sugars are monosaccharides.) Polymers can also be made from different sugar residues, forming heterooligosaccharides or longer heteropolysaccharides. An example of a heterodisaccharide is lactose, the sugar in milk (see top figure, next page).

Lactose also equilibrates in α and β forms.

Nonreducing Disaccharides. A special class of disaccharides is formed by combining two sugar residues through their hemiacetal hydroxyl groups. The most familiar example is **sucrose,** which contains residues of fructose and glucose. Fructose is a ketose isomer of glucose (see second figure, next page).

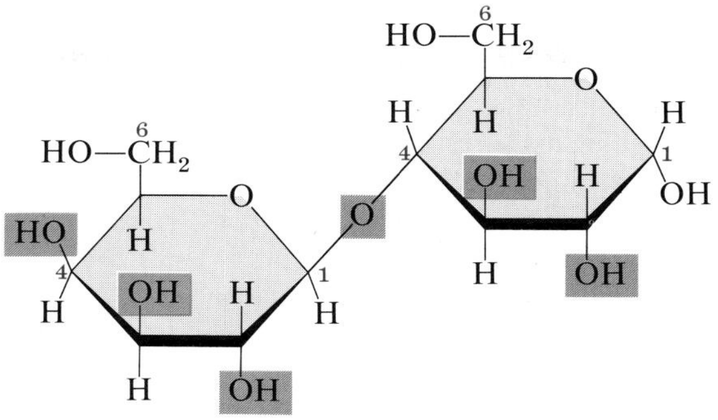

HO—$^{6}CH_2$ HO—$^{6}CH_2$
HO H O H O H H OH H 1 OH
4 OH H 1 H H OH
H H OH

α-lactose
β-D-galactosyl-(1→4)-α-D-glucose

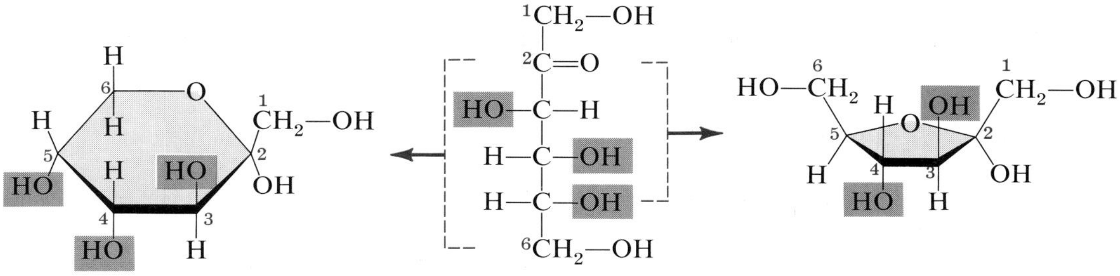

H 6 O 1 H H CH_2—OH 5 H HO 2 HO OH 4 3 HO H

α-D-fructopyranose

$^{1}CH_2$—OH
$^{2}C{=}O$
HO—C—H
H—C—OH
H—C—OH
$^{6}CH_2$—OH

D-fructose
(open-chain form)

6 1 HO—CH_2 H OH CH_2—OH O 5 2 H 4 3 OH HO H

α-D-fructofuranose

Forming a linkage between C-1 of the pyranose form of glucose and C-2 of the furanose form of fructose creates sucrose:

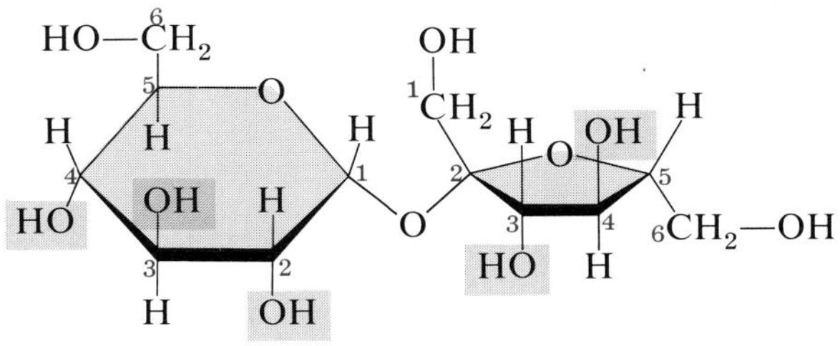

HO—$^{6}CH_2$ OH
5 O $^{1}CH_2$ H
H H H H OH
4 1 2 O 5
HO OH H O 3 4 $^{6}CH_2$—OH
3 2 HO H
H OH

sucrose
α-D-glucopyranosyl-β-D-fructofuranoside

Sucrose is synthesized by plants as a transportable form of carbohydrate; it is a device for moving fuel from the leaves to other parts of the plant, analogous to lactose as a device for moving fuel from mother to infant.

Sucrose has become an important dietary fuel for humans, accounting for over 15 per cent of the total energy consumption of Americans.

Chemically, sucrose is distinctive because both of the potential carbonyl groups are locked in acetal linkage. The rings cannot open, and sucrose does not mutarotate or exist as anomers. Sucrose, therefore, does not exist transiently as a free aldehyde or ketone. The free carbonyl group in other sugars has a reactivity that is accentuated by the neighboring hydroxyl groups; they are readily oxidized by relatively mild reagents, such as alkaline cupric ion solutions. Hence, they are termed **reducing**

sugars because they reduce cupric ion to cuprous ion, whereas sugars such as sucrose are said to be nonreducing sugars. (Boiling samples of urine with blue cupric reagents to detect glucose used to be a familiar part of laboratory examinations as a screening test for diabetes. Automated quantitative analysis for blood glucose concentration and some qualitative tests still hinge on the reducing power of glucose.)

Another nonreducing disaccharide of general interest is **trehalose,** which is α-glucose-(1→1)-α-glucoside. Trehalose is the storage and transport fuel in insects. The general use of disaccharides for transport arises from the need to separate this function from the more general functions of the monosaccharides.

DISACCHARIDES

reducing	maltose = O-α-D-glucopyranosyl-(1→4)-D-glucopyranose
	lactose = O-β-D-galactopyranosyl-(1→4)-D-glucopyranose
nonreducing	sucrose = β-D-fructofuranosyl-α-D-glucopyranoside*
	trehalose = α-D-glucopyranosyl-α-D-glucopyranoside*

*The formal names end in pyranoside, rather than pyranose, to indicate that the combination involves C-1; that is, the compound is a glycoside of both residues.

Higher Polymers

The combination of sugar residues through acetal formation can be extended almost indefinitely. For example, glucose residues can be combined, first to form homooligosaccharides, and then longer homopolysaccharides:

HAWORTH FORMULA

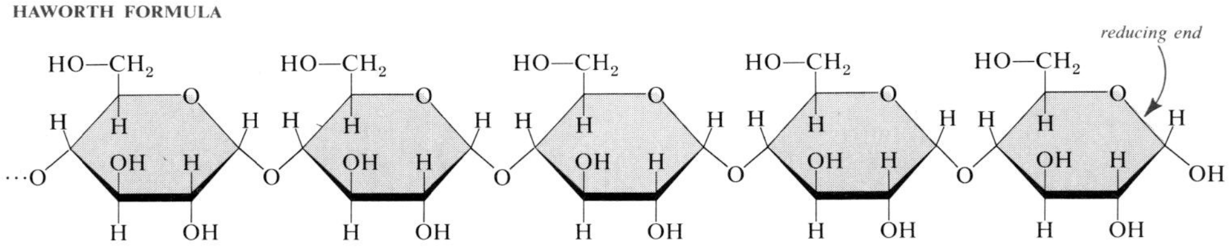

reducing end

CONFORMATIONAL FORMULA

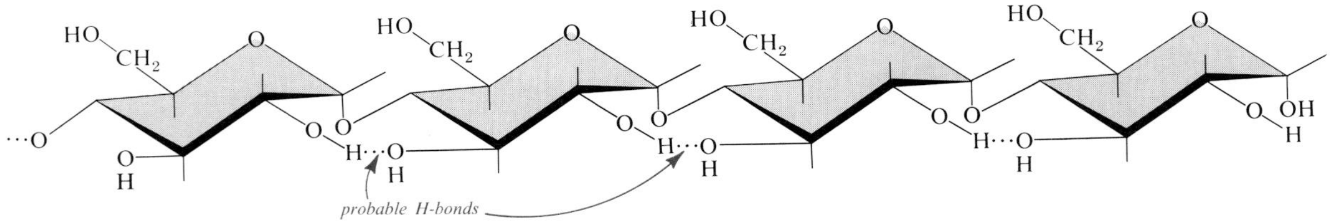

probable H-bonds

amylose chain
α-Glc-(1→4)-Glc

This particular polymer is known as amylose; similar combinations of glucose residues by α(1→4) linkages are important structural features in the stored fuels, glycogen in animals and starch in plants. Only one residue at the end of these polymeric chains is able to mutarotate and behave as a reducing sugar. Various kinds of carbohydrate residues may be combined to form heterooligosaccharides and heteropolysaccharides, and it is in these compounds that one finds important structural components.

Abbreviations. Description of the carbohydrate polymers is greatly aided through the use of abbreviations for the carbohydrate residues:

Aldohexoses	Gal =	D-galactose
	Glc =	D-glucose
	Man =	D-mannose
Ketohexoses	Fru =	D-fructose
Osamines	GalN =	D-galactosamine
	GalNAc =	N-acetyl-D-galactosamine
	GlcN =	D-glucosamine
	GlcNAc =	N-acetyl-D-glucosamine
Uronates	GlcUA =	D-glucuronate
	IdoUA =	L-iduronate
Deoxysugars	Fuc =	L-fucose
	Rha =	L-rhamnose
Neuraminates	Neu =	D-neuraminate
	NeuNAc or NAN =	N-acetyl-D-neuraminate
	Sia =	sialate; unspecified substituted neuraminate (usually N-acetyl-neuraminate in humans)

The linkage between the groups is then specified in the ordinary way. For example, amylose, the glucose polymer illustrated on the facing page, is:

$$\alpha\text{-Glc-}(1\rightarrow4)\text{-}\alpha\text{-}(\text{Glc})_n\text{-}(1\rightarrow4)\text{-Glc}$$

Lactose is β-(Gal)-(1→4)-Glc; sucrose is β-(Fru)-(2→1)-α-Glc, and so on.

GLUCANS: HOMOPOLYMERS OF GLUCOSE

There are many theoretical polymers of glucose; condensation is possible with the hydroxyl group on any of the carbon atoms and may involve either the α or the β anomers. Only a few of the possibilities are known to occur in nature, and let us examine why.

Permissible Configurations

Theoretical analysis of polysaccharide structure is not as advanced as it is for polypeptides, but it is now possible to perceive the structural features of importance and relate them to function. Most of the carbohydrate polymers have an acetal bridge between two pyranose rings, much as we saw in maltose. Since the motion of atoms within the rings is restricted, major alterations in conformation of such polymers are limited to rotations around the hemiacetal ether oxygen. That is, the configuration is mainly fixed by the relative positions of successive rings in the polymer. These positions may be described by a pair of dihedral angles, analogous to the Ramachandran angles used to describe polypeptide conformation. The preferred conformations of lowest energy content cluster near certain angles.

The favored angles in polysaccharides vary with the kind of sugar, the anomer that is involved, and the carbon atoms that are connected because the hydroxyl groups must be oriented so as to favor formation of hydrogen bonds. It now appears that there are four possible secondary structures in polymers of aldopyranoses:

Type A polymers are extended ribbons, such as are seen in cellulose. Cellulose is a polymer of glucose with β-(1→4) linkage, and the type A ribbon is the conformation of minimum energy content for such polymers. It is the ideal conformation for making fibers, because parallel polysaccharide chains can readily be linked by hydrogen bonds (Fig. 10–3). The β-configuration causes adjacent rings in a chain to be rotated 180° relative to each other.

Chitin, a polymer of N-acetylglucosamine residues, has a similar structure. Chitin is the fibrous component of the exoskeleton of arthropods: crabs, spiders, insects.

Type B polymers occur as a helix of variable dimensions, usually left-handed and with an open core (Fig. 10–4). This is the preferred conformation of amylose and of some heteropolysaccharide chains occurring in connective tissues. In each of these cases, there are too many easily interconvertible forms available to these polymers, some involving extensive bonding with water, for them to be useful in making rigid structures, but they are suitable for use as fuels and as a matrix in connective tissues. It is interesting to note the drastic differences in properties between an α(1→4) glucan such as amylose and a β(1→4) glucan such as cellulose.

Type C polymers have a crumpled structure, somewhat like a folded bellows. There are no known natural examples.

Type D polymers have an open structure created by linking sugars through 1→6 bonds. Since an additional degree of freedom is created by rotation around these bonds (Fig. 10–5), such polymers would probably exist in an open extended form in crystals and be in a random form in solution. No pure examples are known, but 1→6 chains are a part of the structure of dextrans, which are polysaccharides made by some bacteria.

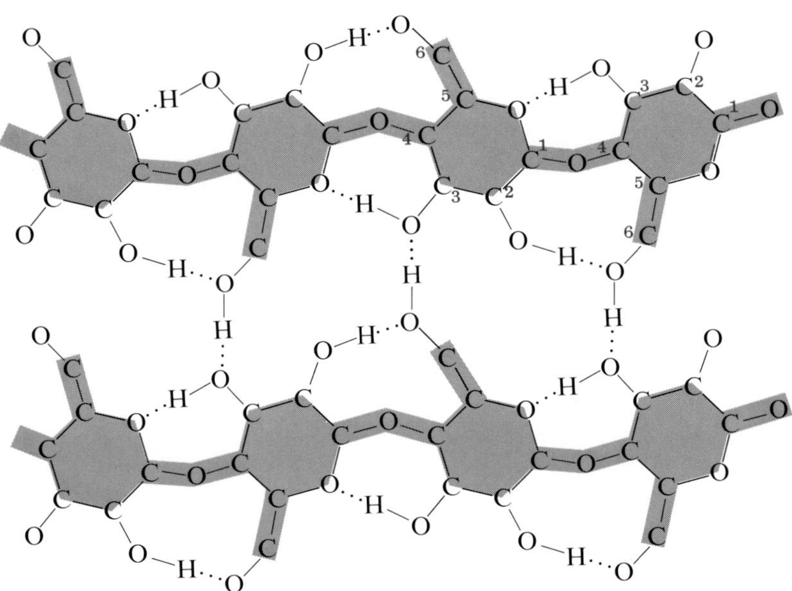

FIGURE 10–3 Cellulose is made from parallel ribbons of glucose residues linked by β-(1→4) bonds. The arrangement is stabilized by hydrogen bonds involving all of the oxygen atoms except those in the ether bridges between rings. The rings are being viewed face-on in this drawing; it is *not* a Haworth projection. It is based on the arrangement shown in K.H. Gardner and J. Blackwell (1974). The Structure of Native Cellulose. Biopolymers 13:1975.

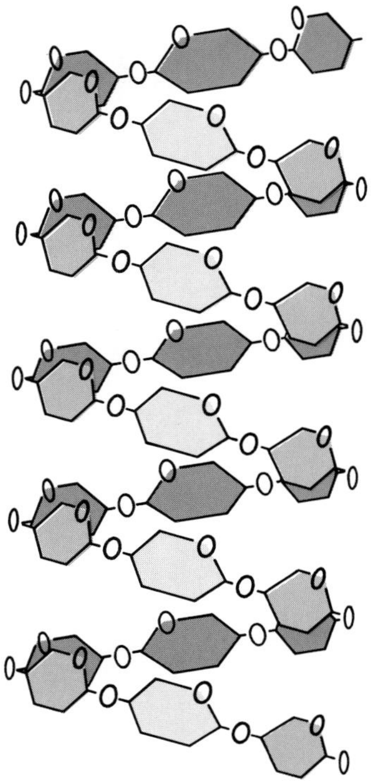

FIGURE 10–4

Amylose chains form open-cored helixes, probably left-handed, of varying sizes. One possible arrangement is shown here.

The Natural Glucans

Glycogen is a polymer that acts as a reservoir of glucose residues. The fundamental structure of glycogen is the amylose chain, with glucose residues joined in an α-(1→4) linkage, but it differs from amylose in having 1→6 branches at every fourth glucose residue in the interior of the molecule (Fig. 10–6). (Longer branched sequences occur at the periphery.) The branches prevent any formation of a helix, and the result is a tree-like structure, sketched in cross section in Figure 10–6*C*. (The role of glycogen as a fuel is discussed in Chapter 27.)

Starch is the storage carbohydrate of plants. It consists of two discrete kinds of molecules, with a minor fraction composed of pure amylose chains and the bulk made of amylopectin, which has a structure similar to that of glycogen but less highly branched. Starch is the major fuel in the diet of most humans.

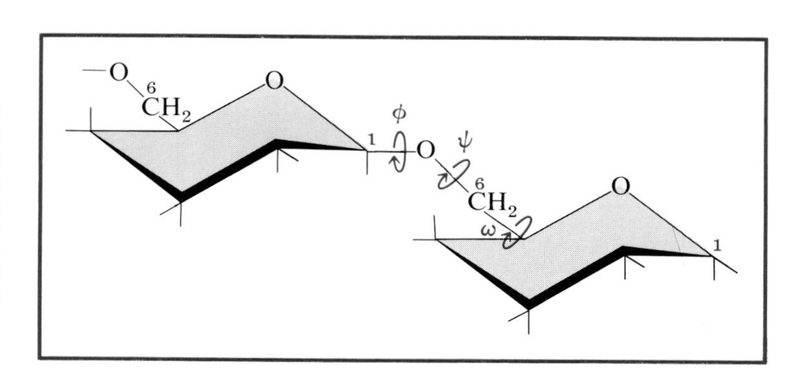

FIGURE 10–5

Glucose residues that are linked by 1→6 bonds can rotate relative to each other around the C-6 to C-5 bond, and three angles must be specified in order to describe their relative position. The greater freedom of motion permits 1→6 polymers to assume open configurations not available to polymers linked at other positions.

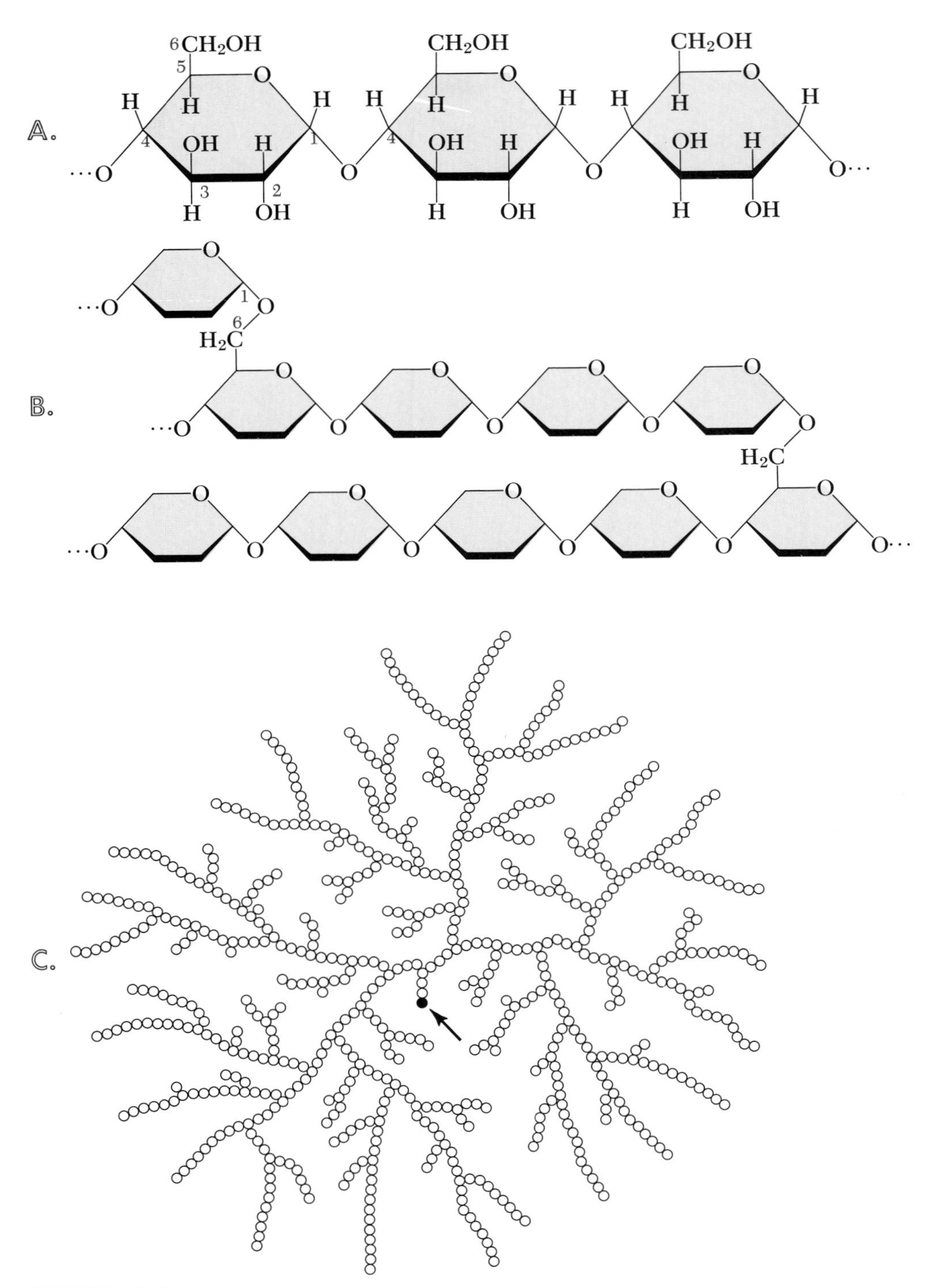

FIGURE 10–6

A. Glycogen contains linear amylose chains, made by linking carbons 1 and 4 of glucose through oxygen. *B.* Branches occur in the amylose chains where carbon 6 of a residue is also linked through oxygen to the C-1 terminal of another chain segment. *C.* A cross section through glycogen showing the tree-like structure created by branched amylose chains. The short inner segments are part of branches extending above and below the cross section. The circles represent glucose residues; only one residue (*arrow*) may assume an open-chain form and mutarotate because it is the only residue that does not have C-1 attached to another residue through an ether linkage.

Cellulose occurs in a few invertebrates but mainly in plants as a structural component. Mammalian cells lack the ability to break the β-(1→4) bonds of cellulose to yield glucose; only ruminants and other animals that harbor a special bacterial flora capable of attacking cellulose are able to utilize it as a dietary fuel. However, the very indigestibility of cellulose makes it important in providing bulk to the intestinal contents of humans.

Dextrans are gelatinous polymers that coat some bacteria. They have an α-(1→6) backbone with varying numbers of 1→3 and 1→4 cross-links, creating a three-dimensional network that is freely accessible to water. They are widely employed as molecular sieves in the separation of proteins and other large molecules by column chromatography because the degree of cross-linking determines the size of the molecule that can penetrate the gels. They are also used to restore blood plasma volume on those infrequent occasions when blood cannot be used. Some species of bacteria that populate the teeth have the ability to convert sucrose, but not glucose, to dextrans, forming plaques. The sticky dextrans aid in fixing the bacteria to the teeth, where they promote decay. The child with the sweet tooth is indeed more likely to find it rotting.

PROTEOGLYCANS

Many compounds are made by covalent combination of proteins and polysaccharides, with a spectrum of composition ranging from mostly protein to mostly polysaccharides. At one end of the spectrum are the **proteoglycans,** also called **proteinpolysaccharides,** in which the protein components are such a minor part of the total mass that their existence was overlooked for many years. These proteoglycans have long heteropolysaccharide chains covalently attached to a protein core, much like bristles on a brush.

Mucous secretions owe their viscous lubricating properties to proteoglycans and glycoproteins, and the proteoglycans formerly were called **mucopolysaccharides.** The major occurrence of proteoglycans is in the matrix in which collagen, elastin, and bone minerals are embedded. The character of connective tissue depends upon the relative proportions of fiber and matrix, as well as upon the nature of the two kinds of components; the intercellular cement is mostly ground substance with a few fibers, whereas a tendon is mostly fibers with minimal ground substance filling the spaces between them.

Chondroitin Sulfates

The most abundant proteoglycans in cartilages and arterial walls contain a heteropolysaccharide, chondroitin, esterified with sulfate groups. Let us examine the important features of these chondroitin sulfates as typical examples of proteoglycans.

The Carbohydrate Chains. Chondroitin is mainly composed of alternating residues of N-acetyl-D-galactosamine and D-glucuronate and therefore may be written as a repeating disaccharide unit:

$$(\text{GalNAc-GlcUA})_n$$

Other proteoglycans also consist of repeating disaccharide units containing an osamine, with the other residue often being a uronate. (We shall later see that the

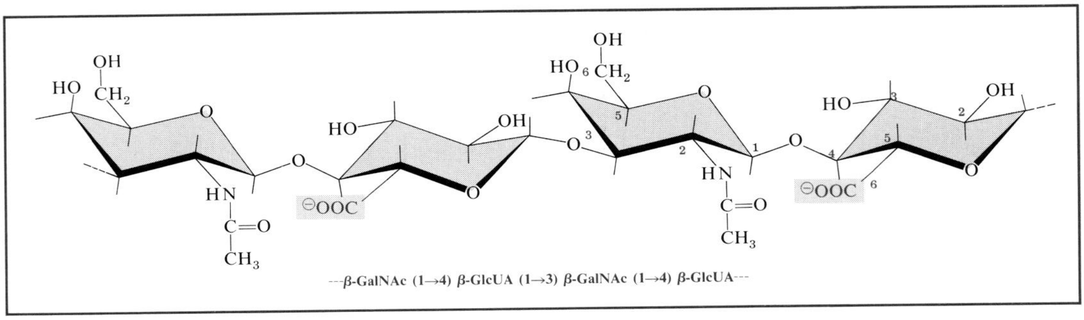

---β-GalNAc (1→4) β-GlcUA (1→3) β-GalNAc (1→4) β-GlcUA---

FIGURE 10–7 **The main chain of the chondroitin sulfates contains alternating residues of N-acetylgalactosamine and glucuronate. Notice that the β-linkages between residues cause alternate residues to be twisted by half-turns. By linking the glucuronate residues through C-4 and the N-acetylgalactosamine residues through C-3, an all-equatorial conformation is achieved, which is similar to the conformation of an amylose chain. Like amylose, the chondroitin chain tends to form helixes.**

carbohydrates are built one residue at a time, not by combining preformed disaccharide units.)

Chondroitin occurs as a straight chain, unlike the branched structure of glycogen. All of the residues are joined in the β-configuration at C-1, and the glycosidic bonds are alternately 1→3 and 1→4:

$$[(1\rightarrow3)\text{-}\beta\text{-GalNAc-}(1\rightarrow4)\beta\text{-GLcUA}]_n$$

The result is a structure in which all of the linkages between residues are equatorial (Fig. 10–7). Chondroitin chains, like those of the other proteoglycans, tend to assume a helical type B configuration in the solid state and are presumably even more open when in free contact with water.

The linkage between the chondroitin chain and the core protein is made by a special sequence of carbohydrate residues containing D-galactose and the pentose **D-xylose** attached to seryl side chains in the polypeptides (Fig. 10–8).

Sulfate groups are present on C-4 of the galactosamine residues in one kind of

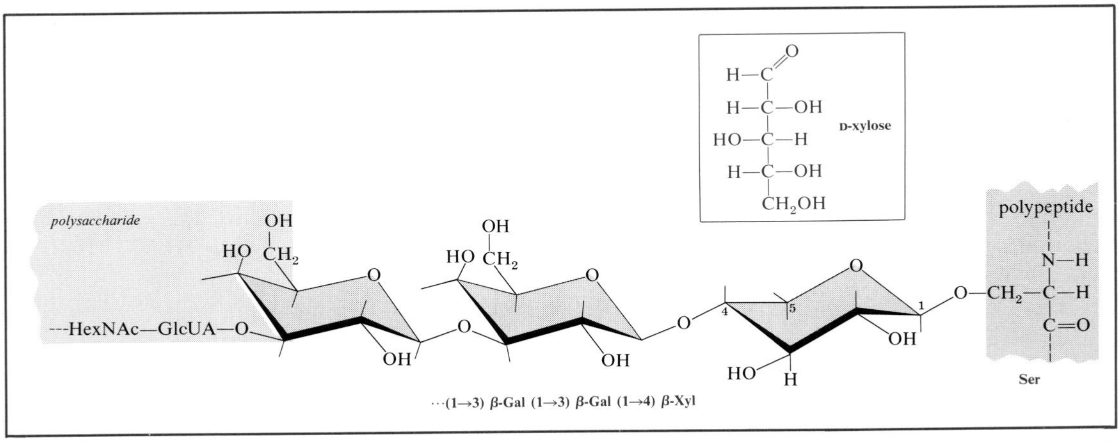

···(1→3) β-Gal (1→3) β-Gal (1→4) β-Xyl

FIGURE 10–8 **In the chondroitin sulfate proteoglycans, the polysaccharide chain is joined to the polypeptide chain through a terminal sequence containing residues of galactose and xylose. (The open-chain formula of xylose is shown in brown.) The xylose in turn is linked to the hydroxyl group of a serine residue in the polypeptide.**

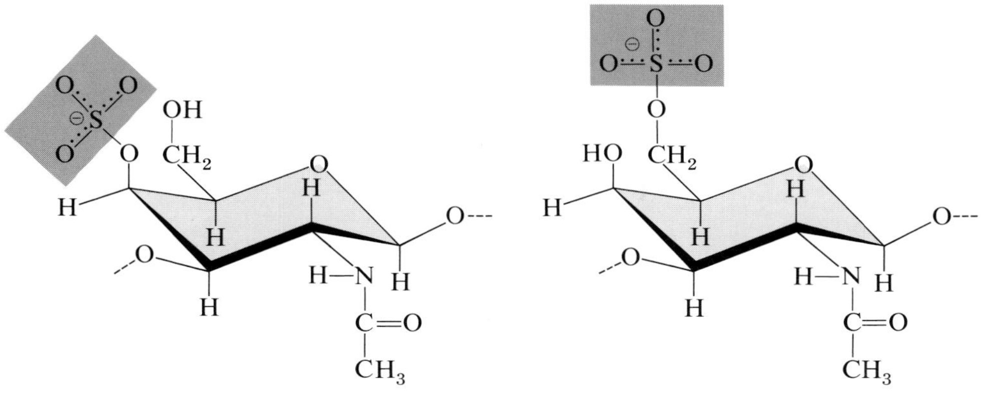

N-acetylgalactosamine-4-sulfate residue

N-acetylgalactosamine-6-sulfate residue

FIGURE 10–9 **Sulfate groups are linked to C-4 of N-acetylgalactosamine residues in some chondroitin sulfates, and to C-6 in others. The result in either case is the anionic form of an ester of sulfuric acid, contributing extra negative charges to the polysaccharide chain.**

chondroitin and on C-6 in another (Fig. 10–9). They add negative charge to the chains in much the same way that phosphate groups add negative charge to polypeptide chains. The proportions of chondroitin 4-sulfate and chondroitin 6-sulfate vary from one tissue to another.

Composition. Typical protein-chondroitin sulfate molecules will have compositions in this range:

20 to 60 polysaccharide chains per molecule; with
20 to 50 disaccharide residues in each chain, attached to
a core protein with a molecular weight of 110,000 to 140,000, giving
a total molecular weight of 2×10^5 to 2×10^6.

Properties. The chondroitin sulfates, like other proteoglycans, have an abundance of negative charges on already hydrophilic carbohydrate chains. Charge repulsion keeps the chains extended from the protein core and separated from each other, but they are free to sweep through the surrounding solution so that the volume occupied by a molecule in solution is much greater than the partial specific volume of the dehydrated solid. Schubert and Hamerman express it beautifully*:

> Statistically such a domain may have a fairly well defined size and shape. It can be visualized somewhat like the definitely shaped head of a tree, a black oak or a lombardy poplar, with branches extending throughout many cubic yards of space though the wood of its branches may occupy only a few per cent of the volume of the head. Small birds can easily fly through the head of the tree, but not through the wood of its branches. Small molecules can easily swim through the domain of a proteinpolysaccharide molecule, larger ones may encounter frequent obstructions, and very large ones could not even enter.

Here we have substances with the ability to exclude large molecules and let small ones through. They have a large negative charge that will attract cations. In short, they may act as molecular sieves and as cation exchangers. It is attractive to

---

*Schubert, M. and D. Hamerman (1968). A Primer on Connective Tissue Biochemistry. Lea and Febiger. Quoted by permission.

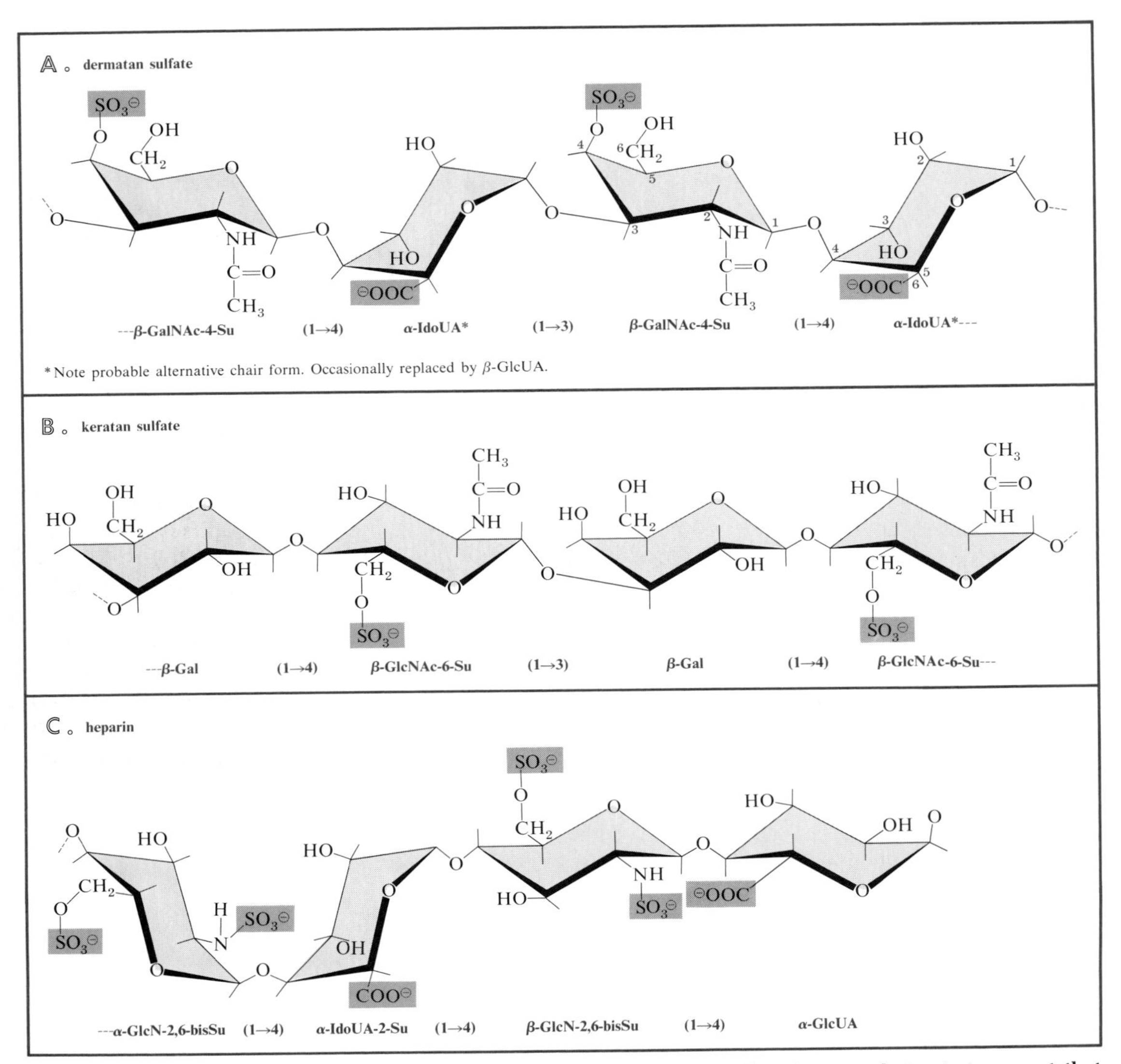

FIGURE 10–10 A. Dermatan sulfate closely resembles chondroitin 6-sulfate in general structure, except that configuration is reversed on C-5 of some of the uronate residues, so that they become residues of iduronate.

B. Keratan sulfate is made by polymerizing galactose with N-acetylglucosamine residues, followed by esterification with sulfate. The molecule is drawn here with a configuration resembling that of chondroitin sulfates, but this is conjectural.

C. Heparan sulfate in connective tissues is believed to resemble the heparin of mast cells in composition and configuration. Heparin contains both iduronate and glucuronate residues, along with N-acetylglucosamine residues. A speculative interpretation of a six-residue repeating unit is given here, with one residue of iduronate not shown at either end. Heparin, and presumably heparan sulfate, unlike other heteropolysaccharides shown here, contains axial links between residues as well as the more common equatorial bonds. These create a more looping conformation.

assume that the proteoglycans have important functions as conduits for the selective transport of materials, but the degree to which this is true is still conjectural.

One clear function of the chondroitin sulfates is to act as a waterbed around cells and fibers, creating a gel with the resilience to disperse shocks without permanent deformation. It seems likely that they interact in specific ways with the collagen fibers from which the structures gain rigidity, but the study of these complex molecules has not progressed to the point where more positive statements can be made.

Connective tissues contain several other sulfated heteropolysaccharides, in addition to the chondroitin sulfates, which differ in the nature of the constituent residues (Fig. 10–10). The proportion of the different compounds varies from tissue to tissue, and we make the obvious inference that appropriate properties are gained by these differences, but we are not yet able to make specific statements on what these properties are or the contribution made by specific heteropolysaccharides. The composition of the proteoglycan fraction can be changed by altering the proportions of two or more proteoglycans, each carrying only one kind of heteropolysaccharide, or it may be changed by adding more than one kind of heteropolysaccharide chain to the same protein core. Both mechanisms appear to be used.

Dermatan sulfates resemble chondroitin sulfates, except that α-L-iduronate residues occur in place of many, but not all, of the β-D-glucuronate residues. The change occurs by an epimerization of the residues after formation of the polysaccharide. The H and COO^- are interchanged on C-5 of many of the glucuronate residues, so that dermatan sulfate contains both of the hexuronates. Owing to the peculiarities of nomenclature, the interchange also changes the designation of the anomer from β-D to α-L, even though the configuration on C-1 has not been altered. The carbohydrate chains are linked to the protein by the same Gal-Gal-Xyl-Ser sequence seen in chondroitin sulfate.

The dermatan sulfates are especially common in skin, hence the name, but they also occur in other tissues.

Keratan sulfates occur as minor constituents along with chondroitin sulfates and are perhaps bound to the same protein cores. They are known to be present in cartilage, the cornea of the eye, and the pulpy nucleus of intervertebral discs. Keratans contain D-galactose, rather than a uronate, in combination with N-acetyl-D-glucosamine:

$$[(1\rightarrow3)\text{-}\beta\text{-Gal-}(1\rightarrow4)\text{-}\beta\text{-GlcNAc}]_n$$

A sulfate group is present on C-6 of the osamine, so the keratan sulfates are also polyanions, but with approximately half of the charge of chondroitin or dermatan sulfates, which also have uronate groups.

Heparan sulfates are more complicated heteropolysaccharides that contain both glucuronate and iduronate residues, along with alternating glucosamine residues that are only partially acetylated. Heparan sulfate layers are sandwiched on both sides of the type IV collagen core in basement membranes. A connecting protein binds the adjacent cells to the heparan sulfate. This protein is **fibronectin,** a two-chain fibrous molecule from fibroblasts; **laminin,** a cross-shaped fibrous molecule from endothelial cells; or **chondronectin,** a more globular molecule from chondrocytes.

Heparan sulfates are related to a more widely studied compound, **heparin,** which has a similar structure with more sulfate groups attached and no acetyl groups on the glucosamine residues.

Heparin is not a constituent of connective tissues; it occurs within the mast cells that line arteries, especially in the liver, lungs, and skin. A repeating unit of six carbohydrate residues has been proposed:

[(1→4)-α-IdoUA-(1→4)-α-GlcN-(1→4)-α-IdoUA-(1→4)-β-GlcN-
a a e e a a e e

(1→4)-α-GlcUA-(1→4)-α-GlcN-]$_n$
e e a e

a = axial *e = equatorial*

in which each glucosamine residue contains sulfate groups on both C-6 (ester linkage) and the nitrogen atom (amide linkage), and each of the iduronate residues contains sulfate in ester linkage on C-2, making a total of eight sulfate groups per six carbohydrate residues. The chains are relatively short with only about 50 carbohydrate residues.

An important property of heparin, and presumably of heparan sulfates, is that some of the residues are linked by α-axial bonds, rather than the β-equatorial bonds common in other proteoglycans. The result is a more looping chain (Fig. 10–10*C*).

Heparin inhibits clotting of blood (p. 348), and commercial preparations are widely used for this purpose. It is used in the management of patients with pulmonary embolism (dissemination of clots to the lung) and deep venous thrombosis. It is also used during surgery on the heart or blood vessels. A low dose of heparin is frequently used as a prophylactic measure prior to some surgical procedures with a high risk of pulmonary embolism. It is sometimes used for the less dramatic purpose of preventing obstruction to access tubes in blood vessels.

Hyaluronate

Hyaluronic acid, occurring as a polyanion, probably qualifies as a proteoglycan by containing a small amount of protein, but the bulk of the molecule is made of a long heteropolysaccharide chain containing some 5000 carbohydrate residues, with a molecular weight of 1 to 3 million. The chain has no sulfate groups and contains alternating residues of D-glucuronate and N-acetyl-D-glucosamine:

[(1→3)-β-GlcNAc-(1→4)-β-GlcUA]$_n$

The long chain of hyaluronate gives it even greater conformational mobility than is seen in the chondroitin and other sulfates. With negative charges only on alternating residues, there is somewhat less charge repulsion. Hyaluronate molecules therefore sweep through very large domains, with approximately 1000 times the volume of the anhydrous molecule. That is, one gram of hyaluronate in excess water will exclude other large molecules from about one liter of space.

Hyaluronate is an effective lubricant as well as being a resilient buffer against mechanical damage. It is therefore an important constituent of the synovial fluid in joints, and it also is a major component of the vitreous humor of the eye, of arterial walls, of the umbilical cord, and a variety of other connective tissues.

The most important functions of hyaluronate may yet be undiscovered. Complexes of hyaluronate with chondroitin sulfates and dermatan sulfates have been isolated from cartilage and from cell cultures. A small **link protein** tightly associates 1:1 with the proteoglycan, and this complex in turn binds hyaluronate. It is not known what the final aggregate does.

GLYCOPROTEINS

The glycoproteins are proteins that contain attached carbohydrates, but they differ from the proteoglycans in that the carbohydrates are not polymers of repeating units. They are in shorter chains, often highly branched. Glycoproteins may contain only a few carbohydrate chains or so many that they amount to more than half the mass of the molecule.

The attachment of carbohydrate residues to polypeptide chains does not convey a particular type of function. Glycoproteins include enzymes, hormones, antibodies, structural proteins, and so on. Despite the diversity of function, most of the glycoproteins are found in a few kinds of locations: in the extracellular fluids, the lysosomes within cells, and in the plasma membrane surrounding cells.

The purpose of the carbohydrate is not always clear. It gives a lubricating property to the carbohydrate-rich proteins of mucous secretions and protects some circulating proteins from removal by the liver. Distinctive carbohydrates on the surface of cells act as receptors for binding specific compounds. Beyond this, we must lean mainly on conjecture. Since many secreted proteins have attached carbohydrate, it has seemed reasonable to suppose that the carbohydrate in some way aids the secretory process, although many proteins are secreted very well without attached carbohydrates. In some cases, the presence of the carbohydrate groups has been shown to be necessary for biological function; they maintain proper conformation, or bind the protein at appropriate sites.

The diverse functions are discussed later. Our purpose now is to describe some common elements in the structure of the attached carbohydrates.

Types of Carbohydrate Linkage

The carbohydrate groups in mammalian glycoproteins are linked to the hydroxyl groups of serine, threonine, and hydroxylysine residues, or to the amide nitrogen of asparagine residues. These two types of linkages are associated with quite different kinds of carbohydrates.

Serine/threonine-linked carbohydrates are frequently simple. We have already noted that type I collagens contain a few sugar residues; typically, there may be one or two β-galactosyl groups and one or two α-glucosyl-(1→2)-β-galactosyl groups attached to hydroxylysyl groups in each 1000 amino acid residues:

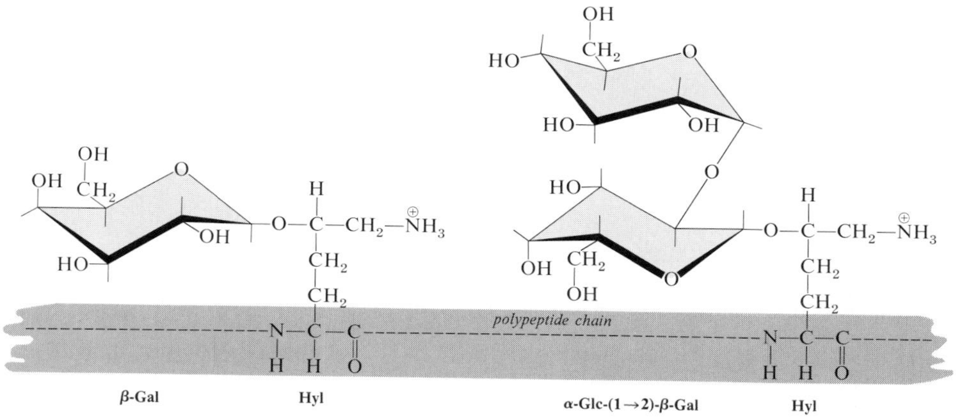

polypeptide chain

β-Gal Hyl α-Glc-(1→2)-β-Gal Hyl

At the other extreme, the submaxillary glycoprotein that has been characterized from sheep contains over 800 disaccharide units—one every six amino acid residues, on the average—that amount to ~40 per cent of its mass. However, nearly all of these units are made of N-acetylneuraminate and N-acetylgalactosamine residues, some attached to serine residues and some to threonine:

$$\alpha\text{-NeuNAc-}(2\rightarrow6)\text{-}\alpha\text{-GalNAc-Ser/Thr}$$

Some carbohydrate groups linked to serine or threonine residues are quite complex branched structures. The heterooligosaccharide in red blood cell membranes that conveys group A specificity is shown in Figure 10–11 as an example.

Asparagine-linked carbohydrate side chains fall into two classes. One is composed only of mannose and N-acetylglucosamine residues, which are frequently branched. Figure 10–12 illustrates typical examples that appear in thyroglobulin, the protein in the thyroid gland in which the hormone thyroxine is generated. The figure also shows a common characteristic of the asparagine-linked glycoproteins—they frequently contain sheared versions of the characteristic side chains that lack one or more carbohydrate residues.

The second class of asparagine-linked polysaccharides contains a wider variety

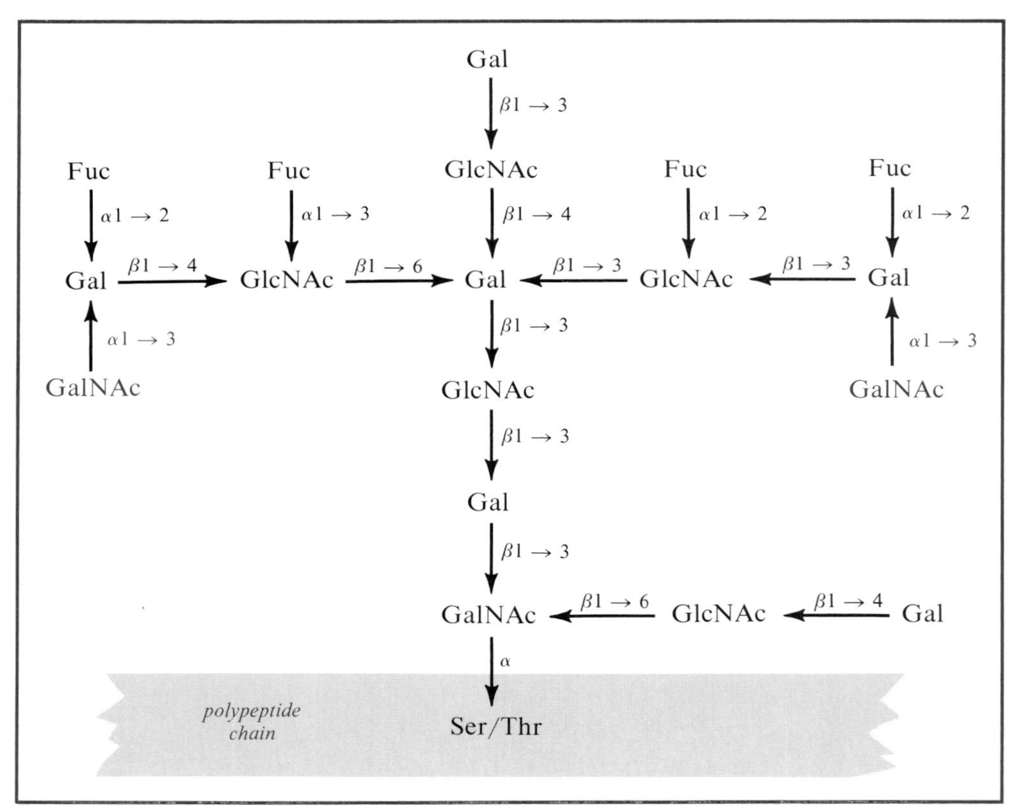

FIGURE 10–11 Red blood cells contain oligosaccharides that are responsible for blood-group specificity. The carbohydrate responsible for type A specificity, shown here, can be linked to seryl or threonyl groups on a polypeptide chain. Individuals with type B blood differ in having galactose residues in place of two N-acetylgalactosamine residues (*brown*). The structure is shown here with the maximum number of side-chain fucose residues, all of which do not occur in every individual.

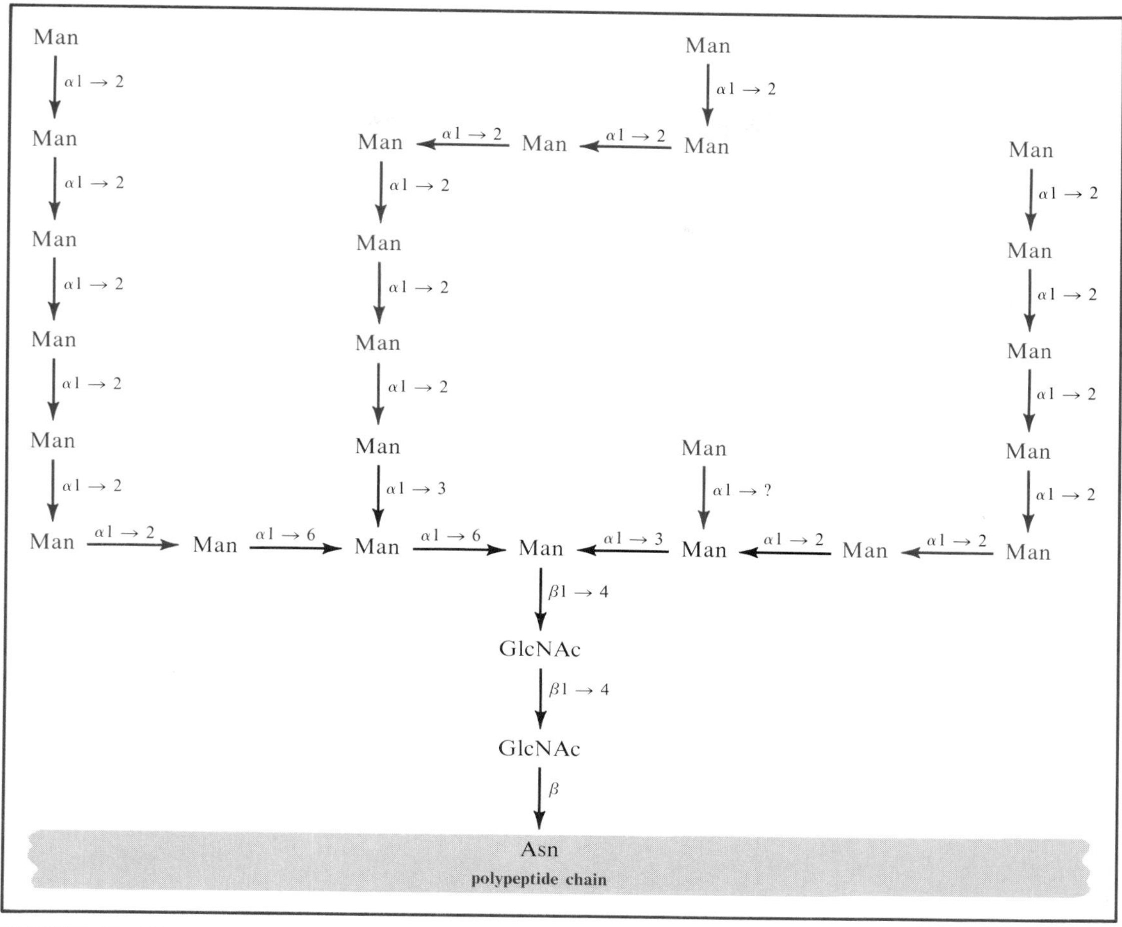

FIGURE 10–12 Some of the carbohydrate groups found in thyroglobulin are made of branched mannose polymers linked to an asparagine residue of the polypeptide chain through two N-acetylglucosamine residues. Mannose is commonly present at branch points in the carbohydrates of glycoproteins. The residues shown in black always occur; varying numbers of those shown in brown are missing in some chains.

of carbohydrate residues in branched structures. Many of the complex side chains are capped by residues of either N-acetyl-D-neuraminate or L-fucose. The function of fucose is not known, but the presence of N-acetyl-neuraminate apparently prevents recognition by sites on the membrane of liver parenchymal cells. If the sialate residue is removed, which sometimes happens as a protein ages in the circulation, the liver takes it up for destruction.

Some glycoproteins contain more than one of these types of carbohydrate branches. Human thyroglobulin contains all three: serine-linked, asparagine-linked and mannose-rich, asparagine-linked and with complex composition.

FURTHER READING

W. Pigman and D. Horton, eds. (1972). The Carbohydrates. Academic Press. A yet incomplete multivolume work. The introductory chapters in vol. 1A are a readable introduction to carbohydrate stereochemistry.

R. Gedden, J.D. Harvey, and P.R. Wills (1977). The Molecular Size and Shape of Liver Glycogen. Biochem. J. 163:201.

D.A. Rees (1977). Polysaccharide Shapes. Halsted Press. Summary of types of polysaccharide conformation by pioneer in field.

V.C. Hascall (1981). Proteoglycans: Structure and Function. p. 1 in V. Ginsburg and P. Robbins, eds. Biology of Carbohydrates, vol. 1. Wiley.

W. Lennarz, ed. (1980). Biochemistry of Glycoproteins and Proteoglycans. Plenum.

U. Lindahl and M. Hook (1978). Glycosaminoglycans and their Binding to Biological Macromolecules. Annu. Rev. Biochem. 47:385.

R.L. Lundblad et al., eds. (1980). Chemistry and Biology of Heparin. Elsevier. Symposium papers.

G.M. Oosta et al. (1981). Multiple Functional Domains of the Heparin Molecule. Proc. Natl. Acad. Sci. U.S. 78:829.

R.G. Spiro (1973). Glycoproteins. Adv. Protein Chem. 27:349. Excellent review.

R. Kornfeld and S. Kornfeld (1976). Comparative Aspects of Glycoprotein Structure. Annu. Rev. Biochem. 45:217. A good snappy summary.

W.D. Comper and T.C. Laurent (1978). Physiological Function of Connective Tissue Polysaccharides. Physiol. Rev. 57:313.

H.K. Kleinman, R.J. Klebe, and G.R. Martin (1981). Role of Collagenous Matrices in the Adhesion and Growth of Cells. J. Cell Biol. 88:473. Review. Includes discussion of nectins and proteoglycan interface.

CHAPTER 11

LIPIDS AND MEMBRANES

The lipids are the waxy, greasy, and oily compounds of the body. They repel water, and this hydrophobic nature is used as a tool for a variety of purposes. Some lipids—the fats—are important fuels, and their coalescence into nearly anhydrous droplets creates a reserve of potential energy that is a much lighter burden to carry than an equal reserve of waterlogged carbohydrate. Still other lipids are major structural components, and their ability to associate so as to exclude water and other polar compounds makes complex organisms possible through the formation of membranes. Membranes separate cells within tissues and organelles within cells, creating compartments with separate chemistries to permit distinct organization and regulation; each compartment becomes an individual part of a more complex whole.

The basic principle of membrane structure has been known for many decades:

Plasma membranes and intracellular membranes are made with a hydrophobic core and polar surfaces.*

*In discussing membranes, we are concerned with those bounding cells and organelles and not with structural sheets, such as basement membranes, or more complex serous membranes, such as the peritoneum.

NATURE OF LIPIDS

Fatty Acids

The fundamental building blocks of stored fats and many structural lipids are straight chain aliphatic carboxylic acids, which may be saturated or may contain one or more double bonds:

$$H_3C{-}(CH_2)_n{-}\overset{O}{\overset{\|}{C}}{-}OH \qquad H_3C{-}(CH_2)_m{-}\left(CH_2{-}\overset{H}{\overset{|}{C}}{=}\overset{H}{\overset{|}{C}}\right)_x{-}(CH_2)_n{-}\overset{O}{\overset{\|}{C}}{-}OH$$

saturated fatty acid — ***cis*-unsaturated fatty acid**

When double bonds are present, they are nearly always in the *cis* configuration, and if there is more than one, they are spaced at three-carbon intervals.

Nomenclature. Both the systematic and trivial names of fatty acids of metabolic or structural importance are listed in Table 11–1. The predominant components of structural lipids contain 16 or more carbon atoms and are shown in boldface type. Trivial names are frequently used for these and for the shortest-chain compounds.

Since lipids usually contain an assortment of different fatty acids, shorthand designations have been devised for indicating composition in terms of the number of carbon atoms and the number of double bonds. Thus 16:0 is **palmitic acid,** with 16 carbon atoms and no double bonds; it is the most abundant saturated fatty acid. 18:1 designates the most abundant unsaturated fatty acids (**oleic** and **vaccenic** acids), with 18 carbon atoms and one double bond. The position of the double bonds is indicated in parenthesis. The important unsaturated fatty acids are shown in Figure 11–1.

Fatty Acids Are Amphipathic. Fatty acids have both hydrophobic and hydrophilic regions. This duality of response to water is the key to the function of biological lipids in general. The hydrocarbon tails try to agglomerate so as to expose a minimum surface, while the polar carboxyl groups attempt to maintain contact with the watery world around them. The length of the hydrocarbon chains determines the

TABLE 11–1 ALKYL CARBOXYLIC ACIDS

Components	Trivial Name	Systematic Name
1	formic	
2:0	acetic	
3:0	propionic	
4:0	butyric	
5:0	valeric	pentanoic
6:0	caproic	hexanoic
8:0	caprylic	octanoic
10:0	capric	decanoic
12:0	lauric	dodecanoic
14:0	myristic	tetradecanoic
16:0	**palmitic**	hexadecanoic
16:1(9)	**palmitoleic**	*cis*-9-hexadecenoic
18:0	**stearic**	octadecanoic
18:1(9)	**oleic**	*cis*-9-octadecenoic
18:1(11)	**vaccenic**	*cis*-11-octadecenoic
18:2(9,12)	**linoleic**	*all cis*-9,12-octadecadienoic
18:3(9,12,15)	**linolenic**	*all cis*-9,12,15-octadecatrienoic
20:4(5,8,11,14)	**arachidonic**	*all cis*-5,18,11,14-eicosatetraenoic
24:0	**lignoceric**	tetracosanoic

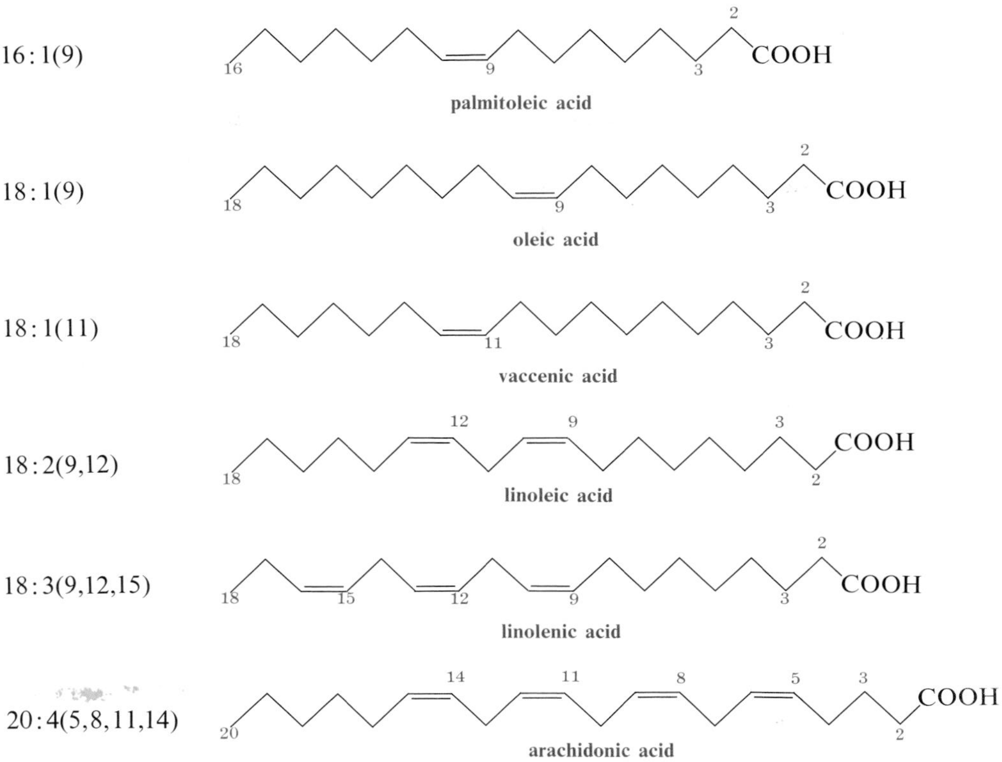

16:1(9) palmitoleic acid

18:1(9) oleic acid

18:1(11) vaccenic acid

18:2(9,12) linoleic acid

18:3(9,12,15) linolenic acid

20:4(5,8,11,14) arachidonic acid

FIGURE 11–1 **Some common unsaturated fatty acids. The configuration is *cis* around each double bond, and multiple double bonds are spaced at three-carbon intervals.**

dominant behavior; hydrophobic interactions are so strong in palmitic acid (16:0) that it is only soluble to the extent of eight parts per million in an acidic solution at 30° C.

Micelles. If palmitic or another long-chain fatty acid is exposed to a neutral solution, it ionizes to form anions, or soaps, with a negatively charged carboxylate group that has a much stronger tendency to associate with water. The sodium or potassium salts of long-chain fatty acids are many times more soluble than the undissociated acids, especially at elevated temperatures. (A true solution of sodium palmitate can reach one millimolar at 60° C, whereas a solution of palmitic acid is saturated at less than 5 micromolar.)

The tendency of the charged head groups to remain in contact with water is so great that the soaps do not precipitate when their solubility is exceeded; instead they form small clusters, or micelles. The simplest form of micelle is a sphere in which the hydrocarbon tails are grouped in the center, and the polar head groups are associated with water and counterions on the surface (Fig. 11–2). Other shapes occur as the ratio of soap to water is increased, or other components are added.

We shall see that formation of micelles with other lipids is aided by the amphipathic character of fatty anions; their detergent action* is an important tool used in the digestion and absorption of lipids.

* A detergent is an amphipathic compound promoting the dispersion in water of substances otherwise incapable of forming micelles, emulsions, or other stable mixtures.

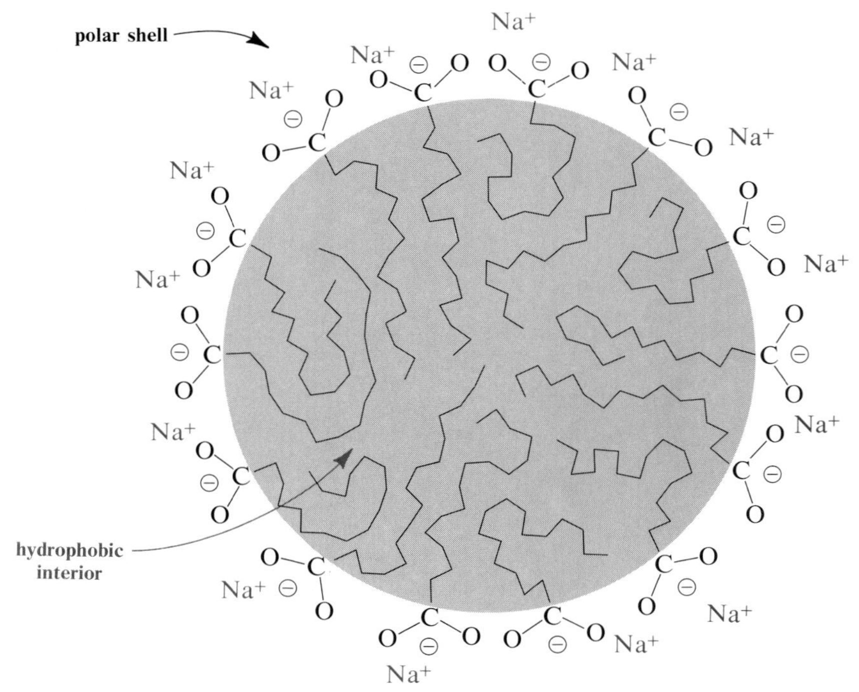

FIGURE 11-2 The anions of long-chain acids, the soaps, form micelles in water.

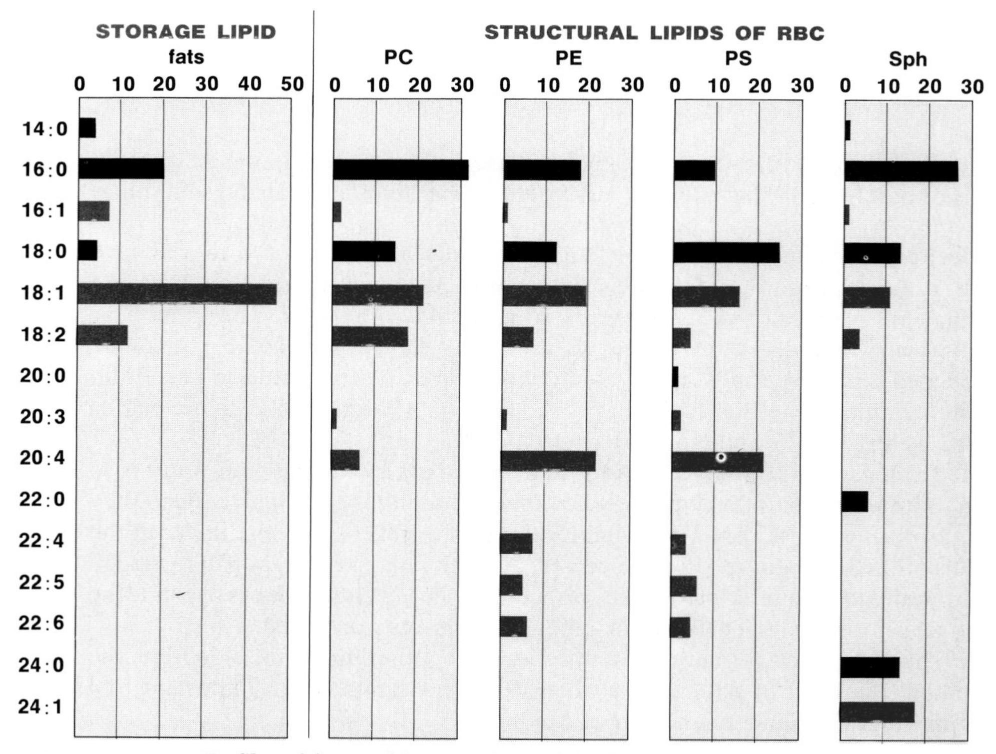

FIGURE 11-3 Profiles of fatty acid composition of lipids. Saturated fatty acids are shown in black, unsaturated in brown. PC, PE, PS are phosphatidylcholine, phosphatidylethanolamine, and phosphatidylserine, respectively. Sph is sphingomyelin.

Effects of Fatty Acid Composition. The nature of lipids built from fatty acids is determined by the length of the fatty acid chains and the number of double bonds in the chains. Alterations in these characteristics are used to control lipid behavior, and fatty acid composition depends upon the purpose for which a lipid is being made. Some idea of the usual variations can be gained by comparing the fatty acids stored as fats in human adipose tissue with the fatty acids used to make structural lipids in human red blood cells (Fig. 11–3). Adipose tissue mainly uses 16:0 and 18:1 fatty acids, whereas the lipids of the red blood cells are built according to more complex directions, with different kinds of structural lipids also differing in fatty acid composition.

What is gained by these variations? Chain length determines the volume of the hydrophobic phase, and double bonds introduce kinks that give a favorable geometry for some purposes, but there is another all-important property that is determined by the fatty acid recipe, and this is the **transition temperature,** or **melting point,** for the lipid. It is imperative that the lipids do not crystallize into rigid structures within the cells. The globules of stored fat must be freely accessible upon demand, and the membranes must permit motion of material through them. Organisms adjust their fatty acid composition to fit the environmental temperatures to which they are exposed.

Figure 11–4 illustrates the principle. The melting points of all but the shortest fatty acids increase in a regular way with increasing chain length. (The shortest have anomalous behavior because their crystal structures are heavily dependent on hydrogen bond formation.) One way to maintain a liquid state would therefore be to use relatively short-chain fatty acids in building lipids. This, however, would limit the hydrophobic character of the lipids, and gaining hydrophobicity is a major purpose of making lipids.

The melting point can be lowered without sacrificing hydrophobic character by introducing *cis*-unsaturated fatty acids. Insertion of one double bond near the middle of an 18-carbon fatty acid makes it safely fluid. (A *trans* double bond or a *cis* double bond near either end of the chain is not nearly as effective.) The effects on transition temperature of chain length and unsaturation depend upon the particular type of lipid, but each lipid is maintained in a fluid state, while also attaining the necessary molecular size and shape, through insertion of particular proportions of the various fatty acids.

Natural fatty acids contain an even number of carbon atoms. The only important exceptions are some short-chain metabolic intermediates. The preference for an even number in the longer-chain compounds is satisfied by forming and degrading the fatty acids in two-carbon units, as will be discussed in detail later. We can see the fundamental reason by examining the melting points again (Fig. 11–4). Fatty acids with an even number of carbon atoms consistently melt at higher temperatures than those with an odd number. This indicates that their hydrocarbon chains pack better. The even-numbered fatty acids are therefore more suitable for creating compact structures. Put in another way, associations of compounds made from the even-numbered acids will be somewhat more stable, and their formation was favored during the original reshuffling of the organic components of the earth by which organisms were formed.

Triglycerides

The fats are esters of glycerol and three molecules of fatty acids. These triglycerides, or neutral fats, are named and numbered from top to bottom when the formula

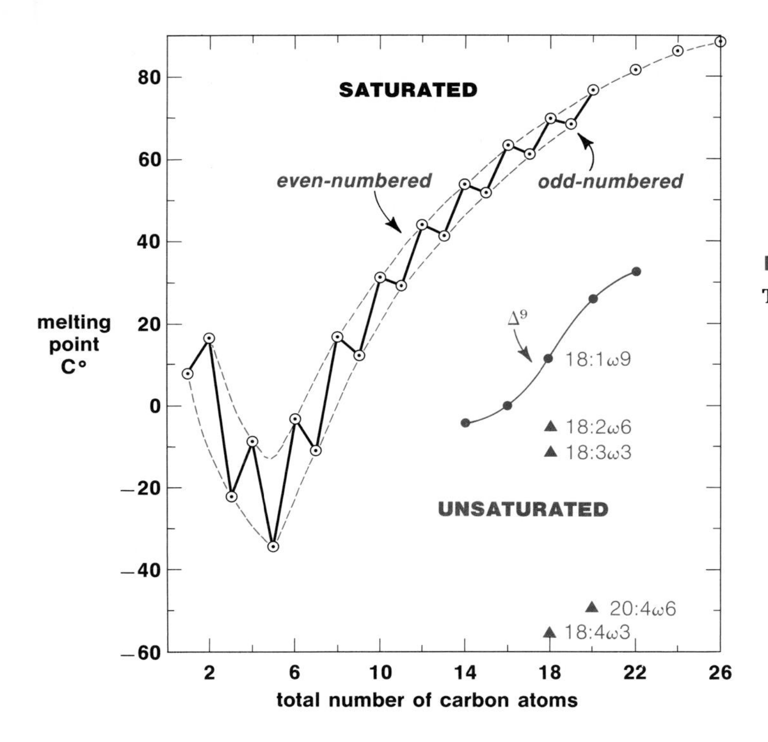

FIGURE 11–4
The melting points of fatty acids.

is drawn so as to have a conventional L-configuration. Thus, 1-palmitoyl-2-oleoyl-3-stearoyl glycerol is:

1-palmitoyl-2-oleoyl-3-stearoyl glycerol

A saturated fatty acid residue is usually present at position 1, and an unsaturated residue at position 2; either may be present at position 3.

There is no inherent reason for numbering glycerol from one end or the other, according to systematic organic nomenclature. The rule described above is a convention, and its application is sometimes indicated by the letters sn, for stereospecific numbering, as in sn-1-palmitoyl-sn-2-oleoyl, etc., but this is really not necessary if the basis for the convention is understood. Describing the isomer as L removes all doubt. If you think there is no structural difference between the two ends of glycerol, mentally try to superimpose opposite ends of two identical molecules without reversing the configuration of the central carbon atom.

The fats occur as oily droplets in cells. The polar contribution of the three ester linkages is relatively small compared to the hydrophobic character of the three long hydrocarbon chains in most fats, so the fats are only slightly soluble in water, and do not form stable micelles by themselves.

A reason why approximately 50 per cent of the fatty acid residues in most fats are unsaturated, and why 16:0 is the prevalent saturated acid, may be deduced from an examination of the melting points (Table 11–2). A triglyceride made of two 16:0 residues and one 18:1 residue melts at 35 or 36° C, barely below body temperature, whereas a triglyceride containing two 18:0 residues and one 18:1 residue melts at

TABLE 11–2 The melting points of triglycerides with various fatty acids esterified with the three carbons of glycerol. These values are for synthetic compounds, presumably DL mixtures of the asymmetric forms. Brown shading indicates unsaturated residues.

Fatty Acid Residue			
	Position		
1	*2*	*3*	**Melting Point**
6:0	6:0	6:0	−25
8:0	8:0	8:0	10
10:0	10:0	10:0	31.5
12:0	12:0	12:0	44
14:0	14:0	14:0	55
16:0	16:0	16:0	66
16:0	18:1(9)	16:0	36
18:1(9)	16:0	16:0	35
16:0	18:1(9)	18:1(9)	18
18:1(9)	16:0	18:1(9)	18
18:0	18:1(9)	18:0	42
18:0	18:1(9)	18:1(9)	23.5
	*[18:2(9.12)]$(16:0)_2$		27
	*$[18:3(9,12,15)]_2$ (16:0)		−3

*Position not specified.

42° C and would be solid at body temperature. A safety factor is provided by mixing in some triglycerides containing two 18:1 fatty acids, which have melting points well below body temperature.

Phospholipids

In order to make useful structures from lipids, these structures must include more strongly amphipathic molecules than the triglycerides so that they can present a greater surface to the aqueous phase. Increased affinity for water is obtained by incorporating charged phosphate ester groups into the lipids. The phospholipids are indispensable components of membranes and are also used as detergents to coat fat droplets for transport within the body.

Phosphatidyl Compounds. Many of the phospholipids are derivatives of phosphatidic acid, which contains glycerol that is esterified with fatty acids on two carbon atoms and with phosphoric acid on the third:

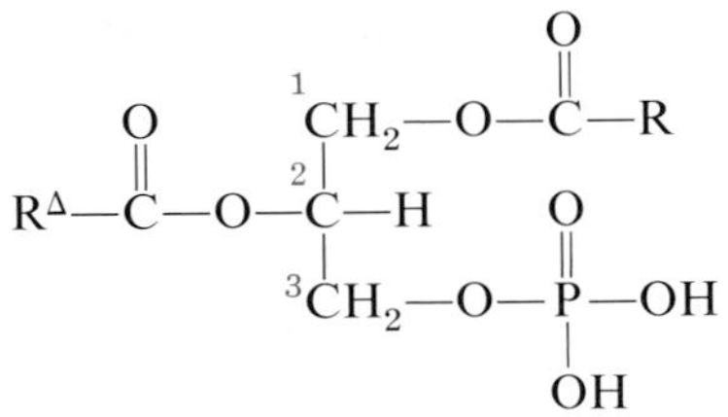

```
                          O
                          ‖
          O      ¹CH₂—O—C—R
          ‖      2|
RΔ—C—O—C—H         O
                  |          ‖
                 ³CH₂—O—P—OH
                              |
                             OH
```

phosphatidic acid
(R^{Δ} = unsaturated chain)

One example is shown here, but the fatty acid composition varies. However, most phosphatidyl compounds have a saturated fatty acid linked to C-1 and an unsaturated fatty acid linked to C-2, as shown.

The phosphatidic acids, like other phospholipids, ionize at physiological H^+

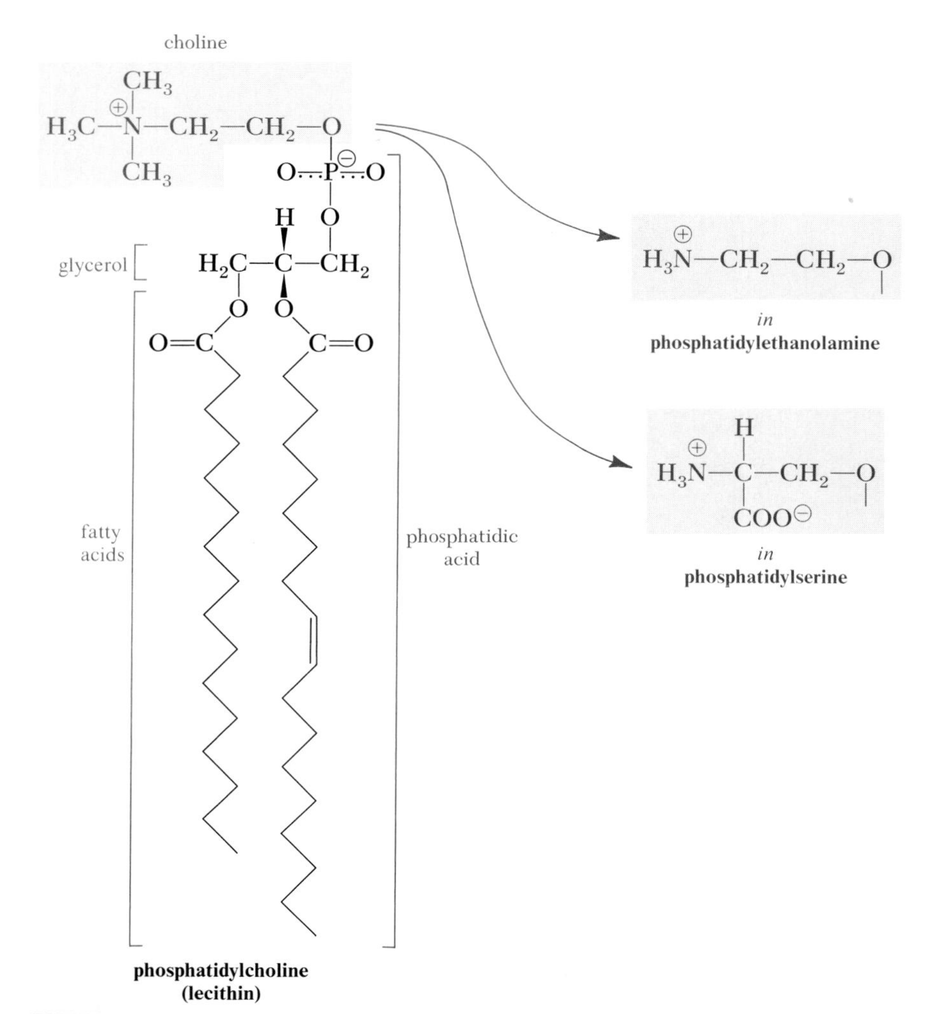

FIGURE 11-5

Major phospholipids in animals include phosphatidyl derivatives of choline, ethanolamine, or serine (*brown shading*). These phospholipids contain both positive and negative charges, and therefore have strongly hydrophilic heads.

concentrations; they carry between one and two negative charges in neutral solution, and we shall refer to them as the phosphatidates.

Phosphatidylcholine, phosphatidylethanolamine, and **phosphatidylserine** are structurally related compounds (Fig. 11-5) that carry both positive and negative charges. We speak of them as if they were single compounds, but there are many phosphatidylcholines and related compounds that vary in fatty acid composition.

TABLE 11-3 FATTY ACID COMPOSITION OF PHOSPHATIDYLCHOLINES IN HUMAN RBC* (listed as mole percent)

Component	16:0	16:1	18:0	18:1	18:2	18:3	20:3	20:4	20:5	22:6
Position 1	61		24	10	0.6					
Position 2	9	1.8		26	35	1.0	4	12	0.5	1.6

*Means of values from several laboratories.

Table 11–3 shows the fatty acid residues found at each position of the phosphatidylcholines in human erythrocytes.

These structures with highly charged heads and hydrophobic tails have a strong detergent action. The phosphatidylcholines, also known as **lecithins,*** are frequently incorporated into processed foods to keep the lipids dispersed.

Phosphatidyl inositols are esters of a cyclic derivative of glucose, with a high proportion of 1-stearoyl-2-arachidonyl species (18:0) (20:4):

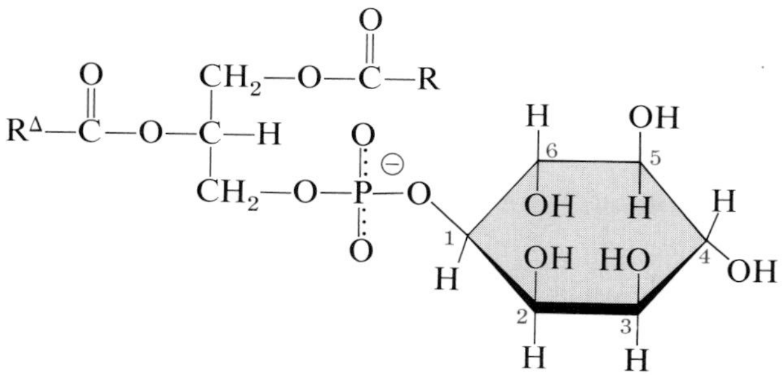

phosphatidylinositol

Phosphatidylglycerol and **diphosphatidylglycerols (cardiolipins)** are esters of phosphatidates with an additional molecule of glycerol:

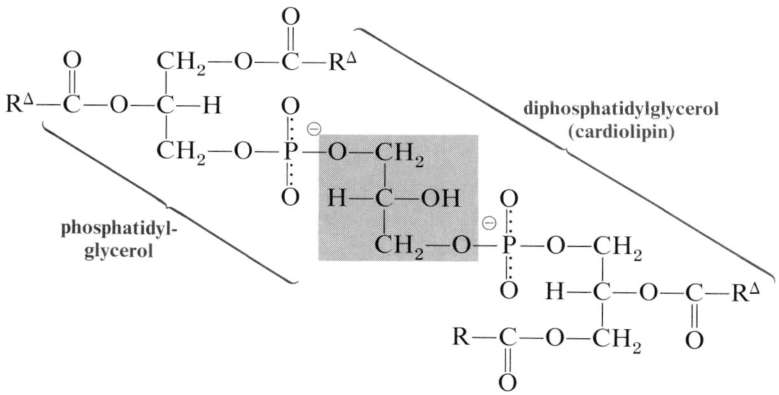

diphosphatidylglycerol (cardiolipin)

phosphatidyl-glycerol

These compounds are rich in polyunsaturated residues; linoleic acid [18:2(9,12)] accounts for 70 to 80 per cent of the fatty acid residues in visceral diphosphatidyl glycerol.

Lysophosphatidate derivatives are compounds lacking a fatty acid residue on C-2 of glycerol:

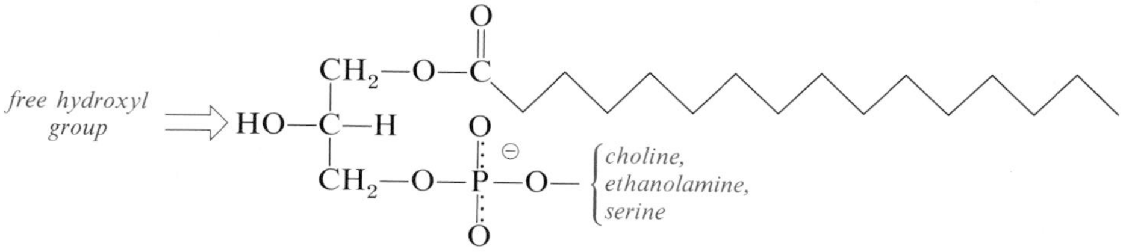

free hydroxyl group

choline, ethanolamine, serine

lysophosphatidyl compound

*Phosphatidylethanolamines were formerly named cephalins.

Thus, there are lysophosphatides, lysophosphatidylcholines, and so on. (The name was coined because the lysophosphatidylcholines, or lysolecithins, are especially effective detergents that cause the lysis of red blood cells.) **Phosphatidal compounds** (phosphatidalcholine, etc.), also known as **plasmalogens,** are relatives of the phosphatidyl compounds in which the fatty acid residue on C-1 is replaced by a long-chain aldehyde:

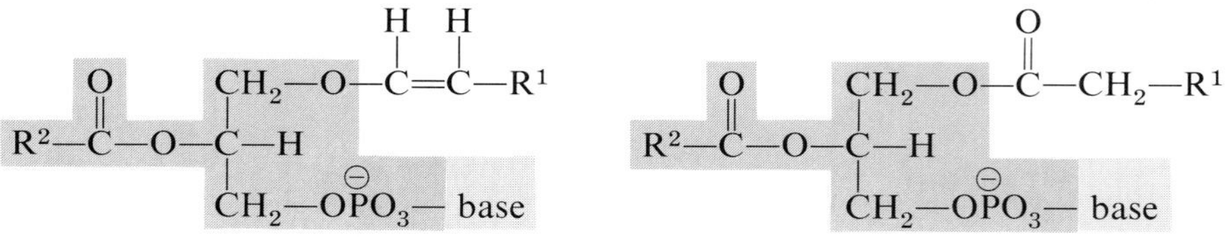

$$R^2\text{—}\overset{\overset{\displaystyle O}{\|}}{C}\text{—O—}\underset{\underset{\displaystyle CH_2\text{—}\overset{\ominus}{O}PO_3\text{— base}}{|}}{\overset{\overset{\displaystyle CH_2\text{—O—}\overset{\overset{\displaystyle H}{|}}{C}\text{=}\overset{\overset{\displaystyle H}{|}}{C}\text{—}R^1}{|}}{C}}\text{—H}$$

phosphatid_al_ compounds

$$R^2\text{—}\overset{\overset{\displaystyle O}{\|}}{C}\text{—O—}\underset{\underset{\displaystyle CH_2\text{—}\overset{\ominus}{O}PO_3\text{— base}}{|}}{\overset{\overset{\displaystyle CH_2\text{—O—}\overset{\overset{\displaystyle O}{\|}}{C}\text{—}CH_2\text{—}R^1}{|}}{C}}\text{—H}$$

phosphatid_yl_ compounds

(base = choline, ethanolamine, or serine)

The phosphatidal compounds may be further reduced to the corresponding alkoxy compounds.

Sphingomyelins are closely similar in shape to phosphatidylcholines (Fig. 11–6) but are of quite different chemical structure, with the polar backbone and one of the hydrocarbon tails contributed by a dihydroxy amine, sphingosine. Sphingosine, therefore, replaces glycerol and one of the fatty acids seen in the phosphatidyl lipids. The other hydrocarbon tail comes from a fatty acid bound by amide linkage to the amino group of sphingosine.

The sphingomyelins contain a high proportion of saturated fatty acid residues, some with unusually long chains. They are the only phospholipids known to contain 24:0 and 24:1 fatty acids (see Fig. 11–3).

Surfactants in the Lung. The detergent action of phospholipids is absolutely necessary for the normal function of the lungs. The lungs are a collection of huge numbers of small bubbles, the alveoli, generating a total of 80 to 100 m^2 of surface area in an adult. (Compare the typical 1.5–2.0 m^2 area of an adult's skin.) The alveolar walls are not inherently strong enough to maintain their shape against the surface tension of water, and the surface tension is reduced by secreting an unusual phosphatidylcholine containing two 16:0 chains, together with lesser amounts of sphingomyelins, into the lung chamber. The phospholipid surfactant enables the lung to assume its functional shape.

Since a fetus is bathed in amniotic fluid and obtains oxygen through the maternal circulation, it has no need for the surfactant until exposure to air at delivery. It forms little of the dipalmitoyl phosphatidylcholine before the 30th week of gestation, and analysis of the amniotic fluid shows a higher concentration of sphingomyelin than of the phosphatidylcholine. Synthesis of the phosphatidylcholine increases rapidly thereafter so that its concentration is twice that (or more) of the sphingomyelins by the 35th week, and the lungs are ready for emergence from the aqueous environment.

Infants that are born before the 35th week may not have sufficiently active surfactant in their lungs for normal respiratory function, and consequently develop the acute respiratory distress syndrome, with rapid, shallow breathing and cyanosis (bluish tinge to the skin). This is the leading cause of morbidity and a major cause of death in premature babies.

A simple test for susceptibility to the respiratory distress syndrome can be made by shaking samples of the amniotic fluid in test tubes with 1, 1.3, and 2 volumes of ethanol, respectively, and examining the tubes for the persistence of bubbles after several minutes. Dipalmitoyl phosphatidylcholine makes a very stable foam under

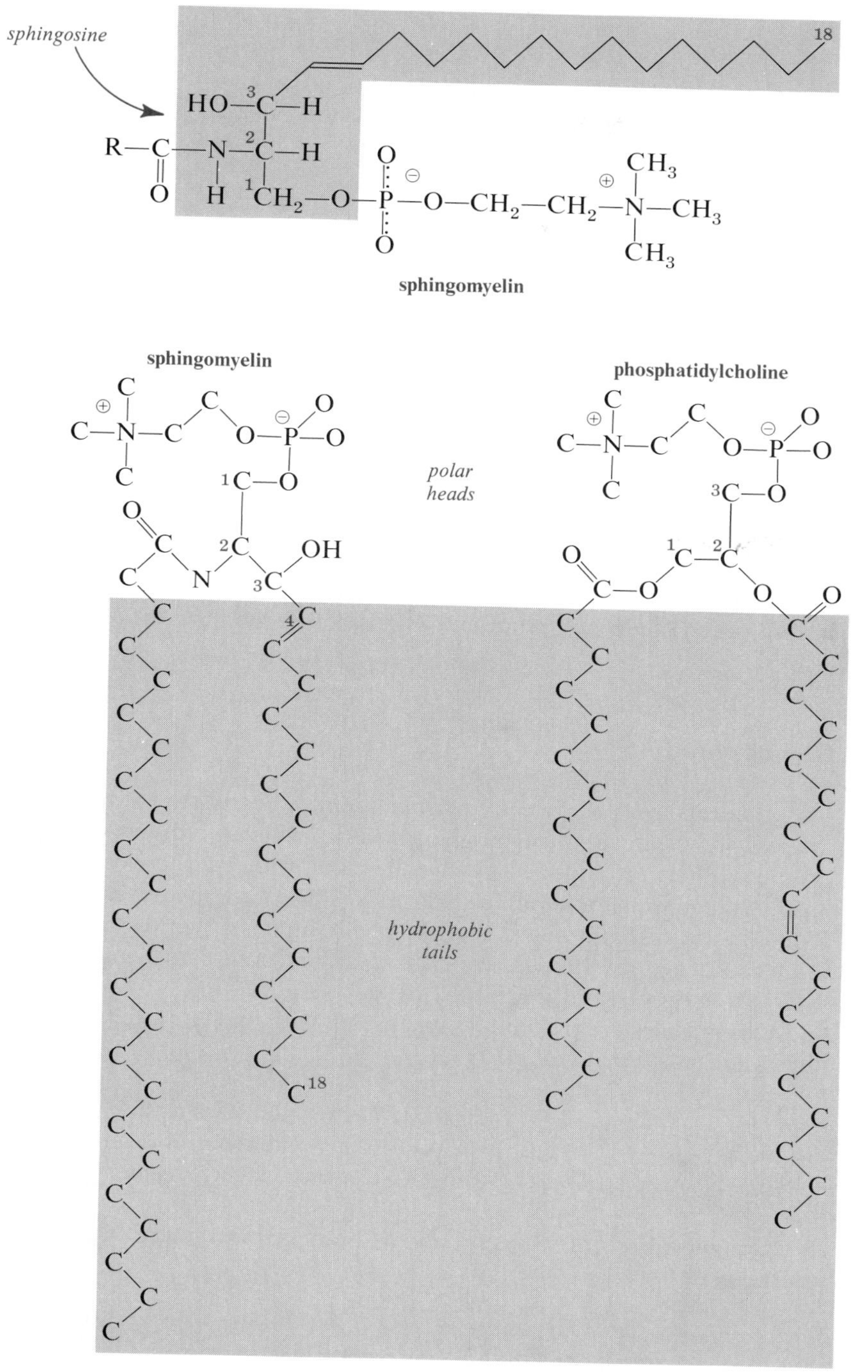

sphingosine

sphingomyelin

sphingomyelin

phosphatidylcholine

polar heads

hydrophobic tails

FIGURE 11-6

Sphingomyelins are phospholipids containing sphingosine, a long-chain amino alcohol (*top, brown shading*), with a fatty acid attached to the amino group in amide linkage. The 1-hydroxyl group is attached to phosphate and choline by ester linkage. The conventional formula does not accurately convey the actual geometry of either sphingomyelin or of phosphatidylcholine, which closely resemble each other. A projection of the location of some of the atoms in atomic models is indicated in the bottom drawing. Sphingomyelin frequently contains a much longer (24-carbon) fatty acid than is common in the phosphatidyl compounds.

ceramide (Cer)

glycosyl groups

glycolipid
(glycosyl ceramide)

FIGURE 11–7 **Glycolipids contain carbohydrate chains attached to a hydroxyl group of a ceramide (Cer).**

these conditions, whereas those containing unsaturated fatty acids and the sphingomyelins do not. Fluid bathing mature lungs will generate persistent bubbles even with the highest concentration of alcohol; the presence of bubbles at only the lowest concentration of alcohol indicates a high risk that the infant will develop respiratory distress.

Glycolipids

The external surface of plasma membranes is studded with carbohydrate groups. Some are supplied by glycoproteins, which we discussed in the preceding chapter. Others are supplied by glycolipids, in which carbohydrates are attached to an **acylsphingosine,** or **ceramide** (Fig. 11–7). The carbohydrates are similar to those found in the glycoproteins, ranging from monosaccharide residues to complex branched structures. Indeed, the carbohydrate chains conveying blood group specificity are mainly found in glycolipids.

Cerebrosides are the simplest examples, with one or more monosaccharide residues attached to ceramide (Cer):

$$\beta\text{-Glc-Cer} = \text{glucocerebroside},$$
$$\beta\text{-Gal-Cer} = \text{galactocerebroside},$$
$$\beta\text{-Gal-}(1\rightarrow 4)\text{-}\beta\text{-Glc-Cer} = \text{lactocerebroside}.$$

Gangliosides are variants of a branched structure containing N-acetylneuraminate residues:

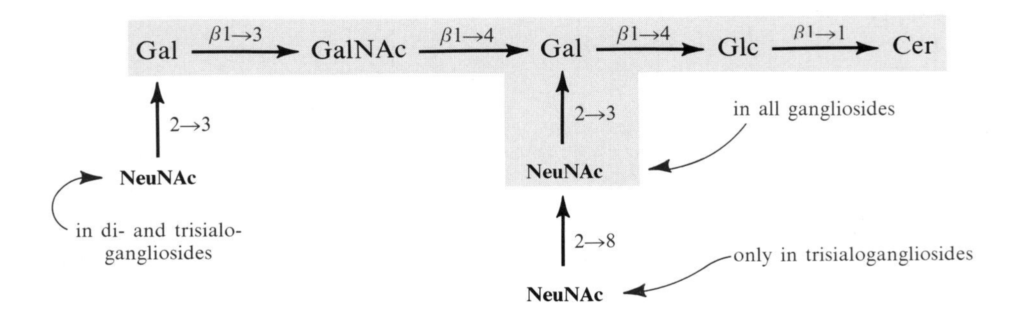

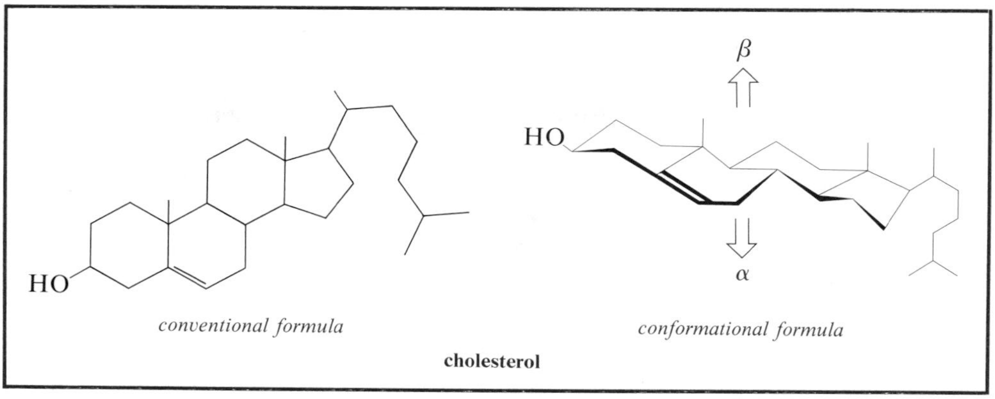

HO
conventional formula
HO
β
α
conformational formula
cholesterol

FIGURE 11–8 Cholesterol is a steroid containing one hydroxyl group and a branched aliphatic side chain. The puckers of the nearly planar polycyclic steroid ring system are better appreciated in a conformational formula (*right*). Groups oriented toward the top as drawn are said to be in the β-configuration; those toward the bottom are in the α-configuration.

Proteolipids

A few membrane proteins are soluble in organic solvents and contain covalently bound fatty acid residues, probably attached in ester linkage to serine or threonine hydroxyl groups. The functional significance is unknown.

Cholesterol

Animal membranes contain a steroid alcohol, cholesterol (Fig. 11–8). We shall have much more to say about the steroids because they include many hormones, but cholesterol is the progenitor of them all in the human body, as well as serving as an important structural component. The critical feature of cholesterol from a structural standpoint is the relatively rigid steroid ring, a nearly planar structure with the polar hydroxyl group at one end, two methyl groups on one side, and a freely movable hydrocarbon tail at the other end.

MEMBRANES

The essential ingredients of membranes are lipids (primarily phospholipids) and proteins. Lipids generate a hydrophobic core with polar faces, and proteins carry out biological functions associated with the membrane, such as catalysis of reactions, or selective transport of compounds from one compartment to another. **Membranes are asymmetric;** the two sides are different. Plasma membranes, which are the cell boundaries, have the carbohydrate chains of glycoproteins and glycolipids exposed on the exterior surface. Some proteins are firmly embedded in one surface; others pass through so as to be exposed on both surfaces; still others are less tightly bound to the surfaces. The membrane may also be attached to intracellular structural proteins such as contractile filaments and microtubules.

The nature of membranes differs widely from organelle to organelle within a cell, from tissue to tissue within an organism, and from organism to organism. All membranes have the hydrophobic core and polar surface with embedded proteins, but the nature of the core, the proteins, and the surface is not the same.

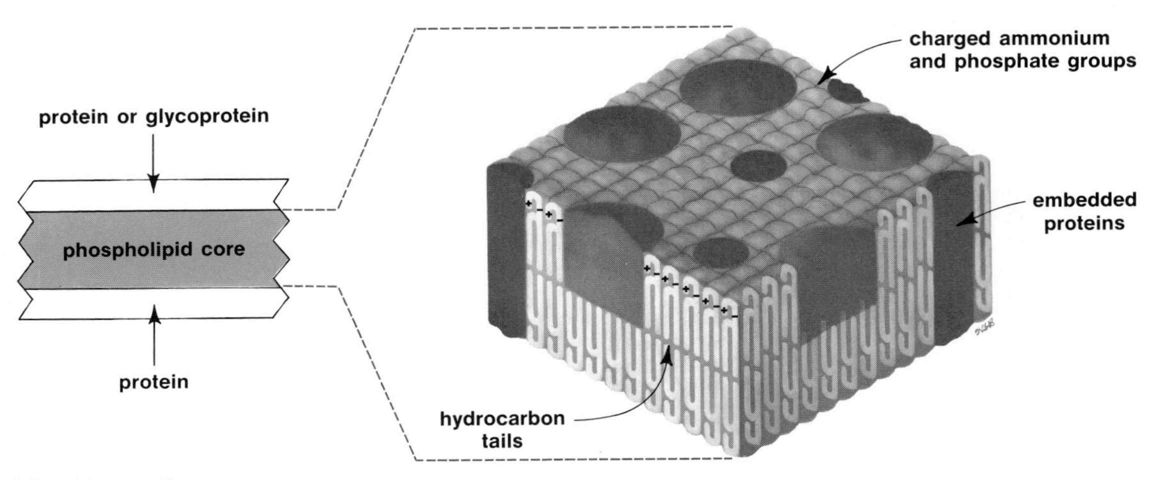

FIGURE 11–9

Generalized structure of membranes, with a phospholipid core and polar faces. Some of the embedded protein molecules are exposed on both sides of the core, others only on one side. The drawing is highly schematic; the actual shapes of the proteins are not known.

The Lipid Core

The core structure of membranes is a double layer of phospholipid molecules, each molecule oriented with its hydrocarbon tails toward the central plane and its polar head toward the surface (Fig. 11–9). Large areas of such double layers cannot survive even small motions in water; they break up into patches, which close upon themselves to create water-filled balloons within the solution. Homogenization of phospholipids in dilute salt solutions creates vesicles, most of them of a uniform microscopic size that can be isolated as a clear suspension—these are the prototypes of membrane-bound biological structures.

Such dispersed preparations were first created by A.D. Bangham in England. His vesicles, known in some circles as bangosomes, are walled by several bilayers of lipid like a hollowed-out onion, resembling the myelin sheath of nerves. Ching-hsien Huang discovered how to make vesicles with only one enclosing bilayer, and these huangosomes are widely used in model studies of membrane phenomena. They are also under intensive study as possible carriers for drugs that otherwise might be destroyed before reaching their target organs or are unable to penetrate cells. The generic term for such experimental particles is **liposome.**

Composition. There is no pat recipe for making the lipid core of a membrane. The core must have a certain degree of structural rigidity but be fluid enough to permit lateral movement of both the constituent lipids and those embedded proteins that are not fixed to some components outside of the membrane. (By contrast, exchange of lipid components from one side of the membrane to the other is very slow.) Fluidity is maintained in animal membranes by the incorporation of phospholipids containing 18:1 fatty acid residues, along with some more unsaturated residues. Most of the lipid core is probably in a liquid crystalline phase, with the polar head groups still associated in a relatively ordered array, but with the fatty acid chains moving about in a more random way. However, there may be patches of more stable gel-like phases embedded in the bulk lipid, with their fatty acid groups in more restrictive parallel alignment. The lipid core is certainly not homogeneous.

The core must also conform to, and perhaps help determine, the curvature of

TABLE 11-4

THE STRUCTURAL LIPIDS

Phospholipids	
phosphatidates	glycerol; saturated fatty acid on C-1, unsaturated fatty acid on C-2, phosphate on C-3
phosphatidylcholine, phosphatidylethanolamine, phosphatidylserine, phosphatidylinositol, di- or monophosphatidyl glycerol	choline, ethanolamine, serine, inositol, or glycerol attached in ester linkage to phosphate of phosphatidate
lysophosphatidyl compounds	phosphatidyl compounds lacking the fatty acid group on C-2
plasmalogens, or phosphatidal compounds	same as phosphatidyl compounds except fatty aldehyde bound to C-1 of glycerol by vinyl ether linkage
alkyl phospholipids	same as phosphatidyl compounds except fatty alcohol bound to C-1 of glycerol by ether linkage
sphingomyelin	sphingosine; saturated fatty acid on N atom; phosphate-choline on C-1
Glycolipids	
ceramide	sphingosine; saturated fatty acid on N atom; same as sphingomyelin without phosphate-choline
cerebrosides	mono- or oligosaccharide; ceramide
gangliosides	N-acetylneuraminate branch on oligosaccharide; ceramide
fucose-containing	resemble blood-group glycoprotein oligosaccharide; ceramide
Cholesterol	steroid ring with one hydroxyl group, two methyl group branches, and hydrocarbon tail

the membrane surface, which may vary from the relatively broad sweep of a large cell envelope to the tight reversals of direction found in the inner membrane of mitochondria. The particular requirements may be met by alterations in the relative proportions of different lipid components (Fig. 11–10) or by alterations in the fatty acid composition of particular kinds of lipids.

Some trends can be noted in the composition of membranes in mammalian cells. Glycolipids are found only on the plasma membrane. The content of sphingomyelin tends to decrease as one goes from the plasma membrane through the interior of the cell to the nucleus, as does the content of cholesterol. The content of diphosphatidylglycerol (cardiolipin) tends to increase in internal membranes; the mitochondrial inner membrane is especially rich in the compounds.*

Functions of the Lipids. We have clues as to the functions of some major components in mammalian plasma membranes, especially the red blood cell membrane, but we have to confess total ignorance of the roles of the tantalizing minor

---

*This similarity in mitochondrial and bacterial membranes bolsters the notion that mitochondria evolved from symbiotic bacteria resident in early eukaryotic cells. The idea was originally advanced to explain the resemblances of mitochondrial and bacterial ribosomes in size and sensitivity to antibiotics. It has now lost favor but is by no means dead.

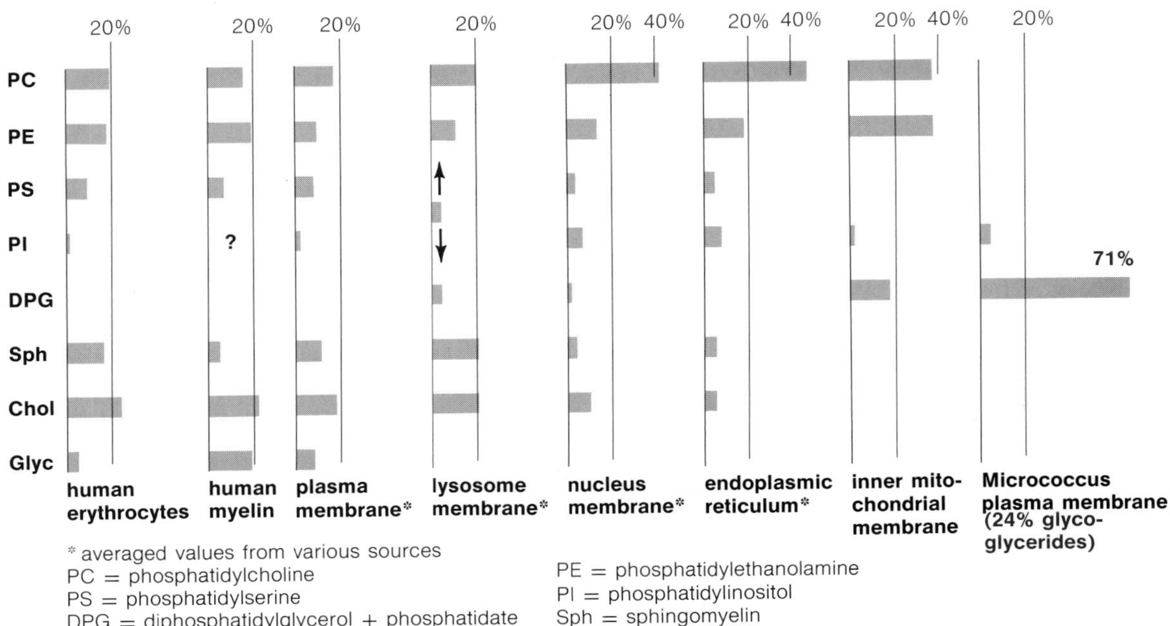

FIGURE 11–10

The proportions of different phospholipids differ from one membrane to another. The inner structures of a cell tend to have membranes with a higher content of diphosphatidylglycerol and a lower content of sphingomyelin than does the plasma membrane.

ingredients—the plasmalogens, lysophosphatidates, and so on, which are not likely to be present because of mere whim.

The relative proportions of phosphatidyl amines and sphingomyelin appear to depend upon the radius of curvature of the membrane. Phosphatidylcholine and sphingomyelin have larger head groups than do phosphatidylethanolamine and phosphatidylserine and therefore tend to occur on the outer surface of the lipid core where there is more area per residue (Fig. 11–11).

Cholesterol appears to fit on end between the hydrocarbon chains of the core, with its hydroxyl group protruding into the polar surface region, where it can form hydrogen bonds with water or the polar head groups (Fig. 11–12). Cholesterol creates a state intermediate between the liquid crystal and gel phases in the surrounding lipids, perhaps by close fit between the steroid and particular lipid chains.

The lipids also interact with the hydrophobic surfaces of proteins in the membranes. Some kinds of lipids associate preferentially with the proteins, creating a

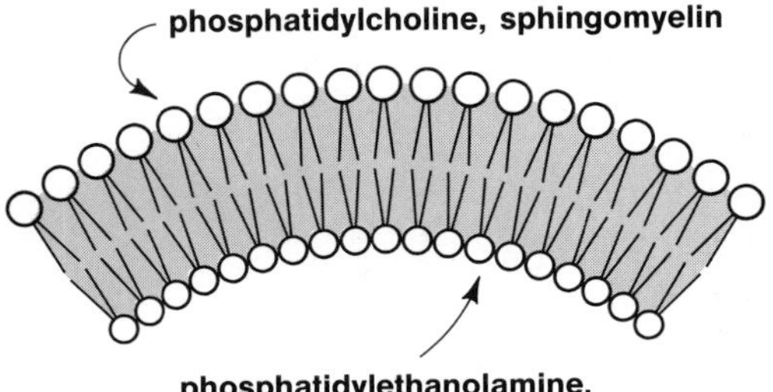

FIGURE 11–11

Since there is more area in the outer surface of a curved membrane than in the inner surface, phospholipid molecules with large head groups will tend to appear preferentially in the outer surface; these include phosphatidylcholine and sphingomyelin.

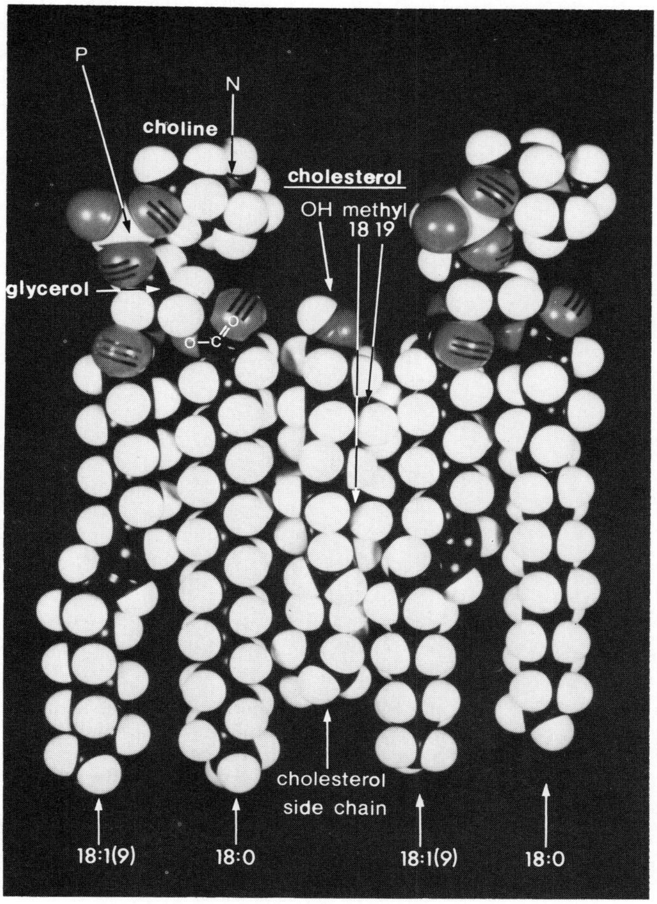

FIGURE 11–12

A molecular model showing how cholesterol might fit between two molecules of phosphatidylcholine. The top half of the cholesterol model is the ring system viewed edge-on; the bottom half is the hydrocarbon side chain. The polar hydroxyl group of cholesterol is at the nonpolar-polar interface. The model demonstrates how the two methyl groups on the right side of the steroid ring might be accommodated in a snug fit by the kink in the unsaturated fatty acid chain of phosphatidylcholine. (Photograph supplied by Prof. Ching-hsien Huang.)

region of different lipid composition in their neighborhood. This region may be less fluid than the bulk lipid phase, aiding in the sealing of the membrane around the protein. Indeed, there is experimental evidence that association with particular lipids is necessary for the biological activity of some membrane-bound proteins.

Anesthetics and the Lipid Core. Loss of consciousness or of local sensation can result from any kind of disturbance of neural transmission; if the disturbance disappears when the cause is removed and there is no permanent damage to the tissues, the causative agent may find application as an anesthetic, either general or local. Most anesthetics are believed to function through effects on membranes.

The common characteristic of a wide variety of anesthetic agents is a preferential solubility in lipids. Indeed, the solubility in olive oil is a direct indicator of the anesthetic potency for many compounds acting as general anesthetics, including several in clinical use (Fig. 11–13).* The chemical nature of these compounds is so diverse that it compels one to conclude that anesthetic action and lipid solubility both hinge on the same property. Attention first focused on the formation of simple solutions in the lipid core, perhaps altering the volume, which would explain why

*An important factor in selecting anesthetics is inflammability. To avoid blowing up all hands in the operating room, anesthesiologists welcomed the introduction of halogenated hydrocarbons, but these also have their perils in the form of potential liver damage upon repeated exposure.

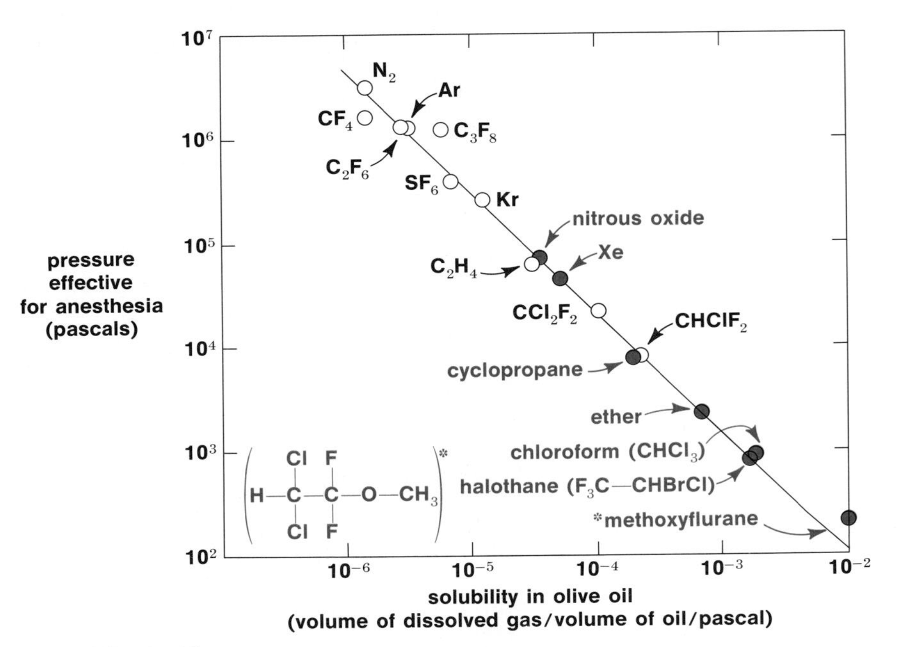

FIGURE 11–13

The effectiveness of general anesthetics administered in the vapor phase is linearly proportional to their solubility in olive oil, which is used as an analogue of the lipid core of membranes. The effectiveness is indicated here as the pressure of gas necessary to achieve a certain level of anesthesia in 50 per cent of the mice used to test the agent. 10^5 pascals = 750 torr. (Modified from E.B. Smith *in* M.J. Halsey, R.A. Millar, and J.A. Sutton, (1974) Molecular Mechanisms in General Anesthesia. Churchill Livingston.)

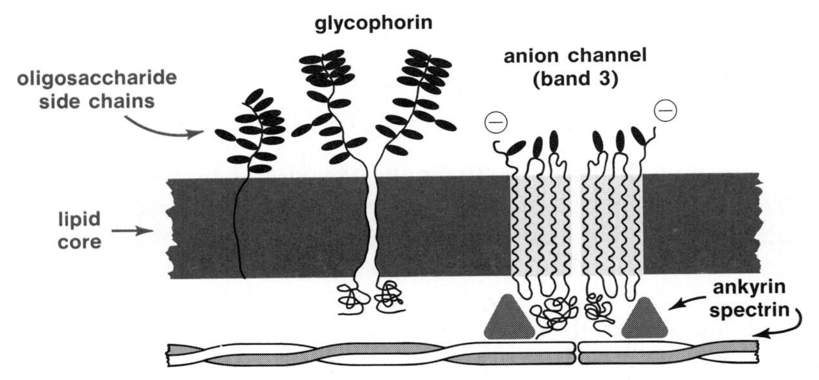

FIGURE 11–14

Highly schematic view of the erythrocyte plasma membrane. The carbohydrate residues of glycoproteins are on the outside. Some proteins terminate in the lipid core (*left*); others, such as the monomers of glycophorin, extend through the core (*center*). The monomers of the anion channel protein are much longer and may pass through the core as many as five times, probably in α-helixes (*right*). The helixes may form a cylindrical channel, shown opened out here. The inside of the membrane is lined with a network of spectrin, which is linked to the anion channel protein by another protein, ankyrin.

increasing the atmospheric pressure reverses anesthesia. This notion is not sustained by further studies, but there are many other possibilities for derangement hinging upon simple hydrophobicity, such as disturbance of protein-lipid interfaces.

Membrane Proteins

Each type of membrane has specific biological functions due to the presence of particular proteins; membranes differ in function because they contain different proteins. Despite these differences, there seems to be a general kind of structure in that some proteins are embedded in one or the other face of the lipid core while others pass through the core and are exposed on both sides of the membrane to aqueous phases (Fig. 11–14). Still other proteins may be more loosely bound to one face, or both, depending upon the membrane. Figure 11–14 sketches many of these features as they occur in the erythrocyte membrane, which is more rigid than the usual plasma membrane. A stabilizing network of a protein, **spectrin,** enables it to survive longer in its passage through the circulation. A hereditary deficiency in the synthesis of one of the two polypeptide chains in spectrin has been shown to cause **spherocytosis,** a defect in which the normally disk-shaped erythrocytes look more like bags (Fig. 11–15).

Little is known about the structure of most membrane proteins, but reasonable assumptions about their nature are supported by some examples at hand. It is likely that the proteins are associated with the lipid core because they have hydrophobic surfaces that cannot otherwise be shielded from water. A collection of hydrophobic groups at one end of a molecule would cause that end to become buried in the lipid core, whereas a hydrophobic middle section of sufficient length would cause a molecule to span the core. This is a simple extension of the principles that govern conformation of proteins in aqueous solution; the exclusion of water from hydrophobic groups is achieved in some proteins by placing the groups in a bulk lipid phase instead of wrapping them into the center of the protein.

The meager information that is at hand on the nature of membrane proteins

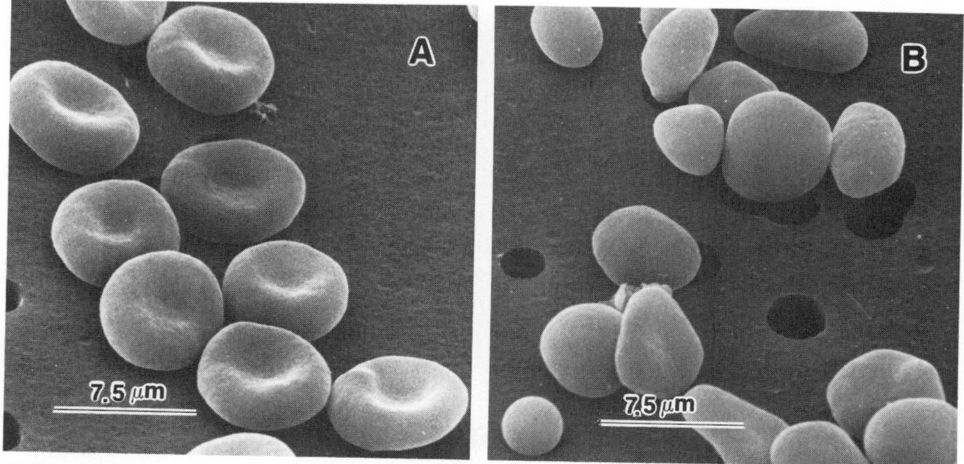

FIGURE 11–15 Scanning electron micrograph of red blood cells. (*A*) Cells from a woman heterozygous for a defect in spectrin synthesis. (*B*) Cells from her homozygous daughter with hereditary spherocytosis. (From P. Agre, E.P. Orringer, and V. Bennett (1982) N. Engl. J. Med. 306:1155. Reproduced by permission.)

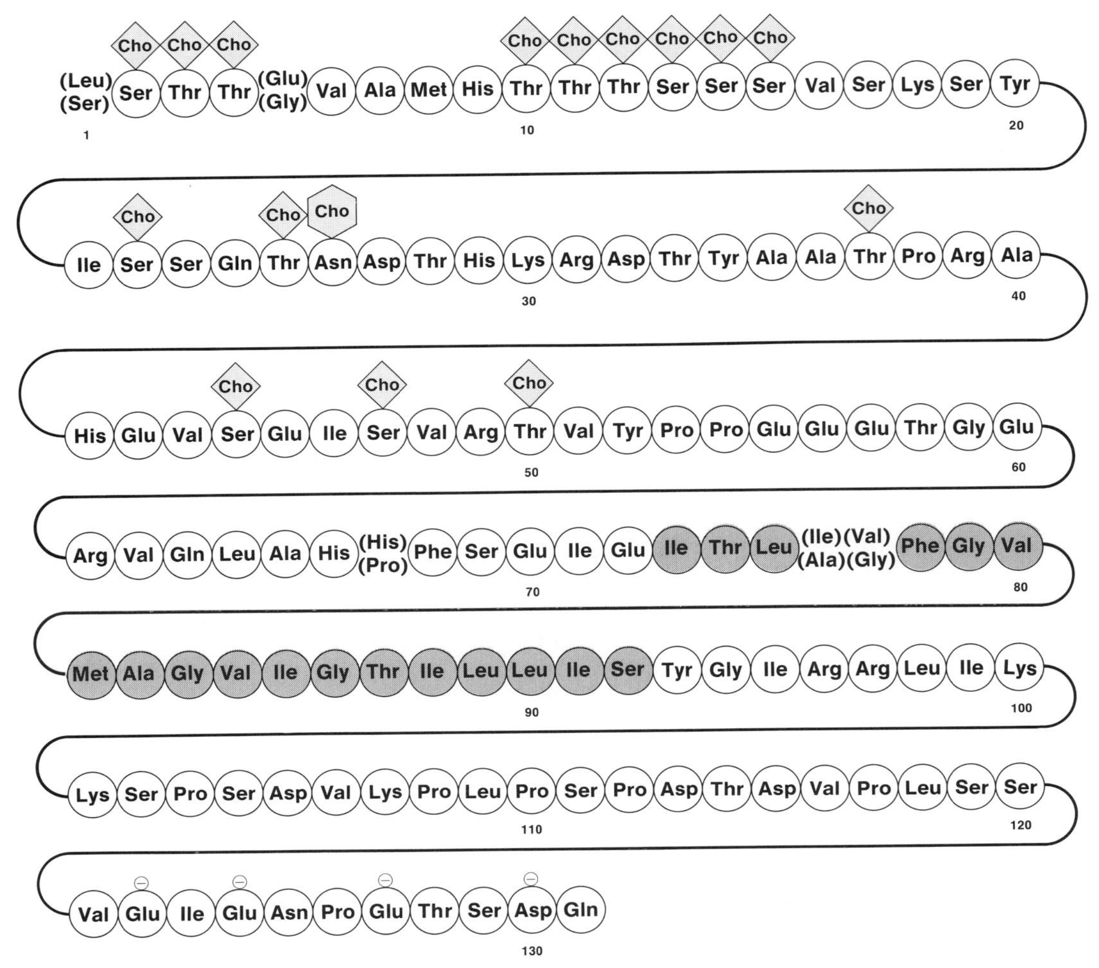

FIGURE 11–16

The primary structure of glycophorin A, a protein that crosses the red blood cell membrane, includes a sequence of hydrophobic amino acid residues that is long enough to span the lipid core. The N-terminal segment appearing on the outside of the cell has many attached hydrophilic carbohydrate groups, while the C-terminal segment on the inside is rich in negatively-charged residues that can bind calcium ions within the cell. (Modified from a drawing in V.T. Marchesi, H. Furthmayr, and M. Tomita (1976) The Red Cell Membrane, Annu. Rev. Biochem. 45:667. Reproduced by permission.)

supports the predictions on structure. Two proteins involved in electron transfers in the endoplasmic reticulum—cytochrome b_5 and cytochrome b_5 reductase—have polypeptide chains in which one end takes care of the business of the protein and the other end is loaded with hydrophobic amino acid residues to be buried in the lipid core. The proteins are catalytically active if the hydrophobic tail is experimentally removed, but they cannot attach to membranes. The tail serves as a binder for the active end.

The amino acid sequence is known for the protein glycophorin, which spans the red blood cell membrane (Fig. 11–16). Here the hydrophobic residues are concentrated in the center of the peptide chain; the N-terminal portion has many attached carbohydrate groups, and the C-terminal portion is rich in negatively charged groups.

The membrane has a channel for passage of anions that is made from a much larger protein, probably occurring as a dimer. The complete amino acid sequence is not known, but it appears that the segments passing through the membrane have primary structures favoring the formation of helixes such that they may associate, perhaps to make a barrel with a hydrophobic surface and polar interior. (This is an inversion of the distribution of polar groups in proteins destined for solution in water.)

The contributions of proteins to the structural character of membranes is yet to be determined. Natural membranes with the proteins in place are stronger and can have larger radii of curvature than single phospholipid bilayers. The contribution of the proteins to function will be touched upon repeatedly in our later discussions.

Membrane Receptors

Many of the proteins and carbohydrates exposed on the surface of membranes are there to form bonds with compounds from the environment. This is especially evident with the external surface of the plasma membrane. For example, cell-to-cell adhesion appears to involve the binding of specific ganglioside structures on one cell by proteins on an adjacent cell.

This kind of binding is often used at receptor sites for identification of molecules to be taken across the plasma membrane into the cell. For example, many extracellular proteins are studded with polysaccharide chains ending in sialyl groups; if the chain loses the sialyl groups, the underlying carbohydrate residues are recognized by receptors on liver cells, which bind the damaged protein and remove it from the circulation.

Specific carbohydrate-protein binding is not always beneficial. Some toxins, for

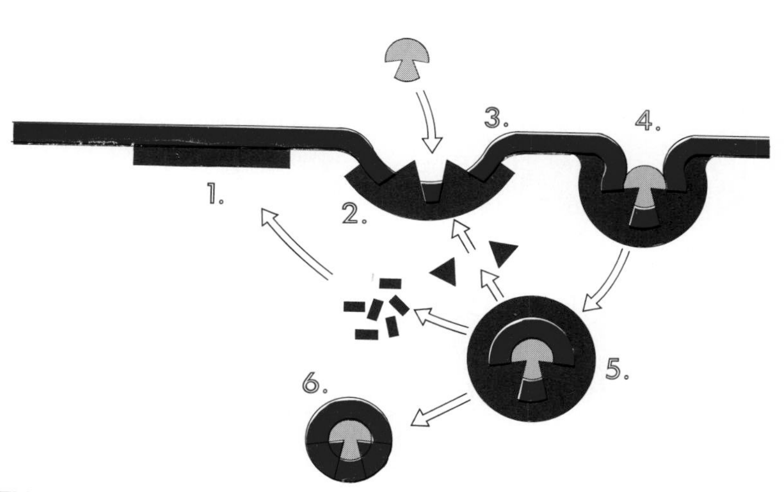

FIGURE 11–17

Endocytosis. *1.* Molecules of clathrin form a circular plaque made of a lattice of hexagons (not shown face on). *2.* The clathrin lattice may trap specific receptor molecules, shown as pyramids; some pentagons form between the lattice hexagons making the clathrin and attached membrane dimple, forming a coated pit. *3.* A molecule from the environment contacts the receptors and becomes bound. *4.* The coated pit puckers around the bound molecule. *5.* A coated vesicle is formed. *6.* The coated vesicle rapidly loses its clathrin and receptor coats, which are incorporated into new coated pits (not shown). The vesicle discharges its contents, probably in lysosomes.

example **cholera toxin** and **ricin** (found in castor beans), contain a domain specific for a plasma membrane ganglioside. The toxin is bound to the ganglioside on intestinal mucosal cells, which then absorb it with a regrettable result.

Coated Pits. Small molecules can cross membranes through specific channels; we shall examine examples later. Larger molecules and aggregated particles must be engulfed by the process of **endocytosis,** in which a portion of the plasma membrane forms a pouch around the molecule, which is then pinched off at the neck to create an internal membrane vesicle (Fig. 11–17).

The receptors at which molecules are bound for transport by endocytosis are indentations coated on the inside with a particular protein, **clathrin** (*clathri* = lattice), which forms a cage around the engulfed molecule or particle. After absorption, the vesicle loses the clathrin coat and fuses with other vesicles. The contents are discharged, usually to lysosomes in the cell. Meanwhile, the clathrin is recycled to the plasma membrane along with the specific receptor molecules from the emptied vesicles to reform the coated pits.

Lectins are proteins that also bind to specific carbohydrate groups but differ from receptor proteins in that they are multivalent; that is, one lectin molecule can bind several carbohydrate molecules. The first known examples were recognized at the beginning of the century when it was found that extracts of some seeds caused red blood cells to agglutinate. It is now known that the agglutinating proteins, called lectins, specifically bind carbohydrate residues occurring on the plasma membrane. For example, the lectin from castor beans—the first recognized—binds galactose or N-acetylgalactosamine residues, while **concanavalin A,** a lectin from the jack bean (*Canavalia ensiformis*), binds terminal mannosyl, glucosyl, or N-acetylglucosaminyl groups in the α-configuration. The specificity of the lectins makes them valuable tools for identifying attached carbohydrate groups.

FURTHER READING

S-I. Hakomori (1981). Glycosphingolipids in Cellular Interaction, Differentiation, and Oncogenesis. Annu. Rev. Biochem. 50:733.

B.M.F. Pearse and M.S. Bretscher (1981). Membrane Recycling by Coated Vesicles. Annu. Rev. Biochem. 50:85.

S.H. Barondes (1981). Lectins: Their Multiple Endogenous Cellular Functions. Annu. Rev. Biochem. 50:207.

I.E. Liener (1976). Phytohemagglutinins (Phytolectins). Annu. Rev. Plant Physiol. 27:291.

P.M. Farrell and M.E. Avery (1979). Hyaline Membrane Disease. Am. Rev. Resp. Dis. 111:659.

B.R. Fink, ed. (1980). Molecular Mechanisms of Anesthesia. Raven.

S.E. Lux (1979). Dissecting the Red Cell Membrane Skeleton. Nature 281:426.

J. Holmgren (1981). Actions of Cholera Toxin and the Prevention and Management of Cholera. Nature 292:413.

D.M. Branton, C.M. Cohen, and J. Tyler (1981). Interaction of Cytoskeletal Proteins on the Human Erythrocyte Membrane. Cell 24:24.

B.E. Ryman and D.A. Tyrell (1980). Liposomes—Bags of Potential. Essays Biochem. 16:49.

P. Agre, E.P. Orringer, and V. Bennett (1982). Deficient Red-Cell Spectrin in Severe, Recessively Inherited Spherocytosis. N. Engl. J. Med. 306:1155.

V. Bennett, J. Davis, and W.E. Fowler (1982). Brain Spectrin, a Membrane-Associated Protein Related in Structure and Function to Erythrocyte Spectrin. Nature 299:126.

A.N. Martonosi (1982). Membranes and Transport. Plenum. A collection of short reviews.

CHAPTER 12

ANTIBODIES: DEFENSIVE PROTEINS

The world swarms with alien species, many of predatory bent. The vertebrates have markedly improved their chances for survival through the development of defenses against actual invasion of the flesh. Before a microorganism can gain entrance, it must pass sturdy mechanical barriers, such as the skin, the filters in the nose, and the constantly sweeping cilia, and survive exposure to destructive chemical agents, like H^+ in the stomach, fatty acids in the skin, and the lysozyme in tears. (Lysozyme is an enzyme that catalyzes hydrolysis of some bacterial cell walls.)

An invader that circumvents the primary barriers mobilizes a second line of defense in the immune system, attracting specialized cells and substances to the site of the breach. The usual result is quiet and efficient disposal of the alien organisms, unless the invasion is massive. In addition to acting as a linebacker for the primary defensive screens, the immune system also may eliminate any aberrant cells that develop within the body.

The Immune Reaction

The initial pattern of the immune response is largely stereotyped and nonspecific; such diverse materials as splinters, foreign serum proteins, and bacterial toxins, in addition to intact microorganisms, will cause the same general effects. A visible inflammation may mark the site of substantial encounters between the obtruding substance and phagocytic cells, lymphoid cells, and the specialized proteins known as antibodies with which this chapter is mainly concerned.

Antigens and Antibodies. The nonspecific natural immune system is backed up by a remarkably specific response in which resistance to a particular foreign substance develops after an initial exposure to it. A substance causing this kind of response is an antigen.

Two kinds of specific responses develop upon exposure to an antigen. In one, specific protein reagents, the antibodies, are synthesized and released from lymphocytes and plasma cells. Antibodies are proteins constructed to combine with antigens; their structure and properties can be determined in detail by the same methods used with other proteins.

The second kind of specific immune response involves a direct mediation of immune function by cells, predominantly lymphocytes. This system of cellular immunity is less accessible to detailed analysis and is beyond the scope of our discussion from a chemical point of view, but it ought to be remembered that immunity results from the harmonious interplay of all of the immune systems, specific and nonspecific.

Our attention will be directed toward the structure, reactions, and synthesis of the antibodies, those diverse proteins that are selected for synthesis in response to the intrusion of an antigen. The ability to make specific immune responses is essential for survival. Paradoxically, it can at times be detrimental not only to recipients of a transplanted kidney, heart, or liver but also to some unfortunates in whom immune reactions destroy their own tissues. Specificity of response means that a particular antigen causes an individual to synthesize only a few kinds of antibody molecules from his entire repertory, but these few are capable of interacting with the antigen.

ANTIGENS

There is no common chemical feature that defines an antigen. Proteins, polysaccharides, nucleic acids, lipids, synthetic polymers—all include antigens. Antigens may be as small as 1,000 daltons or as large as millions of daltons. The structural components of one individual normally do not act as antigens within him, but will provoke an immune response in nearly anyone else except an identical twin.

Antigenic Determinants

Although a molecule must have a certain size to act as an antigen and cause the formation of antibodies, each antibody can bind only to a relatively small exposed portion of the surface of the antigen; this reactive portion is said to be the antigenic determinant for the antibody. That is, an antibody is constructed to fit a particular exposed arrangement of chemical groups in the antigen, and it will not fit other regions of the antigen's surface. One antigen molecule often contains several antigenic determinants because several areas are sufficiently exposed and reactive to cause the formation of antibodies. The result is that there may be several kinds of antibodies reacting with one kind of antigen, each binding at a different site.

Determinants in Myoglobin. The nature of the determinants has been studied in myoglobin from sperm whales. (It is abundant in whale muscles, hence its isolation from this source.) The structure of myoglobin and the sequence of the 153 amino acid residues in its single polypeptide chain are known.

Rabbits and goats can be immunized to sperm whale myoglobin, meaning that the protein can be injected in a form such that it will act as an antigen in either species and provoke the formation of antibodies. Study of the interaction of the

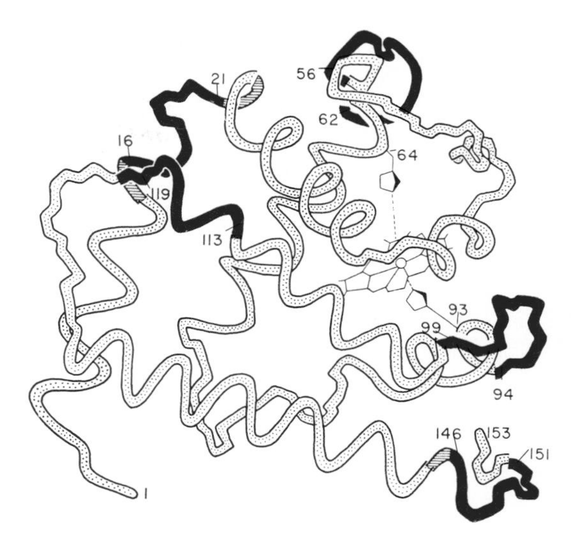

FIGURE 12–1

Antigenic determinants in myoglobin. Those portions of the polypeptide backbone bearing determinant side chains are shown in brown; those in striped brown are sometimes also involved. (Modified from M.Z. Atassi (1975) Immunochemistry 12:423. Copyright by Pergamon Press. Reproduced by permission.)

resultant antibodies and myoglobin showed that either species produced a mixture of antibodies interacting with the same five distinct regions on the myoglobin molecule, which are shown in Figure 12–1. Each of these regions consisted of no more than six or seven amino acid residues. However, each of the regions in itself behaved as several antigenic determinants. Some antibodies interacted only with four particular residues in the region, others with more, but in no case exceeding the seven-residue limit. Estimates of the maximum determinant size in various antigens range from $0.6 \times 0.6 \times 1.5$ to $0.7 \times 1.2 \times 3.4$ nanometers.

A picture develops: Given an exposed area with several amino acid side chains, one antibody may be built to fit a certain combination of these groups, while another antibody is built to fit a different combination. It is the combination of groups that is the antigenic determinant: one group on the antigen may participate in several determinant combinations, each best fitting a different antibody. Put another way, different antibodies may fit different areas on an antigen, and some of the areas overlap.

A substance must have regions of stable geometry if it is to behave as an antigen. Only a few groups are needed to act as a determinant, but the groups must remain in nearly the same relative positions if they are to provoke the formation of a closely fitting antibody. Small molecules are not effective antigens. However, once an antibody has been generated by a molecule sufficiently large to be an antigen, a small molecule bearing the same groups occurring in the determinant region may also interact with the antibody; it need assume the right configuration only transiently in order to be bound. Such small molecules are said to be **haptens** (haptein = to fasten) and have been useful reagents for studying antigen-antibody interactions.

For example, if a protein substituted with dinitrophenyl groups is used as an antigen, some of the resultant antibodies will bind to determinants including the dinitrophenyl groups. These antibodies may also bind free dinitrophenol, even though the free phenol will not act as an antigen by itself. Dinitrophenyl compounds are therefore haptens for these antibodies, and the nature of the hapten-antibody interaction can be studied without the distracting influence of the remainder of the original large antigen molecule.

Vaccines are devices for preventing serious infectious diseases by bringing the immune system to a primed state of readiness.* They have some of the antigenic characteristics of a pathogenic organism or toxin but lack the deleterious biological activity. Some are made by disabling the pathogen, for example, by heat or chemical reactions; others contain less harmful relatives of the pathogen. In either case the objective is to have an antigen that is mildly harmful at the worst but still contains determinants present on the pathogen so that some of the antibodies raised to it will bind to the pathogen.

Specificity of Antibodies

It is not known how many antibodies a mature individual can make; educated guesses range from millions to billions, each differing in the antigen determinant with which it combines. The discrimination of antibodies is often exquisite. Proteins of like function in closely related species usually differ in only a few amino acid residues, but these differences may cause sufficient change in the surface conformation so that antibodies reacting with one have markedly different affinity for the other.

Use of Antibodies for Assay. The discriminatory power of antibodies makes them valuable analytical reagents, used to detect small molecules as well as specific proteins. For example, it is sometimes desirable to determine the blood concentrations of digoxin or digitoxin (Fig. 12–2), drugs commonly used in the treatment of congestive heart failure. The toxic effects of these drugs mimic the symptoms and signs of heart failure so that it is difficult to know if the patient has too much or too little of the drug. However, the drug concentrations can be measured with specific antibodies, which are made by injecting rabbits with a protein to which digoxin or digitoxin has been covalently coupled so that the free drugs then act as haptens. The antibodies are so specific that they can discriminate between digoxin and digitoxin,

*Smallpox has apparently been eradicated by an intensive worldwide search for the remaining cases and immunization of potential contacts with vaccinia virus. (Vaccinia has an uncertain geneology; it may be a modified smallpox or cowpox virus.) The only known smallpox virus extant on the planet is in a few laboratories.

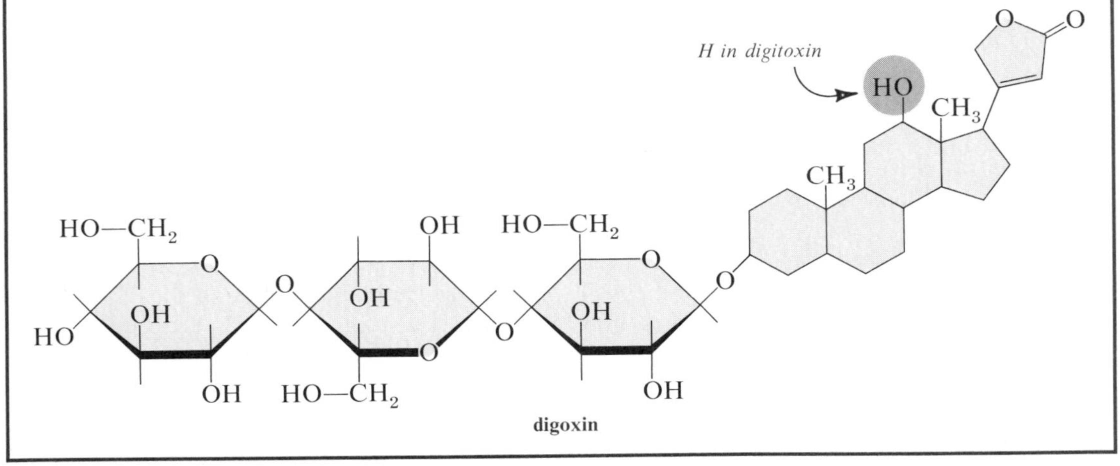

FIGURE 12–2 Structure of digoxin and digitoxin.

which differ only in the presence of a 12-hydroxyl group on the steroid ring. Antibodies prepared with one of the drugs will bind it at only one hundredth the concentration required for equal binding to the other drug.

The actual analysis involves **radioimmunoassay;** the central feature is competition for antibody binding between a radioactively labeled antigen and its unlabeled counterpart in the unknown sample. A complex of antibody with a radioactive drug is added to the unknown sample. The unlabeled drug in the sample displaces part of the radioactive drug from the complex; the measured amount of displaced radioactivity at equilibrium is an index of unlabeled drug concentration. Results from the assay are available in short enough time to make it a useful clinical procedure.

Monoclonal Antibodies

The heterogeneity of the immune response has been a major obstacle in the study of antibodies and their application. Not only is a single antigen likely to have many recognizable determinants, but the characteristics of the antibodies produced vary from individual to individual, and vary within one individual.

The problem of isolating a single antibody for study from the myriads present in a normal individual was solved by studying multiple myeloma. This disease results from a malignant proliferation of plasma cells, typically arising from a single cell, that synthesize and release an extraordinarily large amount of one antibody. Much of our knowledge of antibody structure and function was gained by studying these homogeneous proteins. We also learned an important principle: **one cell produces one antibody.** There are exceptions to be noted later, but normal antibody-forming cells in general behave like multiple myeloma cells in making only one antibody, and there is a separate clone of cells in the body for each antibody found.

MULTIPLE MYELOMA: A NEOPLASTIC PROLIFERATION OF PLASMA CELLS

Incidence—approximately 1 to 4 per 100,000
Sex—males and females equally
Usual age at onset—40 to 60
Clinical expression—
- **(a) bone pain, infection, anemia, elevated serum immunoglobulin with homogeneous peak on electrophoresis;**
- **(b) characteristic punched-out lesions in bone on x-ray;**
- **(c) plasmacytosis (bone marrow contains increased numbers of plasma cells or related cells);**
- **(d) urine may contain large amounts of light chains referred to as Bence Jones protein;**

Treatment—usually alkylating agents and prednisone (a synthetic hormone similar in action to cortisol).

Hybridomas. Techniques have been perfected in recent years for fusing cells and propagating the resultant hybrids in culture. Occasional successes occur even with cells from distantly related species. The application of the methodology to antibody-producing cells now makes it possible to produce a specific antibody to any potent antigen in useful quantities at high purity.

The hybrids are made from myeloma cells and normal spleen lymphocytes. The myelomas cells provide the ability to reproduce rapidly in culture and the spleen cells provide the ability to form antibodies. Some tricks are necessary to get useful results. One is to arrange it so that only the hybridoma cells—the combined myeloma and spleen cells—can grow. Since spleen cells won't grow in the culture medium, this can be done by starving the myeloma cells for purines.

Cells can be starved for purines by not supplying any preformed purines and preventing them from making their own. (Purine metabolism is discussed in detail in Chapter 35.) In this case, mutant myeloma cells were isolated that had lost an enzyme necessary for using hypoxanthine as a source of purines. Both the myeloma cells and the spleen cells can be prevented from making their own purines by adding a certain inhibitor, **aminopterin** (p. 694), to the culture medium.

Here we have it. If a mixture of spleen cells, mutant myeloma cells, and their hybrids is placed in a medium containing hypoxanthine and aminopterin, the spleen cells won't grow anyway; the myeloma cells won't grow because they can't use the hypoxanthine and the aminopterin prevents them from synthesizing purines (Fig. 12–3). The hybrids of spleen and myeloma cells will grow because the genes from the spleen cell provide the enzyme enabling synthesis of purine nucleotides from hypoxanthine. (Deoxythymidine is also included in the medium because aminopterin also blocks its synthesis.)

Another trick was to select myeloma cell mutants that were not only unable to use preformed purines but were also unable to make normal antibody. This was necessary to prevent the hybridomas from making an antibody programmed by the myeloma genes in addition to the one programmed by genes from a spleen cell. Armed with these mutant myeloma cells, it is now possible to prepare hybridomas

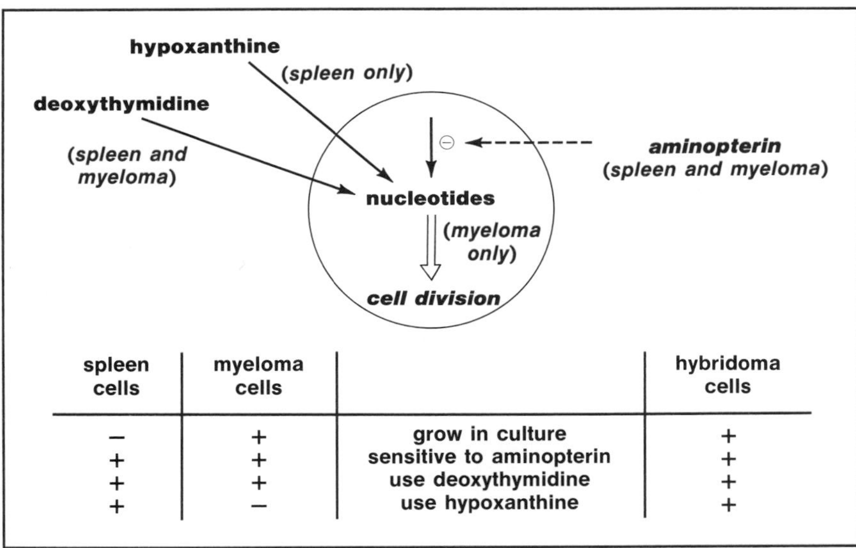

spleen cells	myeloma cells		hybridoma cells
−	+	grow in culture	+
+	+	sensitive to aminopterin	+
+	+	use deoxythymidine	+
+	−	use hypoxanthine	+

FIGURE 12–3 Selective culture of lymphocyte-myeloma hybridomas. Lymphocytes isolated from the spleen cannot grow in culture, and a strain of myeloma cells that cannot use hypoxanthine from the medium is selected. Both kinds of cells are prevented from synthesizing purine nucleotides from amino acids by incorporating aminopterin in the medium. (This also prevents them from making deoxythymidine as a nucleotide, so it is supplied in the medium.) Only the hybridoma cells have the capability of dividing freely, inherited from the myeloma cells, and the ability to use hypoxanthine to sustain nucleic acid synthesis, inherited from the spleen cells.

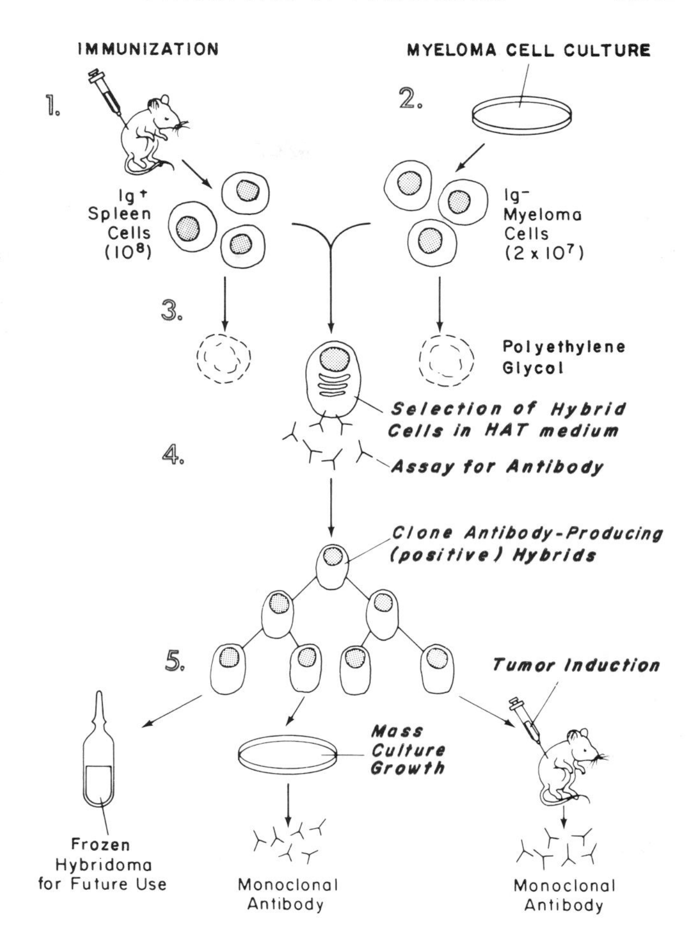

FIGURE 12-4

Outline of monoclonal antibody formation. (*1*) A mouse is immunized with the antigen of interest; its spleen is removed at a time of maximum content of new antibody-forming cells. (2) Myeloma cells deficient in the ability to use hypoxanthine and to produce immunoglobulin are maintained in culture. (3) The two kinds of cells are mixed in the presence of polyethylene glycol, which promotes fusion of plasma membranes. Only a few of the hybrid cells formed are of the desired kind. (*4*) The cells are transferred to a medium containing hypoxanthine, aminopterin, and deoxythymidine. Only the hybridoma cells survive; these are plated out and the separate clones tested for production of the antibody of interest. (5) Those producing an antibody to the antigen of interest are further cultured. They may be frozen for storage, grown in large cultures or injected into the peritoneal cavities of mice for formation of antibody. (Reproduced by permission from B.A. Diamond, D.E. Yelton, and M.D. Scharff (1981) N. Engl. J. Med. 304:1344.)

that synthesize any one of a wide variety of antibodies. The only restriction is that the antigen to which the antibody is to be made be one that provokes a good immune response in the mouse spleen cells. The general procedure is summarized in Figure 12-4.

Monoclonal antibodies are already in wide use as laboratory reagents, especially in the identification and isolation of specific proteins. Commercial preparations will probably supplant currently available crude antisera for many serological tests. It remains to be seen if their promise as therapeutic agents will be fulfilled.

STRUCTURE OF ANTIBODIES

Antibodies react with specific antigens because they are built with combining sites to fit the antigen. The interactions between antibody and antigen depend upon the same forces involved in other biological functions of proteins and knowledge of structure in antibodies clarifies many general features of protein function.

Antibodies are referred to as **immunoglobulins (Ig),** or as **gamma globulins,** a term derived from the behavior of the most abundant class, IgG, upon electrophoresis (p. 21). The gamma globulins of blood serum are mostly antibodies; antibodies are also found in the beta-globulin region, and to a considerably lesser extent in the alpha globulins.

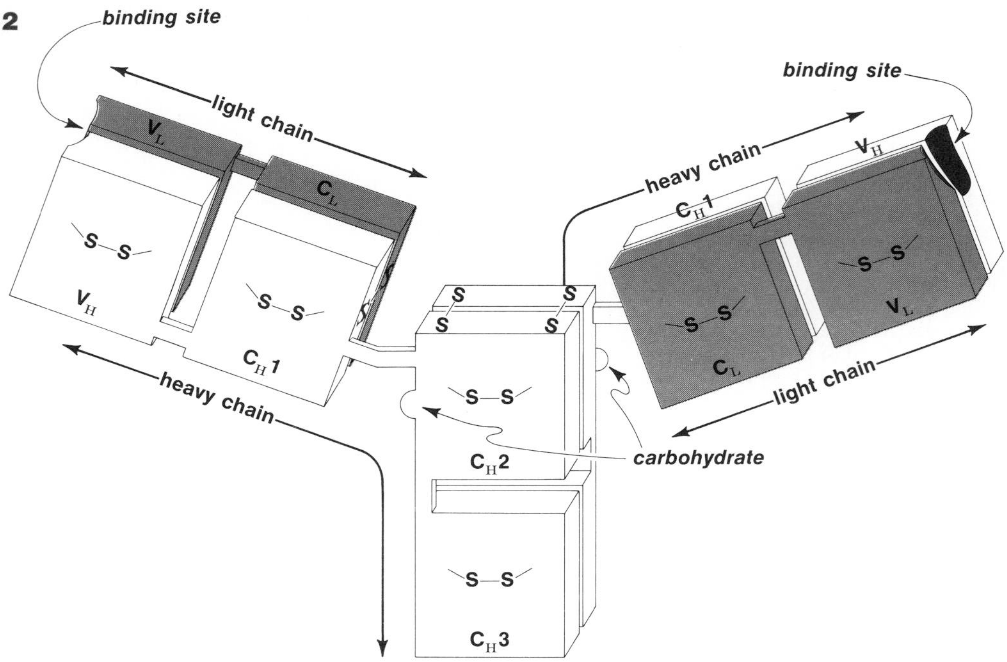

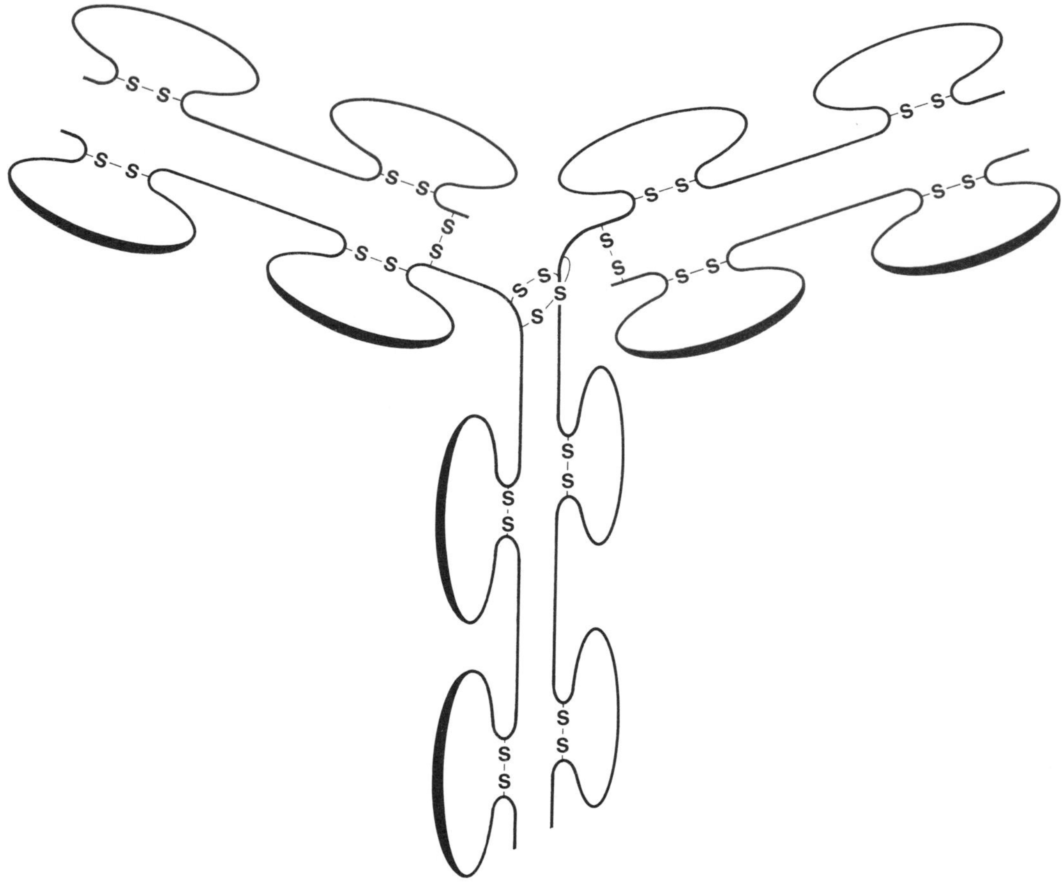

FIGURE 12–5

Schematic outline of antibody structure. Two light chains (*brown*) and two heavy chains (*black*) generate a T- or Y-shaped structure with an axis of symmetry. The chains are organized into domains (*top*). The domains are partially created by disulfide bonds, and disulfide bonds are also used as links between chains, but not always at the positions shown here (*bottom*).

Antibodies Are at Least Divalent. The fundamental unit of an antibody is a Y- or T-shaped molecule with an identical site at the end of each arm for binding an antigen (Fig. 12–5); the antibody can therefore bind two antigen molecules simultaneously. The antibody molecule has an axis of symmetry; rotating one half of the molecule 180° superimposes it on the other half.

Heavy and Light Chains. An antibody contains equal amounts of two kinds of polypeptide chains, heavy and light, with the heavy at least twice the size of the light. One pair of identical heavy chains and one pair of identical light chains combine in this way: Each arm of the molecule contains the entire length of a light chain bound to a nearly equal-sized portion of the amino terminal sequence of a heavy chain. The remaining carboxyl terminal portions of the heavy chains combine to form the stem. The union of the chains is usually stabilized by disulfide bonds.

Domains

Each arm of an antibody is composed of two globular masses of folded polypeptide chain connected by more extended segments. The stem is made from at least two, and sometimes more, globular masses of similar size, also connected to the arms by extended segments and to each other. Here we have an illustration of an important feature of many proteins—the antibodies are composed of domains that are distinct but connected. A domain is a folded compact structure that behaves as an architectural unit. Figure 12–5 showed the domains in schematic form; they can be seen in more detail for one kind of antibody in Figure 12–6.

The Immunoglobulin Fold. The globular domains of antibodies are quite similar in size because each is constructed according to the same architectural plan. Each is made by folding parts of two polypeptide chains: one heavy and one light chain in the arms and two heavy chains in the stem. The folds in each chain are constructed in a strikingly similar pattern, even though their functions differ. (Some refer to the characteristic folds in individual chains as the domains; in this usage, the globular regions we call domains would be regarded as pairs of domains.) Each involves a sequence of approximately 110 amino acid residues, of which 50 to 60 per cent is in two layers of antiparallel pleated sheet arranged somewhat like a sandwich, with the two slices of bread linked by a stabilizing intrachain disulfide bond (Fig. 12–7). Hydrophobic amino acids are invariably present around the S—S bond to further stabilize the structure.

The Binding Site

Variable and Constant Regions. Different antibodies combine with different antigens because they have different amino acid sequences at the ends of their arms. These terminal domains are therefore said to contain the variable regions of the light and heavy chains (V_L and V_H). The remainder of the domains may be identical or nearly identical in many antibodies and are therefore said to be composed of constant regions.

The light chain therefore has one variable region and one constant region (V_L and C_L), while a heavy chain has one variable region (V_H) and three to five constant regions, one in the arm (C_HI) and the others in the stem.

The Variable Region. The known examples of antibody binding sites are irregularly shaped cavities formed by portions of the variable domains from both light and heavy chains. The cavity walls are complementary to an antigenic determinant.

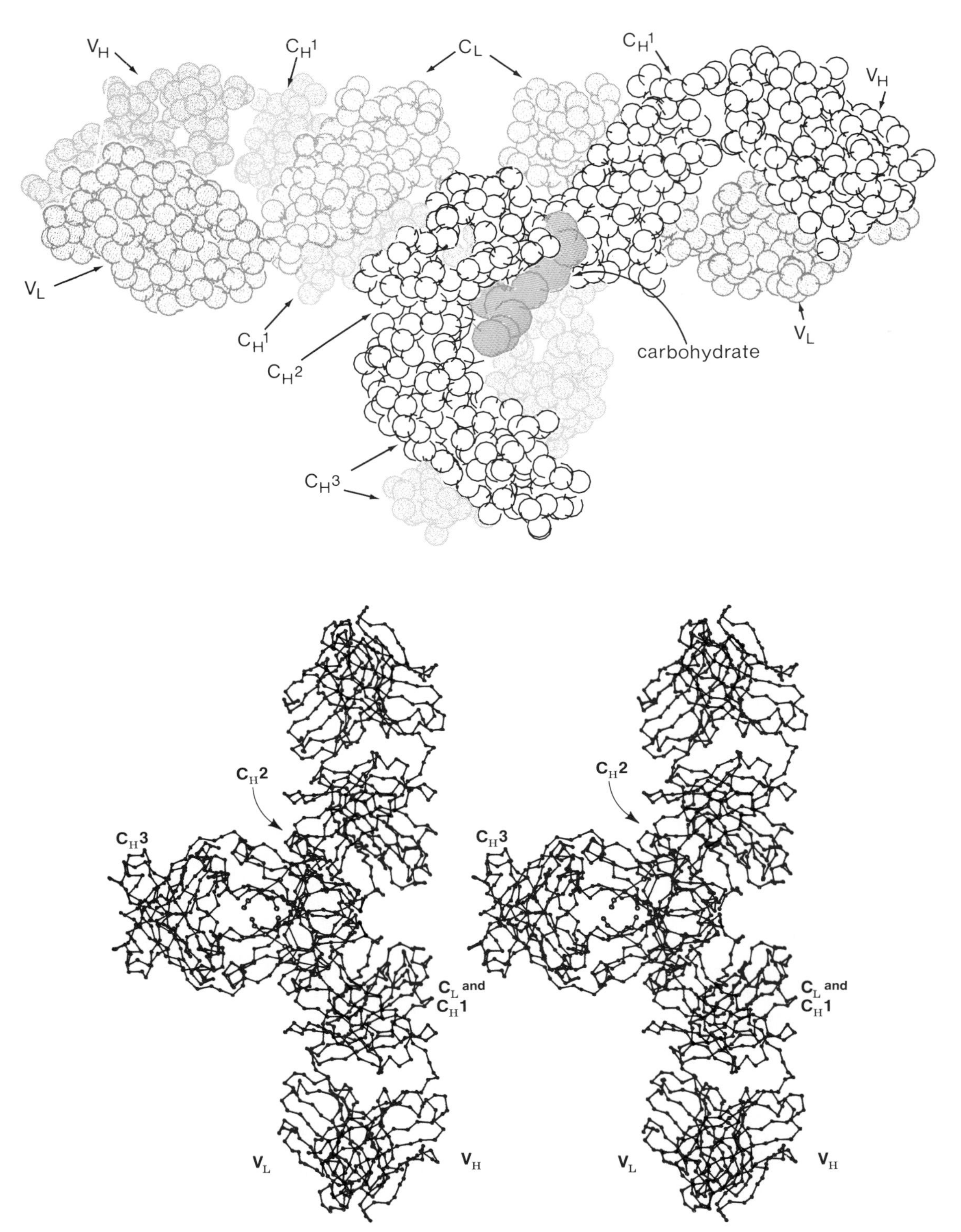

FIGURE 12–6 *See legend on opposite page*

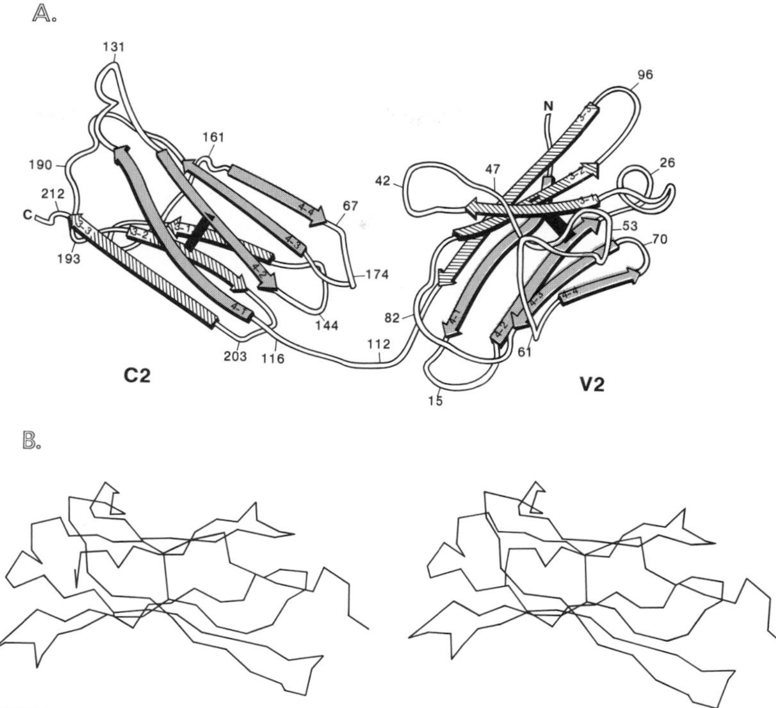

FIGURE 12–7 *A.* Outline of chain folding in a Bence Jones protein containing the variable and constant domains of a light chain. Segments of pleated sheet are indicated by *arrows*. Each domain contains two layers of pleated sheet, one shaded *brown* and the other *striped black*. Note the similarities of the folding in the two domains. (Reprinted with permission from A.B. Edmundson, et al., Rotational Allomerism and Divergent Evolution of Domains in Immunoglobulin Light Chains. (1975) Biochemistry 14:3593. Copyright by the American Chemical Society.)

B. Stereo drawing of the detailed course of the polypeptide chain in a constant domain. Lines connect α-carbon atoms. (Modified from D.R. Davies, E.A. Padlan, and D.M. Segal. Reproduced with permission from the Annual Review of Biochemistry, Vol. 44, p. 654. Copyright 1975 by Annual Reviews, Inc.)

Much of each variable domain does not participate in antigen binding but provides a framework within which the cavities are formed. Antigens are bound in the variable region by three **complementary determining regions** in the light chain and three in the heavy chain. It is the variation of structure within these regions that determines the specificity of antibody for antigen; hence the amino acid composition in these

---

FIGURE 12–6

Detailed structure of an antibody molecule. *Top.* The location of individual residues is indicated by circles; light chains are in *brown*, and heavy chains are in *black*, with one of the chains shaded. Carbohydrate residues are shown by *heavy black shading. Bottom.* Stereo view of the course of the polypeptide chains in the same molecule. Circles locate α-carbon atoms. The top view is modified from E.W. Silverton, M.A. Navia, and D.R. Davies, (1977) Proc. Natl. Acad. Sci. U.S.A., 74:5140. Reproduced by permission. The bottom stereo view was kindly generated by Dr. Navia for this work.

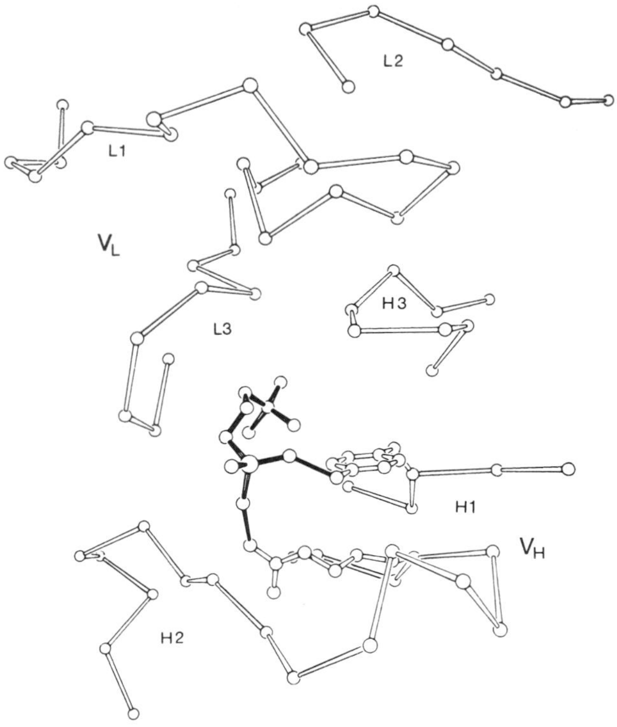

FIGURE 12–8

Outline of the path of the polypeptide chains around an antibody binding site. This site fixes phosphocholine (*heavy black*) as a hapten, but the natural antigen is not known. (Modified from D.R. Davies, E.A. Padlan, and D.M. Segal. Reproduced with permission from the Annual Review of Biochemistry, Vol. 44, p. 663. Copyright 1975 by Annual Reviews, Inc.)

regions differs considerably from one antibody to another. The complementary determining regions are these **hypervariable regions,** generally located at positions 24 to 35, 51 to 57, and 90 to 98 of light chains and positions 31 to 35, 50 to 66, and 99 to 111 of heavy chains. The larger remainder of the polypeptide chains in the variable domain folds so as to bring the hypervariable sequences near each other at the end of an antibody arm, creating a binding site.

An example of a portion of a binding site is shown in Figure 12–8. The site fits O-phosphocholine as a hapten, although the natural antigen for this antibody is not known. The site appears to be $0.6 \times 0.6 \times 0.7$ nm in dimensions; two sides of the entrance are formed by the first and third hypervariable sequences of the light chain, and the other two sides are formed by the second and third hypervariable sequences of the heavy chain.

CLASSES OF IMMUNOGLOBULINS

Antibodies exist in general structural types that appear to be related to different functional roles in the body. Five such classes are known in humans: **A, D, E, G,** and **M. IgG,** the major class in the circulation, is largely confined to the intravascular and extracellular space; **IgA** is the predominant class secreted at mucosal surfaces; **IgE** binds to the plasma membrane of mast cells and basophils; and **IgM** occurs as a cell surface receptor and in the intravascular space. The primary structural difference in the classes is in the kind of heavy chain; these chains designated by the corresponding Greek letters, α, δ, ϵ, γ, and μ. The light chains in any of the five classes are one of two kinds, κ or λ. For example, the chain composition of an IgG molecule may be $\kappa_2\gamma_2$ or $\lambda_2\gamma_2$.

PROPERTIES OF HUMAN IMMUNOGLOBULIN CLASSES TABLE 12–1

Class	IgA	IgD	IgE	IgG	IgM
Heavy chain (H)	α	δ	ϵ	γ	μ
Estimated M. W.					
complete	55K	62K	70K	50K	70K
without carbohydrate	48K	56K	58K	49K	58K
Oligosaccharide chains/heavy chain	2–3		5	1	5
Chain composition	L_2H_2 or L_4H_4	L_2H_2	L_2H_2	L_2H_2	$L_{10}H_{10}$
Extra components	SC, J				J
Approximate total M.W.	162K, 390K	178K	188K	146K	880K
Normal adult serum concentration					
mg/100 ml	200	3	0.05	1,000	120
Range	90–450			800–1,800	60–275

The types of heavy chains differ in the nature of their constant domains, and this leads to further differences between them, which are summarized in Table 12–1. IgA associates into dimers and IgM into pentamers, in which a further single small polypeptide of about 15,000 daltons, the J piece, is incorporated. The IgA dimers include still another polypeptide chain, a secretory component of 70,000 daltons that appears to facilitate secretion through mucous membranes.

The classes also differ in their content of carbohydrate, which makes up from 2.5 to 10 per cent of the weight and is commonly linked from N-acetylglucosamine to asparagine in the constant domains of the heavy chains.

BINDING OF ANTIBODIES

The function of an antibody begins with its binding of an antigen. The specificity is not absolute; the binding site of an antibody will often “recognize” determinants of closely related structure with at least a partial fit into the site, but in general the tightest combination of antibody and antigen is achieved only when the fit is very good and complete. Antigen and antibody must be close friends, not mere acquaintances, for the best union.

Insoluble Antigen-Antibody Complexes

Agglutination. Antibodies can cause cells to agglutinate into large, readily sedimentable, clumps. The cells may be microorganisms, erythrocytes, or nucleated cells because the only requirements for agglutination are an antigen with multiple determinant sites and multivalent antibodies. The multivalent antibodies physically cross-link the polyvalent antigens (Fig. 12–9). Some monoclonal antibodies raised against cellular antigens will not agglutinate the cells, whereas combinations of monoclonal antibodies to different determinants will. Presumably there are too few of some determinants on a cell. For example, if a determinant only occurs in one place, then one divalent antibody molecule specific to that determinant may link two cells, but there is then no further site for another antibody molecule with the same specificity to link another cell.

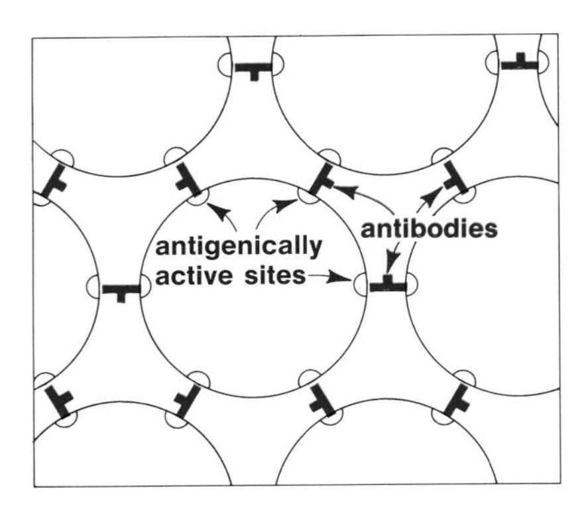

FIGURE 12–9

Cells and other large antigens usually contain many identical antigenic sites. The divalent antibodies may therefore link the antigens into a three-dimensional lattice.

Agglutination of antigen in vivo is probably extremely rare, but the use of agglutination reactions in vitro to type blood is a critical key to the success of transfusions. Their use is extremely important in the diagnosis of some infectious diseases, and in the general identification of many bacteria. The antigenic determinants causing agglutination frequently involve the carbohydrate side chains of glycoproteins in the outer cell membrane. At least nine major blood group systems have been identified in the human erythrocyte on the basis of variations in glycoprotein structure (p. 200). The recipient of transfused red blood cells must have the same determinants found in the donor.

Precipitin Reaction. Mixtures of antibodies raised to a soluble multivalent antigen, such as a particular protein, may cross-link the antigen to form precipitates (Fig. 12–10). This reaction is used as a precise quantitative tool to determine amounts of antibody and antigen and is also useful in the analysis of the structural requirements for specificity of antibodies. The precipitin reaction, like cell agglutination, does not occur with many monoclonal antibodies raised against soluble antigens. The smaller antigens are even less likely than whole cells to have two of the same determinants.

Biological Effects

If there is any protective value to the individual in the agglutination or precipitin reactions, it is yet to be demonstrated. Indeed, adverse consequences follow when mismatched blood is transfused. However, the appearance of antigen-antibody complexes has many powerful consequences that are for the most part

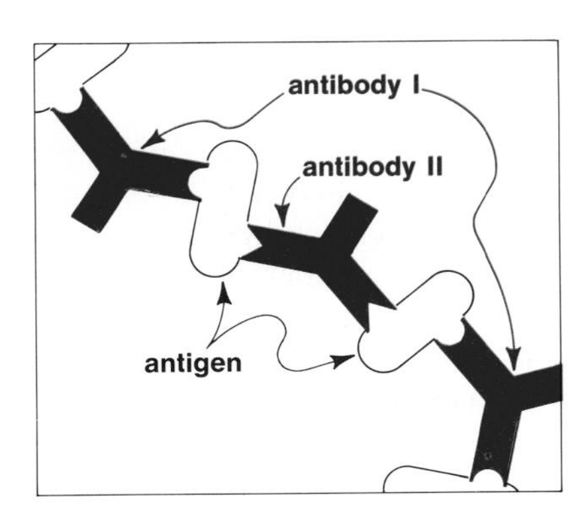

FIGURE 12–10

A precipitin reaction may result from the formation of a lattice with soluble antigens that contain different antigenic determinants through combination with different antibodies.

purposeful and lead to the inactivation and removal of foreign materials. The signal for initiation of these events appears to be alterations in the conformation of the stem of the antibody, caused by binding of antigen to the arms.

Activation of Complement. The complement system is a group of at least 11 serum proteins that are activated in sequence following the union of antigens with some antibodies, and we shall say more about this cascade mechanism in Chapter 18. The end result is lysis of a foreign cell. Activation is especially efficient and rapid upon combination of IgM with erythroctyes or certain bacteria; one molecule of bound IgM can initiate the lysis of one complete cell.

Earlier stages in the activation of complement act as signals for other biological events, such as migration of macrophages to the locale of the antigen, increased phagocytosis of the foreign material, and even visible inflammation.

Deleterious Effects. The effects of antigen-antibody combination are not always good; inflammation may cause extensive damage to basement membranes, with a resultant severe malfunction in the involved organs. Illness, mild to fatal, may result from the activity of IgE antibodies. All normal individuals have IgE, which usually binds strongly to the plasma membranes of mast cells and basophils through its stem without detectable effects. Subsequent binding of antigen to the arms causes these cells to release the potent amines, histamine and serotonin (Chapter 33), and a derivative of arachidonic acid, **leukotriene C-1,** also known as the **slow-reacting substance** (p. 750), which contributes to anaphylactic shock. The effects in unduly sensitive individuals appear as hay fever, asthma, and anaphylactic reactions.

FORMATION OF ANTIBODIES

Antibodies are made in the same way as other proteins. Information in DNA for their amino acid sequence occurs in segments; after transcription, the segments are spliced together to make messenger RNAs.

Multiple Genes—One Polypeptide Chain. The mechanism for making myriads of different antibodies from a minimum of coding information employs multiple genes for the production of a single polypeptide chain, light or heavy.

Three genes code for a light chain. One, the **V gene,** codes for an N-terminal segment, about 96 amino acid residues long, that includes three framework regions and two hypervariable regions along with a portion of a third hypervariable region. A **J (joining) gene** codes for the remainder of the third hypervariable region and a fourth framework region, which are in residues 97 to 109 of the finished polypeptide. A **C (constant) gene** codes for the constant domain.

Four genes apparently code for a heavy chain. A **V gene** codes for three framework and two hypervariable regions. A **D gene,** so named because it adds to the diversity of antibodies, codes for the third hypervariable region, and a **J gene** codes for the fourth framework region. **C genes** code for the constant domains in the various classes of immunoglobulins. (There are several genes for each class, producing subclasses and allotypes within them.) Additional C genes are present to add another heavy chain domain (C_HV) for making a μ chain or δ chain in some cells—those that incorporate IgM into their plasma membranes. These genes cause the formation of tails rich in hydrophobic residues, which undoubtedly are buried in the lipid core of the membranes.

Estimates of the number of variable-region genes that exist within the genome have been determined largely for mice. There appear to be at least four each of J_H and J_L, several hundred V_κ, and at least eight D genes. In humans, five J_κ, about 20 V_κ, and 23 heavy chain genes have been detected.

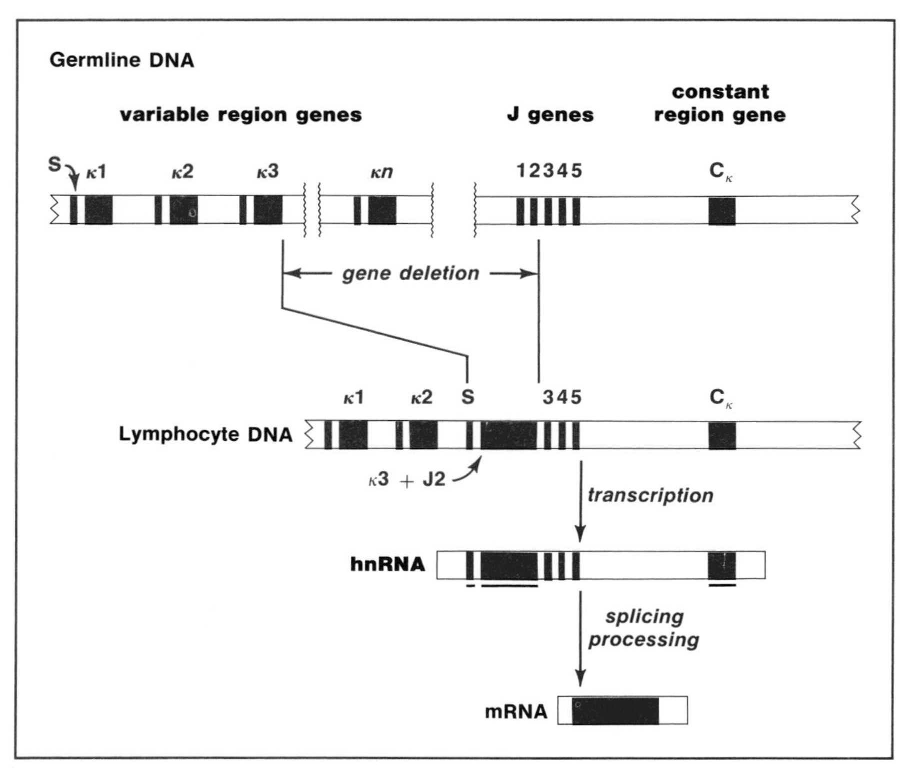

FIGURE 12–11 Processing of genetic information for formation of a specific kappa chain in mice (not to scale). The germline DNA contains several kappa chain variable region and J genes. Each variable region gene is preceded by information for a signal sequence. During maturation of a lymphocyte, one of the kappa chain genes and its accompanying signal sequence is joined to one of the J genes. The DNA containing the joined genes and the following constant region gene is transcribed into hnRNA. Only the joined variable gene and its signal sequence are spliced to the kappa constant region during processing to make a finished light chain messenger RNA. (Not shown) The hydrophobic leader translated from the signal sequence is removed when the light chain polypeptide crosses the endoplasmic reticulum.

The synthesis of antibody polypeptide chains requires processing of information in both DNA and hnRNA (Fig. 12–11.) Maturation of a cell forming an antibody involves combination of a V_L gene and a J_L gene to make a particular light chain gene. A similar recombination of V_H, and D_H, and J_H genes occurs to make a heavy chain gene. Recombination between the end of one gene and the beginning of the downstream gene depends upon the recognition of rather tricky signals in the intervening segment. According to present information, these signals consist of a pair of nucleotide sequences, a heptameric palindrome CACTGTG and the sequence GGTTTTTGT. These sequences occur as such beyond the 3′ end of the sense strand (the downstream end) of the V genes, and in inverted form before the 5′ ends of the sense strand in the J genes. (They will be in the same sequence read upstream in the anti-sense strand from the J genes.) The trick is that the pairs of sequences are separated by a different number of nucleotides in the various places they occur, by a 12 ± 1 base pair spacer beyond a V gene, and by a 23 ± 1 base pair spacer before a J gene. Recombination only occurs by matching a 12-base separation with a 23-base separation according to the present hypothesis on the way the proper gene ends are matched:

$$\text{V gene---CACTCTCN}_{12}\text{G}_2\text{T}_5\text{GT---}$$
$$\text{---TGT}_5\text{G}_2\text{N}_{23}\text{CTCTCAC---J gene}$$

In any event, recombined V_L-J_L or V_H-D_H-J_H genes are transcribed as hnRNA, which is then spliced to remove the intervening segments and to combine it with a separate transcript of a constant region gene, thus making a complete messenger RNA for production of a specific antibody.

By splitting the genetic information and then creating the antibody from various

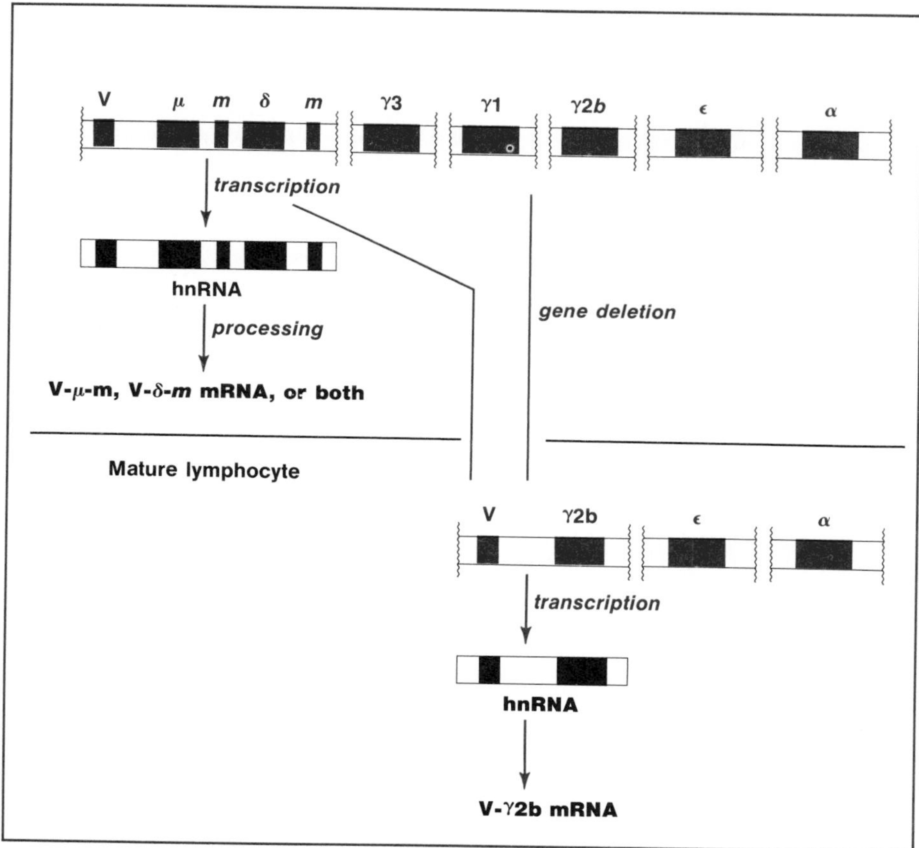

FIGURE 12–12 Processing of genetic information for formation of immunoglobulin heavy chains in mice (not to scale). In the young lymphocyte, a specific joined variable region gene is followed by a succession of genes for the constant regions of the various classes and subclasses of heavy chains. Only the variable and the first two constant region genes are transcribed, however, into a single hnRNA molecule. This is processed into mRNAs in which the variable region is combined with either the μ or the δ constant region genes, along with the extra (m) gene segment for the hydrophobic tail added for binding to membranes. In more mature lymphocytes, constant region genes may be deleted. In the example here, the variable region gene is brought into proximity to a gene for a γ-chain subclass, so the product of transcription and RNA processing is a messenger RNA for an IgG heavy chain.

combinations of the segments it is possible to amplify the number of genes into a much larger number of antibodies. Different binding sites can be made by different combinations of V_L, J_L, V_H, D_H, and J_H. Suppose each made an equal contribution to variability. Then 500 genes (100 of each kind) could generate 100^5 (10^{10}) different antibodies. They do not make equal contributions, but even so, it does not require a major fraction of the total genome to generate millions of different antibodies. In addition, there is a good possibility that some diversity in antibody-forming cells occurs through somatic mutation.

The various classes of immunoglobulins are generated by combining the information for a given heavy chain variable region with information for the various kinds of constant domains. This explains how some multiple myeloma patients can have two populations of myeloma cells forming antibodies in different classes (IgA and IgM, for example) but with one of the variable regions identical in the two populations. It is not certain whether the combination of information occurs at the level of DNA by gene splicing, or by splicing of RNA transcripts, or both. Lymphocytes go through a maturation process during which the constant-region genes being expressed change, but the same variable-region genes are used. They first make IgM with its hydrophobic tail for membrane binding and then shift to IgD, but sometimes they make IgM and IgD simultaneously, which indicates that the messenger RNA is made by combining the variable region transcript sometimes with a transcript of one constant region and sometimes with another. On the other hand, maturation to form one of the other classes of immunoglobulins, such as IgG, appears to be an irreversible event caused by deletion through recombination of DNA (Fig. 12–12).

FURTHER READING

G. McBride (1976). Antibodies Yield Their Secrets and Display Therapeutic Versatility. J.A.M.A. 235:583. Useful introduction.

M.A. Atassi (1975). Antigenic Structure of Myoglobin. Immunochemistry 12:423. Review.

E.W. Silverton, M.A. Navia, and D.R. Davies (1977). Three-Dimensional Structure of an Intact Human Immunoglobulin. Proc. Natl. Acad. Sci. (U.S.) 74:5140.

D.R. Davies, E.A. Padlan, and D.M. Segal (1975). Three-Dimensional Structure of Immunoglobulins. Annu. Rev. Biochem. 44:639.

A.B. Edmundson et al. (1975). Rotational Allomerism and Divergent Evolution of Domains in Immunoglobulin Light Chains. Biochemistry 14:3953.

R.H. Kennett, T.J. McKearn, and K.B. Bechtol, eds. (1980). Monoclonal Antibodies: Hybridomas; A New Dimension in Biological Analyses. Plenum.

B.A. Diamond, D.E. Yelton, and M.D. Scharff (1981). Monoclonal Antibodies. N. Engl. J. Med. 304:1344. Descriptive review.

D.E. Yelton and M.D. Scharff (1981). Monoclonal Antibodies: A Powerful New Tool in Biology and Medicine. Annu. Rev. Biochem. 50:657. More technical review.

E.A. Kabat (1978). The Structural Basis of Antibody Complementarity. Adv. Prot. Chem. 32:1.

M. Robertson (1981). Genes of Lymphocytes I: Diverse Means to Antibody Diversity. Nature 290:625.

J.L. Marx (1981). Antibodies: Getting Their Genes Together. Science 212:1015.

P.A. Hieter, J.V. Maizel, Jr., and P. Leder (1982). Evolution of Human Immunoglobulin κ J Region Genes. J. Biol. Chem. 257:1516.

CHAPTER 13

BINDING FOR TRANSPORT: HEMOGLOBIN AND OXYGEN

Substances frequently are moved from one place to another in the body by binding to a protein carrier. The transport of oxygen by hemoglobin and the nature of this carrier will illustrate many general principles.

Let us first ask why a carrier is necessary. Why isn't simple diffusion of oxygen from the atmosphere to the cells sufficient? Rapid diffusion requires a large difference in concentration over a short distance. Molecular oxygen, **dioxygen,** is a hydrophobic compound, and its low solubility in water limits the concentration differences that can be achieved, so the only way to get a steep gradient is to keep the distance short. However, the distance from cell surface to atmosphere is sufficiently short for effective diffusion in only the simplest organisms, and oxygen is delivered in several stages in the larger organisms (Fig. 13–1). Animals as low in the evolutionary scale as the annelids (for example, earthworms) became possible only by the development of hemoglobin and related proteins as carriers for oxygen, together with a pumped circulation to move the carrier rapidly between surfaces exposed to the atmosphere and the more deeply buried cells.

Individual cells with very high demands for oxygen, such as those in some striated muscles and the heart, also acquired an internal hemoglobin, myoglobin, to ensure a steady supply to the mitochondria, the organelles consuming most of the

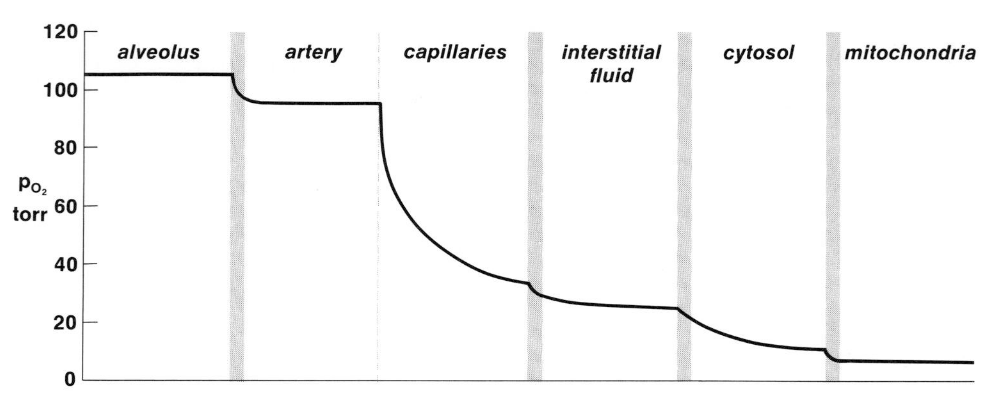

FIGURE 13–1 The transport of oxygen in higher organisms requires the presence of concentration gradients at several sites, with the oxygen tension progressively falling from the lung alveoli to the mitochondria in the peripheral tissues where oxygen is consumed.

oxygen. Short as the distance is, diffusion is not sufficient to sustain the necessary rapid flux of oxygen within those cells.

Magnitude of the Task

How fast must oxygen be transported? Under basal conditions (rest, overnight fast), a typical rate of oxygen consumption by 22-year-old Americans of median size is:

12 millimoles (260 ml) per minute by males (177 cm, 73 kg)
9 millimoles (200 ml) per minute by females (163 cm, 58 kg).

Consumption is increased many times over the basal rate by exercise, reaching a maximum of over 200 mmoles min^{-1} in the male of median size who is a trained athlete. (The maximum consumption of an untrained male of that size is about $\frac{2}{3}$ as great.) Much of the increased rate of transport is accounted for by an increased rate of blood circulation; the hemoglobin carrier is shuttled between lungs and tissues in less than 0.2 of the time required at rest. The remainder of the increase comes from greater release of oxygen by the hemoglobin; the amount of oxygen released per hemoglobin molecule on each pass through a working muscle can be three times the value in a resting muscle.

OXYGEN BINDING SITE

Prosthetic Group. We noted in Chapter 3 that the binding of oxygen by hemoglobin involves the action of two components—heme (iron(II)-porphyrin), and a polypeptide chain wrapped around it. The oxygen molecule is carried in a crevice between these two. The use of heme illustrates a general principle. The performance of some biological functions requires chemical groups that are not present in the 20 amino acids from which proteins are made. Additional needs for chemical reactivity are met by building the necessary groups into some low molecular weight compound, and then constructing a polypeptide chain to associate strongly with this compound. The result is a protein composed in part of polypeptide, known as the **apoprotein,** and in part of another compound, the **prosthetic group** (from *prosthesis,* an additional part). We shall encounter many kinds of apoproteins and their associated prosthetic groups.

The word hemoglobin is a contraction meaning blood globulin, and when the protein was found to contain a porphyrin bound to iron in addition to polymerized amino acids, it was a logical extension to call the iron porphyrin heme and the apoprotein **globin.**

The iron atom in hemoglobin is like many metallic ions that participate in biological processes by forming complexes with proteins or with other components of the cell. Some complexes, such as those of magnesium and calcium, are easily dissociated, so that a given atom of metal is rapidly exchanged from one complex to another. Other ions, such as copper(II), cobalt(II), zinc, and manganese(II), are bound more tightly, often so tightly that the metal is for all practical purposes a fixed part of the structure as long as it exists. This is also the case with the iron(II) ion in heme; it remains in the porphyrin until the porphyrin is destroyed.

Free heme is unstable in contact with oxygen; the Fe(II)-porphyrin is oxidized to Fe(III)-porphyrin, known as hemin. However, little heme exists in free form in cells, because they construct sufficient apoproteins, such as globin, to combine with most of the heme and protect it from oxidation. When oxygen is bound to the iron atom in hemoglobin, a shift of an electron does occur. The result is the conversion of heme to hemin (Fe(III)-porphyrin) with a net positive charge and the reduction of dioxygen to a superoxide anion, O_2^-. However, these oppositely charged ions remain associated in the hemoglobin crevice; it is only the uncharged dioxygen in equilibrium with the superoxide ion that is free to leave, and when it does so, the iron reverts to the Fe(II) state:

$$\text{His—Fe(II)}\ O_2 \rightleftharpoons \text{His—Fe(III)}^{\oplus\ominus}O_2$$

MECHANISM OF TRANSPORT

Binding for transport differs from binding by antibodies in an important way: Transport implies not only a picking up of the substance at the point of supply but also a letting go at the point of demand, and it is the release that antibodies are not built to do. We saw in the antibodies how binding is achieved through specific arrangement of reactive groups; now let us consider what must be done in order to release a bound molecule.

Some transported molecules are acquired by cells in a rather drastic way; the entire carrier complex is engulfed by pinocytosis and then degraded within the cell so as to detach the load. Such a mechanism is prohibitively expensive for transport of something in as high a demand as is oxygen. If hemoglobin had to be destroyed in order to give up its oxygen, our hard-working median man would consume nearly 50 grams of this protein per second.

We noted before (p. 40) that the internal hemoglobin in muscle, myoglobin, is a single polypeptide chain wrapped around a heme molecule, while circulating hemoglobin contains four subunits, two α chains and two β chains, each similar in size to myoglobin. The transport behavior of these two kinds of hemoglobin is quite different; the difference deserves close examination not only because of its inherent importance but also because similar differences in the biological function of proteins will be seen in catalysis by enzymes.

Transport by Myoglobin

The Myoglobin Dissociation Curve. Myoglobin may act as a buffer to minimize changes in oxygen concentration and as a vehicle for transport within muscle cells. One possibility for transport down a concentration gradient is through simple equilibration of the carrier with the transported molecule. Myoglobin could carry oxygen in this way down a relatively steep concentration gradient within cells:

$$\text{Mb} + \text{O}_2 \rightleftharpoons \text{MbO}_2$$

Where the concentration of oxygen is high, the carrier binds more; where the concentration is low, the carrier binds less and will release the oxygen that it brought to the region of low concentration (Fig. 13–2). The carrier is **facilitating diffusion** in a simple way.

The behavior of myoglobin toward oxygen can be described by the ordinary equilibrium expression:

$$\frac{[\text{MbO}_2]}{[\text{Mb}][\text{O}_2]} = K_{eq}$$

Given a fixed supply of myoglobin within a cell, we can calculate the fraction of the total myoglobin that will be oxygenated at varying oxygen concentrations from the above equation and obtain a curve with the hyperbolic shape shown in Figure 13–3.

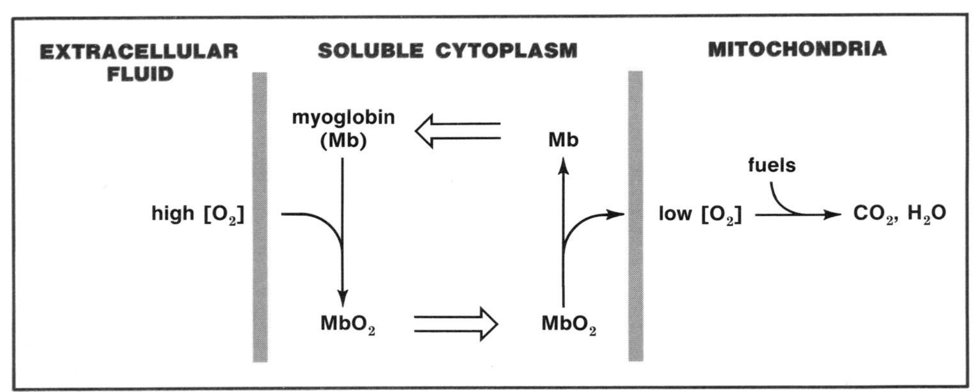

FIGURE 13–2 The presence of myoglobin can facilitate diffusion within a cell. If the concentration gradient of O_2 is in a range over which the concentration of oxygen greatly affects the degree of its binding to myoglobin, much more of the myoglobin will be oxygenated on the high side of the gradient. Effective concentration gradients will therefore be established for oxymyoglobin (MbO_2) in one direction and for deoxymyoglobin (Mb) in the opposite direction, so that molecules of the carrier will go to and fro across the concentration gradient, picking up O_2 on the high side, and releasing it on the low.

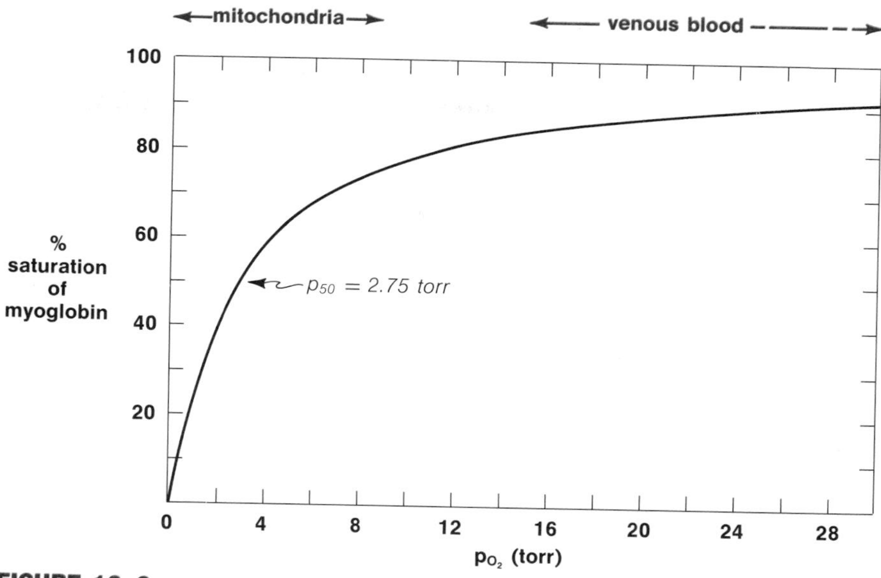

FIGURE 13–3

The effect of oxygen concentration upon the oxygenation of myoglobin. This association curve is a plot of the percentage of myoglobin molecules binding oxygen (per cent saturation) as a hyperbolic function of the $[O_2]$, given as the partial pressure of O_2 gas in equilibrium with the solution. Half of the total myoglobin is oxygenated, that is, $[Mb] = [MbO_2]$, when the p_{O_2} is 2.75 torr. (This is termed the p_{50} for 50 per cent saturation, or the $p_{0.5}$ for half-saturation.) Typical ranges for the measured p_{O_2} of venous blood, and the estimated range in mitochondria are shown at the top. (p_{50} from E. Antonini (1965) Physiol. Rev. 45:123.)

A NOTE ON UNITS

Concentration of oxygen in solution, the **oxygen tension,** is frequently given as the partial pressure, p_{O_2}, of a gas phase in equilibrium with the solution. The customary unit has been the **torr** (mm of mercury). Some now express pressures in **millibars** (dynes cm^{-2}), or **pascals** (newtons m^{-2}). The pascal is the unit of the International System. One torr = 1.333 millibars = 133.3 pascals. One millibar = 100 pascals = 0.750 torr.

The problem with myoglobin, or with any carrier functioning by simple mass-action equilibrium, is that it can be efficient only when it operates over a large concentration gradient; the ratio of oxygen concentrations at the points of uptake and release must be high. We can see this by comparing the approximate oxygen concentrations required for saturating myoglobin to varying extents:

Saturation*	$[O_2]$
0.90	24.75 torr
0.75	8.25
0.50	2.75
0.25	0.92
0.10	0.31

*Saturation is the fraction of the myoglobin molecules containing bound oxygen.

The oxygen concentration must fall to one ninth of its original value, or below, for two molecules of myoglobin to release one molecule of oxygen. (Compare values at 0.75 saturation and 0.25 saturation, for example.)

This is not an impossible requirement; such conditions may occur within some working muscle cells, in which the gradient of oxygen tension may range from 10 torr at the cell surface to perhaps 1 torr at the mitochondria. Myoglobin would indeed facilitate the diffusion of oxygen under those circumstances. (An oxygen molecule can move much faster than the bulky myoglobin molecule, but the solubility of oxygen is so low that there are far fewer molecules of it in solution in the muscle.)

Transport by Hemoglobin

Hemoglobin discharges oxygen at a partial pressure that is sufficiently high to provide the concentration gradient necessary for transport from the blood vessels into the recipient tissues. The construction of hemoglobin from multiple subunits is a critical device for generating release of oxygen at a relatively high pressure.

Cooperative Interaction. The four subunits in hemoglobin, each with its heme, behave much like myoglobin when separated in the laboratory, but they do not act as independent entities in the complete molecule in red blood cells. Hemoglobin behaves differently because the subunits interact in a cooperative way; although the tetramer has a lower affinity for oxygen than do its separate monomers or myoglobin, the binding of oxygen by one subunit in the tetramer enhances further binding of oxygen by the others. Similarly, the loss of one molecule of oxygen from the

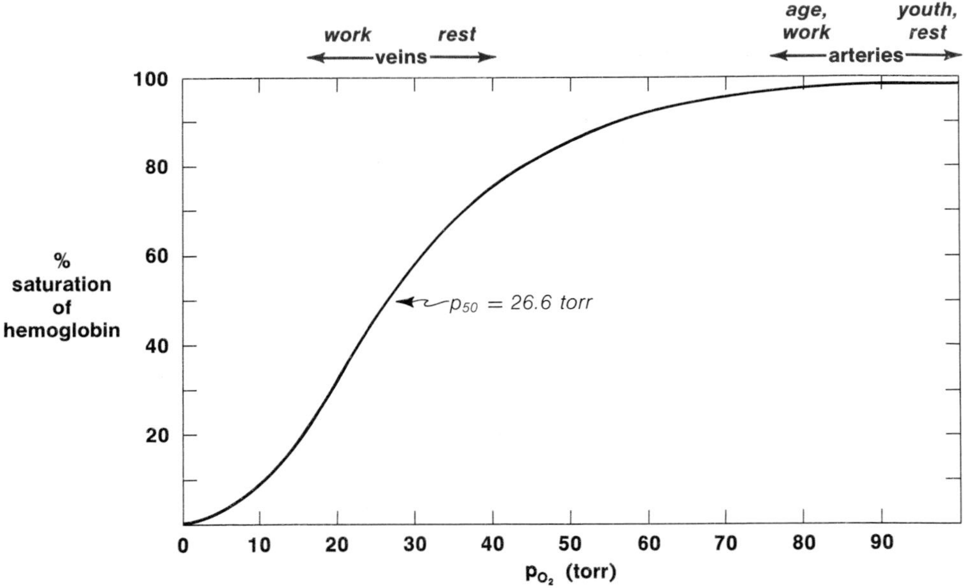

FIGURE 13–4 The oxygen association curve for hemoglobin in red blood cells. The sigmoidal shape is characteristic for multiple binding sites that interact cooperatively. The p_{50} differs by $\pm$ 2 torr in individuals; 26.6 torr is a "standard" value for intact red blood cells. The pO_2 in the venous circulation at given fluxes of oxygen is determined by this value. The value in arterial blood also depends upon the transport capacity of the lungs, which declines with age, and is severely impaired by many pulmonary diseases. ("Standard" association curve in J.W. Severinghaus (1966) J. Appl. Physiol. 21:1108.)

saturated carrier, $Hb(O)_4$, makes it easier to lose additional molecules, and the final oxygen molecule begins to depart at what is still a relatively high oxygen tension. A plot of oxygen binding as a function of tension therefore has a complex sigmoid shape (Fig. 13–4). Hemoglobin is said to exhibit **positive cooperativity** because a change in binding by one subunit makes it easier for a similar change to occur in the other subunits.

Events in oxygen transport can be deduced from Figure 13–4. Exposure of hemoglobin molecules to oxygen tensions near 100 torr in the lungs causes them to become approximately 97 per cent saturated, with an average of $0.97 \times 4 = 3.88$ molecules of oxygen per hemoglobin. Little oxygen is released until the erythrocytes encounter lower oxygen tensions in the capillaries of peripheral tissues. This tension will be 30 to 40 torr in tissues of moderate oxygen consumption, and only one quarter of the transported oxygen will be released, roughly one molecule per molecule of hemoglobin. However, in tissues with high rates of consumption, such as the heart or rapidly working skeletal muscles, the further drop in oxygen tension toward 15 torr will cause hemoglobin to release 80 per cent of its oxygen, or an average of 3.1 molecules; a fall to 10 torr causes a 90 per cent release.* Hemoglobin is therefore constructed so as to provide a large reserve of oxygen-carrying capacity for use during periods of high demand, without a drastic fall in oxygen tension. (Contrast the concentration ratio of 10 required for 90 per cent delivery by hemoglobin with the concentration ratio of 100 required for 90 per cent delivery by myoglobin.) Upon return of the blood to the lungs through the venous circulation, the whole process begins over again.

The Hill Coefficient. If hemoglobin bound oxygen one molecule at a time without cooperative interaction, the combination could be expressed by the same simple association equation developed for myoglobin: $[HbO_2]/[Hb][O_2] = K$, in which Hb represents one subunit. However, if the binding goes through n successive stages, each with a different affinity between hemoglobin and oxygen, the association equation will approach this form:

$$\frac{[Hb(O_2)_n]}{[Hb][O_2]^n} = K$$

in which all of the bound oxygen is assumed to be present as $Hb(O_2)_n$. This equation can be recast logarithmically:

$$n \log[O_2] = \log\frac{[Hb(O_2)_n]}{[Hb]} - \log K$$

According to this equation, a plot of the logarithm of oxygen tension against the logarithm of the ratio of oxyhemoglobin to deoxyhemoglobin will give a straight line with a slope of n, the Hill† coefficient. The Hill coefficient is useful in describing the action of proteins with cooperative interactions, because it is an indication of the minimum number of sites that influence each other. The value for hemoglobin is 2.8, indicating at least three sites of subunit contact that change upon oxygen binding, whereas the value for myoglobin is 1.0, indicating no subunit interaction. (It takes at

*The heart adjusts the flow of blood through itself so as to extract a nearly constant fraction of the oxygen transported through the coronary circulation. The saturation of hemoglobin in the venous drainage from the left ventricle is therefore relatively low, even in an individual at rest.

†Hill, Archibald Vivian (1886–1976), English muscle physiologist and Nobel Laureate.

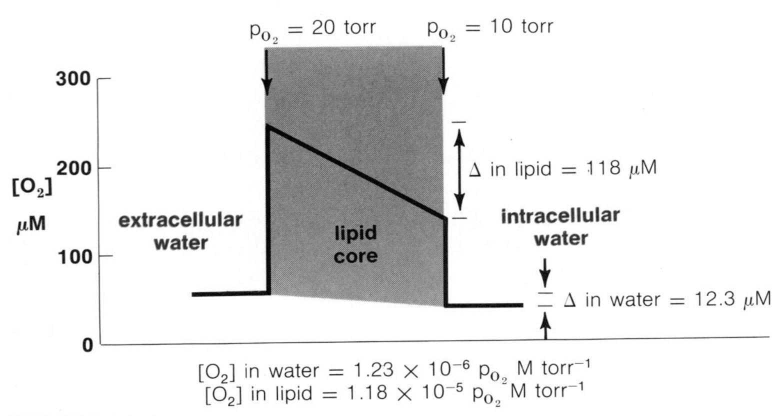

FIGURE 13–5

Demonstration of accelerated diffusion of oxygen in lipid. In this exaggerated example, a pO_2 of 20 torr is assumed at the extracellular water-lipid interface (*left*) and 10 torr at the lipid-intracellular water interface (*right*). The resultant concentration gradient of oxygen is plotted. Because of its higher solubility in the lipid, the concentrations at the two interfaces, and therefore the gradient, is much higher in the lipid than in the aqueous phases. The resultant accelerated diffusion in real cells keeps the pO_2 values nearly equal on the two sides of a membrane.

least two sites to interact.) However, the Hill coefficient is not a theoretically rigorous quantitative description.

Membranes Are No Barrier To Oxygen

Dioxygen, like dinitrogen and the noble gases, has a higher solubility in the lipid core of membranes than it does in the surrounding aqueous phases. The result is that it diffuses more rapidly through a membrane than it does on either side, because any difference in concentration on the two sides of the membrane will be magnified by the higher solubility into a steeper gradient. Figure 13–5 demonstrates the principle with an exaggerated example.

MODIFICATION OF TRANSPORT

Allosteric Effects

A red blood cell is not a packet of pure hemoglobin solution; it contains other compounds, some of which are of critical importance in adjusting the amount of oxygen that is released at given tensions. The components of interest in humans are H^+, CO_2, and the phosphate ester, **2,3-bisphosphoglycerate:**

$$\begin{array}{c} COO^{\ominus} \\ | \\ H—C—O—PO_3^{2-} \\ | \\ CH_2—O—PO_3^{2-} \end{array}$$

2,3-bisphosphoglycerate

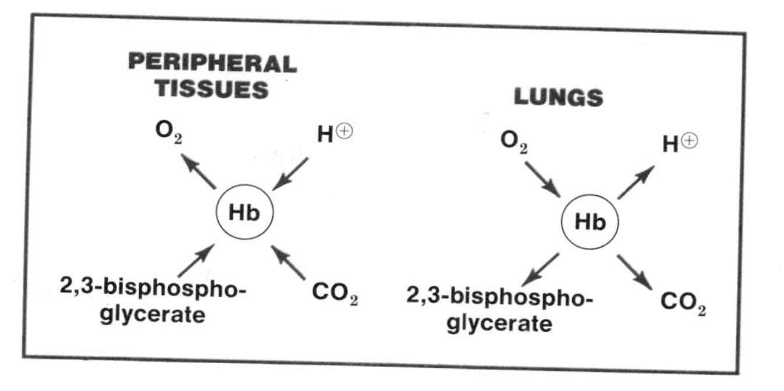

FIGURE 13–6

Oxygen is effectively pushed off hemoglobin by H^+, CO_2, or 2,3-biphosphoglycerate (*left*). Oxygen, in turn, diminishes the binding of all three when blood is returned to the lungs, where its higher concentration forces it onto hemoglobin (*right*).

(2,3-Bisphosphoglycerate is sometimes abbreviated DPG for its older name, 2,3-diphosphoglycerate.) Each of the three components diminishes the affinity of hemoglobin for oxygen, much as if it competes for a place on the carrier (Fig. 13–6). The major physiological function of these effects is to alter the amount of oxygen liberated in the peripheral circulation. Increasing the concentration of any one lowers the amount of oxygen that can be retained by hemoglobin.

Here we have the basis for the regulation of oxygen transport within an individual and also between species. The oxygen tension at the tissues, given a certain demand, is determined by the affinity of hemoglobin for oxygen. A certain base line is established for a species, or for developmental stages of an individual, through the genetically determined construction of a particular hemoglobin. In humans, and in many other animals, the hemoglobin is made with inherently too strong an oxygen affinity for proper function. Adjustments to properly weaken the binding are made through changes in the environment of the hemoglobin molecule—rapid alterations in the concentrations of H^+, CO_2, and slower changes in the concentration of 2,3-bisphosphoglycerate.

Definitions. A change in the biological behavior of a protein caused by compounds not directly involved in its function is an **allosteric effect** (*allo* = other).* Thus H^+, CO_2, and 2,3-bisphosphoglycerate are allosteric **effectors,** which modulate the binding of oxygen by hemoglobin. The term has been modified by usage, as we shall see when we discuss the regulation of enzymes; it is especially pertinent to regulatory responses caused by interactions between multiple binding sites, including the response of hemoglobin to oxygen, itself. **Allosteric effects can frequently be recognized by a sigmoidal plot of response to changing reactant concentrations.**

The magnitude of allosteric effects in human hemoglobin is large. Hemoglobin that has been stripped of CO_2 and phosphate compounds in the laboratory retains a sigmoid response to oxygen tension, but its affinity for oxygen becomes much closer to that of myoglobin (Fig. 13–7). However, addition of physiological concentrations of CO_2 and 2,3-bisphosphoglycerate to a hemoglobin solution at constant H^+ concentration makes the oxygen saturation curve nearly identical to that of intact red blood cells, showing that most of the allosteric effects in the complete system arise from these compounds (at constant pH). The cooperative effects change the association so that hemoglobin will release oxygen at a higher tension in the capillaries. Let us now summarize the action of the individual effectors.

Effect of H^+—the Bohr† Effect. An acidification of blood causes a discharge of oxygen from the red blood cells. The greater the amount of H^+ being poured into the

---

*The term was introduced by Jacques Monod and Francois Jacob, French Nobel Laureates.

†Christian Bohr. Danish pioneer respiratory physiologist. His son Niels had some success as a physicist.

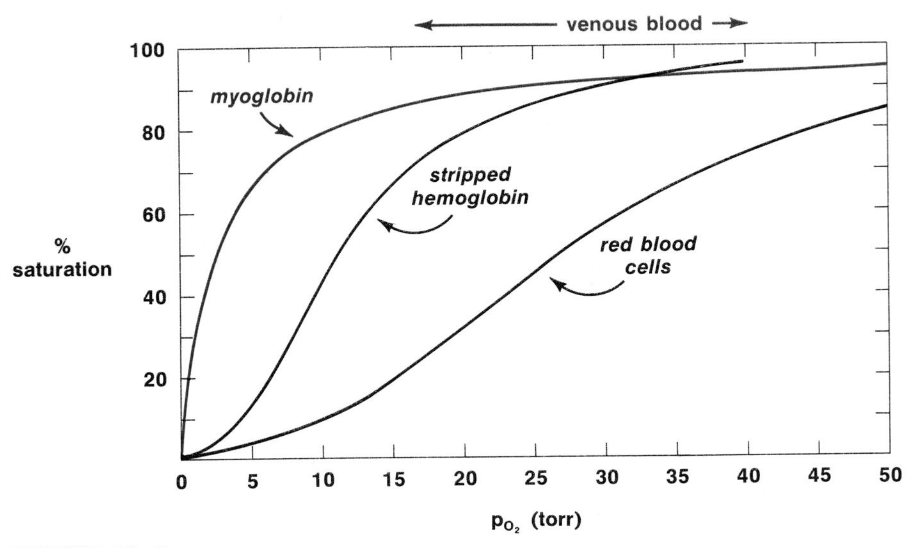

FIGURE 13–7

Solutions of hemoglobin that are stripped of organic phosphates and CO_2 have a much lower affinity for oxygen than do red blood cells. The stripped carrier still has the sigmoid association curve, but the p_{50} is well below the normal range of pO_2 in venous blood, and approaches the p_{50} of myoglobin. (Data from O. Brenna et al. (1972) Adv. Exp. Biol. Med. 28:19.)

blood from the tissues, the greater the release of oxygen in the capillaries. We shall see (Chapter 38) that there is some acidification of venous blood from CO_2 produced by the tissues, but the most important effects occur when there is an unusually high demand for oxygen, as in strenuously exercising muscles, or when the oxygen supply is impaired, as in circulatory defects. Under such circumstances, tissues produce large amounts of H^+, primarily in association with lactate formation. For example, the plasma pH of blood passing through normal hard-working muscles falls from 7.4 to 7.2 or lower. (The pH within erythrocytes is approximately 0.2 unit lower.) The consequences of such a change are shown in Figure 13–8; when H^+ concentration increases (lower pH), hemoglobin can retain less oxygen at a given oxygen tension. The additional delivery is equivalent to 10 per cent saturation (0.4 molecule of O_2 per molecule of hemoglobin) at the 20 torr partial pressure common in the venous drainage of trained working muscles. (The tension is lower in muscles of the flabby who are made to work hard.)

If H^+ decreases the affinity for oxygen, why doesn't it impair the uptake of oxygen by hemoglobin in the lungs? It does, but the saturation curve is so nearly flat at the higher oxygen tensions that the effect is small; hemoglobin is 1 to 2 per cent less saturated at pH 7.2 than it is at pH 7.4 in the lungs. In sum, the Bohr effect provides a built-in delivery of more oxygen in response to demand for oxygen, as manifested by increased H^+ formation.

Effect of CO_2. An important part of the decreased affinity for oxygen of hemoglobin in blood is due to the high concentration of CO_2. The total CO_2, which includes not only the dissolved gas but also carbonic acid and the bicarbonate ion in equilibrium with it, is 25 millimolar in adult arterial blood. The tissues discharge between 0.7 and 1.0 mole of CO_2 for each mole of O_2 consumed, causing a maximum increase of approximately 25 per cent in the total CO_2. Similarly, the baseline CO_2 concentration will also rise if there is any impairment in pulmonary ventilation, and

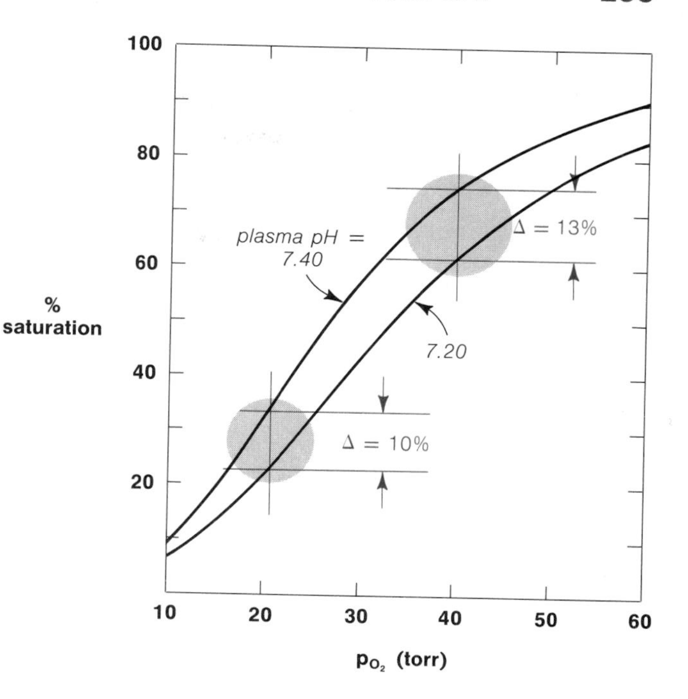

FIGURE 13–8

The Bohr effect is the decrease in affinity of hemoglobin for O_2 at higher H^+ concentrations (lower pH). A fall in the measured plasma pH from 7.4 to 7.2, such as seen in blood draining from hard-working muscles, causes corresponding changes within the erythrocytes, which result in a further 10 per cent discharge of oxygen while holding p_{O_2} at 20 torr—a typical value during heavy exercise. The discharge is even greater at 40 torr (13 per cent), showing that acidosis—a general increase in $[H^+]$ throughout body fluids—would tend to maintain a higher p_{O_2} in the tissues.

this rise further facilitates the release of oxygen to compensate for decreased uptake in the lungs.

Effect of 2,3-Bisphosphoglycerate. This compound, like H^+, diminishes the oxygen affinity of hemoglobin. Changes in 2,3-bisphosphoglycerate concentration are used to create longer-term adjustments of oxygen affinity in response to environmental changes. The phosphate ester is retained within the erythrocyte, and significant changes in its concentration requires hours.

For example, the concentration of 2,3-bisphosphoglycerate increases in erythrocytes when the delivery of oxygen is impaired. (The actual signal apparently is a rise in pH caused by excessive ventilation in the lungs, which causes abnormal losses of CO_2.) Exposure to high altitude is one cause. For example, light exercise during experimental exposure to low O_2 pressure corresponding to an altitude of 4,500 meters causes formation of an additional 0.8 to 1.1 molecules of 2,3-bisphosphoglycerate per molecule of hemoglobin within 24 hours. 2,3-Bisphosphoglycerate also increases when there is a loss of functional hemoglobin; this may result from actual loss of total hemoglobin (anemia), or by impairment of the function of existing hemoglobin through combination with carbon monoxide, which occurs in smokers and in those constantly exposed in other ways.

Structural Basis

What happens to a hemoglobin molecule when it binds oxygen? We have direct evidence from x-ray crystallography for the following:

The iron atom shifts. The iron atom in deoxyhemoglobin, lacking a ligand for the sixth coordination position, has a high spin state, and lies out of the plane of the porphyrin ring. Binding of an oxygen molecule stabilizes a low spin state, with the iron atom in the plane of the ring. There is a corresponding movement of the histidine residue attached to the iron (see p. 42), and a conformational change occurs in

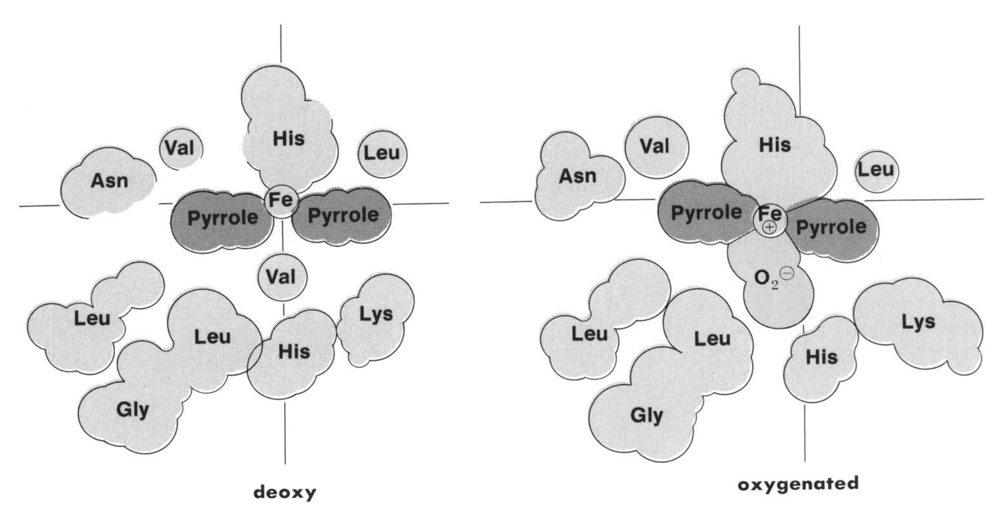

FIGURE 13–9

Changes in the geometry of the β-chain heme pocket upon oxygenation. The areas occupied by some groups in a cross-section through the iron atom are shown. (The porphyrin side chains are omitted.) The coordinates mark the position of the center of the iron atom in deoxyhemoglobin. Note the shift to the side and slight rotation of the Fe-porphyrin, the enlargement of the heme pocket, and the absence of the Val side chain below the iron atom in oxyhemoglobin. (Modified from J. Baldwin and C. Clothia (1979) J. Mol. Biol. 129:175.)

the globin polypeptide chain. (The lower net spin occurs because the strong ligand field of oxygen forces more electrons into paired orbitals in the iron atom.) Figure 13–9 shows some of the changes in the vicinity of the heme pocket caused by the change in position of the iron atom and its attached histidine residue.

The heme pocket changes shape. It is narrow in all of the subunits of deoxyhemoglobin, and oxygen's access to the iron atom in a β subunit is further blocked by a protruding valine side chain. In oxygenated subunits, the pocket opens and deepens, so the heme groups slide farther into the crevice; in the β subunits this moves the iron atom clear of the blocking valine side chain.

How do these differences come about? One can think of oxygen forcing its way into the heme pockets of deoxyhemoglobin, pushing it open, but this is an anthropomorphic picture of molecular behavior. No one knows what happens in detail, but a satisfying rationale can be developed in this way: Think of hemoglobin as breathing, that is, undergoing rapid transient local changes in structure from the most stable configuration. Large changes are rare, and it is only occasionally that the heme pocket will open enough for oxygen to slip in. However, once it does so, it blocks the polypeptide chain from resuming the deoxy configuration, and the chain shifts to a more stable shape conforming to the hemin-superoxide complex.

Similarly, it is only occasionally that hemin-superoxide rearranges back to heme and free O_2, with the O_2 escaping from the pocket. When this does happen, the open pocket is no longer stable; the heme group moves toward the surface, forming the stable deoxy configuration, and the walls of the pocket close in on it.

The subunits of hemoglobin shift position relative to each other. When the conformations of the individual subunits shift upon oxygenation, the orientation of amino acid side chains at the contacts between subunits changes. We can describe these changes most readily by labeling the subunits, α^1, α^2, β^1, and β^2, even though there is no difference betwen the two α chains or the two β chains. There is strong bonding between one α chain and one β chain, and each of the $\alpha\beta$ pairs formed this way move nearly as a unit upon oxygenation; they rotate and slip a little sideways with respect to each other, weakening the contacts between the $\alpha^1\beta^1$ pair and the $\alpha^2\beta^2$ pair:

$$\text{deoxyhemoglobin} + 4O_2 \longrightarrow \text{oxyhemoglobin}$$

It is possible to make reasonable interpretations of the allosteric behavior of hemoglobin in terms of these conformational shifts.

Tense and Relaxed Conformations. In the original Monod-Jacob formulation of allosteric effects, deoxyhemoglobin was said to have a tense conformation that impeded combination with oxygen and fully oxygenated hemoglobin to have a relaxed conformation with no impediments to oxygen binding. According to this view, there is an inherent equilibration of relaxed and tense forms, which favors the tense form of deoxyhemoglobin and the relaxed form of oxyhemoglobin (Fig. 13–10*A*). The tense form has little affinity for oxygen, while the relaxed form has a high affinity, so combination with oxygen stabilizes the relaxed form, making addition of more oxygen molecules easier.

Current experimental evidence favors an induced fit theory, originally intro-

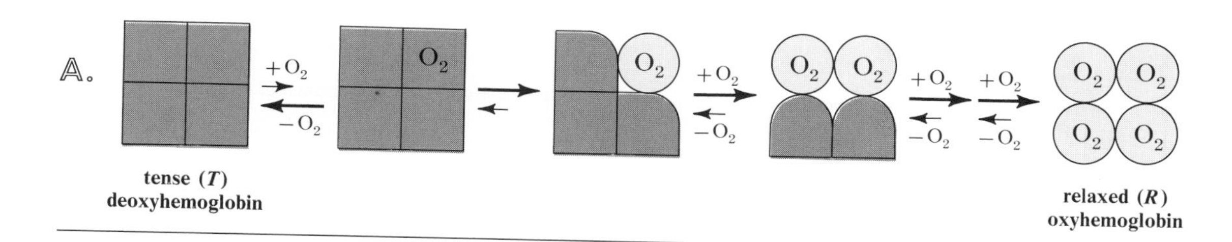

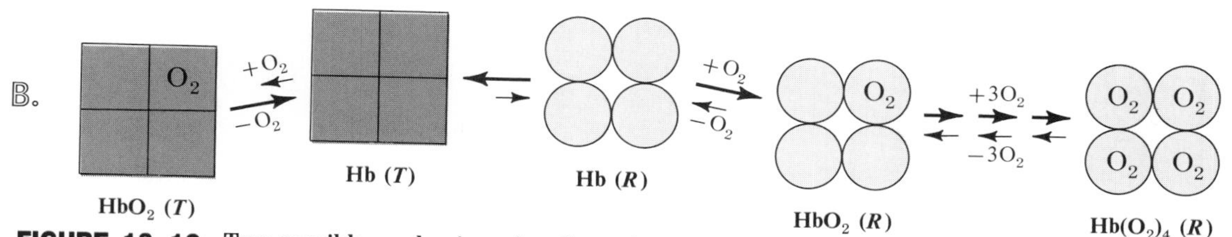

FIGURE 13–10 Two possible mechanisms for allosteric effects. *A.* In the induced fit model, an initial binding occurs with the tense conformation, but the equilibrium is unfavorable. However, binding of one molecule (*far left*) causes a shift in conformation of the subunit to a relaxed state, which modifies the bonds of the adjacent subunits so that they can take on additional O_2 more readily to form the fully oxygenated molecule (*far right*). Either mechanism could involve intermediate steps with distinct intermediates, or all of the oxygen molecules may be bound in a rush, once one is in place. *B.* The classic Monod-Jacob model invoked a constant equilibration of the tense (*brown squares*) and relaxed (*gray circles*) forms, even in the absence of oxygen, but with an equilibrium favoring the tense form (*second and third compounds from the left*). However, the R form has the higher affinity for O_2; therefore raising the pO_2 shifts the equilibrium in favor of the R conformation, which is then readily available for binding more O_2.

duced by Koshland*, in which addition of an oxygen molecule to one subunit alters the conformation of one or more adjacent subunits so as to increase its affinity for oxygen (Fig. 13–10*B*).

Origin of the Bohr Effect. The displacement of oxygen by protons and vice versa is not a direct substitution of one for the other; it is the result of differences in the acidic dissociation constants of some groups in the deoxy and oxygenated forms of hemoglobin, and it is not an all-or-none effect involving one group. Approxi-

*Daniel E. Koshland, Jr. (1920–). American biochemist.

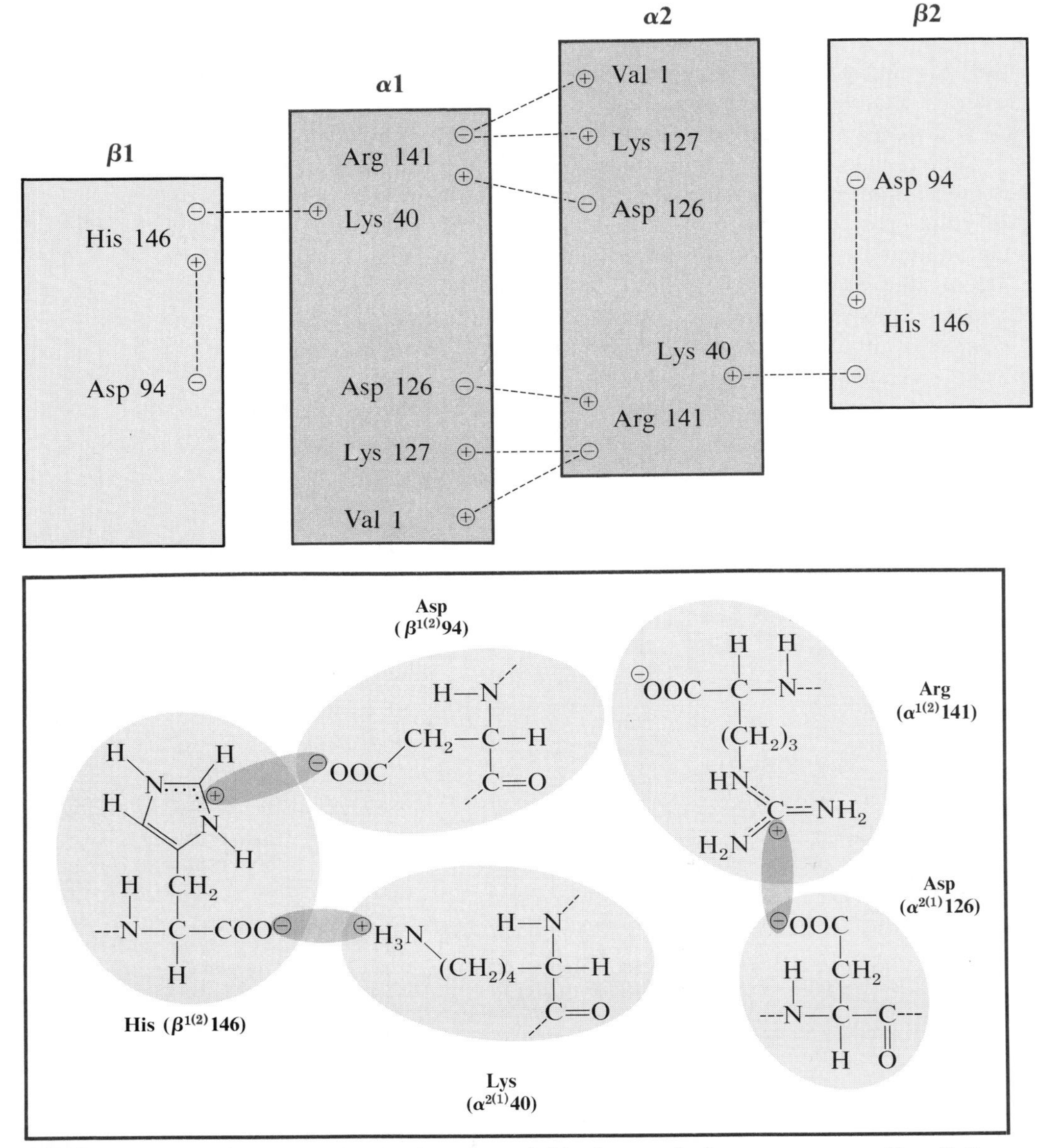

FIGURE 13–11 *Top.* Salt bridges in deoxyhemoglobin that are broken upon oxygenation. *Bottom.* Structures of typical examples of the three general types of electrostatic bonds represented: carboxylate-imidazolium bonds (*upper left*), carboxylate-ammonium bonds (*lower left*), and carboxylate-guanidinium bonds (*right*).

mately 20 per cent of the Bohr protons at neutral pH come from the terminal ammonium group of α chains (Val 1α), 40 per cent from the imidazolium group of a histidine residue in the β chain (His 146β), 15 to 20 per cent from another imidazolium group in the α chain (His 122α), and the balance perhaps from weak changes in several groups.

Some of these changes are caused by a disruption of salt bridges between the two $\alpha\beta$ pairs upon oxygenation. At least four, and perhaps five, such bridges are broken within each pair (Fig. 13–11). The carboxylate groups involved in these bridges are weak enough bases to have little affinity for protons; they retain nearly full negative charge whether free or in a salt bridge. However, the terminal ammonium group and the imidazolium group lose significant amounts of charge when free. Therefore, when oxygenation of hemoglobin causes rupture of salt bridges, these liberated cationic groups will release protons. Contrariwise, if the interior of an erythrocyte becomes more acidic, the increased proton concentration will increase the proportion of positively charged groups, favoring the formation of salt linkages and stabilizing the deoxy (tense) conformation:

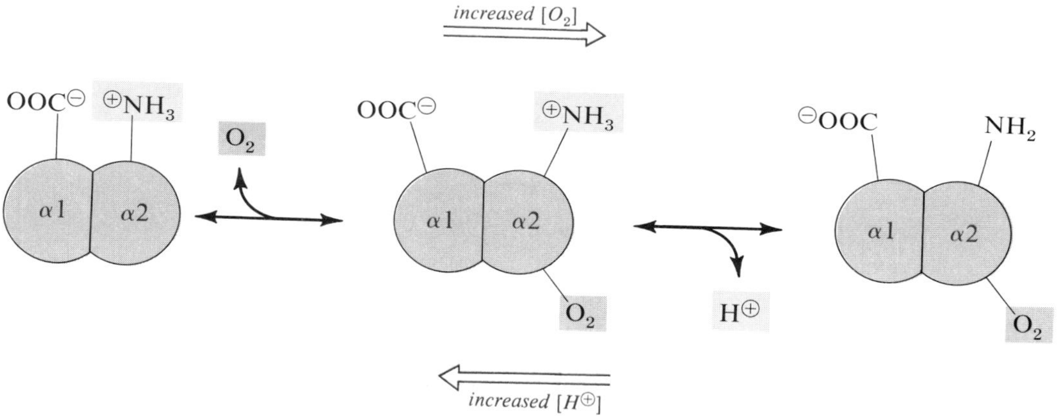

increased $[O_2]$

$OOC^{\ominus}$ $^{\oplus}NH_3$ — $\alpha 1$ $\alpha 2$ ⇌ (O_2) $OOC^{\ominus}$ $^{\oplus}NH_3$ — $\alpha 1$ $\alpha 2$ — O_2 ⇌ ($H^{\oplus}$) $^{\ominus}OOC$ NH_2 — $\alpha 1$ $\alpha 2$ — O_2

increased $[H^{\oplus}]$

Origin of the Bisphosphoglycerate Effect. Eight positively charged groups line the cleft between β subunits. The groups are so positioned that they will form strong electrostatic bonds with the negative charges on 2,3-bisphosphoglycerate when the molecule is in the tense conformation (Fig. 13–12). The bonds are weakened when the hemoglobin molecule is in the relaxed conformation with oxygen bound. The bisphosphoglycerate therefore tends to lock the molecule in the tense conformation and diminish the affinity for oxygen. When bisphosphoglycerate leaves, the space between the β chains is partly filled by one or more unidentified anions, probably sulfate or phosphate.

Origin of the CO_2 Effect. Carbon dioxide combines with terminal amino groups of the polypeptide chains to form carbamates:

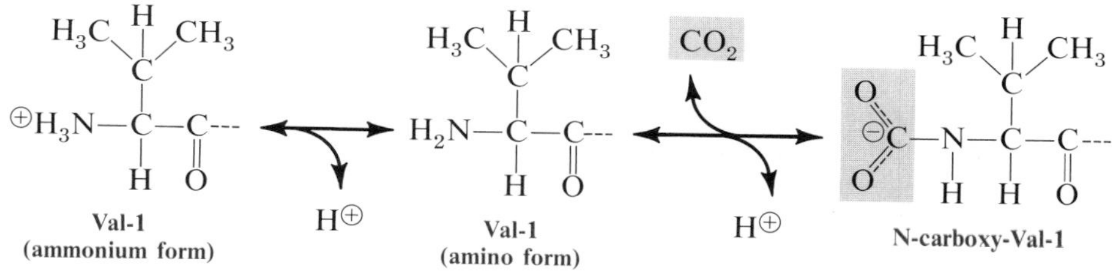

$$^{\oplus}H_3N{-}CH(CH(CH_3)_2){-}C(=O){-}\cdots \rightleftharpoons H_2N{-}CH(CH(CH_3)_2){-}C(=O){-}\cdots + H^{\oplus}$$

$$H_2N{-}CH(CH(CH_3)_2){-}C(=O){-}\cdots + CO_2 \rightleftharpoons {}^{\ominus}OOC{-}NH{-}CH(CH(CH_3)_2){-}C(=O){-}\cdots + H^{\oplus}$$

Val-1 (ammonium form) — **Val-1 (amino form)** — **N-carboxy-Val-1**

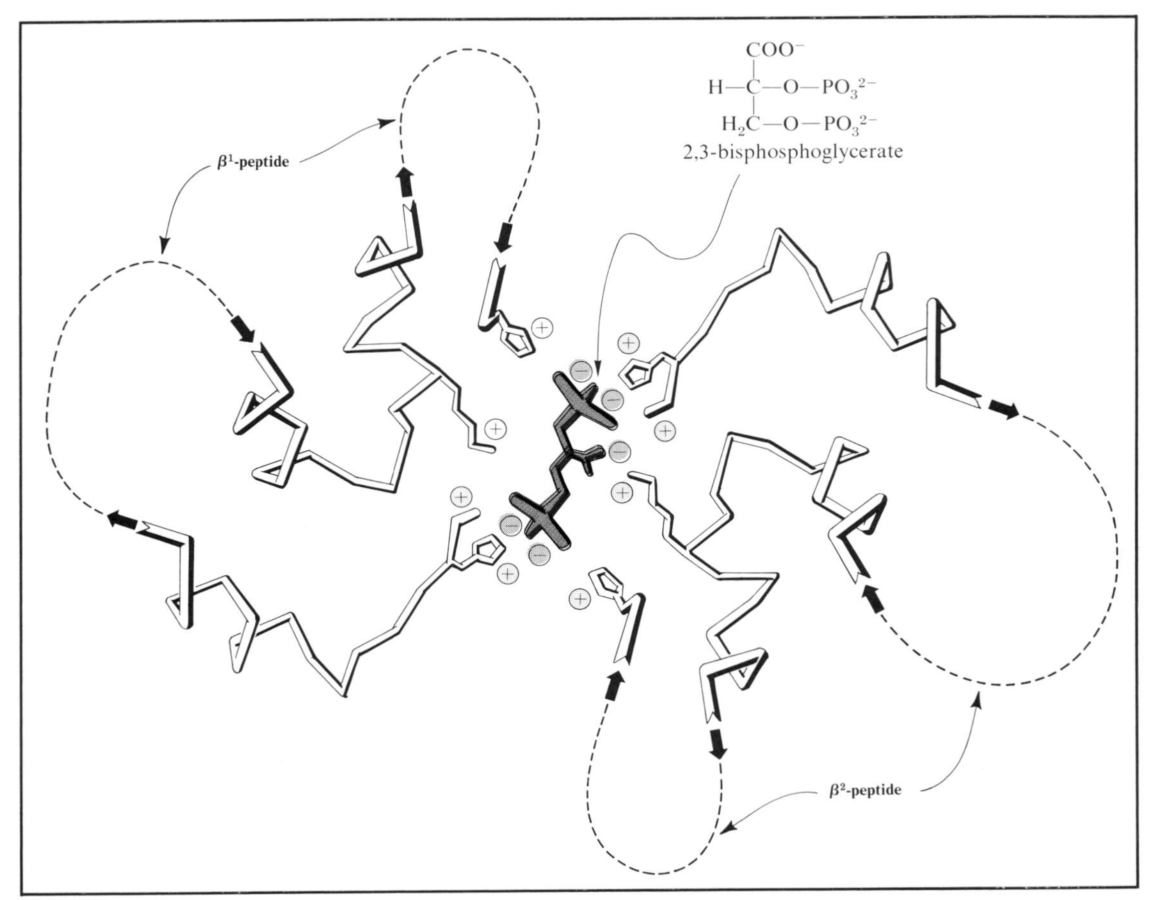

FIGURE 13–12 **2,3-Bisphosphoglycerate, with its five negative charges, combines with positively charged side chains in the cleft between β-subunits of hemoglobin, keeping them separated in the tense, deoxygenated conformation. The compound therefore diminishes the affinity of hemoglobin for oxygen. (Adapted from A. Arrone (1972) Nature 237:146.)**

Negative groups are now present on the formerly positive chain terminals, causing the terminals to shift position and form salt linkages with some of the remaining positively charged groups that ordinarily combine with bisphosphoglycerate. Carbamylation with CO_2 therefore mimics combination with bisphosphoglycerate; the tense conformation is stabilized and the affinity for oxygen is diminished. Because the same sites are involved, the effects of CO_2 and bisphosphoglycerate are not completely additive; one displaces the other to some extent.

STRUCTURAL VARIATIONS

Temporal Variations

Embryos grow so quickly that simple diffusion soon becomes inadequate for oxygen transport, and a carrier must be supplied. However, the conditions *in utero* are quite different than those encountered after birth; hence the need for more than one hemoglobin. Humans contain genes for producing at least seven different hemoglobin subunits, which are expressed at different times during development. The seven are designated α, β, ${}^{A}\gamma$, ${}^{G}\gamma$, δ, ϵ, and ζ. The ζ chains act in place of α chains; the others are all substitutes for β chains. A general chain formula for the human hemoglobins will therefore be:

$$\begin{array}{r|l} & \beta_2 \\ \alpha_2 & \text{or } {}^{A}\gamma_2 \\ \text{or } \zeta_2 & \text{or } {}^{G}\gamma_2 \\ & \text{or } \delta_2 \\ & \text{or } \epsilon_2 \end{array}$$

Embryonic hemoglobin is characterized by the presence of ζ and ϵ chains in place of the α and β chains found in adult hemoglobins. Details of the function of the hemoglobin are yet to be determined.

Fetal hemoglobins, Hb F, contain α chains that are the same as those found in adults, paired with either ${}^{A}\gamma$ or ${}^{G}\gamma$ chains. These γ chains differ from each other only in the presence of alanyl or glycyl groups at position 136.

Adult hemoglobin is mostly Hb A, the $\alpha_2\beta_2$ tetramer we have discussed at length. This is the chain composition for 97 to 98.5 per cent of the hemoglobin found in a sample of adults (mean 97.5 per cent). The remaining hemoglobin is mostly Hb A_2, which has δ chains in place of β chains, and is therefore $\alpha_2\delta_2$.

The mRNA for δ chains is relatively unstable, and this may explain why Hb A_2 is only 1.5 to 3.2 per cent of the total hemoglobin in adults (mean 2.48 per cent). The reason for the existence of this minor component is unknown.

The early maturation sequence is not worked out in detail, but a likely succession is shown in Figure 13–13. After initial growth, the content of Hb F rises; it is half of the total hemoglobin by the time (2 months) the embryo has grown to 30 mm length (crown to rump), and 90 per cent of the total at 50 mm length (2.5 months). The formation of adult hemoglobin, Hb A, begins about the 8th week of gestation but does not exceed 10 per cent of the total until the 30th week, and is 10 to 30 per cent at term. Thereafter, the contents of Hb A and Hb A_2 rise, and the content of

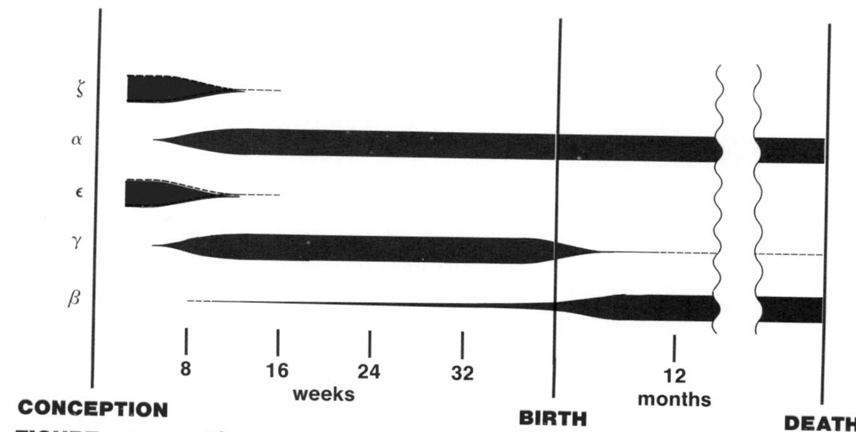

FIGURE 13–13 The sequence of development of hemoglobins. The extent of formation of hybrids such as $\zeta_2\gamma_2$ and $\alpha_2\epsilon_2$ during early life is not known. Hb F, $\alpha_2\gamma_2$, is the principal hemoglobin throughout the remainder of fetal life. The formation of the chains of Hb A is not significant until term and shortly thereafter, when the rate of Hb F synthesis falls rapidly. The formation of the δ chains is not shown—it would make a very thin line, at best—but it presumably parallels that of β chains.

Hb F falls until it is less than 10 per cent of the total at 6 months after birth and less than 2 per cent at 12 months. For some reason the γ chain genes are still turned on in precursors of some erythrocytes, but most adult cells contain no detectable Hb F. The Hb F is 0.03 to 0.7 per cent of the total hemoglobin in adults.

The necessity for a different hemoglobin before birth is clear; the oxygen tension available to the fetal blood in the placenta is lower than the oxygen tension available to the maternal blood in the lungs. Oxygen is transferred from maternal erythrocytes through the placenta to fetal erythrocytes. Therefore, fetal erythrocytes must have a higher affinity for oxygen than adult cells. The oxygen association curve of Hb F is much like that of Hb A; the important distinction is that Hb F binds 2,3-bisphosphoglycerate less tightly and therefore binds oxygen more tightly.

Post-translational Modification. Approximately 10 per cent of the molecules of Hb F are altered by acetylating the terminal amino group of α chains. Blocking the amino groups destroys the reaction with CO_2 and weakens the binding of 2,3-bisphosphoglycerate so that the affinity for oxygen is stronger. (Acetylation is also seen in adult felines; it is strange that the oxygen transport in cats differs from that in all other adult mammals examined.)

Hemoglobin spontaneously reacts with glucose and glucose 6-phosphate, forming several minor components in normal blood. Some have altered electrophoretic mobilities; these were detected first and assigned names*: Hb A_{1a1}, A_{1a2}, A_{1b}, A_{1c}, but there are others that are not separated by electrophoresis. Hb A_{1a1} and A_{1a2} are formed when one or two molecules of glucose 6-phosphate enter the cleft between chains in place of 2,3-bisphosphoglycerate. The most fully characterized and most abundant of these named components is A_{1c}, which is formed by a reaction of glucose with the β-chain terminal valine amino groups, first forming a Schiff's base and then an aminoketone by rearrangement:

```
    H—C=O                                  H—C=N—(Val)Hb                          H—N—(Val)Hb
      |                                      |                                      |
    H—C—OH                                 H—C—OH                                 H—C—H
      |                                      |                                      |
   HO—C—H      + H2N—(Val)Hb  ——→         HO—C—H      —Amadori rearrangement→     C=O
      |                                      |                                      |
    H—C—OH                                 H—C—OH                                HO—C—H
      |                                      |                                      |
    H—C—OH                                 H—C—OH                                 H—C—OH
      |                                      |                                      |
     CH2—OH                                 CH2—OH                                H—C—OH
                                                                                    |
                                                                                   CH2—OH
   glucose                               Schiff's base                          amino ketone
                                          (aldimine)
```

The concentrations of the A_1 hemoglobins increase two- to threefold in individuals with poorly controlled diabetes mellitus, owing to the high average concentration of glucose. The proportion of hemoglobin modified by combination with glucose or glucose phosphate is determined by the time of exposure and the carbohydrate concentrations, other things being constant. This makes the amount of the minor components an effective indicator of the integral of the glucose concentration over the average life span of an erythrocyte (120 days). The concentration of Hb A_{1c} is therefore being used as a measure of the severity of diabetes mellitus and the effectiveness of the control of the blood glucose level.

*The major adult hemoglobin is now usually designated Hb A_0 or simply Hb A, but was formerly called Hb A_1.

HUMAN HEMOGLOBINS*

Component	Chain Composition	M.W. (Anhydrous)	Comments
A	$\alpha_2\beta_2$	64,450	also minor A_1's
A_2	$\alpha_2\delta_2$	64,564	unknown function
^{A}F	$\alpha_2{}^A\gamma_2$	64,734	differ only in Ala or Gly at #136
^{G}F	$\alpha_2{}^G\gamma_2$	64,706	part of Hb F's is acetylated
Gower-2	$\alpha_2\epsilon_2$	?	known only in embryos
Portland-1	$\zeta_2\gamma_2$	?	known only in embryos
embryonic (Gower-1)	$\zeta_2\epsilon_2$		

Occurrence: Adult blood: 97.5–98.5% Hb A + A_1's
1.5–3.0% Hb A_2
0.03–0.7% Hb F
Fetal blood: >90% Hb F from ~10 weeks to 30 weeks gestation
$^G\gamma/^A\gamma = 0.7$ in newborn
p_{50} ~ 26.6 torr in adult, 22.7 in newborn (whole blood)
Concentration: 2.0–2.5 mM (13–16 g/100 ml) in adult blood (8–10 mM to heme)
lower in females than in males
$4.2\text{–}5.0 \times 10^{-16}$ moles/erythrocyte ($27\text{–}32 \times 10^{-12}$ g)

Subunits	Amino Acid Residues	M.W. (Including heme)
α	141	15,742
β	146	16,483
$^A\gamma$	146	16,625
$^G\gamma$	146	16,611
δ	146	16,540
ϵ	146	16,687†
myoglobin	153	17,669

Effectors (all decrease affinity for O_2):

H^+:
0.44 H^+ taken up per O_2 released (Haldane coefficient)
$\Delta \log p_{50}/\Delta \text{pH} = -0.48$ (Bohr coefficient)
60% taken up on α1Val, β146His; 15–20% on 122αHis

2,3-bisphosphoglycerate:
0.9 molecule/molecule of hemoglobin
bound in cleft between β subunits
40% bound in arterial blood; 90% would be bound if all O_2 discharged
content rises in conditions causing O_2 deprivation

CO_2:
1.3 molecules/molecule of hemoglobin in venous blood
bound as carbamate to terminal amino groups

*Note: Quantitative values are typical for adult blood and vary between individuals.
†Tentative

INDIVIDUAL VARIATIONS

Human blood is probably more readily accessible for study than any other experimental tissue, and samples are drawn each year from millions of individuals. Human hemoglobin therefore is excellent for studying genetic variation of a single protein in a single species, each variation representing an alteration in those DNA molecules affecting the synthesis of the protein.

Most mutations are deleterious and will gradually be eliminated, the rate depending upon the severity of the resulting reproductive disadvantage. Most hemoglobin variants therefore occur in only a few individuals, but some occur in many,

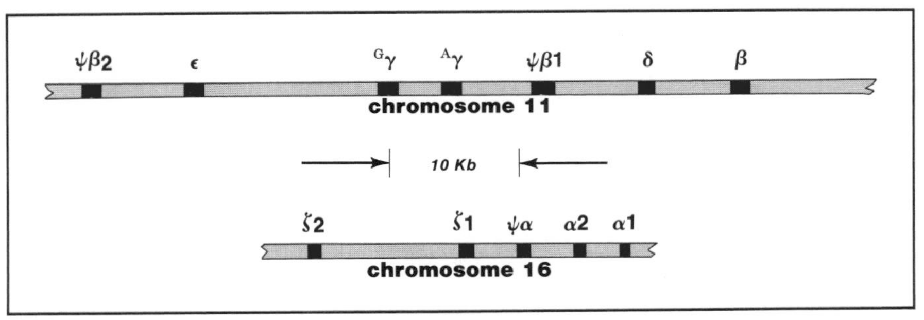

FIGURE 13–14 **The organization of globin genes on human chromosomes 11 and 16. These clusters include some pseudogenes ($\psi\alpha$ and $\psi\beta$), which resemble active genes in base sequence, but are not transcribed. The map is to scale, but the introns within the genes are not indicated.**

even though they are life-threatening. We must look in those cases for a compensatory advantage conveyed by the mutation.

Thalassemias. Impairments of the formation of a single kind of hemoglobin chain result in thalassemias, so-called because the conditions are common in Mediterranean countries (*Thalassa* = sea), and are designated α- or β-thalassemias to show the specific deficiency. They are also prevalent in Southeastern Asia, with an occurrence of some 1 in 100 in Thailand!

Thalassemias could in theory result from any mutation affecting transcription or translation of a globin gene, and they are heterologous, especially the β-thalassemias.

It is easier to understand the thalassemias if the pertinent gene distributions are kept in mind (Fig. 13–14). The genes for α and ζ chains are in a cluster on chromosome 16, while those for β, γ, δ, and ϵ chains are clustered on chromosome 11. Alterations of one cluster will have no direct effect on the other.

α-CHAIN THALASSEMIAS. Most humans have two α-chain genes on each chromosome-16, for a total of four genes. The genes are usually identical; the upstream flanking regions are not, but they are similar. The strong homology probably makes unequal crossovers easy in these regions. Should such a crossover occur, one daughter chromosome ought to have one α-chain gene, with three α-chain genes on the other. Both of these anomalies have been found in human populations.

Whatever the mechanism, gene deletion is the cause of most α-thalassemias. The effects depend upon the number of genes remaining. Total loss of α-chain genes prevents the formation of Hb F, as well as A and A_2. Fetuses survive for a time by making increased amounts of embryonic hemoglobin ($\zeta_2\epsilon_2$) and Hb Portland ($\zeta_2\gamma_2$), but they commonly die before term or shortly after delivery. (The condition is named **hydrops fetalis.**)

Loss of α-chain genes does not prevent the β-like genes from being turned on in the normal way during development. Since there are insufficient α-chains to combine with the β-like polypeptides, the accumulated excess forms homotetramers—β_4, ${}^A\gamma_4$, ${}^G\gamma_4$, or δ_4, and perhaps ϵ_4. Two have been given names: γ_4 is Hb Bart's, and β_4 is Hb H. (Bart's is the very English way of designating St. Bartholomew's Hospital in London.) These homotetramers are unstable and tend to form precipitates **(inclusion bodies)** within red blood cells; not only is there less normal hemoglobin produced, but the life of the cells is shortened.

Loss of only one α-chain gene results in a silent carrier state, α-thalassemia-2 (α-thal-2). It causes no discernible symptoms, but it is detectable by the presence of one to two per cent Hb Bart's in umbilical cord blood or by gene mapping with a restriction endonuclease. The gene composition is $\alpha-$ on one chromosome-16 and $\alpha\alpha$ on the other ($\alpha-/\alpha\alpha$).

Loss of two genes causes **α-thalassemia trait** (α-thal-1), in which the erythrocytes are small (microcytic), but there is little anemia. The loss may occur in cis ($--/\alpha\alpha$), as is common in Southeast Asia, or in trans ($\alpha-/\alpha-$), as is found in Africa and the Mediterranean countries.

Loss of three genes causes **hemoglobin H disease,** with a moderately severe inclusion body anemia. One quarter of the hemoglobin in fetal cord blood may be Hb Bart's and 4 to 30 per cent in adult blood will be Hb H in this condition. The population must contain a double deletion in cis ($--/\alpha\alpha$), as well as a single deletion ($\alpha-$), for the $--/\alpha-$ chromosome composition to occur; therefore Hb H disease, as well as hydrops fetalis, is rare in Africa but common in Southeast Asia. An estimated 25 per cent of the Thai have one or more missing α-chain genes, and hydrops fetalis is the leading cause of stillbirth among them.

β-THALASSEMIAS are caused by two principal types of defects. In one, β^0, there is a nearly complete loss of β-chain production; this is seen in Northern Italy and Southeast Asia. In the other, β^+, there is only partial loss; this occurs in Cyprus and tropical West Africa. Since there are only two genes for β-chain production in most individuals, there is only one heterozygous state possible for each type, and the effects of having only one β-chain gene are usually minor. The condition may only be discovered upon examination of the cells, or after development of anemia during stresses such as pregnancy or infection. Homozygotes or double heterozygotes (β^0/β^+) develop frank anemias, with precipitation of excess α chains. The genes for γ and δ chains are not affected in most cases, so that even heterozygous adults are likely to have increased amounts of Hb F and A_2 in their blood.

Each type of β-thalassemia is heterologous, and only a few of the genetic lesions have been clearly established. Some β^0-thalassemias are due to gene deletions, and the deletion in rare instances may encompass both the β and δ genes ($\delta\beta$-thalassemia) or various other segments of the β-gene cluster. Other β^0-thalassemias result from point mutations, either creating a terminator gene in the middle of one of the exons or changing the sequence at the junction of an exon and the major intron (p. 76) so as to prevent proper splicing of the putative mRNA. One form of β^+-thalassemia is known to result from a similar mutation but in the middle of the smaller intron, where it impedes but does not completely prevent mRNA production.

Serious attempts are being made to reduce the incidence of β-thalassemia in some locations through examination of fetal blood and abortion of homozygotes. The rate in Sardinia is now 1 in 800 births, compared to a former 1 in 250. The situation in Cyprus was even worse, with 1 out of 6 people a heterozygote and 1 out of 135 babies homozygous. Treating the affected children used nearly half of the local blood bank's supply. Repeated transfusions require use of a drug to remove excess iron (p. 672), and this alone consumed 6 per cent of the Ministry of Health's budget. As a result of counseling and screening, the incidence is now down to less than a third of its expected value on Cyprus, and there have been no new cases in London's Cypriot community in two years.

THALASSEMIAS AND MALARIA. Why should potentially lethal genes persist in such abundance in particular regions? The best explanation is that heterozygotes in some way gain partial protection against *Plasmodium falciparum,* the most dangerous malaria parasite, which spends part of its life cycle in red blood cells. This rationale appears most valid with β-thalassemias and the better known sickle cell trait, discussed later. They demonstrate classic cases of balanced polymorphism, with

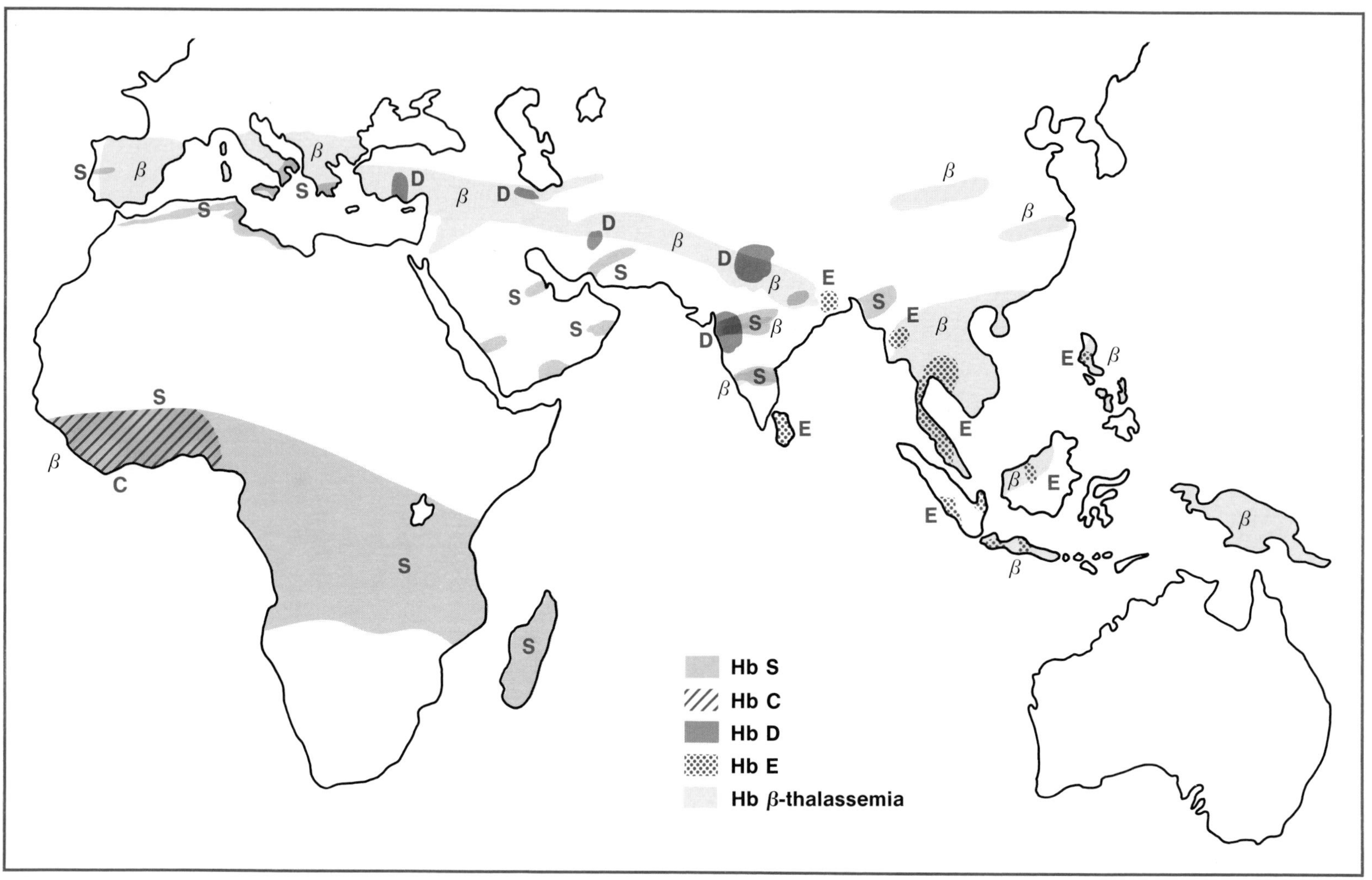

FIGURE 13–15 The geographical distribution of the common human hemoglobin mutants. The incidence of α-thalassemia has not been surveyed extensively, but it is known to be common in parts of Africa and southeastern Asia. The combined location of these mutants roughly outlines the regions of risk for falciparum (malignant) malaria. (This general sketch is based on information from H. Lehmann and R.G. Huntsman, (1974) Man's Hemoglobins, J.B. Lippincott.)

the incidence of the deleterious genes increasing toward levels at which the increase in deaths from their deleterious effects just balances the decrease in the death rate from malaria.

The case for a direct effect of α-thalassemia and some other point mutation genes on malaria is less clear, but assuming a beneficial association of some kind seems reasonable from their prevalence in the falciparum malaria belt (Fig. 13–15).

Coding Variants. Now let us consider mutations that change the amino acid sequence in one or more hemoglobin subunits. We discussed Hb Lepore and Miyada, in which parts of the genes for β and δ chains are combined, and Hb Wayne, created by a frame-shift mutation, in Chapter 7. The more common mutations alter a single base in DNA so as to substitute an amino acid at one position. The effect depends upon the amino acid and its location. Many such variants have substantially different properties (oxygen affinity, cooperativity, stability, solubility), but even variants with apparently normal properties may be synthesized at a slower rate, or destroyed faster, so that a heterozygote will have detectably lower concentrations of the mutant hemoglobin and an increased amount of Hb F. Homozygotes are then likely to be frankly anemic.

COMMON HUMAN HEMOGLOBIN MUTANTS

Thalassemias are impairments of gene transcription.
- α-thalassemias: one or more defective α-chain genes out of a total of four genes.
  - Impair formation of Hb A, A_2. F, embryonic.
  - *All genes affected* (– –/– –)—hydrops fetalis, with death *in utero* or shortly after birth. Accumulate $\zeta_2\gamma_2$ (Portland), γ_4 (Bart's).
  - *Three genes affected* (– –/α–)—Hb H disease, with 4–30 per cent β_4 and some γ_4 in adults. Inclusion bodies, abnormal red cells, mild anemia.
  - *Two genes affected* (α–/α– or – –/$\alpha\alpha$)—thalassemia trait, with 5 per cent Hb H in cord blood at birth, some mild anemia, and altered red cell morphology in adults.
  - *One gene affected* (α–/$\alpha\alpha$)—detectable γ_4 in cord blood at birth, no discernible effects in adults.
  - Common in S.E. Asia.
- β-thalassemias: one or two defective β-chain genes, with partial (β^+) or total (β^0) impairment of affected gene.
  - Impaired formation of Hb A_1.
  - *One gene affected*—thalassemia minor, if gene totally defective. Increased formation of Hb A_2.
  - *Both genes affected*—Cooley's anemia, thalassemia major, if both genes totally defective. 20–80 per cent Hb F in adults, with total hemoglobin falling to 0.3–0.5 mM (2–3 g/100 ml).
  - Common in many parts of the malaria belt.
  - Incidence among U.S. blacks: trait 15 (5–20) per 1,000 births; thalassemia major, 0.06 (0.006–0.1) per 1,000 births.

Hemoglobin variants with amino acid substitutions caused by point mutations. (All incidence figures are for U.S. blacks.*)
- S–S. Sickle cell anemia. β6Glu→Val. Deoxy form less soluble than is A_1. Precipitates in venous blood causing sickling and early cell destruction. Painful crises and short life. 1.6 (0.9–4.9) per 1,000 births; 0.5 (0.3–1.6) per 1,000 all ages.
- S–A. Sickle cell trait. No effect except when impairment of oxygen supply causes [deoxy Hb S] to exceed solubility. 80 (60–140) per 1,000 births.
- C–C. (β6Glu→Lys); D–D (β121Glu→Gln); E–E (β26Glu→Lys). Hb C, D, or E disease. Mild anemia in most instances with little impairment. 0.2 (0.02–0.5) of C–C per 1,000 births.
- C–A. No detectable effects. 30 (10–46) per 1,000 births.
- S–C. Causes sickling (S–C disease). 1.2 (0.3–3.2) per 1,000 births; 0.8 (0.2–2.1) per 1,000 all ages.
- D–A, E–A. No detectable effects.
- S–β-thalassemia. Causes sickling. 0.6 (0.15–1.4) per 1,000 births; 0.3 (0.08–0.7) per 1,000 all ages.
- C–β-thalassemia. 0.2 (0.02–0.5) per 1,000 births.
- Persistent Hb F trait. Failure of Hb F production to fall to normal adult levels. 1 per 1,000 births.
- Persistent Hb F homozygote. 0.00025 per 1,000 births.
- S-persistent Hb F. Causes sickling. 0.04 (0.03–0.07) per 1,000 births.
- C-persistent Hb F. 0.015 (0.005–0.02).
- β-thalassemia-persistent Hb F. 0.008 (0.0025–0.01) per 1,000 births.

*Incidence figures adapted from A.O. Motulsky: (1973) *Frequence of Sickling Disorders in U.S. Blacks*. N. Engl. J. Med., 288:31.

Humans with a long ancestry in the malarial belt commonly have one of four variant hemoglobins—**Hb C, D Los Angeles** (or Punjab), **E,** or **S.** All are so common that it is difficult to regard them as abnormal, although their effects are not innocuous. As many as 30 per cent of the inhabitants of some areas contain a gene for one or more of these variants. All are believed to ameliorate malaria in heterozygotes. All are created by substitution of one residue in each β chain of Hb A.

For example, the first abnormal hemoglobin discovered, Hb S, has a valine residue at position 6 in place of the glutamate residue found in Hb A. It may informatively be designated as $\alpha_2\beta_2^{6Glu \rightarrow Val}$ (read as Glu becoming Val), or more simply as $\alpha_2\beta_2^{6Val}$, or $\alpha^A{}_2\beta^S{}_2$.

SICKLE CELL ANEMIA AND TRAIT. In sickle cell hemoglobin, Hb S, the substitution of valine for glutamate residues at β6 on the surface provides a site for hydrophobic interaction between molecules. This is sufficient in deoxy Hb S to cause agglomeration of the molecules into long tubules. When this occurs, the usual discoid erythrocyte puckers into a sickle shape. Circumstances causing this are critical for the individual because some of the sickled cells are so severely distorted that they cannot recover, and are destroyed. The average lifetime of the cells in a typical homozygote is reduced to 17 days from the normal 120 days.

Since it is the deoxy form that precipitates, the occurrence of sickling depends upon the concentration of both Hb S and oxygen. In a typical homozygote, Hb S will begin to precipitate at oxygen tensions near 50 torr, and most cells will sickle after long exposures to 40 torr. Sickling is slow, however, and cells usually rush back to the lungs for reoxygenation before damage occurs. If there are sluggish spots in the circulation, sickling can begin within them and obstruct the flow further until the oxygen supply for the surrounding tissues is lost. Ulcers on the front of the lower leg (pretibial ulcers) are common in sickle cell patients because of the propensity of the young for banging their shins.

It is imperative for the sickle cell patient to maintain a normal oxygen supply to the lungs. Living at high altitudes, flight in unpressurized aircraft, anesthesia, any condition causing fluid to accumulate in the lungs—all can be life-threatening.

Production of Hb F is increased in sickle cell anemia, and its presence in some way inhibits sickling. Indeed, patients homozygous to Hb S may be relatively free from symptoms if they are among the few who also have some condition causing increased formation of Hb F, such as some forms of homozygous β-thalassemia, or a hereditary inability to suppress γ-chain formation during development (hereditary persistence of fetal hemoglobin).

The concentration of Hb S in heterozygotes—those with sickle cell trait—is usually too low to cause significant sickling under physiological conditions at sea level, and the extent to which it may be a problem under other circumstances is not settled. Trained athletes with the trait had no problems at 3400 meters in the Olympics at Mexico City. On the other hand, four black soldiers died after vigorous exercise in El Paso, elevation 1200 meters, and two trainees collapsed at the Air Force Academy, elevation 1800 meters, all without apparent cause except for being heterozygous to Hb S. (The resultant exclusion of blacks with the trait from some military pursuits is now being challenged in court.) What is clear is that the presence of the trait does mean death for the malaria parasite. The parasite lowers the pH of the cells in which it is growing to the point where they sickle, and this kills the intruders.

RARE POINT MUTATIONS. Over 350 variants in human hemoglobins have been described, most of them point mutations. They have so far been discovered in all of the known subunits except ϵ and ζ. Most are rare and known only in the heterozygous condition, but they have been useful in understanding the structural basis for hemoglobin function.

Mutations that change the lining of the heme pocket, particularly those causing the introduction of polar residues, often make the heme susceptible to oxidation, creating methemoglobins in which at least two of the groups are oxidized to hemin (Fe III). Such hemoglobins are designated **Hb M**: for example, M Osaka ($\alpha_2^{58His\rightarrow Tyr}\beta_2$) or M Milwaukee ($\alpha_2\beta_2^{67Val\rightarrow Glu}$). (The histidyl group altered in M Osaka is near the site at which oxygen is bound; the valyl group altered in M Milwaukee is in contact with the porphyrin ring.) These hemoglobins behave like myoglobin, and it would not be possible for a homozygote to survive.

Other instructive mutations have aided in pinpointing the location of bonds involved in cooperative interactions and the sources of the protons liberated in the Bohr effect.

Variations between Species

The number of different hemoglobins and their transport properties differ among organisms; it is frequently possible to rationalize the alterations in terms of the physiological requirements of the particular species. In general, the larger the animal, the smaller the flux of oxygen required by a given amount of tissue and therefore the lower the oxygen tension required to maintain that flux. Large mammals have hemoglobins with a higher affinity for oxygen so that the tension drops to lower levels before oxygen is released; their hemoglobins also have a smaller Bohr effect.

Some primitive organisms have monomeric hemoglobin units that generate cooperative interactions through reversible associations rather than through the permanent interactions within tetramers seen in mammalian hemoglobins.

Evolution of Hemoglobins. Amino acid sequences are known for hemoglobins from a wide variety of organisms, and it is possible to trace the evolution of the different chains by comparison of the sequences. The shorter the time since the divergence of any two species from a common ancestor, the fewer the changes that will have occurred in the amino acid composition of a given protein.

For example, the number of differences at comparable positions between human subunits and those in other species are:

Subunit	Chimpanzee	Gorilla	Rhesus monkey	Pig	Cow
α-chain	0	1	4	18	17
β-chain	0	1	8	23	25

The evidence from hemoglobin composition (and from protein composition in general) is that the chimpanzee is a close relative of man, the rhesus monkey satisfyingly more distant, and the cow can be eaten with a clear conscience.

Maximizing the Homology. Mutations may cause deletion or insertion of amino acids into a polypeptide, in addition to substitutions. In order to compare two chains, they must be matched in such a way as to allow for the differences in chain length. This is done by aligning the chains with the maximum number of similar amino acids in comparable positions. It is relatively easy to do this with the hemoglo-

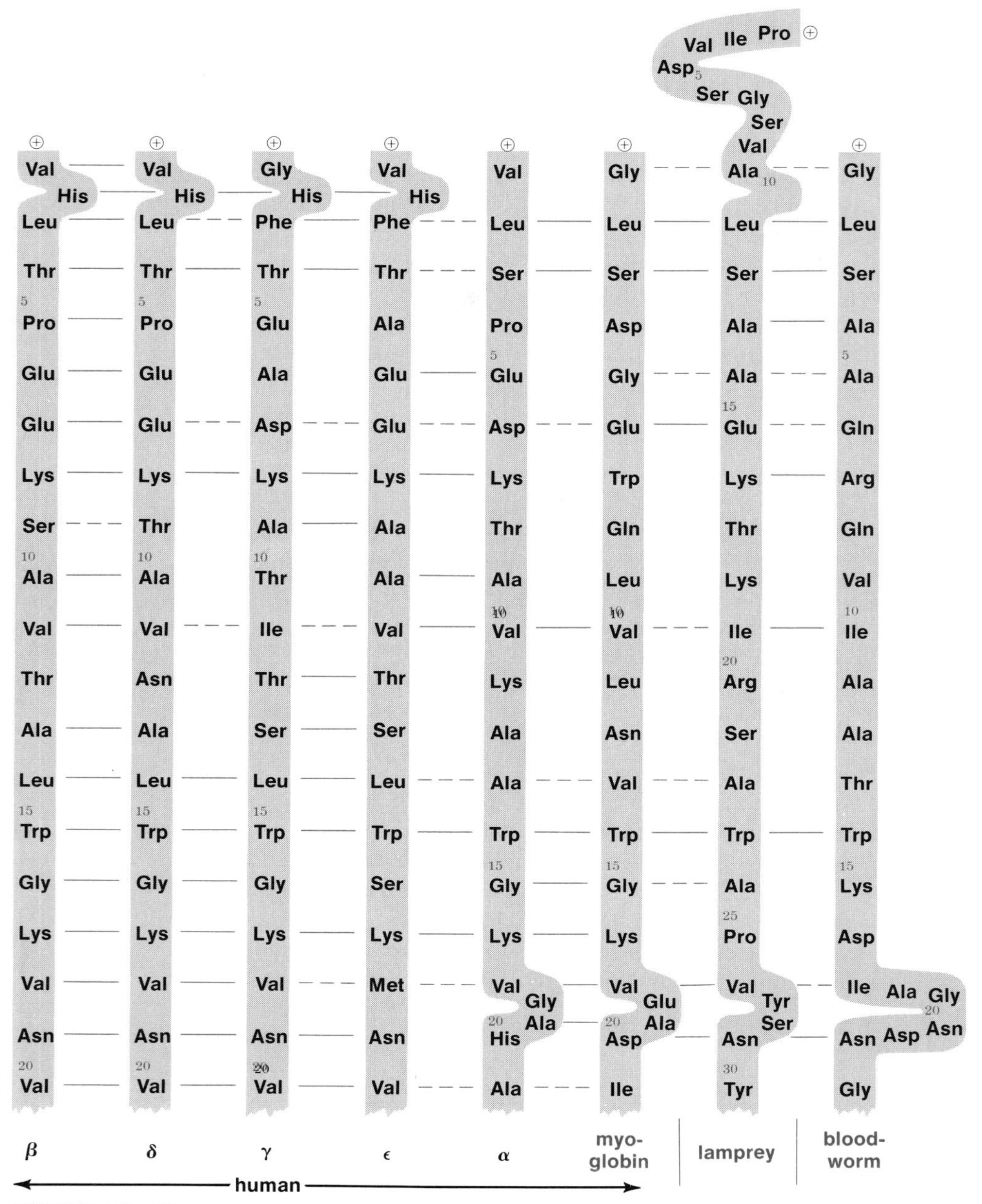

FIGURE 13–16

The amino acid sequences of the amino-terminal portions of different hemoglobin chains are arranged here so as to align residues of comparable function ("maximizing the homology"). Identical residues in neighboring chains are connected by *solid brown lines;* residues of similar character (hydrophobic, H-bonding, etc.) are connected by *dashed brown lines.* Analysis of the relationships allows some deductions to be made about the evolution of the chains. For example, the few differences between β and δ chains indicates a relatively recent divergence, but the common occurrence of 2-His in these and the γ chains indicates they are at least kissing cousins. The additional residues present at α-18,19, and at the corresponding positions in myoglobin, and in the lamprey and bloodworm hemoglobins, are not in the β, δ, or γ chains, indicating that the latter three and the α chain had separate ancestors.

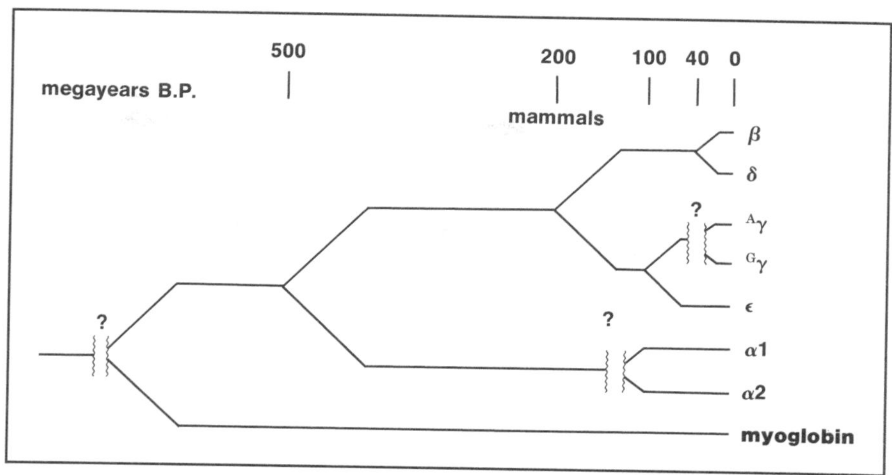

FIGURE 13–17 The order of evolution of various hemoglobin subunits. No indication of relative times is intended. The positions of ε and ζ chains on this family tree are not known.

bin chains. Figure 13–16 shows how the known sequences from the various human chains can be matched with those from a primitive vertebrate and an even more primitive invertebrate. The lines on the figure show where identical or functionally similar amino acids have been aligned in neighboring chains.

Information like that in Figure 13–16 is also a guide to the evolution of the different hemoglobin subunits. The β and δ chains are more like each other than like the γ chain, but the three of them have more resemblances to each other—the insertion of 2-His, for example—than they have to α chains or to myoglobin. Both the α chains and myoglobin have two additional residues (18 and 19) similar to the primitive lamprey hemoglobin, and the bloodworm has four additional residues in a comparable location. Study of the complete sequences along these lines indicates that the hemoglobins have evolved as shown in Figure 13–17.

FURTHER READING

Structure and Function

J. Baldwin and C. Clothia (1979). Haemoglobin: The Structural Changes Related to Ligand Binding and Its Allosteric Mechanism. J. Mol. Biol. 129:175.

R.P. Cole (1982). Myoglobin Function in Exercising Skeletal Muscle. Science 216:523.

Genetics

T. Maniatis et al. (1980). The Molecular Genetics of Human Hemoglobins. Annu. Rev. Genet. 14:145.

A. Efstratiadis et al. (1980). The Structure and Evolution of the Human β-Globin Gene Family. Cell 21:653.

R. Lewin (1981). Evolutionary History Written in Globin Genes. Science 214:426.

R.E. Gale, J.B. Clegg, and E.R. Huehns (1979). Human Embryonic Haemoglobins Gower 1 and Gower 2. Nature 280:162.

Genetic Diseases

E.J. Benz, Jr. and B.G. Forget (1982). The Thalassemic Syndromes: Models for the Molecular Analysis of Human Disease. Annu. Rev. Med. 33:363.

D.J. Weatherall and J.B. Clegg (1981). The Thalassemia Syndromes, 3rd ed. Blackwell.
A.M. Dozy et al. (1979). α-Globin Gene Organization in Blacks Precludes the Severe Form of α-Thalassemia. Nature 280:605.
S.M. Hanash and D.L. Rucknagel, eds. (1980). Clinical Implications of Recent Advances in Hemoglobin Disorders. Med. Clin. N. Am. 64:775.
S.H. Orkin and D.G. Nathan (1981). The Molecular Genetics of Thalassemia. Adv. Hum. Genet. 11:233.
T.H. Maugh (1981). A New Understanding of Sickle Cell Emerges. Science 211:265, 468.
C. Holden (1981). Air Force Challenged on Sickle Trait Policy. Science 211:257.
A.V. Orjih et al. (1981). Hemin Lyses Malaria Parasite. Science 214:666. Possible rationale for thalassemias.

CHAPTER 14

THE ENZYMES: THEIR NATURE

A living organism is a magnificent assembly of chemical reactions, an entity rather than chaos because most of its components make an orderly contribution to existence of the whole. Many of these reactions occur at significant rates only because specific proteins—the enzymes—are present to catalyze them. The enzymes are constructed to control both the kinds of reaction and the rates at which they occur, and it is the existence of these proteins that creates a complex harmony of chemical function.

Enzymes contain the same amino acids found in other proteins, and they have a three-dimensional structure that represents the conformation of least energy content, as do other proteins. All of our present information supports the view that enzymatic catalysis involves the same kinds of bonds that appear during ordinary organic reactions, using structures supplied by amino acid residues and by prosthetic groups. The binding of prosthetic groups is itself determined by conformation of amino acid residues, so the study of the nature of enzymes is in many ways only an extension of the principles of protein structure that we have already developed.

Catalytic Site. Why then do some proteins act as very effective catalysts—more effective than any other known compounds in making some reactions go faster? The answer lies in the ability of proteins to be molded into a variety of shapes. Proteins acting as enzymes are built with clefts that function as catalytic sites. Many compounds can wander in and out, but when one with a proper fit diffuses in, the protein closes around it (Fig. 14–1) and brings to bear chemical groups in a particular geometric conformation. Here is the basis for enzyme function: The catalytic effective-

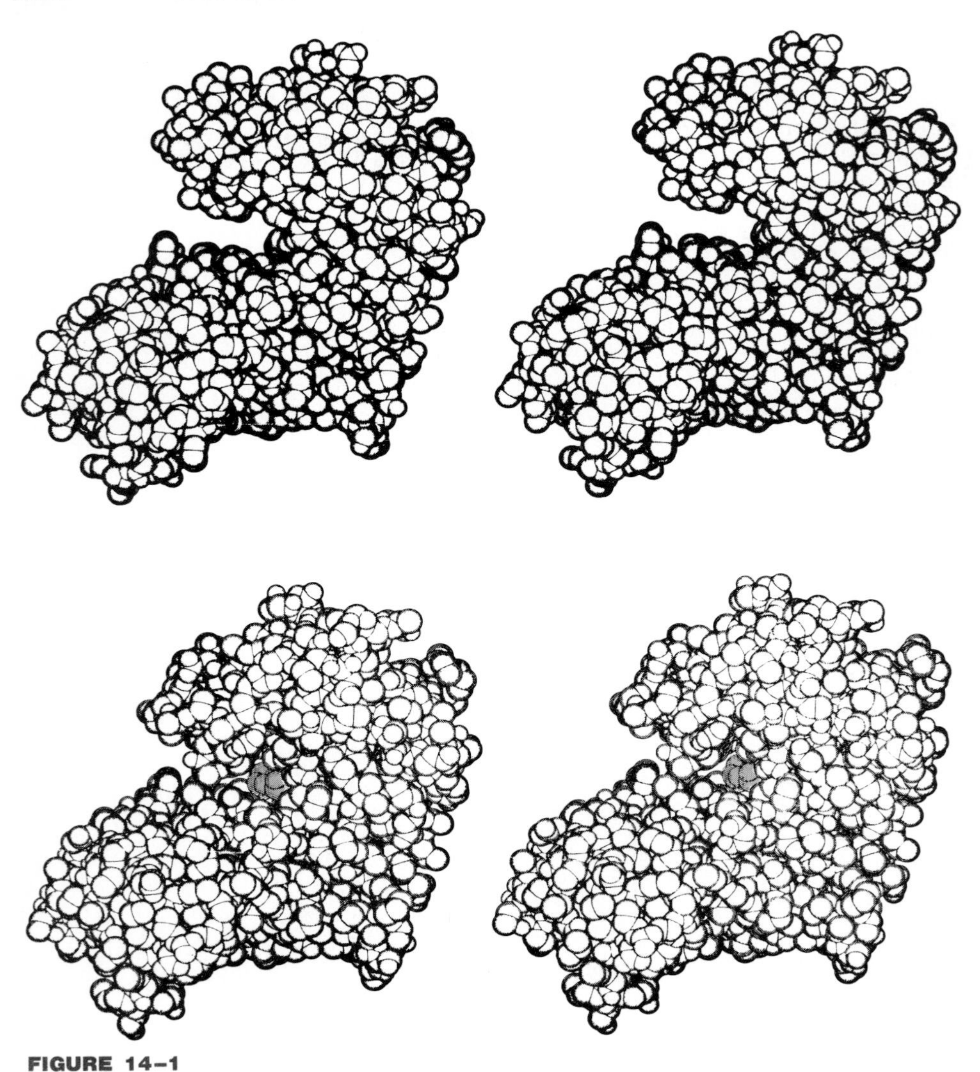

FIGURE 14–1

Shift in conformation of an enzyme upon binding substrate. *Top.* Stereo view of the enzyme with an open cleft accessible to substrates. *Bottom.* Upon entry of substrates (*color*), the enzyme closes around it. The enzyme in this case is hexokinase (p. 310), and the substrates are glucose and ATP. (Color added to drawing from C.M. Anderson, F.H. Zucker, and T.A. Steitz (1979) Science 204:375. © Copyright by the American Association for the Advancement of Science and reproduced by permission.)

ness of amino, carboxyl, and other bonding groups is increased by several orders of magnitude by placing them in fixed spatial arrangements where they can lock onto the affected compound; yet it is possible to stretch or otherwise deform particular bonds through relatively small changes in conformation of the protein.

We shall see that the use of proteins as catalysts conveys other advantages. They can be constructed with separate domains, making rapid multimolecular reactions feasible in some cases, or providing binding sites for allosteric effectors with which to regulate the rates of reactions.

Specificity. The **catalytic site** of enzymes is a part of a **binding site.** Enzymes, like antibodies, are frequently highly specific in binding particular compounds, because the amino acid residues in the binding site closely fit only a few compounds, perhaps one. To use the technical nomenclature, an enzyme may have a high specificity for its substrates.

Substrate is an old term, firmly embedded in the biochemical literature, for a compound whose reaction is catalyzed by an enzyme:

$$\text{substrate} \xrightarrow{\textit{enzyme}} \text{product}$$

Rephrasing the concept of specificity, one enzyme may catalyze the reaction of only a few different substrates. Put still another way, only a few compounds may be capable of acting as substrates for an enzyme.

The specificity of individual enzymes is an important feature of biological control because changes in the activity of particular enzymes will affect only a few compounds. However, there is another aspect of specificity that concerns catalytic ability itself, and this is how it comes about. A selectivity toward a few compounds—high specificity—implies a close molding of the binding site to a particular geometry, with participation of several side chains from the enzyme to create a much

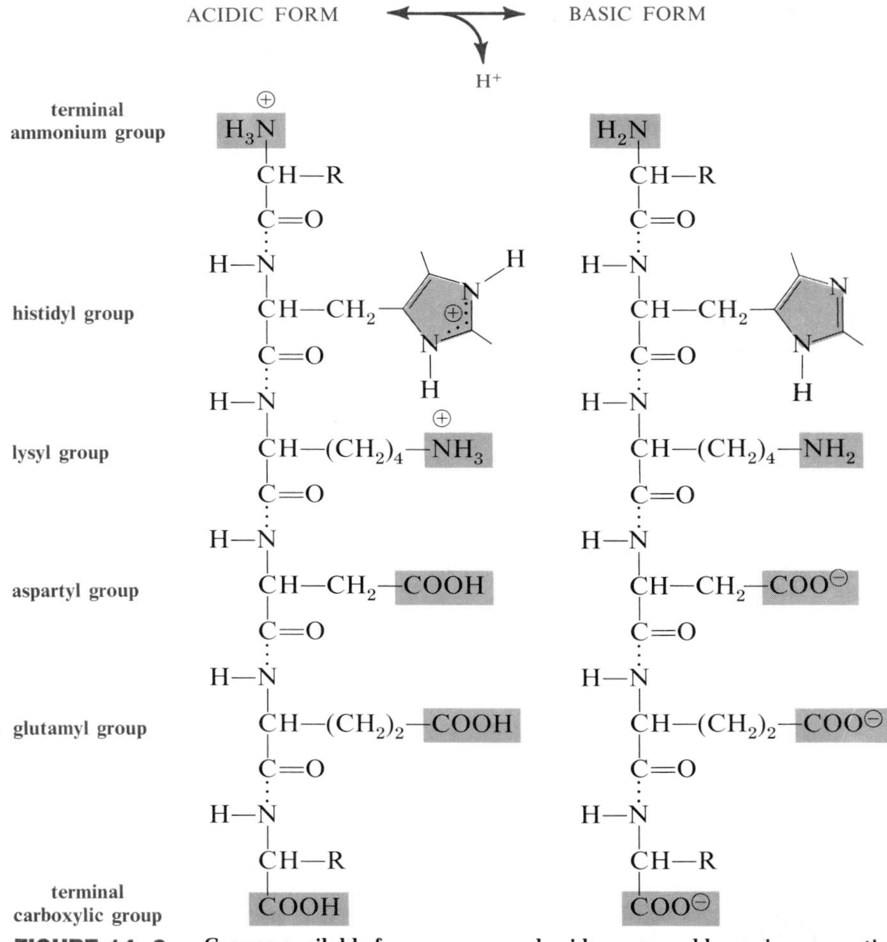

FIGURE 14–2 Groups available for use as general acids or general bases in enzymatic reactions include side chain substitutents and the terminal ammonium or carboxylate groups. Each may in some instances act as an acid and in others as a base; the two forms are shown.

stronger force with "good" substrates than with other compounds. The existence of this stronger force is an important factor in easy deformation of the substrate to break old bonds and create new ones. Enzymes differ from antibodies in that tight binding may change the bound compound.

Catalytic Groups. Another clue to the catalytic mechanism comes from the presence in proteins of groups known to be effective catalysts for ordinary organic reactions. For example, many amino acid side chains are capable of donating or accepting protons so as to act as **general acid** or **general base catalysts** (Fig. 14–2). Other groups, such as hydroxyl or amino groups, may act as **nucleophiles,** and sur-

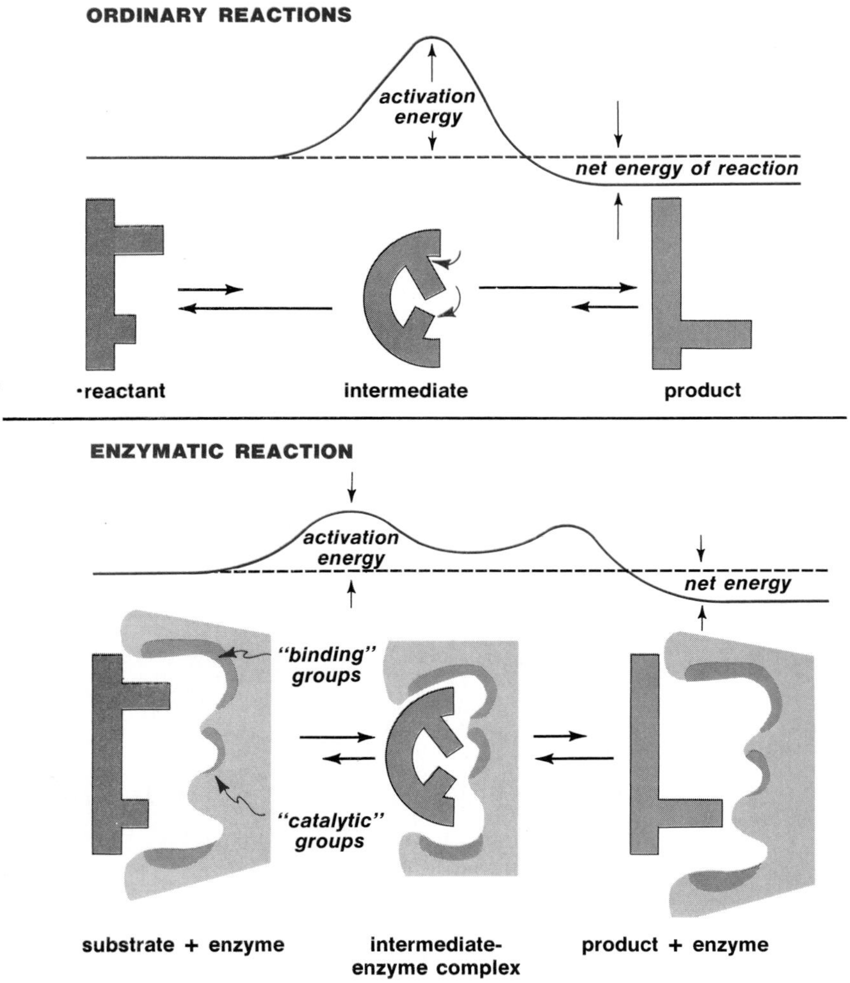

ORDINARY REACTIONS
activation energy
net energy of reaction
reactant
intermediate
product

ENZYMATIC REACTION
activation energy
net energy
"binding" groups
"catalytic" groups
substrate + enzyme
intermediate-enzyme complex
product + enzyme

FIGURE 14–3

Top. **In ordinary chemical reactions, a molecule must assume some unstable intermediate form before it changes into a product. The probability of occurrence of the intermediate is low as shown by the arrows, which is equivalent to saying that it represents a state of high energy content. A molecule must cross this activation energy barrier before reaction occurs, even if the overall process has a substantial net loss of energy.**

Bottom. **In enzymatic reactions, the binding and catalytic groups of the enzyme combine with the substrate in such a way as to make the intermediate-enzyme complex have a lower energy content than the complexes of substrate or product with the enzyme. The activation energy for formation of the intermediate-enzyme complex is therefore much lower than the activation energy for the reaction in the absence of enzyme.**

rounding structures in an enzyme are often designed to reinforce their catalytic properties. (Nucleophiles in effect donate electrons to bond other groups lacking filled electron orbitals.)

General Catalytic Mechanism. Enzymes owe their peculiar catalytic effectiveness to a combination of specific binding and the presence of catalytic groups. In an ordinary chemical reaction, a compound must contort into an unfavorable configuration before it changes into something else (Fig. 14–3); the formation of the less probable intermediate state represents an activation energy barrier that must be overcome before reaction can occur.

Enzymes change this situation by being built so that their combination with the reacting compound is most stable when the compound is in a "high-energy" intermediate form. An enzymatic reaction proceeds through at least three stages—the formation of a complex between enzyme and substrate, the conversion of this complex to an enzyme-intermediate complex, and the further conversion to a complex between enzyme and product that can dissociate:

$$\mathbf{E + S \rightleftharpoons ES \rightleftharpoons EI \rightleftharpoons EP \rightleftharpoons E + P}$$

The groups on the enzyme are arranged to make the total energy of the enzyme intermediate complex less than the total energy in the enzyme-substrate or enzyme-product complexes. The activation barriers in converting enzyme-substrate to enzyme-intermediate, or enzyme-intermediate to enzyme-product are substantially lower than the barrier without enzyme.

The specific binding of substrate to enzyme therefore serves not only to confine particular reactions to a few compounds but also to create a complex that spontaneously favors the formation of a more reactive configuration. The enzyme does several things to achieve this end. It provides catalytic groups in a single molecule, eliminating the need for collision with two or more molecules during the course of the reaction **(entropy effect);** it aligns the substrate at favorable angles relative to the catalytic groups so that the interaction with these groups is facilitated **(orbital steering),** and it provides binding groups at positions that stabilize a reactive intermediate **(propinquity effect).**

Perhaps the easiest way to get a feel for what enzymes do and how they go about it is to examine a few of them in detail, not because these details are something that ought to be committed to memory, but so we can develop some general principles by example. Enzyme mechanisms are rarely "seen" in the sense that they can be frozen for inspection by x-ray crystallography; they are deduced through the use of techniques such as molecular probes that react with active sites, destruction of selected amino acids, spectral changes, and more recently by neutron diffraction; hence our concentration on the principal features rather than the more conjectural fine points of mechanism.

AN ENZYME SAMPLER

Pancreatic Ribonuclease A

Many different enzymes are present in the body to catalyze the hydrolysis of compounds. Some of these hydrolases are secreted into the gastrointestinal tract to split dietary fats, proteins, and polysaccharides into their components for absorption. Other hydrolases are present within tissues to aid in the mobilization of stored fuels or to participate in the constant cycle of degradation and synthesis through which the tissues remain rejuvenated. Among these important enzymes are **phosphodiesterases,** hydrolyzing one of two ester linkages on a single phosphate group, such as occur in phospholipids and in nucleic acids.

The enzyme examined here is one secreted by the pancreas of cattle and other ruminants to catalyze the hydrolysis of RNA; it is a **ribonuclease.** It is believed to be useful to cattle because bacteria in the rumen convert part of the host's nitrogen supply to RNA, which must be degraded to recover the nitrogen. The enzyme is relevant to students of human biochemistry because this relatively small protein, containing only 124 amino acid residues in a known geometry, has been intensively studied as a model for the action of other enzymes.

The Overall Reaction. Ribonuclease A catalyzes the hydrolysis of 5′-phosphate ester bonds in RNA in which the phosphate group is also attached to the 3′-carbon of a pyrimidine nucleoside, such as cytidine or uridine. The 5′-bond may be on any nucleoside, but the enzyme is most active when it is on adenosine:

pyrimidine (5′, 3′) — adenine (5′, 3′)

—P(O⊖O)—O— ribose —O—P(O⊖O)—O— ribose —O—P(O⊖O)—

↓ H_2O, *ribonuclease*

pyrimidine: —P(O⊖O)—O— ribose —P(O⊖O)—OH

adenine: HO— ribose —O—P(O⊖O)—

Continued action of ribonuclease A hydrolyzes a polynucleotide into fragments, all except the terminal pieces bearing a free hydroxyl group at the 5′ end and an ester phosphate group at the 3′ end.

The catalytic site is a cleft in which amino acid side chains and amide groups in the polypeptide backbone are used to identify the nucleic acid substrate and bond it in the proper position for catalysis. However, let us concentrate only on three groups involved in the catalytic mechanism (see top figure, opposite page).

One **histidine** side chain (His 119) bonds in its **acidic imidazolium** form to the substrate; another (His 12) bonds in its **basic imidazole** form.* These groups act as general acids and general bases in the catalytic mechanism. A **lysine** side chain is also involved, and may aid the proper siting of the substrate by its attraction for the negatively charged phosphodiester group.

Electrons shift so that His 12 gains a proton, becoming an acid (see bottom figure, opposite page).

Two oxygen atoms of the pyrimidine nucleoside are now linked in phosphate ester bonds, forming a pentavalent intermediate, which is stabilized by the neighboring lysine side chain. His 119 in its acidic form disrupts the intermediate by donating a proton. This breaks the polynucleotide backbone, and one of the two pieces can now leave (see top figure, p. 278).

The other piece of RNA, the one that must contain a pyrimidine nucleoside,

*The double bond is not incorrectly placed in the imidazole ring of His 12. It sometimes is on the τ-N of histidine residues in proteins.

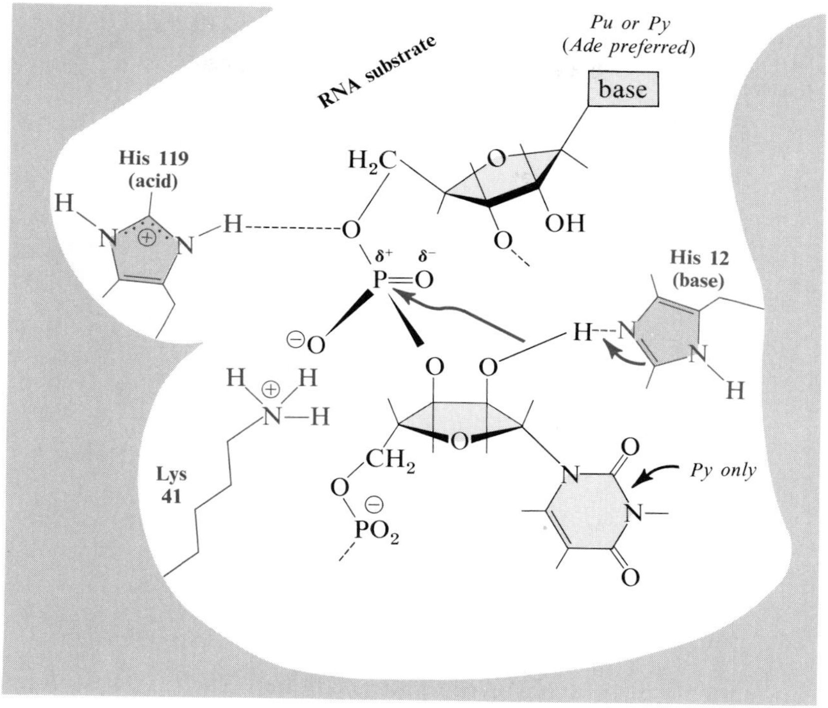

Pu or Py
(Ade preferred)
RNA substrate
base
His 119
(acid)
H₂C
OH
His 12
(base)
Lys
41
CH₂
PO₂
Py only

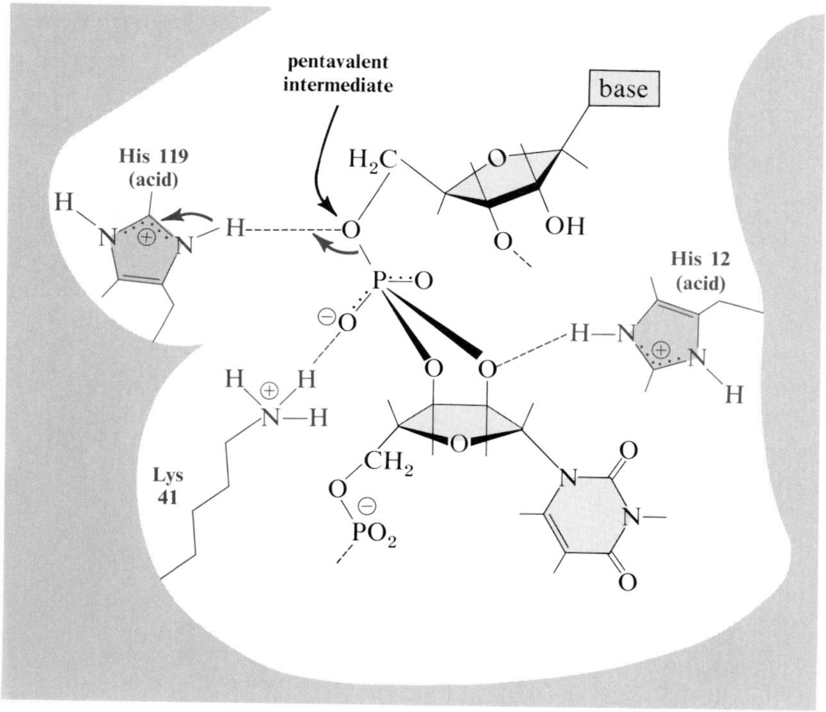

pentavalent
intermediate
base
His 119
(acid)
H₂C
OH
His 12
(acid)
Lys
41
CH₂
PO₂

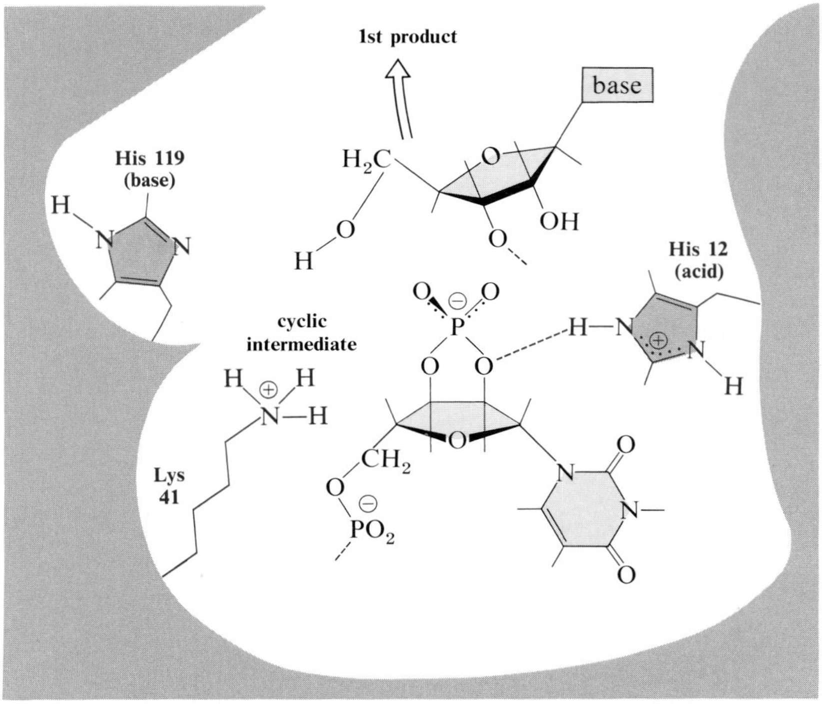

now terminates in a 2′,3′-cyclic phosphodiester. The acid-base forms of the two histidine residues are reversed from those in the original enzyme. These groups return to their original state by catalyzing the cleavage of the cyclic intermediate with water. The steps are essentially a reversal of those that have gone before, using an oxygen atom from water in place of an oxygen atom from ribose. They begin with abstraction of a proton from water by His 119:

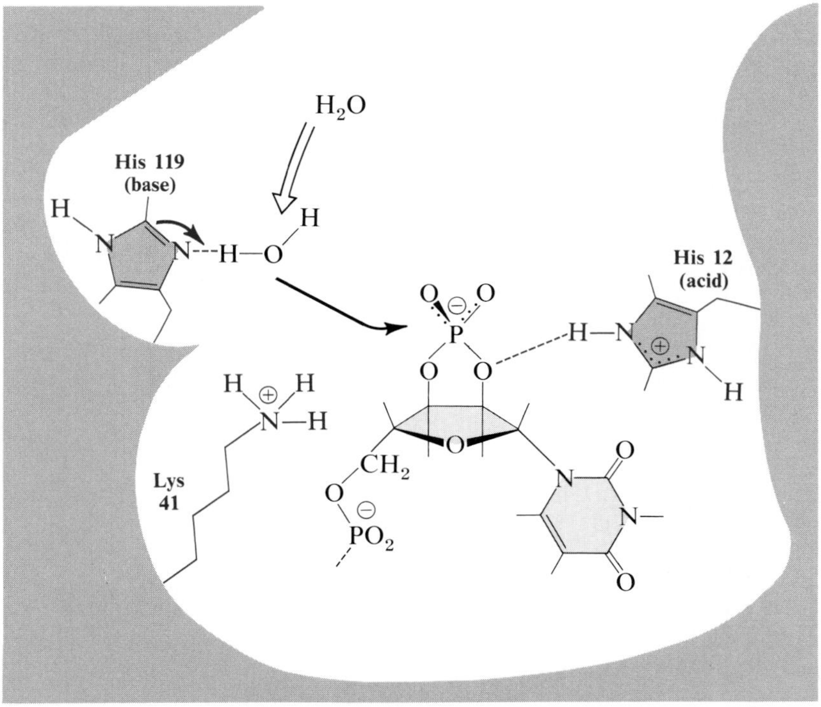

The new pentavalent intermediate is disrupted by donation of a proton from His 12:

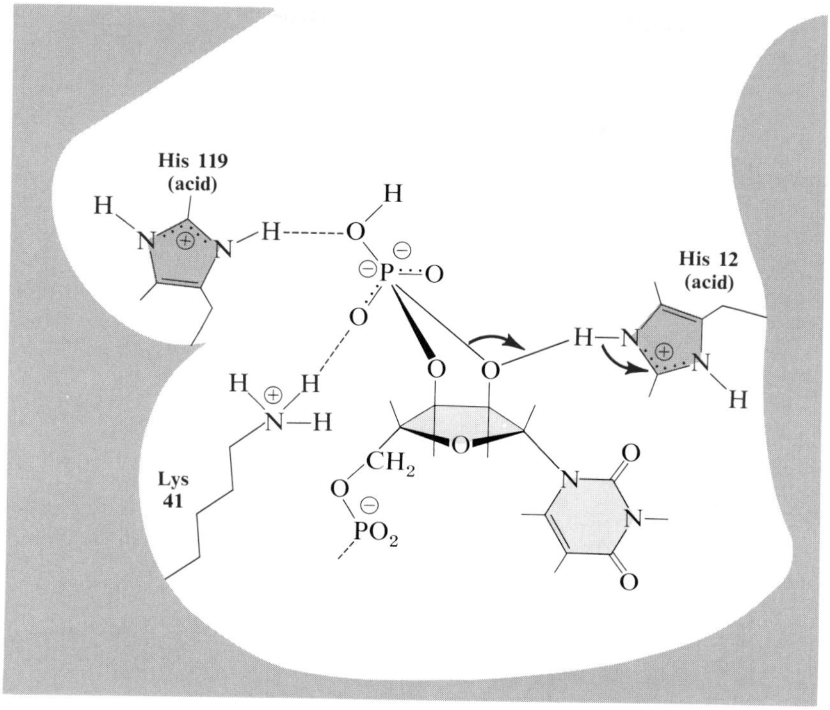

This forms a terminal phosphate group on the remaining portion of RNA, which is now free to leave:

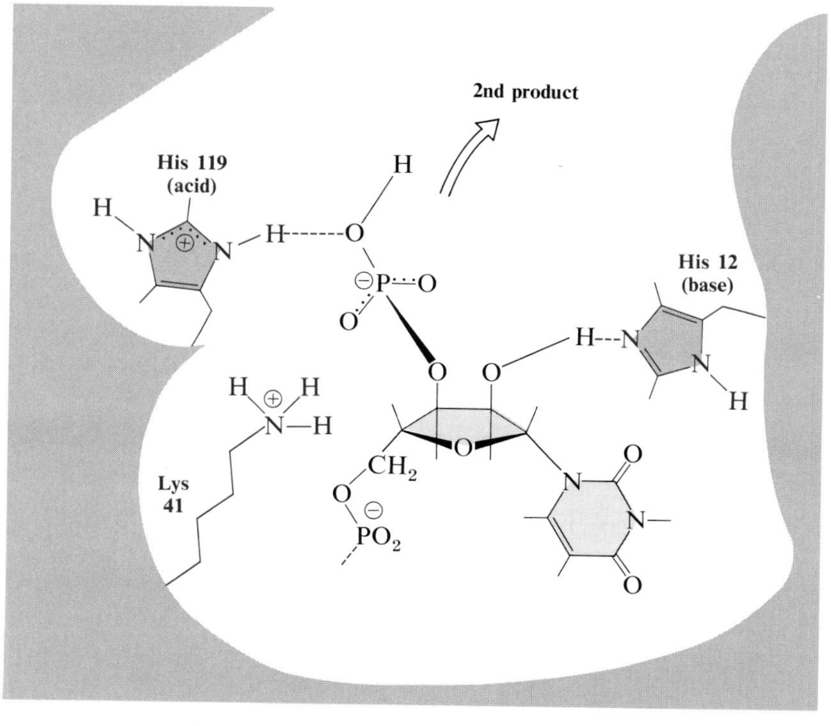

Note how the product-enzyme complex resembles the original substrate-enzyme complex. Indeed, there is nothing in the depicted mechanism to prevent the reaction from running backward and joining two pieces of RNA with a loss of water. However, the position of equilibrium for this reaction strongly favors hydrolysis of RNA into pieces, and the reverse reaction is negligible. (The position of equilibrium for hydrolysis is discussed in Chapter 19.)

Carboxypeptidase A

Peptidases are enzymes hydrolyzing peptide bonds; those hydrolyzing terminal peptide bonds are **exopeptidases.** Carboxypeptidase A is an exopeptidase secreted by the pancreas to hydrolyze the peptide bond of certain carboxyl terminal amino acid residue:

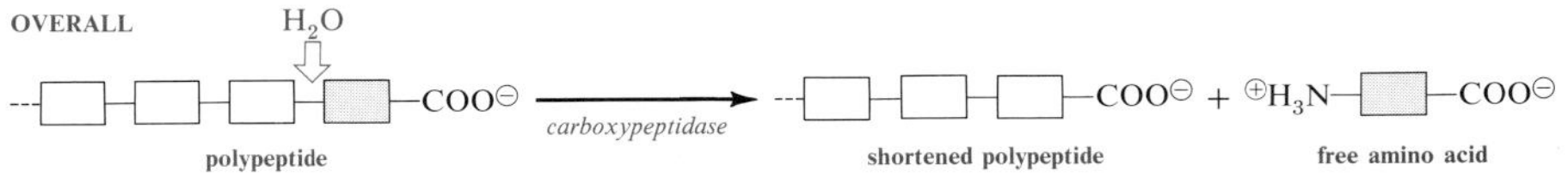

OVERALL

H_2O

polypeptide —COO⊖ → (carboxypeptidase) shortened polypeptide —COO⊖ + ⊕H_3N— free amino acid —COO⊖

The enzyme breaks the peptide bond and transfers the shortened polypeptide chain to a **glutamate** side chain on the enzyme. The acid anhydride bond that is formed is then cleaved with water:

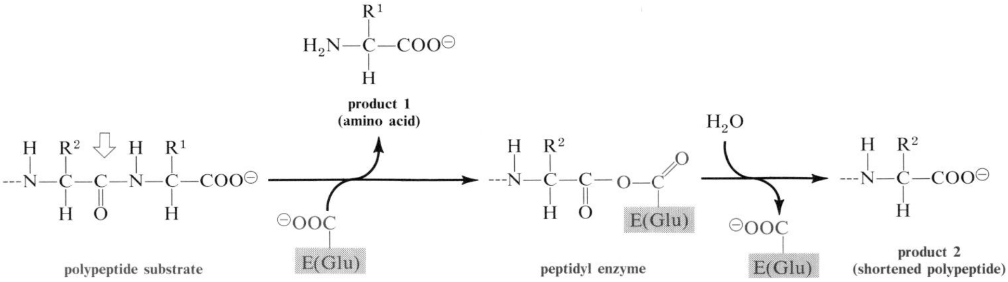

product 1 (amino acid)

polypeptide substrate

E(Glu)

peptidyl enzyme

H_2O

product 2 (shortened polypeptide)

Here we see an example of the use of a nucleophilic group on the enzyme to temporarily form a covalent bond with part of the substrate. In addition, carboxypeptidase is an example of an enzyme with a **metallic ion** in its catalytic site, in this case a **zinc** ion:

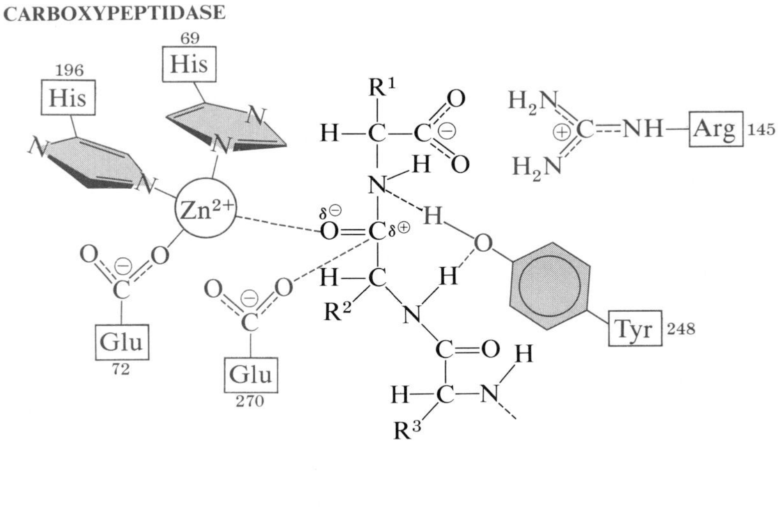

CARBOXYPEPTIDASE

His 196, His 69, Zn^{2+}, Glu 72, Glu 270, Arg 145, Tyr 248

The zinc is held in place through chelation by two histidine side chains and an additional glutamate side chain. Other points of interest: A positively charged **arginine** side chain holds the negatively charged terminal carboxylate group of the substrate. This makes the enzyme specific for the terminal residue. A **tyrosine** phenolic hydroxyl group forms a hydrogen bond with the substrate.

Serine Proteases

Endopeptidases are enzymes catalyzing hydrolysis of interior bonds in polypeptide chains:

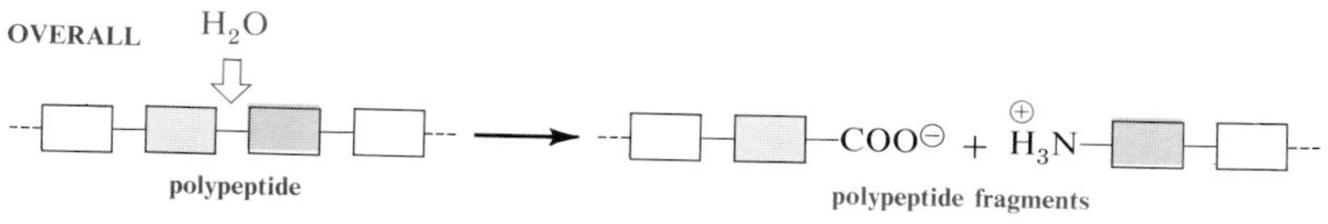

OVERALL H_2O

polypeptide → $COO^{\ominus}$ + $^{\oplus}H_3N$—

polypeptide fragments

They are commonly called **proteases.** Many have identical mechanisms in which the peptide acyl group is transferred to a serine residue. The intermediate acyl-enzyme ester is then cleaved by water:

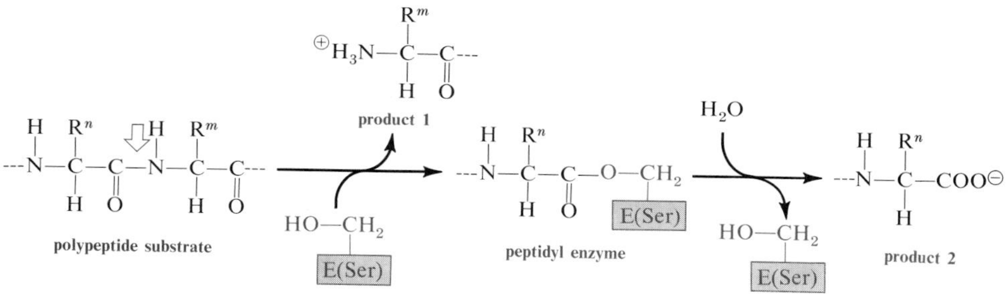

polypeptide substrate

$HO—CH_2$ E(Ser)

$^{\oplus}H_3N—C(R^m)(H)—C(=O)$---

product 1

peptidyl enzyme

H_2O

$HO—CH_2$ E(Ser)

---N(H)—C(R^n)(H)—$COO^{\ominus}$

product 2

These serine proteases contain a **triad** of **aspartate, histidine,** and **serine** residues at the catalytic site (see top figure, following page). The negative aspartate side chain augments the tendency of the basic histidine side chain to pick up a proton and become positively charged. This in turn makes the adjacent serine hydroxyl group a more potent nucleophile because it can more readily lose a proton.

Only three steps of the mechanism are shown in the diagram:

(A) The basic form of the histidine side chain picks up a proton from the serine hydroxyl group, forming a bond between the serine residue and a carbonyl carbon atom in the polypeptide substrate.

(B) The acidic form of the histidine side chain forms a hydrogen bond with the amide nitrogen of the substrate. A shift of electrons then breaks the amide bond and returns the histidine to its basic form.

(C) One piece of the original polypeptide is now covalently bound as a serine ester; the other piece leaves and is replaced by water, shown in place on the enzyme.

Succeeding steps involve a reversal of those shown, with formation of a new tetrahedral intermediate in which the OH of water replaces the NH of the original polypeptide, followed by breakage of the acyl serine bond through a reversal of step (A).

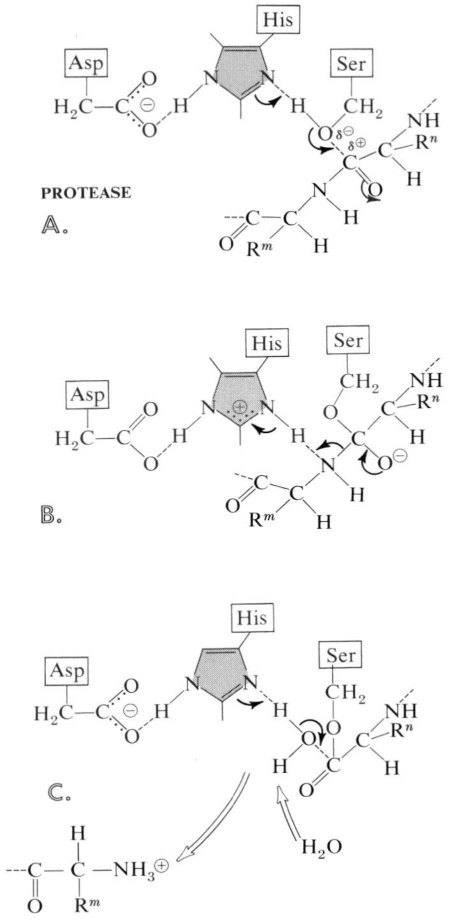

PROTEASE

A.

B.

C.

Fructose Bisphosphate Aldolase

One of the reactions in glucose metabolism is the readily reversible interconversion of a six-carbon sugar chain and two three-carbon pieces, all in the form of phosphate esters:

OVERALL

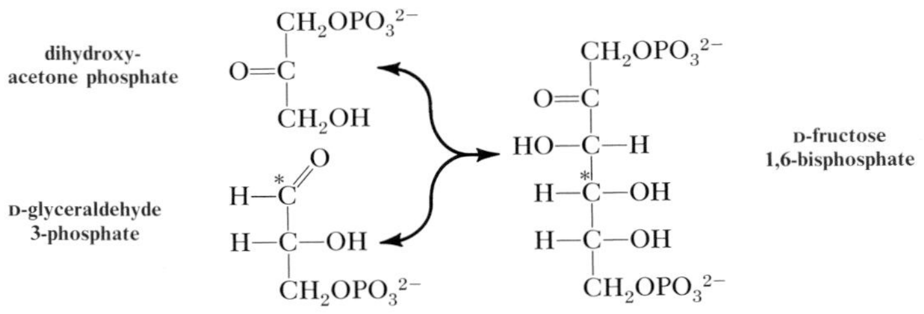

dihydroxy-acetone phosphate

D-glyceraldehyde 3-phosphate

D-fructose 1,6-bisphosphate

Condensation occurs during glucose synthesis, and the reverse cleavage is required for use of glucose as a fuel; we shall consider these metabolic processes in detail later. The point of interest in this reaction is that the enzyme uses a lysine residue to react with a ketose substrate, which is the phosphate ester of either dihydroxyacetone or fructose:

D-fructose 1,6-bisphosphate

*The asterisk identifies a particular C atom for orientation only.

The lysine residue supplies an amino group on a tail of carbon atoms waving out some distance from the remainder of the peptide, and in this case it acts as a nucleophile to attack the carbonyl carbon atom of the ketoses. The substrate becomes covalently bound as a **ketimine (Schiff's base)** to the enzyme, labilizing the adjacent carbon atom for condensation or cleavage. (A histidine residue, not shown, acts as a proton carrier to catalyze these covalent reactions, much as it does in the serine proteases.)

The reaction is catalyzed by an aldolase, so-called because the overall reaction is a variant of an aldol condensation:

$$R^1\text{—CHO} + R^2\text{—CH}_2\text{—CO—X} \rightleftharpoons R^1\text{—CH(OH)—CH}(R^2)\text{—CO—X}$$

PROVISION OF REACTIVE GROUPS BY COENZYMES

Many enzymes differ from those we have discussed previously in that effective catalysis requires the presence of chemical groups that do not occur in the side chains of the common amino acids. These chemical groups must be added in the form of an additional compound, in other words, as a **prosthetic group** or **coenzyme.** In such cases the enzyme consists of the coenzyme and the associated polypeptide chains, which are called the **apoenzyme:**

$$\text{Apoenzyme} + \text{coenzyme} = \text{enzyme.}$$

The polypeptide chains in such enzymes have a third function. They must provide groups not only to bind the substrate and to perform catalysis but also to bind the coenzyme. Let us first discuss an example of an important group of enzymes, the **aminotransferases,** to illustrate these functions.

Aspartate Aminotransferase

The Reaction and the Coenzyme. A critical reaction in all organisms is the transfer of an amino group from one carbon skeleton to another by the process of **transamination.** Frequently, the most active exchange is between glutamate and aspartate:

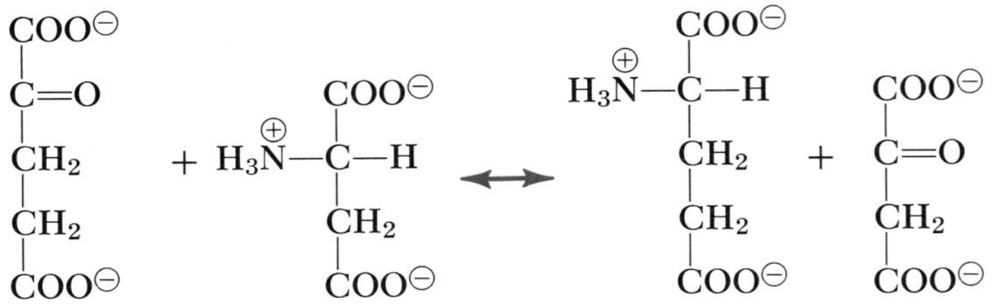

$$\underset{\alpha\text{-ketoglutarate}}{^{\ominus}OOC{-}C({=}O){-}CH_2{-}CH_2{-}COO^{\ominus}} + \underset{\text{L-aspartate}}{^{\ominus}OOC{-}CH(\overset{\oplus}{N}H_3){-}CH_2{-}COO^{\ominus}} \longleftrightarrow \underset{\text{L-glutamate}}{^{\ominus}OOC{-}CH(\overset{\oplus}{N}H_3){-}CH_2{-}CH_2{-}COO^{\ominus}} + \underset{\text{oxaloacetate}}{^{\ominus}OOC{-}C({=}O){-}CH_2{-}COO^{\ominus}}$$

(α-Ketoglutarate is named as 2-oxoglutarate in approved nomenclature, in which the English have a dominant voice. The English use of oxo for keto is spreading in this country, but is not yet common.) This reaction is catalyzed by **aspartate aminotransferase** (also known as glutamate-oxaloacetate transaminase, glutamate-aspartate transaminase, and similar permutations).

The aminotransferases contain **pyridoxal phosphate,** a coenzyme frequently required by enzymes attacking free amino acids. Pyridoxal phosphate is fixed by polypeptide chains of the aminotransferases in such a way that its aldehyde group is brought into proximity with a lysine side chain, with which it reacts to form an aldimine, the reactive structure in the enzyme (Fig. 14–4).

The various groups in pyridoxal phosphate have not been added for decoration. The ionized phenolic group and the positively charged nitrogen of the pyridine ring enhance the reactivity of the aldimine, while the phosphate ester and methyl substituents serve to fix the coenzyme at a specific site. Pyridoxal phosphate illustrates a general property of coenzymes, which is to have a small part of the structure, perhaps a single group, actively participating in the reaction, with the remainder of the molecule serving to accentuate that reactivity and to provide a framework for specific binding to the apoenzyme.

The Steps in Transamination. We shall detail the individual covalent changes in the transamination reaction to show how the coenzyme contributes to catalysis but

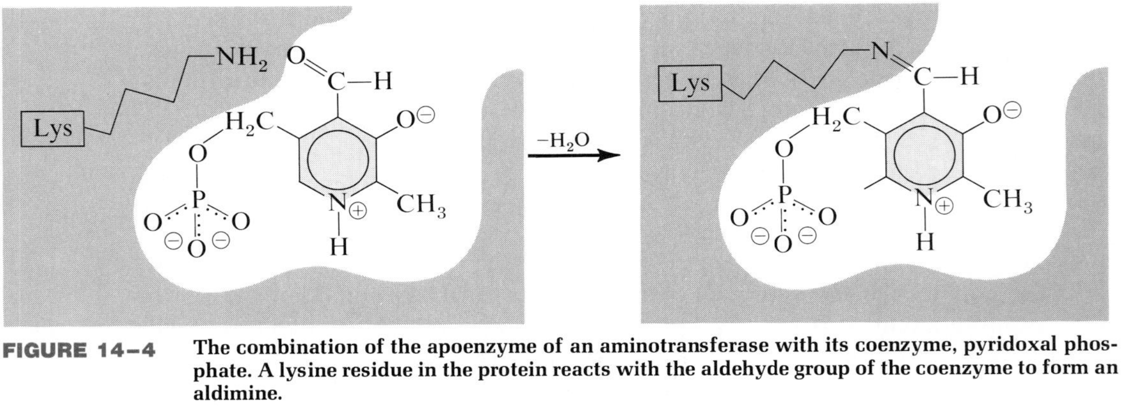

FIGURE 14–4 The combination of the apoenzyme of an aminotransferase with its coenzyme, pyridoxal phosphate. A lysine residue in the protein reacts with the aldehyde group of the coenzyme to form an aldimine.

without giving the actual mechanism, which probably involves acid-base catalysis by histidine side chains (see figure on next page).

(A) A molecule of L-aspartate collides with the enzyme and is bound, probably by its carboxylate groups, so that its amino group is near the aldimine structure on the enzyme.

(B) The double bond shifts from the lysine N atom to the aspartate N atom, releasing the lysine amino group, and forming the aldimine of aspartate. It is much easier to form an aldimine by exchange of groups with a preexisting aldimine than it is by *de novo* combination of amine and aldehyde.

(C) The double bond of the aldimine shifts, forming a ketimine. This tautomerization is the rate-limiting step of the sequence.

(D) The ketimine is hydrolyzed to form oxaloacetate and **pyridoxamine phosphate.** The oxaloacetate is free to dissociate, but the pyridoxamine phosphate is still held to the enzyme by other groups on the coenzyme.

(E) A molecule of α-ketoglutarate (2-oxoglutarate) combines with the protein, also by attachment of carboxylate groups.

(F) A ketimine forms between α-ketoglutarate and pyridoxamine phosphate.

(G) The double bond again shifts, now forming the aldimine of L-glutamate and pyridoxal phosphate.

(H) The aldimine double bond shifts from the nitrogen of L-glutamate to the lysine residue on the protein, leaving glutamate free to dissociate and restoring the original condition of the enzyme so that it can begin the whole process over again with a new molecule of aspartate.

The mechanism can and does operate in the exact reverse of the steps given so that one can start with glutamate + oxaloacetate and arrive at aspartate + α-ketoglutarate.

Coenzymes and Vitamins

In addition to being the first example of a coenzyme we have discussed, pyridoxal phosphate is also the first example of a compound for which a precursor must be supplied in the human diet because the entire compound cannot be synthesized by human tissues.

Coenzymes frequently contain structures not found in ordinary metabolites and behave as catalysts in the sense that they are eventually recovered in their original form. Therefore, they need not be present in high concentrations in the tissues, any more than the apoenzyme need be. We have noted the widespread occurrence of transamination reactions among plants and animals. Most other kinds of enzymatic reactions, with the necessary coenzymes, also have a wide occurrence.

Putting these facts together, it is easy to rationalize the evolutionary truth—organisms whose principal nutrition is supplied by tissues from other organisms frequently have lost their ability to synthesize the structural elements of coenzymes. If the structure is unusual, it need not be made for the quantitatively important metabolic reactions, in which moles of compounds may be handled per day, and its sole function may be in the coenzyme. But only a small amount of coenzyme is required, and if the structure is constantly appearing in dietary compounds, its synthesis becomes unnecessary. Therefore, mutations deleting some of the reactions

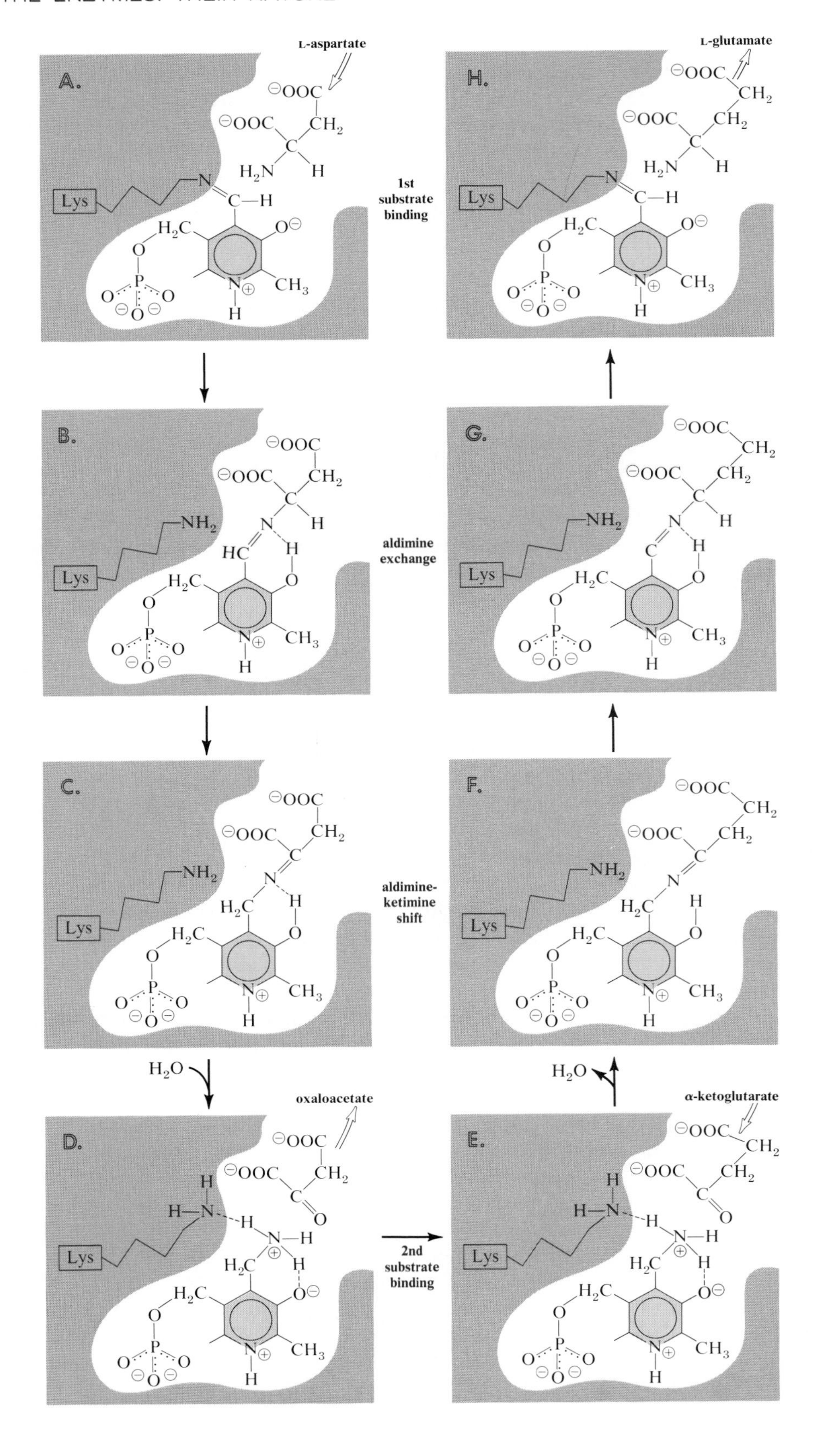
L-aspartate
A.
1st substrate binding
B.
aldimine exchange
C.
aldimine-ketimine shift
D.
oxaloacetate
2nd substrate binding
E.
α-ketoglutarate
F.
G.
H.
L-glutamate
Lys

peculiar to the formation of coenzymes will not cause the death of the animal, and may even give it some advantage by making room in the cell for increased amounts of the enzymes that are absolutely necessary to form other compounds.

When such a deletion has occurred, and small amounts of the necessary organic structure must be supplied in the diet, the dietary compound is a **vitamin.** The formation of all coenzymes does not require vitamins, but many vitamins for which the biological function has been exactly established are known to be required because they are used to form coenzymes.

The substituted pyridine ring of pyridoxal phosphate is an example of a structure that cannot be synthesized by vertebrates, and its dietary precursors are lumped together as **vitamin B_6,** which includes **pyridoxal, pyridoxamine,** and the corresponding alcohol, **pyridoxine:**

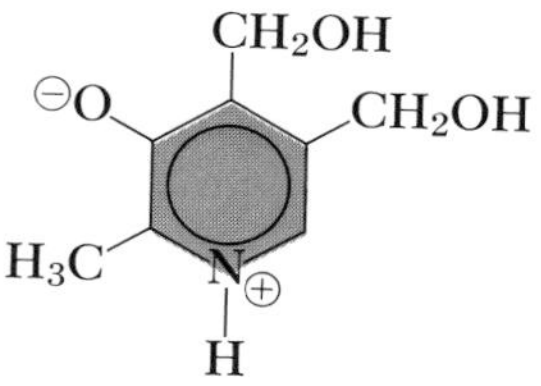

CH_2OH
$^{\ominus}O$ CH_2OH
H_3C $N^{\oplus}$
H

pyridoxine

We can also rationalize the complexity of coenzymes from the example at hand. As shown, the binding of pyridoxal phosphate and the basis for the transamination reaction depend upon a simple reaction between an aldehyde and an amine. There are many aldehydes involved in metabolism. Suppose that one of these many aldehydes came in contact with the lysyl group at the active site after the formation of the peptide chains making up the apoenzyme of transaminase. The formation of an aldimine might well occur. However, it would eventually dissociate, and even though similar accidents might occur, eventually a molecule of pyridoxal phosphate would collide with it. Once this happened, dissociation would become infrequent, because the other binding groups on the molecule fit the particular peptide configuration and hold the coenzyme in place. Two things have been gained. The combination of coenzyme and apoenzyme has been stabilized by the use of multiple binding groups. Just as importantly, a variety of aldehydes can be used in the general metabolism without any significant disruption of the transamination reaction, because only the one very specific aldehyde structure will fit the apoenzyme.

The same kind of an advantage applies to the lysine residue. We shall see that these residues are employed in other enzymes to bind completely different coenzymes, participating in reactions bearing little resemblance to transamination. Each of these kinds of coenzymes has a unique configuration, not capable of being confused with any of the others, and each apoenzyme can't bind the wrong coenzyme to its lysyl groups because it is built to conform to the complex structure of the proper coenzyme.

This represents a beginning on an important concept in metabolism: *The kinds of structures used in biological compounds and the types of reactions they undergo are relatively few.* The close regulation of the complex assembly of reactions that we lump together as metabolism doesn't depend upon each compound's having a unique kind of structure. It depends upon the compound's having a particular *combination* of structures, and the matching of the combination by a configuration built into only a few, perhaps one, of the hundreds of enzymes made by the cell.

THE NATURE OF OXIDATIONS AND REDUCTIONS

Let us briefly review oxidation-reduction reactions before considering a biological example. Three kinds of oxidations are important for our purposes:

(1) Addition of Oxygen. Combination with elemental oxygen is an oxidation and this was the original meaning of the word. A compound is said to be oxidized when its relative content of oxygen is increased, but oxidations are not to be confused with hydrations, in which water is added.

This is an oxidation of an organic compound:

$$\begin{array}{c}\text{H}\quad\text{H}\\ |\quad\; |\\ -\text{C}-\text{C}-\\ |\quad\; |\\ \text{H}\quad\text{H}\end{array} + \frac{1}{2}\,\text{O}_2 \longrightarrow \begin{array}{c}\text{H}\quad\text{H}\\ |\quad\; |\\ -\text{C}-\text{C}-\\ |\quad\; |\\ \text{HO}\quad\text{H}\end{array}$$

This is a hydration and not an oxidation:

$$\begin{array}{c}\text{H}\quad\text{H}\\ |\quad\; |\\ -\text{C}=\text{C}-\end{array} + \text{H}_2\text{O} \longrightarrow \begin{array}{c}\text{H}\quad\text{H}\\ |\quad\; |\\ -\text{C}-\text{C}-\\ |\quad\; |\\ \text{HO}\quad\text{H}\end{array}$$

(2) Dehydrogenations. A compound may also be oxidized by the removal of hydrogen, including both the proton and its associated electron. These dehydrogenations are not to be confused with simple ionizations of acids.

This is an oxidation:

$$\begin{array}{c}\text{H}\quad\text{H}\\ |\quad\; |\\ -\text{C}-\text{C}-\\ |\quad\; |\\ \text{H}\quad\text{H}\end{array} \longrightarrow \begin{array}{c}\text{H}\quad\text{H}\\ |\quad\; |\\ -\text{C}=\text{C}-\end{array} + \text{H}_2$$

This is an oxidation:

$$\begin{array}{c}\text{H}\quad\text{H}\\ |\quad\; |\\ -\text{C}-\text{C}-\\ |\quad\; |\\ \text{H}\quad\text{H}\end{array} + \frac{1}{2}\,\text{O}_2 \longrightarrow \begin{array}{c}\text{H}\quad\text{H}\\ |\quad\; |\\ -\text{C}=\text{C}-\end{array} + \text{H}_2\text{O}$$

(Notice that this reaction creates the starting material for the hydration given above—the compound with the double bond is already as oxidized as is the corresponding alcohol.)

This is also an oxidation:

$$\begin{array}{c}\text{O}\\ \|\\ \text{H}-\text{C}-\text{OH}\end{array} + \frac{1}{2}\,\text{O}_2 \longrightarrow \text{CO}_2 + \text{H}_2\text{O}$$

This is an ionization, and not an oxidation:

$$H-\overset{\overset{O}{\|}}{C}-OH \longrightarrow H-C\overset{O}{\underset{O}{\lesssim}}\ominus + H^{\oplus}$$

(Only a proton is liberated here, and the corresponding electron is retained in the organic anion.)

(3) Electron Transfer. A compound may be oxidized by a simple removal of electrons with a concomitant increase in positive charge. There must be another compound present to accept electrons, and this compound is said to be **reduced** by the gain of electrons. (The name came from the practice of reducing metallic oxide ores to the free metals.)

This is an oxidation-reduction:

$$Fe^{3+} + Cu^{+} \longrightarrow Fe^{2+} + Cu^{2+}$$

In this case, the cuprous ion is being oxidized by the ferric ion. The ferric ion is being reduced by the cuprous ion. We can say that the ferric ion is an oxidizing agent, which becomes reduced when it oxidizes other compounds. The cuprous ion is a reducing agent, which becomes oxidized when it reduces other materials.

The idea that every oxidation must be accompanied by a reduction, and vice versa, has been extended from the simple example we have here to other cases in which it is not obvious. Thus, we say that oxygen is reduced when it adds to organic compounds in the way illustrated in example (1) above, or hydrogen is being reduced in the first reaction given under example (2), while the carbon skeleton is being oxidized. The terminology is somewhat of a formalism in these cases, but the value of the practice will become more apparent as we become familiar with the actual sequences of biological oxidations, in which similar final results are sometimes achieved by the use of the differing types of oxidations.

Lactate Dehydrogenase—An Oxidoreductase

Enzymes that catalyze oxidation-reduction reactions nearly always require a coenzyme, prosthetic group, or bound metallic ion for activity. For example, one of the most common types of reaction is the equilibration of alcohols and carbonyl compounds:

$$H-\overset{|}{\underset{|}{C}}-OH \longleftrightarrow \overset{|}{\underset{|}{C}}=O + 2H^{\oplus} + 2e^{\ominus}$$

The carbonyl compound may be an aldehyde or a ketone. In order for this reaction to proceed, there must be some acceptor (or donor, in the reverse direction) of electrons, and the electron carrier for equilibrating alcohols and ketones (or aldehydes) is usually **nicotinamide adenine dinucleotide (NAD),** or a close relative. The name implies its structure in which nucleotides containing nicotinamide and adenine are linked through a pyrophosphate bond:

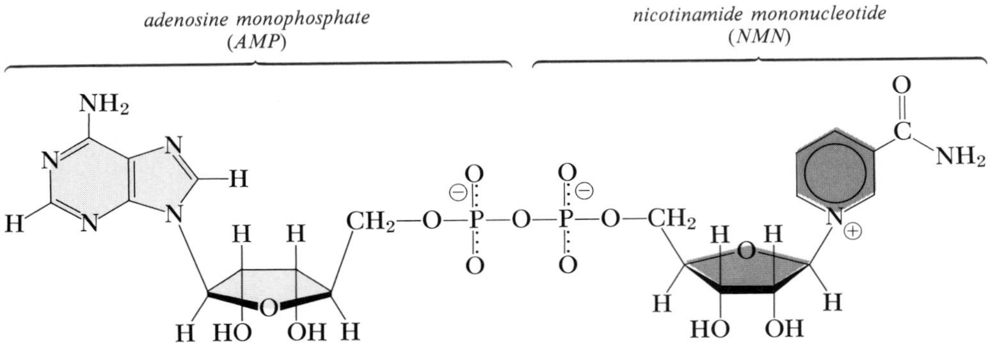

adenosine monophosphate (AMP) *nicotinamide mononucleotide (NMN)*

nicotinamide adenine dinucleotide (NAD)

The pyridine ring of nicotinamide is the part of the molecule that changes during oxidation and reduction. The reduction of NAD requires the addition of a **hydride ion**—a proton with two electrons—and the nucleotide loses a positive charge in the process:

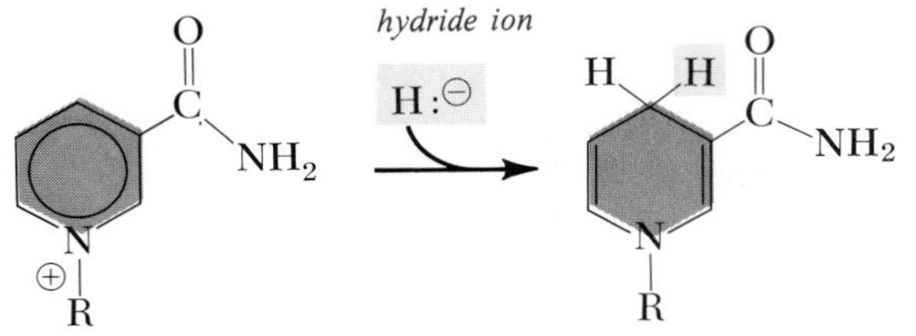

hydride ion $H:^{\ominus}$

$NAD^{\oplus}$ (nicotinamide adenine dinucleotide) **NADH (reduced nicotinamide adenine dinucleotide)**

(The reaction probably goes through free radical intermediates that are not shown.)

A typical example of the use of NAD is in the equilibration of **lactate** and **pyruvate,** which are important products of the degradation of glucose:

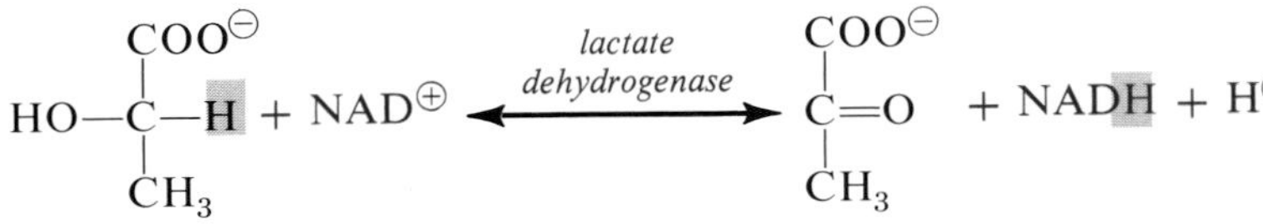

$$HO{-}C(COO^{\ominus})(CH_3){-}H + NAD^{\oplus} \xrightleftharpoons{\text{lactate dehydrogenase}} (COO^{\ominus})(CH_3)C{=}O + NADH + H^{\oplus}$$

L-lactate **pyruvate**

The abbreviations, NAD^+ and NADH, are used in reactions to show only the change in charge, not the total charges on the molecules. When used in text, the charge is omitted from the abbreviation.

The reaction is catalyzed by a lactate dehydrogenase, more particularly by a lactate:NAD oxidoreductase. The name is misleading because the enzyme may catalyze the oxidation of lactate or the reduction of pyruvate, depending upon the relative concentrations of reactants.

Although NAD is consumed during the oxidation of lactate, it is said to be a coenzyme for lactate dehydrogenase because the same molecules of NAD can be used repeatedly to oxidize many molecules of lactate, provided that there is some additional reaction in which NADH is converted back to the oxidized NAD. The semantics here are a little cloudy. A molecule of NAD is not so tightly bound to a molecule of lactate dehydrogenase that it "belongs" to it. It may, and it does, move from one molecule of apoenzyme to another. Indeed, NAD is a second substrate for lactate dehydrogenase, and the only justification for making a distinction between it and the other substrate, lactate, is that a given molecule of NAD has a long lifetime in the cell, being used over and over, whereas molecules of substrates like lactate are rapidly converted to products that migrate out of the cell. Our previous example of a coenzyme, pyridoxal phosphate, differed in that it was tightly bound to apoenzyme and went through a complete cycle of reaction on the one protein molecule.

Domains in Enzymes. Many enzymes are built with distinct domains to perform different functions. One may contribute to the catalytic site, another bind a coenzyme or other reactant, and still another bind allosteric effectors with which to regulate the enzyme. Lactate dehydrogenase contains four subunits, each enzymatically active, and each subunit is composed of two domains. One of these domains is wrapped around the coenzyme. The other contributes groups for the catalytic site, which lies in a cleft between the two domains.

A striking feature of NAD-coupled dehydrogenases is their close structural resemblance in the coenzyme-binding domain. This domain is composed of two very similar parts, one for binding the adenosine phosphate portion of the coenzyme and the other for binding the nicotinamide mononucleotide portion. The general plan of each of these portions involves a twisted pleated sheet lined with segments of α-helix, and was used as an example in Chapter 3 (p. 34). Some other enzymes, catalyzing totally different types of reaction, are also built according to this plan.

The Binding Sites (Fig. 14–5). As with many other types of reactions, catalysis of this oxidation-reduction involves the use of a histidine residue as a general acid–general base (*far left*). An impressive array of side chains is used to hold the coenzyme in place. Note especially the use of a positively charged arginine side chain to bind the negatively charged phosphate group in the coenzyme. Many enzymes catalyzing reaction of phosphorylated substrates have arginine residues appropriately located for this purpose. Here an additional arginine side chain is used to fix the negatively charged carboxylate group of the substrate.

ENZYME NOMENCLATURE

Until the past decade, the creation of names for enzymes was mainly the responsibility of the discoverer, subject to modification by later investigators and to the taste of the various editors of the scientific journals. Decision between alternative names was a matter for the marketplace; those that weren't used disappeared. The only agreement was that all enzyme names ought to have the suffix *-ase,* but this might be preceded by the name of a substrate, the name of a product, or the type of reaction catalyzed.

After 1956, an effort was made by a commission of the International Union of Biochemistry (IUB) to create a more systematic nomenclature. The proposals, like the systematic nomenclature for organic compounds, had a mixed reception in terms of actual usage, even after revision in 1964. A new revision was made in 1972 that faced reality by making the common trivial names the recommended names, wherever possible, but grouped within a systematic framework for classifying enzymes. Readers of this book should be familiar with some of the principles of the systematic classification, if only to enable intelligent exercise of their power to decide the accepted nomenclature through usage. Perhaps the Commissions, like the rest of us, tend to oscillate too wildly between strict regimentation and unrestrained chaos.

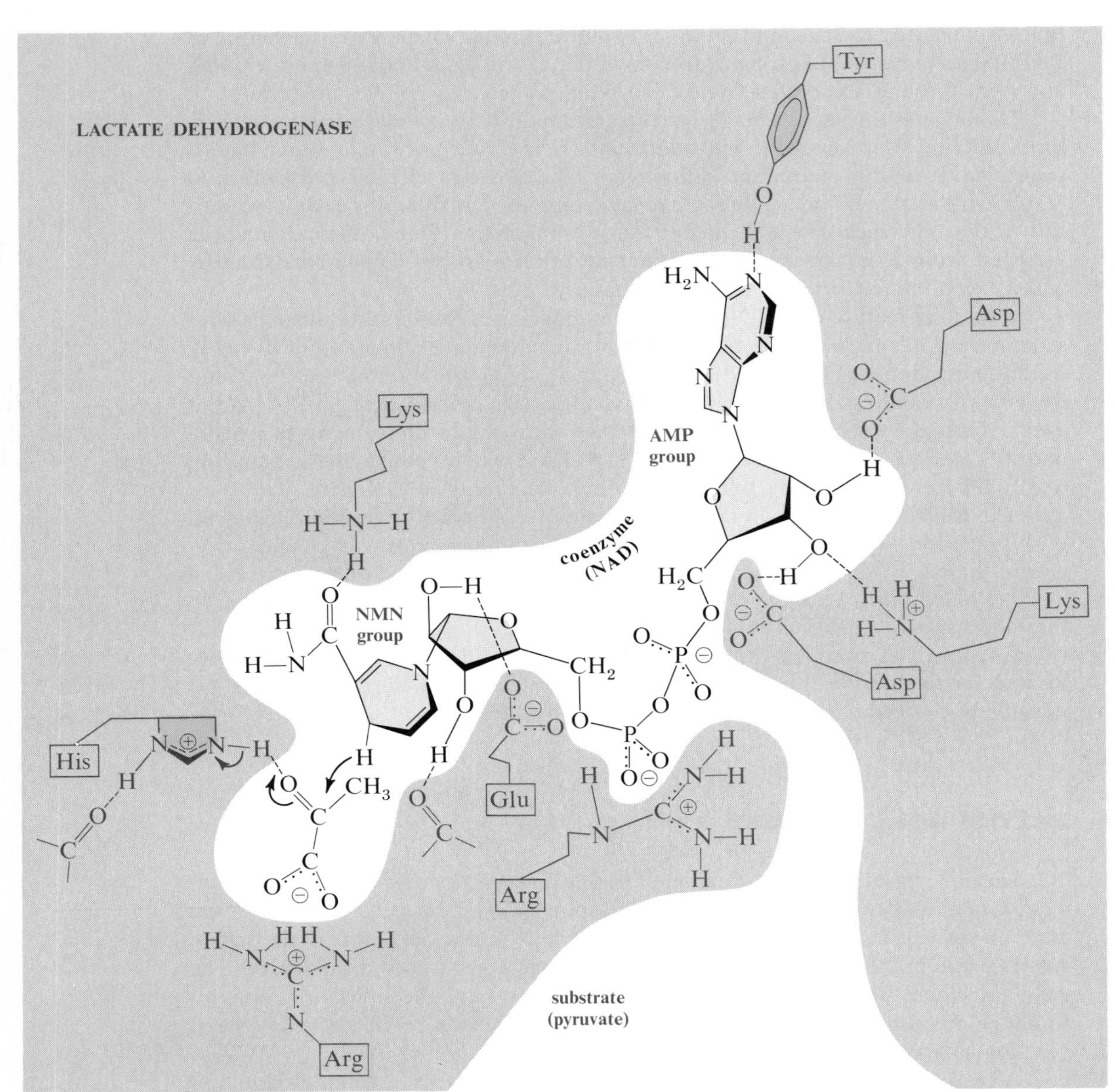

FIGURE 14–5 The binding sites of lactate dehydrogenase for the coenzyme, NADH, and the substrate, pyruvate. (The particular structure is for the muscle enzyme, which usually catalyzes the reduction of pyruvate to lactate, rather than the reverse reaction.) (Based on figure by J.J. Holbrook, A. Liljas, S.J. Steindel, and M.G. Rossman (1975) Lactate dehydrogenase, 11:240, in P.D. Boyer, ed., The Enzymes, 3rd ed. © 1975 by Academic Press. Reproduced by permission.)

General Classes

The IUB nomenclature uses six type designations as the final part of an enzyme name:

1. Oxidoreductases. These are enzymes catalyzing all of the reactions in which one compound is oxidized and another compound is reduced:

dehydrogenases
reductases
oxidases
peroxidases
hydroxylases
oxygenases

2. Transferases. These are enzymes catalyzing reactions not involving oxidation and reduction in which a group containing C, N, P, or S is transferred from one substrate to another:

transferases
trans—ases (such as transaminase, transketolase, transaldolase, transmethylase)

3. Hydrolases. These are enzymes catalyzing hydrolytic cleavages or their reversal. (They do not include enzymes catalyzing the addition or removal of water from one compound, such as fumarase or enolase.) This includes all of the enzymes named as:

esterases
amidases
peptidases
phosphatases
glycosidases

4. Lyases. These are enzymes catalyzing the cleavage of C—C, C—O, C—N bonds, etc., without a hydrolysis or oxidation-reduction:

decarboxylases
aldolases
synthases (but not properly named synthetases)
cleavage enzymes (such as citrate or 3-hydroxy-3-methylglutaryl CoA cleavage enzymes)
hydrases or hydratases or dehydratases
deaminases
nucleotide cyclases

5. Isomerases. These are enzymes catalyzing intramolecular rearrangements not involving a net change in the concentration of compounds other than the substrate:

isomerases
racemases
epimerases
mutases

6. Ligases. These are enzymes catalyzing all of the reactions involving the formation of bonds between two substrate molecules that are coupled to the cleavage of a pyrophosphate bond in ATP or another energy donor. They include all of the enzymes now named as synthetases, except in cases where this name has been misapplied. The name synthetase was originally defined to include those enzymes now being defined as ligases, but it was later applied to enzymes catalyzing the formation of compounds by other mechanisms. Thus, one sometimes sees glycogen synthetase or δ-aminolevulinate synthetase or thymidylate synthetase even though the reactions catalyzed by these enzymes do not involve the cleavage of a high-energy phosphate bond. The term *synthase* was later coined to cover these enzymes, which are lyases in the IUB nomenclature.

Individual Enzymes

Within each type of enzyme, further subdivisions are made and given group numbers preceded by EC for Enzyme Commission. The groups are divided into numbered subgroups, and the individual enzymes within a subgroup are given another number. Thus, each enzyme is designated by four numbers. Some examples will make it clear, and also show advantages and disadvantages of the systematic nomenclature. The recommended trivial name is shown in italics.

1. Oxidoreductases
   1. Acting on the CH—OH group of donors
   1. With NAD or NADP as acceptor
   27. L-lactate:NAD oxidoreductase
       EC 1.1.1.27 (*lactate dehydrogenase*)

2. Transferases
   4. Glycosyltransferases
   1. Hexosyltransferases
   18. α-1,4-Glucan:α-1,4-glucan 6-glycosyltransferase
       EC 2.4.1.18 (1,4-α-*glucan branching enzyme*)
   7. Transferring phosphorus-containing groups
   1. Phosphotransferases with an alcohol group as acceptor
   2. ATP:D-glucose 6-phosphotransferase
      EC 2.7.1.2. (*glucokinase*)

3. Hydrolases
   4. Acting on peptide bonds
   21. Serine proteases
   4. No systematic name given
      EC 3.4.21.4 (*trypsin*)

4. Lyases
   2. Carbon-oxygen lyases
   1. Hydro-lyases
   13. L-Serine hydro-lyase (deaminating)
       EC 4.2.1.13 (*serine dehydratase*)

5. Isomerases
   1. Racemases and epimerases
   3. Acting on carbohydrates and derivatives
   1. D-Ribulose 5-phosphate 3-epimerase
      EC 5.1.3.1 (*ribulosephosphate 3-epimerase*)

6. Ligases
   2. Forming C—S bonds
   1. Acid-thiol ligases
   3. Acid: CoA ligase (AMP-forming)
      EC 6.2.1.3 (*acyl CoA synthetase*)

FURTHER READING

M.L. Bender and L.J. Brubacher (1973). Catalysis and Enzyme Action. McGraw-Hill. Succinct and clear review of general principles of catalysis.

P.D. Boyer, ed. (1971-). The Enzymes, 3rd ed. Academic Press. This multivolume work includes detailed discussions of many mechanisms.

W. P. Jencks (1975). Binding Energy, Specificity, and Enzymatic Catalysis: The Circe Effect. Adv. Enzymol. 43:219. Lengthy review of relation between binding and catalysis.

D.S. Sigman and G. Mouser (1975). Chemical Studies of Enzyme Active Sites. Annu. Rev. Biochem. 44:889.

D.C. Phillips and F.M. Richards (1973). Atlas of Molecular Structures in Biology. I. Ribonuclease S. Oxford.

C.A. Deakyne and L.C. Allen (1979). Role of Active Site Residues in the Catalytic Mechanism of Ribonuclease A. J. Am. Chem. Soc. 101:3951.

A.A. Kossiakoff and S.A. Spencer (1981). Direct Determination of the Protonation States of Aspartic Acid-102 and Histidine-57 in the Tetrahedral Intermediate of the Serine Proteases: Neutron Structure of Trypsin. Biochemistry 20:6462.

I.A. Rose (1975). Mechanism of the Aldose-Ketose Isomerase Reaction. Adv. Enzymol. 43:491.

C.Y. Lai, N. Nakai, and D. Chang (1974). Amino Acid Sequence of Rabbit Muscle Aldolase and the Structure of the Active Center. Science 183:1204.

G.C. Ford, G. Eichele, and J.N. Jansonius (1980). Three-Dimensional Structure of a Pyridoxal Phosphate–Dependent Enzyme, Mitochondrial Aspartate Amino Transferase. Proc. Natl. Acad. Sci. (U.S.) 77:2559. Notable example of dimeric enzyme with each active site generated with groups from both subunits.

D.E. Metzler (1980). Tautomerism in Pyridoxal Phosphate and in Enzymatic Catalysis. Adv. Enzymol. 50:1. Includes discussion of aminotransferase mechanism.

A.E. Martell (1982). Reaction Pathways and Mechanism of Pyridoxal Catalysis. Adv. Enzymol. 53:163.

C.M. Anderson, F.H. Zucker, and T.A. Steitz (1979). Space-Filling Models of Kinase Clefts and Conformation Changes. Science 204:375.

Enzyme Nomenclature (1978). Elsevier. This revised report of the 1972 recommendations of the IUB is a valuable, even though Anglicized, reference work for more general purposes than nomenclature.

CHAPTER 15

RATES OF ENZYMATIC REACTIONS

We dissected enzymatic reactions into individual steps and made qualitative analyses of the mechanisms in the previous chapter. Now we want to understand what is accomplished by the action of enzymes. The metabolic economy is composed of individual chemical reactions, but it is a mistake to assume that memorization of sequences of reactions is enough for understanding the economy; we must also understand quantitative balance.

In short, we cannot begin to understand the physiological function of enzymes until we have some idea of the factors influencing reaction rates, and this idea requires numbers as well as qualitative facts. (Some make the equally serious error of presuming that facility in symbolic manipulations can totally substitute for knowledge of factual detail, but we shall have no trouble in avoiding this mistake in our later discussions.)

The Kinetic Parameters

We saw in the previous chapter that even an enzymatic reaction involving a single substrate and a single product will proceed in at least four steps: The enzyme and substrate combine to form a complex, **ES; ES** is transformed by internal catalysis to an intermediate **transition state** complex, **EI; EI** changes into an enzyme-product complex, **EP,** which decomposes into the enzyme and product. Summarizing these reactions, and the reverse reactions that can also occur, we have:

$$\mathbf{E + S \rightleftharpoons ES \rightleftharpoons EI \rightleftharpoons EP \rightleftharpoons E + P.}$$

There are eight different reactions in this example, which is much simpler than many enzymatic reactions actually are, and each of these reactions will involve a separate mathematical description of its rate.

V_{max} **and** K_M. The kinetic analysis of enzymatic reactions has been simplified by combining the constants for individual steps into two parameters, which describe the overall process in a useful way and can be estimated from routine laboratory determinations. One of these parameters is the **maximum velocity, V_{max}**, which is the theoretical limit for the rate of reaction under defined conditions when the substrate concentration is so high that the active site is constantly occupied by substrate. V_{max} is an inherent property of a given enzyme, and it only depends upon the amount of the enzyme that is present in a given solution. It is a theoretical limit because it can be reached only when the substrate is present at infinite concentration, and no product is present to occupy the active site. It is the velocity of the reaction when the enzyme is **saturated** with substrate.

The second useful parameter is the **Michaelis* constant, K_M**, which is defined as the substrate concentration at which the actual velocity is ½ of the maximum velocity with no product present:

$$K_M = [S], \text{ when } \nu = \tfrac{1}{2} V_{max}, \text{ and } [P] = 0.$$

With these two parameters in mind, we are in a position to describe much of the physiologically important behavior of enzymes.

The Michaelis-Menten Equation

The major complication in analyzing the rate of any enzymatic reaction is the reverse reaction. The mathematics can be greatly simplified if the reverse reaction is negligible compared to the forward reaction. We shall discuss how this ideal is approached under laboratory conditions, but for the moment let us assume that we are dealing with a steady state condition, in which substrate is constantly supplied to maintain its concentration at a fixed level, and the product concentration is kept near zero by whisking it away as fast as it is formed. In this **steady state,** the rate of formation of the intermediate complexes, **ES, EI,** and **EP,** will just balance the rate of their removal so that their concentrations remain fixed.

Under these arbitrary conditions, the velocity of the reaction of a single substrate is described by the Michaelis-Menten equation:

$$\nu = \frac{V_{max} S}{K_M + S}$$

in which ν is the velocity, V_{max} is the maximum velocity with the enzyme concentration and the conditions used, and S is the concentration of substrate. The mathematical form of the relationship between velocity and substrate concentration was recognized long before much was known about the mechanisms of enzymatic reactions, and the equation is valid even though it was originally derived in 1913 using erroneous assumptions.

According to the equation, when $S = K_M$,

$$\nu = \frac{V_{max} K_M}{K_M + K_M} = \frac{V_{max}}{2}$$

which is as it must be, since we defined the Michaelis constant as the substrate concentration at which the velocity is half of the maximal.

*Leonor Michaelis (1875–1949), German (later American) biochemist.

Relation of Substrate Concentration and Rate. We can calculate the effect of substrate concentration on rate:

When $\frac{S}{K_M}$ =	0.1	0.2	0.4	1.0	2.0	4.0	10.0	20.0	50.0	100.0
$\frac{v}{V_{max}}$ =	0.09	0.17	0.29	0.50	0.67	0.80	0.91	0.95	0.98	0.99

An important point emerges. More efficient use of the enzyme without sacrificing control of rate is obtained if the range of substrate concentrations is centered roughly equal to K_M. Within that range, a substantial fraction of the enzyme's catalytic capability is being used, and it is still capable of sensitive response in rate to changes in substrate concentration. Outside of that range either velocity of reaction or its control will suffer. **Most enzymes are built with affinities for their substrates such that their K_M values will lie within an order of magnitude of the physiological concentrations of their substrates.**

Effect of the Reverse Reaction

The simple kinetic analysis we have given depends upon the assumption that the reverse reaction can be neglected, but the concentration of the product is rarely negligible in real biological systems. Since enzymes catalyze attainment of equilibrium from either direction, what we call "product" can also act as a substrate for the reverse reaction. It follows that there will be kinetics of a similar mathematical form when "product" reacts to yield "substrate," complete with its own Michaelis constant and maximum velocity, but operating in the reverse direction.

This gives an additional dimension to the evolutionary design of an enzyme. In effect, there is a competition between what we call substrate and product for the active site on an enzyme. How much time the site is occupied by one or the other will depend upon the relative affinity of the enzyme for the two compounds and their concentrations. The greater the affinity (the lower the K_M) and the higher the concentration, the greater the fraction of V_{max} that can be realized in one direction.

However, the V_{max} and K_M values for forward and reverse reactions are not independent entities. They are related to the equilibrium constant for the reaction according to the Haldane* equation:

$$\frac{V_{max(F)}}{K_{M(F)}} \div \frac{V_{max(R)}}{K_{M(R)}} = K_{eq}$$

in which F and R designate the forward and reverse reactions, respectively.

Let us use the Haldane equation to make two points. Firstly, an enzyme will use the largest fraction of its catalytic ability if the relative affinities for substrate and product are adjusted so that V_{max} is the same in both directions. Secondly, it is not always desirable to catalyze a reversible reaction equally well in both directions; therefore, the enzymes may be built to favor one direction or the other. Now, it is important to remember that the direction in which a reaction can go is determined by the concentrations of reactants relative to the position of equilibrium inherent in

*J.B.S. Haldane (1892–1964). English biochemist and geneticist: also known as far-left politicizer turned disillusioned exile. Not to be confused with his father, J.S. Haldane (1860–1936), respiratory physiologist and source of the Haldane coefficient for H^+ liberation from hemoglobin.

the reaction. **A catalyst can't change the equilibrium constant; it can only make the reaction go faster in whatever direction it is capable of going.** However, an enzyme is a catalyst that can be constructed to drag its feet when concentrations favor reaction in one direction, while happily going to work when concentrations shift so as to favor reaction in the other direction. This property is gained only at some sacrifice of the inherent catalytic potential.

We can develop the principle from a simple example. Imagine a reversible reaction:

$$\mathbf{S} \rightleftharpoons \mathbf{P}$$

for which $K_{eq} = 1$.

The velocity of such a reaction obeying Michaelis-Menten kinetics with both substrate and product present can be calculated from this equation:

$$v = \frac{V_{max(F)}\dfrac{S}{K_{M(F)}} - V_{max(R)}\dfrac{P}{K_{M(R)}}}{1 + \dfrac{S}{K_{M(F)}} + \dfrac{P}{K_{M(R)}}}$$

Let us imagine two different enzymes catalyzing this reaction, but with both having the same catalytic efficiency, that is, the same highest value of V_{max}. Now assume that one enzyme has equal values of K_M and V_{max} in both the forward and reverse directions, but that the K_M values, and therefore the V_{max} values, differ by a factor of 10 in the forward and reverse directions for the second enzyme:

enzyme 1: $K_{M(F)} = K_{M(R)}$; $V_{max(F)} = V_{max(R)} = \mathbf{V}$

enzyme 2: $K_{M(F)} = 10\ K_{M(R)}$; $V_{max(F)} = 10\ V_{max(R)} = \mathbf{V}$

If we now calculate the rates of reaction when the concentration of substrate is 10 × the concentration of product, so that the reaction must be proceeding forward, we find:

enzyme 1 catalyzes the reaction at 43% of **V**
enzyme 2 catalyzes the reaction at 30% of **V**

Enzyme 2 is less efficient even though its V_{max} for the reverse reaction is 0.1 that of enzyme 1.

What would be the point of building something like enzyme 2? We can see why by imagining a set of circumstances in which the reaction runs backward; that is, the concentration of the product is 10 × that of the substrate, the two being just the reverse of the values in the preceding example. Here we find for the reverse reaction:

enzyme 1 catalyzes the reaction at 43% of **V**
enzyme 2 catalyzes the reaction at 8.1% of **V**

Enzyme 1 has no inherent control over the relative velocities; it catalyzes the reaction equally well in both directions. Enzyme 2 doesn't make as efficient use of an identical mechanism, but it will catalyze the forward reaction much faster than it will the reverse reaction, thus acting as a check valve when changing concentrations try to make the reaction run backward from the desired direction. Both kinds of enzymes have their uses.

Isozymes

An organism need not be limited to one particular enzyme for catalyzing a given reaction. Indeed, the same reaction may be catalyzed by completely different proteins that differ in mechanism as well as in structure, a circumstance frequently true when the same reaction occurs in mitochondria and in the soluble cytoplasm. However, in some cases, cells may contain structurally related enzymes utilizing the same mechanism but with different kinetic parameters to fit particular requirements. Such enzymes are said to be isozymes, or isoenzymes.

The lactate dehydrogenases of mammals are good examples. These enzymes are tetramers made from polypeptide chains occurring in two principal forms, designated H for heart and M for muscle. (A third kind is in spermatozoa, but it is of minor importance in the total fuel economy, whatever its other consequences may be.) These polypeptide chains are so similar that they are interchangeable in the tetramer, giving five theoretical possibilities for the major forms of lactate dehydrogenase: H_4, H_3M, H_2M_2, HM_3, and M_4. All occur in tissues. H_4 is predominant in tissues of high oxidative capacity, and it has kinetic parameters that favor catalysis of the oxidation of lactate to pyruvate over the reduction of pyruvate to lactate, and this is appropriate for utilization of lactate as a fuel. Other tissues, such as striated muscle fibers built for quick twitches (white fibers), contain mostly the M_4 isozyme, which has no particular preference for the direction of catalysis. We shall see that the concentrations of reactants usually favor the reduction of pyruvate to lactate in those tissues.

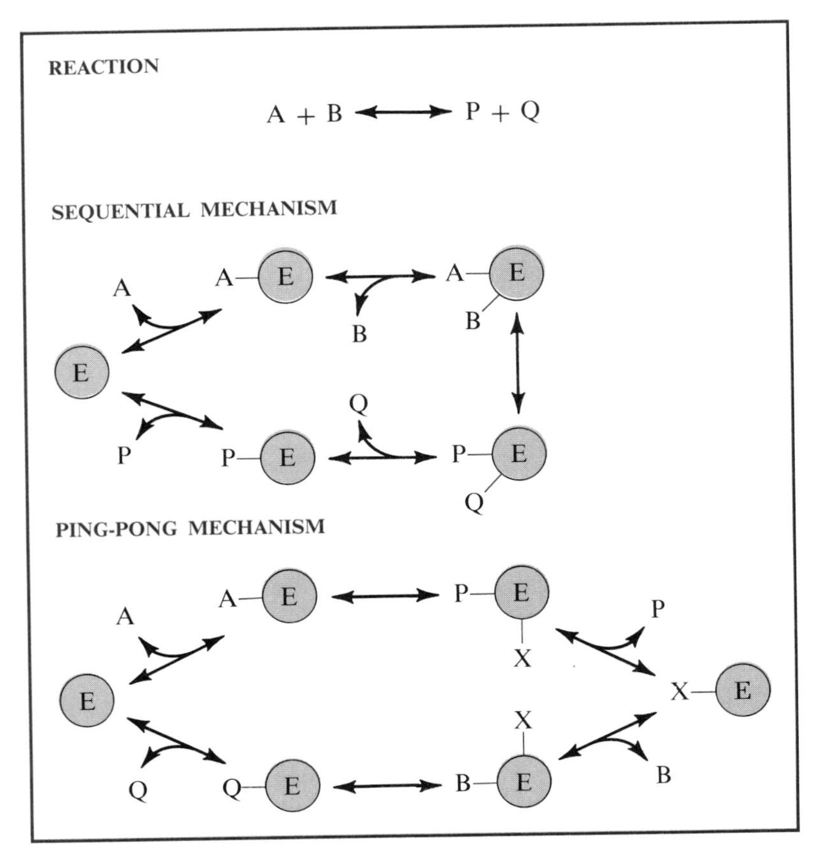

FIGURE 15–1

Reactions involving two substrates may proceed by a sequential mechanism (*top*) in which the substrates successively add to the enzyme before a reaction forming the enzyme-products complex occurs. The products then dissociate, one at a time, from the enzyme. Alternatively, the reactions may proceed by a ping-pong mechanism (*bottom*), in which one substrate reacts on the enzyme to transfer a functional group, X, to it. The first product, P, then leaves before the second substrate, B, is added. The functional group X then is transferred from the enzyme to B, forming the second product, Q.

Reactions with Two or More Substrates

Many enzymatic reactions involve more than one substrate:

$$\mathbf{A} + \mathbf{B} \xrightarrow{E} \mathbf{P}, \text{ or } \mathbf{A} + \mathbf{B} \xrightarrow{E} \mathbf{P} + \mathbf{Q},$$

and so on. The velocity of these reactions depends upon the concentrations of all of the components, and the kinetic equations are more complex. Furthermore, the form of the kinetic equations depends upon the mechanism of the reaction.

For example, two substrates may react in a **sequential** fashion (Fig. 15–1), with both becoming attached to the enzyme before any products are formed. Lactate dehydrogenase and other NAD-coupled dehydrogenases function in this way. The mechanism is said to be **ordered** if the substrates add and the products leave in a defined way (**A** adds before **B, P** leaves before **Q,** for example) and **random** if it makes no difference which substrate is bound first by the enzyme.

In other cases, substrates react by a **ping-pong** mechanism in which one substrate reacts with the enzyme so as to transfer a functional group onto the enzyme, leaving one product. The second substrate then reacts to pick up the functional group, forming a second product. We saw such a mechanism with the aminotransferases, in which the amino group passes from an amino acid substrate to the pyridoxal phosphate coenzyme, followed by its transfer to a keto acid substrate.

MEASUREMENT OF KINETICS

Measurements of the rates of enzymatic reactions are commonly done for two purposes. One is to determine the kinetic parameters of an enzyme for later use. The other is to apply the knowledge of the parameters to assays of the concentration of the enzyme in unknown samples. Such assays are made every day in clinical pathology laboratories; the results are expressed as velocities of reaction catalyzed per amount of sample, and serve as indicators of the progress of a variety of diseases.

Determination of Kinetic Parameters

The Michaelis constant and maximum velocity may be determined for an enzyme most easily by making measurements of the actual reaction velocity at various substrate concentrations, under circumstances in which the rate of the reverse reaction is negligible. Some substrate must be consumed and some product formed if there is to be a reaction velocity to be measured, but it is desirable to have the concentrations change as little as possible, so that the velocity is nearly constant. This is facilitated when a precise and sensitive method is available for measuring the concentration of the product of the reaction. The product obtained per unit time will then be an estimate of the initial velocity of the reaction. With luck, this velocity will be the rate of the enzymatic reaction after steady state is reached but before the overall reaction is slowed down by changes in the product and substrate concentration. (The time required for reaching steady state is usually less than the time required for mixing substrate and enzyme in routine analyses.)

Surprisingly, this approach frequently gives useful information. If a series of determinations of initial velocity are made at varying substrate concentrations, a graph of the sort shown in Figure 15–2 can be constructed from the results. This is the typical hyperbolic curve obtained when Michaelis-Menten kinetics are obeyed.

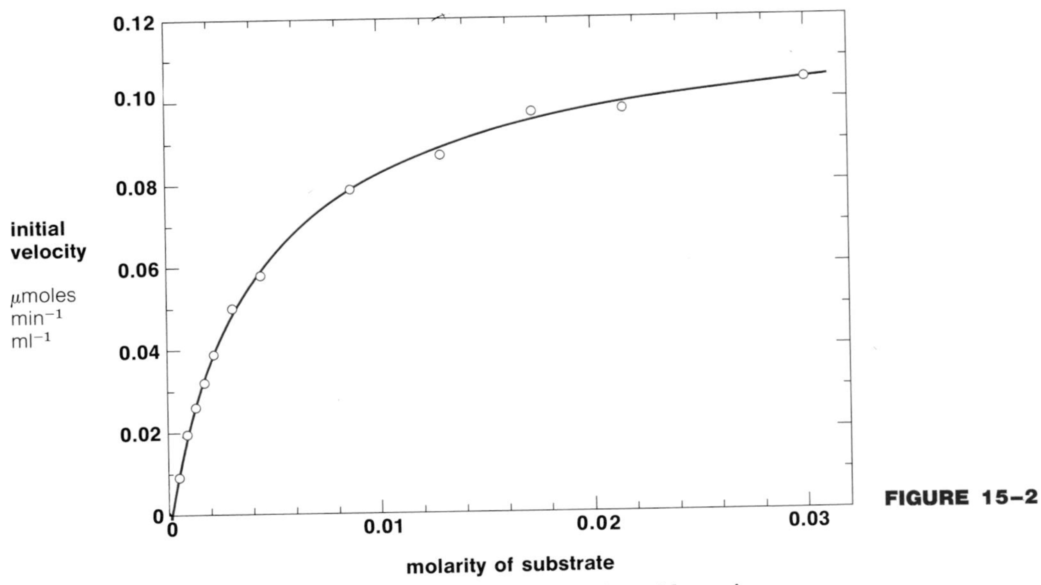

FIGURE 15–2

Measured values of the initial velocity of an enzymatic reaction with varying concentrations of substrate. The scatter apparent at the higher substrate concentrations amounts to 2 per cent of the highest velocity, which is quite good for measurements of most enzymes with standard techniques.

The data can be analyzed in two ways. One is through algebraic analysis, and computer programs have the advantage that they can take into consideration the changing product concentrations as well as the changing substrate concentrations. The other method, almost universally used in the past, is through graphical analysis that neglects the product concentration.

One such analysis is based on inversion and rearrangement of the Michaelis-Menten equation:

$$\frac{1}{\mu} = \frac{K_M QS}{V_{max} S} = \frac{K_m}{Vmax} \times \frac{1}{S} + \frac{1}{V_{max}}$$

If we now take $1/v$ and $1/S$ as our variables, we have a simple linear equation of the type $y = ax + b$ which can be solved either by a linear regression analysis or by graphical analysis.

The reciprocals of the initial velocities are plotted as functions of the reciprocals of substrate concentrations. If Michaelis-Menten kinetics describe the reaction, the result will be a straight line, such as is shown in Figure 15–3, which was derived from the same data as shown in Figure 15–2. These reciprocal plots are known as **Lineweaver-Burk* plots,** after the men best known for developing this type of analysis.

As is indicated on the figure, the intercept on the ordinate is the reciprocal of V_{max} and the intercept on the abscissa is the negative reciprocal of K_M, so both of these constants can be estimated from the plot.

When a reaction involves two or more substrates, the respective Michaelis constants can be estimated with the same sort of procedure by adding a large excess of one, effectively saturating the enzyme, and varying the concentration of the other across its K_M value.

---

*Hans Lineweaver and Dean Burk, American biochemists, now retired.

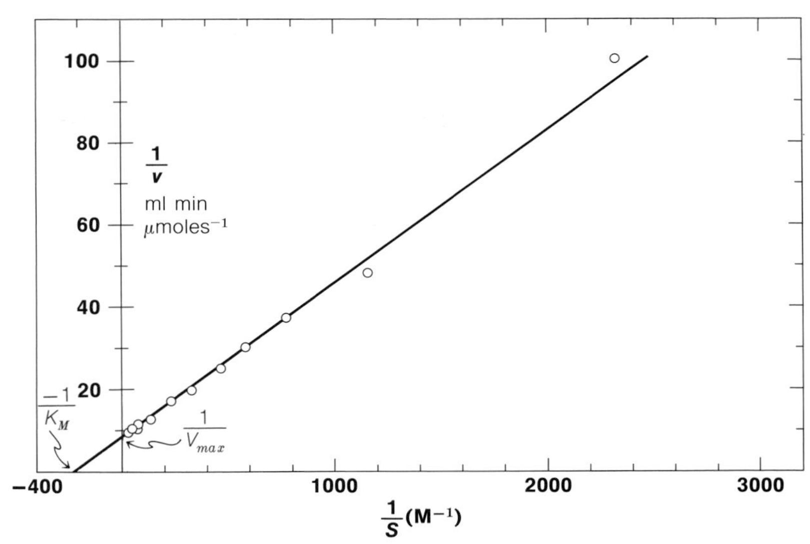

FIGURE 15–3

A Lineweaver-Burk plot of the data from Fig. 15–2. Note the gross scatter evident in values obtained at low substrate concentrations (high values of $1/S$), which were not evident in the steeply rising portion of the regular plot given in Fig. 15–2. The intercepts yield the following values:

$$\text{abscissa} = -\frac{1}{K_M} = -230\ \text{M}^{-1};\ K_M = 0.0043\ \text{M}$$

$$\text{ordinate} = \frac{1}{V_{max}} = 8.3\ \text{ml min}\ \mu\text{moles}^{-1};\ V_{max} = 0.120\ \mu\text{moles min}^{-1}\ \text{ml}^{-1}$$

Other rearrangements of the Michaelis-Menten equation can be used to generate linear plots, but a discussion of their advantages is not necessary for our purpose. (See Further Reading for sources.)

Assay of Enzymes

The assay of enzyme concentrations by kinetic measurements is a common procedure in many kinds of laboratories—clinical, industrial, and experimental. The most common clinical tests involve assay of enzymes in blood, because conditions causing damage to specific tissues usually result in the release of abnormally large amounts of characteristic enzymes, but the concentration of particular enzymes is also determined in tissue samples or in cultured cells, especially to diagnose genetic abnormalities.

It is not yet practical to make a full mathematical analysis of kinetics for routine assays, which frequently involve single incubations of one sample under fixed conditions, so the problem in devising useful procedures is one of minimizing the effect of changing substrate and product concentrations while the enzyme is acting. One wants half as much enzyme to produce half as much product in a given time; one third as much to produce one third as much, and so on. Suppose we fix conditions so that the initial substrate concentration is 10 times the K_M ($S = 10K_M$), and adjust the amounts of sample used so that the richest sample will convert only 0.1 of the substrate to product. How much will the rate change during the assay, assuming the product concentration is too little to have a significant effect?

initial $S = 10\ K_M$ velocity $= 0.91\ V_{max}$

final $S = 9K_M$ velocity $= 0.90\ V_{max}$

Now suppose another sample contains only half as much enzyme. Its V_{max} will therefore be only half the V_{max} of the richer sample, and it will therefore convert less than 0.05 of the substrate to product in the same time:

$$\text{initial } S = 10\ K_M \qquad \text{velocity} = 0.909\ V_{max} = 0.45\ \mathbf{V}$$

$$\text{final } S = 9.5\ K_M \qquad \text{velocity} = 0.905\ V_{max} = 0.452\ \mathbf{V}$$

in which **V** is the V_{max} for the first sample, the sample with twice as much enzyme. It is plain that half the amount of enzyme comes within 1 per cent of producing half the amount of product under the specified conditions (0.452 **V** against 0.900 **V**); this is therefore a good linear assay of enzyme concentration with only a single determination of product concentration for each sample. This is the ideal circumstance the enzyme assayist strives for, but it is not always easy to obtain.

Figure 15–4 shows the time course of reactions that have identical kinetic parameters for the forward reaction but that differ in equilibrium constant, and therefore in kinetic parameters for the back reaction. *Curve A* illustrates the type of reaction beloved by those who assay enzymes, with a high equilibrium constant and low affinity for the product creating nearly linear kinetics for a large part of the reaction course. The amount of product obtained in a single determination is directly proportional to the amount of enzyme present over a wide range with reactions of this type.

Curve D shows a kind of reaction that would be almost impossible to tackle in the ordinary way. The affinity of the enzyme is so much higher for the product than it is for the substrate that the formation of only a little product causes a large drop in reaction rate. (Note that this is true even though the maximum velocity in the forward direction is 1,000 times greater than that in the reverse direction.) Fortunately, enzymes with kinetics like this are very rare, if they ever occur. They wouldn't be very useful catalysts in an organism. *Curve B* illustrates the action of an equally inefficient enzyme, but the position of equilibrium is so far in favor of product, that the catalysis still may be useful.

Curve C shows a more typical time course for a readily reversible reaction. One must either have a sensitive analytical procedure for the product, or take steps to remove the product as fast as it is formed. The latter procedure is preferable, and it may be accomplished by adding an excess of another enzyme to catalyze an additional reaction of the product, or by adding some reagent that selectively reacts with the product and converts it to another compound. Either of these methods also makes it possible to use enzymes for analyzing the concentrations of compounds in samples, even when the equilibrium is not favorable.

For example, the equilibrium of lactate dehydrogenase favors the reduction of pyruvate to lactate:

$$\text{pyruvate} + \text{NADH} + \text{H}^+ \longrightarrow \text{lactate} + \text{NAD}^+$$

NADH has a peak absorption of light at 340 nm, and NAD does not, so the course of the reaction can be followed by noting the decreased absorption of light at this wave length. The changes in absorbance with time may be followed to assay lactate dehydrogenase, or excess enzyme may be added to an unknown sample, and the reaction allowed to go to completion to measure the amount of pyruvate in the sample.

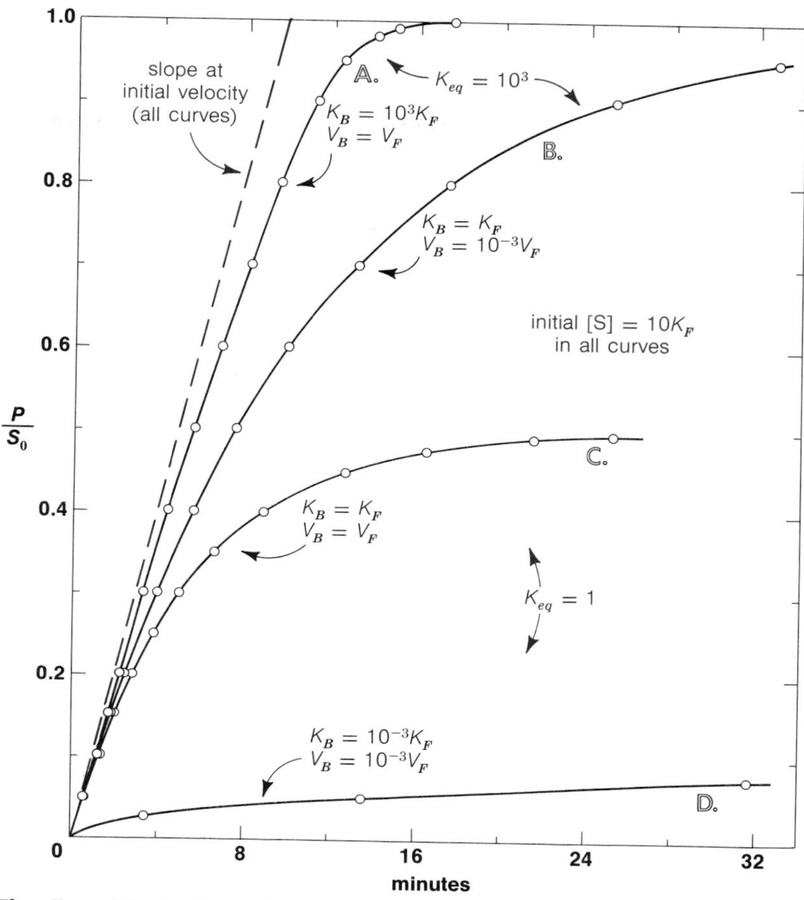

FIGURE 15–4

The effect of the back reaction and the equilibrium constant on the time course of enzymatic reactions. The fraction of the initial substrate that has been converted to product (P/S_0) is plotted as a function of time of reaction with four different enzymes. The relative kinetic parameters for forward and back reactions are given by each curve. All enzymes are assumed to have the same V_{max} in the forward direction; note that all of the reactions would follow the dashed line if substrate was continually added to maintain its concentration, and if the product was removed as fast as it formed.

The reaction cannot be used as such to determine lactate because the equilibrium is unfavorable for the formation of pyruvate, but this determination becomes possible if hydrazine is included in the reaction mixture to remove pyruvate. The overall equilibrium then favors lactate oxidation:

$$\underset{\textbf{L-lactate}}{HO{-}\overset{\displaystyle COO^{\ominus}}{\overset{|}{C}}{-}H \atop {|\atop CH_3}} \xrightleftharpoons[K'_{eq} = 3.4 \times 10^{11}]{NAD^{\oplus} \;\rightarrow\; H^{\oplus} + NADH} \underset{\textbf{pyruvate}}{\overset{\displaystyle COO^{\ominus}}{\overset{|}{C}}{=}O \atop {|\atop CH_3}} \xrightleftharpoons[K'_{eq} \cong 2.4 \times 10^{14}]{\underset{\textbf{hydrazine}}{H_2N{-}NH_2} \;\rightarrow\; H_2O} \underset{\textbf{pyruvate hydrazone}}{\overset{\displaystyle COO^{\ominus}}{\overset{|}{C}}{=}N{-}NH_2 \atop {|\atop CH_3}}$$

$$K'_{eq} \cong 7 \times 10^2$$

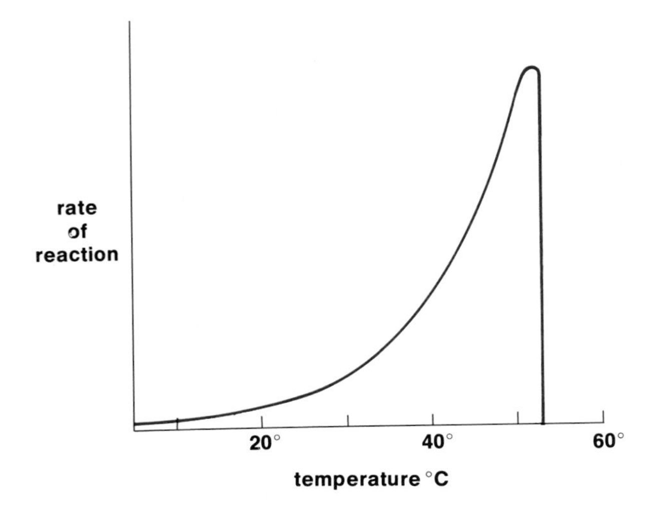

FIGURE 15–5

Enzymatic reactions are like most reactions in that their rates accelerate as the temperature rises above the freezing point. However, unlike most reactions, the rate abruptly falls to low values when the temperature is reached that causes disruption of the conformation of the enzyme.

Units of Enzyme Activity. The common enzyme unit (U) for expressing activity of enzymes has the dimensions of micromole of substrate consumed per minute. It is as much a measure of quantity of enzyme as is mass or other property. Enzyme concentrations are given in units per volume. (See the back endpaper for examples.) These values can be translated into estimates of actual moles of enzyme if its turnover number is known under the same conditions. The turnover number is the number of reaction cycles one molecule of enzyme catalyzes per unit of time (minute or second).

Effect of Temperature

Reactions catalyzed by enzymes are like other reactions in that the rate is increased by increasing temperatures—up to a point. Beyond a critical range, the transition temperature, the activity of an enzyme declines sharply (Fig. 15–5). Not only enzymes, but nearly all proteins lose conformation readily above certain temperatures. When thermal energy becomes great enough to cause the rupture of a few bonds, the neighboring bonds are weakened, and the whole molecule becomes unzippered. This gross loss of conformation is known as denaturation of the protein.

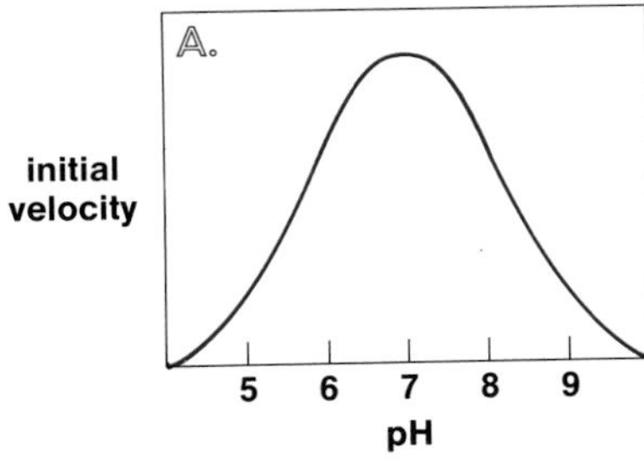

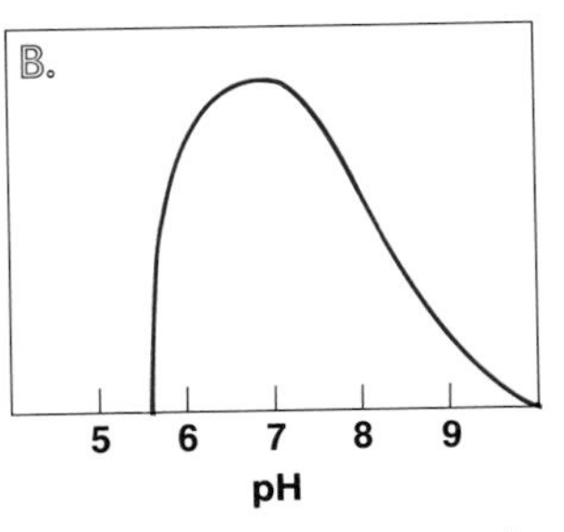

FIGURE 15–6

The effect of pH on the rate of enzymatic reactions. *A.* The enzyme is stable over the pH range tested, and the mechanism of catalysis requires both acidic and basic groups. *B.* The mechanism is similar to that of the above enzyme, but the transition temperature for disruption of the enzyme's structure falls at pH 5.7 below the temperature used for assay.

It happens that most of the proteins in an organism become denatured at temperatures only 10 or 15° C above the usual cellular temperatures. It may be that the margin of safety is small because a certain flexibility of motion is necessary to carry out enzymatic catalysis, or other functions depending upon conformational shifts. The tiger may have to be tickled to a rippling alertness that is not far removed from a destructive frenzy.

Effect of pH

A typical enzyme catalyzes its reaction most effectively at a specific concentration of H^+ (Fig. 15–6), not only because the mechanism depends upon acidic and basic groups being present in optimum proportions but also because the conformation of the protein is also dependent upon the state of charge of bonding groups in general. Severe changes in pH cause substantial lowering of the transition temperature for disruption of conformation.

FURTHER READING

H.L. Segal (1959). The Development of Enzyme Kinetics, vol. 1, p. 1; and Alberty, R.A. (1959). The Rate Equation for an Enzymic Reaction, vol. 1, p. 143. In P.D. Boyer, H.A. Lardy, and K. Myrback, eds., The Enzymes, 2nd ed. Academic Press. These articles remain among the best of the introductory treatments of kinetics.

W.W. Cleland (1970). Steady State Kinetics, vol. 2, p. 1. In P.D. Boyer. ed., The Enzymes, 3rd ed. Academic Press. A pioneer in the development of computer analyses of kinetics treats the subject more intensively.

CHAPTER 16

REGULATION OF ENZYME ACTIVITY

Life hinges upon the ability to adjust the rate of enzymatic reactions. Man has learned to add some of his own adjustments to these physiological controls in some cases; therapeutic drugs are often devices for changing the rate of specific reactions. Intensive use is being made of basic knowledge of enzyme mechanisms to design even more effective compounds.

There are three general ways of changing the rate of an enzymatic reaction:

(1) alter the **substrate or product concentrations,**

(2) alter the **catalytic activity** of individual enzyme molecules through the use of effectors or by covalent modification of the protein, and

(3) alter the **number of enzyme molecules** by changes in the rate of their synthesis or destruction.

Concept of the Rate-Limiting Step

Independent metabolic processes rarely consist of one enzymatic reaction. More often they have several reactions in sequence, in which the product of one reaction is the substrate for the enzyme catalyzing the next reaction, and so on:

$$A \xrightarrow{enzyme\ 1} B \xrightarrow{enzyme\ 2} C \xrightarrow{enzyme\ 3} D \xrightarrow{enzyme\ 4} E \ldots$$

The rate of reaction sequences like this is often governed at one step, which is referred to as the rate-limiting step. Since the rates of each reaction in a sequence must be identical if there is not to be a gross gain or loss of some intermediate compound, it may seem strange to speak of one step as being rate-limiting. The rate-limiting step is not one that goes faster or slower than the others in the sequence, but the one in which small changes in the limiting parameter will cause the largest changes in the overall velocity of the complete sequence. Water flows through a constricted pipe at the same rate at all points throughout its length, but it is the size of a constriction that determines what the rate is at a given pressure head.

WHERE TO REGULATE

The body contains a network of reaction sequences in which the product of one reaction may be a substrate for several others, and vice versa. Which of the many enzymes is the proper choice as a site for control, either through biological evolution or human artifice? The enzyme that is made rate-limiting ought to catalyze **a reaction peculiar to the metabolic process** over which control is desired.

Consider cholinergic nerves. These are neurons that discharge acetylcholine to stimulate a muscle fiber or another neuron. An enzyme catalyzes the synthesis of acetylcholine:

$$\text{acetyl donor} + \text{choline} \xrightarrow[\textit{acetyltransferase}]{\textit{choline}} \text{acetylcholine},$$

which is stored until a nerve impulse causes it to be discharged from the cell. In order to stop the stimulation and recock the recipient cell for another triggering event, the released acetylcholine is rapidly hydrolyzed by another enzyme:

$$\text{acetylcholine} + H_2O \xrightarrow{\textit{cholinesterase}} \text{acetate} + \text{choline}$$

Should the formation of acetylcholine be regulated by changing the supply of the acetyl donor or choline, or the removal of acetate? No, because the acetyl donor is a compound of major importance for many purposes, as we shall see, and we already know that choline is a component of critical phospholipids. Therefore, any attempt at regulating neural transmission in cholinergic nerves by altering the supply of acetyl groups or choline is likely to disturb a variety of other cellular processes. The proper place for control is one of the enzymes peculiar to use of the neurotransmitter, such as choline acetyltransferase or acetylcholinesterase. (We shall consider neurotransmitters in Chapter 33.)

Inherent Regulation by Enzymes. Does the selection of a step as a primary control point mean that the concentrations of other reactants are left free to swing wildly? Not at all, because most enzymes tend to buffer changes in substrate concentration. Enzymes catalyzing readily reversible reactions are usually present in relatively high concentrations. Since they are catalyzing both the forward and the reverse reactions, it takes a lot of enzyme to make the difference in the two rates significant. The net reaction is only a small fraction of V_{max}, so any rise in substrate concentration can cause a relatively large increase in the rate of its removal.

HOW TO REGULATE

Limitation by Substrate Supply

A reaction can obviously be made rate-limiting by shutting off its supply of substrate, but this requires regulation of those reactions supplying the substrate, and they would be the true rate-limiting steps if they occurred in the same location. However, there are some cases in which a compound is made in one place and used as a substrate in another, and its supply is indeed rate-limiting for its utilization. This circumstance is most often seen in fuel metabolism. A prime example is in the use of glucose from the blood by the liver. The enzyme catalyzing the first reaction of glucose metabolism in the liver is built to have a K_M such that it is responsive to changes in the glucose concentration. The reaction is a transfer of phosphate from ATP to glucose:

$$\text{ATP} + \text{glucose} \longrightarrow \text{ADP} + \text{glucose 6-phosphate}$$

It is catalyzed by enzymes called **hexokinases** because they are not very specific for the kind of hexose. The suffix *-kinase* was invented to suggest that the reaction results in the activation of glucose for metabolism. As is implied by the single arrow, the reaction is essentially irreversible under the conditions prevalent in tissues.

Four hexokinases occur in tissues. The types predominant in brain and liver parenchymal cells have been most intensively studied; the measured Michaelis constants are as follows:

	K_M (ATP)	K_M (glucose)
brain hexokinase (type I)	4×10^{-4} M	5×10^{-5} M
liver hexokinase (type IV)	1×10^{-4} M	2×10^{-2} M

The concentration of ATP in these tissues is held near 10^{-3} M or greater. This is substantially greater than K_M for either enzyme, and ATP is therefore not limiting in either case.

Now, the concentration of glucose in the blood supply to the brain and other peripheral tissues usually ranges around 5×10^{-3} M, although it may increase to around 9×10^{-3} M after consumption of a large amount of carbohydrate and may drop to 3×10^{-3} M during fasting or heavy muscular work. The liver operates under different conditions. It receives blood from the intestine through the portal circulation as well as from the arterial circulation, and the portal blood has substantially higher concentrations of glucose during absorption of a carbohydrate-containing meal.

All of the values for glucose concentration we have mentioned are nearly two orders of magnitude higher than the K_M for glucose with the brain hexokinase, a range in which changes in concentration have only small effects on the rate of reaction (see p. 298). This provides a considerable margin of safety for the brain, which is almost completely dependent upon the glucose supply for its major fuel. The observed facts are consistent with this if we allow for a gradient of glucose concentration between the blood and the interior of the neurons, because the concentration of glucose in blood must drop to nearly 2×10^{-3} M before disturbances of the central nervous system begin to be noticed, and even lower concentrations are needed to produce unconsciousness.

On the other hand, the K_M of the liver hexokinase is greater than the concentration of glucose usually found in arterial blood. Therefore, the phosphorylation of glucose catalyzed by this enzyme will be very sensitive to glucose concentration, and this is consistent with the physiological role of the liver. The liver ordinarily produces glucose rather than consuming it, and it only removes more glucose from the blood than it produces during temporary overloads after eating. The balance between production and utilization occurs with blood concentrations near 0.006 M.

A rise in glucose concentration will accelerate the liver hexokinase reaction because the enzyme is operating far below its saturating substrate concentration, whereas the brain hexokinase is always nearly saturated and is relatively insensitive to fluctuations in glucose concentration.

Regulation of Irreversible Reactions

The rate-limiting steps controlled through alteration of an enzyme's activity or of its concentration are usually irreversible reactions—those in which the position of equilibrium is so far in favor of the product that a net back reaction is not feasible in any real sense. (Having a net formation only in one direction does not mean that the back reaction doesn't occur; it means only that the forward reaction is always faster.) The advantage of an effectively irreversible reaction is that a greater range of control is possible without approaching equilibrium. The reaction may be slowed and accumulate substrates and their precursors. It may be accelerated to deplete substrates and accumulate products without any risk of making the reaction go in the opposite direction. The reaction that is the site of control can often be recognized by such crossovers in concentration in response to a regulatory stimulus. For example, the metabolism of glucose includes the following sequence of reactions:

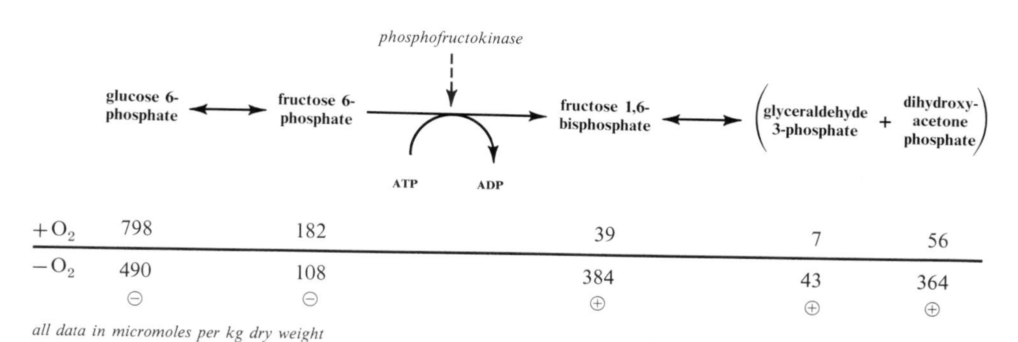

	glucose 6-phosphate	fructose 6-phosphate	fructose 1,6-bisphosphate	glyceraldehyde 3-phosphate	dihydroxyacetone phosphate
$+O_2$	798	182	39	7	56
$-O_2$	490	108	384	43	364
	⊖	⊖	⊕	⊕	⊕

all data in micromoles per kg dry weight

Listed below the compounds in this sequence are their concentrations in rat hearts perfused with and without a supply of oxygen.* The central reaction, catalyzed by the enzyme **phosphofructokinase,** is irreversible, as indicated by the arrow pointing in one direction. Deprivation of oxygen caused a depletion of the compounds supplying that reaction and an increased appearance of the compounds formed by that reaction. The data demonstrate that oxygen deprivation caused an acceleration of the phosphofructokinase reaction, which was the regulated rate-limiting step in the sequence shown.

REGULATION BY EFFECTORS

Enzymes are often regulated through combination with some other compound, which may block the active site or alter the conformation of the enzyme so as to change its catalytic activity. Compounds altering rate in this way are said to be **effectors,** or modulators, of the activity; they may be **activators** that increase the activity or **inhibitors** that decrease the activity. Rate-limiting enzymes at key control points are frequently constructed so that they have distinctive binding sites for the effectors, even though the effector does not participate in the catalytic reaction. This is analogous to the presence of a bonding site for 2,3-bisphosphoglycerate in hemoglobin.

In general, activations are devices used to mobilize stored compounds; fuels are made available for combustion in response to effectors that act as signals of the need for increased energy production. Inhibitions are frequently used to stop the formation of a surfeit of some body constituent.

---

*Data from J.R. Williamson (1965) in B. Chance, ed.: Control of Energy Metabolism. p. 335, Academic Press. Data such as these are often more difficult to interpret than is implied here because a large fraction of compounds present in low amounts is bound to enzymes, and the true concentrations in solution are not measured. This is inherent in the materials and in no way implies lack of skill by the analyst.

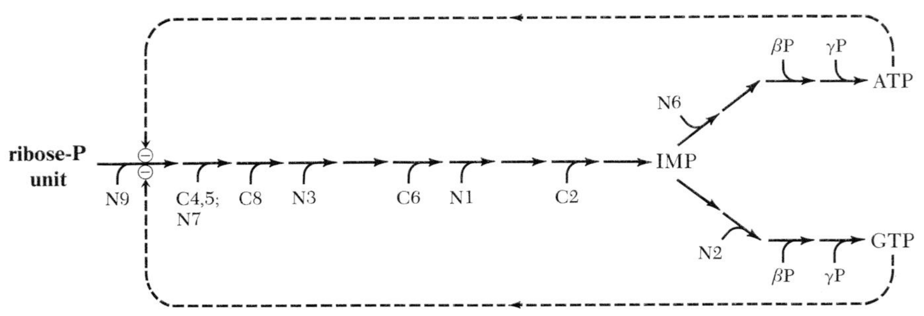

FIGURE 16–1 The purine component of purine nucleotides is assembled one piece at a time on a ribose phosphate unit. The numbers of the successively added atoms indicate their positions in the completed molecule. This long sequence is regulated through feedback inhibition of the first reaction by the final products, thereby preventing wasteful accumulation of the many intermediate compounds.

Negative Feedback. The synthesis of components of the organism frequently involves a succession of intermediates that are not used for any other purpose. For example, the formation of the purines incorporated into nucleotides does begin with common metabolites used for many purposes, but the synthesis proceeds by adding one group at a time to form intermediates that are good for nothing except making purine nucleotide. Each of the steps in the formation requires an enzyme. The problem is how to regulate the rate of a long sequence of reactions so as to produce enough purine nucleotides, but not too much, and without ever accumulating the otherwise useless intermediate compounds. Rather than regulating each reaction involved in such long sequences, the problem is solved by making the ultimate products of the sequence, the purine nucleotides themselves, inhibitors of the first reaction peculiar to purine synthesis (Fig. 16–1). (The reactions are described in Chapter 35.) When ATP and GTP are in short supply, the initial enzyme is not inhibited, and a flow of substrate begins through the long series of enzymatic reactions that ultimately result in the formation of more ATP and GTP. (The activity of all intermediate enzymes is high, so they are not rate-limiting.) When the supply is adequate, the initial enzyme is inhibited, and all of the intermediate reactions also slow down because the enzymes run out of substrates. The general process by which a compound inhibits its own formation is called "negative feedback" after a term borrowed from electronics by Arthur Pardee.

Competitive Inhibitors

An enzymatic reaction can be slowed by any compound occupying the active site in competition with the substrate or an intermediate in the reaction mechanism. Many reactions are physiologically controlled in this way, and a number of drugs are useful because they competitively inhibit particular enzymes. The inhibitor, like the substrate or reaction intermediate with which it competes, frequently will fit only a few enzymes tightly. The action of the drug can be quite specific, as we shall see in many later examples. For now, let us examine some less common applications.

Ethylene glycol, a common constituent of antifreeze preparations, is consumed by some individuals with the expectation of transient or permanent intoxication. The results are frequently disastrous because this compound can be attacked by alcohol dehydrogenase in the liver, with subsequent formation of a variety of organic acids, and a devastating increase in H^+ concentration (acidosis). Problems are also created by precipitation of the calcium salt of oxalic acid in the urinary tract.

$$\underset{\textbf{ethylene glycol}}{\begin{array}{c}HO{-}CH_2\\ |\\ HO{-}CH_2\end{array}} \xrightleftharpoons[\textit{alcohol dehydrogenase}]{NAD^{\oplus} \quad H^{\oplus} + NADH} \underset{\textbf{glycolaldehyde}}{\begin{array}{c}O{=}C{-}H\\ |\\ HO{-}CH_2\end{array}} \xrightarrow[\textit{steps}]{\textit{many}} \underset{\text{organic anions}}{H^{\oplus} +}$$

Ethylene glycol poisoning is treated by administering ethanol, which competes with the glycol for the alcohol dehydrogenase. Here is a neat reversal of the usual principle, with a physiological substrate, ethanol, administered as an inhibitor of the metabolism of a less common compound. (Ethanol is much more abundant than ethylene glycol in natural foodstuffs.)

To cite another example, a competitive inhibitor of hexokinase has been used to chart the metabolic activity of the human brain; 2-fluoro-2-deoxyglucose will replace glucose as a substrate. It is phosphorylated by the enzyme but cannot undergo subsequent reactions, and accumulates at any site where glucose is being metabolized. By using radioactive fluorine (^{18}F), which is a positron emitter, and a suitable array of detectors with computed analysis of the results, it is possible to see where glucose is most actively metabolized within an intact person's brain, and what changes occur in its distribution with various neural stimuli.

The examples given here involve competition for the substrate. With the increasing knowledge of reaction mechanisms, more effort is currently being made to devise competitive analogs of the transition state intermediates. An example is cited later.

Kinetics of Competitive Inhibition. If a substrate (or transition state intermediate) competes with an inhibitor for binding, then raising the concentration of one will diminish the binding of the other. Indeed, this is the hallmark of competitive inhibition. Given a fixed concentration of inhibitor, increasing the concentration of substrate will make the velocity approach the same V_{max} seen in the absence of the inhibitor.

Recognition of competitive inhibition by synthetic compounds has been valuable in deducing the nature of active sites in enzymes, since they frequently are bound by the same groups as a substrate. If an enzyme with Michaelis-Menten kinetics combines with a competitive inhibitor, I, the initial velocity of the reaction is described by:

$$v = \frac{V_{max} S}{S + K_M(1 + I/K_I)}$$

in which $K_I = \frac{[\mathbf{E}][\mathbf{I}]}{[\mathbf{EI}]}$, and $I = [\mathrm{I}]$.

The equation can be rearranged in terms of reciprocals of initial velocity and substrate concentration, as we did with the simple Michaelis-Menten equation, but the essential features are apparent as written. Mathematically, the presence of inhibitor has the effect of increasing K_M in the equation by the factor $\left(1 + \frac{I}{K_I}\right)$, *but it does not change the value of* V_{max}. This is what would be expected, because an infinite concentration of substrate should effectively displace all of a competitive inhibitor.

The equation also shows us that K_I is equal to the concentration of inhibitor that doubles the *apparent* value of K_M. Now if the inhibitor doesn't change the maximum velocity but does alter the apparent value of the Michaelis constant, differing concentrations of inhibitor should give a series of straight lines intersecting at $1/V_{max}$ on the ordinate, when Lineweaver-Burk plots are made of the effect of substrate concentration on rate of reaction.

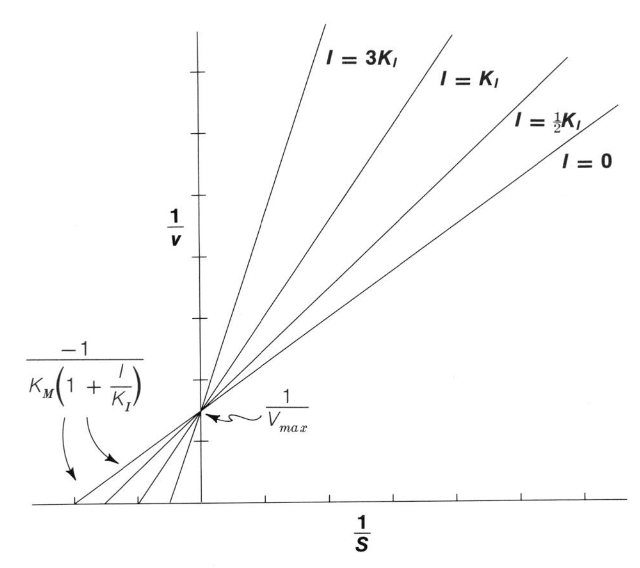

FIGURE 16–2

Lineweaver-Burk plots obtained with varying concentrations, I, of a purely competitive inhibitor. There is a single intercept on the ordinate (vertical axis) because V_{max} is the same in all cases, but the intercept on the abscissa (horizontal axis) changes.

The value of K_I can be estimated from the value of the apparent K_M in the presence of inhibitor, and this value is calculated from the negative reciprocal of the intercept on the abscissa. Figure 16–2 shows an example.

Noncompetitive Inhibitors

Compounds may inhibit an enzymatic reaction by combining with an enzyme in such a way that they are not displaced by the substrate but prevent the enzymatic reaction from occurring.

Inhibitors of Seryl Group-Dependent Hydrolases. We saw that several proteases (endopeptidases) act through a mechanism in which serine side chains are acylated by the substrate. Other hydrolases have similar mechanisms, and acetylcholinesterase is a notable example. Many inhibitors of acetylcholinesterase have been developed that behave like substrates up to the point at which the serine side chain is acylated, but they donate an acyl group that the enzyme is unable to remove. The enzyme mechanism is in effect jammed in the middle. The most effective agents are in two classes: phosphoric acid anhydrides or esters, and carbamic acid esters. An especially effective example is a mixed anhydride of phosphoric acid and hydrofluoric acid:

$(H_3C)_2CH\text{—O—}P(=O)(F)\text{—O—}CH(CH_3)_2$

diisopropylphosphofluoridate

+ **seryl group in protein** (protein—CH_2—OH)

→ $(H_3C)_2CH\text{—O—}P(=O)(\text{—O—}CH_2\text{—protein})\text{—O—}CH(CH_3)_2 + H^{\oplus} + F^{\ominus}$

diisopropylphosphate ester of the protein

The particular agent shown, usually known as diisopropylfluorophosphate (DFP), is an effective tool for demonstrating the participation of seryl groups in an enzymatic reaction because the formation of the phosphate triester prevents the completion of the hydrolytic mechanism. The inhibition is irreversible because even high concentrations of a substrate will not displace the covalently bound inhibiting group.

The effectiveness of esters of phosphofluoridic acid was discovered by accident. Chemists preparing the compounds for the first time in the laboratory of Dr. Willie Lang in Germany developed mental confusion, a sense of constriction of the larynx, and a painful loss of accommodation of the eye. These effects are now known to be characteristic of inhibitors of the enzyme acetylcholinesterase.

The original workers concluded on empirical grounds that the compounds might have value as insecticides. This suggestion was ignored, but with the approach of the Second World War, it was realized that they could provide very potent war gases. A variety of volatile derivatives, many still not discussed in public, were prepared in several of the belligerent countries during the war and after it.

The original suggestion that the compounds ought to be good insecticides was resurrected in the postwar period and proved to be correct. The problem with insecticides is one of diminishing hazards to humans without sparing the bugs. The compound must be modified to lower volatility while retaining the small aliphatic groups so as to permit access to the enzyme cleft. The better compounds are still dangerous, but **malathion,** a sulfur analog that is converted to the oxy compound in tissues, is thought to be safe enough to permit general sale and the spraying of large blocks of California real estate to rid them of Mediterranean fruit flies. Its formula is:

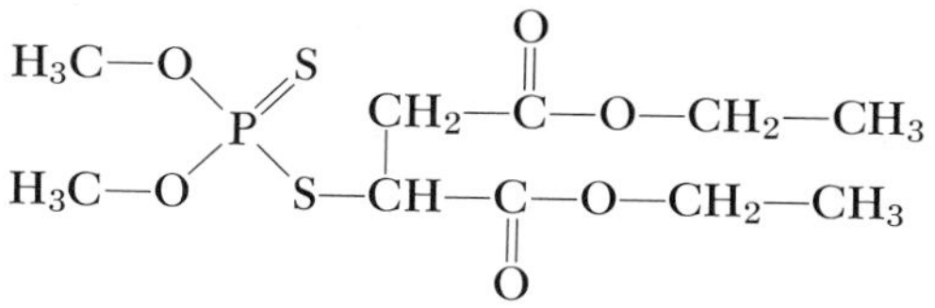

$$(H_3C{-}O)_2P(=S){-}S{-}CH(CH_2{-}C(=O){-}O{-}CH_2{-}CH_3){-}C(=O){-}O{-}CH_2{-}CH_3$$

malathion

Since the most effective phosphate derivatives are dangerous for routine use in the laboratory, other anhydrides such as phenylmethylsulfonyl fluoride:

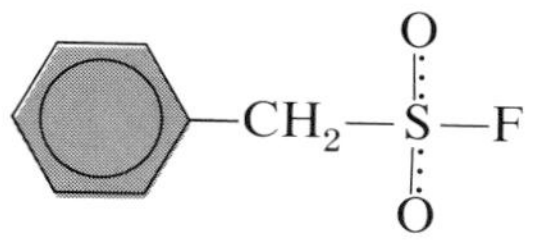

$$C_6H_5{-}CH_2{-}S(=O)_2{-}F$$

phenylmethylsulfonyl fluoride

have been developed to inhibit serine proteases. They are commonly employed during the isolation of proteins that do not themselves contain active seryl groups, in order to inhibit the proteases that are also present in the source tissues. These proteases frequently damage the desired proteins if they are left unchecked.

Carbamic acid esters that block acetylcholinesterase in the same way include insecticides such as **carbaryl (Sevin)** and drugs such as **neostigmine,** which are used in treating **myasthenia gravis** and in counteracting the effect of still other drugs that block transmission in cholinergic nerves. (Myasthenia gravis is a disease assumed to be caused by the formation of antibodies to the acetylcholine receptors. Muscular weakness is a result, which may progress to inability to breathe.)

carbaryl (Sevin)

neostigmine

Still other compounds have been developed to counteract the effect of acetylcholinesterase inhibitors, especially the phosphoric acid derivatives. One, pralidoxime chloride (PAM-2), is a prescription drug for use in cases of insecticide poisoning:

pralidoxime (PAM-2)

blocked seryl group on enzyme

Kinetics of Noncompetitive Inhibition. The effect of these inhibitors is to remove enzyme from the solution. V_{max} is lower, but the remaining enzyme displays the same kinetics as does a more dilute solution of the enzyme with no inhibitor added. In other words, K_M will not be affected, and Lineweaver-Burk plots at varying inhibitor concentrations will give a series of lines that intersect on the abscissa (Fig. 16–3).

We now have two straightforward cases that can be recognized from Lineweaver-Burk plots of the kinetics at different inhibitor concentrations: straight lines intersecting on the ordinate indicate classic competitive inhibition and straight lines intersecting on the abscissa indicate purely noncompetitive inhibition.

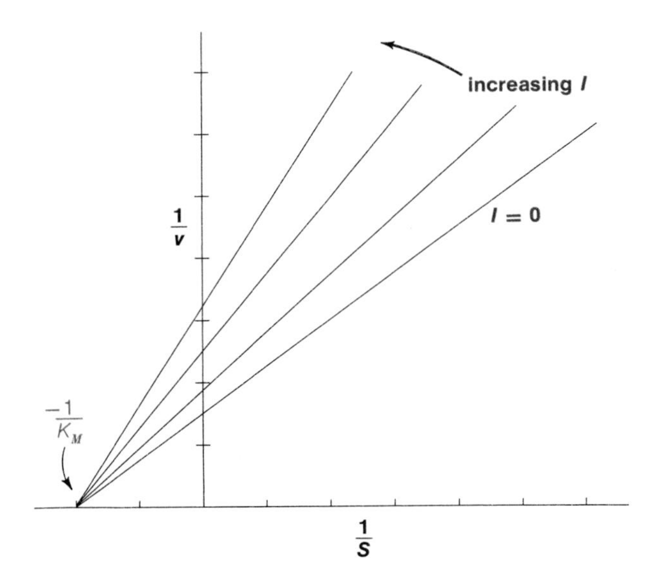

FIGURE 16–3

Lineweaver-Burk plots obtained with varying concentrations of a purely non-competitive inhibitor. K_M is constant in all cases, so there is a single intercept on the abscissa. The inhibitor has the same effect as removal of part of the enzyme, so V_{max} and the intercepts on the ordinate change.

However, these cases do not by any means exhaust the possibilities for inhibition. An inhibitor may react more readily with the intermediate enzyme-substrate complexes than it does with the free enzyme. The inhibitor-enzyme complex may be catalytically active, but with altered values of K_M or V_{max}. The result of the various possibilities may be that varying inhibitor concentrations will produce Lineweaver-Burk plots that are straight lines, but which do not all intersect at one point, so that the inhibition is "mixed" rather than being competitive or noncompetitive. In other cases, Lineweaver-Burk plots produce curved, rather than straight, lines. Permutations of these various kinds of effects can be analyzed in as much detail as patience permits, but such extensive analysis is not required for our purposes. More complete treatment can be found in the references cited at the end of the chapter.

Regulation with Interacting Subunits

Homotropic Kinetics. Some enzymes are built like hemoglobin, with multiple subunits that interact cooperatively so as to create more complex responses to changes in substrate concentration. When the interacting subunits are identical, each with its catalytic site, the resultant kinetics are said to be homotropic. If the binding of one or more molecules of substrate stabilizes a new conformation in which the affinity for the substrate is increased at the remaining binding sites, the cooperativity is said to be positive. With **positive cooperativity,** the kinetic response to changing substrate concentration is *sigmoidal,* much like the oxygen association curve of hemoglobin (Fig. 16–4). (Some enzymes with multiple subunits do not show cooperative interactions and have Michaelis-Menten kinetics. Lactate dehydrogenase is an example.)

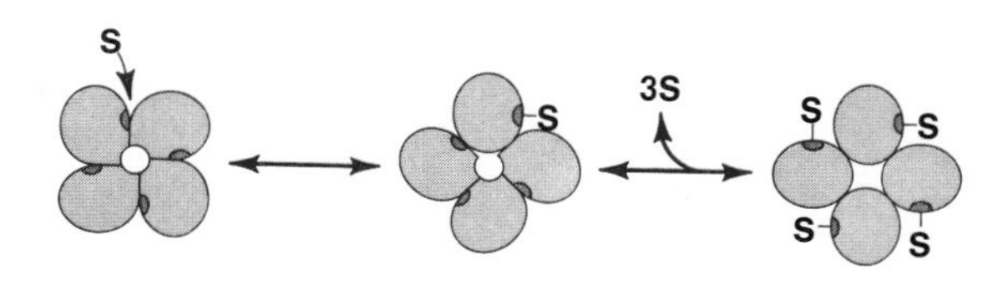

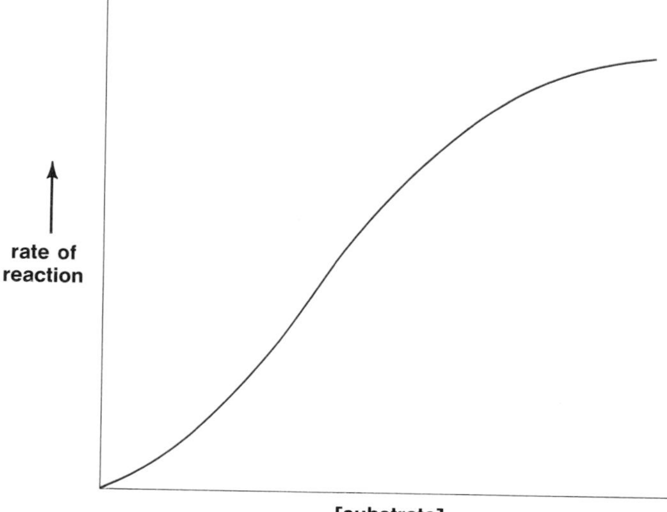

FIGURE 16–4

Enzymes that are constructed from multiple subunits may exhibit complex kinetics, such as the positive cooperativity shown here. In this example, the free tetramer has a conformation in which access to the catalytic sites is impeded in some way (*upper left*). When the substrate concentration is elevated enough to stabilize a conformation with one of the subunits in a more open form (*top center*), the remaining subunits can more readily shift to the fully active conformation that combines with substrate (*upper right*). The result is a sigmoidal response of rate of reaction to substrate concentration.

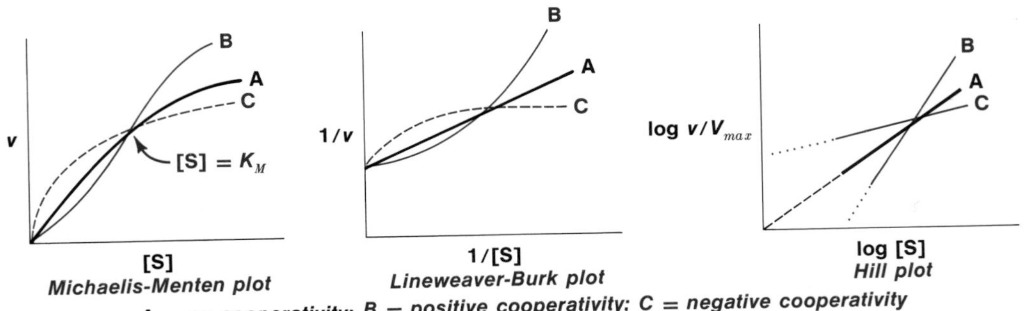

FIGURE 16–5 The presence or absence of cooperativity gives characteristic kinetic plots. In the ordinary Michaelis-Menten plot of initial velocity against substrate concentration (*left*), positive cooperativity is easily distinguished by its sigmoid character (*curve B*), but it is difficult to distinguish Michaelis-Menten kinetics (*curve A*) from kinetics of negative cooperativity (*curve C*) by inspection. The differences become more obvious in a Lineweaver-Burk plot (*center*) in which positive cooperativity gives a curve that is concave upward, while negative cooperativity gives a curve that is concave downward, in contrast to the straight line of Michaelis-Menten kinetics. A logarithmic Hill plot (*right*) also distinguishes the mechanisms by the position of the projected intercept, which is on the abscissa for positive cooperativity and on the ordinate for negative cooperativity, but the greatest value of the Hill plot is that the slope of the central part of the curve gives the Hill coefficient, *n*, which is the minimum number of interacting sites. (The ends of the Hill plot, not shown, cannot be used because they always curve to approach a slope of 1.)

Cooperativity can also be **negative,** with the binding of one substrate molecule hindering the binding of additional molecules. The presence of cooperative interactions and their identification as positive or negative can be recognized by characteristic kinetic plots (Fig. 16–5).

PHYSIOLOGICAL IMPLICATIONS. An enzyme with positive homotropic interactions cannot effectively catalyze its reaction until the substrate concentration has built up to a level approaching the steep portion of the velocity-concentration curve. This is a desirable device for preventing the depletion of a substrate when it undergoes several reactions catalyzed by different enzymes, or when it is irreversibly converted to a product. The rate of reaction falls rapidly toward zero at substrate concentrations well above zero.

Heterotropic Kinetics. Some enzymes that contain cooperatively interacting subunits can also be regulated by binding effectors at sites other than the catalytic

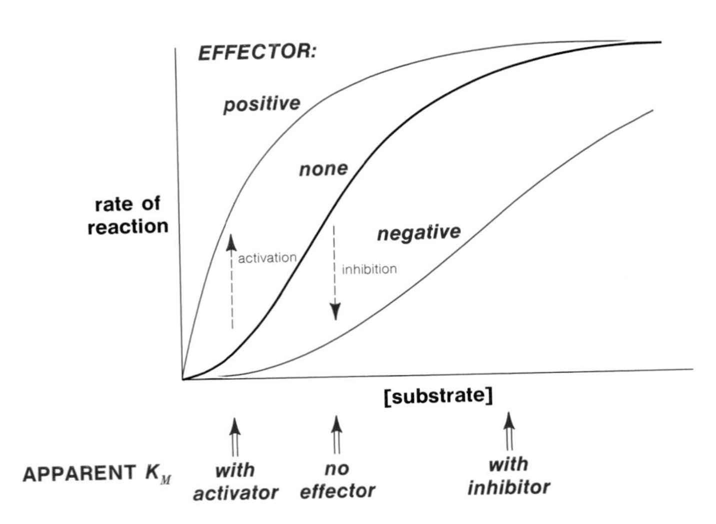

FIGURE 16–6

Kinetics with heterotropic effectors. The plotted examples alter the apparent K_M of the enzyme without substantial effect on the V_{max}. The positive effector in this example eliminates homotropic interactions so that the enzyme has simple Michaelis-Menten kinetics. The negative effector exaggerates the interactions.

sites. Conformational shifts are then caused by changes in concentration of the controlling effector, as well as the substrate.

A heterotropic effector changes the sigmoidal response of velocity to substrate concentration. A frequent effect is to alter the apparent K_M for a substrate without substantial change in V_{max} (Fig. 16–6), but in some cases, the converse is true. When the effector activates an enzyme, it may do so by diminishing internal constraints on catalysis, so that each subunit of the enzyme acts as an isolated molecule with simple Michaelis-Menten kinetics. If the effector inhibits the enzyme, it may do so by stabilizing the interactions between subunits, so that the kinetics become more sigmoidal, and higher concentrations of substrate are required to overcome the interactions.

The binding site for the effector need not even be on the same subunit as the active catalytic site. Many examples are known of enzymes that are aggregates of separate catalytic and regulating subunits, all interacting (Fig. 16–7).

Let us now consider two examples of heterotropic regulation.

PHOSPHOFRUCTOKINASE AND GLUCOSE COMBUSTION. The combustion of glucose residues, derived either from glucose itself or from stored glycogen, involves several key control points. We have already noted one, phosphofructokinase, which catalyzes the transfer of a phosphate group from ATP to fructose 6-phosphate:

fructose 6-phosphate + ATP ⟶ fructose 1,6-bisphosphate + ADP

The purpose of the combustion is to generate usable energy, and we shall later see that the state of the energy balance is signaled by the relative concentration of AMP, which is not a substrate or product of the reaction. When the energy supply is low, the concentration of AMP increases. (We shall look at this important signal in detail in Chapter 25.) An increase in the concentration of AMP therefore indicates a need for more fuel, and phosphofructokinase catalyzes an irreversible reaction on the route by which fuel is supplied. This enzyme is constructed so that AMP is an activating effector, which changes the sigmoidal kinetics of the enzyme to a simple Michaelis-Menten form through loss of cooperative interaction (Fig. 16–8). The V_{max}

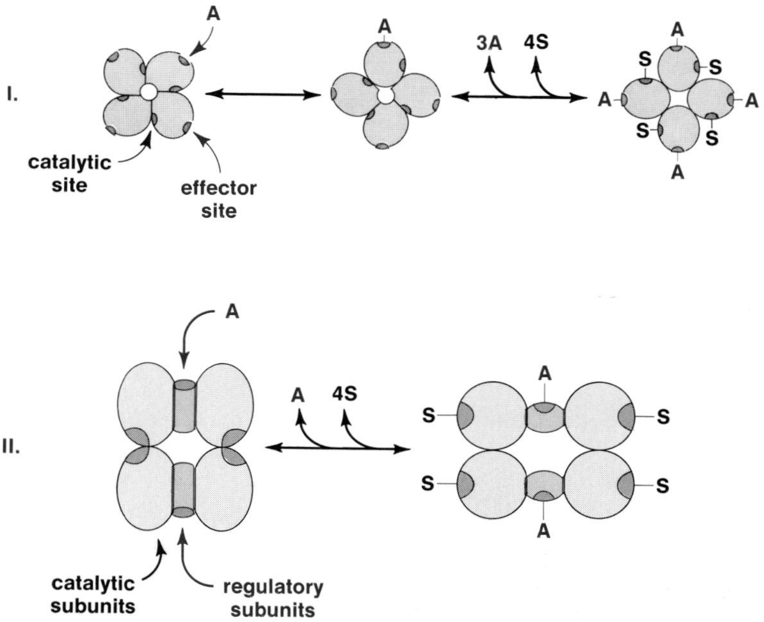

FIGURE 16–7

I. Effector sites may be on the same polypeptide subunits containing the catalytic sites with binding of the effector altering the conformation of the subunits by stabilizing a more reactive form.

II. Effector sites also may be present on separate regulatory polypeptide subunits which interact with the catalytic subunits so as to fix them in a particular conformation. Combination of an activating effector with the regulatory subunits then modifies the interaction so as to permit the catalytic subunits to assume an active conformation. As is indicated here, the number of regulatory sites need not be equal to the number of catalytic sites when they occur on separate polypeptide chains.

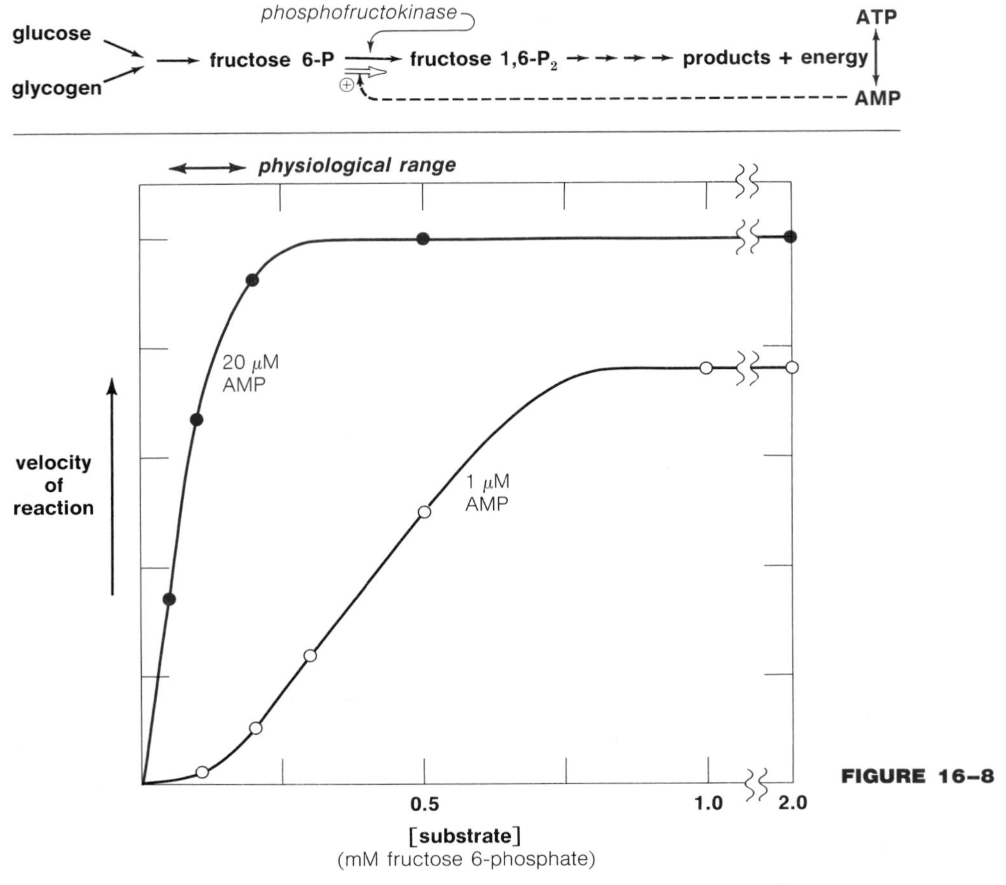

FIGURE 16–8

Regulation of phosphofructokinase by AMP. (Other effectors are not shown.) The activation of phosphofructokinase by AMP is indicated by the broad arrow and plus sign in the reaction sequence at the top. AMP has its effect by changing the homotropic sigmoidal kinetics of the enzyme to simple Michaelis-Menten kinetics, in which the K_M is now in the physiological range of substrate concentration, making the enzyme many-fold more active. (Replotted from data of K. Tornheim and J.M. Lowenstein (1976), Control of Phosphofructokinase from Rat Skeletal Muscle. J. Biol. Chem. 251:7322.)

of the enzyme is also increased to some extent, but the important result is that the affinity of enzyme for its substrate is greatly increased; the K_M is shifted into the normal physiological range of concentration for fructose 6-phosphate.

Amidophosphoribosyl Transferase and Purine Synthesis. The initial step in purine synthesis at which control by negative feedback occurs (Fig. 16–1) involves the displacement of an attached pyrophosphate ester group from a ribose residue by transfer of an amide nitrogen from glutamine:

glutamine + 5-phosphoribosyl pyrophosphate ⟶
glutamate + 5-phosphoribosylamine + pyrophosphate

We shall discuss these compounds in detail in Chapter 35, but the point of interest now is the nature of the control. The enzyme has cooperative interaction only in the presence of one of the purine nucleotides. Therefore, an increase in the concentra-

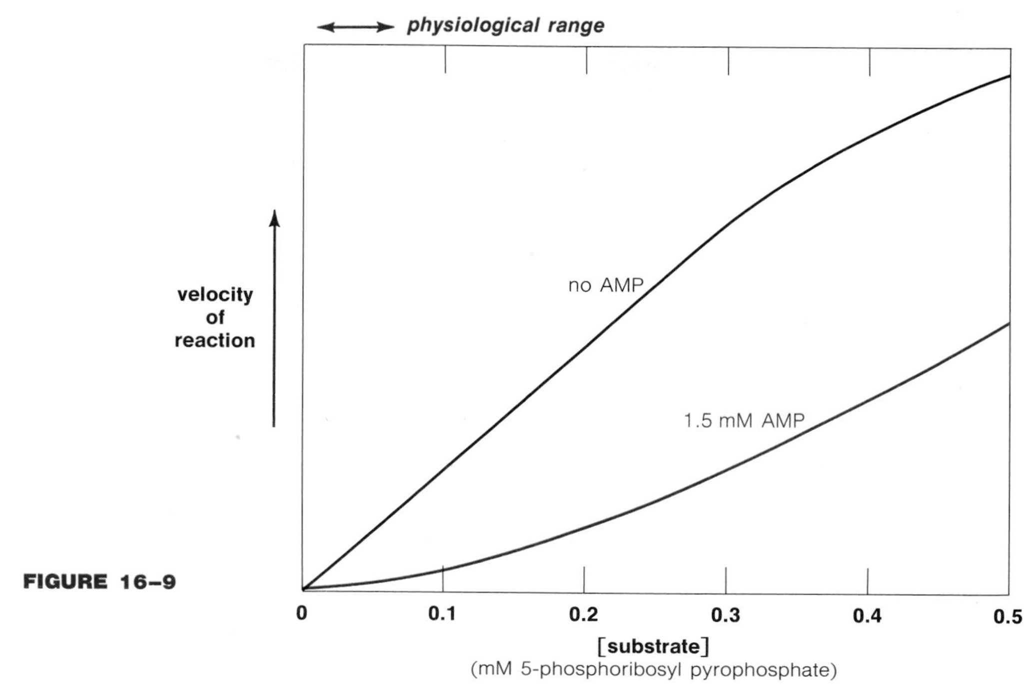

FIGURE 16-9

Regulation of the initial step in the formation of purine nucleotides (see Fig. 16-4). The regulation in this case is a shift by the effector from Michaelis-Menten kinetics to homotropic kinetics, displacing the K_M of the enzyme far above the physiological range of concentrations of the ribose-containing substrate. (Replotted from data of E.H. Holmes, et al. (1973), Human Glutamine Phosphoribosyl Pyrophosphate Amidotransferase. J. Biol. Chem. 248:144.)

tion of any of the adenosine or guanosine phosphates will cause a shift from Michaelis-Menten kinetics to sigmoidal homotropic kinetics and increase the K_M for the ribosyl substrate to a value much above the physiological concentration range. (Fig. 16-9 illustrates the action of AMP as an example; ATP and GTP are probably the important effectors in cells.) The action of purine nucleotide on this enzyme is therefore the opposite of the action of AMP on phosphofructokinase, which makes another important point. The same compound may act as an effector on several enzymes; it may activate some and inhibit others.

COVALENT MODIFICATIONS

Regulation by Phosphorylation

The physiological control of many enzymes involves covalent modifications. We shall consider examples of control by **partial hydrolysis** in the next chapters. Another common method of physiological control is through **phosphorylation of hydroxyl groups** in the enzyme (Fig. 16-10). These groups are in side chains of residues of serine, threonine, or tyrosine in the polypeptide chain. The transfer of phosphate groups to the regulated enzyme from ATP is in itself catalyzed by another enzyme, a **protein kinase.** The phosphate groups are removed to restore the original state by still another enzyme, a **phosphoprotein phosphatase.**

CH_2-OH (—Ser— enzyme) + ATP → (protein kinase) → $CH_2-O-PO_3^{2-}$ (—Ser— enzyme) + ADP

$CH_2-O-PO_3^{2-}$ (—Ser— enzyme) + H_2O → (phosphoprotein phosphatase) → CH_2-OH (—Ser— enzyme) + P_i

FIGURE 16–10

Regulation of enzymes through phosphorylation. A protein kinase catalyzes addition of phosphate groups, and a hydrolase catalyzes their removal.

The affected enzymes resemble other enzymes with cooperative interaction by existing in both active and inactive conformations. The difference is that the shift from one conformation to another is stabilized by the presence or absence of the phosphate ester group on the enzyme (Fig. 16–11). The effect of phosphorylation may be an activation or an inhibition of the activity; we shall encounter examples of each.

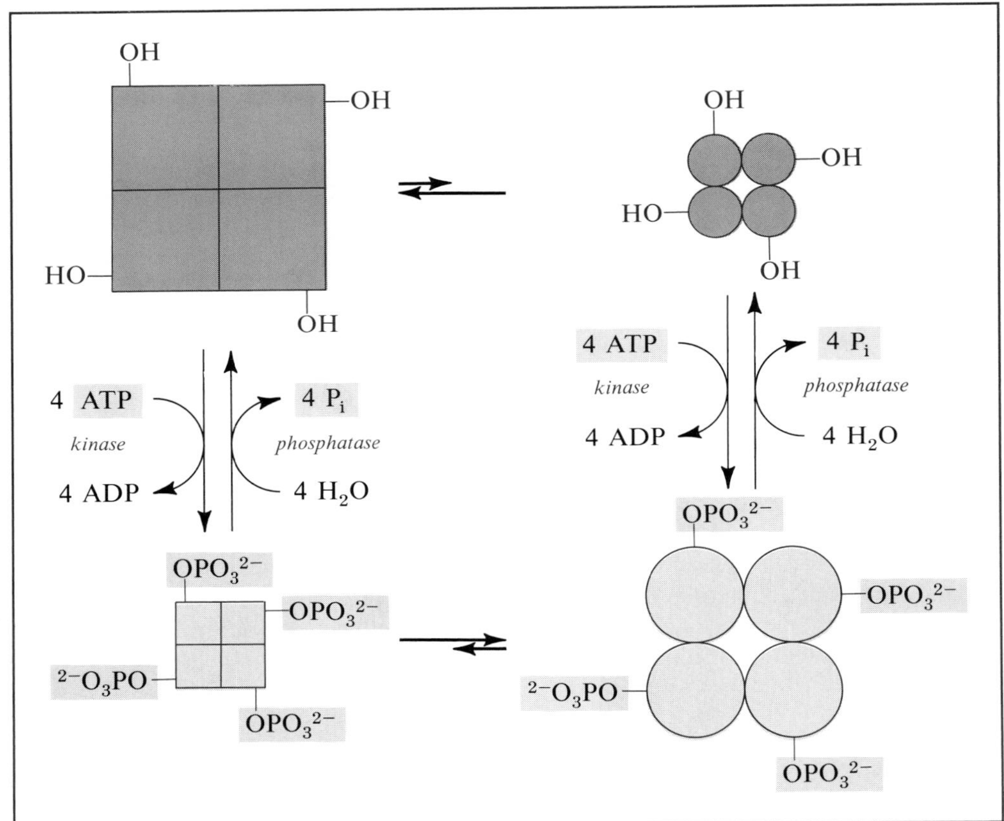

FIGURE 16–11 **Enzymes are probably regulated by phosphorylation through the stabilization of an otherwise unfavorable conformation. In the mechanism illustrated, a tetrameric enzyme exists predominantly in one conformation (*upper left*) when it is not phosphorylated. (Concentration of a conformation is indicated by relative size of the drawing.) Either the favored or the unfavored conformation may be phosphorylated by the transfer of phosphate groups from ATP, the transfer being catalyzed by a kinase. The presence of the phosphate groups shifts the equilibrium between the two conformations, so that the other form (*lower right*) is now favored. When the phosphate groups are removed by hydrolysis (catalyzed by a phosphatase), the conformational shift is reversed. Although the diagram shows the predominant free form in the conventional squares indicating the less reactive species (*upper left*), and the predominant phosphorylated form in the circles indicating the more reactive molecule, this is not true of all enzymes regulated by phosphorylation. Some enzymes are less active when phosphorylated.**

The device of regulation through addition and removal of phosphate groups adds another dimension to control. Whether these groups will be present or absent on the regulated enzyme depends upon the relative activities of the protein kinase and the phosphoprotein phosphatase that catalyze the competing reactions. Here we have enzymes regulated by other enzymes, which are now in turn subject to regulation by effectors. The important difference between this and other effector-mediated mechanisms we have been discussing is that **changes in effector concentration are amplified** by the intervening enzymes.

For example, consider the activity of **glycogen phosphorylase,** which is the enzyme controlling the breakdown of glycogen for use of its glucose residues as fuel.

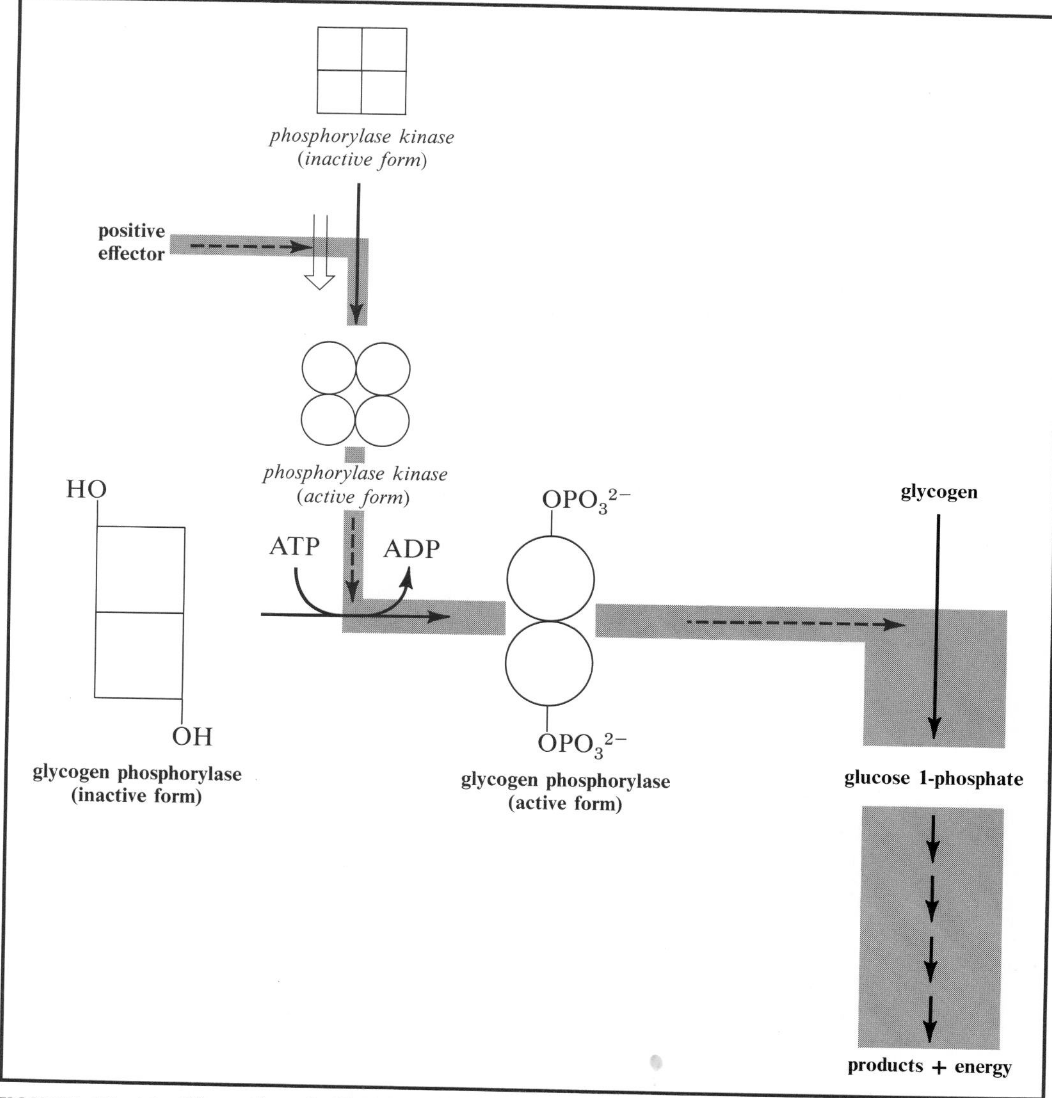

FIGURE 16–12 **The action of effectors is magnified when they influence the phosphorylation of an enzyme. In the example shown here, a positive effector increases the activity of phosphorylase kinase (*top center*), which then accelerates the phosphorylation of glycogen phosphorylase (*center left*). Each molecule of the active kinase can create many active molecules of glycogen phosphorylase. Each of the active phosphorylase molecules in turn catalyzes the cleavage of many glucose residues from glycogen (as glucose 1-phosphate). The magnification of the effector action is indicated by brown shading.**

We shall later see that it is under complex control. The critical feature is a stabilization of the active conformation of the enzyme when the enzyme is phosphorylated (Fig. 16–12). The transfer of phosphate groups to glycogen phosphorylase is catalyzed by another enzyme, a kinase. Activation of the kinase by an effector enables the phosphorylation of many molecules of the glycogen phosphorylase. Each of these many molecules of active enzyme can then catalyze the breakdown of many glucose residues. Without the intervening kinase, a molecule of effector would have to be provided for each of the molecules of glycogen phosphorylase to be activated, but the kinase has multiplied the action of the effector. We shall later see that this **cascade** multiplication of effector action is the basis for the potency of several hormones.

Suicide Substrates

Many organic reagents will react with polar side chain groups in proteins and modify or destroy their function. A new type of reagent is now being developed that is specifically bound in the catalytic cleft of an enzyme and then irreversibly reacts with a group in the cleft. The enzyme is caught by a chemical sting operation, accepting the inactivating reagent as a true substrate. The reagent might, for example, undergo a Michael addition with a lysine amino group in the cleft:

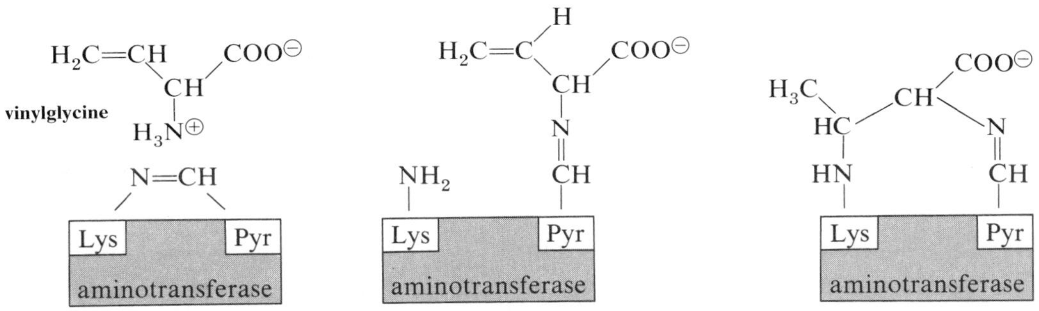

vinylglycine

$H_2C{=}CH$, $COO^{\ominus}$, CH, $H_3N^{\oplus}$, $N{=}CH$ — Lys, Pyr — aminotransferase

H, $H_2C{=}C$, $COO^{\ominus}$, CH, N, CH, NH_2 — Lys, Pyr — aminotransferase

H_3C, $COO^{\ominus}$, CH, HC, N, HN, CH — Lys, Pyr — aminotransferase

There is hope that especially specific drugs can be developed by using mechanisms of this sort.

INDUCTION OR REPRESSION OF ENZYME SYNTHESIS

The ultimate method of regulating the rates of enzymatic reactions is to control the number of enzyme molecules present. Occasionally this may be done by regulating the rate at which particular enzymes are destroyed, but the more common method of control is to regulate the rate of synthesis. We have already discussed in Chapter 5 how this kind of regulation is achieved in prokaryotic organisms through operon control of gene transcription, but the mechanism for comparable control in eukaryotes is not known. However, the results are quite real. We shall frequently have occasion to refer to metabolic adaptations in which the amount of enzyme is adjusted to fit new environmental circumstances. Many hormones are active because of their effects on synthesis of particular enzymes, and more will be said about their mechanism and effects in Chapter 37.

FURTHER READING

P.J. Roach (1980). Principles of the Regulation of Enzyme Activity. 4:203 in L. Goldstein and D.H. Prescott, eds., Cell Biology. Academic Press.
T. Spector and W.W. Cleland (1981). Meanings of K_I for Conventional and Alternate Substrate Inhibitors. Biochem. Pharmacol. 30:1.
M.E. Phelps, D.E. Kuhl, and J.C. Mazziotta (1981). Metabolic Mapping of the Brain's Response to Visual Stimulation: Studies in Humans. Science 211:1445. Includes positron computed tomographs obtained with 1-fluoro-2-deoxyglucose.
A.A. Patchett et al. (1980). A New Class of Angiotensin Converting Enzyme Inhibitors. Nature 288:280. Examples of transition state inhibitors of possible therapeutic value.
M.F. Parry and R. Wallach (1974). Ethylene Glycol Poisoning. Am. J. Med. 57:143. Review.
T.I. Kalman, ed. (1979). Drug Action and Design: Mechanism-Based Inhibitors. Elsevier.
N. Seiler, M.J. Jung, and J. Koch-Weser, eds. (1978). Enzyme-Activated Irreversible Inhibitors. Elsevier. Includes discussions of suicide substrates.
C.J. Peterson et al. (1981). Ethylene Glycol Poisoning. Pharmacokinetics during Therapy with Ethanol and Hemodialysis. N. Engl. J. Med. 304:21.
E.R. Wilson (1959). Molecular Complementarity and Antidotes for Alkylphosphate Poisoning. Fed. Proc. 18:752.
T.P. Singer and R.N. Ondarza, eds. (1981). Molecular Basis of Drug Action. Elsevier. Symposium articles.

CHAPTER 17

HYDROLASES: DIGESTION

Enzymes that catalyze hydrolysis are ubiquitous because destruction, as well as synthesis, is a necessary part of the biological scheme of things. Intracellular hydrolases, discussed later in the book, are used in a constant reshaping of cellular composition that is part of adaptation to changing circumstances. Our concern here is with the enzymes of the gastrointestinal tract that attack the major classes of foodstuffs: the carbohydrates, fats, and proteins. These hydrolases break the hemiacetal bonds of glycosides, the ester bonds of triglycerides, or the amide bonds of polypeptides; they are representatives of the general classes of glycosidases, esterases, and amidases:

$$H_2O + \underset{\text{(ring O)}}{R^1{-}O{-}R^2} \xrightarrow{glycosidase} \underset{\text{(ring O)}}{R^1{-}OH} + HO{-}R^2$$

$$H_2O + R^1{-}\overset{O}{\overset{\|}{C}}{-}O{-}R^2 \xrightarrow{esterase} R^1{-}COO^{\ominus} + H^{\oplus} + HO{-}R^2$$

$$H_2O + R^1{-}\overset{O}{\overset{\|}{C}}{-}\overset{H}{\overset{|}{N}}{-}R^2 \xrightarrow{amidase} R^1{-}COO^{\ominus} + H_3\overset{\oplus}{N}{-}R^2$$

They are introduced at this point to provide an example of the way in which separate enzymatic reactions are designed as parts of a larger process, digestion in this case.

ORGANIZATION OF THE GASTROINTESTINAL TRACT

The gastrointestinal tract is arranged so as to make the food accessible to attack by hydrolases, to secrete hydrolases at the proper time and place, to absorb the products of hydrolysis, to recover components of the digestive secretions for reuse, and to dispose of the indigestible remains as feces (Fig. 17–1).

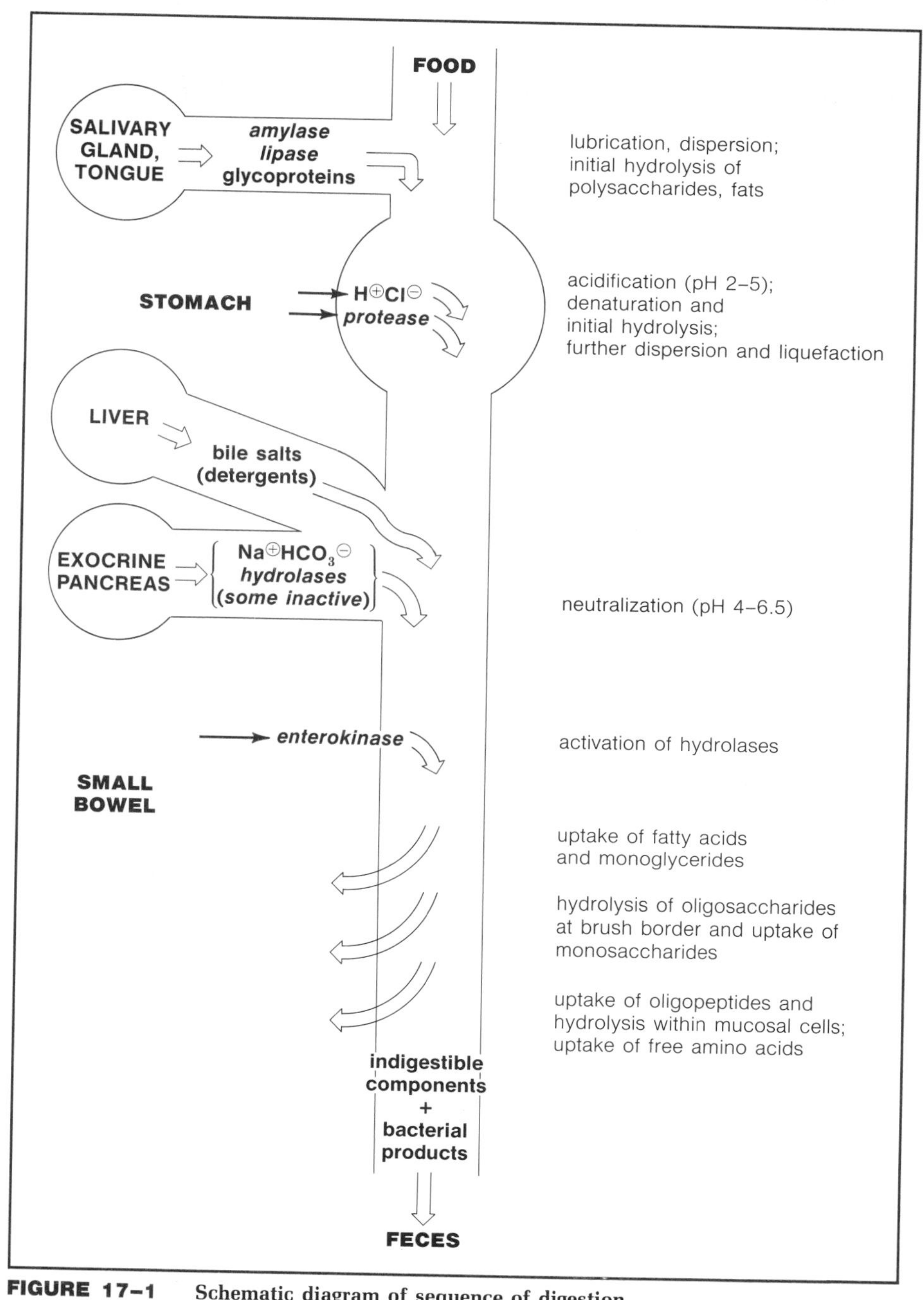

FIGURE 17–1 Schematic diagram of sequence of digestion.

Both mechanical and chemical actions begin in the mouth. The food is dispersed and mixed with saliva, which is rich in lubricating glycoproteins. The saliva also contains **amylase,** a hydrolase attacking glucans. A **lipase** attacking triglycerides is secreted by Ebner's glands in the tongue.

In the stomach, the secretion of **gastrin** by some cells in response to neural stimulation and stretching of the stomach wall causes other cells to secrete high concentrations of HCl, to which the food is thoroughly exposed by mechanical agitation. (The high acidity kills many of the microorganisms in the food.) The action of salivary amylase and the lingual lipase will continue until the H^+ concentration becomes too great. The lipase, in particular, is built to withstand mild acidity, and it works best at pH 5.4 to 6.5. As gastric secretion continues and the contents become more acid, the proteins in the food unfold and become more accessible to attack by hydrolases, that is, they are denatured. The stomach also secretes **proteases** to begin this attack.

The now quite fluid mixture of food and secretions—the **chyme**—passes into the duodenum, where it is exposed to secretions from the pancreas and the liver. The pancreas contributes proenzymes of hydrolases that attack all of the major classes of foodstuffs. Activation of these proenzymes hinges upon exposure to another hydrolase, **enterokinase,** secreted by the small intestine. The activated enzymes function best near pH 7, and the acidic chyme is neutralized by bicarbonate ions. The pancreatic secretion supplies much of the bicarbonate; the remainder is supplied in the bile by the liver, along with bile salts.

Hormonal Regulation of Secretion. It would be costly for the liver and pancreas to be continually secreting bile and enzymes, and hormonal controls exist to provide these materials only when required. Contact with acid chyme causes the intestinal mucosa to release **secretin** into the blood. This polypeptide hormone stimulates the pancreas and the liver to secrete bicarbonate ions. The partially hydrolyzed fats and proteins stimulate the mucosa to release **cholecystokinin,** another polypeptide hormone, which stimulates contraction of the gallbladder to discharge bile and stimulates the pancreas to discharge its hydrolase-loaded secretory vesicles.*

The intestinal mucosa has the dual function of augmenting and continuing the hydrolytic attack on foodstuffs and of absorbing the products in usable form. The mucosa is constructed so as to have a huge surface area. It is extensively folded into flap-like villi, and in addition, the surface of each cell in a villus is further magnified by the formation of the **brush border,** or **microvilli,** which is like the pile of a closely woven carpet on the lumen face of the cell.

Passage through the Unstirred Layer. Hydrolysis of foodstuffs, for good reason, goes through two phases, an incomplete although extensive hydrolysis within the lumen and a final hydrolysis within or on the brush border. The water in contact with the brush border is relatively quiescent during the peristaltic waves that mix and advance the intestinal contents. The products of hydrolysis in the lumen can only reach the brush border for absorption by diffusing through this unstirred layer and an underlying mucous coating. A given mass of incompletely hydrolyzed oligomers diffuses faster than does a comparable quantity of the constituent monomers because the total cross-sectional area of an oligomer is less than the combined areas of the hydrolyzed monomers. If a disaccharide molecule diffused at half the rate of a monosaccharide molecule, it would be carrying fuel from the lumen to the brush border at an equal rate because each disaccharide carries the equivalent of two monosaccharides. However, a disaccharide diffuses at greater than half the rate of a monosaccharide; the diffusion of one molecule is slower, but not proportionately slower.

The task of hydrolyzing the food is nicely balanced between lumen and brush border so that the advantages of faster diffusion of mass in the larger molecules are balanced by the advantages of faster handling of smaller molecules by the brush border itself. (Only part of a large molecule can be in actual contact with the brush border.)

---

*A synthetic analog of cholecystokinin is available to stimulate emptying of the gallbladder for diagnostic purposes. It has only 10 amino acid residues compared to the natural hormone's 33, including an essential sulfate ester of tyrosine.

DIGESTION OF CARBOHYDRATES

Dietary carbohydrates are composed mainly of the polysaccharide starch, with lesser amounts of glycogen and the disaccharides sucrose and lactose. For absorption, fuels are converted to the monosaccharides—glucose, fructose, and galactose.

Starch and glycogen are converted to glucose by the action of several enzymes (Fig. 17–2). Amylase from the salivary glands and the pancreas can only cause hydrolysis of 1,4-α-glucoside bonds if they are no closer than two residues to an outer chain terminal or a 1,6 branch. That is, amylase can attack a polysaccharide at every other 1,4 bond, except outermost bonds and those next to the branches. The result is

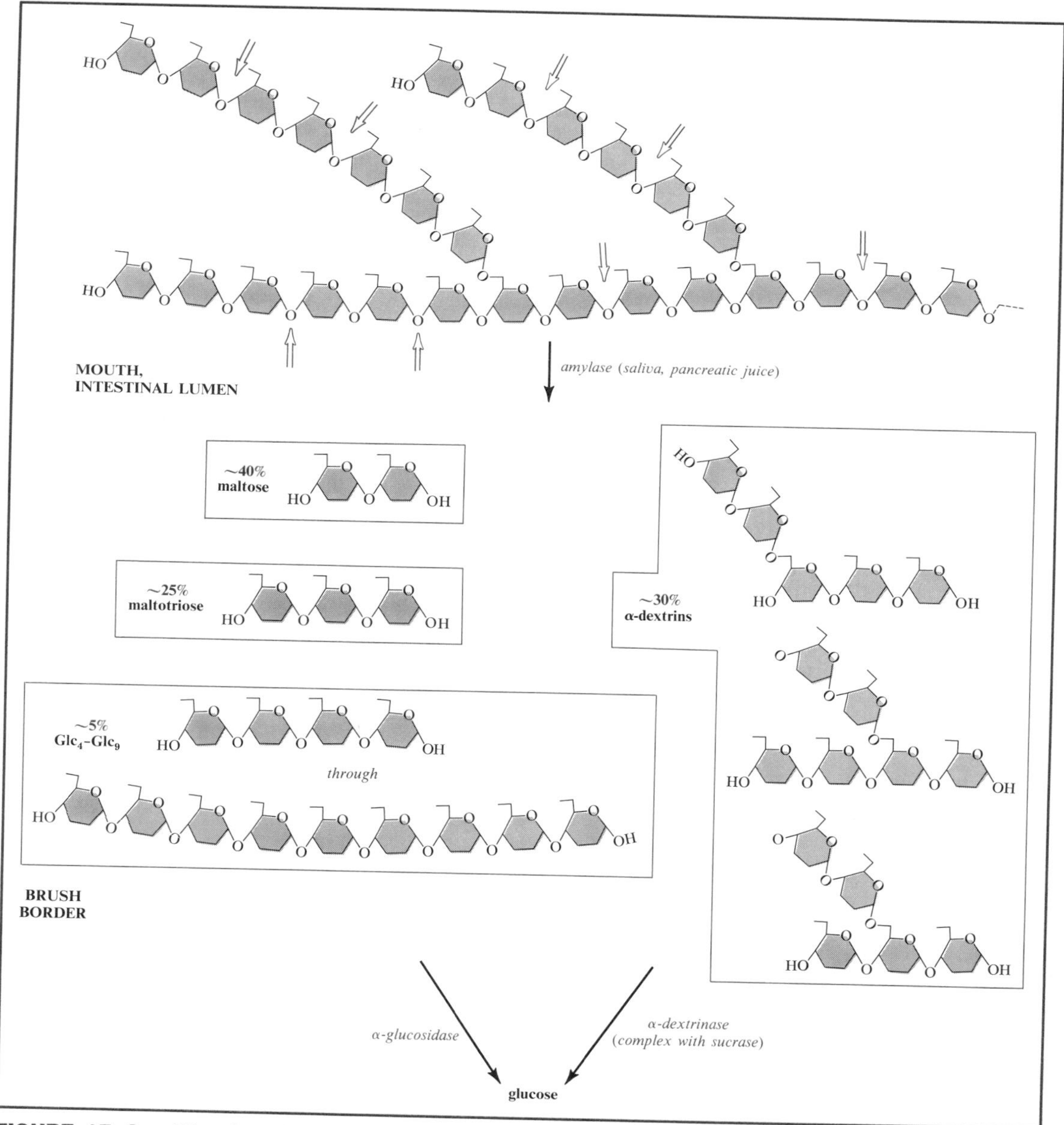

FIGURE 17–2 Digestion of the branched polysaccharides, starch, and glycogen.

hydrolysis of the polysaccharide into a mixture of linear and branched oligosaccharides. Most of the linear compounds are the disaccharide **maltose** and the trisaccharide **maltotriose;** there are small amounts of longer polymers up to Glc_9. The 1,6 branches are in a mixture of Glc_5 to Glc_9 compounds known as **α-dextrins.** These are the components presented to the intestinal mucosa.

Embedded in the brush border membrane of the mucosa are several hydrolases attacking oligosaccharides. Among these are an **α-glucosidase,** which strips off one glucose residue at a time from the linear oligosaccharides, and an **α-dextrinase,** also known as isomaltase, which can break 1,6 bonds as well as 1,4 bonds and therefore releases glucose from the branched oligosaccharides. These enzymes complete the hydrolysis of starch to glucose, which is absorbed.

Additional hydrolases exist in the brush border for attacking other oligosaccharides (Fig. 17–3). These include a **β-galactosidase,** or **lactase;** a **sucrase,** or **β-fructofuranosidase;** and a **trehalase.** The functions of the lactase and sucrase are obvious, but the role of the trehalase is more titillating. Trehalose occurs widely in insects as a storage and transport sugar, and insects may have been of more importance in the primate diet than one might assume from current efforts against cockroaches and the like. More fastidious people encounter trehalose in mushrooms.

Sucrase and α-dextrinase occur as a single complex in the brush border. These two enzymes are first synthesized as one polypeptide chain with a highly hydrophobic amino terminal sequence, which is buried in the brush border membrane to act as a stalk. Unlike the hydrophobic signal sequence of other secreted proteins, the stalk is not removed by hydrolysis. Instead, the remainder of the polypeptide is exposed on the surface of the brush border to the action of pancreatic proteases, which cleave the chain between the two active sites. (If the pancreatic duct is experimentally ligated, the two sites remain as a single chain.) Presumably this is a device

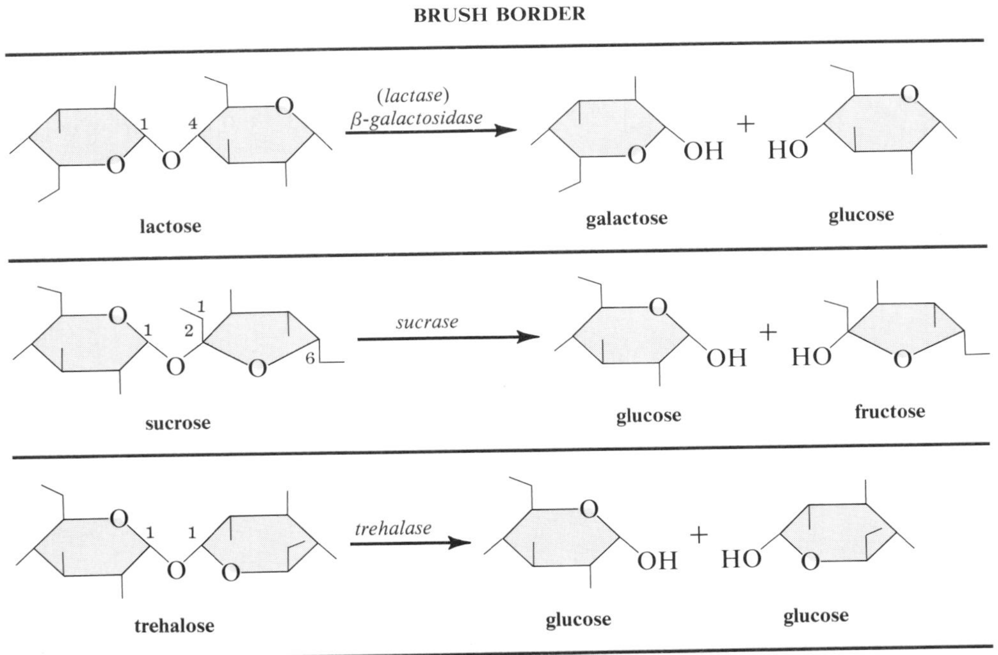

BRUSH BORDER

lactose —(lactase) β-galactosidase→ galactose + glucose

sucrose —sucrase→ glucose + fructose

trehalose —trehalase→ glucose + glucose

FIGURE 17–3 Some dietary disaccharides are hydrolyzed at the intestinal brush border for absorption. Specific enzymes are present to attack lactose, sucrose, and trehalose.

α-Gal-(1→6)-α-Gal-(1→6)-α-Gal-(1→6)-α-Glc-(1→2)-β-Fru	verbascose
α-Gal-(1→6)-α-Gal-(1→6)-α-Glc-(1→2)-β-Fru	stachyose
α-Gal-(1→6)-α-Glc-(1→2)-β-Fru	raffinose

FIGURE 17–4 Some pulses, such as beans, peas, and soybeans, contain oligosaccharides with (1,6)-linked galactose residues that cannot be hydrolyzed for absorption. They may be regarded as sucrose with one, two, or three galactose residues attached.

enabling transport and fixation to the plasma membrane of two activities for the price of one.

Indigestible Carbohydrates. Many polysaccharides cannot be attacked in the primate small bowel. These include cellulose, inulin (a fructose polymer from the Jerusalem artichoke), agar (a heteropolysaccharide from seaweeds), and various other heteropolysaccharides of plant origin, including some pentose polymers from *Psyllium* seeds that are widely sold to add fecal bulk. In most diets, polysaccharides are the most important component of indigestible fiber, and considerable emphasis is placed today on promoting easy passage of feces through the large bowel by increasing their intake.

However, some carbohydrates that are denied to the host as a fuel provide a banquet for the bacterial flora of his gut, and here lies a problem. Some bacteria wear out their welcome when they are given a plentiful supply of usable carbohydrate, through the production of large volumes of gas and lactic acid, along with other products that irritate the intestines. Audible rumblings (given the impressive name of borborygmi) and frank farts add social distress to the physical discomfort of cramps and diarrhea.

Otherwise normal individuals may have this problem through ingestion of foods containing indigestible oligosaccharides. The pulses (beans, peas, and the like) are notorious offenders; these seeds have relatively high concentrations of tri-, tetra-, and pentasaccharides containing α-galactosyl units attached by 1,6 bonds to a sucrose residue (Fig. 17–4), which are not attacked by the intestinal mucosa but are readily hydrolyzed and used by many bacteria.

Similarly, many adults become ill when they drink milk because they have lost lactase from their intestinal brush border as they have grown older. Orientals and Negroes are more likely to be forced to put aside childish foods owing to the resultant lactose intolerance. In Americans of various ancestries, 19 out of 20 Orientals, 14 out of 20 Negroes, and only 2 out of 20 Caucasians developed symptoms after eating 50 grams of lactose, the amount in a quart of cow's milk.

LACTOSE INTOLERANCE*

Occurrence

Common: usually seen in adults
Prevalence varies from 65 per cent in blacks. Mexican-Americans, Orientals, Ashkenazic Jews, and Eskimos to 5 to 15 per cent in Northern Europeans
Rarely congenital

Symptoms

Onset 1/2 to 4 hours after as little as one glass of milk
Abdominal distention, cramps, pain, diarrhea, increased flatulence

*T.M. Bayless, et al.: (1975) Lactose and Milk Intolerance: Clinical Implications. N. Engl. J. Med., 292:1156. Also see N. Engl. J. Med., 294:1057 (1976).

Later studies confirmed that intolerance is widespread among people of varied ancestries. In short, milk is not a very good food for many adults, and it has been pointed out that shipping dried milk to impoverished lands may cause more griping guts than gratitude. The potential magnitude of the problem can be appreciated from one patient, who, upon drinking two liters of milk on each of two days, had 141 passages of flatus per day, including 70 in four hours. His output of 346 ml of gas per hr, mostly H_2 and CO_2, exceeded the previous record of 168 ml per hr achieved by subjects ingesting 50 per cent of the caloric intake as baked beans. (Recognition was sought from the Guinness Book of World Records.)

There is a more serious side to the problem. We shall see later that the pulses are among the better sources of protein readily available in the world, but the pesky carbohydrates that come along with the protein inhibit their more widespread use. Attempts are being made to selectively breed out the undesirable characteristic.

DIGESTION OF FATS

Fats often constitute even more of the potential energy in human diets than do carbohydrates. Fatty acids can be liberated by simple hydrolysis of the ester bonds in triglycerides, but the insolubility of the triglycerides presents a problem; digestion must occur at a phase boundary.

The first step is dispersion of the dietary fat into small particles with sufficient exposed surface area for rapid attack by a hydrolase. Detergent action and mechanical mixing do the job, with the detergent effect being supplied by several components, both in the diet and in the digestive juices, but especially by partially digested fats (fatty acid soaps and monoglycerides) and by bile salts.

Bile Salts. Bile acids, which are steroid derivatives with a carboxyl-containing side chain, are converted to powerful detergents, the bile salts, by forming amide linkage with either glycine or taurine (Fig. 17–5). The principal bile acids are **cholic** and **chenodeoxycholic** acids, either of which may be conjugated with glycine or taurine. For example, cholic acid may form glycocholic or taurocholic acids, occurring as the anions. In addition, some of the compounds occur as the 3-sulfate esters, and are dibasic anions.

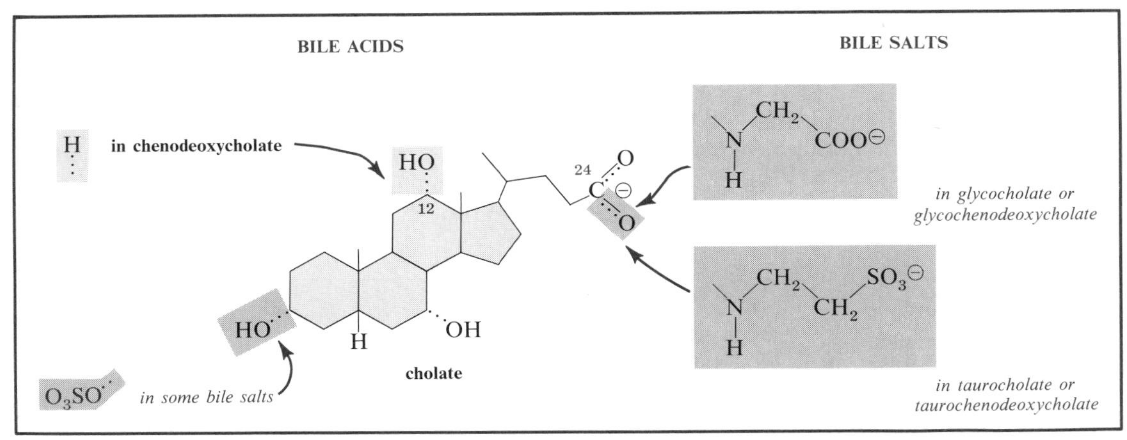

FIGURE 17–5 Bile salts are detergents made by linking bile acids (*black*) with glycine or taurine (*brown*) in amide linkage. The predominant bile acids are cholic acid or chenodeoxycholic acid, which differ in the presence or absence of a 12-hydroxyl group; either may be conjugated with glycine or taurine in bile salts. Some bile salts are further modified by forming sulfate esters with the 3-hydroxyl group on the steroid ring.

The detergent character of the bile salts comes from the presence of hydroxyl or sulfate ester groups on only one side of the molecule. The steroid ring is a puckered plane; the polar groups are on one side, giving a hydrophilic face; the other face is hydrophobic. The hydrocarbon side chain, culminating in a C-24 carboxyl group, adds to the mixture of hydrophobic and hydrophilic character.

Action of Lipase. The lingual lipase can cause a significant hydrolysis of dietary triglycerides, but most of the lipase activity comes from the pancreas. This pancreatic lipase by itself cannot interact with triglyceride droplets even in the presence of bile salts, which complex both the enzyme and the substrate (Fig. 17–6). However, the pancreas secretes an additional small protein, a colipase, that penetrates the surface of the fat droplet more readily and fixes the lipase so as to make triglycerides accessible to its active site. (Either colipase or bile salts also can act to stabilize the lipase at the interface.) Lipase, like other proteins, tends to denature at oil-water boundaries where a force exists to separate its hydrophobic and polar side chains.

Once brought into contact with the triglycerides, pancreatic lipase causes hydrolysis of either the 1 or the 3 ester bonds, but not the bond in the central 2 position (Fig. 17–7). Complete digestion by this enzyme therefore causes a conversion of a triglyceride to two fatty acids and a monoacyl glycerol (monoglyceride).

Absorption. Bile salts form micelles with the fatty acids and monoglycerides liberated at the surface of a fat droplet. (Monoglycerides, with only one hydrocarbon

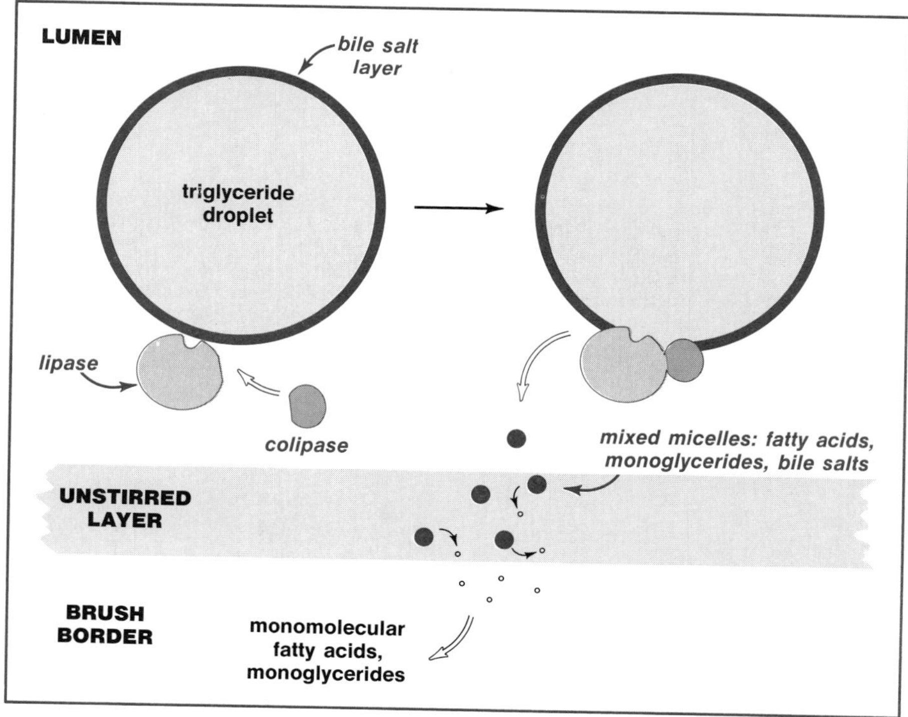

FIGURE 17–6 Triglyceride droplets become coated with a layer of bile salts, which prevents pancreatic lipase from making contact. (The lipase also complexes with bile salts.) An additional small protein, colipase, combines with lipase, bile salts, and the triglyceride droplet in such a way as to bring the lipase active site near the triglycerides, permitting hydrolysis to begin, and maintaining the lipase in its native conformation. The products of hydrolysis leave as mixed micelles, which diffuse readily through the unstirred water layer on the surface of the brush border.

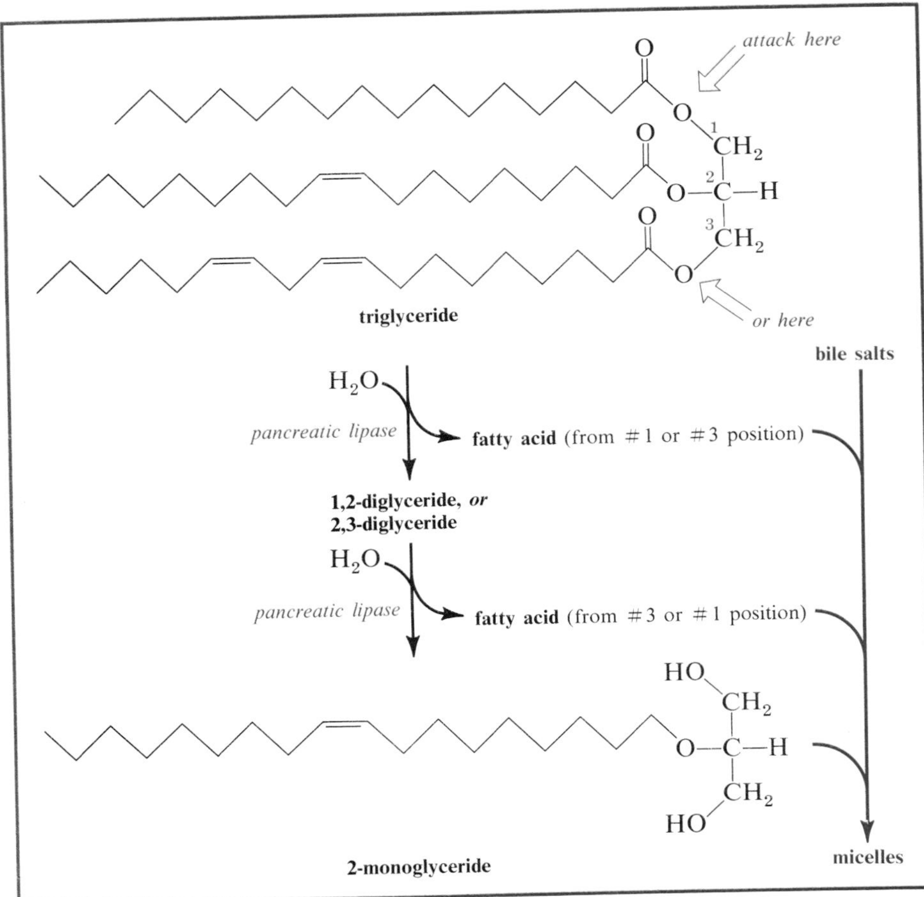

FIGURE 17–7 Digestion of triglycerides begins with an attack on the C-1 or C-3 ester bond by pancreatic lipase. The resultant 1,2- or 2,3-diglycerides are further attacked by the hydrolase to leave the 2-acyl-glycerol, or monoglyceride. The liberated free fatty acids and monoglycerides form micelles with bile salts, which readily diffuse into proximity with the intestinal mucosa.

chain and two free hydroxyl groups, have good detergent action.) The small micellar particles diffuse relatively rapidly and can penetrate the unstirred layer on the surface of the intestinal mucosa at a reasonable rate. The soaps (fatty acid anions) and monoglycerides in true solution as single molecules constantly equilibrate with those in the micellar complexes. The compounds are liberated from the micelles to the unstirred layer so close to the brush border that the diffusion time is short, even though the concentration in true solution is low.

To recapitulate the sequence, a fatty acid residue in dietary fats is dispersed into small droplets, appears either in free fatty acids or in monoglycerides at the surface of the oil droplets, moves into micelles in association with bile salts, moves out of the micelles into free solution, and then passes into the intestinal mucosa through the brush border.

As food passes through the small bowel, the content of fat droplets steadily drops, and the content of micelles rises. These micelles are at first rich in fatty acids and monoglycerides, but they liberate these components to the mucosa as transit continues, with most rapid absorption occurring in the jejunum. In the ileum, the fat droplets are gone, and the micelles are mostly bile salts. (The ileum is constructed to absorb bile salts, and we shall discuss this circulation of steroids in Chapter 36.)

If there is a deficiency of bile salts, either because of failure in production by the liver or failure in delivery due to obstruction, normal transport cannot occur. Other detergents cause enough dispersion of fats to permit the action of lipase, but the formation of micelles is decreased. Oil droplets mainly consisting of free fatty acids still persist in the ileum and go into the feces. Deficiency of bile salts or of lipase therefore results in **steatorrhea**—foul-smelling fatty stools, with a concomitant loss of other lipid soluble compounds, including some vitamins.

DIGESTION OF PROTEINS

The objective of protein digestion is to convert foreign polypeptide chains into their constituent amino acids, which can be used to construct proper tissue components. Hydrolysis of the amide bond between amino acid residues seems straightforward, but getting it done is not so simple.

The variety of combinations of amino acids creates one problem. With 20 amino acids, there can be 400 different combinations of side chains in the two residues immediately adjacent to the peptide bond that is to be hydrolyzed. It is possible to construct a hydrolase that will ignore the adjacent structures and concentrate only on the amide linkage. Such enzymes will hydrolyze almost any peptide; some bacteria make proteolytic enzymes like this, and their lack of specificity makes them very valuable laboratory tools for degrading proteins; subtilisin, from *B. subtilis,* is a commercially available example. (Proteolytic enzyme is a synonym for protease, an endopeptidase.) However, the weak bonding by these enzymes results in relatively poor catalysis with low affinity for the substrate compared to other enzymes. Such broad specificity therefore would require large amounts of enzyme to hydrolyze food proteins in a given length of time. On the other hand, very narrow specificity, with a particular enzyme attacking only one, or a few, combinations of amino acids, would require producing dozens, perhaps hundreds, of different enzymes.

The general plan of protein digestion is a compromise (Fig. 17–8). Four powerful types of endopeptidases and two of exopeptidases are supplied by the digestive juices. These enzymes have specificities broad enough to convert most proteins to a mixture of small oligopeptides and some free amino acids, but the bonding is specific enough to have a reasonably high substrate affinity (low K_M) and catalytic effectiveness (high turnover number). The products are small enough to be absorbed readily, and once absorbed, the oligopeptides can be attacked by a wide battery of intracellular exopeptidases, which need not be produced in the large quantities that would be necessary to maintain activity within the small bowel under attack by the powerful endopeptidases.

Let us now examine some of the components of the general plan. The readily available enzymes from the cow and pig have been studied in detail, but enzymes from other sources, including the human, appear to be quite similar.

Pepsins are secreted by the gastric mucosa. These endopeptidases are constructed to catalyze hydrolysis under acidic conditions, using side-chain carboxyl groups in the mechanism. More than one enzyme is secreted, but they appear to be similar in general properties. The best known favors the hydrolysis of peptides with aromatic or carboxylic groups in the side chains.

Trypsin, chymotrypsin, and **elastase** are three serine proteases secreted by the pancreas. All have quite similar structures, and we discussed their catalytic mechanism earlier (p. 281). They differ in specificity because the catalytic site is in a crevice

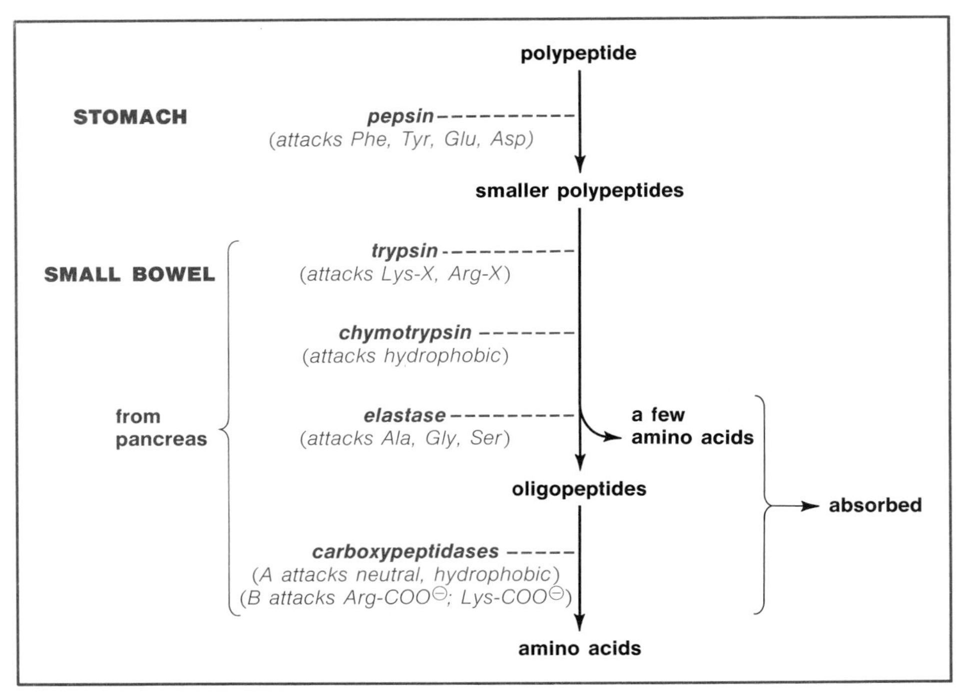

FIGURE 17–8 **Protein digestion involves attack on polypeptide chains by endopeptidases with differing specificities. Pepsin, trypsin, chymotrypsin, and elastase cause hydrolysis at enough different sites along the chains so as to convert dietary proteins (along with proteins from the digestive juices) to a mixture of oligopeptides. The oligopeptides are further shortened by the action of carboxypeptidases, which release the carboxyl terminal residues. The ultimate result is a mixture, mostly of di- and tripeptides along with some free amino acids, all of which are absorbed into the brush border.**

lined with other bonding groups that differ in the three kinds of enzymes. The cleft in trypsin has an aspartyl group with a negative charge and therefore attacks peptides of arginine or lysine, which will bond in the crevice through their positive charges.

The cleft in chymotrypsin is lined with hydrophobic amino acid side chains, which admit segments of polypeptide substrates with similar hydrophobic groups, especially aromatic amino acids, but which exclude those segments containing residues such as Glu, Asp, Gln, Asn, Arg, or Lys, with their polar side chains.

Elastase, named because it once was thought to attack elastin preferentially, differs from chymotrypsin in having valyl and threonyl groups in place of glycyl groups near the entrance to the catalytic crevice. These larger groups will admit only relatively small residues, such as those of alanine, glycine, or serine.

Pancreatic carboxypeptidases are exopeptidases built to attack the peptide bond adjacent to the carboxyl terminus, releasing a free amino acid. (We discussed the mechanism on p. 280.) Two types are known, one hydrolyzing neutral or aromatic residues, the other hydrolyzing arginine or lysine residues. Carboxypeptidases will attack large polypeptides, removing one residue at a time, but they are much more effective as digestion proceeds because the endopeptidases split the original chains into small pieces, each with its carboxyl terminus. The endopeptidases effectively increase the substrate concentration for the carboxypeptidases.

The exopeptidases of the intestinal mucosa attack oligopeptides after absorption and convert them to free amino acids for transfer into the blood. Some of these enzymes occur in the brush border, but most are in the interior cytoplasm. Not all have been described. Those that are known are mainly **aminopeptidase**—enzymes that remove the amino terminal residue by hydrolysis. (For example, the most thoroughly studied is a leucine aminopeptidase, an enzyme attacking amino terminal leucyl groups, or similar residues.) Some of the exopeptidases are dipeptidases attacking oligopeptides with only two residues; others are tripeptidases attacking those with three residues. There is also a **prolidase,** a carboxypeptidase specific for carboxyl terminal proline residues. In any event, the result is that nearly all of the dietary protein is converted to free amino acids before reaching the blood.

Zymogen Activation

Proteases are a menace to any cell. In order to protect the cells that make them, they are synthesized as **proenzymes,** or **zymogens,** which are secreted from the cell before conversion to the active form. The stomach is in less danger, because pepsin is not active at the more neutral intracellular pH, but the pancreas is a loaded bomb, rich in potentially active enzymes.*

A dramatic illustration of the potency of proteases is seen in some people with acute inflammation of the pancreas. This is usually associated with gallbladder disease or alcoholism, which appear to obstruct pancreatic secretion. A particularly virulent form of this disease with a high mortality rate occurs when there is massive activation of the proteases. These people present with severe pain, vomiting, abdominal rigidity, fever, shock, and coma. In the fatal cases the pancreas may be completely liquefied; blood vessels are commonly destroyed (hence the name **acute hemorrhagic pancreatitis** that is applied).

Fortunately, the exocrine pancreas functions placidly in most people, and the secret lies in effective control of zymogen activation. These proenzymes, like other proproteins, are synthesized with extra-long polypeptide chains. The additional residues prevent effective catalysis, either by obstructing access to the active site or by altering the conformation of the protein. Activation involves hydrolysis of the proenzyme, itself, to remove the inhibiting segment of polypeptide.

Pepsinogen has 41 more amino acid residues than does pepsin. It has some proteolytic activity as it is made, sufficient for one molecule of pepsinogen to attack another and convert it to pepsin by hydrolyzing the extra pieces from the molecule. However, there is an important qualification. Pepsinogen is active only below pH 5. Furthermore, although the redundant segment of the molecule is split into pieces, a fragment some 29 residues in length still remains bound to pepsin and inhibits its activity until the pH falls toward 2. Molecules of active pepsin created by accident within the mucosa do nothing more until exposed to the acidity of the gastric contents, not only because they are built to be most active in acid solution but also because the inhibitory peptide clings to them.

Once the proteins are secreted, a few molecules of pepsin can rapidly activate many molecules of pepsinogen through proteolytic attack. The activation is autocatalytic; each molecule of pepsinogen that is converted to pepsin will then attack more pepsinogen, and so on.

The pancreatic peptidases are also secreted as proenzymes. Exposure to secretions from the intestinal mucosa triggers their activation after passage from the pancreatic duct. These secretions include a highly specific protease, known as

*It is not known if Ralph Nader is aware of this common menace.

enterokinase, constructed to remove a particular sequence of residues in **trypsinogen,** the proenzyme form of trypsin (Fig. 17–9). Trypsinogen begins with one to three nonpolar residues, depending upon the species, followed by the distinctive negatively charged Asp-Asp-Asp-Asp-Lys, with the next residue Ile or Val. Enterokinase hydrolyses the bond between Lys and Ile (or Val) in this sequence. The negative charges in the sequence hold the molecule in an inactive conformation; once enterokinase breaks the bond, the remainder of the chain can swing into a new position, creating the active trypsin conformation.

When active trypsin appears, it attacks the other zymogens—**proelastase, chymotrypsinogen,** and **procarboxypeptidases** as well as trypsinogen itself—to convert them to their active forms. For example, a single cleavage of chymotrypsinogen by trypsin permits it to assume a fully active conformation (Fig. 17–10) even though disulfide bonds hold the molecule together. (Further cleavages by chymotrypsin molecules of each other can cause the release of small dipeptide fragments without important modification of activity; still further cleavages destroy the enzyme.)

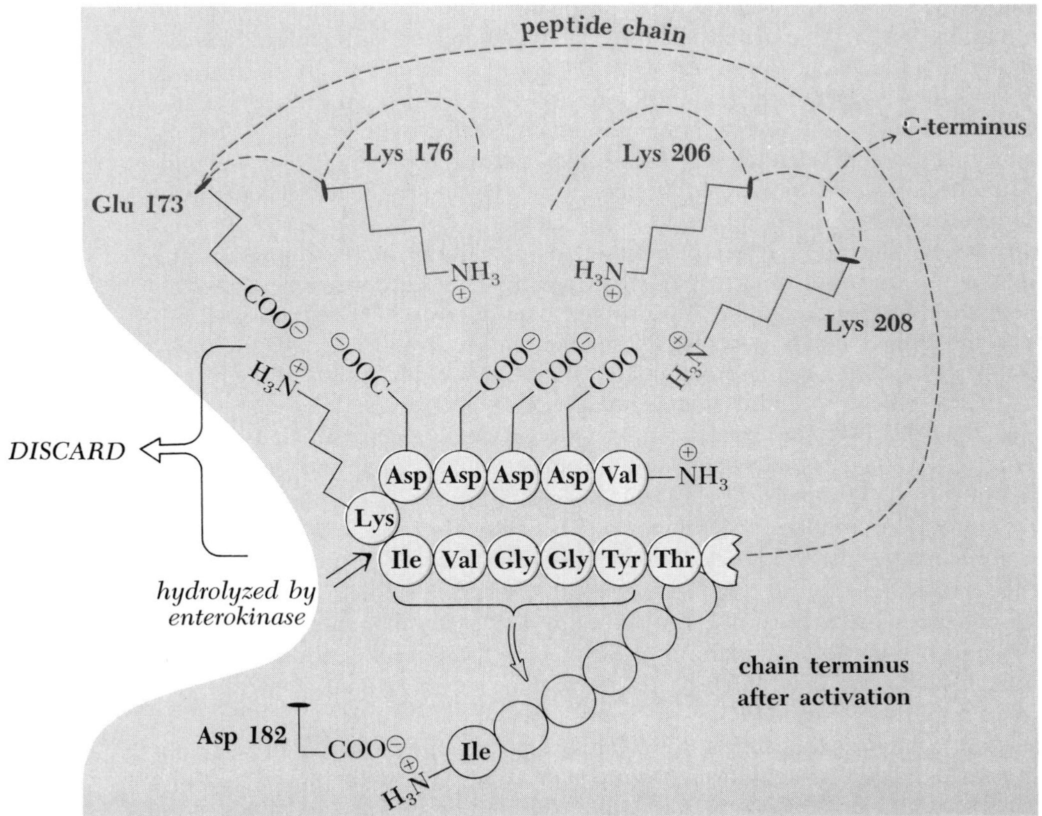

FIGURE 17–9 The activation of trypsinogen by enterokinase involves the hydrolysis of a specific lysyl peptide bond. The segment preceding that bond contains four aspartyl groups in succession, held by their negatively charged chains to positively charged lysyl groups in the remainder of the trypsinogen molecule. These electrostatic bonds (along with another involving the lysyl group being attacked) hold trypsinogen in an inactive strained conformation. When enterokinase breaks the attachment of this sequence to the remainder of the molecule, the new N-terminal segment shifts position to create a more relaxed and active conformation. Trypsin will also attack the same bond in trypsinogen molecules, once activation begins.

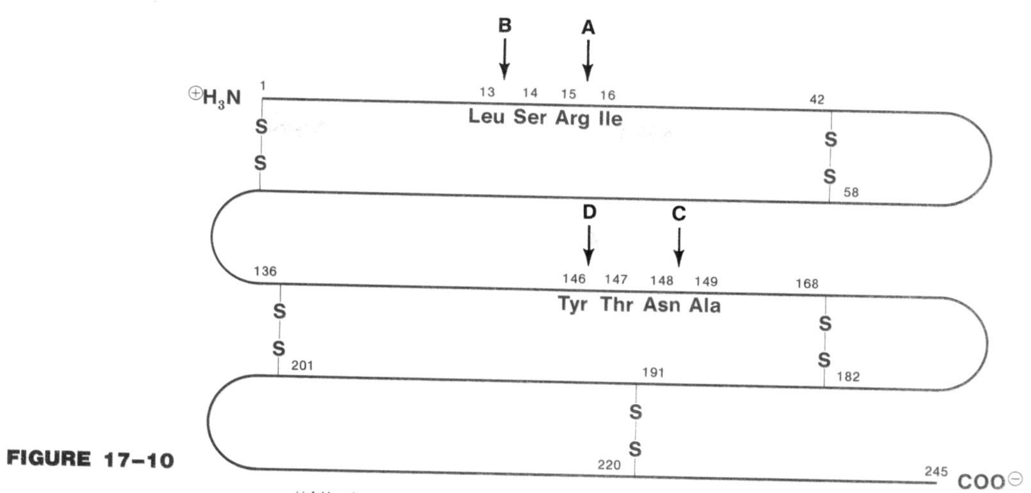

FIGURE 17–10

The topology of chymotrypsinogen activation. The molecule contains five disulfide bridges. Initial activation by trypsin at point A releases the molecule from its strained inactive conformation, but the peptide segments are still held together by the disulfide bonds. Active chymotrypsin will cause further cleavages in chymotrypsinogen molecules, as indicated by B, C, and D.

The key in all of this is the activation of trypsinogen by enterokinase. Once a little active trypsin appears, autocatalytic activation of trypsinogen will create more trypsin, and full activation of the other zymogens rapidly follows. The thing that saves the pancreas is the absence of active trypsin until the juice has safely entered the small bowel. Even so, it is a chancy thing. Trypsinogen, like pepsinogen, has some propensity to assume an active conformation, and exposure to active proteases from other sources is conceivable.

A critical safety factor is provided by the synthesis within the pancreas of another small protein that binds tightly to the active site of trypsin, but that cannot be hydrolyzed. This protein is therefore a **trypsin inhibitor,** which prevents any accidentally activated molecules from creating still more activated molecules. Only when irritation of the pancreas causes active trypsin to be formed in excess of the available supply of inhibitor will disaster occur.

FURTHER READING

Carbohydrate Digestion

G.M. Gray (1975). Carbohydrate Digestion and Absorption. N. Engl. J. Med. 292:1225. Clear review.

G. Semenza (1981). Intestinal Oligo- and Disaccharidases. p. 425 in P.T. Randle, D.F. Steiner, and W.J. Whelan, eds. Carbohydrate Metabolism and Its Disorders, vol. 3. Acadmic Press.

K.W. Simpson et al. (1981). Intestinal Diffusion Barrier: Unstirred Water Layer or Membrane Surface Mucous Coat? Science 214:1241.

H. Sjostrom et al. (1980). A Fully Active, Two-Active-Site, Single-Chain Sucrase-Isomaltase from Pig Small Intestine. J. Biol. Chem. 255:11332.

H.L. Sipple and K.W. McNutt, eds. (1974). Sugars in Nutrition. Academic Press. Includes discussion of gas-forming oligosaccharides in foods.

M.D. Levitt et al. (1976). Studies of a Flatulent Patient. N. Engl. J. Med. 295:260. A case of lactose intolerance.

Fat Digestion

K. Rommel and R. Bohmer, eds. (1976). Lipid Absorption: Biochemical and Clinical Aspects. University Park Press. Useful review articles.

M. Hamosh (1979). A Review: Fat Digestion in the Newborn. Role of Lingual Lipase and Preduodenal Digestion. Pediatr. Res. 13:615.

J.S. Patton and M.C. Carey (1979). Watching Fat Digestion. Science 204:145. Microscopic examination of the phase changes during simulated fat digestion.

P.B. Nair and D. Kritchevsky, eds. (1971). The Bile Acids. Plenum. Three-volume treatise on structure and function.

M. Semeriva and P. Desnuelle (1979). Pancreatic Lipase and Colipase. An Example of Heterogeneous Biocatalysis. Adv. Enzymol. 48:319.

J.W. Riley and R.M. Glickman (1979). Fat Malabsorption—Advances in Our Understanding. Am. J. Med. 67:980. Review.

Protein Digestion

P.D. Boyer, ed. (1971–). The Enzymes, 3rd ed. Academic Press. Articles on specific enzymes. B.S. Hartley and D.M. Shotton. Pancreatic Elastases, 3:349 is especially valuable.

B. Kassell and J. Kay (1973). Zymogens of Proteolytic Enzymes. Science 180:1022.

D.M. Mathews (1975). Intestinal Absorption of Peptides. Physiol. Rev. 55:537.

M. Laskowski, Jr. and I. Kato (1980). Protein Inhibitors of Proteinases. Annu. Rev. Biochem. 49:593.

A. Hershko and A. Ciechanover (1982). Mechanisms of Intracellular Protein Breakdown. Annu. Rev. Biochem. 51:335.

General

D.H. Alpers and B. Seetharam (1977). Pathophysiology of Diseases Involving Intestinal Brush-Border Proteins. N. Engl. J. Med. 296:1047.

E.P. DiMagno, V.L.M. Go, and W.H.J. Summersill (1973). Pancreatic Enzyme Outputs and Malabsorption in Pancreatic Insufficiency. N. Engl. J. Med. 288:813. Demonstration of large-excess production of hydrolyases by normal pancreas.

CHAPTER 18

HYDROLASES: BLOOD CLOTTING AND COMPLEMENT

The life of mammalian cells is absolutely dependent upon proximity to flowing blood as a means of fetching and removing compounds. This traffic is so vital that it is protected and regulated through a number of interdependent devices built into the blood itself and the vessels containing it. Some of these devices depend upon responses by cells and by organs that are beyond the scope of our inquiry, but among the devices is a powerful biochemical tool—the **enzymatic cascade.** When enzymes act in cascade, one enzyme attacks another, which in turn attacks a third, and so on; each step magnifies the effect of the preceding step. Two kinds of cascades are known: successive **phosphorylations** of enzymes, and successive **hydrolysis** of proenzymes.

We are concerned here with four specific but connected phenomena that involve hydrolytic cascades: the clotting of blood, the lysis of clots, the formation of agents (kinins) that dilate vessels and increase permeability, and the lysis of cells upon attachment of antibodies to their plasma membranes. The mechanisms of these events have these common features (Fig. 18–1):

1. They involve limited proteolysis by specific hydrolases in the blood, most of which are serine endopeptidases attacking Arg-X bonds.

2. The active hydrolases are created from proenzymes by attack with still other endopeptidases of similar character.

3. The initial signals for the events trigger a cascade of proenzyme activation in which one enzyme partially hydrolyzes the proenzyme of another, making an active enzyme that in turn activates still another proenzyme, and so on.

4. Inhibitors are present to combine with accidentally activated enzymes and to prevent premature triggering of the cascade.

5. The status quo is restored through constant removal of the activated enzymes, mainly by the liver.

6. Most of the components in these mechanisms are glycoproteins.

We can see in all of this considerable extension of the processes by which pancreatic hydrolases are activated while limiting the risk of premature appearance of functional enzymes.

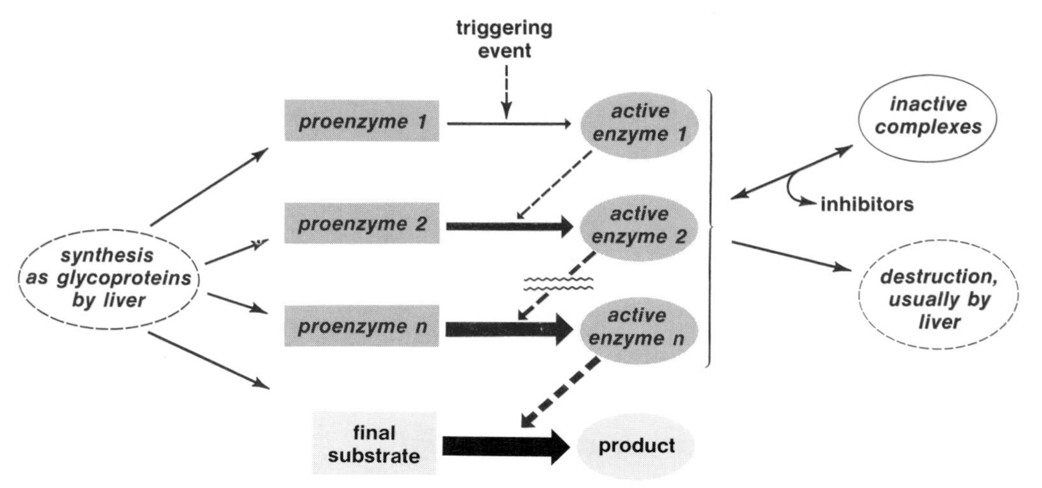

FIGURE 18–1 General outline of cascade activation of endopeptidases in the blood.

BLOOD CLOTTING

The formation of clots to plug defects in blood vessels is one of the more important defense mechanisms of the body. We readily perceive the loss of blood when it pours from gaping wounds, but the potential loss through constant minor injuries to the small vessels is only appreciated when we see the multiple tiny hemorrhages (petechiae) in people deficient in platelets or the large localized hematomas that occur in people with a defective clotting mechanism. These injuries can result from the simple stresses of motion and the accompanying contacts with physical objects, from ordinary chewing, from the normal movement and loss of cells in the gastrointestinal tract, and so on.

Hemostasis—cessation of bleeding—is accomplished by three events. One is **vasoconstriction** of the injured vessel to reduce immediately the flow through the break. Another is **clumping of platelets** in the blood at the site of injury so as to plug the opening temporarily. This may be sufficient to seal small breaks. The third event is **aggregation of a protein, fibrin,** into a clot—a stable three-dimensional lattice that is strong enough to seal the damaged vessel while repairs are being made.

The mechanism may go awry and cause clots to form within an intact vessel. Such a clot is called a **thrombus,** and the occlusion of the vessel is a **thrombosis,** as in coronary thrombosis.

The Platelet Plug

The blood platelets, or thrombocytes, are disk-shaped cells about 1 × 3 microns in size that lack nuclei but contain mitochondria, other cytoplasmic organelles, and distinctive granules, or dense bodies. The blood normally contains about one platelet for each 20 erythrocytes. They arise from megakaryocytes in the bone marrow. The platelets have the property of adhering tightly to collagen fibers. So long as the endothelial lining of blood vessels is intact, platelets and collagen don't meet, but if the lining is disrupted by overt injury or pathological change, platelets stick to the exposed collagen fibers.

Platelets adhering to collagen undergo a remarkable morphological change. They become spiny spheres. The dense bodies disappear from the cytoplasm, and their contents, notably adenosine diphosphate (ADP) and the amine, serotonin (p. 654), appear outside the cell (release reaction). The ADP in some unknown way causes platelets to be sticky, so there is an autocatalytic effect; adhesion of platelets causes release of ADP, which causes still more platelets arriving in the blood to become sticky and in turn to release their dense granules. The growth of the platelet plug continues as long as flowing blood brings more platelets or until the formation of the fibrin lattice, described next, engulfs the plug.

The agglomeration of platelets has another function. Lipoproteins exposed in the platelet plasma membrane are important initiating and accelerating agents in clot formation.

Formation of Fibrin

Let us first consider how a clot forms and then go back through the earlier events that caused its formation. Clotting occurs because a soluble blood plasma protein, **fibrinogen,** is partially hydrolyzed to form fibrin. Fibrin can associate into a three-dimensional lattice, but fibrinogen cannot, even though it has a large surface (45 nm length, 9 nm maximum diameter, MW = 340,000). The extra polypeptide segments in fibrinogen carry negative charges so that electrostatic repulsion aids in keeping the molecules apart.

The chain formula of fibrinogen is $\alpha_2\beta_2\gamma_2$, and the two sets of three unlike chains are linked head to head and side by side by disulfide bonds (the disulfide knot) near the amino terminals in a swollen center (Fig. 18–2). The tails of the chains

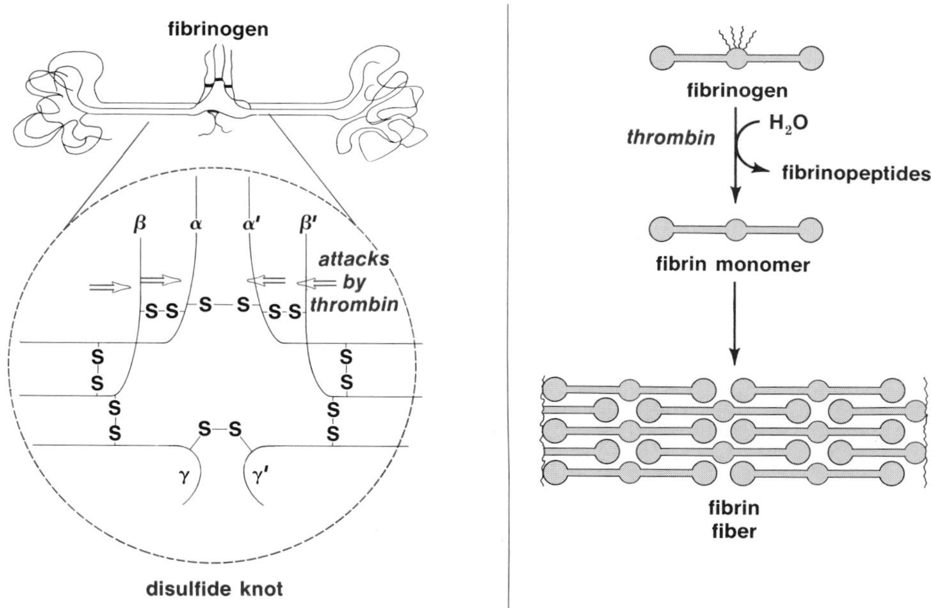

FIGURE 18–2

The fibrinogen molecule (*left*) is made from two identical sets of three unlike chains joined in the center region by a "disulfide knot" shown in enlarged schematic view. Conversion of fibrinogen to fibrin involves hydrolysis of the aminoterminal segments (*brown arrows*) of four of the six chains. (*Right*) Fibrinogen has two terminal globular domains linked by shafts of coiled α-helixes to a central globular domain. The diagram is roughly to scale in broad outline except for the terminal fibrinopeptides removed by hydrolysis to create fibrin monomers (*right center*). The monomers align end-to-end and side-by-side to create a fibrin lattice.

consist of other globular regions linked by coiled coils of α-helix. Fibrinogen is converted to fibrin monomer by hydrolytic removal of amino terminal segments from the α and β chains. Once this is done, the fibrin monomers can associate side by side and end to end to form fibers in an open lattice.

The hydrolysis of fibrinogen to form fibrin is catalyzed by **thrombin,** which is a serine endopeptidase very similar in topology to the digestive endopeptidases. Thrombin specifically attacks Arg-Gly bonds in each of the two α and β chains of fibrinogen, releasing approximately 3 per cent of the original molecule in the form of four polypeptide fragments. Ca^{2+} must be present for polymerization of fibrin monomers, and clotting can be prevented in blood samples by the addition of chelating agents such as citrate, oxalate, or ethylene diamine tetraacetate (EDTA).

The aggregation of fibrin monomers is too weak to serve as a semipermanent plug until final repairs are made. The hydrogen bonds, hydrophobic interactions, and similar forces between monomers are strengthened by the formation of end-to-end covalent bonds (Fig. 18–3). They are formed by the action of a **transglutaminase,** which links the glutamyl portion of a glutamine residue with a lysine side chain. These cross-links are amide bonds between side chains of glutamic acid and lysine residues, the same structures found in hair keratins (p. 172). First, a single cross-link is made between γ-chains in adjacent fibrin molecules, and then two cross-links between α-chains are made more slowly. The cross-links give rigidity to

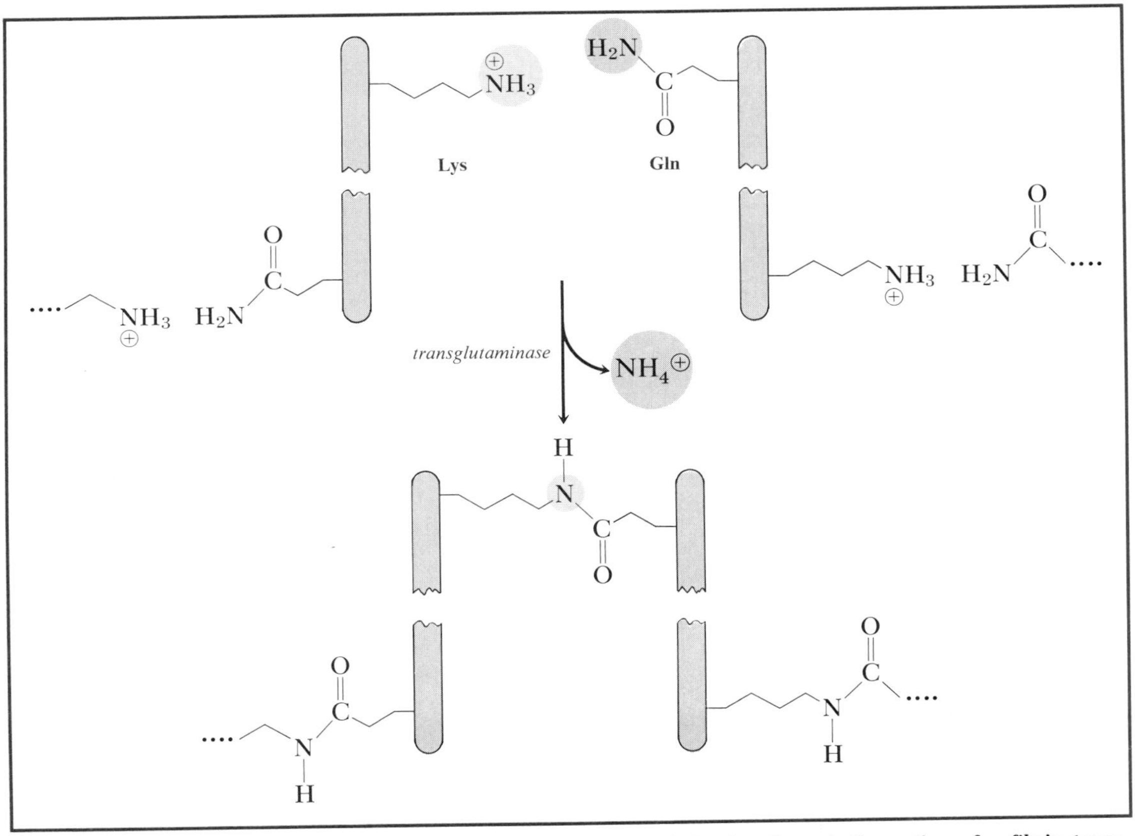

FIGURE 18–3 Fibrin monomers are linked end-to-end in a fibrin clot through the action of a fibrin transglutaminase. This enzyme transfers the glutamyl portion of a glutamine side chain to the nitrogen atom of a lysine side chain on an adjacent fibrin monomer. The amide nitrogen of the original glutamine is released as an ammonium ion. The linked monomers are shown side-by-side to save space, but they are actually aligned end-on.

the clot, making a lattice in which varying numbers of platelets, erythrocytes, and leukocytes are trapped. The strength is amazing for the amount of material involved; a fibrin gel will behave as a rigid solid in the test tube if it contains as little as 0.05 per cent protein. The final step is retraction of the clot into a hard mass. (The semen of rodents also coagulates into a rigid vaginal plug, stabilized by the formation of cross-links by transglutaminase.)

Formation of Thrombin

Fibrinogen is always present in the blood, but normal blood does not clot because thrombin is only present as a proenzyme, which must be activated by partial hydrolysis. Once activated, thrombin not only attacks fibrinogen to make fibrin, it also hydrolyzes specific Arg-Gly bonds in a protransglutaminase to create active transglutaminase.

Before going further, we should note that the components of the clotting mechanism have both trivial names and numbers. Thus, protransglutaminase is also called Factor XIII; an *a* is added to designate the activated forms, as in Factor XIII*a* for the active transglutaminase. These names and some properties are summarized in Table 18–1. In common usage, fibrinogen, prothrombin, and transglutaminase are the preferred terms.

Thrombin is formed from prothrombin by two hydrolyses, both catalyzed by **Factor X*a***, another serine endopeptidase. The first hydrolysis removes an amino-terminal segment by cleaving an Arg-Thr bond, and the second opens a disulfide loop by cleaving an Arg-Ile bond, creating a two-chain thrombin molecule held together by the disulfide bond. (The activation of many of the clotting factors is reminiscent of the conversion of proinsulin to insulin (p. 150).) Factor X*a* can attack prothrombin only when the two are brought together in a complex with Ca^{2+} and phospholipid micelles, along with still another protein, Factor V*a*. Factor V*a* has no enzymatic properties; it only creates a favorable environment for effective union of prothrombin as substrate and Factor X*a* as enzyme. Their combination with phospholipids is a device for bringing together proteins ordinarily present in quite low concentrations in blood plasma. The phospholipids, mainly phosphatidylserine and phosphatidylinositol, are available in activated platelets.

FACTORS IN BLOOD CLOTTING **TABLE 18–1**

Factor* Designation	Trivial Name	M. W. $\times 10^{-3}$	Concentration g/liter	Function of Active Form
I	**fibrinogen**	340	1.5–3.5	precursor of fibrin lattice
II	**prothrombin**	69	0.1–0.2	hydrolyzes fibrinogen
III	tissue factor, thromboplastin	220	–	hydrolyzes Factor VII
IV	$\mathbf{Ca^{2+}}$	0.040	0.085–0.105	cationic cofactor
V	proaccelerin	335	?	protein cofactor for Factor X*a*
VII	proconvertin	50	0.0005–0.002	hydrolyzes Factor X
VIII	anti-hemophiliac factor	?	0.005–0.010	protein cofactor for Factor IX*a*
IX	plasma thromboplastin component	57	0.01–0.02	hydrolyzes Factor X
X	Stuart factor	58.9	0.01–0.02	hydrolyzes prothrombin
XI	plasma thromboplastin antecedent	124	?	hydrolyzes Factor IX
XII	Hageman factor	74†	?	hydrolyzes Factor XI
XIII	**protransglutaminase**	320	?	forms cross-links in fibrin lattice

*Preferred name in bold face
†Bovine; data otherwise for human

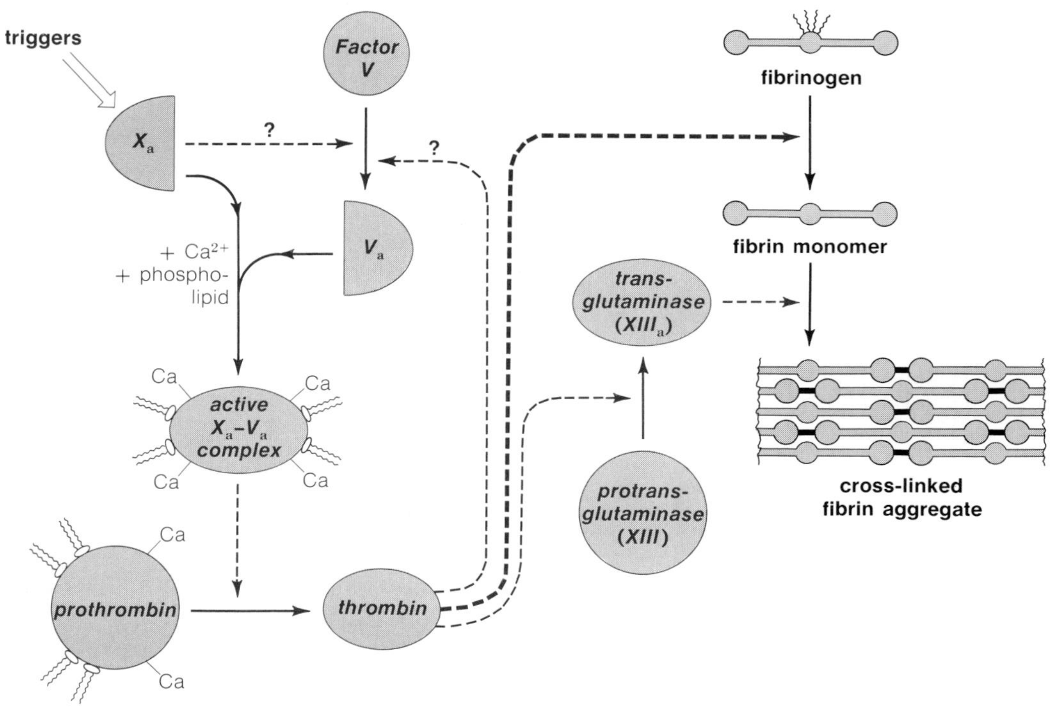

FIGURE 18–4 The terminal events in blood clotting involve partial hydrolysis of several proteins. *Solid lines* indicate the reactions involved; *dashed lines* indicate the responsible enzyme.

Let us summarize events at this point (Fig. 18–4). Clotting Factor V is converted to V*a* by limited proteolytic attack. Factor V*a* then combines with Ca^{2+}, phospholipids, and Factor X*a* to make an active enzyme complex, which converts prothrombin to thrombin by release of a peptide fragment. Thrombin then converts fibrinogen to fibrin and protransglutaminase to transglutaminase, which forms covalent bonds between polymerized fibrin monomeric units.

The activation of transglutaminase illustrates a more general regulatory mechanism. The enzyme in the blood is composed of two pairs of chains, $\alpha_2\beta_2$. The catalytic sites are in the α chains; the β chains act only as regulatory peptides to hinder activation by thrombin. However, once the α chains in the tetramer are hydrolyzed by thrombin, they lose their affinity for β chains, which are now released to reveal the fully active enzyme:

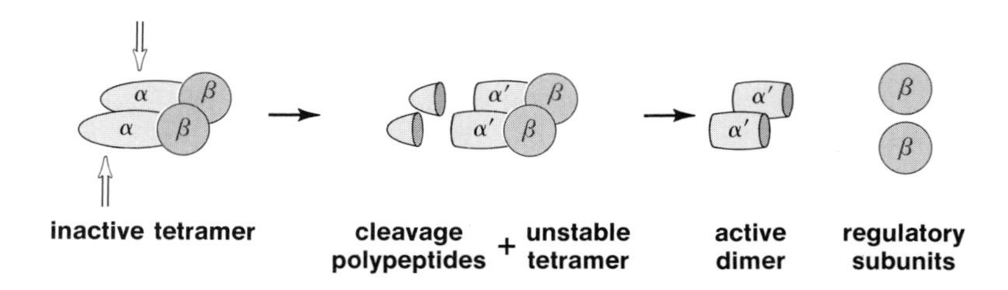

The same α-chains occur in platelets, but without the regulatory β chains. When α_2 is released from adhering platelets, it is attacked by thrombin and converted to the active enzyme much more rapidly than is the $\alpha_2\beta_2$ tetramer found in the blood plasma.

The Triggers for Clotting

The blood clotting mechanism can be triggered *in vivo* by two independent routes, one known as extrinsic because it clearly involves an extravascular factor, the other as intrinsic because all of the components are within the blood. Each pathway generates its own endopeptidase to activate Factor X by partial hydrolysis (Fig. 18–5).

Extrinsic activation is the route for rapid coagulation. It involves the interaction of a **lipoprotein tissue factor (thromboplastin),** which is exposed when the endothelium or other tissues is damaged, with **Factor VII,** a protein in blood plasma. Factor VII can be activated by a number of serine proteases *in vitro,* including Factors X*a*, XI*a*, XII*a*, thrombin, and kallikrein (discussed later), but the route of physiological activation is not established. Factor VII*a* is an endopeptidase that activates Factor X.

Intrinsic activation involves a cascade of at least three other endopeptidases, with the final one activating Factor X. These three are Factors XII, XI, and IX. An auxiliary protein Factor VIII, calcium ions, and platelet phospholipids are also required. The initial event in this sequence is the exposure of Factor XII to the negatively charged surface of collagen or activated platelets. Hydrolysis then converts it to Factor XII*a*, which converts Factor XI to XI*a*. Factor XI*a* in turn activates Factor IX in the presence of Ca^{2+}.

Finally, Factor IX*a*, in the presence of an auxiliary protein, Factor VIII (perhaps modified to be VIII*a*), platelet phospholipid, and Ca^{2+}, will activate Factor X.

People are known who have a hereditary absence of Factor XII, the first component of the sequence, but they usually have little if any trouble from bleeding. This is puzzling, since there is ample evidence for the ability of Factor XII to trigger clotting in the test tube.*

---

*Indeed, Factor XII was discovered because the blood of an individual failed to clot when drawn into a glass test tube; glass activates Factor XII.

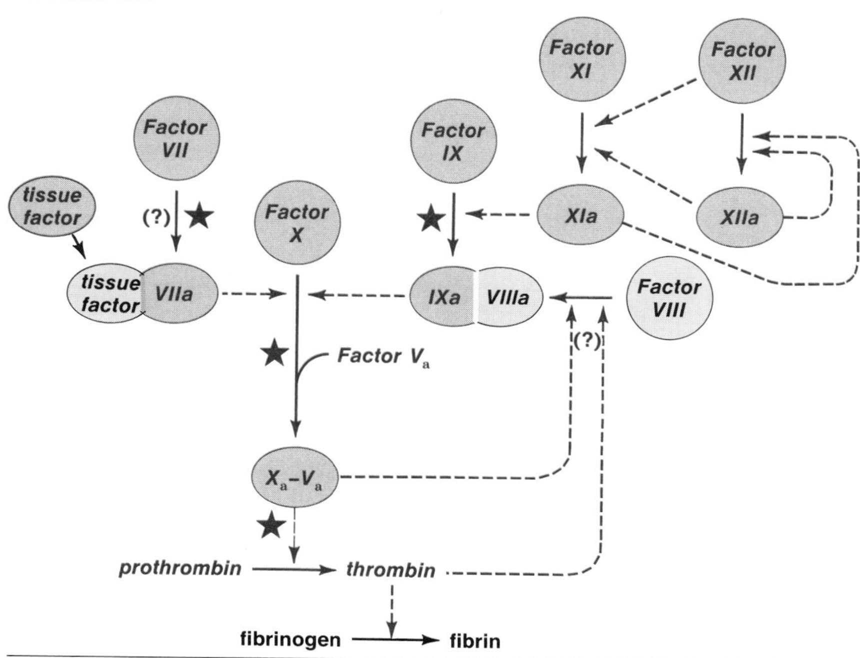

FIGURE 18–5

The formation of Factor X*a* may be caused by either of two sets of reactions. The extrinsic pathway involves uncharacterized proteases from tissues. The intrinsic pathway begins with an activation of Factor XII (see p. 352).

The importance of the intrinsic pathway is best demonstrated by a congenital absence of normal Factor VIII, causing classic **hemophilia.** The affected gene is on the X chromosome and is defective in roughly one in 10,000 infants. Hemophilia therefore appears at this rate in males and is very rare in females (calculated incidence of one in 100,000,000).

"The first record of hemophilia appeared around A.D. 500 in the writings of the Talmudic rabbis. They stated that a third male child would be excused from the ritual of circumcision if two previous male siblings had died from bleeding after their circumcision. By the 12th century, Maimonides stated if a woman had two sons who had bled from circumcision, she could postpone circumcision of future sons whether they be from the same husband or from another husband; thus, it was recognized that the disease was transmitted by the asymptomatic mother."* Excessive bleeding results from relatively trivial causes in these patients. Not only is there a loss of blood, there is also damage to tissues in which bleeding occurs. Hereditary deficiencies of other components in the intrinsic pathway cause similar problems.

Inhibitors of Clotting

Over 10 per cent of the proteins in blood plasma are inhibitors of the enzymes involved in blood clotting and of other proteases. The best understood is **antithrombin III,** a protein that reacts with thrombin, Factor X*a*, or Factor VII*a* to form inert enzyme-substrate intermediates.

There is more antithrombin III in blood than prothrombin; blood is able to clot only because the reaction of the inhibitor with thrombin is much slower than the action of thrombin on fibrinogen so that clotting can occur at locations where the prothrombin to thrombin conversion takes place rapidly.

The reactivity of antithrombin III is greatly increased by combination with the sulfated heteropolysaccharide heparin (p. 196). Heparin therefore prevents clotting. The binding constant for heparin and antithrombin III is in the range of 10^7 to 10^8 M, so low concentrations are effective. The effect of heparin can be counteracted by a protein, platelet factor 4, that is released along with the contents of the dense granules.

The amount of circulating heparin is normally very low; since there is more antithrombin than prothrombin, addition of excess heparin can cause rapid and complete inhibition of clotting, even if all of the prothrombin becomes converted to thrombin. This is the basis of the use of heparin in preventing blood samples from clotting. One micromolar heparin is ample.

An autosomal dominant deficiency of antithrombin III has been reported in several families. This condition, the antithesis of hemophilia and similar disorders, results in severe thrombosis (clotting *in vivo*). Women using **oral contraceptives** frequently have diminished antithrombin III activity in the blood. They have an increased incidence of **thromboembolism** (obstruction of blood vessels by detached clots) in vessels that are rarely affected in other women. (However, immunoassay shows no diminution of the antithrombin III protein.)

Anti-trypsin is a major protease inhibitor in the blood, with a normal concentration of 1.8 to 2.5 g per liter in plasma. It is active against a variety of proteases, despite the name, but its function is uncertain. Some 24 variants of the inhibitor are

---

*Quotation from Harvey R. Gralnick (1977). Factor VIII, Ann. Int. Med. 86:598.

known, and one designated as the Z form results in only 10 to 15 per cent of the normal activity in homozygotes. It is of clinical interest because of an association with pulmonary disease and degeneration of the liver (cirrhosis) in infants. The Z gene frequency is 0.026 in Sweden.

Why doesn't all the blood clot once the cascade mechanism is activated? Part of the reason is the entrapment of the activated factors in the clot through binding to phospholipid and fibrinogen. Another part may be the action of another serine protease known as **protein C.** This protease, like the others, occurs as a proenzyme that is activated by partial hydrolysis, catalyzed in this case by thrombin. Its physiological function is not established, but it has been shown to destroy Factors V*a*, VIII*a*, and X*a* *in vitro*. Some patients with Factors V and VIII deficiencies actually lack an inhibitor of protein C.

Vitamin K and Clotting

Prothrombin, protein C, and Factors VII, IX, and X become activated only in the presence of Ca^{2+}. Calcium is bound to these proenzymes and their activated form because their amino terminal segments contain pairs of glutamate residues that have been modified by carboxylation, creating dicarboxylate side chains with a high affinity for Ca^{2+} (Fig. 18–6). These segments are quite similar and probably have a common evolutionary origin.

The covalent attachment of carboxyl groups to the precursors of these clotting factors by the liver requires the quinol form of vitamin K and oxygen. (An endoperoxide of the vitamin is the other product, and it is reduced first to the quinone and then to the quinol for reuse.) By definition, a compound has vitamin K activity when it enables the synthesis of functional clotting factors. Such compounds are substituted naphthoquinones, of which menaquinone is a typical example (Fig. 18–7). They are hydrophobic compounds that are absorbed from the mixed micelles formed during fat digestion. Any defect in fat absorption will also cause a failure in absorption of vitamin K, and prolonged blood clotting time. Patients with obstruction of the bile duct are routinely given vitamin K before surgery for this reason. The nutritional aspects of the vitamin are discussed in Chapter 41.

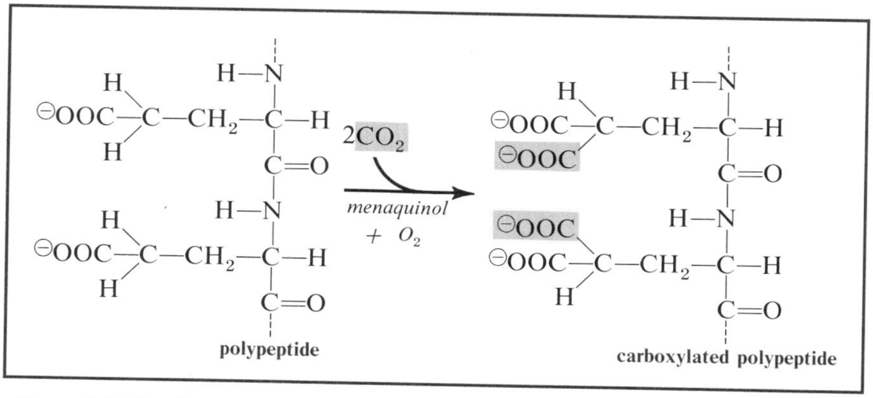

$2CO_2$ → *menaquinol* + O_2

polypeptide → carboxylated polypeptide

FIGURE 18–6

Several clotting factors are carboxylated on C-4 of specific pairs of glutamate residues in the liver before being released into the blood. The carboxylation reaction cannot occur unless menaquinol, or similar reduced derivatives of vitamin K, are available in the liver.

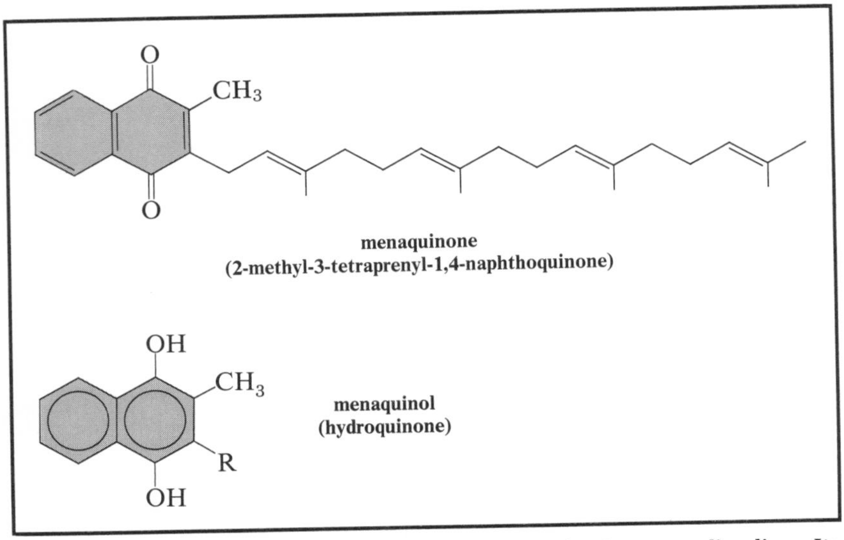

menaquinone
(2-methyl-3-tetraprenyl-1,4-naphthoquinone)

menaquinol
(hydroquinone)

FIGURE 18–7

Menaquinone is the common form of vitamin K occurring in mammalian liver. Its biological activity involves reduction to the corresponding hydroquinone, menaquinol.

Pathological Clotting

Inappropriate formation of thrombi is a common cause of disability and death. Stroke and heart attack are in the common language. A red thrombus containing red blood cells and other cells trapped in a lattice of fibrin may form and grow in some people's veins without any known provocation. Approximately 50,000 people die each year in the United States and another 300,000 are hospitalized as a result of venous thromboembolisms. The formation of clots on a foreign surface is also a major unsolved problem in the use of artificial organs, even with such relatively simple prostheses as artificial heart valves.

White thrombi grow in more compact form along arterial walls, with fewer included erythrocytes, and are major contributors to **myocardial infarction** (loss of blood supply to parts of the heart muscle), **stroke** (loss of blood supply to parts of the brain), and **renal damage.**

We have already discussed the use of heparin in the management of clotting (p. 348). Its great value is that its effects are immediate. Slower and longer-lasting inhibitors of clotting are available in antagonists of vitamin K, and herein lies a tale. Cattle eating spoiled sweet clover develop a hemorrhagic disease, which was shown to be caused by the formation of **bis-hydroxycoumarin** during fermentation of the clover (Fig. 18–8). This compound competes with vitamin K and prevents the formation of prothrombin and other vitamin K–dependent blood clotting factors. It and related compounds are now used in patients prone to abnormal venous clot formation. Since response depends upon loss of already existing prothrombin, it takes two or more days for a useful effect to appear. Heparin is used initially for prompt results.

One of the coumarin derivatives, **warfarin,** is used both clinically (under the trade name Coumadin) and as a rodenticide. It remains an effective rat poison even after continued use because rats fail to associate the delayed deaths of their companions from internal bleeding with the original cause. However, there are disturbing reports of the appearance of rats with genetic resistance to warfarin.

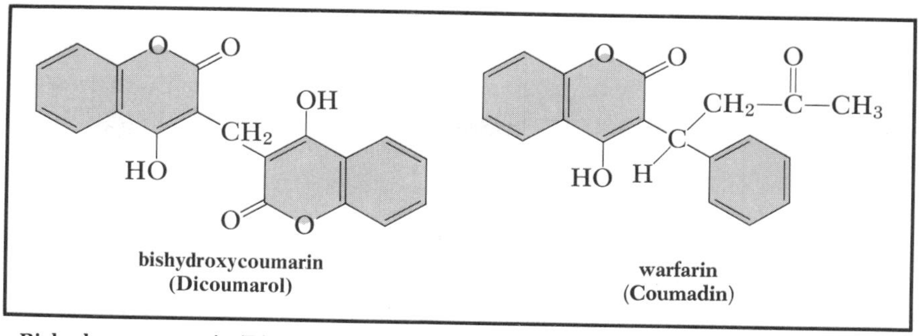

bishydroxycoumarin
(Dicoumarol)

warfarin
(Coumadin)

FIGURE 18-8

Bishydroxycoumarin (Dicoumarol) is an antagonist of vitamin K function. It and its derivatives are used therapeutically to slow clotting. One such derivative, warfarin, is also widely used to poison rats and other small animals.

Abnormal arterial clotting is prevented by drugs acting on platelets. Ordinary aspirin is sometimes used for this purpose.

BRADYKININ, KALLIKREINS, AND LYSIS OF CLOTS

Bradykinin is a small peptide with only nine amino acid residues; it is produced in the blood, where it dilates the vessels and increases their permeability. It is a very potent physiological effector; doses as low as 200 nanograms per kg body weight can cause a measurable fall in blood pressure in small animals. It is also a smooth muscle constrictor and has the added effect of causing intense peripheral and visceral pain. Indeed, production of bradykinin is a mechanism for stimulation of pain receptors upon injury of tissues. It is formed in a variety of inflammatory conditions.

Bradykinin is synthesized by hydrolyzing a particular sequence out of the middle of relatively large precursors, the **kininogens** (Fig. 18–9). (Two sizes of kininogens are known, with 50,000 and 250,000 MW.) The enzymes catalyzing these specific hydrolyses are **kallikreins.** All kallikreins split kininogens at the carboxyl end of the bradykinin sequence, but the position of the amino terminal cleavage varies with the kallikrein. The enzyme from plasma releases bradykinin; glandular tissues cleave kininogen one residue earlier, releasing Lys-bradykinin. A still longer Met-Lys-bradykinin also appears, perhaps owing to cleavages by leukocytes. The extra N-

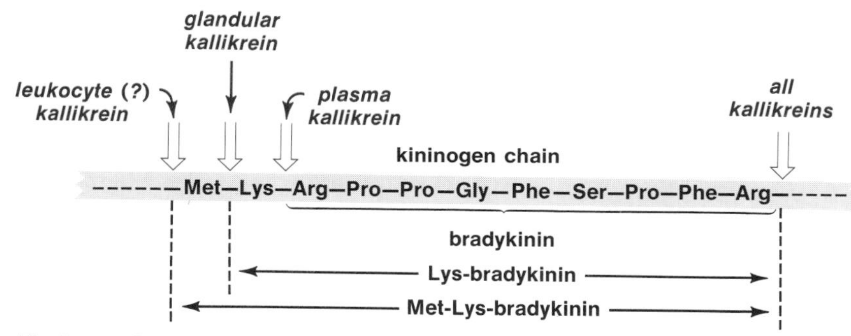

FIGURE 18-9

The long polypeptide chains of kininogens in the blood and tissues contain a specific sequence of amino acids that is released by partial hydrolysis to form bradykinin. The partial hydrolyses are catalyzed by kallikreins.

terminal residues in these derivatives are removed by aminopeptidases in the liver and lungs, forming bradykinin itself.

The events triggering the activation of kallikrein also cause the formation of **plasmin,** an enzyme that hydrolyzes fibrin into soluble pieces (Fig. 18–10). Exactly what happens is somewhat conjectural, but one current hypothesis is probably not far from the mark. It states that everything hinges upon the surface activation of Factor XII. This factor forms a complex on a surface with the high molecular weight kininogen; the complex is enzymatically active. It then attacks prekallikrein, other Factor XII molecules, and Factor XI, converting them all to their activated forms. Some of these, and perhaps activated protein C, may be involved in triggering the conversion of plasminogen to plasmin.

Correct control of sequence is obviously necessary in these complex events. Dilation of the blood vessel by bradykinin and dissolution of the fibrin clot by plasmin must not undo hemostatic constriction and clot formation immediately after injury. The action of kallikrein is blocked by at least three inhibitors in blood plasma. Bradykinin must be formed continuously to maintain its effect because 10 per cent is destroyed in one pass of the blood through the lungs, where a peptidase

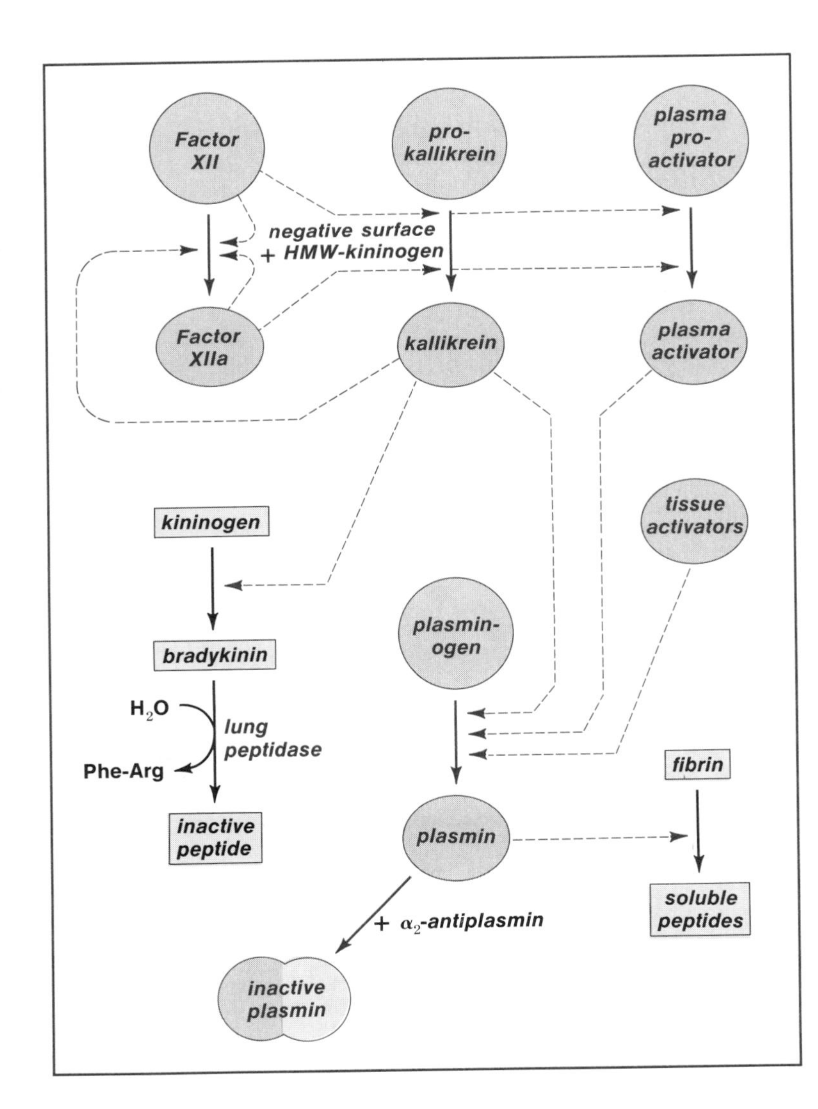

FIGURE 18–10

The formation of bradykinin (*left center*) and the dissolution of fibrin clots (*lower right*) may be triggered by the same events. Kallikrein (*upper center*) catalyzes the activation of plasminogen as well as the formation of bradykinin. Clotting Factor XII*a* will catalyze the formation of kallikrein and of a distinct plasminogen activator (*upper right*) in addition to triggering the intrinsic pathway of blood clotting. Factor XII*a* is formed by at least three autocatalytic routes: Factor XII itself is active in the presence of high molecular weight kininogen (HMW-kininogen) when in contact with a negatively charged surface, and will catalyze the activation of prekallikrein as well as its own activation. Kallikrein will also catalyze the activation of Factor XII, and the resultant Factor XII*a* will also catalyze its own formation in the presence of high molecular weight kininogen and a negatively charged surface.

Bradykinin is destroyed by the action of a lung peptidase, and plasmin is inactivated by contact with a protein inhibitor.

removes the C-terminal Phe-Arg groups. Blood also contains several plasmin inhibitors. Which of these are physiologically important remains to be seen.

COMPLEMENT

The complement system is a collection of some 18 proteins in which interactions occur as a result of partial hydrolyses. The system may be triggered by contact with antibody-antigen complexes or by clotting Factor XII*a*, causing the formation of a protein complex that ruptures cell membranes. Complement is a system designed to destroy foreign cells and to otherwise aid in the removal of foreign materials. Its complexity may stem in part from the necessity of preventing false stimulation that would destroy the host's own cells.

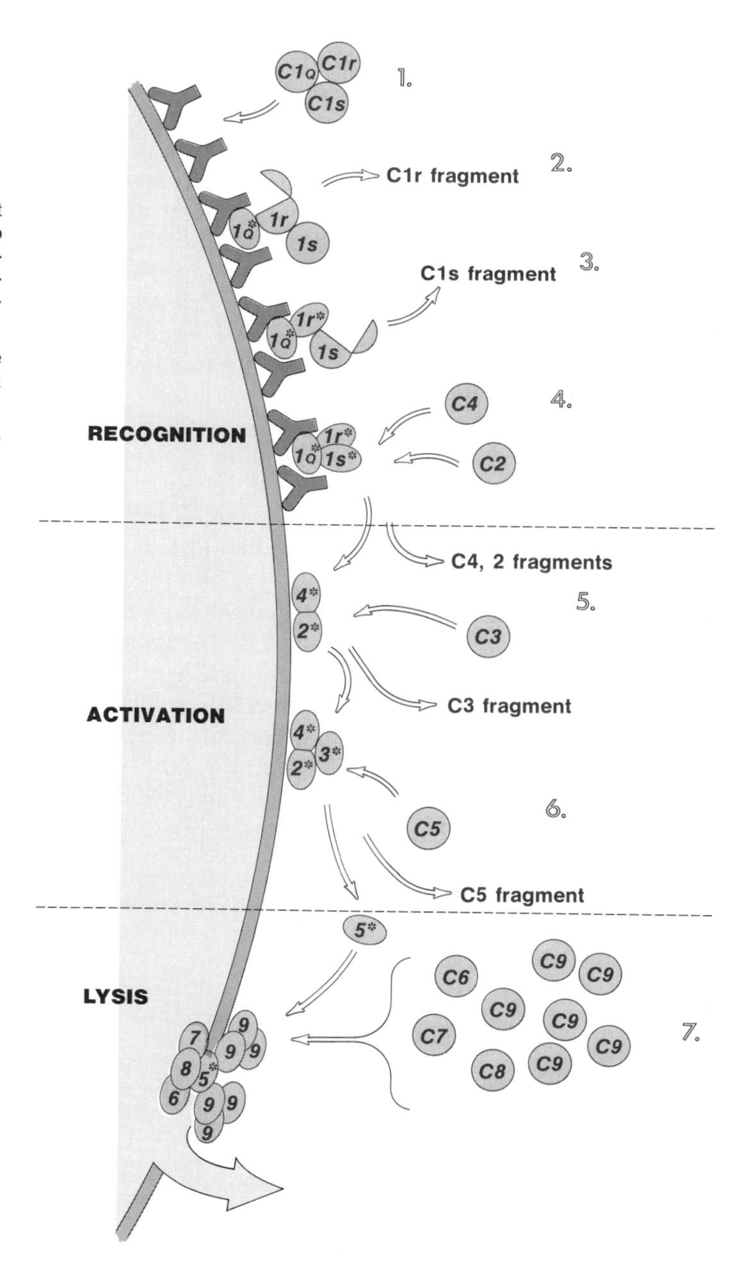

FIGURE 18–11

(*1*) The first component of complement binds to antibodies that are attached to antigens; the C1q subunit of the first component recognizes the altered conformation in the antibody stem caused by reaction with antigen.

(*2*) The C1q subunit becomes an active endopeptidase upon attachment, and partially hydrolyzes the C1r subunit.

(*3*) The modified C1r subunit is an active endopeptidase that attacks the C1s subunit, partially hydrolyzing it.

(*4*) The modified C1s subunit is also an active endopeptidase. It attacks the C4 and C2 components of complement.

(*5*) The partially hydrolyzed C4 and C2 components combine to make still another active endopeptidase, which attacks C3, releasing a fragment from it that acts as a signal to attract macrophages and to cause the release of histamine from neighboring mast cells.

(*6*) The larger portion of the hydrolyzed C3 combines with the C2-C4 complex to create a new endopeptidase activity that will attack C5. The smaller fragment released from C5 also is a signal for macrophages and for histamine release.

(*7*) The larger modified C5 fragment combines with one molecule each of components C6, C7, and C8, and with six molecules of C9 to form a complex that combines with the plasma membrane of the foreign cell, weakening it and causing discharge of its contents.

It is not necessary to examine the functions of complement in great detail for our purposes. The system, like the others examined in this chapter, contains several specific endopeptidases in proenzyme form, which are activated by a cascade mechanism. However, it has the novelty that the action of these enzymes hinges not only upon partial hydrolysis but also upon association of the successive components of the cascade mechanism into complexes.

For example, activation is triggered by combination of a subunit, Clq, in component 1 with either IgG or IgM bound to antigen. The combination of antibody and antigen alters the conformation of the antibody in a way that promotes binding between Clq and the second constant domain of the antibody. The binding causes Clq to become an active endopeptidase that now partially hydrolyzes another subunit, Clr. Clr gains endopeptidase activity with which to attack still another subunit, Cls, which in turn becomes an active endopeptidase (Fig. 18–11).

The Cls subunit of the active complex then attacks two other components, C2 and C4, which associate with each other. The C42 complex (read as C4 + C2) is in turn an endopeptidase that attacts a C3 component, releasing a small fragment (C3b). The larger remainder of C3 combines with active C42 and in turn gains endopeptidase activity to attack C5. C5 is also split to release a smaller fragment. The larger component combines with the remaining components, C6, C7, C8, and six molecules of C9 to make the final membrane-damaging complex. (It is common practice to designate the products of activating hydrolyses as component 3a or component 5b, for example. This practice has been avoided here because it is completely inconsistent with usage in similar systems, such as clotting factors.)

The details are mentioned because the peptide fragments released from C3 and C5 are not merely discarded remnants. They are the compounds responsible for the chemotactic summoning of phagocytic cells to the location of a foreign organism and for the release of histamine, with consequent contraction of smooth muscles and increased vascular permeability.

Component Clq has an interesting architecture. It is composed of six identical subunits, each composed of three polypeptide chains. Most of the amino-terminal portions of the chains form typical collagen triple helixes, except for a kink-forming interruption. The entire protein has 5 hydroxyproline, 2 hydroxylysine, and 18 glycine residues in each 100 residues, and contains 9.8 per cent glucosylgalactose side chains. The carboxy-terminal portions of the chains form a more globular structure. The six collagen-like stems bind at the end into a bundle, bending outward at the kinks like a partially opened six-petaled flower to form individual stalks for the globular regions at the end of the petals.

There is a second pathway for activating complement, which interacts with the classical pathway at C3; its activation is less well characterized, and it is not discussed here.

FURTHER READING

G. Murano (1981). A Basic Outline of Blood Coagulation. Semin. Thromb. Hemostas. 6:140. Clear outline for students.

A. Schafer (1979). Role of Platelets in Thrombotic and Vascular Disease. Prog. Cardiovasc. Dis. 22:31.

E.W. Davie et al. (1979). The Role of Serine Proteases in the Blood Coagulation Cascade. Adv. Prot. Chem. 48:277. Reviews properties of individual clotting factors.

D.A. Walz and L.E. McCoy, eds. (1981). Contributions to Hemostasis. Ann. N.Y. Acad. Sci. vol. 370. Compendium of technical papers.

W.E. Fowler et al. (1981). Cross-Linked Fibrinogen Dimers Demonstrate a Feature of the Molecular Packing in Fibrin Fibers. Science 211:287.

W.H. Kane and P.W. Majerus (1[illegible]81). Purification and Characterization of Human Coagulation Factor V. J. Biol. Chem. 256:1002.

J.A. Penner (1980). Hyperc[illegible]gulation and Thrombosis. Med. Clin. N. Am. 64(4):743.

R.L. Heimark et al. (1980) [illegible]urface Activation of Blood Coagulation, Fibrinolysis, and Kinin Formation. Nature 286:456.

D.M. Kerbiriou, B.N. [illegible]ma, and J.H. Griffin (1980). Immunochemical Studies of Human High Molecular Weight Kin[illegible]ogen and of Its Complexes with Plasma Prekallikrein or Kallikrein. J. Biol. Chem. 255:3952.

R.I. Lundblad et al. (19[illegible]. Chemistry and Biology of Heparin. Elsevier.

F.J. Walker (1980). Re[illegible]tion of Activated Protein C by a New Protein. J. Biol. Chem. 255:5521.

P.M. Gallop, J.B. Lian[illegible] P.V. Hauschka (1980). Carboxylated Calcium-binding Proteins and Vitamin K. N. Engl. J. Me[illegible]1460.

M.A. Ondetti and D[illegible]shman (1982). Enzymes of the Renin-Angiotensin System and Their Inhibitors. Annu. Rev. B[illegible] 51:283.

D. Collen (1980[illegible] Regulation and Control of Fibrinolysis. Thromb. Haemost. 43:77. Review.

B.V. Stadel (19[illegible]Contraceptives and Cardiovascular Disease. N. Engl. J. Med. 305:612,672.

R.A.G. Smith [illegible]). Fibrinolysis with Acyl-enzymes: A New Approach to Thrombolytic Therapy. Nature [illegible]

P.J. Gaffne[illegible]rombosis: A Molecular Approach to Therapy. Nature 290:445.

R.R. Porte[illegible] Reid (1979). Activation of the Complement System by Antibody-Antigen Complex[illegible]cal Pathway. Adv. Prot. Chem. 33:1. Includes structure of Clq.

K.B.M. [illegible] Porter. (1981). The Proteolytic Activation Systems of Complement. Annu. Rev. Bi[illegible] Includes discussion of second pathway of activation.

CHAPTER 19

BIOCHEMICAL ENERGETICS

FREE ENERGY

In our discussion of metabolism, we shall constantly be concerned with de the status of a reaction, or a set of reactions, in relation to the position of equilib **No matter how effective an enzyme may be, it cannot make a reaction go in a di tion that will displace concentrations away from the equilibrium values.**

As reactions go toward equilibrium they can do work. The farther from equili rium that they start, the greater the amount of work that can be obtained from eac mole of reacting compound. A convenient quantitative expression is the free energy change that occurs during the reaction, since this is by definition the capacity to do work under the conditions of constant temperature and pressure occurring in organisms.

Exergonic and Endergonic Reactions. Real reactions always liberate free energy and are said to be exergonic. A reaction that takes up free energy is an endergonic reaction, but reactions that do this don't exist as independent entities. Why do we waste time defining something that exists only in theory? We do so because it turns out to be a handy device for reasoning about some complex reactions that can be imagined to be made up of at least two components, one exergonic and the other endergonic. We treat many real biological reactions as the sum of a highly exergonic reaction, producing lots of free energy, and an endergonic reaction, which consumes part, but not all, of that free energy. Enough free energy must be liberated so that the overall result is at least somewhat exergonic, and it therefore can occur as a real reaction.

This brings us onto more familiar ground, because we are taught from an early age that we burn fuels not only for the energy with which to move but also for the energy necessary for the building of tissues and the like. This isn't too far from the truth; the major defect of the notion is its tendency to equate the concept of energy with a concrete physical entity that can be transferred in a bundle from one reaction to another. The concept is a very useful abstraction, but it is still an abstraction, not a substance.

Biological Examples. What are these exergonic and endergonic reactions, or rather the exergonic and endergonic components of biological reactions? Here are some conspicuous examples:

Exergonic (energy-producing)	Endergonic (energy-consuming)
Oxidation of fuels (carbohydrates, fats, and proteins)	Mechanical movement
Photosynthesis	Synthesis of cellular constituents
Fermentations	Creation of concentration gradients
	Storage of fuels

Anabolism and Catabolism. The sum of the exergonic fermentations and oxidations of fuels is sometimes referred to as catabolism—the disruptive component of metabolism. Synthetic endergonic processes are sometimes referred to as anabolism.

Driving Endergonic Reactions with Exergonic Reactions

Those reactions listed as exergonic are so far from equilibrium that they represent a sort of chemical overkill in which the reaction can be unnecessarily complete for biological purposes. For example, we, like most organisms, completely oxidize glucose:

$$\text{glucose } (C_6H_{12}O_6) + 6\ O_2 \longrightarrow 6\ CO_2 + 6\ H_2O$$

This reaction as given will continue until only one molecule each of glucose and oxygen remain in a sphere of CO_2-saturated water with a diameter 50 times that of the solar system!

On the other hand, the reactions listed as endergonic have such unfavorable positions of equilibrium in water that they will not proceed to even a small extent at any reasonable concentrations of reactants. Consider the formation of a small polypeptide chain by simple union of amino acids:

$$100 \text{ amino acids} \longrightarrow \text{polypeptide} + 99\ H_2O$$

If we start with a generous 0.01 M concentration of amino acid, which is a high value for most intracellular constituents, we find that only one molecule of polypeptide would be formed in an entire universe of solution by the simple reaction given. Most of the amino acids remain there unchanged. Indeed, we noted in discussing proteases that the equilibration of polypeptides with water proceeds toward their hydrolysis, not toward their creation. Something else must be done if the formation of polypeptides from amino acids is to be feasible.

Coupling Endergonic Reactions to Exergonic Reactions. The biological problem is one of somehow combining wildly exergonic reactions, such as the oxidation of glucose, with impossibly endergonic reactions, such as the combination of amino acids into a peptide. It isn't a simple matter of having the two sets of reactions going in the same solution. We could make a solution of glucose, oxygen, and amino acids, along with appropriate catalysts, and find carbon dioxide being vigorously evolved from the oxidation of the sugar while the amino acids sit there essentially unchanged. Once more, energy is not a concrete thing that is automatically transferred from compounds of high chemical potential to those of low potential.

The way in which the oxidation of glucose is made to drive endergonic reactions such as the synthesis of proteins is to modify the oxidation so that it creates ATP and GTP from the nucleoside diphosphates and monophosphates, and inorganic phosphate (P_i). The endergonic reactions are then modified so as to cause the cleavage of the ATP and GTP, as we have already seen in the case of polypeptide synthesis:

$$\text{amino acid} + \text{ATP} + \text{tRNA} \longrightarrow \text{aminoacyl-tRNA} + \text{AMP} + \text{PP}_i$$
$$n(\text{aminoacyl-tRNA} + 2\,\text{GTP}) \longrightarrow \text{polypeptide} + 2n(\text{GDP} + \text{P}_i)$$

It is the exergonic cleavage of the nucleotides that converts what would be an endergonic formation of peptides into an exergonic combination. If the ATP and GTP being consumed for peptide synthesis are now remade in conjunction with the oxidation of glucose, the synthesis of peptides will be effectively driven by the oxidation of glucose, and this is exactly what happens.

We can make a general picture that applies to all metabolic processes:

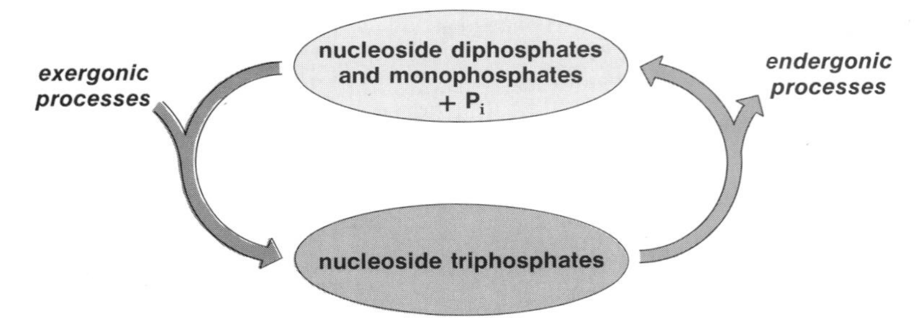

What would be highly exergonic processes, such as the oxidation of glucose, are modified so that nucleoside diphosphates and monophosphates are phosphorylated to create nucleoside triphosphates. The combined reaction is much less exergonic because it now involves the formation of ATP and GTP.

The nucleoside triphosphates are then used in otherwise endergonic reactions, and they are cleaved in these reactions to the corresponding diphosphates or monophosphates so that the reaction becomes exergonic.

The overall effect is as if there were a direct transfer of energy, but the real mechanism is the modification of the exergonic reactions so as to create products of higher chemical potential and to use these products as reactants in the formerly endergonic reactions. The exergonic and endergonic processes became coupled through the involvement of the same nucleotides in both processes, which therefore have the function of energy carriers.

Quantitative Values

Students of metabolism use numerical values of free energy change (ΔG, for Gibbs* energy) for two purposes: as an index of displacement of a reaction from equilibrium and as a device for comparing the potential of different reactions for generating energy under the same conditions.

Displacement from Equilibrium

Given two reactions in which the ratio of reactants to products is equally displaced from equilibrium, the reactions will have equal free energy changes per mole

---

*J. Willard Gibbs (1839–1903): American physical chemist who single-handedly created a large portion of chemical thermodynamics.

reacting, no matter what the equilibrium constant may be. Expressed mathematically, if for

$$A \longrightarrow B,\ [B]/[A] = Q_a,\ \text{with } K_{eq} = K_a,\ \text{and}$$
$$X \longrightarrow Y,\ [Y]/[X] = Q_x,\ \text{with } K_{eq} = K_x,\ \text{then}$$

$\Delta G_{A \to B} = \Delta G_{X \to Y}$ when $K_a/Q_a = K_x/Q_x$.

Indeed, the numerical value of the free energy change can be calculated from this ratio of the equilibrium constant to the actual concentration ratios; this is one of our most important equations in reasoning about energetics:

$$(1) \qquad \Delta G = -RT \ln \frac{K_{eq}}{Q} = -5{,}706 \log \frac{K_{eq}}{Q} \text{ joules mole}^{-1} \text{ at } 25\,^\circ\text{C}$$

in which:

ΔG = Gibbs free energy change, which is negative for spontaneous reactions in which energy is liberated,
R = gas constant = 8.314 joules per mole per degree, or 1.987 calories per mole per degree,
T = absolute temperature,
Q = product of [products]/product of [reactants]. That is, for

$$A + B + C \cdots + n \longrightarrow P + Q + R \cdots + z$$

$$Q = \frac{[P][Q][R] \cdots\cdot [z]}{[A][B][C] \cdots\cdot [n]}.$$

Standard free energy changes for comparing the energy-generating potential of different reactions are related to the equilibrium constants by the expression:

$$\Delta G^0 = -RT \ln K_{eq}.$$

This may be regarded as the free energy change under the hypothetical conditions when all reactants and products are at unit activity (thermodynamic concentration).

It is handy to have in mind a rough idea of the relationship between numerical values for standard free energy changes and equilibrium constants. Figure 19–1 plots the relationship at 0°, 25°, and 38° C (273°, 298°, and 311° K). Looking at the figure we see that a reaction with $\Delta G^0 = -1$ joule per mole is likely to be freely reversible, whereas a value of $\Delta G^0 = -40{,}000$ joules per mole means that the equilibrium lies so far in one direction that it would be difficult to measure the reverse reaction. But what about intermediate values? Here we have to be careful because the relative position of equilibrium depends upon the number of reactants and products. Whether one uses ΔG^0 or K_{eq}, there is frequently no substitute for actual calculation of the equilibrium position in making judgments about the feasibility of a reaction or the equilibrium concentrations of its components. Let us compare three important physiological reactions:

	$\Delta G^{0'}$	K'_{eq}
(1) fumarate + H_2O $\longrightarrow$ malate	−3,670	4.4
(2) polypeptide + H_2O $\longrightarrow$ 2 peptide fragments	−1,880	2.1 M
(3) 2 triose phosphates $\longrightarrow$ fructose 1,6-bisphosphate	−22,850	1.01×10^4 M^{-1}

(As is conventional, the equilibrium expressions for the first two reactions do not include the concentration of water; all pertain to the ionic forms of the reactants at neutrality, and the free energy changes are calculated accordingly.)

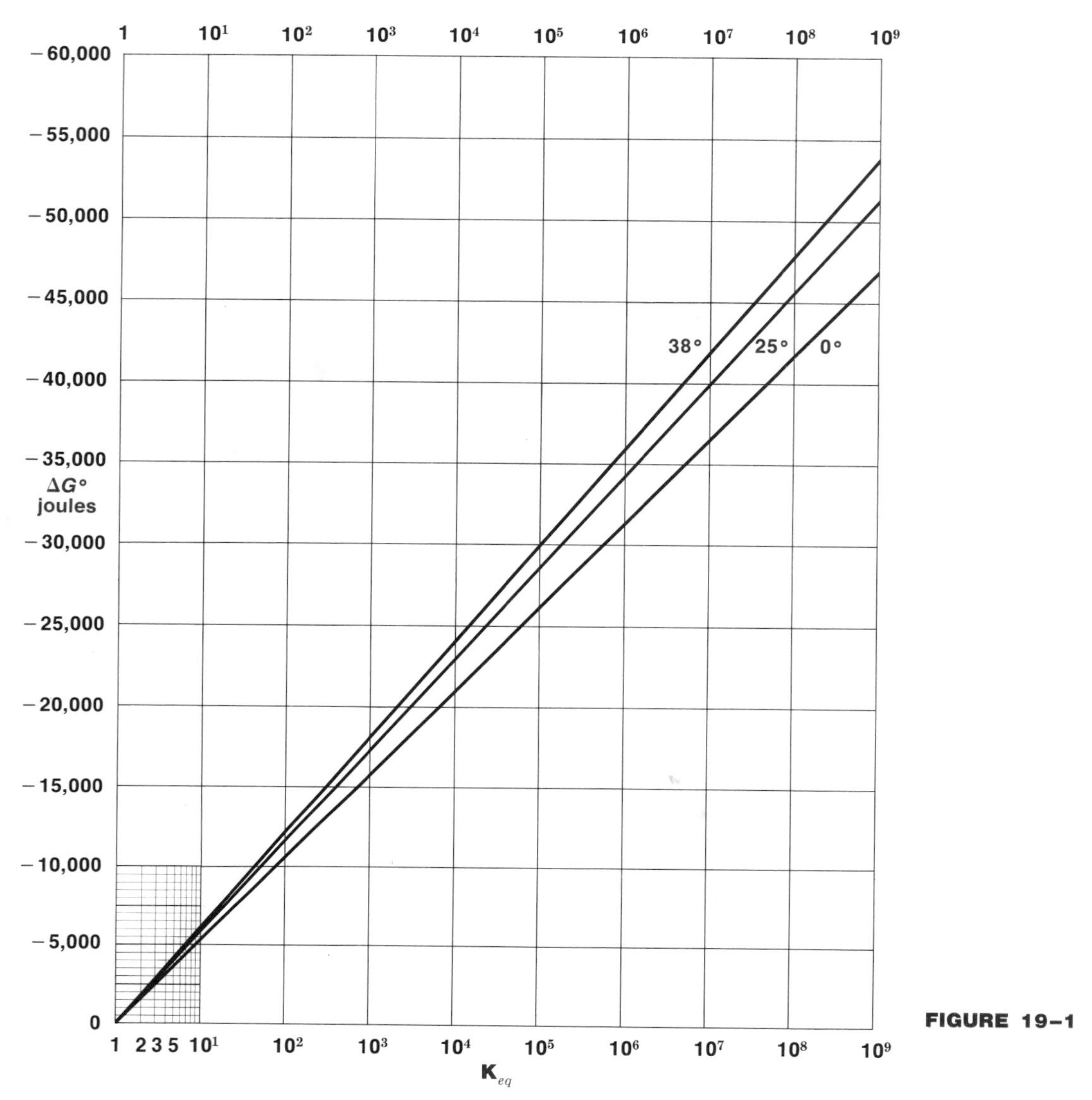

FIGURE 19–1

The relation of standard free energy change and equilibrium constant at three temperatures. The ΔG^0 scale is linear; the K_{eq} scale is logarithmic. The same plot relates positive free energy changes and the reciprocal of the corresponding equilibrium constant.

The first of these reactions is a metabolically important hydration; the second is a hydrolysis by an endopeptidase; the third is the fructose bisphosphate aldolase reaction, which we used as an example of enzyme mechanism (p. 282). Looking only at the numerical values of the standard free energy changes or equilibrium constants, it is easy to assume that the third reaction will go much farther toward completion than the others and that the second reaction will be least complete of the three at equilibrium. What are the facts?

Let us assume that these reactions begin with 0.001 M of each substrate (except water), and calculate the amount, x, of each product formed, and the amount of substrate remaining, $0.001 - x$:

with fumarate, $\frac{x}{0.001 - x} = K_{eq} = 4.4$

$$x = 0.81 \times 10^{-3} \text{ M}$$
$$0.001 - x = 0.19 \times 10^{-3} \text{ M}$$

with polypeptide, $\frac{x^2}{0.001 - x} = 2.1 \text{ M}$

$$x = 0.99952 \times 10^{-3} \text{ M}$$
$$0.001 - x = 0.00048 \times 10^{-3} \text{ M}$$

with triose phosphates, $\frac{x}{(0.001 - x)^2} = 1.01 \times 10^4 \text{ M}^{-1}$

$$x = 0.73 \times 10^{-3} \text{ M}$$
$$0.001 - x = 0.27 \times 10^{-3} \text{ M}$$

The actual positions of equilibrium are in the reverse order of what might be assumed from the numerical values. The hydrolysis, which has the lowest standard free energy change and equilibrium constant, is the most complete and the most difficult to reverse. There is a very important lesson to be learned from this: Standard free energy changes and equilibrium constants are a basis for calculation, not a direct scale of relative position of equilibrium. The distinction between the three reactions we examined that makes the position of equilibrium so different is the relative dimensions of the equilibrium constants. One has no dimensions, the second has a molarity dimension, and the third a reciprocal molarity dimension. The position of equilibrium of the latter two will depend upon the actual concentration of reactants.

There is one rule of thumb that is useful. Most hydrolyses are effectively irreversible under physiological conditions. The low concentrations of reactants found in real tissues favor cleaving one compound into two, even though the equilibrium constant and standard free energy change are low.

Free Energy Changes Are Additive. The principal value of free energy changes in our discussion of metabolic reactions comes from their logarithmic character, which makes the overall change in free energy for any series of reactions the sum of the free energy changes in the individual reactions. Free energy is therefore an especially useful concept for reasoning about the complex series of reactions that constitute metabolism.

The additive nature of free energy changes can be demonstrated by assuming three consecutive reactions, A → B → C → D:

(1) A ⟶ B $\quad K_{eq_1} = \frac{[B_{eq}]}{[A_{eq}]} \quad -\Delta G_1^0 = RT \ln K_{eq_1}$

(2) B ⟶ C $\quad K_{eq_2} = \frac{[C_{eq}]}{[B_{eq}]} \quad -\Delta G_2^0 = RT \ln K_{eq_2}$

(3) C ⟶ D $\quad K_{eq_3} = \frac{[D_{eq}]}{[C_{eq}]} \quad -\Delta G_3^0 = RT \ln K_{eq_3}$

(4) (SUM) A ⟶ D $\quad K_{eq_\omega} = \frac{[D_{eq}]}{[A_{eq}]} \quad -\Delta G_\omega^0 = RT \ln K_{eq_\omega}$

If we solve the first equilibrium expression for $[B_{eq}]$, substitute this value in the equation for the second reaction, solve this for $[C_{eq}]$, and substitute it in the third, we arrive at this expression:

$$K_{eq_1} K_{eq_2} K_{eq_3} = \frac{[D_{eq}]}{[A_{eq}]}.$$

This ratio of concentrations also gives the equilibrium constant, K_{eq_ω}, for the overall reaction. Therefore, the product of the individual equilibrium constants is the overall equilibrium constant:

$$K_{eq_1} K_{eq_2} K_{eq_3} = K_{eq_\omega}$$

Taking the logarithm of the equation, and multiplying by $-RT$, we have:

$$-RT \ln (K_{eq_1} K_{eq_2} K_{eq_3}) = -RT \ln K_{eq_\omega}.$$

Since the logarithm of a product is the sum of the individual logarithms, this is equivalent to saying:

$$\Delta G_1^0 + \Delta G_2^0 + \Delta G_3^0 = \Delta G_\omega^0$$

and this is what we set out to show.

Not only is the additive nature of free energy changes useful in analyzing metabolic processes, but it can also be used to calculate the free energy change for a reaction by simply adding the values for two completely independent reactions whose theoretical sum happens to be the reaction of interest.

For example, it is important to have some idea of the free energy change for the reaction:

(1) $$ATP + H_2O \longrightarrow ADP + P_i.$$

The equilibrium for this reaction lies so far to the right that it is technically difficult to measure the remaining concentrations of reactants with sufficient accuracy for reasonable estimates of the equilibrium constant. However, values can be obtained from two other enzymatic reactions having equilibrium constants low enough to permit reasonable measurements of equilibrium concentrations from which their standard free energies can be calculated:

(2) ATP + glucose $\longrightarrow$ ADP + glucose 6-phosphate

$$\Delta G^{0'} = -23{,}000 \text{ J mole}^{-1}$$

(3) glucose 6-phosphate + H_2O $\longrightarrow$ glucose + P_i

$$\Delta G^{0'} = -13{,}800 \text{ J mole}^{-1}$$

If we add reactions (2) and (3), the overall result is reaction (1), and we can find its free energy change:

$$\Delta G_2^{0'} + \Delta G_3^{0'} = \Delta G_1^{0'} = -36{,}800 \text{ J mole}^{-1} \text{ (at pH 7 and 25° C)}.$$

The prime symbols are used to designate free energy changes at defined conditions, for example, constant pH in this case. When H^+ is a reactant or a product, changes in its concentration would obviously change the free energy and the position of equilibrium for a reaction. Its fixed concentration is already taken into consideration when the prime symbol is a part of the standard free energy designation. This is a convenience for biochemical purposes, because the pH of tissues does not change greatly.

ENERGY-RICH PHOSPHATES

The use of nucleoside triphosphates and similar compounds for energy exchange hinges upon the liberation of relatively large amounts of free energy when these compounds are hydrolyzed. If the synthesis of some compound is to be made feasible by combination with the cleavage of ATP, then the cleavage must liberate more free energy than the synthesis consumes so that the combination of the two processes into one reaction also releases free energy instead of consuming it.

Let us examine the characteristics of those phosphate bonds that have a relatively high free energy of hydrolysis.

Phosphate Anhydrides and Esters. ATP can be hydrolyzed through several routes involving the cleavage of P—O bonds, some of which are summarized in Figure 19–2. The corresponding standard free energy changes and equilibrium constants for these reactions at 25° C are as follows:

Reaction	$\Delta G^{0\prime}$ (joules/mole)	K'_{eq}
(A) $ATP + H_2O \rightarrow ADP + P_i$	−36,800	2.8×10^6 M
(B) $ADP + H_2O \rightarrow AMP + P_i$	−36,000	2.0×10^6 M
(C) $ATP + H_2O \rightarrow AMP + PP_i$	−40,600	1.3×10^7 M
(D) $PP_i + H_2O \rightarrow 2P_i$	−31,800	3.7×10^5 M
(E) $AMP + H_2O \rightarrow A + P_i$	−12,600	1.6×10^2 M

$\mathbf{P_i}$ and $\mathbf{PP_i}$ are abbreviations for inorganic orthophosphate and pyrophosphate, respectively. The corresponding ionic forms that are most abundant at pH 7.4 are HPO_4^{2-} and $HP_2O_7^{3-}$.

The prime marks on the symbols for the standard free energy change and the equilibrium constant signify that these values apply to some empirical conditions; in this case, they were determined at a concentration of H^+ of $10^{-7.4}$ M (pH 7.4), and a concentration of Mg^{2+} of 10^{-4} M. The concentration of magnesium ion is important because it alters the apparent equilibrium by forming complexes with the phosphates. The empirical equilibrium constants are calculated with the total nucleotide concentrations, including all ionic forms and the magnesium complexes.

The reactions are given here, in terms of the free compounds, even though 0.9 of the ATP exists as the Mg complex. This is done in order to simplify the reactions. It is rather difficult to see what is going on when all of the ionic species are specified; for example, here is a more complete approximation of the hydrolysis of ATP at pH 7.4 and 0.5 mM Mg^{2+}:

$$0.88\ MgATP^{2-} + 0.014\ HATP^{3-} + 0.106\ ATP^{4-} + H_2O \longrightarrow 0.40\ MgADP^{-} + 0.06\ HADP^{2-} + 0.54\ ADP^{3-} + 0.04\ MgHPO_4 + 0.10\ H_2PO_4^{-} + 0.86\ HPO_4^{2-} + 0.44\ Mg^{2+} + 0.85\ H^{+}.$$

The low standard free energy change shows that there is something different about the hydrolysis of the phosphate of AMP, which forms free adenosine. If we look at the structures shown in Figure 19–2, we see that it alone represents the hydrolysis of a phosphate ester, a compound formally made from an alcohol (adenosine) and an acid (phosphoric acid). The other hydrolyses all represent the cleavage of an oxygen bridge between two phosphorus atoms, in other words, the cleavage of a **pyrophosphate** bond.

Formally, pyrophosphoric acid is an acid anhydride; it represents the combination of two molecules of phosphoric acid with the loss of water. More molecules of phosphoric acid may be added with the further loss of water to make extended chain polyphosphoric acids, which are polyanhydrides:

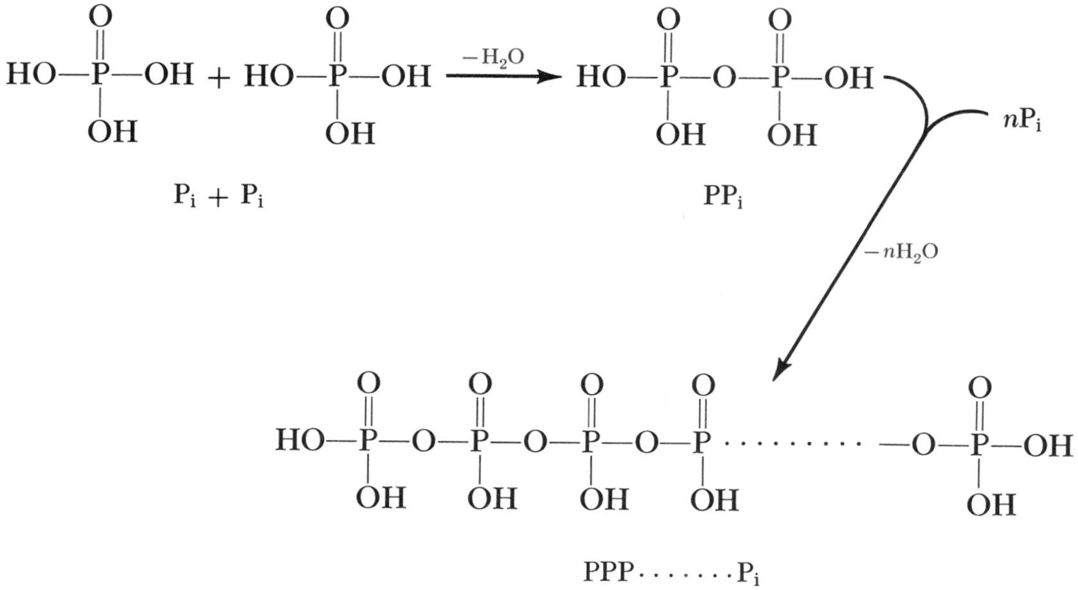

$$\underset{P_i}{HO-\overset{\overset{O}{\|}}{\underset{\underset{OH}{|}}{P}}-OH} + \underset{P_i}{HO-\overset{\overset{O}{\|}}{\underset{\underset{OH}{|}}{P}}-OH} \xrightarrow{-H_2O} \underset{PP_i}{HO-\overset{\overset{O}{\|}}{\underset{\underset{OH}{|}}{P}}-O-\overset{\overset{O}{\|}}{\underset{\underset{OH}{|}}{P}}-OH}$$

$$\xrightarrow[-nH_2O]{nP_i} \underset{PPP\cdots\cdots P_i}{HO-\overset{\overset{O}{\|}}{\underset{\underset{OH}{|}}{P}}-O-\overset{\overset{O}{\|}}{\underset{\underset{OH}{|}}{P}}-O-\overset{\overset{O}{\|}}{\underset{\underset{OH}{|}}{P}}-O-\overset{\overset{O}{\|}}{\underset{\underset{OH}{|}}{P}}\cdots\cdots\cdots-O-\overset{\overset{O}{\|}}{\underset{\underset{OH}{|}}{P}}-OH}$$

High-Energy, or Energy-Rich, Phosphates. We see by comparing the table and Figure 19–2 that the standard free energy is quite highly negative in each case in which a pyrophosphate bond is hydrolyzed. For this reason, such bonds and the compounds that contain them are said to be energy-rich or high-energy. (Some physical chemists become quite upset by this usage, believing the terms imply that the liberated free energy is concentrated in the bond or that the compounds are activated to unusual energy states, and neither of these conditions occurs. Most people who use the terms understand that they simply designate bonds or compounds with large standard free energies of hydrolysis, and this free energy change is the result of the changes in chemical potential of all components of the system, including water and the hydrogen ion.)

Bonds are sometimes designated with a squiggle symbol if their cleavage by hydrolysis has a large negative standard free energy change. Thus, ATP might be designated as A—P∼P∼P, because the final two bonds are of the energy-rich anhydride type, while the bond between adenosine and phosphorus is of an ordinary ester type. (We shall later see that there are other kinds of groups in addition to phosphate groups that are attached by energy-rich bonds in the sense that their hydrolysis has a large equilibrium constant.)

The Phosphate Pool

Adenine nucleotides are the ones most commonly used to couple endergonic and exergonic processes. We shall see that the oxidation of fuels and related exergonic processes involve the simultaneous formation of ATP from ADP, whereas the formation of many compounds involves the simultaneous cleavage of ATP to ADP and P_i, or to AMP and PP_i.

Other nucleotides are also used as energy sources for synthesis. Guanosine triphosphate (GTP) drives peptide synthesis and the movement of ribosomes. It is the cleavage of the pyrophosphate bonds in the precursor nucleoside triphosphates that drives nucleic acid synthesis. Cytidine triphosphate (CTP), uridine triphosphate (UTP), and even deoxythymidine triphosphate (dTTP) drive some syntheses.

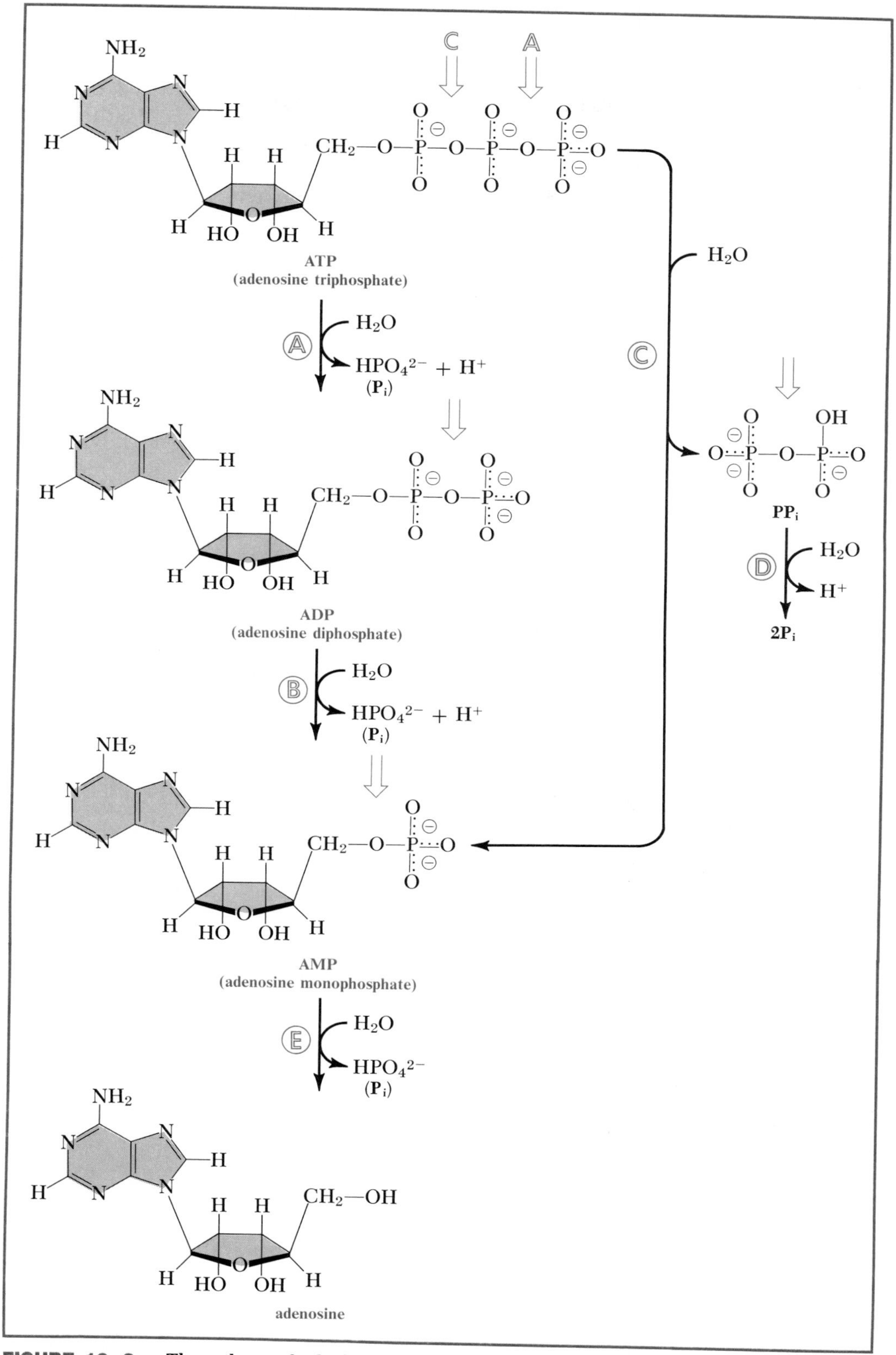

FIGURE 19–2 The pathways for hydrolysing phosphate bonds in ATP. The reactions are listed in the text by the same letters. The large open arrows indicate the bonds broken by hydrolysis in the lettered reactions.

REGENERATION OF TRIPHOSPHATE BY NUCLEOSIDE DIPHOSPHOKINASE. When CTP, GTP, UTP, and other triphosphates are used to drive reactions, they are converted to their diphosphates or monophosphates. The monophosphates are also released when nucleic acids are destroyed. It is therefore necessary that there be some mechanism for converting the mono- and diphosphates back to their triphosphates for reuse, if metabolism is to continue. Most of the conversions hinge upon a transfer of phosphate from ATP. For example, there is a relatively nonspecific nucleoside diphosphate kinase that catalyzes the reversible transfer of a phosphate group from ATP to various nucleoside diphosphates:

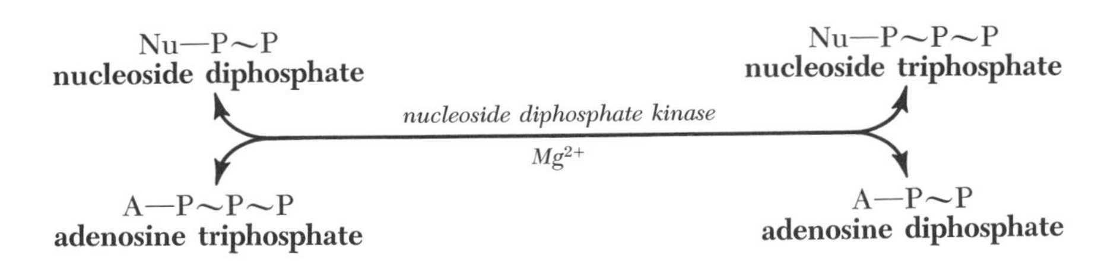

The effect of the action of this enzyme is to make the ratio of concentrations of the triphosphates and diphosphates nearly the same for all the nucleotides.

NUCLEOSIDE MONOPHOSPHATE KINASES. Similarly, nucleoside monophosphate kinases catalyze the transfer of a phosphate group from ATP to various nucleoside monophosphates:

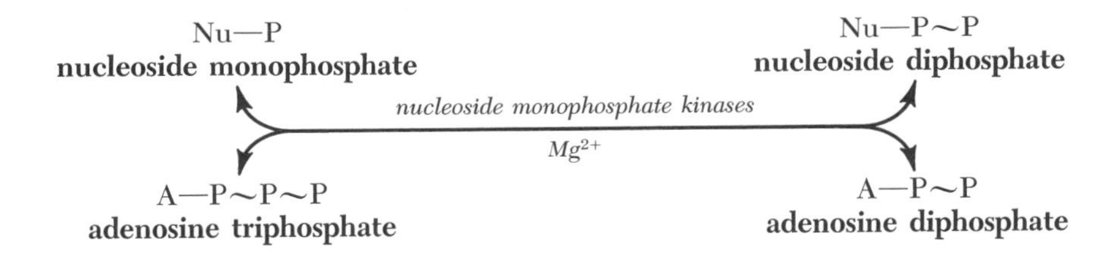

Each nucleoside monophosphate reacts with a specific enzyme (an AMP kinase, a GMP kinase, a CMP kinase, and so on). The AMP kinase is of particular importance because the adenosine phosphates participate in so many processes. It is the most active of the monophosphate kinases and catalyzes the reaction:

$$\text{ATP} + \text{AMP} \longleftrightarrow 2\ \text{ADP}.$$

We see that it would require two phosphorylations to convert the resultant ADP completely to ATP. Therefore, we can make a mental reminder that **any reaction forming AMP from ATP is effectively spending 2 moles of high-energy phosphate per mole of AMP formed.**

Complete Interrelationships. The interplay between these various high-energy phosphates is summarized in Figure 19–3. Since ATP is used in rephosphorylation of the other kinds of nucleotides, expenditure of any other high-energy phosphate for endergonic purposes, as represented on the right of the figure, is equivalent to the conversion of ATP to ADP. The ultimate result in each case is the replenishment of the high-energy phosphate by highly exergonic processes coupled to the phosphorylation of ADP, as shown on the left.

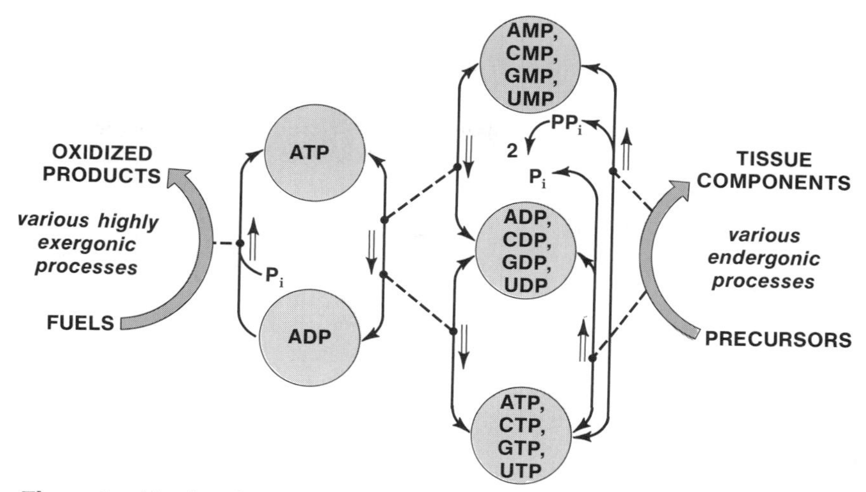

FIGURE 19–3

The nucleoside phosphate pool. Exergonic processes, such as the oxidation of fuels, drive the phosphorylation of ADP to form ATP (*left*). Part of the ATP is used to phosphorylate nucleoside monophosphates to form nucleoside diphosphates, and to phosphorylate nucleoside diphosphates to form nucleoside triphosphates. The other nucleoside triphosphates, as well as ATP, may then be used to drive various endergonic processes, such as the formation of tissue components from precursors. In doing so, either a phosphate or a pyrophosphate group is split from the nucleoside triphosphate. (Note that double-headed arrows are used to indicate potentially reversible reactions. The large arrows alongside indicate the direction of physiological flow.)

Phosphagens. The concentrations of the nucleoside di- and triphosphates represent only a small reserve for the energy requirements of cells. These compounds are intermediates, not stored fuels. Typical concentrations range around several millimoles of ATP, tenths of a millimole of ADP, and 0.01 (or less) of a millimole of AMP per kilogram of active tissue (excluding gross structural components such as cartilage in animals, extracellular structures in protists, and woody tissue in plants). The concentrations of the guanine, cytosine, and uracil nucleotides are typically a tenth or less of that of the adenine nucleotides. The lack of reserves of high-energy phosphate is no problem for cells that require only relatively slow acceleration of ATP-consuming processes, because the ATP-producing reactions can be accelerated at the same pace. However, the muscles of animals represent a special case, in which the tissues have evolved to give rapid responses. Quick contraction of muscles represents a prompt demand on high-energy phosphate of greater magnitude than can be met by any reasonable metabolic capacity for generating ATP. For example, a man can expend as much as 6 mmoles of ATP per kg of muscle per second in a sudden burst of activity, but the maximum production by fuel metabolism, which only occurs after several seconds' delay, is 1 mmole $kg^{-1}s^{-1}$. The quick contraction can occur only at the expense of high-energy phosphate already at hand.

If the necessity for high-energy phosphate were the only consideration, the need could be met by simply increasing the ATP concentration in the tissue. However, this would affect the rate of all of the processes in which ATP is involved, and cause an intolerable disruption of the general metabolism. Hence, different compounds have been evolved to serve as a high-energy phosphate store. They are known collectively as the phosphagens, and are substituted phosphoguanidinium compounds (Fig. 19–4). (These compounds are named both as the phosphoguanidines and as the guanidine phosphates. For example, phosphocreatine and creatine phosphate are the same thing.)

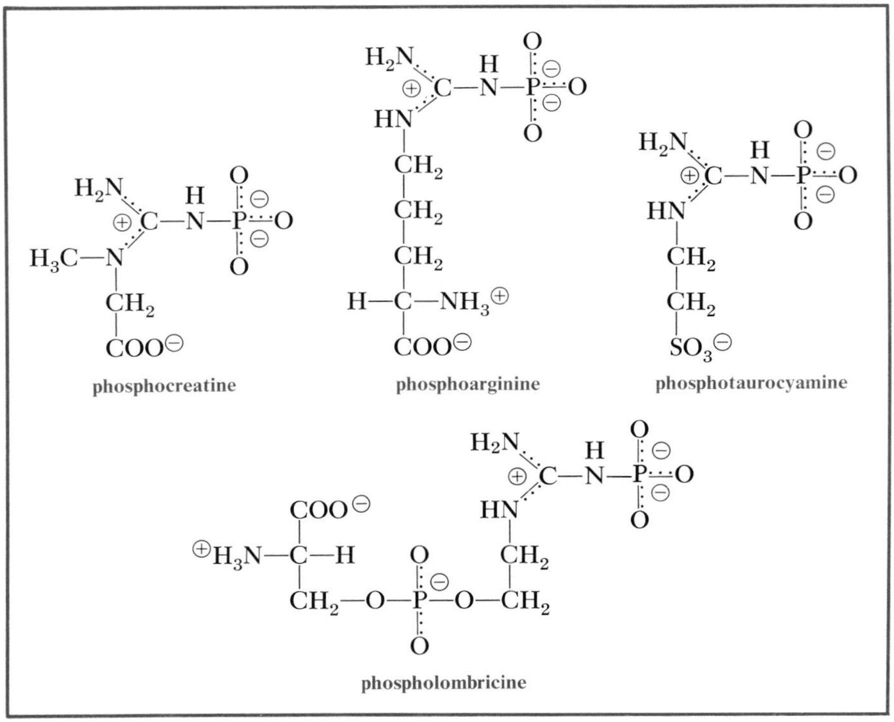

phosphocreatine

phosphoarginine

phosphotaurocyamine

phospholombricine

FIGURE 19–

The phosphagens are phosphorylated derivatives of guanidinio compounds used to store high-energy phosphate, especially in muscles. The compounds used as phosphagens differ in various phyla.

Phosphocreatine is the phosphagen found in most vertebrates and in some invertebrates. Phosphoarginine is found in many other invertebrates. The other compounds listed occur in fewer phyla. The phosphagens typically occur in tens of millimoles per kilogram of muscle. They also occur in brain, presumably to protect this vital tissue from an accidental transitory lack of ATP. Their function as energy stores hinges upon enzymes that catalyze the equilibration with ADP, such as the creatine kinase found in mammalian muscles and brain:

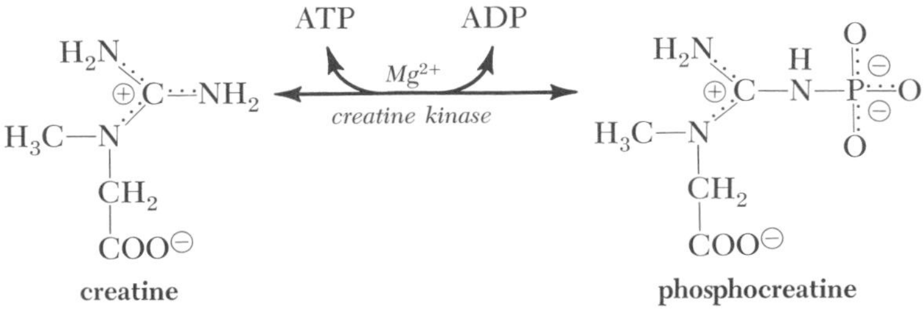

creatine + ATP ⇌ phosphocreatine + ADP (Mg^{2+}; *creatine kinase*)

This reaction effectively expands the supply of readily available high-energy phosphate, and the supply can be rebuilt at rest for use in the next rainy day.

Role of Mg^{2+}. The kinases mentioned here, like most enzymes catalyzing the transfer of phosphate groups, require Mg^{2+} for activity. (This ion is required for some other enzymes, as well.)

FURTHER READING

P.D. Boyer, H.A. Lardy, and K. Myrbäck, eds. (1960). The Enzymes, 2nd ed., Academic Press. The following articles are of interest: M.J. Johnson, Enzyme Equilibria and Thermodynamics, vol. 3, p. 407. Written by one of the pioneers in applying clear thinking to the subject. Directed toward the beginning professional student. R.M. Bock, Adenine Nucleotides and Properties of Pyrophosphate Compounds, vol. 2, p. 3. Detailed summary.

F. Lipmann, (1941). Metabolic Generation and Utilization of Phosphate Bond Energy. Adv. Enzymol. 1:99. The classic introduction of the concept.

Interunion Commission on Biothermodynamics (1976). Recommendations for Measurement and Presentation of Biochemical Equilibrium Data. Quart. Rev. Biophys. 9:439.

CHAPTER 20

USE OF HIGH-ENERGY PHOSPHATE: MUSCULAR CONTRACTION

High-energy phosphate is used for three general purposes: mechanical work, the generation of concentration gradients, and chemical syntheses. All are important, but of the three, mechanical work represents by far the greatest peaked load in most animals. A human working with a power output that can be continued for several hours may be cleaving pyrophosphate bonds in his skeletal muscles at six or more times the rate at which they are being cleaved in all other tissues; a sudden burst of activity, as in a jump, may push the rate of cleavage in muscles toward 100 times that in other tissues. This chapter is concerned with the mechanism by which the conversion of ATP to ADP and P_i is transduced into rapid application of force over a distance, i.e., work.

STRUCTURE OF MUSCLES

Muscles contract because they contain sets of filaments that are forced to slide over each other by a cleavage of ATP into ADP and P_i. This is the basic arrangement in all kinds of muscles, but there are variations on this theme. Smooth muscle cells, which contract relatively slowly and maintain tension over long periods of time without much additional expenditure of energy, are packed with filaments. Striated muscle cells, themselves organized into fibers, are subdivided into **myofibrils.** The myofibrils are made by lengthwise repetition of a fundamental arrangement, the **sarcomere.** A sarcomere is the longitudinal segment of a myofibril between crosswise divisions known as **Z-bands,** which are rigid, thin circular plates of protein (Fig. 20–1). The Z-band is an anchor between sarcomeres; a set of **thin filaments** bristles from each side into adjacent sarcomeres; these filaments are polymers of the protein **actin.** Imagine a coin on edge with a loose bundle of fibers glued end-on to either face, making a long cylinder of fibers with the coin in the middle. Now imagine several of these cylinders spaced end to end so that there are gaps between the fiber bundles. The space between coins, including the gap, is a sarcomere. The gap is filled with a bundle of another kind of filament, the **thick filament,** which intermeshes end to end with the thin filament bundles. Thick filaments are composed of a protein

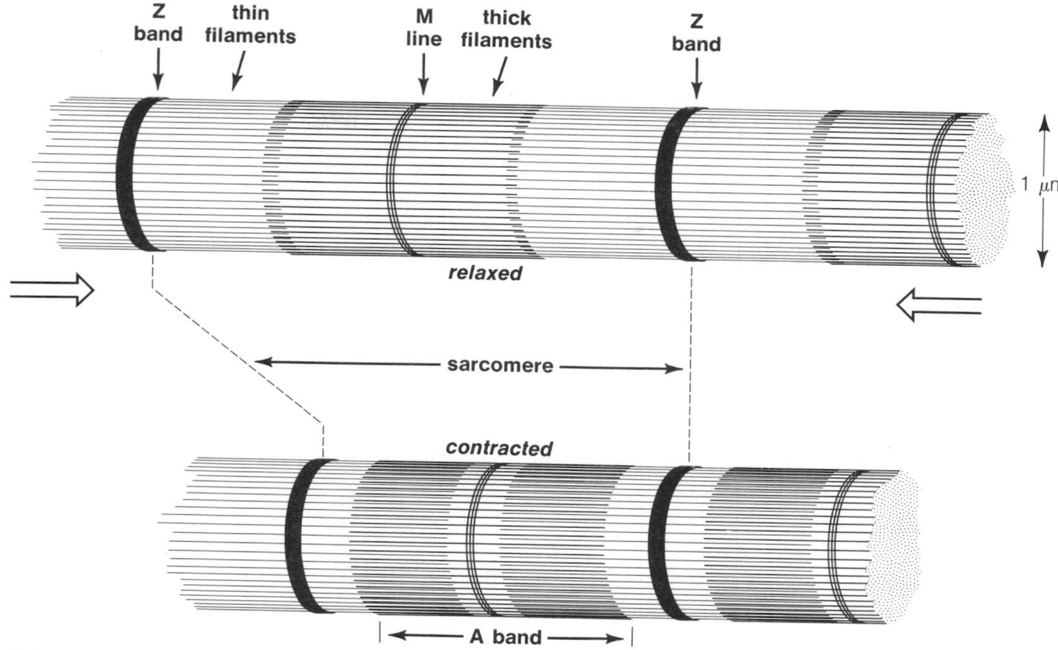

FIGURE 20–1 Structure of striated muscles.

known as **myosin;** they are held in a bundle by central protein cross-links known as the **M-line.**

The striated muscle fibril therefore can be pictured as bundles of thick filaments alternating with paired bundles of thin filaments, pushed together so that the ends of the thick filaments and thin filaments overlap. The separated thick and thin filaments and their overlap create microscopically visible striations in the myofibrils of striated muscle.

In sum, a striated muscle fiber is a bundle of a few thousand myofibrils that run its length. Each myofibril is in turn an end-to-end stack of a few thousand sarcomeres. Each sarcomere is a bundle of a few thousand filaments, which are the actual contractile elements of the muscle.

Types of Striated Muscles. Although we shall emphasize their common features, not all striated muscles are alike. The fibers of the heart (cardiac fibers) differ from those in skeletal muscles. Skeletal muscles are composed of at least three types of fibers in many animals. Two of the types are red in color, compared to a relatively white third type. The **white fibers** utilize glycogen or glucose almost exclusively. The **red fibers** have a high capacity for using oxygen and readily burn fats as well as carbohydrates. The red color comes from higher concentrations of myoglobin and other hemoproteins used for oxygen transport and consumption. One of the red fiber types contracts slowly and steadily (slow-twitch) so as to enable maintenance of tension over longer periods. The other red fibers and white fibers contract rapidly (fast-twitch), enabling swift development of high tension. Fiber composition depends upon the use of the muscle. Migratory birds such as ducks have mostly red fibers in their flight muscles, whereas ground-living birds such as chickens have mostly white fibers. However, chickens are constantly running and have mostly red fibers in their thigh muscles. Hares run in bounding bursts, and their hind leg muscles are rich in white fibers. Muscles adapt. If those rich in white fibers are experimentally stimulated at a steady low frequency over extended periods, the fibers become predominantly the slow-twitch type.

Human muscles are made of mixtures of fiber types, with the proportions varying with the muscle and the way in which the individual uses it. As might be expected, quick bursts of effort promote development of fast-twitch fibers. Occasional intense effort, as in sprinting, develops muscles rich in white fibers; repeated strenuous effort, as in most heavy manual labor or athletic training regimens, leads to a greater content of fast-twitch fibers of high oxidative capacity; long bouts of steady effort cause more slow-twitch fibers to develop.

Thin Filaments

The thin filaments are double-stranded chains of beads, in which each bead is a molecule of actin (Fig. 20–2*A*). Actin has a single polypeptide chain. The two strands are twisted together with one full turn in about 13 beads. The individual beads within a strand cohere strongly enough to sustain contractile tension, even though there are no covalent bonds between them. The polymerization of actin beads to form a strand involves the cleavage of one ATP molecule per bead, and the resultant ADP remains attached to each bead in the polymerized strand. (This fixed ADP is not to be confused with ADP liberated during the contractile process discussed below.)

The thin filaments also contain two regulatory proteins. **Tropomyosin** is a long, thin molecule made from two α-helical chains twisted into a coiled coil. Molecules of tropomyosin lie end to end near the groove on each side of the thin filament. Their

A.

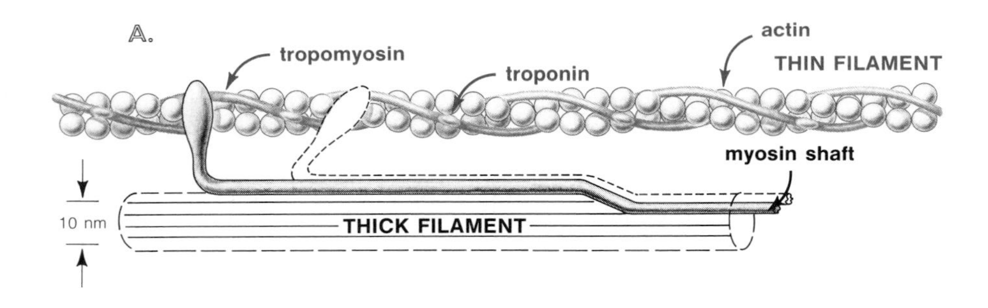

B.

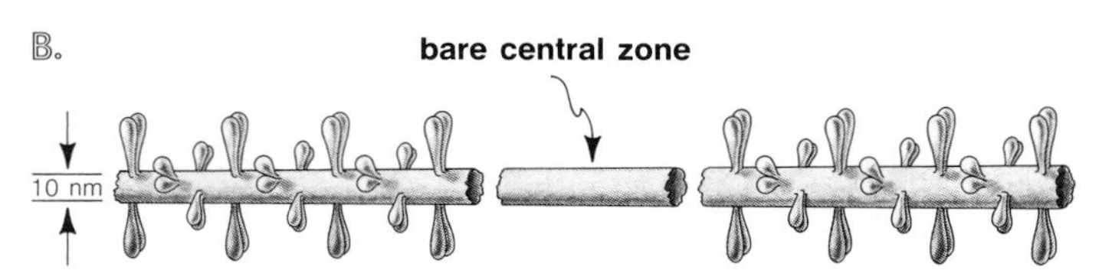

FIGURE 20–2

A. The thin filaments (*brown*) are polymers of globular actin molecules in two twisted strands like chains of beads. Long molecules of tropomyosin lie near the groove of the strands on each side. Each molecule of tropomyosin, which covers seven actin monomers, is bound with a molecule of troponin, a more globular protein. The thick filaments (*black*) are bundles of myosin molecules. Myosin is a protein with a long shaft, made of coiled α-helixes, and a pair of identical globular heads. The heads protrude at right angles in a relaxed muscle. When a muscle contracts, the heads flex toward the shaft (dashed lines), the end of which lifts from the thick filament. The flexure moves the thin filament toward the center of the thick filament.

***B.* Myosin molecules are spaced within the thick filaments so that the heads protrude in a spiral arrangement. The heads are oriented toward the ends of the filaments with the shafts toward the center, so there is a zone in the middle devoid of protruding heads.**

relative position with respect to the groove is controlled by another protein, **troponin,** which is composed of three different polypeptide chains. (Their functions are described below.) There is one molecule of troponin per tropomyosin molecule.

Thick Filaments

The thick filaments are built from stacked molecules of myosin (Fig. 20–2*A* and *B*). Myosin is a large protein with twin globular heads attached to a single long shaft. The shafts are stacked into bundles from which the heads protrude (Fig. 20–2*B*); adjacent molecules overlap in such a way that successive heads appear at 60° angles around the shaft and farther toward the ends. The positions of the heads therefore make spirals around the bundled shafts. The molecules are stacked from the center of the thick filaments toward their ends so that a portion of the middle of the filament has no heads, but the two ends do. A cross-linking protein connects the central bare segments in the M-line.

Myosin is made from six polypeptide chains. Two of these are identical and much heavier than the other chains. These heavy chains are used to construct the shaft (156 × 2 to 3 nm) and part of the two identical heads of a myosin molecule. The shaft portion is formed by folding most of the carboxyl-terminal end of the heavy chains into α-helixes and further twisting the two coils into a long coiled coil. A much shorter length from the amino-terminal ends has a globular conformation and is incorporated into the heads. The remainder of each identical head is constructed from two different light chains.

Myosin

- 460,000 MW
- Combined to form thick filaments, 1,600 × 14 nm
- Long shaft with two globular heads at one end
- Heavy chains (two, identical):
  - C-terminal α-helix paired in coiled coil to form 156 × 2 to 3 nm shaft
  - N-terminal portion in separate globular heads, 19 × 4.2 nm
  - Some lysine and histidine side chains are methylated
- Light chains (four, two each of two kinds; different myosins have different kinds):
  - One of each kind in each globular head
  - One chain phosphorylated
- Catalyzes hydrolysis of ATP to ADP and P_i in isolated form (ATPase activity)

Actin

- 43,000 MW globular protein, 4–5 nm diameter
- Polymerize into double chains, like twisted strands of beads, to make thin filaments
- Polymerization involves cleavage of ATP, and ADP remains on each monomeric unit

Tropomyosin

- 66,000 MW coiled coil of two α-helical chains
- 41 × 2 nm length covers seven actin molecules
- Draped lengthwise on each side of thin filaments
- Lateral position determines if contraction can occur

Troponin

- Regulates position of tropomyosin
- 78,000 MW globular molecule
- One molecule per tropomyosin molecule
- Three subunits
  - T subunit binds tropomyosin
  - C subunit binds Ca^{2+}
  - I subunit also required to regulate contact of actin and myosin by shifting attached tropomyosin

---

The heavy chains have identical primary structures in different striated muscle fiber types, but the light chains differ. One of the kinds of light chains is phosphorylated on a serine residue. This is critical to the action of smooth muscles, which must stay contracted for long periods, because removal of the phosphate latches the molecule into the contracted conformation, but no important function of the phosphate groups has been demonstrated for myosin in striated muscles.

Myosins can differ in another way. Some of the lysyl and histidyl groups in the heavy chain are modified by methylation of their side chains, and the number of added methyl groups is determined by fiber type. Trimethyllysine residues occur in both cardiac and skeletal muscle myosin, but monomethyllysine is found only in the skeletal muscles. It is not known how these changes affect the properties of the molecule.

THE CONTRACTILE EVENT

The fundamental action of contraction is a binding of the actin chains by the globular head of myosin, which then bends on its base, pulling the actin chain into a new position closer to the center of the sarcomere. In order to accommodate this movement, the shaft of myosin also can bend at another location indicated by a dashed outline in Figure 20–2*A*. Since the bundles of thin filaments overlap opposite ends of one bundle of thick filaments, the effect of the movement is to pull both sets of thin filaments toward each other over the thick filaments, which do not move. The sarcomere is shortened without altering the length of either the thin or the thick filaments. The displacement is accompanied by a release of ADP and P_i.

Calcium Ions Trigger Contraction. Muscles are prevented from contracting in the absence of stimulation by draping one molecule of tropomyosin across seven actin beads in such a way as to prevent them from making effective contact with myosin heads. Tropomyosin assumes this position because of its interaction with an attached troponin complex. One of the three troponin subunits has a high affinity for tropomyosin. A second subunit makes the troponin act as an inhibitor of contraction; presumably it combines with the actin in such a way as to push the attached tropomyosin into the blocking position.

The third subunit of troponin can bind Ca^{2+}. When it does, the troponin conformation changes so as to move the attached tropomyosin within the thin filament groove. Effective contact between actin and myosin can then occur.

In a resting fiber, the 0.1 μM concentration of Ca^{2+} is well below the half-saturation level of troponin (5 μM), and troponin holds tropomyosin out of the groove. When a neuron stimulates the fiber through the motor endplate, Ca^{2+} is discharged from the sarcoplasmic reticulum surrounding the myofibrils, raising the concentration to 10 μM, or more, and causing troponin to move tropomyosin out of the way.

Sequence of Events. The major events during muscle contraction appear to be these:

Contracted myosin heads are in a low-energy conformation, which is converted to the high-energy relaxed conformation by reacting with ATP (as the Mg complex) (Fig. 20–3*A*). The reaction probably involves phosphorylation of the myosin head, but the MgADP that is produced also remains bound to the myosin, along with phosphate, upon creation of the high-energy conformation (Fig. 20–3*B*). (When isolated myosin reacts with ATP, the intermediate complex reacts with water to release both ADP and P_i. Free myosin therefore acts as an ATP hydrolase, or ATPase, and this action is accelerated, if actin is present, to form an actomyosin complex.)

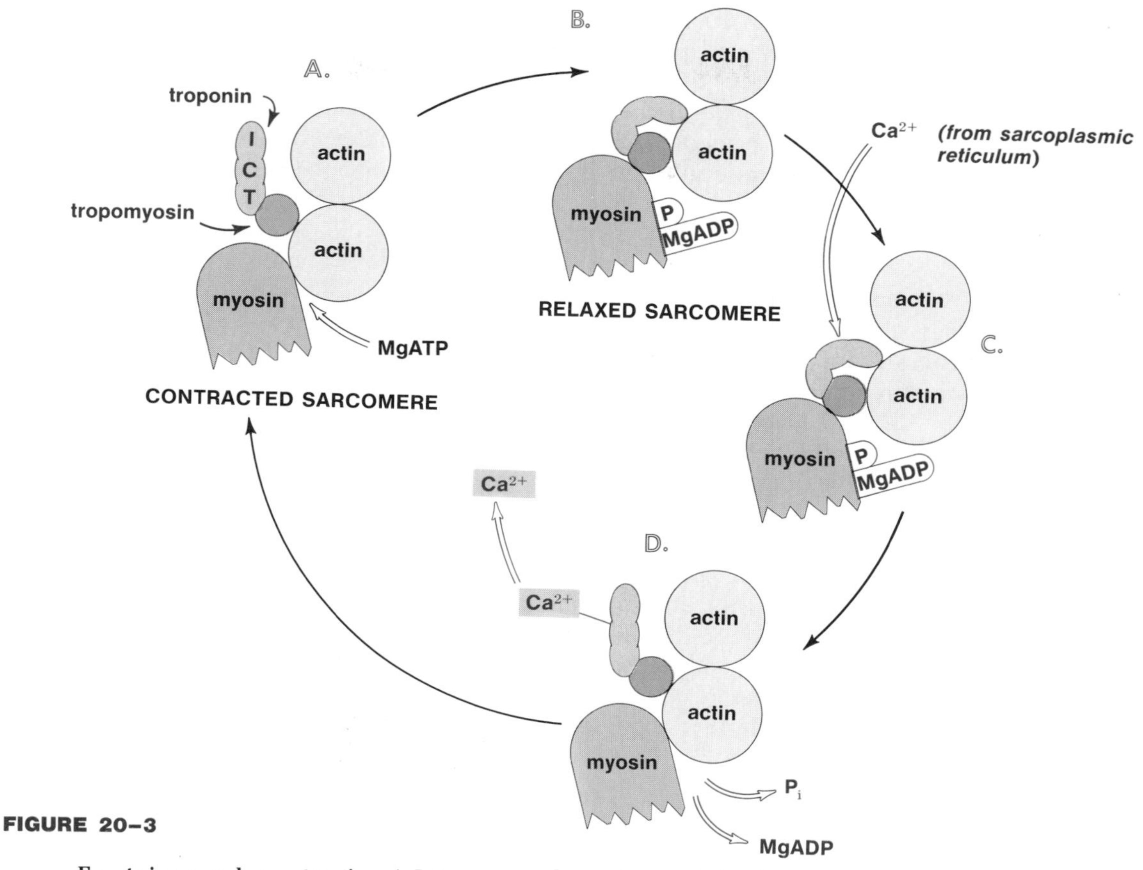

FIGURE 20–3

Events in muscular contraction. *A*. In a contracted muscle, myosin heads are in contact with actin monomers. The tropomyosin molecule, seen in cross section, does not interfere with the contact. Relaxation begins with a binding of ATP as the magnesium chelate.

***B*. ATP is cleaved on myosin to create a high-energy relaxed conformation in which the protein is phosphorylated and MgADP is still bound. Troponin binds to actin as a result, dragging the attached tropomyosin between actin and myosin so as to prevent their interaction.**

***C*. Contraction is stimulated by a binding of Ca^{2+}, released from the sarcoplasmic reticulum, to the C component of troponin.**

***D*. The binding of Ca^{2+} to troponin changes its conformation so that it no longer holds tropomyosin in place between actin and myosin. When actin and myosin interact, myosin reverts to its low-energy contracted conformation, with loss of the phosphate group and MgADP.**

Ca^{2+} is released from the sarcoplasmic reticulum upon excitation and is bound by the troponin-tropomyosin complex, which moves so that actin can interact with myosin (Fig. 20–3*C*). The combination of the myosin heads with actin displaces the phosphate group as P_i and the bound ADP (Fig. 20–3*D*). The loss of phosphate and ADP causes the myosin heads to shift to their bent, or low-energy, configuration, pulling the attached actin along and contracting the muscle. The twin heads of a myosin molecule presumably act in some concerted mechanism with one actin bead, but details are not known. After contraction, Ca^{2+} leaves, and MgATP is once more bound to initiate a new cycle of relaxation and contraction, which will repeat so long as the $[Ca^{2+}]$ remains high.

Not all of the myosin heads are in position to contract at once. When a thin filament is moved relative to a thick filament by attachment of one head to one bead, it brings other beads into proximity to other heads, both on the same and on other thick filaments so that waves of interaction can occur. This arrangement also prevents slippage when a contracted head again reacts with MgATP and releases a bead so as to move back to its relaxed conformation, ready to go through the entire sequence again. Some of the heads in the thick filament are always hanging on during contraction. As the sarcomere shortens, more and more heads are in contact, and tension increases until the opposing actin filaments begin to overlap. Tension falls abruptly with further overlap, presumably because of mutual interference to the binding by myosin.

When the nerve quits firing, the sarcoplasmic reticulum recovers its discharged calcium; the cytoplasmic concentration falls too low for effective binding by troponin, and the troponin-tropomyosin complex moves back to its inhibitory position. Actin and myosin no longer interact, and the myosin heads remain in their relaxed conformation.

The existence of two identical heads on myosin, both presumably moving in the same direction, presents an interesting topological problem that has not been addressed. We can move both extended arms toward our face in similar ways because they are mirror images. If the two halves of our body were identical, as in myosin, one would expect the arms to move in opposite directions.

Contraction of smooth muscles involves a somewhat different sequence of events. They are triggered to contract by entry of extracellular calcium through receptor-mediated channels in the plasma membrane. New drugs are now available for relaxing the smooth muscles in arterial walls by blocking the entry of calcium through the plasma membrane. They are proving to be useful in the management of heart disease and other conditions in which undesirable arterial spasms occur.

HIGH-ENERGY PHOSPHATE BALANCE

Each time a myosin head advances along an actin filament and pulls it back, a molecule of ATP is hydrolyzed to ADP and P_i. The power of a muscle fiber—the **rate** at which it can apply force over a distance—therefore depends upon the rate at which ATP can be supplied to drive contraction, as well as upon the properties of the contractile apparatus. Muscles are built to convert the ADP and P_i back to ATP in a short time so as to enable contraction to continue even though ATP is present in limited concentration. The regeneration of ATP is accomplished in part by the combustion of fuels and by the conversion of glycogen to lactate, as we shall recount in ensuing chapters, but the most immediately available source is the store of **phosphocreatine** (or other **phosphagens** in lower animals) contained within the muscle cells. Phosphocreatine is not used for any purpose other than the phosphorylation of ADP, and it has a relatively low affinity for Mg^{2+} or Ca^{2+}, so its concentration can be varied over wide ranges without disturbing other metabolic processes or grossly changing the divalent cation concentration.

Creatine Kinase Equilibrium

Phosphocreatine is a store of high-energy phosphate that can be used to create more ATP when ATP is being hydrolyzed by muscular contraction and can be regenerated when the muscle is at rest (Fig. 20–4). There is sufficient creatine kinase in striated muscle to keep its reaction near equilibrium, except perhaps in the most violent bursts of activity.

Phosphocreatine is a compound designed to keep the concentration of ATP relatively constant until a muscle nears exhaustion. It can do this only because the

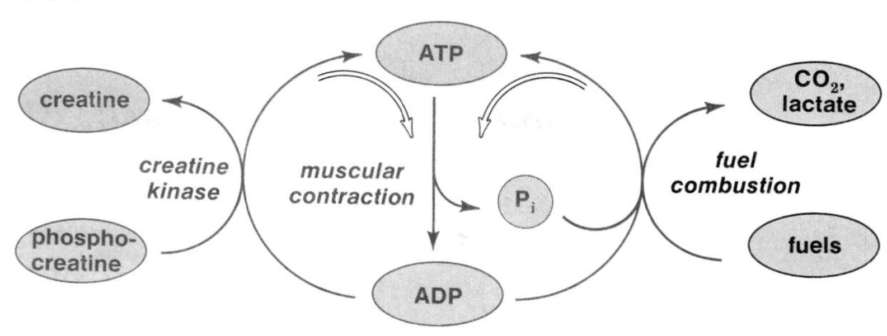

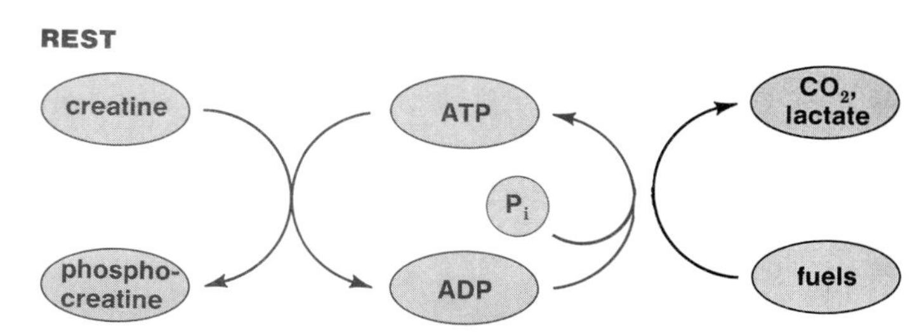

FIGURE 20–4

The controlling event in a working muscle is the hydrolysis of ATP to ADP and P_i by muscular contraction (*center top*). The resultant increase in concentration of ADP will cause phosphate to be transferred from phosphocreatine by creatine kinase (*left top*), and will also cause increased fuel utilization to convert the ADP and P_i back to ATP (*right top*). Both processes serve to provide more ATP for continued muscular contraction.

At rest, the utilization of fuels (*bottom right*) can create a high concentration of ATP relative to ADP, since ATP is not being hydrolyzed rapidly by muscular contraction. The resultant shift in the creatine kinase equilibrium causes phosphate to be transferrea from ATP to creatine, thereby rebuilding the store of phosphocreatine.

equilibrium constant for ATP formation by the creatine kinase reaction is greater than one:

$$\text{ADP} + \text{phosphocreatine} \longrightarrow \text{ATP} + \text{creatine} \qquad K'_{eq} = 54.$$

(The equilibrium constant hinges upon the H^+ and Mg^{2+} concentration; the quoted value is typical for intracellular conditions, with pH = 6.9 and $[Mg^{2+}] = 0.5\ \mu M$.)

At first glance, the high equilibrium constant may seem to make phosphocreatine a less effective energy store. In resting muscles, only two thirds of the total creatine is present as phosphocreatine; one third still remains as free creatine even though the ratio [ATP]/[ADP] is 100 or more. It is precisely this disparity of concentration ratios that enables the maintenance of nearly constant ATP concentrations. We can calculate from the known equilibrium constants and typical values for concentrations in resting skeletal muscles what ought to happen when 90 per cent of the high-energy phosphate stored in phosphocreatine is discharged.

	Millimolar Concentrations				
	[PCr]	[Cr]	[ATP]	[ADP]	[ATP][creatine] / [ADP][phosphocreatine]
resting muscle	20	10	5.94	0.055	54
working muscle	2	28	4.52	1.17	54

We see that ATP remains at 76 per cent of its original concentration, even though the phosphocreatine concentration has fallen to 10 per cent of its original value. (The sum of the ATP and ADP concentrations is slightly less in working muscle because of the AMP kinase reaction discussed below.)

Figure 20–5 plots these concentrations as the total high-energy phosphate store in a muscle falls. The drop in total store is a guide to the power being developed by the muscle.

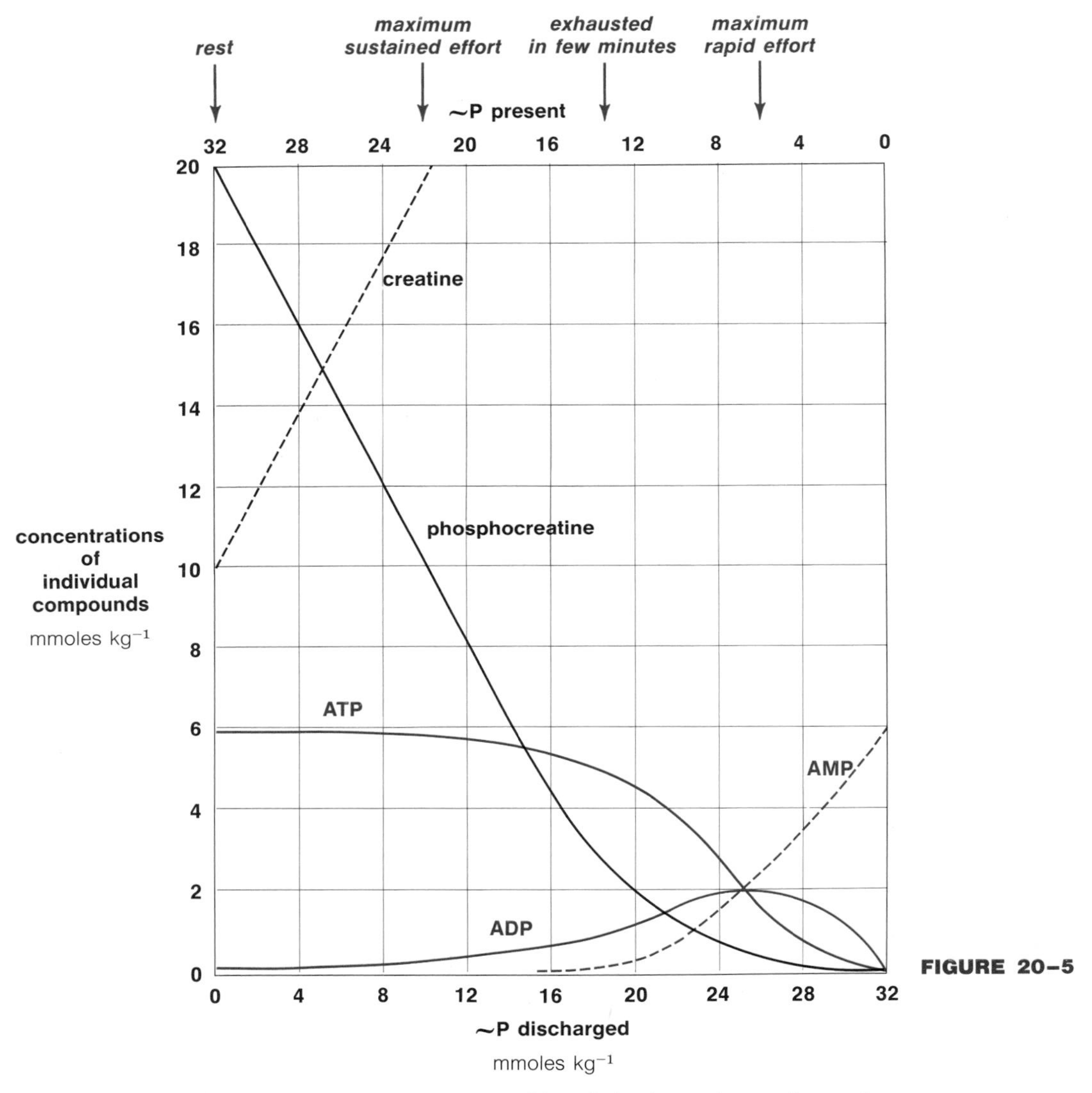

FIGURE 20–5

Changes in the high-energy phosphate pool of working skeletal muscles, as increasing amounts of high-energy phosphate are depleted. The concentration of total high-energy phosphate represents a steady-state balance between the rate of depletion by the muscle power output and the rate of production by utilization of fuels. The greater the work load, the lower the steady-state concentration. Estimates of the concentrations corresponding to various loads are indicated by brown arrows at the top.

Most of the depletion of high-energy phosphate is due to a fall in phosphocreatine concentration even at quite heavy work loads. Marked drops in ATP concentration, with concomitant rises in the concentration of ADP and AMP, occur only when the muscles are working at rates causing exhaustion in minutes, or less.

Note that the use of some compound with a standard free energy of hydrolysis similar to that of ATP for an energy store would not do. Assume a compound X, which is phosphorylated to form P—X by a reaction with $K'_{eq} = 1$:

$$X + ATP \longleftrightarrow P\text{—}X + ADP$$

When [ATP]/[ADP] = 100, as at rest, the concentration of P—X will also be 100 times that of X. However, in order to use half of the stored P—X, the concentration of ATP must also fall to half of its original value. This explains why phosphagens with seemingly unfavorable equilibrium have evolved.

Calculated values were used to make the point because ordinary analyses of muscle samples cannot distinguish between the small concentration of free ADP, which determines the equilibrium, and the larger amount bound to actin and other proteins. Only recently has nuclear magnetic resonance been applied to the measurement of the high-energy phosphate compounds in live and kicking humans; the results to date fully vindicate this application of theory to deduce real physiological events.

The blood concentration of creatine kinase is used clinically as an index of muscle degeneration. It is elevated, for example, after myocardial infarction. An important use is in the early detection of **Duchenne's dystrophy.** This X-linked recessive condition causes deterioration of the muscles and mental retardation in young boys, who die before adulthood. The delayed development and difficulty in moving are often overlooked or dismissed for several years, with the unsuspecting parents conceiving additional afflicted infants. Measurement of the creatine kinase level in all infants who fail to walk by 18 months of age has been proposed as a way to minimize further grief for the parents.

AMP Kinase Equilibrium

Striated muscles contain high concentrations of AMP kinase, catalyzing the reaction:

$$ATP + AMP \xrightarrow{Mg^{2+}} 2\ ADP \qquad K'_{eq} = 1.0$$

so the three adenosine phosphates are kept nearly at equilibrium.

The AMP kinase provides a means of using the second high-energy phosphate bond in ATP for muscular contraction, like this:

$$2\ ATP \xrightarrow{muscular\ contraction} 2\ ADP + 2\ P_i$$

$$2\ ADP \xrightarrow{AMP\ kinase} ATP + AMP$$

$$\text{SUM:} \quad ATP \longrightarrow AMP + 2\ P_i$$

In order for this reaction to contribute a significant amount of high-energy phosphate, the ADP concentration must be substantially elevated. We saw in the preceding section that this can occur only when the phosphocreatine supply is nearly exhausted. Therefore, AMP only becomes a major fraction of the total adenosine phosphates when a muscle is working near its limit, and this is also shown in Figure 20–5.

AMP Deaminase. The amount of high-energy phosphate that can be drained for extreme exertion is increased in another way. An enzyme is present in muscles to catalyze the hydrolytic removal of the amino group in the purine ring of AMP,

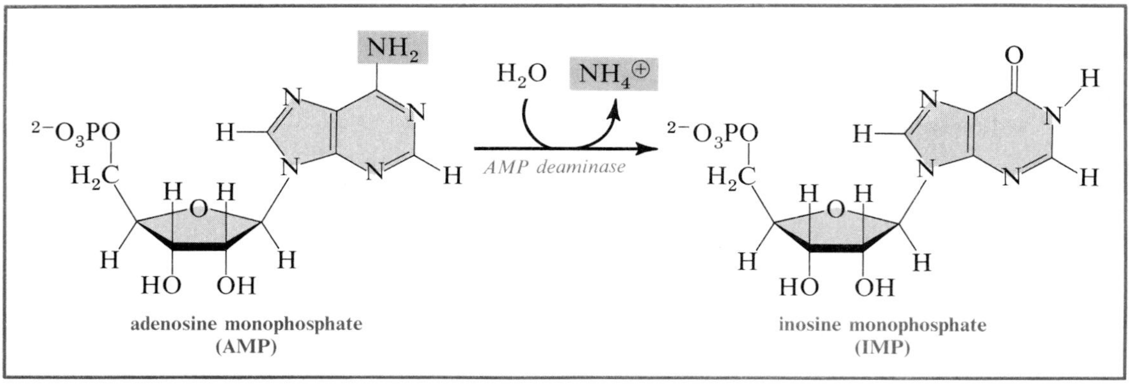

FIGURE 20–6 **AMP deaminase catalyzes the hydrolytic removal of the amino group from adenosine monophosphate as ammonium ion, forming inosine monophosphate.**

thereby converting AMP to inosine monophosphate (Fig. 20–6). This removes one of the products of the AMP kinase reaction and therefore shifts the equilibrium position so as to use more of the high-energy phosphate remaining in ADP. However, this is a rather drastic measure because the adenine ring must be rebuilt by energy-consuming steps, which we will discuss later (Chapter 35).

Microfilaments

Actin and myosin occur within all of the body's cells. The intracellular actin, and perhaps the myosin, differs from that in sarcomeres. Much of it is organized into microfilaments, forming a meshwork through the cytoplasm, but some is in monomeric form. These microfilaments give form and motion to the cell, but the mechanism of motion is not known, nor is the function of the lesser amount of intracellular myosin clear.

FURTHER READING

A. Elliott and G. Offer (1978). Shape and Flexibility of the Myosin Molecule. J. Mol. Biol. 123:505.

K.A. Taylor and L.A. Amos (1981). A New Model for the Geometry of the Binding of Myosin Cross-bridges to Muscle Thin Filaments. J. Mol. Biol. 147:297.

R.S. Adelstein and E. Eisenberg (1980). Regulation and Kinetics of the Actin-Myosin-ATP Interaction. Annu. Rev. Biochem. 49:921.

J. Squire (1981). The Structural Basis of Muscular Contraction. Plenum.

P. Milvy, ed. (1979). The Marathon. Ann. N.Y. Acad. Sci. vol. 301. Contains many informative articles on muscular function. See especially the discussion of fiber types by B. Saltin.

G.A. Sleep and S.J. Smith (1981). Actomyosin ATPase and Muscle Contraction. Curr. Top. Bioenerg. 11:239.

P.F. Dillon et al. (1981). Myosin Phosphorylation and the Cross-Bridge Cycle in Arterial Smooth Muscle. Science 211:495.

M.O. Aksoy, R.A. Murphy, and K.E. Kamm (1982). Role of Ca and Myosin Light Chain Phosphorylation in Regulation of Smooth Muscle. Am. J. Physiol. 242:C109.

D.G. Gadian and G.K. Radda (1981). NMR Studies of Tissue Metabolism. Annu. Rev. Biochem. 50:69.

B.R. Ross et al. (1981). Examination of a Case of Suspected McArdle's Syndrome by ^{31}P Nuclear Magnetic Resonance. N. Engl. J. Med. 304:1338. Example of application of in vivo analysis by NMR to a clinical problem.

S. Rehumnen and M. Harkonen (1980). High-Energy Phosphate Compounds in Human Slow-Twitch and Fast-Twitch Muscle Fibers. Scand. J. Clin. Lab. Invest. 40:45. Example of use of muscle biopsies from volunteers.

D.E. Crisp, F.A. Ziter, and P.F. Bray (1982). Diagnostic Delay in Duchenne's Muscle Dystrophy. J.A.M.A. 247:478.

CHAPTER 21

USE OF HIGH-ENERGY PHOSPHATE: ION TRANSPORT ACROSS MEMBRANES

(Na^+ + K^+)-ADENOSINETRIPHOSPHATASE (ATPase)

We shall discuss in this chapter how the cleavage of ATP to ADP and P_i can generate ionic concentration gradients across cell membranes and some of the uses to which these gradients are put. A prime example of such gradients is the difference in concentration of Na^+ and K^+ across the plasma membrane (Fig. 21–1). **The extracellular environment is rich in Na^+, and the interior of the cell is rich in K^+.** This is true for every cell in the body.

If the plasma membrane were freely permeable to these ions, K^+ would move out of the cell, and Na^+ in, until the ratios of their concentrations inside and outside of the cell became identical. The concentration gradients represent a displacement from equilibrium, and free energy must be expended to create them. This is done by coupling the formation of the gradients with the hydrolysis of ATP to ADP and P_i. A substantial part of the fuel consumption at rest is used to generate high-energy phosphate for this purpose.* Once the gradients are created, they also represent a store of free energy, and we shall see how the discharge of these gradients can be used to drive otherwise endergonic processes.

Sodium and potassium ions are pumped in opposite directions with a simultaneous hydrolysis of ATP by a protein that spans the plasma membrane. This protein is named (Na^+ + K^+)-adenosine-triphosphatase, sometimes shortened to Na,K-ATPase, because isolated preparations hydrolyze ATP, but only in the presence of Na^+ and K^+. As with many membrane proteins, its action has been difficult to study in detail since it must be associated with phospholipid to maintain its activity. Furthermore, x-ray diffraction patterns from which the structure could be deduced are not attainable because lipid-protein complexes do not form rigidly ordered arrays. Much of the surface of the protein components is shielded from combination with reagents that are useful as probes in deducing the structure of more hydrophilic proteins.

*Some estimate that more than half of the energy output of the body at rest is used to maintain the cation concentration gradients. Specialists tend to appropriate most of the ATP for their pet processes, and all such claims add up to much more than the total ATP available, so skepticism is warranted. However, the demand for ATP to maintain ion concentration gradients is clearly not trivial.

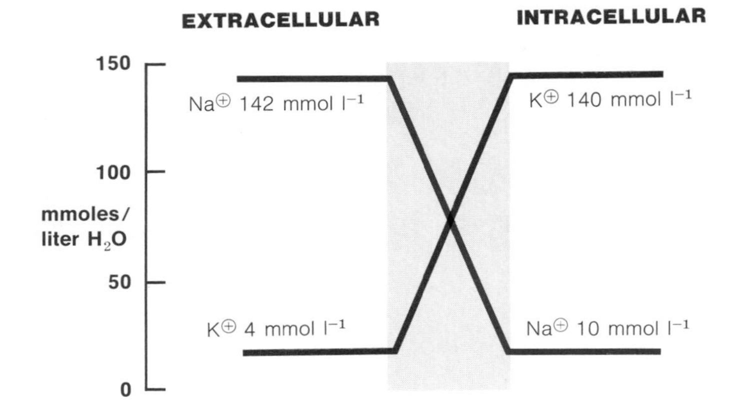

FIGURE 21–1

Large gradients in Na^+ and K^+ concentrations exist across the plasma membranes of cells. The sodium ion concentration is high outside the cells, and the potassium ion concentration is high inside. The cited values are typical for mammalian cells.

These difficulties are dwelt upon because they create similar problems in understanding the action of other membrane-bound systems that we shall encounter. However, in this case and in the others, the important results are clear, even though the detailed mechanism is not.

The overall reaction involves the transport of more Na^+ out than K^+ in, with the ratio approaching 3:2 for each ATP hydrolyzed:

$$3\,Na^{\oplus}_{(in)} + 2\,K^{\oplus}_{(out)} + ATP + H_2O \longrightarrow 3\,Na^{\oplus}_{(out)} + 2\,K^{\oplus}_{(in)} + ADP + P_i.$$

There is an important consequence of the unequal distribution of cations in the two directions: The concentration gradient is used to make a typical cell interior become 50 to 90 millivolts more negative than the external environment. The energy expended by the cation pump is therefore used to create both a chemical potential and an electrical potential across the plasma membrane.

An important feature of the ATPase is that it spans the plasma membrane. It appears to be a tetramer made from two kinds of subunits ($\alpha_2\beta_2$). One kind of subunit has an ATP-binding site exposed within the cell where ATP is available from fuel metabolism for driving the pump. The second kind of subunit is a glycopeptide exposed on the outside surface of the cell. **Digoxin** and similar cardiac glycosides (p. 228) combine with this subunit, and some of their effects may be due to an inhibition of Na^+ and K^+ exchange. (One of the glycosides, ouabain, is frequently used as an experimental tool to block the exchange.) The arrangement of the subunits is otherwise unknown.

The operation of the pump involves a phosphorylation of one of its protein constituents; phosphate is transferred from ATP to an aspartate side chain carboxylate group on the protein, creating a high-energy phosphate-carboxylate anhydride bond (Fig. 21–2).

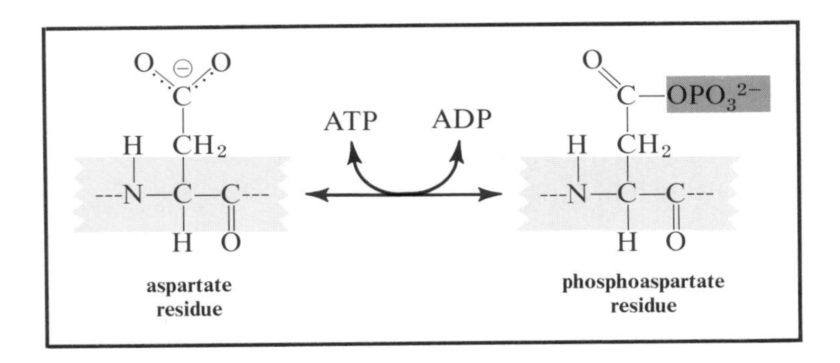

FIGURE 21–2

The action of ($Na^+ + K^+$)-adenosine-triphosphatase (ATPase) involves phosphorylation of an aspartate side chain in the enzyme by ATP. The phosphoaspartate residue formed is an anhydride of a carboxylic acid and phosphoric acid; its free energy of hydrolysis is comparable to that of ATP.

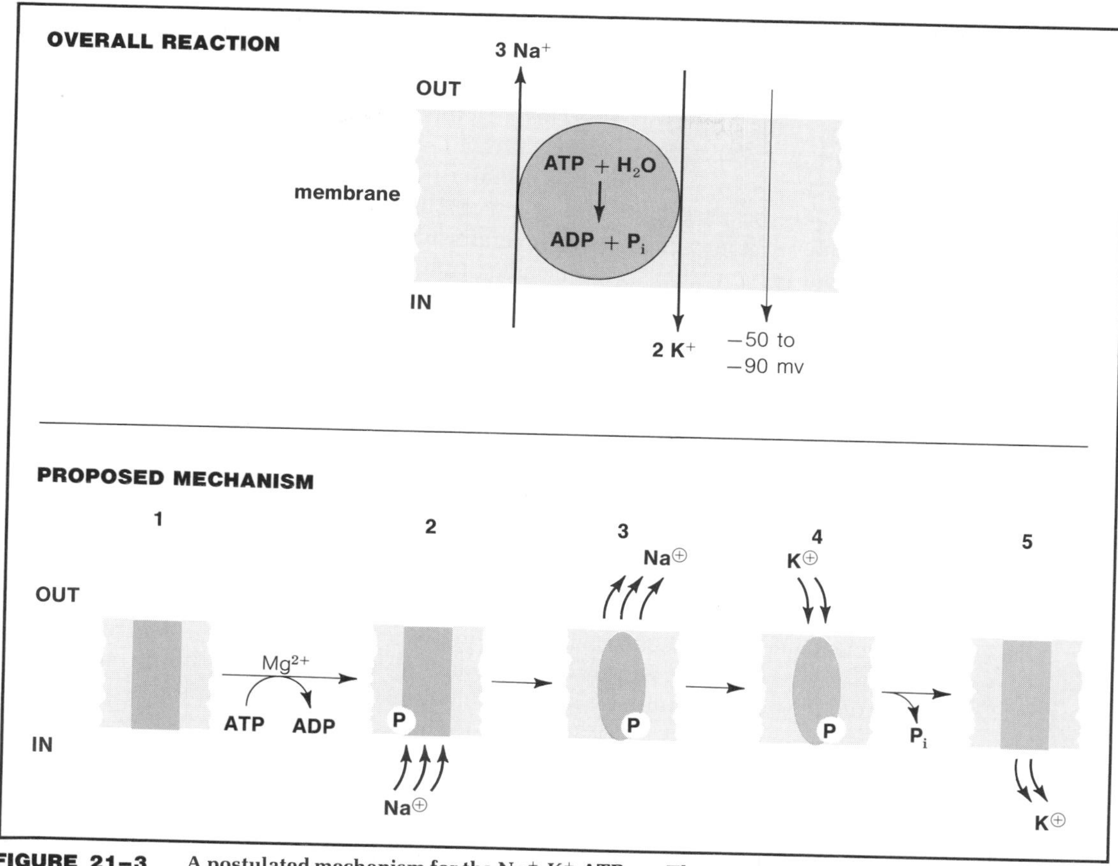

FIGURE 21–3 A postulated mechanism for the Na^+, K^+-ATPase. The overall reaction is shown at the top. When the ATPase (*1, bottom*) binds MgATP in the presence of sodium ions, a phosphate group is transferred from the MgATP to the enzyme (*2*). This causes a conformational shift that exposes the sodium ions to the outside surface and weakens their binding by the enzyme (*3*), while increasing the affinity of the enzyme for potassium ions (*4*). Binding of K^+ causes a release of the phosphate group, and a reversion to the original conformation (*5*). In this conformation, the cation binding site is exposed to the interior of the cell, and has a low affinity for the cation, which is released.

The phosphorylation is associated with a binding of intracellular Na^+ for export. Cleavage of the phosphate group on the protein to form P_i is associated with the final release of K^+ within the cell. In any event, some conformational change must occur upon addition and removal of the phosphate group that changes the affinity of the proteins for the cations, and forces them to move against the electrochemical gradient. Many suggestions have been advanced for the nature of the intervening steps; one is shown in Figure 21–3.

UTILIZATION OF THE Na^+, K^+ GRADIENT

The chemical or the electrical potential differences created by the cation gradient can be used to drive other processes. In order to use the chemical potential, one or both cations must move so as to restore equilibrium concentrations, and the movement must occur in association with the desired effect. When Na^+ does move back into the cell or K^+ escapes, they do so largely by combination with specific transport proteins. In effect, cations that have been pumped uphill by the ATPase now flow downhill through these transport proteins with the energy of the flow made to serve some additional purpose. The phospholipid core of the membranes prevents any sizeable accidental leakage of cations across it.

Nature of the Transport Proteins. Before examining specific examples, let us consider the general nature of the carriers. We are dealing here with facilitated diffusion—the use of a protein to accelerate transport down a concentration gradient, much like the postulated use of myoglobin to move oxygen within a cell. In theory, transport might be aided in several ways (Fig. 21–4). A protein might bind a molecule and move with it from one side of the membrane to the other. The protein might span the membrane and rotate within it so as to expose the binding site, first on one side of the membrane and then on the other. The necessary motions for these two possibilities have been excluded on theoretical grounds and by physical measurements in many instances, leaving a third possibility as the likely choice: A protein may span the membrane and undergo reversible conformational changes that expose a binding site to one side of the membrane or to the other. This is much like the mechanism proposed for the Na^+, K^+-ATPase, except that there is no energy input to favor one conformation or another. Additional speculative possibilities can be devised for selective binding by proteins that will move compounds through the membrane. Such a protein is, in effect, creating a channel, or pore, through the membrane, except that it has the characteristic specificity of binding for particular compounds that we associate with protein action.

Facilitated diffusion can be distinguished from passive diffusion through a truly open hole in the membrane by its specificity and on kinetic grounds. Like enzymes, transport proteins approach saturation with increasing substrate concentrations until further increases cause little change in the rate of transport, whereas the rate of passive diffusion will increase linearly (Fig. 21–5). Active transport, such as that catalyzed by the Na^+,K^+-ATPase, can often be distinguished from facilitated diffusion by its ability to create concentration gradients, that is, active transport can change concentrations or electrical potentials across a membrane away from the equilibrium values.

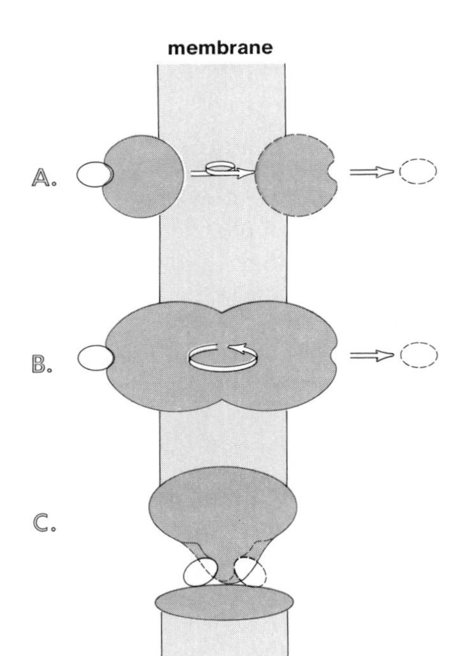

FIGURE 21–4

A transport protein (*brown*) might move a bound molecule through a membrane in any one of several ways. *A.* The protein may move through the lipid core while rotating to expose the binding site on the opposite face. *B.* The protein may extend through the membrane so that simple rotation moves the binding site from one side to the other. *C.* The protein may represent a closed pore or channel that exists in two conformations. A shift from one conformation to the other exposes the binding site to the opposite side of the membrane. Most transport proteins are now believed to function in this or similar ways involving conformational shifts without gross physical movement of the entire protein complex. (The sketch is diagrammatic.)

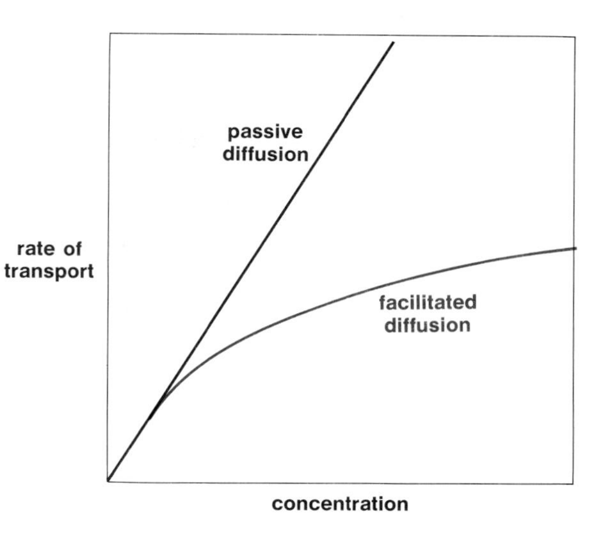

FIGURE 21–5

The use of a transport protein to facilitate diffusion can be distinguished from passive diffusion through an open pore by kinetic measurements. A protein can be saturated, so an increased concentration will cause the rate of transport to approach a maximum velocity, like the rate of an enzymatic reaction. Passive diffusion through an open pore will show a linear increase of velocity with an increasing concentration.

Excitable Tissues

One important use of the Na^+,K^+ gradient is to support excitability in nerves and muscle fibers. Cells in those tissues have specific channels through which the ions can move to discharge the gradients, but the channels are **gated.** That is, their permeability to the ions is altered by a signal. The signal can result from an increased concentration of a specific neurotransmitter (Chapter 33) or from a change in electrical potential across the membrane near the channel. It is the response to the changing potential that makes it possible for a signal to travel down a nerve.

Transmission depends upon the diminution of the electrical potential across a nerve membrane—the membrane is said to be **depolarized**—by allowing Na^+ to flow through specific gated Na^+ channels. In order to understand depolarization, it is necessary to understand the origin of the potential. The existence of different concentrations of ions across a membrane is not enough in itself for the generation of a potential; the ions must also move through the barrier. We are dealing here with **diffusion potentials**—electrical gradients created by the movement of ions down concentration gradients. When an ion moves from one side of the membrane to the other, it is carrying charge and creating a potential, but in order for this to happen, the membrane must be permeable to the ion (Fig. 21–6). The membrane of neurons is constructed so that channels for K^+ are slightly open in the resting state, but the channels for Na^+ are nearly completely closed. The result is that K^+ tends to move out of the neuron at a rate that creates a potential to just balance the concentration gradient for K^+, thereby making the interior of the nerve approximately 85 millivolts negative.

When the nerve is stimulated, the gate of the Na^+ channel adjacent to the site of stimulation opens wide, and the rush of these cations toward the interior of the cell now creates a local potential that is some 30 millivolts negative on the outside of the cell, completely overcoming the potential created by the small current of K^+ in the opposite direction. The shift in potential, the action potential, in some way causes the gates of neighboring Na^+ channels to open. In the ordinary course of propagation, preceding gates will already be open, so the wave of increased conductance goes in one direction. (Propagation may involve displacement of Ca^{2+} near the gates to act in a manner resembling its effect on the contractile apparatus of muscles.)

After the action potential peaks, the Na^+ channel closes, and the gate of a K^+ channel opens wide. K^+ now rushes through to restore the original membrane poten-

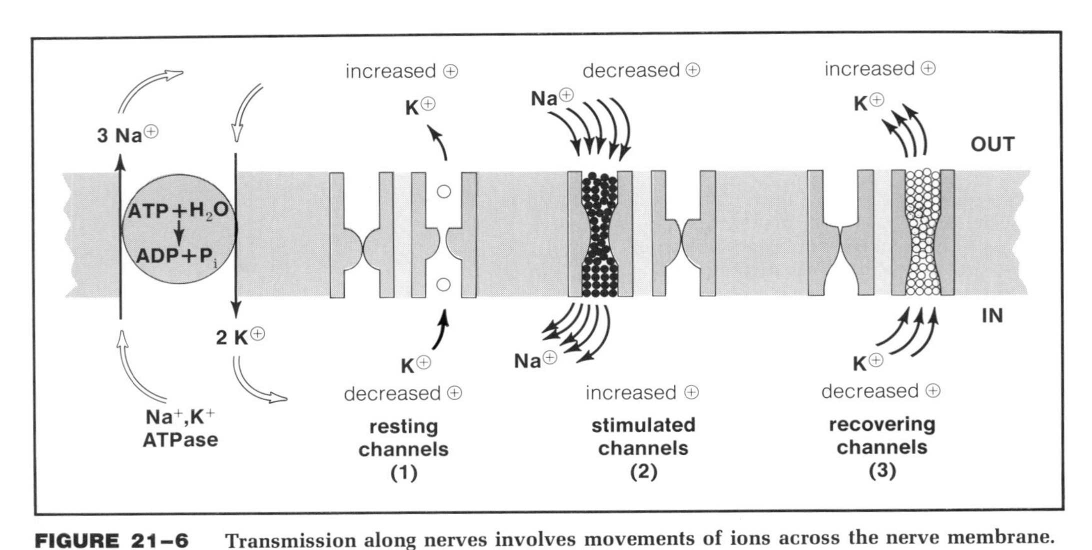

FIGURE 21–6 Transmission along nerves involves movements of ions across the nerve membrane. Energy for transmission is obtained from the concentration gradients for Na^+ and K^+ created by the Na^+, K^+-ATPase (*left*). The nerve contains gated channels through which these gradients can be discharged. (*1*) In a resting nerve, the channel for K^+ is partially open, creating a potential across the membrane that is negative on the inside and positive on the outside. (*2*) When a stimulus is received, the channel for Na^+ is thrown wide open for a brief period. The resultant flux of positive charges from outside to inside creates a sharp potential pulse of opposite polarity—positive inside, and negative outside. (*3*) This is immediately followed by an opening of the K^+ channel and a closing of the Na^+ channel. The rate of flow of K^+ is not as great as the rate of the preceding flow of Na^+, but it is still sufficient to restore the original electrical potential (positive outside, and negative inside) in a relatively short time.

tial. The maximum flux of K^+ is still much less than the maximum flux of Na^+, so recovery is slower than the initial activation: this is an important feature for unidirectional propagation. Preceding channels cannot refire until some time after the passage of the wave of stimulus.

The total displacement of ions during the cycle of stimulation and recovery is only a small fraction of the concentration gradients. If the ATP supply is stopped so that a nerve cannot replenish the Na^+,K^+ gradient by action of the ATPase, it is still capable of firing a large number of times before the existing gradient is exhausted.

Active Transport

It is often desirable for cells to accumulate compounds in excess of the concentrations found in the external environment. For example, the intestinal mucosa absorbs glucose and some neutral amino acids from the lumen even when the concentration of these compounds is well below the concentration in the cell. This permits the recovery of the greatest possible amount of these important metabolites before they are destroyed by the large bacterial populations in the lower bowel or are discharged in the feces.

Creating these concentration gradients requires energy, and the energy is supplied by a simultaneous discharge of the Na^+ concentration gradient. This gradient is created by the ATPase pump on the serosal side, away from the lumen, of the mucosal cells. The two processes are coupled by constructing transport proteins that interact both with Na^+ and with the desired compound (glucose or amino acid) in

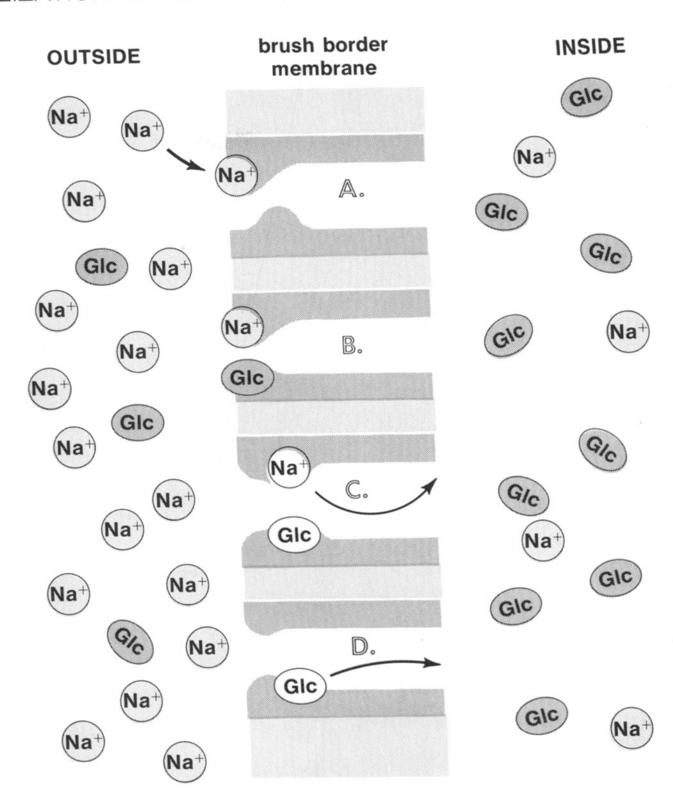

FIGURE 21–7

A possible mechanism whereby the concentration gradient for Na^+ across the brush border membrane in the intestinal mucosa is used to pump glucose into the cell from the intestinal lumen. A transport protein is used that binds both glucose and Na^+. *A.* Binding of Na^+ is not sufficient for transport of the cation. *B.* The binding of Na^+ increases the affinity of the protein for glucose, which is bound to a second site. *C.* The binding of both Na^+ and glucose facilitates equilibration with a conformation in which the binding sites are exposed to the other side of the membrane, where the $[Na^+]$ is low. *D.* Loss of Na^+ weakens the affinity for glucose, which also leaves the protein, even though its concentration is relatively high in the neighboring solution.

The necessary ionic concentration gradient is generated by the Na^+, K^+-ATPase in the basal-lateral membrane of the mucosal cells (not shown). Glucose can cross that membrane by facilitated diffusion, since its concentration inside the cell has been made higher than the concentration in blood.

such a way that one crosses the cell membrane only slowly without the other (Fig. 21–7). The equilibrium position for transport by this mechanism is achieved only when the potential gradient (chemical + electrical) for glucose (or amino acid) is equal and opposite to the potential gradient for Na^+.

We see here an example of **symport**—a simultaneous transport of two compounds in one direction. We will see in later chapters examples of **antiport,** in which the movement of a compound in one direction drives the movement of a second compound in the opposite direction (Fig. 21–8), but here, too, one concentration gradient is being used to create the other. (Transport devices for a single compound are sometimes called **uniporters.**)

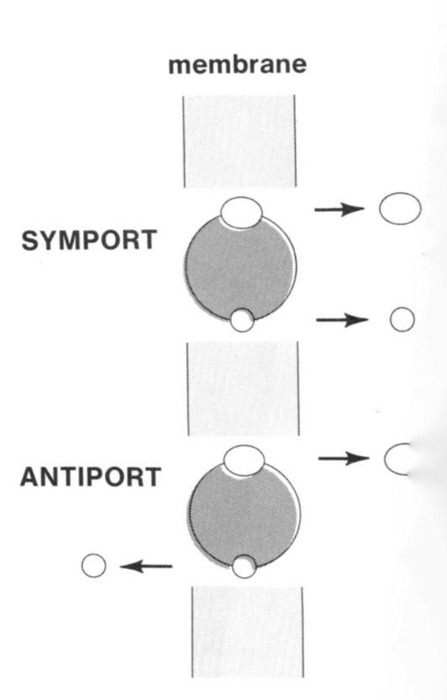

FIGURE 21–8

A concentration gradient for one compound can be made to create a concentration gradient for another compound in two ways. Symporters will transport both compounds in the same direction through a membrane, with the flow of one down its pre-existing gradient carrying the other up a gradient. The Na^+-coupled transport of glucose into the intestinal mucosa is an example of symport. Antiporters similarly couple transport of one compound to transport of the other, but in the opposite direction.

The association between intestinal absorption of glucose and sodium ions led to the development of a technique for the management of **cholera.** This infection causes diarrhea so severe that the fecal stream issues from the anus like water from a faucet; the depletion of water and salts may be lethal. However, a constant infusion of glucose into the gut by means of a tube through the mouth causes a reabsorption of salt and associated water with a dramatic reduction in the amount of intravenous fluid required to maintain water and salt balance.

The kidney utilizes the same kind of mechanism for recovery of glucose and some neutral amino acids from filtered blood plasma as it passes through tubules in the organ. Some specialized cells have a brush border in the lumen; the other sides of these cells next to the blood vessels have a highly active Na^+,K^+-ATPase creating a gradient for Na^+ from the cells to the blood stream, and from the lumen of the tubules into the cells (Fig. 21–9) in the same way that similar cells do in the intestine. Recovery of Na^+ from the tubules can therefore occur by facilitated diffusion, utilizing a Na^+ carrier. (We shall discuss regulation of salt and water balance in Chapter 38.) Specific symporters use the Na^+ gradient for recovery of glucose and amino acids from the filtrate.

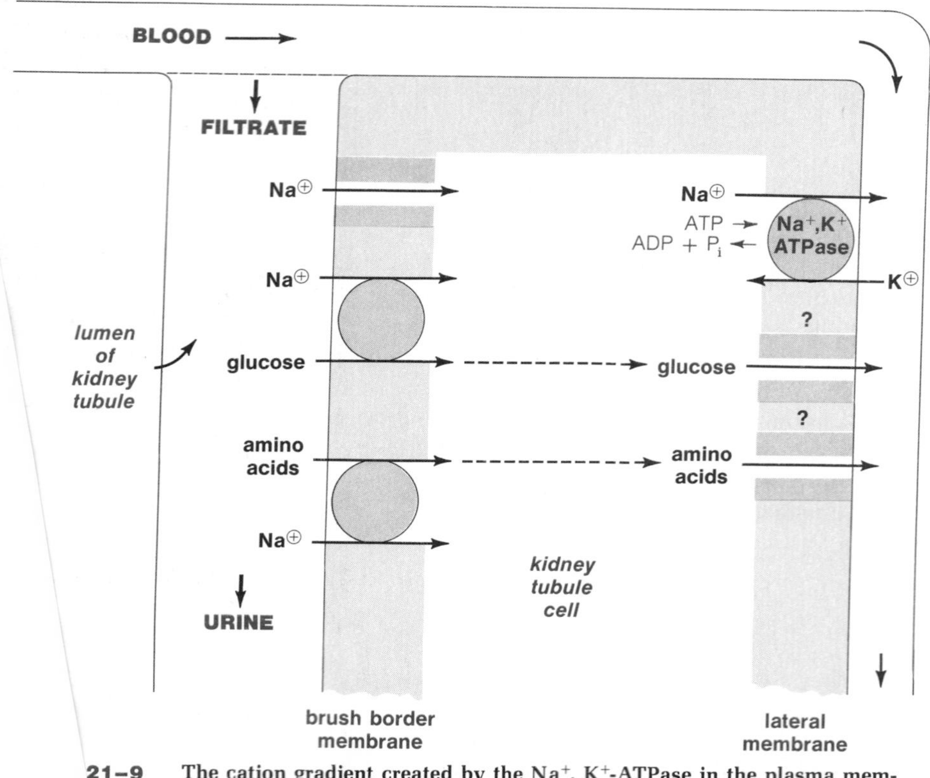

21–9 **The cation gradient created by the Na^+, K^+-ATPase in the plasma membrane is used by kidney cells during the conversion of a blood filtrate into urine. The filtrate (*left*) passes over the brush border membranes of cells in kidney tubules. Since it has a higher Na^+ concentration than the interior of the cells, this gradient can be used to remove glucose and amino acids from the filtrate by simultaneous transport of Na^+, using a symporter. Additional Na^+ can be recovered by facilitated diffusion through uncharacterized channels.**

The low Na^+ concentration within the cells is created by Na^+, K^+-ATPase in the basal and lateral membranes. These same membranes have carriers for passage of glucose and amino acids into the extracellular space, and thence to the blood, by facilitated diffusion.

Associated Anion Gradients

We have concentrated on the movement of cations, but each side of a membrane must have negative charges almost equal in number to the positive charges. The electrical potentials represent only a minute quantitative imbalance, and it follows that any net movement of cations must be associated with an equivalent movement of anions. This movement is acomplished through specific channels for Cl^- and HCO_3^-. The membranes of the cell are nearly impermeable to many other anions, such as phosphate esters, polycarboxylate compounds, and the like. Such exchanges of these compounds as do occur involve relatively slow symport and antiport mechanisms.

(Ca^{2+} + Mg^{2+})-ADENOSINETRIPHOSPHATASE

An essential feature of muscular contraction is the storage of Ca^{2+} in the sarcoplasmic reticulum at concentrations much greater than those in the cytoplasm around the contractile filaments. Stimulation of contraction involves opening channels so that these calcium ions can pour out rapidly. The recovery phase of contraction involves pumping the Ca^{2+} back into the sarcoplasmic reticulum. This is accomplished by an ATP-driven mechanism that in many respects resembles the mechanism for pumping Na^+ and K^+. Ca^{2+} is pumped inside the reticulum and Mg^{2+} is pumped out, and the apparent stoichiometry is:

$$2\ Ca^{2+}_{(out)} + ATP \longrightarrow 2\ Ca^{2+}_{(in)} + ADP + P_i.$$

The pump consists of two components; one reacts with ATP outside the reticulum and the other is a protein-lipid component that may serve as a channel through the lipid core of the membrane to the inside of the reticulum. The mechanism of the transfer involves phosphorylation of an aspartyl residue, as in the Na^+,K^+-ATPase.

The storage of Ca^{2+} within the sarcoplasmic reticulum is made easier by the presence of **calsequestrin,** a calcium-binding protein on the interior surface of the membrane. This protein is very rich in carboxylate side chains (Glu and Asp), which chelate Ca^{2+} and diminish its effective concentration while keeping it readily available for use through dissociation.

FURTHER READING

S.R. Levinson (1981). The Structure and Function of the Voltage-Dependent Sodium Channel. p. 315 in T.P. Singer and R.N. Ondarza, eds. The Molecular Basis of Drug Action. Elsevier.

R.L. Post (1981). The Sodium and Potassium Ion Pump. p. 299 in Singer and Ondarza.

K.J. Sweadner and S.M. Goldin (1980). Active Transport of Sodium and Potassium Ions: Mechanism, Function, and Regulation. N. Engl. J. Med. 302:777. Review.

L.C. Cantley (1981). Structure and Mechanism of the (Na,K)-ATPase. Curr. Top. Bioenerg. 11:201. Review.

A.A. Martonosi, ed. (1980). Calcium Pumps. Fed. Proc. 39:2401. Symposium papers.

U. Hopfer and R. Groseclose (1980). The Mechanism of Na-Dependent D-Glucose Transport. J. Biol. Chem. 255:4453.

CHAPTER 22

OXIDATIONS AND PHOSPHORYLATIONS

We have shown in preceding chapters how the raw materials of metabolism—oxygen and food—are made available. Now we turn to their reactions. For what purposes is oxygen used?

(1) **Most of the oxygen is used to burn fuels for the generation of ATP from ADP and P_i.** A single kind of enzyme complex in the inner membrane of mitochondria catalyzes this reaction of oxygen as an ultimate electron acceptor. The mitochondria have aptly been termed the powerhouses of the cell.

(2) **Oxygen is consumed to alter compounds for structural purposes.** The alteration may be an oxygenation, such as the conversion of proline residues in collagen

precursors to hydroxyproline residues (p. 162), or a dehydrogenation, such as the introduction of double bonds into fatty acid chains, as we shall see later (p. 534). Different enzymes catalyze the various reactions of this type; many are found in the endoplasmic reticulum, but some are in the mitochondrial membranes as well as other organelles.

(3) **Oxygen is consumed to alter compounds for protective purposes.** For example, white blood cells (leukocytes) use oxygen to kill ingested bacteria. In other cases, potentially dangerous compounds are converted to innocuous derivatives through oxidations. The leukocyte enzymes are located on the plasma membrane; the detoxifying enzymes, of which there are many, are mostly on the endoplasmic reticulum.

Almost all of the oxygen-consuming reactions occur on membranes. There are some exceptions, but in general the enzymes using oxygen, a hydrophobic substrate, are associated with lipid.

OXIDATIVE PHOSPHORYLATION

The formation of ATP by phosphorylation of ADP is a concomitant of the combustion of fats, of carbohydrates, and even of protein. The total oxidations of a typical fat and of glucose go like this in the laboratory:

$$\text{(fat)}\quad C_{55}H_{102}O_6 + 77\tfrac{1}{2}\,O_2 \longrightarrow 55\,CO_2 + 51\,H_2O$$

$$\text{(glucose)}\quad C_6H_{12}O_6 + 6\,O_2 \longrightarrow 6\,CO_2 + 6\,H_2O.$$

Cells modify these reactions so as to cause the phosphorylation of a large number of molecules of ATP per molecule of fuel consumed; the overall biological process can be appoximated by these equations:

$$\text{(fat)}\quad C_{55}H_{102}O_6 + 77\tfrac{1}{2}\,O_2 + 437\,(ADP + P_i) \longrightarrow 55\,CO_2 + 437\,ATP + 488\,H_2O$$

$$\text{(glucose)}\quad C_6H_{12}O_6 + 6\,O_2 + 36\,(ADP + P_i) \longrightarrow 6\,CO_2 + 36\,ATP + 42\,H_2O.$$

How is this accomplished? Each of these processes must represent the summation of many individual reactions, because hundreds of molecules simply don't collide simultaneously in one grand reaction. The combustions proceed in small steps, with the formation of many successive intermediate compounds catalyzed by individual enzymes. However, all of the associated generation of high-energy phosphate is accomplished by the action of only four different enzymes. Three of these catalyze **substrate-level phosphorylations,** that is, the generation of high-energy phosphate by reactions of the fuel metabolites themselves, without direct participation of oxygen. These are discussed in the next chapters. The fourth enzyme, with which we are now concerned, is the one responsible for **oxidative phosphorylation,** that is, the generation of pyrophosphate bonds coupled to the reduction of oxygen. This enzyme forms over 85 per cent of the high-energy phosphate consumed in the course of a day by the entire body.

The Role of Membranes. The process of oxidative phosphorylation is localized in membranes in all kinds of cells in which it occurs. Oxidative phosphorylation occurs in the inner membranes of mitochondria in animals and in other eukaryotes, including fungi and the higher plants. In bacteria, it occurs in the cytoplasmic membrane, the innermost of the envelopes enclosing the cell.

MITOCHONDRIAL MEMBRANES. A mitochondrion consists of two sacs, one inside the other. The outer membrane is relatively unwrinkled, but the inner mem-

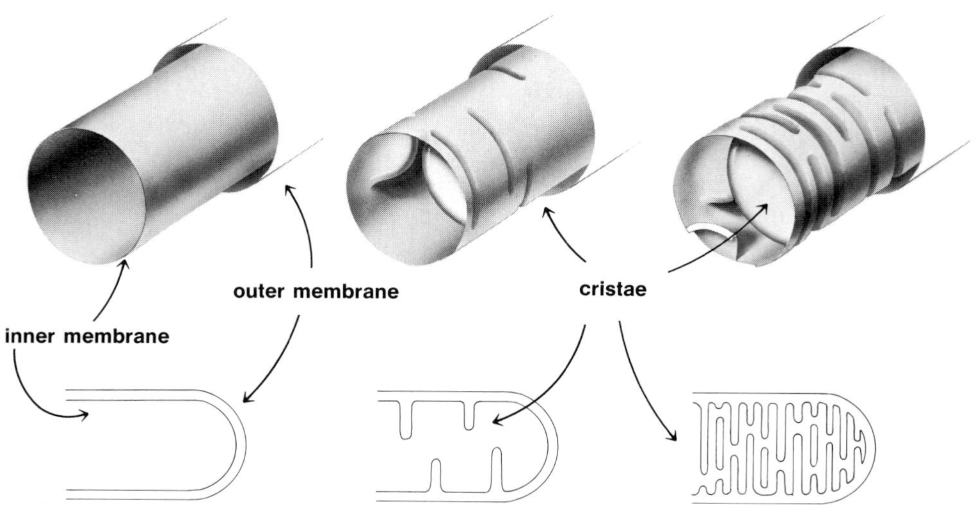

FIGURE 22–1

Idealized gross structure of mitochondria, shown in cutaway view and in longitudinal section. The structure is made from two concentric bags, and the membrane comprising the inner bag may be indented into ridges, tubes, or pouches, called cristae. As shown, there may be few cristae or there may be so many that they nearly fill the interior, with cross sections appearing to have nearly even laminations.

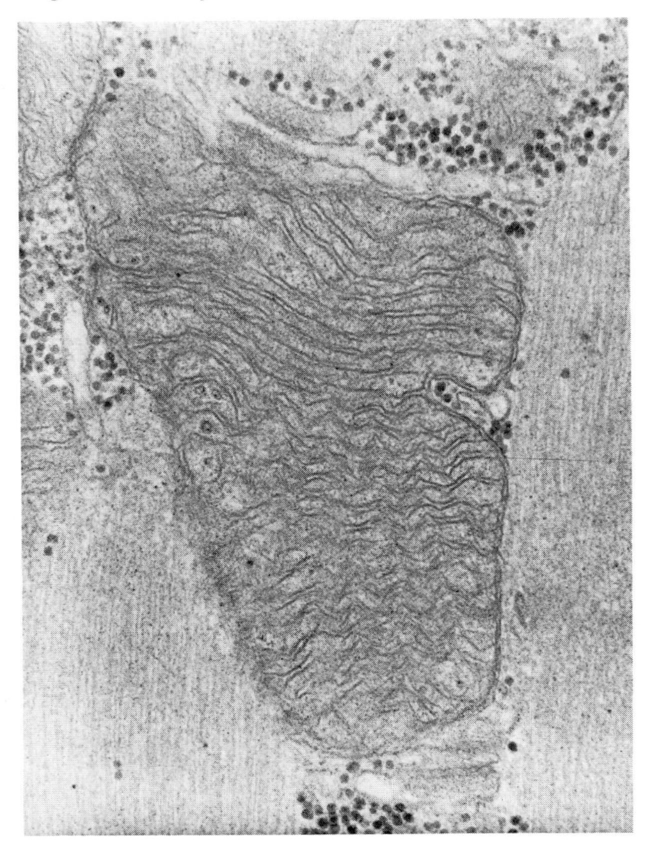

FIGURE 22–2

A mitochondrion in rat thigh muscle. The mitochondrion is enclosed within the outer membrane, and is filled with cristae of the inner membrane. The mitochondrion is intimately associated with muscle fibers to either side. More mitochondria lie in a band above and below the one seen here. The large surface of the inner membrane is characteristic of mitochondria in tissues having a high oxygen consumption, such as muscles built for extended periods of work. Compare the mitochondria in liver (p. 501). Magnification 60,000×. (Courtesy of Dr. Carlo Bruni.)

brane has bulges into the center of the mitochondrion, which greatly extend its entire surface. The number and shape of these bulges, which are known as cristae, vary from one kind of a cell to another, even in differing cells of the same organism (Figs. 22–1, 2), but in all cases the membrane is continuous. The material enclosed by the inner membrane is the matrix of the mitochondrion.

The outer membrane is freely permeable to molecules up to about 400 M.W. and has a relatively small number of associated enzymes. A few other enzymes are located in the intermembrane space, but most of the catalytic activity of mitochondria is found within the inner membrane or its enclosed matrix. The inner membrane contains the enzymes responsible for transferring electrons to oxygen and for the associated phosphorylation of ADP to create ATP. The matrix is rich in other enzymes, many of which are necessary for the catabolism of fuels.

General Mechanism of Oxidative Phosphorylation

Oxidative phosphorylation is associated with the appearance of a proton concentration gradient across the membrane where it occurs. The easiest explanation for this association, and the one most widely accepted, is that the reduction of oxygen to water creates the proton concentration gradient, which represents a store of potential energy, and this energy is converted (transduced) into the form of high-energy phosphate by moving protons back down the gradient to the region of lower concentration (Fig. 22–3).

The gradient is frequently referred to as **electrochemical** because it contains potential energy in two forms, the proton concentration difference and the electrical potential also created by displacement of these positively charged ions. In mitochondria, oxygen consumption pumps protons out of the matrix toward the soluble cytoplasm (cytosol), making the matrix more negative and alkaline than the cytosol. The generation of ATP diminishes the differences in potential and H^+ concentration because it requires proton movement from the cytosol side toward the matrix side to drive it.

A CAVEAT. It is not clear if protons must be pumped completely across a membrane or if they only shift position within it for phosphorylation to occur, with the external changes in pH being a secondary phenomenon. For our purposes it makes no difference because the proton gradient and high-energy phosphate are *de facto* interchangeable; a depletion of one is equivalent to a depletion of the other, and it simplifies the story to regard a complete transfer of H^+ as an obligatory step between oxygen reduction and high-energy phosphate generation.

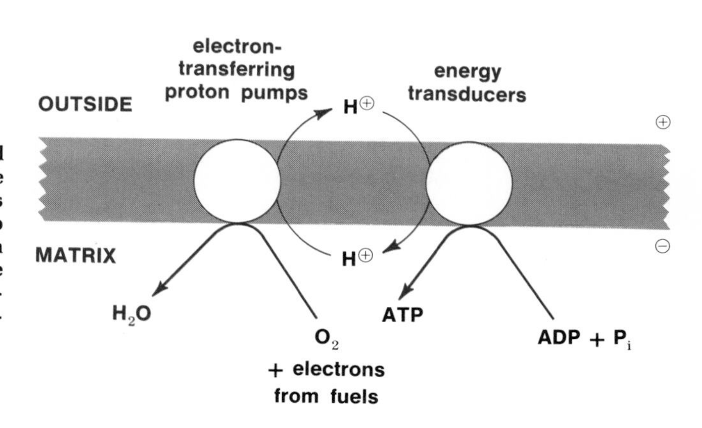

FIGURE 22–3

Energy transduction in the mitochondrial inner membrane. Enzyme complexes use the energy of reduction of oxygen by electrons from fuels to pump protons from the matrix to the cytosol surface. The protons return through another enzyme complex that uses the energy of their passage from the high-concentration, positive potential outer side to synthesize ATP from ADP and P_i.

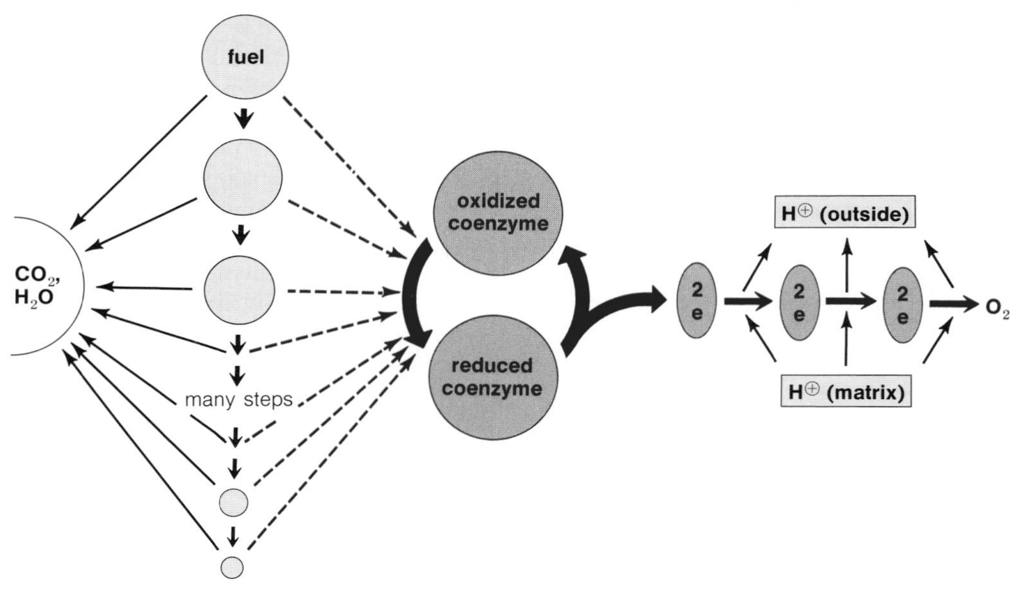

FIGURE 22–4

Many different fuels are attacked by the same oxidized coenzymes. The fuels are chipped away, two electrons at a time, with their substance gradually appearing as CO_2 and H_2O. The coenzymes become reduced, and their oxidized forms are regenerated by transfer of electrons to molecular oxygen through as many as three intermediate enzyme complexes. These transfers are highly exergonic by themselves, and each drives the pumping of protons from the matrix across the inner membrane, creating an electrochemical gradient.

The oxidation of a typical substrate by oxygen liberates enough free energy to sustain the formation of two or three molecules of ATP for each atom of oxygen consumed. The transfer of electrons from the substrate so as to reduce oxygen to water is therefore broken up into a maximum of three steps. Here we have another example of the beautiful economy of design often achieved by evolution. Three proton-pumping enzyme complexes capture energy from the reduction of oxygen; these suffice even though electrons for the reduction are obtained from many different fuel metabolites (Fig. 22–4). Each complex pumps enough protons across the membrane to drive the formation of one molecule of ATP from ADP and P_i per pair of electrons transferred. (Reduction of oxygen to water requires two electrons per oxygen atom.)

The degradation of fuels is likewise broken into many steps so that electrons are supplied a pair at a time to feed the sequence of proton pumps.

The Electron Transfers

The three proton-pumping enzyme complexes catalyze these successive electron transfers:

$$\text{NAD} \longrightarrow \text{ubiquinone} \longrightarrow \text{cytochrome c} \longrightarrow O_2.$$

We discussed NAD (nicotinamide adenine dinucleotide) earlier (p. 290), and ubiquinone and cytochrome c are described later. Each of these transfers increases the electrochemical gradient enough to generate one ATP from ADP and P_i.

Electrons are fed into this sequence by oxidizing compounds derived from dietary fuels. These oxidations are catalyzed by dehydrogenases, each specific for a few, perhaps only one, of the fuel metabolites. Some use NAD as the oxidizing agent, while others use ubiquinone directly. The reason for bypassing NAD is as follows: The reaction of some substrates with oxygen liberates only enough energy for the formation of two molecules of ATP. Dehydrogenases attacking these substrates bypass one of the three proton pumps by transferring their electrons directly to ubiquinone (Q):

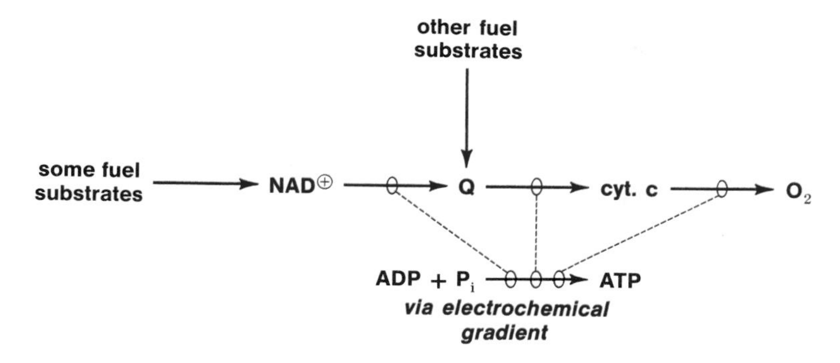

These sequences of electron flow are nothing more than a series of oxidation-reduction reactions, each of which can be represented as:

$$\textbf{reduced A + oxidized B} \longrightarrow \textbf{oxidized A + reduced B.}$$

This is a typical reaction of the sort we discussed on pages 288 to 290. Let us go through it with a specific example, the oxidation of malate to oxaloacetate (Fig. 22–5).

We shall see in the next chapter that the oxidation of all kinds of fuels results in the formation of malate, which is then oxidized to oxaloacetate. (The structures are not important for the moment.) The oxidation of malate by oxygen:

$$\text{malate} + \tfrac{1}{2}\,O_2 \longrightarrow \text{oxaloacetate} + H_2O$$

liberates enough energy to form three molecules of ATP per malate oxidized (six per molecule of oxygen consumed). Efficient capture of this energy therefore requires

$COO^{\ominus}$–CHOH–CH_2–$COO^{\ominus}$ L-malate; NAD; NADH; QH_2; Q; 2 (cytochrome c); 2 (reduced cytochrome c); H_2O; $\tfrac{1}{2}O_2$; $COO^{\ominus}$–C=O–CH_2–$COO^{\ominus}$ oxaloacetate; H⊕ (matrix); H⊕ (outside)

FIGURE 22–5 **The oxidation of malate.**

that the electrons from malate first be transferred to NAD on their way to oxygen. A malate dehydrogenase in the mitochondrial matrix catalyzes this reaction:

$$\text{malate} + \text{NAD}^{\oplus} \xrightleftharpoons{\textit{malate dehydrogenase}} \text{oxaloacetate} + \text{NADH} + \text{H}^{\oplus}$$

This is a freely reversible reaction, and all of these reactants and products are free to move about in the matrix. The NADH released will sooner or later encounter the first of the proton-pumping complexes on the inner membrane. When it does, its electrons are transferred to ubiquinone, which is dissolved in the lipid core of the membrane:

$$\text{NADH} + \text{H}^{\oplus} + \text{Q} \longrightarrow \text{NAD}^{\oplus} + \text{QH}_2$$

$$\text{H}^{\oplus}\ (\text{matrix}) \rightarrow \text{H}^{\oplus}\ (\text{outside})$$

Q = ubiquinone, and QH_2 = ubiquinol, its reduced form. (Note that we are doing separate bookkeeping on the protons required to balance the oxidation-reduction reaction and those pumped across the membrane.) The transfer regenerates the limited supply of NAD necessary for the original oxidation of malate. Ubiquinol dissolved in the lipid core of the membrane is accessible to the second proton-pumping enzyme complex, which transfers electrons from ubiquinol to cytochrome c. Each molecule of cytochrome c can carry only one electron, so two molecules are required per pair of electrons transferred:

$$\text{QH}_2 + 2\,\overset{(\text{Fe(III)})}{(\text{cytochrome c})} \longrightarrow \text{Q} + 2\left(\overset{(\text{Fe(II)})}{\text{reduced cytochrome c}}\right) + 2\,\text{H}^{\oplus}$$

$$\text{H}^{\oplus}\ (\text{matrix}) \rightarrow \text{H}^{\oplus}\ (\text{outside})$$

The third proton-pumping enzyme complex transfers electrons from reduced cytochrome c to oxygen. Four electrons are required per molecule of oxygen reduced:

$$4\left(\overset{(\text{Fe(II)})}{\text{reduced cytochrome c}}\right) + \text{O}_2 + 4\,\text{H}^{\oplus} \longrightarrow 4\,\overset{(\text{Fe(III)})}{(\text{cytochrome c})} + 2\,\text{H}_2\text{O}$$

$$\text{H}^{\oplus}\ (\text{matrix}) \rightarrow \text{H}^{\oplus}\ (\text{outside})$$

In all of this sequence, only malate and oxygen are consumed, except for the associated pumping of protons across the mitochondrial inner membrane. All of the other components are coenzymes or proteins available only in fixed amounts within the membrane, and they must be constantly recycled from the oxidized form to the reduced form and back to the oxidized form to handle the continual flow of fuel metabolites passing through the mitochondria. (The amounts of these membrane components do change, but hours are required for significant effects compared to the many turnovers per second that occur during fuel metabolism.) The overall reaction therefore is:

$$\text{malate} + \tfrac{1}{2}\,\text{O}_2 + n\,(\text{protons inside}) \longrightarrow \text{oxaloacetate} + \text{H}_2\text{O} + n\,(\text{protons outside})$$

The pumped protons can then be used to generate ATP:

$$n\,(\text{protons outside}) + 3\,(\text{ADP} + \text{P}_i) \longrightarrow n\,(\text{protons inside}) + 3\,\text{ATP} + 3\,\text{H}_2\text{O}$$

This makes the overall reaction for the oxidation of malate become:

$$\text{malate} + \tfrac{1}{2}\,\text{O}_2 + 3\,(\text{ADP} + \text{P}_i) \longrightarrow \text{oxaloacetate} + 4\,\text{H}_2\text{O} + 3\,\text{ATP}.$$

The **P:O ratio** for this reaction, that is, the moles of high-energy phosphate generated per atom of oxygen consumed, is 3.0. This ratio is commonly used to express the effective energy yield from oxidations; that is why the reactions are cast per atom of oxygen consumed (one-half molecule of O_2). The practice also simplifies bookkeeping in terms of fuel consumption.

Yield of Pumped Protons. It is not known exactly how many protons are moved across the inner membrane per pair of electrons passing through the enzyme complexes. Some experts insist four H^+ are moved at each of the three proton-pumping sites when a pair of electrons moves through them, and that four H^+ move back inside the mitochondrion for each ATP generated, which makes the P:O ratio equal 3.0 for the oxidation of malate and of other metabolites from which electrons are transferred to NAD. We shall use these values as a basis for further discussion. To summarize, we shall assume:

(1) 4 H^+ are pumped from the matrix side to the cytosol side of the mitochondrial inner membrane:
- (a) each time NADH is oxidized by ubiquinone,
- (b) each time ubiquinol is oxidized by 2 cytochrome c, and
- (c) each time 2 reduced cytochrome c are oxidized by oxygen.

(2) 4 H^+ move from the cytosol side to the matrix side of the inner mitochondrial membrane each time ADP and P_i combine to form ATP. If the proton gradient is discharged in any other way, then the movement of each H^+ toward the matrix is equivalent to the loss of 0.25 potential ATP.

Now let us consider the components of this system in more detail.

NADH Dehydrogenase

The transfer of electrons from NADH in the mitochondrial matrix to ubiquinone in the membrane core, and the accompanying pumping of protons, is catalyzed by a highly organized enzyme complex, NADH dehydrogenase. This complex includes a **flavoprotein** and **iron sulfide proteins** as electron carriers. The mechanism of proton pumping is not known for this, or the other electron-transferring complexes, but it presumably requires some specific organization of the components in the complexes.

The Flavoprotein. Flavoproteins contain derivatives of **riboflavin,** a vitamin, as electron-carrying coenzymes. The oxidized form of riboflavin has an intense yellow color (flavus means yellow), which is characteristic of flavoproteins. The reduced form is colorless.

In general, the flavin coenzymes are tightly bound to their apoproteins so that a cycle of reduction and oxidation must be completed by each enzyme molecule before it can begin to repeat its catalytic mechanism. (Contrast the behavior of any dehydrogenase catalyzing the reduction of NAD in the mitochondrial matrix; the NADH that is formed wanders off to be oxidized by another enzyme, and the dehydrogenase can in the meantime react with a different molecule of NAD.)

Flavoproteins often function as bridges between two-electron and one-electron carriers because they can form quite stable **semiquinones,** which are free radicals

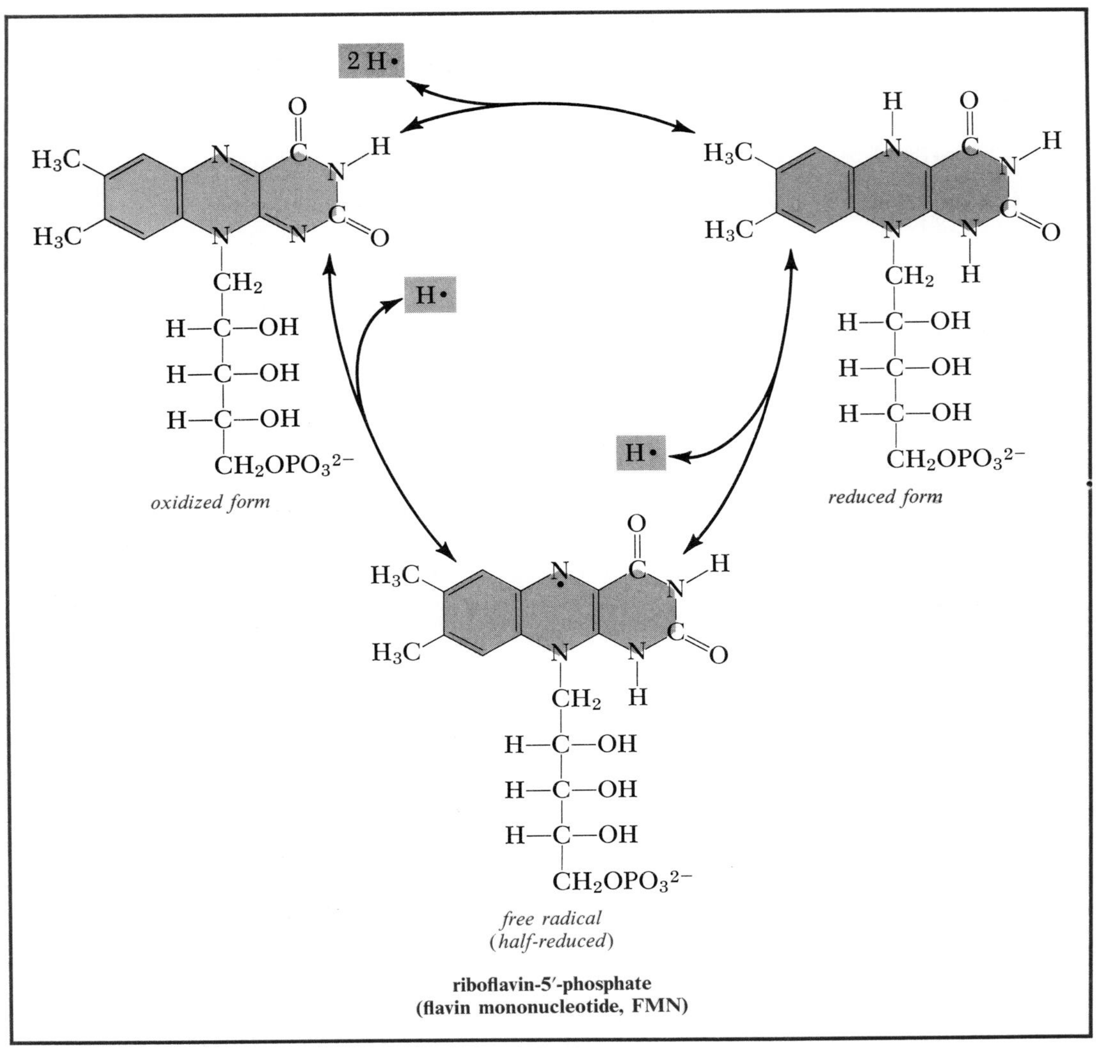

FIGURE 22–6 **The flavin coenzymes may transfer either two electrons (*top*), or one electron. The transfer of one electron involves an intermediate free radical, or semiquinone form (*bottom*), and it may occur in either of two ways (*right* or *left*).**

with flavin reduced by only one electron, as well as the fully reduced form, which carries two electrons (Fig. 22–6). In NADH dehydrogenase, the flavoprotein is exposed on the matrix surface of the membrane where it can oxidize NADH to NAD while the flavin becomes fully reduced by a pair of electrons. These electrons can then be transferred one at a time to iron-sulfide proteins by intermediate formation of the semiquinone.

FIGURE 22–7 **Iron-sulfide proteins occur in two principal forms, one with two atoms each of iron and sulfide (*top*), and one with four atoms each (*bottom*). Each iron atom is also bonded to the sulfur of cysteine side chains in the protein, making a total of four sulfur atoms linked to each iron atom. These structures are established for bacterial and plant iron-sulfide proteins (ferredoxins), but the structures of those from mitochondria are not known.**

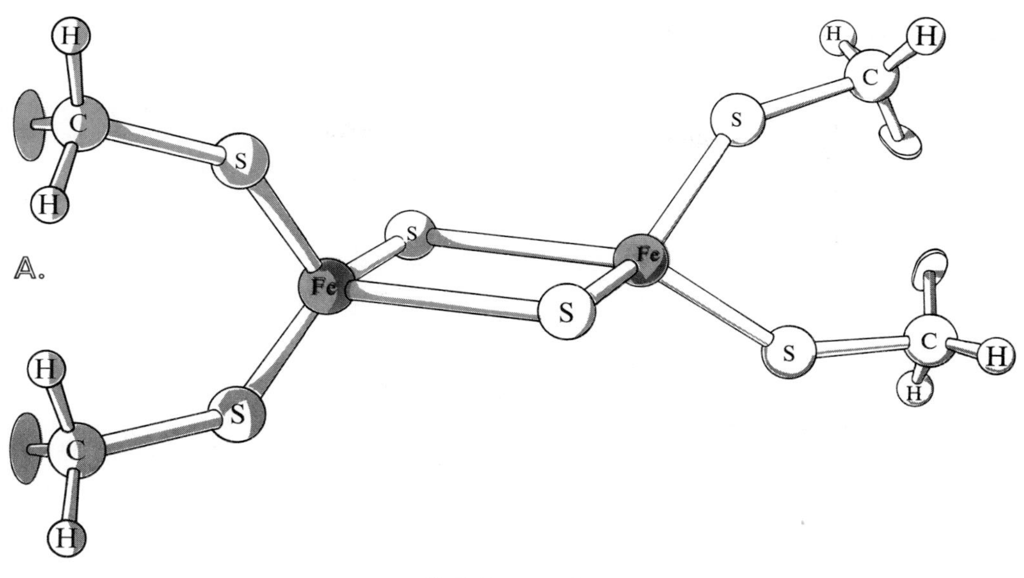

$Fe_2S_2Cys_4$ complex

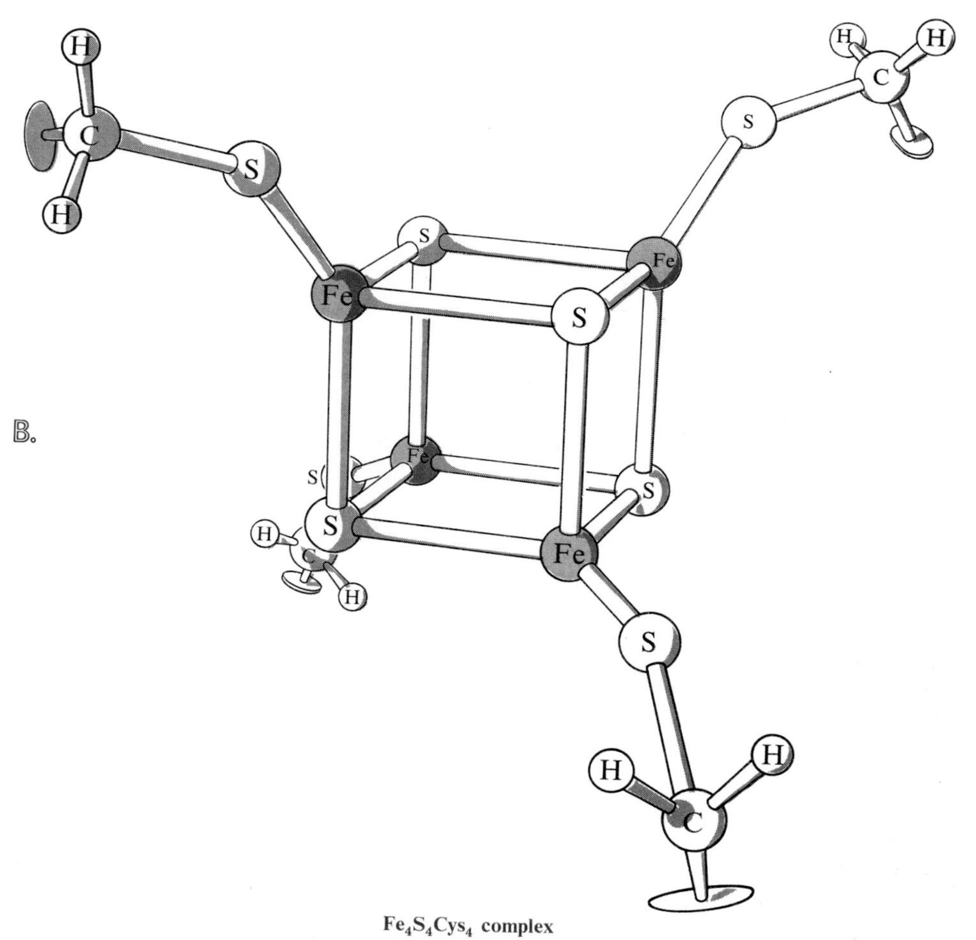

$Fe_4S_4Cys_4$ complex

FIGURE 22–7 *See legend on opposite page.*

The Iron-sulfide Proteins. NADH dehydrogenase contains at least four different proteins with attached lattices of iron and sulfur atoms that are used to transfer single electrons. The exact structures of these lattices in NADH dehydrogenase are not known. However, the structures of related proteins are known. Among these are the ferredoxins from bacteria and plants, which differ in the number of iron atoms involved (Fig. 22–7). In all of these, the iron atoms are chelated with sulfur atoms, which are in part supplied by cysteinyl groups in the associated protein and in part as inorganic sulfide ions.

The entire lattice carries only one electron. Iron-sulfide proteins were referred to in the older literature as **nonheme iron.** The important feature of the iron-sulfide proteins is that their relative affinity for electrons can be varied over a wide range by changing the nature of the polypeptide chain. Some are relatively strong oxidizing agents; others are powerful reducing agents—even stronger than NADH.

Ubiquinone (formerly known as **coenzyme Q**) is actually a group of compounds, all containing the same quinone structure but substituted with a side chain composed of varying numbers of **prenyl** groups linked head to tail (Fig. 22–8). The most common version in mammalian tissues has 10 prenyl groups and is designated as Q_{10}.

Quinones resemble the flavins in forming relatively stable semiquinones, which enables them to handle electrons in pairs or one at a time. NADH dehydrogenase probably fully reduces ubiquinone to ubiquinol, but by only one electron at a time, with protons pumped at each transfer.

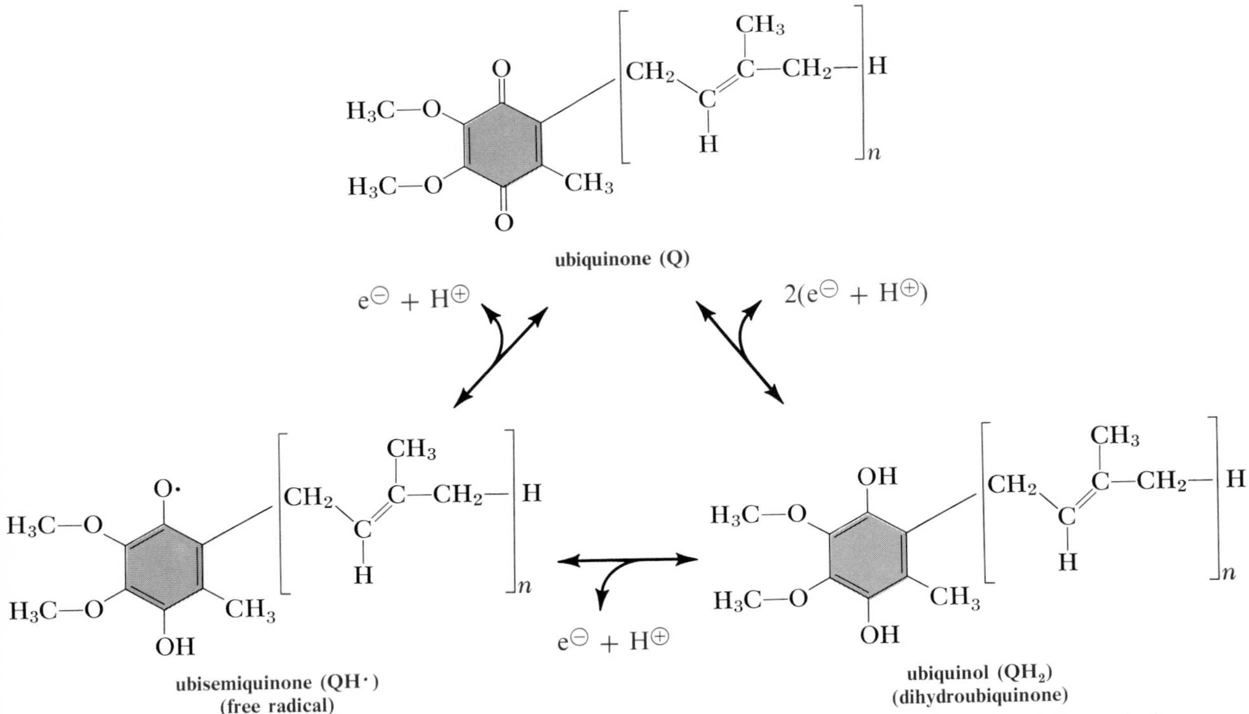

ubiquinone (Q)

$e^{\ominus} + H^{\oplus}$

$2(e^{\ominus} + H^{\oplus})$

$e^{\ominus} + H^{\oplus}$

ubisemiquinone (QH·)
(free radical)

ubiquinol (QH_2)
(dihydroubiquinone)

FIGURE 22–8 **The ubiquinones have a characteristic quinone structure with hydrocarbon tails made from varying numbers of prenyl groups. The most abundant ubiquinone in animal mitochondria has 10 prenyl groups in its side chain (Q_{10}). Like other quinones, the ubiquinones may be reduced one electron at a time through the semiquinone free radical, or may be reduced directly to the quinol (the dihydroquinone) by two electrons.**

Other Flavoprotein Dehydrogenases

Those substrates whose oxidation cannot sustain the generation of three molecules of ATP are attacked by specific dehydrogenases in the mitochondrial membrane. These dehydrogenases also have flavoprotein prosthetic groups and transfer electrons through iron-sulfide proteins to ubiquinone. Unlike NADH dehydrogenase, they do not pump protons across the membrane and therefore can catalyze oxidations that are energetically less favorable. Typical substrates for these flavoproteins are saturated acyl compounds, which are oxidized to the corresponding unsaturated derivatives. An important example is **succinate dehydrogenase,** which catalyzes the oxidation of succinate to fumarate:

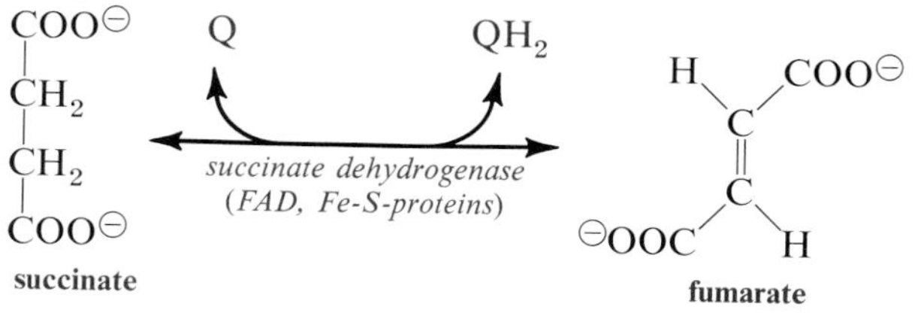

$$\text{COO}^{\ominus}\text{—CH}_2\text{—CH}_2\text{—COO}^{\ominus} \xrightleftharpoons[\textit{succinate dehydrogenase (FAD, Fe-S-proteins)}]{Q \rightarrow QH_2} {}^{\ominus}\text{OOC—CH=CH—COO}^{\ominus}$$

succinate **fumarate**

This dehydrogenase is a complex containing three iron-sulfide proteins; the flavin prosthetic group is **flavin adenine dinucleotide (FAD)** rather than the riboflavin 5′-phosphate found in NADH dehydrogenase:

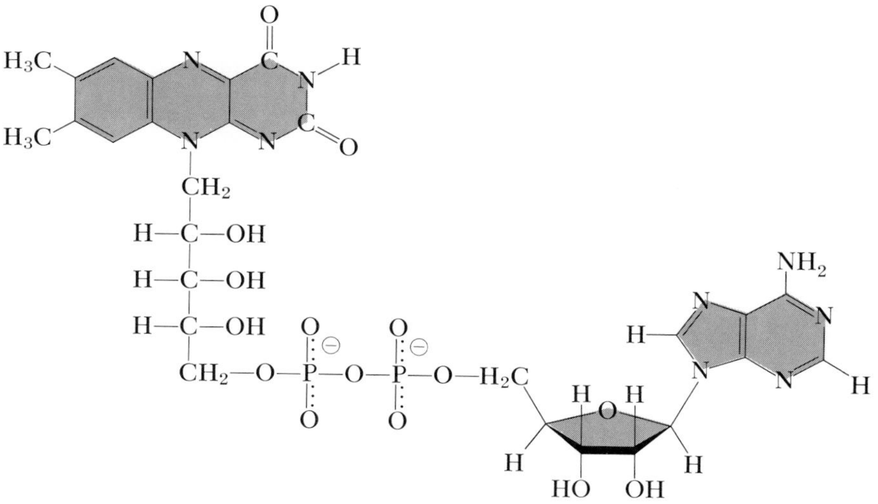

H_3C, H_3C, N, N, N, N–H, C=O, C=O, CH_2, H—C—OH, H—C—OH, H—C—OH, CH_2—O—P(O)(O$^{\ominus}$)—O—P(O)(O$^{\ominus}$)—O—H_2C, H, H, H, H, HO, OH, NH_2, N, N, N, N, H, H

flavin adenine dinucleotide (FAD)

The bc_1 Complex

The enzyme complex catalyzing the oxidation of ubiquinol contains an iron-sulfide protein of the Fe_2S_2 type and two types of cytochromes. The cytochromes are hemoproteins like hemoglobin, but, unlike hemoglobin, their iron atoms are oxidized and reduced to transfer electrons between other compounds. The different types of cytochromes vary in the nature of the heme group and its mode of attachment to the apoprotein (Fig. 22–9). Each type is given a letter designation—a, b, c, and so on—that was assigned in the days when the cytochromes were mainly known as changing absorption bands in the spectrum of light passed through working insect

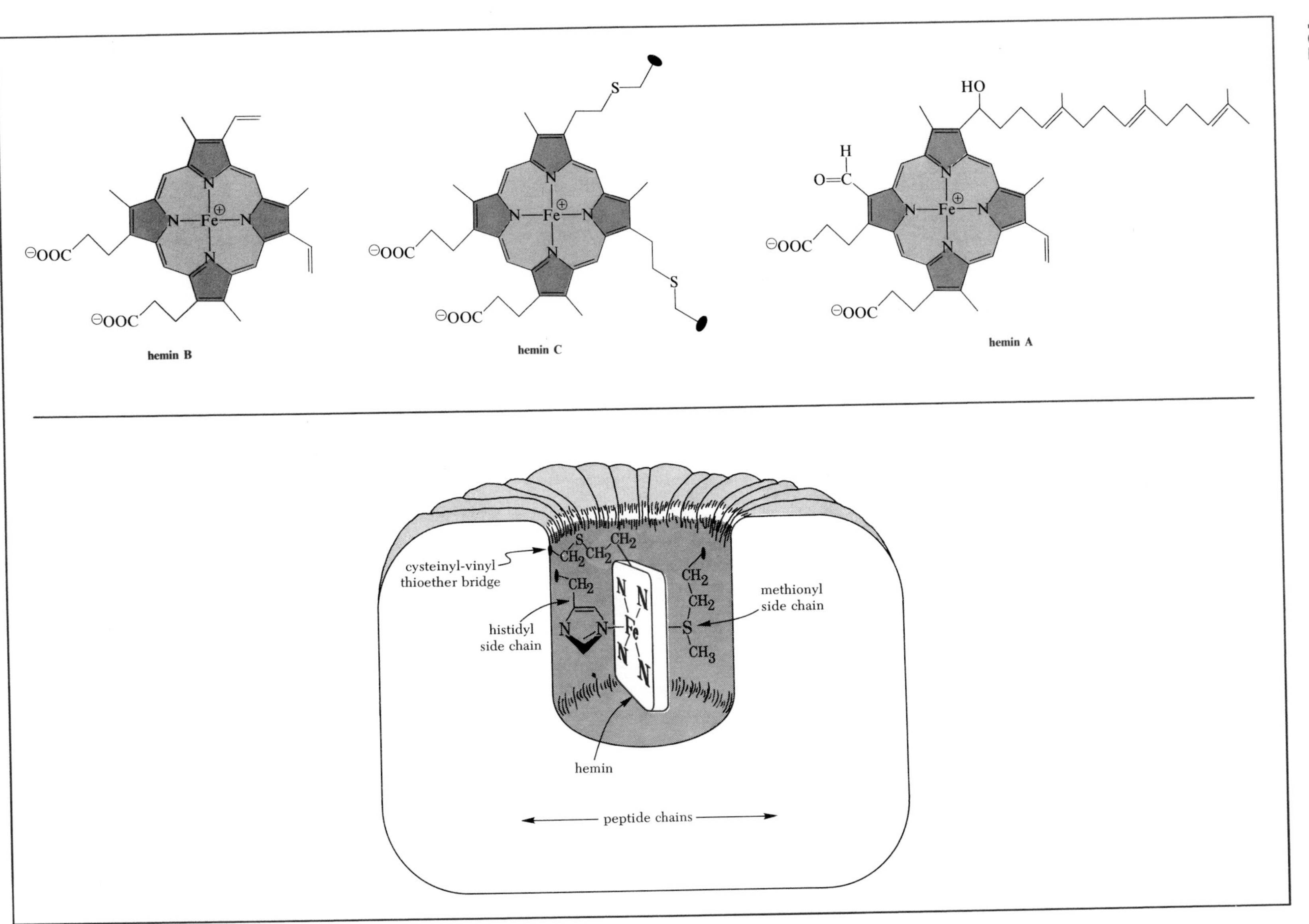

FIGURE 22–9 *See legend on opposite page*

muscles,* with the unknown compound absorbing at the longest wavelength called cytochrome a, that absorbing the next longest wavelength called cytochrome b, and so on. Unfortunately, the order of wavelength does not correspond to the physiological sequence in which they function. Subscript numbers were added as new individual cytochromes within the same type were found, that is, those with similar prosthetic groups but with different apoproteins. The cytochromes in the complex that oxidizes ubiquinol are b and c_1 cytochromes; this bc_1 complex transfers its electrons to cytochrome c, which is a protein different from cytochrome c_1. Cytochrome c is the electron-accepting substrate, while cytochrome c_1 is a part of the proton-pumping enzyme complex catalyzing the electron transfer.

The cytochromes b have the same iron-protoporphyrin IX complex found in hemoglobin, with no covalent bonds between hemin and polypeptide. The cytochromes c and c_1 also have the same prosthetic group, except that the vinyl side chains of the porphyrin are added to cysteinyl side chains, creating covalent thioether bonds between hemin and polypeptide (Fig. 22–9, *bottom*). The cytochromes a contain a different iron-porphyrin, hemin A, with side chains including a formyl group and a hydrophobic polyprenyl tail.

The structure of cytochrome c is known from x-ray crystallography; its hemin group is in a deep crevice, with the coordinating position of the iron atom filled by histidine and methionine side chains. The exact conformations of the other cytochromes are not known.

Since each cytochrome can carry only one electron on its heme groups, two molecules of cytochrome c must be reduced for each molecule of substrate, such as malate or succinate, that is oxidized by the mitochondria. The bc_1 complex is arranged across the inner membrane, with electrons removed from ubiquinol and ubisemiquinone near the matrix side, while the cytochrome c to which the electrons are transferred is attached to the cytosol side of the inner membrane. It is not known how this transmembrane arrangement facilitates the pumping of protons that accompanies the electron transfer.

The aa_3 Complex: Cytochrome Oxidase

The third proton-pumping complex transfers electrons from cytochrome c to oxygen. It spans the inner membrane, reacting with reduced cytochrome c on the cytosol side, and with oxygen on the matrix side. Most of the oxygen consumed in the body is reduced by this enzyme complex. The mechanism of electron transfer is undoubtedly complex, since four molecules of reduced cytochrome c must donate their single electrons so as to reduce a molecule of oxygen to water without the release of any partially reduced intermediates. (We shall see later that partially reduced oxygen is dangerously reactive.) In order to accomplish this, the complex

*Our knowledge of electron transfer is in many ways a triumph of spectroscopic detection. Changes in visible light absorption have been used to discover cytochromes and follow their behavior as well as that of flavoproteins. Ultraviolet spectroscopy is used to follow reduction of NAD. Many iron-sulfide proteins were discovered as bands in electron paramagnetic spectrograms at low temperatures.

FIGURE 22–9 **(*Top*) The cytochrome hemins. Hemin B is identical to the hemin in methemoglobin. Hemin C is similar except for the addition of cysteine sulfhydryl groups to the vinyl side chains. Hemin A has one methyl group oxidized to an aldehyde and a polyprenyl tail. (*Bottom*) Schematic view of hemin binding in cytochrome c.**

contains two cytochromes, a and a_3, each containing a molecule of hemin A, and two different proteins with a bound atom of Cu(II) on each. Enough is known about the individual behavior of the four electron carriers during the reduction of oxygen to enable formulation of hypothetical schemes, but the mechanism is still too uncertain to warrant close study for our purposes. The mechanism of proton pumping accompanying the electron transfers is even more uncertain.

The Complete Electron Transfer Chain

It was formerly common to draw detailed flow diagrams for mitochondrial electron transfer like this:

substrate → NAD → FMN-protein → Fe-S protein → Q → cyt b →
Fe-S protein → cyt c_1 → cyt c → cyt a → Cu(II) → cyt a_3 → O_2.

The emphasis is now on the enzyme complexes as units, with the definitive stages of transfer being those we discussed initially:

substrate → NAD → Q → cyt c → O_2.

We can better understand the reasoning in terms of the free energy changes that are involved. These changes in oxidation-reduction reactions are frequently expressed in terms of the electrical potential generated by the reaction. Let us review the basis for these numbers.

Reduction Potentials

An electrical cell consists of two electrodes, with an oxidation occurring at one and a reduction at the other. It is convenient to separate the reaction occurring at each electrode and express its potential individually, although this cannot be done in an absolute way. For purposes of comparison, the potential of an electrode at which hydrogen gas (1 atmosphere) and protons (1 M) are at equilibrium is taken as zero. Therefore, when one sees a value of −0.315 volts for the reaction:

$$NAD^{\oplus} + H^{\oplus} + 2\,e^{\ominus} \longrightarrow NADH$$

it is really the potential for a cell in which the other electrode reaction is:

$$H_2 \longrightarrow 2\,H^{\oplus} + 2\,e^{\ominus}$$

and the overall reaction becomes:

$$NAD^{\oplus} + H_2 \longrightarrow NADH + H^{\oplus}.$$

Unless otherwise stated, reduction potentials are given in which the oxidized form of the specified compound gains electrons from its electrode, and electrons are obtained from H_2 at the other electrode of the cell. If the reaction runs in the opposite direction, an equal oxidation potential of opposite sign is created.

Reduction potentials are like free energies in that they depend upon the concentrations of the cell components; they are usually cited as standard potentials, $E^{0\prime}$,

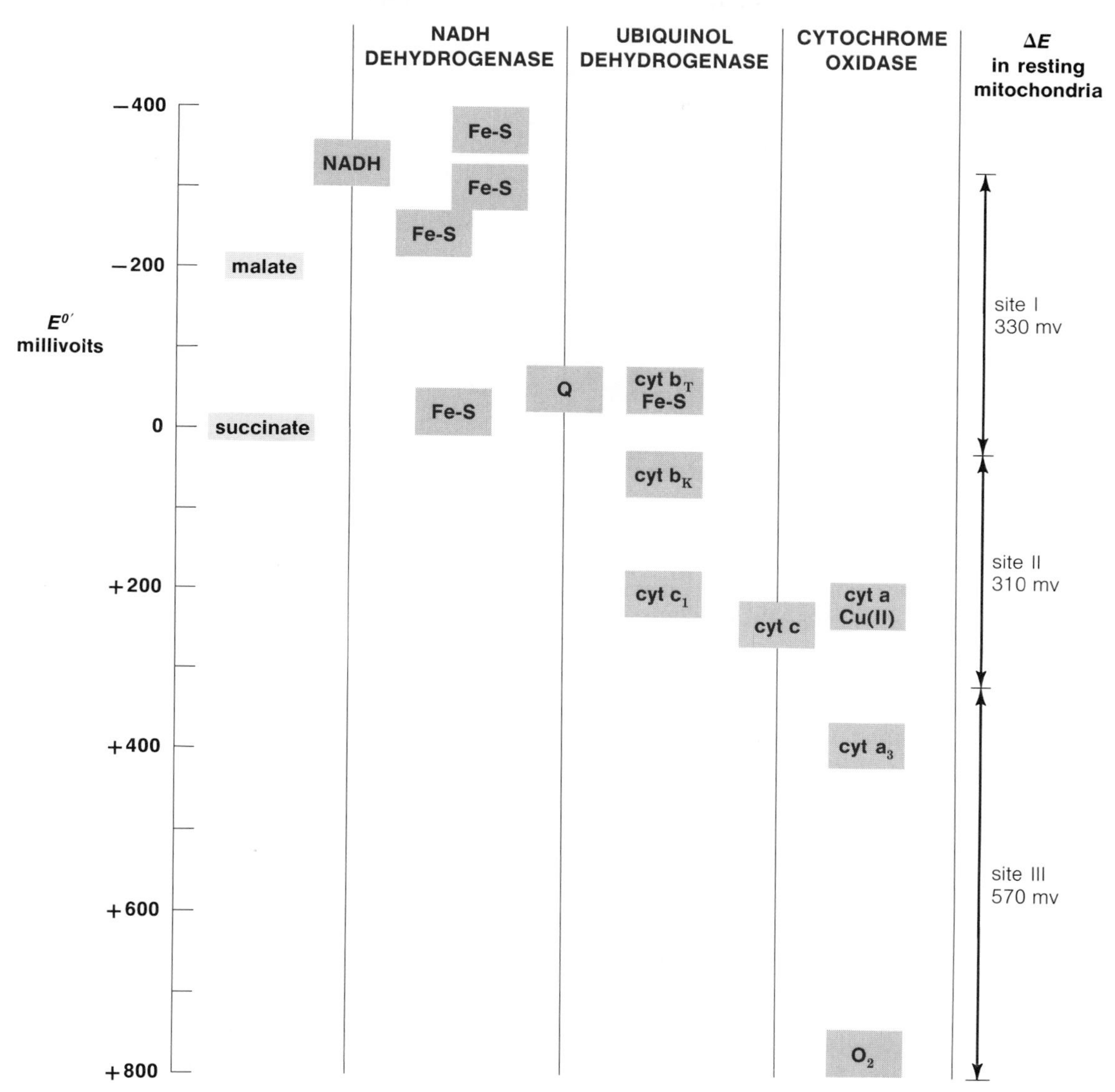

FIGURE 22–10 A plot of the components of the electron transfer chain on a standard reduction potential scale. Beginning with the reduced form of the substrates (malate or succinate), each successive complex of electron-transferring proteins contains at least one component matching in potential the electron donor from which it will gain electrons, and one component matching the electron acceptor to which it will transfer electrons. The gap in potential between these components provides the energy to generate high-energy phosphate.

The right column (*black*) shows estimates of the actual potential drop (not standard potential) across the three major stages in mitochondria of resting muscles when the high-energy phosphate concentration is near maximum.

analogous to standard free energies. $\Delta G^{0\prime}$, in which the concentration of H^+ is fixed at a specified value around the electrode in question, and all other reactants except H_2O, including the H^+ around the hydrogen electrode, are taken to be at 1 molar concentration. The standard reduction potential, $E^{0\prime}$, is related to the change in standard free energy by the equation:

$$\Delta G^{0\prime} = -nFE^{0\prime}$$

in which n = number of electrons transferred in the reaction, and F = Faraday constant = 96,487 joules per mole per volt.

Cell potentials can be added, so the standard potential for the reaction between NADH and oxygen can be obtained from the individual electrode standard potentials at 25° C, pH 7.0:

$$\begin{array}{lll} & \text{NADH} + \text{H}^{\oplus} \longrightarrow \text{NAD}^{\oplus} + \text{H}_2 & E^{0\prime} = 0.315 \text{ volts} \\ & \tfrac{1}{2}\,\text{O} + \text{H}_2 \longrightarrow \text{H}_2\text{O} & E^{0\prime} = 0.815 \text{ volts} \\ \hline \text{SUM:} & \text{NADH} + \tfrac{1}{2}\,\text{O}_2 + \text{H}^{\oplus} \longrightarrow \text{NAD}^{\oplus} + \text{H}_2\text{O} & E^{0\prime} = 1.130 \text{ volts} \end{array}$$

The standard free energy change for the oxidation of NADH by O_2 at pH 7 and 25° C can be calculated from that value:

$$-2 \times 96.487 \times 1.130 = -218.060 \text{ J mole}^{-1}.$$

Carrier Potentials. A pattern emerges when the potentials of the electron carriers are plotted (Fig. 22–10). Many of the carriers within a proton-pumping enzyme complex have potentials clustered near the same value, but others differ more markedly. The range of potentials spanned within one complex overlaps the range spanned by the next. The current interpretation is that some carriers in each complex are "tuned" to match the carriers donating electrons to and accepting electrons from the complex. For example, many of the iron-sulfide carriers in NADH dehydrogenase have potentials near that of NAD, whereas at least one has a potential near that of ubiquinone. This arrangement is believed to facilitate efficient transport of electrons between the complexes.

COUPLED PHOSPHORYLATION

ATP Synthase. The potential energy generated as an electrochemical gradient across the inner mitochondrial membrane by the electron transfers is converted to potential energy in the form of high-energy phosphate by a complex of enzymes exposed on the inner surface of the membrane. This complex comprises a globular headpiece attached through a stalk to a base piece buried in the lipid core (Fig. 22–11). Current speculations lean toward the notion that energy transduction is accomplished through conformational changes in the proteins of the ATP synthase complex that involve shifts in the acidic dissociation constants of groups toward both the cytosol and the matrix sides of the complex. All efforts to detect covalent intermediates in the process have failed; indeed there is evidence that the synthesis of ATP, like the cleavage of ATP by myosin during muscular contraction, involves the formation of an energy-rich complex of protein, ADP, and P_i without overt phosphorylation of the protein.

When the ATP synthase complex is isolated from mitochondria, it behaves like an ATPase, a hydrolase cleaving ATP. However, when it is incorporated into recon-

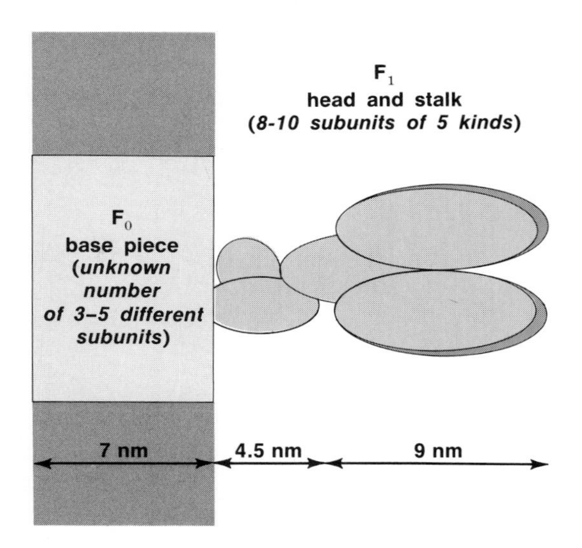

FIGURE 22–11

Schematic representation of the energy-transducing complex in the mitochondrial inner membrane. An ATP-synthesizing head (ATP-synthase) is attached by a stalk to a proton-conducting base-piece that is exposed to both sides of the membrane. The components of the complex are isolated and studied in two parts, F_0 and F_1.

stituted lipid vesicles, it unequivocally acts as a reversible ATP-coupled proton pump. That is, experimental creation of a difference in H^+ concentration across the membrane will enable the complex to synthesize ATP from ADP and P_i. If ATP is added to the preparation, it will be hydrolyzed while a proton concentration gradient is created, clearly demonstrating the easy interconversion of energy in the forms of electrochemical gradient and pyrophosphate bonds.

Mitochondrial Transport of ATP, ADP, and P_i

The major consumption of high-energy phosphate occurs outside of the mitochondria, but the major production is inside. To link the two processes ADP and P_i constantly enter the mitochondrial matrix, and ATP leaves (Fig. 22-12). These polar compounds are moved across the inner membrane by transporters. The entrance of P_i is coupled to the entrance of H^+, and the entrance of ADP is coupled to the exit of ATP. Both of these transporters are discharging the electrochemical gradient. The **P_i-H^+ symporter** is electroneutral, that is, it causes no net movement of electrical charge, but it is effectively transporting protons back into the matrix. The **ADP-ATP antiporter,** on the other hand, is moving more negative charges out than it is bringing in, so it is effectively discharging the outside-positive electrical potential across the membrane. The sum of these two processes is therefore discharging both the electrical and the chemical concentration components of the electrochemical gradient. The electron transfer pumps must move one H^+ from the matrix to the cytosol to generate the energy expended by the movements of one molecule each of ADP, P_i, and ATP.

Since we are somewhat arbitrarily assigning a total requirement of 4 H^+ pumped across the inner membrane to replace each ATP hydrolyzed, it follows that three out of the four are used by ATP synthase to make the high-energy phosphate bond within the mitochondria, and the fourth H^+ is used to move the ADP and P_i in and the ATP out. (There is more than a little sophistry in estimates of this kind. Those who like to believe only 2 H^+ need be moved by ATP synthase per ATP generated conclude that each of the electron transfer pumps must move 3 H^+, one to drive the antiporters and two to make the high-energy phosphate bond. Everyone's

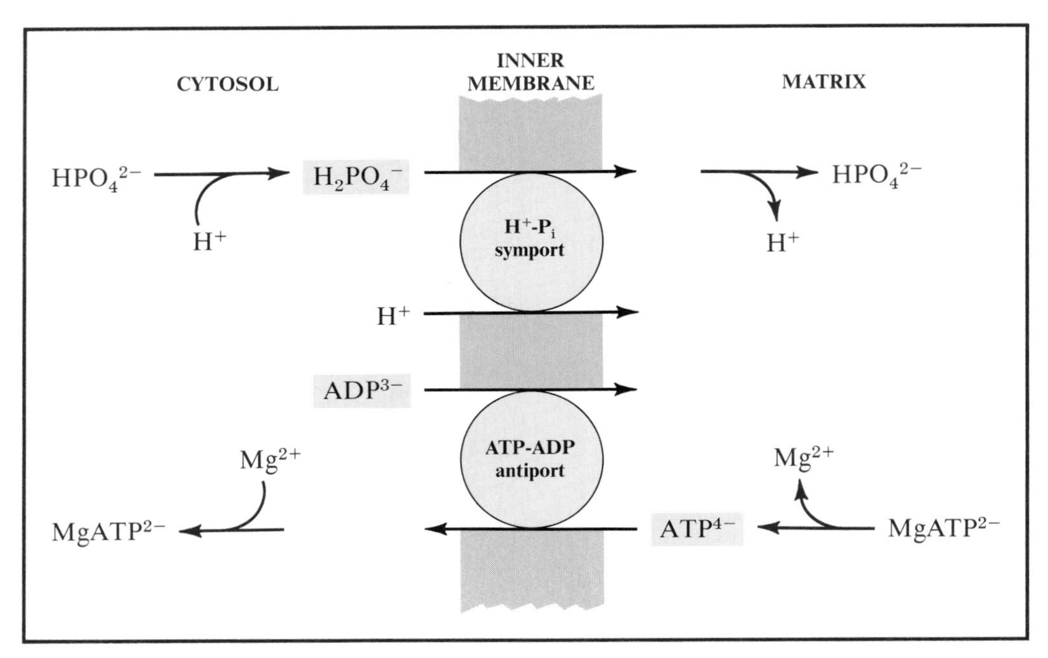

FIGURE 22–12 **Transport of substrates and products of the phosphorylation process across the inner mitochondrial membrane by transporters. A symporter carries a proton and monobasic P_i; it is driven by the H^+ concentration gradient. An antiporter exchanges ADP^{3-} and ATP^{4-} and is driven by the electrical potential across the membrane. There is no discharge of the proton gradient from the release of a proton by the transported $H_2PO_4^-$ ion, because there is a counterbalancing uptake when the P_i is used to make ATP:**

$$ADP^{3-} + HPO_4^{2-} + H^+ \longrightarrow ATP^{4-} + H_2O$$

ledgers always balance in this kind of bookkeeping. To make this true in our case, any process causing the movement of one H^+ across the inner membrane will be deemed equivalent to the loss of 0.25 high-energy phosphate.)

Transport as Phosphocreatine in Muscles

Maximum effort by striated muscle fibers requires a high flux of high-energy phosphate from the inner mitochondrial membrane to the myofibrils—a greater flux than can be achieved by simple diffusion of ATP. To meet the demand, phosphocreatine is used not only as a high-energy phosphate buffer (p. 376), but also as a carrier (Fig. 22–13). Creatine kinase occurs in the space between the inner and outer mitochondrial membranes as well as in the sarcoplasm. Phosphocreatine is much smaller than ATP and in higher concentration; each molecule can move faster, and there are more of them. When phosphocreatine is used to regenerate ATP near a twitching fiber, more phosphocreatine diffuses from the mitochondria to replace it, and a countercurrent of creatine moves back to the mitochondria for phosphorylation by the mitochondrial creatine kinase.

Efficiency of Phosphorylation

Since the exact mechanism of energy transfer is not known, we cannot state with certainty that one high-energy phosphate bond can be produced whenever two elec-

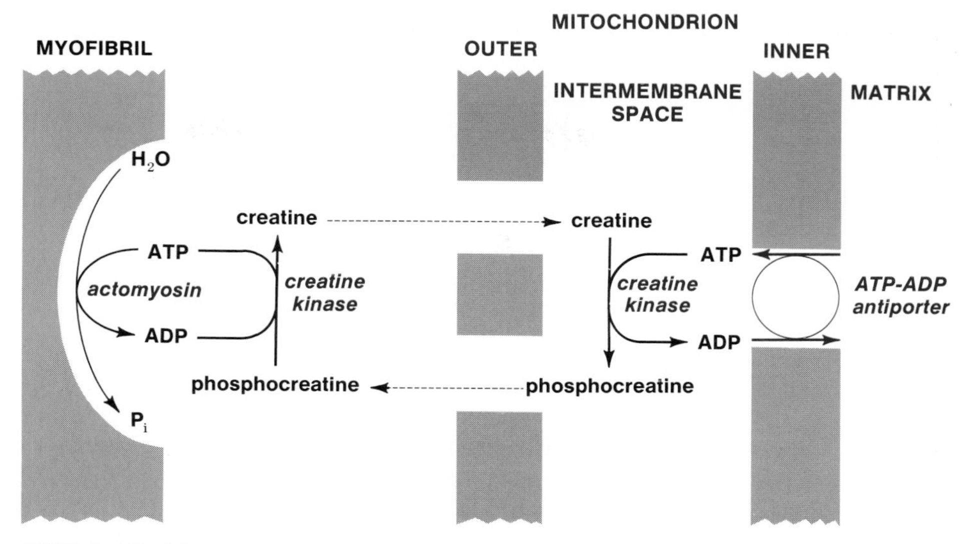

FIGURE 22–13 Transport of high-energy phosphate as phosphocreatine in myofibrils.

trons pass through one of the proton-pumping complexes. We can get some reassurance by a theoretical assessment of the possible yield.

Phosphate Potential and ATP Yield. Remember that the Gibbs energy change for a reaction depends not only upon the inherent character of the reactants, as expressed in the standard free energy change, but also upon the actual concentrations of reactants and products. The free energy change required to form ATP is:

$$\Delta G = +36{,}800 \text{ J/mole} + RT \ln [\text{ATP}]/[\text{ADP}][\text{P}_i].$$

This actual free energy change is referred to by some as the phosphate potential, ΔGp, although, strictly speaking, phosphate potential is the equivalent electrode potential. The intent in any case is a quantitative index of the unfavorable concentrations overcome by electron transfers.

The electron transfers generate a fixed amount of potential energy from the oxidation of a given substrate. It follows that in theory more moles of ATP could be generated when the ratio $[\text{ATP}]/[\text{ADP}][\text{P}_i]$ is low than there could when it is high.

Let us test the two possibilities. We can phrase the first question this way: When the ATP and ADP concentrations are equal, what is the theoretical yield of high-energy phosphate from the oxidation of malate:

$$\text{malate} + \tfrac{1}{2}\,\text{O}_2 \longrightarrow \text{oxaloacetate} + \text{H}_2\text{O} \qquad \Delta G^{0\prime} = -192{,}000 \text{ J mole}^{-1}?$$

Since the standard free energy of formation of ATP is only $+36{,}800 \text{ J mole}^{-1}$, one might think it should be possible to make $192/36.8 = 5.2$ moles of ATP per mole of malate oxidized. However, we have already learned that standard free energies cannot be used for this kind of reasoning when there are molarity dimensions in the equilibrium constants, and the actual equilibrium conditions must be calculated.

Let us see what happens if we assume that all reactant concentrations, except those of ATP and ADP, are fixed at values similar to those found in living cells:

$$[\text{oxaloacetate}]/[\text{malate}] = 0.01; \; [\text{P}_i] = 0.01 \text{ M}; \; [\text{O}_2] = 0.001 \text{ atm}.$$

The calculation is laborious, but the result is important, and let us rapidly skim through the method.

Assume that x moles of ATP are generated per atom of oxygen consumed. The stoichiometry then becomes:

$$\text{malate} + x(\text{ADP} + \text{P}_i) + \tfrac{1}{2}\,\text{O}_2 \longrightarrow \text{oxaloacetate} + x\text{ATP}$$

$$\Delta G^{0'} = 36{,}000x - 192{,}000 \text{ J mole}^{-1}$$

and the equilibrium constant is:

$$K'_{eq} = e^{-\Delta G^0/RT} = \left(\frac{[\text{oxaloacetate}]}{[\text{malate}]}\right) \times \left(\frac{[\text{ATP}]}{[\text{ADP}]}\right)^x \times \left(\frac{1}{[\text{P}_i]}\right)^x \times \left(\frac{1}{[\text{O}_2]}\right)^{1/2}.$$

If we set the concentrations of ATP and ADP to be equal and solve for x, we find $x = 4.04$ moles of ATP per atom of oxygen consumed. This means that as many as four molecules of ATP could be generated per molecule of malate oxidized if the cell will tolerate the concentration of ATP never being greater than the concentration of ADP.

Let us now ask the second question. What would be the theoretical ratio of ATP and ADP concentrations that could be generated by the oxidation of malate if the yield of ATP is held at 3.0 moles per atom of oxygen consumed (P:O = 3.0)? Substituting $x = 3.0$ in the above equation, and solving for the [ATP]/[ADP] ratio, we find that equilibrium is not reached until [ATP]/[ADP] = 858.

Evolution has made a choice between these extremes of high molar yield and high concentration ratios. In our discussion of high-energy phosphate stores in muscle, we noted that [ATP]/[ADP] had to be over 100 in order to charge the phosphocreatine store to the levels actually present. If the ratio is 150, then only 3.2 molecules of ATP can be made per molecule of malate oxidized (P:O = 3.2). It is apparent that the choice has been made for a high ratio of [ATP]/[ADP], and a P:O ratio of 3.0 means that phosphorylation is proceeding at nearly its maximum efficiency under existing steady state concentrations. This choice appears to be an absolute one, inherent in the mechanism used; that is, the P:O ratio is not increased when the ATP/ADP ratio is experimentally maintained at a low level.

The Regulation of Fuel Combustion

Electron transfers and phosphorylations are usually tightly coupled in mitochondria. It therefore follows that the oxidation of fuels hinges upon an available supply of ADP and P_i. We can reason this way:

1. Fuels are oxidized by NAD or flavin coenzymes.

2. The oxidations convert NAD and flavin coenzymes to their reduced forms.

3. The reduced forms of the coenzymes must be converted back to the oxidized forms if fuel combustion is to continue. (For example, there is a limited supply of NAD in the mitochondria. If the NADH rises, there will be less NAD available.)

4. Conversion of the reduced coenzymes to their oxidized form requires transfer of electrons through the proton-pumping enzyme complexes. There is a limit to how steep the electrochemical gradient can be made, and electron transfer stops if the limit is reached.

5. Diminution of the electrochemical gradient to permit continued electron transfer is mainly accomplished by conversion of ADP and P_i to ATP.

6. The availability of ADP and P_i hinges upon the rate at which ATP is converted to ADP and P_i by other processes, such as muscular contraction, the Na,K-ATPase, and chemical syntheses.

Therefore, the rate of ATP consumption directly controls the rate of fuel combustion and mitochondrial oxygen consumption.

Adenylate Energy Charge. The critical point is that combustion of fuels is regulated by changes in the ratio $[ATP]/[ADP][P_i]$, which can be expressed quantitatively as changes in the phosphate potential. Another measure of this effect, the adenylate energy charge, has been used, but it has flaws. Essentially, adenylate energy charge is an indication of the fraction of the possible high-energy phosphate bonds that actually are present in the adenine nucleotides. This fraction will be 1.0 when ATP is the only adenine nucleotide present, and zero when AMP is the only one present. Since ADP has only one high-energy bond and a full charge is two high-energy bonds per adenine nucleotide,

$$\text{Adenylate energy charge} = \frac{2[\text{ATP}] + [\text{ADP}]}{2([\text{ATP}] + [\text{ADP}] + [\text{AMP}])}.$$

Unfortunately, the numerical values are not a useful guide to the metabolic state. When the [ATP]/[ADP] ratio is very high, as in resting muscle, very large changes in phosphate potential, and therefore in the rate of oxygen consumption, can occur with relatively small absolute changes in the ADP concentration, but the calculated adenylate energy charge changes only little. (We already noted that large changes in the ATP concentration do not occur in muscle until it approaches exhaustion.)

INHIBITORS OF OXIDATIVE PHOSPHORYLATION

Since oxidative phosphorylation supplies most of the ATP that is required for so many vital functions, any compound severely interfering with the process is surely lethal. We might expect such compounds to have their major application outside of the laboratory as pesticides for killing animals. We also might expect them to be toxic to all animals if they can enter the tissues. This is indeed the case. However, some of the compounds have important uses inside the laboratory as experimental tools, and some are even employed therapeutically in sublethal doses. Let us consider some examples.

Inhibitors of Electron Transfer

Cyanides are among the poisons better known by the general public. They are not extraordinarily potent (the minimum lethal dose for humans is estimated to be of the order of 1 to 3 millimoles), but the small HCN molecule rapidly enters tissues, so that a sufficient quantity may be lethal within a few minutes. It is this quick effect, forestalling effective countermeasures in the absence of advance preparation, that has gained the cyanides so much respect, and has led to their use as pesticides against such diverse enemies as rats and insects in ships, moles in lawns, and murderers in some states.

His lawyer probably doesn't realize it, but the felon executed in a gas chamber is having his mitochondrial electron transport blocked at its terminal reaction. The cyanide ion, CN^-, combines tightly with cytochrome oxidase ($K_i \sim 8\text{–}9 \times 10^{-8}$ M). (Cyanide also combines with other iron porphyrins, especially in the ferric state, but with much less affinity than it has for cytochrome oxidase.) The consequence is the cessation of transfer of electrons to oxygen. The previous electron carriers in the chain accumulate in their reduced state, and the generation of high-energy phosphate ceases.

The effect of cyanide is as fundamental as deprivation of oxygen, and like the latter, causes rapid damage to the brain. Indeed, there have been clinical trials of sublethal doses of cyanide as a means of causing corrective disturbances in the brains of psychotics, much as with electric or insulin shocks used for the same purpose. No advantage appeared to counterbalance the disadvantage of the dangerously small margin of safety in dosage.

Cyanide is also encountered in other circumstances. **Cassava root,** an important foodstuff for many people, contains glycosides that liberate dangerous quantities. Cyanide also is liberated from a glycoside in apricot pits, presently notorious as **laetrile.** It also may be a hazard in the therapeutic use of sodium ferrinitrosocyanide, $Na_2Fe\ NO(CN)_5$, **(sodium nitroprusside)** for hypertensive crisis.

Hydrogen sulfide will also combine with cytochrome oxidase, and few realize that it is as toxic as HCN. Its bad odor gives more warning, but there have been fatalities from only a few inspirations at high concentrations of the gas. The gas is a particular problem for workmen on oil-drilling platforms at sea. Hydrogen sulfide often occurs in oil-bearing strata; it is a lethal menace in all drilling operations, but especially in the confined quarters of the platforms. (In the test tube, 0.1 mM sulfide inhibits cytochrome oxidase more than does 0.3 mM cyanide—96 per cent against 90 per cent.)

The **antimycins** are antibiotics produced by species of streptomyces. They strongly associate with mitochondria and block the passage of electrons from cytochrome b to cytochrome c_1; 0.07 micromole of antimycin A_1 per gram of mitochondrial protein is effective. As laboratory tools, they permit experimental distinction between events in the earlier and later parts of the electron transfer chain. Since they kill the host as well as invaders, they are not therapeutic agents. However, they have been employed to kill fish in small lakes for restocking. Only one microgram per liter is required and the compound disappears within a day.

Piericidin A, another antibiotic produced by streptomyces, has an action similar to that of rotenone. **Rotenone** is a compound extracted from the roots of tropical plants (*Derris elliptica, Lonchoncarpus nicou*) that complexes avidly with NADH dehydrogenase; only 30 nanomoles per gram of mitochondrial protein are effective. It is a valuable tool for distinguishing routes of electron flow beginning with NAD-coupled dehydrogenases from those beginning with flavoproteins since it does not affect the latter. It acts between the iron-sulfide proteins and ubiquinone. The results of its effect on oxidative metabolism have long been exploited by primitive people. Rotenone is relatively nontoxic to mammals because it is absorbed poorly, although exposure of the lungs to the dust is a little more dangerous. However, the compound readily passes into the gills of fish and the breathing tubes of insects, and is intensely toxic to these animals. Fish-eaters apply preparations of the appropriate plant roots to ponds, collect the floating fish whose mitochondrial NAD remains reduced, and eat them with impunity. Rotenone was in favor as a relatively safe insecticide, with low toxicity to land vertebrates and a short lifetime, but it is now frowned upon by monitors of public safety using more rigorous standards.

Barbiturates also block NADH dehydrogenase, but much higher concentrations are required. The sedative actions of these compounds appear to depend on other actions on neural membranes, but inhibition of respiration may augment the effect.

Inhibitors of Oxidative Phosphorylation

Oligomycins. These antibiotics from various streptomyces inhibit the transfer of high-energy phosphate to ADP. Therefore, they also inhibit electron transfers coupled to phosphorylation, but they have no effect on oxidation-reduction reactions that are not coupled. They are widely used as experimental tools for discriminating between the two kinds of reactions. They combine with a protein that seems to be in the stalk of the phosphorylating particles and block any conversion of ADP to ATP, as well as the use of ATP for creating ionic concentration gradients, but they have no effect on the direct formation of ionic concentration gradients by electron transfer.

Atractyloside. This compound from *Atractylis gummifera,* a plant native to Italy, attracted renewed attention when several children were poisoned by eating rhizomes of the plant, resulting in three deaths. Atractyloside blocks oxidative phosphorylation by competing with ATP and ADP for a site on the ADP-ATP antiport of the inner membrane. It therefore prevents renewal of the ATP supply in the cytosol. In the laboratory, it is used as a device for separating intra- and extramitochondrial changes in the ATP balance.

Bongkrekate is a toxin formed by a bacterial species (Pseudomonas sp.) in a Javan coconut preparation ("bongkrek"). It also blocks the ATP-ADP antiport and has caused human fatalities. Two micromoles per gram of mitochondrial protein suffice.

Uncouplers of Oxidative Phosphorylation

If electron transfers in mitochondria can somehow be dissociated from phosphorylation, the supply of ATP will be impaired as effectively as if phosphorylation were inhibited. This uncoupling can be achieved, in effect, by causing a constant discharge of the high-energy state that is ordinarily used to generate ATP. Uncoupling can be distinguished from inhibition in this way: Uncoupling causes an increased oxygen consumption in the absence of increased utilization of ATP. Inhibition of phosphorylation, or inhibition of the ATP-ADP antiport, diminishes oxygen consumption in normal coupled mitochondria.

2,4-Dinitrophenol acts as an uncoupler at a concentration of 10 micromolar. It does so because both the phenol and the corresponding phenolate ion are significantly soluble in the lipid core of the inner membrane. The phenol diffuses through the core toward the matrix, where it loses a proton; the phenolate ion then diffuses back toward the cytosol side, where it picks up a proton to repeat the process. Other acids with lipid-soluble anions, such as long-chain fatty acids, can also act as uncouplers by the same mechanism.

Valinomycin is an example of an ionophore, a compound that binds ions for transport. This particular example, produced by streptomyces, is an antibiotic with a ring structure that closely fits K^+, but with a sufficiently hydrophobic exterior to permit passage through the inner mitochondrial membrane. It combines with K^+ from the cytosol side of the membrane, and the charged complex is driven by the electrical potential to the inside surface where K^+ can be released. This in turn dissipates part of the potential, permitting more H^+ to be pumped out of the mitochondria. The result is a constant exchange of K^+ for H^+ across the membrane, which also discharges the high-energy state, thereby causing increased consumption of oxygen.

EXTRAMITOCHONDRIAL OXYGEN CONSUMPTION

Oxidation of Xenobiotics by Monooxygenases

A monooxygenase is an enzyme introducing one atom of oxygen from O_2 into a compound while reducing the other atom of oxygen to water. The endoplasmic reticulum, especially in liver cells, has several monooxygenases of broad specificity, which attack a wide variety of compounds. Some of the compounds affected are normal physiological intermediates: many others are alien to the programmed routes of metabolism, and are therefore called **xenobiotics** (xeno = foreign). The xenobiotics include many products of the organic chemist's art that are hardly likely to have been encountered in the past evolutionary history. Why then has the ability to oxidize them been developed? The best explanation is that the monooxygenases of broad specificity are protective devices. Many exotic compounds were encountered in the environment long before the advent of humans; some as products of plants and bacteria, others produced by spontaneous chemical reactions, especially in fires. Some are deleterious through direct interference with enzymatic reactions or neural transmission; others are potent carcinogens. Addition of oxygen atoms to these compounds often destroys the deleterious effects or makes it easier to excrete them. In some cases, the oxidation converts a relatively innocuous compound to a toxic one; this is especially the case with some aromatic hydrocarbons that become carcinogens only upon oxidation, but on balance it seems reasonable that the monooxygenases remove a threat more often than they make one, since they have persisted.

Since the monooxygenases reduce one atom of O_2 to water while transferring the other to a substrate, they require a source of electrons for the reduction. These are obained from a reduced coenzyme, commonly reduced nicotinamide adenine dinucleotide phosphate, NADPH, which is a reduced derivative of NAD:

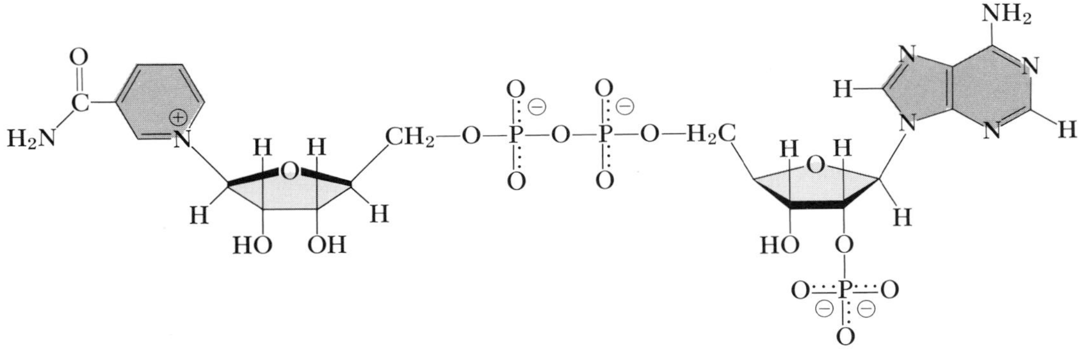

nicotinamide adenine dinucleotide phosphate (NADP)

The additional phosphate group on the ribose in its adenosine moiety enables some enzymes to react specifically with this coenzyme in its oxidized or reduced forms, and not with NAD or NADH even though both NADP and NAD are nearly identical in their oxidation-reduction potentials. We shall see that a portion of the fuel supply is expended to generate reducing power in the form of NADPH, especially in the cytosol, rather than for the generation of high-energy phosphate:

$$\text{fuel metabolites} + NADP^+ \longrightarrow \text{oxidized metabolites} + NADPH + H^+.$$

The amount of fuel diverted in this way is not a large fraction of the total, and the

NADPH that is formed is used for chemical syntheses as well as in the action of monooxygenases that we are now considering.

Two kinds of broadly specific monooxygenases are known. One is a flavoprotein in which FAD directly transfers electrons from NADPH to O_2, while adding oxygen to amine nitrogen or thioether sulfur atoms in almost any compound:

$$R(R')N{-}H \text{ or } R(R')S \xrightarrow[\text{NADPH} + H^{\oplus} \;\rightarrow\; \text{NADP}^{\oplus} + H_2O]{O_2,\ \textit{flavin monooxygenase}} R(R')N{-}OH \text{ or } R(R')S{\rightarrow}O$$

The other type consists of several **cytochromes P-450** (named for a distinctive absorption band). These enzymes have somewhat overlapping specificities; one is always present, another is induced to appear in the endoplasmic reticulum by exposure to compounds such as phenobarbital, and still another appears after exposure to polycylic aromatic hydrocarbons. They contain the same hemin group found in methemoglobin, except that it is bound to the protein through the sulfur of a cysteine residue.

The cytochromes P-450 are only active in association with another enzyme, an **NADPH-cytochrome P-450 reductase,** which is a flavoprotein necessary for transfer of electrons from NADPH to the monooxygenase:

$$\text{NADPH} \xrightarrow[\textit{(FAD-FMN)}]{\textit{NADPH-cytochrome P-450 reductase}} \text{cytochrome P-450}; \quad O_2 \rightarrow H_2O; \quad \text{(substrate)-H} \longrightarrow \text{(substrate)-OH}$$

Other monooxygenases are used during the synthesis and breakdown of normal cellular components. These are much more specific in the kind of substrates attacked, and will be discussed when we encounter the processes in which they are involved.

Dioxygenases

These enzymes insert both atoms of dioxygen into the substrates attacked. For example, some cleave compounds containing double bonds, with the formation of a pair of carbonyl groups:

$$R{-}CH{=}CH{-}R' + O_2 \longrightarrow R{-}CH{=}O + O{=}CH{-}R'.$$

Many of these dioxygenases contain only an atom of Fe(II) as a prosthetic group; others have hemin attached, but not in a pocket as in the cytochromes. They are used in specific metabolic processes discussed later, and most occur in the cytosol.

Flavoprotein Oxidases

Peroxisomes, or **microbodies,** are organelles of poorly defined function. They contain a few flavoprotein oxidases that use molecular oxygen and produce hydrogen peroxide, H_2O_2, along with other enzymes that use the hydrogen peroxide as an oxidizing agent. For example, they contain a **D-amino acid oxidase,** which catalyzes the destruction of D-amino acids:

$$H_2O + H-\underset{R}{\overset{COO^{\ominus}}{C}}-NH_3^{\oplus} \xrightarrow[\text{D-amino acid oxidase}]{O_2 \;\; H_2O_2} \underset{R}{\overset{COO^{\ominus}}{C}}=O + NH_4^{\oplus}$$

Many antibiotics contain D-amino acids, and presumably compounds of similar toxicity might be formed in the body if the precursors were not removed.

The peroxidases of the peroxisomes can use the hydrogen peroxide to attack a variety of compounds; for example, they will convert primary alcohols to aldehydes:

$$\underset{\textbf{alcohol}}{R-CH_2-OH} + H_2O_2 \xrightarrow{peroxidase} \underset{\textbf{aldehyde}}{R-CH=O} + 2\,H_2O$$

but their physiological substrates are not known. They are all hemin-containing enzymes. One of them, **catalase,** is one of the most effective catalysts known. This enzyme decomposes higher concentrations of hydrogen peroxide to oxygen and water:

$$2\,H_2O_2 \longrightarrow 2\,H_2O + O_2$$

and this reaction protects tissues during topical application of germicidal concentrations of H_2O_2, but it is probable that the action of catalase as a peroxidase is more significant.

TOXIC METABOLITES OF OXYGEN AND LEUKOCYTE FUNCTION

The reduction of dioxygen can involve a variety of intermediates differing by one in electron content:

Superoxide, O_2^-, is the anion of the perhydroxyl radical, HO_2, and has one more electron than dioxygen.

Hydrogen peroxide, H_2O_2, is the undissociated form of the dianion, O_2^{2-}, which has two more electrons than dioxygen.

Hydroxyl radicals, $\cdot OH$, are equivalent to half of a hydrogen peroxide molecule, but are much more reactive. (They are formed when ionizing radiation passes through oxygen-saturated water, and are responsible for much of the resultant oxidations.) Addition of one electron to each hydroxyl radical results in the formation of hydroxide ions, OH^-, the completely reduced form of oxygen.

Many of the enzymatic reactions in which dioxygen is consumed proceed by mechanisms in which superoxide or peroxides are formed. For example, Figure 22–14 illustrates a postulated mechanism for cytochrome P-450 reactions, We have also noted that superoxide is probably the bound form of oxygen in oxyhemoglobin (p. 245). There is inevitably some leakage of the intermediates from these various reactions. Superoxide is especially dangerous. It does not itself react readily with

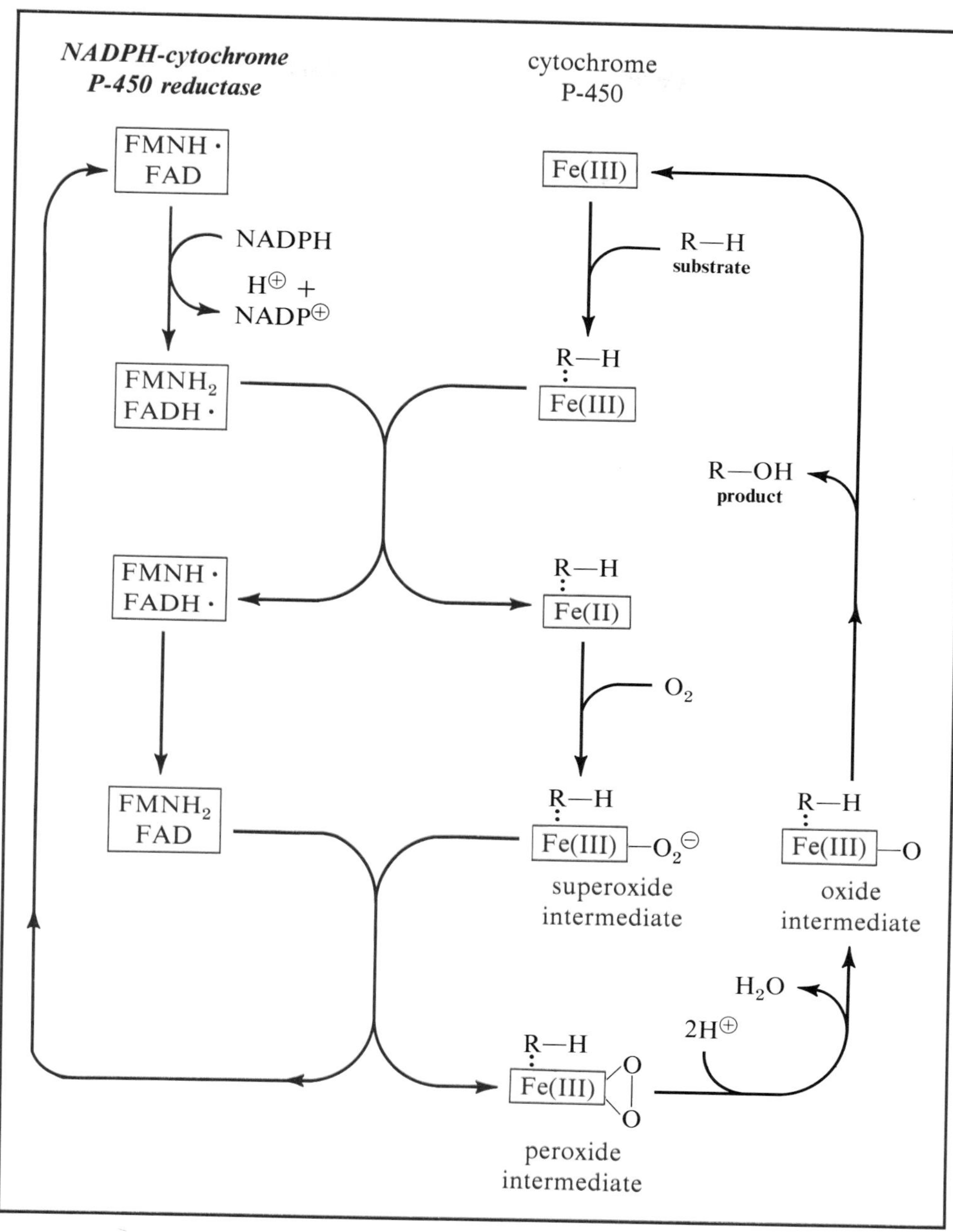

FIGURE 22–14 A postulated mechanism for catalysis of substrate hydroxylations by the cytochromes P-450. Advantage is taken of the ability of flavoproteins to transfer one or two electrons (*left*). Two electrons are accepted from NADPH and then dispensed one at a time, first to reduce the hemin in cytochrome P-450 so that it will convert oxygen to bound superoxide, and then to reduce the superoxide to a bound peroxide, which decomposes to form the bound oxygen atom that is transferred to the substrate. (Condensed from more extensive versions in J.L. Vermillion et al: (1981), J. Biol. Chem. 256:266 and R.E. White and M.J. Coon (1980), Annu. Rev. Biochem. 49:315.)

most cellular constituents, but it will spontaneously combine with peroxides to form hydroxyl radicals and singlet oxygen, which are disruptively reactive:

$$O_2^{\ominus} + H_2O_2 \longrightarrow HO\cdot + {}^1O_2 + OH^{\ominus}$$

(Singlet oxygen is a form of dioxygen in which the electrons are in less stable orbitals with antiparallel spins. The Δ-form, with one pair of orbitals unfilled, is the principal kind of singlet oxygen.)

Almost without exception, cells that can survive exposure to dioxygen contain **superoxide dismutase** to protect themselves from this dangerous sequence of reactions by destroying superoxide:

$$2\,H^{\oplus} + O_2^{\ominus} + O_2^{\ominus} \longrightarrow H_2O_2 + O_2$$

Mitochondria and bacteria have a dismutase with Mn(III) bound to the protein; the cytosol of eukaryotic cells has another kind of dismutase with both Cu(II) and Zn(II) attached. (This is another argument for similar evolutionary origin of mitochondria and modern bacteria.) The enzyme is called a dismutase because it has the effect of transferring electrons from one molecule of superoxide to another; one molecule is reduced to oxidize the other. The enzyme, like catalase, is extremely efficient. In its presence, the rate of superoxide dismutation approaches the theoretical limit set by diffusion. It must be efficient if it is to be protective because it forms hydrogen peroxide, the other component of the reactions by which superoxide generates hydroxyl radicals and singlet oxygen. The superoxide concentration must be kept extremely low.

Effective as it may be, superoxide dismutase cannot cope with excessive loads. **Paraquat** is a herbicide that is also highly toxic to animals and bacteria because it is reduced by cellular enzymes to a free radical that immediately reacts with dioxygen to form superoxide and the original paraquat molecule. Paraquat therefore acts as a catalyst for the formation of superoxide, which leads to cell destruction. (It was used to destroy marijuana plants in Mexico, and there was some alarm about possible triggering of untoward chains of free radical reactions in potheads peddled the salvaged products.)

Phagocytic leukocytes—those that engulf alien objects in an envelope of plasma membrane—are exceptions. These cells deliberately produce both superoxide and hydrogen peroxide to destroy engulfed microorganisms. Macrophages may also destroy tumor cells in this way. When the plasma membrane of these cells is perturbed, a **respiratory burst** follows within 20 to 70 seconds. Much of the extra oxygen consumption, if not all, can be accounted for by the action of two enzymes catalyzing these reactions on the plasma membrane:

$$NADPH + 2\,O_2 \longrightarrow NADP^{\oplus} + H^{\oplus} + 2\,O_2^{\ominus}$$

$$NADH + H^{\oplus} + O_2 \longrightarrow NAD^{\oplus} + H_2O_2\ (\sim 85\%) + O_2^{\ominus}\ (\sim 15\%)$$

The reduced nicotinamide nucleotides being consumed in these reactions are rapidly regenerated through oxidation of fuels, especially glucose. (Macrophages contain phosphocreatine, perhaps to tide them over during this transient diversion of fuels from oxidative phosphorylation.)

After engulfment of the intruder, the resultant phagosome is fused with a lysosome. The phagocyte lysosomes are rich in a **myeloperoxidase.** (The green color of pus comes from the color of this hemin-containing enzyme.) This enzyme catalyzes the formation of hypochlorite from hydrogen peroxide and chloride ion. The resultant bath of Chlorox not only subjects the alien object to the action of hypochlorite as such, but also to attack by singlet oxygen produced from it:

$$H_2O_2 + Cl^{\ominus} \longrightarrow H_2O + OCl^{\ominus}$$
$$H_2O_2 + OCl^{\ominus} \longrightarrow H_2O + {}^1O_2 + Cl^{\ominus}$$

CHRONIC GRANULOMATOUS DISEASE

Results from defective oxidative metabolism in phagocytic leukocytes.

Frequent infections in childhood with formation of granuloma.

Commonly a defect in superoxide-forming NADPH oxidase.

Infection by bacteria containing catalase more common. (Phagocytes destroy those lacking catalase by oxidizing halide with accumulating H_2O_2 through the action of myeloperoxidase.)

FURTHER READING

General Reviews

A. Tzagoloff (1982). Mitochondria. Plenum.

A.N. Martonosi, ed. (1982). Membranes and Transport. Plenum. Collection of short reviews, including many on mitochondrial function.

Proton-Pumping Electron Transfer Complexes

C.L. Stayman, ed. (1982). Electrogenic Ion Pumps. Curr. Topics Membr. Transp., vol. 16. Includes discussion of mitochondrial proton pumps.

M. Wikstrom, K. Krab, and M. Saraste (1981). Proton-Translocating Cytochrome Complexes. Annu. Rev. Biochem. 50:623.

M. Gutman (1980). Electron Flux through the Mitochondrial Ubiquinone. Biochem. Biophys. Acta 594:53. Review.

B.L. Trumpower (1981). New Concepts on the Role of Ubiquinone in the Mitochondrial Respiratory Chain. J. Bioenerg. Biomemb. 13:1. Review.

M.V. Sweeney and J.C. Rabinowitz (1980). Proteins Containing 4Fe-4S Clusters—An Overview. Annu. Rev. Biochem. 49:139.

G.v. Jagow and M. Sebald (1980). b-Type Cytochromes. Annu. Rev. Biochem. 49:281.

M. Erecinska and D.F. Wilson (1978). Cytochrome c Oxidase: A Synopsis. Arch. Biochem. Biophs. 188:1.

Oxidative Phosphorylation

D.F. Wilson, M. Erecinska, and P.L. Dutton (1974). Thermodynamic Relationships in Mitochondrial Oxidative Phosphorylation. Annu. Rev. Biophys. Bioeng. 3:203.

R.L. Cross (1981). The Mechanism and Regulation of ATP Synthesis by F_1-ATPases. Annu. Rev. Biochem. 50:681.

R.H. Fillingame (1980). The Proton-Translocating Pumps of Oxidative Phosphorylation. Annu. Rev. Biochem. 49:1079.

J.J. Lemasters and W.H. Billica (1981). Nonequilibrium Thermodynamics of Oxidative Phosphorylation by Inverted Membrane Vesicles of Rat Liver Mitochondria. J. Biol. Chem. 256:12949. Evidence for differing ATP yields from the three proton-pumping complexes.

S.J. Ferguson and M.C. Sorgato (1982). Proton Electrochemical Gradients and Energy-Transduction Processes. Annu. Rev. Biochem. 51:185.

Cyanide

E.E. Robin (1977). Dysoxia. Arch. Intern. Med. 137:905. Contains discussion of cyanide poisoning.
R.F. Palmer and K.C. Lasseter (1975). Sodium Nitroprusside. N. Engl. J. Med. 292:294. Discusses liberation of cyanide.

Extramitochondrial Oxygen Consumption

B.G. Malmström (1982). Enzymology of Oxygen. Annu. Rev. Biochem. 51:21.
R.E. White and M.J. Coon (1980). Oxygen Activation by Cytochrome P-450. Annu. Rev. Biochem. 49:315.
N.E. Tolbert (1981). Metabolic Pathways in Peroxisomes and Glyoxysomes. Annu. Rev. Biochem. 50:133.
J.A. Badwey and M.L. Karnovsky (1980). Active Oxygen Species and the Functions of Phagocytic Leukocytes. Annu. Rev. Biochem. 49:695.
H.M. Hassan and I. Fridovich (1979). Paraquat and *Escherichia coli*. Mechanism of Production of Extracellular Superoxide Radical. J. Biol. Chem. 254:10846. Fridovich has been the leader in uncovering the role of superoxide and superoxide dismutase.
B.M. Babior and W.A. Peters (1981). The O_2^--Producing Enzyme of Human Neutrophils. J. Biol. Chem. 256:2321.
P.D. Lew et al. (1981). A Variant of Chronic Granulomatous Disease: Deficient Oxidative Metabolism Due to a Low-Affinity NADPH Oxidase. N. Engl. J. Med. 305:1329.

CHAPTER 23

THE CITRIC ACID CYCLE

GENERAL DESCRIPTION

The combustion of fuels can be sketched very simply if we use only bold strokes (Fig. 23–1). Each kind of major fuel is converted to **acetyl** groups, which are handled by attachment to a particular coenzyme known as **coenzyme A.** The acetyl groups, regardless of source, can then be oxidized to CO_2 and H_2O.

Simple as this sketch is, it describes the major part of metabolic processes in tissues such as muscles and nerves, in which metabolism is mainly geared to the production of ATP for physical movement or for transmission of neural impulses. We shall see later that the oxidation of acetyl groups accounts for roughly two thirds of the total oxygen consumption and ATP production in most animals.

The sequence of reactions by which acetyl groups are oxidized is known as the citric acid cycle. The reactions function as a unit to remove electrons from acetyl groups within mitochondria and to feed the electrons into the oxidative phosphorylation pathway, with the ultimate transfer to molecular oxygen. The prime purpose of the complete citric acid cycle is the complete oxidation of acetyl groups in a way that will cause the formation of ATP.

The citric acid cycle gets its name because it begins with the formation of **citrate,** the anionic form of citric acid. (The cycle is also known as the Krebs cycle after Sir Hans A. Krebs,* its discoverer, and as the tricarboxylic acid cycle.) Citrate has six carbon atoms, and it is formed by the condensation of the two-carbon acetyl group removed from acetyl coenzyme A with a four-carbon compound, **oxaloacetate** (Fig. 23–2). The sequence proceeds with an oxidation that causes the loss of one carbon atom as CO_2, leaving the five-carbon compound, **α-ketoglutarate (2-oxoglutarate).** A

*H.A. Krebs (1900–1981). One of the outstanding scientists of the century. German-born and trained, he moved to England as a refugee.

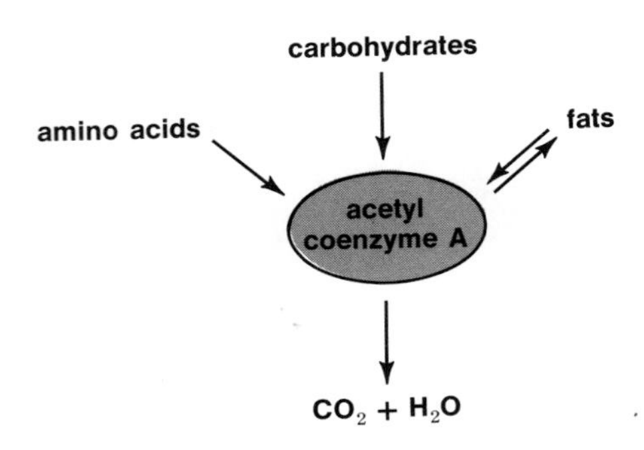

FIGURE 23–1

Acetyl coenzyme A is formed from carbohydrates, fats, and amino acids. Whatever the source, a large fraction of it is oxidized to CO_2 and H_2O, although any excess may be used to form fats for storage.

second carbon atom is then lost as CO_2 in another oxidation. Two more oxidations finally transform the four-carbon remainder into oxaloacetate, which replaces the oxaloacetate originally consumed to complete the cycle.

The dehydrogenases responsible for the four oxidations catalyze the transfer of hydrogen atoms, including the associated electrons, to NAD or directly to ubiquinone. In either case, the electrons are used for oxidative phosphorylation with an ultimate consumption of oxygen. Coupling with oxidative phosphorylation is possible because the enzymes involved in the cycle are found on the inner membrane of mitochondria or in the matrix. The result of all of this is the complete combustion of the acetyl group.

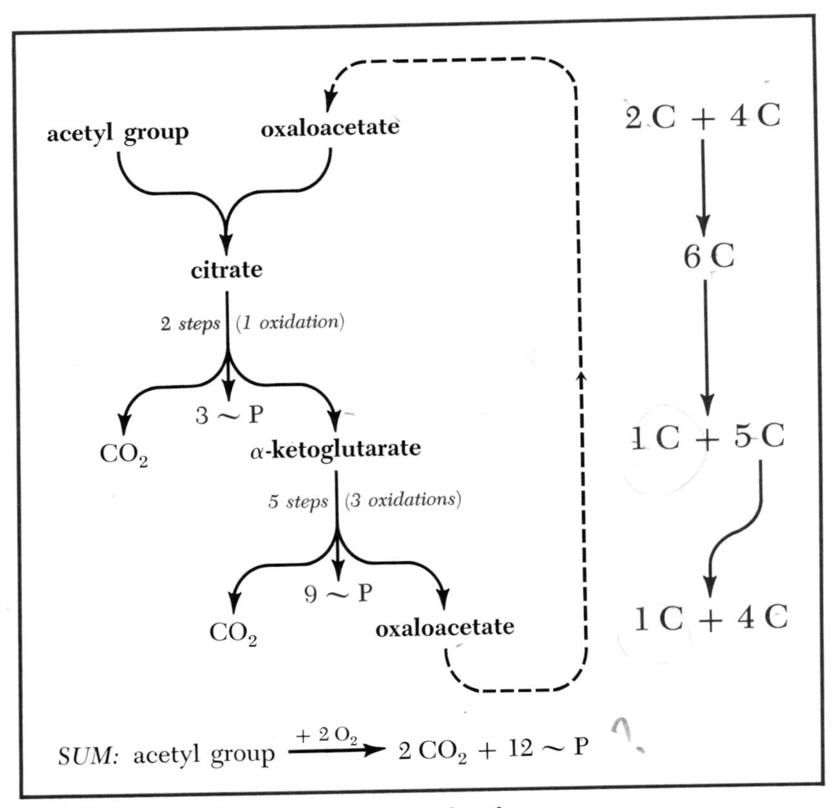

FIGURE 23–2 **Outline of the citric acid cycle.**

ACETYL COENZYME A: THE COMPOUND

The Structure of Coenzyme A

Coenzyme A contains one sulfhydryl group that is the reactive part of the coenzyme, combining with acyl groups to form thiol esters. It is therefore commonly abbreviated as CoA-SH. Using this shorthand description, acetyl coenzyme A is:

$$CH_3-\overset{\overset{\displaystyle O}{\|}}{C}-S-CoA$$

This handy way of abbreviating the structure properly emphasizes the acyl group being transported and its linkage as a thiol ester, but coenzyme A is actually quite complex. The reactive sulfhydryl group is at one end of a molecule comprising an adenine nucleotide and the three other compounds named in Figure 23–3. We have already noted that coenzymes frequently contain unusual structures—unusual in the sense that animals have lost the ability to form them, and they must be supplied in the diet. Coenzyme A has such a structure, contained in the vitamin, **D-pantothenic acid.** Animals lacking pantothenate die, but a natural deprivation of pantothenate is very rare because coenzyme A is present in all living cells and therefore in the natural foods of all animals.

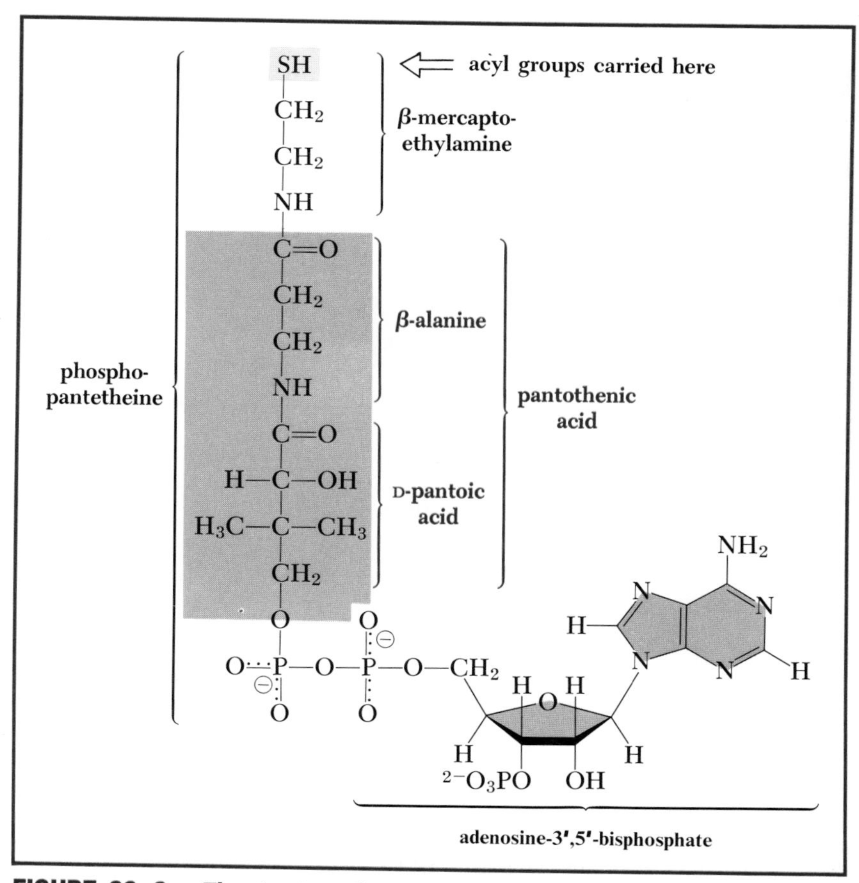

FIGURE 23–3 The structure of coenzyme A and the names of its constituents.

Energetics of Thiol Esters

Thiol esters such as acetyl coenzyme A are high-energy compounds, with a standard free energy for hydrolysis of −35,700 joules per mole at pH 7 and 38°. This means that the cleavage of a thiol ester, like the cleavage of a pyrophosphate bond, can be used to drive other reactions.* (It also follows that these thiol esters can be made only by coupling their formation to some highly exergonic processes, as we shall see in the following chapters.)

REACTIONS OF THE CITRIC ACID CYCLE

Let us now look at the individual reactions of the citric acid cycle in some detail, not only because of their importance but also because each and every one is an

---

*Thiol esters formed between glutamyl and cysteinyl side chains have recently been discovered in the C3 and C4 components of complement and in α_2-macroglobulin, a plasma protease inhibitor. The esters apparently are buried until the proteins are activated by partial hydrolysis. Upon exposure, the groups react with other proteins to form covalent bonds.

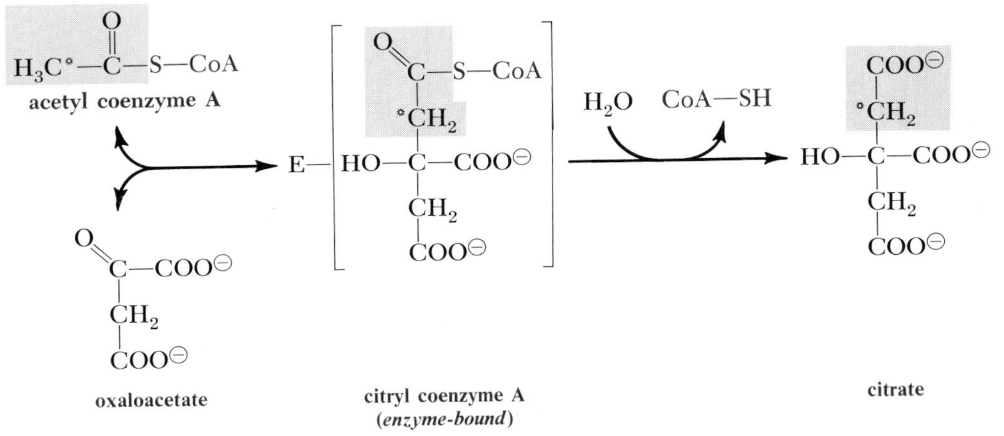

$H_3C^*—C(=O)—S—CoA$
acetyl coenzyme A

$O=C(—COO^{\ominus})—CH_2—COO^{\ominus}$
oxaloacetate

$E—[HO—C(—^*CH_2—C(=O)—S—CoA)(—COO^{\ominus})—CH_2—COO^{\ominus}]$
citryl coenzyme A
(*enzyme-bound*)

H_2O CoA—SH

$HO—C(—^*CH_2—COO^{\ominus})(—COO^{\ominus})—CH_2—COO^{\ominus}$
citrate

GENERAL TYPES OF REACTIONS

(1) Cleavage of a thiol ester of coenzyme A to drive a synthesis:

$$(A)—S—CoA + H—(B) \longrightarrow (A—B) + HS—CoA$$

(2) Condensation of carbonyl group and methylene group:

$$O{=}C< + H—\overset{|}{\underset{|}{C}}—H \longrightarrow HO—\overset{|}{\underset{|}{C}}—\overset{|}{\underset{|}{C}}—H$$

FIGURE 23–4 Citrate synthase catalyzes a condensation of acetyl coenzyme A and oxaloacetate to form citryl coenzyme A as an enzyme-bound intermediate. The asterisk indicates the position of a particular carbon atom during the reaction.

example of an important type of reaction that we shall repeatedly encounter. Gaining a good grasp now will save much later effort. Furthermore, the individual reactions of the cycle and the compounds involved in them are also parts of other metabolic processes that we shall examine later.

Formation of Citrate (2C + 4C → 6C). The cycle begins with the condensation of acetyl coenzyme A and oxaloacetate to form citrate (Fig. 23–4), catalyzed by the enzyme, **citrate synthase.** As the figure shows, the reaction proceeds in two parts on the enzyme surface; the initial condensation results in the formation of citryl coenzyme A (the coenzyme A thiol ester of citric acid). The citryl coenzyme A is then hydrolyzed to release free citrate and coenzyme A into solution.

The formation of a carbon-to-carbon bond by the condensation of a carbonyl group with an activated methylene carbon is a common type of reaction in metabolism. These are familiar reactions to the organic chemist, and are variants of the classical aldol condensation, which we examined in connection with the fructose bisphosphate aldolase reaction (p. 283).

It so happens that free acetate can be made to condense with oxaloacetate, so what is the point of evolving a reaction that utilizes the coenzyme A ester, since it is destroyed before the reaction is over? The answer lies in the position of equilibrium of the reaction. The condensation of acetate and oxaloacetate by themselves has an equilibrium position such that only relatively high concentrations of the starting materials will result in significant formation of citrate. However, making acetyl coenzyme A the starting material adds the highly exergonic hydrolysis of the coenzyme A ester to the reaction. The sum of these two processes makes the formation of citrate an almost irreversible reaction, so much so that it can readily proceed even with an intramitochondrial concentration of oxaloacetate of the order of 10 micromolar.

$$K' = \frac{[\text{citrate}][\text{CoA}]}{[\text{acetyl CoA}][\text{oxaloacetate}]} = 2.24 \times 10^6.$$

Formation of α-Ketoglutarate (6C + NAD → 5C + CO_2 + NADH)

Citrate → Isocitrate. The metabolism of citrate begins by converting it into a compound that can be oxidized. Citrate has a tertiary alcohol group, which could not be attacked without breaking a carbon bond, but rearrangement into the isomer, isocitrate, creates a secondary alcohol group that can be oxidized. This rearrangement is achieved by removing water and adding it back (see figure at top of next page). (The formula of citrate is rotated from that in the previous sketch to show the reacting end at the top.) The dehydration and hydration involved in the rearrangement are catalyzed by the same enzyme, **aconitase.** When aconitase catalyzes the addition of water to the double bond of *cis*-aconitate, the hydroxyl group may be added to either carbon; in one case citrate is formed and in the other, isocitrate. The enzyme will also catalyze the removal of water from either citrate or isocitrate to form *cis*-aconitate as an intermediate bound on the enzyme. Aconitase is an example of a **hydratase,** an enzyme catalyzing the reversible hydration of double bonds. (It is unlike many enzymes of this type, which usually require no cofactors, in that it contains an iron-sulfide complex. This is baffling, because there is no transfer of electrons during the reaction.) The interconversion of alcohols and unsaturated compounds by addition or removal of water is a common type of biochemical reaction, and the standard free energy change is usually small enough so that the reaction can go in either direction to a biologically significant extent. In this case the equilibrium

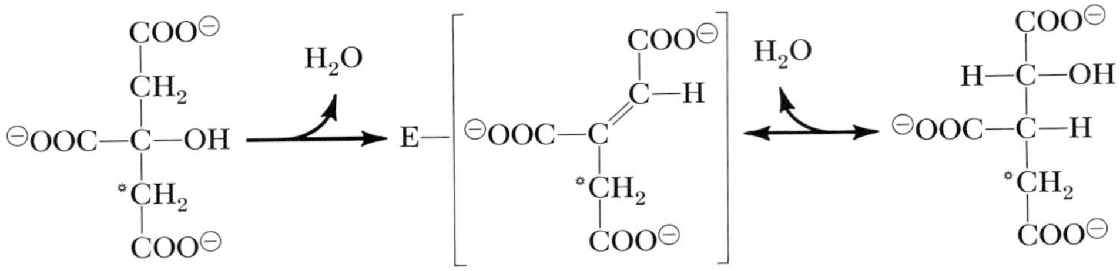

$$\begin{array}{c} COO^{\ominus} \\ | \\ CH_2 \\ | \\ {}^{\ominus}OOC-C-OH \\ | \\ {}^{*}CH_2 \\ | \\ COO^{\ominus} \end{array} \xrightarrow{\nearrow H_2O} E-\left[\begin{array}{c} COO^{\ominus} \\ | \\ C-H \\ \| \\ {}^{\ominus}OOC-C \\ | \\ {}^{*}CH_2 \\ | \\ COO^{\ominus} \end{array}\right] \overset{H_2O}{\longleftrightarrow} \begin{array}{c} COO^{\ominus} \\ | \\ H-C-OH \\ | \\ {}^{\ominus}OOC-C-H \\ | \\ {}^{*}CH_2 \\ | \\ COO^{\ominus} \end{array}$$

citrate — ***cis*-aconitate (*enzyme-bound*)** — **isocitrate**

> **GENERAL TYPE OF REACTION**
>
> **Interconversion of alcohols and unsaturated compounds by hydration and dehydration:**
>
> $$\begin{array}{cc} | & | \\ -C- & C- \\ | & | \\ H & OH \end{array} \longrightarrow H_2O + \begin{array}{c} | \quad | \\ -C=C- \end{array}$$

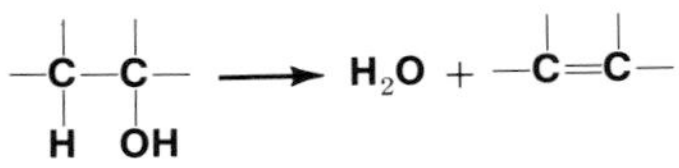

mixture contains about 91 per cent citrate, 3 per cent *cis*-aconitate, and 6 per cent isocitrate. Since the intramitochondrial concentration of citrate is greater than 1 millimolar, the equilibrium concentration of isocitrate would be of the order of 0.1 mM, a reasonable level for metabolic intermediates.

Under ordinary circumstances, the reactions go through the sequence as given: citrate → (aconitate) → isocitrate. There is nothing peculiar about aconitase to cause this: it only catalyzes the attainment of equilibrium. The net flow of metabolites goes in one direction because citrate is constantly being formed by the irreversible citrate synthase reaction, and isocitrate is constantly being removed by the next reaction, in which isocitrate is oxidized.

Oxidative Decarboxylation of Isocitrate. The alcohol group of isocitrate is oxidized by NAD in a reaction catalyzed by an isocitrate dehydrogenase:

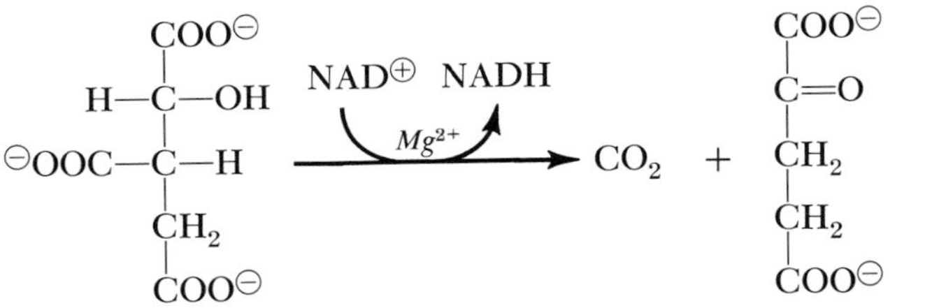

$$\begin{array}{c} COO^{\ominus} \\ | \\ H-C-OH \\ | \\ {}^{\ominus}OOC-C-H \\ | \\ CH_2 \\ | \\ COO^{\ominus} \end{array} \xrightarrow[Mg^{2+}]{NAD^{\oplus} \quad NADH} CO_2 + \begin{array}{c} COO^{\ominus} \\ | \\ C=O \\ | \\ CH_2 \\ | \\ CH_2 \\ | \\ COO^{\ominus} \end{array}$$

isocitrate — **α-ketoglutarate**

> **GENERAL TYPE OF REACTION**
>
> **Oxidative decarboxylation of a 3-hydroxy acid:**
>
> $$\begin{array}{c} OH \\ | \\ -C- \\ | \\ H \end{array}\!\!\begin{array}{c} | \\ -C-COO^{\ominus} \\ | \end{array} + NAD^{\oplus} \longrightarrow \begin{array}{c} O \\ \| \\ -C- \end{array}\!\!\begin{array}{c} | \\ -C-H \\ | \end{array} + CO_2 + NADH$$

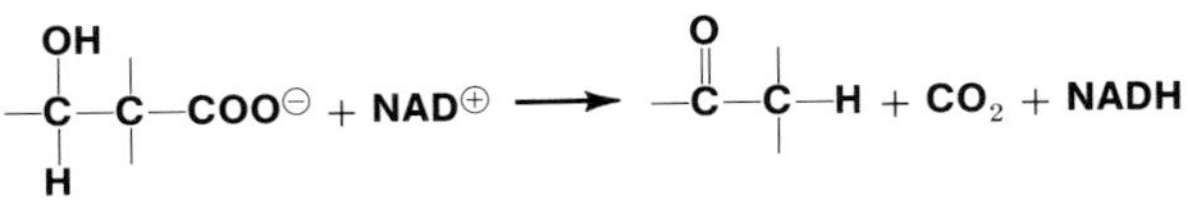

This reaction results not only in the conversion of the alcohol to the corresponding ketone, but also in the loss of one of the carboxyl groups as CO_2 to form α-ketoglutarate. The reaction is therefore an example of an **oxidative decarboxylation.** It is quite common, although not obligatory, for the enzymatic oxidation of the hydroxyl group in a 3-hydroxyl carboxylic acid to cause the simultaneous loss of CO_2. In the example at hand, this not only occurs, but it does so without even a transitory appearance of the intermediate β-keto compound. The concentration of CO_2 is low, owing to its constant removal through the lungs, and this makes the reaction effectively irreversible.

The enzyme we are discussing apparently catalyzes both the oxidation and the decarboxylation in a concerted way, but there are related enzymes that catalyze the two parts separately. When this happens, the initial oxidation of isocitrate by NAD causes the temporary formation of the corresponding ketone known as **oxalosuccinate** on the enzyme:

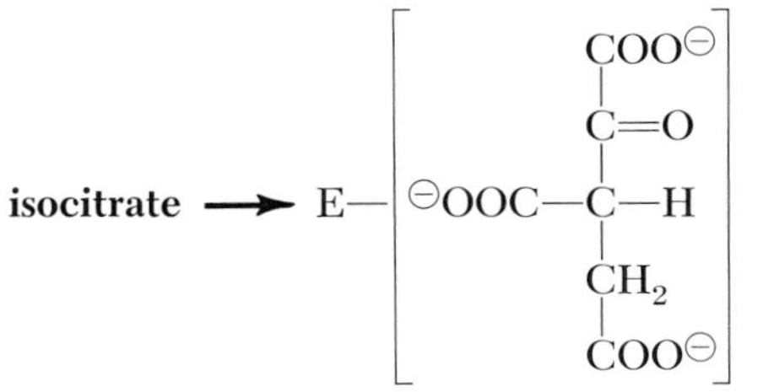

$$\textbf{isocitrate} \longrightarrow E-\left[{}^{\ominus}OOC-\underset{\underset{\displaystyle COO^{\ominus}}{|}}{\underset{\underset{\displaystyle CH_2}{|}}{\overset{\overset{\overset{\displaystyle COO^{\ominus}}{|}}{\displaystyle C{=}O}}{\overset{|}{C}}}}-H \right]$$

This is a straightforward alcohol dehydrogenase type of reaction, but notice that the carbonyl group of the intermediate oxalosuccinate is in the β-position relative to the middle carboxylate group. (Of course, it is also in the α-position relative to the top carboxylate group, but that is not relevant to our point.) Now, we know from organic chemistry that β-ketoacids are readily decarboxylated:

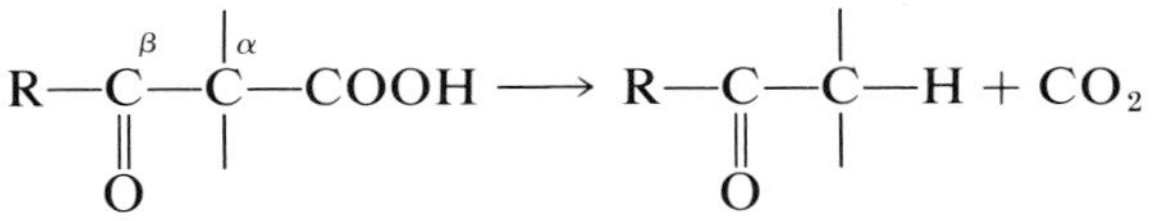

$$R-\underset{\underset{\displaystyle O}{\|}}{\overset{\beta}{C}}-\underset{|}{\overset{|\,\alpha}{C}}-COOH \longrightarrow R-\underset{\underset{\displaystyle O}{\|}}{C}-\underset{|}{\overset{|}{C}}-H + CO_2$$

Therefore, we might expect the intermediate oxalosuccinate to readily lose CO_2, and it does.

Role of Mg^{2+}. Isocitrate dehydrogenase requires Mg^{2+} for activity like many enzymes that catalyze β-decarboxylations. Enzymes catalyzing dehydrogenations without decarboxylation do not require metallic ions for activity.

Oxidative Decarboxylation of α-Ketoglutarate (5C + NAD → 4C + CO_2 + NADH)

The next reaction of the citric acid cycle is an oxidation of α-ketoglutarate by NAD in such a way that the first carboxyl group appears as CO_2. By transferring the four-carbon succinyl group onto coenzyme A, part of the energy of the oxidation is captured as the energy-rich thiol ester, succinyl coenzyme A (next page).

α-Ketoglutarate is an example of a 2-ketocarboxylate, and it is a convenient formalism to regard the oxidation of these compounds as the sum of two organic reactions. One is a simple decarboxylation to produce CO_2 and an aldehyde:

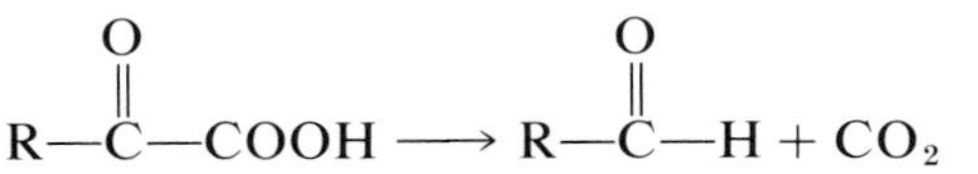

$$R-\overset{\overset{\displaystyle O}{\|}}{C}-COOH \longrightarrow R-\overset{\overset{\displaystyle O}{\|}}{C}-H + CO_2 \quad \Delta G^{0\prime}\ -23{,}000\ \text{J mole}^{-1}$$

$$\begin{matrix} COO^{\ominus} \\ | \\ C{=}O \\ | \\ CH_2 \\ | \\ CH_2 \\ | \\ COO^{\ominus} \end{matrix} + NAD^{\oplus} + CoA{-}SH \xrightarrow[\alpha\text{-ketoglutarate dehydrogenase complex}]{CO_2} \begin{matrix} O \\ \| \\ C{-}S{-}CoA \\ | \\ CH_2 \\ | \\ CH_2 \\ | \\ COO^{\ominus} \end{matrix} + NADH$$

α-ketoglutarate **succinyl coenzyme A**

GENERAL TYPE OF REACTION

Oxidative decarboxylation of 2-keto acid to form thiol ester:

$$R{-}\overset{O}{\overset{\|}{C}}{-}COO^{\ominus} + NAD^{\oplus} + CoA{-}SH \longrightarrow R{-}\overset{O}{\overset{\|}{C}}{-}S{-}CoA + CO_2 + NADH$$

and the other is the oxidation of the aldehyde:

$$R{-}\overset{O}{\overset{\|}{C}}{-}H + NAD^{\oplus} + H_2O \longrightarrow R{-}COO^{\ominus} + NADH + 2H^{\oplus}$$
$$\Delta G^{0'} \sim -50{,}000 \text{ J mole}^{-1}.$$

These steps could proceed independently, and do so in some microorganisms, but the two steps combined have a negative standard free energy change that is ample for the creation of a high-energy bond; the reaction is modified to provide for the formation of the coenzyme A ester, rather than the free anion of the carboxylic acid product.

The formation of the high-energy thiol ester bond is quite independent of the concomitant formation of high-energy phosphate by mitochondrial electron transport; it serves to tap the extra free energy available from the oxidative decarboxylation of the 2-ketocarboxylate, which is substantially greater than the free energy released during the oxidation of alcohols.

The immediately preceding isocitrate dehydrogenase reaction is also an oxidative decarboxylation, but of a 3-hydroxycarboxylate. The standard free energy change for that reaction is only −400 J per mole, far too little to support the formation of an extra high-energy bond even at the low concentrations of CO_2 occurring in tissue.

α-Ketoglutarate Dehydrogenase Complex. We can see that there are at least three things going on during the conversion of α-ketoglutarate to succinyl coenzyme A: decarboxylation, oxidation, and formation of the thiol ester, and all of this is too much to be accomplished by a single enzymatic reaction. Instead, cells construct a marvelous example of an integrated enzyme complex in which a sequence of discrete reactions is carried out in a coordinated way. This complex is a distinctive arrangement of different protein molecules. Each kind of protein catalyzes a separate part of the reaction sequence, but the association into a single complex makes it possible to link the reactions together without release of intermediates into solution.

The decarboxylating dehydrogenase is the component of the complex that initially reacts with α-ketoglutarate; it contains a coenzyme, **thiamine pyrophosphate,** to carry an aldehyde intermediate.

Thiamine is another of the essential compounds whose synthesis has been deleted during the evolution of mammals, and it is therefore a vitamin. It is used in

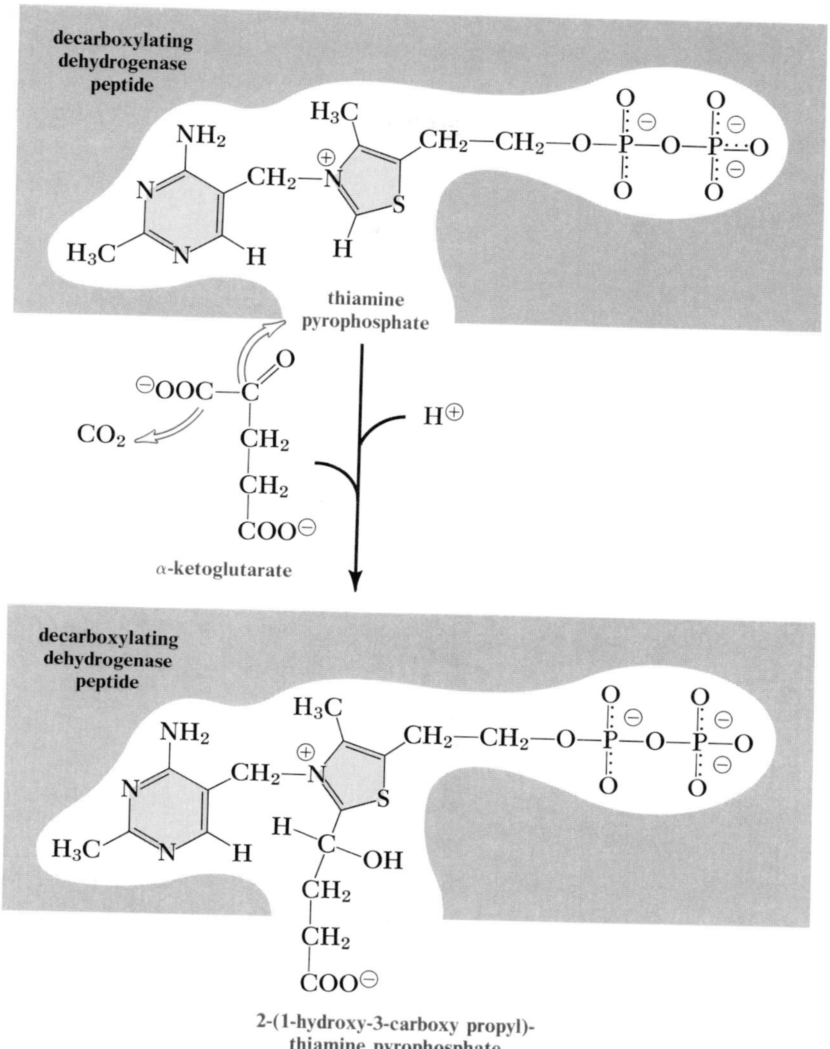

decarboxylating
dehydrogenase
peptide

H_3C NH_2 CH_2—N⊕ CH_2—CH_2—O—P—O—P—O S N N H_3C H H

thiamine
pyrophosphate

$^{\ominus}OOC$—C=O

CO_2

CH_2

CH_2

$COO^{\ominus}$

α-ketoglutarate

$H^{\oplus}$

decarboxylating
dehydrogenase
peptide

H_3C NH_2 CH_2—N⊕ CH_2—CH_2—O—P—O—P—O S N N H_3C H

H—C—OH

CH_2

CH_2

$COO^{\ominus}$

2-(1-hydroxy-3-carboxy propyl)-
thiamine pyrophosphate

FIGURE 23-5

Thiamine pyrophosphate, attached to the decarboxylating dehydrogenase component of α-ketoglutarate dehydrogenase complex, accepts a potential succinic semialdehyde group from α-ketoglutarate, with a simultaneous loss of CO_2.

other reactions than the one at hand for carrying potential aldehyde groups. In the present reaction (Fig. 23-5), the carbonyl carbon of α-ketoglutarate is bonded to thiamine during the accompanying cleavage of the carboxyl group.

The bound intermediate would be succinic semialdehyde if it were released. Instead, the potential succinaldehydate group is oxidized by transfer to a bound coenzyme, **lipoic acid,** on another protein in the complex, **transsuccinylase:**

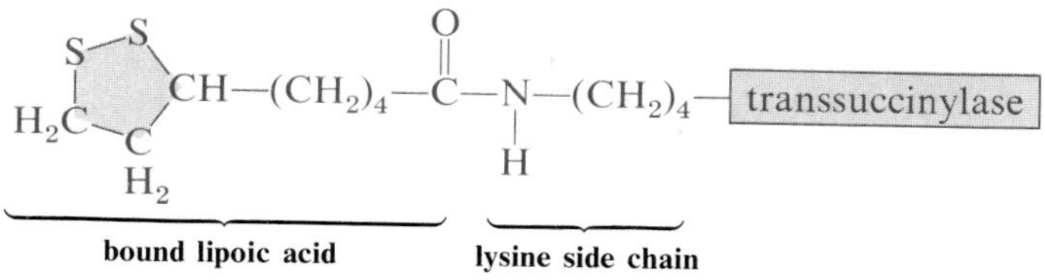

S—S H_2C C H_2 CH—$(CH_2)_4$—C(=O)—N(H)—$(CH_2)_4$— transsuccinylase

bound lipoic acid **lysine side chain**

The reactive part of lipoic acid is a disulfide group, which acts as both an oxidizing agent for the substrate "aldehyde" and as a receptor for the resultant succinyl group. Two things are happening in this single reaction: The lipoyl group is being reduced to a dihydrolipoyl group, to which the succinyl group is attached as a thiol ester.

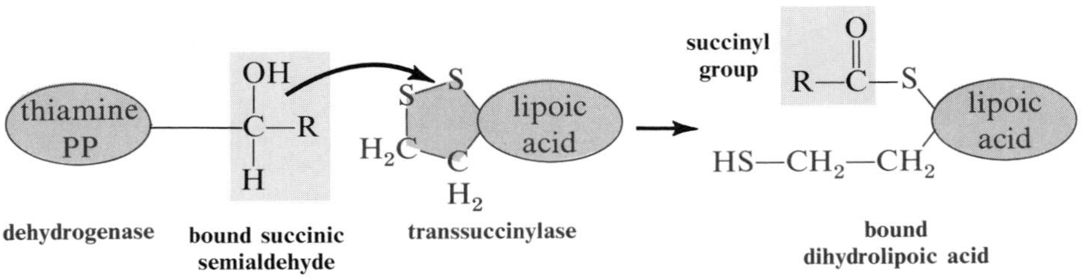

dehydrogenase — bound succinic semialdehyde — transsuccinylase → succinyl group — bound dihydrolipoic acid

TRANSFER OF SUCCINYL GROUP TO COENZYME A. The transsuccinylase component carries lipoyl groups as oxidizing agents, but the protein is also an enzyme, and its function is to convert the high-energy succinyl thiol ester to a usable form by transferring the succinyl group to an incoming molecule of coenzyme A, enabling it to leave as succinyl coenzyme A:

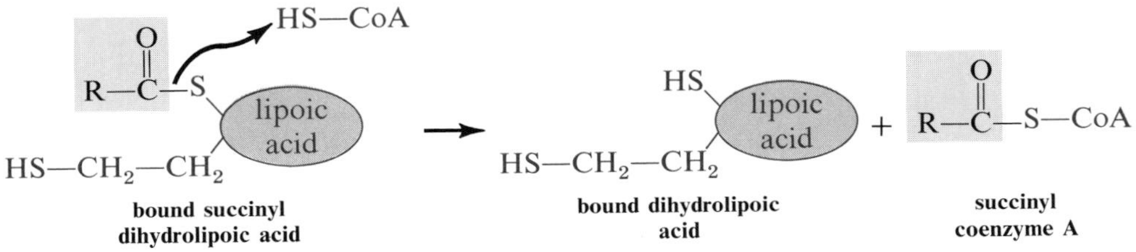

bound succinyl dihydrolipoic acid + HS—CoA → bound dihydrolipoic acid + succinyl coenzyme A

DIHYDROLIPOAMIDE DEHYDROGENASE. The action of transsuccinylase completes the oxidative decarboxylation of α-ketoglutarate and the capture of part of the energy as a thiol ester. However, the electrons removed from the substrate are now present in the two thiol groups of a dihydrolipoic acid residue. The disposal of these electrons is handled by the third kind of protein in the enzyme complex, dihydrolipoamide dehydrogenase. This dehydrogenase catalyzes the oxidation of dihydrolipoyl groups by NAD, completing the overall reaction of the complex:

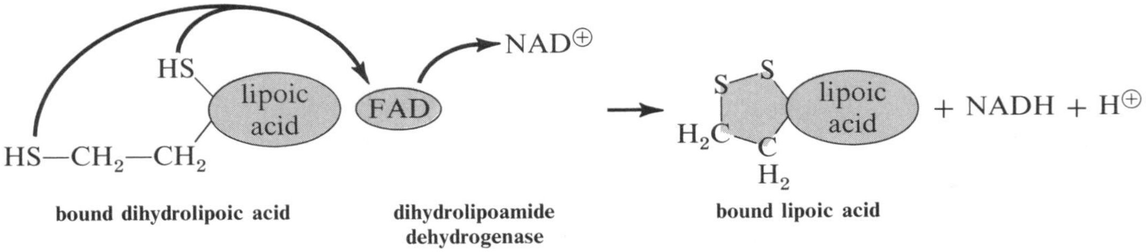

bound dihydrolipoic acid — dihydrolipoamide dehydrogenase (FAD), $NAD^{\oplus}$ → bound lipoic acid + NADH + $H^{\oplus}$

The two results are the final restoration of the transucсinylase to its original form, and the formation of NADH, which can be used for oxidative phosphorylation in the inner membrane of the mitochondria.

Dihydrolipoamide dehydrogenase is a flavoprotein. It seems odd for a flavoprotein to be used in generating NADH, since most flavoproteins are stronger oxidizing agents than NAD, but the transfer of electrons from two thiol groups to NAD is apparently a special case, because other reactions of this sort also are catalyzed by

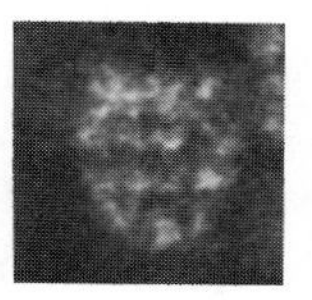

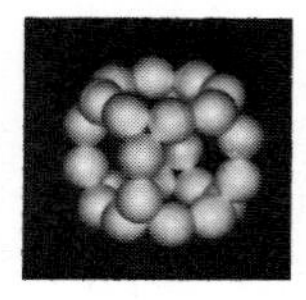

FIGURE 23-6

The α-ketoglutarate dehydrogenase complex is built around a transsuccinylase core. (*Left*) An electron micrograph of one molecule of complete complex as isolated from bovine kidney. From such seemingly indistinct images of the complex and its components, it is possible to deduce a model for the core (*right*). (Both photographs courtesy of Prof. Lester J. Reed.)

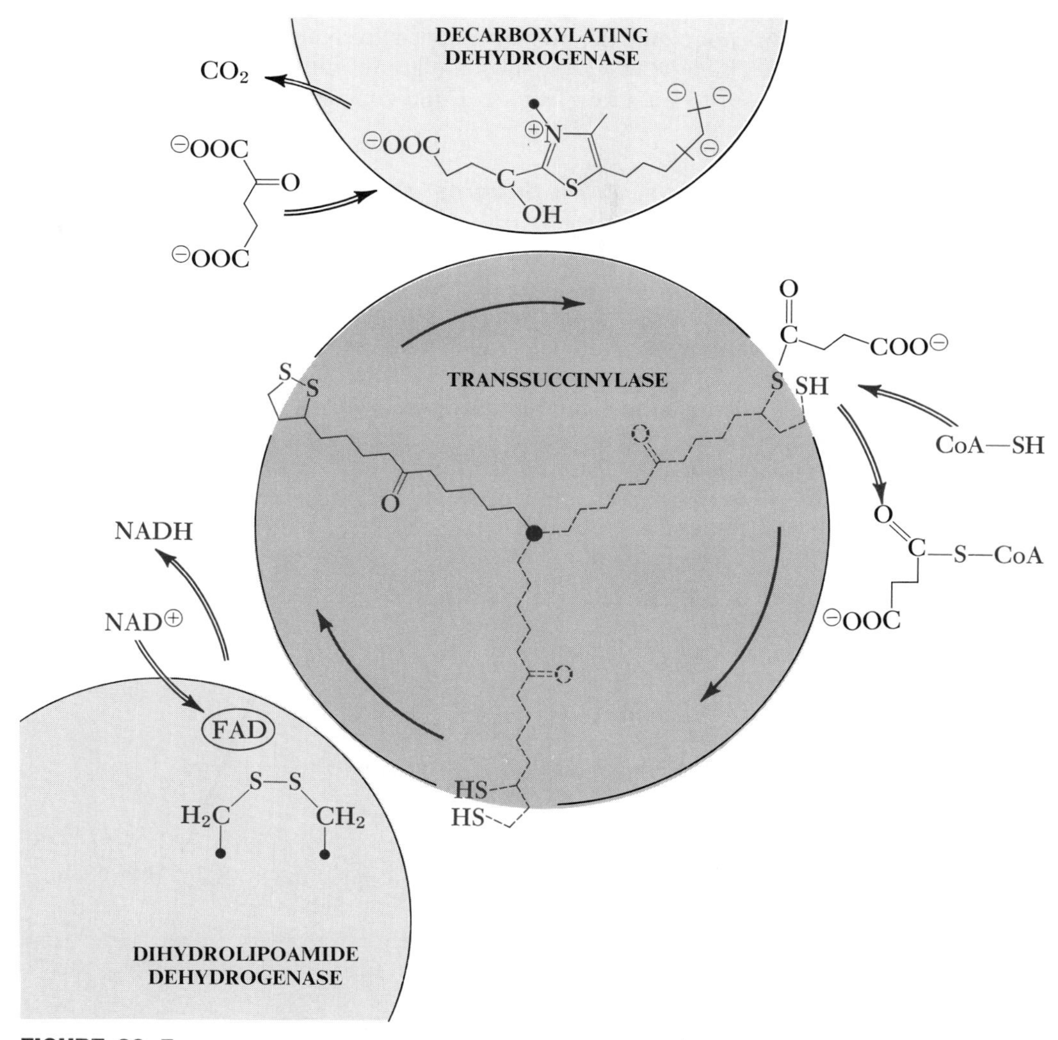

FIGURE 23-7

Integrated action of the α-ketoglutarate dehydrogenase complex is believed to depend upon the wide area that can be covered by the long lipoyl-lysyl chains (*center*) on the transsuccinylase polypeptides. The substrate, α-ketoglutarate, is shown entering at the upper left with the loss of CO_2 and attachment of the resultant succinic semialdehyde to thiamine pyrophosphate on the decarboxylating dehydrogenase component of the complex. The subsequent transfer to a lipoyl group and then to coenzyme A is shown at *center right*. The reoxidation of a dihydrolipoyl group by a separate dehydrogenase component is shown at *lower left*, with transfer of the electrons to NAD, forming NADH.

similar flavoproteins. The reaction involves the cooperative action of the FAD coenzyme and a residue of cystine in the enzyme; the mechanism is not clear, but it may involve a covalent combination between an SH group and the flavin so that a pair of electrons is, in effect, shared between them.

GEOMETRY OF THE DEHYDROGENASE COMPLEX. The α-ketoglutarate dehydrogenase complex is a huge molecule, comparable in size to ribosomes. The transsuccinylase components, some 24 identical polypeptide subunits, are in its core (Fig. 23–6), with the decarboxylating dehydrogenase and dihydrolipoamide dehydrogenase components arranged on the periphery.

Lipoate is attached to the core transsuccinylase by forming an amide bond with lysine side chains. This places the reactive disulfide group at the end of a long flexible chain and some 12 atoms removed from the polypeptide backbone. The ability of the chain to swing the disulfide group into contact with the different proteins is believed to be a key feature of the enzyme complex (Fig. 23–7).

Recovery of Energy From Succinyl Coenzyme A

The next step in the citric acid cycle is cleavage of succinyl coenzyme A to drive a synthesis of GTP from GDP and P_i (Fig. 23–8). This recovers the potential energy of the thiol ester bond as a high-energy phosphate. Here we have an example of a **substrate-level phosphorylation,** in which a high-energy phosphate is directly created by reactions of a metabolite, rather than by electron flow on the mitochondrial inner membrane. The reaction is catalyzed by **succinyl CoA synthetase,** and it involves the formation of an intermediate succinyl phosphate. The phosphate is transferred, first onto the imidazole side chain of a histidine residue in the enzyme and then onto guanosine diphosphate, creating guanosine triphosphate. Why is GTP a product

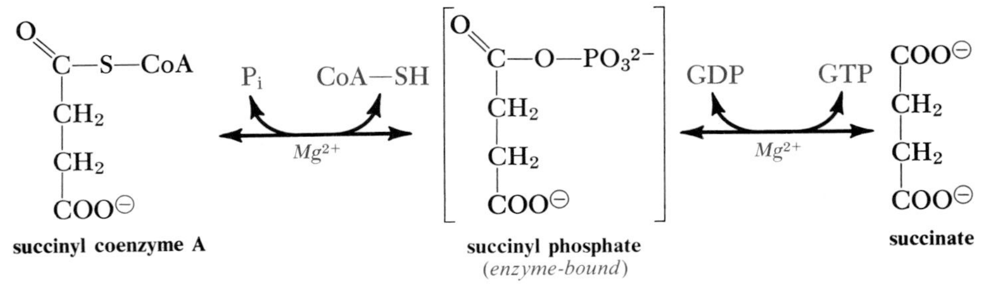

$$\underset{\textbf{succinyl coenzyme A}}{O{=}C(-S-CoA)-CH_2-CH_2-COO^{\ominus}} \xrightarrow[Mg^{2+}]{P_i \;\; CoA{-}SH} \underset{\substack{\textbf{succinyl phosphate}\\ (enzyme\text{-}bound)}}{\left[O{=}C(-O-PO_3^{2-})-CH_2-CH_2-COO^{\ominus}\right]} \xrightarrow[Mg^{2+}]{GDP \;\; GTP} \underset{\textbf{succinate}}{COO^{\ominus}-CH_2-CH_2-COO^{\ominus}}$$

GENERAL TYPE OF REACTION

Reversible synthesis of high-energy phosphate by cleaving a thiol ester:

$$R{-}\overset{O}{\overset{\|}{C}}{-}S{-}CoA + P_i + NDP \longrightarrow R{-}COO^{\ominus} + NTP + HS{-}CoA$$

FIGURE 23–8 Succinyl coenzyme A is cleaved by P_i to form an intermediate enzyme-bound succinyl phosphate. The high-energy phosphate is then transferred to GDP, forming GTP.

rather than ATP, which is just as likely a possibility on energetic and mechanistic grounds? We can speculate that GTP has a dual use—a signal for regulation of metabolism in addition to its participation in the high-energy phosphate pool. In any event, we already noted in Chapter 19 that GTP and ADP are equilibrated to form GDP and ATP by a nucleoside diphosphokinase. Mitochondria also contain a GTP:AMP phosphotransferase, so the high-energy phosphate created by the cleavage of the succinyl coenzyme A is fully available for all reactions in which ATP is used.

Synthetases. The enzyme is named synthetase to be consistent with the nomenclature of other enzymes of this type, because the reaction is freely reversible, and synthetases are enzymes catalyzing the formation of compounds at the expense of high-energy phosphate.

Conversion of Succinate to Oxaloacetate

The remainder of the reactions in the citric acid cycle are concerned with the regeneration of oxaloacetate from succinate, with a concomitant use of the electrons for oxidative phosphorylation.

Oxidation of Succinate. Succinate is oxidized to fumarate by succinate dehydrogenase (Fig. 23–9). We have already noted (p. 401) that this dehydrogenase is a complex of a flavoprotein and iron-sulfide proteins on the inner mitochondrial membrane. (Flavin adenine dinucleotide is the coenzyme.) The dehydrogenase transfers

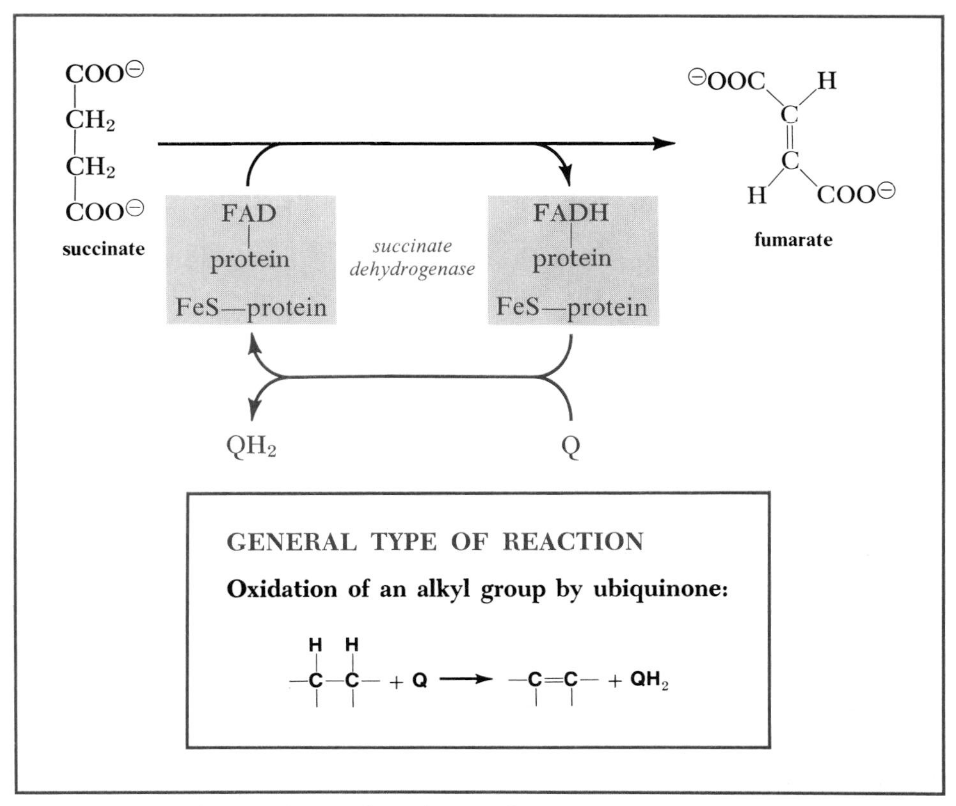

FIGURE 23–9 **The oxidation of succinate to fumarate.**

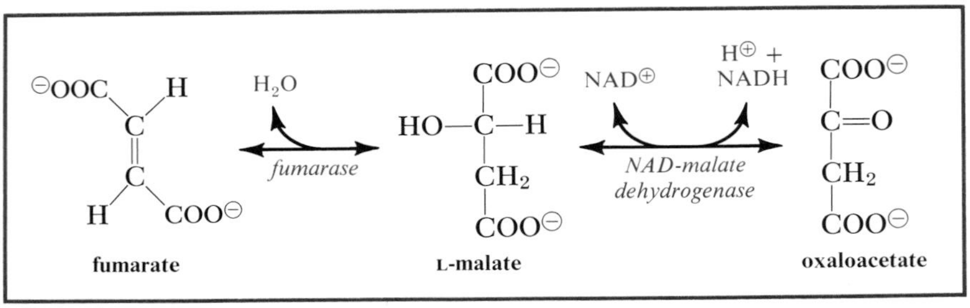

fumarate ($^{\ominus}OOC$–CH=CH–$COO^{\ominus}$) + H_2O ⇌ (*fumarase*) L-malate ($COO^{\ominus}$–CH(OH)–CH_2–$COO^{\ominus}$); L-malate + $NAD^{\oplus}$ ⇌ (*NAD-malate dehydrogenase*) oxaloacetate ($COO^{\ominus}$–C=O–CH_2–$COO^{\ominus}$) + $H^{\oplus}$ + NADH

FIGURE 23–10 **Fumarate is equilibrated with oxaloacetate by a hydration to form malate, followed by an oxidation with NAD.**

electrons directly to ubiquinone, bypassing the first phosphorylation site in the electron transfer scheme. It follows that the overall transfer of electrons from succinate to oxygen will result in the generation of only two high-energy phosphates.

Regeneration of Oxaloacetate. Water is added across the double bond of fumarate to form L-malate, which then undergoes the final oxidation of the citric acid cycle to become oxaloacetate (Fig. 23–10). We have here typical hydratase and alcohol dehydrogenase reactions. The NADH produced by the dehydrogenase is available for oxidative phosphorylation, and the oxaloacetate replaces the molecule originally used to create citrate.

THE COMPLETE CYCLE

Total Stoichiometry

Now that we have gone over the individual steps, let us survey the complete cycle as it is portrayed in Figure 23–11. If the cycle is really oxidizing the acetyl group of acetyl coenzyme A to CO_2 and H_2O, the sum of all the equations for the individual reactions must add up to the formal stoichiometry:

$$(1)\quad H_3C{-}\overset{\overset{\displaystyle O}{\|}}{C}{-}S{-}CoA + 2\,O_2 \longrightarrow 2\,CO_2 + CoA{-}SH + H_2O$$

after any high-energy phosphate that is produced has been utilized and converted back to the starting materials. If we add all of the reactions in the cycle, we arrive at the sum:

$$(2)\quad H_3C{-}\overset{\overset{\displaystyle O}{\|}}{C}{-}S{-}CoA + 3\,NAD^{\oplus} + Q + GDP + P_i + 2\,H_2O \longrightarrow 2\,CO_2 + CoA{-}SH + 3\,NADH + QH_2 + GTP.$$

We see here the production of two molecules of CO_2 and the four transfers of pairs of electrons necessary for the oxidation. Neglecting ATP production, electrons will be transferred eventually to oxygen according to the equation:

$$(3)\quad 3\,NADH + QH_2 + 2\,O_2 \longrightarrow 3\,NAD^{\oplus} + Q + 4\,H_2O.$$

The concomitant pumping of protons and generation of ATP is neglected because, as

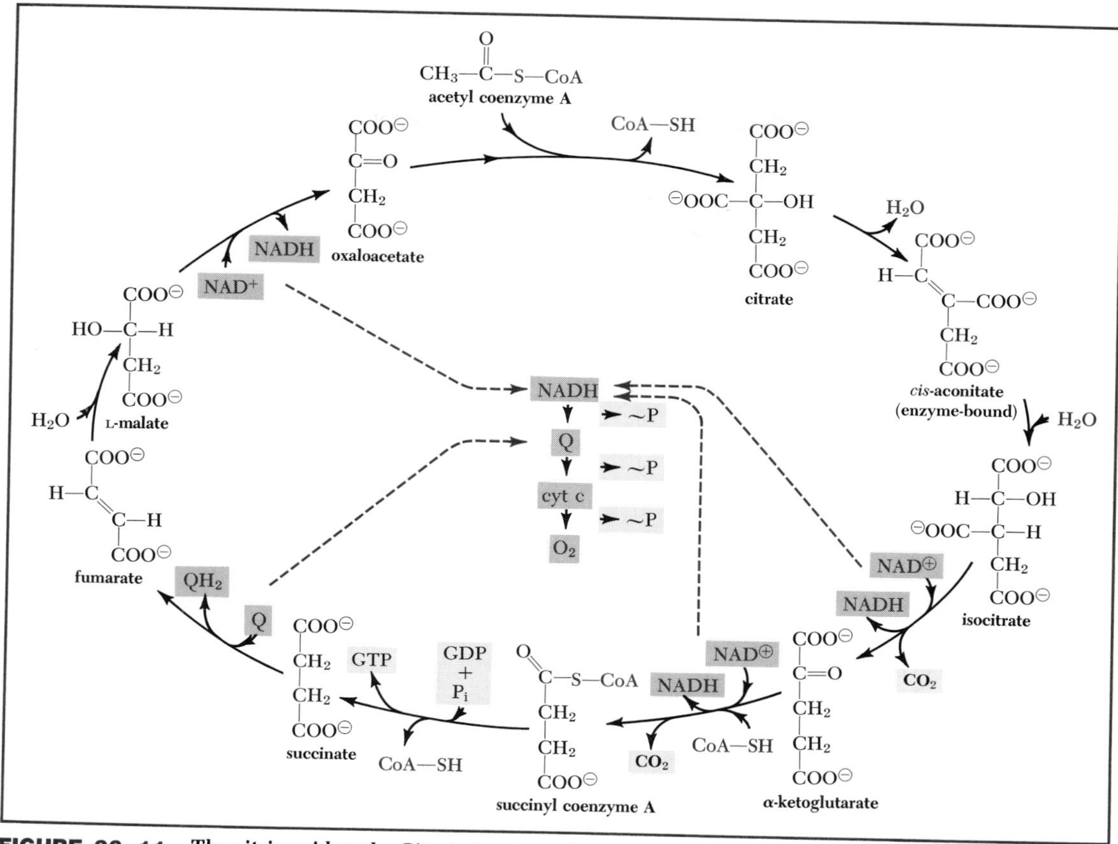

FIGURE 23–11 The citric acid cycle. Citrate is oriented so as to correspond atom-for-atom with *cis*-aconitate and isocitrate; when pictured this way, the atoms in citrate derived from acetyl coenzyme A are at the bottom of the formula. The coenzymes involved in electron transfers are shown in brown shaded boxes. Components of the high-energy phosphate pool, and also the CO_2 produced by the cycle, are shown in gray shaded boxes.

discussed later, the potential energy of the electrochemical gradient and the phosphate potential is first dissipated elsewhere before the citric acid cycle can operate to regenerate the energy supply. The result is maintenance of the status quo rather than accumulation of pumped protons or of high-energy phosphate insofar as the overall reaction in the whole body is concerned. For the same reason, when GTP is utilized in other reactions, the effect will be:

$$(4) \quad GTP + H_2O \longrightarrow GDP + P_i.$$

If we add equations (*3*) and (*4*) to the sum for the cycle, equation (*2*), we indeed have the formal stoichiometry for the oxidation of acetyl groups.

In addition to showing the oxidation of the acetyl group, there is another important lesson in the stoichiometry. **The citric acid cycle does not involve the net production or consumption of oxaloacetate or of any other constituent of the cycle itself.** The reactions of the citric acid cycle do not provide a route for making additional oxaloacetate from acetyl groups. Failure to appreciate this point has led intelligent men into serious error in the past, and it must be kept firmly in mind to avoid making the same mistake in the future.

High-Energy Phosphate Yield. A grand total of 12 molecules of ATP can be formed during complete oxidation of the acetyl group if there is perfect coupling between oxidation and phosphorylation. Three molecules of NADH are formed. The oxidation of each of these by O_2 results in the formation of three molecules of ATP from ADP and P_i, for a total of nine molecules of ATP. In addition, one molecule of ubiquinol is directly formed and the oxidation of this by O_2 results in the formation of two more molecules of ATP. Finally, the molecule of GTP formed in the succinyl CoA synthetase reaction is equivalent to an additional molecule of ATP:

$$\text{GTP} + \text{ADP} \longrightarrow \text{GDP} + \text{ATP}.$$

Since 2 O_2 are consumed in the citric acid cycle, the P:O ratio, the ratio of high-energy phosphates produced to atoms of oxygen consumed, is 12:4 or 3.00. We later shall have occasion to use this number in assessing the relative energy production obtained from various fuels.

The Main-Line Sequence

Much of the citric acid cycle is made from a sequence of reactions that occurs again and again in biochemistry, sometimes complete and sometimes as fragments.

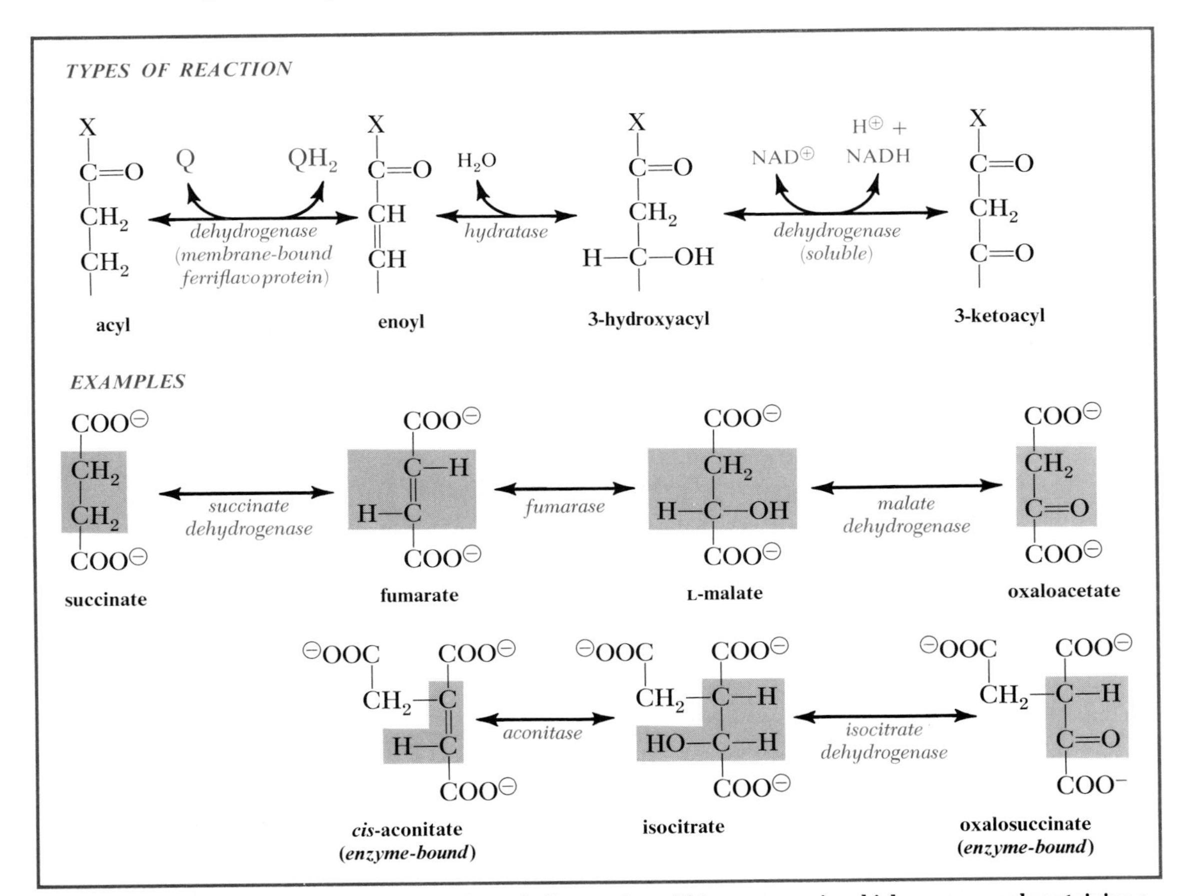

FIGURE 23-12 The mainline sequence of metabolic reactions. This sequence, in which a compound containing a segment of saturated hydrocarbon chain (acyl group) is successively converted to the unsaturated analogue (enoyl group), an alcohol (hydroxyacyl group), and then to a carbonyl compound (ketoacyl group) is common in metabolism. The citric acid cycle is mainly composed of reactions of this sort, as shown in the examples.

This sequence begins with the anion or ester of a saturated carboxylic acid, which is successively modified by oxidation to the unsaturated compound, followed by hydration of the double bond to form an alcohol group, and subsequent oxidation of this group to form a keto compound (Fig. 23–12). The figure shows how this sequence occurs in the citric acid cycle, with the formulas of the compounds arranged to match the general scheme.

Regulation of the Citric Acid Cycle

Since the citric acid cycle is in the major routes of fuel combustion in many cells, there must be some control of the rate at which it proceeds (Fig. 23–13). It wouldn't do to have the oxidative machinery going full tilt like a runaway boiler at times of rest nor would it do to have it only sluggishly responsive when there is an immediate demand for ATP.

Stoichiometric Regulation by ADP. Since a major consequence of the action of the citric acid cycle is the conversion of ADP and P_i to ATP through the mechanism of oxidative phosphorylation, it follows that the cycle won't function unless there is

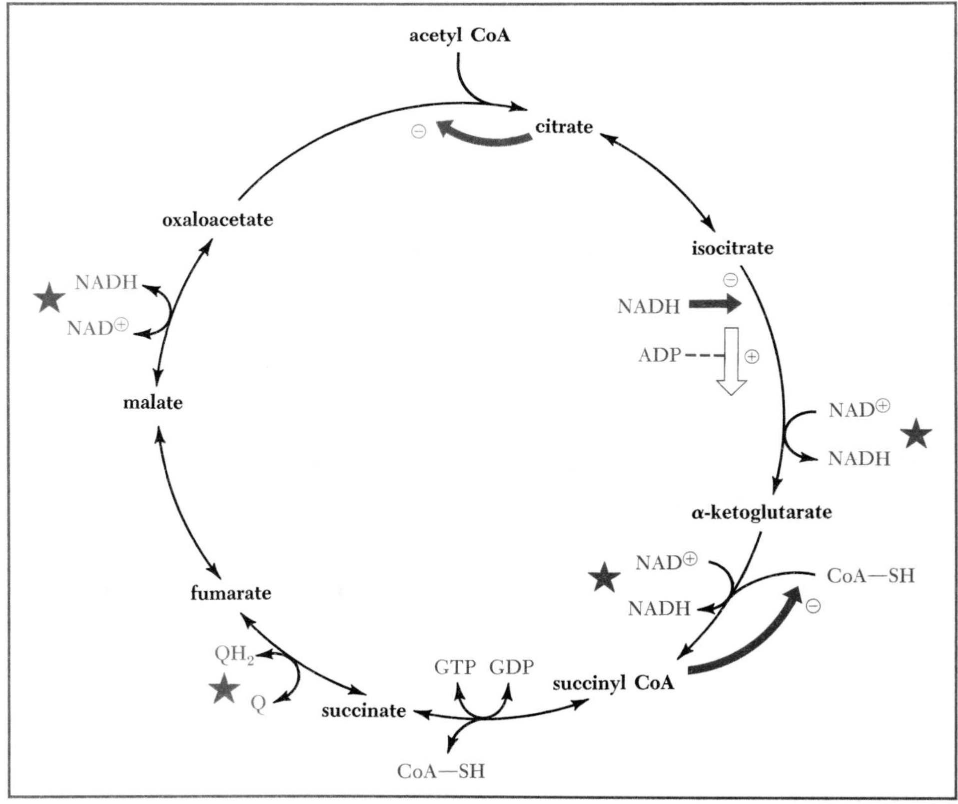

FIGURE 23–13 Regulation of the citric acid cycle. The starred reactions require oxidized coenzymes, and the ratio of oxidized to reduced coenzyme is governed by the availability of ADP and P_i for oxidative phosphorylation. Other specific controls prevent irreversible steps in the cycle from depleting the supply of cofactors and cycle intermediates. The action of negative effectors is indicated by ⊖; the action of positive effector (⊕) is indicated by an open arrow next to the reaction arrow.

a supply of ADP and P_i. It is the concentration of ADP that is usually limiting in animal tissues. Electrons can't be transferred to O_2 unless the cells are doing something that requires ATP and causes the production of ADP (such as moving, synthesizing proteins, or transporting ions). Oxygen consumption diminishes in resting cells, and the electron carriers accumulate in their reduced form. To be specific, a lack of ADP at rest prevents NADH from being reoxidized back to NAD. The resultant lack of NAD will slow the oxidations of isocitrate, α-ketoglutarate, and L-malate. Ubiquinol will also accumulate because it cannot be reoxidized to ubiquinone. Not only will the lack of ubiquinone slow the oxidation of NADH, but it will also slow the oxidation of succinate.

The rate of exchange of ATP and ADP across the inner membrane also is increased by a rise in external ADP concentration. Since changes in the rate of oxidative phosphorylation tend to be exponentially proportional to changes in ADP concentration (p. 477), only small shifts in the internal ATP-ADP balance of the mitochondrion suffice to control the rate of the oxidative metabolism.

We have emphasized ADP concentration as the normal regulating factor, but it may be that the muscular weakness or impaired cardiac performance sometimes seen in patients with a low P_i concentration (hypophosphatemia) is due to P_i becoming limiting. (The concentration may by lowered by prolonged use of antacids, which cause passage of insoluble phosphates through the bowel, for example.)

Regulation by Effectors. Some of the reactions in the citric acid cycle require individual regulation because they are essentially irreversible under physiological conditions. There is always the danger that such reactions will continue until they have consumed the available supply of substrate or coenzyme.

Consider, for example, the **citrate synthase** reaction. Its position of equilibrium is so far toward the formation of citrate that all available acetyl coenzyme A or oxaloacetate could be converted to citrate, even when the remainder of the citric acid cycle is ticking along at the resting rate. This is prevented by making the rate of the reaction critically dependent upon the oxaloacetate concentration. The K_M for oxaloacetate is in the physiological concentration range; furthermore, citrate is a competitive inhibitor for oxaloacetate on the enzyme. The effect is double-barreled. An accumulation of citrate raises its concentration as an inhibitor, but it also lowers the concentration of oxaloacetate as a substrate. Why? Because the complete cycle must function at the same rate to restore the oxaloacetate consumed in the first step. Any accumulation of intermediates in the cycle represents a depletion of oxaloacetate.

Now consider **isocitrate dehydrogenase.** It is especially important because it is the first committed reaction in the citric acid cycle—it has no other function than to participate in the cycle, and it is irreversible at intracellular concentrations of its reactants. (Citrate synthesis is used in fatty acid synthesis [Chapter 28], as well as in the cycle.) Two important allosteric effects are used here. ADP is a specific activator, which lowers the K_M for isocitrate. A rise in ADP concentration is a signal of a need for more high-energy phosphate, and the response to this signal includes an accelerated injection of substrate into the citric acid cycle.

Isocitrate dehydrogenase is also inhibited by NADH at an allosteric site. Any slowing of oxidative phosphorylation, which leads to an accumulation of NADH, therefore slows the enzyme in two ways. If the [NADH] is high, the [NAD] must be low, and NAD is required as a substrate for the reaction. The NADH also combines with the allosteric site to make the enzyme less active. (Any control of isocitrate dehydrogenase also tends to control citrate synthase because changes in isocitrate concentration are accompanied by changes in citrate concentration. Aconitase rapidly equilibrates the two compounds.)

The **α-ketoglutarate dehydrogenase** reaction represents a threat to the supply of coenzyme A for other reactions. Indeed, 70 per cent of the coenzyme A supply in some tissues is present as succinyl coenzyme A under some conditions, even though the enzyme is regulated. The compound tends to accumulate at rest when the phosphate potential is high, owing to the lack of GDP and P_i for its subsequent use. The extent of accumulation is controlled by constructing the α-ketoglutarate dehydrogenase so that its product, succinyl coenzyme A, is a competitive inhibitor for one of its substrates, coenzyme A. Here again is a double-barreled effect. A rise in succinyl coenzyme A concentration in itself inhibits, but it also represents a depletion of coenzyme A, causing still more effective inhibition.

FURTHER READING

J.M. Lowenstein (1967). The Tricarboxylic Acid Cycle, vol. 1, p. 147, in D. E. Greenberg, ed.: Metabolic Pathways, 3rd ed. Academic Press. A detailed discussion by one of the more lucid writers.

P.A. Srere (1975). The Enzymology of the Formation and Breakdown of Citrate. Adv. Enzymol. 43:57.

L.J. Reed (1974). Multi-enzyme Complexes. Accts. Chem. Res. 7:40.

P.D. Boyer, ed. (1976–). The Enzymes, 3rd ed. Academic Press. This multivolume work includes detailed discussions of several of the enzymes of the citric acid cycle.

J. Fothergill (1982). From CoA to Complement: Thioesters as the Spring in the Molecular Mouse Trap. Nature 298:705.

CHAPTER 24

THE OXIDATION OF FATTY ACIDS

Fatty acids are major fuels for animals, both as a primary supply from the diet and as a secondary supply created from other dietary components. They are stored within cells as fats (triglycerides) and later released through the blood stream to meet the demands of many tissues, especially muscles. The importance of the oxidation of fatty acids is not limited to the obese or to devotees of greasy foods; it is a critical part of the metabolic economy in the lean as well as the lardy.

The complete combustion of fatty acids to CO_2 and H_2O occurs in the mitochondria, where the transfer of electrons from the fatty acids to oxygen can be used to generate ATP. The combustion occurs in two stages; the fatty acid is sequentially oxidized so as to convert all of its carbons to acetyl coenzyme A. The acetyl coenzyme A is oxidized by the reactions of the citric acid cycle. Both stages generate ATP by oxidative phosphorylation.

INTRACELLULAR TRANSPORT OF FATTY ACIDS

Free fatty acids appear within a cell by absorption from the extracellular fluid, or by intracellular hydrolysis of triglycerides. Those used as fuels are mostly long-chain compounds with 16 or 18 carbon atoms and from zero to two double bonds. Whatever the disposition of a fatty acid is to be—formation of a structural component, storage as fat, or combustion as a fuel—it is first combined with coenzyme A to form the highly polar thiol ester, which is easily soluble in the aqueous phases of the cell.

Formation of Acyl Coenzyme A. The acyl coenzyme A compounds with which we are now concerned are identical to acetyl coenzyme A in structure, except that they are made from long-chain fatty acids instead of the two-carbon acetic acid. They are like acetyl coenzyme A in being high-energy thiol esters, with a standard free energy of hydrolysis comparable to that of ATP, and they are made from the free fatty acids at the expense of high-energy phosphate:

$$\underset{\text{fatty acid}}{H_3C{-}(CH_2)_n{-}COO^{\ominus}} \xrightarrow[\text{ATP} \rightarrow \text{AMP} + \text{PP}_i,\ Mg^{2+}]{\text{CoA—SH},\ \textit{acyl CoA synthetase}} \underset{\text{acyl coenzyme A}}{H_3C{-}(CH_2)_n{-}\overset{\overset{\displaystyle O}{\|}}{C}{-}S{-}CoA}$$

$$PP_i + H_2O \xrightarrow{\textit{inorganic pyrophosphatase}} 2\ P_i$$

The formation of acyl coenzyme A is readily reversible in the test tube, but the reaction is made to go nearly to completion by the action of an **inorganic pyrophosphatase.** This enzyme hydrolyzes PP_i and keeps its concentration very low. The result is the hydrolysis of an additional high-energy phosphate bond. **Any synthetic reaction in which ATP is cleaved to AMP and PP_i is in effect being driven by the hydrolysis of two high-energy phosphate bonds,** owing to the accompanying hydrolysis of PP_i.

Long-chain fatty acids in the cytosol may be converted to acyl coenzyme A at three locations. Liver cells have acyl CoA synthetases on peroxisome membranes, the endoplasmic reticulum, and the outer mitochondrial membrane. **Peroxisomes** use the acyl groups as a fuel for hydrogen peroxide formation as discussed later. The **endoplasmic reticulum** makes structural components and triglycerides for storage from fatty acyl groups. The **mitochondria** use the acyl groups for fuel in oxidative phosphorylation. This occurs in the liver, but it is especially important in red muscle fibers, where fatty acids are the principal fuel for long-term work. Liver cells contain a small binding protein to transport fatty acids; its function is not clearly defined, but it may preferentially supply the endoplasmic reticulum and peroxisomes. Myoglobin binds fatty acids to its surface as well as oxygen to its heme group, and therefore may serve to supply both fuel and the oxygen to consume it to mitochondria within red muscle cells.

Transport as Acyl Carnitine. The mitochondrial inner membrane is nearly impermeable to coenzyme A and its derivatives, permitting separate regulation of the concentrations of these compounds within and without mitochondria. Acyl coenzyme A therefore cannot move as such from the cytosol to the mitochondrial matrix. Instead, the acyl groups are moved across the inner membrane by transferring them from coenzyme A to another compound, **carnitine,** at the outer surface of the inner membrane (Fig. 24–1). An antiporter within the inner membrane then exchanges acyl carnitine on one side of the membrane for carnitine on the other. The acyl carnitine that appears on the inner surface of the inner membrane is once more equilibrated with acyl coenzyme A, this time utilizing coenzyme A from the mitochondrial matrix. The fatty acids appear as acyl coenzyme A within the matrix without compromising the independent controls within the separate cellular compartments.

Genetic Defects in Acyl Transfer. Some people are unable to use long-chain fatty acids in one or more tissues because of defects in a **carnitine palmitoyl transferase** or some defect in the synthesis or transport of carnitine. The severity of the consequences depends upon the number of processes affected. Those in whom only one of the two carnitine palmitoyl CoA transferases in red muscle fibers is affected get along surprisingly well. Triglycerides accumulate in the blood, but acute symp-

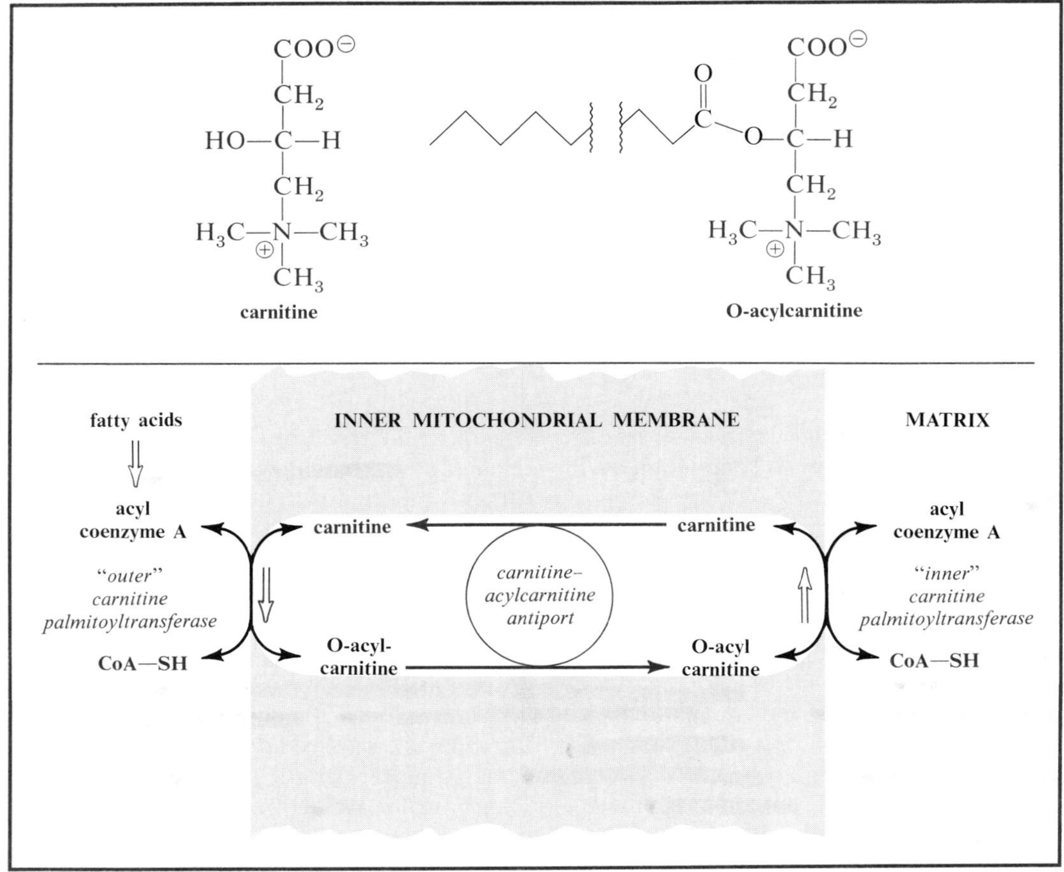

FIGURE 24–1 Acyl groups are transferred across the inner mitochondrial membrane by combination with carnitine to form O-acylcarnitine. Acyl groups are transferred from coenzyme A to carnitine on the outer surface of the membrane, and the O-acylcarnitine formed is moved to the inner surface by exchange with free carnitine using an antiport mechanism. The acyl group is then transferred from carnitine to coenzyme A within the mitochondrion, creating acyl coenzyme A that can be attacked by the acyl coenzyme A dehydrogenases exposed on the inner surface.

toms may not appear for years, when a bout of strenuous exercise causes a demand for ATP beyond the ability to supply it. The muscle cells then break down, with an appearance of myoglobin (a relatively small protein) in the urine, and an elevated concentration of larger muscle proteins, such as creatine kinase, in the blood.

Defects in the supply of carnitine to cells, which may be due to a failure of its synthesis (p. 657) or of its transport into cells, may be life-threatening in infancy. The results include lethargy, enlargement of the liver and heart, and damage to the muscles from excessive lipid storage. If the defect is in synthesis, it is treatable by feeding carnitine.

Transport of Shorter Chain Fatty Acids. The bulk of the fatty acids used as fuels contain 16 or more carbon atoms. However, shorter fatty acids do occur, and are used as fuels even more readily than the longer-chain compounds. Those with 10 carbon atoms or fewer are sufficiently soluble in water that no special provision is necessary for moving them. They diffuse rapidly through membranes, including the inner mitochondrial membrane. An enzyme in the mitochondrial matrix, **butyryl CoA synthetase,** acts on any fatty acid with 4 to 12 carbon atoms; it, too, synthesizes the coenzyme A derivatives by splitting ATP to AMP and PP_i. Acetate represents a special case. Because of the central role of acetyl coenzyme A in metabolism, any free acetate that appears is promptly converted to acetyl coenzyme A by a synthetase found in both the cytosol and the mitochondrial matrix. This synthetase will also convert propionate to propionyl coenzyme A:

$$CH_3-CH_2-\overset{\overset{\displaystyle O}{\|}}{C}-S-CoA$$

propionyl coenzyme A

The AMP generated during formation of the shorter-chain coenzyme A derivatives is removed through phosphorylation by GTP:

$$GTP + AMP \longleftrightarrow GDP + ADP.$$

The ADP can be handled by oxidative phosphorylation and the GDP by the citric acid cycle.

In addition, any long-chain fatty acids appearing in the mitochondrial matrix can be converted to the coenzyme A derivative by a synthetase that uses GTP in a reversible reaction:

$$\text{fatty acid} + \text{CoA-SH} + GTP \longrightarrow \text{acyl coenzyme A} + GDP + P_i.$$

However, we do not understand what special purpose the guanine nucleotide serves in these reactions.

FORMATION OF ACYL COENZYME A

By cleaving ATP to AMP and PP_i:
- **acyl coenzyme A synthetase (C12 and longer) on**
  - (1) peroxisome membrane—provides fuel for peroxide formation
  - (2) endoplasmic reticulum—for use in fat storage
  - (3) mitochondrial outer membrane—provides fuel for oxidative phosphorylation
- **butyryl coenzyme A synthetase (C4 to C12)**
  - only in mitochondrial matrix
- **acetyl CoA synthetase (C2 or C3)**
  - (1) cytosol—function unknown
  - (2) mitochondrial matrix—enables combustion of acetate

By cleaving GTP to GDP and P_i:
- **acyl CoA synthetase (GDP-forming)**
  - only in mitochondrial matrix

OXIDATION OF SATURATED CHAINS

When the molecules of acyl coenzyme A reach the inner surface of the inner mitochondrial membrane, oxidation can begin. The overall scheme is a beautiful example of economy of design. As is illustrated for palmitoyl coenzyme A in Figure 24–2, the scheme is essentially repetitious of the main-line sequence we discussed in the last chapter, with each repetition concluded by a thiolysis. A saturated acyl derivative (acyl coenzyme A) is oxidized to the unsaturated derivative (enoyl coen-

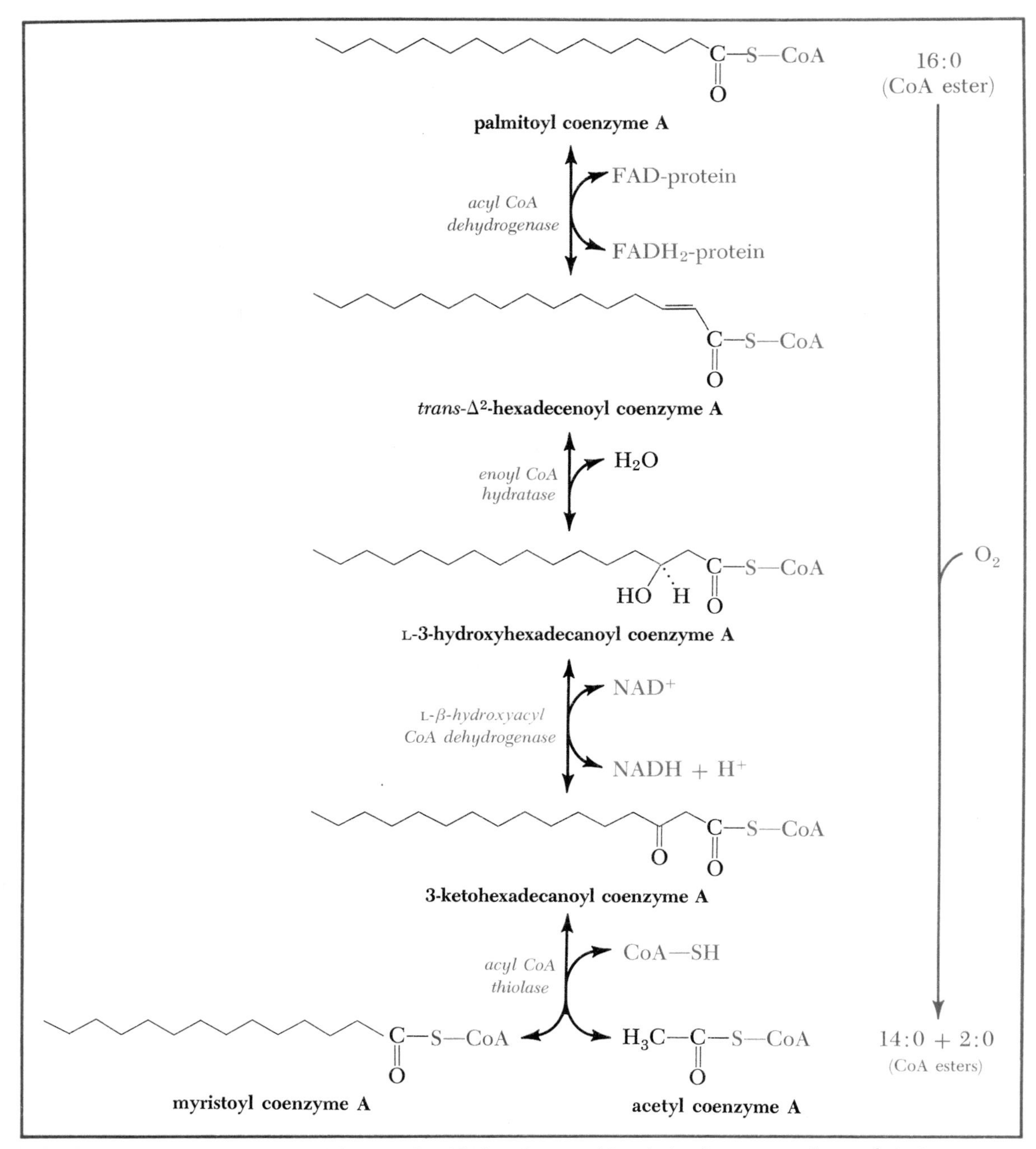

FIGURE 24–2 **The reactions of fatty acid oxidation shown with palmitoyl coenzyme A as substrate.**

zyme A) by a flavoprotein, followed by hydration to form an alcohol group (3-hydroxyacyl coenzyme A), which is oxidized by NAD to a ketone (3-ketoacyl coenzyme A). The 3-ketoacyl coenzyme A is then cleaved with another molecule of coenzyme A, liberating acetyl coenzyme A and a new acyl coenzyme A compound that is two carbon atoms shorter than the original substrate.

The marvelous thing is that the shorter acyl coenzyme A can then undergo the same sequence of reactions, clipping off another two-carbon unit as acetyl coenzyme

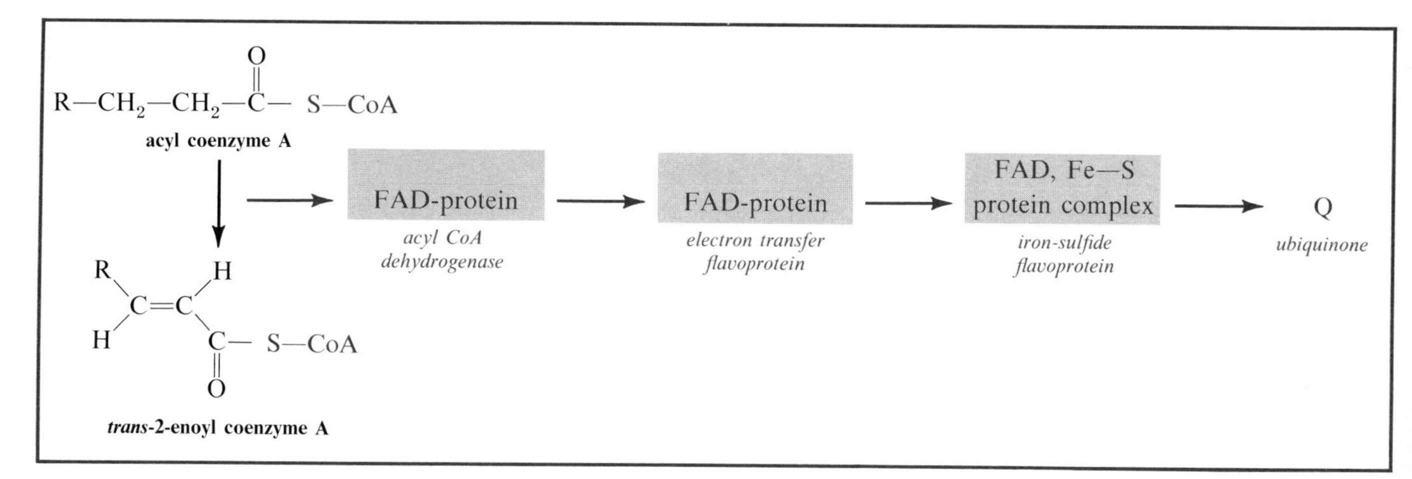

FIGURE 24–3 The oxidation of acyl coenzyme A compounds involves passage of electrons through a series of flavoproteins in the inner mitochondrial membrane before they finally reach ubiquinone.

A, with the whole process repeated until the entire long-chain acyl group has been chopped into two-carbon acetyl groups, which can be oxidized by the citric acid cycle.

Let us now examine some of the details of this elegant and simple process.

Acyl coenzyme A dehydrogenases, found in the inner mitochondrial membrane, are flavoproteins because the oxidation of an acyl coenzyme A, like the oxidation of succinate in the citric acid cycle, does not release enough free energy to support the formation of 3 ATP. Therefore, it is necessary to bypass the first proton-pumping complex and transfer electrons to ubiquinone. The dehydrogenases remove two hydrogen atoms from acyl groups to form *trans* isomers of enoyl groups. Three of these enzymes are known, with different specificities for substrate chain length. One works best on short-chain compounds (C_4 to C_8), another on intermediate-size compounds (C_6 to C_{16}), and another on long-chain compounds (C_6 to C_{18} with a peak at C_{14}).

Acyl coenzyme A dehydrogenases contain no iron and transfer electrons to ubiquinone by a route that first involves another flavoprotein, the **electron-transfer flavoprotein (ETF),** and then still another flavoprotein and iron-sulfide complex (Fig. 24–3). Why these extra flavoproteins are used for fatty acid oxidation is not clear, since the end result is similar to that of succinate oxidation: Electrons are introduced into the mitochondrial electron transport chain at the level of ubiquinone with two moles of high-energy phosphate generated per mole of acyl coenzyme A oxidized.

Hydration and Dehydrogenation. The next two reactions occur in the mitochondrial matrix. The hydration of unsaturated acyl coenzyme A compounds to form the corresponding L-3-hydroxyacyl derivatives is straightforward and quite similar to the hydration of fumarate in the citric acid cycle, but it is catalyzed by a different enzyme, **enoyl CoA hydratase.**

The oxidation of the alcohol group by NAD also has no unusual features. There is no possibility of an accompanying decarboxylation such as we saw in the oxidation of isocitrate because there is no free carboxylate group—it is fixed as a coenzyme A ester. NAD for this reaction comes from the same pool supplying all NAD-coupled dehydrogenases in the matrix, and the NADH that is produced is oxidized with the generation of 3 ATP.

Cleavage and Coenzyme A. Esters of 3-keto acids are subject to C—C scission through attack by a nucleophilic reagent, both in the test tube and in biological systems. The biological nucleophilic reagent frequently is coenzyme A, and the cleavage therefore is a **thiolysis**—a splitting by a thiol aided by enzymatic catalysis.

The release of acetyl coenzyme A by this reaction is highly favored at intracellular concentrations, while generating a new fatty acyl coenzyme A ester from the remaining carbon atoms.

The complete sequence (Fig. 24–4) involves a repetition of oxidation, hydration, oxidation, and cleavage until the final four-carbon remnant (butyryl coenzyme A)

GENERAL TYPE OF REACTION

Cleavage of a 3-ketoacyl coenzyme A compound with coenzyme A, or their synthesis by the reverse reaction.

$$\mathrm{R{-}\overset{O}{\overset{\|}{C}}{-}\overset{|}{\underset{|}{C}}{-}\overset{O}{\overset{\|}{C}}{-}S{-}CoA + HS{-}CoA \longrightarrow R{-}\overset{O}{\overset{\|}{C}}{-}S{-}CoA + H{-}\overset{|}{\underset{|}{C}}{-}\overset{O}{\overset{\|}{C}}{-}S{-}CoA}$$

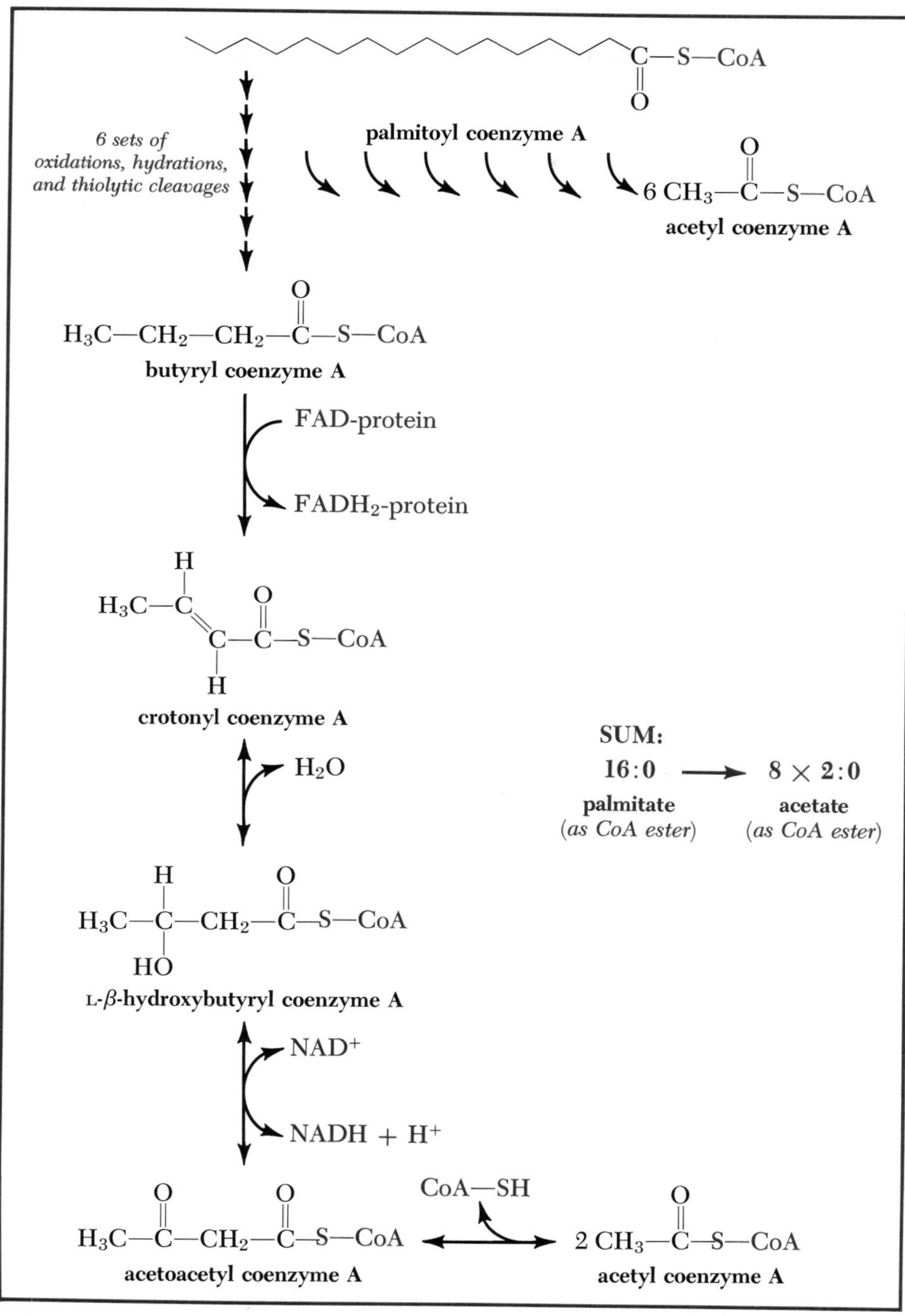

FIGURE 24–4 Repetitions of the main-line sequence of oxidations followed by thiolytic cleavage finally converts the last four carbons of a fatty acid to butyryl coenzyme A, with the others having been converted to acetyl groups in acetyl coenzyme A. The repetition of the sequence with butyryl coenzyme A completes the total conversion of the fatty acid carbons to acetyl groups.

has been oxidized and cleaved, at which point all of the carbon atoms in the original fatty acids will have been converted to acetyl groups.

ODD-CHAIN FATTY ACIDS. What happens if a fatty acid with an odd number of carbon atoms is introduced into mitochondria? There are such fatty acids in nature, although they are much less common than those with an even number of carbon atoms. The sequence of reactions proceeds just as before. (We expect this because it is asking too much to expect an enzyme to be specific for substrates with chains of, say, 14, 16, or 18 carbon atoms without the enzyme also binding those of 15 or 17 carbon atoms.) Acetyl coenzyme A units are successively cleaved from these odd-numbered chains until the very end of the sequence, when the three-carbon remainder is in the form of **propionyl coenzyme A.**

This compound is not oxidized directly in the citric acid cycle, but it has a metabolism that is important in the breakdown of amino acids and will be considered in connection with them (p. 598). The odd-chain fatty acids are only a small fraction of the total, and only the terminal three carbons appear as propionyl coenzyme A. The metabolism of propionyl coenzyme A is therefore not of quantitative significance in fatty acid oxidation.

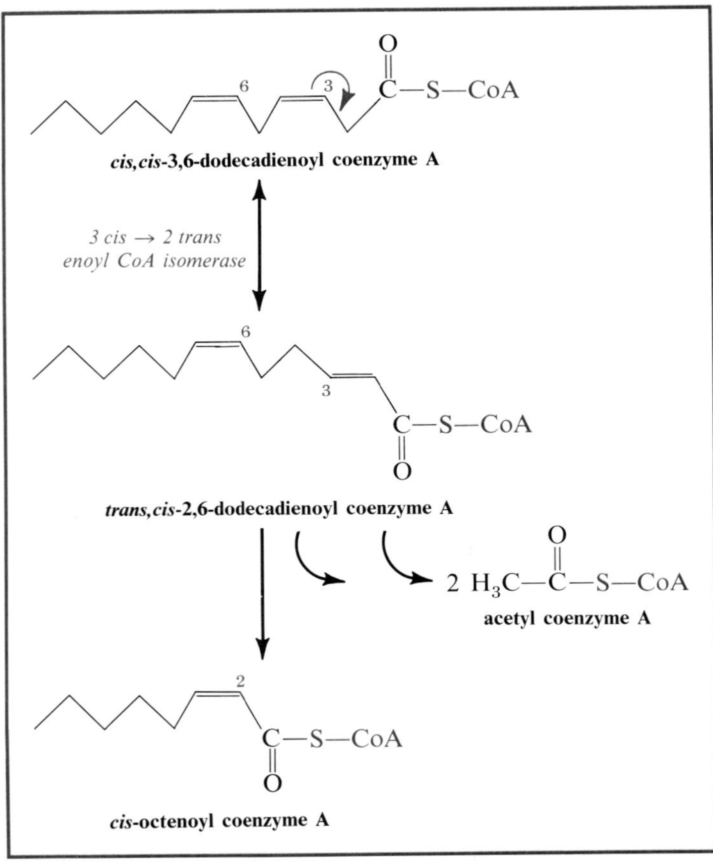

FIGURE 24–5 The 3-*cis* unsaturated coenzyme A esters formed by the oxidation of unsaturated fatty acids are converted to 2-*trans* isomers by the action of an isomerase. Since the 2-*trans* isomer is an intermediate in the main-line sequence, degradation can proceed in the usual way until another double bond is encountered.

OXIDATION OF UNSATURATED FATTY ACIDS

Since over half of the fatty acid residues in body lipids are unsaturated, a large part of the high-energy phosphate generated in many tissues must come from the oxidation of this type of compound. Oleic, 18:1(9), and vaccenic, 18:1(11), acids are the most abundant of the unsaturated fatty acids, accounting for approximately half of all residues in a typical triglyceride, but others with more double bonds are not trivial. The general route includes the reactions already described for oxidation of saturated acids, with the addition of two more.

As an example, let us consider the metabolism of the linoleic acid anion—*all cis*-$\Delta^{9,12}$ octadecadienoate. Its metabolism illustrates all of the reactions involved. The double bonds are well removed from the carboxylate end, so it will begin to be metabolized as if it were an ordinary saturated fatty acid, beginning with the formation of the coenzyme A ester, which is then carried three times through the regular sequence of oxidation, hydration, oxidation, and cleavage to liberate acetyl coenzyme A:

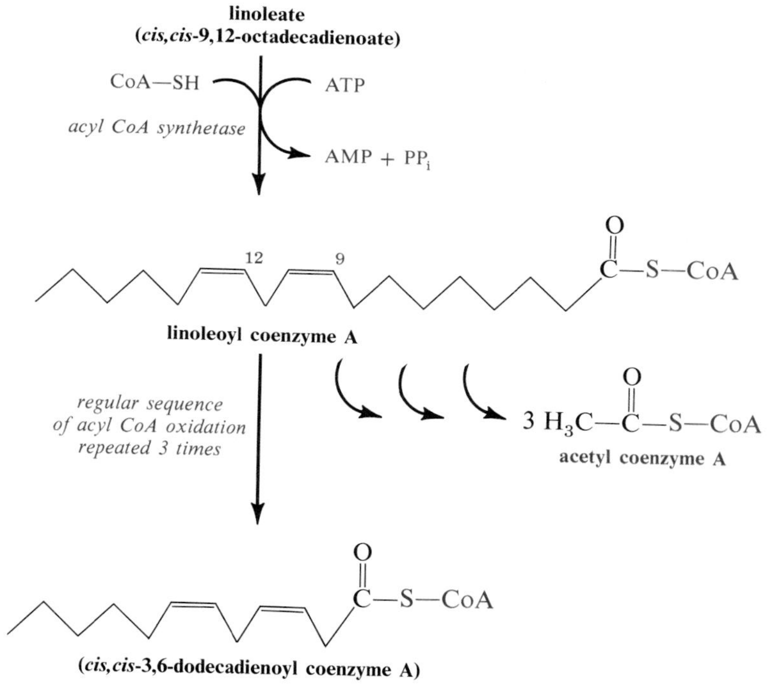

linoleate
(*cis,cis*-9,12-octadecadienoate)

CoA—SH → ATP
acyl CoA synthetase
→ AMP + PP_i

12 9 C(=O)—S—CoA

linoleoyl coenzyme A

regular sequence of acyl CoA oxidation repeated 3 times

$3\ H_3C—C(=O)—S—CoA$
acetyl coenzyme A

C(=O)—S—CoA

(*cis,cis*-3,6-dodecadienoyl coenzyme A)

The sequence can proceed until three molecules of acetyl coenzyme A have been cleaved, but no farther, because the chain is then shortened so that the double bond that was originally between C9 and C10 of linoleic acid is now between C3 and C4 in the remaining chain. Further oxidation is blocked, because the acyl CoA dehydrogenase attacks C3 and C2.

Isomerization of Double Bond. One of the additional enzymes then comes into play. This enzyme catalyzes a migration of the double bond from the third to the second carbon atom, and at the same time changes the configuration from *cis* to *trans* (Fig. 24–5). Equilibrium for this reaction is reached with about seven eighths in the *trans* form, so no energy donors or receptors are involved. After migration, the double bond is in the same position and has the same configuration as the regular intermediates of fatty acid oxidation, and the compound once more enters the general pathway, beginning with a hydration.

Hydration of *cis*-Enoyl CoA. After two more molecules of acetyl coenzyme A are cleaved from the chain, the second of the original double bonds in linoleic acid is now between C2 and C3, but it is still in its *cis* configuration. However, the enoyl hydratase that catalyzes the addition of water during the regular sequence is not specific for configuration, and it will also cause the addition to the *cis* form.

The rub is that hydration of a *cis* Δ^2-enoyl coenzyme A by this enzyme results in the formation of a D-3-hydroxyacyl CoA, rather than the L-isomer formed in the regular sequence, and D-3-hydroxyacyl coenzyme A derivatives are not substrates for the L-3-hydroxyacyl CoA dehydrogenases used in the regular route. At this point the second of the additional enzymes, a racemase, brings the D-3-hydroxy compounds into the regular route by catalyzing the interconversion of the D- and L-isomers. As the L-isomer is removed by the action of the hydroxyacyl CoA dehydrogenase, the racemase converts more of the D-isomer to the L, so the effect is to divert all of the D-compound into the regular pathway of fatty acid oxidation (Fig. 24–6).

Here is another example of the beautiful economy of organization of metabolism. The introduction of two additional types of enzymes, an enoyl coenzyme A isomerase and a 3-hydroxyacyl coenzyme A racemase, makes it possible to handle any combination of double bonds found in an unsaturated chain through the same route used for saturated fatty acids. (It is an interesting exercise in mental gymnastics

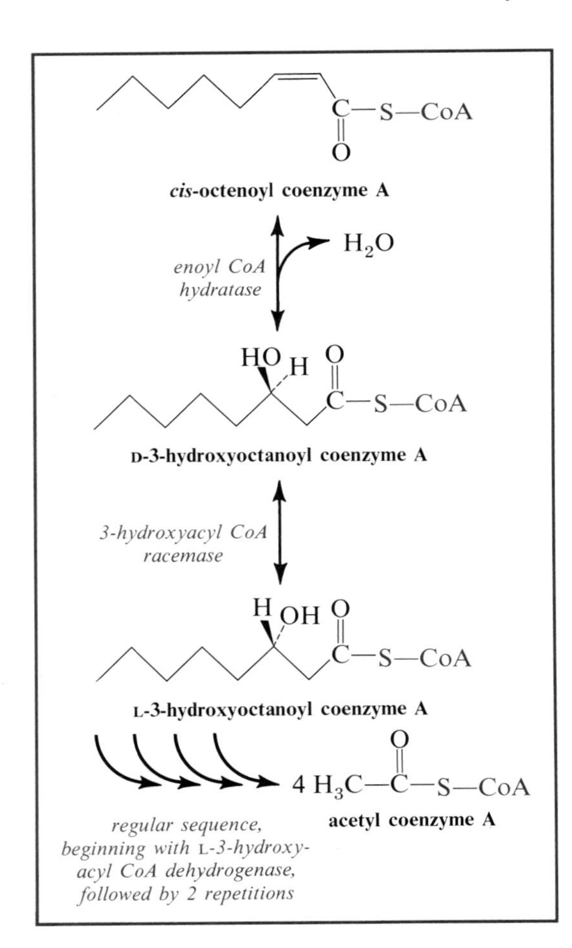

FIGURE 24–6

The action of a 3-hydroxyacyl CoA racemase.

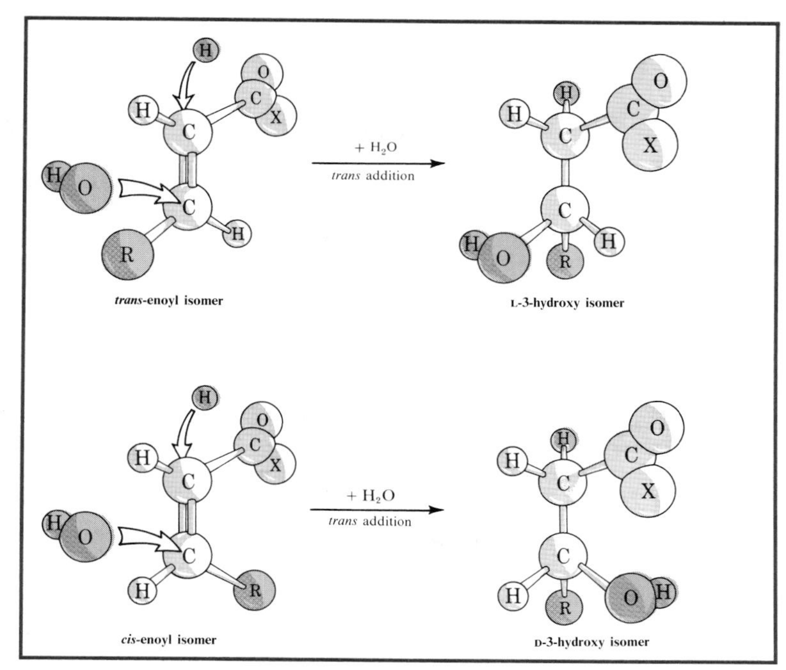

FIGURE 24–7 **If an enzyme catalyzes *trans* addition of the elements of water to a double bond, that is, addition of the H behind and OH in front of the bond as shown, in such a way that a *trans*-unsaturated compound is converted to an L-hydroxy compound, then the enzyme must catalyze the conversion of the *cis*-unsaturated isomer to a D-hydroxy compound. (Assuming specificity permits the enzyme to react with both isomers.)**

to test this on a random assortment of double bonds in a long chain, except that the structure, —C=C=C—, is forbidden. Allenes are rare among natural compounds, being known only as products of the metabolism of a few kinds of microorganisms.)

Mechanism of Hydration. Why do the *cis* and *trans* isomers yield different stereoisomers upon hydration? Hydratases are like other enzymes in that they are constructed to bind substrates in a particular way. If the addition of the elements of water to a *trans* bond create an L-isomer of the resultant hydroxy compound, then addition of water in a similar way to a *cis* bond must create a D-isomer. Why this is so is shown in Figure 24–7.

We noted in the previous chapter that hydration of *cis*-aconitate produces D-isocitrate and hydration of fumarate, which is a *trans* isomer, produces L-malate. The hydration of *cis* to form D and of *trans* to form L is a handy mnemonic in these cases, but it is more of an accident than a general mechanism, and we shall see an important exception to this "rule" when we consider fatty acid synthesis.

OXIDATION OF FATTY ACIDS VIA ACETOACETATE

We have outlined the major route for using fatty acids as fuels: The fatty acids are oxidized and cleaved to acetyl coenzyme A, followed by combustion of the acetyl groups through the citric acid cycle in the same cell. In other words, this route involves the total combustion of a fatty acid by single cells. It is especially important in muscles, including the heart.

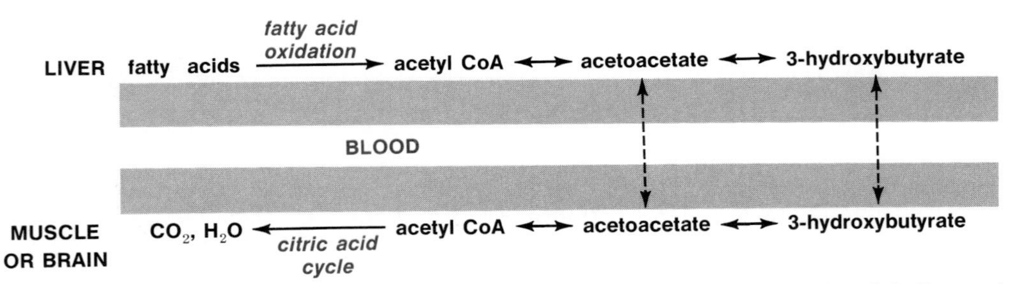

FIGURE 24–8 Fatty acids can be oxidized to CO_2 and H_2O by the cooperative action of the liver and the muscles or brain, involving intermediate formation of a mixture of acetoacetate and D-3-hydroxybutyrate.

However, there is another way in which fatty acids can be oxidized that involves the generation of acetyl groups in the liver or kidneys and their combustion in other tissues (Fig. 24–8). The acetyl groups are transported through the blood in the form of **acetoacetate** (3-oxobutyrate in systematic nomenclature) or **D-3-hydroxybutyrate:**

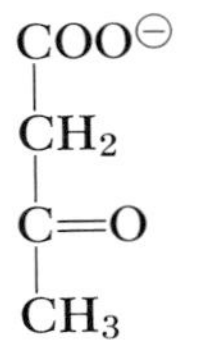

$COO^{\ominus}$–CH_2–C(=O)–CH_3
acetoacetate

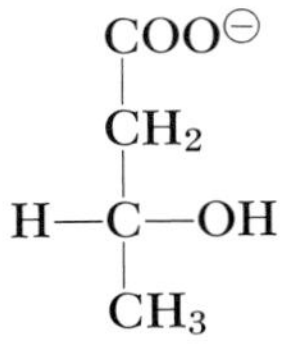

$COO^{\ominus}$–CH_2–CH(OH)–CH_3
D-3-hydroxybutyrate

Offhand, it might appear that this circuitous route offers no advantages to the organism over the direct oxidation of fatty acids in the receptor tissues. The important difference is that acetoacetate and D-3-hydroxybutyrate, unlike the long-chain fatty acids, can be used as a fuel by nervous tissue in place of the glucose ordinarily used. Conditions in which the supply or utilization of glucose is impaired, such as starvation or diabetes mellitus, cause an increased production of these compounds.

Formation of Acetoacetate. Acetoacetate is produced in the liver and kidneys by a simple two-step process (Fig. 24–9), in which acetyl coenzyme A and acetoacetyl coenzyme A first condense with the release of one molecule of coenzyme A to form 3-hydroxy-3-methylglutaryl coenzyme A, which is then cleaved at a different point

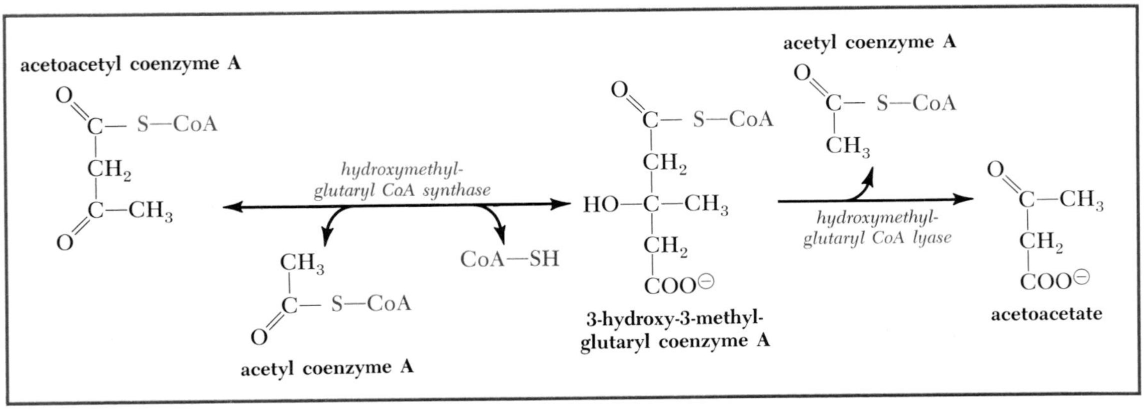

acetoacetyl coenzyme A + acetyl coenzyme A —*hydroxymethylglutaryl CoA synthase*→ 3-hydroxy-3-methylglutaryl coenzyme A + CoA—SH

3-hydroxy-3-methylglutaryl coenzyme A —*hydroxymethylglutaryl CoA lyase*→ acetoacetate + acetyl coenzyme A

FIGURE 24–9 The formation of acetoacetate.

to yield free acetoacetate and acetyl coenzyme A in an irreversible reaction. If we think of acetoacetyl coenzyme A as being analogous to oxaloacetate, we see that the condensation reaction is exactly analogous to the formation of citrate in the first step of the citric acid cycle (p. 424). (However, it is catalyzed by a quite different enzyme, 3-hydroxy-3-methylglutaryl CoA synthase.)

What is the effective result? Since acetyl coenzyme A and acetoacetyl coenzyme A are in equilibrium because of the reaction catalyzed by acetoacetyl coenzyme A thiolase (p. 446), all of the carbons of 3-hydroxy-3-methylglutaryl coenzyme A can be formed from acetyl coenzyme A:

acetyl CoA + acetyl CoA	⟶	acetoacetyl CoA + CoA-SH
acetyl CoA + acetoacetyl CoA	⟶	3-hydroxy-3-methylglutaryl CoA + CoA-SH
SUM: 3 acetyl CoA	⟶	3-hydroxy-3-methylglutaryl CoA + 2 CoA-SH

When 3-hydroxy-3-methylglutaryl coenzyme A is cleaved, one of the molecules of acetyl coenzyme A is recovered:

3-hydroxy-3-methylglutaryl CoA ⟶ acetyl CoA + acetoacetate

so the overall result is the formation of acetoacetate from two molecules of acetyl coenzyme A.

Formation of D-3-Hydroxybutyrate. Part of the acetoacetate formed in the liver is converted to D-3-hydroxybutyrate in the mitochondrial cristae by an NAD-coupled D-3-hydroxybutyrate dehydrogenase, which is a typical alcohol dehydrogenase catalyzing the reversible reaction:

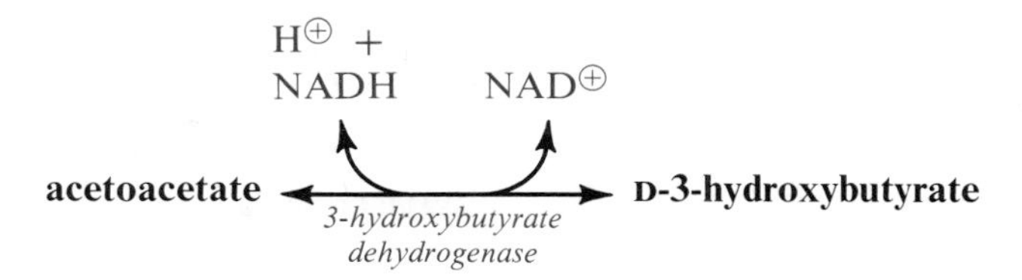

(Why is the dehydrogenase built to form the D-3-hydroxy isomer? Presumably so as to differentiate the 3-oxybutyrate route from the usual route of fatty acid oxidation, which involves the L-3-hydroxy derivatives. No good rationalization has been given for the enzyme's occurrence in tight association with phospholipids in the inner membrane.) The amount of conversion of acetoacetate to D-3-hydroxybutyrate is determined by the ratio of NADH to NAD in the mitochondria. Indeed, the relative amounts of the two oxybutyrates produced by the liver is used as an index of the state of reduction of NAD in mitochondria. The formation of hydroxybutyrate is, in effect, withdrawing electrons from the mitochondria to make a more reduced substrate, and later oxidation of the compound in peripheral tissue can produce more high-energy phosphate than does the oxidation of acetoacetate.

Utilization of 3-Oxybutyrates. Acetoacetate and D-3-hydroxybutyrate are transported into skeletal muscles, cardiac muscle, and the brain. The mitochondria of these tissues also contain the NAD-coupled 3-hydroxybutyrate dehydrogenase catalyzing the conversion of this compound to acetoacetate. However, the muscle mitochondria also have another enzyme, **acetoacetate—succinate CoA transferase,** that catalyzes the transfer of coenzyme A from succinyl coenzyme A to acetoacetate (Fig. 24–10). The reaction is reversible, but the constant removal of acetoacetyl coenzyme A to form acetyl coenzyme A makes it proceed in one direction physiolog-

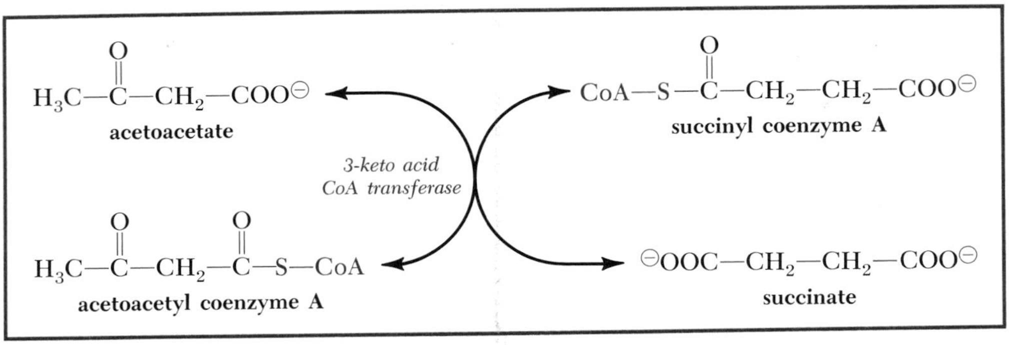

$H_3C-CO-CH_2-COO^{\ominus}$ (acetoacetate) + $CoA-S-CO-CH_2-CH_2-COO^{\ominus}$ (succinyl coenzyme A) ⇌ $H_3C-CO-CH_2-CO-S-CoA$ (acetoacetyl coenzyme A) + $^{\ominus}OOC-CH_2-CH_2-COO^{\ominus}$ (succinate)

3-keto acid CoA transferase

FIGURE 24–10 **Muscles and brain convert acetoacetate to acetoacetyl coenzyme A for use as a fuel by transferring coenzyme A from succinyl coenzyme A.**

ically. The uptake of acetoacetate catalyzed by this enzyme is, in effect, at the expense of one mole of high-energy phosphate. This is so because it involves the conversion of succinyl coenzyme A to succinate, and GTP would otherwise be obtained by this conversion through the succinyl CoA synthetase reaction in the citric acid cycle (p. 432).

In sum, the muscles and brain have a mechanism for converting D-hydroxybutyrate to acetoacetate, and the carbons of acetoacetate obtained in this way and by direct diffusion from the blood are injected into the citric acid cycle in the form of acetyl coenzyme A. Two moles of acetyl coenzyme A are obtained from one mole of acetoacetate (Fig. 24–11).

Ketonemia and Ketosis. Acetoacetate is constantly undergoing spontaneous decarboxylation to form **acetone.** The reaction is slow, but if the concentration of acetoacetate becomes high, enough acetone may be formed to make its odor detectable in the breath. This is part of the reason that acetoacetate, D-3-hydroxybutyrate, and acetone were collectively called "the **ketone bodies**" by early investigators even though acetone is a minor part of the total. The term now seems quaint, but it is still in use, and an increase in blood concentrations of the compounds is called ketonemia.

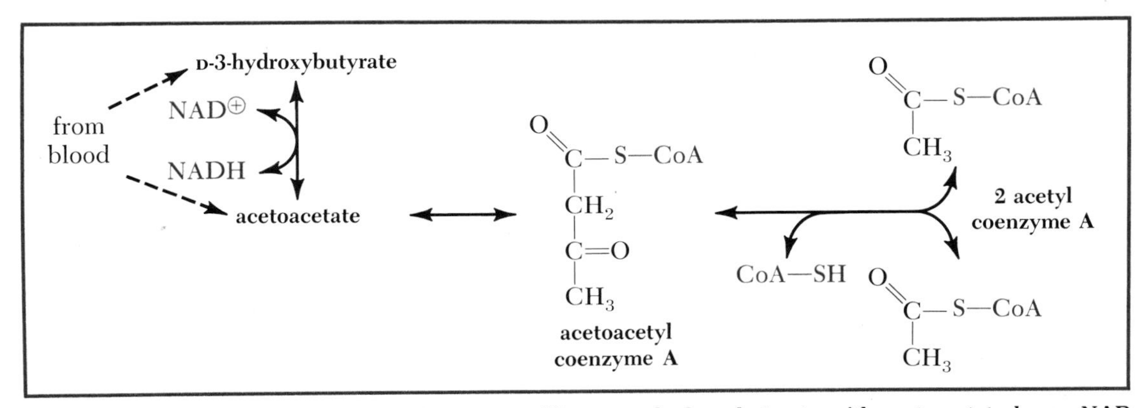

FIGURE 24–11 **Muscles and brain also can equilibrate D-3-hydroxybutyrate with acetoacetate by an NAD-coupled dehydrogenase, and can therefore use both acetoacetate and D-3-hydroxybutyrate obtained from the blood, with the formation of two molecules of acetyl coenzyme A from either compound.**

If the formation of ketone bodies is so rapid that large concentrations begin to appear in the urine, an individual is said to have ketonuria, and to be in a state of ketosis. (Since H^+ is produced along with the oxybutyrates, ketosis is frequently accompanied by **acidosis.**)

The concentrations of the ketone bodies in a resting individual are a sensitive indicator of this dependence on fatty acids as a fuel. Here are some concentrations in resting normal individuals before breakfast and after a further 7-day fast (J. Clin. Invest. 45:1751, 1966):

Time of Fast	Blood Concentrations (mM)		Daily Urinary Excretion (mmoles)	
	AcAcO$^-$	*3-OH-Bu*	*AcAcO*$^-$	*3-OH-Bu*
overnight	0.013	0.016	0.049	0.027
180 hours	1.1	4	10.9	77.1

In another experiment in which medical students were fasted for 3 days, it was shown that 18 per cent of their CO_2 expired at rest came from the oxidation of acetoacetate and D-3-hydroxybutyrate. Even with the sharply increased CO_2 output upon exercise, 10 per cent of it still came from these compounds.

Typical values in the blood of a diabetic entering an emergency ward with acidosis would be 3 mM acetoacetate and 10 mM D-3-hydroxybutyrate. Before the introduction of insulin, higher values were commonly seen in patients with diabetic ketoacidosis.

YIELD OF HIGH-ENERGY PHOSPHATE FROM FATTY ACID OXIDATION

Yield from Direct Combustion. As we saw, each cleavage of acetyl coenzyme A from a saturated acyl chain is preceded by a pair of oxidations, one catalyzed by a flavoprotein and the other by an NAD-coupled alcohol dehydrogenase. Electrons from the flavoprotein are inserted into the mitochondrial electron transport system at the ubiquinone level and therefore yield two high-energy phosphate bonds per molecule of substrate oxidized. Electrons from NADH go through the complete electron transport mechanism and yield three high-energy phosphate bonds per molecule of substrate oxidized. The sum of the two reactions is therefore five high-energy phosphate bonds generated per acetyl coenzyme A unit cleaved.

It takes only two cuts to divide a string into three pieces. Similarly, the conversion of palmitate (16:0) as palmitoyl coenzyme A to eight molecules of acetyl coenzyme A requires only seven pairs of oxidations, which produce 35 molecules of ATP through oxidative phosphorylation. A further 12 molecules of high-energy phosphate will be generated upon the subsequent oxidation of each acetyl group in the citric acid cycle.

We must now take into account the two high-energy phosphate bonds expended in forming the original coenzyme A ester, and also allow for the fact that each double bond already present in the original molecule eliminates the necessity for one oxidation catalyzed by an acyl CoA dehydrogenase and therefore diminishes the high-energy phosphate yield by two.

If we go through the necessary arithmetic, we arrive at equations such as these for the complete oxidation of palmitate, which is the most abundant saturated fatty acid in many plant and animal tissues, and for the complete oxidation of oleate, which is the most abundant of all fatty acid residues in the higher animals:

Palmitate:

$$C_{16}H_{31}O_2 + 23\ O_2 + 129\ (ADP + P_i) \longrightarrow 16\ CO_2 + 129\ ATP$$

Oleate:

$$C_{18}H_{33}O_2 + 25\tfrac{1}{2}\ O_2 + 144\ (ADP + P_i) \longrightarrow 18\ CO_2 + 144\ ATP$$

We see that the P:O ratios are 129/46 and 144/51, or 2.80 and 2.82, respectively.

Yield from Combustion via 3-Oxybutyrates. The equations just given are those that apply to the complete oxidation of a fatty acid within the mitochondria of one cell. How is the picture changed if the fatty acids are oxidized to acetoacetate or to D-3-hydroxybutyrate in the liver, and these compounds are transported to the peripheral tissues for oxidation? The total balance is changed very little, but the oxidations, and therefore the production of high-energy phosphate, are distributed between the liver and the peripheral tissues. When one mole of palmitate is oxidized *via* acetoacetate, 0.30 of the required total oxygen consumption and 0.26 of the total ATP production occur in the liver; when it is oxidized *via* 3-hydroxybutyrate, 0.22 of the oxygen consumption and 0.17 of the ATP production occur in the liver. In either case the P:O ratio for oxidation of the ketone body in the peripheral tissues is 2.89, so the peripheral tissues have a higher yield of ATP per molecule of O_2 consumed in oxidizing the ketone bodies than they do when they completely oxidize palmitate. Conservation of the oxygen supply may be another reason why the brain is adapted to use ketone bodies but not fatty acids.

REGULATION OF MITOCHONDRIAL FATTY ACID OXIDATION

The oxidation of the fatty acids to acetyl coenzyme A, like the subsequent oxidation of acetyl coenzyme A via the citric acid cycle, requires ADP to be available for coupled oxidative phosphorylation. If there is no demand for high-energy phosphate, then there is no production of ADP, no electron transport, and no fatty acid oxidation. Very simple.

However, given a demand for high-energy phosphate, the primary regulation in animals appears to hinge on the amount of substrates available. The 3-oxybutyrates are preferentially used by peripheral tissues when they are available, and high concentrations of the free fatty acids promote the formation of the ketone bodies in the liver. We shall have more to say about the relative utilization of fatty acids after we examine the other kinds of fuels available.

FATTY ACID OXIDATION IN PEROXISOMES

The fatty acids undergo the same sequence of oxidation, hydration, oxidation, and thiolysis in peroxisomes that they do in mitochondria (Fig. 24–12). However, the enzymes are all different. The most important change for peroxisomal function is the use of an FAD-containing acyl CoA oxidase, which forms hydrogen peroxide by transferring electrons directly from long-chain acyl coenzyme A to oxygen. The proteins catalyzing the subsequent reactions also differ from those in mitochondria. There are no enzymes attacking the shorter-chain acyl coenzyme A's in peroxisomes, so the process ends with the formation of octanoyl coenzyme A. Palmitoyl coenzyme

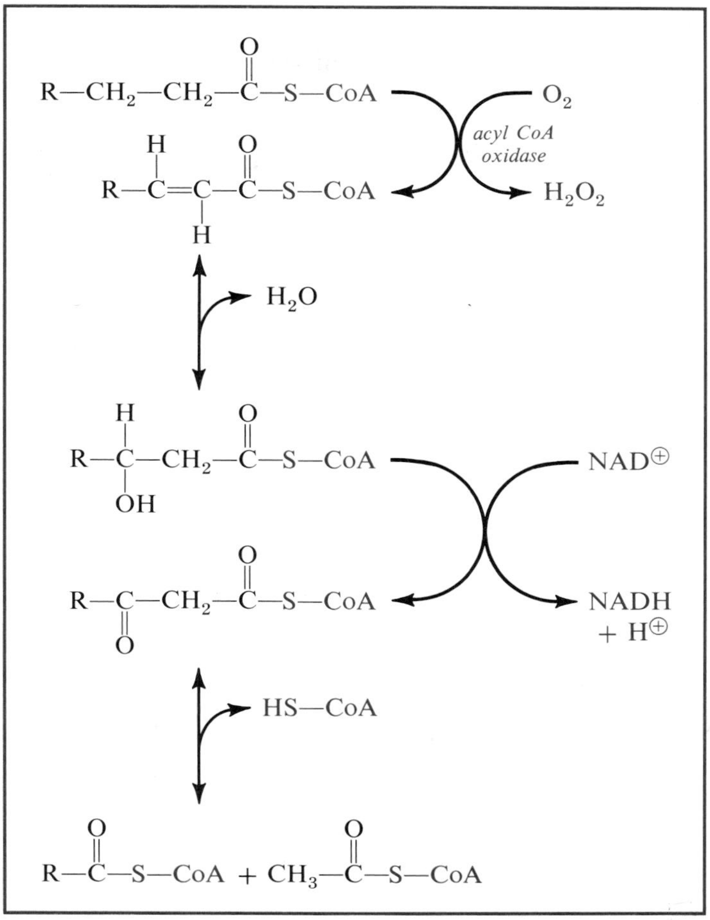

$$\mathrm{R{-}CH_2{-}CH_2{-}\overset{O}{\overset{\|}{C}}{-}S{-}CoA} \xrightarrow[\text{acyl CoA oxidase}]{O_2 \;\to\; H_2O_2} \mathrm{R{-}\overset{H}{\overset{|}{C}}{=}\underset{H}{\underset{|}{C}}{-}\overset{O}{\overset{\|}{C}}{-}S{-}CoA}$$

$$\updownarrow \; H_2O$$

$$\mathrm{R{-}\overset{H}{\overset{|}{\underset{OH}{\underset{|}{C}}}}{-}CH_2{-}\overset{O}{\overset{\|}{C}}{-}S{-}CoA} \xrightarrow{NAD^{\oplus} \;\to\; NADH + H^{\oplus}} \mathrm{R{-}\underset{O}{\underset{\|}{C}}{-}CH_2{-}\overset{O}{\overset{\|}{C}}{-}S{-}CoA}$$

$$\updownarrow \; HS{-}CoA$$

$$\mathrm{R{-}\overset{O}{\overset{\|}{C}}{-}S{-}CoA + CH_3{-}\overset{O}{\overset{\|}{C}}{-}S{-}CoA}$$

FIGURE 24–12
The pathway of oxidation of fatty acids in peroxisomes is like that in mitochondria except that the first oxidation is catalyzed by an oxidase that reduces oxygen to hydrogen peroxide.

A, for example, is converted to one molecule of octanoyl coenzyme A and four molecules of acetyl coenzyme A in peroxisomes. The organelles also contain an octanoyl carnitine transferase and an acetyl carnitine transferase, so the products of fatty acid oxidation are exported as octanoyl carnitine and acetyl carnitine, both of which can enter mitochondria for further oxidation. No high-energy phosphate is generated in the peroxisomes as a result of the oxidation. We shall see later (p. 469) that the NADH generated can indirectly be reoxidized by mitochondria, so the major loss of energy caused by the use of fatty acids in peroxisomes comes from the four nonproductive oxidase reactions. In addition, the amount of fatty acid consumed by the peroxisomes is small compared to the amount oxidized in mitochondria. All in all, the price for sustaining the protective functions of peroxisomes is not excessive.

FURTHER READING

S.J. Wakil, ed. (1970). Lipid Metabolism. Academic Press. Detailed summaries of several aspects.
P.B. Garland et al. (1969). Interactions between Fatty Acid Oxidation and the Tricarboxylic Acid Cycle. Pages 163–212 in J.M. Lowenstein, ed. Citric Acid Cycle, Academic Press.
C.L. Hall and H. Kamin (1975). The Purification and Some Properties of Electron Transfer Flavoprotein

and General Fatty Acyl Coenzyme A Dehydrogenase from Pig Liver Mitochondria. J. Biol. Chem. 250:3470.

F.J. Ruzicka and H. Beinert (1975). A New Iron-sulfur Flavoprotein of the Mitochondrial Electron Transfer System. Biochem. Biophys. Res. Comm. 66:622.

P. Harris, J. Gloster, and B.J. Ward (1980). Transport of Fatty Acids in the Heart. Arch. Mal. Coeur 73:593. Short review in English.

E.O. Balasse, F. Fery, and M.-A. Neuf (1978). Changes Induced by Exercise in Rates of Turnover and Oxidation of Ketone Bodies in Fasting Man. J. Appl. Physiol. 44:5.

J.D. McGarry and D.W. Foster (1980). Regulation of Hepatic Fatty Acid Oxidation and Ketone Body Production. Annu. Rev. Biochem. 49:395.

Carnitine

J.A. Idell-Wenger (1981). Carnitine:acylcarnitine Translocase of Rat Heart Mitochondria. J. Biol. Chem. 256:5597.

M.E. Tripp et al. (1981). Systemic Carnitine Deficiency Presenting as Familial Endocardial Fibroelastosis. N. Engl. J. Med. 305:385.

B.M. Patten et al. (1979). Familial Recurrent Rhabdomyolysis Due to Carnitine Palmityl Transferase Deficiency. Am. J. Med. 67:167.

P.R. Chapoy et al. (1980). Systemic Carnitine Deficiency—A Treatable Inherited Lipid-Storage Disease Presenting as Reye's Syndrome. N. Engl. J. Med. 303:1389.

R.A. Frenkel and J.D. McGarry, eds. (1980). Carnitine Biosynthesis, Metabolism, and Functions. Academic Press.

Peroxisomes

N.E. Tolbert (1981). Metabolic Pathways in Peroxisomes and Glyoxysomes. Annu. Rev. Biochem. 50:133.

CHAPTER 25

THE OXIDATION OF GLUCOSE

Glucose is a major fuel for most tissues and is especially important for the brain. Much of the glucose is derived from starch, which accounts for over half of the fuel in the diets of most humans, although less than this in the United States and in other highly developed countries. Glucose is also produced from other dietary components by the liver and, to a lesser extent, by the kidneys.

The combustion of glucose is conveniently considered in parts (Fig. 25–1): (1) the transport of glucose into cells, followed by the formation of glucose 6-phosphate; (2) the transformation of glucose 6-phosphate into a form that can be split to yield two triose phosphates; (3) the conversion of the triose phosphates into pyruvate; and (4) the oxidation in mitochondria of pyruvate to acetyl coenzyme A, which can be further oxidized by the citric acid cycle.

The extramitochondrial part of the sequence, the conversion of glucose to pyru-

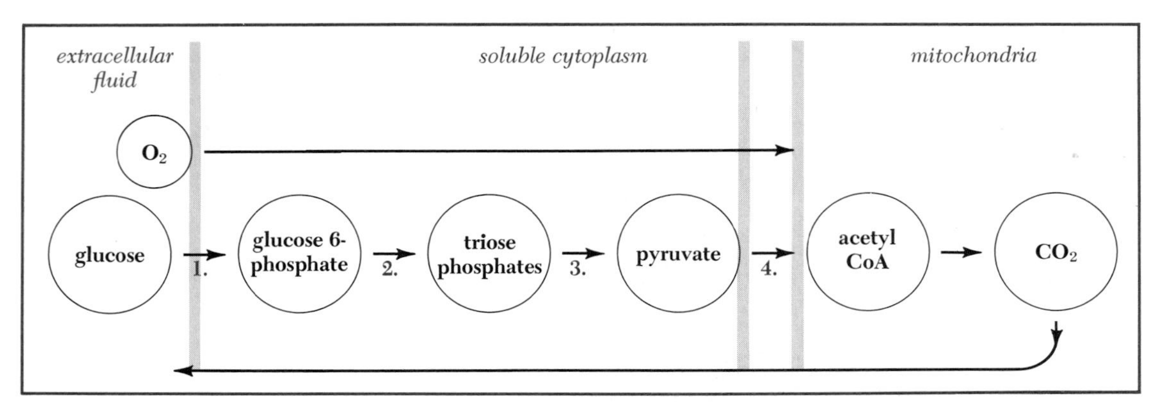

FIGURE 25–1 General outline of the oxidation of glucose. The numbered segments of the scheme are discussed in the text.

vate, is frequently called the **Embden-Meyerhof* pathway,** and it involves enzymes that are in solution in the cytoplasm, or loosely attached to membrane surfaces. High-energy phosphate is generated by substrate level phosphorylation in this pathway, as well as by oxidative phosphorylation during the subsequent mitochondrial oxidations.

CELLULAR UPTAKE OF GLUCOSE

The driving force for the passage of glucose into most cells is a concentration gradient across the plasma membrane. However, the polar glucose molecules cannot freely cross the membrane, and their passage is facilitated by specific carrier proteins in the membranes.

The concentration gradient that enables cells to obtain glucose from the blood by facilitated diffusion is not a gift; it represents a store of energy that must have been obtained somewhere. Part comes from expenditure of high-energy phosphate to maintain a relatively high concentration of glucose in the blood, between 4 and 6 mM in most humans after an overnight fast (M.W. of glucose is 180). The concentration may go as low as 3 mM or as high as 9 mM at other times in normal individuals, depending upon diet and activity, but the range is small in relation to the amount of the compound that is metabolized each day and the intervals between meals. We have already seen that ATP is expended in the cells of the small bowel and kidney in order to take up glucose by Na^+-linked active transport (p. 387), and this accounts in part for the ability to maintain the blood glucose concentration. We shall see in the next two chapters how high-energy phosphate is also expended to regenerate glucose from other compounds within the body, and this is the way that the blood glucose concentration is maintained in the periods between meals.

The carriers for facilitated diffusion in many cells have no known controls other than the concentration gradient itself. Cells with modest demands for glucose, such as red blood cells, which have lost most of their organelles, and those in the lens of the eye and in bones, have an unregulated equilibration of internal and external glucose concentrations. This is also true of the liver; it is an organ that both uses and

---

*Gustave Embden (1874–1933): German biochemist; one of the great pioneers in the study of metabolism. Otto Meyerhof (1884–1951): Another outstanding German biochemist, who received the Nobel prize in 1922, and who sought refuge in the United States in 1938.

produces glucose. Both the rate and direction of transport change as the balance of internal and external concentrations shifts.

The brain cells represent a special case. They have a high demand for glucose, and their capacity for transport is greater than the demand under most circumstances. The carriers are not saturated with glucose, but are still able to supply glucose at two to three times the rate of demand (0.26 micromoles per min per g of tissue) when blood glucose concentrations are within the normal range. However, transport does become limiting at concentrations below a calculated limit of 1.44 mM. Symptoms of deprivation, such as confusion or hallucinations, may appear in some at even higher concentrations (2.5 mM, or more), whereas others may tolerate concentrations approaching 1 mM for brief periods without effect. The damage to the brain from sustained deprivation can be so severe as to cause death or a vegetative existence.

Facilitated diffusion of glucose into the muscles (cardiac, skeletal, or smooth) and adipose tissue (fat cells) is governed in part by the hormone **insulin.** Little glucose will move into fat cells or resting muscle cells unless insulin is present. It has been shown that fat cells maintain a reservoir of glucose-receptor protein on the endoplasmic reticulum, and insulin causes the protein to move to the plasma membrane. One effect of insulin therefore is to increase the number of transport receptors, not to change the activity of an individual receptor. It is yet to be proved if insulin acts on muscle fibers in the same way.

The pancreas increases its output of insulin in response to a higher blood glucose concentration, enabling muscles to obtain more for their needs from the ample supply. At lower glucose concentrations, much of this fuel is reserved for the nervous system, which is not dependent upon insulin for glucose uptake. In other words, glucose utilization in resting muscles or fat cells becomes limited by the rate of transport when the insulin level is low.

Working muscles are not totally dependent upon the action of insulin for a glucose supply. Repeated contraction in some way increases the rate of entry of glucose if a small amount of insulin is present. (This is in part why adults with diabetes mellitus are encouraged to exercise regularly. Juvenile diabetics with a nearly total absence of insulin secretion are not benefited in this way.)

Phosphorylation of Glucose

The metabolism of glucose begins by transferring a phosphate group to it from ATP in a reaction catalyzed by a hexokinase:

$$\text{HO—CH}_2\text{-glucose} \xrightarrow[\textit{hexokinase}]{\text{ATP} \;\; Mg^{2+} \;\; \text{ADP}} {}^{2-}\text{O}_3\text{P—O—CH}_2\text{-glucose}$$

α- or β-D-glucose

α- or β-D-glucose 6-phosphate

A route for generating high-energy phosphate begins by spending a molecule of high-energy phosphate! However, this is not an accidental quirk of evolution. ATP is being expended to trap glucose as glucose 6-phosphate, which does not pass freely through the membranes. The reaction is essentially irreversible at physiological concentrations of reactants and products because the pyrophosphate bond of ATP that is being cleaved has a much higher standard free energy of formation than does the phosphate ester bond in glucose 6-phosphate. It is this irreversibility that enables the intracellular concentration of glucose to be maintained at a low level so that glucose will flow into the cell without further expenditure of energy. Indeed, the concentration in the brain cells is kept so low that it is difficult to measure. Recall that the brain hexokinase has half maximum velocity at a glucose concentration of 50 micromolar, the K_M (p. 310).

THE FORMATION OF TRIOSE PHOSPHATES

The conversion of glucose 6-phosphate to two molecules of triose phosphates involves three reactions (Fig. 25–2). Glucose 6-phosphate (an aldose phosphate) is converted to fructose 6-phosphate (a ketose phosphate), which can be phosphorylated on C1. The resultant fructose 1,6-bisphosphate is then cleaved in the middle to create two molecules of triose phosphates. One is glyceraldehyde 3-phosphate (an aldose phosphate) and the other is dihydroxyacetone phosphate (a ketose phosphate). These two triose phosphates are equilibrated by another reaction. Each of these reactions is an example of a type that occurs repeatedly in metabolism and it is worth looking at them in some detail.

Glucose 6-Phosphate → Fructose 6-Phosphate. This freely reversible interconversion involves shifting the potential carbonyl group of the glucose residue from C1 to C2. It is an isomerization of the aldohexose into a ketohexose in the form of the phosphate ester, catalyzed by **glucose phosphate isomerase** through the intermediate formation of an enediol:

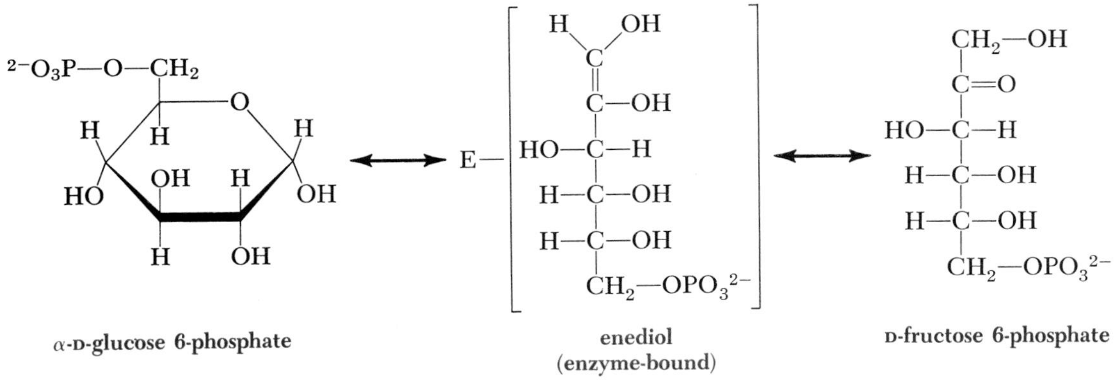

$^{2-}O_3P—O—CH_2$ (pyranose ring: H, H, O, H, HO, OH, H, OH, H, OH) ⟷ E— [H, OH / C ‖ C—OH / HO—C—H / H—C—OH / H—C—OH / $CH_2—OPO_3^{2-}$] ⟷ $CH_2—OH$ / C=O / HO—C—H / H—C—OH / H—C—OH / $CH_2—OPO_3^{2-}$

α-D-glucose 6-phosphate | **enediol (enzyme-bound)** | **D-fructose 6-phosphate**

We shall later see similar interconversions of other aldose and ketose phosphates by isomerases specific to them. An interesting sidelight of the enzyme at hand is that it binds the α-pyranose form of glucose 6-phosphate, but the open-chain form of fructose 6-phosphate. The α- and β-pyranose forms equilibrate, so all of the glucose 6-phosphate is available to the enzyme. It is true that a larger fraction of the ketose phosphate exists in the open chain form than is the case with the free sugars because the only possible ring closure is a less-favored five-membered furanose ring, but even so, fructose 6-phosphate exists mainly in the ring form at equilibrium.

Fructose 6-Phosphate → Fructose 1,6-Bisphosphate. When fructose 6-phosphate is formed, a new primary alcohol group appears on C1, and the next step in the Embden-Meyerhof pathway is the phosphorylation of this group by transfer from ATP. The action of **phosphofructokinase** commits the cell to metabolizing glu-

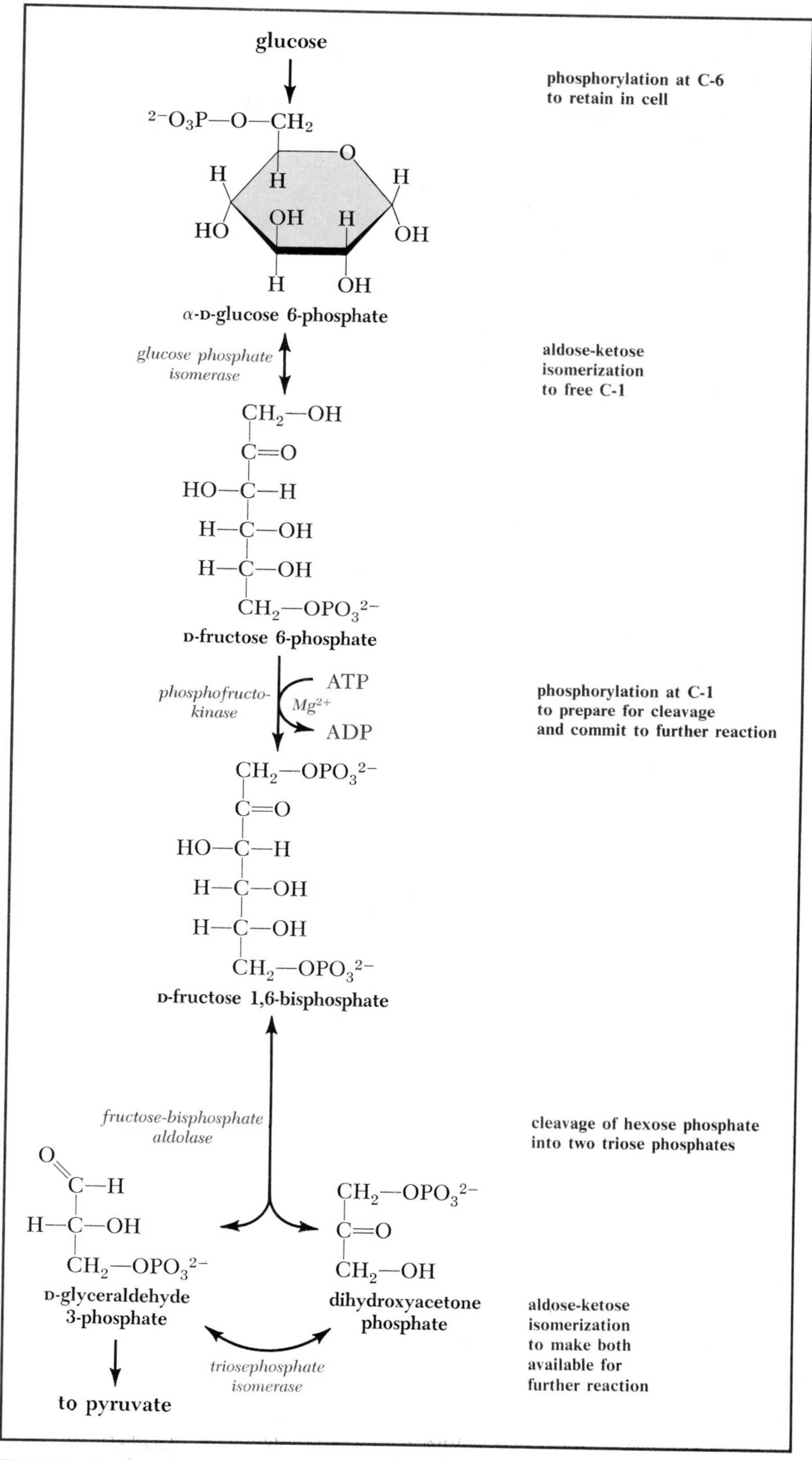

FIGURE 25–2 The conversion of glucose 6-phosphate to two molecules of triose phosphate.

cose rather than storing it or converting it to some other hexose, and we shall see that this enzyme is also a key site of regulation. The reaction catalyzed by phosphofructokinase, like the phosphorylation of glucose catalyzed by hexokinase, is essentially irreversible under physiological conditions.

Fructose 1,6-Bisphosphate → Triose Phosphates. Since fructose bisphosphate is a molecule with phosphate groups on both ends, it can be split in the middle to form the two isomeric triose phosphates—the aldose, D-glyceraldehyde 3-phosphate, and the ketose, dihydroxyacetone phosphate. We discussed the mechanism of the **fructosebisphosphate aldolase** that catalyzes this reaction in Chapter 14 (p. 283).

Glyceraldehyde 3-Phosphate ↔ Dihydroxyacetone Phosphate. The triose phosphates formed in equal amounts by the aldolase reaction are aldose-ketose isomers, and they are interconverted by a **triose phosphate isomerase** in the same way that glucose and fructose phosphates are interconverted by the glucose phosphate isomerase.

The existence of triose phosphate isomerase makes it possible to divert all of the carbons of glucose into any pathway utilizing one of the triose phosphates. We shall see in the next section that glyceraldehyde 3-phosphate is converted to pyruvate; indeed, all of the carbons of glucose can be converted to pyruvate owing to the presence of triose phosphate isomerase. This is true even though the concentration of dihydroxyacetone phosphate is much greater than the concentration of glyceraldehyde phosphate at equilibrium. Figure 25-3 represents the pools of compounds in a tissue as tanks, connected by reactions in the form of pipes. Fructose bisphosphate and the triose phosphates are equilibrated freely. We can see from the diagram that conversion to pyruvate has the immediate effect of lowering the concentration of glyceraldehyde 3-phosphate, but this will cause replacement of glyceraldehyde 3-phosphate from the dihydroxyacetone phosphate equilibrated with it, which is

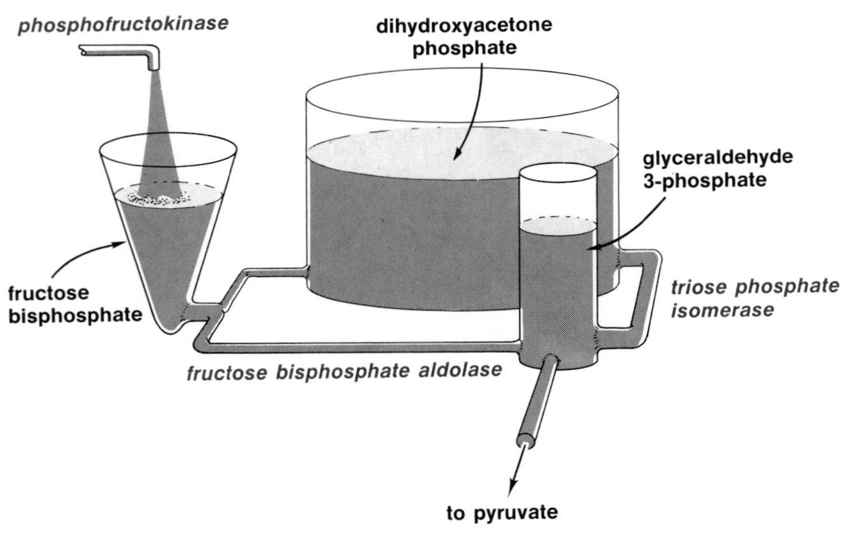

FIGURE 25-3

Fructose bisphosphate and the triose phosphates are maintained near equilibrium concentrations, with dihydroxyacetone phosphate present in the highest concentration. The pool of these three compounds is fed by the nearly irreversible formation of fructose bisphosphate. The removal of glyceraldehyde 3-phosphate as the direct precursor of pyruvate causes part of it to be replaced from dihydroxyacetone phosphate in order to maintain equilibrium. The result is that all of the hexose carbons flow toward pyruvate to an equal extent. (The diagram shows the proportion of fructose bisphosphate increasing when the total concentration of the compounds rises, since it is one molecule in equilibrium with two.)

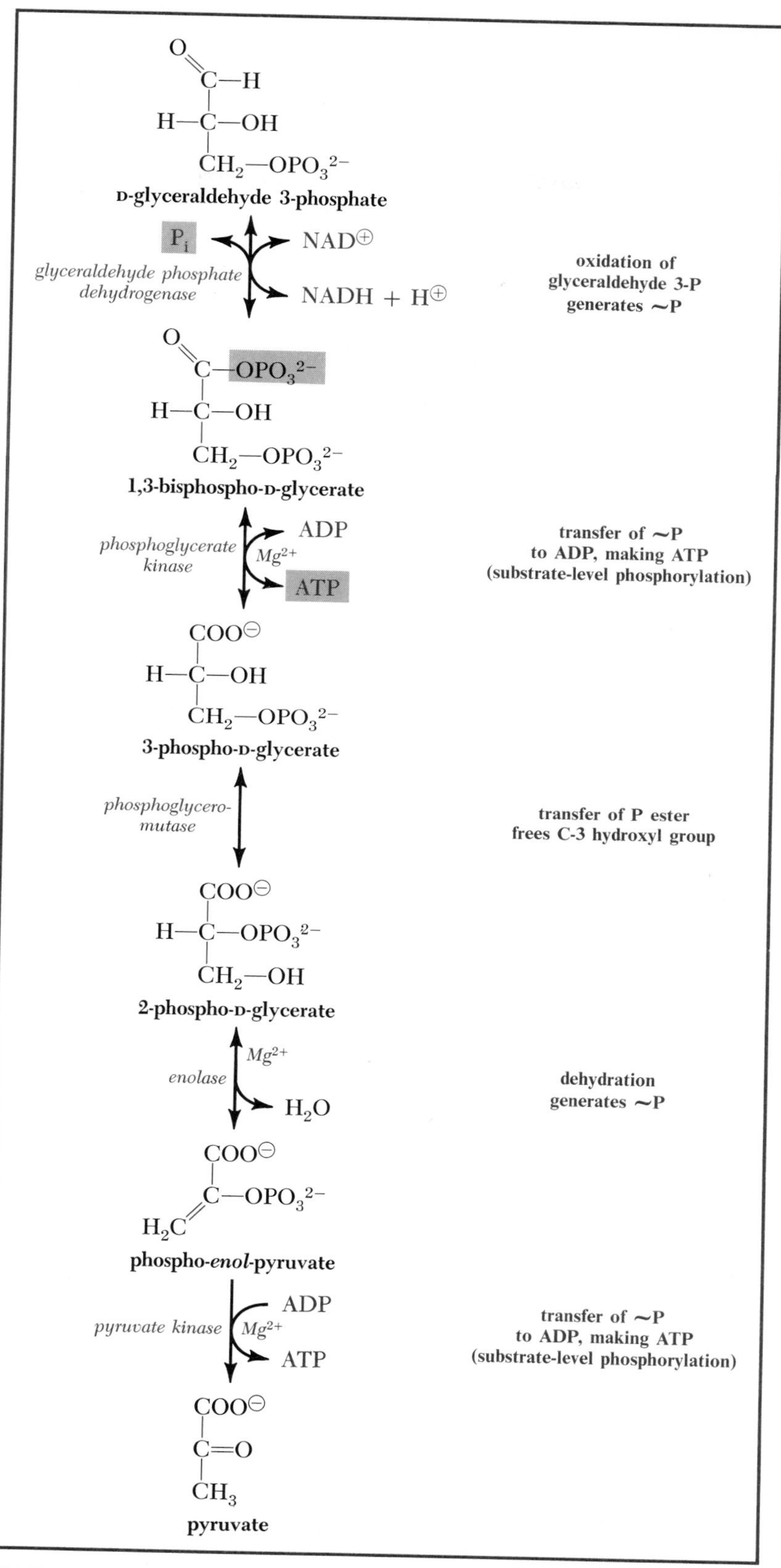

FIGURE 25–4 The conversion of glyceraldehyde 3-phosphate to pyruvate. For each molecule of pyruvate produced, two molecules of ATP are formed by substrate-level phosphorylations.

therefore also being consumed. Both halves of the fructose phosphates are converted to pyruvate in equal amounts.

THE FORMATION OF PYRUVATE FROM TRIOSE PHOSPHATES

The conversion of the triose phosphates to pyruvate in the soluble cytoplasm of cells initiates the actual recovery of energy from the metabolism of glucose. Since both triose phosphates are involved, two molecules of pyruvate are formed from each molecule of glucose 6-phosphate consumed. Essentially, the sequence involves an oxidation of glyceraldehyde 3-phosphate with capture of the energy as high-energy phosphate, followed by rearrangements that result in the formation of still further high-energy phosphate (Fig. 25–4). The high-energy phosphate is used to form ATP from ADP.

Glyceraldehyde 3-Phosphate ↔ 3-Phosphoglycerate. The aldehyde group on glyceraldehyde 3-phosphate is oxidized by NAD to form a high-energy phosphate bond in 1,3-bisphosphoglycerate. The high-energy phosphate is then transferred to ADP, releasing 3-phosphoglycerate. Oxidation of an aldehyde normally forms a carboxylic acid, 3-phosphoglyceric acid in this case. When we discussed the oxidation of α-ketoglutarate to succinyl coenzyme A in the citric acid cycle (p. 427), we noted that such oxidations have a standard free energy change sufficiently negative to support the formation of a high-energy bond at physiological concentrations. In the example at hand, the energy of oxidation of glyceraldehyde 3-phosphate is captured by causing the simultaneous formation of an anhydride bond between the carboxyl group and phosphate. The products of the reaction catalyzed by **glyceralde-**

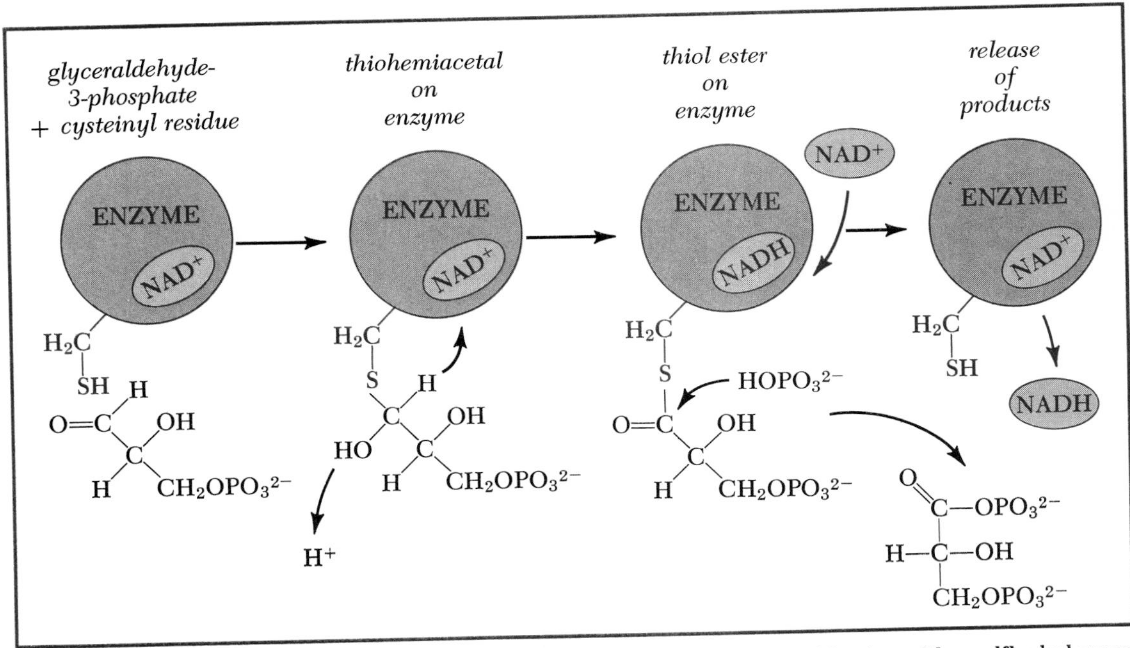

FIGURE 25–5 The oxidation of glyceraldehyde 3-phosphate involves its combination with a sulfhydryl group on the dehydrogenase, forming a thiohemiacetal. This structure is oxidized by NAD, also on the enzyme, to form a thiol ester, which is cleaved by inorganic phosphate to produce 1,3-bisphosphoglycerate. The NADH produced on the enzyme is displaced by a molecule of NAD from the solution.

hyde-3-phosphate dehydrogenase are NADH and 1,3-bisphosphoglycerate, and the 1-phosphate anhydride is a high-energy bond. Disposal of the NADH is discussed later.

The mechanism of the glyceraldehyde-3-phosphate dehydrogenase is somewhat like that of α-ketoglutarate dehydrogenase. In both cases, the energy of the oxidation is first captured in the form of an energy-rich thiol ester bond, which is then cleaved by inorganic phosphate to make a carboxylic acid-phosphoric acid anhydride bond (Fig. 25-5). However, a cysteine residue, not a dihydrolipoyl group, is the source of the thiol group in this case.

Glyceraldehyde-3-phosphate dehydrogenase is susceptible to inhibition by reagents reacting with thiols. This provided a powerful tool for early investigations in carbohydrate metabolism because treatment of an intact tissue with iodoacetate

$$\text{peptide}\!-\!SH + I\!-\!CH_2\!-\!COO^{\ominus} \longrightarrow \text{peptide}\!-\!S\!-\!CH_2\!-\!COO^{\ominus} + H^{\oplus} + I^{-}$$

stopped metabolism of the triose phosphates at this point. The resultant failure of ATP regeneration made it possible to show that high-energy phosphate is depleted during muscular work.

Phosphoglycerate kinase catalyzes the transfer of the high-energy phosphate from C1 of 1,3-bisphosphoglycerate to ADP, forming ATP and releasing 3-phosphoglycerate.

3-Phosphoglycerate → Pyruvate. The remaining reactions in the formation of pyruvate rearrange the three-carbon compounds in such a way as to capture the free energy difference between 3-phosphoglycerate and pyruvate in the form of ATP.

PHOSPHOMUTASES. The phosphate group is first transferred from C3 to C2 of glycerate. Enzymes catalyzing intramolecular transfers of phosphate are called phosphomutases, phosphoglyceromutase in this case. The mechanism of phosphoglyceromutase is typical of many (Fig. 25-6). The active enzyme contains a phosphorylated histidine residue, and the phosphate is transferred to either 3-phosphoglycerate or 2-phosphoglycerate to form a transient intermediate, **2,3-bisphosphoglycerate,** on the enzyme surface.

The small concentrations of the intermediate 2,3-bisphosphoglycerate necessary in the solution to prime the reaction are made by a bisphosphoglycerate mutase in the cytoplasm, which catalyzes a transfer of the phosphate from the first to the second carbon of 1,3-bisphosphoglycerate:

$$\underset{\textbf{1,3-bisphospho-D-glycerate}}{O\!=\!C(OPO_3^{2-})\!-\!CH(OH)\!-\!CH_2OPO_3^{2-}} \xrightarrow[\text{bisphosphoglycerate mutase}]{Mg^{2+}} \underset{\textbf{2,3-bisphospho-D-glycerate}}{COO^{\ominus}\!-\!CH(OPO_3^{2-})\!-\!CH_2OPO_3^{2-}}$$

This priming reaction ought not to be confused with the major pathway we are discussing, in which most of the 1,3-bisphosphoglycerate is used to form ATP by the phosphoglycerate kinase reaction discussed earlier. This mutase also supplies the 2,3-bisphosphoglycerate used in erythrocytes to regulate the affinity of hemoglobin for oxygen (p. 250).

ENOLASE. The 2-phospho-D-glycerate created by the mutase reaction is next dehydrated by the action of an enzyme, enolase, to form phospho-*enol*-pyruvate.

PYRUVATE KINASE. The final reaction of the Embden-Meyerhof pathway to pyruvate is the transfer of the high-energy phosphate from phospho-*enol*-pyruvate to

phosphohistidyl group + 3-phosphoglycerate ⇌ histidyl group + 2,3-bisphosphoglycerate ⇌ phosphohistidyl group + 2-phosphoglycerate

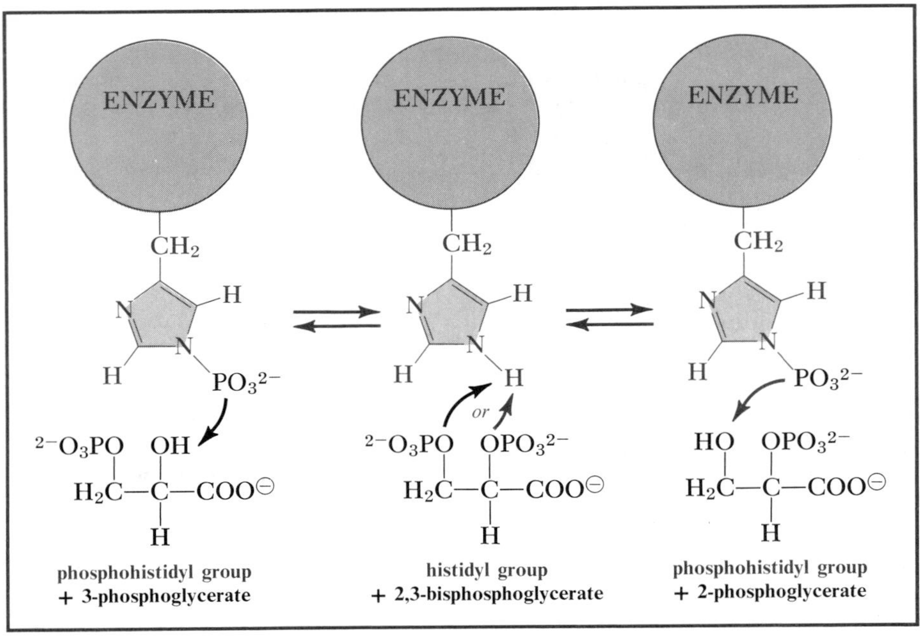

FIGURE 25–6 Mutases that catalyze intramolecular transfer of phosphate frequently have mechanisms involving phosphorylation of a histidine residue in the enzyme by a doubly phosphorylated form of the substrate. As is shown in the center, either of the phosphate groups on 2,3-bisphosphoglycerate may be transferred to a histidyl group. Therefore, either 3-phosphoglycerate or 2-phosphoglycerate may be formed. Since the reactions are reversible, the enzyme will catalyze the interconversion of the 2- and 3-phosphoglycerates. A small amount of 2,3-bisphosphoglycerate is required to prime the enzyme with phosphate groups.

ADP, thereby forming ATP. The equilibrium position of the reaction is far toward ATP formation ($K'_{eq} = 6{,}500$ at pH 7.4, 30° C).

Phospho-*enol*-pyruvate is a high-energy phosphate compound because release of the enol by cleavage of the phosphate allows it to revert spontaneously to the keto form:

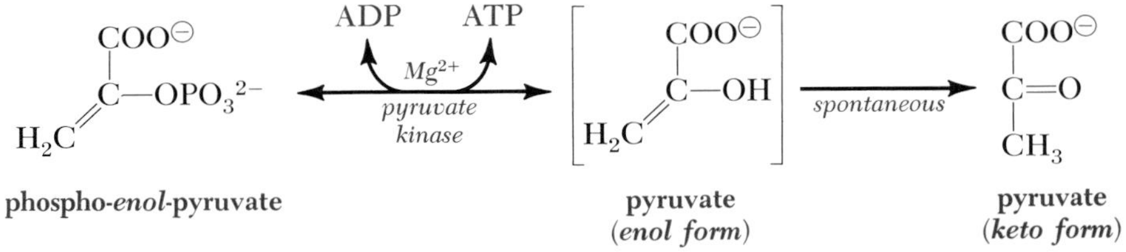

phospho-*enol*-pyruvate ⇌ (ADP → ATP, Mg^{2+}, pyruvate kinase) [pyruvate (*enol form*)] → (spontaneous) pyruvate (*keto form*)

The position of equilibrium is far in the direction of the ketone form of pyruvate, and the large amount of free energy released by this equilibration is added to the free energy of phosphate ester hydrolysis.

THE COMPLETE EMBDEN-MEYERHOF PATHWAY

If we begin with a molecule of glucose and add all of the reactions involved in its conversion to two molecules of pyruvate according to the Embden-Meyerhof scheme, the result is:

$$\text{glucose} + 2\,\text{NAD}^{\oplus} + 2\,(\text{ADP} + \text{P}_i) \longrightarrow 2\,\text{pyruvate}^{\ominus} + 2\,\text{NADH} + 2\,\text{ATP}.$$

(You should convince yourself that this is so, but note that this, and the equations that follow, are not balanced for H^+ and H_2O.) Three things happen simultaneously: glucose is oxidized to pyruvate, NAD is reduced to NADH, and ADP is phosphorylated to form ATP. There can be no Embden-Meyerhof pathway without all three events, which means that NAD, ADP, and P_i, as well as glucose, must be present.

Oxidation of Cytoplasmic NADH. Since the Embden-Meyerhof pathway consumes NAD and produces NADH, it cannot proceed without some means of regenerating the oxidized nucleotide. The NADH of eukaryotic cells cannot enter mitochondria to be oxidized by molecular oxygen through the apparatus of oxidative phosphorylation, and it therefore must be handled in the cytosol. The only reactions of significance by which this is done involve the reduction of carbonyl compounds to alcohols:

$$\text{C}{=}\text{O} + \text{NADH} + \text{H}^{\oplus} \rightleftharpoons \text{H}{-}\text{C}{-}\text{OH} + \text{NAD}^{\oplus}$$

The equilibrium position of most reactions of this type favor the formation of the alcohol group.

But what to do with the accumulating alcohols? They could leave cells as end products in metabolism. We shall discuss the way some cells use this method of handling the problem in the next two chapters; but many cells, particularly those of the striated muscles and nervous tissue, remove the resulting alcohol by transporting it into mitochondria where it can be oxidized by molecular oxygen. This effectively transports electrons from NADH in the cytosol to oxygen in mitochondria. Let us look at two important examples.

THE GLYCEROL PHOSPHATE SHUTTLE. The cytosol contains a dehydrogenase that equilibrates glycerol 3-phosphate and dihydroxyacetone phosphate with NAD and NADH (Fig. 25–7). This is a straightforward equilibration of secondary alcohol and the corresponding ketone. (Dihydroxyacetone phosphate is an intermediate in

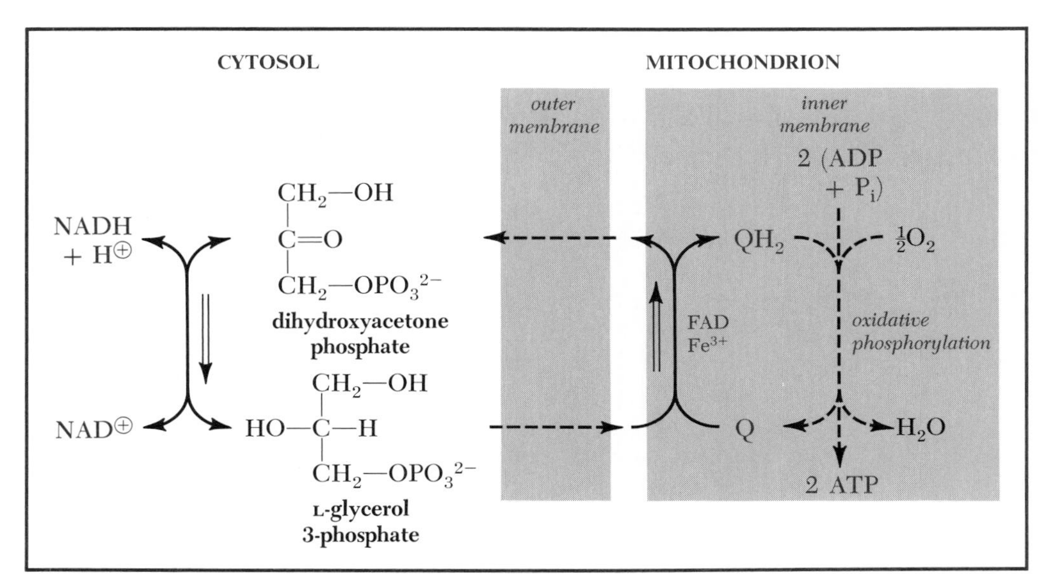

FIGURE 25–7 The glycerol phosphate shuttle of electrons.

the conversion of glucose to pyruvate and is also in equilibrium with glyceraldehyde 3-phosphate, but we are here talking about an entirely different function of the compound.) When the concentration of NADH tends to rise owing to the action of the Embden-Meyerhof pathway, there will be a shift from dihydroxyacetone phosphate to glycerol 3-phosphate that oxidizes part of the increased NADH to NAD.

Now, glycerol 3-phosphate can cross the outer mitochondrial membranes and become exposed to the action of a dehydrogenase present on the outer surface of the inner membrane. This dehydrogenase also catalyzes the interconversion of glycerol 3-phosphate and dihydroxyacetone phosphate; however, it is an iron-containing flavoprotein similar to succinate dehydrogenase, and the use of this stronger oxidizing agent makes the equilibrium strongly favor the formation of the ketone. In addition, the electrons removed are directly transferred to ubiquinone, and the re-oxidation of reduced ubiquinone by O_2 results in the generation of two high-energy phosphate bonds per glycerol 3-phosphate oxidized.

Here we have it. Dihydroxyacetone phosphate acts as a catalytic carrier for electrons in the glycerol phosphate shuttle, much in the way that oxaloacetate acts as a catalytic carrier for acetyl groups in the citric acid cycle. An increase in [NADH] raises the ratio [glycerol 3-P]/[dihydroxyacetone P] in the cytosol and the mitochondrial oxidation lowers the same ratio in the mitochondria. Therefore, there is a concentration gradient for glycerol 3-phosphate into mitochondria and for dihydroxyacetone phosphate out of mitochondria.

Since there is no net consumption or production of dihydroxyacetone phosphate in the shuttle, this function does not detract from utilization of the triose phosphate in the Embden-Meyerhof pathway. The glycerol phosphate shuttle is especially important in white muscle fibers of the higher animals and the flight muscles of insects.

The overall reaction of the glycerol phosphate shuttle is nothing more than:

$$\text{NADH}_{cytosol} + \tfrac{1}{2}\,\text{O}_2 + 2\,(\text{ADP} + \text{P}_\text{i}) \longrightarrow \text{NAD} + 2\ \text{ATP}.$$

When the two molecules of NADH produced in the Embden-Meyerhof pathway are oxidized by the shuttle, the effective result is the oxidation of glucose to pyruvate by O_2:

$$\text{glucose} + \text{O}_2 + 6\,(\text{ADP} + \text{P}_\text{i}) \longrightarrow 2\ \text{pyruvate}^{\ominus} + 6\ \text{ATP}.$$

Four of the six ATP are produced by oxidative phosphorylation as a result of the shuttle, and two are produced by substrate-level phosphorylations in the Embden-Meyerhof pathway itself.

THE MALATE-ASPARTATE ELECTRON SHUTTLE. Red skeletal muscles, heart muscle, and the brain transport electrons into mitochondria by a mechanism that involves amino acids and the malate dehydrogenase reaction. We encountered malate dehydrogenase in our discussion of the mitochondrial citric acid cycle, but the same reaction also occurs in the cytosol:

$$\text{malate}^{2-} + \text{NAD}^{+} \longleftrightarrow \text{oxaloacetate}^{2-} + \text{NADH} + \text{H}^{+}.$$

The complete system is sketched in Figure 25–8; let us reason how it works, numbering the steps as shown in the figure.

1. Accumulation of NADH in the cytosol will shift the malate dehydrogenase reaction to the left, regenerating NAD by reducing oxaloacetate (the ketone) to malate (the alcohol).

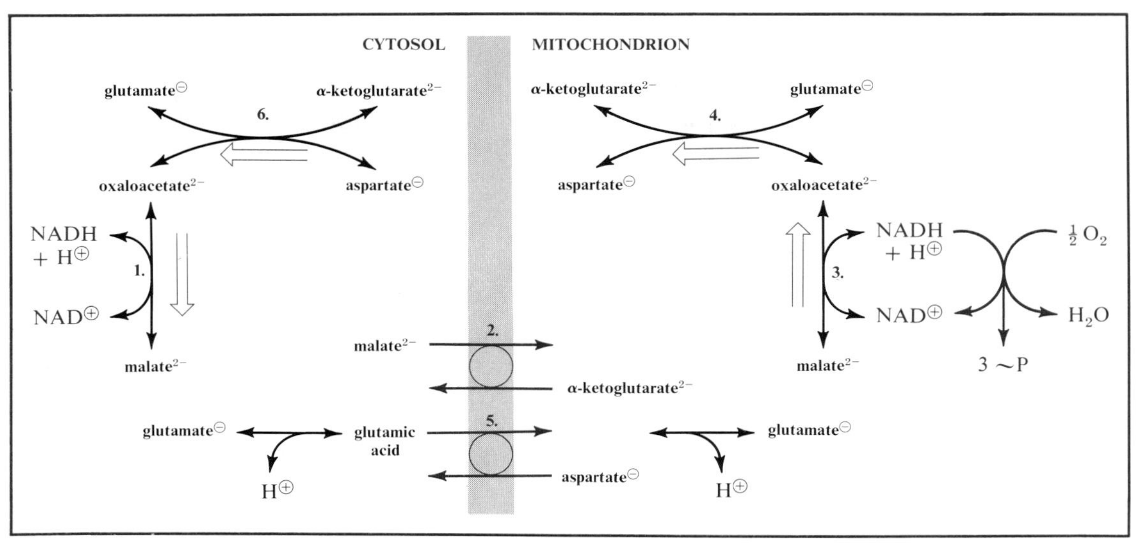

FIGURE 25–8 **The malate aspartate shuttle of electrons. The reactions are numbered in the sequence discussed in the text.**

2. Malate moves into the mitochondrial matrix by exchange for α-ketoglutarate; this exchange is catalyzed by a specific antiporter protein and is driven by the concentration differences between the cellular compartments; no additional source of energy is required.

3. When malate appears within the mitochondrial matrix, it is again equilibrated with oxaloacetate through oxidation by NAD. The NADH that is formed is in the mitochondrial matrix, and its electrons are used to generate 3 ATP while reducing oxygen.

4. In order for the exchange of malate to occur, α-ketoglutarate must be supplied within the mitochondrial matrix. This is done by transamination of glutamate with the oxaloacetate. We discussed this transamination reaction as an example of enzyme mechanism (p. 284), and it causes the formation of α-ketoglutarate and aspartate.

5. The transamination reaction requires glutamate to be supplied and aspartate to be removed within the mitochondria. This is done by an antiport exchange. However, this antiporter is specific for neutral glutamic acid and the anionic form of aspartate.* It is, therefore, moving negative charges out of the mitochondria and dissipating the electrochemical gradient created by electron transport in the inner membrane. This is the same as saying that the antiporter is driven by part of the energy of oxidation in the mitochondrial inner membrane. To the extent that the high-energy state is dissipated in this way, oxidative phosphorylation will be diminished.

6. Aspartate aminotransferase, like malate dehydrogenase, also occurs in the cytosol. Therefore, the aspartate pumped out of mitochondria will react with the α-ketoglutarate that moves out of mitochondria to regenerate oxaloacetate and glutamate in the cytosol. This completes the sequence. The overall result is an oxidation

---

*The antiporter may separately bind glutamate and H^+ rather than glutamic acid, but the net reaction is the same.

of NADH to NAD in the cytosol and a reduction of NAD to NADH in the mitochondrial matrix. The transfer of electrons is made possible even though the [NADH]/[NAD] ratio is much lower in the cytosol than in the mitochondrial matrix, because energy is expended to move the intermediates across the inner membrane against this concentration gradient. The amount of energy used cannot be stated exactly, but a minimal value is equivalent to the synthesis of 1/4 ATP.

THE OXIDATION OF PYRUVATE

The final steps in the total combustion of glucose by plants and animals involve the transport of pyruvate from the cytosol into the mitochondria. Pyruvate is actively transported into the matrix through a simultaneous transport of H^+ in a symport mechanism. Since mitochondrial electron transport maintains a higher external H^+ concentration than is present in the matrix, the movement of H^+ down this gradient can pump transport of pyruvate into the matrix. Once inside the mitochondrion, pyruvate is oxidized to acetyl coenzyme A by a pyruvate dehydrogenase complex, and then to CO_2 and H_2O by the citric acid cycle:

$$\underset{\textbf{pyruvate}}{COO^{\ominus}-C(=O)-CH_3} \xrightarrow[\textit{pyruvate dehydrogenase complex}]{NAD^{\oplus},\ CoA-SH \ \rightarrow\ NADH,\ CO_2} \underset{\textbf{acetyl coenzyme A}}{CH_3-C(=O)-S-CoA} \xrightarrow[\textit{citric acid cycle}]{} CO_2, H_2O$$

These mitochondrial oxidations account for over 80 per cent of the total ATP obtained from the complete oxidation of glucose.

The Pyruvate Dehydrogenase Complex

Pyruvate is an α-ketocarboxylate, which is oxidized by the same sort of mechanism as is α-ketoglutarate (p. 427). A decarboxylase polypeptide in the pyruvate dehydrogenase complex catalyzes the removal of CO_2 through combination of the remaining two carbons with an attached molecule of thiamine pyrophosphate as a hydroxyethyl group (Fig. 25–9). A dehydrogenase polypeptide then catalyzes the transfer of the hydroxyethyl group, which is equivalent to acetaldehyde in oxidation state, to a **lipoyllysyl** side chain on still another peptide. The transferred group has been oxidized to an **acetyl** group, combined with what is now a dihydrolipoyllysyl side chain. Finally, the acetyl group is transferred to coenzyme A, and the remaining dihydrolipoyllysyl group is reoxidized to its disulfide form by transferring electrons through **FAD** to NAD.

The pyruvate dehydrogenase differs from α-ketoglutarate dehydrogenase in animal tissues in having the decarboxylase and dehydrogenase activities on separate polypeptide chains.

Disturbances in Pyruvate Oxidation

Thiamine Deficiency. Thiamine must be supplied in the diet as a precursor for thiamine pyrophosphate. (If you deduce that the coenzyme is formed by a transfer of the pyrophosphate group from ATP, leaving AMP, you are absolutely right. There is

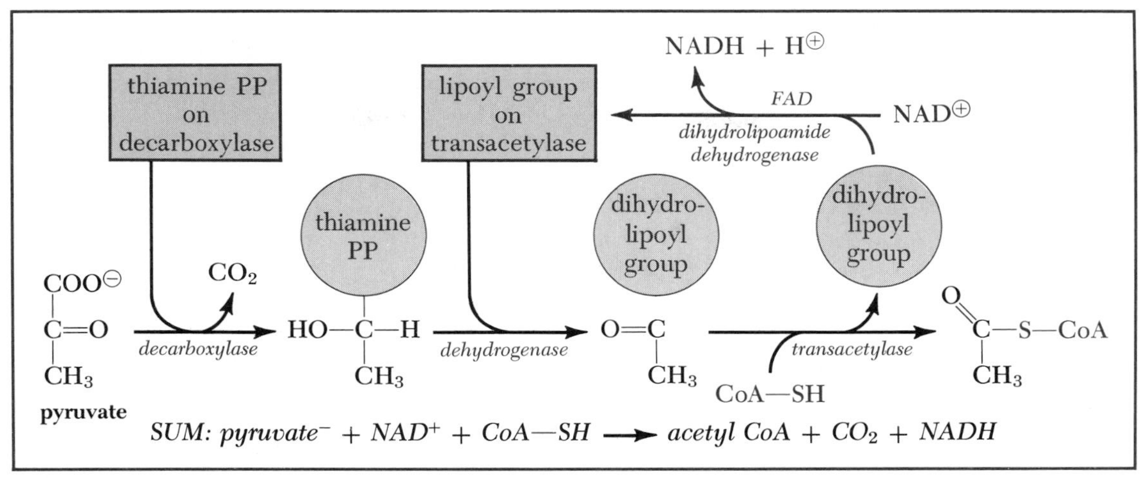

FIGURE 25–9 **The oxidative decarboxylation of pyruvate uses the same kind of mechanism seen with α-ketoglutarate dehydrogenase, except that the enzyme has separate decarboxylase and dehydrogenase polypeptide chains.**

a small amount of a thiamine pyrophosphokinase in animal tissues to catalyze this reaction.) If the supply of thiamine is restricted, then one or more enzymes requiring thiamine pyrophosphate will also be deficient.

Thiamine pyrophosphate is required for the oxidation of pyruvate into acetyl coenzyme A and for the oxidation of α-ketoglutarate in the citric acid cycle, as well as for other reactions that we shall encounter later. A defect in the citric acid cycle will impair the metabolism of fatty acids and amino acids, as well as glucose; yet all of the evidence we have at hand links the deficiency primarily to disturbances of carbohydrate metabolism, especially in the brain. For example, the thiamine requirement of a human is dependent upon carbohydrate intake. On typical mixed diets in this country, an adult will need about 1.5 micrograms of thiamine per gram of carbohydrate ingested (0.8 micromole per mole) to avoid symptoms of deficiency. Perhaps this is an indication of a greater loss of coenzyme from the intermediate states of actively working enzymes.

Concentrations of pyruvate and α-ketoglutarate were measured in the blood of normal and of thiamine-deficient individuals during fasting, and following ingestion of 100 grams of glucose (data from Metabolism, 14:141 [1965]):

	Normal		Thiamine-deficient	
	Fasting	*After glucose*	*Fasting*	*After glucose*
[pyruvate]	31 μM	42 μM	49 μM	115 μM
[α-ketoglutarate]	6 μM	7 μM	11 μM	14 μM

The α-ketoglutarate concentration is doubled in the thiamine-deficient, whether starved and using fatty acids as fuels, or using glucose as fuel. The pyruvate concentration is more than doubled in the thiamine-deficient, but only after feeding glucose.

There have been, and still are, ample opportunities for studying the gross effects of these enzymatic disabilities in humans. This may seem strange in view of the fact that thiamine pyrophosphate is an obligatory coenzyme for all of the organisms from which natural human foods are derived, and therefore ought to be a constant constituent of the diet. Indeed, a primary thiamine deficiency, as opposed to a deficiency secondary to a starvation diet, would be a rare thing were it not for two human traits. The first of these is a dislike of coarse food. Hence, humans developed the technique of removing the hard outer layers from seeds, leaving the soft, starchy interior for cooking directly or making into flour. Unfortunately, most of the thiamine pyrophosphate is in the cells of the outer layers. Therefore, those people heavily dependent upon seeds for food are liable to a deficiency of thiamine, and this is most acute in the rice-consuming areas of the Orient. The deprivation is not complete, so that the deficiency develops relatively slowly in many cases; the resultant illness is **beriberi.**

The second human trait resulting in thiamine deficiency is the desire to diminish unpleasant stimulations through the use of ethanol as a depressant. In plain English, a drunk may live only with his bottle and eat so little food that an acute deficiency of thiamine rapidly develops. The manifestations differ from those of beriberi, and the illness is known as **Wernicke's disease.** In addition to its occurrence in alcoholics, it was well known in Japanese prisoner camps during the Second World War and has developed quite rapidly in elderly patients maintained on intravenous glucose solutions without vitamin supplementation.

The first symptoms of beriberi are usually abnormal sensations in the limbs. The early disturbance of nervous function may grow worse, finally resulting in paralysis and wasting of the limbs (dry beriberi). This is consistent with the major dependence of the nervous system on glucose metabolism as a source of energy. However, the nerves of other individuals may successfully compete for the limited thiamine supply, and the cardiovascular system may be caught short. This may be manifested by congestive heart failure with **edema**—seepage of liquid into the tissues so that they become puffy (wet beriberi). There may be a combination of neural and cardiac symptoms, in which the neurological symptoms can be a life-saver by keeping the patient bedridden so that he doesn't overload his damaged heart.

Infantile beriberi is a leading cause of death in some areas, causing vomiting and a peculiar aphonia—a soundless crying.

The more sudden deficiency of Wernicke's disease primarily appears as defects of the central nervous system, with acute mental disturbances and failures of motor control, but even in these cases the mental impairment sometimes masks effects on the circulation that can result in sudden cardiac failure.

Here, then, are different clinical entities with deprivation of thiamine as a common fundamental cause, but with the tissue predominantly affected depending upon the rate of deprivation. We have dwelt upon this deficiency at some length to illustrate how difficult it is to recognize all of the factors that influence the rates of a few enzymatic reactions in a real animal.

Arsenite and Mercuric Ion Poisoning. During the action of pyruvate dehydrogenase, dihydrolipoyl groups are formed, which have a closely spaced pair of sulfhydryl groups. Metallic ions with a high affinity for sulfhydryl groups, such as mercuric or arsenite ions, are bound much more tightly to a pair on one molecule than they are to two sulfhydryl groups on separate molecules. For example, trivalent arsenic was formerly in common use by murderers and suicides, essentially because the formation of a dihydrolipoyl–arsenite chelate prevents the reoxidation of the dihydrolipoyl group necessary for continued activity of the α-keto acid dehydrogenase complexes. Indeed, arsenite poisoning mimics some effects of thiamine deficiency on the nervous system and causes an accumulation of pyruvate in the blood. An American ambassadress to Rome had her pyruvate dehydrogenase inhibited in this way by her bedroom wallpaper, which had a patrician design partially created with the mellow green of cupric arsenite.

The high affinity of arsenite for compounds containing neighboring thiol groups means that it accumulates in the protein keratin. This constituent of hair, skin, horns, and the like contains a high concentration of disulfide bonds. Some of the parent sulfhydryl groups tightly bind arsenite during maturation of the protein. Enough of the hair and fingernails survived dissolution in the otherwise well-aged corpse of Charles Francis Hall, an explorer who died in Northern Greenland in 1871 after a two-week illness, to enable proof in 1968 that he had been murdered by his associates, because the hair and nails grown during his last days of life contained high levels of arsenic and older portions did not.*

Treatment of acute poisoning by either arsenite or mercuric ions uses 2,3-dimercaptopropanol, also known as British Anti-Lewisite, BAL, because it was originally developed as an antidote for that arsenical war gas. BAL, with its adjacent sulfhydryl groups, can compete with dihydrolipoyl residues for binding with the metallic ions, forming a soluble chelate that is excreted in the urine:

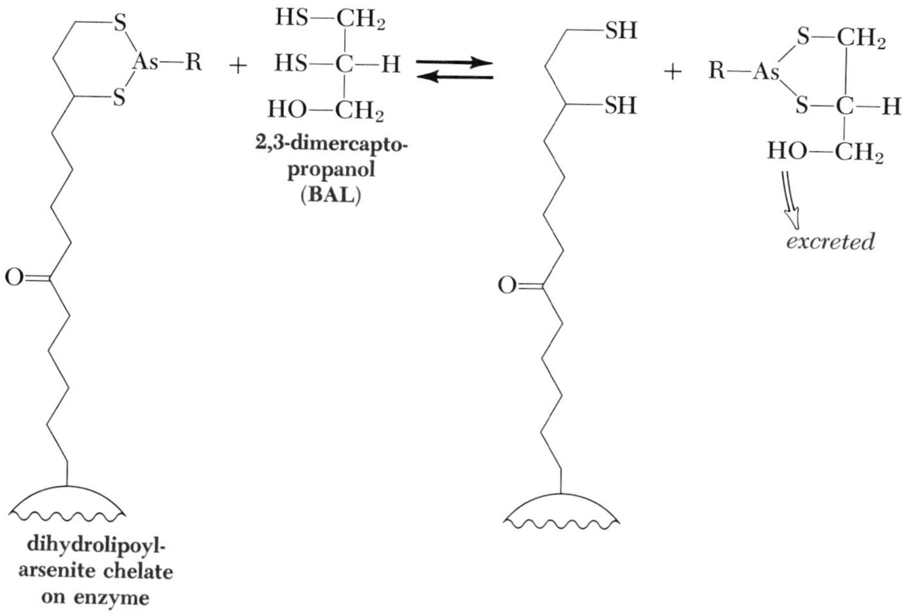

As—R + HS—CH₂ / HS—C—H / HO—CH₂ ⇌ SH / SH + R—As(S—CH₂ / S—C—H) / HO—CH₂

dihydrolipoyl-arsenite chelate on enzyme

2,3-dimercapto-propanol (BAL)

excreted

Congenital Deficiencies and Ataxia. Ataxia is defective muscular coordination; it is caused by a number of acquired and hereditary disturbances of neural function. Approximately half of those with ataxia have less than normal levels of pyruvate dehydrogenase activity. At least some result from genetic defects in the synthesis of one of the components of the pyruvate dehydrogenase complex. (There are many possibilities; the defect may be in the formation of one kind of polypeptide chain, in the availability or attachment of one of the coenzymes, and so on.) If 35 to 50 per cent of the activity remains, no symptoms may develop until ages 5 to 15 years, when slowly progressive ataxia appears. The patient is said to have **spinocerebellar ataxia.** If the deficiency is at the severe end of the spectrum, with less than 15 per cent of the pyruvate dehydrogenase activity remaining, the symptoms will appear in early infancy (before 6 months of age), and will be rapidly progressive. The infant will have severe mental retardation, microcephaly (small head), ataxia, spasticity, and other neurological dysfunctions.

*The wife of a patient hospitalized at the University of Virginia with suspected arsenic poisoning was seen briskly cutting his hair after being advised that samples were needed for analysis.

THE COMPLETE PROCESS

We have now seen all of the reactions necessary for the total oxidation of glucose: the oxidation of glucose to pyruvate in the cytosol, the transfer of the resultant electrons from the cytosol to mitochondria, the oxidation of the pyruvate to acetyl coenzyme A in mitochondria, and finally the total oxidation of acetyl coenzyme A by the citric acid cycle in mitochondria. What is the result?

If we add the individual equations when the glycerol phosphate shuttle is used, we arrive at an overall stoichiometry:

$$\text{glucose} + 6\ O_2 + 35.5\ (\text{ADP} + P_i) \longrightarrow 6\ CO_2 + 35.5\ \text{ATP}.$$

Nearly 36 molecules of high-energy phosphate are generated per molecule of glucose burned, and the P:O ratio is approximately 3.0. Slightly less oxygen is consumed to generate a given quantity of ATP when glucose is burned than is consumed when fatty acids are the fuel (P:O ratio approximately 2.8).

Stoichiometry of High-Energy Phosphate

Let us examine exactly how the figure of 35.5 high-energy phosphates produced per glucose consumed was obtained. It is important to remember that two moles of triose phosphates are produced from each mole of glucose so that the stoichiometry for all of the reactions beginning with the oxidation of glyceraldehyde 3-phosphate must be multiplied by 2.

1. One high-energy phosphate is consumed to phosphorylate glucose:	−1
2. One high-energy phosphate is consumed to phosphorylate fructose 6-phosphate:	−1
3. Two high-energy phosphates are produced by oxidative phosphorylation in mitochondria for each of the two pairs of electrons transported from glyceraldehyde 3-phosphate via NADH and the glycerol phosphate shuttle:	+4*
4. One high-energy phosphate is gained by transfer to ADP from each of two molecules of 1,3-bisphosphoglycerate:	+2
5. One high-energy phosphate is gained by transfer to ADP from each of two molecules of phospho-*enol*-pyruvate:	+2
6. One H^+ is discharged across the inner mitochondrial membrane during the transport of each pyruvate from the cytosol to the mitochondrial matrix. This is equivalent to the loss of $\frac{1}{4}$~P per H^+ discharged (perhaps more):	−0.5
7. Three high-energy phosphates are produced by oxidative phosphorylation in mitochondria from reoxidation of each of the two molecules of NADH formed by oxidation of pyruvate:	+6
8. Twelve high-energy phosphates are produced by the complete oxidation of each of the two molecules of acetyl coenzyme A in the citric acid cycle:	+24
TOTAL	+35.5

* This value is higher in tissues utilizing the malate-aspartate shuttle for electron transfer. Oxidation of each NADH created in the mitochondria by the shuttle produces three ATP, for a total of six; however, some potential ATP is dissipated in the shuttle itself, amounting to at least 0.5 mole per mole of glucose oxidized. The total yield from oxidation of glucose in these tissues will therefore be no more than 37 ATP. This yield is close to the maximum that can be obtained when the [ATP]/[ADP] ratio is near 100, while still permitting the $[P_i]$ to fall below 1 mM, as it does in some organisms. (This is true when $[CO_2]/[O_2] = 100$ and the [glucose] = 1 mM.)

REGULATION OF GLUCOSE OXIDATION

The metabolism of glucose is so central to the life of the organism that several controls are built into it so as to balance the many and varying demands for the finite supply. These controls are easiest to understand in the muscles, in which the combustion of glucose may increase as much as 200-fold upon going from rest to maximum power output. They are summarized in Figure 25–10 so that the function of each control mechanism we shall discuss can be placed in relation to the overall process of glucose oxidation. Two things must happen if muscles are to compensate for an increased rate of work: (1) Those reactions causing a synthesis of ATP from ADP and P_i must accelerate so as to supply ATP as fast as it is being consumed, and (2) enough fuel must be supplied to maintain the ATP-synthesizing reactions.

The primary signal for these adjustments is the increase in the concentration of ADP that occurs whenever the rate of ATP hydrolysis, as in muscular contraction, exceeds the rate of its synthesis. We noted that the supply of phosphocreatine prevents large changes in the concentration of ATP only because the [ATP]/[ADP] ratio is high (p. 379). It only takes a small absolute change in the low concentration, of ADP in resting muscle to make a very large relative change in its concentration, and it is this large relative change that acts as a signal in two ways: (1) the increasing concentration of ADP directly accelerates those reactions using ADP as a substrate to make ATP, and (2) the increased concentration of ADP causes an even more magnified increase in the concentration of AMP through the AMP kinase reaction, and AMP acts as an allosteric effector to make more glucose available for combustion (Fig. 25–11).

Acceleration of ATP synthesis by a rising ADP concentration in the cytosol involves three reactions:

CYTOSOL *(triose P ⟶ pyruvate, Fig. 25–11*D):

$$\text{1,3-bisphosphoglycerate} + \text{ADP} \longleftrightarrow \text{ATP} + \text{3-phosphoglycerate}$$
$$\text{phospho-}\textit{enol}\text{-pyruvate} + \text{ADP} \longrightarrow \text{ATP} + \text{pyruvate}$$

CYTOSOL-MITOCHONDRIAL ANTIPORT *(not shown in Fig. 25–11)*:

$$\text{ADP}_{\text{cytosol}} + \text{ATP}_{\text{mitochondria}} \longrightarrow \text{ADP}_{\text{mitochondria}} + \text{ATP}_{\text{cytosol}}$$

The ADP-ATP antiporter is believed to be rate-limiting for oxidative phosphorylation, and therefore for pyruvate combustion, in mitochondria. When it is accelerated by the increasing ADP supply in the cytosol, there is a corresponding rise in the ADP concentration in the mitochondrial matrix, which in turn accelerates consumption of pyruvate and oxygen.

MITOCHONDRIA *(pyruvate ⟶ CO_2, H_2O, Fig. 25–11*F,G):

$$\text{NADH} + \text{Q} + \text{ADP} \longleftrightarrow \text{ATP} + \text{QH}_2 + \text{NAD}^+$$
$$\text{QH}_2 + 2(\text{cyt c}) + \text{ADP} \longrightarrow \text{ATP} + 2(\text{reduced cyt c}) + \text{Q}$$
$$2(\text{reduced cyt c}) + \tfrac{1}{2}\text{O}_2 + \text{ADP} \longrightarrow \text{ATP} + \text{H}_2\text{O} + 2(\text{cyt c})$$

Does the concentration of ADP have to rise twofold to get a twofold increase in the rate of pyruvate formation in the cytosol and pyruvate combustion in the mitochondria? No, because the net reaction velocity will vary as the square of the substrate concentration when two sequential reactions utilizing the same substrate are near equilibrium; when three such reactions are in sequence, the velocity will vary as the cube of the substrate concentration. The two ADP-consuming and ATP-producing kinase reactions in the cytosol and the three

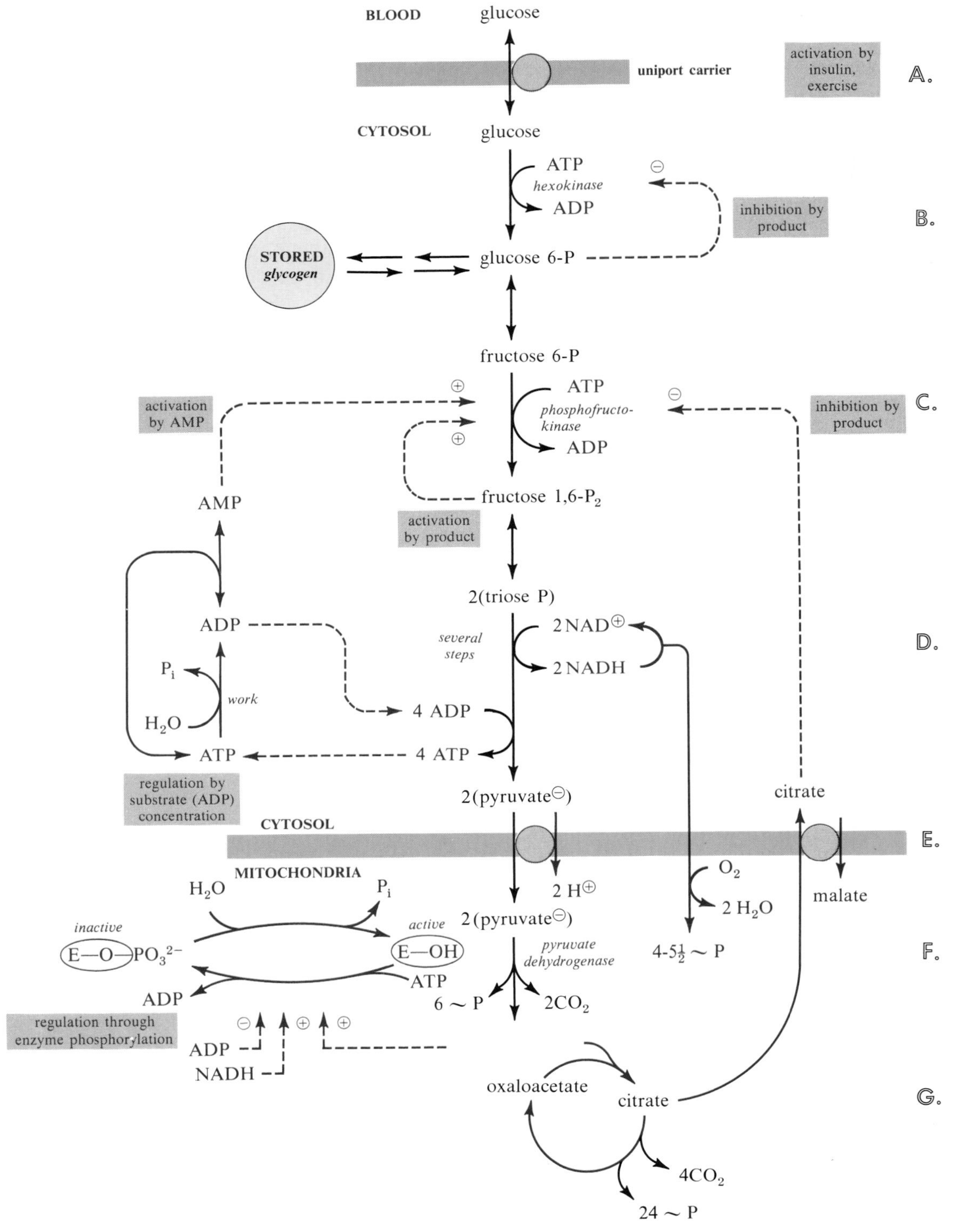

FIGURE 25–10 *See legend on opposite page*

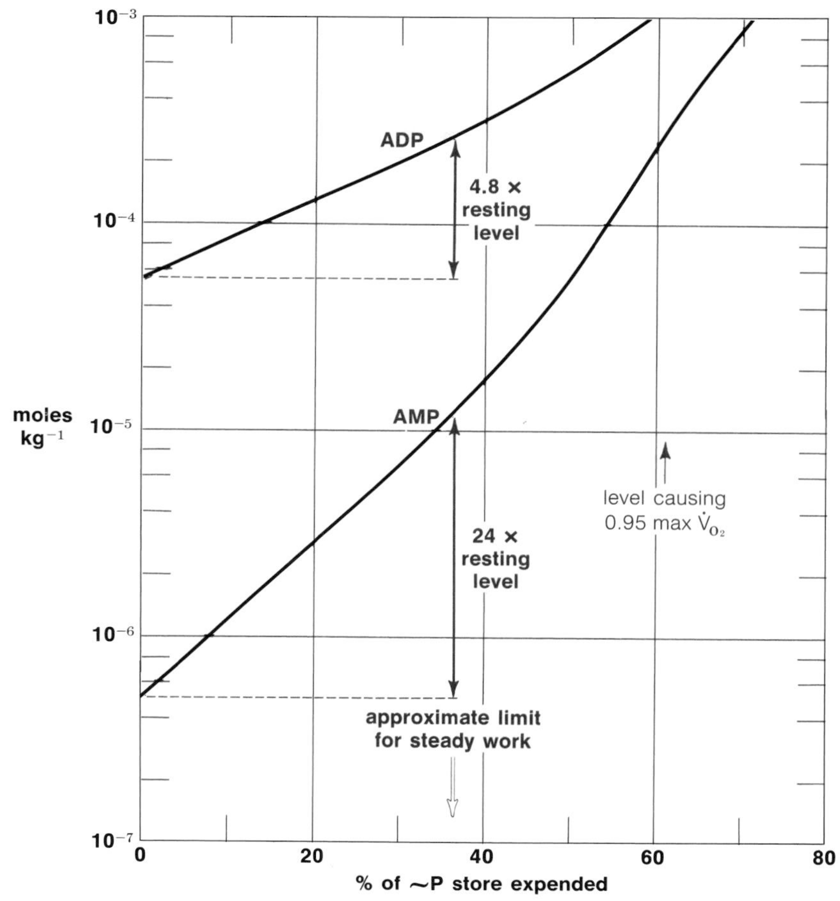

FIGURE 25–11 The changes in concentrations of ADP and AMP in skeletal muscles as the total high-energy phosphate stores are discharged. The extent of discharge is a measure of the power output of a muscle. Although the absolute concentrations are low with light activity, the relative changes are very sensitive to changes in power output, especially in the case of AMP.

FIGURE 25–10

Regulation of glucose oxidation. *A.* Activation of glucose transport into muscle by insulin or exercise.

B. Inhibition of hexokinase by its product, glucose 6-phosphate.

C. Phosphofructokinase is inhibited by ATP (not shown); the inhibition is relieved by AMP and relief is augmented by the product of the reaction, fructose 1,6-bisphosphate. The enzyme is inhibited by citrate, which enters the cytosol from mitochondria by an antiport mechanism.

D. The conversion of triose phosphates to pyruvate depends upon availability of ADP, which is also a substrate. Unless ATP is being utilized, the ADP concentration will fall and triose phosphate oxidation will slow.

E. The extent of regulation of pyruvate exchange across the inner mitochondrial membrane is not known.

F. The pyruvate decarboxylase component of pyruvate dehydrogenase is inactivated by phosphorylation. Phosphorylation is accelerated by NADH or acetyl coenzyme A, and is inhibited by ADP.

G. The citric acid cycle, as well as the electron transport shuttles, hinges upon the availability of ADP to maintain oxidative phosphorylation (not shown).

proton-pumping electron transfer complexes in the mitochondria are not all close to equilibrium, but even so, they affect each other sufficiently to make the rate of the combined reactions accelerate exponentially in response to a rising ADP concentration.

Regulation of the Fuel Supply. The acceleration of pyruvate formation in the cytosol and its combustion in the mitochondria by a rising ADP concentration will continue until the new rate of ATP synthesis just balances the rate of ATP utilization. A new steady state is then reached in which the [ATP]/[ADP] (and therefore the [phosphocreatine]/[creatine]) remains at some lower value compared to the ratios in resting muscles. However, this acceleration can occur only if there is an ample supply of triose phosphates from which to make pyruvate. The muscles must be designed so that the enzymes converting glucose to triose phosphate can supply sufficient fuel for maximum sustained effort. That is, the V_{max}'s must be high, and yet the enzymatic reactions must be nearly stopped at rest. Consider only the conversion of glucose 6-phosphate to triose phosphates:

$$\textbf{glucose 6-phosphate} \xrightarrow[\text{3 steps}]{\text{ATP} \curvearrowright \text{ADP}} \textbf{2 triose phosphates}$$

This is an irreversible process. Left to itself with fully active enzymes, it would rapidly consume glucose and ATP until one or the other was used up. (Question: What would happen in an intact cell with an unregulated complete Embden-Meyerhof pathway and mitochondrial oxidations?) This undesirable event is prevented by close regulation of phosphofructokinase, which catalyzes the one irreversible step in the sequence.

Phosphofructokinase is almost totally inhibited by physiological concentrations of ATP. Of course, ATP is a substrate for the enzyme, and the reaction won't go without it, but there are also allosteric sites to bind ATP and make the enzyme inactive. The secret of control is to relieve this inhibition whenever more triose phosphates are needed. This is done by binding AMP in competition with ATP. The result is an activation of phosphofructokinase by AMP, which we used as an example in Chapter 16 (p. 319), but it is more accurately described as a relief of the inhibition by ATP.

AMP as an Effector. Several enzymes respond to AMP, rather than ADP or ATP, as an allosteric effector because it is a more sensitive indicator than ADP or ATP of the demand for high-energy phosphate. This is a consequence of the AMP kinase reaction:

$$2\,\text{ADP} \longleftrightarrow \text{ATP} + \text{AMP}; \qquad K'_{eq} = \frac{[\text{ATP}][\text{AMP}]}{[\text{ADP}]^2} \cong 1.$$

Dividing both numerator and denominator by $[\text{ATP}]^2$ and rearranging, we see that:

$$\frac{[\text{AMP}]}{[\text{ATP}]} \cong \left(\frac{[\text{ADP}]}{[\text{ATP}]}\right)^2.$$

Since the concentration of ATP changes very little until muscles are nearly exhausted, the AMP concentration will be varying nearly as the square of the ADP concentration, and AMP is therefore a magnified signal of demands for high-energy phosphate, as is shown in Figure 25–11.

Phosphofructokinase is also regulated by other effectors. Citrate inhibits, and this is reasonable, since an accumulation of citrate means that the citric acid cycle in the muscle is being supplied with excess substrate, and the use of glucose residues as a source of acetyl coenzyme A ought to be slowed.

The enzyme is also activated by its own product, fructose 1,6-bisphosphate, when AMP is present. Here we have a rare example of positive feedback with an acceleration of the reaction tending to accelerate it still further. This is evidently a device for magnifying the activating effects. (Remember that the activity of this enzyme must be regulated over a several hundredfold range in order to meet the fuel demand as muscles using glucose go from rest to maximum work.)

Phosphorylation of Pyruvate Dehydrogenase

Total combustion of glucose requires that pyruvate be oxidized to acetyl coenzyme A in mitochondria as fast as it is formed in the cytosol, but not so fast that it is completely removed, because it has other uses. In addition, muscles also make acetyl coenzyme A from fatty acids, and there is no point in consuming pyruvate if there is already a good supply of acetyl groups. Control is necessary, and **the rate of the pyruvate dehydrogenase reaction is regulated by phosphorylating the enzyme to make it inactive.**

This is the first specific example we have encountered of regulation through enzyme phosphorylation (Fig. 25–12). The decarboxylase and dehydrogenase polypeptides in the pyruvate dehydrogenase complex occur together as $\alpha_2\beta_2$ tetramers, and it is the decarboxylase component that is phosphorylated, making it inactive. This phosphorylation is catalyzed by another enzyme, a protein kinase, that is present in the dehydrogenase complex. Two to three molecules of the **pyruvate decarboxylase kinase** are present per dehydrogenase complex, depending upon the tissue. A

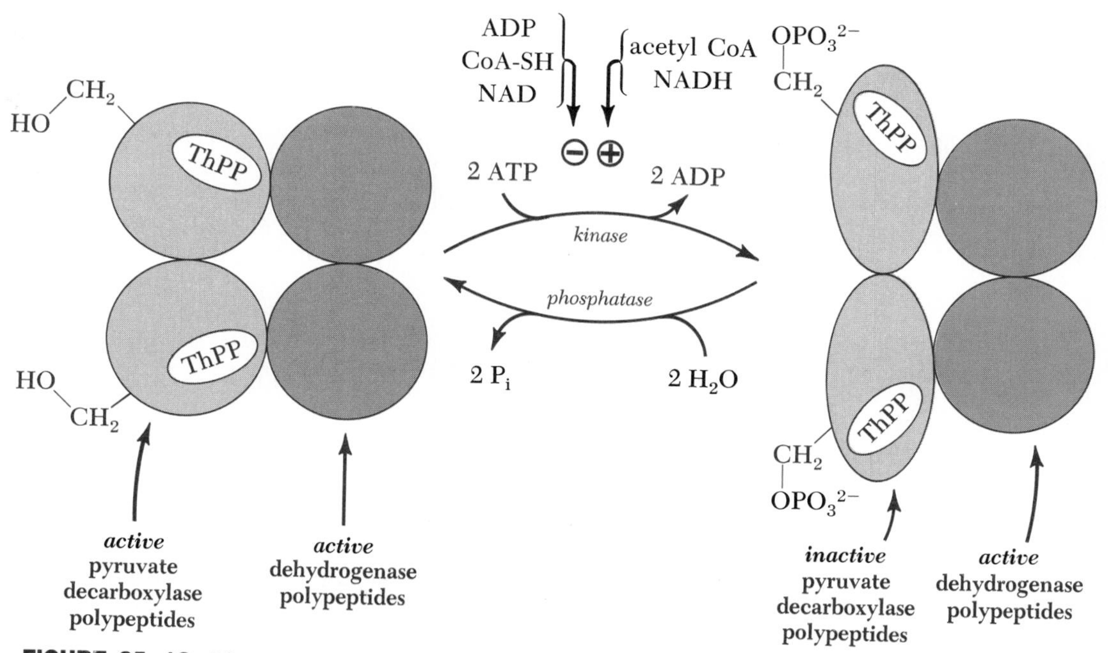

FIGURE 25–12 The pyruvate dehydrogenase complex contains a regulatory kinase that catalyzes the phosphorylation of a serine residue in the decarboxylase component making it inactive, and a phosphatase that removes the inactivating phosphate group.

phosphoprotein phosphatase, also present in the complex, restores activity to the enzyme by hydrolyzing the phosphate ester groups. The overall activity of pyruvate dehydrogenase therefore depends upon the relative activity of the kinase and the phosphatase within it.

COMPOSITION OF MAMMALIAN PYRUVATE DEHYDROGENASE COMPLEX

pyruvate dehydrogenase component:
30 per complex, each is tetramer containing 2 polypeptides with decarboxylase site and 2 polypeptides with dehydrogenase site.

transacetylase component:
60 polypeptide chains per complex, each bearing lipoyl group; forms core of complex to which other components are bound.

lipoamide dehydrogenase component:
6 dimers of identical chains per complex, one bound FAD molecule per chain; each serves several of the other components because the lipoyl groups in transacetylase can pass electrons and acetyl groups between themselves to reach the fewer lipoamide dehydrogenase chains.

pyruvate decarboxylase kinase component:
3 per complex in kidney; 2 (?) in heart; catalyzes rapid phosphorylation of one serine side chain in each decarboxylase polypeptide, which inactivates it.

pyruvate decarboxylase phosphatase:
number per complex not known; maximum activity less than that of kinase; removes phosphate group from decarboxylase polypeptide, which activates it.

The kinase is the regulated enzyme, with the phosphatase having a nearly constant and substantially smaller V_{max}. **The kinase is activated by acetyl coenzyme A and NADH; it is inhibited by coenzyme A, NAD, and ADP.** If the citric acid cycle becomes overloaded, the [acetyl CoA]/[CoA] rises; this activates the kinase and causes phosphorylation of the pyruvate dehydrogenase component. The phosphorylation inhibits the oxidation of pyruvate, blocking formation of still more of the excess acetyl coenzyme A. Similarly, if the electron transfer complexes are overloaded, either through an oversupply of oxidizable substrates or an undersupply of ADP and P_i, the [NADH]/[NAD] rises and activates the kinase so as to shut off further oxidation of pyruvate that would aggravate the oversupply. Both of these controls can be overriden by an increase in the concentration of ADP, which says in effect that no matter what else is wrong, more high-energy phosphate is needed.

Regulation of the Glucose 6-Phosphate Supply

The hexokinase of muscles and brain is regulated in two ways. It is inhibited by its own product, glucose 6-phosphate. Unless the glucose trapped within the cell by the hexokinase reaction is being used, the accumulating glucose 6-phosphate prevents more from being taken up. Also, hexokinase in brain, and perhaps in muscles, is reversibly bound to mitochondria. The bound enzyme is markedly less sensitive to inhibition by glucose 6-phosphate; it has a lower K_M for ATP, and very likely is in an environment with a higher [ATP]/[ADP], owing to the nearby operation of the mitochondrial ATP:ADP antiporter. The fraction of the total enzyme remaining free in the cytosol is determined by the glucose 6-phosphate and glucose concentrations, which diminish mitochondrial binding. Since glucose 6-phosphate both inhibits the enzyme and increases the amount that is sensitive to inhibition, the binding phenomenon amplifies the end-product inhibition and sensitizes regulation of the rate of glucose uptake to changes in the rate of glucose combustion.

FURTHER READING

F. Dickens, P.J. Randle, and W.J. Whelan (1968). Carbohydrate Metabolism and Its Disorders, vol. 1, chaps. 1,2,3. Academic Press. Contains detailed discussions of processes in animal tissues.

C.M. Veneziale, ed. (1981). The Regulation of Carbohydrate Formation and Utilization in Mammals. University Park Press. Contains reviews on hexokinase, phosphofructokinase, and 2-ketoacid dehydrogenase complexes.

J. Wahren (1977). Glucose Turnover During Exercise in Man. Ann. N.Y. Acad. Sci. 301:45.

H. Lund-Andersen (1979). Transport of Glucose from Blood to Brain. Physiol. Rev. 59:305.

J.E. Wilson (1980). Brain Hexokinase, the Prototype Ambiquitous Enzyme. Curr. Topic Cell Regul. 16:1.

L.J. Reed (1981). Regulation of Mammalian Pyruvate Dehydrogenase Complex by a Phosphorylation-Dephosphorylation Cycle. Curr. Topic Cell Regul. 18:95.

K. Uyeda (1979). Phosphofructokinase. Adv. Enzymol. 48:193.

J.B. Blass (1979). Disorders of Brain Metabolism. Neurol. 29:280. A review.

M. Prick et al. (1981). Pyruvate Dehydrogenase Deficiency Restricted to Brain. Neurology 31:398. Suggestive, but not definitive.

P.D. Boyer, ed. (1975–). The Enzymes. Academic Press. Detailed discussions of individual enzymes may be found in this multivolume work.

CHAPTER 26

GLYCOLYSIS AND GLUCONEOGENESIS

Glycolysis literally means the splitting of glucose. Glucose can be cleaved to form two molecules of lactic acid according to the formal stoichiometry

$$\underset{\text{D-glucose}}{C_6H_{12}O_6} \longrightarrow 2\ \underset{\text{L-lactate}}{HO{-}C(COO^{\ominus})(CH_3){-}H} + 2\,H^+$$

The diversion of glucose to lactate and H^+ sometimes occurs in place of the complete combustion of glucose to CO_2 and H_2O. The formation of lactate generates ATP without using O_2.

Glycolysis is nothing more than the Embden-Meyerhof pathway for the conversion of glucose residues to pyruvate with the added action of lactate dehydrogenase:

$$\underset{\text{pyruvate}}{CH_3{-}C(=O){-}COO^{\ominus}} + NADH + H^{\oplus} \xleftrightarrow{\textit{lactate dehydrogenase}} \underset{\text{L-lactate}}{HO{-}C(COO^{\ominus})(CH_3){-}H} + NAD^{\oplus}$$

$$K' = \frac{[\text{lactate}][\text{NAD}]}{[\text{pyruvate}][\text{NADH}][H^+]} = 1.11 \times 10^{11}\ (\text{pH } 7.0,\ 38°).$$

The cytosol of most, if not all, cells of the mammalian body contain enough of this enzyme to maintain the reaction near equilibrium. The direction in which the reaction goes in a particular cell depends upon the relative concentration of the reactants. The determining factor usually is the relative efficiency of electron shuttles in oxidizing NADH in the cytosol and transferring the electrons to oxygen in mitochondria. The [NAD]/[NADH] ratio determines the [pyruvate]/[lactate] ratio at equilibrium. If NADH is formed faster than the electron shuttles can oxidize it back to NAD, more pyruvate will be converted to lactate. Of course, pyruvate must also be available, but the principal source of NADH in the cytosol of most cells is the conversion of glucose residues to pyruvate, so cytosolic NADH and pyruvate tend to be formed together.

The circumstances causing formation of lactate are of profound physiological and clinical importance, so let us consider them in detail.

GLYCOLYSIS IN MUSCLES

Most muscles make lactate at times during their normal function, whereas other tissues do not. The Embden-Meyerhof pathway in muscles often makes NADH and pyruvate in the cytosol faster than they can be handled by mitochondria. One cause is a deprivation of oxygen. Isometric contraction (straining against an unmoving resistance) or working in tight quarters can temporarily occlude local blood vessels as a part of everyday life. The resultant lack of oxygen prevents mitochondrial oxidations of NADH and pyruvate, so the lactate concentration rises. An experiment was done in which normal young men isometrically contracted leg muscles to exhaustion after the occlusion of the circulation in the legs. Small biopsy samples were taken from the muscles before and after the occlusion and exercise and were found to contain the following:

	Measured				Calculated
	mmoles/kg muscle				
	lactate	pyruvate	[lact]/[pyru]	pH	[NAD]/[NADH]
resting	0.71	0.07	10	7.09	713
fatigued	24.5	0.14	178	6.56	140

(Data from Sahlin, Harris and Hultman (1975). Biochem. J. 152:173. Recalculated for fresh muscle, assuming 77% water, from dry weight concentrations.)

The rise in lactate concentration without a corresponding rise in pyruvate concentration attests to the accumulation of NADH and the corresponding decline in NAD concentration. The accumulation of H^+ accompanying the accumulation of lactate is discussed later.

Values obtained in muscle samples are cited here because they more accurately reflect the NAD/NADH balance, but the accumulating lactate also crosses the plasma membrane on an anion transporter and appears in the blood. Pyruvate also moves into the blood, but there is much less of it.

Aerobic Glycolysis

Most skeletal muscles in humans produce lactate when they are working very hard, even though the circulation is unimpeded and they are consuming oxygen at a high rate. The extent to which this occurs depends upon the enzymatic constitution of the muscle, as well as the power output. Red fibers with a high mitochondrial content produce little or no lactate, whereas white fibers with few mitochondria readily form lactate upon stimulation. Even the white fibers consume some oxygen,

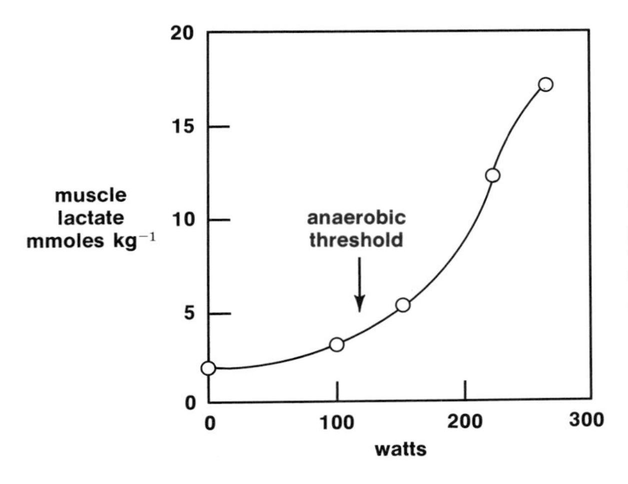

FIGURE 26–1

Variation of muscle lactate concentration with work load. Typical values in leg muscle biopsies after three-minute periods of work on a bicycle ergometer. The power output is expressed in watts. Composite data from various studies with untrained young men.

and the balance between oxidation and glycolysis in them depends upon their power output. Muscles have a so-called **anaerobic threshold,** the rate of work at which any further increase in the work load will cause a steep increase in lactate formation (Fig. 26–1). (It is anaerobic only in the sense that the increased glucose catabolism is not matched by a proportional increase in the already high oxygen consumption.) The blood lactate concentration has risen to 4 mM at the threshold. The large muscles in most people reach this threshold when their oxygen consumption ($\dot{V}_{O_2}$) is 0.6 of the maximum their mitochondria can achieve. Long-distance runners, with a smaller proportion of white fibers, reach the threshold only when their $\dot{V}_{O_2}$ is 0.9 of the maximum.

Yield of ATP. When glucose residues from stored glycogen are converted to lactate as the end product, ATP is formed only by the reactions of the Embden-Meyerhof pathway. No further ATP is made during the disposal of NADH by reducing pyruvate to lactate:

$$\text{glucose in glycogen} + 3\,(\text{ADP} + \text{P}_i) + 2\,\text{NAD}^{\oplus} \longrightarrow 2\,\text{pyruvate}^{\ominus} + 3\,\text{ATP} + 2\,\text{NADH} + 4\,\text{H}^{\oplus}$$

$$2\,\text{pyruvate}^{\ominus} + 2\,\text{NADH} + 2\,\text{H}^{\oplus} \longrightarrow 2\,\text{lactate}^{\ominus} + 2\,\text{NAD}^{\oplus}$$

$$\text{SUM:}\quad \text{glucose} + 3\,(\text{ADP} + \text{P}_i) \longrightarrow 2\,\text{lactate}^{\ominus} + 2\,\text{H}^{\oplus} + 3\,\text{ATP}^*$$

This contrasts with the much greater yield of ATP from complete combustion of stored glucose residues:

$$\text{glucose in glycogen} + 6\,\text{O}_2 \longrightarrow 6\,\text{CO}_2 + 36.5\text{–}38\ \text{ATP}.$$

Origin of the Anaerobic Threshold. Pyruvate is formed by the Embden-Meyerhof pathway during both lactate formation and complete combustion:

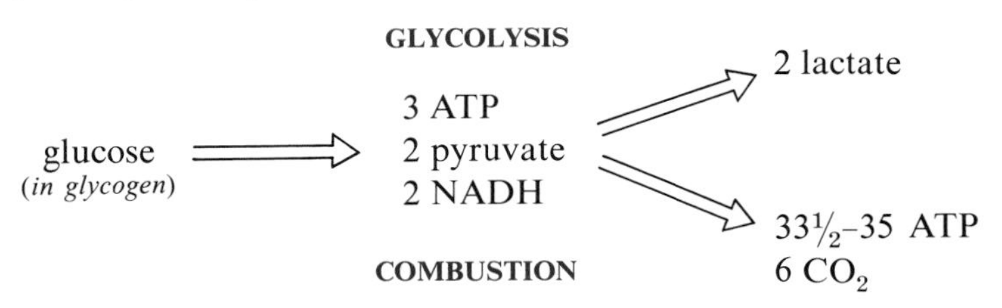

*This equation does not include the fractional stoichiometry for H^+ taken up during ATP generation, which is already balanced by H^+ released during ATP utilization. Utilization precedes generation (p. 411).

In order to produce a given amount of ATP, much more pyruvate must be formed when lactate is the product than when the pyruvate is oxidized to CO_2 and H_2O. This is the reason for the abrupt rise in lactate formation at the anaerobic threshold.

Consider first muscles that begin working at a steady pace, one that can be maintained for hours. The ADP concentration rises with the beginning of contractions, and accelerates both the Embden-Meyerhof pathway and mitochondrial oxidations. Generation of ATP by mitochondrial oxidations is sufficient to prevent more than small rises in the cytosolic pyruvate and NADH concentrations. When ATP generation matches ATP consumption, there is no further rise in ADP concentration and a steady state is achieved in which only a small part of the pyruvate is reduced to lactate in a given time. Most is oxidized.

Now consider what happens as the rate of working is increased. The ADP concentration rises in both white and red fibers. The white fibers have fewer mitochondria but a higher concentration of cytosolic enzymes catalyzing the Embden-Meyerhof pathway. As the ADP concentration rises, the mitochondria will be made to work at capacity and still not be able to meet the demand for ATP, so the ADP concentration will continue to rise and accelerate the Embden-Meyerhof pathway until it can generate ATP as fast as it is consumed. This surge in pyruvate and NADH formation is the cause of the surge in lactate formation. Beginning with glycogen, 12 to 13 glucose residues must be converted to lactate to form the same amount of ATP produced by converting one glucose residue to CO_2 and H_2O.

Suppose that a fiber works at a low rate and is hydrolyzing 37 molecules of ATP each microsecond. Both pyruvate formation and pyruvate oxidation will accelerate until one residue of glucose is oxidized completely each microsecond. This will generate ATP as fast as it is being formed, and the ADP concentration will stop rising.

Now suppose that an identical increase of effort occurs at high work loads when the mitochondrial oxidations are nearing full capacity, but the Embden-Meyerhof pathway is not. (A typical human muscle can make pyruvate 25 times faster than it can oxidize it.) The ADP concentration will continue rising until over 12 more glucose residues are being converted to pyruvate each microsecond, since only 3 ATP are generated per residue. Most of the extra pyruvate will be converted to lactate because the mitochondrial oxidations are operating near their V_{max}. We see that an increase in power output at high work loads will cause an acceleration of pyruvate formation up to 12 times greater than will the same increase at a low work load.

The advantage of aerobic glycolysis is the high power output it will support. Being able to make pyruvate 25 times faster than it can be oxidized means that ATP can be made twice as fast by converting glycogen to lactate as it can by oxidizing glycogen completely: $25 \times 3 = 75$ ATP compared to 36.5 to 38 ATP in the same period of time.

The disadvantage of glycolysis is the high rate of glycogen consumption; a given power output can only be sustained for one twelfth as long by glycolysis as it can by combustion of a given amount of glycogen.

Regulation of the Fuel Supply for Glycolysis

At its maximum, the rate of the Embden-Meyerhof pathway in working muscles can be one thousand times that of resting muscles. This creates a problem. How can there be enough of the enzymes to catalyze at the maximum rate necessary and still prevent wasteful loss of stored fuel at rest? Keeping the reactions between fructose bisphosphate and lactate in check is no problem, because they will not proceed without a supply of ADP and P_i, but a large amount of the glycogen could be converted to fructose bisphosphate at rest if there was even a slow leak through the phosphofructokinase reaction.

Role of the Fructose Phosphate Cycle. The regulatory problem is solved by providing white muscle fibers with an enzyme that slowly hydrolyzes fructose bisphosphate back to fructose 6-phosphate in resting fibers:

$$\text{D-fructose 1,6-bisphosphate} + H_2O \xrightarrow{\text{fructose bisphosphatase}} \text{D-fructose 6-phosphate} + P_i$$

The phosphatase is inhibited by AMP, whereas phosphofructokinase is activated by AMP. The maximum activity of phosphofructokinase is much greater than that of fructose bisphosphatase. However, the concentration of AMP in resting muscle is so low that phosphofructokinase is only slightly activated and fructose bisphosphatase is only slightly inhibited. The kinase has only a small fraction of its potential activity, and this is balanced by the nearly full activity of the phosphatase. This creates a slow futile cycle, in which fructose bisphosphate is hydrolyzed nearly as fast as it is formed, with the net result being a hydrolysis of ATP (Fig. 26–2). This prevents any accumulation of fructose bisphosphate.

When the muscle begins to work, the sharp increase in AMP concentration (p. 378) completely inhibits the phosphatase, while accelerating the phosphofructokinase and providing a net formation of fructose bisphosphate from which to make

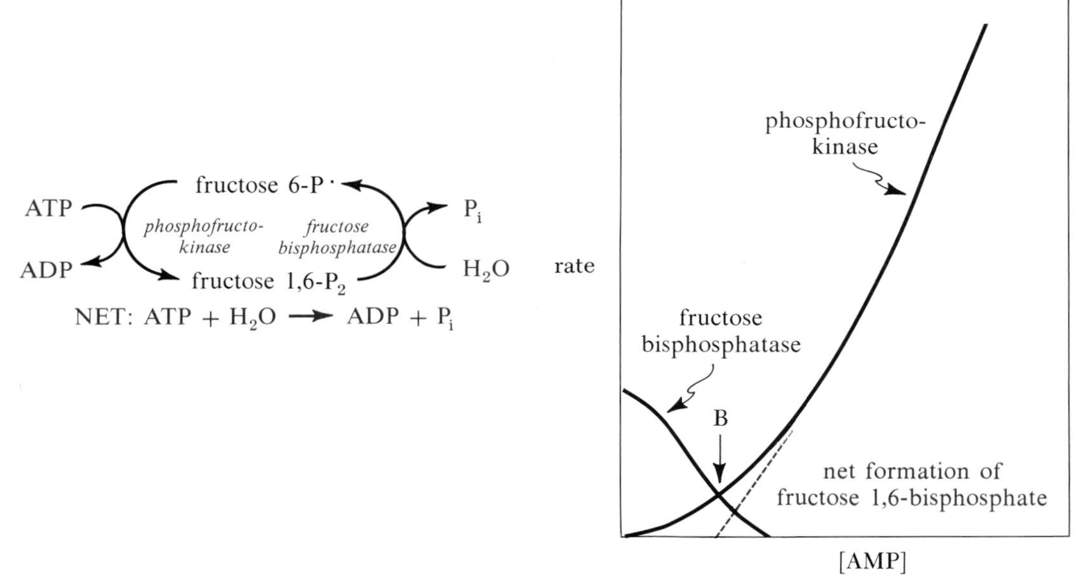

FIGURE 26–2

The presence of fructose bisphosphatase in white skeletal muscle fibers creates the potential for a futile cycle (*left*) in which the net result is hydrolysis of ATP. However, phosphofructokinase is activated, and fructose bisphosphatase is inhibited, by AMP. At the low concentrations of AMP in resting muscle, the phosphatase is more active than the kinase. Upon beginning contraction, the rise in concentration of AMP beyond point B results in a rapidly accelerating net formation of fructose 1,6-bisphosphate (*dashed line*), owing to activation of the kinase and inhibition of the phosphatase.

pyruvate. The presence of the phosphatase gives a positive on-off character to regulation of fructose bisphosphate formation.

LACTATE FORMATION AND THE OXYGEN SUPPLY

The capacity for transporting oxygen to muscles is ordinarily near the maximum demand. Those who move to a higher altitude are no longer able to provide enough oxygen to meet the demand, and their production of lactate is accelerated at more modest work loads than are required to cause rapid glycolysis at sea level.

Acute interruptions of the oxygen supply also cause lactate production in organs that normally consume the compound rather than producing it, such as the heart, brain, or liver. The interruption may be localized as in obstruction of a blood vessel, or it may be more general, as in pulmonary edema (accumulation of fluid in the lungs), obstructions of the airway, exposure to carbon monoxide, and so on. The formation of lactate may extend the life of the affected organs under these conditions. On the one hand, a lot of lactate must be produced to generate the same energy as does the normal oxidative metabolism, but on the other, a little extra ATP may make the difference between potential recovery and irreversible damage.

Glycolysis in Other Animals

Not all air-breathing animals live on the land; those that lack gills and dive under water close their access to oxygen, and the small store they can carry with them is not adequate to sustain their activity through oxidative phosphorylation for more than a few minutes in most cases; the diver does not differ from the surface-dweller in this respect.

Since the diver is deprived of oxygen, his muscles produce lactate from glycogen. Diving reptiles are extreme examples; they produce large quantities of lactate and are built to tolerate the associated massive acidosis. The blood pH of a diving turtle may fall to 6.7 from its normal 7.4 as a result of glycolysis, but the ATP generated in this way is sufficient to permit survival for as long as 12 hours at 22° without oxygen. Even the lizards and snakes can survive as long as 45 minutes under these conditions.

Whether it be mammal, reptile, or bird, the diving animal has a brain that is as dependent upon oxidative phosphorylation as is ours, and it can stay below only as long as there is oxygen in the blood circulating through its brain. The animals are able to achieve long dives because they conserve the small amount of oxygen carried with them for use by the vital tissues through a reflex response to immersion. The circulation to the skeletal muscles is partially or completely closed, and the heart rate drastically slows. Adjustments of this sort occur to a lesser extent in trained human divers. Indeed, nearly all of us were divers when we entered the world; the temporary deprivation of oxygen during delivery produces a similar reflex in the fetus.

Ordinarily, loss of the oxygen supply for more than four minutes causes permanent damage to the human brain. It has recently been noted that immersion in cold water may slow brain metabolism sufficiently so that it will survive much longer. Resuscitation has been successful on apparently dead victims of drowning in cold water.

Glycerol Phosphate in Insects. The flight muscles of insects lack lactate dehydrogenase and therefore do not carry out glycolysis. However, incubation of the muscles in the absence of oxygen causes an accumulation of glycerol 3-phosphate and pyruvate, and it is sometimes said that these are normal end-products of glucose metabolism in insects. However, this is wrong. The flight muscles have a very active oxidative metabolism, including an efficient glycerol phosphate shuttle. Insects couldn't afford to lift the large quantities of fuel that would be required to sustain a long flight by glycolysis and therefore dropped lactate dehydrogenase. The appearance of more than small amounts of glycerol 3-phosphate and pyruvate is an experimental artifact caused by the lack of oxygen. (The same compounds would appear in mammalian muscles in the absence of oxygen if the muscles lacked lactate dehydrogenase: they don't appear because the equilibria are such as to favor the formation of lactate from glycerol 3-phosphate and pyruvate.)

Fermentations

Mechanisms similar to glycolysis are used by those microorganisms that have evolved to fill ecological niches in which there is a supply of carbon compounds but little or no oxygen. These organisms are properly named **anaerobes.** They may be **obligatory** and use only anaerobic pathways, or they may be **facultative,** able to switch from aerobic to anaerobic metabolism when needed. The overall anaerobic fuel metabolism of microorganisms is known as fermentation.

Some bacteria produce lactate as the primary product of glucose metabolism by the same route used in mammalian muscles, but the best known fermentation is the cleavage of glucose by some yeasts to form ethanol and CO_2. These yeasts also use the reactions of the Embden-Meyerhof pathway for the formation of pyruvate, but they lack lactate dehydrogenase. Instead, they contain a pyruvate decarboxylase that converts pyruvate to acetaldehyde and CO_2. (The enzyme utilizes thiamine pyrophosphate as a coenzyme, and its mechanism is the same as that of the pyruvate decarboxylase component of pyruvate dehydrogenase in animals.) Acetaldehyde is then reduced by NADH to ethanol, through an equilibration catalyzed by alcohol dehydrogenase (Fig. 26–3).

There are many other and more complicated fermentations known in microorganisms, producing mixtures of products such as butanol, butyrate, glycerol, acetone, and acetate. Hydrogen gas is also a frequent product.

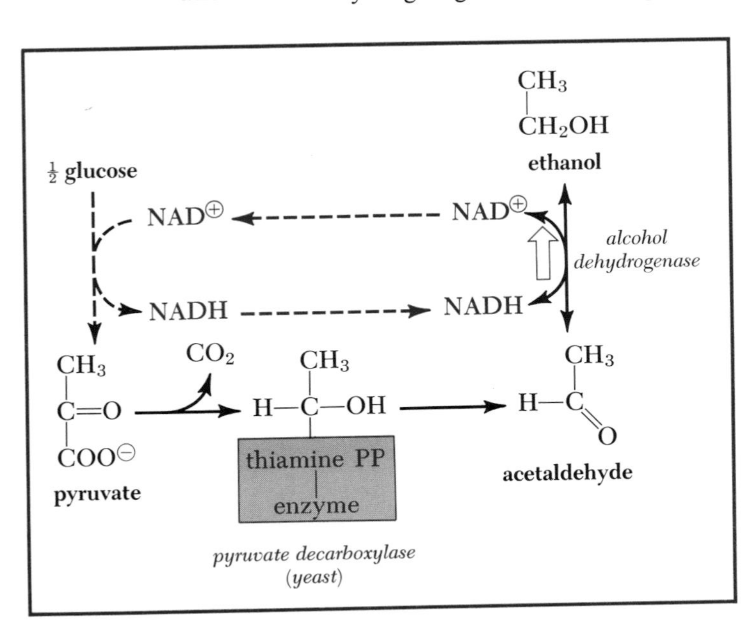

FIGURE 26–3

Yeasts form ethanol from glucose by first making pyruvate and NADH through the Embden-Meyerhof pathway. The pyruvate is decarboxylated to acetaldehyde, and the NADH is used to reduce acetaldehyde to ethanol.

THE UTILIZATION OF LACTATE: OXIDATION OR GLUCONEOGENESIS

The lactate and lesser amounts of pyruvate transported into blood from rapidly contracting muscles represent incompletely oxidized glucose residues and are potential fuels in themselves. Animals improve the overall efficiency of fuel metabolism by using lactate and pyruvate in other tissues.

Combustion of Lactate

One way of disposing of lactate is to oxidize it to CO_2 and H_2O, and this occurs in red muscle fibers, the heart, and the brain. We have been emphasizing the formation of lactate from pyruvate by the lactate dehydrogenase in rapidly contracting muscles, but the actual direction of the pyruvate-lactate interconversion depends upon the concentrations of the compounds involved. Tissues with a high oxidative capacity, such as the red muscle fibers, heart, and brain, can maintain a ratio of [NAD]/[NADH] high enough to favor the oxidation of lactate to pyruvate, which is then further oxidized in mitochondria in the usual way. Lactate is therefore a substitute for glucose as a fuel for combustion in those tissues. Lactate may pass directly from white to red fibers within a muscle.

We noted before (p. 300) that tissues with high oxidative capacity contain an isozyme of lactate dehydrogenase that is fully effective only when concentrations favor the oxidation of lactate, rather than its formation.

Efficiency. The conversion of glucose to lactate in muscles followed by the combustion of the lactate in other tissues is as efficient as the direct combustion of glucose within a single cell:

Muscle		~P Yield
Skeletal		
glucose (in glycogen) ⟶	2 lactate	+3
Heart		
2 lactate + 2 NAD^+ ⟶	2 pyruvate + 2 NADH	0
2 NADH (cytosol) + O_2 (mitochondria) ⟶	2 NAD^+ (cytosol) + 2 H_2O	+5.5* (using malate-aspartate shuttle)
2 pyruvate (cytosol) ⟶	2 pyruvate (mitochondria)	−0.5*
2 pyruvate + 5 O_2 ⟶	6 CO_2 + 6 H_2O	+30
Sum		
glucose in glycogen + 6 O_2 ⟶	6 CO_2 + 6 H_2O	38

*This value presumes that movement of 4 H^+ across the inner mitochondrial membrane is equivalent to one high-energy phosphate.

The only price paid for the rapid supply of energy from glycolysis is a temporary acidification.

Gluconeogenesis: Conversion of Lactate to Glucose

Only part of the lactate formed during vigorous exercise is burned in other tissues. Much of the remainder is converted back to glucose residues, or, in times when the glucose supply is ample, stored as fat (Chapter 28). **The formation of glucose from noncarbohydrate precursors such as lactate is known as gluconeogenesis.**

The conversion of glucose to lactate by the Embden-Meyerhof pathway is irreversible; otherwise it wouldn't have much value as a means of generating ATP. Some different reactions must be used to convert lactate to glucose.

Remember that in the Embden-Meyerhof pathway, two ATP are used to convert glucose to a pair of triose phosphates:

(*1*) glucose + 2 ATP ⟶ 2 triose phosphates + 2 ADP.

The subsequent conversion of the triose phosphates to lactate generates four ATP, and it is the combination of the two conversions that makes a net yield of two ATP formed per glucose converted to lactate:

(*2*) 2 triose phosphates + 4 ADP + 2 P_i ⟶ 2 lactate + 4 ATP
SUM: glucose + 2 ADP + 2 P_i ⟶ 2 lactate + 2 ATP.

The overall glycolytic reaction is irreversible, which means glucose can't be made from lactate by hydrolyzing two ATP to ADP and P_i. Furthermore, we know that glycolysis is irreversible beginning with glycogen, even though the ATP yield is increased to three. This means glucose 6-phosphate also cannot be made from lactate, even by cleaving three ATP. Something must be done to spend more high-energy phosphate if glucose is to be made from lactate, and this will require some modification of the reactions of the Embden-Meyerhof pathway.

Possible Gluconeogenesis in White Skeletal Muscle Fibers. There is increasing evidence that a substantial part of the lactate generated within white skeletal muscle fibers is converted back to glycogen within them. Not all accept this interpretation, but we can make a beginning on understanding the mechanics of gluconeogenesis by examining the possibility.

Since four ATP are generated by converting 2 triose phosphates to 2 lactate, it would require the cleavage of four ATP to reverse the process and make triose phosphates out of lactate. Is this energetically feasible? The answer is yes, barely. When the [ATP]/[ADP] ratio is brought back toward the value of 150 in resting muscle, this segment of the glycolytic pathway will be near equilibrium, and therefore reversible.*

The stumbling block for reversal of glycolysis is that the kinase reactions between glucose and the triose phosphate use ATP (reaction (*1*), above); making these irreversible steps go backward would require them to generate ATP, which is energetically impossible under physiological conditions. These reactions must be bypassed, and we have already seen one such bypass in white skeletal muscle fibers.

These fibers contain fructose bisphosphatase, which hydrolyzes fructose bisphosphate to fructose 6-phosphate. This is an irreversible reaction in the right direction for making glucose 6-phosphate from lactate. The presence of the phosphatase may be enough to enable white fibers to remake part of their expended glycogen when they recover from repeated contractions (Fig. 26–4). If so, they can use their low oxidative capacity during rest to recharge the fuel supply for the next heavy burst of effort.

Hepatic Gluconeogenesis and the Cori Cycle. The conversion of lactate to glucose 6-phosphate, and thus to glycogen, that may occur in white muscle fibers re-

---

* My unpublished calculations.

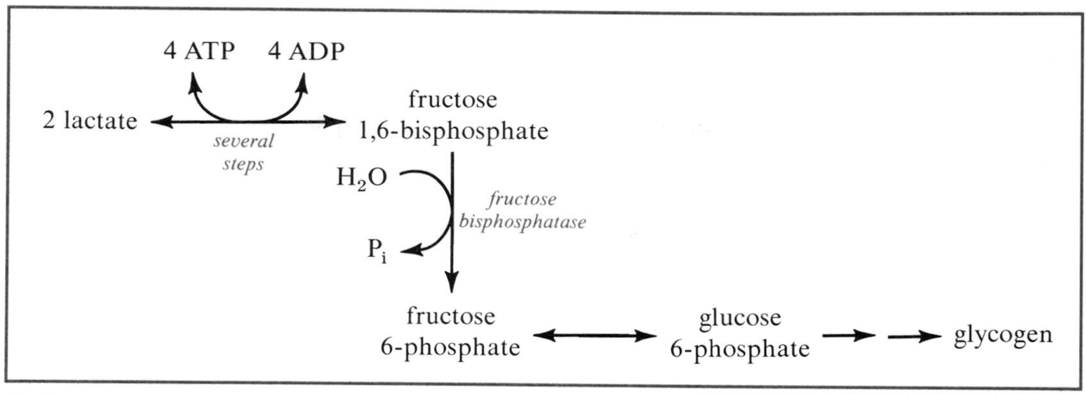

FIGURE 26–4 **Possible route for gluconeogenesis in white skeletal muscle fibers. When the lactate and ATP concentrations are high, and fructose bisphosphatase is activated in the fibers recovering from heavy work, some of the fructose bisphosphate that is hydrolyzed may be replaced from lactate by reversal of part of the Embden-Meyerhof pathway.**

quires high lactate concentration and high [ATP]/[ADP] ratios. The liver has an [ATP]/[ADP] ratio ranging near 10. Yet, it, and to a lesser extent the kidneys, carry out gluconeogenesis even at low concentrations of lactate or pyruvate. It does so by two devices. One is the use of a phosphatase to bypass the hexokinase reaction and hydrolyze glucose 6-phosphate:

$$\text{glucose 6-phosphate} + H_2O \longrightarrow \text{glucose} + P_i.$$

Glucose 6-phosphatase occurs in the endoplasmic reticulum of liver and kidneys, but not muscles. It occurs as a complex spanning the membrane. A transporter component moves glucose 6-phosphate from the cytosol side into contact with a hydrolase component on the lumen side. It is the presence of this enzyme that enables the liver, but not muscles, to release most of the glucose stored as glycogen.

The second device used to drive hepatic gluconeogenesis is the cleavage of additional high-energy phosphate during the conversion of lactate to triose phosphates:

$$\text{2 lactate} + 6 \sim\text{P} \longrightarrow \text{2 triose phosphates.}$$

This makes the formation of triose phosphates clearly irreversible.

Driving Gluconeogenesis by Carboxylation-Decarboxylation. During glycolysis, phospho-*enol*-pyruvate is converted to pyruvate by pyruvate kinase, with the formation of ATP. During gluconeogenesis in the liver, but not in muscles, pyruvate is converted back to phospho-*enol*-pyruvate by two enzymes, with the cleavage of one ATP and one GTP. The extra high-energy phosphate is put into the conversion by carboxylating pyruvate to oxaloacetate:

$$\text{pyruvate} + CO_2 + \text{ATP} \longrightarrow \text{oxaloacetate} + \text{ADP} + P_i.$$

This is an irreversible reaction, not occurring in muscles. The oxaloacetate is then decarboxylated by transferring a phosphate from GTP:

$$\text{oxaloacetate} + \text{GTP} \longrightarrow \text{phospho-}\textit{enol}\text{-pyruvate} + CO_2 + \text{GDP.}$$

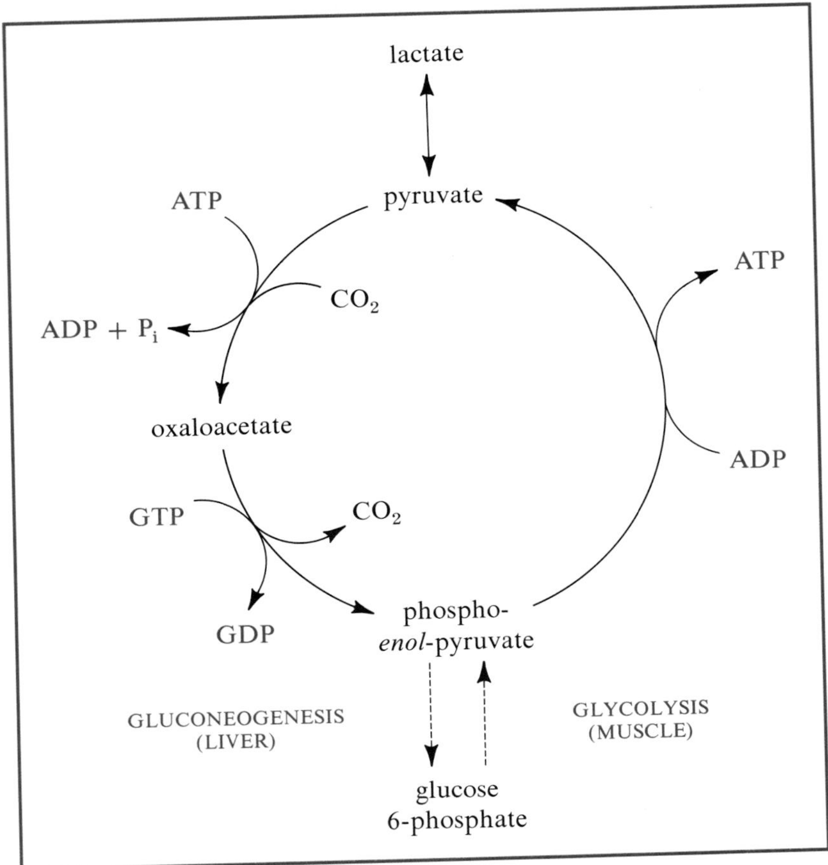

FIGURE 26–5 **Gluconeogenesis in the liver involves an irreversible conversion of pyruvate to phospho-*enol*-pyruvate through a carboxylation-decarboxylation sequence (*left*). In skeletal muscles, it is the conversion of phospho-*enol*-pyruvate to pyruvate that is favored because it does not involve an additional high-energy phosphate compound.**

The effect is to add the energy of decarboxylation of oxaloacetate in order to push the unfavorable conversion of pyruvate to phospho-*enol*-pyruvate. Figure 26–5 contrasts this two-step conversion with the single pyruvate kinase reaction in glycolysis. **Similar sequences of ATP-driven carboxylation followed by decarboxylation are used in other instances to drive biological synthesis,** so let us examine the two reactions in more detail.

Pyruvate → Oxaloacetate. The formation of phospho-*enol*-pyruvate begins with carboxylation of pyruvate at the expense of ATP to form oxaloacetate:

$$\underset{\textbf{pyruvate}}{\begin{matrix} CH_3 \\ | \\ C{=}O \\ | \\ COO^{\ominus} \end{matrix}} + CO_2 + ATP + H_2O \xrightarrow[Mg^{2+},\ Mn^{2+},\ K^{+}]{\textit{pyruvate carboxylase}} \underset{\textbf{oxaloacetate}}{\begin{matrix} COO^{\ominus} \\ | \\ CH_2 \\ | \\ C{=}O \\ | \\ COO^{\ominus} \end{matrix}} + ADP + P_i$$

Dissolved CO_2 is in equilibrium with carbonic acid and the bicarbonate ion; the bicarbonate ion is the form used by pyruvate carboxylase.

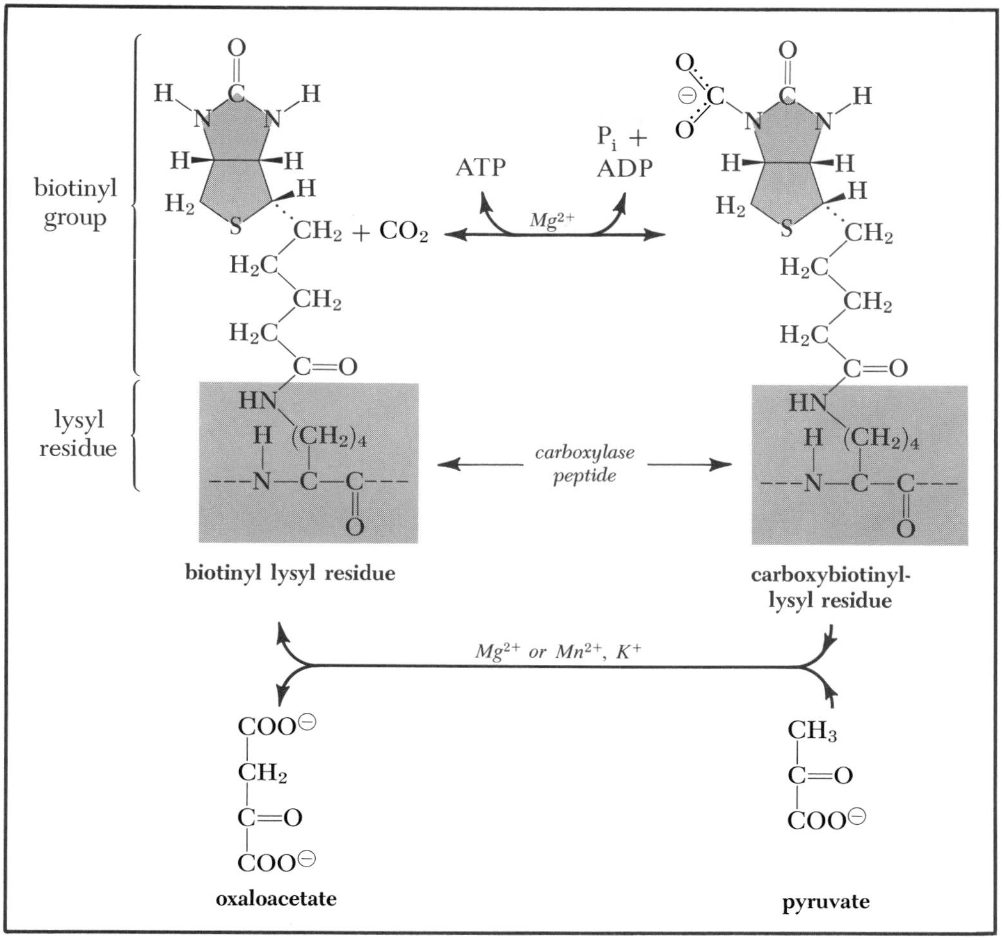

biotinyl group

lysyl residue

ATP

P_i + ADP

Mg^{2+}

CO_2

carboxylase peptide

biotinyl lysyl residue

carboxybiotinyl-lysyl residue

Mg^{2+} or Mn^{2+}, K^+

oxaloacetate

pyruvate

FIGURE 26–6 **The mechanism of pyruvate carboxylase involves a biotinyl group attached to a lysine residue. The biotinyl group is carboxylated at the expense of high-energy phosphate.**

Carboxylation reactions of this type involve the participation of biotin, a coenzyme attached to lysine residues of the specific enzymes. Biotin serves as a carrier for the "active" CO_2 that is transferred as a substrate (Fig. 26–6). Biotin cannot be synthesized by animals and is required in the diet.

Pyruvate carboxylase has an unusual metal requirement; in addition to a requirement for a monovalent cation, which is satisfied by K^+ within the cell, it also requires a divalent cation, and this requirement appears to be met by either Mg^{2+} or Mn^{2+} *in vivo*. The enzyme isolated from calf liver, as well as from tissues of lower animals, contains both of these ions. However, the enzyme from chickens fed a manganese-deficient diet has mainly Mg^{2+}, and it works equally well.

Pyruvate carboxylase also has a nearly absolute requirement for acetyl coenzyme A as an activator. This compound is not directly involved in the reaction, but the rate of conversion of pyruvate to oxaloacetate is governed by the extent to which acetyl coenzyme A has accumulated. Why would acetyl coenzyme A accumulate? We can at this point visualize two possible circumstances: There might be a shortage of oxaloacetate to combine with the acetyl coenzyme A and form citrate. Increased activity of the pyruvate carboxylase, thereby causing the formation of more oxaloacetate, primes the citric acid cycle in this case.

Acetyl coenzyme A may accumulate because enough is being formed from fatty acids to supply the energy requirements of the cell. Increasing the conversion of pyruvate to oxaloacetate then diverts the carbons of pyruvate to the formation of glucose.

Oxaloacetate → Phospho-*enol*-pyruvate. Oxaloacetate is converted to phospho-*enol*-pyruvate by phosphorylation with GTP, accompanied by a simultaneous decarboxylation:

$$\underset{\text{oxaloacetate}}{\begin{array}{c} COO^{\ominus} \\ | \\ C{=}O \\ | \\ CH_2 \\ | \\ COO^{\ominus} \end{array}} + GTP \underset{Mn^{2+},\ K^{+}}{\xrightleftharpoons{\text{phosphopyruvate carboxykinase}}} \underset{\text{phospho-}enol\text{-pyruvate}}{\begin{array}{c} COO^{\ominus} \\ | \\ C{-}OPO_3^{2-} \\ \| \\ CH_2 \end{array}} + GDP + CO_2$$

This reaction effectively discharges the chemical potential gained at the expense of ATP in the preceding carboxylation of pyruvate. This discharge drives the energetically unfavorable transfer of phosphate from GTP to the pyruvate moiety.

The **phosphopyruvate carboxykinase** catalyzing this reaction differs from pyruvate carboxylase in mechanism. The reaction does not involve biotin, and CO_2 is not "activated." The enzyme is present in the mitochondria of some species and in the cytosol of others; it is in both cellular compartments of human liver. The result in either case is the appearance of phospho-*enol*-pyruvate in the cytosol. No expenditure of energy is required to transport phospho-*enol*-pyruvate from mitochondria to cytosol.

The Cori Cycle.* The white muscle fibers and the liver can cooperate in a cyclical turnover of glucose and lactate:

(*1*) *Muscle:* glycogen ⟶ lactate (*into blood stream*)
(*2*) *Liver:* lactate (*from blood stream*) ⟶ glucose (*into blood stream*)
(*3*) *Muscle:* glucose (*from blood stream*) ⟶ glycogen

Energy Cost of the Cori Cycle. Gluconeogenesis isn't free. Fuel must be expended to generate the ATP hydrolyzed during the conversion of lactate to glucose, and the Cori cycle is especially expensive. The reactions are tabulated in Table 26–1.

A total of 6.5 moles of high-energy phosphate is consumed in the liver for each

*Carl F. Cori (1896–) and Gerty T. Cori (1908–1955): Austrian-born American biochemists and Nobel laureates.

TABLE 26–1 STOICHIOMETRY OF GLUCONEOGENESIS FROM LACTATE (H^+ IS OMITTED)

2 lactate⁻ + 2 NAD^+	⟶	2 pyruvate⁻ + 2 NADH(*cytosol*)
2 pyruvate⁻(*cytosol*) + 0.5 ~ P*	⟶	2 pyruvate⁻(*mitochondria*)
2 pyruvate + 2 CO_2 + 2 ~ P	⟶	2 oxaloacetate(*mitochondria*)
2 oxaloacetate + 2 ~ P	⟶	2 phospho-*enol*-pyruvate(*mitochondria* or *cytosol*)
2 phospho-*enol*-pyruvate + 2 NADH + 2 ~ P	⟶	fructose bis-P + 2 NAD^+(*cytosol*)
fructose bis-P + 2 H_2O	⟶	glucose + 2 P_i
SUM: 2 lactate⁻ + 6.5 ~ P	⟶	glucose

*Assuming discharge of 4 H^+ across inner membrane is equivalent to loss of one ATP.

mole of glucose produced from 2 moles of lactate. How is the liver to provide this high-energy phosphate? It does so by oxidizing some fuel, which is equivalent to the loss of part of the glucose. The calculations will be summarized in the next chapter for those with especial interest, but it is possible to arrive at the approximation that for each 100 millimoles of glucose residues in glycogen converted to lactate by glycolysis in the muscles, only 80 millimoles could come back and be stored again as glycogen. However, the muscle originally obtained 300 millimoles of ATP from glycolysis, and only 20 millimoles of glucose have been lost in the end, so the net yield is 15 moles of ATP gained to drive the muscles for each mole of glucose consumed. This is a lot better than the 2 moles obtained when lactate is the final product, as it is in fermenting bacteria. To summarize:

1. Energy-yielding glycolysis in muscle:

$$100 \text{ glucose (as glycogen)} \longrightarrow 200\,(\text{lactate}^{\ominus} + \text{H}^{\oplus}) + 300 \sim \text{P}$$

2. Energy-consuming resynthesis of glucose in liver and storage in muscle, using part of lactate as fuel:

$$200 \text{ lactate} + 120\ O_2 \longrightarrow 80 \text{ glucose (as glycogen)} + 120\ CO_2$$

$$\text{SUM: } 20 \text{ glucose (as glycogen)} + 120\ O_2 \longrightarrow 120\ CO_2 + 300 \sim \text{P}$$

$$\sim\text{P/glucose} = 15$$

This yield is much less than the 38 ATP per glucose residue disappearing that is obtained when lactate is oxidized in other tissues, so why isn't all of the lactate burned rather than being converted back to stored fuel? Probably several factors are involved. One is the lessened demand for fuel, even by the heart, during recovery from strenuous work. Another is the conservation of carbohydrate as a fuel for emergencies.

Even though 20 per cent of the potential ATP is expended for each lactate that is processed in the liver rather than being directly oxidized, the total loss is not as great as it seems. Maximum lactate production is used only for brief periods, of the order of a minute. Even in our civilized times, being able to go all-out for a minute is sometimes life-saving, and the waste of what might have been another two minutes worth of ATP, to take the worst possible case, isn't a great price. (After study of the next chapter, those interested may calculate the net high-energy phosphate yield per glucose consumed when the conversion of lactate to glycogen occurs within the white muscle fiber in which the lactate has been formed. It is 25 ATP formed per glucose residue consumed.)

LACTIC ACIDOSIS

Transitory increases in lactate concentration, and an associated increase in H^+ concentration, are a normal consequence of exercise. Values up to 17 mM lactate in blood are common, and may be higher in some individuals after extreme exertion. Even excitement will cause more lactate to be formed, although at much lower levels.

However, some conditions cause a more persistent elevation in lactate concentration; the underlying metabolic disturbance, as well as the change in acid-base balance (Chapter 38), is frequently a cause for concern. Any concentration of lactate greater than 2 mM after resting is considered abnormal.

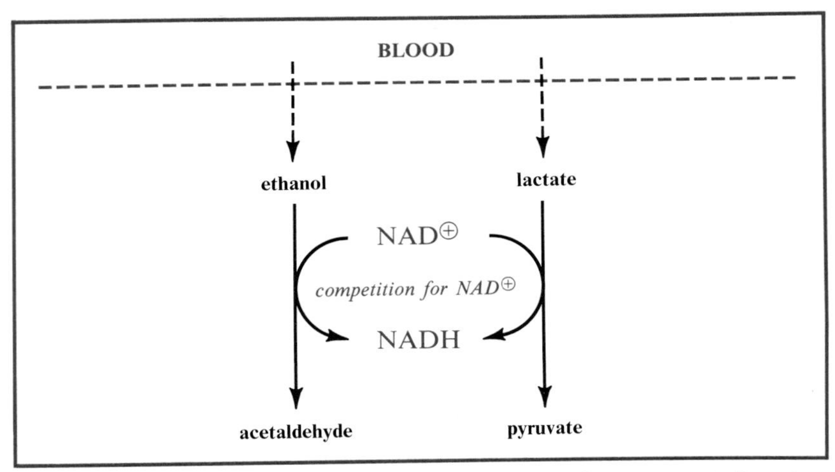

FIGURE 26–7 Ingestion of ethanol tends to prevent gluconeogenesis from lactate in the liver because the oxidation of ethanol competes for the NAD necessary for the oxidation of lactate.

The lactate concentration may rise because of increased production from glucose. We have already noted that any impairment of the oxygen supply will lead to a general increase in lactate production. This may occur during anesthesia. A similar increase is seen with any inhibitor of the oxidative processes, so long as glucose is available. For example, cyanide poisoning can cause a massive lactic acidosis.

The lactate concentration also may rise because of decreased gluconeogenesis in the liver. A congenital absence of any of the necessary isozymes that are peculiar to the liver (or other gluconeogenic tissues) would cause lactate accumulation. Infants have been found with persistent lactate levels as high as 20 mM who lacked the hepatic fructose bisphosphatase, but not the enzyme in muscle.

Gluconeogenesis can also be impaired by drinking ethanol. The metabolism of ethanol begins by oxidation with NAD in the liver. The resultant increase in the [NADH]/[NAD] ratio makes the equilibrium for lactate dehydrogenase shift toward lactate (Fig. 26–7).

TABLE 26–2 LACTIC ACIDOSIS

Characteristics
- blood lactate concentration >2 mM without exercise
- blood pH low to variable extent
- [lactate]/[pyruvate] ratio often increased over resting

Causes and Resultant Lactate Concentrations*
1. low blood [O_2] (3–4 mM)
2. severe anemia (<2 mM)
3. congenital absence of enzymes necessary for gluconeogenesis (e.g., hepatic fructose bisphosphatase 3–23 mM)
4. leukemia (12–25 mM)
5. cardiovascular shock (2–35 mM)
6. surgical bypass of the heart (2–8 mM)
7. anesthesia
8. diabetes mellitus (10–31 mM)
9. ethanol ingestion (2–8 mM)
10. phenformin intoxication (10–31 mM)

*Based on Oliva (1970), Am. J. Med., *48*: 209.

Lactic acidosis also accompanies a variety of physiological and pathological states (Table 26–2). Phenformin, a drug formerly used for the oral treatment of diabetes mellitus, sometimes provoked lactic acidosis in an unknown way.

FURTHER READING

M.J. Rennie and R.H.T. Edwards (1981). Carbohydrate Metabolism of Skeletal Muscle and Its Disorders. 3:1 in P.J. Randle, D.F. Steiner, and W.J. Whelan, eds. Carbohydrate Metabolism and Its Disorders. Academic Press. Excellent review.

P.R. Moret et al. (1980). Lactate. Springer-Verlag. Informative discussions of physiological and pathological events. See especially the discussion of intracellular gluconeogenesis by L. Hermansen and O. Vaage, p. 46.

R. Margaria (1976). Biomechanics and Energetics of Muscular Exercise. Clarendon Press. Includes discussion of lactate production.

R.W. McGilvery (1975). Use of Fuels for Muscular Work. p. 12 in H. Howald, and J. Poortmans, eds.: Metabolic Adaptation to Prolonged Physical Exercise. Birkhauser Verlag Basel. An extension and amplification of the treatment given here. (Neglects propulsion of mitochondrial ADP-ATP antiport by membrane potential, however.)

J. Keul, E. Doll, and D. Keppler (1972). Energy Metabolism of Human Muscle, vol. 7 of E. Jokl, ed.: Medicine and Sport. University Park Press. Contains good and quite readable reviews.

H.T. Andersen (1966). Physiological Adaptations in Diving Vertebrates. Physiol. Rev. 46:212.

H. Siebke et al. (1975). Survival After 40 Minutes' Submersion Without Cerebral Sequelae. Lancet 1:1275.

R.W. McGilvery (1975). Biochemical Concepts. W.B. Saunders. Chapter 16 is a more extensive discussion of fermentations. Not bad—not bad at all.

C.M. Veneziale, ed. (1981). The Regulation of Carbohydrate Formation and Utilization in Mammals. University Park Press.

K. Sahlin, R.C. Harris, and E. Hultman (1975). Creatine Kinase Equilibrium and Lactate Content Compared with Muscle pH in Tissue Samples Obtained after Isometric Exercise. Biochem. J. 152:173.

K. Sahlin (1978). Intracellular pH and Energy Metabolism in Skeletal Muscle of Man. Acta Physiol. Scand. Suppl. 455.

G.A. Brooks and G.A. Gaesser (1980). End Points of Lactate and Glucose Metabolism after Exhausting Exercise. J. Appl. Physiol. 49:1057.

F. Plum, ed. (1973). The Threshold and Mechanism of Anoxic-Ischemic Brain Injury. Arch. Neurol. 29:385. Symposium papers.

H. Bossart and C. Perret, eds. (1978). Lactate in Acute Conditions. S. Karger. Symposium papers.

L. Stolberg et al. (1982). D-Lactic Acidosis Due to Abnormal Gut Flora. N. Engl. J. Med. 306:1344. Some bacteria form D-lactate from glucose, and surgical tinkering with the small bowel may encourage their growth.

CHAPTER 27

STORAGE OF GLUCOSE AS GLYCOGEN

The demand for high-energy phosphate is not transient, and fuels must always be available to supply the processes generating high-energy phosphate. In order for an animal to interrupt its food intake—to eat meals rather than small snacks and to sleep for long intervals—it must store fuels at times of excess and be able to recover them at times of deprivation. These stores include a polymer of glucose: glycogen (p. 192). (Plants use starch in a comparable way.)

Glycogen appears in muscle cells as discrete particles (Fig. 27–1), with masses up to 2×10^7 daltons and more. Liver contains both individual particles and covalently linked aggregates as large as 10^9 daltons. The many hydroxyl groups make glycogen quite polar, and the particles contain 1.1 grams of bound water per gram of polysaccharide. Another 1 to 3 grams of water may interact with the particles. The

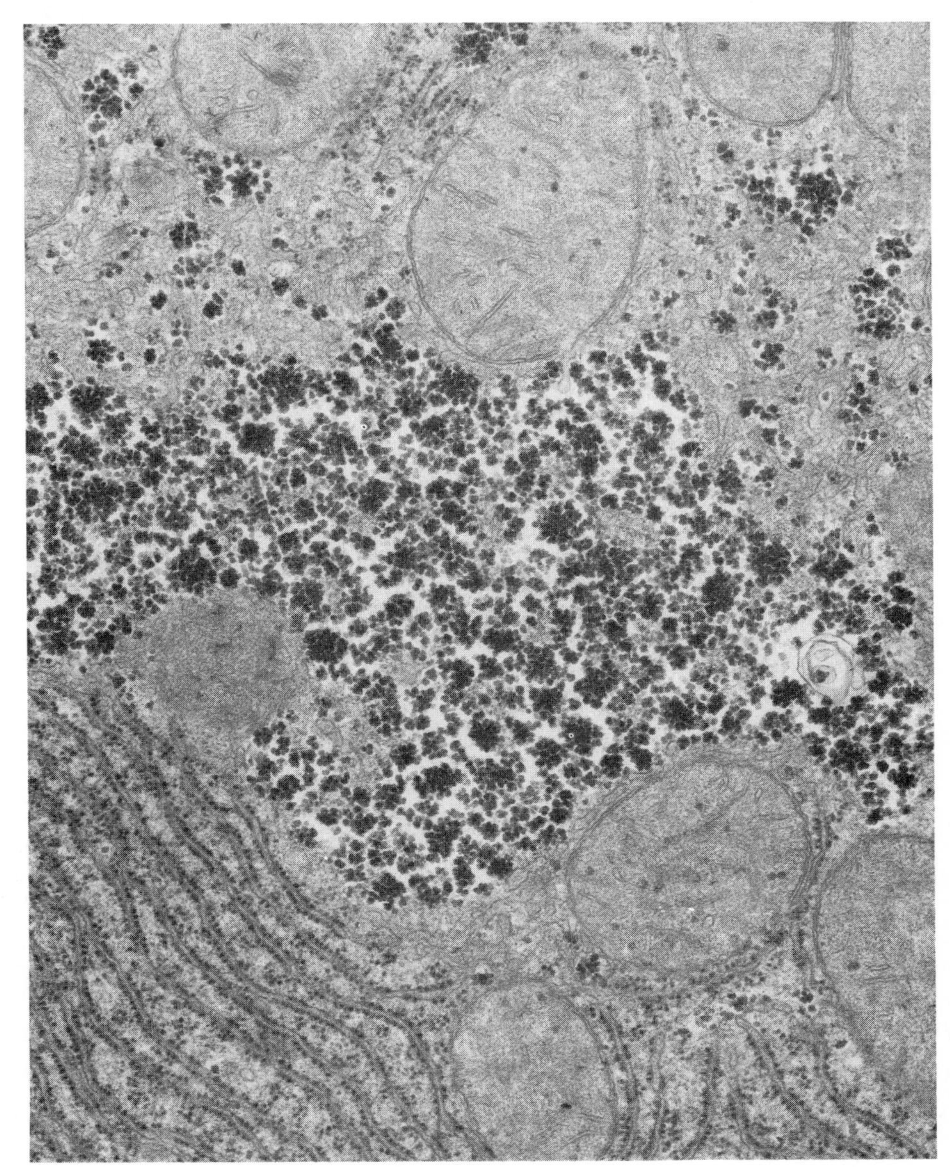

FIGURE 27–1 Dense clumps of glycogen fill the *center* of this view of a liver cell in a rat fed a high-carbohydrate diet. The large organelles are mitochondria. Rough endoplasmic reticulum and associated ribosomes are at the *lower left;* smooth endoplasmic reticulum is at the *upper right.* Magnification 31,000×. (Courtesy of Dr. Robert R. Cardell, Jr.)

particles also contain the enzymes responsible for the synthesis and later breakdown of glycogen.

The amount of glycogen varies widely, not only among different tissues but also within a tissue, depending upon the supply of glucose and the metabolic demand for energy. Most of the body's glycogen is in the liver and muscles. We shall mainly be concerned with these two tissues, but glycogen is also important to the economy of other tissues. In general, the content of glycogen in the liver varies greatly with the diet, while that in the muscles varies greatly with exercise.

Much of our knowledge about the glycogen concentration of human tissues

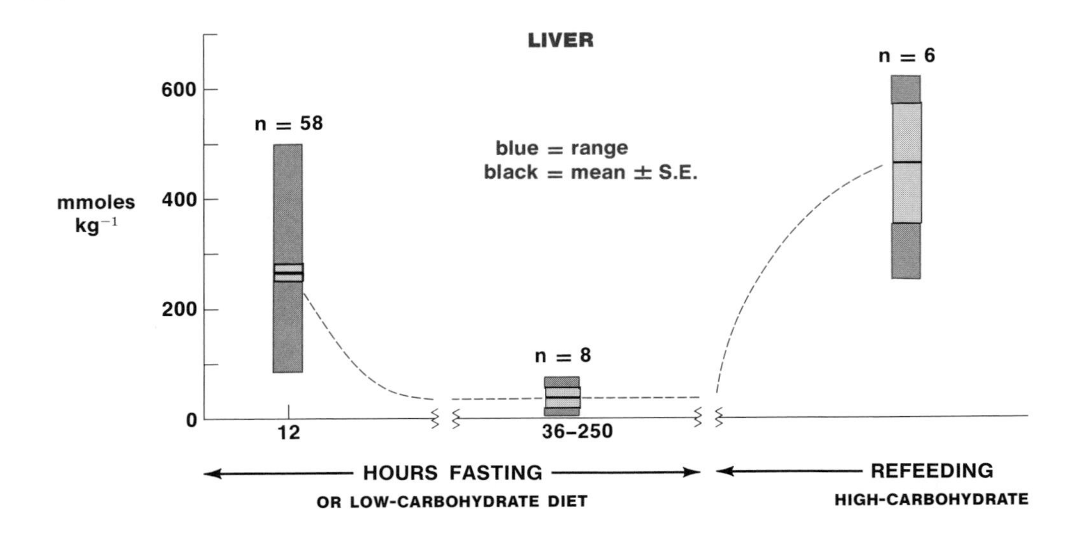

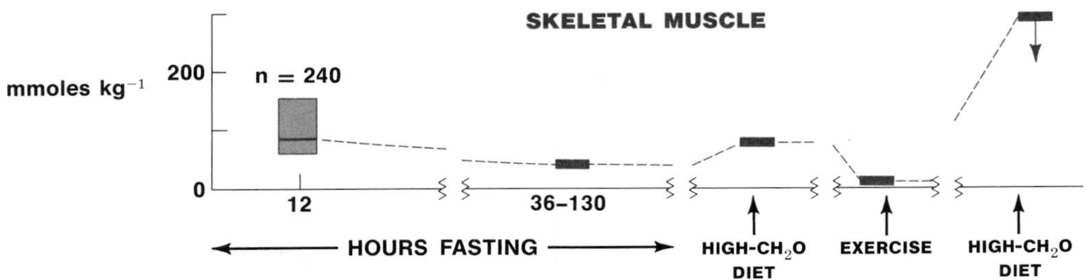

FIGURE 27–2

Changes in the glycogen content of human liver and skeletal muscles with fasting and high-carbohydrate diets. Changes are also shown for skeletal muscles with exercise. This is a composite of data from several sources. Solid bars indicate values obtained from a small number of subjects. Note that the glycogen content of muscle responds to glucose feeding more sharply after exercise.

Values based on data from the following:
Nilsson, L.H:S. (1973). *Liver Glycogen Content in Man in the Postabsorptive State.* Scand. J. Clin. Lab. Invest. 32:317.
Nilsson, L.H:S., and E. Hultman (1973). *Liver Glycogen in Man—the Effect of Total Starvation or a Carbohydrate-poor Diet Followed by Carbohydrate Refeeding.* Scand. J. Clin. Lab. Invest. 32:325.
Hultman, E. (1967). *Studies on Muscle Metabolism of Glycogen and Active Phosphate in Man with Special Reference to Exercise and Diet.* Scand. J. Clin. Lab. Invest., 19 (suppl. 94).
Hultman, E., J. Bergstrom, and A.E. Roch-Norlund (1971). *Glycogen Storage in Human Skeletal Muscle.* Adv. Expt. Biol. Med. 11:273.
Saltin, B., and J. Karlsson (1971). *Muscle Glycogen Utilization during Work of Different Intensities.* Adv. Exp. Biol. Med. 11:289.

comes from analysis by Swedish investigators of biopsy specimens from normal volunteers. (We are all indebted to university students and members of the fire department in Stockholm for contributing pieces of themselves to satisfy our curiosity.) The data have some limitations, because only readily accessible regions of the organs can be explored in living subjects, but the values are consistent with other quantitative information on human fuel metabolism. Results under various conditions from several laboratories are summarized in Figure 27–2.

A typical value for the liver glycogen content in a well-fed human is near 400 millimoles of glucosyl residues (65 grams dry weight) per kilogram of tissue; there is less after fasting and more upon eating a high-carbohydrate meal.

The skeletal muscles typically have 85 millimoles of glucosyl residues (14 grams) per kilogram of tissue, which does not change much with overnight fasting or with a high-carbohydrate diet. However, the content falls to 1 millimole per kilogram, or lower, after an hour or two of heavy work. After depletion in this way, high-carbohydrate diets on subsequent days can cause the level to rise as high as 300 millimoles per kilogram.

Even though the liver usually has a higher concentration of glycogen than do the skeletal muscles, the muscles contain the bulk of the total glycogen store, owing to their large mass. A man weighing 70 kilograms will have about 28 kilograms of skeletal muscle, but only about 1.6 kilograms of liver. Given typical contents per kilogram of tissue, 400 millimoles in the liver and 85 millimoles in the muscles, the total store is 0.6 moles in the liver and 2.4 moles in the muscles. The total in the body, considering all organs, will therefore range somewhat over 3 moles in a well-fed individual and near 3 moles after an overnight fast (the postabsorptive state).

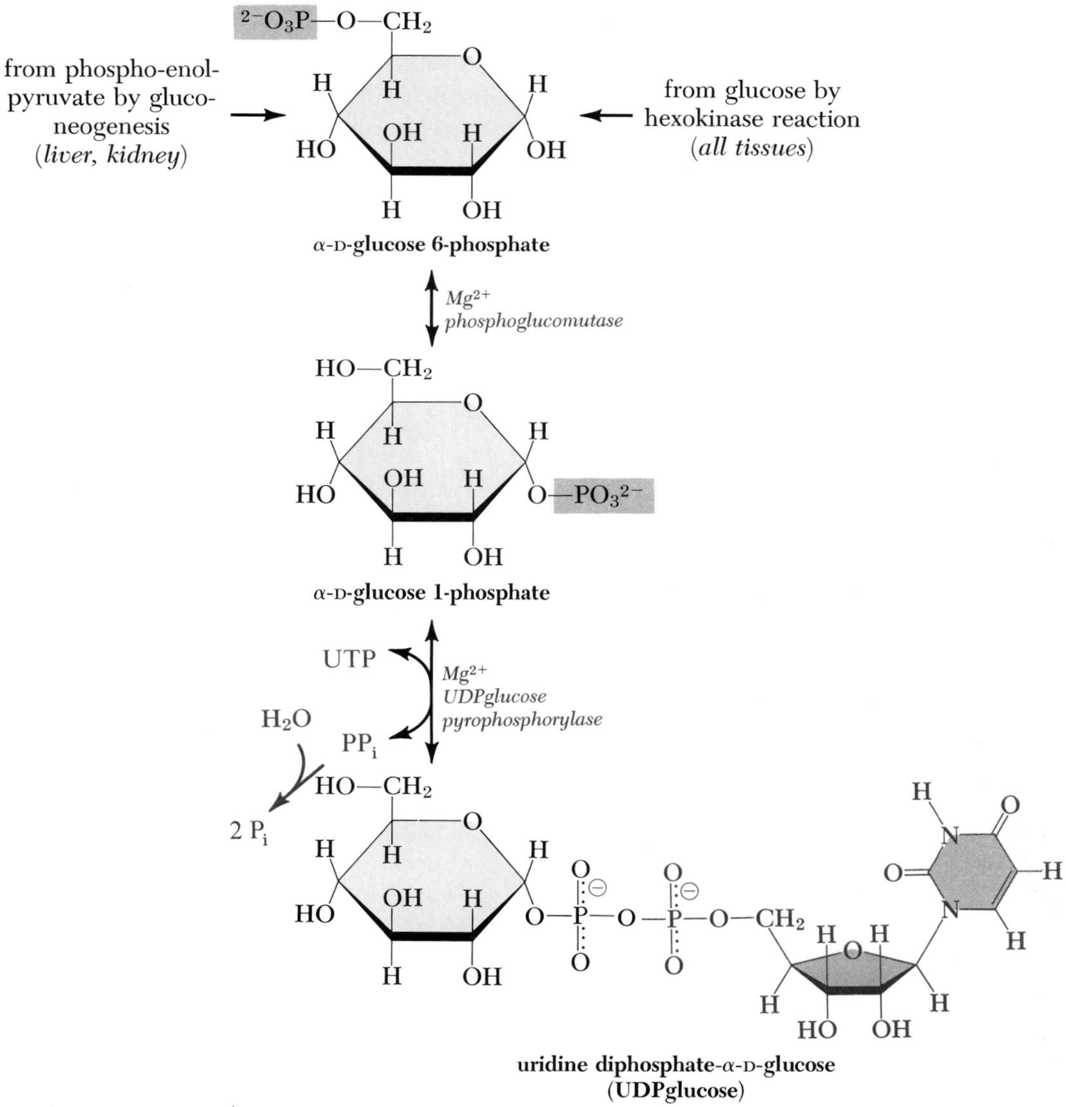

from phospho-enol-pyruvate by gluco-neogenesis (*liver, kidney*) → $^{2-}O_3P$—O—CH_2 ← from glucose by hexokinase reaction (*all tissues*)

α-D-glucose 6-phosphate

↕ *Mg^{2+} phosphoglucomutase*

HO—CH_2 … O—PO_3^{2-}

α-D-glucose 1-phosphate

UTP → PP_i ; *Mg^{2+} UDPglucose pyrophosphorylase*

H_2O → 2 P_i

uridine diphosphate-α-D-glucose (UDPglucose)

FIGURE 27–3

Storage of excess glucose residues as glycogen involves the formation of uridine diphosphate glucose (UDP-glucose).

MECHANISM OF GLYCOGEN STORAGE

Glycogen is made by adding one glucosyl residue at a time to an existing glycogen molecule, creating amylose chains that are then rearranged to make branches. The overall process is conveniently thought of in three stages: (1) the conversion of glucose 6-phosphate to **uridine diphosphate glucose (UDP-glucose);** (2) the transfer of glucosyl units from UDP-glucose onto residues in glycogen chains, so as to make growing amylose chains composed of α-1,4-linkages; and (3) the creation of branches by shifting portions of the chains onto the C6 hydroxyl groups of adjacent chains.

The Formation of UDP-Glucose. Since it is the C1 of glucose that will later be attached to glycogen, the sequence begins by transferring the phosphate group of glucose 6-phosphate to form glucose 1-phosphate (Fig. 27–3). This freely reversible reaction is catalyzed by **phosphoglucomutase,** which utilizes the corresponding bisphosphate compound, glucose 1,6-bisphosphate, in small concentrations as an intermediate (compare phosphoglycerate mutase, p. 467).

Glucose 1-phosphate then reacts with UTP to form UDP-glucose and inorganic pyrophosphate. As is the case with other reversible reactions involving pyrophosphate, hydrolysis by inorganic pyrophosphatase keeps the pyrophosphate concentra-

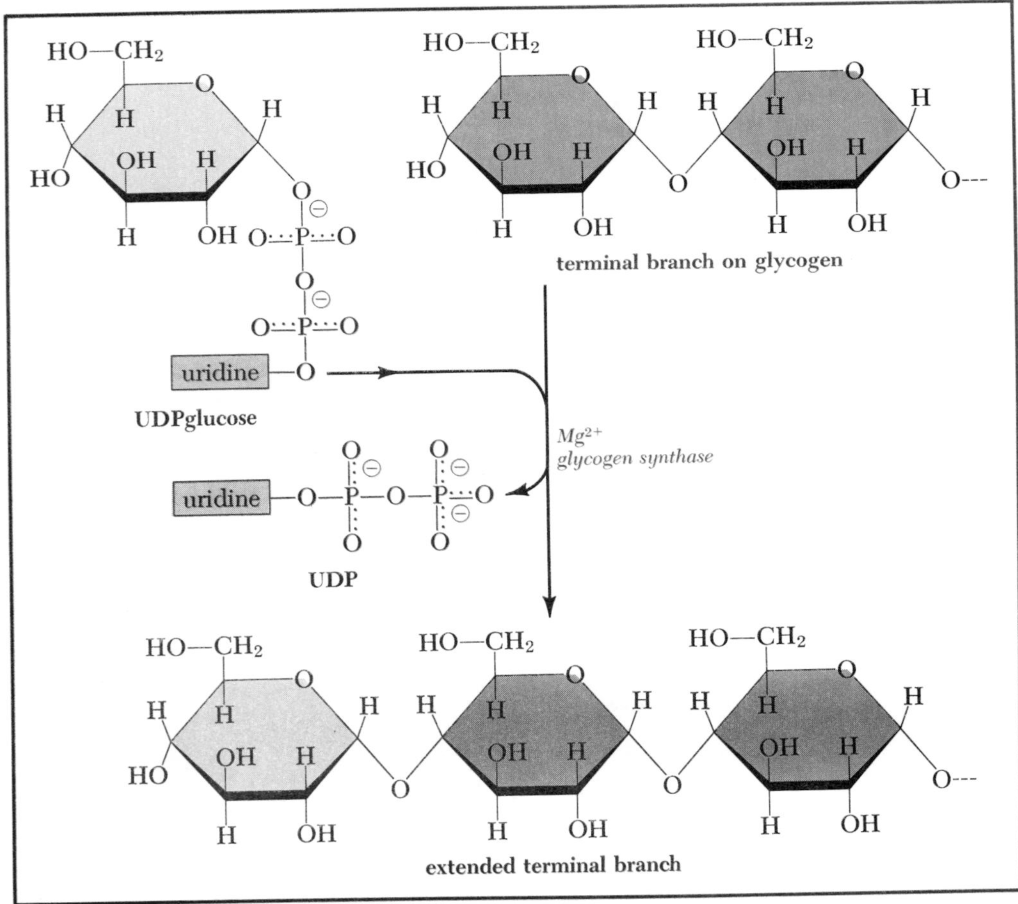

FIGURE 27–4 Terminal amylose chains on glycogen are extended by transfer of glucose residues from UDP-glucose to the C-4 hydroxyl groups of the chains.

tion low, and pulls the reaction in the direction of UDP-glucose formation. The UTP utilized in this reaction is generated by a **nucleoside diphosphokinase** reaction:

$$\text{ATP} + \text{UDP} \longleftrightarrow \text{ADP} + \text{UTP}$$

so the ultimate source of the high-energy phosphate is ATP.

Formation of Amylose Chains. UDP-glucose donates glucose residues used for extending the terminal branches of glycogen in a reaction catalyzed by **glycogen synthase.** The reaction is specific for the hydroxyl group on C4 of the glycogen residues, so the new glucosyl residue simply extends the 1,4-chain (Fig. 27–4). Since the nature of the chain is not changed by extension, the **glycogen synthase** reaction can be repeated indefinitely. A molecule of UDP is released for each glucosyl residue added.

Formation of Branches. If nothing else happened, the result of the glycogen synthase reaction would be long amylose chains composed only of 1,4-bonds. However, cells storing glycogen also have a branching enzyme, a **glycosyl-4:6-transferase,** that catalyzes the transfer of a segment of amylose chain onto the C6 hydroxyl of a neighboring chain (Fig. 27–5). This enzyme moves a block of seven 1,4-residues

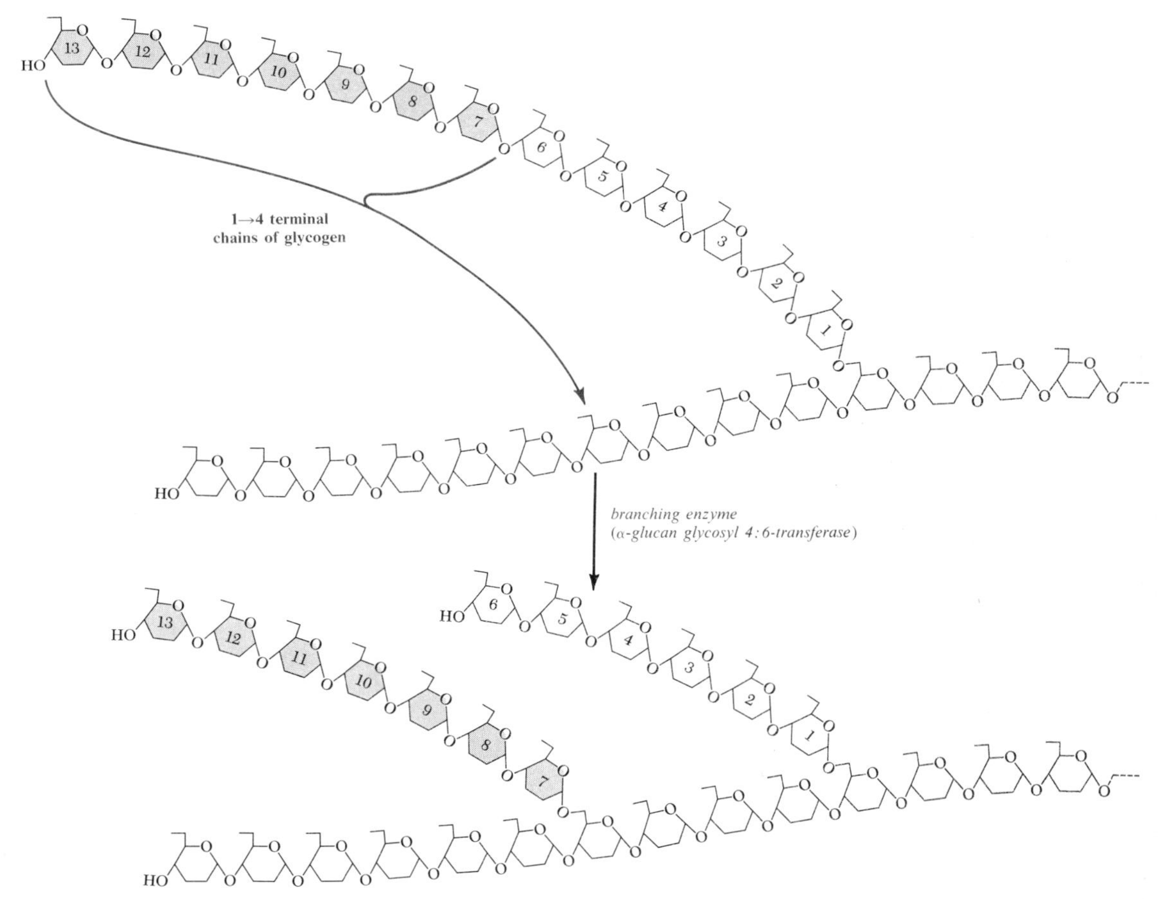

FIGURE 27–5 Branches are created in glycogen by transferring seven-residue segments of amylose terminal chains to hydroxyl groups on carbon 6 of glucose residues that are four residues removed from existing branches. A terminal branch must be at least 11 residues long before a segment is transferred from it.

from a chain at least 11 residues in length and transfers it onto another segment of amylose chain at a point four residues removed from the nearest branch. (Transfer of seven residues is favored, but specificity is not absolute, and this number is emphasized for clarity.) The new branch is therefore usually seven residues in length, and the remaining stub from which it was removed is at least four residues in length, but is more commonly six to nine residues long. 1,6-Glucosides have about 4,800 joules per mole less standard free energy content than do 1,4-glucosides, so the equilibrium of reaction favors branching.

Further Growth. The new branch and the remaining stub from which it was obtained can grow in length by addition of more 1,4-glucosyl residues from UDP-glucose through the action of glycogen synthase. When enough have been added to extend them to at least 11 residues beyond the branch, a further transfer by the branching enzyme may create still another branch, which also can grow in length. What limits the size of the molecule and the number of particles is not known. There is little information to justify speculation, although large glycogen particles appear to be made of subunits of approximately the theoretical limit for tightly packed spheres. What is certain is that skeletal muscle does not accumulate glycogen indefinitely, even with a high concentration of blood glucose. The content of glycogen in the liver can rise markedly under heavy dietary loading, with the formation of large aggregates, but even in this tissue glucose is absorbed slowly enough with more usual conditions so that an increased fraction of the excess supply is used for making fat (next chapter) after the glycogen content rises above 50 to 60 grams per kilogram of tissue. The mashed livers of force-fed Strasbourg geese, from which the pâte de foie gras favored by epicures is made, owe their special properties to high contents of both glycogen and fat.

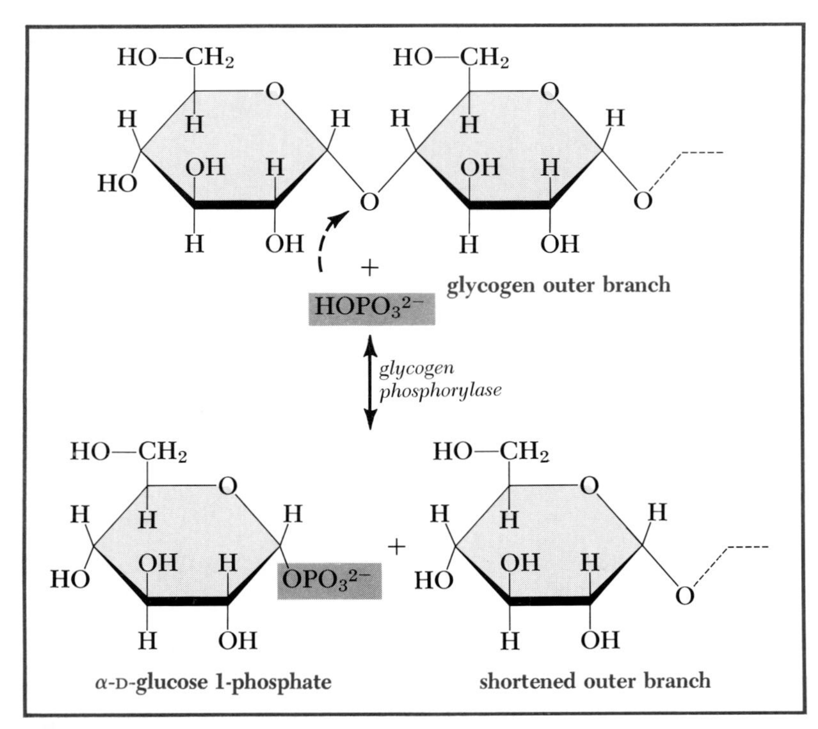

FIGURE 27–6 **The recovery of the glucose residues stored as glycogen begins with phosphorolysis of terminal glucosidic bonds to form glucose 1-phosphate.**

Priming Glycogen Synthesis. Glycogen storage is mainly an extension of already existing polysaccharide chains, but there must be some way of making the core molecules that are to be enlarged. It is possible that reactions yet to be discovered put priming carbohydrate residues on a protein core. There is some evidence that this is not necessary, that the synthetic reactions used to enlarge glycogen will also make new starter molecules, although slowly.

UTILIZATION OF GLYCOGEN AS FUEL

Phosphorolytic Cleavage. When glycogen is mobilized, all of the glucose residues appear as glucose 1-phosphate, except for one molecule of free glucose released from each 1,6 branch. The primary reaction is a simple phosphorolysis of 1,4 bonds at the end of each branch (Fig. 27–6). This reaction, which is catalyzed by **phosphorylase,** shortens an outer branch by one residue, which appears as glucose 1-phosphate. The reaction is freely reversible, with the equilibrium ratio of P_i to glucose 1-phosphate concentrations near 3.5. However, the concentration ratio in animal tissues is always greater than that, so the reaction always goes in the direction of formation of glucose 1-phosphate.

When a glucose residue is removed by phosphorolysis, the shortened 1,4-branch

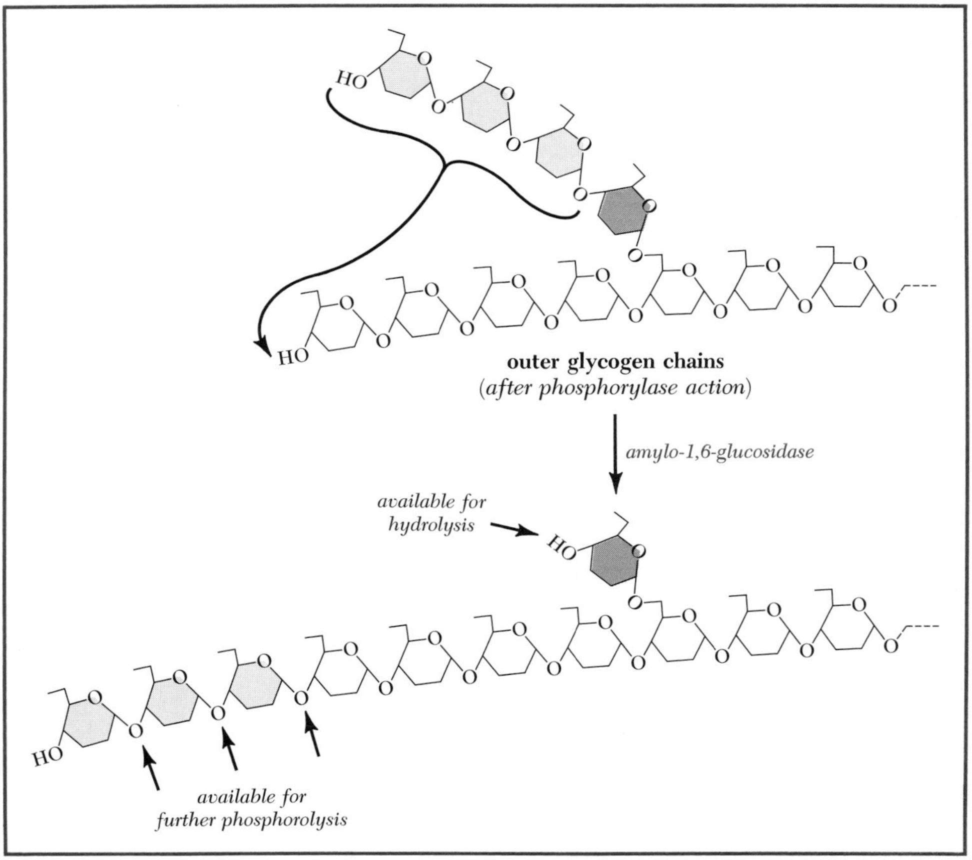

FIGURE 27–7 A transferase active site in amylo-1,6-glucosidase catalyzes the transfer of three residues from the stubs of glycogen branches to the C4 hydroxyl groups of amylose stubs. Another active site on the same enzyme then catalyzes the hydrolysis of the exposed 1,6-linked residue. Further attack by phosphorylase on the extended 1,4-amylose chain can then remove the three added glucose residues.

on the glycogen molecule can be attacked again to remove another residue, followed by further attack on the remainder. However, phosphorylase by itself will not catalyze the removal of all of the terminal residues. In anthropomorphic terms, it won't go near the branches. Expressed more mechanistically, the specificity of the enzyme is such that the 1,4-bond attacked must be at least four residues removed from a 1,6-branch. Therefore, phosphorylase alone will not make the outer branches any shorter than four glucosyl residues. The function of this specificity is not clear. Glycogen completely sheared by phosphorylase to the four-residue stubs is called the phosphorylase limit dextrin, or ϕ-dextrin. (Dextrin is a term borrowed from starch chemistry, describing partially hydrolyzed glucose homopolymers, and it is widely used for describing intermediates obtained during chemical investigation of structure, but this does not imply the occurrence of these discrete types of molecules in cells.)

Removal of Branches. Further degradation of glycogen requires a second enzyme, **amylo-1,6-glucosidase.** This enzyme catalyzes two successive reactions. In the first reaction, it acts as a **glucosyl transferase** akin to the branching enzyme functioning during glycogen synthesis, but differing in that it transfers three glucosyl residues from a shortened branch onto another chain terminus so as to lengthen it (Fig. 27–7).

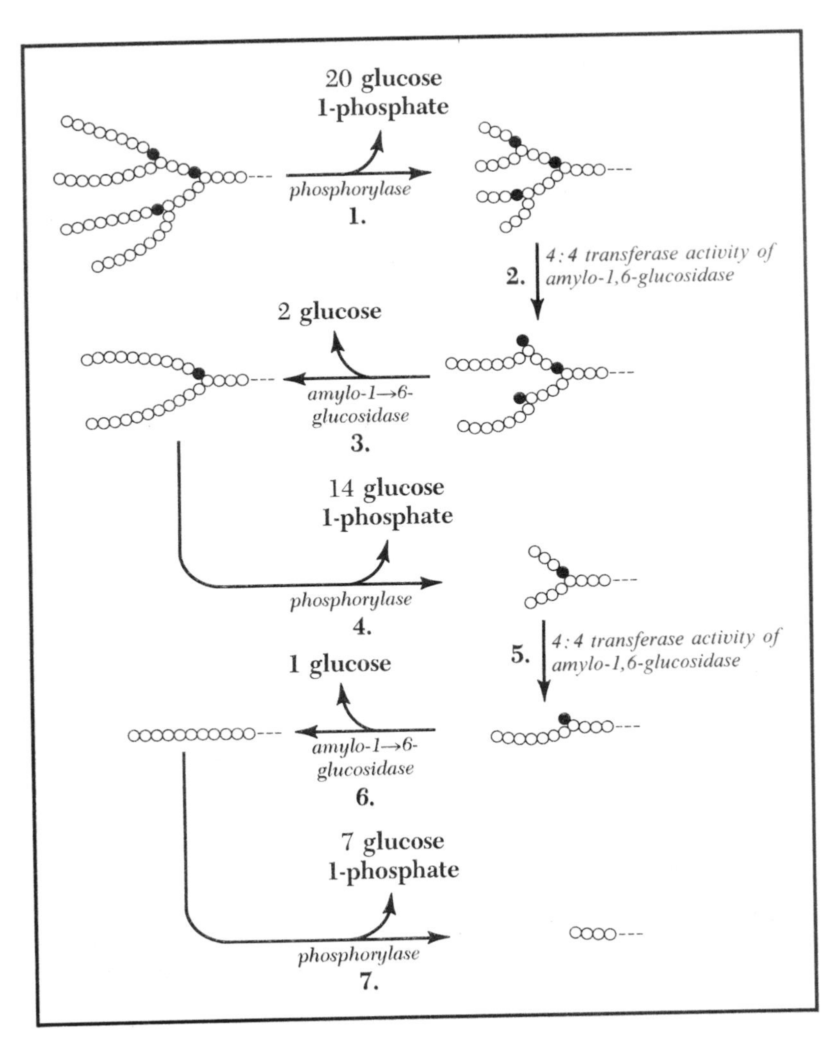

FIGURE 27–8

Summary of a sample sequence during glycogen breakdown. The diagram shows the fate of four outer chains and the next two tiers of branches from which they arise. Each chain is shown with nine residues beyond the outermost branch. The residues linked by 1,6 bonds and released as free glucose are indicated by *solid circles.*

***SUM:* Out of the 44 glucosyl residues removed, three have appeared as free glucose and 41 as glucose 1-phosphate.**

Suppose that adjacent chains are shortened as much as possible at a branch by phosphorylase. Each will then be four glucosyl residues in length. The glucosyl-4:4-transferase activity in amylo-1,6-glucosidase catalyzes the removal of three of the residues of the stub of the branch, leaving a single residue attached to C6. The enzyme transfers the three-residue package to the end of the other stub, which now has seven residues in 1,4-linkage and can therefore be attacked again by phosphorylase.

The amylo-1,6-glucosidase then catalyzes its second reaction: hydrolysis of the single residue remaining attached by a 1,6-linkage at the branch to yield free glucose. The hydrolytic activity is absolutely specific for a 1,6-linkage to a single residue; this protects longer branches on glycogen from internally disruptive attack by the enzyme.

Removal of the single branched residue exposes a straight chain of 1,4-linked residues down to the next branch, so phosphorylase can catalyze the removal of more residues as glucose-1-phosphate before reaching its limit of four residues next to a branch. What is the result? The whole sequence (Fig. 27–8) causes a continual stripping of glucosyl residues as glucose 1-phosphate, except for the single residues at branches that are removed as free glucose. The degree of branching is such that 11 to 14 molecules of glucose 1-phosphate are formed for each molecule of free glucose released.

REGULATION OF GLYCOGEN STORAGE

The objectives of control over glycogen metabolism are to store glucose at times of plenty and mobilize the supply at times of need, while preventing the reactions of both storage and mobilization from occurring at the same time. If both sets of reactions are proceeding simultaneously, the result is a futile cycle accomplishing nothing more than the hydrolysis of ATP. The critical reactions are these:

Synthesis

(1)	glucose 1-phosphate + UTP	$\longrightarrow$	UDP-glucose + PP_i
(2)	$PP_i + H_2O$	$\longrightarrow$	$2\ P_i$
(3)	UDP-glucose	$\longrightarrow$	UDP + glycogen chain
(4)	UDP + ATP	$\longrightarrow$	UTP + ADP

Mobilization

(5)	glycogen chain + P_i	$\longrightarrow$	glucose 1-phosphate
SUM	$ATP + H_2O$	$\longrightarrow$	ADP + P_i

The strategy of control is to make glycogen phosphorylase and glycogen synthase respond to similar signals, but in opposite directions. When the glycogen phosphorylase becomes active, the synthase becomes inactive, and *vice versa.* In addition, UDP-glucose is made a competitive inhibitor of its own formation so that it will not accumulate when the synthase is inactive.

The nature of the signals regulating glycogen metabolism and the responses to them differ among tissues. In the **muscles,** the objective is to mobilize glycogen as a fuel for contraction and to store it at rest if the glucose supply is adequate. Consequently, the signals are the **intracellular concentrations of Ca^{2+} and AMP,** which rise when contraction begins, and the **blood concentrations of adrenaline and insulin.** Adrenaline signals a state of alert, perhaps requiring fuel expenditure, while insulin signals that glucose is available for storage.

The liver regulates glycogen supply so as to buffer the blood glucose concentration. After glucose is absorbed by the small bowel, it first goes to the liver *via* the portal vein and part is stored as glycogen. The liver can also store glucose residues synthesized by gluconeogenesis. When the blood glucose concentration falls, the liver breaks down glycogen and releases glucose into the blood to maintain the concentration. Important signals for glycogen metabolism in the liver include the **glucose concentration** within hepatic cells and the blood concentrations of two pancreatic hormones, **insulin** to indicate an adequate glucose supply and **glucagon** to indicate a need for more glucose. In addition, the liver responds to stimulation of its **sympathetic nerve supply,** and to other hormones to be considered later.

Many of these signals act to regulate both glycogen phosphorylase and glycogen synthase. These enzymes are regulated in two ways: by combination with allosteric effectors and by covalent modification through phosphorylation and dephosphorylation.

CONTROL BY PHOSPHORYLATION

Glycogen synthase and glycogen phosphorylase each exists in active and inactive conformations. Allosteric effectors and phosphorylation regulate these enzymes by stabilizing one conformation. Phosphate groups are added by transfer from ATP in reactions catalyzed by kinases, and they are removed by hydrolysis in reactions catalyzed by phosphatases:

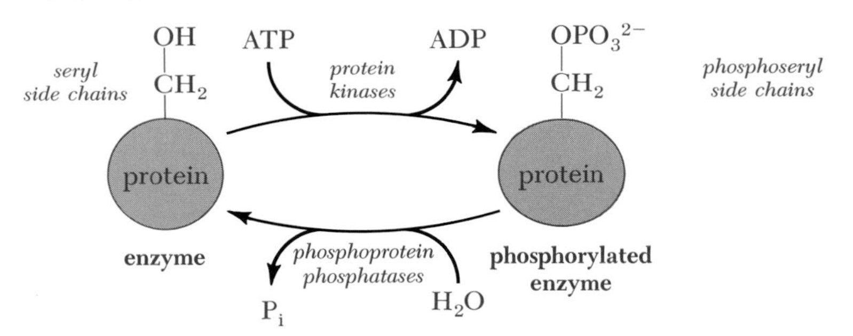

The kinases and phosphatases affecting the enzymes of glycogen metabolism also act on enzymes responsible for other metabolic processes that are discussed later. In general, **phosphorylation activates catabolic enzymes and inactivates anabolic (synthetic) enzymes.** This is true for the enzymes of glycogen metabolism. Phosphorylation of glycogen phosphorylase locks it in the active conformation. That is the P-form (phosphorylated form) is glycogen phosphorylase a **("a" for intrinsically active),** but phosphorylation locks glycogen synthase in the inactive conformation, designated glycogen synthase b:

glycogen synthase a is *not* phosphorylated
glycogen synthase b is phosphorylated
glycogen phosphorylase a is phosphorylated
glycogen phosphorylase b is *not* phosphorylated.

The mechanisms for regulating the interconversion of active and inactive conformations of glycogen phosphorylase and glycogen synthase by the combination of allosteric effectors and phosphorylation-dephosphorylation reactions are considerably more complex than anything we have encountered before. Let us first dissect the responses in mammalian skeletal muscles.

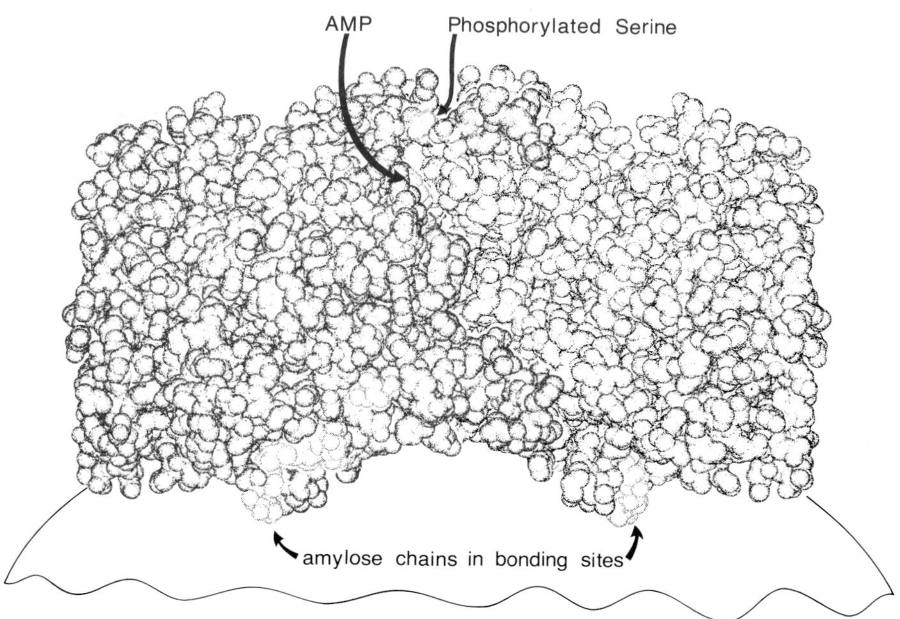

FIGURE 27–9 A molecule of phosphorylase as viewed from the side. In this computer drawing, the concave catalytic surface is facing down on a glycogen particle. The catalytic cleft is out of view, but part of the glycogen chains in the bonding sites of the two identical subunits are visible. The convex control surface is facing up; one phosphorylated serine residue and part of a bound AMP molecule are visible. (Modified from R.J. Fletterich, S. Sprang and N.B. Madsen (1979). Can. J. Biochem. 57:789. Reproduced by permission. Multicolor drawings from other viewpoints are included in their paper.)

REGULATION OF PHOSPHORYLASE IN MUSCLE

Phosphorylase is an enzyme with multiple binding sites on each of its two identical subunits (Fig. 27–9). The enzyme clings to a glycogen particle like a barnacle on a rock, wrapping itself around a terminal amylose chain. This holding site has no catalytic function. Another crevice on the same concave face fits the terminal amylose chain to be cleaved with inorganic phosphate. This catalytic site has a deeply buried molecule of **pyridoxal phosphate.** Pyridoxal phosphate is usually employed in reactions involving combination of its aldehyde group with amino groups on substrates (p. 284), but here it is its phosphate group that is activated, probably to form a transient pyrophosphate-like intermediate, and no other example of this function is known.

The opposite face of glycogen phosphorylase, the convex face, is the control face. It has sites for binding AMP and for phosphorylation.

AMP activates glycogen phosphorylase. Only the active conformation of the enzyme tightly binds AMP. Phosphorylase a, the P-form, is already in the active conformation, so combination with AMP does not affect its activity. Most of glycogen phosphorylase b, the dephospho form, is in the inactive conformation in the absence of AMP, but a rise in the concentration of AMP pulls the equilibrium toward the active conformation, even though the enzyme is not phosphorylated (see figure at top of next page). Since a rising AMP concentration is a sensitive indicator of a fall in high-energy phosphate concentration, which will accelerate the Embden-

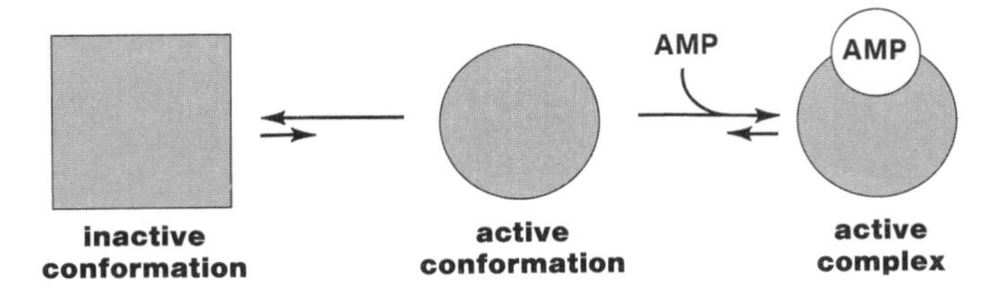

Meyerhof pathway (p. 488), its effect on glycogen phosphorylase is an important means of matching fuel supply and demand.

Calcium ions indirectly activate glycogen phosphorylase. They affect the activity of another enzyme, phosphorylase kinase, which in turn catalyzes the phosphorylation of glycogen phosphorylase:

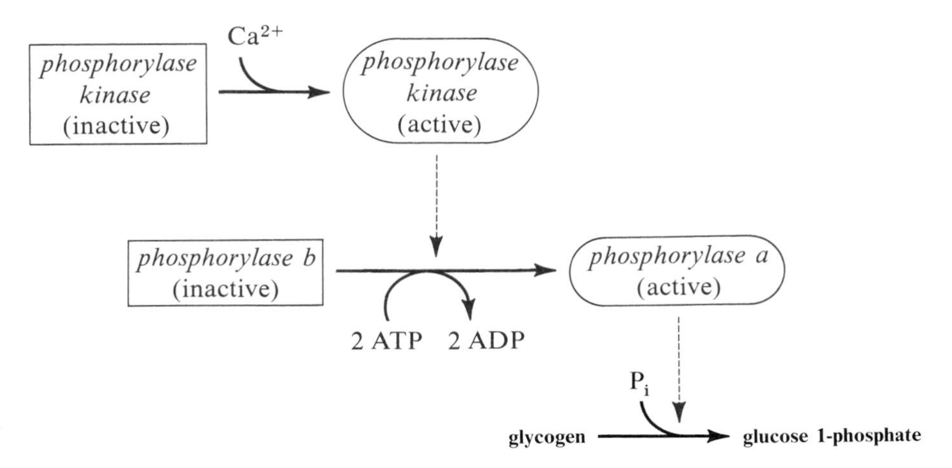

Phosphorylase kinase has the subunit composition $\alpha_4\beta_4\gamma_4\delta_4$. It is unusual in that the δ subunit is **calmodulin,** a protein that also exists separately as an important regulatory protein in many kinds of cells.

Calmodulin binds four calcium ions per monomer, and its active conformation only appears when all the binding sites are filled. The activation of calmodulin therefore depends upon the fourth power of the calcium ion concentration:

$$\text{calmodulin} + 4\,Ca^{2+} \longrightarrow \text{calmodulin-}Ca_4\text{ complex}$$

$$K = \frac{[\text{calmodulin-}Ca_4]}{[\text{calmodulin}][Ca^{2+}]^4}$$

This has the effect of creating a concentration threshold; only a small increase in $[Ca^{2+}]$ above the threshold is necessary to fully activate calmodulin.

When the calmodulin subunits within phosphorylase kinase are activated, the entire molecule shifts to an active conformation. The kinase then catalyzes the conversion of glycogen phosphorylase b to glycogen phosphorylase a, fixing it in its active conformation, and this is the way in which the release of calcium ions in a stimulated muscle fiber causes an accelerated breakdown of glycogen.

White and red skeletal muscle fibers have different isozymes of phosphorylase kinase. Both contain calmodulin subunits, but the white muscle isozyme also binds additional active calmodulin from the sarcoplasm to further augment its catalytic activity. This makes the white fibers sensitive to lower $[Ca^{2+}]$ than are red fibers; they mobilize glycogen more promptly upon stimulation, as is consonant with their role in rapid development of high tension for high power output.

Adrenaline and Cyclic AMP

When an animal is confronted with an emergency that may demand prompt and strenuous physical activity—flee or fight—it is advantageous for it to mobilize glucose 1-phosphate rapidly from glycogen in its skeletal muscles so as to have the fuel available even before the signal for contraction is given. The advantage is achieved because emergencies of this sort are recognized in the central nervous system, which stimulates the adrenal medulla (the tissue in the core of the adrenal glands) to release adrenaline and the related noradrenaline into the blood stream:

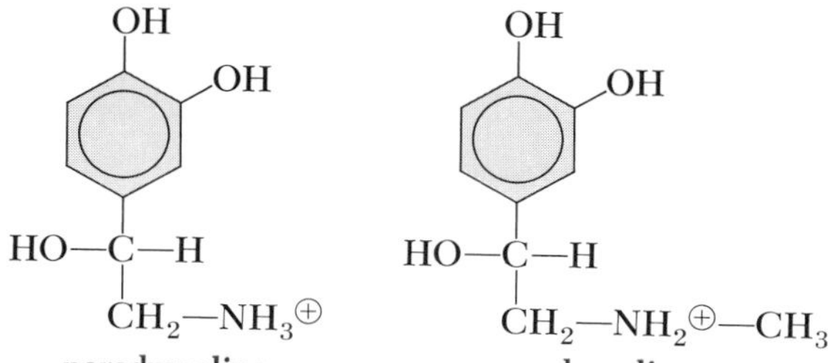

noradrenaline **adrenaline**

The human adrenal medulla releases more adrenaline than noradrenaline, but the reverse is true in some mammals. Both compounds are frequently lumped under the general name of **catecholamines.** They are frequently called **epinephrine** and **norepinephrine** in American medical circles because Adrenalin is a trademark. However, the tissue receptors for catecholamines are called **adrenergic receptors** by all.

Adenylate Cyclase and Cyclic AMP. Adrenaline stimulates the breakdown of glycogen in skeletal muscle by tripping a series of events that causes phosphorylase kinase to be locked in an active conformation by phosphorylation. These events begin with activation of an enzyme, **adenylate cyclase,** in the plasma membrane of the muscle fibers. This enzyme occurs as a complex spanning the plasma membrane, with the adrenaline receptor protein on the outside and the catalytic site on the inside. (Activation involves binding of GTP by a regulator protein in the complex.) Adenylate cyclase catalyzes an intramolecular condensation of ATP to produce **adenosine 3′,5′-monophosphate,** more familiarly known as **cyclic AMP** (Fig. 27–10).

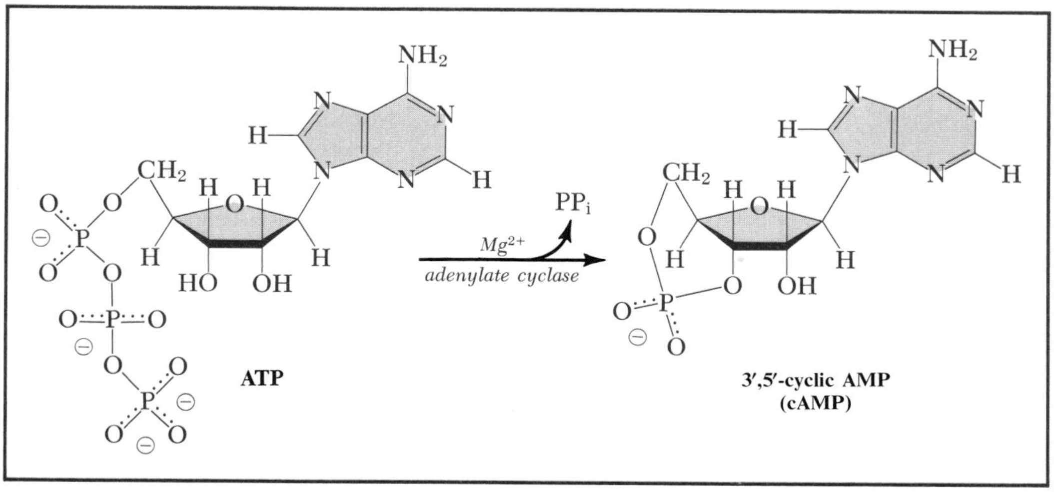

ATP → (Mg^{2+}, *adenylate cyclase*; PP_i) → **3′,5′-cyclic AMP (cAMP)**

FIGURE 27–10 The formation of cyclic AMP from ATP.

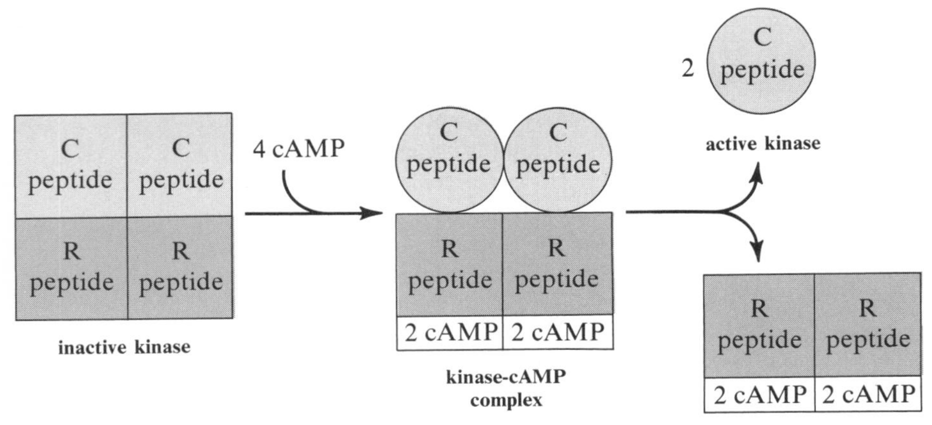

FIGURE 27–11 The activation of the cAMP-dependent protein kinase.

The compound is sometimes designated as **3′,5′-AMP** and sometimes as **cAMP.** Cyclic AMP acts as a **"second messenger,"** conveying the signal received in the form of adrenaline on the plasma membrane to the responsive enzymes within the cell.

Activation of cAMP-Dependent Protein Kinase. The major effect of cyclic AMP in eukaryotic cells is the activation of a particular protein kinase. The activation occurs in this way: The protein kinase consists of two pairs of subunits: R_2C_2, in which R is a regulatory subunit and C is a catalytic subunit. The combined tetramer is inactive (Fig. 27–11). Each of the regulatory subunits will combine with two molecules of cAMP, but when it does, the association with the catalytic subunits is weakened, and they are released as fully active monomers.

Effect of Protein Kinase on Phosphorylase Kinase. The cAMP-dependent protein kinase catalyzes the transfer of phosphate from ATP to eight serine side chains in phosphorylase kinase. Each of the four β subunits has one site that is rapidly phosphorylated, stabilizing the active conformation of the enzyme in the presence of Ca^{2+}. Each α subunit has one site that is less rapidly phosphorylated; the function of this phosphorylation is discussed later.

The phosphorylation of phosphorylase kinase is the final link in a multienzyme cascade by which an elevation in blood adrenaline concentration causes conversion of glycogen to glucose 1-phosphate (and lesser amounts of glucose) (Fig. 27–12):

(1) Adrenaline is bound by a cell-surface receptor protein.

(2) A conformational shift occurs, permitting a neighboring regulator protein to bind GTP.

(3) A further conformational shift exposes the catalytic site of a neighboring adenylate cyclase protein to MgATP in the cytosol, and cyclic AMP is formed.

(4) Cyclic AMP combines with a particular protein kinase complex in the cytosol, causing it to release its catalytic subunits.

(5) The catalytic subunits of the protein kinase react readily with the β subunits, and less readily with the α subunits, of phosphorylase kinase on glycogen particles.

(6) Phosphorylation of the β subunits of phosphorylase kinase fixes the enzyme in its active conformation, even at low calcium ion concentrations, so that it catalyzes transfer of phosphate groups from ATP to glycogen phosphorylase.

(7) Glycogen phosphorylase begins the breakdown of glycogen to glucose 1-phosphate.

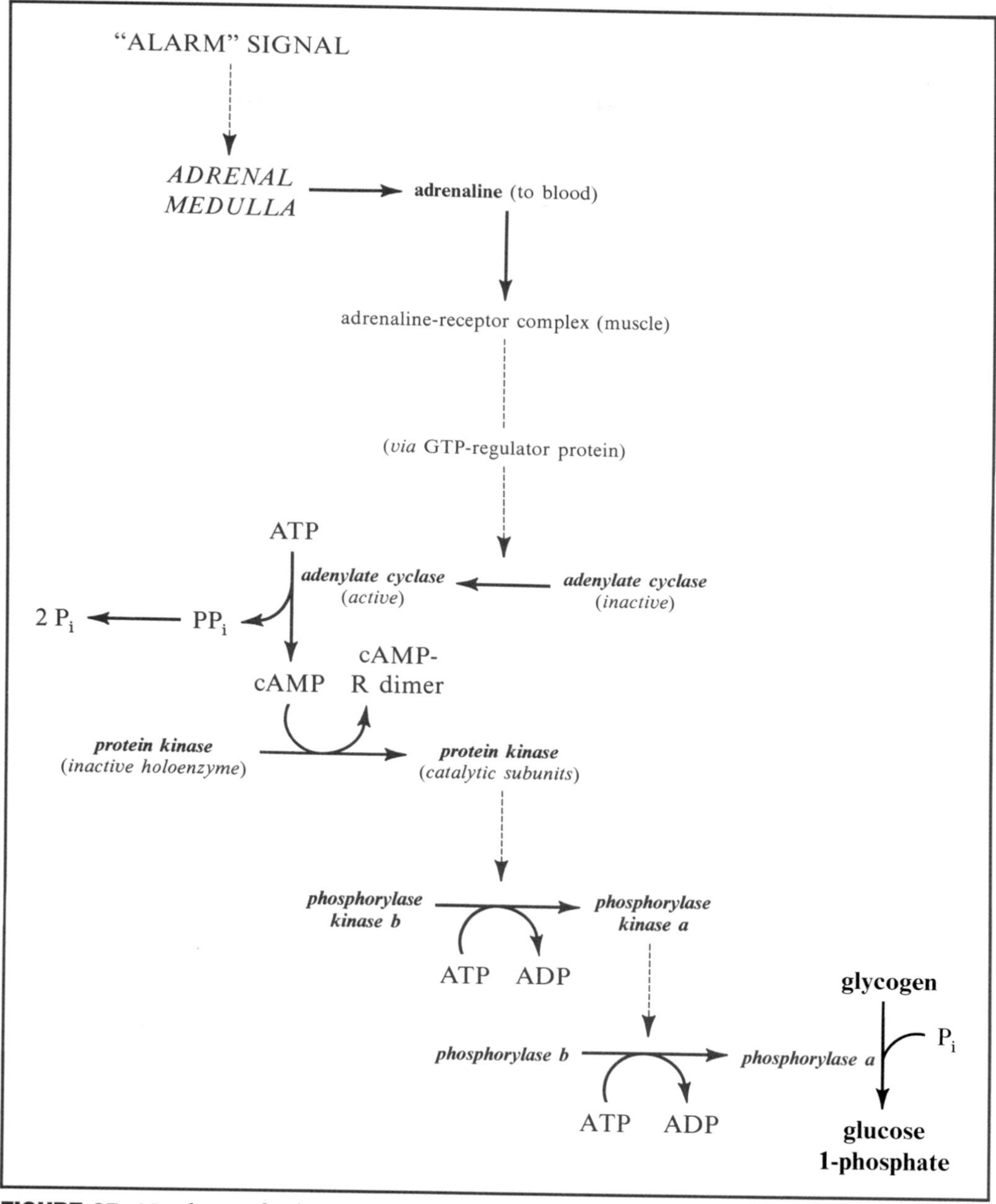

FIGURE 27–12 **A cascade of enzyme activations translates the appearance of adrenaline into an accelerated formation of glucose 1-phosphate from glycogen.**

Since there is a magnification by successive enzyme action, a minute amount of adrenaline has a very large effect, so large that 2 micromoles (364 μg) is a potent dose for humans. The normal blood concentration of adrenaline is in the nanomolar range. It has been estimated that a rise in the concentration of cyclic AMP to 1 micromole per kg of muscle causes the formation of 25,000 times as much glucose 1-phosphate per minute.

An increased calcium ion concentration itself can activate phosphorylase kinase in muscles at rest or during initial contractions. The governing factor is pH. Phosphorylase kinase b, the dephosphorylated form, is not activated by calcium ions unless the pH is higher than 6.8. This value is exceeded in resting fibers, in which the pH is near 7.0, and there is a transient rise to pH 7.2 or more when contraction begins, owing to the net hydrolysis of phosphocreatine. As contractions continue, the pH of the muscle falls as formation of CO_2 and lactate accelerate, and phosphorylase kinase must be phosphorylated to the a form if it is to be active. Phosphorylation does not diminish its requirement for calcium ions, but permits it to be active in a more acidic environment.

REVERSAL OF PHOSPHORYLASE ACTIVATION

When the demand for fuel has passed, the breakdown of glycogen is halted by restoring glycogen phosphorylase to its inactive conformation. A muscle recovering from work will continue accelerated synthesis of ATP until the phosphate potential is restored to the resting level. As this occurs, the AMP concentration declines below the level at which AMP can fix glycogen phosphorylase b in its active conformation. However, glycogen phosphorylase a remains active in the absence of AMP, and several additional events occur in a resting muscle to convert glycogen phosphorylase a back to glycogen phosphorylase b and prevent its reactivation in the absence of further excitation of the muscle. These events include:

(1) Activation of protein phosphatases to hydrolyze the phosphate groups from glycogen phosphorylase a and phosphorylase kinase a;

(2) Inactivation of the cAMP-dependent protein kinase through hydrolysis of cyclic AMP by cyclic nucleotide phosphodiesterases;

(3) Cessation of further formation of cyclic AMP, after the adrenaline concentration falls, through prompt hydrolysis of GTP by the regulatory protein in the adenylate cyclase complex (p. 513);

(4) Rapid destruction of circulating adrenaline to lower its concentration in the absence of continued stimulation of the adrenal medulla. This will be discussed in Chapter 33.

The activation of protein phosphatases also involves the enzyme cascade (Fig. 27–13). Protein phosphatase-1 is an enzyme capable of hydrolyzing phosphate groups on many proteins. It occurs in many tissues, including the muscles, where it will convert glycogen phosphorylase a to glycogen phosphorylase b, and also will remove phosphate groups from the β subunits of phosphorylase kinase, provided that the α subunits are also phosphorylated. This phosphatase is regulated by the phosphorylated form of another protein known as protein inhibitor-1.

Protein inhibitor-1 is phosphorylated by the same cAMP-dependent protein kinase that converts phosphorylase kinase b to phosphorylase kinase a. Using these facts, we can expand the events in the cascade regulation of glycogen breakdown:

(1) The cAMP-dependent protein kinase rapidly phosphorylates not only the β-subunits of phosphorylase kinase, making it active, but also protein inhibitor-1, which combines with protein phosphatase-1 to inactivate it.

(2) The protein kinase slowly phosphorylates the α-subunits of phosphorylase kinase, making it susceptible to attack by active protein phosphatase. During the interval between phosphorylation of the β- and α-subunits, phosphorylase kinase will remain fully active, ensuring some minimum response to the adrenaline signal.

(3) Protein phosphatase-1 can catalyze the removal of the phosphate group from its own inhibitor, and it begins to do so, causing the inhibitor to dissociate.

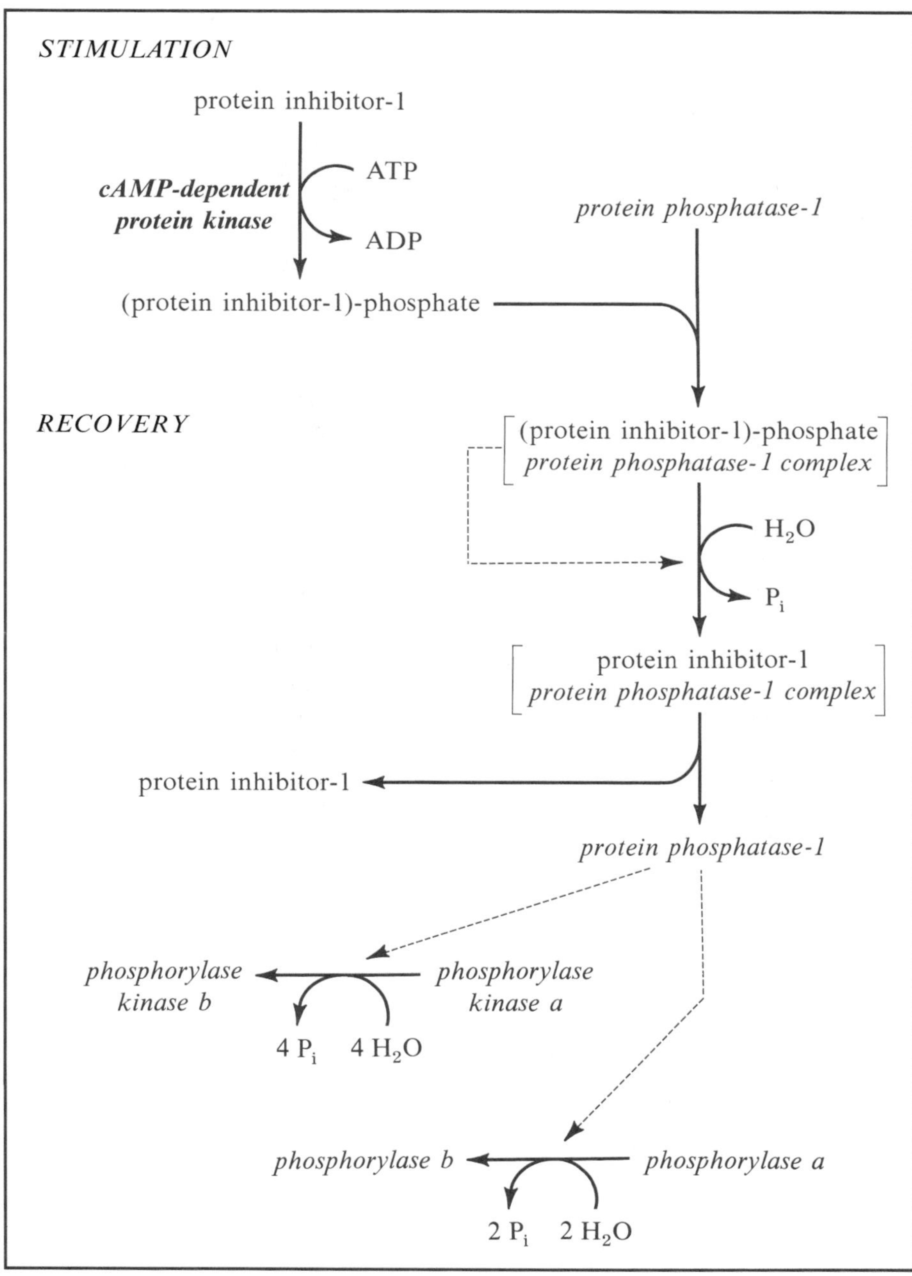

FIGURE 27–13 Regulation of protein phosphatase-1, the principal enzyme inactivating phosphorylase a. (*Top*) Stimulation of the cAMP-dependent protein kinase causes phosphorylation of a protein inhibitor, which then combines with protein phosphatase-1 and prevents its action on other proteins. (*Bottom*) The phosphatase can catalyze hydrolysis of the phosphate on the inhibitor bound to it, causing the protein inhibitor to dissociate. The freed phosphatase can then act on phosphorylase kinase a and phosphorylase a to inactivate them.

(4) As the active protein phosphatase is released, it removes the phosphate groups from the β-subunits of phosphorylase kinase, making it inactive.

(5) The active protein phosphatase-1 will not attack glycogen phosphorylase a that has AMP bound to it. If the AMP concentration is low, then the protein phosphatase will convert glycogen phosphorylase a to glycogen phosphorylase b, which is inactive at low AMP concentrations. A high AMP concentration thus not only activates glycogen phosphorylase b, it also maintains phosphorylase a in its inherently active phosphorylated state, once it is formed.

(6) A protein phosphatase-2 is also present to catalyze the removal of the phosphate groups from the α-subunits of phosphorylase kinase, returning the enzyme to its resting state, ready to go through the activation cycle again.

REGULATION OF GLYCOGEN SYNTHASE IN MUSCLE

Glycogen synthase also has active and inactive conformations. When the enzyme is not phosphorylated, it is mostly in the active conformation. Phosphorylation stabilizes the inactive conformation of this enzyme. Each of the identical subunits in glycogen synthase may be phosphorylated by three kinds of kinases, each acting at different locations on the polypeptide chain. Phosphorylase kinase is the most effective; only one site need be phosphorylated by it to stabilize the inactive conformation (glycogen synthase b). A Ca^{2+}-dependent (cAMP-independent) protein kinase is next most effective, and it will add as many as four phosphate groups to each monomer. The cAMP-dependent protein kinase is least effective; it phosphorylates two sites in each monomer, but without generating full synthase activity. The inactivation of glycogen synthase in muscle is therefore stimulated by the same events that activate glycogen phosphorylase: a rise in the concentrations of adrenaline in the blood, or calcium in the sarcoplasm, or both.

Regulation by Glucose 6-Phosphate. Glucose 6-phosphate activates glycogen synthase b, raising the V_{max} and lowering the K_M for UDP-glucose. It seems likely that this effect would be useful for gathering up glucose 6-phosphate and storing the glucosyl group as glycogen at any time an accumulation of the compound isn't being utilized for energy production. Whether this is true for mammalian muscle is disputed by many, but it so happens that frogs have a glycogen synthase with an absolute requirement for glucose 6-phosphate for activity, even in the free form, and it is therefore difficult to doubt the physiological relevance of the effect to the croakers.

Regulation by Glycogen Concentration. Glycogen inhibits the removal of phosphate from glycogen synthase b by a protein phosphatase. As glycogen accumulates in the muscle, it therefore shuts off its own production by decreasing the conversion of glycogen synthase b to its more active free form. This effect is the most likely explanation of why animals on high carbohydrate diets don't store massive quantities of glycogen in their muscles.

PHOSPHORYLATION OF MYOSIN

Phosphorylation and dephosphorylation are also used to control the contractile apparatus, at least in smooth muscles. The light chains of myosin are partially phosphorylated in all muscles by a specific protein kinase requiring the calcium-calmodulin complex for activity. The physiological function of the phosphorylation in striated muscles is unknown, but it is absolutely necessary for contraction of smooth muscles, in which it provides a locking device to sustain the contracted state without additional expenditure of ATP. There is a corresponding protein phosphatase and inhibitor protein to control the removal of the phosphate groups.

GLYCOGEN METABOLISM IN THE LIVER

The turnover of glycogen in the liver is regulated through the same kinds of mechanism seen in striated muscles. However, several of the enzymes involved are isozymes of those occurring in muscles and are regulated by metabolites in different ways. One of the important modulations of enzyme activity in the liver is a combination of glucose with phosphorylase a, which inactivates the enzyme. This means that an elevated concentration of glucose inhibits the breakdown of liver glycogen to form more glucose. In addition, the glucose-glycogen phosphorylase a complex is a

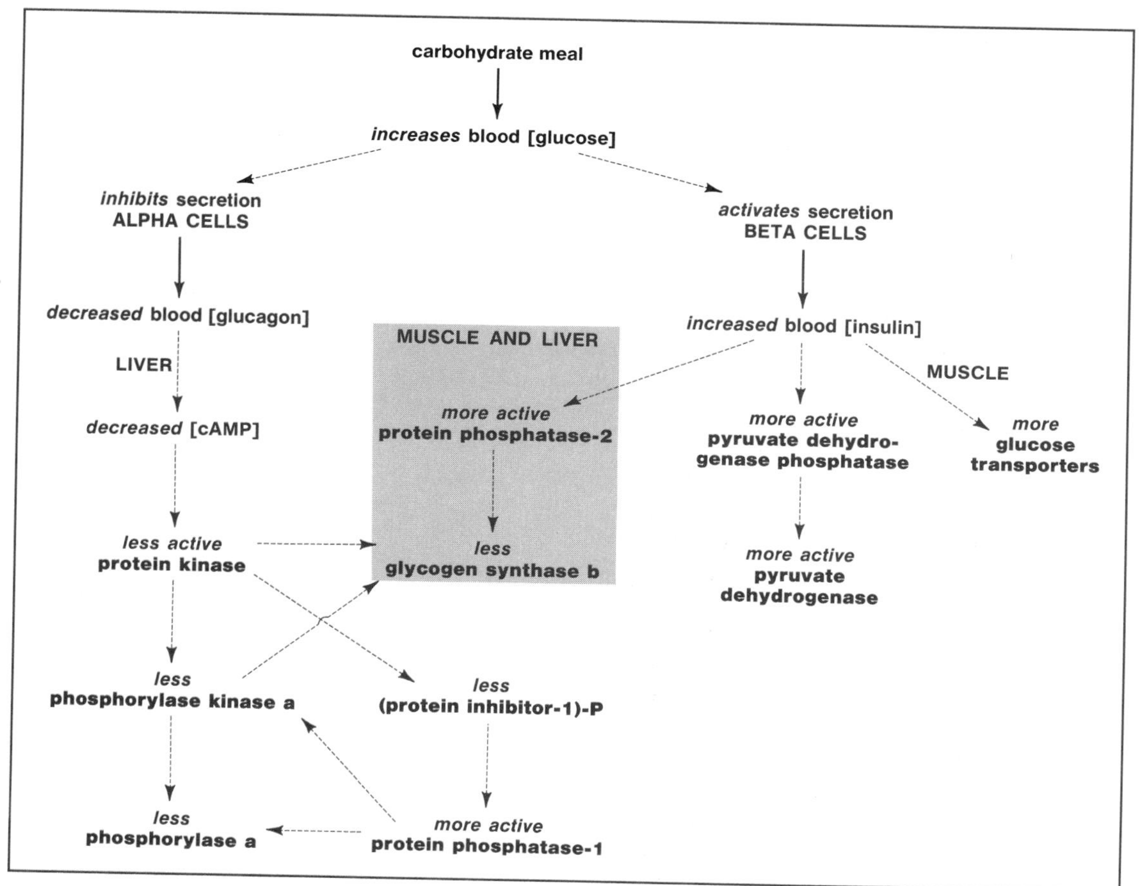

FIGURE 27–14 The regulation of glycogen metabolism by pancreatic hormone secretion. An elevated blood glucose concentration has opposing effects on the secretion of glucagon (*left*) and insulin (*right*). The decrease in glucagon concentration diminishes the cyclic AMP cascade activation of phosphorylase and inactivation of glycogen synthase in the liver, both by diminution of kinase activity and augmentation of phosphatase activity. The increase in insulin concentration increases the activity of glycogen synthase and pyruvate dehydrogenase in both the liver and muscles by activating the protein phosphatases affecting those enzymes. Insulin also causes the appearance of more glucose transporters in the plasma membrane of muscle fibers.

better substrate for protein phosphatase-1 than is glycogen phosphorylase a alone, so the presence of glucose accelerates the conversion of glycogen phosphorylase a to the less active b form.

Glucagon and Insulin (Fig. 27–14)

Secretion by Pancreas. The pancreas of higher animals is both an exocrine organ that elaborates enzymes to digest food and an endocrine organ that secretes hormones into the blood. The pancreas contains organized nests of cells known as the **islets of Langerhans,** which play a major role in regulating metabolism by secreting insulin, and glucagon. We have already encountered insulin, which consists of two polypeptide chains linked by disulfide bonds, as an example of modification of a polypeptide after its synthesis (p. 148); the cells forming insulin are known as **beta cells.** Glucagon is also a polypeptide but has only some 29 residues in the known examples, and it is secreted by different cells in the islets, the **alpha cells.** (More is said about pancreatic endocrine function in Chapter 37.)

These hormones have opposing effects on glucose metabolism. The alpha cells are stimulated to release glucagon when the blood glucose concentration falls, and the beta cells release insulin when the glucose concentration rises.

The regulation of glycogen metabolism in the liver illustrates a general principle: **Adenylate cyclase is activated by different hormones in different tissues. Adenylate cyclase in the liver is complexed with receptors for glucagon, whereas it is not in muscles. Therefore, glucagon secretion accelerates the breakdown of glycogen in liver** through activation of the cAMP-dependent protein kinase, but it has no direct effect on glycogen metabolism in the muscles. When glycogen is being broken down in the liver, most of the resultant glucose 6-phosphate is hydrolyzed to glucose by the action of glucose-6-phosphatase. The glucose moves out of the liver into the blood and is carried to other tissues.

On the other hand, **insulin promotes the storage of glucose as glycogen in both the liver and the muscles.** It is now over 70 years since a graduate student at the University of Chicago, E.L. Scott, demonstrated that the pancreas secreted a factor lowering the blood glucose concentration, and 60 years since an insulin preparation was first used to save a diabetic patient, but we still do not know the details of how insulin works. We know that it causes more glucose transporter proteins to be moved to the plasma membrane of cells in at least some sensitive tissues (p. 739), but this does not explain its internal actions on cellular metabolism. Relief may be in sight. There is evidence that combination of insulin with its receptors causes the release of a small protein from the plasma membrane into the cytoplasm, where it acts as a second messenger. This messenger is not derived from insulin directly; it may be a fragment of a larger protein in the plasma membrane. It is believed to have three effects: an inhibition of the cAMP-dependent protein kinase, an activation of a protein phosphatase-2 that attacks glycogen synthase b (not the one attacking glycogen phosphorylase a), and an activation of the protein phosphatase component of the mitochondrial pyruvate dehydrogenase complex. Therefore, **insulin causes glycogen synthase and pyruvate dehydrogenase to become more active and suppresses activation of glycogen phosphorylase.** The result is increased synthesis of glycogen in the liver and muscles, and an increased conversion of pyruvate to acetyl coenzyme A in these organs. These effects are in addition to the action of insulin in accelerating glucose transport into muscles.

In general, insulin secretion rises when the secretion of glucagon falls, and *vice versa.* Suppose both are being secreted. There is some protection against simultaneous activation of the synthase and glycogen phosphorylase because glycogen phosphorylase a, but not the glucose-glycogen phosphorylase a complex, is a potent inhibitor of the action of protein phosphatase-2 on glycogen synthase b. Glucagon seemingly can override the effects of insulin on glycogen turnover in the liver, provided that the glucose concentration is low.

SUMMARY OF GLUCOSE HOMEOSTASIS

Let us pull all of this together by recapitulating the events that occur when the blood glucose concentration rises and falls.

Excess Glucose

We eat and the blood glucose concentration rises. There are three important direct results:

1. The pancreatic islets put out less glucagon and more insulin. Less glucagon means less cyclic AMP formed in the liver, and less active protein kinase. With less active protein kinase, there is less formation of active glycogen phosphorylase and less inactivation of glycogen synthase. More insulin also results in more active glycogen synthase in both the liver and the muscles and more rapid transport of glucose into the muscles.

2. Glucose combines with glycogen phosphorylase in the liver. This causes increased loss of active glycogen phosphorylase.

3. Increased blood glucose concentration results in the net formation of glucose 6-phosphate in the liver, since the liver glucokinase is not saturated with substrate and therefore can respond to changes in blood glucose concentration.

All of these things taken together mean an elevation in blood glucose concentration is itself a sufficient signal to cause further storage of glucose as glycogen in both the liver and muscles. The increased glycogen synthase activity in the muscles causes increased removal of glucose 6-phosphate: glucose 6-P $\leftrightarrow$ glucose 1-P $\rightarrow$ UDP-glucose $\rightarrow$ glycogen. Since glucose 6-phosphate is an inhibitor of its own formation by the hexokinase reaction (p. 482), its increased removal to form glycogen also promotes increased uptake of glucose in the muscle.

The deposition of glycogen in the muscles will proceed only until a characteristic level is reached even if elevated concentrations of glucose persist. However, the liver will continue to store more glycogen, although at an increasingly slower rate. We shall see in the following chapters that more and more glucose is converted to fat for storage as the glycogen reserves begin to be filled.

Deprivation of Glucose

The blood concentration falls as more time elapses from the preceding meal. Here we have the opposite effects:

1. The pancreatic islets put out more glucagon and less insulin. More glucagon means more cyclic AMP formed in the liver and a more active protein kinase. The protein kinase causes the conversion of glycogen synthase to its inactive phosphorylated form, the conversion of glycogen phosphorylase to its active phosphorylated form, and the phosphorylation of protein inhibitor-1, which inactivates protein phosphatase-1.

2. The fall in the amount of glucose slows the hydrolysis of glycogen phosphorylase a because phosphorylase a without glucose is a poorer substrate for protein phosphatase-1 than is its glucose complex.

3. The low concentration of glucose results in a low rate of the liver glucokinase reaction so that the liver adds glucose to the blood rather than removing it.

What are the results? The changes in enzyme activity in the liver stop the formation of glycogen and accelerate its degradation. The resultant rise in the concentration of glucose 6-phosphate will accelerate the hydrolysis of this compound to form free glucose, which will diffuse into the blood to prevent a further drop in concentration. The maintenance of the blood glucose concentration will enable the brain to continue using this fuel at its normal rate, and the supply for this purpose is conserved by the simultaneous slowing of glucose uptake and of glycogen formation in the muscles, owing to the declining insulin levels.

THE EFFICIENCY OF GLYCOGEN STORAGE

What is the price the organism pays for storage of excess glucose as glycogen? We can show it is only 3 per cent of the potential for generating ATP. Let us put the question in this way: How much ATP will be generated per glucosyl group that disappears during combustion if the glucose has been stored as glycogen, compared to the amount generated when glucose is used directly without storage?

It requires two high-energy phosphates to store one glucose residue as glycogen, according to the balance:

$$\begin{aligned}
\text{glucose} + \text{ATP} &\longrightarrow \text{glucose 6-phosphate} + \text{ADP}\\
\text{glucose 6-phosphate} &\longrightarrow \text{glucose 1-phosphate}\\
\text{glucose 1-phosphate} + \text{UTP} &\longrightarrow \text{UDP-glucose} + \text{PP}_i\\
\text{UDP-glucose} + \text{glycogen} &\longrightarrow \text{glucosyl-glycogen*} + \text{UDP}\\
\text{PP}_i + \text{H}_2\text{O} &\longrightarrow 2\,\text{P}_i\\
\text{UDP} + \text{ATP} &\longrightarrow \text{UTP} + \text{ADP}
\end{aligned}$$

$$\text{SUM: glucose} + 2\,\text{ATP} + \text{glycogen} + \text{H}_2\text{O} \longrightarrow \text{glucosyl-glycogen} + 2\,(\text{ADP} + \text{P}_i).$$

Where is the high-energy phosphate to be obtained? It may be from combustion of any fuel, but we can approximate the net cost by assuming that glucose is the only fuel available, and that part of its supply must be burned immediately in order to store the remainder, and the complete combustion of one glucose molecule will generate 35.5 ATP in white fast-twitch fibers. From this, we can estimate that 5.3 per cent of the glucose will be burned to store the remaining 94.7 per cent as glycogen:

$$.947 \times 2\ \text{ATP consumed} \cong .053 \times 35.5\ \text{ATP produced}$$

When the stored glycogen is later used, approximately 95 per cent of it will be converted to glucose 1-phosphate; the remaining 5 per cent will be released as free glucose from hydrolysis of the 1,6 branches.†

Suppose that there were 100 millimoles of glucose originally available. 94.7

*An extended glycogen chain is designated glucosyl glycogen to distinguish the added residues (s) from the primer.

†These figures are based on the assumption that the proportion of 1,4 bonds is higher in the portion of the glycogen molecule degraded than it is in the entire molecule.

millimoles were stored as glycogen, while 5.3 millimoles were burned to provide the energy for storage. When the 94.7 stored millimoles are recovered, 5 per cent, or 4.7 millimoles, is released as free glucose, and 90 millimoles are obtained as glucose 1-phosphate. Oxidation of glucose 1-phosphate generates 36.5 ATP; therefore $90 \times 36.5 = 3{,}285$ millimoles of ATP will be formed.

Out of the original 100 millimoles of glucose, 4.7 millimoles reappeared as glucose during the breakdown of glycogen; therefore, only 95.3 have disappeared to generate 3,285 millimoles of ATP, for a yield of 34.5 ATP per glucose molecule disappearing. This compares with 35.5 that could be obtained by direct oxidation of glucose. The cost of glycogen storage is therefore only 3 per cent of the potential ATP—a small price for an immediately available supply of rapidly combustible fuel. (This analysis neglects the cost of carrying the extra weight of the stored fuel.)

Efficiency of the Cori Cycle

A similar analysis can be made for the Cori cycle, in which part of the lactate made from glycogen in white muscle fibers is converted to glucose in the liver and returned to the muscles for re-storage as glycogen. This calculation is again a simplification of complex balances in which it is assumed that combustion of lactate by the liver provides the high-energy phosphate for gluconeogenesis. It goes like this:

Skeletal muscles: 100 millimoles of glucose residues in glycogen yield 200 millimoles of lactate plus 300 millimoles of ATP.

Liver: 200 millimoles of lactate arrive in the blood from the muscles.

1. 31.3 millimoles are oxidized to CO_2 to generate 548 millimoles of ATP (17.5 millimoles per lactate, utilizing the malate-aspartate electron shuttle).
2. The remaining 168.7 millimoles of lactate are converted to 84.4 millimoles of glucose, utilizing the 548 millimoles of ATP generated in step 1.

Skeletal muscles: 84.4 millimoles of glucose arrive in the blood from the liver.

1. 94.7 per cent of the glucose or 79.9 millimoles can be stored as glycogen by burning the other 5.3 per cent to provide the energy.
2. This replaces all but 20.1 millimoles of the original glucose residues used in the muscle; therefore, the muscle gained the 300 millimoles of ATP used initially for contraction at the expense of 20.1 millimoles of glucose, for a yield of 14.9 ATP per glucose residue consumed.

This value is 40.9 per cent of the 36.5 ATP generated by total oxidation in the muscle of a glucose residue in glycogen.

GENETIC DEFECTS IN GLYCOGEN METABOLISM

A number of humans have been discovered who are deficient in one of the enzyme activities necessary for glycogen metabolism. The combined incidence of these defects is about 1 in 40,000 births. One case might arise per year in Los Angeles. However, they add to the catalogue of genetic defects that does demand the attention of every pediatrician. In addition, their effects have promoted our understanding of normal carbohydrate metabolism.

A defect may occur in the formation of any of the proteins necessary for either the breakdown or the synthesis of glycogen, so that the result could be either a high

TABLE 27–1 GLYCOGEN STORAGE DISEASES

Deficient Enzyme	Name	Type	Site of Deficiency	Incidence
glucose 6-phosphatase	von Gierke's disease	I	liver, kidneys	1:200,000
lysosomal α-glucosidase	Pompe's disease	II	all tissues	1:200,000
amylo-1,6-glucosidase	limit dextrinosis	III	all tissues	1:200,000
branching enzyme	amylopectinosis	IV	all tissues	very low
glycogen phosphorylase	McArdle's disease	V	skeletal muscle	low
glycogen phosphorylase	Hers' disease	VI	liver	?
phosphofructokinase		VII	skeletal muscle, erythrocytes	very low
adenyl cyclase(?)		VIII	brain, liver	very low
phosphorylase kinase		IX	liver, other tissues (not muscle)	1:100,000
cAMP-sensitive protein kinase		X	liver, muscles	very low

Note: Type numbers VI, VIII, and IX have been assigned to different deficiencies by some authors.

or a low content of glycogen with a normal structure, or the presence of glycogen with abnormalities in the degree of branching. A classification of the resultant clinical conditions is given in Table 27–1. However, it ought to be remembered that genetic errors can affect the formation of a protein in various ways; the rate of synthesis of normal protein may be changed, or mutations of coding may affect amino acid composition at any point in the molecule. The result is that there may be many types of deficiency of a given enzyme, some more severe in their consequences than others, so the rigid classification of the table does not adequately convey the actual spectrum of diseases.

A genetic defect may affect all tissues with an active glycogen metabolism. However, in some cases, a defective gene will disturb glycogen metabolism in only a few tissues because normal genes for different isozymes are still being expressed in other organs. Thus, defects in glycogen phosphorylase may be peculiar to the liver or to the skeletal muscles, because different isozymes are made in these organs.

Two decades ago it was believed that glycogen is synthesized by glycogen phosphorylase, because its reaction is readily reversed in the test tube. A young man was discovered who was weak and who developed severe pain in his muscles after modest exercise, and he was shown to have high concentrations of glycogen and an absence of glycogen phosphorylase in his muscles. This proved that he was not synthesizing glycogen by the phosphorylase reaction and provided strong supporting evidence for the then newly discovered UDP-glucose pathway. (The condition is known as **McArdle's disease,** after its discoverer. A psychiatrist engagingly confesses [Ann. Intern. Med. *62*:412, 1965] that he thought McArdle's patient was displaying classic hysteria owing to an unhappy childhood. We are not told how much the phosphorylase deficiency may have contributed to the unhappy childhood.)

McArdle's disease tells us something else. Impairment of glycogen utilization in skeletal muscles is not fatal. The impairment is very real; there isn't any unsuspected route for utilizing the polysaccharide, and this is easily demonstrated by shutting off the blood supply to an arm with a tourniquet. Clenching of the fist causes a sharp rise in lactate concentration in a normal individual, ending in painful tetany, but there is no rise in a patient with McArdle's disease. However, the importance of carbohydrate metabolism for full efficiency is shown by the weakness of the patient and is supported by the uncommon occurrence of the genetic lesion in the population—this is a mutation that is eliminated rapidly.

This view of glycogen metabolism as a convenience for full efficiency of muscles rather than as an absolute necessity is reinforced by the discovery of individuals with a deficiency of phosphofructokinase in skeletal muscles, who have the same clinical picture found in McArdle's disease. They not only accumulate glycogen because of an inability to use it, they can't even use glucose taken up from the blood stream directly, and they still survive.

Accumulation of normal glycogen in the liver causes massive enlargement of the organ, so much so that it may occupy a large fraction of the abdomen in affected children. This in itself causes surprisingly little difficulty; infants without glycogen phosphorylase in the liver grow to be adults. However, if the failure is due to an absence of glucose-6-phosphatase, which prevents the use of stored glycogen to maintain the blood glucose concentration between meals, the consequences without treatment are grave. The blood glucose concentration falls, while the lactate concentration rises. The brain is usually damaged, perhaps owing in part to the absence of glucose as a fuel and in part to the lactic acidosis. The treatment is repeated feeding of carbohydrate at two- to three-hour intervals night and day. If the infant can be brought through the first four years, his chances for gradual adjustment to a less heroic feeding schedule are good.*

Defects in phosphorylase kinase are interesting because two patterns of inheritance are known. One is an autosomal recessive and the other is an X-linked recessive. If we had only this information, we could infer that the kinase contains at least two polypeptide chains coded by genes in different chromosomes, and we have already seen that the kinase contains some three different chains. In addition, another kind of phosphorylase kinase deficiency has been discovered that affects both the liver and the muscles, indicating that these isozymes have some common subunit.

One of the most damaging storage diseases results from accumulation of glycogen by lysosomes. The enzyme affected is an α-glucosidase that attacks either 1,4 or 1,6 linkages and is not involved in the major routes of glycogen metabolism. It probably is a device for removing glycogen trapped during the normal scavenging function of lysosomes, which become filled to the bursting point when the enzyme is missing. All tissues are affected, but damage to the heart is usually the immediate cause of death, which occurs in infancy.

FURTHER READING

P.J. Roach (1981). Glycogen Synthase and Glycogen Synthase Kinase. Curr. Top. Cell Regul. 20:45.

R.J. Fletterich and N.B. Madsen (1980). The Structures and Related Functions of Phosphorylase a. Annu. Rev. Biochem. 49:31.

S.G. Withers et al. (1981). Evidence for Direct Phosphate-Phosphate Interaction between Pyridoxal Phosphate and Substrate in the Glycogen Phosphorylase Catalytic Mechanism. J. Biol. Chem. 256:10759.

V. Dombradi (1981). Structural Aspects of the Catalytic and Regulatory Function of Glycogen Phosphorylase. Int. J. Biochem. 12:125. A review.

General Aspects of Regulation

F. Huijing (1975). Glycogen Metabolism and Glycogen Storage Diseases. Physiol. Rev. 55:609.

D.A. Hems and P.D. Whitton (1980). Control of Hepatic Glycogenolysis. Physiol. Rev. 60:1.

L. Hue (1981). The Role of Futile Cycles in the Regulation of Carbohydrate Metabolism. Adv. Enzymol. 52:247.

Phosphorylation-Dephosphorylation

P. Cohen, ed. (1980). Recently Discovered Systems of Enzyme Regulation by Reversible Phosphorylation. Elsevier. Includes historical review.

---

*Partial bypass of portal blood flow around the liver by surgery (portacaval shunt) gives some relief in extreme cases.

E.G. Krebs (1981). Phosphorylation and Dephosphorylation of Glycogen Phosphorylase: A Prototype for Reversible Enzyme Covalent Modification. Curr. Top. Cell Regul. 18:401.
O.M. Rosen and E.G. Krebs, eds. (1981). Protein Phosphorylation. 2 vols. Cold Spring Harbor Lab. A collection of many short articles.
E.Y.C. Lee at al. (1980). The Phosphoprotein Phosphatases: Properties of the Enzymes Involved in the Regulation of Glycogen Metabolism. Adv. Cycl. Nucl. Res. 13:95.
J.T. Stull (1980). Phosphorylation of Contractile Proteins in Relation to Muscle Function. Adv. Cycl. Nucl. Res. 13:40.

Hormonal Control

E.M. Ross and A.G. Gilman (1980). Biochemical Properties of Hormone-Sensitive Adenylate Cyclase. Annu. Rev. Biochem. 49:533.
R.H. Unger and L. Orci, eds. (1981). Glucagon—Physiology, Pathophysiology, and Morphology of the Pancreatic A-Cells. Elsevier. The same authors have a shorter review under a similar title in the N. Engl. J. Med. 304:1518,1575 (1981).
J. Larner et al. (1981). Toward an Understanding of the Mechanism of Action of Insulin. p. 1 in C.M. Veneziale. The Regulation of Carbohydrate Formation and Utilization in Mammals. University Park Press. This article from the laboratory leading current research into insulin action also includes a historical note on the discovery of insulin.
R.T. Curnow and J. Larner (1979). Hormonal and Metabolic Control of Phosphoprotein Phosphatases. 6:77 in G. Litwack, ed. Biochemical Action of Hormones. Academic Press.
D.A. Walsh and R.H. Cooper, ibid., p. 1. The Physiological Regulation and Function of cAMP-Dependent Protein Kinases.

Calcium as Effector

R.H. Kretsinger (July, 1980). Structure and Evolution of Calcium-Modulated Proteins. CRC Crit. Rev. Biochem. p. 119. Review by the discoverer of the importance of calcium-binding proteins.
C.B. Klee, T.H. Crouch, and P.G. Richman (1980). Calmodulin. Annu. Rev. Biochem. 49:489.
W.Y. Cheung (1979). Calmodulin Plays a Pivotal Role in Cellular Regulation. Science 207:19.

CHAPTER 28

STORAGE OF FATS

NATURE AND DISTRIBUTION OF FAT STORES

Fat is a fuel that is laid down for long-term storage. Glycogen and starch are fuels for short-term storage or for the maintenance of organisms in the presence of limited amounts of oxygen. The human epitomizes this dual storage; the ordinary adult has only enough glycogen to maintain activity for one day or less, but he can live from his fat for nearly a month. A human being is built for a daily routine in which he oxidizes glucose residues for energy immediately after meals while rebuilding glycogen reserves and converting any excess glucose to fatty acids. As the time of the last meal recedes, and the glycogen supply again becomes depleted, more and more of his energy is obtained by oxidizing fatty acids previously stored as triglycerides. Even the overnight fast is sufficient to cause the amount of oxygen used for the oxidation of fatty acids from fats to be twice that used for the oxidation of glucose from glycogen at rest.

The great advantage of fat as a stored fuel is that it is light in weight, and the initial appearance on earth of organisms with large fat stores evidently coincided with the development of the ability to move over relatively long distances without an intake of food. Salmon and ducks are alike in building up large stores of fat before they begin their long migrations, but vertebrates of more fixed domicile, along with many insects, also can store fat for less dramatic exertion.

Location of Fat. Most people have a general idea of the character of animal fat and give it an equally diffuse anatomical role. Let us take a moment to discuss the sites of deposition so that we can more fully appreciate the biochemistry of these tissues. Fat deposition began to be important with the evolution of vertebrates, and the liver was the initial site of deposition. Modern sharks frequently have massive livers containing cells loaded with triglycerides. With the appearance of bony fish, fat began to be deposited to a greater extent in and around the muscle fibers, creating the oily flesh we see in salmon and sardines. Insects went on another pathway and created a multipurpose organ with many of the functions of the vertebrate liver, but which contains so much fat that it is known as the **fat body.** The advanced vertebrates, beginning with some fish, developed a discrete **adipose tissue** by modifying the same kind of cells from which the blood cells are derived. These **adipocytes** contain globules of triglyceride that may constitute 90 per cent or more of the mass of the cell. Adipose tissue is especially prevalent in subcutaneous tissues, around deep blood vessels, and in the abdominal cavity.

Adipose tissue, although it appears relatively formless and is difficult to handle experimentally because of its high content of fat, is well-organized with an active metabolism appropriate for its important function as an internal larder. Adipose tissue can become the largest in the body, comprising half or more of the total mass of some individuals. Humans can become tubs of lard. Such people are objects of humor, disdain, or concern in our society, but in societies subject to famine they may be happily living on their own fat while burying the last of their formerly trim companions.

Plants also store fat, especially in the tissues surrounding the embryos in seeds. The light weight of this kind of a fuel no doubt aids in the dispersal of small seeds, but fat is also the predominant stored fuel in large seeds, where its hydrophobic character may be of primary importance in protecting the embryo until time for development.

THE ORIGIN OF FATTY ACIDS

The stored fatty acids originate mainly from the diet or by synthesis from glucose, the proportion depending upon the relative amount of fats or glucose ingested. Worldwide, carbohydrates represent the largest source of chemical potential in the human diet, but those with a high fat intake, such as a typical American, may convert little of their glucose to fatty acids for long-term storage. We do most of our conversion of glucose to fatty acids in the liver, and then store the products in our adipose tissue, but the rats and mice commonly used for laboratory studies synthesize fatty acids in both liver and adipose tissue.

Overall Mechanism

When adipocytes or hepatocytes make fatty acids from glucose, they first oxidize the glucose to acetyl coenzyme A. This is done in exactly the same way as it is when the acetyl groups are destined to be burned in the citric acid cycle by muscles (Chapter 25). That is, glucose is oxidized to pyruvate along the Embden-Meyerhof pathway in the cytosol, and the pyruvate is then oxidized to acetyl coenzyme A by the pyruvate dehydrogenase complex in mitochondria. The differences between glucose oxidation and conversion of glucose to fat come after the formation of acetyl coenzyme A. A surfeit of glucose is accompanied by regulatory events that cause acetyl groups to be diverted to fat synthesis in the cytosol rather than being used for combustion or for the formation of 3-oxybutyrates in the mitochondria.

Fatty acids are synthesized from acetyl coenzyme A by a process that is similar to a reversal of fatty acid oxidation. That is, the chain is made longer by successive additions of acetyl groups, each addition being followed by reduction to an increasingly longer fatty acid residue. We shall now consider the nature of these two processes—chain lengthening and reduction to fatty acids.

The Enzymes

Before discussing the mechanism of fatty acid synthesis, let us examine the enzymes involved.

Acetyl-CoA Carboxylase. Rebuilding a fatty acid chain from acetyl groups requires expenditure of energy. The energy for carbon-to-carbon condensation is supplied by a process of carboxylation and decarboxylation of acetyl groups in the cytosol. (This process resembles the carboxylation and decarboxylation of pyruvate that is used to drive the synthesis of glucose from lactate [p. 493]). Each two-carbon unit that is added to the growing fatty acid chain is supplied as malonyl coenzyme A, which is obtained by carboxylating acetyl coenzyme A:

$$\underset{\textbf{acetyl coenzyme A}}{CH_3-C(=O)-S-CoA} \xrightarrow[\substack{\textit{acetyl CoA carboxylase} \\ \textit{(biotin-protein)}}]{CO_2 \quad ATP \quad ADP + P_i} \underset{\textbf{malonyl coenzyme A}}{COO^{\ominus}-CH_2-C(=O)-S-CoA}$$

Acetyl-CoA carboxylase contains biotin to act as a CO_2 carrier in the same way as it does in pyruvate carboxylase. **Malonyl coenzyme A is only used to make fatty acids; therefore, its formation has been made the rate-controlling step in fatty acid synthesis.**

Fatty acid synthase, also occurring in the cytosol, is an unusual enzyme.* It has two identical subunits (220,000 to 240,000 daltons). A subunit is made by translating a single messenger RNA molecule, even though it has seven different catalytic sites to catalyze the many reactions necessary for converting malonyl groups to a fatty acid. **The product is palmitate,** in which the two carbon atoms at the methyl terminal are obtained directly from acetyl coenzyme A and the remaining 14 carbon atoms are obtained, two at a time, from malonyl coenzyme A (see figure at top of next page).

Bacteria contain separate proteins to catalyze the individual reactions of fatty acid synthesis; even the formation of malonyl coenzyme A occurs in two stages (carboxylation of biotin, and transfer of the carboxylate group to acetyl coenzyme

---

*The enzyme is frequently named fatty acid synthetase, but its action does not fit the definition of a synthetase.

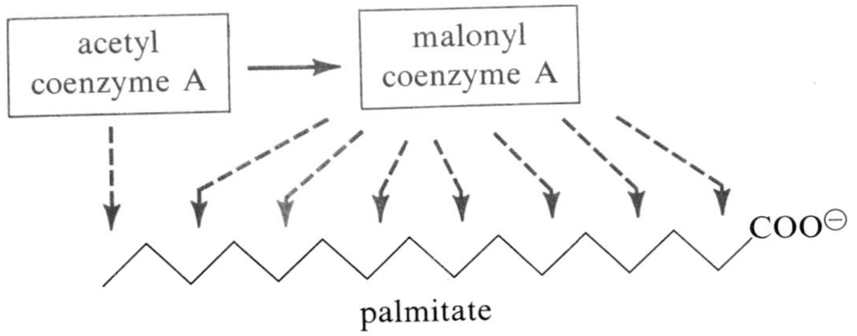

acetyl coenzyme A → malonyl coenzyme A

$COO^{\ominus}$

palmitate

A). This lucky circumstance made it much easier to discover the sequence of reactions, since each reaction could be studied separately. Demonstration that the eukaryotic enzyme complex operated in a similar fashion then followed.

The fatty acid synthase contains two all-important sulfhydryl groups; the growing fatty acid is shuffled between these groups. One is relatively fixed in position because it is on a cysteine residue. It acts as a parking place for acyl groups that are to be lengthened. The other sulfhydryl group carries the extended chain while it undergoes the reactions necessary for reduction to a saturated acyl group, and it also accepts the acetyl and malonyl groups from which the fatty acid is built. This sulfhydryl group can swing across the seven different catalytic sites because it is located in a residue of **phosphopantetheine** (Fig. 28–1). Phosphopantetheine is part of the structure of coenzyme A, but here it is bound to a serine residue in one domain (the acyl carrier domain) of each fatty acid synthase subunit. The phosphopantetheinyl group in the synthase has the same acyl-carrying function during synthesis that it has in coenzyme A during fatty acid oxidation. The important difference is that the group is covalently fixed to one enzyme complex in fatty acid synthase, whereas coenzyme A is in solution and is free to migrate from one enzyme to another.

Assembly of the Carbon Chain

The synthesis of fatty acids begins with the binding of an acetyl group to fatty acid synthase (Fig. 28–2*A*). (The events described here can occur simultaneously on

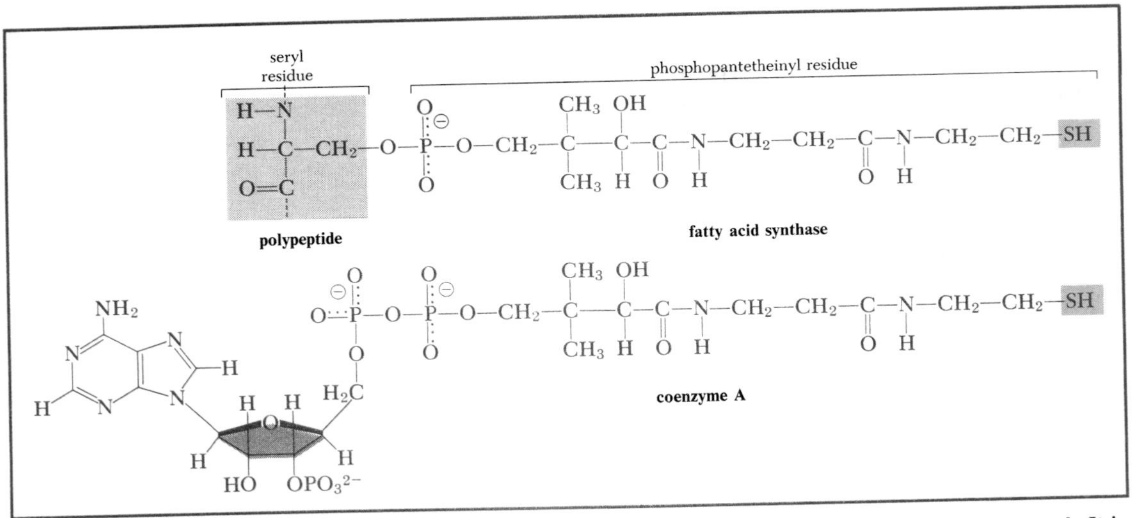

seryl residue

phosphopantetheinyl residue

H—N

H—C—CH_2—O—P—O—CH_2—C—C—C—N—CH_2—CH_2—C—N—CH_2—CH_2—SH

O=C

polypeptide

fatty acid synthase

NH_2

coenzyme A

HO OPO_3^{2-}

FIGURE 28–1 **A residue of phosphopantetheine is found in both fatty acid synthase and in coenzyme A. It is bound to the synthase by formation of a phosphate ester with a serine residue.**

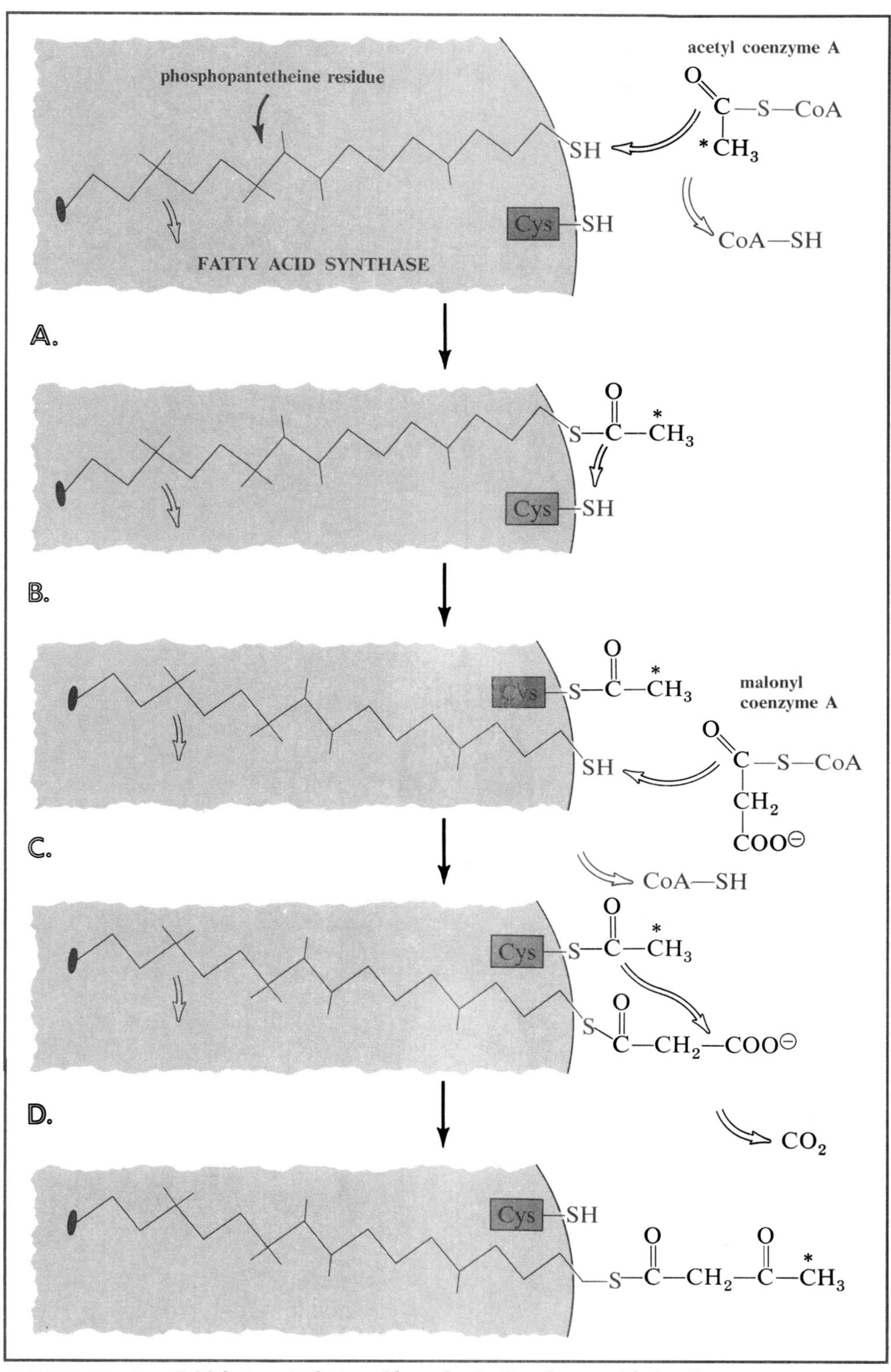

FIGURE 28–2 Initial steps in fatty acid synthesis. See the text for details.

the two identical sets of catalytic sites.) This short-chain acyl group is the tail upon which a molecule of palmitate will be constructed toward the head by addition of two-carbon units. It is obtained by transfer from acetyl coenzyme A, first to the phosphopantetheine sulfhydryl group and then to the cysteine sulfhydryl group for temporary storage (Fig. 28–2*B*).

The next step is a transfer of a malonyl group from malonyl coenzyme A to the now-vacant phosphopantetheine residue (Fig. 28–2*C*). The synthase is thus loaded with an acetyl group and a malonyl group, and the two groups are condensed (Fig. 28–2*D*). The condensation involves transfer of the acetyl group (the tail) to the malonyl group (the new head), which is simultaneously decarboxylated. The loss of free energy upon decarboxylation drives the reaction toward completion. The acetoacetyl product is two carbon atoms longer than the starting acetyl group.

Reduction to an Acyl Group

The next steps in fatty acid synthesis convert the acetoacetyl group to the corresponding saturated butyryl group by a pair of reductions and a dehydration. The first reduction forms a D-3-hydroxybutyryl group (Fig. 28–3*A*). Water is removed to form the unsaturated crotonyl group (Fig. 28–3*B*), which is then reduced to the saturated butyryl group (Fig. 28–3*C*).

Each of the steps in the sequence is catalyzed by a distinct site on fatty acid synthase; the substrate group is moved from one site to the next while bound to a phosphopantetheine residue on the enzyme.

NADPH as the Electron Carrier. The reductions during fatty acid synthesis use electrons from NADPH (p. 414). The reason is that NADP is maintained in a highly reduced state compared to NAD. We shall soon see that NADP is involved in a few energetically favorable oxidations with equilibrium positions far in favor of formation of NADPH. NAD, on the other hand, is used for reactions in which the equilibrium is less favorable for formation of NADH. We can see the result in the calculated ratios of oxidized and reduced forms in liver cytosol from well-fed rats*:

*Data from H.A. Krebs, and R.W. Veech (1969). Adv. Enzym. Regul. 7:397.

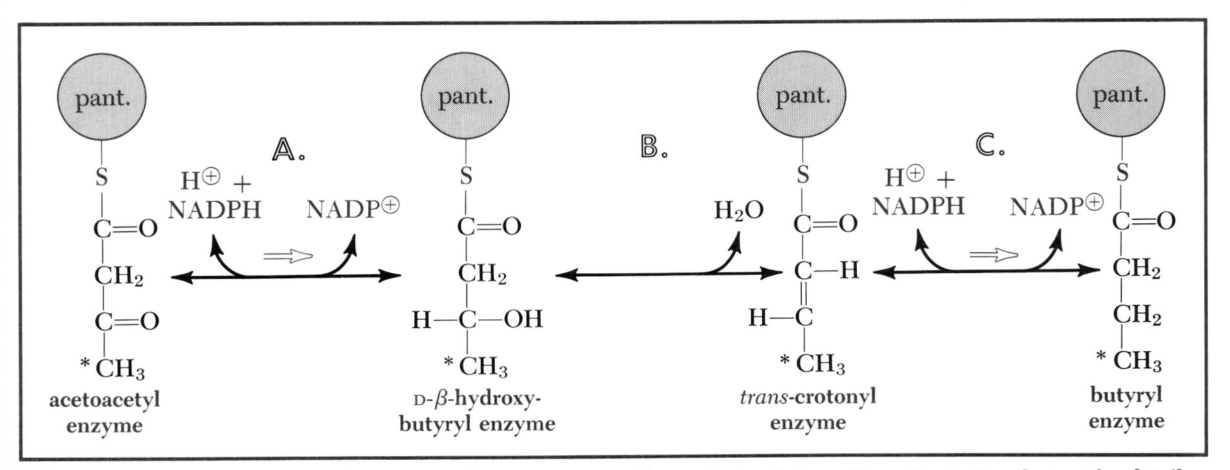

FIGURE 28–3 **Reduction of the 3-ketoacyl group to an acyl group during fatty acid synthesis. See the text for details.**

$$\frac{[\text{NADPH}]}{[\text{NADP}]} = 77 \qquad \frac{[\text{NADH}]}{[\text{NAD}]} = 0.00083.$$

The proportion of the reduced form of NADP is 10^5 that seen with NAD! Given these ratios, reduction of other compounds by NADPH will liberate 30,000 more joules of free energy per pair of electrons, the energetic equivalent of driving each reduction by the hydrolysis of ATP. The high concentration of NADPH is used to push synthetic reactions, of which the reduction of acetyl groups to fatty acids is an example.

Chain Growth

The general sequence of condensation and reduction catalyzed by fatty acid synthase is:

$$\underset{n\ \text{carbons}}{\textbf{acyl group}} \xrightarrow[-CO_2]{+\ \text{malonyl group}} \textbf{3-ketoacyl group} \xrightarrow{\text{NADPH}} \textbf{3-hydroxy-acyl group} \xrightarrow{-H_2O} \textbf{enoyl group} \xrightarrow{\text{NADPH}} \textbf{acyl group}$$

n carbons ←——— *(n + 2) carbons* ———→

In the initial sequence, the acetyl group is the first acyl group, and a butyryl group is the product acyl group. The enzyme then transfers the butyryl group to the cysteine residue. The phosphopantetheine group is again loaded with a malonyl group obtained from malonyl coenzyme A, and the entire sequence is repeated. This will happen over and over, making a growing fatty acid residue in which all but the two carbon atoms at the methyl end have passed through malonyl coenzyme A.

Chain Termination. The chain grows on the fatty acid synthase until a 16-carbon palmitoyl residue is produced. The final step is a hydrolysis to release the chain as free palmitate. It is not known how the chain length is limited to 16 carbon atoms; the fatty acid synthase is evidently constructed to expel the fatty acid once it reaches this size. The mammary glands of some animals, such as the cow, synthesize shorter chain fatty acids, but the synthase isolated from these tissues also makes palmitate in the laboratory. The apparent paradox was resolved when those mammaries were discovered to contain a protein that combines with fatty acid synthase and causes it to hydrolyze shorter chain acyl derivatives. It is as if the additional protein occupies part of a notch for the growing chain in such a way as to activate early hydrolysis.

The stoichiometry for the formation of palmitate from acetyl coenzyme A, allowing for the intermediate formation of seven moles of malonyl coenzyme A, is (neglecting balance of H^+ and H_2O):

(1) 7 acetyl-S-CoA + 7 CO_2 + 7 ATP ⟶
7 malonyl-S-CoA + 7 ADP + 7 P_i

(2) acetyl-S-CoA + 7 malonyl-S-CoA + 14 NADPH ⟶
palmitate + 7 CO_2 + 14 $NADP^{\oplus}$ + 8 CoA-SH

SUM: 8 acetyl-S-CoA + 7 ATP + 14 NADPH ⟶
palmitate + 7 ADP + 7 P_i + 8 CoA-SH + 14 $NADP^{\oplus}$

MODIFICATION OF THE FATTY ACID CHAIN

Fatty acids are modified by elongating the carbon chain and introducing double bonds. Quantitatively, the most important modification is the conversion of palmitate to **oleic acid, 18:1(9),** which is the most abundant fatty acid in animal fat. This involves only the addition of one pair of carbon atoms and the introduction of one double bond. Stored fat contains lesser amounts of **linoleic acid, 18:2(9,12),** and **linolenic acid, 18:3(9,12,15),** which are found in the diet and originally synthesized in plants. Other fatty acids with longer chains and more double bonds are of great qualitative importance as constituents of membrane lipids and as precursors of other compounds. These are ordinarily made by lengthening the chains of linoleic and linolenic acids.

CHAIN ELONGATION

Elongation on Endoplasmic Reticulum. Palmitate is converted to palmitoyl coenzyme A, and the chain is then elongated by the addition of one or more acetyl groups from malonyl coenzyme A. The sequence of reactions is exactly the same as we outlined for the action of fatty acid synthase, except that the elongation enzymes are located on the endoplasmic reticulum. Two carbon units can be added to various saturated or unsaturated fatty acids by this system until the total length of the chain reaches 24 carbon atoms. The elongation system works better with unsaturated fatty acids. It is difficult for it to elongate a saturated fatty acid beyond 18:0, so **stearic acid** is the principal saturated product. Longer chain fatty acids are mainly polyunsaturated.

Chain Desaturation

Both plants and animals contain **desaturases** for introducing double bonds into fatty acid residues. These enzymes are mixed function oxidases that use NADH and cytochrome b_5 to activate dioxygen for attacking an acyl coenzyme A. For example, the Δ^9-desaturase converts stearoyl coenzyme A to oleoyl coenzyme A by this reaction:

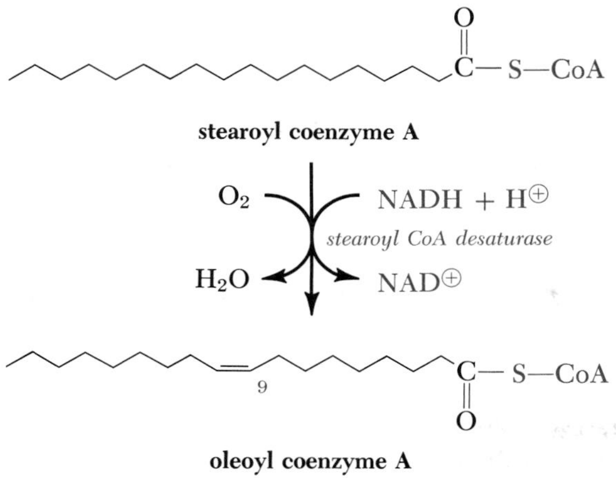

O
‖
C—S—CoA

stearoyl coenzyme A

O_2 → H_2O; NADH + $H^{\oplus}$ → $NAD^{\oplus}$

stearoyl CoA desaturase

9

C—S—CoA
‖
O

oleoyl coenzyme A

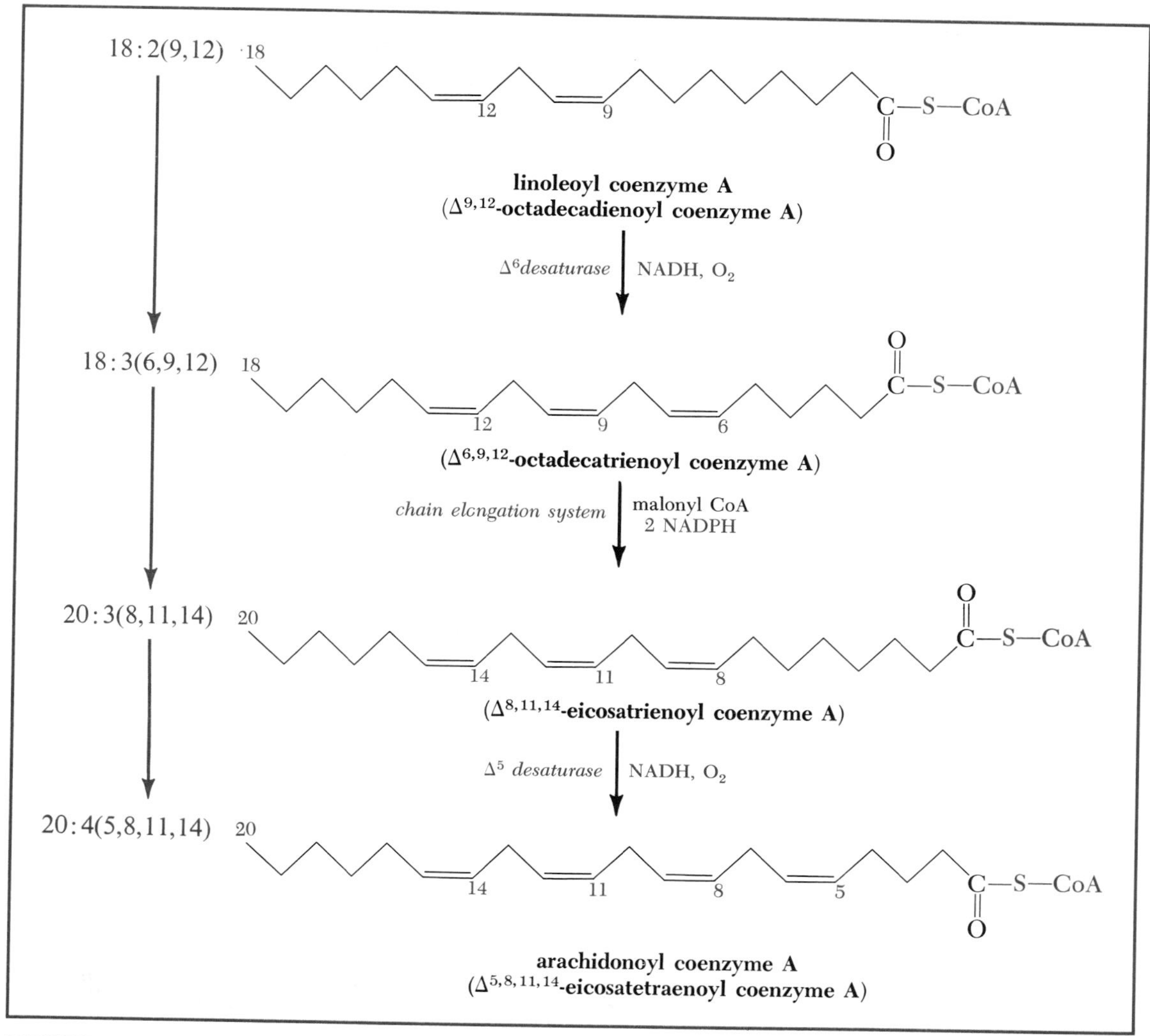

FIGURE 28–4 **The synthesis of arachidonic acid residues from linoleic acid residues involves a combination of chain lengthening and further desaturation toward C1 in the coenzyme A derivatives. All of the illustrated compounds have a 6-carbon tail beyond the final double bond, a feature that identifies long-chain fatty acids formed from dietary linoleic acid because animals cannot introduce double bonds in that portion of the molecule.**

In addition, animals have a Δ^5 and a Δ^6 desaturase, which will introduce additional double bonds toward the carboxyl end of the fatty acid chain. They cannot introduce double bonds beyond C9. Plants, on the other hand, have the ability to introduce double bonds, not only at C9, but also at C12 and C15.

Families of Fatty Acids. Palmitic acid groups are the shortest substrates for the animal Δ^9 desaturase. The resultant **palmitoleic acid** residue therefore has seven carbon atoms beyond the double bond, while oleic acid has nine. If these are elongated (with or without additional double bonds) in animals, there will still be seven or nine carbon atoms beyond the last double bond. **Any polyunsaturated fatty acid residue in the body with fewer than seven carbon atoms beyond the last double bond must be derived from a dietary fatty acid of plant origin.** Thus, dietary linoleate,

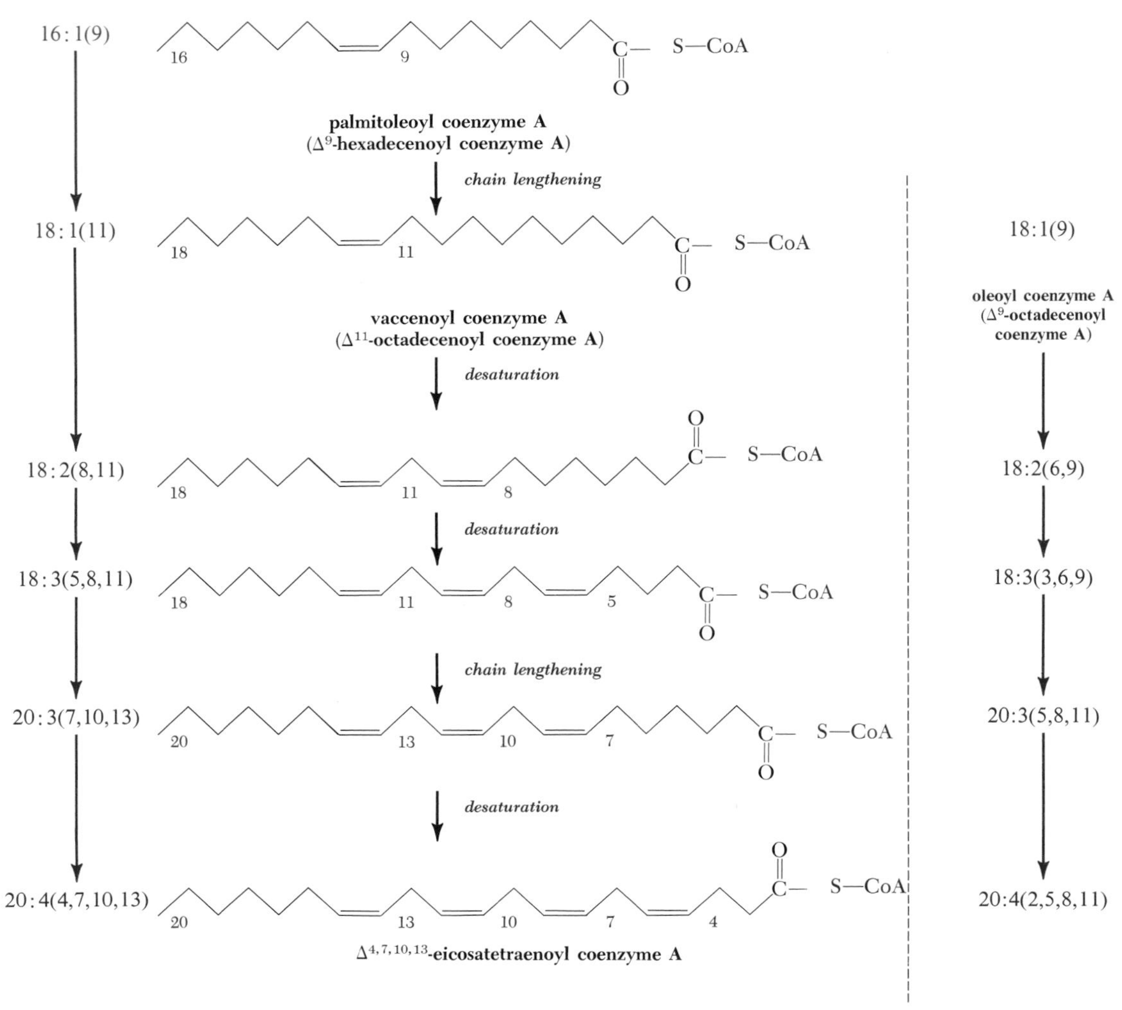

FIGURE 28–5 When the diet lacks linoleic acid as a precursor for fatty acids, substitutes with 7-carbon tails beyond the last double bond may be synthesized in increased amounts for incorporation into membrane lipids. The scheme for synthesizing fatty acids with 9-carbon tails is also indicated.

18:2(9,12), can be converted by a combination of desaturation and chain elongation to **arachidonate,** 20:4(5,8,11,14), the precursor of the prostaglandin hormones, but palmitate cannot (Fig. 28–4). The closest animals can come is to begin with palmitoleoyl coenzyme A, 16:1(9), and successively add carbons and desaturate to create 20:4(4,7,10,13) or to begin with oleoyl coenzyme A and create 20:4(2,5,8,11) (Fig. 28–5). (The pathway from palmitoleoyl CoA also creates residues of vaccenic acid, 18:1(11), which is a common constituent of animal fats, although not as abundant as oleic acid residues, 18:1(9).)

A family of precursor and its products generated in animals through elongation and desaturation can be recognized by subtracting the number designating the last double bond from the total number of carbon atoms. The result is the same within a family. For example, with linoleate, 18:2(9,12), and arachidonate, 20:4(5,8,11,14): $18 - 12 = 6 = 20 - 14$.

ELECTRON SUPPLY IN FATTY ACID SYNTHESIS

The reduction of acetyl groups to fatty acids is accomplished by oxidizing NADPH to NADP, and the NADPH used for this purpose is maintained in a high concentration relative to NADP. The NADPH is generated by three major processes: (1) the oxidative decarboxylation of isocitrate in the cytosol; (2) the oxidative decarboxylation of malate in the cytosol; and (3) the sequential oxidation of glucose 6-phosphate to ribulose 5-phosphate and CO_2.

Most of the NADPH generated by these reactions is used for converting glucose to fatty acids, but it is also an important source of electrons for other purposes, including syntheses and the conversion of methemoglobin to hemoglobin. The reactions deserve detailed attention.

Generation of NADPH by Isocitrate Dehydrogenase

The cytosol, and to a lesser extent mitochondria, have an NADP-coupled isocitrate dehydrogenase:

$$\underset{\text{isocitrate}}{\begin{array}{c} COO^{\ominus} \\ | \\ H-C-OH \\ | \\ ^{\ominus}OOC-C-H \\ | \\ CH_2 \\ | \\ COO^{\ominus} \end{array}} \xrightarrow{NADP^{\oplus} \quad NADPH} CO_2 + \underset{\alpha\text{-ketoglutarate}}{\begin{array}{c} COO^{\ominus} \\ | \\ C=O \\ | \\ CH_2 \\ | \\ CH_2 \\ | \\ COO^{\ominus} \end{array}}$$

It catalyzes the same oxidative decarboxylation of isocitrate that occurs in mitochondria during the citric acid cycle except for the use of NADP rather than NAD as the oxidizing agent. The extra energy liberated by the decarboxylation drives the reaction toward completion, generating a high [NADPH]/[NADP] ratio. This appears to be a major route for generating NADPH in fetal livers, but it is not clear how it is integrated in the general metabolic processes. It is probably of minor importance in adults.

Malate Dehydrogenases and Acetyl Transfer

The cytosol of liver cells contains two malate dehydrogenases. One catalyzes the same readily reversible NAD-coupled oxidation of malate to oxaloacetate that occurs during the citric acid cycle within mitochondria. (The equilibrium favors malate formation.) The other both oxidizes malate and decarboxylates the intermediate oxaloacetate to pyruvate:

$$\underset{\text{L-malate}}{\begin{array}{c} COO^{\ominus} \\ | \\ HO-C-H \\ | \\ CH_2 \\ | \\ COO^{\ominus} \end{array}} \xrightarrow[\textit{NADP—malate dehydrogenase}]{NADP^{\oplus} \quad NADPH} \underset{\text{pyruvate}}{\begin{array}{c} COO^{\ominus} \\ | \\ C=O \\ | \\ CH_3 \end{array}} + CO_2$$

Here again, the decarboxylation liberates additional free energy and drives the reaction toward completion.

The existence of the two malate dehydrogenases provides a neat device for transferring electrons from NADH to form NADPH in the cytosol:

$$\underset{\text{oxaloacetate}}{\begin{array}{c}COO^{\ominus}\\ |\\ C{=}O\\ |\\ CH_2\\ |\\ COO^{\ominus}\end{array}} \xrightleftharpoons[\text{NAD}^{\oplus}]{\text{NADH}} \underset{\text{L-malate}}{\begin{array}{c}COO^{\ominus}\\ |\\ HO{-}C{-}H\\ |\\ CH_2\\ |\\ COO^{\ominus}\end{array}} \xrightarrow[\text{NADP}^{\oplus}\quad CO_2]{\text{NADPH}} \underset{\text{pyruvate}}{\begin{array}{c}COO^{\ominus}\\ |\\ C{=}O\\ |\\ CH_3\end{array}}$$

SUM: NADH + NADP$^{\oplus}$ + oxaloacetate $\longrightarrow$ NAD$^{\oplus}$ + NADPH + pyruvate + CO_2

Malate acts only as a transient intermediate in this process, which is a major source of NADPH for fatty acid synthesis in the liver of adults. (The existence of the two dehydrogenases demonstrates that the oxidation of 3-hydroxycarboxylic acids may or may not involve a simultaneous decarboxylation of the 3-ketocarboxylic acid formed.)

This rather tricky transfer of electrons fits very smoothly into the overall mechanism of fatty acid synthesis from glucose. The NADH generated in the cytosol by converting glucose to pyruvate along the Embden-Meyerhof pathway is used in this way to make NADPH, rather than accumulating to a point at which pyruvate would be reduced to lactate. The NADPH is then consumed to convert pyruvate to palmitate, via acetyl coenzyme A.

The malate dehydrogenase reactions fit into the synthetic scheme in still another way. Acetyl groups formed in the mitochondria by the oxidation of pyruvate need to be transferred to the cytosol for fatty acid synthesis. This is done by an ingenious scheme in which citrate is formed in the mitochondria and transported to the cytosol (Fig. 28–6). Citrate is then cleaved in the cytosol, supplying acetyl groups for the fatty acid chain and also providing the oxaloacetate for driving electron transfer via

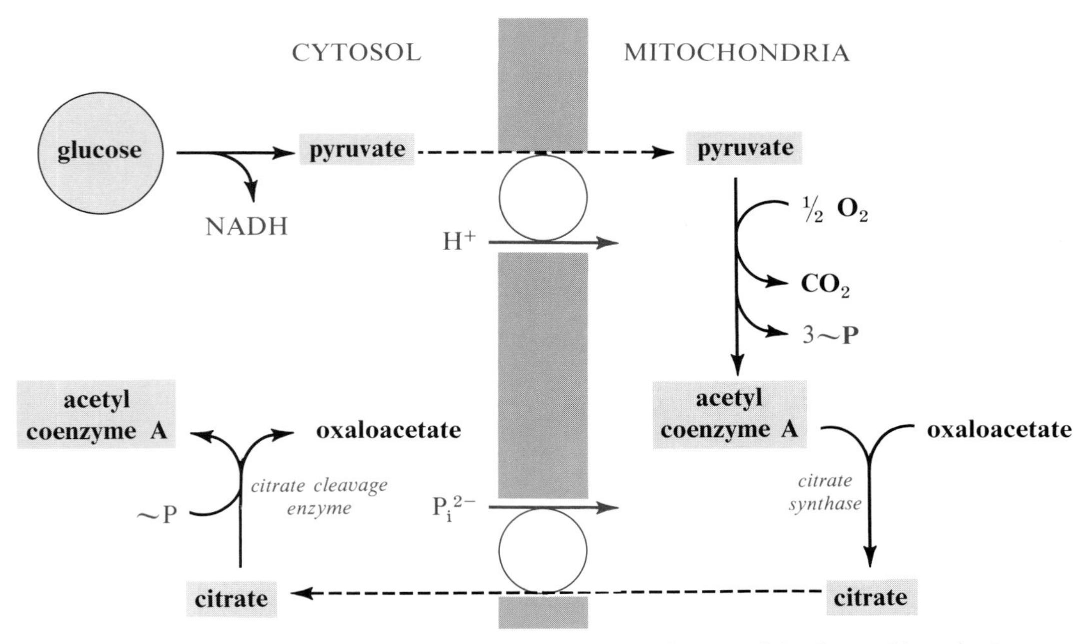

FIGURE 28–6 Use of citrate to transport acetyl groups into the cytosol for fatty acid synthesis.

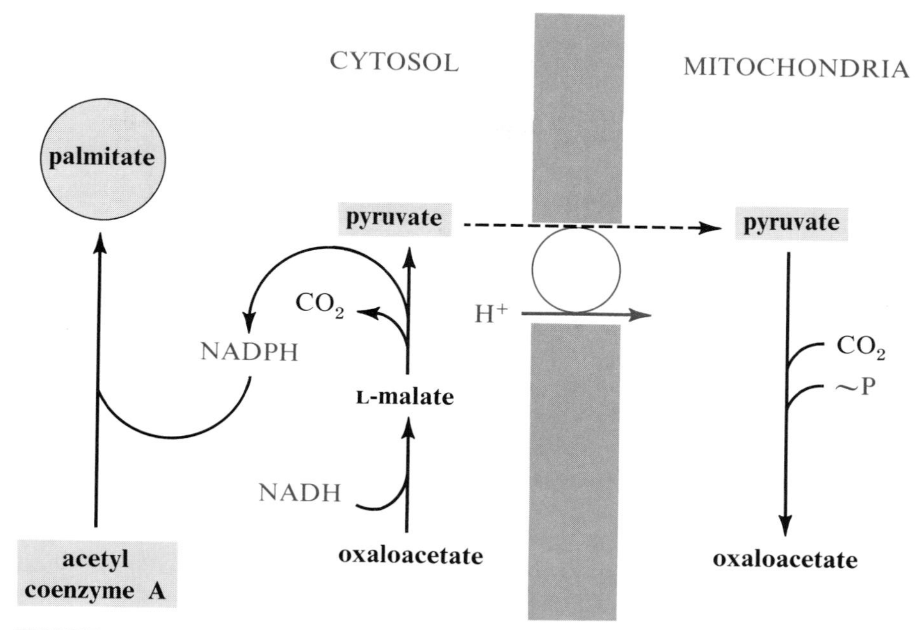

FIGURE 28–7 Coupling of NADPH generation for fatty acid synthesis with the citrate-pyruvate cycle. The conversion to pyruvate of oxaloacetate released from citrate (not shown) in the cytosol is used to transfer electrons from NADH to NADP. Oxaloacetate is regenerated from the pyruvate in the mitochondria for use in citrate synthesis (not shown).

malate. In this scheme, citrate is formed from oxaloacetate and acetyl coenzyme A by the familiar citrate synthase reaction of the citric acid cycle in mitochondria. Instead of being oxidized, it is transported to the cytosol by exchange for P_i or other anions. It is cleaved to form acetyl coenzyme A and oxaloacetate in the cytosol by a separate enzyme, and the cleavage occurs at the expense of ATP.

The Citrate-Pyruvate Cycle. Now comes an even more clever part of the mechanism. The acetyl coenzyme A released in the cytosol by the cleavage of citrate is used to create fatty acid chains. The oxaloacetate that is also released is used by the malate dehydrogenase route for transfer of elecrons from NADH to NADP, and it is converted to pyruvate and CO_2 as a result (Fig. 28–7). The original source of the oxaloacetate was in the mitochondria, where it was used to make citrate. We have seen in our discussion of gluconeogenesis from lactate (p. 495) that oxaloacetate is made in the mitochondria by carboxylation of pyruvate:

$$\underset{\text{pyruvate}}{CH_3\text{–}C(=O)\text{–}COO^{\ominus}} + CO_2 + ATP + H_2O \xrightarrow[Mg^{2+},\ Mn^{2+}]{\textit{pyruvate carboxylase}} \underset{\text{oxaloacetate}}{{}^{\ominus}OOC\text{–}CH_2\text{–}C(=O)\text{–}COO^{\ominus}} + ADP + P_i$$

Therefore, the pyruvate formed in the cytosol by the sequence citrate→oxaloacetate→pyruvate is transported into mitochondria to replace the pyruvate used in the sequence pyruvate→oxaloacetate→citrate!

The citrate-pyruvate cycle simultaneously moves an acetyl group from mitochondria to the cytosol and shifts a pair of electrons from NADH to NADP. Two molecules of ATP are expended per turn of the cycle, one in carboxylating pyruvate and the other in cleaving citrate.

Integration of Citrate–Pyruvate Cycle and Fatty Acid Synthesis. This is a fascinatingly interwoven set of reactions; let us go through it again, this time in a step-by-step analysis:

1. An excess supply of glucose to the liver or adipose tissue causes an accelerated formation of pyruvate and NADH in the cytosol by the Embden-Meyerhof pathway.

2. Pyruvate is transported into mitochondria. Part is oxidized to acetyl coenzyme A and part is carboxylated to form oxaloacetate. Acetyl coenzyme A and oxaloacetate condense to form citrate.

3. Tissues synthesizing fatty acids from glucose have a relatively slow citric acid cycle; therefore, most of the citrate formed is available for exchange into the cytosol. Here we see the first of many examples of the use for other purposes of individual reactions also employed in the citric acid cycle.

4. Citrate is cleaved in the cytosol to form oxaloacetate and acetyl coenzyme A at the expense of ATP. The acetyl coenzyme A is used for fatty acid synthesis.

5. The oxaloacetate released from citrate in the cytosol is reduced to malate by the NADH formed in the Embden-Meyerhof pathway.

6. The malate is oxidatively decarboxylated to form NADPH, pyruvate, and CO_2. The NADPH is used in forming fatty acids.

7. When the pyruvate moves into mitochondria, it is effectively replacing the pyruvate formerly consumed to make oxaloacetate, so the only pyruvate now missing is that which was oxidized to form acetyl coenzyme A.

The sum of all of the reactions involved shows that the formation of 8 molecules of acetyl coenzyme A in the cytosol via pyruvate also forms 8 molecules of NADPH. We have seen that 14 molecules of NADPH are required to form palmitate. Let us now consider how the additional 6 molecules of NADPH are obtained.

PENTOSE PHOSPHATE PATHWAY

The balance of the NADPH necessary for synthesizing fatty acids, as well as the NADPH used in other processes, is made by oxidizing glucose-6-phosphate twice with NADP (Fig. 28–8). **Glucose-6-phosphate dehydrogenase** catalyzes the oxidation of C1 to a carbonyl group. The substrate is a potential aldehyde in the form of its hemiacetal, and the product is a potential acid in the form of its lactone, or inner ester. The next reaction hydrolyzes this inner ester to release **6-phospho-D-gluconate.** (Gluconic acid is the 1-carboxylic analogue of glucose.) This irreversible hydrolysis pulls the preceding dehydrogenation toward completion, so that it will generate a high NADPH concentration.

In the next reaction, 6-phosphogluconate is oxidatively decarboxylated, again using NADP as the oxidizing agent. Here, it is the accompanying decarboxylation that releases the additional free energy necessary to generate a high [NADPH]/[NADP] ratio. The oxidation occurs on C3 of the phosphogluconate, so the product is **D-ribulose 5-phosphate,** the ketose isomer of D-ribose 5-phosphate. The reaction is a typical oxidative decarboxylation of a 3-hydroxycarboxylate without release of the intermediate 3-keto compound. It is therefore of the same type as the oxidative decarboxylations of isocitrate or malate, which are also used to generate NADPH.

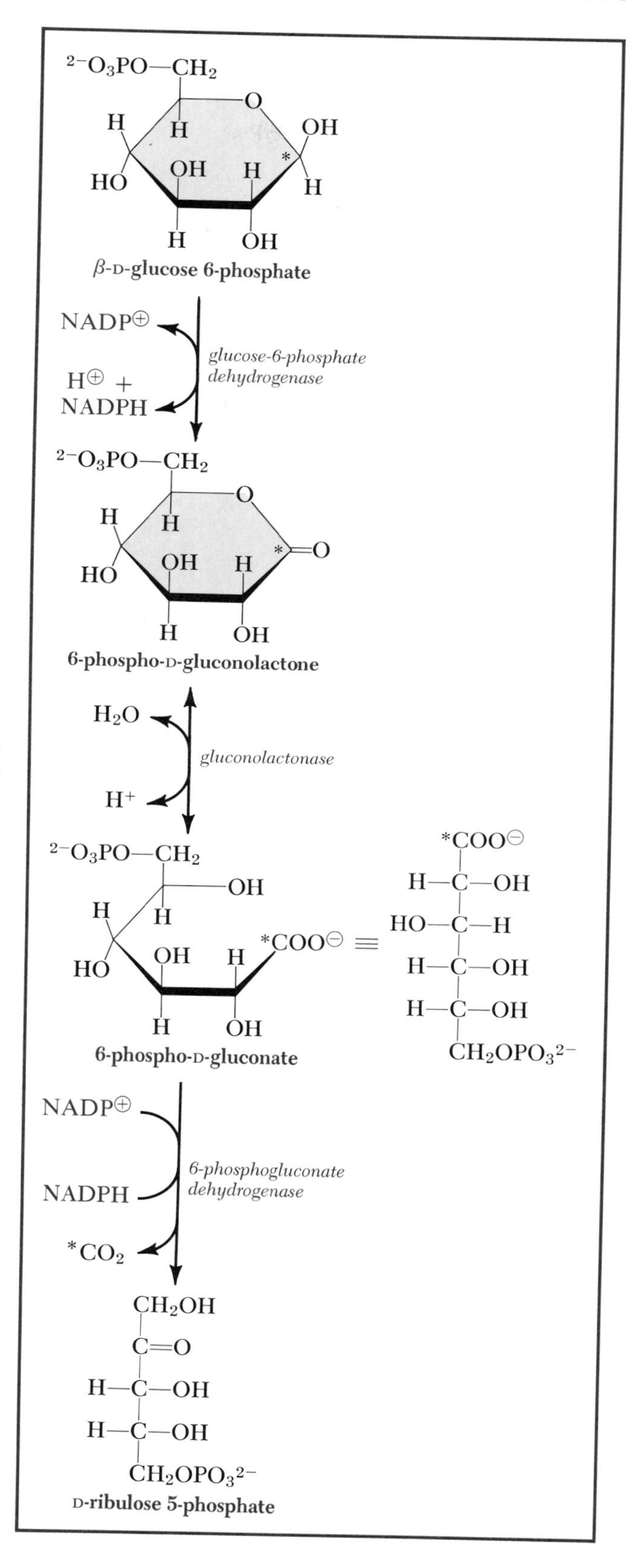

FIGURE 28–8

Oxidative decarboxylation of glucose-6-phosphate to ribulose 5-phosphate generates two molecules of NADPH. (6-Phosphogluconate is shown in both Haworth and linear formulas to illustrate its relationship to the other compounds.)

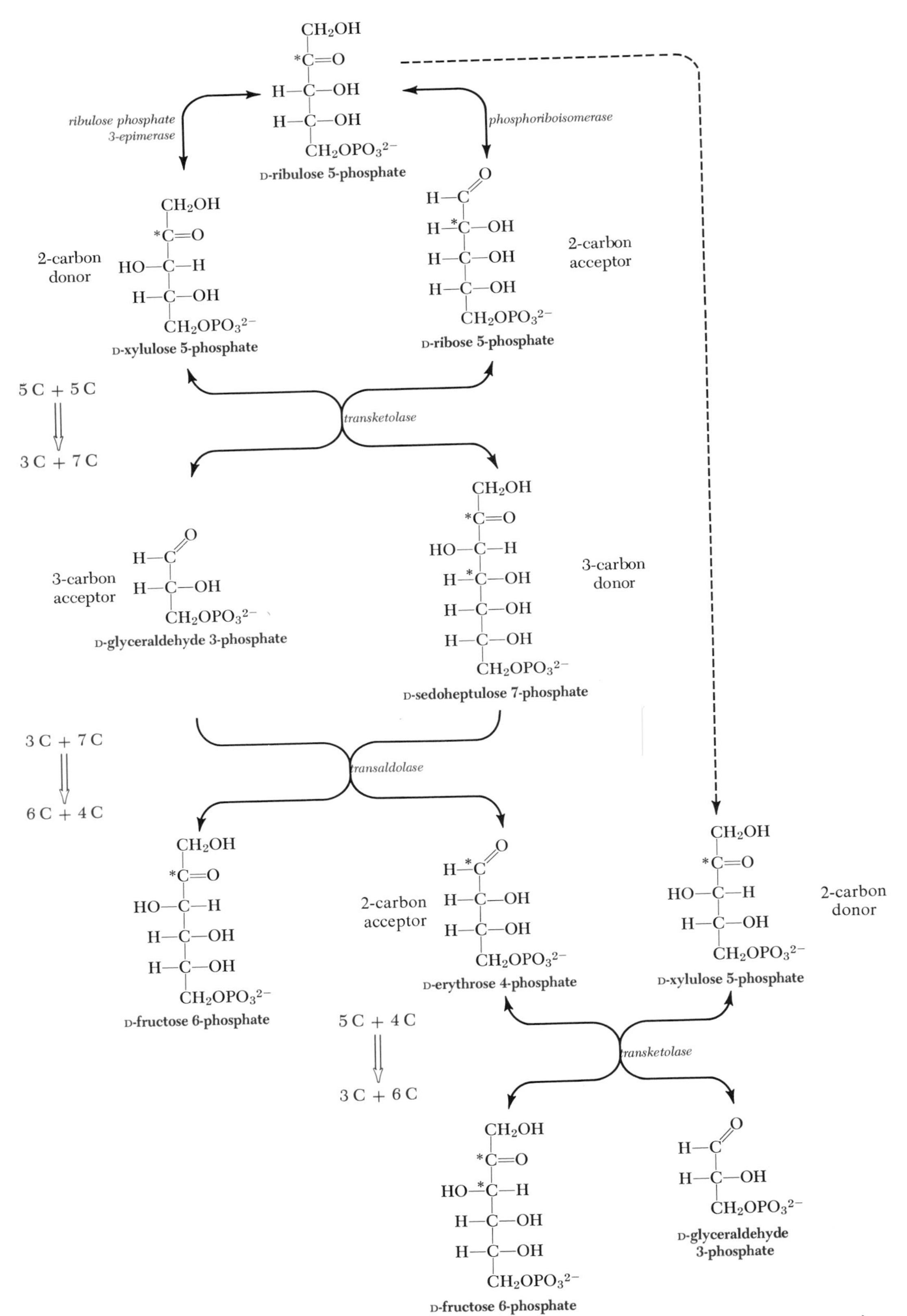

FIGURE 28–9 *See legend on opposite page*

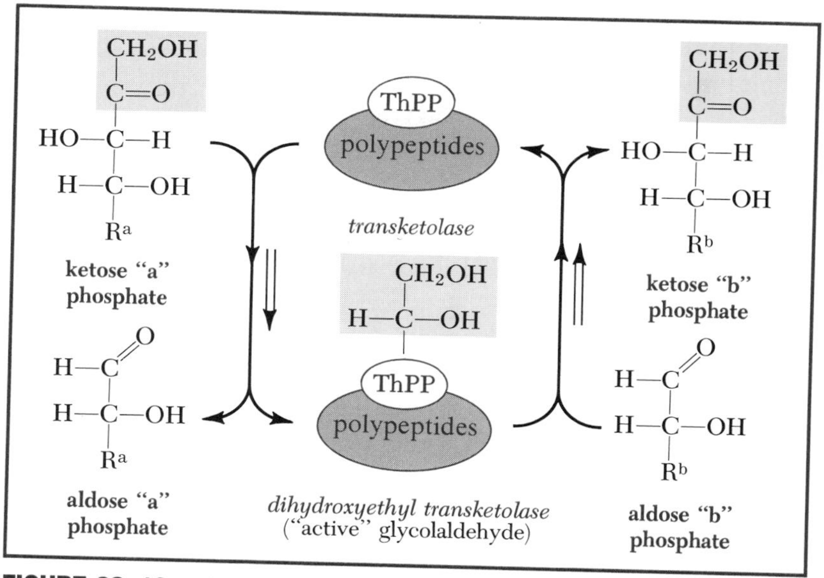

FIGURE 28–10 The action of transketolase. A two-carbon group is transferred from ketose phosphate "a" to thiamine pyrophosphate on the enzyme, leaving the remaining carbons as aldose phosphate "a." The two-carbon group is then transferred to another aldose phosphate ("b") converting it to a new ketose phosphate. The configuration of the asymmetric carbons in the ketose phosphate is said to be D-*threo* because this is the configuration found in the four-carbon aldose, D-threose (p. 181).

This sequence of reactions produces two moles of NADPH for each mole of CO_2. This is the crucial part of the overall economy of the pentose phosphate pathway: one carbon of glucose appears as CO_2 for each pair of NADPH generated. The pentose phosphate, representing five sixths of the original glucose, is recovered for further use by the remaining reactions of the pathway, which we shall now consider.

The metabolism of ribulose 5-phosphate begins with a pair of isomerizations (Fig. 28–9, top). One is a straightforward aldose-ketose isomerization, which equilibrates ribulose 5-phosphate and ribose 5-phosphate. The ribose 5-phosphate can be used to make nucleotides (p. 680), but the large quantity made during active fatty acid synthesis from glucose is handled in another way, and this requires **D-xylulose 5-phosphate,** which is formed by an epimerase that equilibrates the D- and L-configurations on C3 of ribulose 5-phosphate.

The 5-phosphates of the three D-pentoses—ribulose, ribose, and xylulose—are the substrates for a series of reactions in which the carbon atoms are reshuffled so as to wind up with hexose and triose phosphates. The redistribution involves two reactions catalyzed by one enzyme, **transketolase.** Transketolase moves two-carbon units from ketose phosphates to aldose phosphates, and it contains thiamine pyrophosphate as a coenzyme.

FIGURE 28–9 Ribulose 5-phosphate formed during the pentose phosphate pathway is recovered by converting it to fructose and glyceraldehyde phosphates. This is accomplished by successive transfers of two- and three-carbon groups, catalyzed by transketolase and transaldolase. The result is the conversion of 15 carbon atoms in the form of three molecules of ribulose 5-phosphate to two molecules of fructose 6-phosphate and one molecule of glyceraldehyde 3-phosphate.

The first two carbons of the ketose phosphate are transferred to the aldose phosphate with intermediate carriage on the thiamine pyrophosphate (Fig. 28–10). Almost any ketose and aldose phosphate are substrates, so long as the ketose has a D-*threo* configuration on the carbons adjacent to the carbonyl group, and the aldose has a D-configuration on the carbon adjacent to its carbonyl group.

5 + 5 = 3 + 7. D-Xylulose 5-phosphate is a two-carbon donor for transketolase, and D-ribose 5-phosphate is an acceptor. Transketolase therefore forms glyceraldehyde 3-phosphate and sedoheptulose 7-phosphate. (This particular seven-carbon ketose was first found as a polymer in sedums, fleshy-leafed plants; hence the name.)

3 + 7 = 6 + 4. The two products of transketolase action on the pentose phosphates are in turn substrates for **transaldolase,** which is an enzyme that catalyzes the transfer of a three-carbon unit from the ketose phosphate. The products are **fructose 6-phosphate** and **erythrose 4-phosphate.**

What do we have now? Two molecules of the original ribulose 5-phosphate have been consumed, and their ten carbon atoms have been converted to a molecule each of fructose 6-phosphate and erythrose 4-phosphate. Fructose phosphate is a compound of the Embden-Meyerhof pathway and represents full rehabilitation, as it were, of six carbons derived from pentose phosphates. However, four carbons still remain in the form of erythrose 4-phosphate. What do we do with these?

5 + 4 = 3 + 6. Erythrose 4-phosphate is an aldose phosphate and therefore is also an acceptor substrate for transketolase, becoming fructose 6-phosphate after the transfer of a two-carbon unit.

The donor of the two carbons once more is xylulose 5-phosphate, derived from ribulose 5-phosphate, and a molecule of glyceraldehyde 3-phosphate remains after the transfer. This represents the consumption of a third molecule of ribulose 5-phosphate.

Now what do we have? Beginning with three molecules of ribulose 5-phosphate, two molecules of fructose 6-phosphate and one molecule of glyceraldehyde 3-phosphate have been formed by the following transformations:

$$
\begin{array}{lrcl}
 & 5C + 5C & \longrightarrow & 3C + 7C \text{ (transketolase)} \\
 & 3C + 7C & \longrightarrow & 6C + 4C \text{ (transaldolase)} \\
 & 5C + 4C & \longrightarrow & 3C + 6C \text{ (transketolase)} \\
\hline
\text{SUM:} & 5C + 5C + 5C & \longrightarrow & 6C + 6C + 3C
\end{array}
$$

Total Stoichiometry of the Pentose Phosphate Pathway. Each of the ribulose 5-phosphate molecules that we have disposed of so neatly was derived from a molecule of glucose 6-phosphate, with a concomitant generation of CO_2 and two molecules of NADPH. The original problem was to generate six molecules of NADPH so as to complete the formation of palmitate from glucose (p. 533). In order to do this three molecules of glucose 6-phosphate must be oxidized by the pentose phosphate pathway:

$$3\text{ (glucose 6-phosphate)} + 6\text{ NADP}^+ \longrightarrow 3\text{ (ribulose 5-phosphate)} + 3\text{ CO}_2 + 6\text{ NADPH}$$

We have just seen that three molecules of ribulose 5-phosphate are converted to one molecule of glyceraldehyde 3-phosphate and two molecules of fructose 6-phosphate:

$$3\text{ (ribulose 5-phosphate)} \longrightarrow 2\text{ (fructose 6-phosphate)} + \text{glyceraldehyde 3-phosphate}$$

When we add the two equations for the overall process, the result therefore is:

3 (glucose 6-phosphate) + 6 $NADP^+$ $\longrightarrow$
2 (fructose 6-phosphate) + glyceraldehyde 3-phosphate + 3 CO_2 + 6 NADPH

This is the final stoichiometry for the complete pentose phosphate pathway. In effect, the pathway has oxidized only half of a glucose residue to CO_2, with half appearing as triose phosphate. The other two glucose residues are merely converted to fructose residues, just as they would be in the ordinary Embden-Meyerhof pathway.

THE TOTAL STOICHIOMETRY OF FATTY ACID SYNTHESIS

Having examined routes for supplying the necessary NADPH, we are now in a position to cast the complete balance for the conversion of glucose to fatty acids. The result is a beautiful demonstration that metabolism is an array of reactions, ordered in character of the intermediates and in anatomical location of the enzymes, and susceptible to rational, albeit not easy, analysis. Let us go through this one step at a time, not because the process is something to be memorized in detail, but rather so we can see the magnificent harmony it displays.

To begin, one molecule of palmitate is made from eight molecules of acetyl coenzyme A (neglecting ionic charges, water, and H^+) (p. 533).

$$\text{8 (acetyl-S-CoA)} + \text{7 ATP} + \text{14 NADPH} \longrightarrow \text{palmitate} + 7\,(\text{ADP} + \text{P}_i) + 14\,\text{NADP}^+ + 8\,(\text{CoA-SH})$$

The eight molecules of acetyl coenzyme A are provided in the cytosol along with a transfer of hydrogen from NADH to NADP by the pyruvate–citrate cycle (p. 539):

$$8\,(\text{acetyl-S-CoA}_{(mitoch.)}) + 8\,\text{NADH} + 8\,\text{NADP}^+ + 20\,\text{ATP} \longrightarrow 8\,(\text{acetyl-S-CoA}_{(cytosol)}) + 8\,\text{NAD}^+ + 8\,\text{NADPH} + 20\,(\text{ADP} + \text{P}_i)$$

(The high-energy phosphate balance here includes an assessment of 4 ~P for the transport of eight pyruvate and eight citrate molecules that were omitted to simplify an already intricate discussion. This assumes a loss of one H^+ from the mitochondrial gradient for each molecule transported, with the loss of four H^+ equivalent to a loss of one ATP. Eight ~P were used to carboxylate pyruvate, and eight more to cleave citrate.)

These acetyl coenzyme A molecules in the mitochondria are formed by oxidizing pyruvate with a concomitant oxidative phosphorylation (p. 472):

$$8\,\text{pyruvate} + 4\,\text{O}_2 + 24\,(\text{ADP} + \text{P}_i) + 8\,\text{CoA-SH} \longrightarrow 8\,(\text{acetyl-S-CoA}) + 8\,\text{CO}_2 + 24\,\text{ATP}$$

The pyruvate in turn arises from glyceraldehyde 3-phosphate by the Embden-Meyerhof pathway (p. 465):

$$8\,(\text{glyceraldehyde 3-phosphate}) + 8\,\text{NAD}^+ + 16\,\text{ADP} + 8\,\text{P}_i \longrightarrow 8\,\text{pyruvate} + 8\,\text{NADH} + 16\,\text{ATP}$$

One molecule of glyceraldehyde phosphate, along with the remaining six molecules of NADPH that are required, is formed by the pentose phosphate pathway:

$$3 \text{ (glucose 6-phosphate)} + 6 \text{ NADP}^+ \longrightarrow \text{glyceraldehyde 3-phosphate} + 2 \text{ (fructose 6-phosphate)} + 3 \text{ CO}_2 + 6 \text{ NADPH}$$

The other seven molecules of glyceraldehyde phosphate are formed by the usual steps of the Embden-Meyerhof pathway:

$$3\tfrac{1}{2} \text{ (fructose 6-phosphate)} + 3\tfrac{1}{2} \text{ ATP} \longrightarrow 7 \text{ (glyceraldehyde 3-phosphate)} + 3\tfrac{1}{2} \text{ ADP}$$

Of the $3\tfrac{1}{2}$ fructose 6-phosphate required, only 2 are made by the pentose phosphate pathway; the other $1\tfrac{1}{2}$ are made by the usual isomerase reaction:

$$1\tfrac{1}{2} \text{ (glucose 6-phosphate)} \longrightarrow 1\tfrac{1}{2} \text{ (fructose 6-phosphate)}$$

Finally, the total of $4\tfrac{1}{2}$ molecules of glucose-6-phosphate is made from glucose:

$$4\tfrac{1}{2} \text{ glucose} + 4\tfrac{1}{2} \text{ ATP} \longrightarrow 4\tfrac{1}{2} \text{ (glucose 6-phosphate)} + 4\tfrac{1}{2} \text{ ADP}$$

Adding all of these steps together we get a grand total:

$$4\tfrac{1}{2} \text{ glucose} + 4 \text{ O}_2 + 5 \text{ (ADP} + \text{P}_i) \longrightarrow \text{palmitate} + 11 \text{ CO}_2 + 5 \text{ ATP}$$

How about that! Everything works out, and there is even a small production of high-energy phosphate. Essentially, the balance states that adipose tissue consumes 27 carbon atoms as glucose to store 16 carbon atoms as palmitate, with the remaining 11 appearing as CO_2. It does so in such a way that the eight moles of acetyl coenzyme A that must be transported from mitochondria to cytosol exactly equals the number of pairs of electrons that must be transferred from NADH to NADP in the cytosol. The citrate moving out of mitochondria supplies both needs in the cytosol, and pyruvate returning replaces the oxaloacetate used in making citrate, so there is no need for a diffusion of oxaloacetate across the mitochondrial membrane. The entire process is shown as a flow sheet in Figure 28–11.

FORMATION OF TRIGLYCERIDES

Fatty acids, whether obtained by hydrolysis of dietary fat or by synthesis from glucose, are converted to triglycerides for transport to, and for deposition in, adipose tissue. The fatty acids are first converted to their coenzyme A thioesters (p. 441). The acyl groups are then transferred to the hydroxyl groups of glycerol, which is initially provided in the form of dihydroxyacetone phosphate. Let us look at these processes in detail.

The Initial Acylation. There are two routes for placing an initial acyl group on what will be the glyceryl backbone of a triglyceride (Fig. 28–12). In one, the acyl group is transferred to dihydroxyacetone phosphate before reduction to create a glycerol residue. In the other route, reduction of dihydroxyacetone phosphate to glycerol phosphate occurs first, with the acyl group then added. The ultimate source of the glycerol residue is glucose, or some gluconeogenic precursor in either case. The monoacyl glycerol phosphate product is trivially named as a **lysophosphatidate.** (The

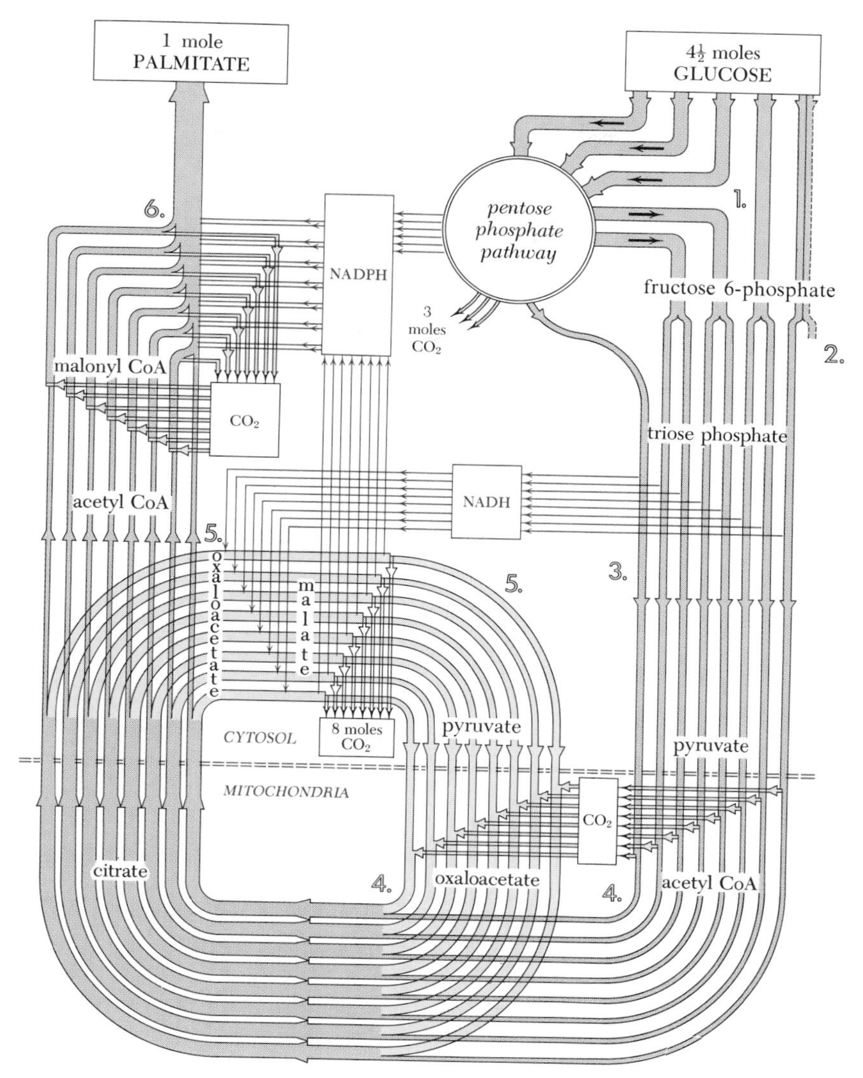

FIGURE 28–11 Flow sheet for the synthesis of palmitate from glucose. The flow of carbons appearing in palmitate is shown by *heavy shading. Lighter shading* indicates the recycling of oxaloacetate via pyruvate from the cytosol to the mitochondria. Transfers of electrons are shown in *brown.*

(*1*) Four and a half moles of glucose are required to make one mole of palmitate; three of these pass through the pentose phosphate pathway to generate six moles of NADPH. The remainder are converted directly to fructose 6-phosphates by the Embden-Meyerhof pathway.

(*2*) Two additional moles of fructose 6-phosphate are formed by the pentose phosphate pathway, making a total of $3\frac{1}{2}$ moles, which are converted to seven moles of triose phosphate. Another mole of triose phosphate is formed by the pentose phosphate pathway.

(*3*) The eight moles of triose phosphate created in step 2 are converted to pyruvate by the Embden-Meyerhof pathway, with the formation of eight moles of NADH.

(*4*) Eight moles of pyruvate are oxidized to acetyl coenzyme A in the mitochondria, and the acetyl coenzyme A combines with oxaloacetate to form citrate (8 moles).

(*5*) The citrate passes from the mitochondria into the cytosol, where it is cleaved to acetyl coenzyme A and oxaloacetate. The oxaloacetate is reduced to malate, using the eight moles of NADH formed in step 2. The malate is then oxidized to pyruvate by the NADP-coupled dehydrogenase to form an additional eight moles of NADPH. The pyruvate moves back into the mitochondria, where it is carboxylated to regenerate oxaloacetate consumed in step 3.

(*6*) The eight moles of acetyl coenzyme A transported into the cytosol as citrate are carboxylated (7 moles) and condensed to form palmitate. The necessary reductions consume the NADPH generated in steps 1 and 4.

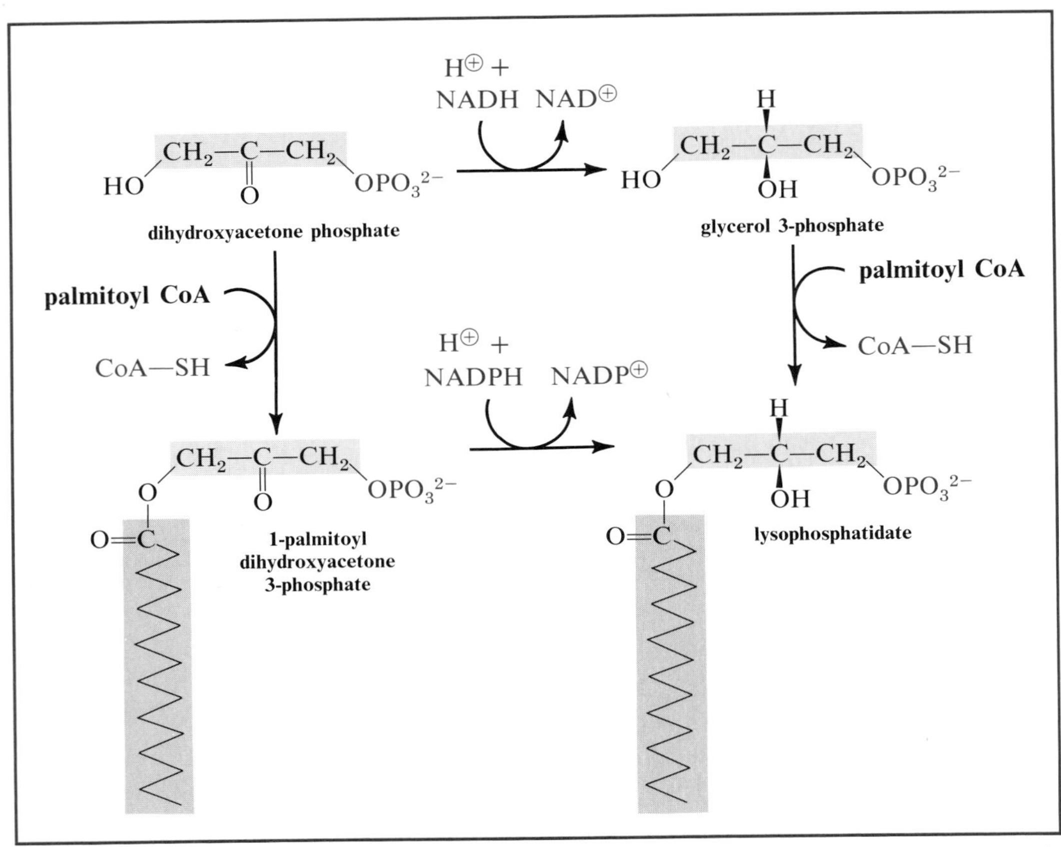

FIGURE 28–12 The first fatty acid incorporated during triglyceride synthesis is usually palmitate. The palmitoyl group may be placed on dihydroxyacetone phosphate before or after it is reduced to glycerol 3-phosphate.

lyso- prefix is given by analogy with lysolecithin, a monoacyl glyceryl phosphate ester of choline, which has the property of causing lysis of cells by its detergent action.)

The initial acylation preferentially involves a saturated fatty acid, and this is true with either dihydroxyacetone phosphate or glycerol 3-phosphate as the acceptor for the acyl group. Figure 28–12 shows a palmitoyl group, the most common at position 1, being transferred.

Since the ultimate precursors and products appear to be the same in the two routes, except for the electron carriers, it is not certain why both routes exist. There is some evidence that the direct acylation of dihydroxyacetone phosphate is preferred at times of glucose deprivation.

The Second and Third Acylations. An unsaturated fatty acid residue is transferred from its coenzyme A derivative to the free hydroxyl group at position 2 of lysophosphatidates, forming **phosphatidates,** the 1,2-diacyl esters of glycerol 3-phosphate (see figure at top of next page). This is true in nearly all tissues, but there are interesting exceptions, including human mammary glands, in which a palmitate residue is transferred. A protein may be present in these tissues to combine with the acyl transferase and modify its specificity during triglyceride synthesis.

The phosphatidate formed by the second transfer is then attacked by a specific phosphatase, exposing the final hydroxyl group of the glyceryl backbone. Still an-

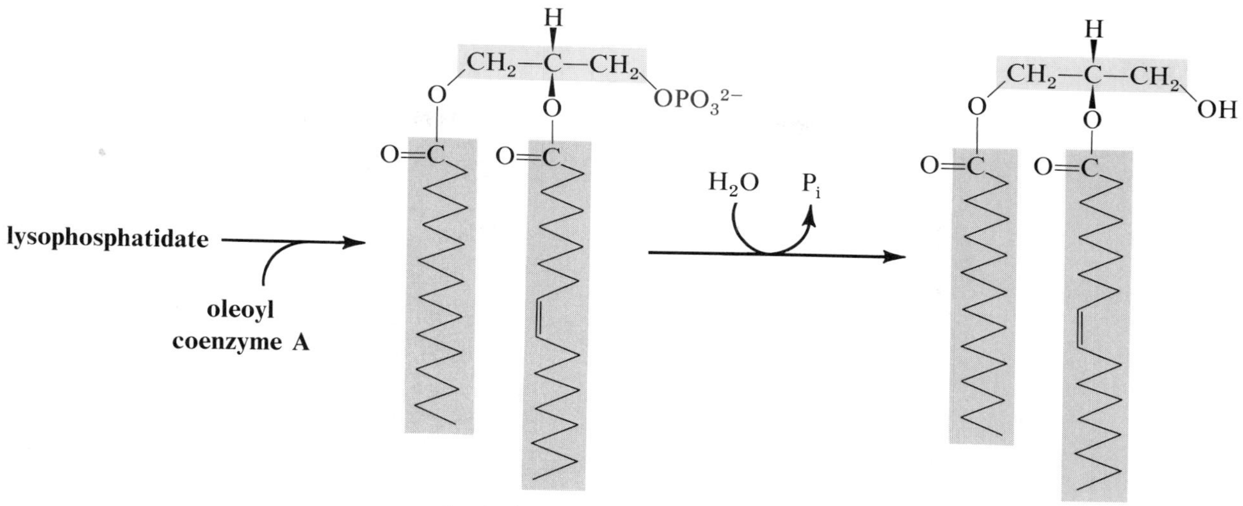

lysophosphatidate
oleoyl
coenzyme A
H
CH2—C—CH2
O
O
OPO3 2−
O=C
O=C
H2O
Pi
H
CH2—C—CH2
O
O
OH
O=C
O=C

phosphatidate 1,2-diglyceride

other acyl transferase, this one specific for a diglyceride acceptor, then adds the final fatty acid residue, which may be either saturated or unsaturated:

diglyceride
oleoyl
coenzyme A
H
CH2—C—CH2
O
O
O
O=C
O=C
O=C

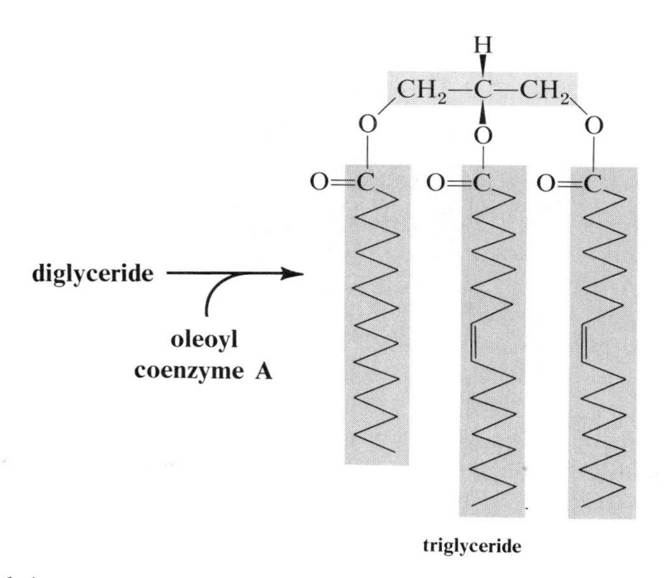

triglyceride

The result in a typical human will be a mixture of triglycerides with the following acyl groups comprising most of the total:

20% 16:0 (palmitoyl)
7% 16:1 (palmitoleoyl)
50% 18:1 (oleoyl and vaccenoyl)
10% 18:2 (linoleoyl)

This is not necessarily a reflection of the relative synthesis of the various types of fatty acids. It should not be forgotten that dietary fatty acids are frequently incorporated into triglycerides without change, and the composition of the diet partly determines the character of the stored fat. This is shown strikingly by experiments in which individuals were fed corn oil, linseed oil, or coconut oil for periods of a year or

TABLE 28–1 COMPOSITION OF TRIGLYCERIDES IN PLANT OILS AND IN ADIPOSE TISSUE OF HUMANS EATING THE OILS*

	Percentage of Total Fatty Acid Residues						
Fatty Acid	*Humans on Random American Diet*	*Corn Oil*	*Humans with 40% of Energy Source as Corn Oil for 3 Years*	*Linseed Oil*	*Humans Eating 83 g of Linseed Oil Daily for 1 Year*	*Coconut Oil*	*Humans Eating 60 g of Coconut Oil for 1.5 Years*
8:0						8	
10:0						10	
12:0	0.7		0.1		0.1	47	14.5
14:0	3.3		0.7		1.2	16	13.9
16:0	19.5	13	15.3	6	14.7	8	17.5
16:1	6.9		2.2		5.8	1	7.6
18:0	4.2	4	2.2	4	5.4	2	2.7
18:1	46.3	29	32.1	22	35.5	6	30.4
18:2	11.4	54	45.2	16	20.5	2	9.3
18:3	0.4			52	13.7		0.1

*Data for adipose tissue from J. Hirsch: (1965) *Handbook of Physiology*. p. 148, section 5. American Physiological Society. Data for oil composition from H. E. Longenecker: (1959) J. Biol. Chem., *130*: 167, and E. Fedeli, and G. Jacini: (1971) Adv. Lipid Res., *9*: 335.

more, and the fatty acid composition of their subcutaneous adipose tissue was compared with that of individuals eating the usual random American diet. The fatty acids of these oils and the resultant changes in the adipose tissue are shown in Table 28–1.

Several things emerge from these data. There is little chain lengthening beyond 18 carbon atoms. The data with a random diet show that the liver isn't very fancy in its selection of fatty acids for storage. The palmitoyl and oleoyl residues make up nearly two thirds of the total and linoleate from the diet is the next most abundant residue. There is little introduction of additional double bonds beyond the first needed to supply the oleoyl and palmitoleoyl residues. The fact that a diet high in 18:2 or 18:3 fatty acids leads to a storage of these residues in adipose tissue suggests that there is no inherent screen against polyunsaturated fatty acids. Eating coconut oil, with its abundance of short chain residues, does lead to an increased storage of 12:0 and 14:0 residues, but most of the 8:0 and 10:0 residues found in the food have evidently been modified or disposed of elsewhere.

REGULATION OF FATTY ACID SYNTHESIS

The control of fuel metabolism in the liver requires decisive choices between major metabolic routes. Shall glucose be converted to pyruvate for fat synthesis, using the Embden-Meyerhof pathway, or shall pyruvate be converted to glucose through gluconeogenesis? Shall acetyl coenzyme A be converted to fatty acids by making malonyl coenzyme A, or shall fatty acids be oxidized to acetyl coenzyme A to provide energy for the liver and perhaps fuel for other tissues in the form of acetoacetate and 3-hydroxybutyrate?

The principal device for switching metabolic routes is phosphorylation and dephosphorylation of key enzymes. These gross changes are supplemented with allosteric effects for fine tuning and further insurance against misuse of the fuel supply.

The overall purpose is to avoid futile cycles, while providing the capability to both synthesize and mobilize stored fat.

Phosphorylations inhibit fatty acid synthesis and promote gluconeogenesis. Glucagon promotes these phosphorylations through activation of the cyclic AMP-dependent protein kinase in the liver. **Catecholamines,** such as **noradrenaline** released by the sympathetic nerves, promote the phosphorylations by activating other protein kinases. The mechanism of this action of catecholamines is unknown; it involves a combination with receptors (α-adrenergic receptors) that are different from β-adrenergic receptors responsible for activation of adenylate cyclase. (Adrenergic receptors are discussed in Chapter 37.)

Insulin promotes fatty acid synthesis and inhibits gluconeogenesis through activation of specific protein phosphatases, thus reversing the effects of glucagon or catecholamines.

Regulation of the Fructose 6-Phosphate/Fructose 1,6-Bisphosphate Cycle. The liver isozymes of phosphofructokinase and fructose bisphosphatase, like the muscle isozymes, are affected by citrate and AMP, but the major effector may be the recently discovered **fructose 2,6-bisphosphate,** which activates phosphofructokinase and inhibits fructose bisphosphatase. It is formed from fructose 6-phosphate:

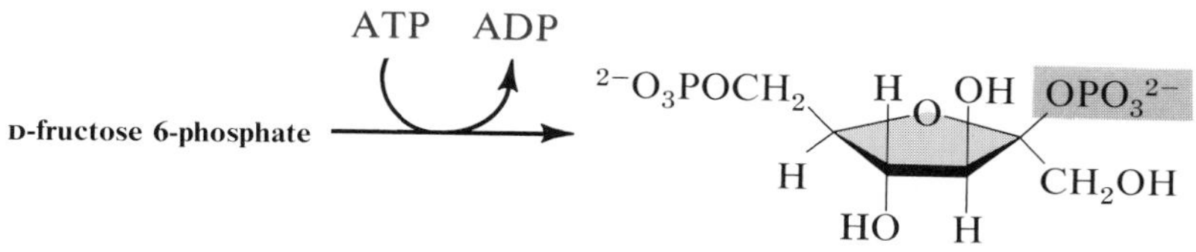

ATP ADP

D-fructose 6-phosphate →

$^{2-}O_3POCH_2$ H O OH OPO_3^{2-}

H CH_2OH

HO H

β-D-fructose 2,6-bisphosphate

Glucagon or catecholamines block its formation. They do so by stimulating protein kinases to phosphorylate the enzyme catalyzing the reaction. (This regulatory enzyme is provisionally named phosphofructokinase-2, to distinguish it from the familiar enzyme of the Embden-Meyerhof pathway, which can be designated more specifically as phosphofructokinase-1.) Phosphorylation inactivates phosphofructokinase-2, so the end result of glucagon or catecholamine action is to promote the conversion of pyruvate to glucose-6-phosphate at the fructose bisphosphatase step. The phosphofructokinase-2 is presumably dephosphorylated by an insulin-responsive phosphatase; if so, insulin blocks gluconeogenesis at this step and promotes fatty acid synthesis from glucose at the phosphofructokinase-1 step, by increasing the formation of fructose 1,6-bisphosphate, therefore of triose phosphates.

Regulation of the Pyruvate/Phospho-*enol*-pyruvate Cycle. Gluconeogenesis in the liver requires the conversion of pyruvate to phospho-*enol*-pyruvate via oxaloacetate (p. 494). Fatty acid synthesis requires the conversion of phospho-*enol*-pyruvate directly to pyruvate by pyruvate kinase, and this is the controlled enzyme. The liver isozyme of **pyruvate kinase** is inactivated through phosphorylations catalyzed by the glucagon- and catecholamine-responsive protein kinases. It is activated through dephosphorylation catalyzed by an insulin-responsive protein phosphatase. The V_{max} of pyruvate kinase is much greater than the opposing maximum rate of gluconeogenesis, and pyruvate kinase is specifically activated by fructose 1,6-bisphosphate as an additional signal that there is an ample carbohydrate supply. Therefore, when the blood insulin concentration is high and the blood glucagon concentration is low, there will be a net formation of pyruvate from phospho-*enol*-pyruvate.

Regulation of Pyruvate Dehydrogenase. The pyruvate dehydrogenase of hepatic mitochondria, which forms the acetyl coenzyme A necessary for fatty acid synthesis, is regulated by phosphorylation and dephosphorylation in the same way as is the muscle enzyme (p. 481); they may be identical. In any event, the enzyme is clearly activated by an insulin-responsive protein phosphatase, but the physiological controls over its inactivation through phosphorylation are less clear.

Regulation of the Acetyl Coenzyme A/Palmitate Cycle. Acetyl CoA carboxylase catalyzes the first committed step in fatty acid synthesis, and regulation of this enzyme controls all of the remaining steps. It, too, is inactivated through phosphorylation by the hormone-responsive protein kinases, and is activated through dephosphorylation by an insulin-responsive protein phosphatase. In addition, the enzyme has an absolute requirement for citrate. Not only is citrate used as the source of acetyl coenzyme A in the cytosol, it also signals the availability of ample supplies of oxaloacetate, and therefore pyruvate. (If acetyl coenzyme A itself were the signal, the carboxylase would be turned on when the liver is already flooded with fatty acids and rapidly oxidizing them.) Acetyl coenzyme A carboxylase is inhibited by malonyl coenzyme A, so it will only function as fast as is necessary to supply fatty acid synthase. It is also inhibited by palmitoyl coenzyme A so as to prevent the synthesis of fatty acids faster than they can be incorporated into triglycerides.

Finally, malonyl coenzyme A inhibits carnitine palmitoyl transferase on the outer face of the inner mitochondrial membrane. Thus, fatty acids will not be sent into the mitochondria for oxidation whenever acetyl coenzyme A carboxylase is turned on for fatty acid synthesis.

GLUCONEOGENESIS OR FATTY ACID SYNTHESIS?

Summary of Regulation by Phosphorylation and Dephosphorylation of Enzymes

Glucagon or catecholamines stimulate protein kinases to phosphorylate.

Insulin stimulates a protein phosphatase to dephosphorylate.

Phosphorylation activates:
- **phosphofructokinase-2 to make fructose 2,6-bisphosphate**
  - **fructose 2,6-bisphosphate inhibits fructose 1,6-bisphosphatase, and therefore inhibits gluconeogenesis**
  - **fructose 2,6-bisphosphate stimulates phosphofructokinase-1 to make fructose 1,6-bisphosphate, and therefore promotes fatty acid synthesis**
    - **fructose 1,6-bisphosphate acts as a signal to stimulate pyruvate kinase**

Phosphorylation inhibits:
- **pyruvate kinase from converting phospho-*enol*-pyruvate to pyruvate, which blocks fatty acid synthesis**
- **pyruvate dehydrogenase from oxidizing pyruvate to acetyl coenzyme A, which blocks fatty acid synthesis**
- **acetyl coenzyme A carboxylase from forming malonyl coenzyme A, which blocks fatty acid synthesis**

Desaturation. The mechanism of control of the rate of introduction of double bonds into the fatty acid chains is unknown, but the effects are clear. More double bonds appear in the stored triglycerides and in the structural lipids when the environmental temperature falls. Since double bonds lower the melting point, this provides a mechanism to prevent crystallization of the lipids in the tissue.

Furthermore, polyunsaturated compounds that can be completely synthesized from acetyl coenzyme A—those with a seven- or nine-carbon tail—are not formed if there is an adequate dietary supply of linoleate and linolenate, or some other precursors with only three- or six-carbon saturated tails. These essential fatty acids are preferentially used for the construction of structural lipids when they are available; when they are not, more of the synthetic substitutes will be formed.

BROWN ADIPOSE TISSUE

Animals can generate extra heat without shivering by stimulating a specialized kind of adipose tissue. This tissue is brown because it is rich in mitochondria; less than half of its mass is triglycerides. Increased concentrations of catecholamines, either from the blood or released by the many sympathetic nerve endings in the tissue, cause a sharp increase in oxidation of fatty acids with a concomitant production of heat. Infants are especially dependent upon brown adipose tissue for temperature regulation. Much of it disappears with age, but significant amounts are still present in many adults (Fig. 28–13).

Increased oxidation without performing additional work represents an effective uncoupling of oxidation and phosphorylation. The mitochondria of brown adipose tissue contain large amounts of a protein not detected in other mitochondria; this protein closely resembles the ATP-ADP antiporter in amino acid sequence, but instead it acts as a proton transporter, undoing the work of the electron transfer system by moving protons from the cytosol back into the mitochondria. This discharge of the electrochemical gradient allows additional electron transfer to occur without formation of high-energy phosphate.

A current notion of the way in which heat generation is regulated hinges upon inhibition of the proton transporter by ATP and relief of the inhibition by AMP. Exposure of the brown adipocyte to noradrenaline causes the ATP concentration to

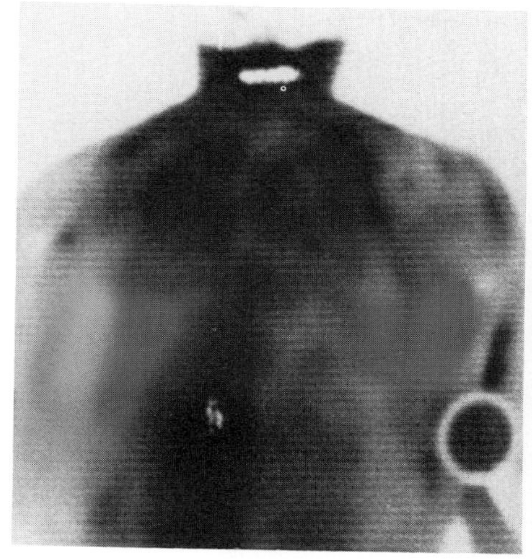

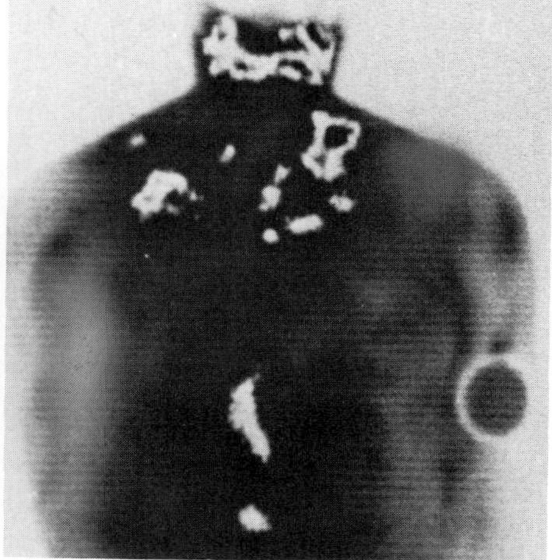

FIGURE 28–13 Location of superficial brown adipose tissue in a normal 36-year-old male. The white patches are areas of the skin with a temperature greater than 34.1° before (*left*) and after (*right*) administration of a catecholamine to provoke thermogenesis. (From N.J. Rothwell and M.J. Stock (1979), Nature 281:31. Reproduced by permission.)

fall and the AMP concentration to rise, thus increasing the leakage of protons back into the mitochondria. How this is done is not certain. Some conjecture that noradrenaline in some way stimulates the Na^+,K^+-ATPase of the plasma membrane, perhaps by a cation leak, and the ensuing consumption of ATP is sufficient to change the concentration. (The brown adipose tissue mitochondria have much less ATP synthase than do other mitochondria.)

FURTHER READING

Routes and Enzymes

K. Stoops and S.M. Wakil (1981). Animal Fatty Acid Synthetase. A Novel Arrangement of the β-Ketoacyl Synthetase Sites Comprising Domains of the Two Subunits. J. Biol. Chem. 256:5128.

M.R. Pollard et al. (1980). The $\Delta 5$ and $\Delta 6$ Desaturation of Fatty Acids of Varying Chain Length by Rat Liver: A Preliminary Report. Lipids 15:690.

T. Okayasu et al. (1981). Purification and Partial Characterization of Linoleoyl CoA Desaturase from Rat Liver Microsomes. Arch. Biochem. Biophys. 206:21.

P. Madvig and S. Abraham (1980). Relationship of Malic Enzyme Activity to Fatty Acid Synthesis and the Pathways of Glucose Catabolism in Developing Rat Liver. J. Nutr. 110:90ff.

W. Blom, S.M.P.F. de Muinck Keizer, and H.R. Scholte (1981). Acetyl-CoA Carboxylase Deficiency: An Inborn Error of De Novo Fatty Acid Synthesis. N. Engl. J. Med. 305:465. Report of infant with severe brain and muscle damage.

Regulation

E. Van Schaftingen, D.R. Davies, and H.G. Hers (1981). Inactivation of Phosphofructokinase 2 by Cyclic-AMP Dependent Protein Kinase. Biochem. Biophys. Res. Comm. 103:362.

S.J. Pilkis et al. (1981). The Role of Fructose 2,6-Bisphosphate in Regulation of Fructose-1,6-bisphosphatase. J. Biol. Chem. 256:11489.

S. Ly and K.-H. Kim. (1981). Inactivation of Hepatic Acetyl-CoA Carboxylase by Catecholamine and Its Agonists through the α-Adrenergic Receptor. J. Biol. Chem. 256:11585.

J.B. Blair (1981). Regulatory Properties of Hepatic Pyruvate Kinase. p. 121 in C.M. Veneziale, ed. The Regulation of Carbohydrate Formation and Utilization in Mammals. University Park Press.

P. Cohen, ed. (1980). Recently Discovered Systems of Enzyme Regulation by Reversible Phosphorylation. Elsevier.

R. Jeffcoat (1979). The Biosynthesis of Unsaturated Fatty Acids and Its Control in Mammalian Liver. Essays Biochem. 15:1.

R.M. Denton and A.P. Halestrap (1979). Regulation of Pyruvate Metabolism in Mammalian Tissues. Essays Biochem. 15:37.

Brown Adipose Tissue

W.P.T. James and P. Trayhurn (1981). Thermogenesis and Obesity. Br. Med. Bull. 37:43. Includes review of brown adipose tissue structure and function.

N.J. Rothwell and M.J. Stock (1979). A Role for Brown Adipose Tissue in Diet-Induced Thermogenesis. Nature 281:31.

CHAPTER 29

TURNOVER OF FATS AND LIPOPROTEINS: THE CHOLESTEROL CONNECTION

Deposition of Fat

Triglycerides are stored directly in the muscle fibers that will use them as fuels, or in the adipose tissue. They are transported to those tissues as chylomicrons or very low-density lipoproteins, which are small droplets coated with a mixture of proteins, phospholipids, and cholesterol.

Chylomicrons are formed by the small bowel; they appear first in its lymphatic drainage and then in the blood via the thoracic duct. These droplets are so large that they give a milky appearance to the blood plasma after a fat-rich meal, and will float to the top of the plasma in a test tube left standing. The triglycerides in chylomicrons are synthesized by the intestinal mucosa, using fatty acids absorbed from digested food. The fatty acids are first converted to acyl coenzyme A derivatives; they may be used for complete synthesis of a new triglyceride (p. 546) or added to absorbed monoglycerides:

$$\text{acyl coenzyme A} + \text{2-monoacylglycerol} \longrightarrow \text{1,2-diglyceride} + \text{CoA-SH}$$

The diglyceride is then converted to a triglyceride in the usual way.

Very low-density lipoproteins transport fat from the liver. These droplets are large enough to make blood plasma opalescent when they are in high concentrations but not so large as to float into a separate layer without centrifugation. The secreted triglycerides may contain fatty acids synthesized by the liver from glucose (or other sources of acetyl coenzyme A), or fatty acids extracted from the blood by the liver.

Fatty acids are removed by the action of lipases from both chylomicrons and very low-density lipoproteins as they pass through capillaries. The principal sites are the muscles and adipose tissue. The capillaries in these tissues are coated with a **lipoprotein lipase,** which hydrolyzes either the 1 or the 3 ester bond of tri- and diglycerides, releasing free fatty acid anions, which are carried through the plasma membrane of the recipient cells by a protein carrier. The remaining monoglyceride is hydrolyzed by a monoacylglycerol lipase, which is free in blood plasma, and is also found in blood platelets and within the recipient cells.

After fatty acids move into the cell, they are converted into the coenzyme A esters and then into triglycerides for storage.

The lipoprotein lipase that initiates fat uptake from the blood is synthesized in the interior of either the capillary endothelial cells or the parenchymal cells and then moves outside, where it adheres tightly to heparan-like polysaccharide chains (p. 197) on the endothelial cell surfaces. This lipase is activated upon contact by a specific protein present in chylomicrons and very low-density lipoproteins. (The lipase can also be activated by heparin, which displaces it from the endothelial cell membranes into the blood. Heparin has long been known to have a fat-clearing action, but it is uncertain if this is among its normal physiological functions.)

Mobilization of Fat

Stored fatty acids are released from adipose tissue by a hormone-sensitive lipase. This enzyme, which is distinct from lipoprotein lipase, is activated through phosphorylation by cAMP-dependent protein kinase. It will hydrolyze any and all of the bonds in triglycerides. The first hydrolysis is rate-limiting, because the enzyme acts 10 times as fast on diglycerides and 4 times as fast on monoglycerides as it does on triglycerides. (It also acts effectively on cholesteryl esters of fatty acids.) Monoglycerides do not accumulate, because the adipocytes also contain a separate monoacylglycerol lipase to help things along.

The fatty acids released from adipocytes are bound to serum albumin for transport in the blood. Serum albumin is the most abundant protein in the blood plasma; 35 to 50 grams are present per liter. It has two sites with a high affinity for fatty acids, and five more sites of moderate affinity. The fatty acid load in blood plasma ranges from 0.5 to 0.8 millimole per liter in an individual at rest after an overnight fast, and half or less of these values after eating. However, these relatively low concentrations do not reflect the true fatty acid traffic in the blood. Turnover is rapid, and the blood concentration is the result of a balance between the rate of release from adipose tissue and the rate of consumption by other tissues, especially working muscles.

Fate of Glycerol. Adipose tissue cannot use the glycerol released by hydrolysis of fats. It passes into the blood and is taken up by the liver, which has a glycerokinase catalyzing the reaction:

$$\text{glycerol} + \text{ATP} \longrightarrow \text{glycerol 3-phosphate} + \text{ADP}$$

The glycerol phosphate may be used to form triglycerides or to add to the dihydroxyacetone phosphate supply, since the two compounds are equilibrated in the cytosol by glycerol phosphate dehydrogenase. Glycerol can, therefore, be converted to glucose or to fatty acids, or it can be used as a fuel.

Regulation of Fat Turnover

Fat metabolism in the various organs is coordinated through hormonal and neural controls. **Insulin promotes fat storage and blocks fat release in adipocytes; it favors synthesis of fats from glucose in the liver, and it permits the use of glucose by the muscles.** The general effect of insulin is therefore to promote the use of glucose and the conservation of fat. Burn the wood and spare the oil, as it were.

Insulin's stimulation of glucose transport into adipocytes in itself promotes fat storage. Glucose is necessary as a source of the glycerol residue on which triglycerides are constructed. Also, more lipoprotein lipase appears in adipose tissue in response to continued elevation of insulin concentration, and this response may be caused by glucose metabolites within the cell.

The site of fat deposition is regulated in part by the nature of lipoprotein lipases. The enzyme in muscle capillaries has a high affinity for the triglyceride-bearing lipoproteins, and therefore can extract fat from the blood lipoproteins even when they are at relatively low concentrations. The enzyme made by adipocytes has a lower affinity (higher K_m), and therefore becomes more effective when the supply of triglycerides is abundant. The engine, not the storage tank, is served first when fuel is limiting.

Fats are mobilized from the adipocytes in response to catecholamines released from the sympathetic nerve endings or arriving in the blood. This stimulus frequently coincides with the onset of muscular work that will require fuel. Catecholamines activate adenylate cyclase in the adipocyte plasma membrane to produce cyclic AMP. The cAMP-dependent protein kinase then phosphorylates adipose tissue lipase, making it active.

Insulin antagonizes the effects of catecholamines on adipocytes, apparently by stimulating the hydrolysis of cyclic AMP in this way: The cyclic nucleotide phosphodiesterase requires calcium-loaded calmodulin for activity, but the calcium concentration in the cell is ordinarily kept very low by an ATP-driven calcium pump in the plasma membrane. Insulin appears to block the action of the pump and permit a constant leakage of calcium ion to raise the internal calcium ion concentration. This would account for the known activation of the cyclic nucleotide phosphodiesterase that follows exposure of adipocytes to insulin, and therefore for the hormone's inhibition of fat mobilization.

The fate of fatty acids depends upon circumstances. A rise in their blood concentration promotes their use by skeletal muscles, so most of the fatty acids released from adipocytes will be burned by the muscles during exercise. The rising concentration also promotes uptake of free fatty acids in the liver. Since mobilization of fatty acids ordinarily occurs when the blood insulin concentration is low and the blood glucagon concentration is high, owing to a diminished glucose supply, the liver will not be synthesizing fatty acids. Fatty acid oxidation will therefore not be blocked by the inhibition of the carnitine transport system with malonyl CoA. The fatty acyl groups will be oxidized in the liver mitochondria, forming acetoacetate and D-3-hydroxybutyrate. These ketone bodies are transported to the muscles and also used as fuel. The blood concentrations of neither the ketone bodies nor the free fatty acids become very high so long as exercise continues. When it ceases, both will transiently rise until stimulation of the adipose tissue also ceases.

At rest after an overnight fast, the liver removes approximately one third of the circulating fatty acids that are slowly being released from the adipose tissue. It sends approximately 60 per cent of the carbon atoms back into the blood—40 per cent as ketone bodies, and 20 per cent as triglycerides in very low-density lipoproteins. The fatty acid oxidation is sufficiently slow at rest to permit the fatty acyl coenzyme A concentrations to rise beyond the level at which triglyceride synthesis occurs.

LIPOPROTEIN TURNOVER

Structure and Composition

Three major kinds of lipoproteins are secreted into the blood: chylomicrons formed by the intestinal mucosa, very low-density lipoproteins formed by the liver, and high-density lipoproteins formed by the liver. All have the same general structure upon maturation in the circulation, with a hydrophobic core of triglycerides and cholesteryl esters surrounded by an amphipathic shell of apoproteins, phospholipids, and free cholesterol. The phospholipids are mainly phosphatidylcholines and sphingomyelins. Appropriate lipids must be available for the synthesis of the lipoproteins, but it is the apoprotein components that enable formation of spherical droplets rather than bilayer vesicles or small micelles. Some apolipoproteins, perhaps all, contain segments of a-helix with polar side chains aligned along one side and nonpolar side chains on the other. This arrangement facilitates formation of interfaces between the hydrophobic core and the polar surroundings.

Many kinds of apolipoproteins are known; they are grouped according to function and occurrence rather than by defined evolutionary kinship. Their nomenclature is volatile, but class designations in common use are **apo A, B, C,** and **E.** Each class may consist of several different proteins, for example, **A-I, II, III, IV** and **C-I, II, III.** (The former D is now A-III.) The liver makes representatives of each class of apoproteins, but the intestinal mucosa makes only A and B apoproteins.

The triglyceride-rich chylomicrons and very low-density lipoproteins are stabilized by B apoproteins. The liver makes one kind of apo B and the intestinal mucosa another, somewhat smaller, kind. When the appropriate phospholipids and cholesterol, as well as a triglyceride supply, are available, the hepatic type of apo B, along with apo E and apo C, coalesce with them to make nascent very low-density lipoproteins, which are secreted. If it is the intestinal type of apo B, along with apo A, that is present, they and the appropriate lipids form into nascent chylomicrons. (The intestinal mucosa also synthesizes some variant very low-density lipoproteins; they contain the intestinal apo B and apo A-IV.)

Metabolism of Lipoproteins

The composition of chylomicrons and very low-density lipoproteins begins to change shortly after they are secreted. They retain their characteristic B apoproteins, but exchange other apoproteins with each other and with the high-density lipoprotein. Indeed, chylomicrons are not attacked by lipoprotein lipases until they gain apo C-II, a specific activator of the enzyme, from either very low-density or high-density lipoproteins. Further exchanges are forced by the shrinkage of the shell that occurs as triglycerides are lost. In addition, the shrinking core is partially replenished with cholesteryl esters of long-chain fatty acids.

The high-density lipoproteins made by the liver act as carriers and reservoirs for lipids and some apoproteins during the metabolism of the triglyceride-rich lipoproteins. They are especially important as transfer agents for cholesterol. The high-density lipoproteins accumulate cholesterol, which an attached enzyme converts into cholesteryl esters of long-chain fatty acids. This enzyme is a **lecithin-cholesterol acyltransferase,** which transfers an acyl group (usually 18:2) from phosphatidylcholines in the high-density lipoprotein shell to cholesterol in the shell (see top, p. 559). **The cholesteryl esters are highly hydrophobic,** and as they accumulate, some are carried to the shrinking chylomicrons and very low-density lipoproteins. There is some evidence that the blood contains a specific protein to carry out cholesteryl ester exchanges. (The lysophosphatidylcholines remaining as the other product of acyl transfer are returned to other tissues for reacylation.)

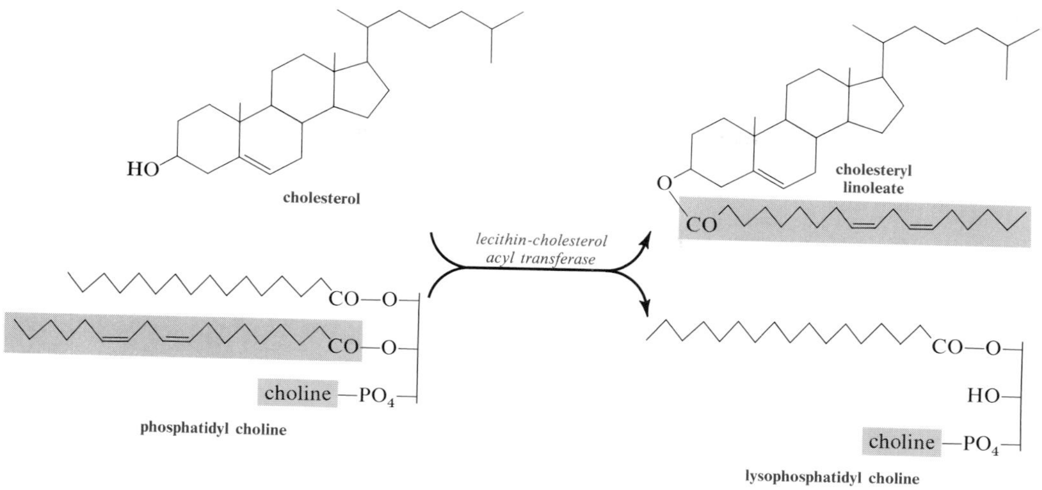

HO

cholesterol

lecithin-cholesterol acyl transferase

CO—O—

CO—O—

choline —PO_4—

phosphatidyl choline

O

CO

cholesteryl linoleate

CO—O—

HO—

choline —PO_4—

lysophosphatidyl choline

The lecithin-cholesterol acyltransferase necessary for the transfer operation is synthesized separately by the liver, but binds to the nascent high-density lipoprotein. It does not become active until the nascent lipoprotein gains a specific apo A-I activator by transfer from chylomicrons. The nascent high-density lipoprotein is first made as a bilayer disc in which phospholipids and some cholesterol coalesce around apo E. Upon gaining the lecithin-cholesterol acyltransferase and its apo A-I activator, the formation of cholesteryl esters begins, first using the lipids originally present, and then those gained by transfer from the other lipoproteins. The cholesteryl ester droplet forms a core like those seen in other lipoproteins, making the molecule more nearly spherical. As the esters accumulate, some are transferred to the other lipoproteins.

The changes in composition of the various lipoproteins during maturation and loss of triglycerides are tabulated in the following:

SUMMARY OF LIPOPROTEIN MATURATION

	Chylomicrons (intestinal mucosa)	VLDL (liver)	HDL (liver)
newly synthesized	apo B apo A-I,II,IV triglyceride core; phospholipid-cholesterol shell	apo B apo C-I,II,III triglyceride core; phospholipid-cholesterol shell	apo E apo A-I,II,III apo C-I,II,III phospholipid-cholesterol bilayer
newly circulating lipoproteins	retains apo B loses some apo A gains C apoproteins gains apo E	retains apo B gains C apoproteins gains apo E	gains apo A loses some apo C loses some apo E cholesterol ⟶ cholesteryl ester
changes during triglyceride uptake	loses triglyceride retains apo B loses apo C loses some apo E gains cholesteryl esters	loses triglyceride retains apo B loses apo C loses some apo E gains cholesteryl esters	gains apo C gains phospholipid + cholesterol, forming cholesteryl esters
fate	destroyed in liver	becomes intermediate-density, then low-density lipoproteins; destroyed in liver and peripheral tissues	fate not clear

Disposal of the Lipoproteins. The chylomicron remnants and low-density lipoproteins are fat-depleted relics, which are removed from the circulation by attachment to specific receptors in coated pits, followed by incorporation into internal vesicles and digestion in lysosomes. They still retain some triglyceride, because the apoproteins lost during their shrinkage include the apo C-II necessary to keep lipoprotein lipases active, but much of the core is now cholesteryl esters. The coated pit receptors for these relic lipoproteins recognize apo E or apo B. Binding is blocked by the presence of the C apoproteins, and this is why less-depleted lipoproteins are not prematurely removed. The liver rapidly clears chylomicron remnants through binding of the remaining apo E, and more slowly removes low-density lipoproteins. Other tissues clear roughly half of the low-density lipoproteins, primarily by binding apo B. (Only a small amount of apo E, and virtually no apo A or apo C, is present in low-density lipoproteins.)

PLASMA LIPOPROTEINS

Property	Chylomicrons	Very Low-density Lipoproteins	Low-density Lipoproteins	High-density Lipoproteins
Typical Per Cent Composition				
fat	87	55	8	0.6
phospholipid	8	20	24	21
free cholesterol	1.5	10	8	5
cholesteryl esters	2	5	35	15
protein	1.7	9	25	55
Density	0.92–0.96	0.95–1.006	1.019–1.063	1.063–1.21
Molecular Weight	5×10^8 (typical)	$5\text{–}100 \times 10^6$	$2.2\text{–}3.5 \times 10^6$	3.2×10^5 (HDL_2) 1.75×10^5 (HDL_3)
Diameter (nm)	75–600	28–75	17–26	10.8 (HDL_2)
Function	triglyceride transport from small bowel	triglyceride transport from liver	cholesterol transport	cholesterol transport

CHOLESTEROL CYCLES

We have seen that cholesterol is a necessary component of membranes and of lipoproteins. It is also the precursor of bile acids and of several hormones. It passes to and from tissues in two cyclical processes, one involving lipoprotein turnover, and the other involving bile acid turnover. **Cholesterol and its products are mainly lost in the feces.** The losses are made up in part from the diet and in part by synthesis from acetyl coenzyme A.

Cholesterol and Lipoprotein Turnover

About two thirds of the cholesterol in the blood is found in low-density lipoproteins. The normal total cholesterol content is 3.1 to 5.7 mM in blood serum, of which roughly one fourth is free cholesterol in the surface shells of various lipoproteins, and the balance is cholesteryl esters in the lipoprotein cores.

Cholesterol is regarded as a nasty word by many laymen and by my collaborator because of its association with lipoproteins. **Anything causing accumulation of the lipoproteins rich in cholesteryl esters—chylomicron remnants, intermediate-density lipoproteins, or low-density lipoproteins—is almost certain to cause severe atherosclerosis.** However, the engulfment of low-density lipoproteins is a normal device for delivery of cholesterol to extrahepatic tissues, in which the cholesteryl esters are hydrolyzed by lysosomal lipases. The cholesterol obtained in this way suppresses internal synthesis of new cholesterol, but if the supply is still in excess of cellular needs the cell exports cholesterol by a mechanism not yet understood, and it is taken up by the high-density lipoproteins, presumably for disposal in the liver. Increased amounts of total cholesterol in high-density lipoproteins and decreased amounts in low-density lipoproteins are therefore associated with increased disposal of peripheral cholesterol and decreased risk of atherosclerotic changes, according to current views.

Cycling of Cholesterol in the Bile

Most of the cholesterol synthesized or absorbed from the diet each day is used to replace the bile acids and cholesterol lost in the feces. Bile contains a micelle of phosphatidylcholines, free cholesterol, and bile salts. The concentration of the bile salts in the bile is 8 to 12 times the concentration of cholesterol, but only 0.2 to 0.4 of the cholesterol in the intestinal lumen is absorbed, whereas nearly all of the bile salts, or their constituent bile acids, are absorbed. The result is that the steroid fecal losses are roughly equally partitioned between cholesterol and the bile salts (and their bacterial degradation products). Typically, an adult must replace 2 millimoles of total steroids per day, of which 1.8 millimoles will be synthesized as cholesterol, and 0.2 millimoles obtained from the diet (assuming the diet is not rich in cholesterol).

This enterohepatic circulation of the bile salts (Fig. 29–1) is strikingly efficient. The liver synthesizes conjugates of cholic and chenodeoxycholic acids, which appear in the bile. Most of these are absorbed as such in the ileum, but some are not absorbed until they reach the lower ileum, where bacteria hydrolyze the attached glycine or taurine, and reduce a sizeable fraction of the free bile acids to deoxycholate or lithocholate. Most of the free acids (as anions) are absorbed.

The absorbed bile salts and bile acids are transported to the liver through the portal blood in the form of complexes with serum albumin. The liver removes them, with up to 95 per cent extracted in a single passage. Both the secondary bile acids (deoxycholic and lithocholic), as well as the primary compounds, are once more conjugated with taurine and glycine for secretion into the bile. The result is that the percentage of the total bile salts present as the various compounds in bile typically is:

glycocholate—24	glycochenodeoxycholate—24
taurocholate—12	taurochenodeoxycholate—12
glycodeoxycholate—16	various lithocholates— 4
taurodeoxycholate— 8	

Roughly 0.3 of the total represents steroids modified by passage through the bowel. Eating stimulates discharge of bile from the gallbladder, and the bile salts recirculate two or three times through the liver per meal.

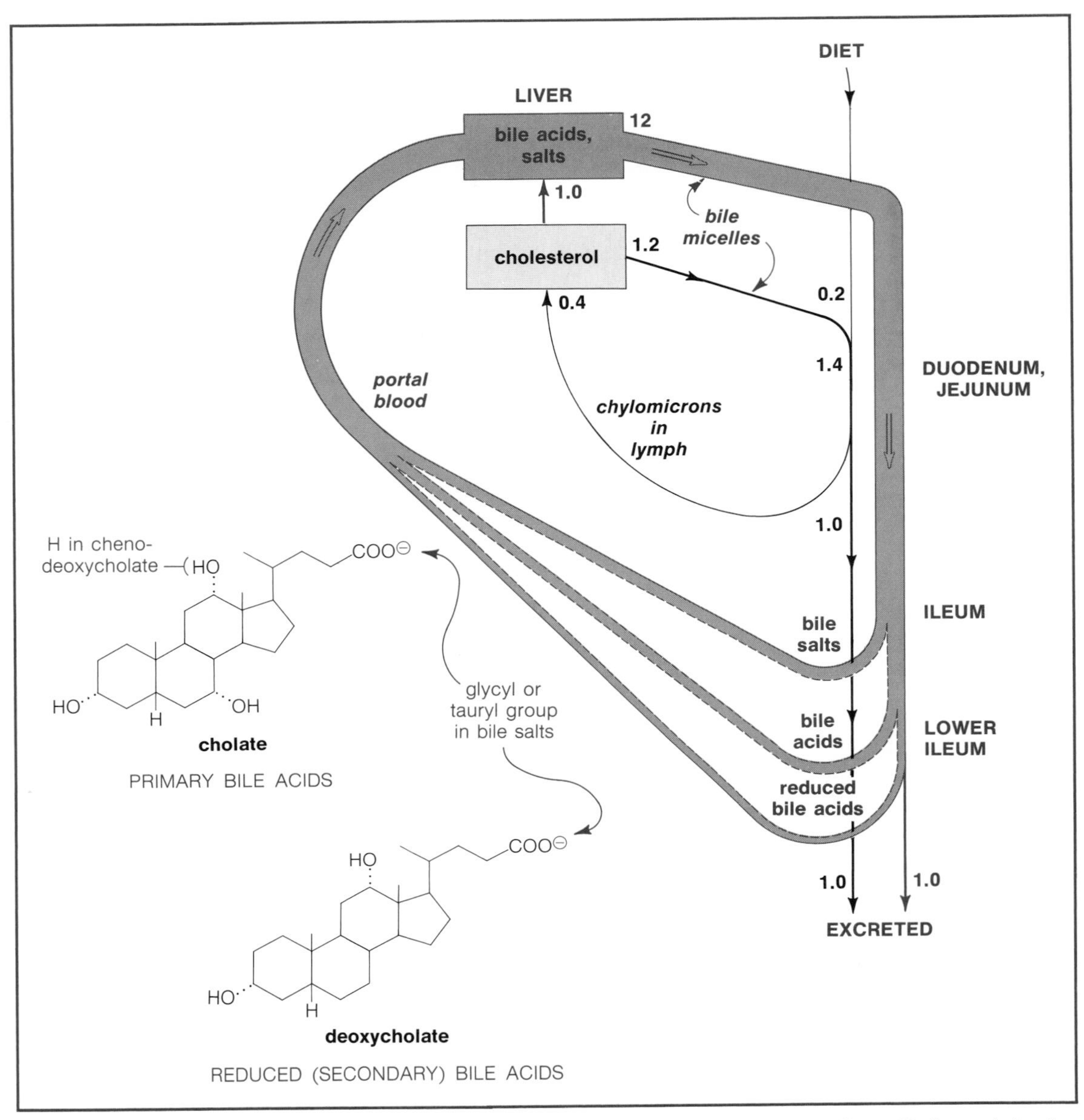

FIGURE 29–1 **Recirculation of bile salts and cholesterol from the small bowel to the liver. Cholesterol circulation shown in *black*, bile salt circulation in *brown*. The numbers indicate approximate millimoles per day.**

Synthesis of Bile Acids and Salts

Conversion of cholesterol to the bile acids, **cholate** and **chenodeoxycholate,** begins with hydroxylation at C7 (Fig. 29–2), and this is the rate-limiting step, which is regulated through an inhibition by chenodeoxycholate as an ultimate product. A series of additional oxygenase and dehydrogenase reactions removes part of the side chain, introduces additional hydroxyl groups, and inverts the configuration of the already existing 3-hydroxyl group.

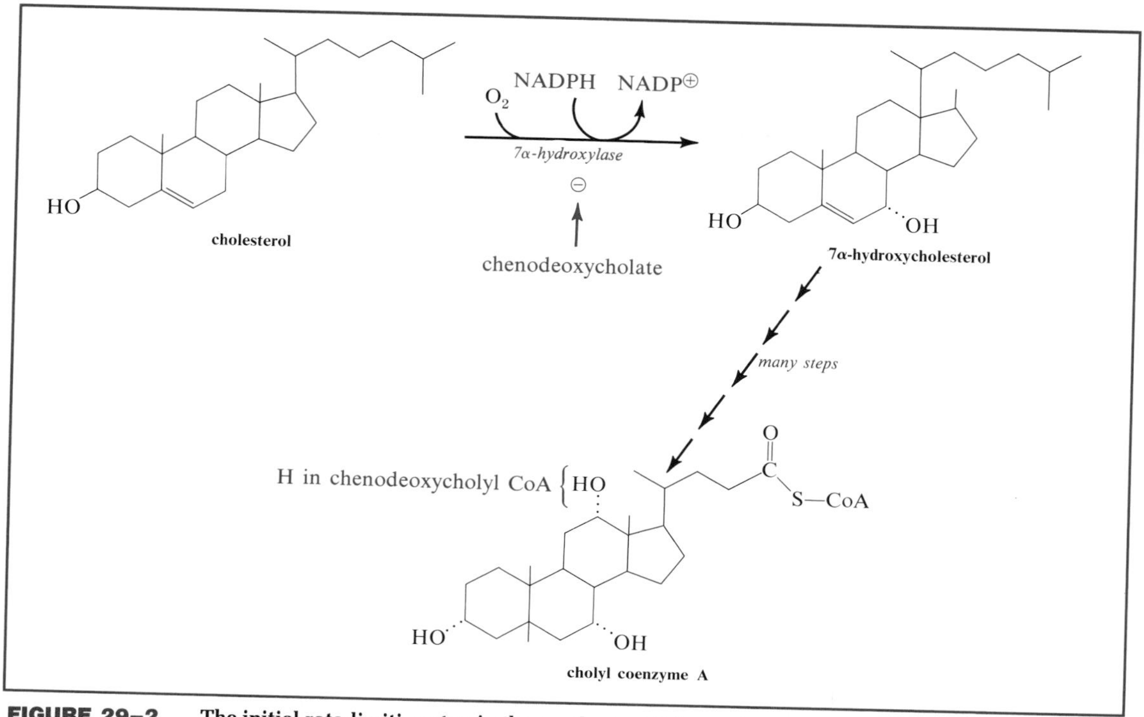

FIGURE 29–2 The initial rate-limiting step in the synthesis of bile acids from cholesterol is a hydroxylation at the 7α position. The bile acids are converted to their coenzyme A derivatives for reaction with glycine or taurine to form bile salts.

Finally, the steroid acids are converted to the coenzyme A derivatives, followed by transfer of the acyl groups to either **taurine** or **glycine,** forming the bile salts (p. 332). Taurine is formed by decarboxylation of cysteic acid, which in turn is made by oxidation of cysteine:

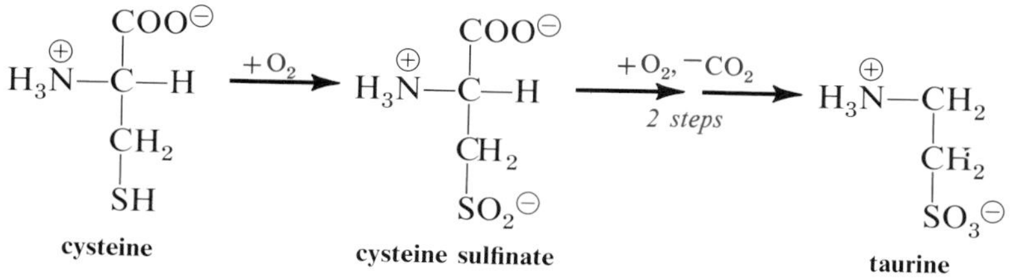

$$\underset{\text{cysteine}}{H_3\overset{\oplus}{N}-CH(COO^{\ominus})-CH_2-SH} \xrightarrow{+O_2} \underset{\text{cysteine sulfinate}}{H_3\overset{\oplus}{N}-CH(COO^{\ominus})-CH_2-SO_2^{\ominus}} \xrightarrow[\text{2 steps}]{+O_2,\ -CO_2} \underset{\text{taurine}}{H_3\overset{\oplus}{N}-CH_2-CH_2-SO_3^{\ominus}}$$

SYNTHESIS OF CHOLESTEROL

The dietary intake of cholesterol often replaces only a small fraction of the excreted steroids. The body content is maintained through synthesis of cholesterol.

The steroid skeleton is made by condensing six prenyl groups, otherwise known as isoprene groups:

$$-CH_2-\overset{CH_3}{\overset{|}{C}}=CH-CH_2-$$

Many other polyprenyl compounds occur in nature. Animals, for example, make the side chain of ubiquinone (p. 400), and plants make a wide variety of compounds, including the carotenes, rubber, and the cyclic terpenoid compounds (camphor, pinene, limonene, and many others).

The precursor of all these compounds is **3-hydroxy-3-methylglutaryl coenzyme A** (Fig. 29–3), which is formed from acetyl coenzyme A in both the cytosol and the mitochondria. It is used for acetoacetate synthesis in mitochondria (p. 452), where it also appears as an intermediate in leucine metabolism. In the cytosol, it is used for steroid synthesis. **The rate-controlling step in cholesterol synthesis** is the initial reduction of 3-hydroxy-3-methylglutaryl coenzyme A to **mevalonate** on the endoplasmic reticulum. Mevalonate is then phosphorylated in three stages, with the third stage creating an intermediate that loses phosphate and is decarboxylated to create an isoprenol (as a pyrophosphate ester).

This **Δ³-isopentenyl diphosphate** is the general donor of prenyl groups for the synthesis of all polyprenyl compounds. The acceptor molecules are Δ²-isopentenyl compounds (Fig. 29–4). The initial acceptor, **dimethylallyl diphosphate,** is made by simple conversion of the Δ³ compound to its Δ² isomer. An enzyme then catalyzes transfer of prenyl groups to it, first making the diprenyl (C_{10}) compound **geranyl diphosphate,** and then the triprenyl (C_{15}) compound **farnesyl diphosphate.** Farnesyl diphosphate is used for cholesterol synthesis; another transferase is present to add an additional prenyl group to make a C_{20} precursor of longer chain polyprenyl compounds. (This reaction is especially important to plants, which use the C_{20} compound to make C_{40} carotenes.)

In order to make the steroid ring, two molecules of farnesyl diphosphate are condensed head-to-head and reduced by NADPH to form **squalene** (Fig. 29–5). This symmetrical 30-carbon hydrocarbon is named for its occurrence in shark liver oil (*Squalus* spp.).* It, and the subsequent intermediates in cholesterol synthesis, are

---

*Konrad Bloch, who received the Nobel Prize for elucidation of the pathway of cholesterol synthesis, combined work and pleasure by traveling to Bimini for collection of shark livers.

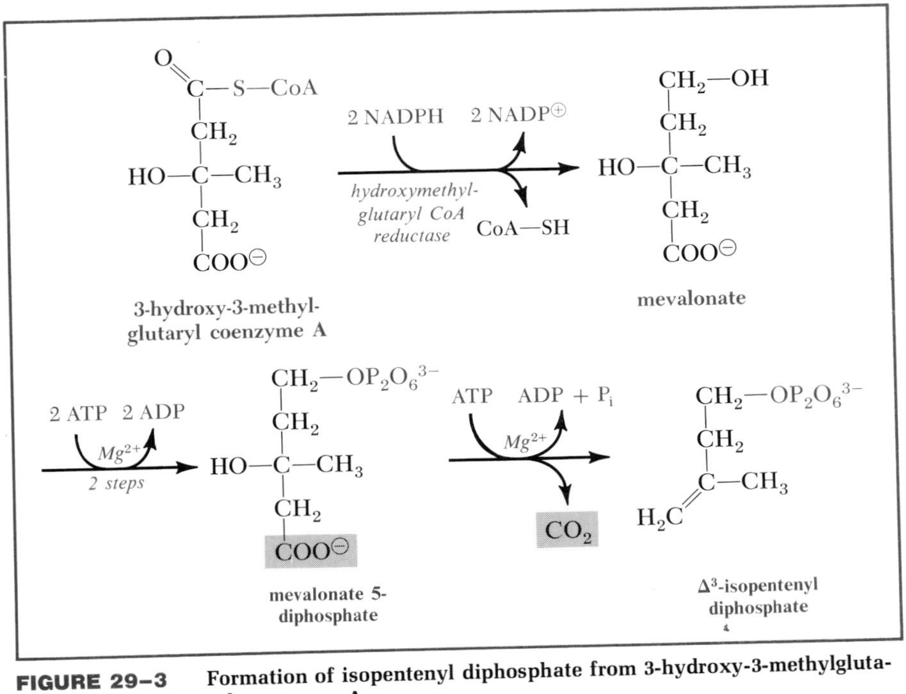

3-hydroxy-3-methylglutaryl coenzyme A: $O{=}C(S{-}CoA){-}CH_2{-}C(OH)(CH_3){-}CH_2{-}COO^{\ominus}$

$\xrightarrow[\text{hydroxymethylglutaryl CoA reductase}]{2\ NADPH \rightarrow 2\ NADP^{\oplus};\ CoA{-}SH}$

mevalonate: $CH_2{-}OH{-}CH_2{-}C(OH)(CH_3){-}CH_2{-}COO^{\ominus}$

$\xrightarrow[\text{2 steps}]{2\ ATP \rightarrow 2\ ADP,\ Mg^{2+}}$

mevalonate 5-diphosphate: $CH_2{-}OP_2O_6^{3-}{-}CH_2{-}C(OH)(CH_3){-}CH_2{-}COO^{\ominus}$

$\xrightarrow{ATP \rightarrow ADP + P_i,\ Mg^{2+};\ CO_2}$

Δ³-isopentenyl diphosphate: $CH_2{-}OP_2O_6^{3-}{-}CH_2{-}C(CH_3){=}CH_2$

FIGURE 29–3 **Formation of isopentenyl diphosphate from 3-hydroxy-3-methylglutaryl coenzyme A.**

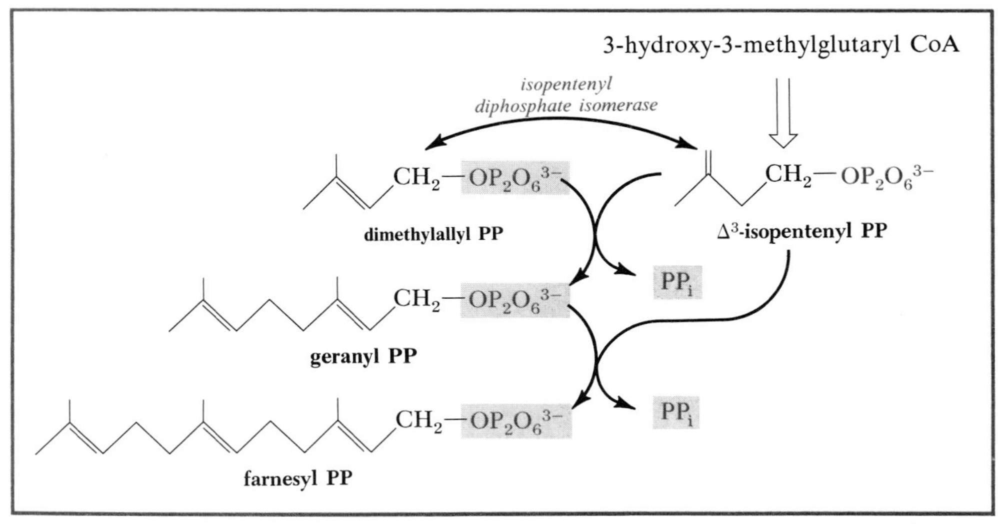

FIGURE 29–4 **Farnesyl pyrophosphate is formed from three molecules of isopentenyl pyrophosphate.**

hydrophobic and require the presence of specific carrier proteins to bring them in contact with the appropriate enzymes on the endoplasmic reticulum. One of these is a mixed-function oxidase that converts squalene to a cyclic oxide, which then undergoes a concerted internal condensation to form **lanosterol,** the parent steroid in animals.

Lanosterol differs from cholesterol in having three additional methyl groups, and in the location of double bonds. The methyl groups are removed by a succession of oxidations (Fig. 29–6) interspersed with relocation of the double bonds by appropriate reductions and oxidations (Fig. 29–7) to form cholesterol.

Regulation of Cholesterol Synthesis. The activity of **hydroxymethylglutaryl CoA reductase,** the rate-controlling enzyme catalyzing the first step in cholesterol synthesis, is regulated in two ways. One is by altering the number of enzyme molecules. Significant changes can occur in a few hours or less because the enzyme turns over rapidly. Less enzyme is synthesized when the product concentrations rise. The actual effectors may be hydroxylated derivatives of cholesterol. Chenodeoxycholate, for example, suppresses formation of the enzyme in the liver. Other hydroxylated cholesterols are formed elsewhere, and may be physiologically important.

The enzyme is also regulated through phosphorylation and dephosphorylation. The phosphorylated form is inactive, as is the case with other regulated enzymes involved in synthetic processes. The kinase catalyzing the phosphorylation is itself phosphorylated by still another kinase before it becomes active. The primary signals stimulating the phosphorylation sequence are not clear.

Cholesterol Gallstones. The bile frequently becomes supersaturated with cholesterol in the gallbladder, and cholesterol crystals begin to grow in some individuals, forming stones. This does not happen in everyone, apparently because some additional components must be present as nucleation centers, but it is still a major problem. A typical patient with gallstones is said to be female, fat, 40, and fertile, with some 15 million women and 5 million men affected in the United States. There are a million new cases and 500,000 cholecystectomies (gallbladder removals) per year.

Attempts have been made to alleviate the problem through administration of chenodeoxycholate, with the expectation that it will inhibit cholesterol synthesis and diminish the amount secreted in the bile. This has proved to be the case, with substantial dissolution of existing stones occurring in a significant fraction of the patients. However, the number requiring surgery was little affected, and the incidence of side effects added some additional risk. Further study is being made with ursodeoxycholate, a bile acid found in bear bile, which acts at lower concentrations.

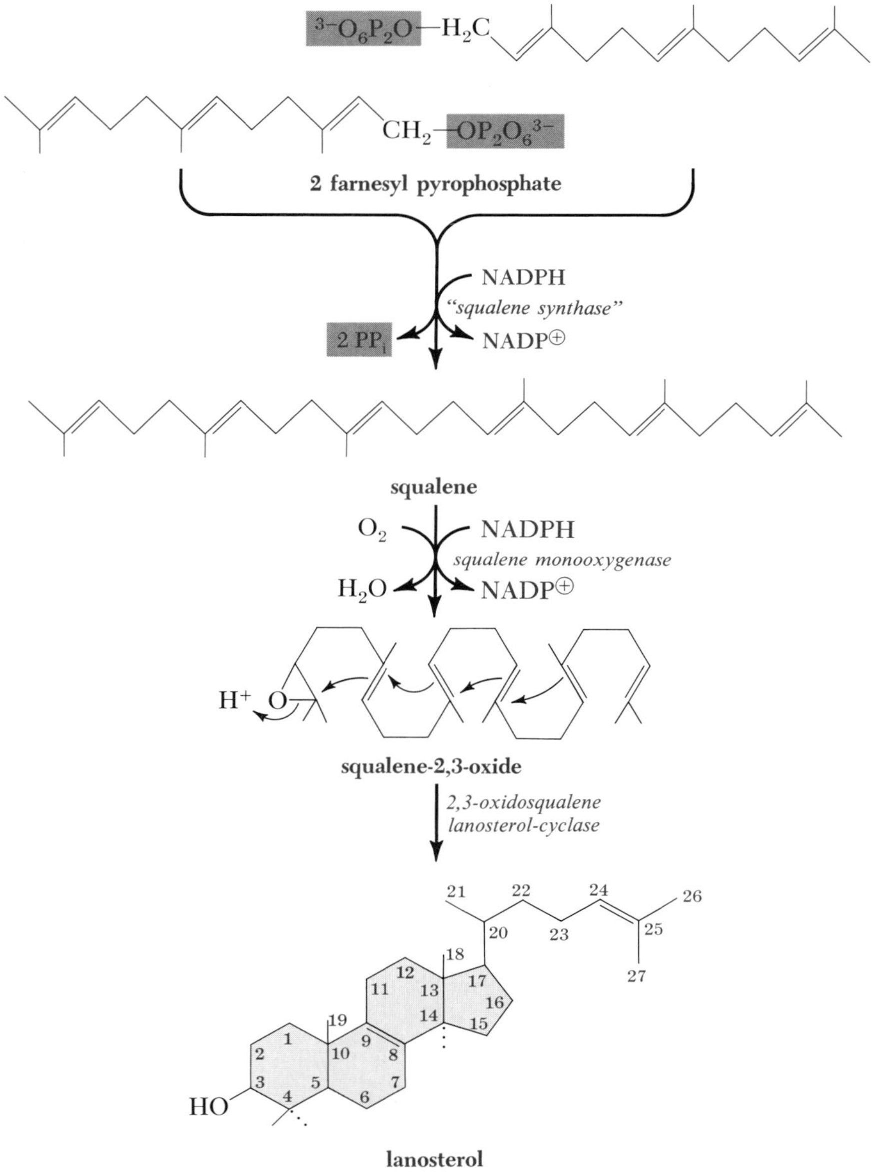

$^{3-}O_6P_2O$—H_2C

CH_2—$OP_2O_6{}^{3-}$

2 farnesyl pyrophosphate

NADPH

"squalene synthase"

2 PP_i

NADP⊕

squalene

O_2

NADPH

squalene monooxygenase

H_2O

NADP⊕

H^+ O

squalene-2,3-oxide

2,3-oxidosqualene lanosterol-cyclase

21 22 24 26 20 23 25 18 12 17 11 13 27 16 14 19 15 1 9 2 10 8 3 5 7 4 6

HO

lanosterol

FIGURE 29–5 **Synthesis of lanosterol, the parent steroid compound in animals, from farnesyl groups.**

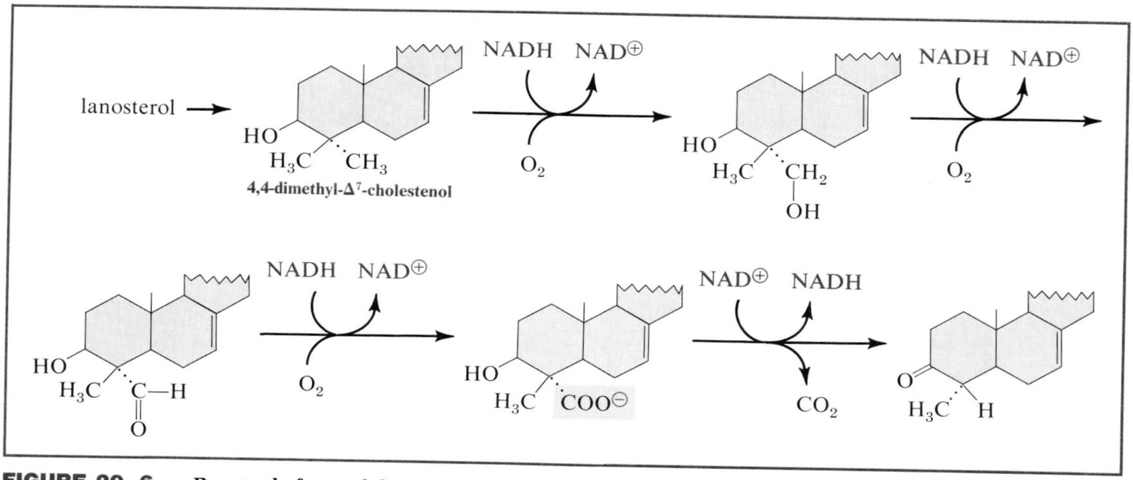

lanosterol → 4,4-dimethyl-Δ7-cholestenol

FIGURE 29–6 Removal of one of the methyl groups from lanosterol. Three successive reactions by an α-methyl sterol oxidase convert the methyl group first to an alcohol, then an aldehyde, and finally a carboxylate group. The carboxylate group is removed by oxidative decarboxylation. (An intermediate 3-ketocarboxylate is not shown.) Two more methyl groups are removed by similar sequences of reactions.

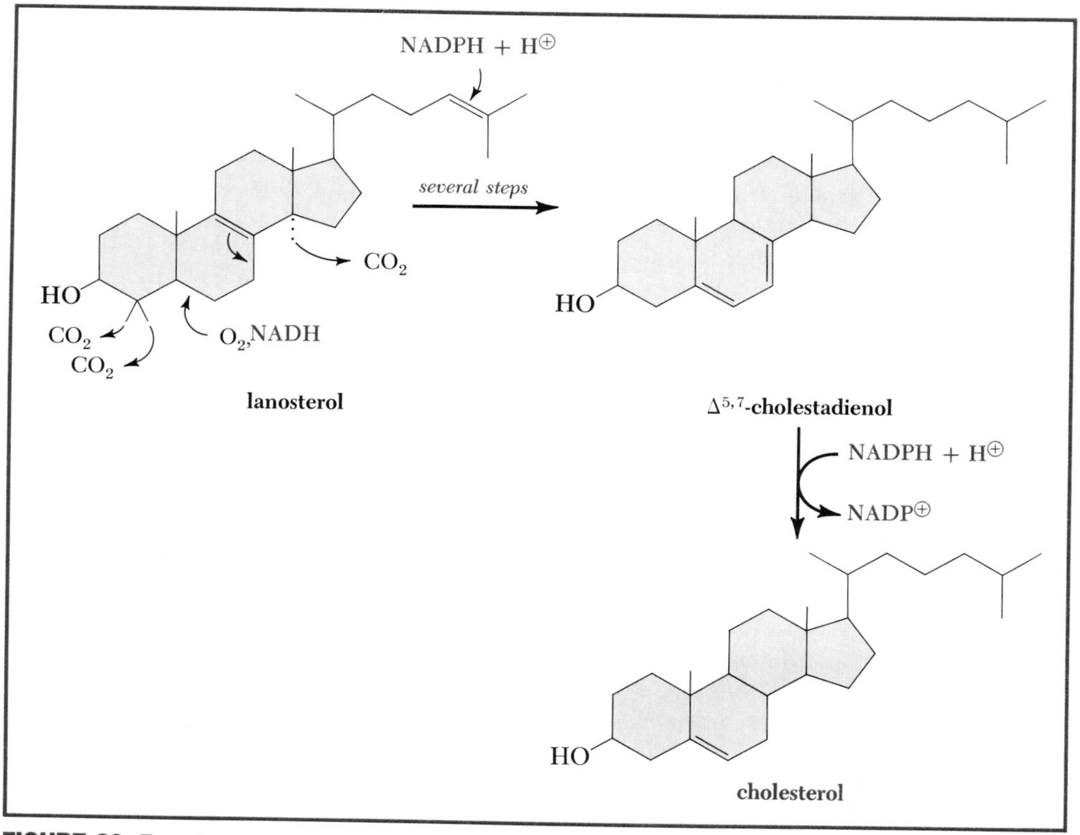

FIGURE 29–7 A summary of the conversion of lanosterol to cholesterol. The final step is a reduction of $\Delta^{5,7}$-cholestadienol to cholesterol.

LIPID TRANSPORT AND DISEASE

Defective Synthesis of Specific Transport Proteins

All of these genetic defects are rare, but have been important in stimulating study of more common clinical entities of less well-defined origin. **Some people are unable to synthesize normal amounts of serum albumin.** Considering the importance of albumin in transporting free fatty acids, as well as other substances, and its important role in maintaining the osmotic pressure of the blood, it is easy to predict that hypoalbuminemia (analbuminemia) is not compatible with life.* Unfortunately, the prediction is totally wrong, and the presence of this deficiency is sometimes detected only as a coincidental result of routine examination in seemingly healthy adults. Intensive study may reveal degenerative changes and abnormal lipid patterns, but it is difficult to reconcile the relative benignity of the condition with the heavy selection pressure that must exist to keep it so rare and with the known multiple functions of albumin.

Apolipoprotein A deficiency causes Tangier disease,† characterized by the development of orange tonsils, pharynx, and rectal mucosa during a frequently benign childhood. Lipid infiltration of the cornea, a peripheral neuropathy, and splenomegaly occur later. The blood triglyceride concentration is high and the cholesterol concentration is low. These various effects can be rationalized as a result of the high-density lipoprotein deficiency caused by lack of apo A. Without these lipoproteins, it is not possible to remove cholesterol from the peripheral cells, nor is a normal exchange of apoproteins possible. Cholesterol and cholesteryl esters accumulate, especially in reticuloendothelial cells and Schwann cells. (Lipid deposits are frequently yellow to red from lipid-soluble pigments, such as the carotenes.) The very low-density lipoproteins are deficient in apo C, including the lipoprotein lipase activator normally obtained in part from the now-missing high-density lipoprotein, and this could explain their retention in the circulation (high triglyceride concentration) and a failure to mature into low-density lipoproteins (causing low total cholesterol concentration).

Apolipoprotein B deficiency results in a loss of ability to make chylomicrons and very low-density lipoproteins. Without very low-density lipoproteins, low-density lipoproteins are not formed. If the loss is total, the consequences are grave, because there is a generalized failure of both triglyceride and cholesterol metabolism. The afflicted infant will be unable to transport fat from the intestinal mucosa; the cells become stuffed with fat droplets while the excess appears in the foul-smelling stools. The erythrocytes become prickled, owing to lack of normal cholesterol exchange in the plasma membrane. There is a general neural and muscular degeneration; lipid imbalance, deficiency of essential fatty acids, deficiency of lipid-soluble nutrients—any or all of these may be responsible.

Apolipoprotein E deficiency causes familial dysbetalipoproteinemia, in which cholesteryl ester–rich chylomicron remnants are not bound to the apo E-sensitive receptors in the liver, and accumulate in the blood. This causes the appearance of **xanthomas** and **atheromas.** A xanthoma is a deposit of lipid in the skin. It may be in the form of a rash or a large flat plaque, or it may be nodular, depending upon its cause. Xanthomas may appear whenever the blood has a persistently high concentration of any lipid, but the composition of the lipid is changed during transport and processing from blood to skin. The xanthomas are usually yellow to orange from lipid-soluble pigments, such as carotenes. They are an unsightly nuisance but not in themselves crippling.

---

*I made such a prediction, but luckily not in print.

†Tangier Island, Virginia.

Atheromas are another matter. These are cholesteryl ester-rich plaques developing in arterial walls and blocking normal blood flow. The rationale of their origin is that injuries to arteries are repaired by proliferation of smooth muscle cells in the arterial wall, and these cells engulf plasma constituents in the site of injury. If these constituents include high concentrations of cholesterol, the cells accumulate cholesteryl esters as insoluble inclusions, preventing normal maturation at the repair site.

Whatever the mechanism, apoprotein E deficiency results in premature atherosclerotic disease.

Enzyme Deficiencies

Familial lecithin-cholesterol acyltransferase deficiency (LCAT-deficiency) prevents the normal transfer of cholesterol as cholesteryl esters. The nascent high-density lipoproteins cannot accumulate cholesteryl esters and mature from their lamellar bilayer form. All of the other circulating lipoproteins are abnormal. Opacities appear in the cornea during childhood. Anemia and kidney damage develop later.

Familial lipoprotein lipase deficiency prevents the clearance of triglycerides from the blood. The accumulation of chylomicrons usually overshadows the accompanying accumulation of very low-density lipoproteins, which become detectable on a fat-free diet. The high concentration of chylomicrons gives a cream of tomato soup appearance to the blood. While xanthomas may develop, the presenting symptom is often abdominal pain, caused by acute inflammation of the pancreas. The condition can be treated by a diet low in fats. The disease may convey some protection against cardiovascular disease.

Familial lysosomal lipase deficiency (Wolman's disease) causes an accumulation of engulfed triglycerides and cholesteryl esters in many tissues. It is fatal in infancy if the deficiency is complete.

LDL-Receptor Deficiency

Familial hypercholesterolemia is an autosomal dominant condition in which the receptors for low-density lipoproteins are defective. At least three different mutants are known in some 325 families, including over 1800 individuals heterozygous to the condition. These patients have been intensively studied because of their propensity for developing cardiovascular diseases. Heterozygotes have total serum cholesterol concentrations in the range of 7 to 14 mM (275 to 550 mg/dl), while homozygotes have concentrations of 17 to 36 mM (650 to 1000 mg/dl). (The normal concentration is 3.1 to 5.7 mM [120 to 220 mg/dl].) Half of the male heterozygotes will have heart attacks before age 50 and half of the females before age 60. The homozygotes develop cardiovascular disease in childhood, with death from myocardial infarction commonly occurring before age 20.

In addition to coronary artery disease, those afflicted also may develop xanthomas and corneal infiltration, the "arcus senilis."

Recent trials of drugs to lower the blood cholesterol concentration in these patients have had promising results. These include a combination of an ingested ion exchange resin that binds bile acids (cholestipol) along with nicotinate (niacin), which inhibits mobilization of fatty acids from adipose tissue. The rationale is that the constant removal of bile acids in the feces will stimulate removal of cholesterol through its conversion to more bile acids, and diminution of the blood free fatty acid concentration will alleviate the known tendency of saturated fatty acids to stimulate cholesterol synthesis.

Other promising drugs include probucol and ML236B, which inhibit cholesterol synthesis directly. However, it remains to be seen if long-term use of any of these agents will diminish the incidence of cardiovascular disease.

Other Inherited Hyperlipidemias

Two conditions of apparently genetic origin are much more common, and much less well understood, than those just discussed. An estimated one out of every 100 people is heterozygous to an autosomal dominant **familial hypertriglyceridemia,** in which the very low-density lipoprotein concentration is elevated. The effects of the condition are variable. Obesity, high blood pressure, high uric acid concentration, and resistance to insulin action (glucose intolerance) may occur in various combinations. The condition may be associated with increased risk of early heart attacks, but this is not settled.

Familial multiple hyperlipoproteinemia is another relatively common condition in which either the very low-density or the low-density lipoprotein concentrations may be elevated at a given time. Ten per cent of the survivors of myocardial infarctions occurring before age 60 are said to come from families bearing this autosomal dominant condition. As is the case with familial hypertriglyceridemia, the exact nature of the mutations causing the condition is not known.

CHOLESTEROL AND CARDIOVASCULAR DISEASE

The turnover of lipoproteins may be altered by dietary changes or as a result of other diseases. Epidemiological studies indicate that changes in cholesterol turnover, whatever the cause, are associated with changes in the incidence of cardiovascular disease. There is a clear correlation between atherosclerotic disease of the coronary arteries and the concentration of total cholesterol in the blood, which mainly reflects the cholesterol content of the low-density lipoproteins. There is an even more clear negative correlation between coronary disease and the concentration of cholesterol in the high-density lipoprotein fraction. That is, the chances of heart disease are increased in those with elevated LDL concentrations and diminished in those with elevated HDL concentrations. For example, consider a 46-year-old male with a systolic blood pressure of 150 torr who does not smoke and has normal glucose tolerance. If his total cholesterol concentration is 165 mg/dl, his probability of developing coronary heart disease in the next six years is 0.024. If his cholesterol concentration is 315 mg/dl, the probability is 0.090. If the concentration of cholesterol associated with high-density lipoproteins is 30 mg/dl, those probabilities should be multiplied by 1.82. If the HDL cholesterol is 65 mg/dl, the probabilities should by multiplied by 0.45.

That is, over the range of concentrations likely to be encountered, variation in LDL-cholesterol concentration changes the associated risk of coronary heart disease by a factor of two, whereas variation in the HDL-cholesterol concentration changes it by a factor of four. There is no association of risk with triglyceride concentration.

Note carefully that we are talking about associations, not about causal relationships. It is to be emphasized that there is no evidence that manipulation of the blood concentration of lipoproteins will in itself change the risk of coronary heart disease. Even if it would, there is cause for reflection. Data are accumulating that a low level of total cholesterol, or a high level of HDL-cholesterol, is associated with an increased risk of cancer of the colon in some groups. The death rate from cancer of the colon in a given cohort is much less than that from coronary heart disease, but it is still substantial. Finally, it must be emphasized that we are speaking of atherosclerotic disease. Myocardial infarctions are at least sometimes caused by vascular spasms in the absence of demonstrable atherosclerosis. The relationship between dietary factors and cardiovascular diseases will be discussed in Chapter 42.

FURTHER READING

Lipoproteins

R. Havel, J.L. Goldstein, and M. Brown (1980). Lipoproteins and Lipid Transport. p. 393 in P.K. Bondy and L.E. Rosenberg, eds. Metabolic Control and Disease, 8th ed. Saunders. Perhaps the best comprehensive review.

S. Eisenberg, ed. (1979). Lipoprotein Metabolism. Prog. Biochem. Pharmacol. vol. 15. Review articles.

A.M. Scanu and F.R. Landsberger, eds. (1980). Lipoprotein Structure. Ann. N.Y. Acad. Sci. vol. 348.

L.D. Grouse (1980). A Medical Misdemeanor: I Harbored Evil Thoughts About the Frederickson Fat Classification. J.A.M.A. 244:2090. The author's explanation of why he as a medical student couldn't take seriously a widely quoted classification of hyperlipidemias is also an explanation of why the classification is not used in this chapter.

Cholesterol and Bile Acid Synthesis; Gallstones

G.J. Schroepfer, Jr. (1982). Sterol Biosynthesis. Annu. Rev. Biochem. 51:555.

E.D. Beytia and J.W. Porter (1976). Biochemistry of Polyisoprenoid Biosynthesis. Annu. Rev. Biochem. 45:113.

M.S. Brown and J.L. Goldstein (1980). Multivalent Feedback Regulation of HMGCoA Reductase, A Control Mechanism Coordinating Isoprenoid Synthesis and Cell Growth. J. Lip. Res. 21:505.

M.S. Brown, P.T. Kovanen, and J.L. Goldstein (1981). Regulation of Plasma Cholesterol by Lipoprotein Receptors. Science 212:628.

A. Sedaghat and S.M. Grundy (1980). Cholesterol Crystals and the Formation of Cholesterol Gallstones. N. Engl. J. Med. 302:1274.

L. Swell et al. (1981). Bile Acid Synthesis in Humans. Cancer Res. 41:3757.

L.C. Schoenfeld et al. (1981). Chenodiol (chenodeoxycholic acid) for Dissolution of Gallstones. The National Cooperative Gallstone Study. Ann. Intern. Med. 95:257.

Tokyo Cooperative Gallstone Study Group (1980). Efficacy and Indications of Ursodeoxycholic Acid Treatment for Dissolving Gallstones. Gastroenterology 78:542.

Genetic Defects

G. Franceschini et al. (1980). A-1_{Milano} Apoprotein. Decreased High-density Lipoprotein Cholesterol Levels with Significant Lipoprotein Modifications and without Clinical Atherosclerosis in an Italian Family. J. Clin. Invest. 66:892.

J.P. Kane et al. (1981). Normalization of Low-Density Lipoprotein Levels in Heterozygous Familial Hypercholesterolemia with a Combined Drug Regimen. N. Engl. J. Med. 304:251.

H. Mabuchi et al. (1981). Effects of an Inhibitor of 3-Hydroxy-3-Methylglutaryl Coenzyme A Reductase on Serum Lipoproteins and Ubiquinone-10 Levels in Patients with Familial Hypercholesterolemia. N. Engl. J. Med. 305:478.

J.V. Weinstock, H. Kawanishi, and J. Sisson (1979). Morphologic, Biochemical, and Physiologic Alterations in a Case of Idiopathic Hypoalbuminemia (Analbuminemia). J.A.M.A. 67:132.

R.A. Norum et al. (1982). Familial Deficiency of Apolipoproteins A-I and C-III and Precocious Coronary Artery Disease. N. Engl. J. Med. 306:1513.

R.S. Lees and A.M. Lees (1982). High-Density Lipoproteins and the Risk of Atherosclerosis. N. Engl. J. Med. 306:1546.

Dietary Lipids and Disease

S.B. Hulley et al. (1980). Epidemiology as a Guide to Clinical Decisions. The Association Between Triglyceride and Coronary Heart Disease. N. Engl. J. Med. 302:1383.

M.F. Oliver (1982). Risks of Correcting the Risks of Coronary Disease and Stroke with Drugs. N. Engl. J. Med. 306:297.

R.K. Shekelle et al. (1981). Diet, Serum Cholesterol, and Death from Coronary Heart Disease. The Western Electric Study. N. Engl. J. Med. 304:65. Also see p. 1168 for a barrage of letters concerning this paper.

B. Peterson, E. Trell, and N.H. Sternby (1981). Low Cholesterol Level as Risk Factor for Noncoronary Death in Middle-Aged Men. J.A.M.A. 245:2056.

R.R. Williams et al. (1981). Cancer Incidence by Levels of Cholesterol. J.A.M.A. 245:247.

M.F. Oliver (1981). Diet and Coronary Heart Disease. Br. Med. Bull. 37:49.

E. Brittain (1982). Probability of Coronary Heart Disease Developing. West. J. Med. 136:86. Simplified presentation of risks calculated from the Framingham study (Ann. Int. Med. 90:85).

CHAPTER 30

AMINO ACIDS: DISPOSAL OF NITROGEN

We opened our discussion of biochemistry by examining the amino acids as constituents of proteins. The use of amino acids as precursors of proteins must constantly be kept in mind, even though amino acid metabolism is also used for other purposes.

It is easy to show that just maintaining a supply of amino acids from which proteins can be made isn't a simple process. We eat proteins, which are hydrolyzed to the constituent amino acids in the gut. At the same time, tissue proteins are also being hydrolyzed to form amino acids, which mix with those derived from food as an **amino acid pool** in the extracellular fluid upon which all tissues may draw for their requirements.

Part of the amino acid is used to rebuild tissue proteins, but the total amount of protein remains relatively constant in most adults from day to day, which means that the quantity of amino acids used to make tissue proteins is not greater than the quantity obtained by the breakdown of tissue proteins. Amino acids aren't excreted in significant quantities. Therefore, the usual adult will have a surplus of amino acids equivalent to the amount he has ingested. The surplus must be disposed of in other ways. In short, there is an active metabolism of amino acids that constantly must be reconciled with maintenance of a supply for protein synthesis. Let us summarize the nature of this metabolism (Fig. 30–1).

The amino acid pool is constantly drained for the synthesis of other nitrogenous constituents. One of these, urea, is deliberately made as a device for disposal of

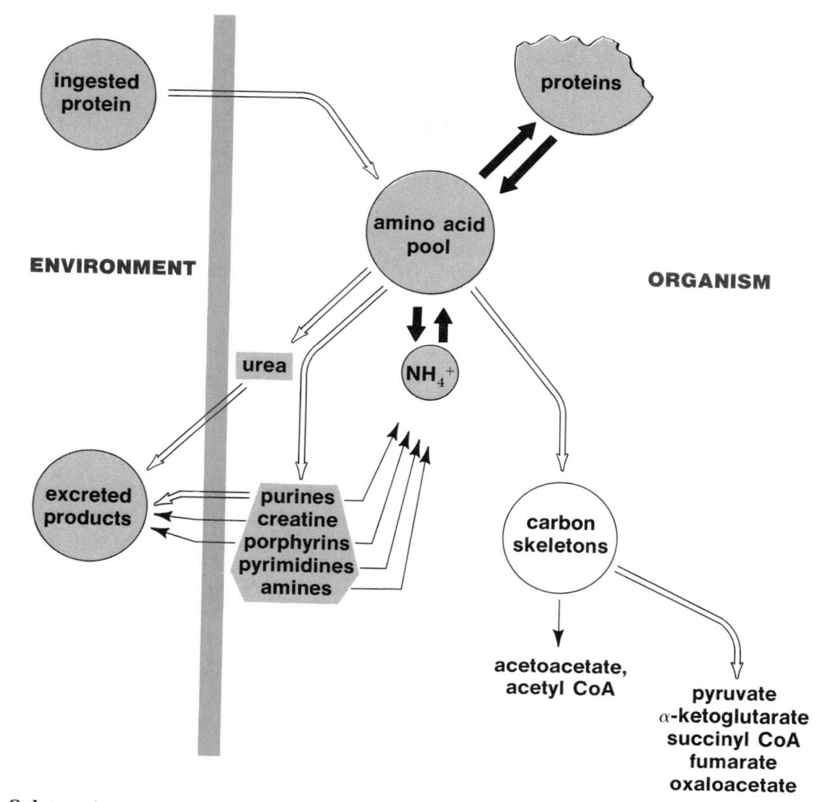

FIGURE 30–1

Schematic outline of the flow of nitrogen through the body. Arrow thicknesses roughly indicate relative magnitudes.

excess nitrogen in the urine. Others are vital tissue components: purines, pyrimidines, porphyrins, amines, creatine, and so on. These compounds are constantly being broken down and re-made, as are the proteins themselves. Part of their nitrogen can be recovered as ammonia, which is a precursor of several amino acids, but part is lost through excretion of other products.

Amino acids are used as fuel. The bulk of the excess amino acid intake is degraded through oxidative pathways associated with the formation of high-energy phosphate. How important a fuel are they? They are usually less significant than fat or carbohydrate but are sometimes of major importance. The extent of the contribution to the energy supply obviously depends upon the composition of the diet. The rib roast eaten by an affluent human contains protein and fat to the extent of one quarter each of its cooked weight, with virtually no carbohydrate. The fish or chicken more likely to be on most tables has three to six times as much protein as fat when purchased raw, but the fat will be increased in the final food to some extent through the cooking oil used in its preparation.

Considering the total intake, most humans obtain only one tenth, rarely more than one fifth, of their high-energy phosphate by oxidizing amino acids, but purely carnivorous animals generate nearly one half of their supply in this way.

Since there are some 20 amino acids commonly metabolized during the utilization of proteins, the description of protein metabolism inevitably becomes more complicated than the description of the major routes of carbohydrate and fat metabolism, and it is tempting to skip over the subject lightly with a few superficial generalities, but the temptation ought to be, and will be, resisted. The metabolism of amino acids is complicated, and it is important.

The very abundance of enzymes involved in amino acid metabolism ensures that pediatricians will encounter infants lacking one of the enzymes, even though a genetic defect in a particular enzyme may be uncommon or even rare. Indeed, pediatrics is to a considerable extent an exercise in practical biochemistry. Furthermore, aberrations of the nitrogen economy accompany many of the most pressing of current medical problems in the adult population.

We shall deal in this chapter with some of the general processes of amino acid metabolism, and discuss the fate of the individual amino acids in the next chapters.

TRANSPORT OF NITROGEN

Uptake of Amino Acids

Use of Transporters. Many cells are capable of concentrating amino acids from the extracellular environment, but the processes of transport have not been explored in detail in most cases. Separate carriers are known to exist for some neutral amino acids, cationic amino acids, and anionic amino acids.

A Na^+-coupled symport mechanism for the absorption of neutral amino acids is present in the brush borders of the intestinal mucosa and the kidney tubules. The passage of Na^+ down its concentration gradient causes simultaneous transport of amino acids against their concentration gradient in the same way that glucose is moved in these cells (p. 387). (Some amino acids are transported into striated muscles by a Na^+-dependent process that is stimulated by insulin.) **Cystinuria,** which causes the formation of stones of the slightly soluble cystine in the urinary tract (p. 15), is a hereditary defect in the ability of the kidney to transport cystine from the lumen back into the cells. Abnormal quantities of arginine, lysine, and ornithine are also excreted, indicating that the hereditary defect also affects one or more carriers for these amino acids as well. Nearly 1 in 40 people is estimated to carry a gene for cystinuria, with 1 in 7000 infants homozygous to this recessive gene. (The incidence varies greatly among different populations.)

There is strong evidence that a known set of reactions involving 5-glutamyl peptides (Fig. 30–2) is used for amino acid transport into many tissues, including the brush borders of the kidney and the small bowel, the brain, and red blood cells. These peptides, commonly known as γ-glutamyl peptides because they involve the carboxyl group attached to the γ-carbon atom, include glutathione (5-glutamylcysteinylglycine). The tissue content of glutathione ranges as high as eight millimoles per kilogram. (Glutathione is also used as a reducing agent, but this function is not involved in amino acid transport.)

The amino acid to be transported is converted to a γ-glutamyl derivative by moving the γ-glutamyl group of glutathione or another γ-glutamyl peptide onto it. The γ-glutamyl transferase is believed to be exposed on the external side of the plasma membrane, with the products released on the internal side. In any event, the γ-glutamyl amino acid is attacked by a cyclotransferase that releases the amino acid inside the cell and forms pyroglutamate (5-oxoproline) from the glutamyl group.

As Figure 30–2 shows, the starting materials can be synthesized from the other product of the amino acid transfer, creating a cycle. When glutathione is used as a glutamyl group donor, the other product is cysteinylglycine, which can be hydrolyzed to its constituent amino acids. The ring in pyroglutamate is opened to form glutamate by a reaction requiring ATP. Glutathione can then be rebuilt from glutamate, cysteine, and glycine by two ATP-requiring synthetases.

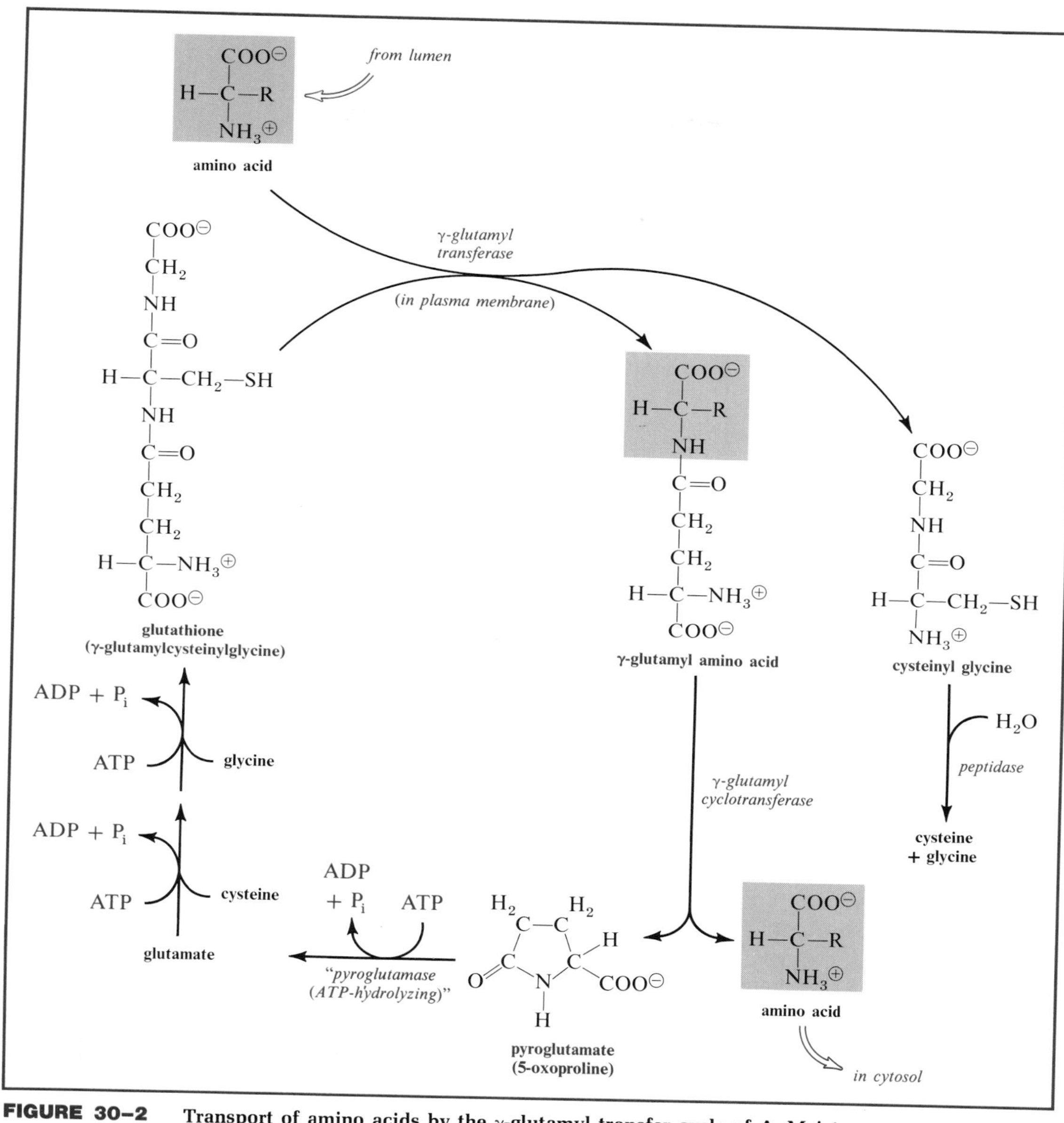

FIGURE 30–2 **Transport of amino acids by the γ-glutamyl transfer cycle of A. Meister.**

According to the mechanism given, the transport of one molecule of amino acid requires the expenditure of three high-energy phosphate bonds, but there are alternative possibilities that are less expensive involving γ-glutamylcysteine itself as the glutamyl group donor. The quantitative importance of these pathways is yet to be assessed. Much of the load on cystine transport in the kidney tubules may normally come from cleavage of glutathione. **Cystinosis,** a rare condition in which cystine is deposited in the lysosomes of many cells, also may result from some defect in glutathione metabolism. (A severe form of cystinosis causes irreversible damage to the kidneys from the crystals forming within it.) Still another rare condition, **pyroglutamic aciduria (oxoprolinuria),** is due to lack of the enzyme synthesizing glutathione from γ-glutamylcysteine. The synthesis of γ-glutamylcysteine is regulated through negative feedback by glutathione, and if glutathione is not formed, the enzyme

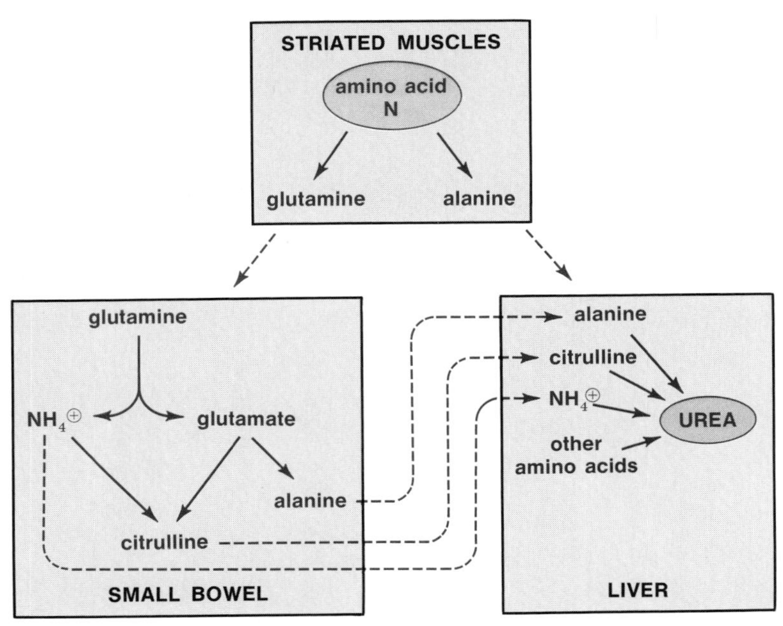

FIGURE 30–3

Transfer of nitrogen through the blood. The nitrogen of amino acids degraded in striated muscles leaves mainly as glutamine and alanine. Much of the glutamine is removed from the blood in the small bowel, reappearing in the portal blood as ammonium ions, alanine, and citrulline. The liver converts the nitrogen of these various compounds to urea for excretion.

functions unchecked, with the product continuously broken down to form large quantities of pyroglutamate (5-oxoproline).

Transport Between Tissues

The skeletal muscles, the intestines, and the liver are particularly important in disposing of excess amino acids, but much of the nitrogen is channeled into only a few compounds for transport between these tissues (Fig. 30–3). For example, the skeletal muscles export nitrogen mainly in the form of **glutamine** and **alanine.** Glutamine is made by combining ammonium ions and glutamate at the expense of ATP:

$$\underset{\text{L-glutamate}}{{}^{\ominus}OOC{-}CH_2{-}CH_2{-}CH(NH_3^{\oplus}){-}COO^{\ominus}} + NH_4^{\oplus} + ATP \xrightarrow{\text{glutamine synthetase}} \underset{\text{L-glutamine}}{H_2N{-}C({=}O){-}CH_2{-}CH_2{-}CH(NH_3^{\oplus}){-}COO^{\ominus}} + P_i + ADP$$

We will see how the nitrogen from many amino acids can appear in glutamate and in ammonium ions, which can then be combined as glutamine for transport from the muscles.

Glutamine is neutral, nontoxic, and crosses plasma membranes readily. It is synthesized as a device for storing and transporting ammonium ions, as well as for incorporation into proteins, and this function is especially important in the brain and the striated muscles. Muscles, with their great mass, are more significant quantitatively.

Most of the glutamine is extracted from the blood by the small intestine, which contains a **glutaminase** catalyzing the hydrolysis to glutamate and NH_4^+. Some of the ammonium ions leave the small bowel as such, but some are converted to citrulline (p. 584) for export. Both the ammonium ions and the citrulline go directly from the bowel to the liver through the portal blood and are used to make urea.

The nitrogen removed from many amino acids in the muscles or the small bowel appears first in glutamate, and then in alanine by transamination. Alanine is the amino acid with the highest concentration in blood. The liver takes up alanine and removes the nitrogen for urea synthesis through the same chain of transamination.

REMOVAL OF NITROGEN ATOMS

Nitrogen atoms are removed from amino acids in several ways; the pathways used depend upon the particular amino acid. Removal makes the nitrogen available for incorporation into other compounds or for excretion. The major part of the excreted nitrogen usually is in the urine as urea:

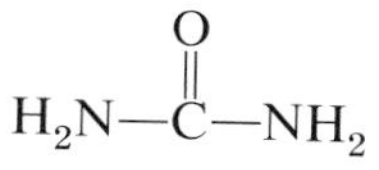

$$H_2N-\overset{\overset{\displaystyle O}{\|}}{C}-NH_2$$

urea

Let us focus for now upon routes by which nitrogen from amino acids finds its way into urea for excretion. **All of the nitrogen in urea is derived from two precursors, ammonium ion and aspartate, one atom from each.** Our initial task is to describe ways in which nitrogen from other amino acids can appear in these two compounds; we shall then see how urea is formed.

Amino Group Transfer

The most general route for removing nitrogen from amino acids is a transfer of the amino group to a keto acid by aminotransferases. The most active of these enzymes is **aspartate aminotransferase,** which catalyzes the equilibration of glutamate and oxaloacetate with α-ketoglutarate and aspartate:

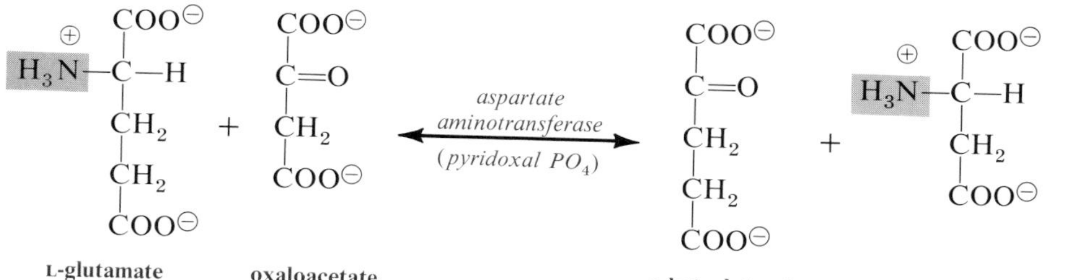

$$\underset{\text{L-glutamate}}{{}^{\oplus}H_3N-CH(COO^{\ominus})-CH_2-CH_2-COO^{\ominus}} + \underset{\text{oxaloacetate}}{{}^{\ominus}OOC-C(=O)-CH_2-COO^{\ominus}} \xrightleftharpoons[\text{(pyridoxal } PO_4)]{\text{aspartate aminotransferase}} \underset{\alpha\text{-ketoglutarate}}{{}^{\ominus}OOC-C(=O)-CH_2-CH_2-COO^{\ominus}} + \underset{\text{L-aspartate}}{{}^{\oplus}H_3N-CH(COO^{\ominus})-CH_2-COO^{\ominus}}$$

We considered the mechanism of this readily reversible reaction earlier (p. 286).

Since aspartate aminotransferase is present in high concentrations in most cells,

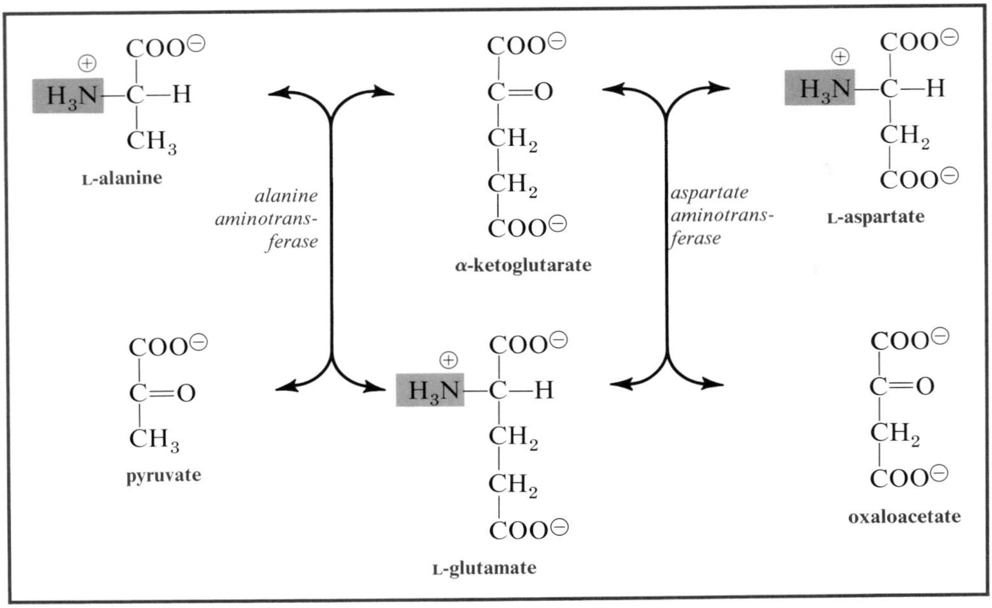

L-alanine ⇄ (alanine aminotransferase) α-ketoglutarate ⇄ (aspartate aminotransferase) L-aspartate

pyruvate, L-glutamate, oxaloacetate

$$\underset{\text{L-alanine}}{^{\oplus}H_3N{-}CH(CH_3){-}COO^{\ominus}} + \underset{\alpha\text{-ketoglutarate}}{^{\ominus}OOC{-}CH_2{-}CH_2{-}C(=O){-}COO^{\ominus}} \rightleftharpoons \underset{\text{pyruvate}}{CH_3{-}C(=O){-}COO^{\ominus}} + \underset{\text{L-glutamate}}{^{\oplus}H_3N{-}CH(COO^{\ominus}){-}CH_2{-}CH_2{-}COO^{\ominus}}$$

$$\underset{\text{L-glutamate}}{^{\oplus}H_3N{-}CH(COO^{\ominus}){-}CH_2{-}CH_2{-}COO^{\ominus}} + \underset{\text{oxaloacetate}}{^{\ominus}OOC{-}C(=O){-}CH_2{-}COO^{\ominus}} \rightleftharpoons \underset{\alpha\text{-ketoglutarate}}{^{\ominus}OOC{-}CH_2{-}CH_2{-}C(=O){-}COO^{\ominus}} + \underset{\text{L-aspartate}}{^{\oplus}H_3N{-}CH(COO^{\ominus}){-}CH_2{-}COO^{\ominus}}$$

FIGURE 30–4 **The action of alanine and aspartate aminotransferases makes it possible for the nitrogen of alanine to appear first in glutamate and then in aspartate.**

the reaction remains near equilibrium, and any removal of aspartate for urea synthesis will be withdrawing nitrogen from glutamate. Now, there are many other aminotransferases that catalyze transfer of amino groups from different amino acids to α-ketoglutarate, forming glutamate. These reactions are also readily reversible, so nitrogen from these amino acids will also appear, first in glutamate and then in aspartate, as aspartate is used in urea synthesis. This is the route through which at least half of the excess nitrogen is converted to urea. Figure 30–4 illustrates this sequence of events with alanine (the alanine enzyme is the second most active aminotransferase), but the same sort of thing occurs with many amino acids.

Clinical literature commonly refers to these aminotransferases as glutamate-oxaloacetate transaminase (SGOT; S for serum), and glutamate-pyruvate transaminase (SGPT).

Formation of Ammonium Ion

Oxidative Deamination. Much of the nitrogen of the amino acids sooner or later appears in the form of glutamate because of the action of aminotransferases, but all of glutamate nitrogen need not be used to form aspartate. Glutamate is also subject to oxidative deamination in a reaction catalyzed by glutamate dehydrogenase:

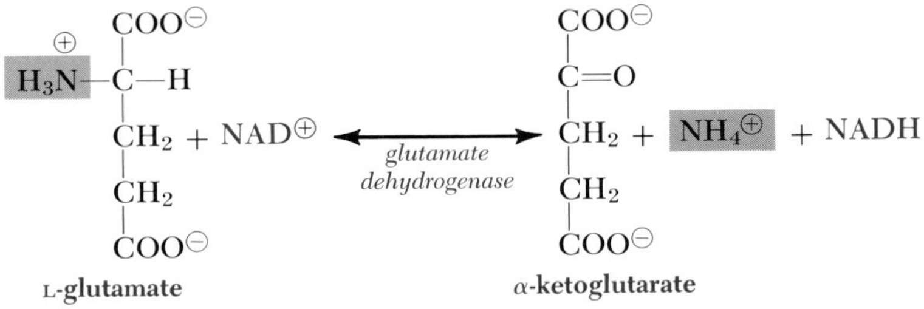

$$\underset{\text{L-glutamate}}{^{\oplus}H_3N{-}CH(COO^{\ominus}){-}CH_2{-}CH_2{-}COO^{\ominus}} + NAD^{\oplus} \xrightleftharpoons[\textit{dehydrogenase}]{\textit{glutamate}} \underset{\alpha\text{-ketoglutarate}}{^{\ominus}OOC{-}C(=O){-}CH_2{-}CH_2{-}COO^{\ominus}} + NH_4^{\oplus} + NADH$$

Glutamate dehydrogenase is present in the mitochondria of most, if not all, tissues. The enzyme will catalyze the reaction with either NAD or NADP, and it is readily reversible. We are now concerned primarily with the oxidation of glutamate and the accompanying release of ammonium ion, but should note that the reaction provides a capability for creating glutamate from ammonia by a reversal of the reaction.

The combined action of glutamate dehydrogenase and the various aminotransferases provides a route for making ammonium ion. Here we get another glimpse of the versatility of metabolic routes; the nitrogen from many amino acids appears as glutamate; the glutamate may be used to generate aspartate by an aminotransferase reaction, or it may be used to generate ammonium ions by the dehydrogenase reaction. Which way the reactions go depends upon the relative balance of the components:

amino N → **glutamate N** → **aspartate N** → **urea**

glutamate N → $NH_4^{\oplus}$ → **urea**

Direct Deamination. Compounds containing amino groups, such as amino acids, could hypothetically lose ammonia by a direct deamination analogous to the dehydration of an alcohol:

$$\overset{\oplus}{H_3N}\text{—}\underset{\underset{\large R}{|}}{\underset{H\text{—}C\text{—}H}{|}}\overset{COO^{\ominus}}{\overset{|}{C}}\text{—H} \longrightarrow \underset{\underset{\large R}{|}}{H\text{—}C}{=}\overset{COO^{\ominus}}{\overset{|}{C}}\text{—H} + NH_4^{\oplus}$$

Such reactions with simple amines or amino acids have an unfavorable position of equilibrium, and they are feasible in aqueous solution only when the character of the product is such as to form a resonating conjugated system. Therefore, only a few of the amino acids are directly deaminated in this way, and only one, histidine, in animals:

$$\text{Histidine (imidazole–}CH_2\text{–}CH(\overset{\oplus}{N}H_3)\text{–}COO^{\ominus}) \xrightarrow{\textit{histidase}} \text{Urocanate (imidazole–}CH{=}CH\text{–}COO^{\ominus}) + NH_4^{\oplus}$$

The product, urocanate, is fully conjugated. (Aspartate, phenylalanine, and tyrosine are also directly deaminated in bacteria or plants.)

Deamination by Dehydration. Serine and threonine have hydroxyl groups in addition to the ammonium group on their carbon chain. In effect, the carbon skeleton of these amino acids is already more oxidized than that of many amino acids.

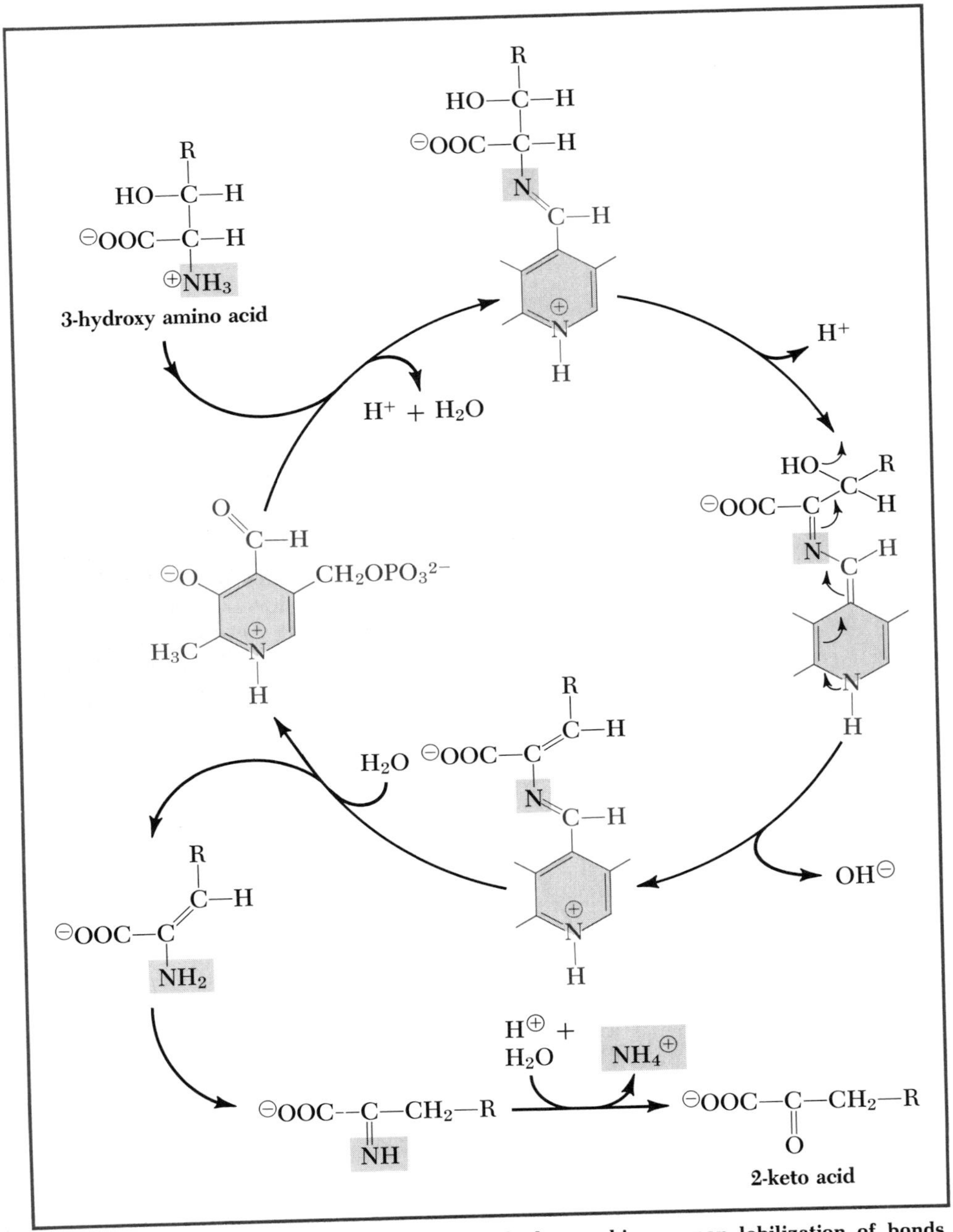

FIGURE 30–5 The mechanism of serine dehydratase hinges upon labilization of bonds around the α-carbon atom of the substrate through combination with the pyridoxal phosphate coenzyme. (Complete structure shown only at left center.)

They are somewhat analogous to glycerate, with its adjacent hydroxyl group, and we saw in the discussion of the Embden-Meyerhof pathway that the dehydration of glycerate to form pyruvate liberates enough free energy to sustain the formation of a high-energy phosphate (p. 467). Similarly, the deamination of serine and threonine to form the corresponding 2-keto compounds liberates enough free energy to be essentially irreversible:

$$\underset{\text{L-serine}}{\overset{\oplus}{H_3N}\text{—}\underset{CH_2OH}{\overset{COO^\ominus}{C}}\text{—H}} \xrightarrow{\textit{serine dehydratase}} \underset{\text{pyruvate}}{\underset{CH_3}{\overset{COO^\ominus}{C}}\text{=O}} + NH_4^{\oplus}$$

$$\underset{\text{L-threonine}}{\overset{\oplus}{H_3N}\text{—C(COO}^\ominus\text{)—H; H—C—OH; CH}_3} \xrightarrow{\textit{serine dehydratase}} \underset{\text{2-ketobutyrate}}{COO^\ominus\text{—C(=O)—CH}_2\text{—CH}_3} + NH_4^{\oplus}$$

The enzyme is called a dehydratase rather than a deaminase, because the reaction proceeds by the initial loss of the elements of water (Fig. 30–5). Serine dehydratase, like the aminotransferases and some other kinds of enzymes involved in amino acid metabolism, contains **pyridoxal phosphate** as a coenzyme (p. 284). This coenzyme, through the formation of a Schiff's base, labilizes all of the bonds around the α-carbon of the amino acid, and which bond is broken depends upon the surrounding catalytic groups contributed by the enzyme protein.

Hydrolytic Deamination. Ammonium ions are released from the amide groups of asparagine and glutamine by simple hydrolysis:

$$\underset{\text{L-glutamine}}{O\text{=C(—NH}_2\text{)—CH}_2\text{—CH}_2\text{—H—C(—}\overset{\oplus}{N}H_3\text{)—COO}^\ominus} \xrightarrow[\textit{glutaminase}]{H_2O \quad NH_4^{\oplus}} \underset{\text{L-glutamate}}{COO^\ominus\text{—CH}_2\text{—CH}_2\text{—H—C(—}\overset{\oplus}{N}H_3\text{)—COO}^\ominus}$$

$$\underset{\text{L-asparagine}}{O\text{=C(—NH}_2\text{)—CH}_2\text{—H—C(—}\overset{\oplus}{N}H_3\text{)—COO}^\ominus} \xrightarrow[\textit{asparaginase}]{H_2O \quad NH_4^{\oplus}} \underset{\text{L-aspartate}}{COO^\ominus\text{—CH}_2\text{—H—C(—}\overset{\oplus}{N}H_3\text{)—COO}^\ominus}$$

A bacterial asparaginase is frequently used in the treatment of childhood leukemia, in which the leukemic cells, unlike normal cells, cannot synthesize asparagine.

Deamination by the Purine Nucleotide Cycle. A major source of ammonium ion in skeletal muscles is the hydrolytic deamination of adenosine monophosphate (Fig. 30–6). The rate of this reaction increases sharply as a muscle approaches exhaustion, and the AMP concentration rises correspondingly sharply (p. 378). Mentioning the reaction at this time is not irrelevant to the point at hand, because we shall see when we discuss purine nucleotide metabolism (p. 684) that the resultant

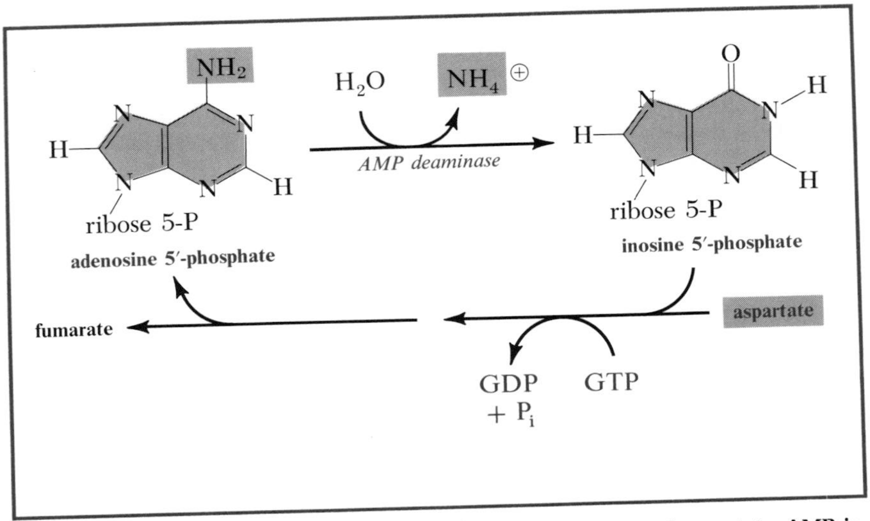

FIGURE 30–6

The purine nucleotide cycle of J.M. Lowenstein for deamination of aspartate. AMP is deaminated by a hydrolysis, and the resultant IMP is converted back to AMP by transfer of nitrogen from aspartate, with the energy for the transfer supplied by cleavage of GTP.

inosine monophosphate is converted back to AMP by transfer of nitrogen from aspartate:

$$\text{IMP} + \text{aspartate} + \text{GTP} \longrightarrow \text{AMP} + \text{fumarate} + \text{GDP} + \text{P}_i$$

The combination of this reaction with the deamination reactions provides a device by which the nitrogen of aspartate, and therefore of other amino acids, can appear as ammonium ion. (Another example of the use of aspartate as a nitrogen donor is given later.)

DISPOSAL OF NITROGEN: UREA SYNTHESIS

One of the fundamental facts about animals is that they are intolerant of even modest concentrations of ammonium ion in the cellular environment. Simple organisms living in water have no problem with nitrogen disposal. Ammonia diffuses out freely and is thereby diluted to a very low concentration. The ionic gradients and basic metabolic processes of amino acid metabolism may have evolved to fit the circumstance of low ammonium ion concentration. Adaptation for more efficient disposal of nitrogen only became necessary with the development of larger size and an enclosed circulation with impermeable skin, which were necessary preludes for movement from marine to terrestrial environments.

Modern animals of the higher phyla have efficient means of maintaining the ammonium ion concentration at very low levels—70 μM is the upper limit for the normal range in human blood sampled from arm veins. Many animals, including the mammals, accomplish this by converting ammonia to urea. (It is common to speak of ammonia and ammonium ion somewhat interchangeably—they are in equilibrium. At the pH of blood, only 1 per cent of the total is present as the base, so it is better to express concentrations as ammonium ion concentrations. Ammonia is the form that passes through cell membranes. The particular ionic form used in many enzymatic reactions has not been determined.)

The liver is the principal site of urea synthesis. Urea is made by a simple hydrolysis of the amino acid arginine catalyzed by the enzyme **arginase:**

$$\underset{\text{L-arginine}}{H_2N{-}C({=}NH_2^{\oplus}){-}NH{-}CH_2{-}CH_2{-}CH_2{-}CH(NH_3^{\oplus}){-}COO^{\ominus}} + H_2O \xrightarrow[Mn^{2+}]{arginase} \underset{\text{urea}}{H_2N{-}\overset{O}{\overset{\|}{C}}{-}NH_2} + \underset{\text{L-ornithine}}{{}^{\oplus}NH_3{-}CH_2{-}CH_2{-}CH_2{-}CH(NH_3^{\oplus}){-}COO^{\ominus}}$$

Urea synthesis therefore requires some mechanism for producing arginine. The other product of the hydrolysis of arginine is the dibasic amino acid, ornithine. (Ornithine is the next lower homologue of lysine, but it is not used for protein formation.) Arginine is rebuilt from ornithine by a cyclical process. The ornithine portion of the molecule is used over and over again, acting only as a carrier for the carbon atom and two nitrogen atoms released as urea. This is analogous to the continual reuse of the oxaloacetate portion of citrate in the citric acid cycle.

We have considered the citric acid cycle earlier because of its central role in metabolism. Historically, the urea cycle became known earlier, being described in the late 1930's by H. A. Krebs (who later discovered the citric acid cycle), and this description of a metabolic process in terms of a cycle was one of the major milestones in biochemistry.

As we mentioned earlier, one of the nitrogen atoms in urea is derived from ammonium ion and the other from aspartate. The carbon is obtained from CO_2. The complete process of urea synthesis is shown diagrammatically in Figure 30–7.

Synthesis of Carbamoyl Phosphate. The first step in rebuilding arginine from ornithine involves the combination of NH_4^+ and CO_2 (as bicarbonate ion) to make **carbamoyl phosphate.*** This is the critical step for lowering the ammonium ion concentration, which is assured in two ways: *First,* the reaction is made irreversible by cleaving two molecules of ATP, even though the product, carbamoyl phosphate, contains a high-energy phosphate bond. *Second,* the enzyme, carbamoyl phosphate synthetase, is present in more than adequate amounts for coping with the ammonia load. It constitutes a major part of the mitochondrial matrix proteins; indeed, its estimated concentration of 0.2 to 1.0 mM in the matrix is higher than that of most substrates. Its K_M is 250 μM for NH_4^+, but the high concentration of the enzyme enables it to handle the load at $[NH_4^+]$ well below this value. Since the physiological concentration of NH_4^+ (*ca.* 70 μM) *is* well below the K_M, any elevation in its concentration will greatly accelerate its removal.

Carbamoyl phosphate synthetase has an absolute requirement for an activator, **N-acetyl-glutamate,** which is not otherwise involved in urea synthesis. This activator is synthesized within mitochondria from acetyl coenzyme A and glutamate as a signal of elevated amino acid concentrations. (Arginine is an especially effective signal.) Eating a high-protein diet causes more acetylglutamate to be made, and it in turn increases the activity of carbamoyl phosphate synthetase, anticipating the need for increased disposal of excess nitrogen.

---

*Carbamoyl is official nomenclature for the —CO—NH_2 group, but carbamyl is sometimes used.

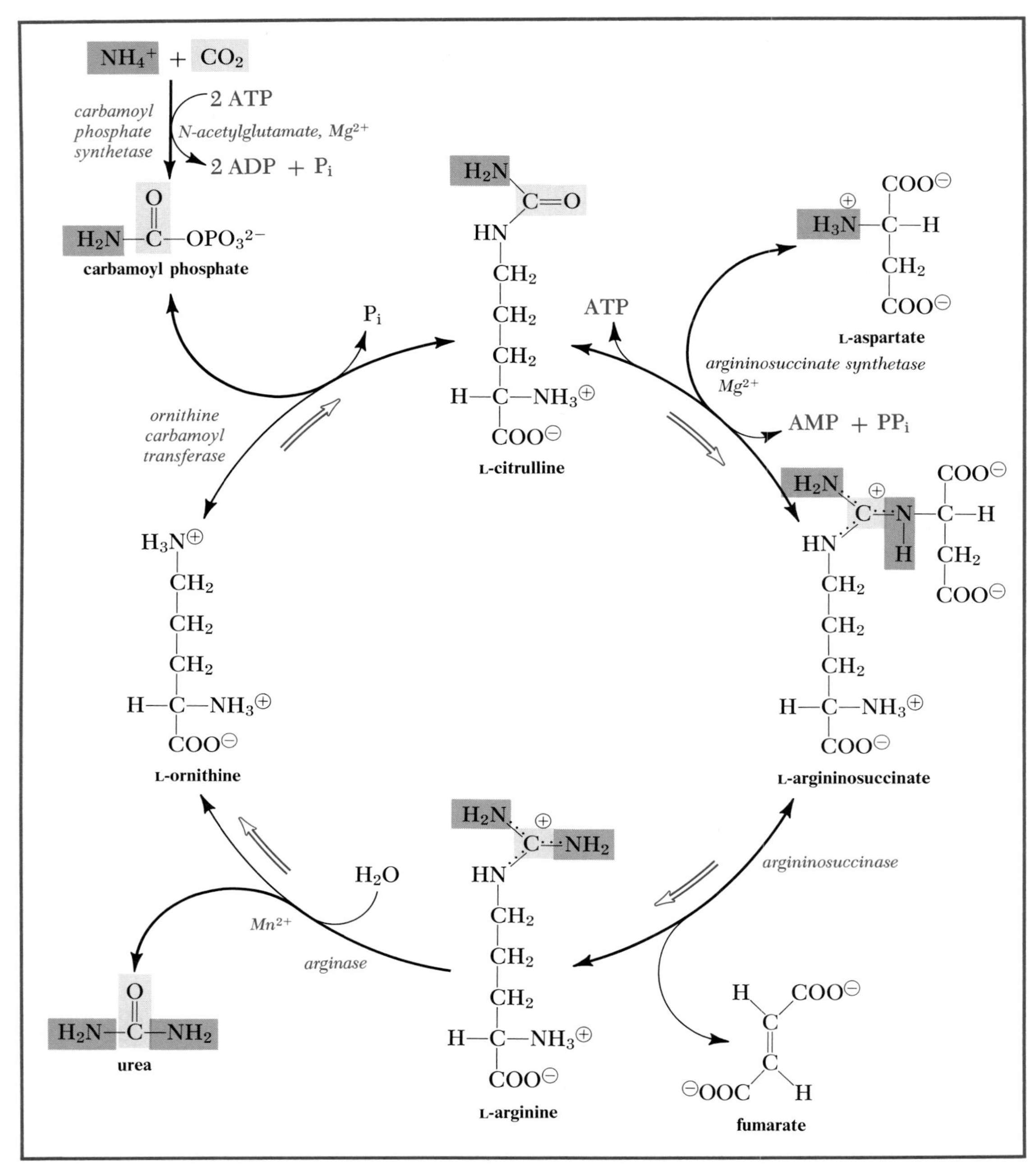

FIGURE 30–7 The synthesis of urea by the Krebs-Henseleit cycle. This is the same H.A. Krebs who later discovered the citric acid cycle. Kurt Henseleit was a student who worked for his M.D. thesis with Krebs.

Formation of Arginine *via* Citrulline. The carbamoyl group of carbamoyl phosphate is transferred to ornithine, forming citrulline. (Citrulline is primarily of importance as an intermediate in arginine synthesis; some residues of the amino acid occur in proteins, for example, in hair, but they are formed by modification of arginine residues, not by insertion during translation.) The reaction is driven by the loss of the high-energy phosphate in carbamoyl phosphate.

Citrulline then gains a nitrogen atom from aspartate to form arginine. The condensation of aspartate and citrulline is driven by the effective cleavage of two high-energy phosphate bonds. (PP_i is an initial product.) The resultant **argininosuccinate** is then cleaved so as to liberate arginine as one product and fumarate as the other. The effect of the sequence is to form a —C=N—C bridge on one side of the central nitrogen atom and cleave it on the other; the condensation and cleavage effectively transfer the nitrogen of aspartate to citrulline.

Complete Cycle. In sum, the formation of one molecule of urea requires the consumption of one molecule each of NH_4^+, aspartate, and CO_2, with a cleavage of four high-energy phosphate bonds:

1. $NH_4^+ + CO_2 + 2\,ATP \longrightarrow$ carbamoyl phosphate $+ 2\,ADP + P_i$
2. carbamoyl phosphate + ornithine $\longrightarrow$ citrulline + P_i
3. citrulline + aspartate + ATP $\longrightarrow$ argininosuccinate + AMP + PP_i
4. AMP + ATP $\longrightarrow$ 2 ADP
5. $PP_i + H_2O \longrightarrow 2\,P_i$
6. argininosuccinate $\longrightarrow$ arginine + fumarate
7. arginine + $H_2O \longrightarrow$ urea + ornithine

SUM: $NH_4^+ + CO_2 + 4\,ATP$ + aspartate $\longrightarrow$ urea + fumarate $+ 4\,ADP + 4\,P_i$

Compartmentation of Urea Synthesis

Nitrogen metabolism also involves transport between cellular compartments. The reactions of urea synthesis are divided between mitochondria and the cytosol in such a way as to facilitate utilization of the carbon skeletons (Fig. 30–8). We now encounter an architectural arrangement that is in many ways even more beautiful and sophisticated than the design of fatty acid synthesis, so much so that the consideration of the nuances still occupies the attention of many subtle minds. The major features are the assignment of citrulline synthesis to the mitochondrial matrix, and arginine synthesis and hydrolysis to the cystosol. Citrulline is made from ammonium ions *via* carbamoyl phosphate in the matrix, and the sources of the ammonium ions are the glutamate dehydrogenase and glutaminase reactions, both of which also occur in the matrix.

Citrulline leaves the mitochondria before it is converted, first to argininosuccinate and then to arginine. Why? The reason is that the transfer of nitrogen from aspartate forms fumarate as the other product, and the reactions for the disposal of fumarate occur in the cytosol. Fumarate equilibrates with malate, and we have already recited how malate may be directly converted to pyruvate for use in fatty acid synthesis during times of glucose plenitude, or converted to phospho-*enol*-pyruvate *via* oxaloacetate as a source of glucose during times of glucose deprivation. After cleavage of arginine to urea and ornithine, transport of the ornithine cation back to the mitochondrial matrix is driven by the electrical potential across the inner membrane. This discharge of potential energy is equivalent to the loss of an additional $0.25 \sim P$ per molecule of urea synthesized.

The message here is that urea synthesis is organized within a liver cell in such a way as to coordinate nitrogen disposal with the other reactions required for generating precursors and disposing of other products. We can see this more clearly by examining the fate of specific amino acids; let us use alanine and glutamate as examples—alanine because it comes to the liver in the highest concentration and glutamate because it is the most active intermediate.

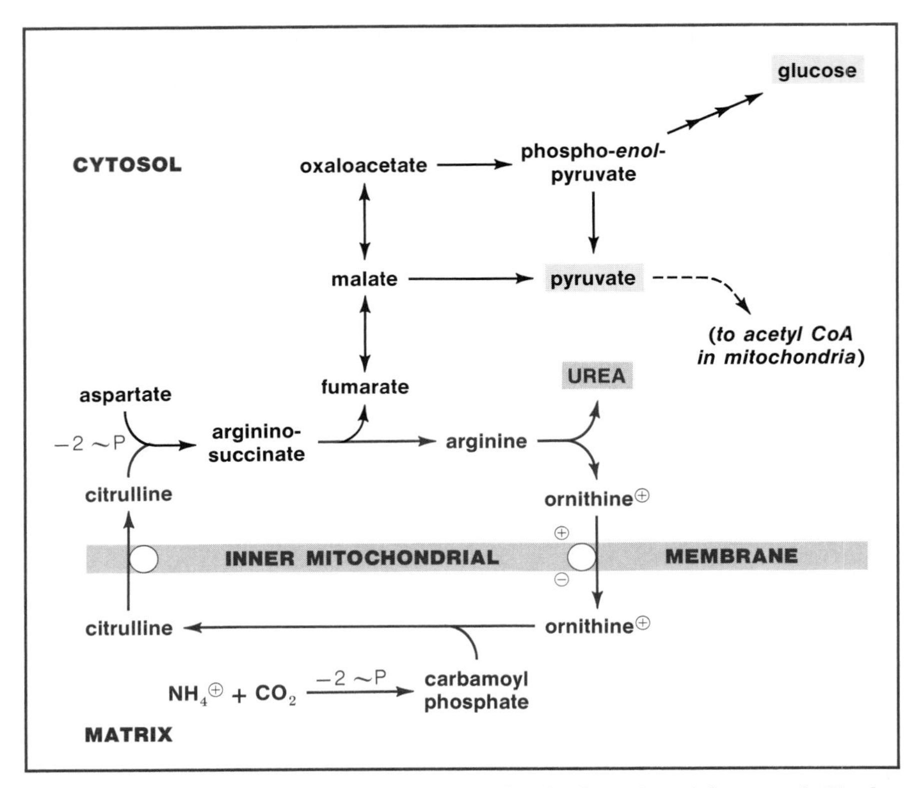

FIGURE 30–8

The partition of urea synthesis between the mitochondrial matrix and the cytosol. Citrulline synthesis occurs in the matrix, and arginine synthesis and hydrolysis occur in the cytosol. After transfer of nitrogen from aspartate, the remaining fumarate can be used as a precursor of pyruvate or glucose. The energy for pumping ornithine into the matrix is supplied by the electrical potential across the inner membrane.

Disposal of Excess Alanine (Fig. 30–9). The nitrogen in excess alanine will appear in glutamate in both the cytosol and mitochondrial matrix of human liver cells, because both compartments contain alanine aminotransferase:

$$\text{alanine} + \alpha\text{-ketoglutarate} \longrightarrow \text{pyruvate} + \text{glutamate}.$$

Alanine can therefore contribute either or both nitrogen atoms of urea because ammonium ion can be generated from glutamate in the mitochondria (by glutamate dehydrogenase), and aspartate can be generated from glutamate in the cytosol (by aspartate aminotransferase, reusing the fumarate formed during urea synthesis as a source of oxaloacetate).

The excess **pyruvate** generated from excess alanine can be removed by the routes of fuel metabolism we discussed in earlier chapters. For example, it can be converted to acetyl coenzyme A for combustion or fatty acid synthesis (not shown in Fig. 30–9), or it may be converted to malate *via* oxaloacetate for transport out of the mitochondria into the cytosol, followed by conversion to glucose *via* phospho-*enol*-pyruvate.

The balance of the overall processes can be shown by tabulating a likely sequence of reactions. The numbers correspond to those in Figure 30–9. (High-energy phosphate, CO_2, NAD, and similar components are omitted for simplicity.*) The

*An even more interesting picture emerges when the reactions are tabulated in detail, and the use of all components including NAD and NADH in the two cellular compartments is accounted for. The results of such lengthy arithmetic are used in Chapter 32.

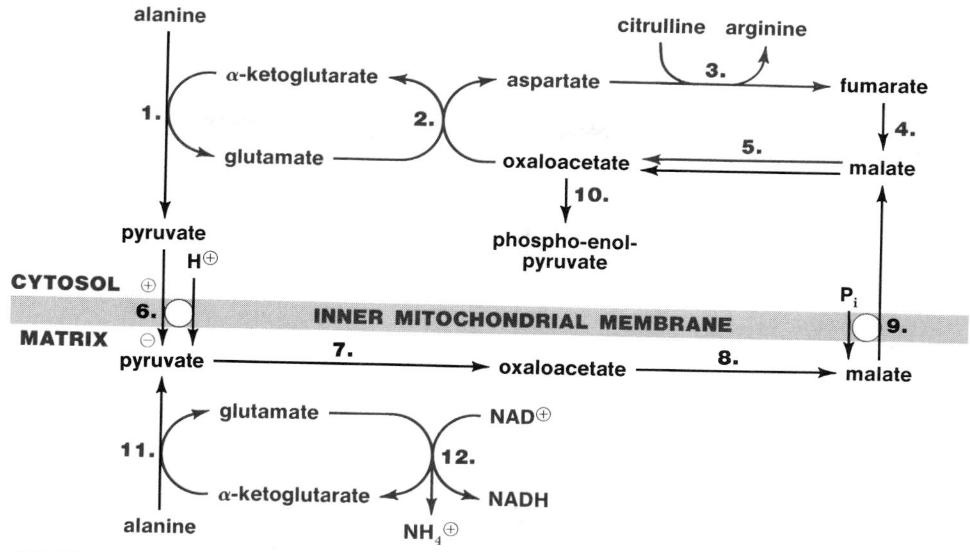

FIGURE 30–9 Coordination of alanine metabolism in the liver with other metabolic processes. Nitrogen may be removed by alanine aminotransferase in either the mitochondrial matrix or in the cytosol to appear as NH_4^+ or aspartate, respectively (brown). In either case, the pyruvate produced can be converted to malate and then to phospho-*enol*-pyruvate as a precursor of glucose (black). Reactions are discussed by number in the text.

reactions are shown in two sets to illustrate how alanine can contribute both nitrogen atoms of urea, one incorporated during arginine synthesis and the other during citrulline synthesis.

I. METABOLISM OF ALANINE DURING ARGININE SYNTHESIS

A. **Use of nitrogen and recycling of four-carbon carrier** (*cytosol*).

1. alanine + α-ketoglutarate ⟶ pyruvate + glutamate
2. glutamate + oxaloacetate ⟶ α-ketoglutarate + aspartate
3. aspartate + citrulline ⟶ arginine + fumarate
4. fumarate ⟶ malate
5. malate ⟶ oxaloacetate

Subtotal IA: alanine + citrulline ⟶ pyruvate + arginine

B. **Use of pyruvate for gluconeogenesis.** (An equally valid conversion to acetyl CoA for combustion or use in fatty acid synthesis is not shown.)

6. $\text{pyruvate}_{cytosol} \longrightarrow \text{pyruvate}_{mitochondria}$ (*An H^+ symporter effectively utilizes the proton concentration gradient to pump pyruvate.*)
7. pyruvate ⟶ oxaloacetate
8. oxaloacetate ⟶ malate
9. $\text{malate}_{mitochondria} \longrightarrow \text{malate}_{cytosol}$ (*An exchange for P_i. Since P_i and malate are equally charged for this exchange, it is not actively pumped.*)

Cytosol:

5. malate ⟶ oxaloacetate
10. oxaloacetate ⟶ phospho-*enol*-pyruvate

Subtotal IB: pyruvate ⟶ phospho-*enol*-pyruvate

TOTAL I: alanine + citrulline ⟶ arginine + phospho-*enol*-pyruvate

II. METABOLISM OF ALANINE DURING CITRULLINE SYNTHESIS

A. **Use of nitrogen** (*mitochondria*).

11. alanine + α-ketoglutarate $\longrightarrow$ pyruvate + glutamate
12. glutamate $\longrightarrow$ α-ketoglutarate + $NH_4^{\oplus}$

not shown: $NH_4^{\oplus} + CO_2 \longrightarrow$ carbamoyl phosphate
carbamoyl phosphate + ornithine $\longrightarrow$ citrulline

Subtotal IIA: alanine + ornithine $\longrightarrow$ pyruvate + citrulline

B. **Use of pyruvate for gluconeogenesis**—same as IB except transfer of pyruvate across inner membrane not necessary.

TOTAL II: alanine + ornithine $\longrightarrow$ phospho-*enol*-pyruvate + citrulline

Disposal of Excess Glutamate (Fig. 30–10). Excess glutamate can be handled in two ways within mitochondria. It may be oxidized by glutamate dehydrogenase to release NH_4^+, or its nitrogen may be transferred to oxaloacetate to form aspartate. In either case, the carbon skeleton of the excess glutamate appears as α-ketoglutarate, and an excess of this compound within mitochondria is oxidized to malate by the reactions of the citric acid cycle. (Here we have another illustration of the extent to which these reactions are used for other purposes within the liver.) The fate of malate depends upon what else is happening. When aspartate is being withdrawn for use in the cytosol, the depletion of oxaloacetate to make aspartate will accelerate the oxidation of malate to oxaloacetate. The carbon skeletons will then appear as malate *via* fumarate within the cytosol, owing to urea synthesis. When glutamate is removed by oxidation, the accumulating malate will pass directly into the cytosol. By either route, malate in the cytosol is available for synthesis of fatty acids or of glucose.

A likely balance can also be cast for glutamate metabolism in the liver; again omitting high-energy phosphate, CO_2, NAD, and NADH:

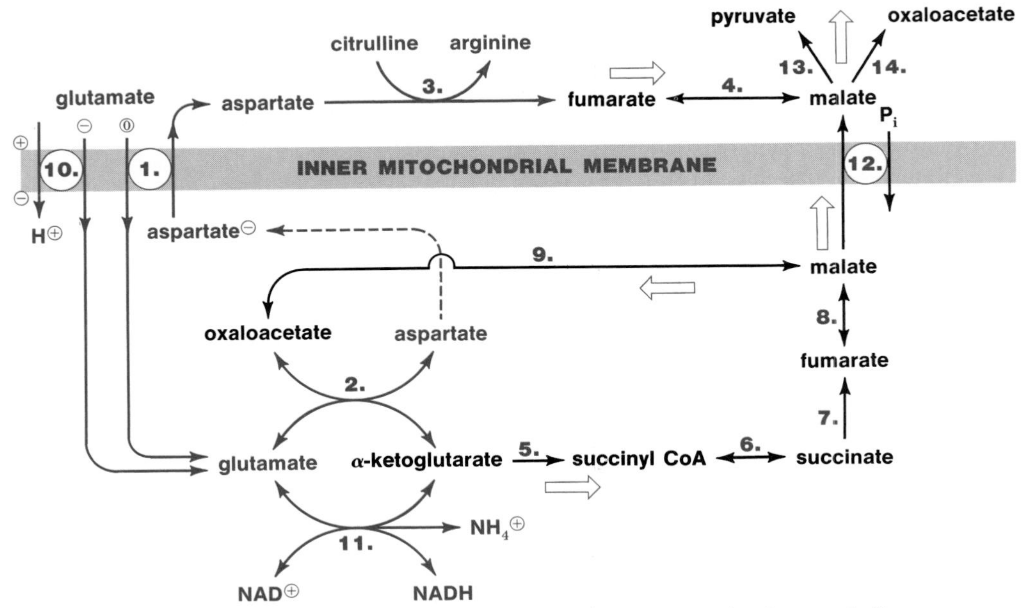

FIGURE 30–10 Coordination of glutamate metabolism in the liver with other metabolic processes. Reactions are discussed by number in the text.

I. USE OF GLUTAMATE IN ARGININE SYNTHESIS

1. glutamic acid$_{cytosol}$ + aspartate$^{\ominus}_{mitochondria}$ $\longrightarrow$ glutamic acid$_{mitochondria}$ + aspartate$^{\ominus}_{cytosol}$

(This transport is driven by the potential gradient across the inner mitochondrial membrane because the carrier will react only with the uncharged glutamic acid and with the aspartate anion. In addition, the transport is effectively discharging the proton concentration gradient, since the glutamic acid released in the matrix will again ionize:

glutamic acid $\longrightarrow$ glutamate$^{\ominus}$ + H$^{\oplus}$.)

Mitochondria:

2. glutamate + oxaloacetate $\longrightarrow$ α-ketoglutarate + aspartate
5–9. α-ketoglutarate $\longrightarrow$ oxaloacetate (reactions of citric acid cycle*)

Cytosol:

3. aspartate + citrulline $\longrightarrow$ fumarate + arginine
4. fumarate $\longrightarrow$ malate
13 or 14. malate $\longrightarrow$ pyruvate (*fat synthesis*) or malate $\longrightarrow$ oxaloacetate (*gluconeogenesis via phospho*-enol-*pyruvate*)

TOTAL I: glutamate + citrulline $\longrightarrow$ arginine + pyruvate (or oxaloacetate)

II. USE OF GLUTAMATE IN CITRULLINE SYNTHESIS

10. glutamate$^{\ominus}_{cytosol}$ + H$^{\oplus}_{cytosol}$ $\longrightarrow$ glutamate$^{\ominus}_{mitochondria}$ + H$^{\oplus}_{mitochondria}$

(This reaction uses the proton concentration gradient to pump glutamate into the mitochondrial matrix.)

Mitochondria:

11. glutamate $\longrightarrow$ α-ketoglutarate + $NH_4^{\oplus}$
not shown: $NH_4^{\oplus}$ + ornithine $\longrightarrow$ citrulline (via carbamoyl phosphate)
5–8. α-ketoglutarate $\longrightarrow$ malate
12. malate$_{mitochondria}$ + $P_{i_{cytosol}}$ $\longrightarrow$ malate$_{cytosol}$ + $P_{i_{mitochondria}}$
13 or 14. malate $\longrightarrow$ pyruvate or oxaloacetate

TOTAL II: glutamate + ornithine $\longrightarrow$ citrulline + pyruvate (or oxaloacetate)

AMMONIA TOXICITY—HEPATIC COMA AND GENETIC DEFECTS

Ammonia is toxic to the brain, and yet half of all of the excess nitrogen that is to be excreted as urea must first appear as ammonia. The ability to handle large quantities of ammonia and yet prevent it from leaking into the peripheral circulation depends upon both the high capacity of the liver for urea synthesis and the partition of metabolic functions among the tissues. Many tissues remove ammonia from the peripheral circulation and form glutamine. The skeletal muscles account for half of this peripheral clearance of ammonia. Using the intestinal mucosa to degrade glutamine ensures that any ammonia produced is delivered directly to the liver through

*The sum of reactions 2 and 5–9 is a mitochondrial oxidation of glutamate to aspartate:

$$\text{glutamate} + 2\,O_2 + 12\,(\text{ADP} + P_i) \longrightarrow \text{aspartate} + CO_2 + 12\,\text{ATP}$$

the portal circulation, along with any ammonia formed in the intestinal lumen by bacterial action. (The bacteria can produce ammonia not only from proteins, but also from urea diffusing into the bowel from the circulation. Thus, the capacity for urea synthesis must be even greater than would be estimated by the daily loss in the urine.) The intestinal mucosa also aids urea synthesis by converting part of the nitrogen to citrulline, which is then delivered to the liver for completion of the synthetic cycle.

Ammonia toxicity occurs in previously normal adults if they sustain liver damage, or if ammonia is introduced directly into the general circulation at a rate greater than it can be cleared by the peripheral tissues. The latter circumstance is uncommon; it has occurred with bacterial infections in the bladder and in patients in whom the drainage from the kidneys was surgically diverted to the lower bowel. In both conditions, there is direct venous return of the resultant ammonia to the heart, bypassing the liver. (Any wastage of the muscles further aggravates these conditions by diminishing the capacity to clear the peripheral blood of ammonia.)

Degeneration of the liver arises from a number of causes, but perhaps the most common is too much alcohol coupled with too little of other dietary constituents, leading to chronic hepatic cirrhosis.* More acute damage is caused by toxic agents, such as halogenated hydrocarbons (e.g., halothane or carbon tetrachloride), or by viral infections. A high concentration of ammonium ions results, and this may be a major factor in causing the coma frequently seen with both chronic and acute damage. (The glial cells in the brain—the cells surrounding neurons—are particularly affected, and assume the same abnormal forms that can be created experimentally by toxic doses of ammonium ions. These cells are apparently responsible for much of the ammonium ion turnover in the brain; they are the sites of glutamine synthesis.)

There have been various reports of improvements in hepatic coma through the administration of glutamate, ornithine, or citrulline to bolster the inadequate rate of ammonium ion removal, but the consequences of hepatic degeneration are usually so manifold that little can be done other than to diminish the sources. The easiest way to do this is to cut down on the protein intake, but this is not without its own hazards, since many of the individuals with chronic hepatic failure have already been on a diet deficient in protein.

Some help is obtained by diminishing the quantity of ammonium ion entering the portal circulation from the bowel. This can be done by giving antibiotics to decrease the intestinal bacteria, because they produce large bursts of ammonium ion after a protein meal and also form a urease that hydrolyzes urea back to ammonium and bicarbonate ions. (A substantial amount of the excess nitrogen in the body normally recirculates through urea in this way.)

Absorption of ammonium ions is also diminished by feeding a nonabsorbable synthetic sugar, lactulose, that is fermented by intestinal bacteria so as to make the contents more acidic. Only neutral ammonia will go across cell membranes, and increasing the $[H^+]$ increases the $[NH_4^+]/[NH_3]$ ratio, thereby diminishing the rate of absorption even though the total ammonia-ammonium ion pool is unchanged.

None of these measures solves the overall problem completely, because half of the nitrogen atoms must pass through ammonium ion on their way to urea, and elimination of all ammonium ion formation would also eliminate all urea formation.

Genetic Defects in Urea Synthesis. One out of every 2500 newborn infants can be expected to have a severe deficiency of one of the six enzymes necessary for

*In older days when the University of Virginia had a certain reputation for bibulosity, an in-house pathologist said that no member of the faculty came to autopsy with a normal liver. Now, not even the Big Ten can boast a more soberly dedicated collection of savants.

normal urea synthesis: carbamoyl phosphate synthetase, acetylglutamate synthetase, ornithine carbamoyl transferase, argininosuccinate synthetase, argininosuccinase, and arginase. Any of the deficiencies causes hyperammonemia and mental impairment. First, it must be emphasized that infants with a complete absence of an enzymatic activity necessary for urea synthesis die promptly after the first feeding of a protein meal, unless aggressive therapy is begun immediately. These total impairments are described as "neonatal" for this reason.

Partial defects permitting continued, albeit often uncertain, survival are known for each of the enzymes involved in urea synthesis. The most damaging are those affecting the first two enzymes, carbamoyl phosphate synthetase or ornithine carbamoyl transferase. Surprisingly, those who survive a partial impairment usually have a normal, or nearly normal, excretion of urea. That seems strange. What, then, is the difficulty? The problem is that the rate of urea synthesis is normal only because the ammonium ion or carbamoyl phosphate concentrations have risen to a higher fraction of the K_M value so that fewer enzyme molecules can work faster. It is the higher ammonium ion concentration (as high as one millimolar), not a failure in urea synthesis, that is responsible for the symptoms, which include lethargy, stupor, vomiting, convulsions, and other indications of central nervous system impairment. Fewer than 100 patients with these two deficiencies have been described.

Partial impairments of the enzymes converting citrulline to arginine sometimes have less severe consequences, probably because the ammonium ion concentrations do not rise to such high levels. Symptoms may appear gradually over weeks or months, but some degree of mental retardation is common. A substantial fraction of the nitrogen in the urine appears as citrulline (argininosuccinate synthetase deficiency), or as argininosuccinate (argininosuccinase deficiency), although urea is still the major product. In these, as with other point mutations, a spectrum of abnormalities is created, depending upon the part of the enzyme molecule affected. It was demonstrated through examination of the enzyme in cultured fibroblasts that the mutation of argininosuccinate synthetase in one patient resulted in an altered K_M for citrulline, from 0.4 to over 10 millimolar, which would in itself lead to an increased accumulation of citrulline before its rate of removal balanced its rate of formation.

Therapy of the genetic defects in urea synthesis includes immediate relief of severe hyperammonemia by peritoneal dialysis, then diminution of the nitrogen load and provision of alternative routes of nitrogen excretion. The protein content of the diet can be diminished if sufficient essential amino acids are supplied (p. 821). The nitrogen content can be lowered further by supplying the 2-keto analogues in place of some of the amino acids. (The aminotransferases convert the 2-keto compounds to the amino acids.) However, this is very expensive therapy to continue for extended periods.

Nitrogen excretion can be increased in different ways. One promising technique is an adaptation of a laboratory exercise to which students in biochemistry courses were frequently subjected in the past. If one eats benzoic acid (as sodium benzoate), it is converted to the glycine conjugate **hippuric acid**, which is excreted in the urine:

COO⊖ (benzoate) + HS—CoA, ATP → AMP + PP_i → C(=O)—S—CoA (**benzoyl coenzyme A**) + **glycine** → C(=O)—N(H)—CH_2—COO⊖ (**hippurate**)

benzoate | HS—CoA | ATP | AMP + PP_i | **benzoyl coenzyme A** | **glycine** | **hippurate**

The benzoate is converted to benzoyl coenzyme A in liver mitochondria, and the benzoyl group is then transferred to glycine. Sodium benzoate has a very low toxicity, and it has been successfully used to increase nitrogen excretion as hippuric acid in an infant with ornithine carbamoyl transferase deficiency. Phenylacetic acid may prove to be even more useful, because it is excreted as the glutamine conjugate, in which two moles of nitrogen are lost:

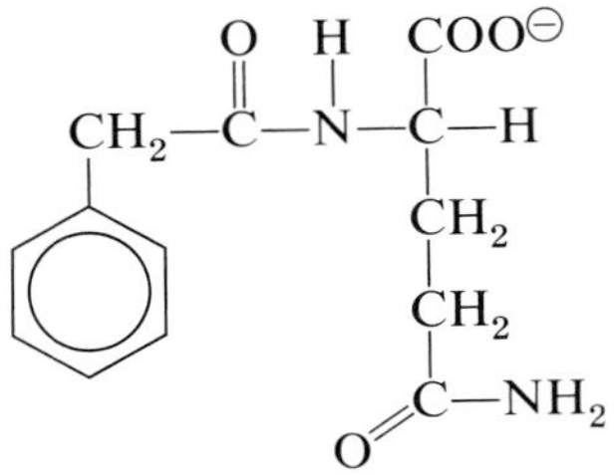

phenylacetylglutamine

Argininosuccinate is itself a relatively nontoxic and readily excreted end product, so infants with a **deficiency of argininosuccinase** can be successfully treated by feeding increased amounts of arginine, not only to ensure an ample supply of this amino acid for protein synthesis, but also to increase the nitrogen excretion. (The excess arginine is cleaved to urea and ornithine. The urea is readily excreted, and the ornithine picks up two more nitrogen atoms upon conversion to argininosuccinate, which is also excreted.) Arginine administration is not adequate therapy for **citrullinemia (argininosuccinate synthetase deficiency),** because only one nitrogen atom can be added to the resultant ornithine, and citrulline is reabsorbed in the kidney, so less is excreted. This condition is therefore treated with sodium benzoate and dietary manipulation.

An infant with a **deficiency of acetylglutamate synthetase** has been successfully treated with N-carbamoylglutamate. Unlike N-acetylglutamate, the natural activator, this compound can enter mitochondria, where it serves as an activator of carbamoylphosphate synthetase.

FURTHER READING

H.G. Windmueller (1982). Glutamine Utilization by the Small Intestine. Adv. Enzymol. 53:201.

A. Meister (1981). On the Cycles of Glutathione Metabolism and Transport. Curr. Top. Cell. Regul. 18:21.

P. Lund (1980). Glutamine Metabolism in the Rat. FEBS Lett. 117 Suppl.:K86

J.D. Butler and S.P. Spielberg (1981). Decrease of Intracellular Cystine Content in Cystinotic Fibroblasts by Inhibitors of γ-Glutamyl Transpeptidase. J. Biol. Chem. 256:4160.

O.W. Griffith (1981). The Role of Glutathione Turnover in the Apparent Renal Secretion of Cystine. J. Biol. Chem. 256:12263.

D. Wellner and A. Meister (1981). A Survey of Inborn Errors of Amino Acid Metabolism and Transport in Man. Annu. Rev. Biochem. 50:911.

M. Walser (1982). Urea Cycle Disorders and Other Hereditary Hyperammonemic Syndromes. p. 402 in J.B. Stanbury et al., eds. The Metabolic Basis of Inherited Disease, 5th ed. McGraw-Hill.

L.E. Rosenberg and C.R. Scriver (1980). Disorders of Amino Acid Metabolism. p. 583 in P.K. Bondy and L.E. Rosenberg, eds. Metabolic Control and Disease, 8th ed. Saunders.

C. Bachman et al. (1981). N-Acetylglutamate Synthetase Deficiency: A Disorder of Ammonia Detoxication. N. Engl. J. Med. 304:543.

M.L. Batshaw et al. (1982). Treatment of Inborn Errors of Urea Synthesis: Activation of Alternative Pathways of Waste Nitrogen Synthesis and Excretion. N. Engl. J. Med. 306:1387.

CHAPTER 31

AMINO ACIDS: DISPOSAL OF THE CARBON SKELETONS

When amino acids are directly degraded without being used to make other cellular constituents, all of the carbon atoms appear as CO_2 or as eight familiar intermediates of fuel metabolism (Fig. 31–1). These are acetoacetate, acetyl coenzyme A, crotonyl coenzyme A, pyruvate, α-ketoglutarate, succinyl coenzyme A, fumarate, or oxaloacetate. How has this come about? It seems reasonable to assume that the 20 amino acids found in proteins were selected from many possible structures during early evolution, not only because of the particular properties they contribute to polypeptides, but also because their metabolism can be integrated easily with the metabolism of glucose and fatty acids.

The ultimate disposition of the carbon atoms from amino acids therefore depends upon what is happening to fatty acids and carbohydrates, with these exceptions: **acetyl groups formed within muscles, and acetoacetate from any source, are oxidized to CO_2 and H_2O at all times.** Within the liver, where most carbon skeletons are handled, they may at some times be used to form glucose and 3-oxybutyrates, and at other times to form fatty acids, as was noted for alanine and glutamate in the preceding chapter. Let us review the choices.

During fasting or times of low carbohydrate intake when fatty acids are the principal fuel, pyruvate and intermediates in the citric acid cycle are converted to glucose:

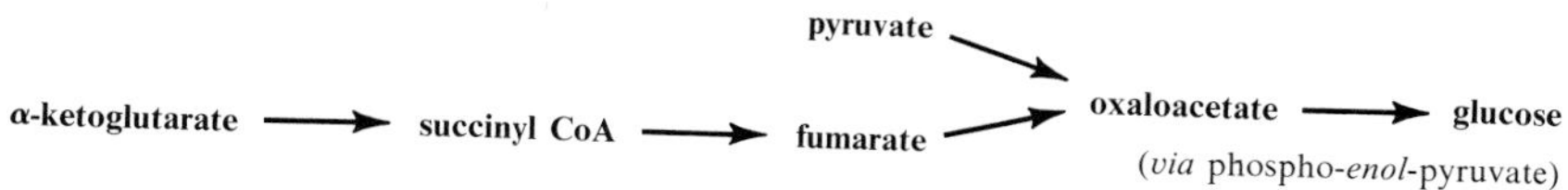

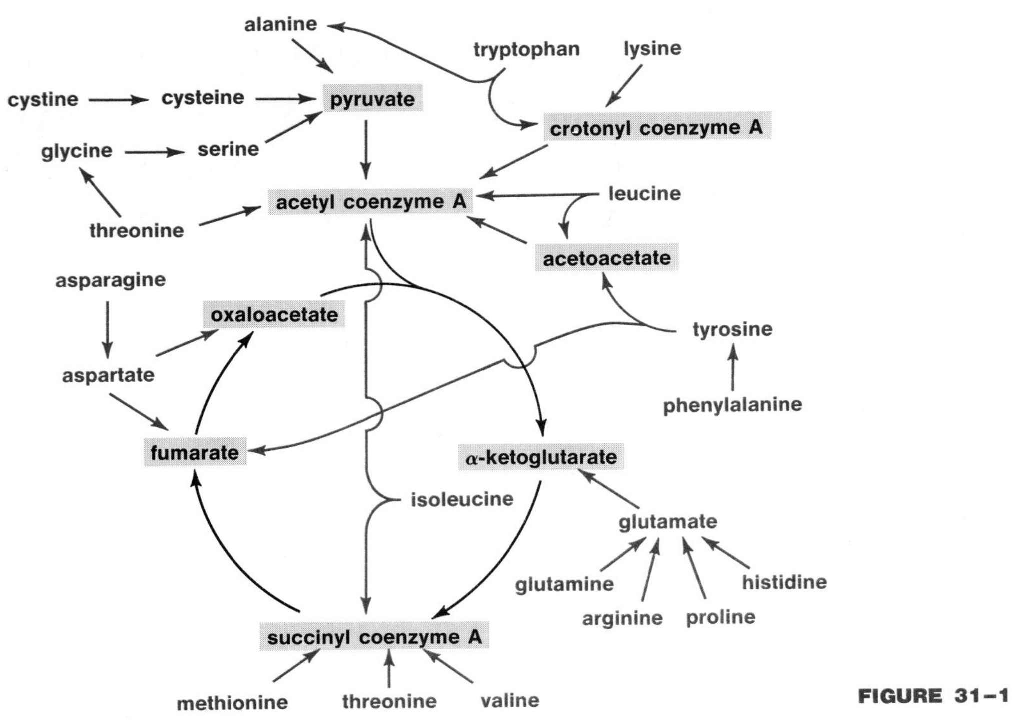

FIGURE 31–1

Outline of the fate of the carbon skeletons of the amino acids when used as fuels.

At the same time, intermediates of fatty acid oxidation (crotonyl coenzyme A and acetyl coenzyme A) are converted to the ketone bodies:

$$\text{crotonyl CoA} \longrightarrow \text{acetyl CoA} \longrightarrow \text{acetoacetate} \longrightarrow \text{D-3-hydroxybutyrate}$$

Glucose and the ketone bodies are then used by the other tissues. Amino acids acting as precursors of glucose under these conditions are said to be glucogenic; those acting as precursors of the ketone bodies are said to be ketogenic.

After a carbohydrate meal when glucose is already present in ample supply, part of the intermediates may be converted first to acetyl coenzyme A and then to fatty acids for storage:

$$\alpha\text{-ketoglutarate} \longrightarrow \text{succinyl CoA} \longrightarrow \text{fumarate} \longrightarrow \text{malate} \longrightarrow \text{pyruvate} \longrightarrow \text{acetyl CoA} \longrightarrow \text{palmitate} \longrightarrow \text{oleoyl CoA} \longrightarrow \text{triglycerides}$$

With this background, let us consider in this and the following chapter some routes by which each amino acid's carbon skeleton is converted to these common intermediates and then correlate the results. (We do not have time to consider all known possibilities, but those we shall discuss are the major alternatives.) Routes will also be listed by which the nitrogen atoms can appear as NH_4^+ or as aspartate for incorporation into urea. **All of the amino acids are glucogenic during starvation unless otherwise specified.**

Aspartate

Nitrogen: Used directly in arginine synthesis, or appears as NH_4^+ in purine nucleotide cycle (pp. 582 and 583).

Carbons: Appear as fumarate in arginine synthesis or in purine nucleotide cycle. Also may appear as oxaloacetate as a result of transamination in peripheral tissues.

GLUTAMATE AND ITS PRECURSORS

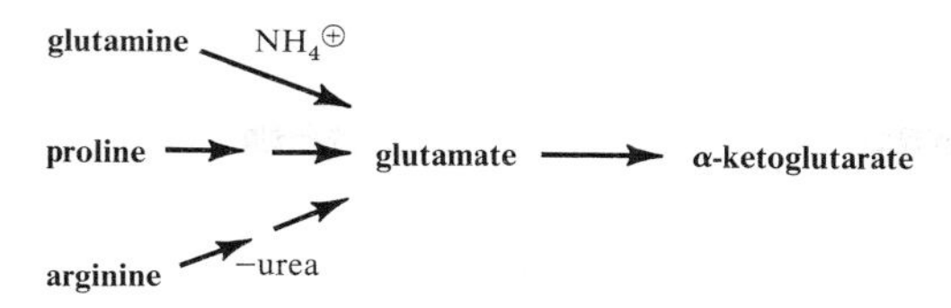

(Histidine also is converted to glutamate, as discussed in the next chapter.)

Glutamate

Nitrogen: Appears directly as NH_4^+ (glutamate dehydrogenase, p. 578) or as aspartate (aspartate aminotransferase, p. 577).

Carbons: Appear directly as α-ketoglutarate. As noted before (p. 588), this compound is oxidized to malate by the reactions of the citric acid cycle, producing high-energy phosphate. The malate passes into the cytosol, where it will be converted to glucose (*via* oxaloacetate and phospho-*enol*-pyruvate), or to fatty acids (*via* pyruvate and acetyl coenzyme A). Therefore, three of the five carbon atoms may appear as glucose, or two as fatty acids, with the remainder appearing as CO_2. (We are speaking here of net formation of products, not of isotope incorporation.)

Glutamine

Nitrogen: Amide nitrogen appears as NH_4^+ (glutaminase, p. 577).

Carbons: Appear as glutamate, which is then metabolized as given above.

Proline

Nitrogen and carbon skeleton: Appear as glutamate.

The transformation of L-proline to L-glutamate is simple. Proline is oxidized by a cytochrome-linked enzyme in mitochondria, forming an unsaturated ring. Although simple hydrolysis of the ring at the double bond spontaneously yields the semialdehyde of glutamate, it is the ring form, pyrroline carboxylate, that is directly oxidized to yield free glutamate (Fig. 31–2).

Arginine

Nitrogens: Two appear directly as urea and one appears in glutamate.

Carbons: One appears directly as urea and five appear in glutamate.

As we have seen, arginine is hydrolyzed to urea and ornithine as a part of the process of urea synthesis. Ornithine is re-used in the urea cycle, but excess arginine derived from the diet or tissue proteins causes the formation of excess ornithine beyond the amount needed to sustain the urea cycle.

The terminal amino group of ornithine is transferred to α-ketoglutarate:

L-arginine ($H_2N{\cdots}C(\oplus){\cdots}NH_2$, NH—CH_2—CH_2—CH_2—H—C—$NH_3^{\oplus}$, $COO^{\ominus}$) + H_2O → (*arginase*, *Mn^{2+}*) → H_2N—C(=O)—NH_2 (**urea**) + L-ornithine ($^{\oplus}NH_3$—CH_2—CH_2—CH_2—H—C—$NH_3^{\oplus}$, $COO^{\ominus}$)

L-ornithine ⇌ (*ornithine aminotransferase*; α-Kg* → Glu*) ⇌ glutamate semialdehyde (H—C=O, CH_2, CH_2, H—C—$NH_3^{\oplus}$, $COO^{\ominus}$) ⟷ Δ¹-pyrroline 5-carboxylate → ($NAD^{\oplus}$ → NADH) → glutamate

L-arginine **L-ornithine** **glutamate semialdehyde**

*α-Kg = α-ketoglutarate
Glu = glutamate

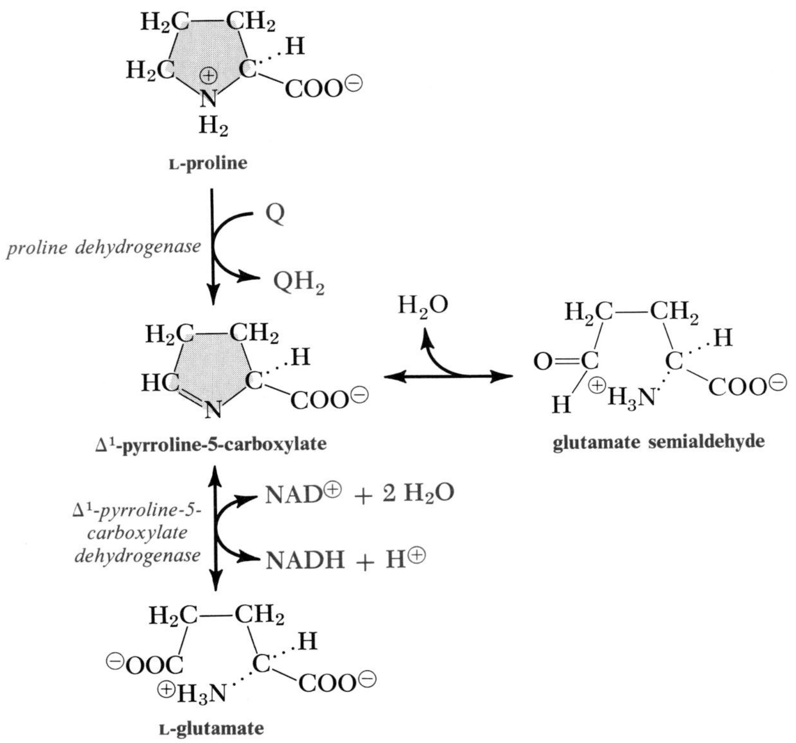

L-proline

proline dehydrogenase

Q → QH_2

H_2O

Δ^1-pyrroline-5-carboxylate ⇌ glutamate semialdehyde

Δ^1-pyrroline-5-carboxylate dehydrogenase

$NAD^{\oplus} + 2\ H_2O$ → $NADH + H^{\oplus}$

L-glutamate

FIGURE 31–2

Oxidation of proline to glutamate in mitochondria.

The product is the semialdehyde of glutamate, which can be oxidized to glutamate after equilibration to Δ^1-pyrroline-5-carboxylate (see under Proline). (Recall that the equilibrium position for the oxidation of aldehydes, and therefore for the equilibrated pyrroline compound, is far in the direction of the acid [p. 428].) A hereditary deficiency of the aminotransferase has been shown to cause atrophy of the retina, resulting in blindness.

PRECURSORS OF PYRUVATE

Alanine

Nitrogen: Appears in glutamate by transamination (p. 578).
Carbons: Appear as pyruvate.

Serine

Nitrogen: Directly deaminated to NH_4^+ by serine dehydratase (p. 580).
Carbons: Appear as pyruvate.

Cysteine and Cystine

Nitrogen: Appears in glutamate by transamination.
Carbons: Appear as pyruvate.
Sulfur: Appears as sulfate.

When we discussed the formation of proteins, we noted that cysteine residues in proteins (sulfhydryl groups) are interconverted with cystine residues (disulfide groups) by exchange reactions (p. 148). This is a general type of reaction within cells so that the cystine present in proteins undergoing degradation is interconvertible with cysteine, which is the form that enters the major pathways of metabolism.

It is clear that the metabolism of cysteine produces pyruvate and inorganic sulfate as the major products, but there is still uncertainty over the routes. In what may be the major pathway, cysteine is oxidized in the cytosol to cysteine sulfinate (Fig. 31–3), which undergoes transamination. The resultant 2-keto compound releases SO_2 as sulfite; sulfite is oxidized to inorganic sulfate, which may be excreted or incorporated into other cellular constituents.

Sulfite oxidase is a complex protein, containing molybdenum and a hemoprotein resembling cytochrome b_5. It is located in the intermembrane space of mitochondria, where its electrons can be transferred to cytochrome c on the outer surface of the inner membrane. As a result, one high-energy phosphate bond is generated per sulfite oxidized. This is a trivial part of the metabolic economy, but the oxidation of sulfite is evidently not at all trivial to the function of the brain. Infants born with a congenital deficiency of sulfite oxidase have such a severe neurological impairment that they shortly become functionally decerebrate.

A variety of other possible routes of cysteine metabolism have been reported, many quantitatively minor and of uncertain purpose. For example, a small fraction of the sulfur of the body is excreted in the urine as **thiosulfate** and **thiocyanate,** which apparently arise by the route shown in Figure 31–4. The carbon skeleton appears as pyruvate in most of these routes.

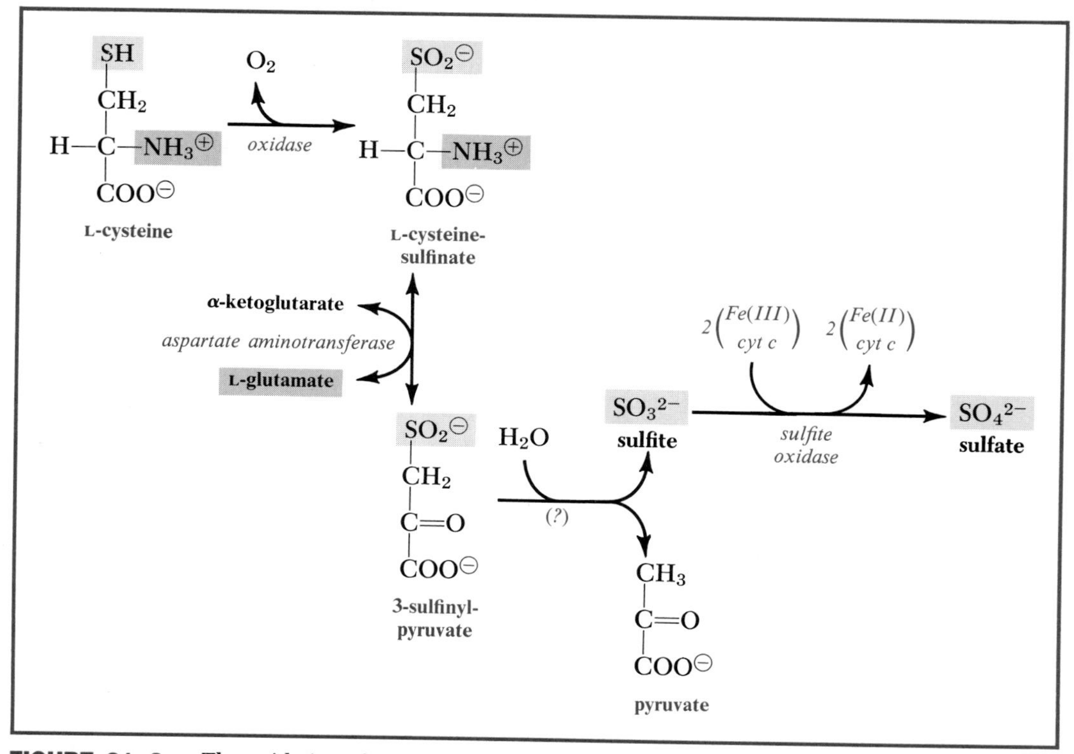

FIGURE 31–3 **The oxidation of cysteine to pyruvate and sulfite.**

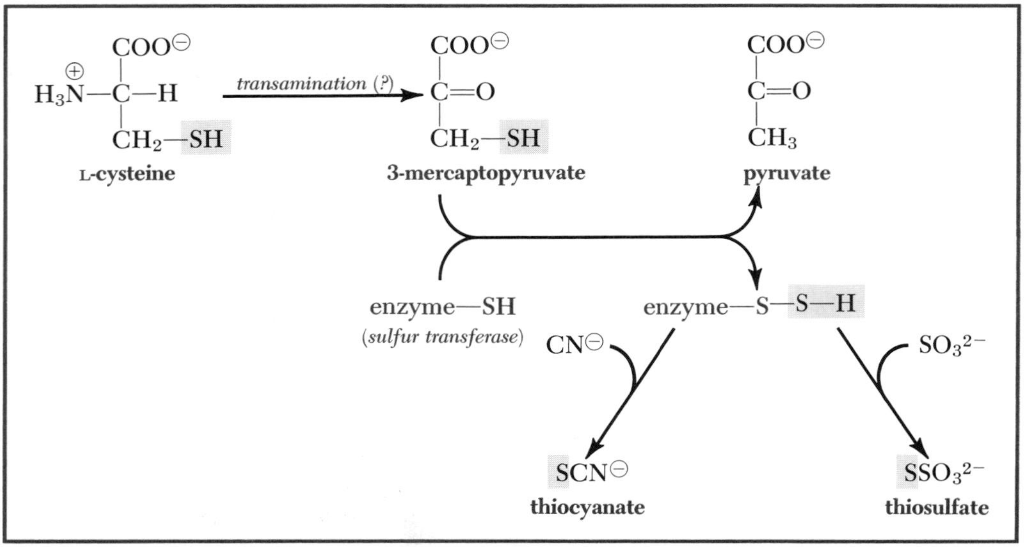

FIGURE 31–4 Minor routes of metabolism of cysteine include the formation of thiocyanate and thiosulfate.

Threonine and Propionyl Coenzyme A Metabolism

(An alternative route of threonine metabolism involving the formation of glycine is discussed in the next chapter.)

Nitrogen: Direct deamination produces NH_4^+ (serine dehydratase).

Carbons: Appear as succinyl coenzyme A *via* propionyl coenzyme A.

The initial deamination of threonine forms NH_4^+ and 2-ketobutyrate (p. 581), which is handled as the next higher homologue of pyruvate. That is, it is oxidatively decarboxylated by pyruvate dehydrogenase complex to form propionyl coenzyme A (Fig. 31–5) in the same way that pyruvate is converted to acetyl coenzyme A. Both the initial deamination and the oxidative decarboxylation are irreversible reactions.

Propionyl coenzyme A is an important intermediate in the metabolism of other amino acids, and it is also produced from the terminal three carbon atoms of fatty acids with odd-numbered chains when they are oxidized. The route for metabolism of this compound is summarized in Figure 31–6. The first reaction is catalyzed by propionyl CoA carboxylase, which contains biotin and acts by the same sort of mechanism seen with acetyl CoA carboxylase and pyruvate carboxylase. The product is **methylmalonyl coenzyme A,** which is an optically active compound. A racemase exists to catalyze the equilibration of the two isomers; the (R) isomer is the substrate for the next reaction. (It isn't at all clear why different isomers are used in this sequence.)

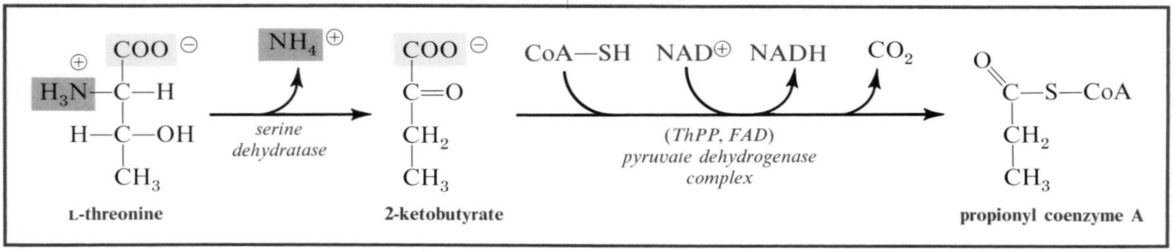

FIGURE 31–5 2-Ketobutyrate arising from threonine is oxidatively decarboxylated to propionyl coenzyme A.

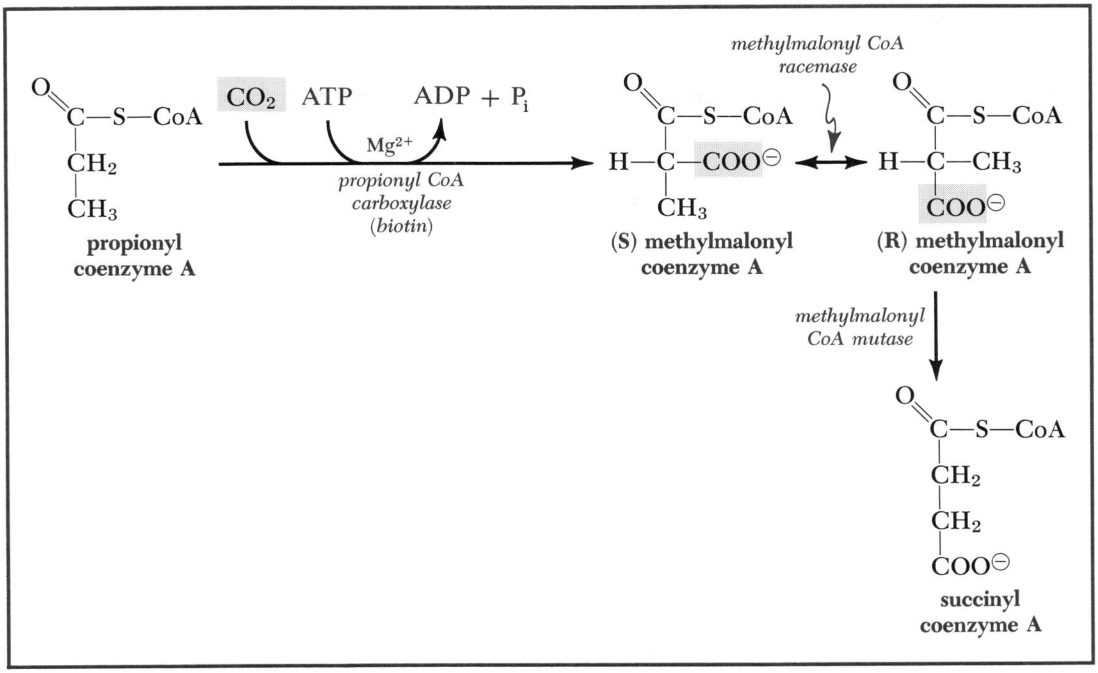

propionyl coenzyme A (CH_3—CH_2—C(=O)—S—CoA) + CO_2 + ATP → ADP + P_i (*propionyl CoA carboxylase (biotin)*, Mg^{2+}) → (S) methylmalonyl coenzyme A (H—C(CH_3)($COO^{\ominus}$)—C(=O)—S—CoA) ⟷ (*methylmalonyl CoA racemase*) (R) methylmalonyl coenzyme A (H—C(CH_3)($COO^{\ominus}$)—C(=O)—S—CoA) → (*methylmalonyl CoA mutase*) succinyl coenzyme A ($COO^{\ominus}$—CH_2—CH_2—C(=O)—S—CoA)

FIGURE 31–6 **Propionyl coenzyme A is metabolized by a carboxylation followed by rearrangements to form succinyl coenzyme A.**

The final reaction is the rearrangement of (R)-methylmalonyl coenzyme A to succinyl coenzyme A, catalyzed by a mutase. Since succinyl coenzyme A is converted to oxaloacetate, it follows that threonine and all other compounds giving rise to propionyl coenzyme A or methylmalonyl coenzyme A are glucogenic.

We have not seen anything comparable to the mutase reaction heretofore. It involves a coenzyme derived from vitamin B_{12}, and let us consider this important compound.

Cobalamin Coenzyme and Vitamin B_{12}

The coenzyme for methylmalonyl CoA mutase is **adenosylcobalamin** (Fig. 31-7)—a very complicated molecule containing **cobalt** bound to nitrogen or carbon on all six coordination positions. The salient part of the molecule is the large tetrapyrrole ring surrounding the cobalt, which resembles a porphyrin ring superficially, but differs in being more saturated and in lacking one methylene bridge. This is known as the **corrin** ring. The cobalamins are corrinoid compounds.

The cobalamin coenzyme has two unusual chemical features. First, it contains a metal-carbon bond, the only known biological example of this linkage. Second, it is made from cob(I)alamin, the most powerful nucleophile known, in which cobalt is present at a univalent oxidation state. When cobalamin combines with an adenosyl carbonium ion (obtained from ATP) to form the cobalt-carbon bond, there is a shift of electrons to form a cobalt(III)-carbanion bond. The detailed mechanism of action of the coenzyme is still unknown, but it involves rupture of one of the cobalt coordination bonds by the substrate, with formation of free radicals and migration of the substituent groups (Fig. 31-8).

What is commonly referred to as vitamin B_{12} is the same as adenosylcobalamin except that the adenosyl group is replaced by a cyanide ion. The valence of cobalt in

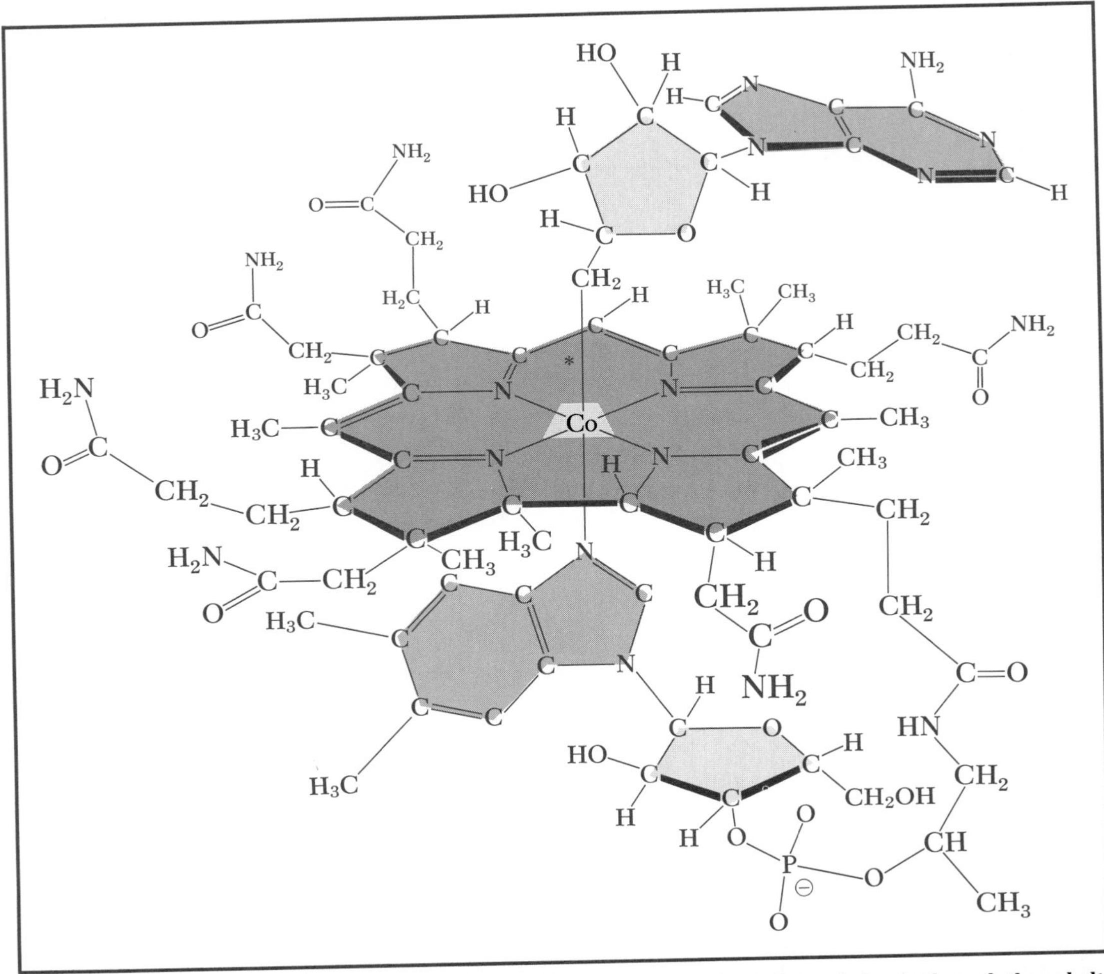

FIGURE 31–7 **Adenosylcobalamin. The adenosyl moiety (*top*) is attached to cobalamin through the cobalt atom (*center*). Cobalamin includes a corrin ring (*center color shading*) and a benzimidazole group (*bottom color shading*). The asterisk designates the bond that is broken during the function of the coenzyme. This bond is between Co(III) and a carbanion (charges not shown).**

cyanocobalamin is +3, so a net charge of +1 remains on the metal. Cyanocobalamin is the form of the compound that was originally isolated during the search for the vitamin. Although we ordinarily think of cyanide as a dangerous poison, it is produced by many microorganisms in small amounts, and the affinity of the cobalamins for cyanide is so great that cyanocobalamin may be created during the handling of the microbial cultures from which it is isolated. Indeed, large doses of hydroxycobalamin have been used in treating cases of cyanide poisoning. Most of the cobalamin in animal tissues is present as the adenosyl or methyl derivatives, which are therefore the natural dietary forms of the vitamin.

The search for the vitamin was originally spurred when it was recognized that humans with **pernicious anemia** were deficient in some dietary factor. In addition to severe anemia, the effects of this condition include serious disturbances of the central nervous system that result in abnormal sensation, motion, behavior, or thought.

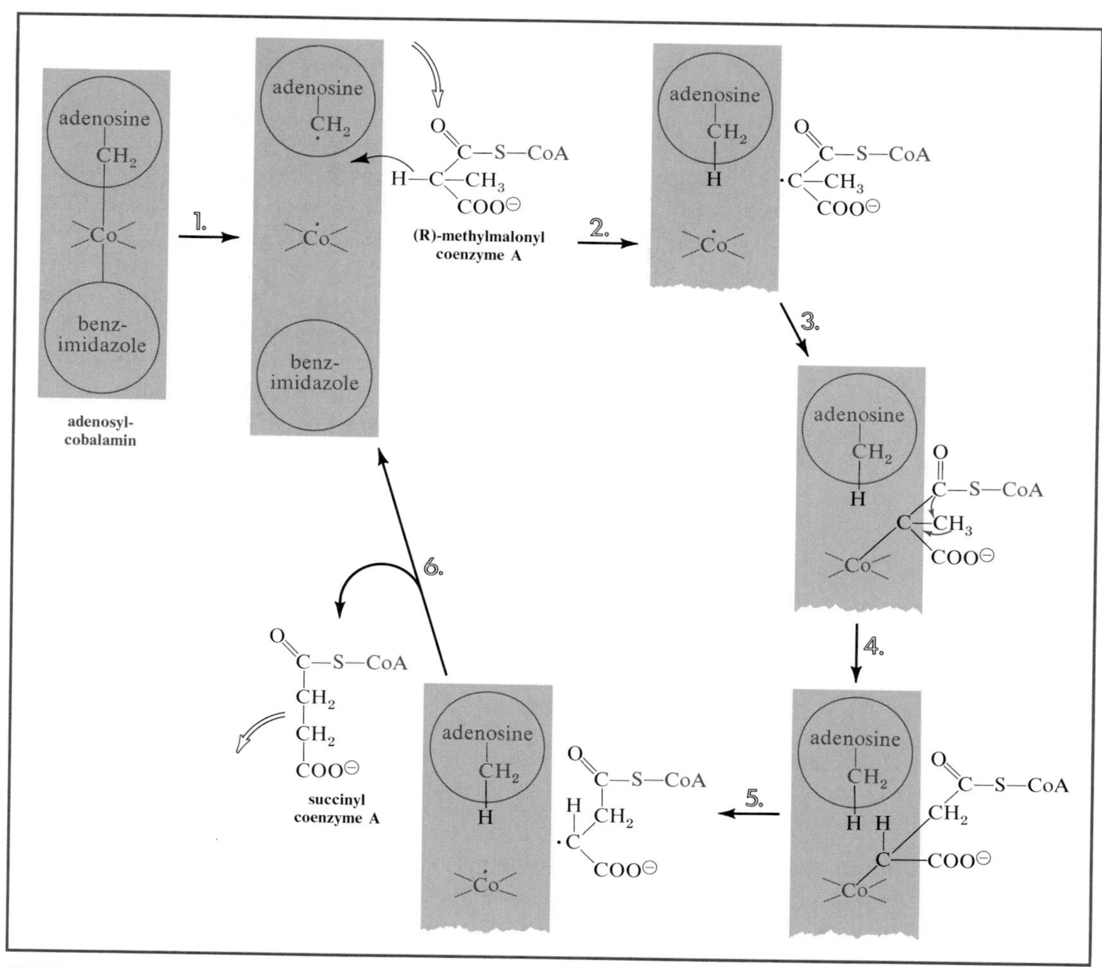

FIGURE 31–8 A postulated mechanism of action for methylmalonyl CoA mutase.

Curiously, pernicious anemia turned out to be a defect of the stomach, rather than a dietary deficiency disease. The absorption of dietary cobalamins depends upon the formation by the gastric mucosa of a carbohydrate-rich protein known as the **intrinsic factor.** People with pernicious anemia do not make this protein. Before purified cobalamins were available, treatment consisted of eating hog stomach preparations to supply the missing intrinsic factor and liver to supply the vitamin (the extrinsic factor).

PERNICIOUS ANEMIA

Described in 1855 by Dr. Thomas Addison and called pernicious because it is a fatal disease. Affects those of Northern European ancestry, including American blacks, but not Africans. Major systems affected are the hematopoietic system and the central nervous system. Degenerative changes are seen in the nervous system. There is no secretion of HCl in the stomach, and there is an increased incidence of carcinoma of the stomach.

The amount of cobalamin required in the diet is very small—about one nanomole a day for a normal individual. Even those with pernicious anemia can absorb enough if the oral dose is raised to the level of one micromole or so. The entire body contains approximately two micromoles. The cobalamins therefore represent some of the most potent biological agents known. The formation of cobalamins is the only known biological function of cobalt, but it is a critical one for animals, and this brings us to a striking paradox. It is almost a truism that the metabolic processes of plants and animals proceed with the same kinds of reactions, with the important exception of photosynthesis. It is therefore startling to discover that the yeasts, the green algae, and all of the higher plants have no need for cobalt and contain no cobalamins, despite the fact that these organisms deal with the same kinds of compounds metabolized by cobalamin-dependent reactions in animals. This drastic evolutionary parting of the metabolic way has few direct consequences over most of the world. However, there are some areas, particularly in Australia, in which the soils have a very low cobalt content. Plants grow, despite their low cobalt content, but some animals that eat the plants don't.

The need for cobalt is especially pronounced in ruminants, and it is the proclivity of the Australians to herd sheep that exposed the unsuspected deficiency in their land. Cud-chewing animals depend upon microbial fermentations in their rumen to break down cellulose in the diet. (Cellulose is a glucose polysaccharide, but unlike starch and glycogen, it is made of β-glucosyl residues and is not attacked by the usual battery of enzymes in mammalian intestines.) The fermentations produce acetate and butyrate, which can serve as sources of fatty acids, but they also convert a large part of the carbohydrate to propionate, and this is the major source of glucose for the animals.

The conversion of propionate to glucose depends upon the same sequence of reactions we have been discussing, with the addition of a thiokinase reaction to form propionyl coenzyme A. Therefore, the ruminants have a rapid propionate metabolism, including an active methylmalonyl mutase and its necessary cobalamin coenzyme. If they ingest enough cobalt, the microorganisms that digest carbohydrate will also synthesize the requisite cobalamin for them.

Defects in Propionyl Coenzyme A Metabolism

The consequences of isolated impairment of propionyl coenzyme A metabolism are shown by several rare genetic defects in the metabolic route. Impairment of propionyl CoA carboxylase causes **propionic acidemia,** and impairment of methylmalonyl CoA mutase (and perhaps the isomerase) causes **methylmalonic acidemia.** These differing conditions present with similar clinical pictures, with repeated bouts of vomiting, ketosis, and high serum glycine concentration. (Methylmalonic acidemia is also a consequence of defective cobalamin supply.) There is no good explanation for the accumulation of glycine in the conditions, but the accompanying massive ketosis and acidosis may in part result from a mimicking of acetyl coenzyme A by propionyl coenzyme A. Citrate synthase uses propionyl coenzyme A to form methylcitrate, which in turn reacts to form methylisocitrate. However, the methyl group blocks any further reaction, so the compound is an inhibitor of isocitrate dehydrogenase. Failure to oxidize acetyl groups would lead to excessive formation of acetoacetate.

BRANCHED CHAIN AMINO ACIDS: LEUCINE, ISOLEUCINE, AND VALINE

Nitrogen: Transferred from all to α-ketoglutarate, forming glutamate.

Carbons: Leucine: Converted to 3-hydroxy-3-methylglutaryl coenzyme A, the

AMINOACIDOPATHIES

Over 122 inherited disorders are known in the metabolism or transport of amino acids. The first genetic defects labeled as inborn errors of metabolism (Garrod, 1923) were aminoacidopathies. These usually recessive conditions can result from any one of a number of different mutations. The effects of different alterations in the same enzyme may be quite different; on the other hand, two patients with similar manifestations of illness may have different enzyme deficiencies. Some mutations affect the supply of a necessary coenzyme or weaken the affinity of an apoenzyme for its coenzyme; these can be corrected in some cases by administering large doses of an appropriate vitamin (cobalamin, pyridoxine, thiamine, biotin). Such therapy should always be tried when an aminoacidopathy is encountered.

precursor of acetoacetate (p. 452).

Isoleucine: Converted to equal amounts of succinyl coenzyme A (*via* propionyl coenzyme A) and acetyl coenzyme A, or to butyryl coenzyme A by a minor route.

Valine: Converted to succinyl coenzyme A *via* propionyl coenzyme A.

The branched chain amino acids, unlike most of the others, are taken up by the striated muscles after a protein meal and are at least partially oxidized in those tissues. The brain also may use significant amounts. Still higher concentrations are taken up by the liver. Whatever the partition may be between the tissues, the general features of the routes seem clear.

The amino groups from all of the branched chain amino acids can be transferred to α-ketoglutarate. Each of the remaining 2-ketocarboxylates then undergoes an oxidative decarboxylation to form the coenzyme A esters (Fig. 31–9).

Some human genetic defects in the enzymes catalyzing these first two steps are illuminating. One rare defect causes **hypervalinemia,** in which there is an accumulation of valine without an accompanying accumulation of leucine or isoleucine. Another rare defect causes an accumulation of both leucine and isoleucine. Evidently there is one aminotransferase for valine and another for leucine and isoleucine in humans.

Approximately one out of 250,000 infants is born with **maple syrup urine disease,** in which both the branched chain amino acids and their 2-keto analogues accumulate. It is caused by a defect in the branched chain 2-ketoacid dehydrogenase complex. (This complex is like α-ketoglutarate dehydrogenase; it contains thiamine pyrophosphate, lipoate, and FAD, and requires NAD and coenzyme A as cofactors.) Since the aminotransferase reactions are readily reversible, an accumulation of a 2-keto acid causes a corresponding accumulation of its precursor amino acid.

Although the name has a certain puckishness, the consequences of the classic form of the condition are grim. Affected infants appear normal at birth, but develop severe neurological damage within a few days after birth, and die within several weeks unless treated. The damage may be due to toxic effects of the accumulating 2-keto compounds, or of the amino acids themselves. Treatment of this, like many other aminoacidopathies, involves rigid dietary restrictions to reduce the intake of the affected amino acids to the minimal level consistent with growth. Continued rigid control occupies much of the family's attention and is expensive.

Let us now go on to consider the further metabolism of the branched chain skeletons and see how variations on familiar types of reactions have been developed to cope with these compounds. The coenzyme A esters produced by oxidative decarboxylation are typical fatty acid derivatives, except for the branches, and they un-

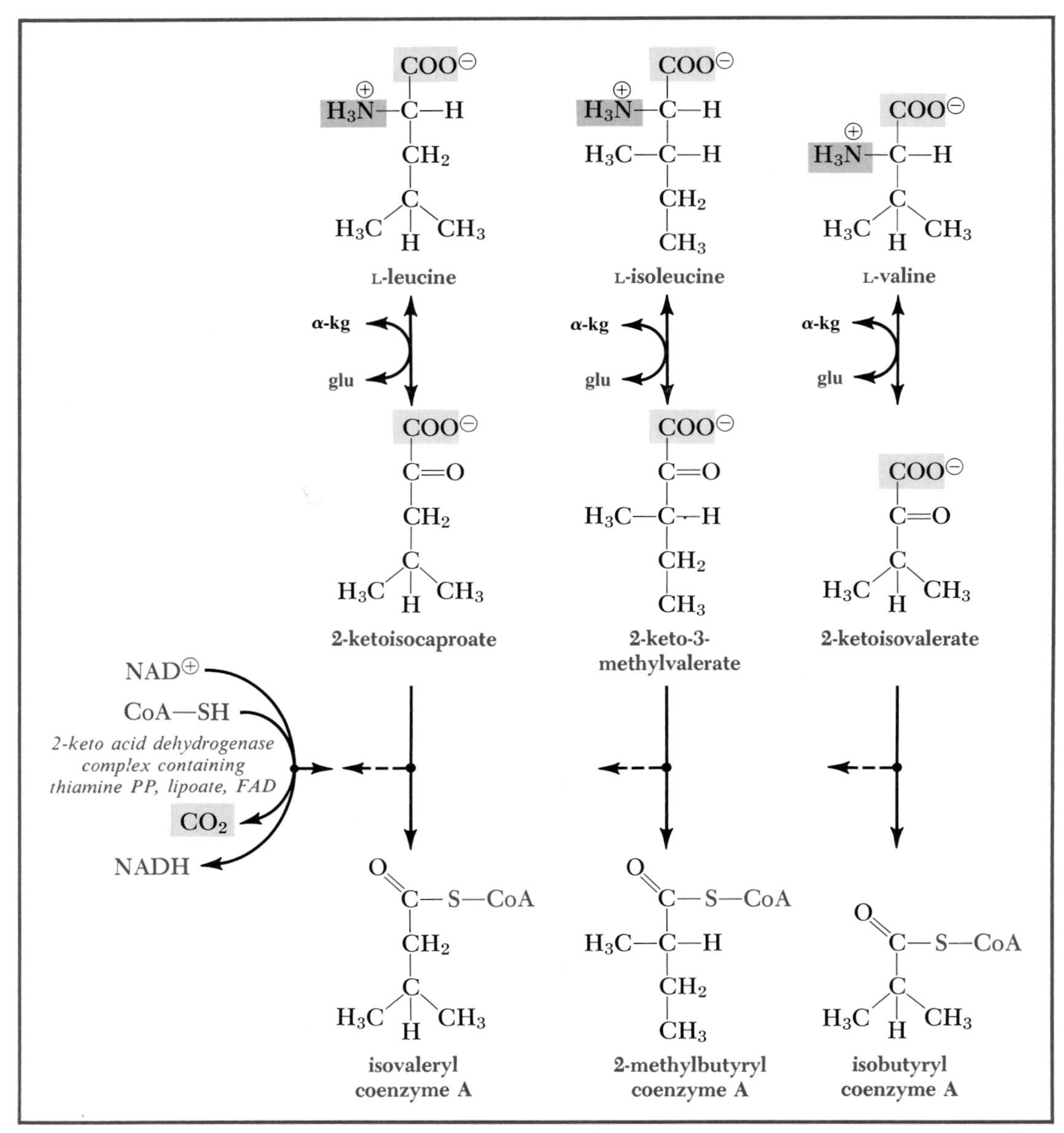

FIGURE 31–9 **The branched chain amino acids are metabolized by similar initial steps, beginning with a transamination. The resultant 2-ketocarboxylates are oxidatively decarboxylated by a branched chain keto acid dehydrogenase complex resembling α-ketoglutarate dehydrogenase. This reaction forms branched chain fatty acids (as the coenzyme A esters) containing one less carbon atom than the original amino acids. The succeeding steps are detailed in the next two figures.**

dergo the same kind of oxidation by flavoproteins seen with other acyl coenzyme A compounds.

The branched chain acyl coenzyme A compounds derived from valine and isoleucine undergo the standard sequence of oxidation, hydration, and oxidation seen with ordinary fatty acid derivatives. The major pathways are shown in Figure 31–10. There is an alternative pathway for handling the carbon skeleton of isoleucine, made necessary by spontaneous formation of an unnatural stereoisomer at the branch.

This occurs because (S)-2-keto-3-methylvalerate formed from isoleucine is in equilibrium with the enol form, and therefore racemizes:

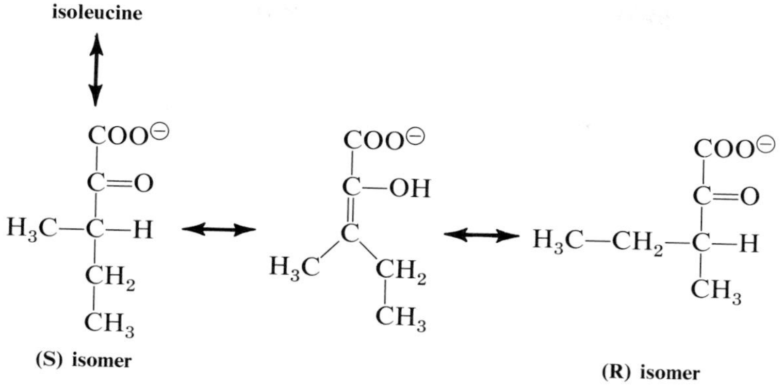

isoleucine

$$\updownarrow$$

$$\begin{array}{c} COO^{\ominus} \\ | \\ C{=}O \\ | \\ H_3C{-}C{-}H \\ | \\ CH_2 \\ | \\ CH_3 \end{array} \longleftrightarrow \begin{array}{c} COO^{\ominus} \\ | \\ C{-}OH \\ \| \\ C \\ H_3C \diagup \quad \diagdown CH_2 \\ | \\ CH_3 \end{array} \longleftrightarrow \begin{array}{c} COO^{\ominus} \\ | \\ C{=}O \\ | \\ H_3C{-}CH_2{-}C{-}H \\ | \\ CH_3 \end{array}$$

(S) isomer (R) isomer

The original (S)-isomer is oxidized on the longer chain at the branch, as shown in Figure 31–10, but the dehydrogenase attacks the shorter chain in the (R)-isomer:

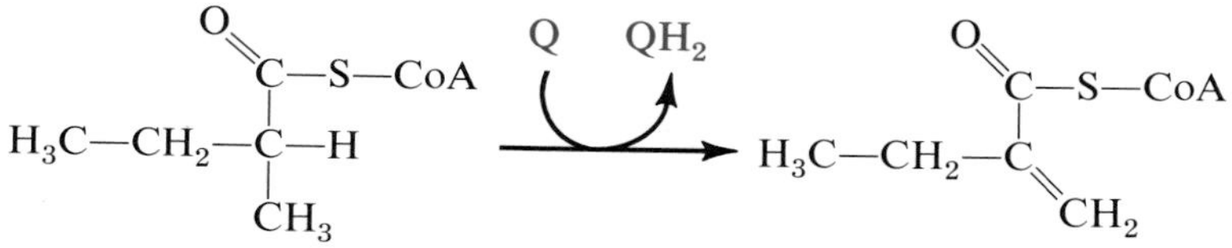

$$H_3C{-}CH_2{-}CH(CH_3){-}C(=O){-}S{-}CoA \xrightarrow{Q \;\to\; QH_2} H_3C{-}CH_2{-}C(=CH_2){-}C(=O){-}S{-}CoA$$

The subsequent reactions are identical to those shown for the valine carbon skeleton, except that each intermediate has one more carbon atom, and the final product is butyryl coenzyme A.

Since the major catabolic routes of valine and isoleucine both give rise to one mole of succinyl coenzyme A *via* propionyl coenzyme A, three of the carbons of these amino acids appear as glucose. Isoleucine also forms a mole of acetyl coenzyme A by the major route, and a mole of butyryl coenzyme A by the minor route, so two or four of its carbons may appear as acetoacetate. There is also a net production of one or two moles of CO_2 from isoleucine and two from valine.

The isovaleryl coenzyme A formed from leucine is handled in a different way. The existence of a few infants with an apparently normal fatty acid metabolism who excrete isovalerate when fed leucine, but not isobutyrate or 2-methylbutyrate when fed valine or isoleucine, shows that there must be a separate isovaleryl CoA dehydrogenase. Some observers say that the affected infants smell like cheese, others say like sweaty feet. They are effectively treated by feeding them glycine. Elevating its concentration causes increased formation of isovaleryl glycine, which is excreted in the urine:

$$\text{isovalerate} \longrightarrow \text{isovaleryl coenzyme A} \xrightarrow{\text{glycine}} \text{isovalerylglycine.}$$

The initial oxidation of isovaleryl coenzyme A forms the corresponding enoyl compound, 3-methylcrotonyl coenzyme A, as expected (Fig. 31–11). If one were dealing with a straight-chain fatty acid derivative, a hydration followed by an oxidation to yield a 3-ketoacyl coenzyme A compound would then be expected. (Recall

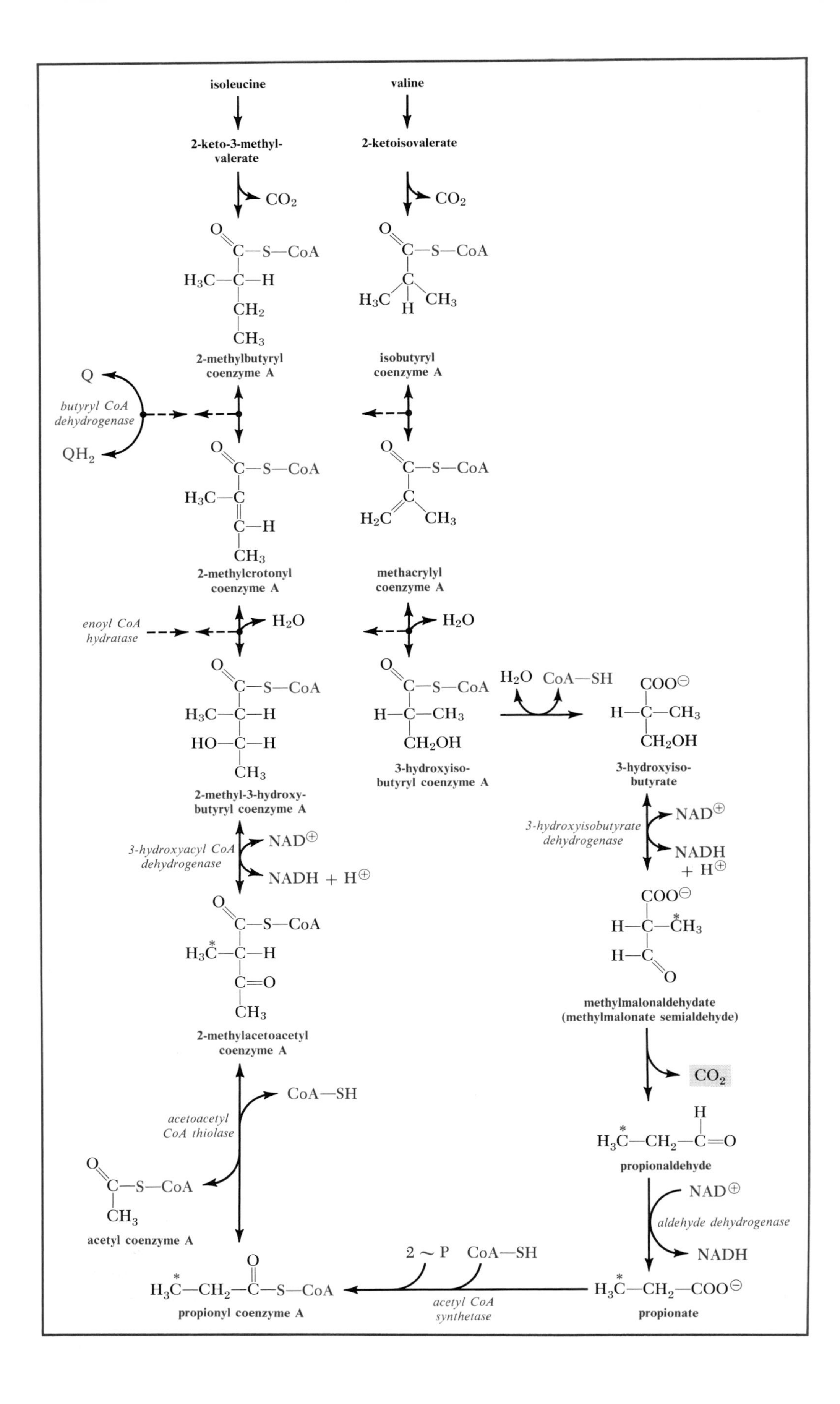

isoleucine
valine
2-keto-3-methyl-valerate
2-ketoisovalerate
CO_2
CO_2
2-methylbutyryl coenzyme A
isobutyryl coenzyme A
Q
butyryl CoA dehydrogenase
QH_2
2-methylcrotonyl coenzyme A
methacrylyl coenzyme A
enoyl CoA hydratase
H_2O
H_2O
2-methyl-3-hydroxy-butyryl coenzyme A
3-hydroxyiso-butyryl coenzyme A
H_2O CoA—SH
3-hydroxyiso-butyrate
3-hydroxyacyl CoA dehydrogenase
$NAD^{\oplus}$
$NADH + H^{\oplus}$
3-hydroxyisobutyrate dehydrogenase
$NAD^{\oplus}$
$NADH + H^{\oplus}$
2-methylacetoacetyl coenzyme A
methylmalonaldehydate (methylmalonate semialdehyde)
CO_2
acetoacetyl CoA thiolase
CoA—SH
acetyl coenzyme A
propionaldehyde
$NAD^{\oplus}$
aldehyde dehydrogenase
NADH
2 ~ P
CoA—SH
acetyl CoA synthetase
propionyl coenzyme A
propionate

FIGURE 31-10

(*Left*) The metabolism of the branched chain coenzyme A esters obtained from isoleucine and valine begins with the same main-line sequence of reactions seen with ordinary fatty acids. An oxidation by a flavoprotein is followed by hydration to a 3-hydroxy derivative, which is oxidized to a 3-keto derivative. However, the metabolism of the valine skeleton deviates in that the coenzyme A is lost before oxidation of the alcohol group. Metabolism of the isoleucine skeleton continues on the main-line sequence with a thiolytic cleavage to form acetyl coenzyme A and propionyl coenzyme A. The methylmalonic semialdehyde derived from valine is decarboxylated and oxidized to propionate, which also appears as propionyl coenzyme A.

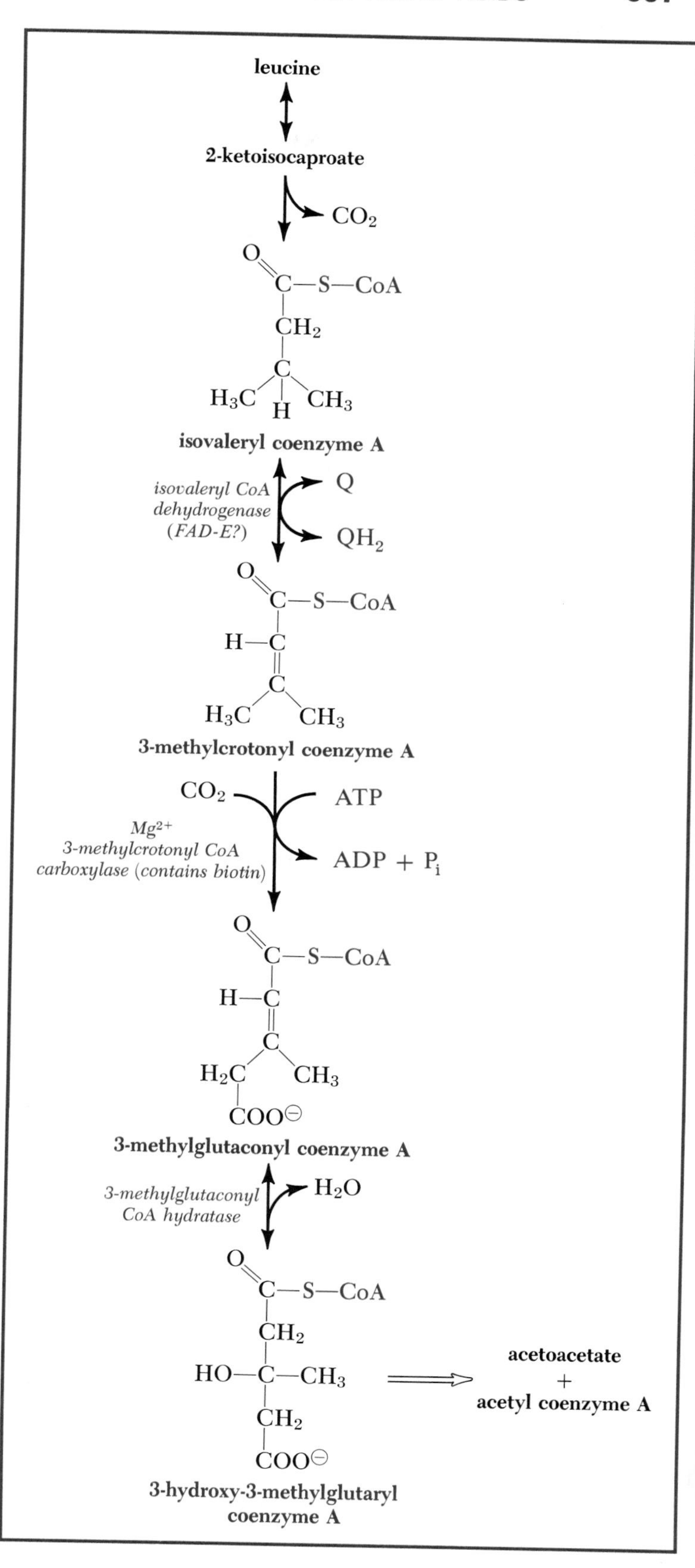

FIGURE 31-11

(*Right*) Isovaleryl coenzyme A obtained from the metabolism of leucine is converted to acetyl coenzyme A and acetoacetate, with the intermediate formation of 3-hydroxy-3-methylglutaryl coenzyme A.

the main-line sequence, p. 436.) However, this sequence cannot be continued here because of the branch on C3, which blocks formation of a 3-keto compound.

A different route has evolved, in which the compound is carboxylated by a typical biotin-containing carboxylase at the expense of a high-energy phosphate. The result is the coenzyme A ester of an unsaturated dicarboxylic acid. This is then hydrated to form 3-hydroxy-3-methylglutaryl coenzyme A, which is also the intermediate used in the production of acetoacetate. Therefore, the compound will be cleaved in either the liver or muscles to yield acetyl coenzyme A and acetoacetate by routes we have already considered. Leucine, therefore, is unlike the amino acids considered earlier. It is a ketogenic, rather than a glucogenic amino acid, and all six carbons may appear as acetoacetate.

AROMATIC AMINO ACIDS

Phenylalanine and Tyrosine

Nitrogen: Transferred to α-ketoglutarate, producing glutamate.

Carbons: Appear as acetoacetate and fumarate.

Phenylalanine is oxidized to form tyrosine, and tyrosine is further metabolized with the ultimate formation of fumarate and acetoacetate. The sequence of oxidations by which the aromatic ring is broken is catalyzed by a number of monooxygenases and dioxygenases localized in the soluble cytoplasm and the endoplasmic reticulum. These reactions therefore do not lead to the generation of high-energy phosphate. Of course, the fumarate and the acetoacetate eventually produced are intermediates of fuel metabolism, and since a mole of each is produced, the potential production of high-energy phosphate in the whole body is considerable, with three carbons convertible to glucose, four appearing as acetoacetate, and two appearing as CO_2.

The conversion of phenylalanine to tyrosine involves an irreversible oxidation by **phenylalanine hydroxylase.** The reaction consumes molecular oxygen and at the same time causes the oxidation of a type of cofactor we have not seen before, **tetrahydrobiopterin** (Fig. 31–12). The mechanism of the reaction probably involves opening of the pterin ring to form a peroxide intermediate. In any event, the reduced pterin must be regenerated by a second enzyme at the expense of NADH. (Tetrahydrobiopterin is synthesized from GTP.)

The major pathway for the metabolism of **tyrosine** begins with a transamination with α-ketoglutarate, catalyzed by a specific aminotransferase (Fig. 31–13). The resultant *p*-hydroxyphenylpyruvate is oxidized by a **dioxygenase,** a kind of enzyme that causes the addition of a full molecule of O_2.

This particular dioxygenase is typical of a subclass with an unusual mechanism. Enzymes of its sort use a second substrate to generate an intermediate peroxide as the active oxidant:

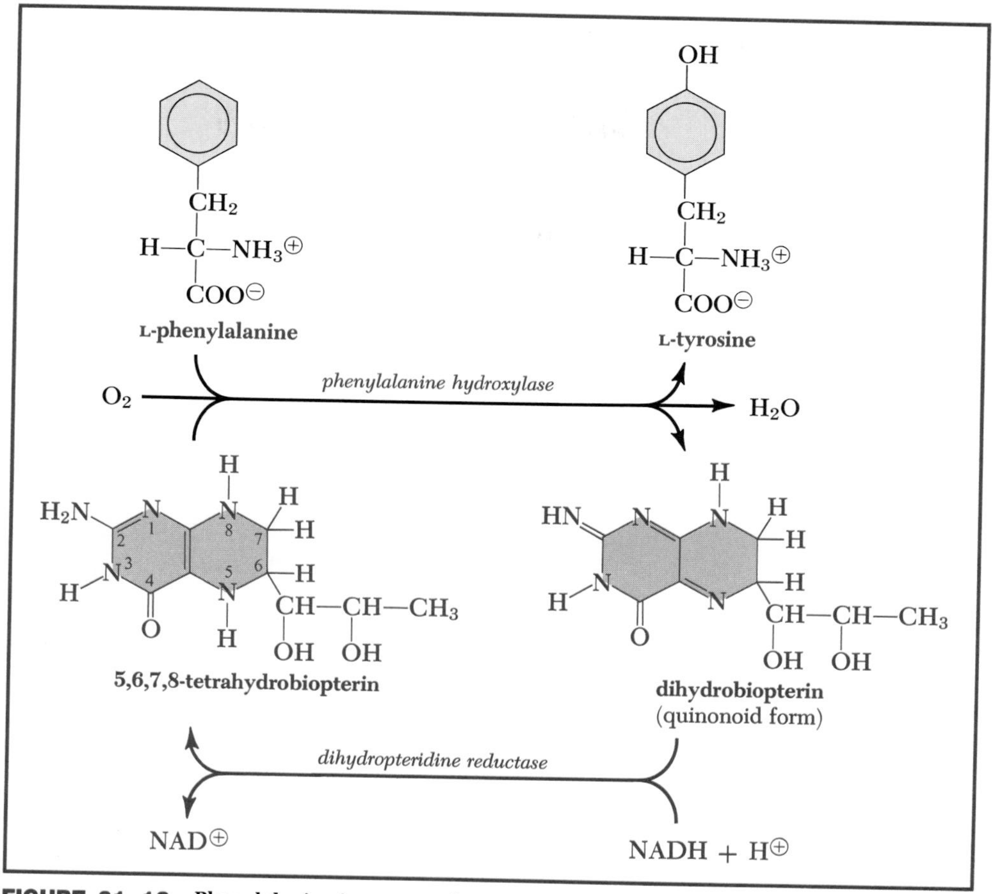

FIGURE 31–12 Phenylalanine is converted to tyrosine by the action of phenylalanine hydroxylase (phenylalanine 4-monooxygenase). This mixed-function oxidase uses tetrahydrobiopterin as an electron-donating coenzyme to activate molecular oxygen.

In the case at hand, the second substrate is the 2-keto side chain on the aromatic ring, but with other enzymes, such as the prolyl hydroxylase that converts proline residues in collagen to hydroxyproline residues, α-ketoglutarate is used as the second substrate.

The dioxygenase causes a hydroxylation of the ring, an oxidative decarboxylation of the side chain, and a shift of the acetyl group on the ring. The resultant 2,5-dihydroxyphenylacetate has the trivial name of **homogentisate.** Homogentisate is attacked by another dioxygenase, which does not require a second substrate. It cleaves the ring with dioxygen to form the *cis*-unsaturated compound C-maleoylacetoacetate. An isomerase converts this to the *trans*-compound, C-fumaroylacetoacetate, and this is hydrolyzed to yield fumarate and acetoacetate.

This is a remarkably complex series of oxidations. Rupture of an aromatic ring so as later to yield the same metabolites encountered in fat and carbohydrate metabolism is in itself somewhat of a stunt. Does the fact that this can take place explain why proteins have phenylalanine rather than phenylglycine or phenylaminobutyrate? Who knows? Furthermore, we shall see that similar kinds of reactions are used to make several other kinds of aromatic compounds in small quantities, so the type of reaction has been evolved for more general purposes than the particular sequence we are considering.

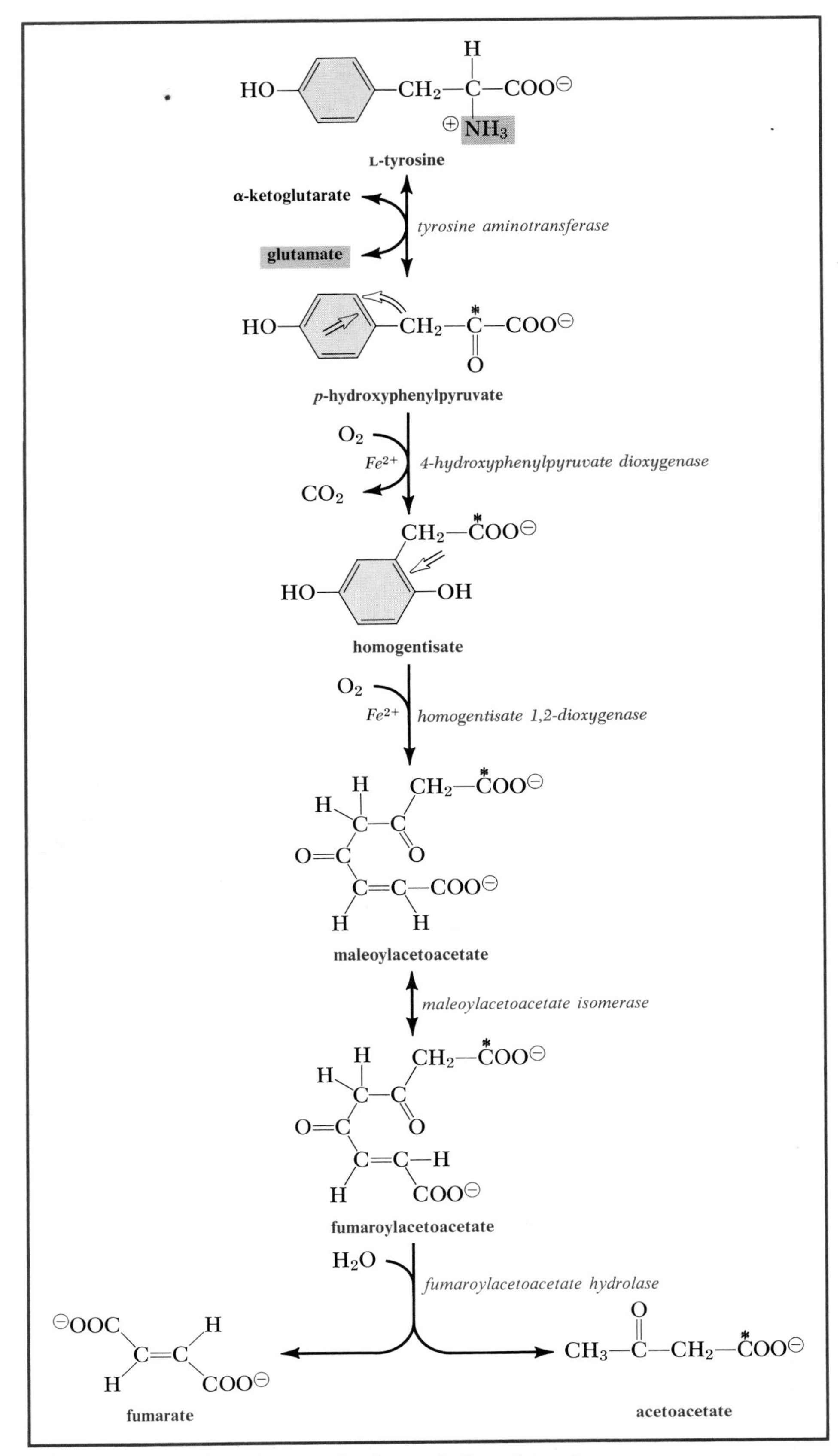

FIGURE 31–13 The major route for the metabolism of tyrosine.

Genetic Defects

Phenylketonuria (hyperphenylalaninemia) results from defects in the oxidation of phenylalanine to tyrosine. Part of the accumulated phenylalanine is transaminated, forming phenylpyruvate. This compound is excreted in the urine, along with phenylalanine, hence the name phenylketonuria. It is also known as phenylpyruvic oligophrenia, because the condition results in early neurological damage preventing normal intellectual development. Treatment consists of early avoidance of high concentrations of phenylalanine in the diet, but the problem is the same as in maple syrup urine disease. The body cannot synthesize phenylalanine, so some must be present to build normal proteins; it is difficult to gain a proper balance, and an expensive diet is required.

PHENYLKETONURIA

Recessive genetic disorders associated with elevated blood phenylalanine concentrations in which serious mental retardation is the major manifestation. Known phenotypes include:

- **PH^0—almost complete absence of phenylalanine hydroxylase activity**
- **PH^-—markedly reduced phenylalanine hydroxylase activity**
- **$DHPR^0$ or $DHPR^-$—absent or markedly reduced dihydropteridine reductase activity**
- **Deficient dihydrobiopterin synthesis**

Screening tests for elevated phenylalanine concentration detect almost all afflicted infants if done 24 hours after the start of milk feedings in full-term infants (5 to 7 days in premature infants).

Phenylalanine hydroxylase deficiencies can be treated with diets containing no more than 200 to 500 mg per day of phenylalanine, continued as long as possible.

Defects in tetrahydrobiopterin formation, much less common, also impair formation of some neurotransmitters (Chapter 33) and require additional measures.

Women treated for phenylketonuria may mature and bear children, but the fetus is at least a heterozygote to the condition, and at risk for damage from the combined maternal and fetal deficiencies in phenylalanine metabolism.

Phenylketonuria is more common than most genetic defects. We earlier discussed it as an example (p. 120). The full blown condition is only seen in homozygotes, yet it occurs once in each 14,000 births, which means that the heterozygotes, detectable only by deliberately loading them with phenylalanine and testing for blood levels of phenylalanine and phenylpyruvate, must be about one in 60 of the population. Such a prevalence makes one immediately suspicious that the prevalence is due to a balanced polymorphism similar to some hemoglobinopathies. But under what circumstances could this condition possibly be advantageous? We don't know. Perhaps the conservation of phenylalanine had an advantage on some peculiar diet.* Perhaps it is still advantageous in some regions of the world where food is frequently in short supply.

---

*The incidence among Celts is especially high—one in 5000 births. Perhaps the condition was a defense against haggis.

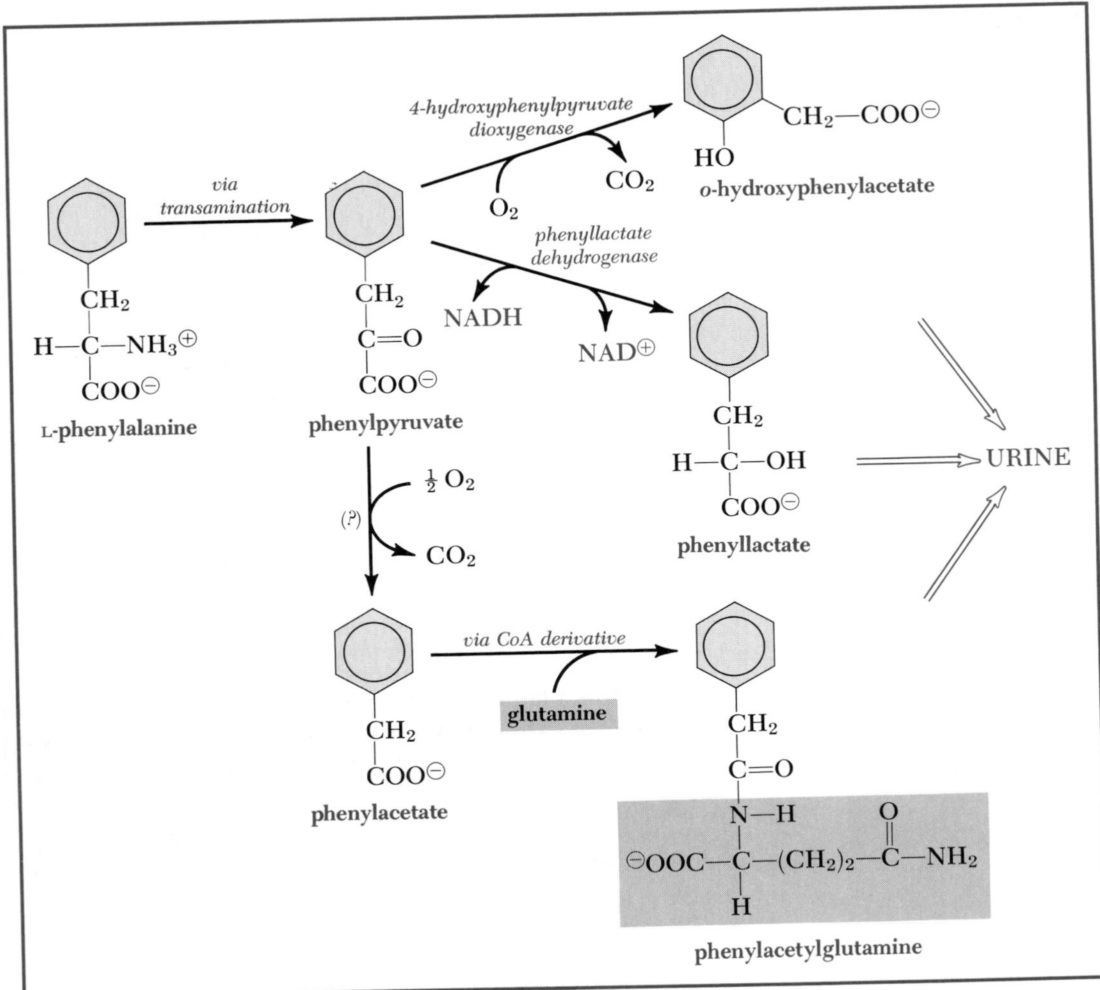

FIGURE 31–14 **Usually trivial routes of phenylalanine metabolism become more important in patients lacking phenylalanine hydroxylase activity, and an increased amount of a variety of products appears in the urine.**

People with phenylketonuria also excrete some other aromatic compounds (Fig. 31–14), representing aberrations of the normal process of metabolism due to the high concentration of phenylpyruvate. The compound is reduced by a phenyllactate dehydrogenase in the liver, forming phenyllactate. Phenylpyruvate is also oxidized in the same way as is the normal *p*-hydroxyphenylpyruvate, except that the product is *o*-hydroxyphenylacetate, rather than the 2,5-dihydroxyphenylacetate that is homogentisate. The monohydroxy compound is not further metabolized and appears in the urine.

The suspicion that the prevalence of phenylketonuria may reflect some occasional advantage to the heterozygotes is deepened by the fact that the other known genetic abnormalities of tyrosine metabolism are much less common, although their consequences are sometimes less grave.

A single clear-cut case of **tyrosinosis** was reported in an adult by Grace Medes in 1927. The facts are consistent with the absence of 4-hydroxyphenylpyruvate dioxygenase in the patient who excreted both tyrosine and the keto acid. The excretion of both increased with phenylalanine feeding, but homogentisate was metabolized at the normal rate. The case is of historical interest in that the metabolism of the aromatic acids was not well understood until painstaking studies were made of the metabolism of this one human. The compounds that did and didn't appear in the urine now appear obvious in light of our knowledge, but much of that knowledge is a result of just those measurements. This condition ought not be confused with **tyrosinemia.** When tyrosinemia is caused by a deficiency of tyrosine aminotransferase, it results in mental retardation and opacity of the eye cornea. Another variety, perhaps due to a fumaroylacetoacetate hydrolase deficiency, results in progressive hepatic and renal damage, with early death.

Another form of tyrosinemia has recently been discovered in which 4-hydroxyphenylpyruvate dioxygenase is defective. A new amino acid, **hawkinsin,** is excreted:

hawkinsin

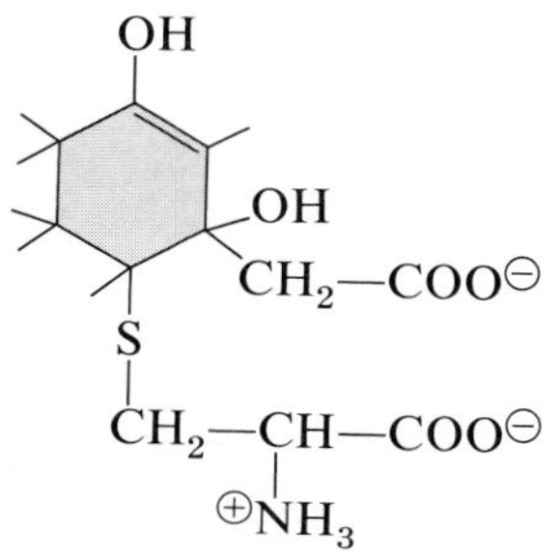

OH

OH

$CH_2—COO^{\ominus}$

S

$CH_2—CH—COO^{\ominus}$

$^{\oplus}NH_3$

It probably is formed by an accumulation of the intermediate epoxide in the dioxygenase mechanism (see earlier discussion), which reacts with glutathione. This would explain the simultaneous excretion of pyroglutamate because the loss of glutathione would cause overproduction of γ-glutamylcysteine and therefore of pyroglutamate (p. 575). The two known infants with hawkinsinuria, after a stormy beginning requiring intravenous supplementation, apparently adapted to the condition and thrived on a normal regimen.*

Finally, in about one in 200,000 births, infants are found whose urine turns black on standing owing to the homogentisate they are excreting because they lack homogentisate oxidase. Homogentisate is a substituted hydroquinone, and these diphenols are notoriously susceptible to auto-oxidation, forming a mixture of highly colored products. The condition is known as **alcaptonuria,** and individuals with it live until well into reproductive age with no difficulty other than whatever esthetic offense the darkening urine may represent. Many in their fourth or fifth decade will develop arthritis. The degeneration of the connective tissue in the joints is apparently associated with a deposition of pigment (ochronosis), presumably resulting from further oxidation of homogentisate in cartilage.

What do these genetic defects teach us? The active life possible with alcaptonuria shows that the oxidation of aromatic amino acids is not imperative for energy production. We have to stretch our imagination to visualize circumstances in which the few grams of available metabolite represented by these compounds might tip the balance. Those with phenylketonuria or tyrosinemia are in more difficulty, but their greater problems evidently arise from the accumulation of metabolites, which in themselves disturb metabolic processes, rather than from a failure to produce some needed compounds or an adequate supply of high-energy phosphate.

---

*One of the infants was said to smell like an Australian swimming pool. It is not clear if swimming pools smell the same in all cultures.

Tryptophan

Nitrogens: One appears as alanine, and one as NH_4^+.

Carbons: Three appear as alanine, four as crotonyl coenzyme A, one as formate, and three as CO_2.

Tryptophan is usually the least abundant of the amino acids in the diet, and is not a major substrate for the generation of high-energy phosphate. However, the unusual indole ring that it contains is used as a precursor for other cellular components, as we shall see later. The formation of these substances is usually satisfied by a small fraction of the dietary tryptophan consumption. The balance of the carbon skeleton is metabolized to CO_2 by way of alanine made from the side chain and crotonyl CoA formed from the ring, which can be converted to glucose and acetoacetate, respectively. Metabolism begins with an irreversible oxidation by O_2, catalyzed by tryptophan 2,3-dioxygenase, an enzyme containing heme (Fig. 31–15). This

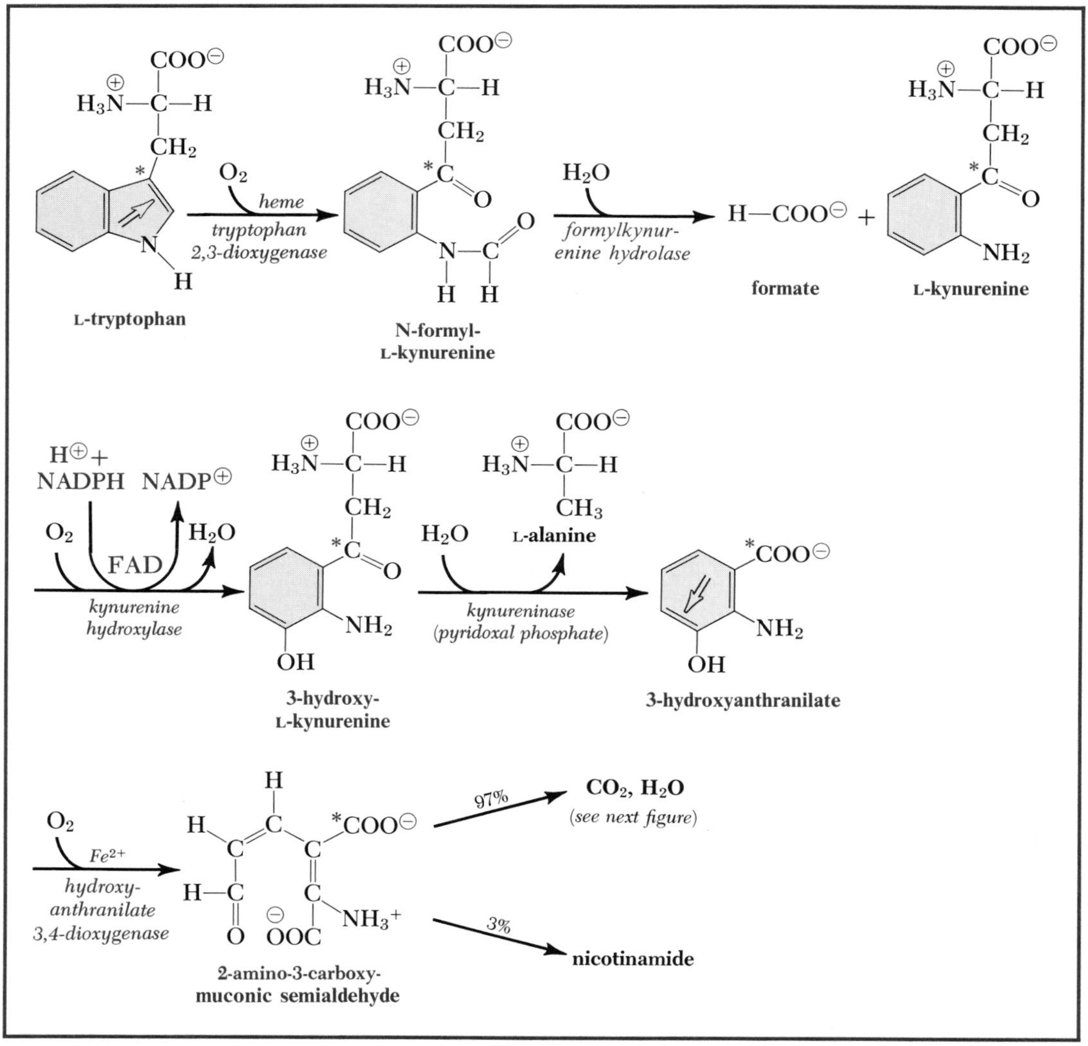

FIGURE 31–15 **The route for the complete metabolism of tryptophan involves dioxygenases that cleave the two rings, with the concurrent formation of formate and alanine. Part of the final intermediate shown here may be used to form nicotinate, but most is further metabolized as shown in the next figure.**

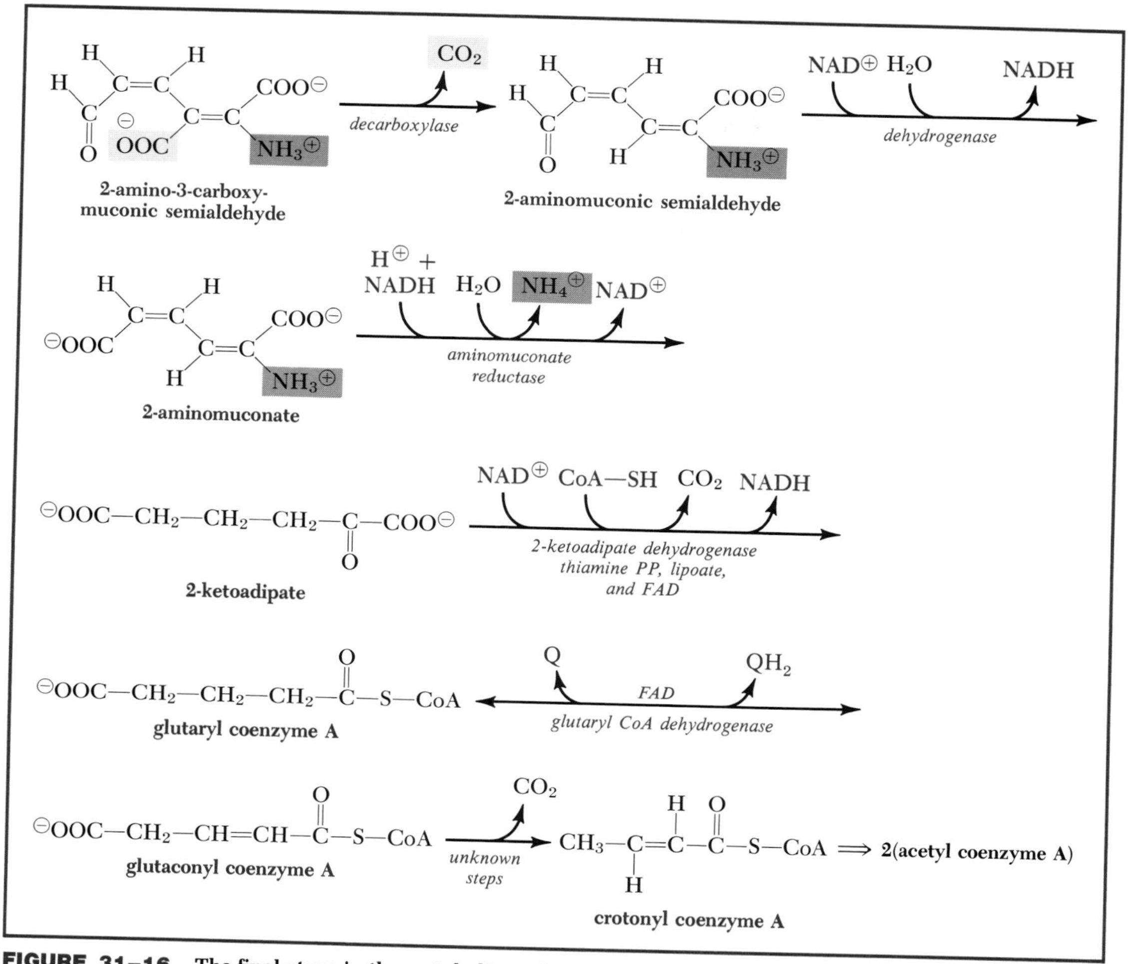

FIGURE 31–16 The final steps in the metabolism of tryptophan. The last three reactions are also involved in the metabolism of lysine.

reaction opens the indole ring. Carbons are removed as formate and alanine. A succeeding oxidation opens the phenyl ring. The oxidations are of the same sort encountered in the metabolism of phenylalanine and tyrosine. We have already considered the disposal of alanine. We shall consider the fate of formate more completely in the next chapter. Suffice it to say now that it can be oxidized to CO_2 and H_2O.

The remaining reactions by which the carbons are converted to CO_2 and crotonyl CoA, which is an intermediate of fatty acid metabolism, involve different compounds than those we have seen, but no particular novelties in the types of reactions (Fig. 31–16).

LYSINE

Nitrogens: Both transferred to α-ketoglutarate, forming glutamate.

Carbons: Two appear as CO_2 and four as crotonyl coenzyme A.

Lysine, like tryptophan, is metabolized *via* 2-ketoadipate, eventually forming crotonyl CoA. However, lysine is a relatively abundant amino acid. The amount

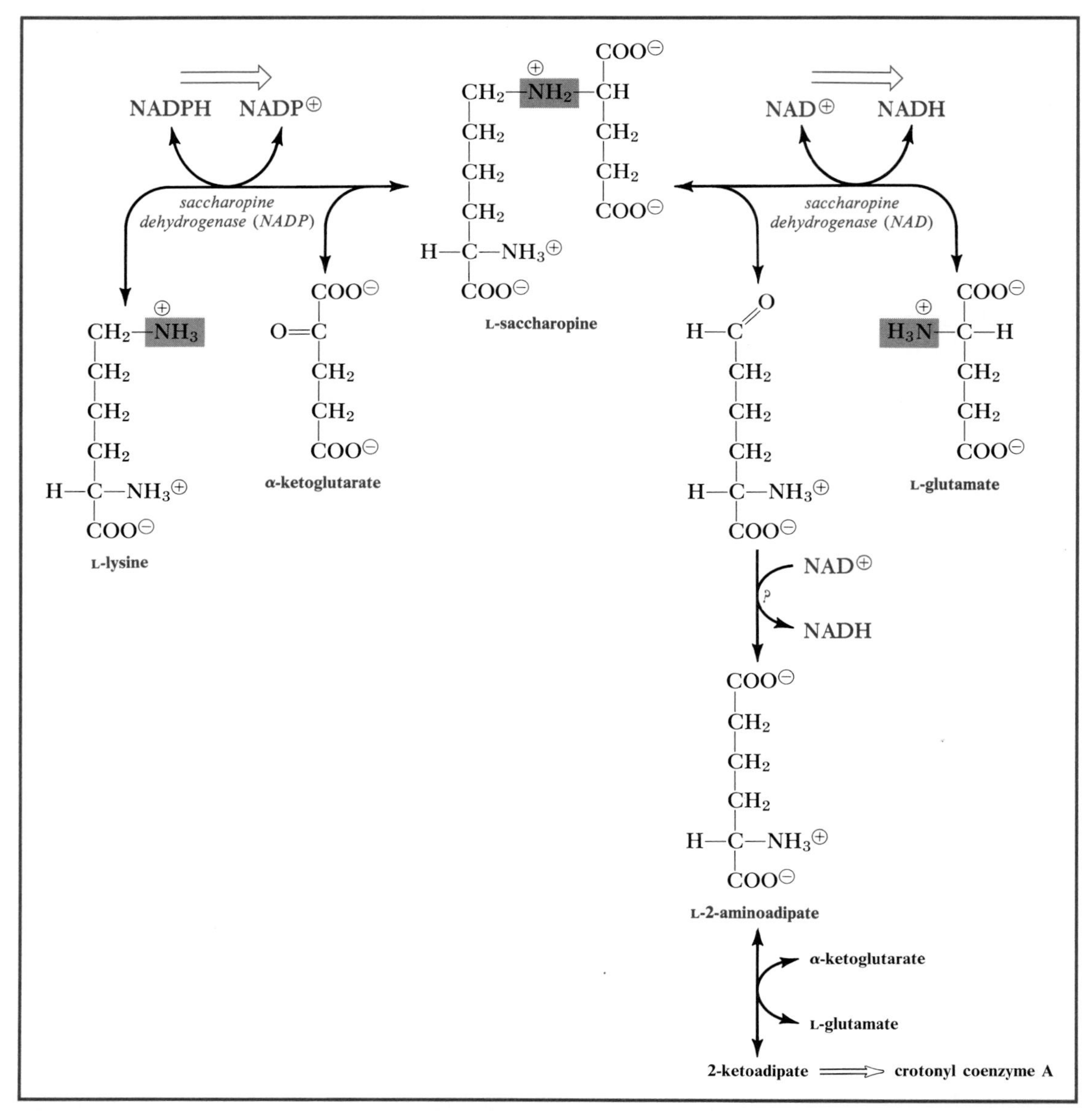

FIGURE 31–17 **The major route for the metabolism of lysine.**

consumed on most diets is comparable to the intake of the branched chain amino acids. It is metabolized in mitochondria with the generation of high-energy phosphate.

The route involves a condensation with α-ketoglutarate and a reduction to form the compound saccharopine, which is then cleaved with an oxidation on the opposite side of the nitrogen bridge so as to release glutamate and the semialdehyde of 2-aminoadipate (Fig. 31–17). (The semialdehyde spontaneously cyclizes much like sugars.) The semialdehyde is then oxidized to form 2-aminoadipate. This is the next higher homologue in the dicarboxylic amino acid series; it does not occur in proteins, but it will transaminate, forming 2-ketoadipate. Here we have a classic example of the differential use of NAD and NADP. The reduction is favored by the high [NADPH]/[NADP] ratio and the oxidation by the high [NAD]/[NADH] ratio. The sequence is irreversible for all practical purposes.

FURTHER READING

E. Adams and L. Frank (1980). Metabolism of Proline and the Hydroxyprolines. Annu. Rev. Biochem. 49:1005.

M. Walser and J.R. Williamson, eds. (1981). Metabolism and Clinical Implications of Branched Chain Amino and Ketoacids. Elsevier.

L.E. Rosenberg and C.R. Scriver (1980). Disorders of Amino Acid Metabolism. p. 583 in P.K. Bondy and L.E. Rosenberg, eds. Metabolic Control and Disease, 8th ed. Saunders.

J.B. Stanbury et al. (1982). The Metabolic Basis of Inherited Disease, 5th ed. McGraw-Hill. Includes several chapters on defects of amino acid metabolism.

D. Wellner and A. Meister (1981). A Survey of Inborn Errors of Amino Acid Metabolism and Transport in Man. Annu. Rev. Biochem. 50:911.

C.R. Scriver and C.L. Clow (1980). Phenylketonuria: Epitome of Human Biochemical Genetics. N. Engl. J. Med. 303:1336,1394.

B. Wilcken et al. (1981). Hawkinsinuria. A Dominantly Inherited Defect of Tyrosine Metabolism with Severe Effects in Infancy. N. Engl. J. Med. 305:865.

A.J.L. Cooper, M.T. Haber, and A. Meister (1982). On the Chemistry and Biochemistry of 3-Mercaptopyruvic Acid and the α-Keto Analog of Cysteine. J. Biol. Chem. 257:816.

L.J. Filler, ed. (1979). Glutamic Acid: Advances in Biochemistry and Physiology. Raven.

D.E. Matthews et al. (1981). Regulation of Leucine Metabolism in Man. Science 214:1129.

S.W. Bailey and J.E. Ayling (1980). Cleavage of the 5-Amino Substituent of Pyrimidine Cofactors by Phenylalanine Hydroxylase. J. Biol. Chem. 255:1174.

CHAPTER 32

AMINO ACIDS: ONE-CARBON POOL AND TOTAL BALANCE

The metabolism of some amino acids involves an associated metabolism of one-carbon units. Groups containing single carbon atoms are transferred for many different purposes, including the introduction of carbon atoms into purine rings and the formation of methyl groups on compounds such as choline, creatine, and the bases of nucleic acids. Of course, we have already seen many examples of the utilization or formation of CO_2 through carboxylations or decarboxylations, but we are now concerned with the transfer of carbon atoms in a less fully oxidized state. It is convenient to lump together the sources of these more reduced carbon groups as the one-carbon pool, but it will become apparent as the story unfolds that the pool is more like a series of interconnected basins.

NATURE OF ONE-CARBON GROUPS

What precisely are we talking about? Single carbon atoms can exist in various oxidation states, which are represented by the series of simplest organic compounds: methane, methanol, formaldehyde, formate, and carbon dioxide. It is possible to incorporate carbon units at each of these oxidation states, except that of methane, into other compounds by the formal elimination of water. The results are given in Figure 32–1. All of the possible structures of one-carbon groups shown in the figure are found in biological compounds. The figure uses the general R-designation for groups bearing one-carbon units, but the actual nature of the groups is limited by their chemical potential in most cases to those containing N, O, or S atoms. Furthermore, the figure includes carbonic acid, the hydrated form of CO_2, for completeness even though compounds at this oxidation state are handled by carboxylation and decarboxylation, not by the reactions of the one-carbon pool.

Figure 32–1 is organized so that each row differs from its neighbors by the equivalent of two electrons; it follows that the rows are interconvertible only by oxidations or reductions. Those groups on the same row are of equivalent oxidation-reduction state and are interconvertible by removal or addition of the elements of water (p. 288).

The division of the one-carbon groups according to oxidation state also divides them according to their typical types of reactions. For example, we have already seen that CO_2 is carried by biotin for the formation of carboxylate groups. Now we shall see a pteridine compound, **tetrahydrofolate,** used for the transfer of single carbon groups equivalent in oxidation state to formate, formaldehyde, and sometimes methanol. That is, tetrahydrofolate carries methylidyne groups, methylene groups, and sometimes methyl groups.

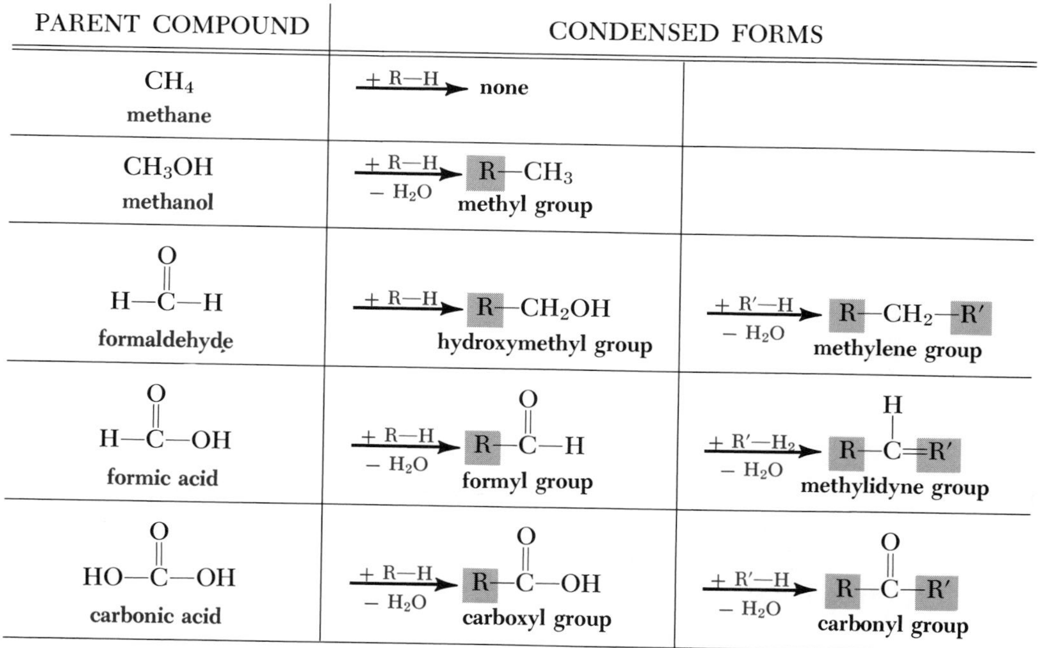

PARENT COMPOUND	CONDENSED FORMS	
CH_4 methane	$\xrightarrow{+\text{R—H}}$ none	
CH_3OH methanol	$\xrightarrow[-H_2O]{+\text{R—H}}$ R—CH_3 methyl group	
H—C(=O)—H formaldehyde	$\xrightarrow{+\text{R—H}}$ R—CH_2OH hydroxymethyl group	$\xrightarrow[-H_2O]{+\text{R}'\text{—H}}$ R—CH_2—R′ methylene group
H—C(=O)—OH formic acid	$\xrightarrow[-H_2O]{+\text{R—H}}$ R—C(=O)—H formyl group	$\xrightarrow[-H_2O]{+\text{R}'\text{—H}_2}$ R—C(H)=R′ methylidyne group
HO—C(=O)—OH carbonic acid	$\xrightarrow[-H_2O]{+\text{R—H}}$ R—C(=O)—OH carboxyl group	$\xrightarrow[-H_2O]{+\text{R}'\text{—H}}$ R—C(=O)—R′ carbonyl group

FIGURE 32–1 One-carbon compounds of various oxidation states and their formal relationships to groups of the same oxidation states. The reverse reactions involving hydrations or hydrolyses are not shown. The methylidyne group is often termed a methenyl group.

Finally, we shall see how methyl groups, the most reduced of the transferrable single-carbon groups, are usually derived from an activated form of the amino acid methionine.

TETRAHYDROFOLATE

Tetrahydrofolate is a reduced pteridine like tetrahydrobiopterin (p. 609), but it has the specific function of carrying one-carbon units and does not participate in mixed-function oxidase reactions. Plants and many protists can combine the pteridine with p-aminobenzoic acid to form pteroic acid (Fig. 32–2), followed by conjugation of the tetrahydro derivative with a molecule of glutamate to make **tetrahydropteroylglutamate,** which is a more systematic name for tetrahydrofolate. Still more molecules of glutamate may be attached through the 5-carboxyl group to make tetrahydropteroyl polyglutamates. Animals, including us, usually depend upon plants for the principal supply, although the protists are sometimes of importance. (Rats, for example, disgust us by eating their own feces, but this enables them to use

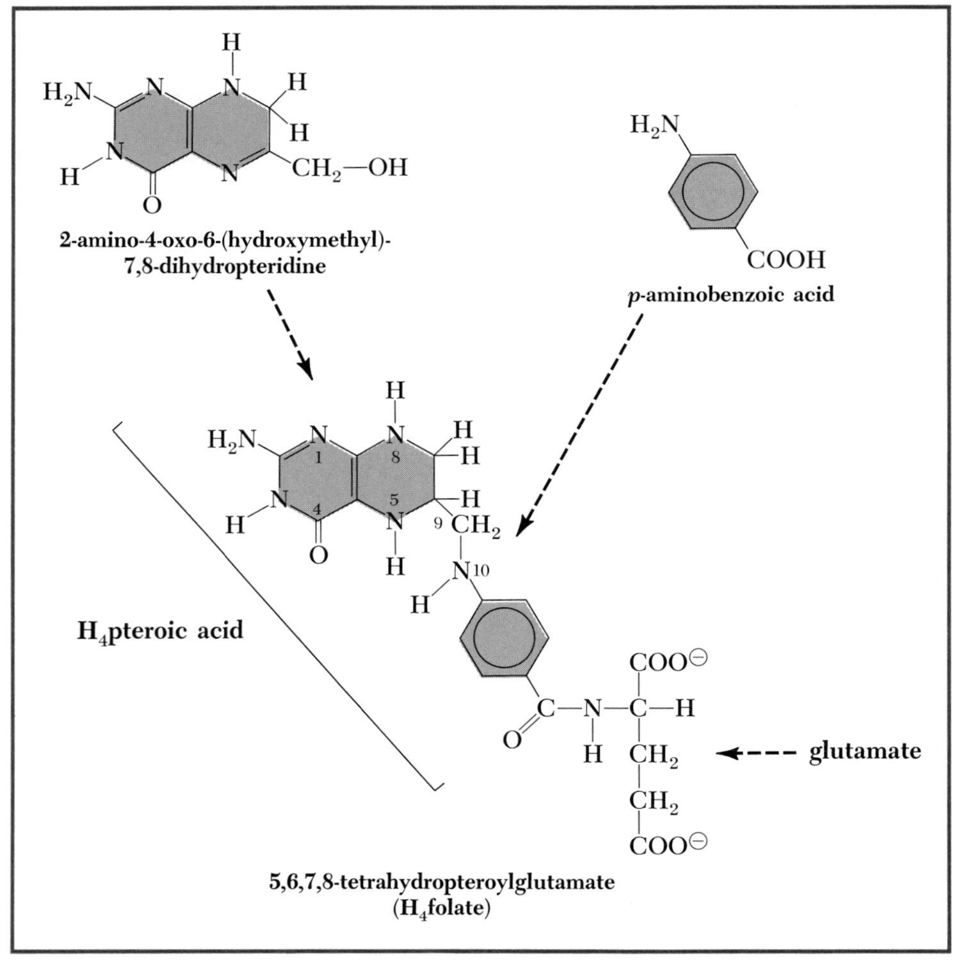

FIGURE 32–2 Tetrahydrofolate (H_4folate) and its precursors. It is more formally named tetrahydropteroylglutamate, abbreviated as H_4PteGlu.

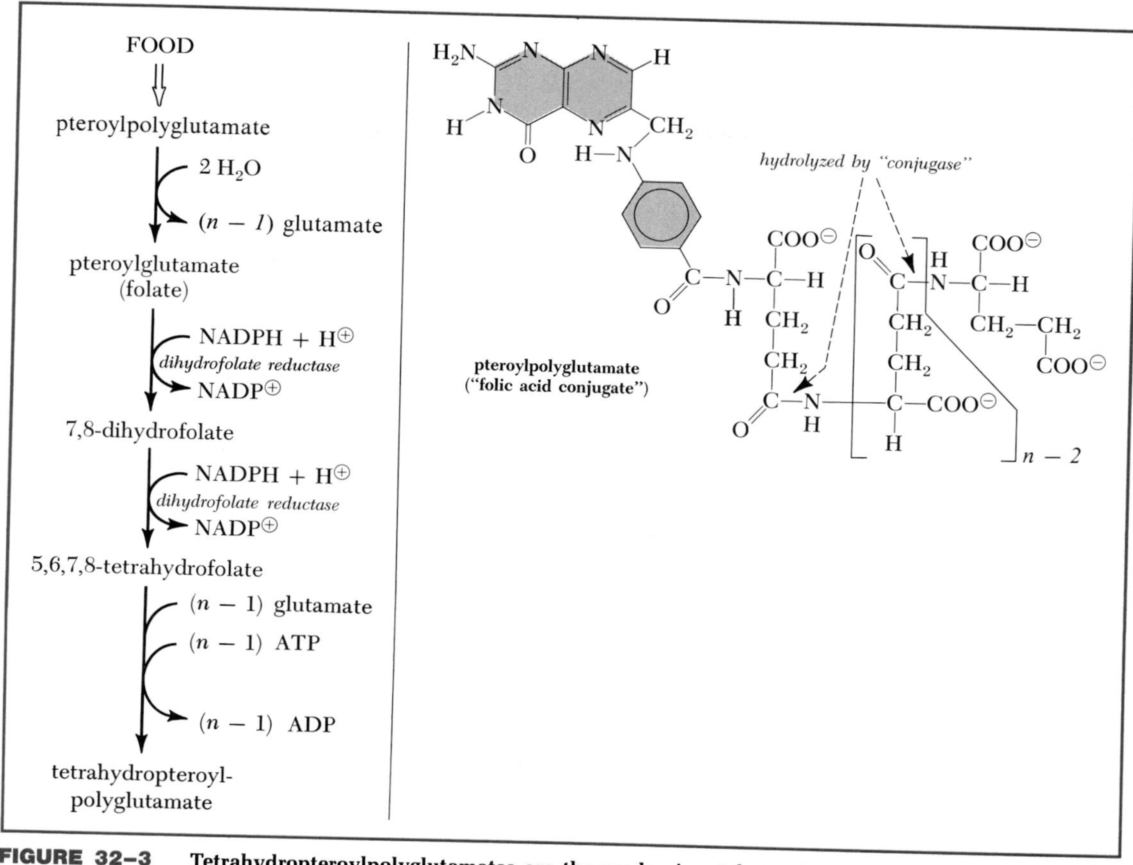

FIGURE 32–3 Tetrahydropteroylpolyglutamates are the predominant forms in raw foodstuffs, but are often oxidized before absorption. The intestinal mucosa removes all but one glutamyl group from the polyglutamates and reduces the resultant folate to its tetrahydro form in two steps. The polyglutamate chain is reconstructed within cells at the expense of ATP. (The plant coenzyme typically has three glutamate residues, whereas the human coenzyme has five or six residues in the polyglutamate chain.)

the folate and other vitamins synthesized by their intestinal bacteria. We aren't built to take advantage of this practice, and it is not recommended, even for the most avid devotee of health foods.) We shall say more about the origin of folate and its turnover when we discuss nutrition (p. 844).

Tetrahydrofolate occurs within cells—plant, animal, and microbial—as polyglutamate derivatives, in which additional molecules of glutamate are attached by amide linkage through their 5-carboxyl groups. The tetrahydro compounds are readily oxidized to folate compounds by atmospheric oxygen during the preparation and digestion of foods. Hydrolases in the intestinal mucosa remove all but one glutamyl group, so it is folate itself that is the principal form that is absorbed. It is reduced after absorption, first to dihydrofolate and then to tetrahydrofolate. The **dihydrofolate reductase** is a different enzyme from dihydrobiopterin reductase, and uses NADPH, not NADH, as the reducing agent. When tetrahydrofolate is taken up by cells, additional glutamyl groups are once more placed on the molecule (Fig. 32–3). Most of the tetrahydrofolate in mammalian tissues has five or six glutamyl groups, and it is these forms that are most active as coenzymes. However, for simplicity, we shall refer to the reduced forms as H_2folate and H_4folate, it being understood that it is the polyglutamyl derivatives that are reacting.

The biological significance of pterins to animals was first noted in an odd way. The pterins were known to occur in animal pigments, for example on the wings of butterflies. The basis for their chemistry was laid in the late 1800's, but the subject remained stagnant until the 1930's when one of the compounds, xanthopterin, was synthesized. This compound was shown to alleviate an anemia produced in rats fed on goat milk—a curious observation. More significantly, salmon in Washington hatcheries fed on diets in which yeast was used as a vitamin supplement also became anemic, but this anemia was corrected by feeding small amounts of xanthopterin. Obviously, the fish needed the insect pigment. With this background, it was much easier to connect the pteridines with the vitamin folic acid, which was isolated in the 1940's, and to show that it was chemically related. So much attention was concentrated on folate derivatives that the simpler pterins were neglected, and it was much later that the use of these compounds in oxidation-reduction reaction, such as the oxidation of phenylalanine, was demonstrated.

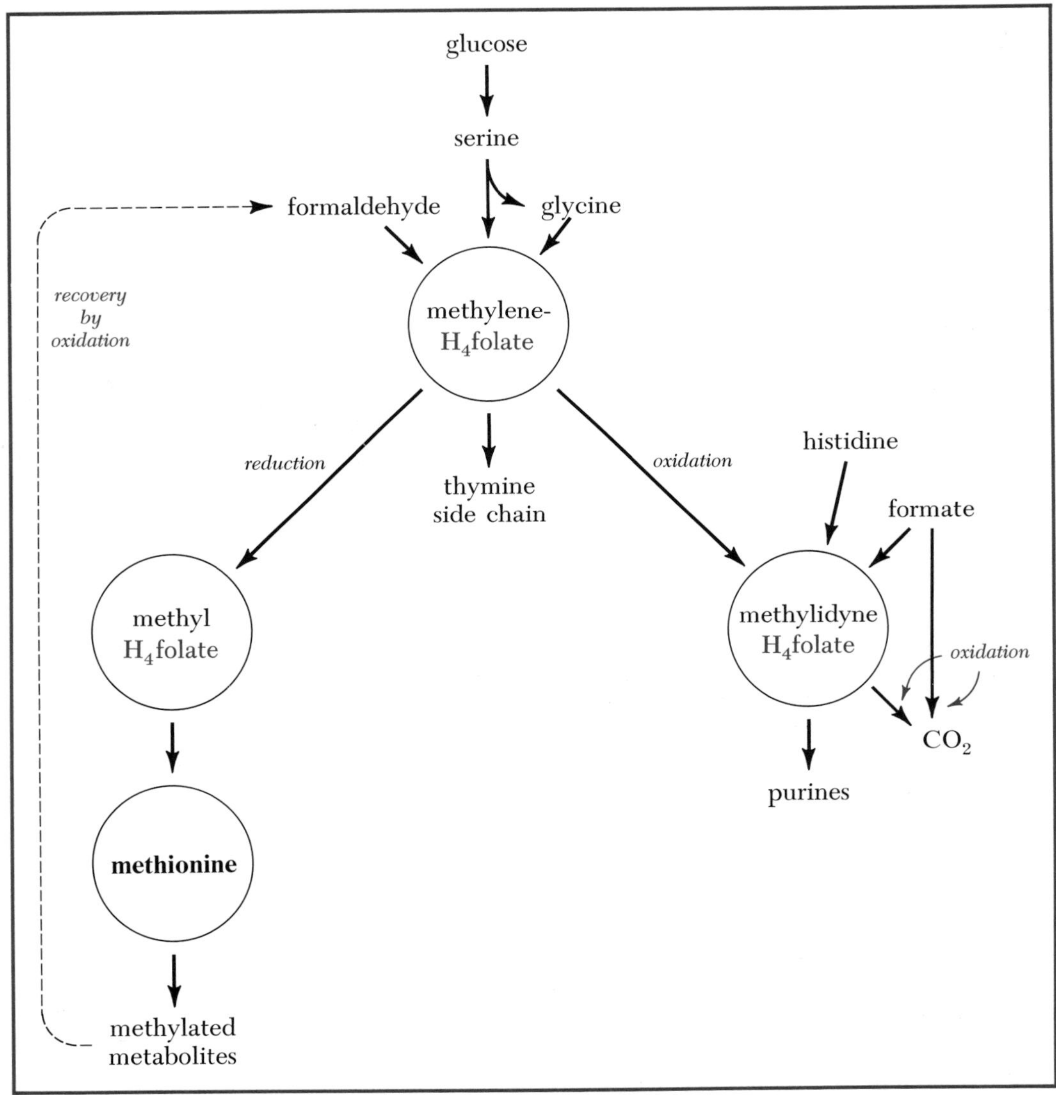

FIGURE 32–4 Overall view of one-carbon metabolism. One-carbon groups are ultimately derived from glucose *via* serine and glycine. The principal uses of the compounds are to form methylated compounds, the side chain of thymine, and carbons in the purine ring. One-carbon groups can be recovered by oxidation to formaldehyde or formate and by direct transfer from histidine. Excess one-carbon groups are removed by oxidation to CO_2.

ONE-CARBON GROUPS ON TETRAHYDROFOLATE

General Outline

The function of tetrahydrofolate is to receive one-carbon groups from various sources, retain them while they are oxidized or reduced from one form to another, if necessary, and then deliver them for the creation of new compounds. The general pathways for doing all of this are outlined in Figure 32–4.

Methylene-H_4folate, the derivative containing a group at the oxidation level of formaldehyde, is of central importance. The methylene group may be derived from formaldehyde itself, but the major sources are the amino acids serine and glycine. Serine donates a methylene group, thereby becoming glycine, and glycine itself is a source of an additional methylene group.

The methylene groups are used to form the side chains of thymine incorporated into DNA (Chapter 35), and are sources of methyl groups (by reduction) and methylidyne groups (by oxidation).

Methylidyne-H_4folate and the equivalent formyl-H_4folate are made by oxidation of methylene-H_4folate and by combination of H_4folate with formate. Formate is not only ingested by ant-fanciers, but it also is a normal intermediate of metabolism. Histidine, too, is a source of a methylidyne group.

Methylidyne and formyl H_4folate are obligatory precursors of the purines.

Methyl-H_4folate is made by reduction of methylene-H_4folate and is used to regenerate methyl groups on methionine, which is the precursor of most methylated metabolites of the sort we shall consider later in this chapter. The degradation of such compounds results in the oxidation of the methyl groups to formaldehyde, which is recovered by combination with H_4folate.

Let us now look at some of the details of transfer of one-carbon units *via* H_4folate.

PRIMARY SOURCES FOR THE ONE-CARBON POOL

Methylene Groups from Serine and Glycine. Serine combines with pyridoxal phosphate on a transferase that transfers the hydroxymethyl group to nitrogen atoms at positions 5 and 10 of H_4folate (Fig. 32–5). The other product is glycine. **The equilibration of serine and glycine in this way within mitochondria is one of the most important reactions in nitrogen metabolism:**

$$\textbf{serine} + \textbf{H}_4\textbf{folate} \longleftrightarrow \textbf{glycine} + \textbf{5,10-methylene-H}_4\textbf{folate}$$

This reaction is rapid enough in most animals to provide their needs for glycine, which is therefore not an essential amino acid for them.

Glycine is degraded within mitochondria by a reaction sequence that is equivalent to an oxidative deamination with an associated decarboxylation:

$$\text{glycine} + \text{NAD}^{\oplus} + \text{H}_4\text{folate} \longrightarrow \text{NH}_4^{\oplus} + \text{CO}_2 + \text{5,10-methylene-H}_4\text{folate} + \text{NADH}.$$

The resultant NADH supplies electrons for oxidative phosphorylation.

A high demand for one-carbon units may be met by converting serine to glycine, by oxidizing glycine, or by both processes. The reactions also provide a means of converting excess glycine to serine in the absence of a demand for one-carbon groups. The resultant serine can then be catabolized *via* pyruvate:

1. glycine + $1/2\ O_2$ + H_4folate $\longrightarrow$ $NH_4^{\oplus}$ + CO_2 + 5,10-methylene-H_4folate + 3 ~ P
2. 5,10-methylene-H_4folate + glycine $\longrightarrow$ H_4folate + serine
3. serine $\longrightarrow$ pyruvate + $NH_4^{\oplus}$

   SUM: 2 glycine + $1/2\ O_2$ $\longrightarrow$ 2 $NH_4^{\oplus}$ + CO_2 + pyruvate + 3 ~ P

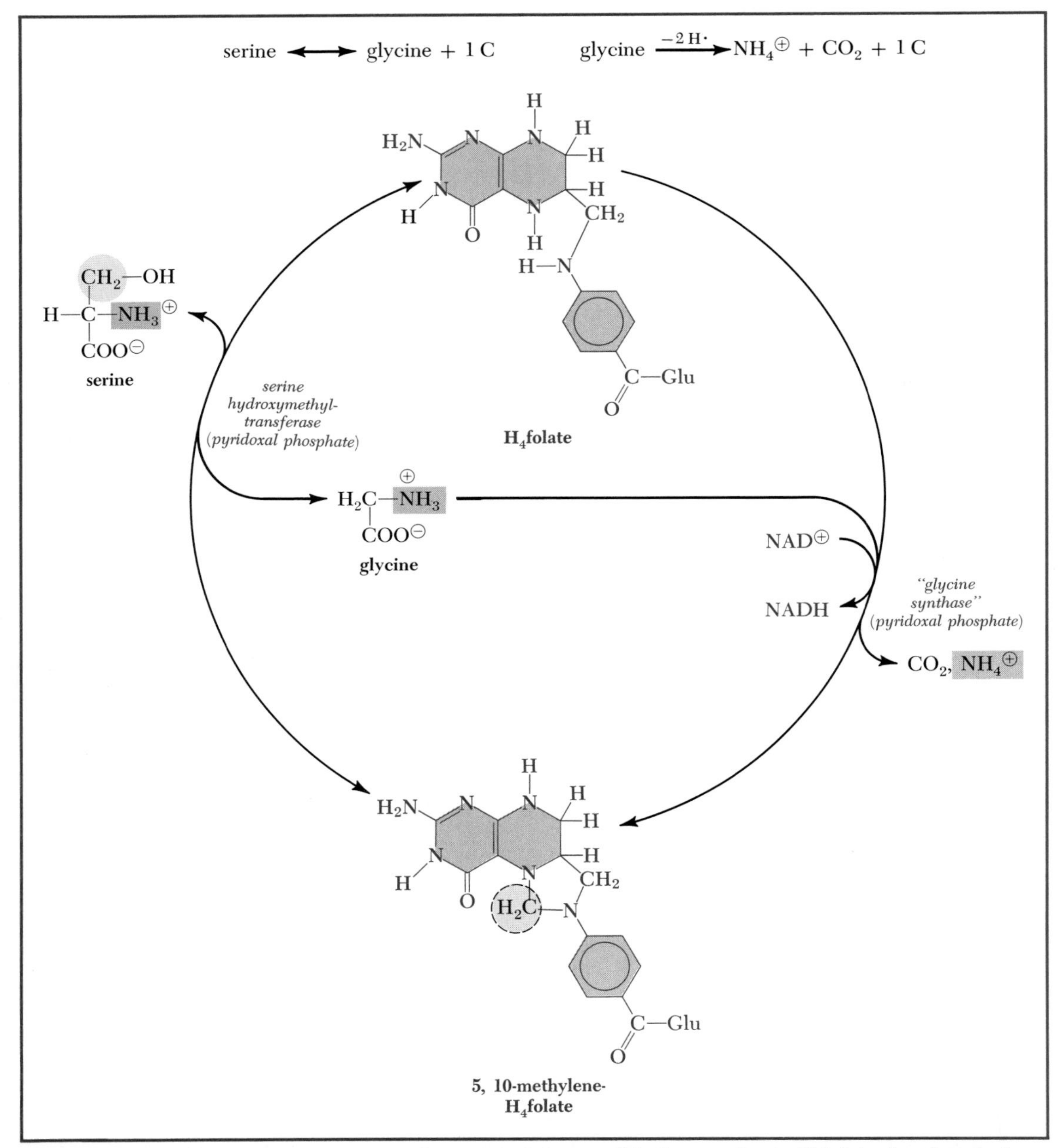

FIGURE 32–5 A methylene group can be transferred reversibly from serine to H_4folate, forming glycine. Glycine can be oxidized to form another methylene group, along with CO_2 and $NH_4^{\oplus}$.

We can therefore state:

Glycine

Nitrogen: Appears as NH_4^+.

Carbons: One fourth (one from every two glycine) appear directly as CO_2, three fourths appear in pyruvate.

Glycine is clearly a glucogenic amino acid during fasting.

Source of Serine. Serine is a metabolically active amino acid that is the precursor of several cellular constituents in addition to being a major component of proteins. Animals have, therefore, preserved the ability to synthesize the amino acid from glucose, using routes that perforce differ from the irreversible deamination used in catabolism and discussed in the preceding chapter.

The immediate precursor of serine is hydroxypyruvate, or phosphohydroxypyruvate:

$$\underset{\substack{\textbf{hydroxypyruvate}\\ \textit{or}\\ \textbf{3-phosphohydroxypyruvate}}}{\begin{array}{l} COO^{\ominus} \\ | \\ C{=}O \\ | \\ CH_2{-}O(H,\ PO_3^{2-}) \end{array}} \xrightleftharpoons[\substack{\textit{serine-pyruvate}\\ \text{or}\\ \textit{phosphoserine}\\ \textit{aminotransferase}}]{\text{amino acid} \quad \text{keto acid}} \underset{\substack{\textbf{serine}\\ \textit{or}\\ \textbf{phosphoserine}}}{\begin{array}{l} \quad\ \ COO^{\ominus} \\ \overset{\oplus}{H_3N}{-}C{-}H \\ \quad\ \ CH_2{-}O(H,\ PO_3^{2-}) \end{array}}$$

Amino groups are transferred from alanine to hydroxypyruvate, or from glutamate to the phosphorylated form.

Phosphohydroxypyruvate can be made directly from glucose, hydroxy pyruvate indirectly *via* glycerol (Fig. 32–6), so methylene groups on H_4folate can also rise

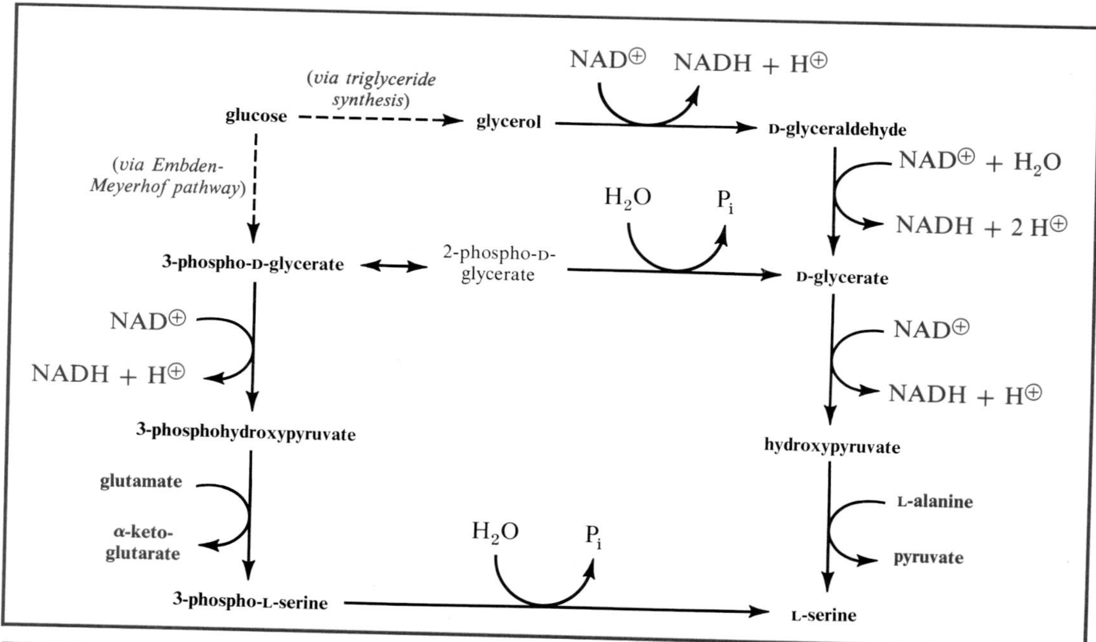

FIGURE 32–6 Serine can be formed from glucose by two independent routes. One involves phosphorylated derivatives in the Embden-Meyerhof pathway, with the necessary nitrogen obtained from glutamate (*left*). The other involves nonphosphorylated compounds derived *via* glycerol, and the nitrogen is obtained from alanine (*top and right*). These are the routes by which carbon from glucose is injected into the one-carbon pool.

from glucose by these routes. The reason for the existence of two routes for synthesis is not clear.

Glycine from Threonine. Excess threonine is partially degraded by serine dehydratase (p. 581) to form 2-ketobutyrate. However, a major fraction is attacked by a dehydrogenase. The 3-keto intermediate is cleaved with coenzyme A into glycine and acetyl coenzyme A:

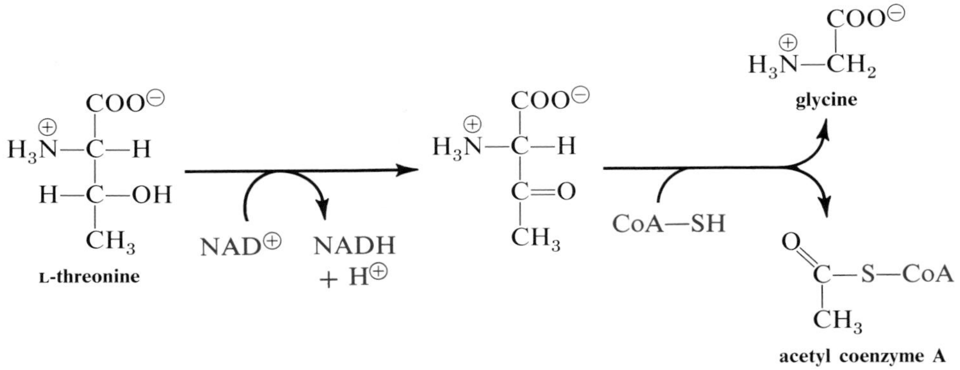

L-threonine + $NAD^{\oplus}$ → 2-amino-3-ketobutyrate + NADH + $H^{\oplus}$; + CoA—SH → glycine + acetyl coenzyme A

To the extent that this route is used, threonine is both glucogenic and ketogenic.

Methylidyne Groups

Tracer studies with ^{14}C-labeled serine and glycine show that carbon atoms from these compounds are rapidly incorporated at the oxidation state of formate into purine nucleotides and the like. This is accomplished most by an oxidation of 5,10-methylene-H_4folate to 5,10-methylidyne-H_4folate by an NAD-coupled dehydrogenase (Fig. 32–7).

Formation from Histidine. We noted in Chapter 31 that the catabolism of histidine begins with a deamination and ends with the formation of glutamate. Intermediate reactions involve opening the imidazole ring, and doing this creates **N-forminoglutamate.** The formimino group is the nitrogen analogue of a formyl

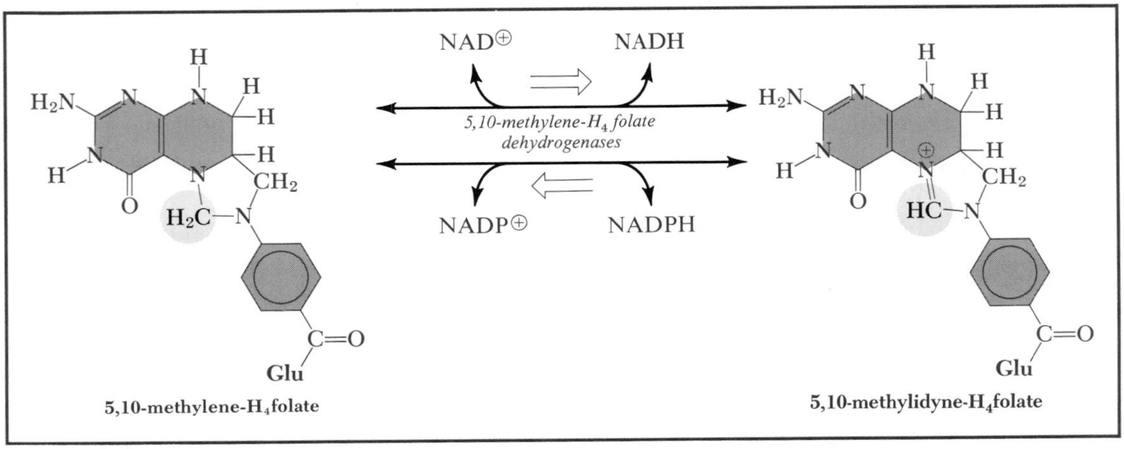

5,10-methylene-H_4folate ⇌ 5,10-methylidyne-H_4folate ($NAD^{\oplus}$ → NADH; NADPH → $NADP^{\oplus}$; 5,10-methylene-H_4folate dehydrogenases)

FIGURE 32–7 **Interconversion of 5,10-methylene-H_4folate, and 5,10-methylidyne-H_4folate.**

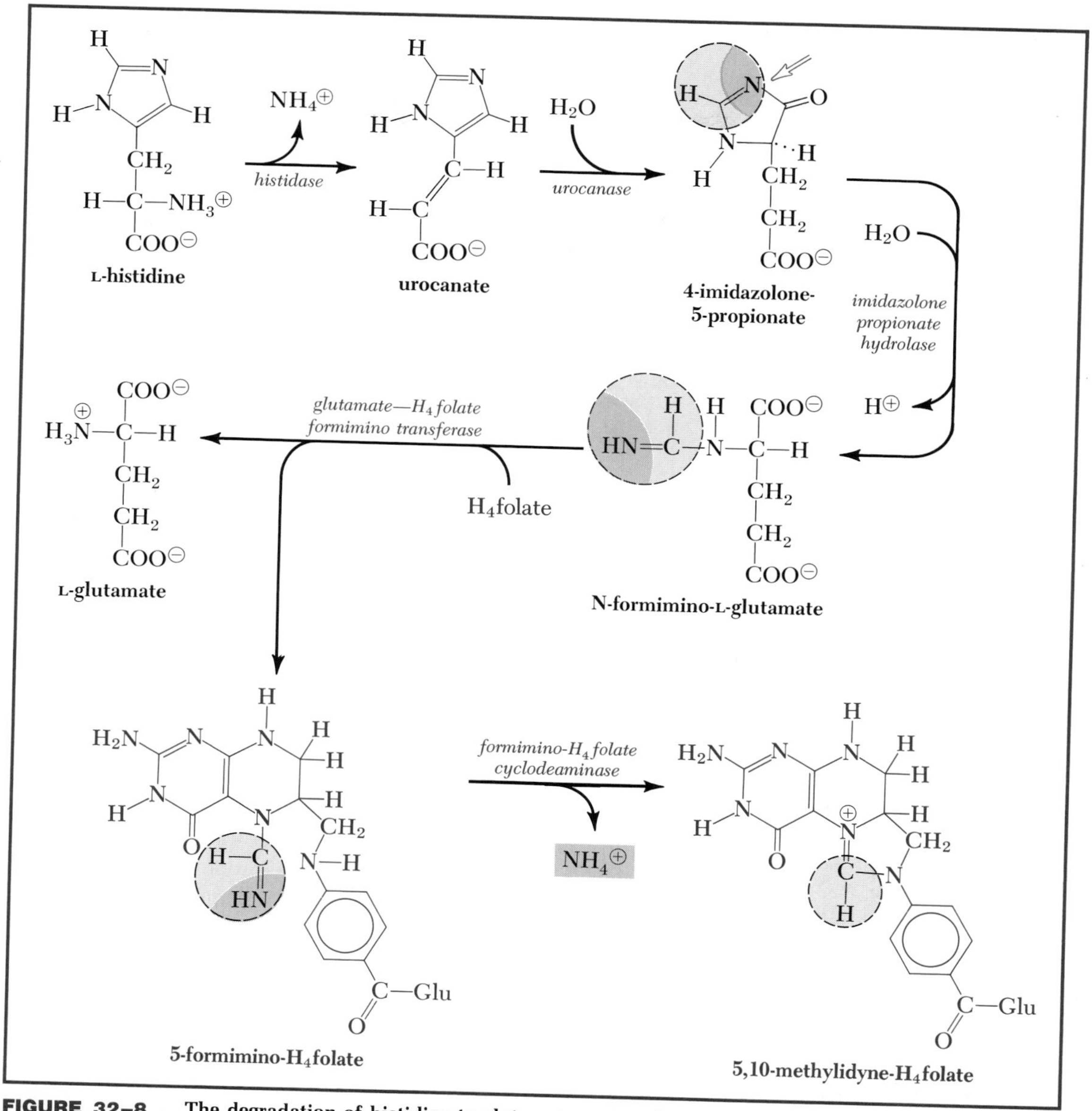

FIGURE 32–8 **The degradation of histidine to glutamate, ammonia, and a methylidyne group.**

group, and it is transferred intact to H_4folate (Fig. 32–8), followed by loss of NH_4^+ and cyclization to form 5,10-methylidyne-H_4folate. To summarize:

Histidine

Nitrogens and carbons: One nitrogen and five carbons appear as glutamate, two nitrogen as NH_4^+, and one carbon as methylidyne-H_4folate.

Since histidine is one of the less abundant amino acids, this route is not of great quantitative importance, but it does reflect the state of the H_4folate pool in this way: The formimino transferase reaction is slowed by a deficiency of H_4folate, which therefore causes an accumulation of **formiminoglutamate (FIGLU)** and its excretion in the urine, especially after ingestion of extra histidine.

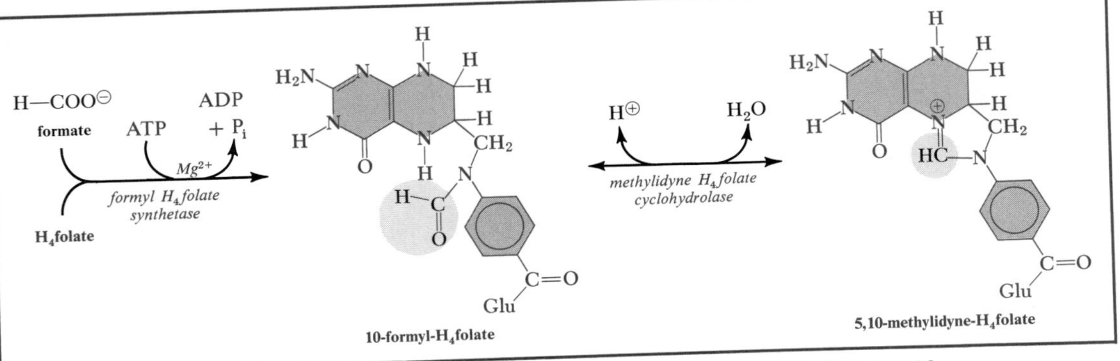

FIGURE 32–9 The interconversion of carbon groups at the oxidation state of formic acid.

Formation from Formate. Free formate is incorporated into the one-carbon pool *via* 10-formyl-H_4folate by the action of a synthetase (Fig. 32–9). A major source of formate is the oxidation of formaldehyde, which we shall shortly see is obtained by oxidation of methyl groups.

METABOLISM OF METHYL GROUPS

Methionine as Methyl Donor

S-Adenosylmethionine. A large number of biological compounds contain methyl groups that have been added to some parent compound, usually by attachment to O or N atoms, but sometimes to C atoms. The usual source of these methyl groups is the methyl thioether group of methionine, which is activated for transfer by forming S-adenosylmethionine (Fig. 32–10). S-Adenosylmethionine is formed by transfer of the adenosyl group from ATP to the sulfur atom of methionine, which makes a **sulfonium** group. It is a high-energy compound with respect to the C—S bonds, and its formation is driven in effect by the hydrolysis of all of the phosphate

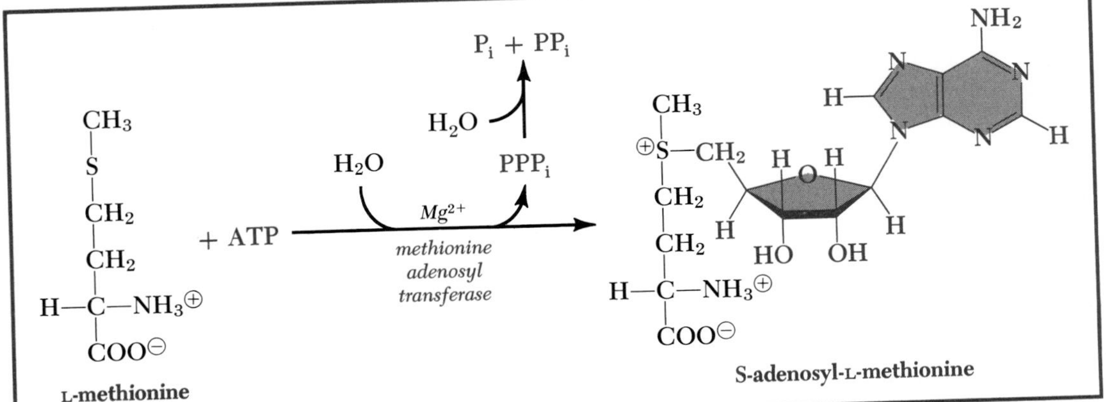

FIGURE 32–10 The formation of S-adenosylmethionine is driven, in effect, by the hydrolysis of two high-energy and one low-energy phosphate bonds, since the concentrations of PPP_i and PP_i are kept at low levels by hydrolases.

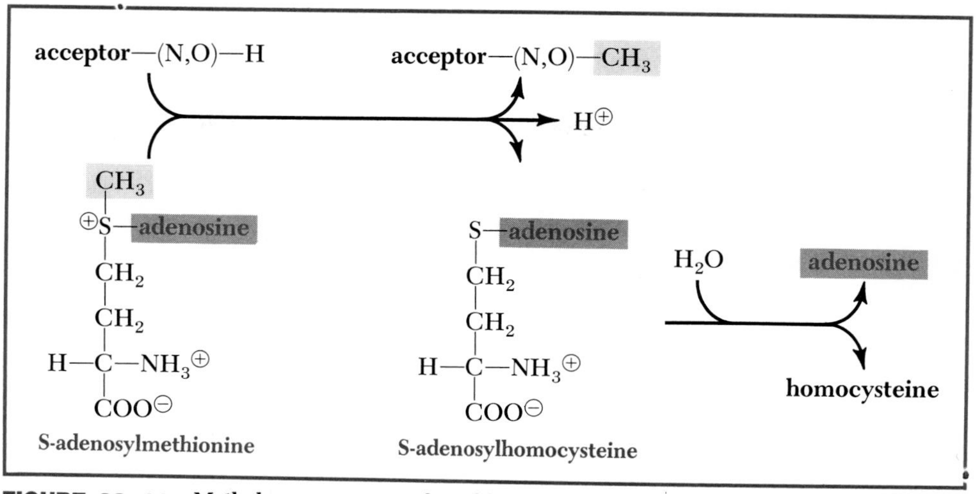

acceptor—(N,O)—H → acceptor—(N,O)—CH_3 + $H^{\oplus}$

CH_3—$^{\oplus}$S—adenosine, CH_2, CH_2, H—C—$NH_3^{\oplus}$, $COO^{\ominus}$

S-adenosylmethionine

S—adenosine, CH_2, CH_2, H—C—$NH_3^{\oplus}$, $COO^{\ominus}$

S-adenosylhomocysteine

H_2O → adenosine + homocysteine

FIGURE 32–11 **Methyl groups are transferred from S-adenosylmethionine to various acceptors, usually groups containing N or O atoms. The S-adenosylhomocysteine formed is a potent inhibitor of the methyltransferases, and it is removed by hydrolysis.**

bonds in ATP, since the initial transfer of the adenosyl group liberates inorganic triphosphate (PPP_i), which is rapidly hydrolyzed to PP_i and P_i.

Transfer of Methyl Groups. When methyl groups are transferred from S-adenosylmethionine to an acceptor, such as an amine or alcohol (Fig. 32–11), the remaining **S-adenosylhomocysteine** no longer has a charged sulfonium group. It is a simple thioether, analogous to methionine, and the resultant loss of free energy makes the methyl transfers essentially irreversible. Let us now consider important examples of these transfers.

Formation of Creatine

Creatine includes fragments of methionine, glycine, and arginine. It is made by transfer of a methyl group from S-adenosyl-methionine to guanidinioacetate (Fig. 32–12). The guanidinioacetate is formed by transfer of an amidinio group from arginine to glycine, leaving ornithine.

The formation of creatine provides a means of storing high-energy phosphate in muscles and nerves, but it also represents a constant drain on the organism's supply of methionine. This is true because phosphocreatine spontaneously cyclizes at a slow rate to form **creatinine,** which is excreted in the urine, and there is no known way of recovering the methyl groups transferred to make creatine. The rate of loss depends only upon the total phosphocreatine content at a given temperature and pH and is therefore quite constant from day to day in a given individual. (A typical adult male excretes about 15 millimoles per day.) The constancy can be used to test the reliability of urine samples—the idea being that a bottle alleged to contain all of a day's urine ought to contain all of a day's creatinine and this can be measured.

Creatinine output can also be used to simplify some comparative clinical measurements by making it unnecessary to collect timed samples of urine. Since the quantity of creatinine in a conveniently collected urine sample is in itself an index of the volume of blood filtered to excrete that sample, one compares the ratio of the urinary output of some other metabolite and the output of creatinine. The ratio will change in proportion to the relative rate of excretion of the metabolite.

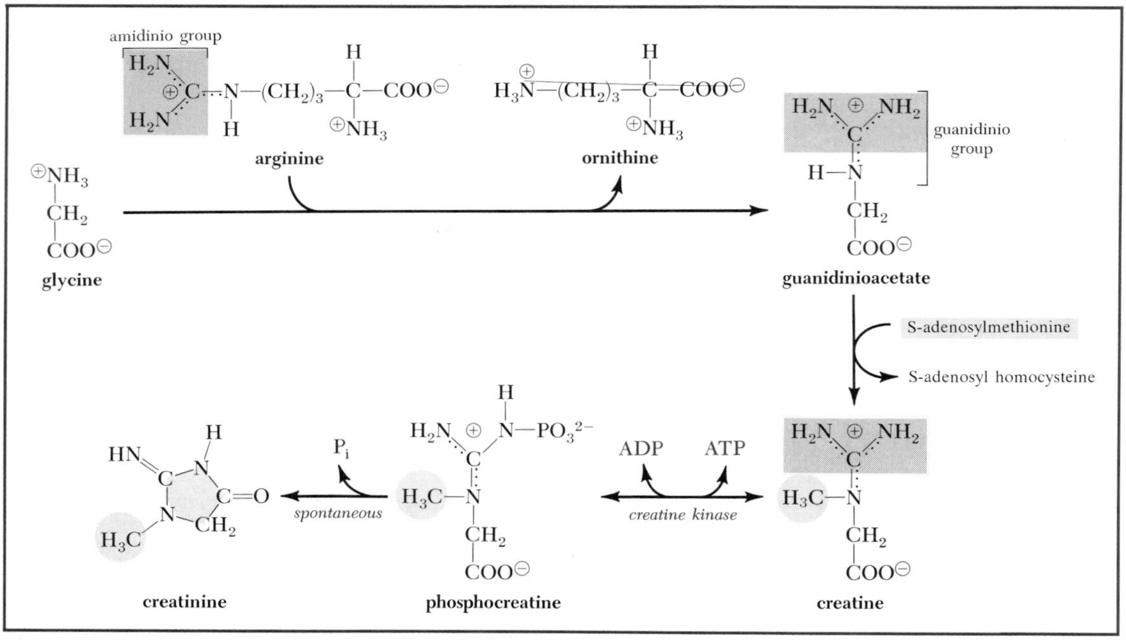

FIGURE 32–12 **Creatine is made by combining glycine, an amidinio group from arginine, and a methyl group from S-adenosylmethionine. There is a continual loss of these groups in the urine as creatinine, which is formed by spontaneous cyclization of phosphocreatine.**

Since creatinine excretion depends upon phosphocreatine content, it can also be used to assess muscle mass. When muscle degenerates for any reason—from paralysis, or from muscular dystrophy—the creatinine content of the urine falls. In addition, any rise in blood creatinine concentration is a useful indicator of kidney malfunction, since it is normally eliminated rapidly, and is less subject to the influence of diet and other factors than is the blood urea concentration.

Choline and Phospholipids

Choline is the most abundant N-methyl compound in the body:

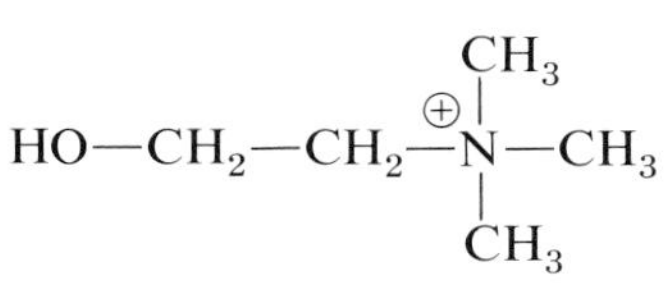

$$HO-CH_2-CH_2-\overset{\oplus}{N}(CH_3)_3$$

choline

Although it is qualitatively important as the precursor of acetylcholine, which is a transmitter in some nerves, choline occurs mainly in phospholipids—phosphatidylcholines, sphingomyelins, and the like. Indeed, choline is synthesized in the form of phosphatidylcholines (Fig. 32–13).

Synthesis of choline involves a transfer of three methyl groups from S-adenosyl methionine. The monomethylethanolamine derivative is made by one enzyme, and the addition of the next two methyl groups is catalyzed by another enzyme. The phospholipids are constantly turning over through hydrolysis, so choline generated in this way is later released.

The ethanolamine moiety is created from serine by an ingenious arrangement (Fig. 32–14). The phosphatidyl group in phosphatidylethanolamine is transferred to serine, creating phosphatidyl serine and releasing free ethanolamine. A second enzyme catalyzes the decarboxylation of phosphatidylserine, leaving phosphatidylethanolamine as a product. Here we have an example of another class of enzymes, the **amino acid decarboxylases,** that utilize the labilizing effect of condensation with pyridoxal phosphate to promote cleavage of one of the C-2 bonds.

The sum of these two enzymatic reactions amounts to a net formation of ethanolamine from serine:

1. serine + phosphatidylethanolamine $\longrightarrow$ phosphatidylserine + ethanolamine
2. phosphatidylserine + H^+ $\longrightarrow$ phosphatidylethanolamine + CO_2

SUM: serine + H^+ $\longrightarrow$ ethanolamine + CO_2

Since serine can be made from glucose and ammonia, it follows that ethanolamine can also be completely synthesized from these compounds, and a dietary supply is not necessary. Now let us see how ethanolamine is converted to phosphatidylethanolamine.

Use of the Free Base. In addition to the ethanolamine and choline obtained from the syntheses we outlined, the pool of the free compounds is also supplemented by

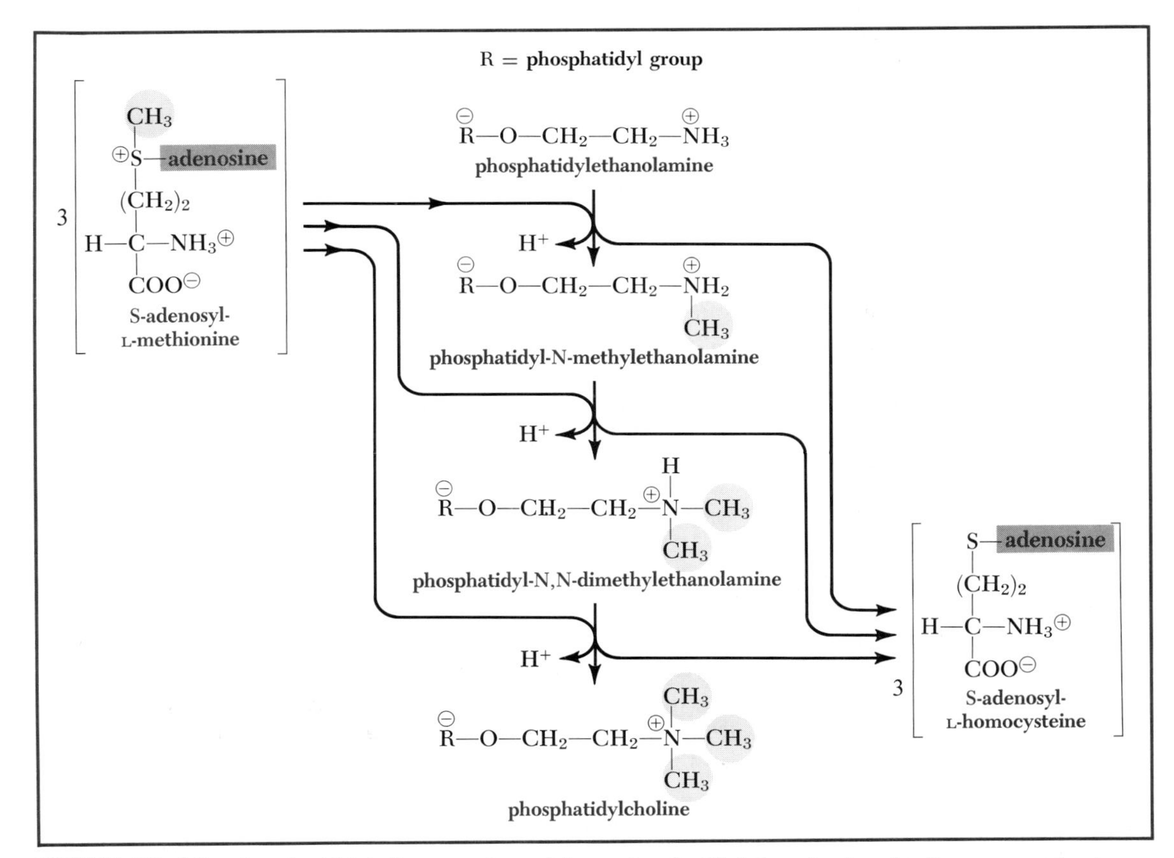

FIGURE 32–13 Phosphatidylcholines are formed from phosphatidylethanolamines by three successive transfers of methyl groups from S-adenosylmethionine.

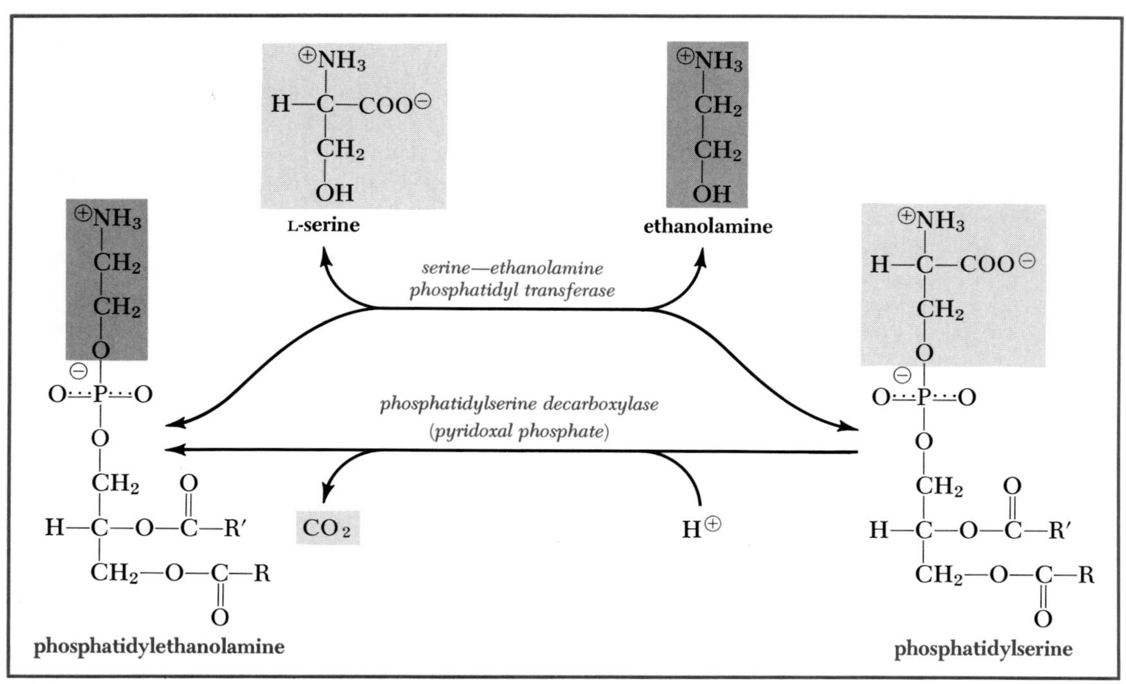

FIGURE 32–14 **Phosphatidylserines are formed from the corresponding phosphatidylethanolamines by exchange of serine for ethanolamine. Some of the phosphatidylserines are decarboxylated to phosphatidylethanolamines, with a net effect of converting serine to ethanolamine. Pyridoxal phosphate is used as a coenzyme by amino acid decarboxylases.**

the diet. (Indeed, the dietary supply can be sufficient to supply most of the needs.) The free compounds are used to make phospholipids through an intermediate conversion to cytidine diphosphate derivatives (Fig. 32–15), at a total cost of 3 $\sim$ P per phosphatidyl base made.

Summarizing the picture at this point, methyl groups from methionine can be used to make choline in the form of phosphatidylcholines, provided that phosphatidylethanolamines are available. However, the demand for synthesis can be alleviated to the extent that choline and ethanolamine are obtained from the diet.

Methylation of RNA

We noted in our discussion of protein synthesis that transfer RNA and ribosomal RNA contain bases other than the usual guanine, adenine, cytosine, and uracil. The distinctive bases frequently are methylated derivatives of the more common compounds (p. 82). All of these methyl derivatives are created by enzymatic transfer of methyl groups from S-adenosylmethionine to the ribonucleates, which originally contain only the nonmethylated bases at the time of assembly on DNA templates. Even the thymine found in RNA is made by methylation of uracil, not by direct incorporation of thymidine at the time of nucleic acid formation. The bases to be methylated appear to be located in particular sequences. For our purposes now, the important thing to recognize is that the formation of these specific ribonucleates represents a drain upon the supply of methyl groups represented by S-adenosylmethionine.

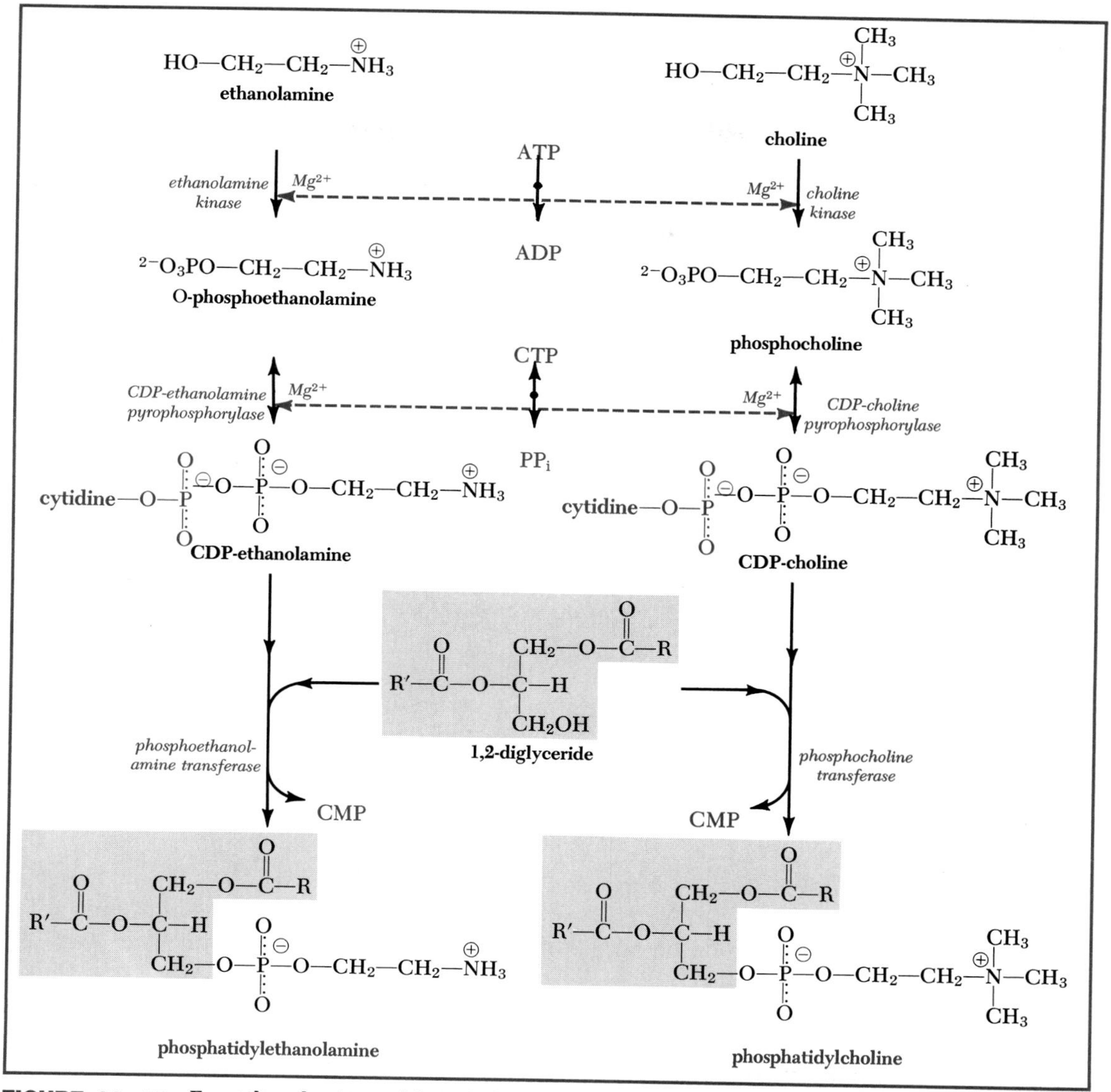

FIGURE 32–15 Free ethanolamine and free choline can be incorporated into phosphatidyl compounds by converting them to the CDP-derivatives, from which the phospho-base moieties are transferred to 1,2-diglycerides.

REGENERATION OF METHIONINE

In these methylation reactions, only the methyl group in methionine is being consumed; the remainder of the molecule is intact as S-adenosylhomocysteine, which is hydrolyzed to release free homocysteine:

$$\text{S-adenosylhomocysteine} + H_2O \longrightarrow \text{adenosine} + \text{homocysteine.}$$

There is no route for making homocysteine in mammals, and it is important to salvage this important molecule. All that is required is methylation of the sulfur atom to regenerate methionine, and there are two ways by which this is done. One involves the recovery in mitochondria of a methyl group from choline; the other involves the generation in the cytosol of a methyl-H_4folate.

Recovery of Methyl Groups from Choline. It is possible to salvage one of the methyl groups from an excessive supply of choline for use elsewhere. This is done by raising the chemical potential of choline through an oxidation of its alcohol group to a carboxylate group (Fig. 32–16). This creates a **betaine** ("bay-tah-een"), an N,N,N-trimethylamino acid.* The loss of free energy upon removing one methyl group from the quaternary nitrogen is sufficient to support a transfer of the group to homocysteine. The tertiary amino group in the remaining dimethylglycine has too little free energy to support additional transfers. However, many N-methyl compounds are catabolized by oxidizing the methyl group to formaldehyde, and dimethylglycine is converted in this way to glycine and two molecules of formaldehyde. In short, one out of the three methyl groups transferred from S-adenosyl methionine to make choline can be recovered as methionine. The other two can also re-enter the one-carbon pool, but as 5,10-methylene-H_4folate *via* formaldehyde. Note, however, the **dietary choline can be used as a source of a methyl group** in lieu of synthetic pathways. Note also that an average yield of 3 ~ P per methyl group is obtained during the degradation of choline.

Formaldehyde in excess can be oxidized to CO_2 and H_2O. It is first oxidized to formate, which is then converted to the H_4folate derivative before the final oxidation. All of these oxidations occur within mitochondria, and therefore contribute their mites to oxidative phosphorylation.

Although H_4folate and its derivatives cross the inner mitochondrial membrane only very slowly, there is interaction between the one-carbon pool inside and outside the mitochondria through transfer of amino acids bearing the groups. The hydroxymethyl transferase equilibrating glycine and serine occurs both in the mitochondria and in the cytosol, so passage of these two amino acids is in effect carrying methylene groups.

***De Novo* Synthesis from Methylene-H_4folate.** The supply of methyl groups obtained from the diet in the form of methionine and choline is supplemented by generating them from the one-carbon pool. 5,10-Methylene-H_4folate is irreversibly reduced to 5-methyl-H_4folate (Fig. 32–17), and the methyl group is then transferred to homocysteine, regenerating methionine.

The transferase catalyzing the reaction requires **methylcobalamin** as a coenzyme, rather than the deoxyadenosylcobalamin necessary for methylmalonyl CoA mutase (p. 599). The methyl group is transferred from the coenzyme to homocysteine, and the coenzyme is recharged from 5-methyl-H_4folate.

Many of the deleterious effects of vitamin B_{12} deficiency are believed to result from a failure of this critical transfer of methyl groups. One important consequence is an accumulation of excessive amounts of 5-methyl-H_4folate. This is ordinarily the most abundant form of H_4folate, and the form that crosses cell membranes, so it is not the increase in concentration of the compound that causes the problem but the depletion of the smaller amounts of the other forms of H_4folate. **The effects of vitamin B_{12} deficiency therefore include an accompanying folate deficiency, in addition to an impairment of methionine synthesis and methylmalonate metabolism.**

---

*Named for its occurrence in *Beta vulgaris,* the common beet. More specifically, it is glycine betaine, distinguishing it from other trimethylamino acids.

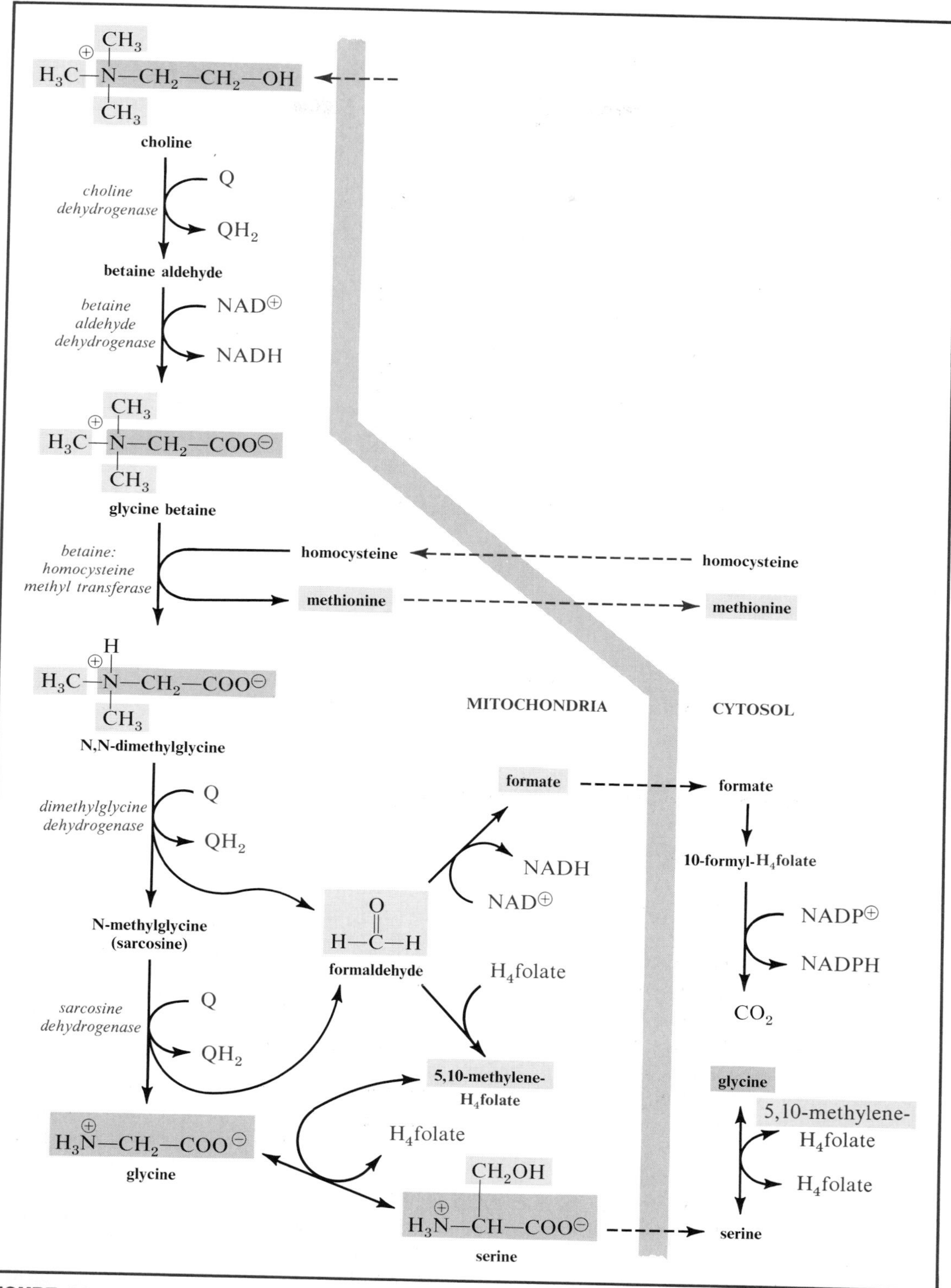

FIGURE 32–16 Choline is constantly degraded in mitochondria by oxidation of its alcohol group to a carboxylate group, followed by removal of the methyl groups. One methyl group is transferred to homocysteine (*center*), and the other two are oxidized to formaldehyde. The one-carbon groups are believed to move from the mitochondria to the cytosol in the form of methionine and serine.

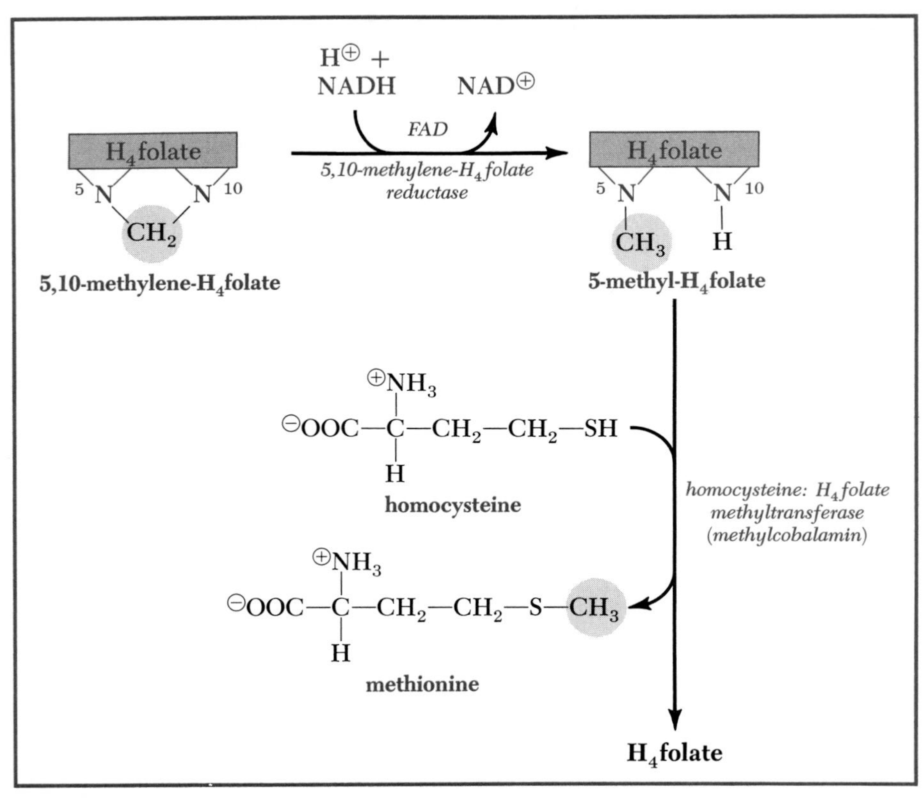

FIGURE 32–17 **The methyl group of methionine can be synthesized by transfer from 5-methyl-H_4folate to homocysteine. The transfer involves methylcobalamin.**

DEGRADATION OF HOMOCYSTEINE

Nitrogen: Released as NH_4^+.
Carbons: Converted to succinyl coenzyme A.
Sulfur: Appears as cysteine.

Although some homocysteine can be recovered as methionine by methylation reactions, part is constantly lost through degradation by an irreversible route in which the sulfur appears as cysteine and the carbon skeleton as 2-ketobutyrate, which we have already seen is metabolized *via* propionyl coenzyme A and CO_2 to succinyl coenzyme A. Therefore, methionine is a source of cysteine and methyl groups and is a glucogenic amino acid.

The actual degradation of homocysteine is a simple two-step process involving the formation of **cystathionine** and its cleavage to form cysteine, 2-ketobutyrate, and ammonium ion (Fig. 32–18). Both of the enzymes involved have pyridoxal phosphate as a coenzyme.

Defects in both these enzymatic steps occur in humans. **Homocystinuria,** the excretion of the oxidized form of homocysteine, occurs in a hereditary absence of cystathionine synthase, and results in mental retardation. **Cystathioninuria,** the excretion of cystathionine, also is associated with mental defects. When one considers the unusually high content of cystathionine in the brains of normal individuals (0.2 to 0.5 mg per g of tissue, as contrasted with 7 to 8 μg per g in other tissues) and the disturbances in the nervous system apparently created by an inability to make or

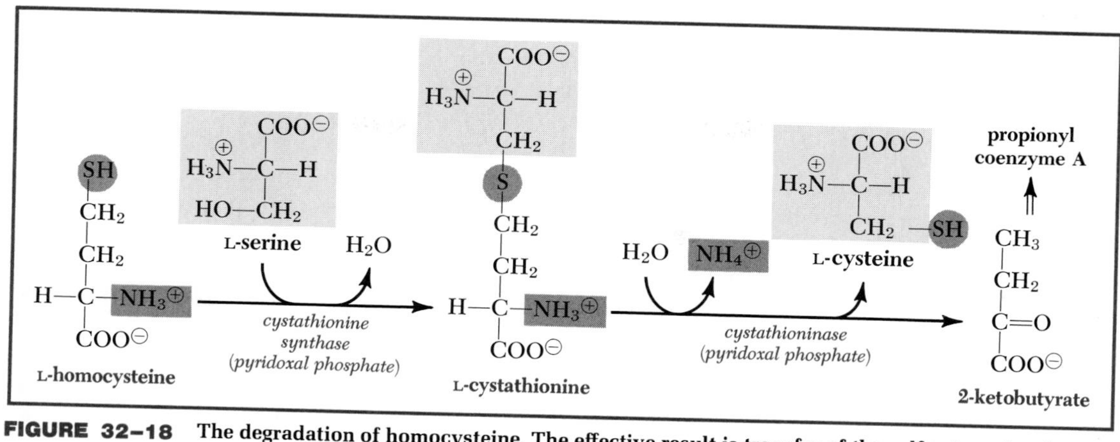

FIGURE 32–18 The degradation of homocysteine. The effective result is transfer of the sulfur to serine, forming cysteine, and a deamination with the carbon skeleton appearing as 2-ketobutyrate.

destroy this amino acid, one suspects it may have some presently unknown function in nerve physiology. Primates have a higher concentration of cystathionine in the brain than do other mammals, whereas birds have much lower levels than do mammals. This is not to imply that the term "hypocerebrocystathioninic" could rationally be substituted for bird-brained as an epithet.

Low levels of cystathionine synthase activity have another consequence of great interest: damage to the endothelial cells in arteries. Indeed, administration of homocysteine has been used as an experimental tool to demonstrate that endothelial cell damage is probably an essential preliminary to the development of atherosclerotic plaques. The mechanism for damage is not clear. Accumulation of homocysteine causes the formation of a mixed disulfide with cysteine; this is the principal form in which homocysteine appears in the blood, and the formation of the compound further diminishes an already low supply of cysteine resulting from the failure to metabolize homocysteine.

The question arises: Is there any connection between coronary artery disease and a disturbance in homocysteine metabolism? Very preliminary results suggest there may be in at least some cases, but this is by no means conclusive. A very, very small sample ($n = 25$) of coronary patients had significantly higher concentrations of the homocysteine-cysteine mixed disulfide than did control patients. In the general population, the incidence of homocystinuria may be as high as 1:50,000, implying that 1 in 100 people are heterozygotes, so it is possible that additional people have less severe impairments in homocysteine metabolism, but sufficient to cause an elevated concentration as the preliminary data suggest.

Before leaving this subject, let us re-emphasize that methionine is an important dietary constituent for three purposes. Most importantly, **it is required as an amino acid for the formation of proteins,** and no other dietary component will substitute for this purpose, since the homocysteine chain cannot be made. As a **source of methyl groups** it is important, but not imperative, since choline in the diet can provide methyl groups, and a further supply is available through the one-carbon pool by the formation of methyl-H_4folate. Finally, it is a **source of sulfur** available for the formation of cysteine. This relieves the necessity of an absolute requirement for cysteine in the diet, but it means that sufficient methionine must be present to provide an excess sulfur supply, if the content of cysteine is low.

TOTAL BALANCE OF AMINO ACID DEGRADATION

Now that we have completed a survey of amino acid degradation, let us take stock. What is the net result of the use of amino acids as fuels? It is not possible to give a single answer to that question because several tissues are involved in ways that are influenced by other metabolic events. However, we can make an approximation by defining conditions. Let us develop a model for the fate of the amino acids in a piece of beefsteak eaten by an individual who has been fasting for one day. Assume that this hungry person has eaten 1000 millimoles of amino acids. The average formula weight of the amino acid residues in beef muscle is near 110, so this quantity of amino acids will be obtained from 110 grams of protein, which is the amount in about 530 grams of raw lean meat. Since some of the amino acids contain more than one nitrogen atom, the total nitrogen content of the protein is near 1390 milliatoms. To make our calculations, we shall specify exact amounts, even though they are not known with this precision for real foods. As a frame of reference it is helpful to know that the quantity we are dealing with is close to the mean total daily intake of protein for young white male adults in the United States.

The calculations that follow are based on the following composition (Orr, M.L., and B.K. Watt: [1957] *Amino Acid Content of Foods.* Home Economics Research Report No. 4, U.S.D.A.), given in millimoles of each residue per 1000 millimoles of all residues:

Ala	82.5	His	28.5	Pro	54.5
Arg	47	Ile	51	Ser	51
*Asx	89	Leu	79.5	Thr	47
Cys	13	Lys	76	Trp	7.5
*Glx	131	Met	21	Tyr	24
Gly	105	Phe	32	Val	60.5

amide nitrogen (=Asn + Gln) = 110

Events In Skeletal Muscles

The branched chain amino acids from proteins eaten by a fasting individual are mainly taken up by the skeletal muscles. The muscles put out nitrogen as glutamine and alanine, for the most part. Assuming that the muscles metabolize the carbon skeletons, and supplying a small amount of glutamate from the liver, the following stoichiometry can be developed:

Consumed By Muscles

51 isoleucine	21 glutamate
79½ leucine	919½ O_2
60½ valine	

Produced By Muscles

53 alanine	634 CO_2
79½ glutamine	4,855 ~P

*Asx = Asn + Asp; Glx = Gln + Glu.

Events in the Small Bowel

Most of the aspartate, asparagine, glutamate, and glutamine from the diet is processed in the intestinal mucosa, and the mucosa also is the tissue mainly responsible for clearing glutamine from the blood for use as a fuel. Much of the ammonium ion released by hydrolysis of asparagine and glutamine appears as such in the portal blood. Alanine is the other predominant nitrogenous product. (Citrulline and other compounds are also produced, but we can simplify the stoichiometry by considering only alanine without serious distortion of the final result.)

Consumed By Small Bowel

from the diet	*from the blood*
89 (aspartate + asparagine*)	79½ glutamine
131 (glutamate + glutamine*)	315¾ O_2

Produced By Small Bowel

299½ alanine	510 CO_2
189½ NH_4^+	1,838 ~P

Events In Liver

The alanine poured out by the muscles and the small bowel and the balance of the dietary amino acids are mainly handled in the liver. The nitrogen appears as urea, the sulfur as sulfate, and the carbon skeletons as CO_2 and glucose in a fasting individual, except for a small amount of acetoacetate. (We shall neglect D-3-hydroxybutyrate formation, since the amount involved is minor.)

Consumed By Liver

435 alanine	54½ proline
47 arginine	51 serine
13 cysteine	47 threonine
105 glycine	7½ tryptophan
28½ histidine	24 tyrosine
76 lysine	189½ NH_4^+
21 methionine	1,166½ O_2
32 phenylalanine	

Produced By Liver

695¾ urea	367½ CO_2
21 glutamate	56 acetoacetate
396 glucose	150 ~P
34 SO_4^{2-}	

The striking thing here is the negligible yield of high-energy phosphate from a large consumption of oxygen. **Amino acid metabolism by itself will not sustain the energy-requiring processes of the liver under these conditions.** Note also that the

*Dietary asparagine + glutamine = 110.

stoichiometry allows a small output of glutamate to replace the amount used in skeletal muscles to make glutamine. This is consistent with the observed events *in situ.*

Events In Brain

To complete the picture, we must account for the glucose and acetoacetate produced by the liver. A reasonable assumption in a fasting man is the use of the fuels by the brain, for which the stoichiometry is:

$$396 \text{ glucose} + 56 \text{ acetoacetate} + 2{,}600\ O_2 \longrightarrow 2{,}600\ CO_2 \times 15{,}346 \sim P.$$

Total Stoichiometry

If we add the balances within the individual tissues, the complete equation for the metabolism of the amino acids in beefsteak becomes:

$$1{,}000 \text{ amino acids} + 5{,}001\tfrac{3}{4}\ O_2 \longrightarrow 695\tfrac{3}{4} \text{ urea} + 4{,}111\tfrac{1}{4}\ CO_2 + 22{,}188 \sim P + 34\ SO_4^{2-}.$$

The overall P:O ratio in the whole body is 2.22.

The inevitable question is whether this is an academic exercise without relevance to real events. We have a clue from old observations: Animals can be caused to excrete glucose in the urine by removing the pancreas so as to create diabetes, or by treating them with phlorhizin, which prevents the reabsorption of glucose from the filtered blood plasma in the kidney. (Phlorhizin is a polyphenolic glycoside found in the bark of apples, cherries, and other rosaceous plants.) When properly performed, the experiments resulted in nearly quantitative recovery of any glucose formed in the animal. The amount of glucose formed by the metabolism of amino acids could then be measured by analyzing the excretion of nitrogen and of glucose. The results were expressed as a D:N ratio, the weight of glucose (dextrose) formed per weight of nitrogen excreted.

Now, the theoretical D:N ratio, according to our stoichiometry, would be 396×180 grams of glucose per $695\tfrac{3}{4} \times 2 \times 14$ grams of excreted N, which calculates out as 3.66. The actual observed values on animals fed meat were 3.63 in dogs, 3.68 in humans treated with phlorhizin, and 3.63 to 3.73 in humans with diabetes!

Another test of theory is through the observed relative value of proteins as fuels, compared to fats and carbohydrates. This will be discussed more extensively in Chapter 39, when we shall use the theoretical values in reasoning about the total metabolic economy.

Origin of the Stoichiometries

A detailed discussion of the calculations made in developing the preceding equations is beyond our scope. Those with special interest might try to reconstruct the reasoning. The model for skeletal muscle metabolism was developed to make the ratio of exported glutamine and alanine 3:2 and minimize the requirement for use of glutamate from the blood.

It is necessary to account for the movement of metabolites within cellular compartments, and the following hints will be useful: (1) Aspartate enters mitochondria by energy-driven glutamate-aspartate antiport. (2) α-Ketoglutarate leaves mitochondria by exchange for malate, and the malate can be returned by exchange for P_i. There is a constant flow of P_i into mitochondria for oxidative phosphorylation. (3) Alanine aminotransferase occurs in both the cytosol and mitochondria, and alanine can move across the inner mitochondrial membrane without exchange. This enables movement of nitrogen atoms from one compartment to the other. (4) Phospho-*enol*-pyruvate can be formed from oxaloacetate in both the cytosol and the mitochondria, and it moves across the inner membrane. (5) Pyruvate moves into the mitochondria together with H^+, making this movement also energy-driven.

SYNTHESIS OF AMINO ACIDS

Animals rely upon the diet for the provision of some amino acids. Indeed, it might be asked why animals synthesize any of the amino acids, since all 20 are obtained from the food. However, we have seen that some of the amino acids have important metabolic roles, which in themselves involve a constant synthesis and degradation of the compounds in quantities much beyond those needed for protein synthesis. Animals have maintained an ability to synthesize these amino acids *de novo,* that is, from glucose as a carbon source and ammonium ion as a nitrogen source.

The premier example is **glutamate,** which can be formed by a reversal of the glutamate dehydrogenase reaction:

$$\underset{\alpha\text{-ketoglutarate}}{{}^{\ominus}OOC{-}CH_2{-}CH_2{-}C(=O){-}COO^{\ominus}} + NH_4^{\oplus} + \text{NADPH or NADH} \xrightleftharpoons{\text{glutamate dehydrogenase}} \underset{\text{L-glutamate}}{{}^{\ominus}OOC{-}CH_2{-}CH_2{-}CH(\overset{\oplus}{N}H_3){-}COO^{\ominus}} + \text{NADP}^{\oplus} \text{ or NAD}^{\oplus}$$

This is the point of entry for ammonia nitrogen into other nitrogenous compounds in animal tissues. The α-ketoglutarate required for the reaction can be made from glucose *via* citrate. (Both the oxaloacetate and the acetyl coenzyme A used for citrate formation can be made from pyruvate.) Since glutamate dehydrogenase can function with either NAD or NADP, some presume that it uses NAD for oxidation and NADPH for the synthetic reduction, but a rational explanation for this discrimination and a demonstration of its occurrence has yet to be made.

Any increase in the concentration of glutamate will automatically cause an increase in the concentration of **aspartate** and **alanine,** because the aminotransferases constantly equilibrate these amino acids and the available supply of glutamate, oxaloacetate, and pyruvate. Therefore, aspartate and alanine need not be supplied in the diet (Fig. 32–19).

Arginine is continually being synthesized by the reactions of the urea cycle, but if arginine is withdrawn from the cycle, ornithine must be provided from some other source to replace it. It is known that ornithine can be made from glutamate by animals, but it is not known how. The reaction of ornithine aminotransferase (p. 595) is reversible, so ornithine can be made from glutamate semialdehyde, but how is glutamate semialdehyde made? The oxidation of glutamate semialdehyde to glutamate is irreversible, like most aldehyde oxidations, and there is no known alternative route from glutamate to the aldehyde in animals (Fig. 32–20).

Proline is also known to be synthesized from glutamate by a process involving the reduction of the cyclic form of glutamate semialdehyde with NADPH, but the

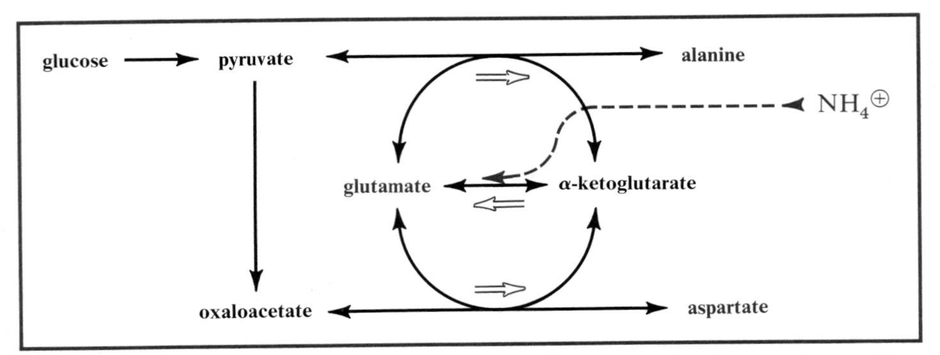

FIGURE 32–19 Carbon from glucose and nitrogen from ammonium ions can be used for total synthesis of alanine, aspartate, and glutamate in animals.

same uncertainty exists over the source of the semialdehyde. Here we also see another example of the selective use of electron carriers. Proline is degraded through the use of the flavoprotein to favor oxidation; it is synthesized through the use of NADPH to favor reduction.

The retention of arginine and proline synthesis by animals reflects their important metabolic roles. Arginine is involved in the synthesis of urea, creatine, and other compounds. Proline is an important fuel for the muscles of insects and other animals; it is also turned over more rapidly than many amino acids in human tissues, although a specific function of its metabolism has not been described.

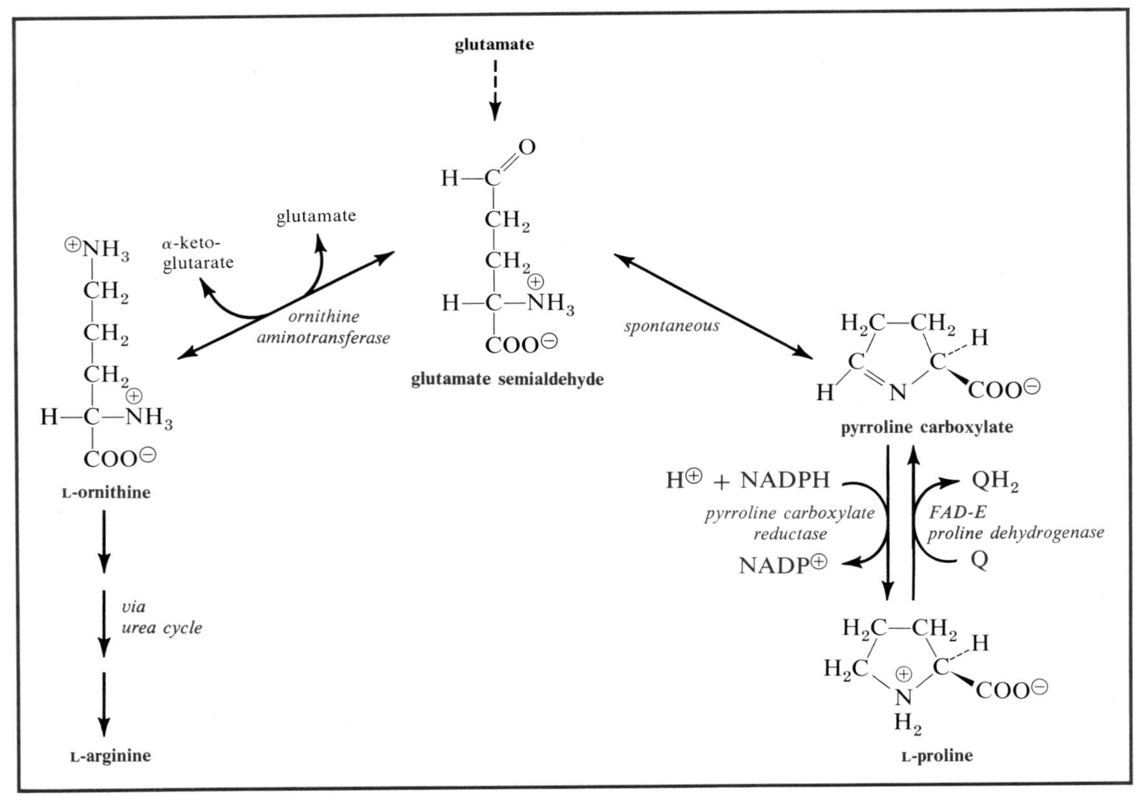

FIGURE 32–20 Arginine and proline can be synthesized from glutamate, but the mechanism of formation of the intermediate glutamate semialdehyde is not known.

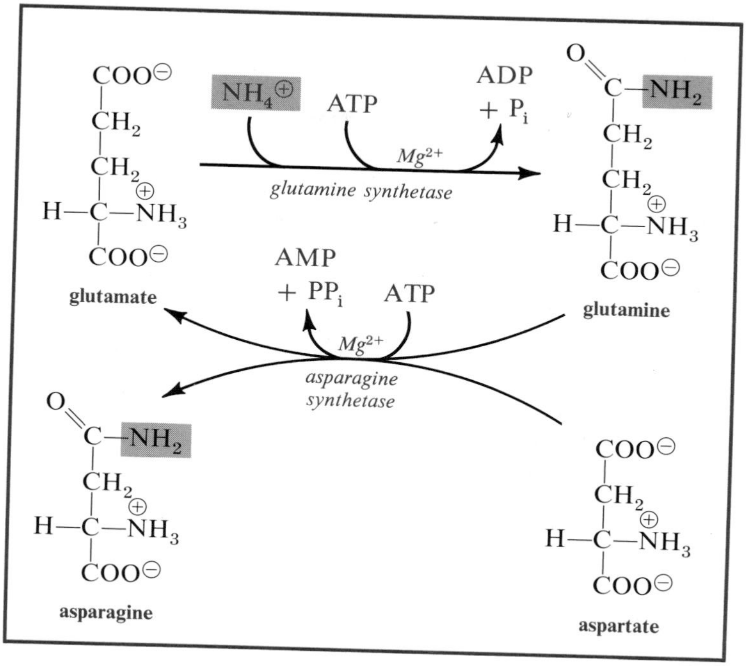

FIGURE 32-21

The amide nitrogen of glutamine is obtained directly from ammonium ion. It may be transferred to aspartate to generate asparagine.

Serine, and therefore **glycine,** can be made from glucose and glutamate through routes described earlier in the chapter. They have a rapid turnover in animals reflecting their participation in the one-carbon pool and in the formation of other nitrogenous compounds. (We shall see later that glycine is required for purine synthesis.)

Glutamine is made from glutamate and ammonium ion at the expense of high-energy phosphate. **Asparagine** is made from aspartate by the transfer of the glutamine amide nitrogen, also at the expense of high-energy phosphate (Fig. 32-21). Here we see the first example of a general function of glutamine: It often acts as a donor of an amino group during the synthesis of other compounds.

None of the other 11 amino acids can be made *de novo* by mammals. However, **tyrosine** can be made from phenylalanine, and **cysteine** can be made from methionine, so these two amino acids are not essential constituents of the diet, so long as the supply of their precursor amino acids is adequate. The practical implications of this are discussed in Chapter 40.

Finally, we come to a hard-core list of nine amino acids that must be present in the human diet:

Essential Amino Acids in Human Diets

histidine	lysine	threonine
isoleucine	methionine	tryptophan
leucine	phenylalanine	valine

Inspection of the routes of degradation of these amino acids will show that they include effectively irreversible steps; animals have no other reactions to bypass these steps. There may be one exception. A route has been proposed for the synthesis of leucine, beginning with a branched chain 3-ketoacid formed from valine (Fig. 32-22). The last step in this route involves an equilibration of β-leucine, the 3-amino

from acetyl coenzyme A; from valine — COO^-, CH_2, $O=C$, CH, H_3C CH_3 — 3-ketoisocaproate; Glu α-KG → "β-leucine" (COO^-, CH_2, H_3N^+-C-H, CH, H_3C CH_3); leucine 2,3-aminomutase, adenosyl cobalamin ⇌ L-leucine (COO^-, H_3N^+-C-H, CH_2, CH, H_3C CH_3)

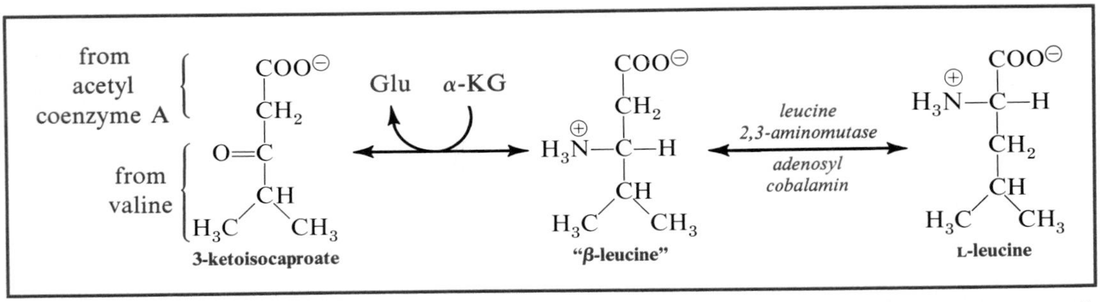

FIGURE 32–22 A postulated route for the synthesis of leucine in humans. The carbon skeleton is constructed from branched chain fatty acids (as the coenzyme A ester) to form the 3-keto, and then the 3-amino analogue of leucine. An enzyme requiring adenosylcobalamin as a coenzyme can interconvert leucine and its 3-amino analogue.

analogue of leucine, with leucine itself. The 2,3-aminomutase catalyzing this reaction is a cobalamin-dependent reaction. The most compelling evidence for leucine synthesis by this route is a decline in the leucine concentration and a rise in the β-leucine concentration in patients with pernicious anemia. Treatment with cyanocobalamin causes a fall in the valine concentration and a rise in the leucine concentration. However, it is clear that this putative route for leucine synthesis in man is not adequate to maintain the supply of the amino acid, and some must be supplied in the diet. We depend upon plants and autotrophic bacteria to maintain our supply of the essential amino acids.

FURTHER READING

D. Mastropaolo et al. (1980). Folic Acid: Crystal Structure and Implications for Enzyme Binding. Science 210:334.

J.O. Schulman et al. (1980). Genetic Disorders of Glutathione and Sulfur Amino-Acid Metabolism. Ann. Int. Med. 93:330.

J.G. Hilton, B.A. Cooper, and D.S. Rosenblatt (1979). Folate Polyglutamate Synthesis and Turnover in Cultured Human Fibroblasts. J. Biol. Chem. 254:8398.

Y. Aoyama and Y. Motokawa (1981). L-Threonine Dehydrogenase of Chicken Liver. Purification, Characterization, and Physiological Significance. J. Biol. Chem. 256:12367.

J.M. Poston (1980). Cobalamin-dependent Formation of Leucine and β-Leucine by Rat and Human Tissue. Changes in Pernicious Anemia. J. Biol. Chem. 255:10067.

CHAPTER 33

AMINES

"Waste not, want not" is an adage that fits the physiological use of amino acids. In addition to their indispensable function as precursors of proteins and their auxiliary use as fuels, some are precursors of a wide variety of small molecules containing nitrogen that have potent physiological actions, which make them more important than their relatively low concentrations would indicate. Many have been studied intensively because of their function in the nervous system; others are hormones or constituents of coenzymes; still others have unknown roles, but their importance is manifested by the consequences of disturbances in their metabolism. In addition, some of the amino acids themselves are of physiological importance in their own right.

NEUROTRANSMITTERS

Transmission of a nervous impulse between cells occurs by diffusion of specific chemical compounds, which act as signals. The junctions at which communication occurs are **synapses.** The synapses at the **neuromuscular junctions** are also known as **motor end-plates.** The basic pattern is that the nerve carrying the information to the next cell contains vesicles loaded with a transmitting substance. These vesicles are concentrated at the nerve terminals. When a wave of depolarization reaches a terminal, the vesicles fuse with the cell membrane and discharge their contents into the gap, or **synaptic cleft,** between the cells. The transmitter diffuses to the adjacent cell, combines with receptors on the cell membrane, and stimulates or inhibits the activity of the receiving cell. One neuron may be surrounded by terminals from several different neurons, and the activity of the receiving neuron is influenced by the summation of the various stimuli it is receiving.

After passage of an impulse, status quo is restored by removal of the released transmitter and restocking of the vesicles. In most cases, the bulk of the transmitter is pumped back into the presynaptic cell through the action of an Na^+-coupled symporter, but we shall see one case in which the transmitter is destroyed to remove it.

Many different compounds are used as neurotransmitters. They include amino acids such as **glutamate** and **glycine,** and small peptides such as the **enkephalins** that cause an opiate-like relief of pain. A discussion of the neurophysiology of these compounds is beyond our scope. We are concerned here with the formation and metabolism of another class of neurotransmitters, the amines.

Acetylcholine

Acetylcholine is a general transmitting agent for efferent impulses, those going out from the central nervous system, between neurons. It is also the transmitting agent at the motor end-plates of skeletal muscle fibers and in terminals of the parasympathetic nervous system.

Acetylcholine is formed by a simple reaction between choline and acetyl coenzyme A in the cytoplasm. Neurons obtain choline by concentrating it from the blood and by internal synthesis via phosphatidylcholine (p. 631). They make acetyl coenzyme A from pyruvate in their mitochondria, and transport it to the cytosol as citrate, in the same way that acetyl groups are transported by other cells for fatty acid synthesis (p. 538). The mechanism by which the newly synthesized acetylcholine is concentrated in vesicles is unknown. (In general, synaptic vesicles contain a mixture of proteins and other compounds of uncharacterized function.)

When acetylcholine is released into the synaptic cleft, it is destroyed by **acetylcholinesterase.** If this enzyme is inhibited, the persisting acetylcholine continues to stimulate its receptors. We noted earlier (p. 314) that the mechanism of acetylcholinesterase involves the side chain of a serine residue and that phosphate insecticides and some war gases act by combining with this seryl group. Inhibitors such as neostigmine (p. 316) can also be used to slow the action of acetylcholinesterase reversibly with a dose-response relationship that makes them useful drugs.

Black widow spider venom also causes overstimulation of cholinergic nerves, but in a different way. The fusion of synaptic vesicles with the presynaptic membrane is probably caused by an influx of Ca^{2+} from the extracellular fluid; this influx is in turn caused by the depolarizing effect of the nerve stimulus arriving at the synaptic region. The spider venom appears to increase the cationic permeability of the membrane bilayer, thereby initiating discharge of acetylcholine.

Disturbances in the function of the acetylcholine receptors cause relaxation of skeletal muscles. **Curare** blocks the combination of acetylcholine and receptor, and it has been used for this purpose by South American Indians and by anesthesiologists of diverse ancestries. (The anesthesiologists now employ tubocurarine, a purified active ingredient, or synthetic derivatives.)

Some pathological conditions have curare-like effects, and indeed cause increased sensitivity to curare. These include **myasthenia gravis** (p. 315), in which receptor function is impaired, perhaps by antibodies. The remaining receptor function is stimulated by neostigmine. **Eaton-Lambert syndrome** occurs as a result of several cancers, and apparently is a defect in the release of acetylcholine from the synaptic vesicles, rather than a defect in receptor function. Patients therefore respond poorly to neostigmine, since there is little acetylcholine in the synaptic cleft, even with inhibition of acetylcholinesterase.

4-Aminobutyrate

Glutamate is decarboxylated at certain inhibitory synapses by a specific enzyme, which contains pyridoxal phosphate as a coenzyme:

$$\text{L-glutamate } (H_3N^{+}-CH(COO^{-})-CH_2-CH_2-COO^{-}) + H^{+} \xrightarrow[\text{(pyridoxal phosphate)}]{\text{glutamate decarboxylase}} \text{4-aminobutyrate } (H_3N^{+}-CH_2-CH_2-CH_2-COO^{-}) + CO_2$$

Here we have another example of the action of the general class of amino acid decarboxylases (p. 631). The resultant 4-aminobutyrate (commonly referred to in neurological literature as GABA, for gamma-aminobutyrate) is concentrated in the synaptic vesicles as a transmitter. Its release increases passage of Cl^{-} through the postsynaptic membrane.

The compound is recovered from the synaptic cleft by active transport. However, it is also metabolized within the neurons by a route involving an initial transamination followed by an NAD-linked dehydrogenation to succinate:

$$\text{4-aminobutyrate } (H_3N^{+}-CH_2-CH_2-CH_2-COO^{-}) + \alpha\text{-ketoglutarate} \underset{\text{(pyridoxal phosphate)}}{\overset{\text{aminobutyrate aminotransferase}}{\rightleftharpoons}} \text{succinic semialdehyde } (OHC-CH_2-CH_2-COO^{-}) + \text{L-glutamate}$$

$$\text{succinic semialdehyde} + NAD^{+} + H_2O \xrightarrow{\text{aldehyde dehydrogenase}} \text{succinate } (^{-}OOC-CH_2-CH_2-COO^{-}) + NADH + 2\,H^{+}$$

On the face of it, the route represents a bypass of the α-ketoglutarate dehydrogenase reaction in the citric acid cycle:

1. glutamate $\longrightarrow$ 4-aminobutyrate + CO_2
2. 4-aminobutyrate + α-ketoglutarate $\longrightarrow$ succinate semialdehyde + glutamate
3. succinate semialdehyde + NAD^{+} $\longrightarrow$ succinate + CO_2 + NADH

SUM: α-ketoglutarate + NAD^{+} $\longrightarrow$ succinate + CO_2 + NADH

Much has been made of this as a major route of glucose metabolism in brain, but optimistic estimates allow no more than 0.1 of the α-ketoglutarate to be handled by this route. Beyond that, it seems purposeless as a bypass in the citric acid cycle, because the generation of GTP ordinarily occurring from succinyl coenzyme A would be lost.

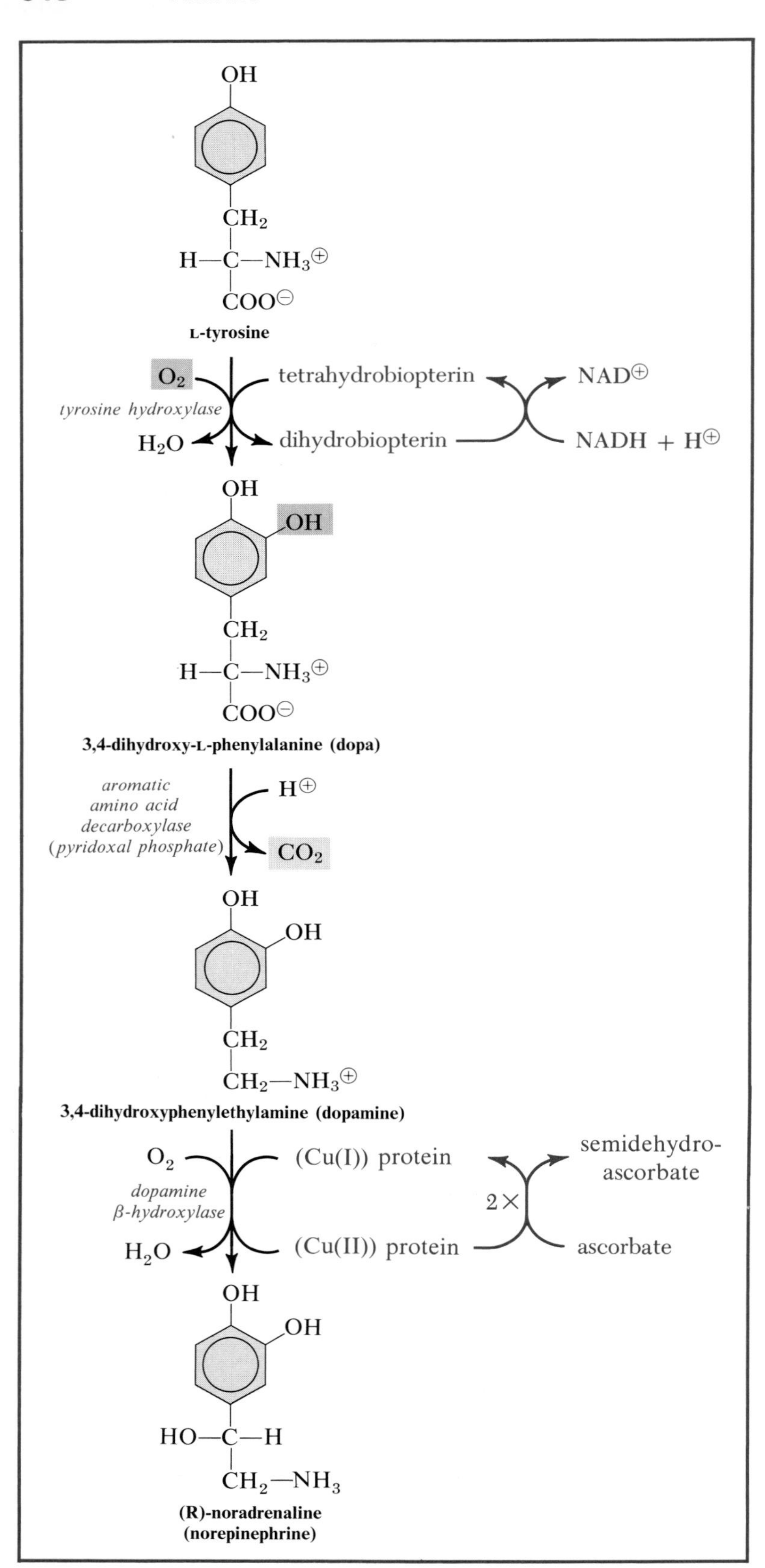

FIGURE 33–1

The formation of noradrenaline from tyrosine. The final hydroxylation is believed to involve successive one-electron transfers to ascorbate.

Adrenaline, Noradrenaline, and Melanin

Most of the tyrosine in the body that is not used for protein synthesis is metabolized to CO_2 and urea by the route beginning with transamination (p. 610). A small part of the tyrosine is diverted to make other components in some tissues by pathways beginning with an oxidation. These tissues are derived from the ectodermal layer of an embryo and include some nerves, the skin, and the adrenal medulla. While the routes are trivial in a quantitative sense, they are of major qualitative importance for proper function.

The oxidation of tyrosine in nerves and the adrenal medulla is catalyzed by a tyrosine hydroxylase, tyrosine 3-monooxygenase, which resembles phenylalanine hydroxylase (p. 608). It activates molecular oxygen with tetrahydrobiopterin so that one atom of oxygen goes onto the aromatic ring of the substrate and the other oxidizes the coenzyme (Fig. 33–1). The product is **3,4-dihydroxyphenylalanine (dopa).** This amino acid is then decarboxylated to form **3,4-dihydroxyphenylethylamine (dopamine),** which is used as a transmitter in some neurons of the brain, for example, in the basal ganglia.

In other nerves and the adrenal medulla, dihydroxyphenylethylamine is oxidized on the β-carbon of the side chain to form noradrenaline. The hydroxylase is a

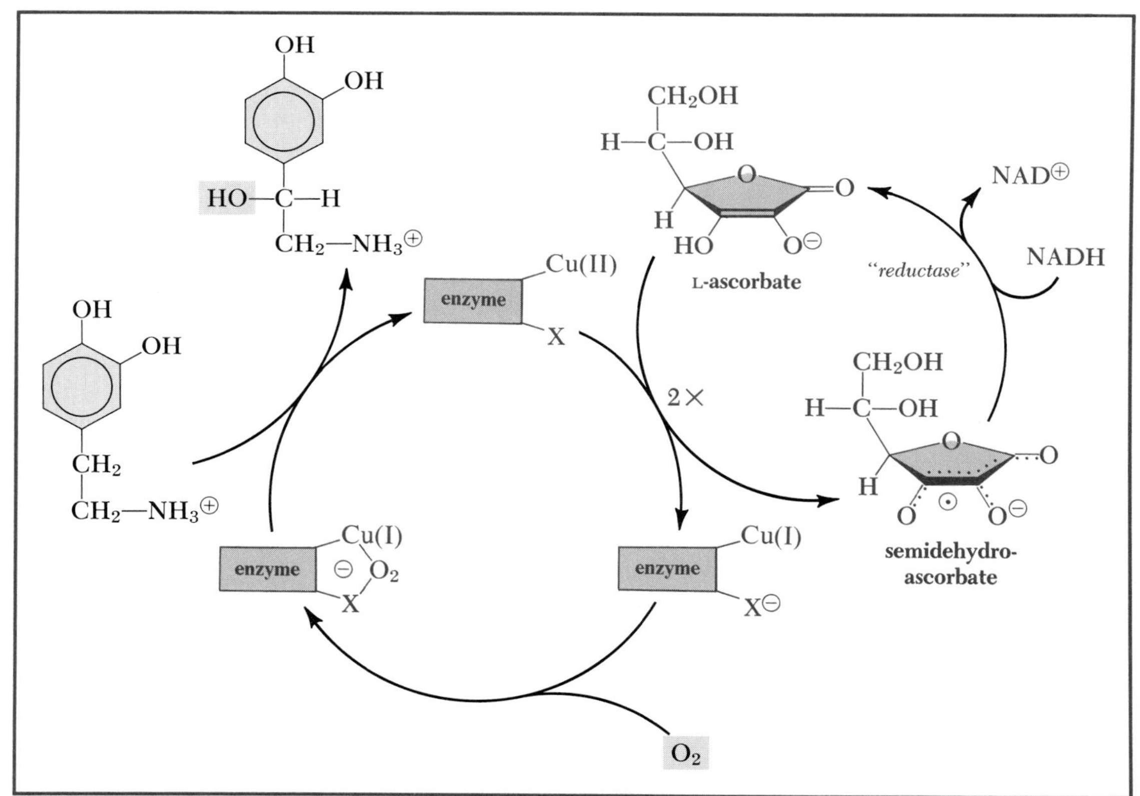

FIGURE 33–2 A postulated mechanism for the hydroxylation of dopamine involves reduction of copper and some unknown group X on the enzyme by successive transfers of single electrons from ascorbate. The reduced enzyme combines with dioxygen and transfers one oxygen atom to dopamine, using the other oxygen atom to re-oxidize its copper and "X" group. The semidehydroascorbate formed from the reaction has the free radical electron and its negative charge delocalized by resonance. It is reduced back to ascorbate by NADH in an uncharacterized reaction.

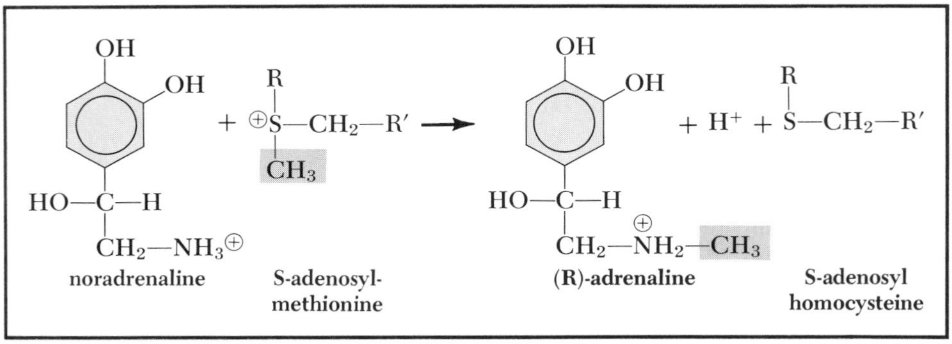

noradrenaline + S-adenosyl-methionine → (R)-adrenaline + H^+ + S-adenosyl homocysteine

FIGURE 33–3 **A noradrenaline methyltransferase uses a methyl group from S-adenosylmethionine to convert noradrenaline to adrenaline.**

copper-protein and is an example of a class of mixed-function oxidases that use ascorbate (vitamin C) as a second reducing agent in reactions with molecular oxygen (Fig. 33–2). Noradrenaline is the transmitter substance in the majority of sympathetic nerve terminals and at some synapses in the central nervous system. It is further methylated in the adrenal medulla to form adrenaline, with S-adenosyl methionine as methyl donor (Fig. 33–3).

The various dihydroxyphenylethylamines, such as dopamine, noradrenaline, and adrenaline, are lumped under the term **catecholamine.**

The word is derived from cathechol, which is a fairly recent shortening of the trivial name pyrocatechol, or 1,2-dihydroxybenzene (catechol was a more complex aromatic compound from which pyrocatechol was obtained by heating). The nomenclature is unfortunate, but it is in common usage.

Regulation. The rate-limiting step in catecholamine synthesis is the conversion of tyrosine to dihydroxyphenylalanine by tyrosine hydroxylase. This enzyme is controlled in neurons, but not in the adrenal medulla, by a calmodulin-sensitive protein kinase. The rise in calcium ion concentration caused by a neural impulse stimulates the kinase to phosphorylate an activator protein. The phosphorylated protein then combines with the tyrosine hydroxylase, causing it to initiate the replenishment of the catecholamines. Defective regulation leading to excessively high concentrations of noradrenaline may be a factor in the development of schizophrenia.

Fate of Catecholamines. After discharge into the synaptic cleft as transmitters, catecholamines are recovered by an Na^+-coupled symporter into the cytosol, from which they can be again concentrated into vesicles. However, catecholamines in the cytosol are subject to attack by a monoamine oxidase in the outer membrane of mitochondria. This enzyme is an important protective device; its activity is high enough to remove amines rapidly. It is a flavoprotein that uses dioxygen and forms hydrogen peroxide as a product:

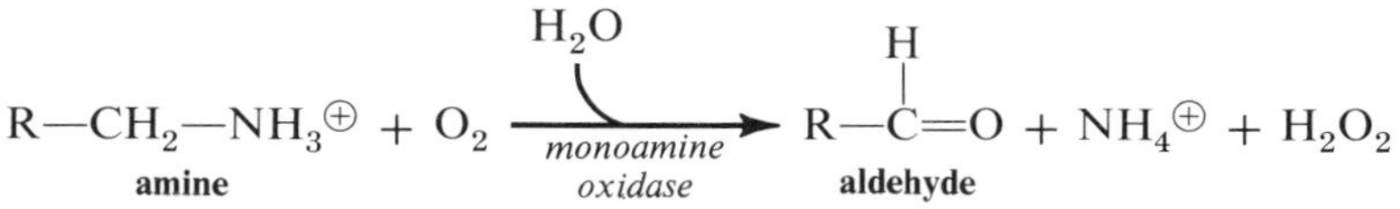

$$\underset{\textbf{amine}}{R{-}CH_2{-}NH_3^{\oplus}} + O_2 \xrightarrow[\textit{monoamine oxidase}]{H_2O} \underset{\textbf{aldehyde}}{R{-}\overset{H}{\overset{|}{C}}{=}O} + NH_4^{\oplus} + H_2O_2$$

A second enzyme in the cytosol methylates the 3-hydroxy group in deaminated catechol compounds. The methylated aldehyde that is formed may either be oxidized to an acid or reduced to an alcohol:

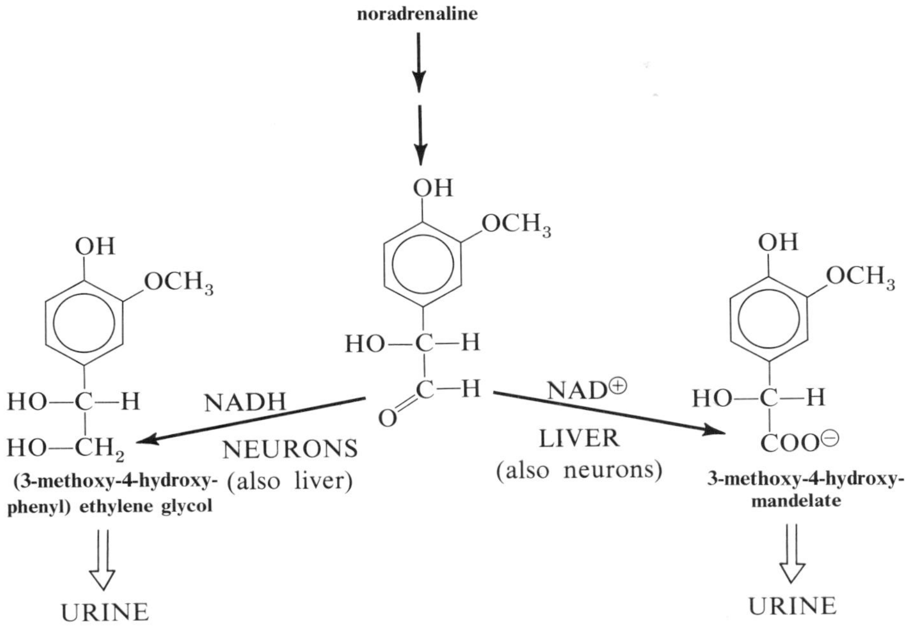

noradrenaline

NADH
NEURONS
(also liver)

$NAD^{\oplus}$
LIVER
(also neurons)

(3-methoxy-4-hydroxy-phenyl) ethylene glycol

URINE

3-methoxy-4-hydroxy-mandelate

URINE

The alcohol (3-methoxy-4-hydroxy-phenylethylene glycol) is the principal product in the central nervous system. It is slowly transported into the blood and excreted in the urine.

However, the principal catechol derivative in the urine is the acid as its anion, **3-methoxy-4-hydroxymandelate,** formed by the liver from catecholamines released by the adrenal medulla and other tissues. The determination of its excretion is a useful guide to the clinical diagnostic management of tumors such as neuroblastoma and pheochromocytoma that form adrenaline, noradrenaline, and dopamine. The compound is known clinically as vanillmandelate or vanillylmandelate, an example of bastard nomenclature that is excruciating to chemists. Its origin lies in the widespread occurrence of fragrant methylated 3,4-dihydroxyphenyl derivatives in plants; among them is vanillin in vanilla. Eugenol in cloves is another example:

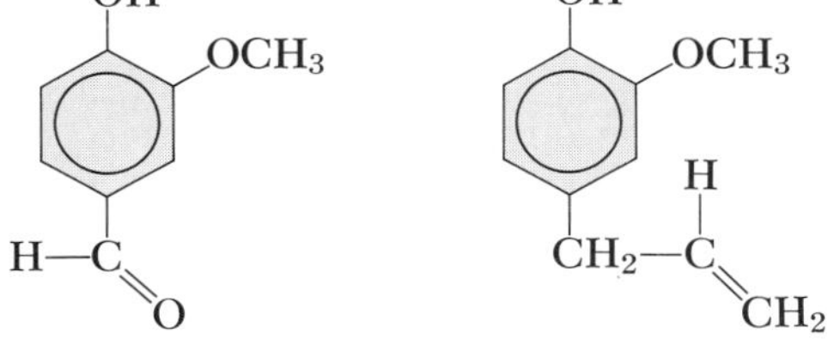

vanillin
(in vanilla)

eugenol
(in cloves)

NEUROBLASTOMA

Malignant neoplasm arising from cells derived from neural crest and sympathetic nervous system

Frequently originates in adrenal medulla

Comprises 15 to 50 per cent of neonatal malignancies and 5 per cent of all childhood cancer deaths

Some produce dopamine; some dopamine and noradrenaline

Serum level of dopamine β-hydroxylase correlates well with level of urinary vanillylmandelic acid

Dopamine excreted as homovanillic acid:

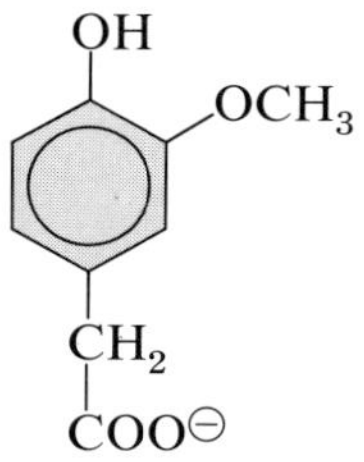

OH, OCH_3, CH_2, $COO^{\ominus}$

homovanillate

PARKINSON'S DISEASE

Clinical manifestations related to deficiency of dopamine in areas of the brain responsible for motor stimulation.

Estimated to affect 1 per cent of the population over 50 years old.

Characteristic features include:
- **expressionless face;**
- **eye blinks at 5 to 10 per minute** (normal 15 to 20);
- **slow voluntary movements** (but will move as fast as anyone in a fire);
- **resting tremor,** including pill-rolling motion of fingers;
- **muscle rigidity,** manifested by a series of cogwheel jerks when a limb is moved by the examiner;
- **interruptions in voluntary movements.**

Treatment is based upon replenishment of dopamine in the central nervous system. Dopamine itself cannot readily enter the neurons and is rapidly destroyed. Its precursor, dopa, can enter the neurons. Large doses are required because dopa decarboxylase in other tissues will rapidly convert it to dopamine unless a decarboxylase inhibitor is also administered. Carbidopa is used for this purpose because it does not enter the neurons:

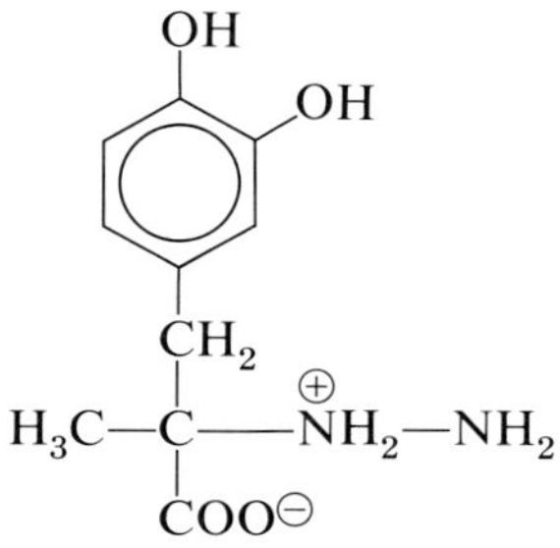

OH, OH, CH_2, $H_3C—C—\overset{\oplus}{N}H_2—NH_2$, $COO^{\ominus}$

carbidopa

The **melanins,** the pigments of skin and hair, are complex polymers in which a major constituent is formed from tyrosine via dihydroxyphenylalanine (Fig. 33–4). Synthesis occurs in **melanosomes,** which are specialized vesicles formed by **melanocytes** and phagocytosed by many neighboring keratinocytes. The initial reaction is an oxidation catalyzed by a phenol monooxygenase peculiar to melanosomes. It is a copper-containing mixed-function oxidase that carries out a tricky sequence. The initial hydroxylation converts tyrosine to dihydroxyphenylalanine, but the oxygen-activating electron donor for the hydroxylation is dihydroxyphenylalanine. The enzyme makes one of its own substrates! The enzyme not only makes dihydroxyphenylalanine, it oxidizes it to a quinone. The dihydroxyphenylanine serves the same purpose during its own formation that tetrahydrobiopterin serves in the tyrosine hydroxylase reaction in nerves; it becomes oxidized to a quinone in the same way that tetrahydrobiopterin is oxidized to dihydrobiopterin by the other monooxygenase.

The dopaquinone formed by the oxidations undergoes a series of fast spontaneous reactions in which an indole ring is formed and the carboxyl group is lost as CO_2. The various intermediates in this series of reactions participate in condensations to form a three-dimensional polymer. Melanins are complex structures of varying and unknown composition. There are at least three different types, including the yellow-brown eumelanins, more reddish pheomelanins, in which some sulfur-containing compounds are also polymerized, and brilliant red trichochromes, which are lower molecular weight compounds also containing sulfur. (Although the trichochromes occur in red hair, they are probably not as important as the pheomelanins in producing the distinctive hue.)

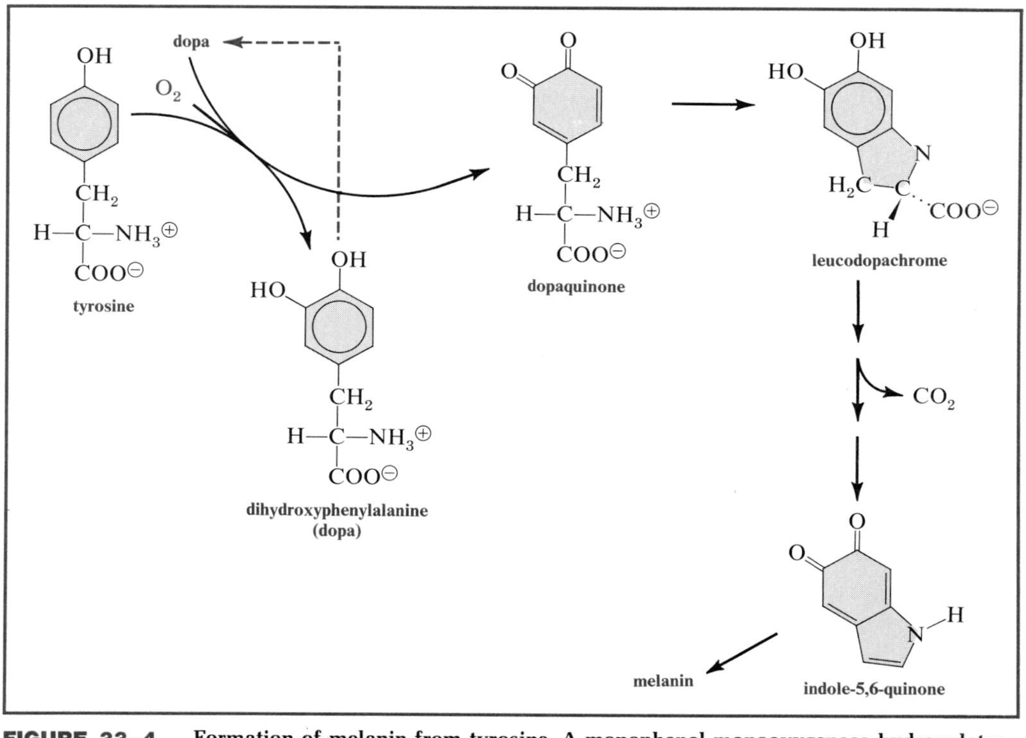

FIGURE 33–4 **Formation of melanin from tyrosine. A monophenol monooxygenase hydroxylates tyrosine to dopa (dihydroxyphenylalanine) while oxidizing dopa to dopaquinone. Dopa is both a product of the reaction and the electron-donating substrate. Dopaquinone undergoes a series of spontaneous reactions to yield quinones that polymerize to form melanins.**

Serotonin (5-Hydroxytryptamine)

Serotonin is a transmitter in some neurons of the central nervous system. These neurons affect many aspects of behavior: appetite, aggression, sleep, sexual activity, and so on. The route of formation is like that of noradrenaline. It begins with hydroxylation of C5 in tryptophan (Fig. 33–5) by a mixed function oxidase using tetrahydrobiopterin as the electron donor. The 5-hydroxytryptophan is decarboxylated to form serotonin by dihydroxyphenylalanine decarboxylase. The tryptophan 5-monooxygenase is activated by a phosphorylated protein in the same way that tyrosine 3-monooxygenase is activated.

As with the catecholamines, released serotonin is recovered by an Na^+-coupled symporter. Any of the compound that escapes packaging in vesicles is removed by the action of monoamine oxidase and aldehyde dehydrogenase, which convert it to the corresponding carboxylate for excretion in the urine.

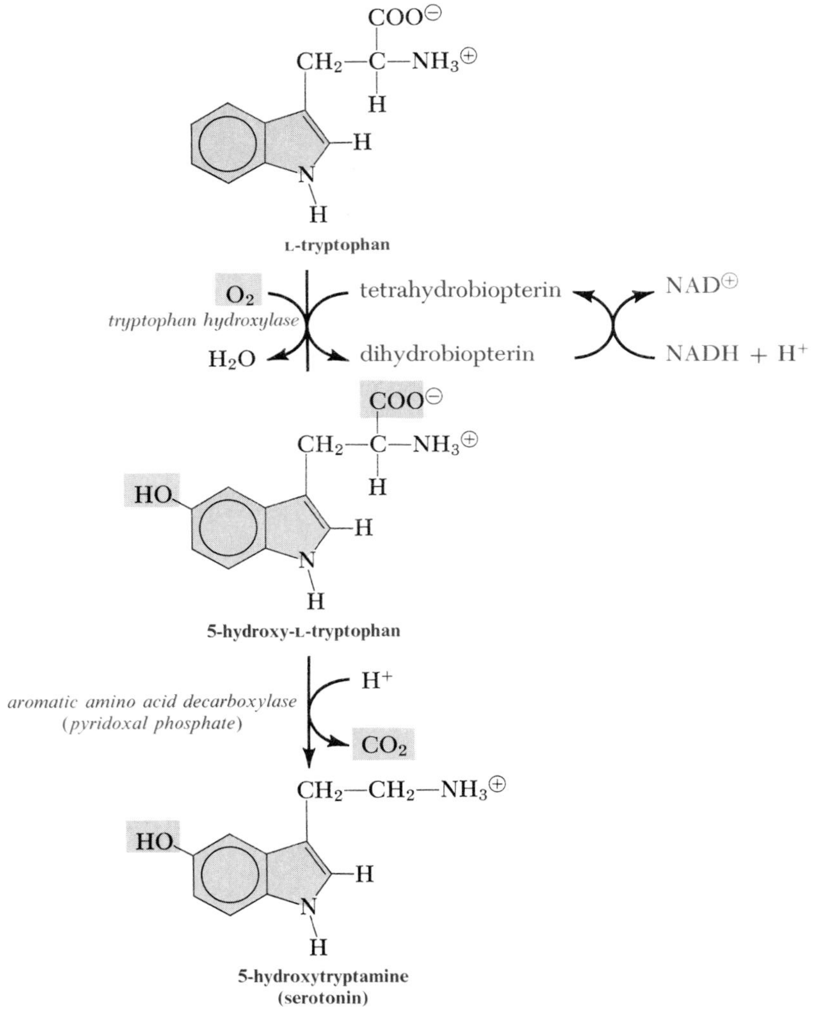

$COO^{\ominus}$
$CH_2—C—NH_3^{\oplus}$
H
—H
N
H
L-tryptophan

O_2 — tetrahydrobiopterin — $NAD^{\oplus}$
tryptophan hydroxylase
H_2O — dihydrobiopterin — $NADH + H^+$

$COO^{\ominus}$
$CH_2—C—NH_3^{\oplus}$
HO
H
—H
N
H
5-hydroxy-L-tryptophan

aromatic amino acid decarboxylase (pyridoxal phosphate)
H^+
CO_2

$CH_2—CH_2—NH_3^{\oplus}$
HO
—H
N
H
5-hydroxytryptamine (serotonin)

FIGURE 33–5 **The formation of serotonin from tryptophan.**

Neuropharmacology

Control of the metabolism and function of neurotransmitters in the human is the basis of action for a number of the most important drugs available to a clinician, and the study of this action is a large part of pharmacology. We can identify the major site of action of drugs by using the neurotransmitter noradrenaline as an example; drugs are known that do the following:

(1) Interfere with noradrenaline synthesis by inhibiting tyrosine 3-monooxygenase: alpha-methyl-*p*-tyrosine.

(2) Act as a false transmitter: alpha-methyl-DOPA is a precursor of such a compound. It is converted to α-methyldopamine and transported into granules, where it is converted to α-methyl-noradrenaline, which displaces noradrenaline and is not a neurotransmitter.

(3) Interfere with transport of the transmitter: reserpine blocks uptake of noradrenaline into vesicles. Imipramine and chlorpromazine interfere with reutilization of noradrenaline released outside of the cell.

(4) Inhibit breakdown of the neurotransmitter: pargyline inhibits monoamine oxidase and potentiates the effects of noradrenaline (and other amines) by blocking their degradation.

(5) Stimulate release of the neurotransmitter: tyramine, the product of decarboxylation of tyrosine, does this and can precipitate a hypertensive crisis. Aged cheese and Chianti wine, which are rich in tyramine, must be avoided by patients taking pargyline for this reason.

(6) Deplete the supply of the neurotransmitter by causing a slow release—too slow to stimulate the receptors but too rapid for replenishment: guanethidine.

(7) Inhibit normal release in response to the action potential: bretylium.

(8) Interfere with the receptor site: phenoxybenzamine.

(9) Mimic the normal transmitter: phenylephrine.

In considering this battery of drugs, it is of interest to note that many of them may be useful in treating psychiatric patients. Those that increase the synaptic concentrations of noradrenaline are antidepressant; those that lower it aggravate depression. Blocking agents for dopamine receptors are useful in more severe psychotic conditions.

IMPORTANT AMINES IN OTHER TISSUES

Histamine

Mast cells, the specialized cells that store heparin (p. 198), also make and store histamine by decarboxylating histidine:

$$\text{L-histidine} + H^+ \xrightarrow[\text{(pyridoxal phosphate)}]{\text{histidine decarboxylase}} \text{histamine} + CO_2$$

L-histidine

histamine

Histamine affects at least two kinds of receptors. **H1 receptors** are present in some smooth muscles, and cause, for example, an expansion of capillaries by dilating arterioles and constricting venules, and constriction of bronchi. **H2 receptors** in the gastric mucosa cause an increased secretion of hydrochloric acid. Drugs have been tailored to affect selectively these different receptors. Antagonists of H1 receptors often have aromatic rings with positively charged side chains and include the familiar over-the-counter antihistaminics that relieve the stuffed-up nose. In contrast, antagonists of H2 receptors such as cimetidine are employed to minimize gastric acid secretion. These contain imidazole rings with polar, but uncharged, side chains.

Histamine, like noradrenaline, may be metabolized by routes involving methylation of the ring or oxidation of the amino group. The methyl derivatives or the imidazoleacetates appear in the urine. The oxidation of the amino group of the parent compound is catalyzed by a relatively nonspecific enzyme, **diamine oxidase,** that also oxidizes a number of aliphatic compounds containing two amino groups, whereas the methylated derivative is a substrate for monoamine oxidase.

Putrescine, Spermidine, and Spermine

Putrescine received its ugly name because it was discovered as a product of the action of bacteria in decaying meat, being formed by decarboxylation of ornithine, which in turn arises from bacterial hydrolysis of arginine in the meat. (The product of decarboxylation of lysine was named **cadaverine** for the same reason.)

The polyamino compounds, spermidine and spermine, were named for their discovery in human semen. Indeed, crystals of spermine phosphate in semen were one of the things noted by van Leeuwenhoek with his newly invented microscope, and their presence is still used as a part of the legal identification of suspect stains, but it was not until 1926 that the structure of the compound was worked out. It was realized only in recent years that these compounds are widespread in tissues and have important functions. These polycations occur in association with the nucleic acids, which are polyanions. They apparently facilitate ribosome formation, and may well be involved in replication, in chromatin formation, and in translation.

Putrescine is formed by the action of a straightforward amino acid decarboxylase on ornithine and is converted to spermidine and spermine (Fig. 33–6). The necessary aminopropyl groups—one in spermidine and two in spermine—are obtained from S-adenosylmethionine. Ornithine decarboxylase has a very short half-life (10–20 minutes in the rat) and must be synthesized equally rapidly. This fast turnover makes it possible to alter the concentration of the enzyme in a short time. A large increase in activity is an early event in tissue growth, preceding the S phase.

We ordinarily think of S-adenosylmethionine only as a methyl group donor. However, Guilio Cantoni noted, and it appears obvious after it has been pointed out, that there is no mechanistic reason that any one of the three substituents on the sulfur atom of S-adenosylmethionine couldn't be the group transferred to another compound. Here we have a case in which it is the amino acid skeleton that is transferred, with an accompanying decarboxylation.

The **methylthioadenosine** formed as a byproduct of polyamine synthesis is split by a phosphorylase into methylthioribose 1-phosphate and adenine. **Methionine is rebuilt** from the former; the route is not known, but some of the ribose carbons are incorporated into the amino acid, along with the methylthio group. **Adenine is salvaged to** rebuild adenine nucleotides (Chapter 35).

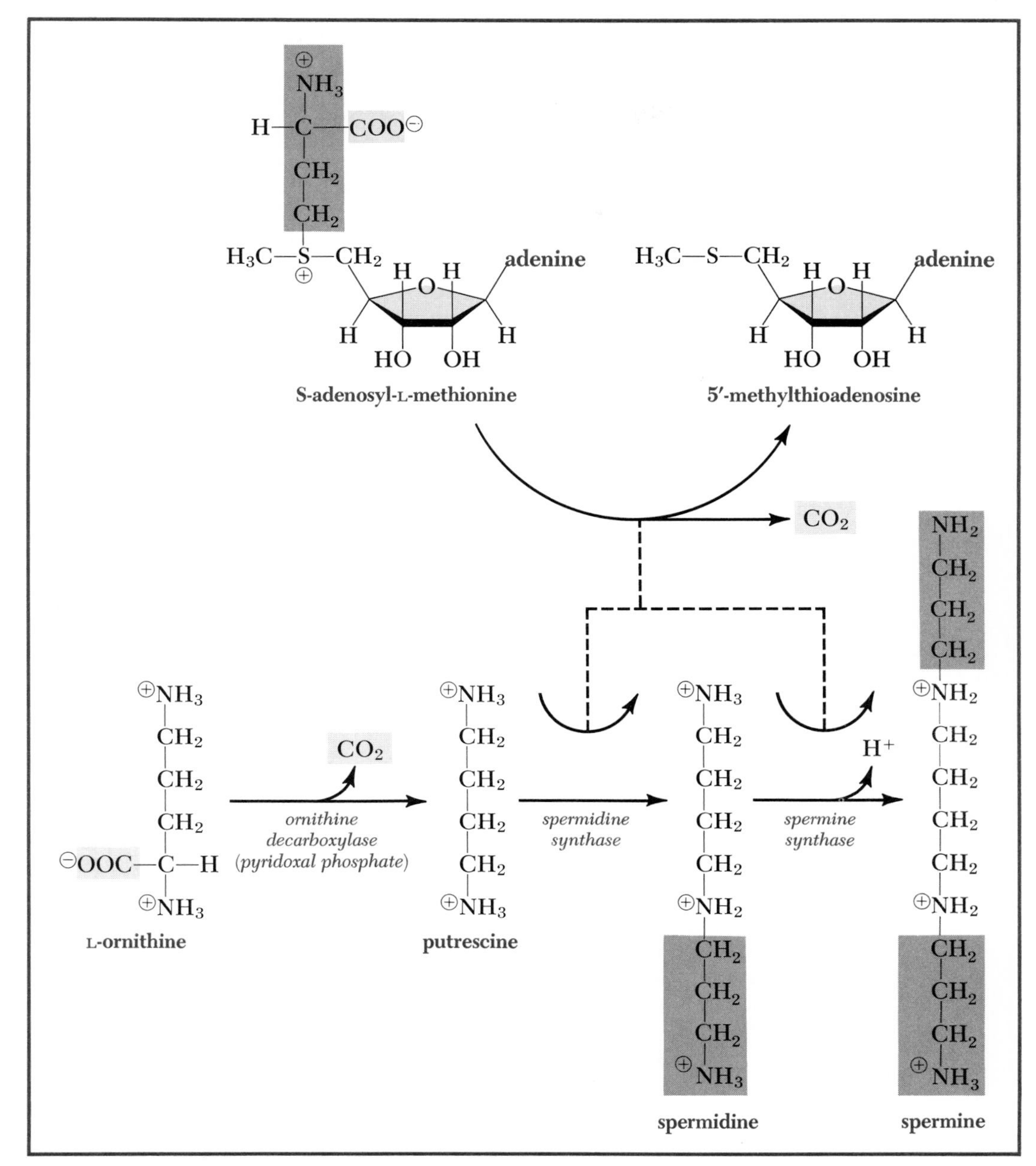

FIGURE 33–6 The formation of putrescine, spermidine, and spermine.

Carnitine

Carnitine is made from N,N,N-trimethyllysine (Fig. 33–7). This amino acid is not made from lysine as such, but is obtained during the normal turnover of those proteins in which some lysine residues are modified by methylation. (Examples include histones and myosin.) The conversion of trimethyllysine to carnitine begins with a hydroxylation on C3. The mixed-function oxidase catalyzing this reaction is

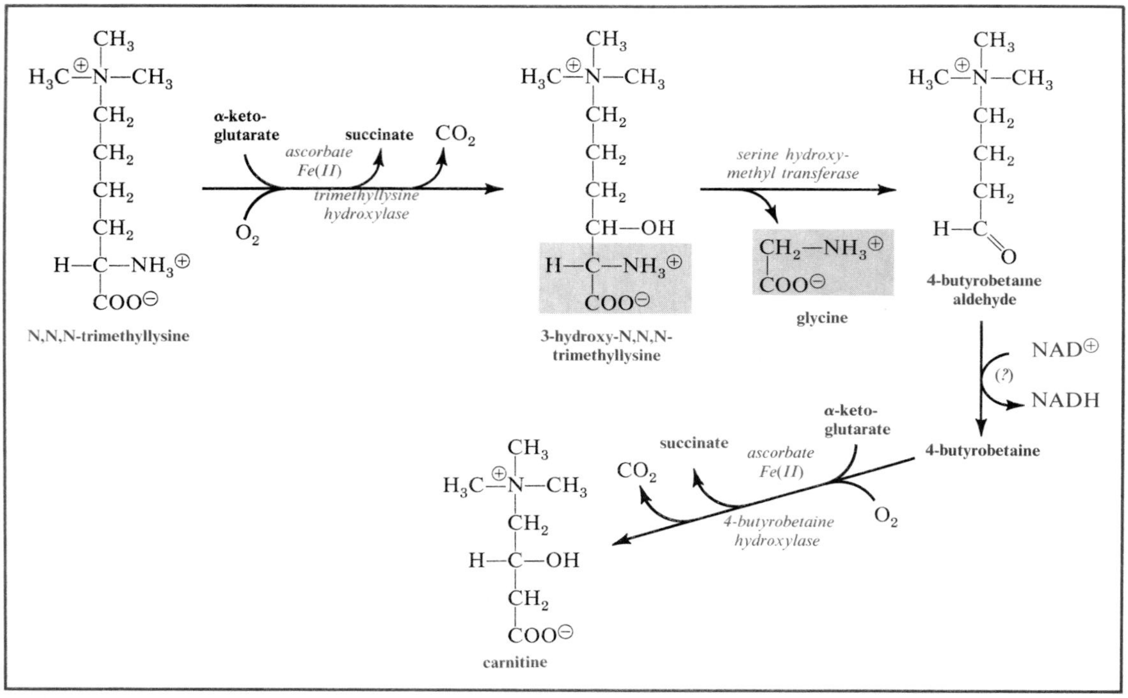

FIGURE 33–7 **The synthesis of carnitine involves the action of two monooxygenases that use α-ketoglutarate as an electron donor to activate dioxygen. Both require Fe(II) and are activated by ascorbate.**

an example of a special class that uses α-ketoglutarate as an electron donor. The mechanism of action includes the formation of an intermediate peroxide, much like the mechanism of 4-hydroxyphenylpyruvate-3-dioxygenase (p. 608). One of the atoms of oxygen then combines with trimethyllysine and the other is used for oxidative decarboxylation of α-ketoglutarate to succinate. (Other examples of dioxygenases requiring α-ketoglutarate as a second substrate are the **prolyl hydroxylase** and **lysyl hydroxylase** that create hydroxyproline and hydroxylysine residues in collagen (p. 162).)

3-Hydroxytrimethyllysine can be regarded as a derivative of serine, and it is attacked by serine hydroxymethyltransferase, splitting off glycine to yield 4-butyrobetaine aldehyde. After oxidation of the aldehyde to the corresponding acid, carnitine is formed by hydroxylation. This last reaction is catalyzed by still another example of a dioxygenase using α-ketoglutarate as an electron acceptor.

Before going on, we ought to note that these oxygenases require Fe(II) for activity. **Their activity is also increased by ascorbate.** Although it is not absolutely required in the test tube, ascorbate is apparently necessary to maintain the enzymes in active form within tissues.

Other Examples

Some compounds are embarrassing because they occur at too high concentrations in specialized tissues to ignore them completely, and yet their functions are unknown. For example, skeletal muscles contain up to 30 millimoles per kilogram of **carnosine,** which is a peptide of β-alanine and histidine:

$$
{}^{\oplus}H_3N{-}CH_2{-}CH_2{-}\overset{O}{\overset{\|}{C}}{-}\overset{H}{\overset{|}{N}}{-}\underset{H}{\underset{|}{\overset{CH_2\text{-(imidazolyl)}}{\overset{|}{C}}}}{-}COO^{\ominus}
$$

carnosine
(β-alanylhistidine)

(The compound is made from β-alanine and histidine at the expense of ATP, which is cleaved to AMP and PP_i. β-Alanine, which also occurs in coenzyme A, is made by decarboxylating aspartate.) Carnosine has also been found in high concentrations in the olfactory bulb of the mouse. Perhaps some general function in specialized excitatory tissues will yet be disclosed.

Another example is **N-acetylaspartate,** which is present at a concentration of 5 millimoles per kilogram of human brain, a level exceeded only by glutamine and glutamate among the amino acid derivatives, but no physiological role has been demonstrated.

Taurine is also present in high concentrations in some tissues, and has some undescribed functions other than serving as a precursor of bile salts (p. 563). Developing brains are rich in taurine, although the concentration drops later. Taurine represents half of the total free amino acids in adult hearts, in which its half-life of 15 days is unusually high for an amino acid.

FURTHER READING

J.K. Bluszajn and R.J. Wurtman (1981). Choline Biosynthesis by a Preparation Enriched in Synaptosomes from Rat Brain. Nature 290:417.

E.J. Diliberto, Jr. and P.L. Allen (1981). Mechanism of Dopamine β-Hydroxylation. J. Biol. Chem. 256:3385.

O. Hornykiewicz (1982). Brain Catecholamines in Schizophrenia—A Good Case for Noradrenaline. Nature 299:484.

A.M. Körner and J.M. Palawek (1982). Mammalian Tyrosinase Catalyzes Three Reactions in the Biosynthesis of Melanin. Science 217:1163.

O. Hayaishi, Y. Ishimura, and R. Kido, eds. (1980). Biochemical and Medical Aspects of Tryptophan Metabolism. Elsevier.

D.R. Morris and L.J. Morton, eds. (1981). Polyamines in Biology and Medicine. Dekker.

C.J. Bacchi et al. (1980). Polyamine Metabolism: A Potential Therapeutic Target in Trypanosomes. Science 210:332.

E.S. Canellakis et al. (1979). The Regulation and Function of Ornithine Decarboxylase and of the Polyamines. Curr. Top. Cell Regul. 15:155.

R.J. Huxtable and H. Pasantes-Morales, eds. (1982). Taurine in Nutrition and Neurology. Adv. Exp. Biol. Med., vol. 139. A symposium.

CHAPTER 34

TURNOVER OF PORPHYRINS AND IRON

Hemoproteins and other iron-containing proteins are constantly synthesized and degraded with a concomitant synthesis and degradation of the associated porphyrins and a reuse of the ligated iron atoms. While more is known about the turnover of the abundant hemoglobin, the other hemoproteins are also important because of the central role of the iron-containing electron carriers in metabolism. The synthesis and degradation of protoporphyrin IX in hemoglobin is a significant part of the nitrogen economy and involves a major part of the economy of iron in the body.

Nomenclature of the Porphyrins

Before going further, it is necessary to say something about the classification of porphyrins. The basic unit of porphyrins is the pyrrole ring, with four of these linked by one-carbon bridges to form the large porphyrin ring. Each of the four pyrrole rings may have two side chains attached, and these side chains may differ among the four pyrrole groups. The name of the porphyrin indicates the kinds of side chains; it does not indicate the arrangement of the particular pyrrole groups. Some porphyrins contain the kinds of pyrrole groups shown in Figure 34–1, and the composition of physiologically important examples may be summarized as follows:

uroporphyrins—each pyrrole group has an **acetate** and a **propionate** side chain.

coproporphyrins—each pyrrole group has a **methyl** and a **propionate** side chain.

protoporphyrins—each of two pyrrole groups has a **methyl** and a **propionate** side chain; each of the other two has a **methyl** and a **vinyl** side chain.

The porphyrins of a given name can vary among themselves in the order in which the pyrrole rings are put together. Suppose we designate the acetate and

FIGURE 34–1

Physiologically important porphyrins. The nature of the constituent pyrroles as given at the top define the type of porphyrin. It is not hard to visualize sequences of decarboxylation and oxidations by which the last two could be formed from the parent uroporphyrins listed first.

	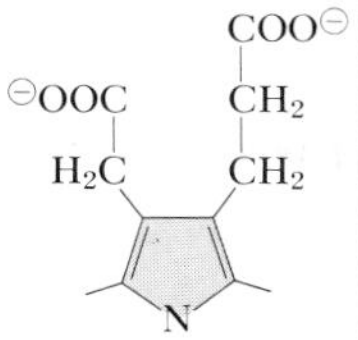 $COO^{\ominus}$, $^{\ominus}OOC$–H_2C, CH_2–CH_2, N (pyrrole)	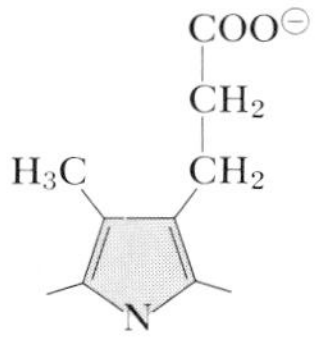 $COO^{\ominus}$, H_3C, CH_2–CH_2, N (pyrrole)	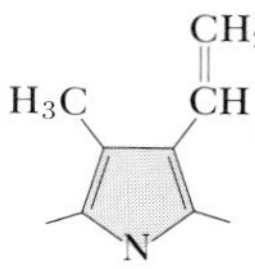 H_3C, CH_2=CH, N (pyrrole)
uroporphyrin	4	—	—
coproporphyrin	—	4	—
protoporphyrin	—	2	2

propionate groups in the uroporphyrins as A and P. There are four possible ways of combining the pyrroles bearing these groups so as to make a porphyrin:

Type I	Type II	Type III	Type IV
A P P A A P P A	A P A P P A P A	A P A A P P P A	A P P P A A A P
(-AP-AP-AP-AP-)	**(-AP-PA-AP-PA-)**	**(-AP-AP-AP-PA-)**	**(-AP-PA-PA-AP-)**

Each of these four uroporphyrins is designated by a Roman numeral. (There are only four uroporphyrins because any other reversal of pyrrole groups beyond those shown is superimposable on one of the four by turning the ring over.)

Now, if two of the groups in uroporphyrin are changed into a third kind of group, which is the circumstance seen in protoporphyrins, then there are 15 possible combinations. Hans Fischer* wrote down the 15 possibilities, and showed that the porphyrin in hemoglobin had the same arrangement as the ninth he had tabulated. Hence, the porphyrin in heme is designated as protoporphyrin IX.

All natural porphyrins are derived from uroporphyrin I, in which there is a regular alternating sequence of groups, as might be expected if the pyrroles are combined head-to-tail, or from uroporphyrin III, which represents a reversal—an isomerization—of one of the pyrrole groups.

PORPHYRIN SYNTHESIS

The complex porphyrin molecule is made from two simple precursors, succinyl coenzyme A and glycine (Fig. 34–2). These compounds are readily available within mitochondria, where they are condensed to form **5-aminolevulinate.** This is the rate-controlling step in porphyrin biosynthesis.

---

*Hans Fischer (1881–1945): German biochemist and Nobel Laureate. Not to be confused with Emil Fischer (1852–1919), also a German biochemist and Nobel Laureate, discoverer of much of the fundamental knowledge of the chemistry of proteins, carbohydrates, and nucleic acids; nor with Emil Fischer's late son, H.O.L. Fischer, a carbohydrate chemist of distinction at Toronto and Berkeley; nor with E.H. Fischer, very much alive at Seattle, and not bad as a biochemist, either. (This list is by no means exhaustive.)

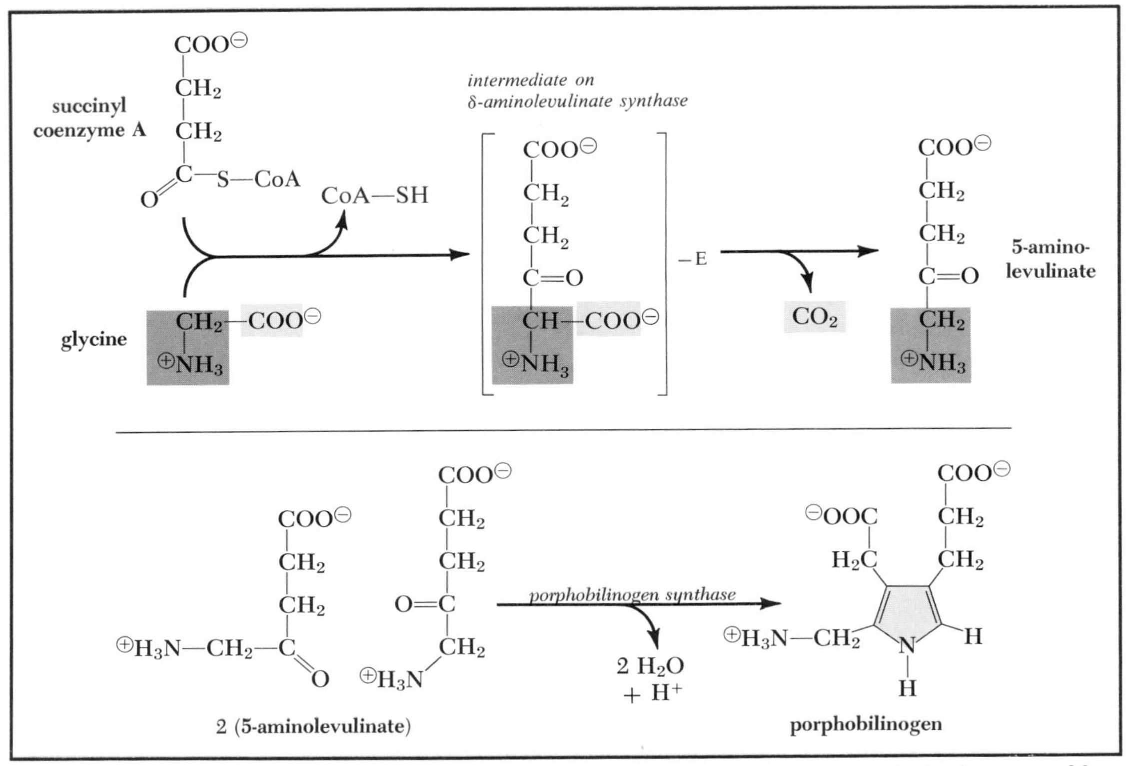

FIGURE 34–2 Porphyrin synthesis begins with two successive condensations. A pyrrole ring is generated from two molecules each of succinyl coenzyme A and glycine.

The reaction involves an intermediate condensation of glycine with pyridoxal phosphate. The mechanism is not shown; one of the H atoms on C2 of glycine leaves after condensation; the resultant carbanion then unites with the electropositive carbonyl carbon of succinyl coenzyme A.

The 5-aminolevulinate passes into the cytosol, where two molecules condense to form **porphobilinogen.** This is the parent pyrrole compound and four molecules of it are combined to make **uroporphyrinogen III** (Fig. 34–3). (The porphyrinogens are porphyrins in which the bridge atoms between pyrrole rings are in the reduced, or methylene, state, in contrast to the methylidyne bridges seen in porphyrins.)

Two proteins are involved in the formation of uroporphyrinogen III. **Uroporphyrinogen I synthase** catalyzes deamination of the porphobilinogen precursor, with an accompanying head-to-tail condensation to make a hydroxymethylbilane (Fig. 34–3). (A bilane is a linear tetrapyrrole.) This bilane will slowly condense spontaneously to make uroporphyrinogen I. However, a second enzyme, **uroporphyrinogen III cosynthase,** rapidly alters the bilane by closure of the porphyrinogen ring, with a simultaneous reversal of the final pyrrole. A postulated intermediate is shown in the figure. The product is uroporphyrinogen III, in which the side chains have the order necessary to be a precursor of protoporphyrin IX.

The remaining steps (Fig. 34–4) involve decarboxylation of the aceto side chains to form methyl groups (coproporphyrinogen III), oxidative decarboxylation of two of the propiono side chains to form vinyl groups (protoporphyrinogen IX),

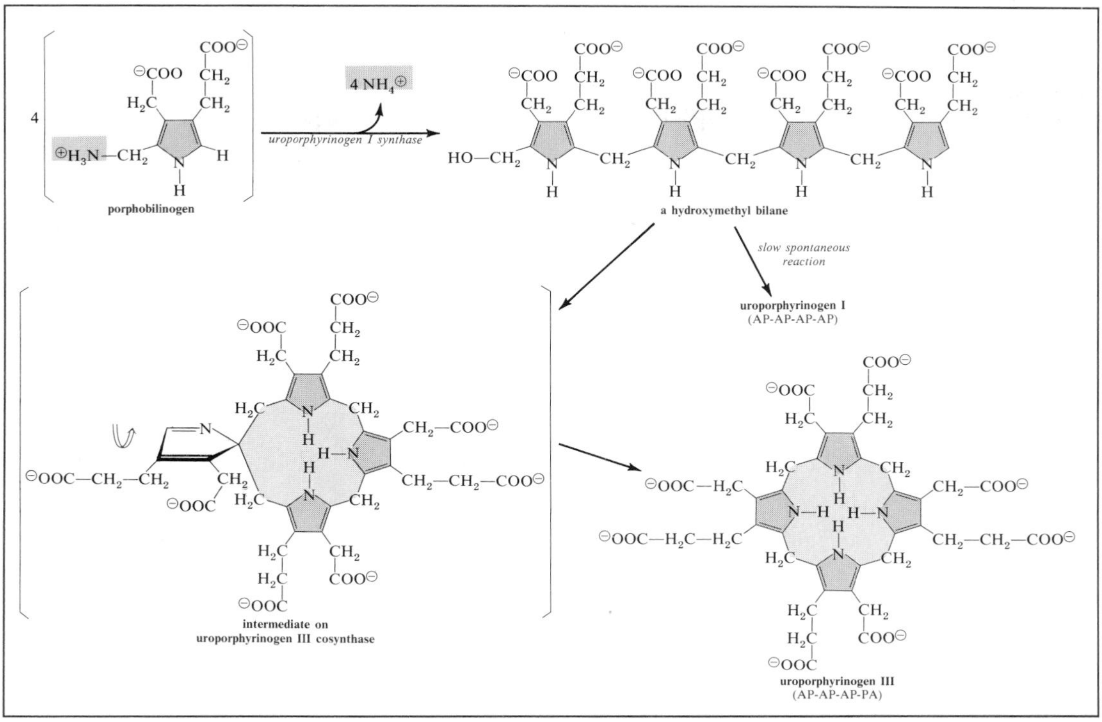

FIGURE 34–3 Uroporphyrinogen is formed from four molecules of porphobilinogen by two associated enzymes. The first forms an intermediate linear tetrapyrrole, or bilane. The second immediately cyclizes the bilane to form an intermediate in which both methylene bridge groups are linked to one carbon atom on the final pyrrole ring. This ring can now be twisted and the bridge shifted so that the order of its side chains is reversed, thus forming uroporphyrinogen III. If the second enzyme is deficient, then the bilane will slowly condense to form uroporphyrinogen I.

and the oxidation of the methylene bridges to methylidyne bridges (protoporphyrin IX). The latter two steps are catalyzed by mitochondria, but it is not known where the enzymes are localized within the organelle. However, the final step of heme synthesis, the addition of Fe(II) to protoporphyrin IX, is catalyzed by an enzyme localized on the inside of the inner membrane, at least in liver.

Regulation of hemoglobin synthesis involves control of both porphyrin and globin formation (Fig. 34–5). The regulating factor is an accumulation of heme, which in the free form is spontaneously oxidized to hemin. The mechanisms of control may differ in the liver and the precursors of red blood cells. Porphyrin synthesis in both tissues is regulated mainly by controlling the activity of δ-aminolevulinate synthase within mitochondria. Hemin represses the formation of the enzyme. There is some evidence that it also may control the enzyme in an unusual fashion. The synthase is made in the cytoplasm on mRNA transcribed from nuclear DNA, and is transported into the mitochondria. Hemin apparently blocks this transport, thus preventing the enzyme from gaining access to its intramitochondrial supply of succinyl coenzyme A. (Glycine is available in the cytosol, as well as in the mitochondria.) The relative physiological importance of this means of controlling porphyrin synthesis is not known. The synthase turns over rapidly, with a half-life of 20 minutes in the rat, so the response can be prompt by either mechanism.

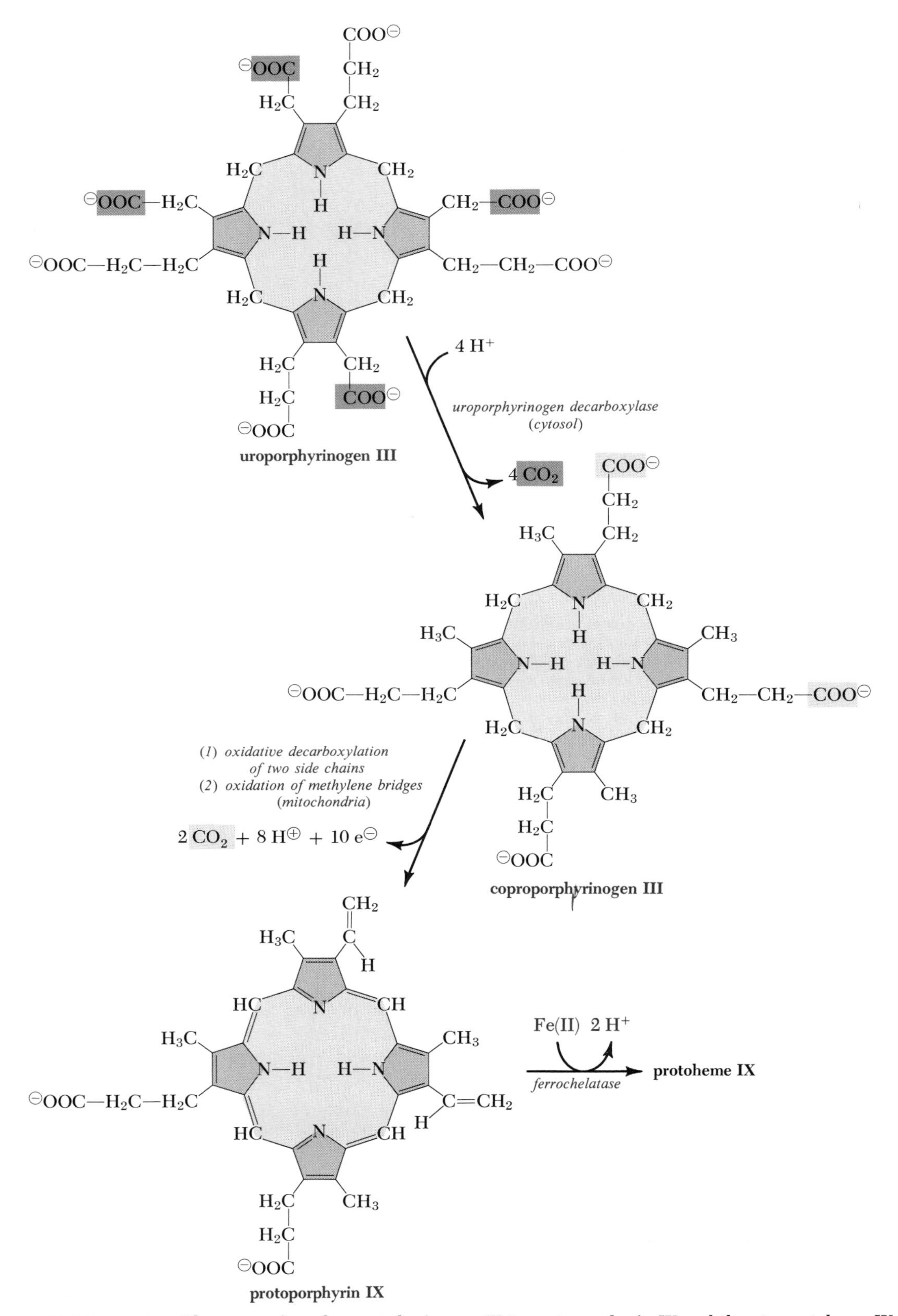

FIGURE 34–4 The conversion of uroporphyrinogen III to protoporphyrin IX and then to protoheme IX.

PORPHYRIAS

Porphyrias are caused by disorders in the synthesis of heme. Since the clinical manifestations of these uncommon disorders are similar to those of many other diseases, and their biochemical basis is only dimly recalled by those lacking the benefits of exposure to this textbook, it is a mark of distinction to suspect and diagnose a case of porphyria when it is encountered.

About 15 per cent of heme synthesis occurs in the liver and nearly all of the remainder occurs in erythrocyte precursors, so the porphyrias are separated into hepatic or erythropoietic types. A majority of the hepatic and all of the erythropoietic porphyrias are inherited. Those in which the biochemical defect appears to be known include the following:

Erythropoietic protoporphyria is an autosomal dominant disease caused by a **deficiency of ferrochelatase.** Exposure to long wavelength ultraviolet light provokes burning and itching of the skin, owing to photochemical reactions resulting from activation of the accumulated porphyrin. Feces, erythrocytes, and blood plasma contain increased amounts of protoporphyrin IX. There is frequently an associated anemia.

Acute intermittent porphyria is an autosomal dominant disease caused by a **deficiency of uroporphyrinogen I synthase** in both the liver and erythropoietic cells. One in 1000 Laplanders is afflicted. Although the biochemical defect is completely expressed, clinically significant sequelae are less common. Symptoms rarely occur before puberty; they include in order of frequency, abdominal pain, vomiting, constipation, paralysis, and psychological symptoms. (The English may blame or thank acute intermittent porphyria for the madness of George III to the degree it may have contributed to the separation of the American colonies.) The urine of the afflicted may have a port wine color from photooxidation of the porphobilinogen excreted, together with 5-aminolevulinate, in large amounts. The clinical disease is activated by any factor increasing the synthesis of δ-aminolevulinate synthase. These include the administration of various drugs. (Barbiturates, alcohol, chloramphenicol, some steroid hormones, are examples.) Administration of hemin (usually as hematin, the hydroxide form) appears to suppress δ-aminolevulinate synthase activity and relieve the condition in some.

Hereditary coproporphyria is an autosomal dominant disease caused by a **deficiency of coproporphyrinogen synthase** in the liver. Since it also causes an accumulation of 5-aminolevulinate and porphobilinogen, it has symptoms similar to those of acute intermittent porphyria, but with the addition of skin photosensitivity caused by the accumulating porphyrins. Coproporphyrinogen III is excreted in the urine.

Porphyria cutanea tarda is the commonest porphyria. The activity of uroporphyrinogen decarboxylase is decreased in both the liver and erythropoietic cells, causing excretion of large amounts of uroporphyrinogen and uroporphyrins, which makes the urine pinkish to brown. The disease is associated with liver damage, especially from alcoholism, and with moderate iron overloads. There is evidence for a genetic susceptibility to loss of decarboxylase activity in those afflicted. Phlebotomy (draining blood from a vein) improves the condition by depleting the iron stores. The major symptoms are those of skin photosensitivity. As may be expected, they are worse in summer and more common in males. The disease frequently becomes apparent during the fourth through sixth decades.

Lead poisoning has multiple effects on the enzymes of heme synthesis. Uroporphyrinogen I synthase is especially sensitive because it contains zinc, which is displaced by lead, as well as other metallic ions. (Administration of zinc is sometimes beneficial in lead poisoning.) Ferrochelatase is also inhibited. Protoporphyrin IX accumulates in *acute* lead poisoning for this reason. However, zinc protoporphyrin accumulates in *chronic* poisoning because there is a concomitant impairment of iron absorption, and the remaining ferrochelatase causes incorporation of zinc rather than the missing iron. The abdominal pain and neurological symptoms of lead posioning are believed to be caused by the resultant porphyria. However, there is also a hemolytic anemia caused by a defect in nucleotide metabolism (next chapter).

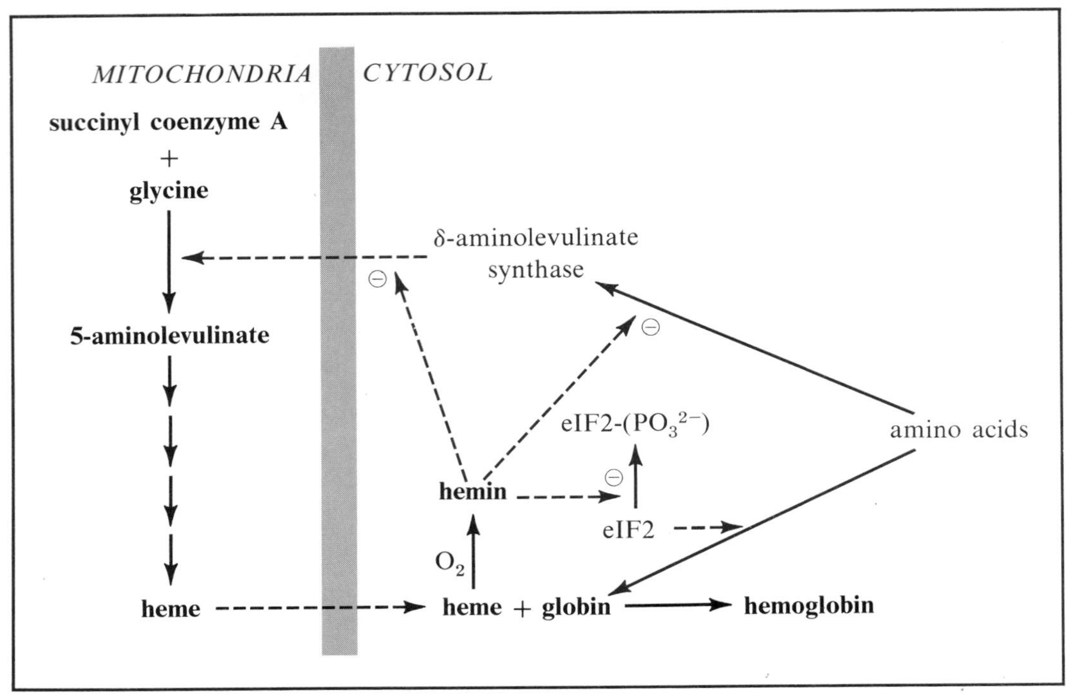

FIGURE 34–5 Regulation of hemoglobin synthesis by the concentration of hemin. If free heme accumulates, it is spontaneously oxidized to hemin, which represses the formation of δ-aminolevulinate synthase and perhaps blocks its transfer from the cytosol into the mitochondrial matrix. Hemin is also an inhibitor of a protein kinase that phosphorylates eIF2 (p. 93) to make it inactive. Hemin therefore permits the formation of more globin to combine with the excess heme.

Hemin activates the synthesis of globin polypeptide chains in erythrocyte precursors in this way: The cells contain a protein kinase that catalyzes the phosphorylation of a serine residue in eIF-2, one of the initiation factors necessary for translation of mRNA. The phosphorylation inactivates the factor and therefore blocks protein synthesis. (The formation of all proteins is apparently inhibited, but the globin polypeptides are the major products in the affected cells.) Hemin combines with the protein kinase and makes it inactive. As is usually the case when proteins are controlled by phosphorylation, a protein phosphatase is also present; the degree of phosphorylation is controlled by the balance of protein kinase and protein phosphatase activities. When the kinase is inhibited by hemin, the phosphatase gains ground and initiation of protein synthesis proceeds. As more globin appears to combine with heme, the kinase will become ascendant and porphyrin synthesis will slow. The result is a nice balance of porphyrin and globin synthesis, neither being allowed to get ahead of the other.

RED CELL SURVIVAL AND METHEMOGLOBIN

The mean lifetime of a red blood cell in adult humans is 120 days, and most of the constituent hemoglobin molecules survive unscathed for that time. The hemoproteins of other tissues turn over much more rapidly, with half-lives of a few days. Even so, the greater mass of hemoglobin present in the body makes it the greater source of degraded porphyrins, contributing about 80 to 85 per cent of the total.

Although hemoglobin is remarkably stable, it is slowly, but continually oxidized to methemoglobin, which contains Fe(III). The processes involved are very slow compared to the enzymatically catalyzed metabolic events we have been considering, but their cumulative effects would be disabling without some means of reducing the methemoglobin back to hemoglobin. The mechanisms of methemoglobin formation probably include the occasional loss of superoxide from the Fe(III)-superoxide complex during normal oxygen transport (p. 245), and attack by extraneous superoxide and hydrogen peroxide formed by the reaction of oxygen and other compounds. (Some aniline dyes have been notorious for catalyzing methemoglobin formation. Infants were poisoned by laundry marks on their diapers in this way before the potential danger was appreciated. Other infants were affected by water rich in nitrate, part of which is reduced to nitrite. Much older infants get methemoglobinemia by sniffing nitrite esters.)

The ultimate source of electrons used to reduce methemoglobin is glucose. Erythrocytes, lacking mitochondria, depend upon glycolysis to generate ATP. Only a small portion of the NADH need be diverted from reduction of pyruvate for methemoglobin reduction.

$$2\,(\underset{\text{methemoglobin}}{\text{Fe(III)-porphyrin-protein}}) + \text{NADH} + \text{H}^{\oplus} \longrightarrow 2\,(\underset{\text{hemoglobin}}{\text{Fe(II)-porphyrin-protein}}) + \text{NAD}^{\oplus}$$

In addition, the erythrocytes have an active pentose phosphate pathway for generating NADPH.

The enzyme catalyzing the reduction of methemoglobin by NADH, and perhaps by NADPH, resembles the cytochrome b reductase found on the endoplasmic reticulum and involved in the desaturation of fatty acids (p. 534). Perhaps it is identical, and the membrane-bound enzyme is somehow solubilized during maturation of the erythrocyte.

Methemoglobin formation may also be minimized by removing superoxide and hydrogen peroxide through the action of superoxide dismutase and a glutathione peroxidase:

$$2\ \text{G-SH} + H_2O_2 \xrightarrow{\textit{glutathione peroxidase}} \text{G-S-S-G} + 2\ H_2O$$

The oxidized glutathione is then reduced with NADPH:

$$\text{G-S-S-G} + \text{NADPH} + \text{H}^{\oplus} \xrightarrow{\textit{glutathione reductase}} \text{2-G-SH} + \text{NADP}^{\oplus}$$

Methemoglobin is not in itself especially deleterious except that it represents a loss of oxygen transport. Those rare individuals with a genetic deficiency of the NADH–methemoglobin reductase activity appear alarmingly blue but seem to thrive, with normal fertility and life spans. (It only takes 1.5 grams of methemoglobin to mimic the color of 5 grams of deoxyhemoglobin.) Somehow they are able to compensate for 20 per cent or more of the heme being oxidized to hemin without impairing the life of the erythrocytes, whereas normal individuals exposed to a chemical causing an equal degree of methemoglobinemia will have extensive destruction of their red blood cells.

The relatively innocuous consequences of a deficiency of the methemoglobin reductase are made even more puzzling by the sometimes drastic effects of a relatively common **deficiency of glucose-6-phosphate dehydrogenase.** This deficiency was discovered when a large number of American blacks developed hemolytic crises after taking primaquine, an antimalarial drug, during their military service. It was shortly found that there are many different variants in this enzyme, much as there are in hemoglobin, with some quite rare and some quite common; some are innocuous, while others are seemingly deleterious. The common mutants were found to be present in the same populations as the common hemoglobin mutants: those that were exposed to falciparum malaria. It is believed that the increased rate of methemoglobin formation somehow protects against malaria. Serious problems occur only with the development of illness from other causes, or upon ingestion of compounds promoting methemoglobin formation. Such compounds occur in nature, for example, in a certain vetch known as the fava "bean" (*Vicia favia*), but they are also found among many commonly used drugs, such as the sulfonamides, and even aspirin, in addition to the antimalarials.

The gene for glucose-6-phosphate dehydrogenase occurs on the X chromosome. All of the red blood cells in an affected male will be deficient, but only part in the heterozygous female. (The condition was used to demonstrate the Lyon hypothesis that only one of the X chromosomes is functional in female somatic cells.) An estimated 13 per cent of American black males and 20 per cent of the females carry a defective gene for glucose-6-phosphate dehydrogenase. The incidence is 50 per cent or greater among Kurdish Jews, 14 per cent in Sardinia, and 2.7 per cent in Malta. (It is only 0.4 per cent in mainland Italy.)

Still more puzzling is the lack of methemoglobinemia or hemolysis in those rare individuals with genetic defects in erythrocyte glutathione synthesis.

PORPHYRIN DEGRADATION

The principal sites of heme catabolism are the spleen and the liver. Which site is used to handle hemoglobin depends upon the circumstances of red cell destruction. Usually, the aged red blood cells are destroyed in the spleen, and heme degradation also begins there.

When hemoglobin is disrupted, some of the heme is oxidized to hemin. However, this makes no difference, because porphyrin degradation begins by reduction of hemin to heme with NADPH, followed by rupture of the methylidyne bridge between the two rings carrying vinyl groups (Fig. 34–6). A sequence of oxidations removes the bridge carbon as carbon monoxide. The change in the structure decreases the affinity for iron, which dissociates. The released linear tetrapyrrole has a green color and was named **biliverdin** because of this. It and related compounds are responsible for the pigmentation of well-developed bruises.

Biliverdin is reduced by NADPH at the center methylidyne bridge to form **bilirubin,** which is named for its red-brown color. The shift in color comes from the loss of part of the resonance between the two halves of the molecule. Bilirubin is transported in the plasma by attachment to that versatile carry-all, albumin, and is taken up by the liver.

Bilirubin is made more soluble in the liver by attaching residues of **D-glucuronate** through glycosidic bonds. Suffice it for now to note that UDP-glucuronate is made by the oxidation of UDP-glucose, and its glucuronate residue may be transferred to acceptors such as bilirubin in the presence of appropriate enzymes. The conversion of bilirubin to its glucuronide enables secretion into the bile without the formation of crystals. Even so, there is enough hydrolysis in the bile that free bilirubin is an important constituent of gallstones.

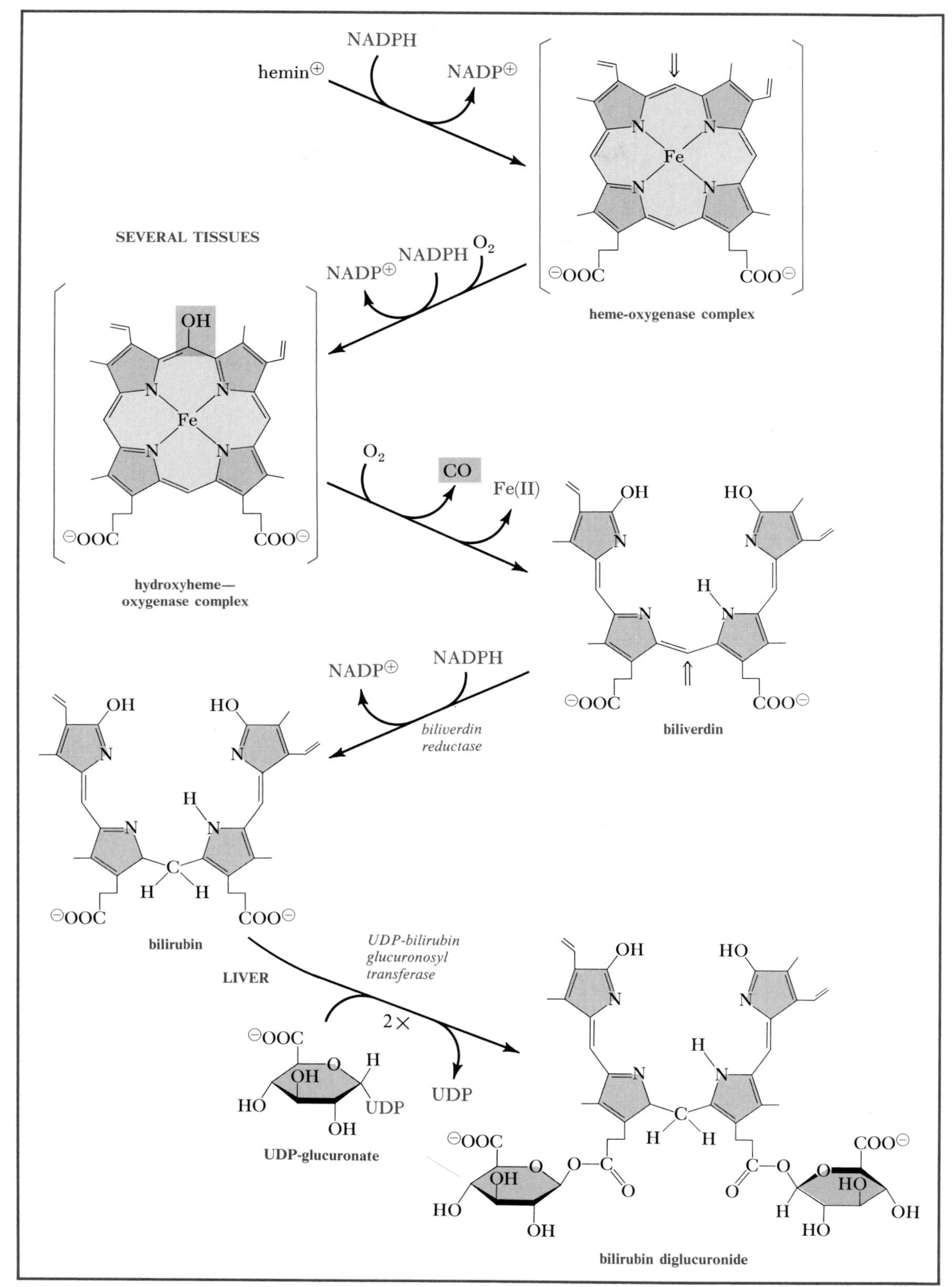

FIGURE 34–6 Degradation of hemin begins by reduction to heme and formation of a complex with an oxygenase system in the endoplasmic reticulum. A mixed function oxidase hydroxylates one of the methylidyne bridges, followed by a dioxygenase reaction in which the bridge is cleaved to liberate carbon monoxide and the iron atom, and form biliverdin. Biliverdin is reduced to bilirubin, which travels through the blood to the liver for conjugation with glucuronosyl groups and passage into the bile.

TABLE 34–1 HAPTOGLOBIN AND HEMOPEXIN

Haptoglobin
Two binding sites for $\alpha\beta$ Hb^+ dimers; each site itself a pair of sites, one for α and one for β subunits
M.W. 100,000; 0.4–1.8 g per liter of blood

Hemopexin
One binding site for hemin
M.W. 57,000; 0.6–0.8 g per liter of blood
Hemin complex taken up by liver

Biliary excretion disposes of bilirubin for all metabolic purposes. However, there are further transitions of esthetic significance. The glucuronyl residues are removed by hydrolysis within the bowel, and the liberated bilirubin undergoes a series of reductions by the microorganisms present, with the formation of **urobilinogens** and **stercobilinogens.** The methylidyne bridges are reduced in all of these compounds; one or both vinyl groups are reduced in the urobilinogens, and the stercobilinogens have two pyrrole rings also reduced. Part of the urobilinogen is reabsorbed and excreted in the urine. The various bilinogens are colorless but are spontaneously oxidized to re-form one of the methylidyne bridges upon exposure to oxygen. The partially oxidized compounds are known as the **urobilins** and **stercobilins.** These are the compounds largely responsible for the brown hue of feces and the yellow color of urine. Obstruction of the bile ducts leads to light-colored feces and dark urine. The skin is tinted yellow by the bilirubin that normally is converted to urobilin in the feces. Jaundice may also result from hepatic damage, hereditary defects in the formation of bilirubin glucuronide in the liver, or rapid destruction of red blood cells beyond the capacity of the liver to form the glucuronide. Premature infants are especially prone to jaundice because their capacity for glucuronide conjugation has not reached adult levels. Brain damage may result. The bilirubin level is diminished by exposing them to light, with a resultant photochemical destruction of the compound. Specific carrier proteins are present in the blood to carry oxidized hemoglobin or hemin, should the erythrocyte be broken elsewhere in the circulation, and minimize the loss through the kidney. **Haptoglobins** carry methemoglobin dimers, and **hemopexins** carry free hemin (Table 34–1). Serum albumin also binds hemin, but at only 1/60 the affinity of hemopexin; however, the concentration of albumin is 50-fold greater, so it has a significant fraction of the free hemin attached. The albumin-hemin complex is not taken up by the liver; it may act as a reservoir, with the hemin transferred to hemopexin for removal.

Turnover of Iron

Iron atoms are not allowed to move about the body unescorted. Specific carriers are employed for extracellular transport and intracellular storage of iron. Even if there were no such carriers, few iron ions could exist in the free form in the body. This is true because these ions have a high affinity for complexing groups containing O, N, or S, and the free ions therefore react readily with a variety of cellular constituents. Indeed, if enough inorganic iron salt is consumed to swamp the capacity of the approved carriers, the excess is acutely toxic. (Children have died after eating their mothers' attractively coated iron tablets; it is claimed to be the second most common cause of accidental poisoning in childhood—aspirin is still number one.)

The known forms in which the **available iron pool**—as opposed to the specific heme proteins and other iron-containing electron carriers—exists are summarized in the following and in Table 34–2.

Iron is transported in the blood between tissues by combination with specific proteins, **transferrins,** which have a high affinity for two Fe(III) ions. This affinity is so high that these proteins are half-saturated at a [Fe(III)] of only 25 atoms of free Fe(III) per liter of blood.

Within cells, Fe(III) is stored by combination with other proteins, apoferritins, which can ligate many atoms of the metal per molecule, so that the iron content of **ferritins** approaches one quarter of the total mass. This is possible because the apoprotein is composed of 24 subunits in a spherical shell that is penetrated by six channels. Iron enters the channels and is deposited in the core as a hydroxy phosphate. These complex ferritin molecules associate to form small granules in the cytoplasm.

As the iron content of a cell represented by ferritin grows, an increased amount of another type of granule, **hemosiderin,** appears. Hemosiderin is an ill-defined complex of ferric ions with hydroxide ions, various polysaccharides, and proteins, with a third of its weight as iron.

It is not known how iron is carried through membranes or within the cell. It may be that the more soluble Fe(II) compounds are the only forms that cross membranes, but this is not established. However, Fe(II) preferentially enters the channels in apoferritin, which acts as a ferroxidase, using O_2 to convert the trapped iron to Fe(III) for deposition. It is not even known how the iron is released from ferritin for use. Reduction to Fe(II) is a good possibility.

Once in the blood, iron is oxidized to Fe(III) by a ferroxidase known as **ceruloplasmin.** (It is bright blue owing to the presence of copper ions.) Fe(III) is tightly bound to transferrin. Cells receiving iron have receptor sites for transferrin, which exchanges between the blood and intercellular fluid. Apotransferrin is used over and over, but it is not known how the iron is removed. It may be reduced at the receptor site, or the receptor-transferrin complex may be engulfed by pinocytosis, followed by reduction within the cell and release of the apoprotein.

Iron losses from the body are usually low, and most of the iron released from degraded hemoproteins is retained within the body. There is no specific secretory mechanism for iron within the kidney, and the amount in the urine is less than 10 μmoles per day. In all, men lose about 20 μmoles per day with moderate activity, of which about two thirds is lost from the bowel. Women lose more through menstruation, the loss ranging from 50 to 15,000 μmoles per period, with an average

TABLE 34–2

Transferrins

Many variants in one individual
Single polypeptide chain; M.W. 81,000, 6 per cent carbohydrate
Two binding sites for Fe(III); no cooperative interaction
Anion binding required for iron binding; bicarbonate or carbonate is usual anion

$$K' = \frac{[\text{Fe chelate}][\text{H}^+]^3}{[\text{apoprotein}][\text{Fe(III)}][\text{HCO}_3^-]} = 10^3\ \text{M}$$

$$\text{Effective } K' \text{ in blood} = \frac{[\text{chelate}]}{[\text{apoprotein}][\text{Fe(III)}]} = 5 \times 10^{23}\ \text{M}^{-1}$$

Ferritins

May be composed of two similar polypeptide chains in varying proportions
24 subunit hollow shell; subunit M.W. 18,500–19,000; 12.4–13.0 nm external dia., 7–8 nm internal dia.
Core with up to 4,000 Fe atoms, mainly $(FeO(OH))_8(FeO \cdot PO_4H_2)$
Small amount of apoferritin in blood from normal tissue turnover; level is a guide to iron store, with 1 μg apoferritin per liter of blood representing *ca.* 8 mg stored iron

of 250 μmoles, which is equivalent to about 9 additional μmoles per day on a long-term basis.

It is obvious that if the absorption of iron exceeds these various small losses, the total iron content of the body will rise with age, and this in fact is a frequent occurrence in males. The only regulation of the quantity of iron in the body is a regulation of absorption from the intestine, although its mechanism is not known.

Iron is most readily absorbed when it occurs in hemin; the absorbed hemin is attacked within the intestinal mucosa by heme oxygenase to liberate iron and biliverdin in the same way that it is attacked in the spleen or liver. Inorganic iron is much less readily absorbed, although the fraction taken in is increased by simultaneously eating meat, for unknown reasons. Much probably depends upon the form of the iron. It may be that only certain chelates can readily enter mucosal cells; this may explain why ingestion of ascorbate with iron salts improves absorption, since the ascorbate would reduce iron to Fe(II), and form a soluble chelate. (Certain chelating agents diminish iron absorption, for example, phytate, an inositol polyphosphate occurring in plants such as spinach.)

Whatever the mechanism of absorption, the fraction of the iron passing through the mucosa into the blood depends upon the amount of iron already in the body. The intestinal mucosal cells form ferritin, and this deposit is lost when the cells slough into the lumen. This may be part of the control mechanism, but there also may be more direct regulation of transit through the cells.

Iron Overload. A chronic accumulation of iron causes excessive deposits of ferritin and hemosiderin to appear within cells, especially within the liver. This in itself is believed to be damaging; there is a high correlation of cirrhosis of the liver with these deposits. (Other nutritional problems may sometimes contribute to the damage.) Three kinds of circumstances are likely to cause a dangerous accumulation. One is a hereditary deficiency in control over iron absorption known as **hemochromatosis.** The deficiency may be in regulation by the intestinal mucosa, or in the regulation of any aspect of hemoprotein synthesis or degradation. (Anemia causes increased iron absorption.) These people absorb a larger quantity of iron than they need when the dietary intake is normal; any increase in the intake magnifies their problem. It takes many years to build these abnormal iron stores; consequently, hemochromatosis is typically a disease of mature men. It takes longer for it to appear in women owing to their larger iron losses.

HEMOCHROMATOSIS

Primary (idiopathic)

Autosomal recessive.

Gene frequency may be as high as 5 per cent in those of British or French ancestry, with disease not fully expressed in all homozygotes. Causes enlarged liver, metallic-gray skin, diabetes, loss of libido.

Hemosiderin deposits especially prominent in liver and pancreas, with iron content up to 100 times normal.

Treated by phlebotomy, removing 500 ml of blood per week, or by therapy with microbial siderophores—compounds with a high affinity for iron secreted by some bacteria.

Secondary

Transfusion hemosiderosis. Caused by constant breakdown and replacement of hemoglobin in chronic anemias, such as thalassemia.

Dietary. Caused by high iron intake. Alcoholics are at greatest risk owing to concomitant liver damage.

Excessive iron may also accumulate from repeated transfusions given to compensate for hemolytic conditions, such as thalassemia. This is perhaps the most common source of excessive iron deposits in children.

Finally, excessive iron may accumulate from excessive intake. The intestinal control over absorption is not perfect. It limits the fraction of dietary iron that is absorbed, but any increase in the absorbable iron content of the diet will cause some increase in the absolute amount taken up. This is classically illustrated by the Bantus of South Africa, who prepare meals and make beer in large iron pots so that the average of 50 μmoles originally in the foodstuffs ingested per day is increased to 2,000 μmoles or more. Many of these people are ill, and the condition is designated as the iron pot syndrome. We need not travel so far afield, because wines sold in the United States have contained as much as 500 μmoles per liter. Finally, a few areas of the world have extraordinary amounts of iron in water used for drinking, and some people in those areas accumulate excessive stores.*

*I once drank beautifully clear and cold water from a Forest Service well in Colorado and then had the disquieting experience of seeing a heated pot of water become opaque from a massive precipitate of iron hydroxide. Much lower concentrations than this would be dangerous upon lengthy exposure.

FURTHER READING

(Also see Chapter 41 for a discussion of iron in nutrition.)

A. Goldberg and M.R. Moore, eds. (1980). The Porphyrias. Clin. Hematol. 9:225. Several review articles.

W.C. Mentzer, ed. (1981). Enzymopathies. Clin. Hematol. 10:1. Reviews genetic defects in erythrocyte metabolism.

K. Yamauchi, N. Hayashi, and G. Kikuchi (1980). Translocation of δ-Aminolevulinate Synthase from the Cytosol to the Mitochondria and Its Regulation by Hemin in the Rat Liver. J. Biol. Chem. 255:1746.

T.S. Hronis and J.A. Traugh (1981). The Use of Porphyrins as Probes to Examine the Requirement for Hemin in the Maintenance of Protein Synthesis in Reticulocyte Lysates. J. Biol. Chem. 256:11409.

B.F. Felsher et al. (1982). Decreased Hepatic Uroporphyrinogen Decarboxylase Activity in Porphyria Cutanea Tarda. N. Engl. J. Med. 306:766,799.

P. Aisen and L. Listowosky (1980). Iron Transport and Storage Proteins. Annu. Rev. Biochem. 49:357.

A. Jacobs and M. Worwood, eds. (1980). Iron in Biochemistry and Medicine II. Academic Press. Review articles.

T.H. Bothwell et al. (1979). Iron Metabolism in Man. Blackwell. General review.

National Research Council Subcommittee on Iron (1979). Iron. University Park Press. Review with emphasis on environmental iron and its toxicity.

A. Jacobs, ed. (1982). Disorders of Iron Metabolism. Clin. Haemotol. 11:239,486.

C.A. Finch and H. Huebers (1982). Perspectives in Iron Metabolism. N. Engl. J. Med. 306:1520.

CHAPTER 35

TURNOVER OF NUCLEOTIDES

We have seen two general functions of nucleotides: to play key roles in many metabolic reactions, and to act as precursors of the nucleic acids. Both the metabolic pool of nucleotides and the ribonucleic acids are constantly being synthesized and degraded. The attrition of these compounds and the associated requirements for synthesis of replacements is most acute in those tissues in which normal function involves loss of entire cells, such as the skin, the intestinal mucosa, and the blood. These tissues also require the constant formation of new DNA.

The turnover of RNA is also rapid in cells that secrete proteins as a major part of their function. Prominent examples are cells of various parts of the gastrointestinal tract that pour out large volumes of digestive juices, rich in enzymes and mucoproteins. In addition to secreting many proteins, the liver also rapidly changes its intracellular protein composition, which imposes an additional load on the formation of RNA. Even the stable cell population in adult muscles and nerves has an internal turnover of RNA although at a slower rate than that of most tissues.

An Overall View

Let us summarize the general features of nucleotide turnover:

(1) Both the pyrimidine and the purine nucleotides are constantly degraded to the free bases and ribose phosphates.

(2) The free bases and ribose phosphates can be used in salvage pathways to rebuild nucleotides.

(3) The base components also can be synthesized *de novo* on ribose phosphate from common metabolic intermediates so as to increase the size of the nucleotide pool or replace any degraded components.

(4) The ribose phosphates can be synthesized from glucose or metabolized by the usual reactions of the pentose phosphate pathway.

(5) There is a constant attrition of bases through degradative reactions:
  a. The pyrimidines are degraded to common metabolic intermediates.
  b. The purines are converted to uric acid, a more oxidized purine that is excreted in the urine.

The low solubilities of uric acid and its monosodium salt will cause our attention to be focused on purine nucleotide turnover; precipitation of these compounds causes trouble. Formation of sodium urate crystals damages the joints, kidneys, and other tissues, causing gout. Uric acid is a common constituent of stones in the urinary tract. A variety of hereditary conditions and disease states cause clinically significant elevations in uric acid production.

DEGRADATION AND SALVAGE

Nucleic acids are hydrolyzed by **endonucleases,** which attack bonds in the middle of the polynucleotide chains, and **exonucleases,** which hydrolyze successive terminal bonds. Many such enzymes have been identified in mammalian cells, some specific for DNA, some for RNA, and some attacking both. Several of them in the nucleus appear to be concerned with repair of DNA or maturation of RNA after initial transcription. Other nucleases occur in the lysosomes, presumably to handle foreign nucleic acids and damaged cellular components. These include a 3′-exonuclease, which forms oligonucleotides with terminal 2′,3′-cyclic phosphodiester bonds, much like the intermediate in the action of bovine pancreatic ribonuclease (p. 277). A cyclic phosphodiester hydrolase then cleaves these bonds. Most of the nucleases form 5′-nucleotides.

The general pathway for degrading 5′-nucleotides begins with hydrolysis of the phosphate ester group to release the nucleosides (Fig. 35–1). The nucleosides are cleaved with inorganic phosphate, releasing the free bases and forming ribose 1-phosphate. Many of the modified bases in tRNA and rRNA are excreted as such, but it is mRNA that turns over most rapidly, and this larger source of unmodified bases contributes to the general nucleotide pool of the cell, which is constantly degraded and rebuilt. Ribose 1-phosphate is interconvertible with ribose 5-phosphate through the action of phosphoglucomutase, the same enzyme acting on glucose phosphates. The ribose 5-phosphate, together with the bases, is then used to rebuild nucleotides, or is used as fuel.

Provision of Ribose Phosphate Groups. When nucleotides are synthesized, either by salvage of existing bases or by *de novo* synthesis, the ribose phosphate moiety is supplied in the form of **5-phospho-α-D-ribose diphosphate,** commonly referred to as **phosphoribosyl pyrophosphate (PRPP).** This compound is made from ribose 5-phosphate, which can be generated from glucose 6-phosphate by the pentose phosphate pathway (p. 542) or from ribose 1-phosphate:

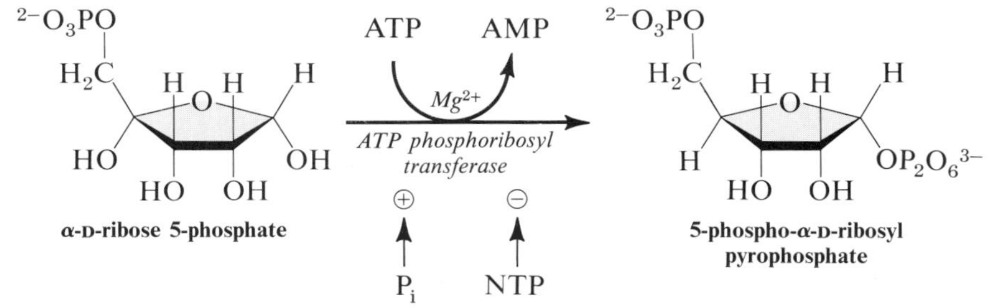

$^{2-}O_3PO$–H_2C (ribose ring; HO, HO, OH, OH)
α-D-ribose 5-phosphate

ATP → AMP (Mg^{2+})
ATP phosphoribosyl transferase
⊕ P_i ⊖ NTP

$^{2-}O_3PO$–H_2C (ribose ring; H, HO, OH, $OP_2O_6^{3-}$)
5-phospho-α-D-ribosyl pyrophosphate

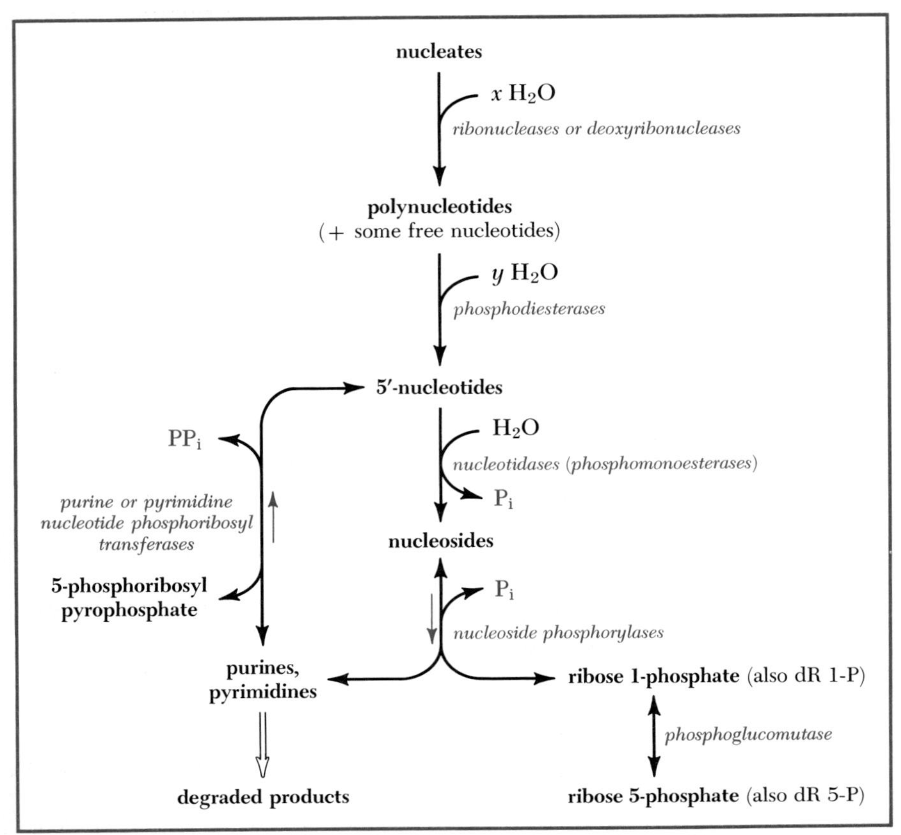

FIGURE 35–1 Outline of the general processes for degradation of nucleic acids and other nucleotides. The purines and pyrimidines arising in this way or obtained from the diet may be re-incorporated into nucleotides by reactions with phosphoribosyl pyrophosphate (*left*), or they may be further degraded. The ribose portion appears as ribose 5-phosphate.

The ATP phosphoribosyl transferase catalyzing this reaction is an important site of regulation of nucleotide synthesis. The enzyme has an absolute requirement for inorganic phosphate; the effect is to increase the synthesis of nucleotides for the metabolic pools whenever their degradation becomes excessive, as manifested by accumulation of P_i. The enzyme is inhibited by nucleotides, thereby preventing their accumulation through either the salvage pathway or *de novo* synthesis.

The salvage pathway involves a condensation of free bases with phosphoribosyl pyrophosphate to form a nucleotide:

$^{2-}O_3PO$ — H_2C … $OP_2O_6^{3-}$ + base (Pu or Py, N–H) $\xrightarrow[\text{phosphoribosyl transferase}]{Mg^{2+}}$ PP_i + nucleotide ($^{2-}O_3PO$ — H_2C …, N–Pu or Py)

5-phospho-α-D-ribosyl pyrophosphate **base** **nucleotide (nucleoside 5′-phosphate)**

PURINE METABOLISM

Degradation (Fig. 35–2)

The concentration of adenine nucleotides is severalfold greater than the concentration of other nucleotides, and their turnover is correspondingly greater. We already noted (p. 581) that AMP is deaminated by a simple hydrolysis to produce inosine monophosphate as part of the purine nucleotide cycle. This is also the first step in degradation of the nucleotide. Inosine monophosphate and guanosine monophosphate are then hydrolyzed to form the nucleosides and cleaved with P_i to liberate hypoxanthine and guanine.

Hypoxanthine is oxidized to xanthine and then to uric acid by molecular oxygen. The enzyme catalyzing these successive reactions, **xanthine oxidase,** occurs in the soluble cytoplasm of the cell, and yet it is a molybdenum-containing iron-sulfide protein. The presence of **molybdenum** in this, as well as the related aldehyde oxidase also present in the cytoplasm, makes it necessary to have traces of the inorganic element in the diet. The molybdenum, ordinarily hexavalent, is transiently reduced to the pentavalent compound during the passage of electrons to oxygen. Oxygen is reduced to hydrogen peroxide as the other product.

Guanine is at the same oxidation level as xanthine, so simple hydrolytic deamination of guanine liberated by degradation of its nucleotides will form xanthine. Xanthine therefore is an intermediate in the degradation of both major purines, and uric acid is the end-product.

Since the first ionization of uric acid has a pK′ near 5.4, the urate concentration is 100 times the uric acid concentration in blood plasma at pH 7.4. It is cleared efficiently by the kidney, but if the urinary pH drops to 5.4, as it commonly does, then the urinary concentration of uric acid equals the concentration of urate.

Most mammals other than the primates degrade urate further, using uricase, a copper-containing enzyme, to generate allantoin as an end-product:

$$\text{urate} + O_2 \xrightarrow{\text{urate oxidase (uricase)}} \text{allantoin} + H_2O_2 + CO_2$$

This compound has a solubility in excess of 5,000 mg per liter. Compare this with the solubilities of the purines (mg per liter):

Compound	Blood Serum	Water	Urine at pH 5
sodium urate	68	1,200	—
uric acid	—	65	150
xanthine	100	?	50
hypoxanthine	1,150	?	1,400
guanine	?	39	?

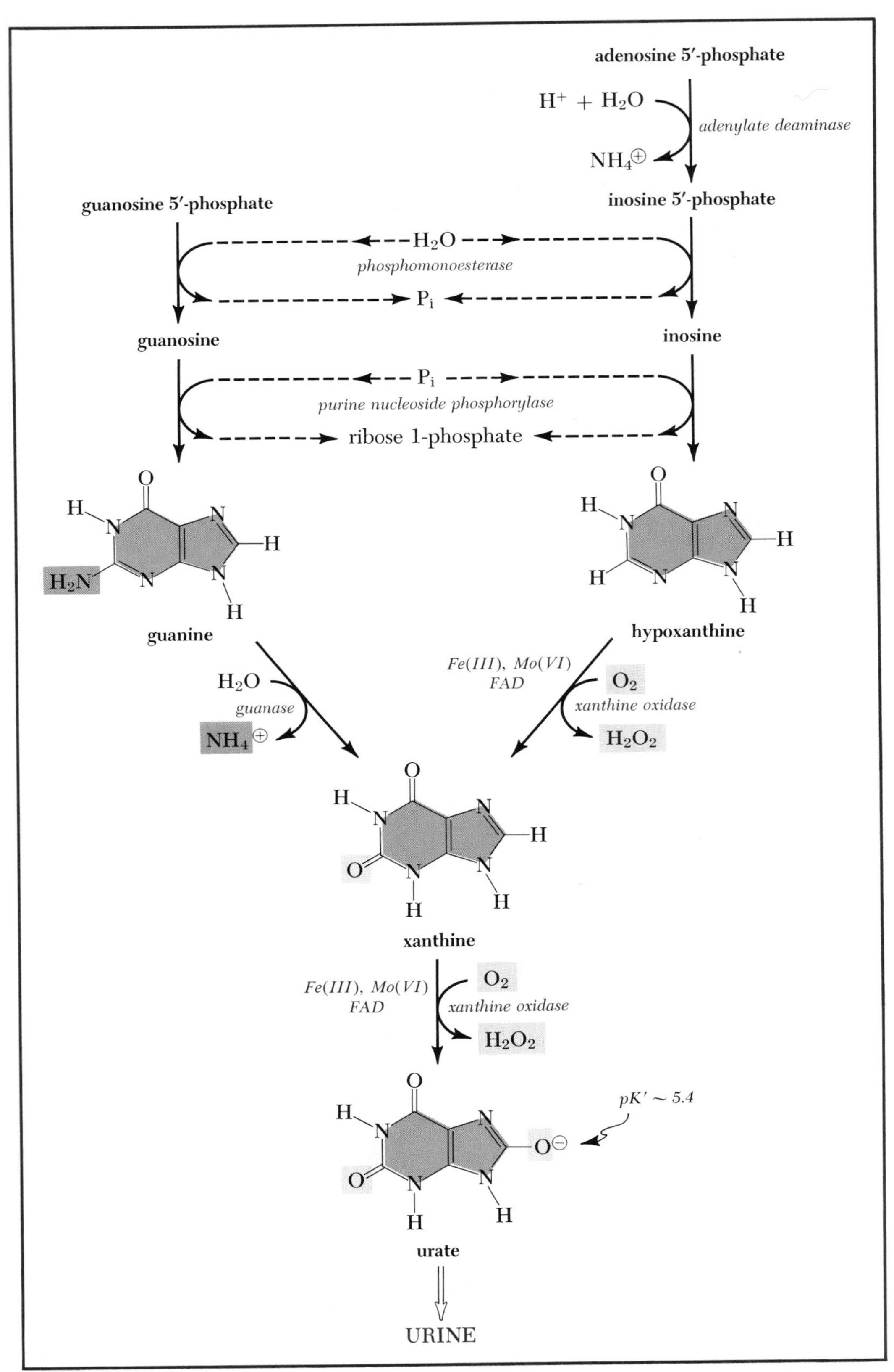

FIGURE 35–2 The degradation of purine residues to urate. The product exists in the urine as a mixture of urate and uric acid (not shown).

We must ask why the primates, including humans, stop purine metabolism with the formation of uric acid, since the solubility product of the sodium salt is so low that concentrations over 70 mg per liter in the sodium-rich extracellular fluids are supersaturated, and the free acid is little more soluble in urine. If some intact purine must be excreted, why not choose hypoxanthine, which is much more soluble?

The answer is clear-cut with birds and reptiles, which conserve water in their closed eggs and during flight by converting nearly all of the waste nitrogen to solid masses of uric acid. That is, they synthesize purines instead of urea to dispose of amino acid nitrogen. Many invertebrates make either uric acid or guanine for the same purpose.

The primates, however, make urea as the principal excretory product of nitrogen metabolism, and dilute it with a constant supply of water. Even so, their purine turnover is rapid enough to make troublesome amounts of uric acid, and we shall return to this point later.

Purine Salvage

Adenosine has a special role. It causes vasodilation, and this is believed to be part of the mechanism for increasing blood flow through the heart and other muscles when their power output is increased. Adenosine is released through the action of a 5′-nucleotidase. In addition, adenosine is transported from the liver, in part within red blood cells, for use in other tissues. In either case, adenosine is converted to inosine within the cells by the action of **adenosine deaminase,** or is phosphorylated to make AMP by an **adenosine kinase.** When inosine is formed, hypoxanthine is liberated from it by a nucleoside phosphorylase.

Purines brought to a cell or formed within it by turnover of its nucleotides are salvaged by two enzymes (Fig. 35–3). A **hypoxanthine phosphoribosyltransferase** is specific for either hypoxanthine or guanine; an **adenine phosphoribosyltransferase** only reacts with adenine.

The hypoxanthine phosphoribosyltransferase is especially active in the central nervous system and other tissues in which the capacity for *de novo* synthesis of purines is limited. These cells depend upon efficient recovery of pre-existing purines for normal function. Since hypoxanthine is formed from adenine nucleotides and guanine from guanine nucleotides, the one enzyme suffices. (The concentration of free adenine is usually low, and only a small activity of adenine phosphoribosyl transferase is necessary to cope with it. Some adenine is formed from thiomethyl-adenosine [p. 656] and some is obtained from the diet.)

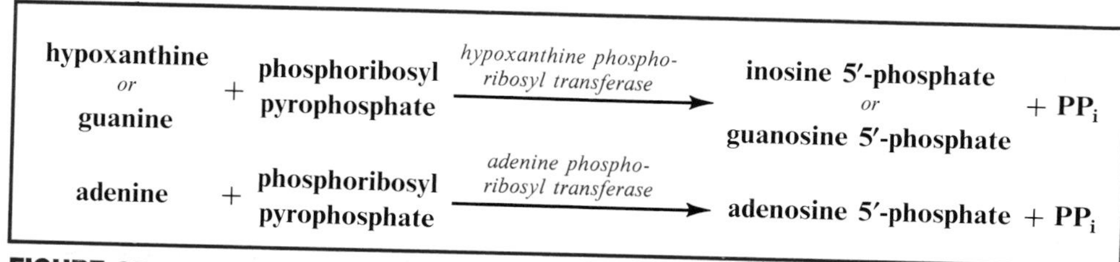

FIGURE 35–3 Salvage of free purines. The enzyme recovering guanine or hypoxanthine is the more active.

Purine Synthesis

De novo synthesis of purines involves a piece-by-piece assembly of the rings on ribose phosphate, and the origin of the ring skeleton is shown in Figure 35–4.

One of the four nitrogen atoms and two of the carbons come from an intact **glycine** molecule. Another nitrogen is contributed by **aspartate,** and two more by **glutamine.** Two of the carbons come from the **one-carbon pool** as tetrahydrofolate derivatives, and the remaining carbon is added as $\mathbf{CO_2}$. The initial product when the ring is assembled is inosine monophosphate.

Inosine monophosphate (IMP) is the precursor of both guanosine and adenosine monophosphates. Simple replacement of the oxygen by an amino group derived from aspartate makes the adenine nucleotide. Oxidation of the ring and replacement of the added oxygen by the amide group of glutamine make the guanine nucleotide.

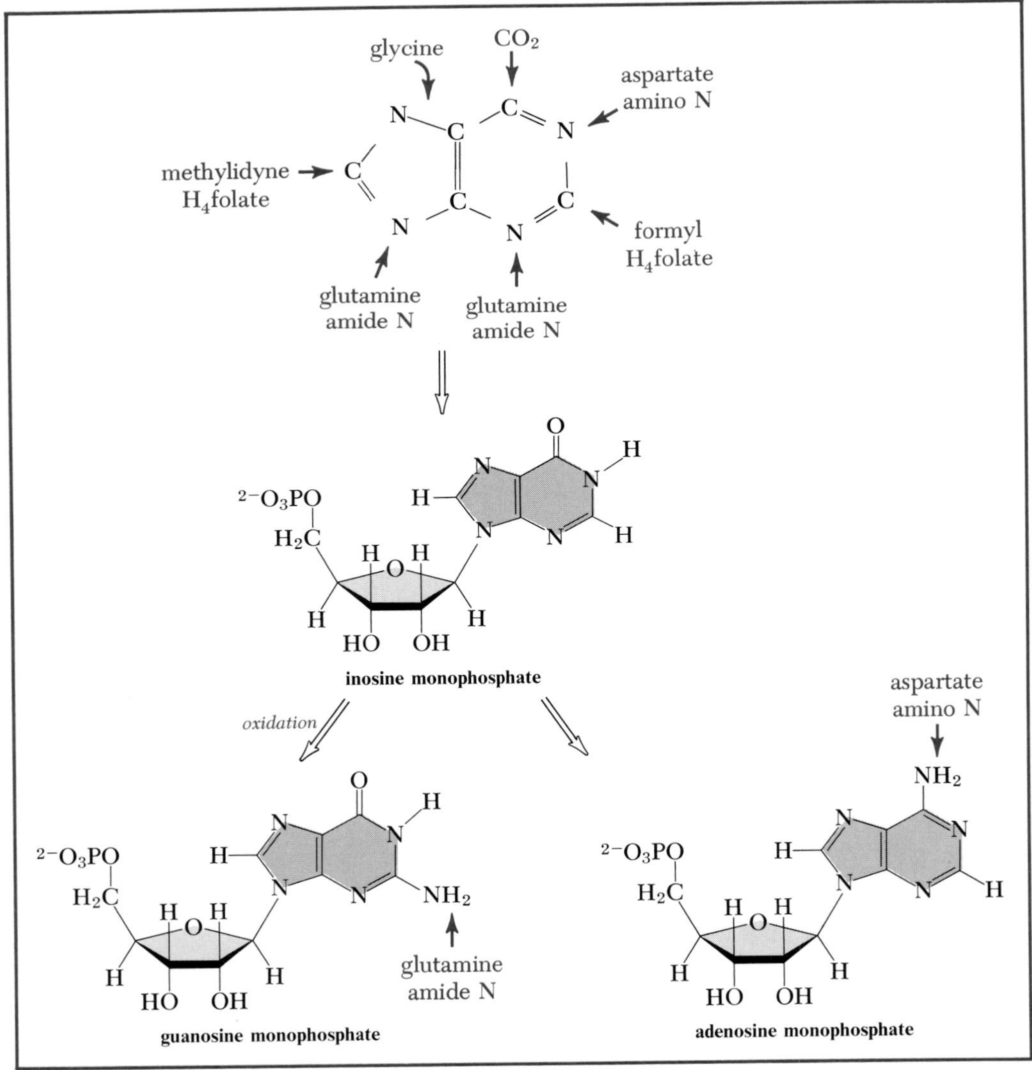

FIGURE 35–4 Sources of the atoms of purines. The rings are assembled on a ribose 5-phosphate residue obtained from 5-phosphoribosyl pyrophosphate (not shown).

The initial reaction is a transfer of the amide group of glutamine to phosphoribosylpyrophosphate forming **phosphoribosylamine** and liberating inorganic pyrophosphate. Since the ribosylpyrophosphate has the α-configuration and the ribosylamine the β-configuration, there is an inversion of configuration during the reaction. The remainder of the purine ring is built around this added nitrogen, so the products will have the characteristic β-ribosyl configuration found in the nucleotides.

A glycyl group is attached intact to the amino group on ribose by a straightforward synthetase reaction in which ATP is cleaved to drive the reaction. The reaction is reversible because of the similar free energies of hydrolysis of the aminoacyl amide and ATP, but the product is constantly removed by the next reaction.

A formyl group is then added to the free amino group of the glycyl residue. This completes the atoms necessary for the five-membered ring of the purines. The formyl donor is 5,10-methylidyne-H_4folate in the one-carbon pool.

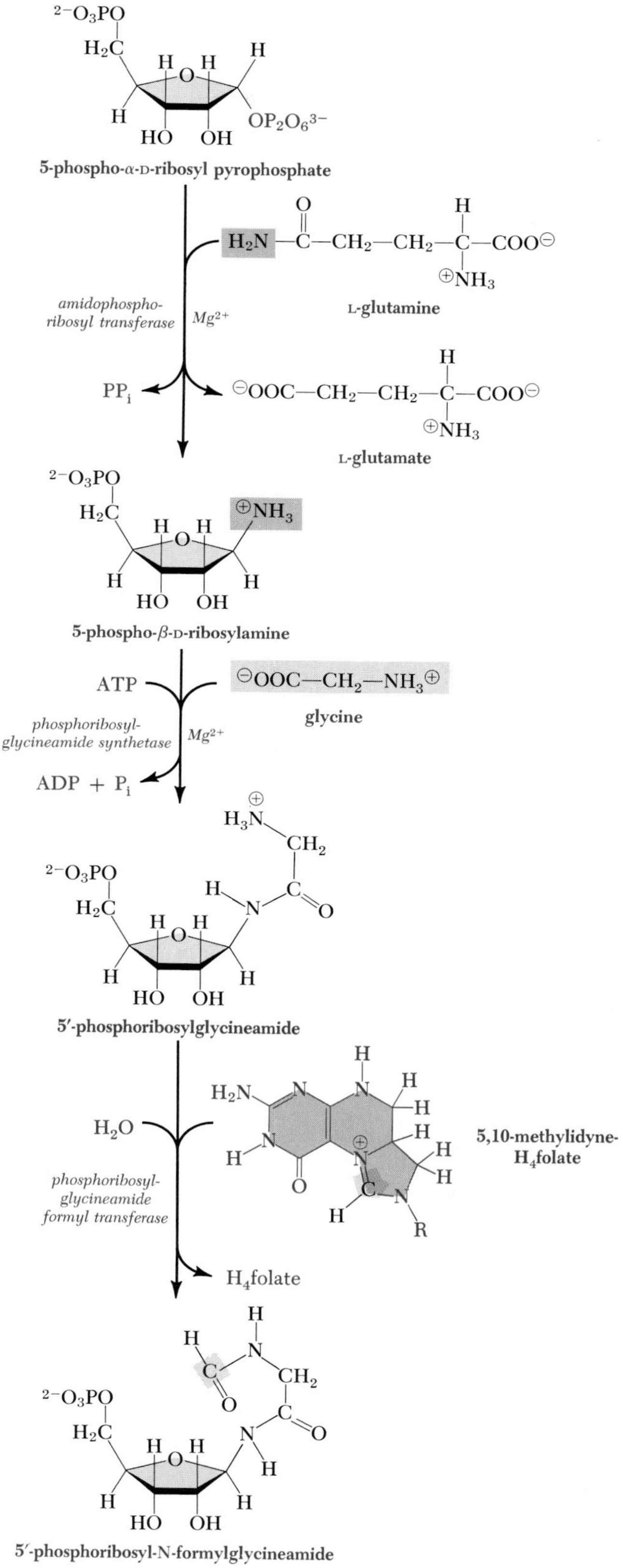

5-phospho-α-D-ribosyl pyrophosphate

L-glutamine

amidophosphoribosyl transferase Mg^{2+}

PP_i

L-glutamate

5-phospho-β-D-ribosylamine

ATP

glycine

phosphoribosylglycineamide synthetase Mg^{2+}

ADP + P_i

5′-phosphoribosylglycineamide

H_2O

5,10-methylidyne-H_4folate

phosphoribosylglycineamide formyl transferase

H_4folate

5′-phosphoribosyl-N-formylglycineamide

ATP
phosphoribosylformylglycineamidine synthetase
Mg^{2+}
$H_2N-C(=O)-CH_2-CH_2-CH(^{\oplus}NH_3)-COO^{\ominus}$
L-glutamine
ADP + P_i
L-glutamate

A further nitrogen is next transferred from the amide group of glutamine, and the reaction is driven by the hydrolysis of ATP. Since the free energy of hydrolysis of ordinary amides, in contrast to aminoacyl amides, is less than the free energy of hydrolysis of ATP, the reaction is essentially irreversible.

5′-phosphoribosyl-N-formylglycineamidine

ATP
phosphoribosylaminoimidazole synthetase
Mg^{2+}
ADP + P_i

The five-membered imidazole ring of the purine is then formed in a reaction driven by ATP. The amino group added in the preceding reaction now sticks out.

5′-phosphoribosyl-5-aminoimidazole

phosphoribosylaminoimidazole carboxylase
CO_2
$H^{\oplus}$

In the next step, CO_2 adds to the imidazole ring. The carboxylase catalyzing this reaction has no biotin, pyridoxal phosphate, or thiamine pyrophosphate attached, and therefore differs in mechanism from most of the enzymes catalyzing addition or removal of carboxylate groups.

5′-phosphoribosyl-5-aminoimidazole-4-carboxylate

ATP
phosphoribosylaminoimidazolesuccinocarboxamide synthetase
Mg^{2+}
$^{\oplus}H_3N-CH(COO^{\ominus})-CH_2-COO^{\ominus}$
L-aspartate
ADP + P_i

In the next pair of reactions, the amino group of aspartate is transferred to form an amide of the carboxyl group created by the preceding reaction. The first step is the formation of the amide between the aminoimidazole carboxylate and the amino group of aspartate, using ATP as an energy donor.

5′-phosphoribosyl-5-aminoimidazole-4-(N-succino)carboxamide

The product then is cleaved on the other side of the connecting nitrogen atom to liberate fumarate and leave the aminoimidazole carboxamide as a product. The pair of reactions is analogous to the mechanism in urea synthesis by which the nitrogen of aspartate is transferred to citrulline, forming arginine (p. 584), with an intermediate N-succino compound formed in both cases. (However, ATP is cleaved to form pyrophosphate in arginine synthesis, with the constant removal of pyrophosphate ensuring that the reaction will be drawn toward arginine.)

adenylosuccinase

fumarate

5′-phosphoribosyl-5-aminoimidazole-4-carboxamide

The final atom necessary for the purine ring is transferred from 10-formyl-tetrahydrofolate by an irreversible reaction.

phosphoribosylaminoimidazole-carboxamide formyl transferase K^+

10-formyl-H_4folate

H_4folate

5′-phosphoribosyl-5-formamidoimidazole-4-carboxamide

With the final atom for the purine in place, the ring is formed by an enzyme catalyzing the simple removal of the elements of water to produce the purine nucleotide, inosine 5′-phosphate.

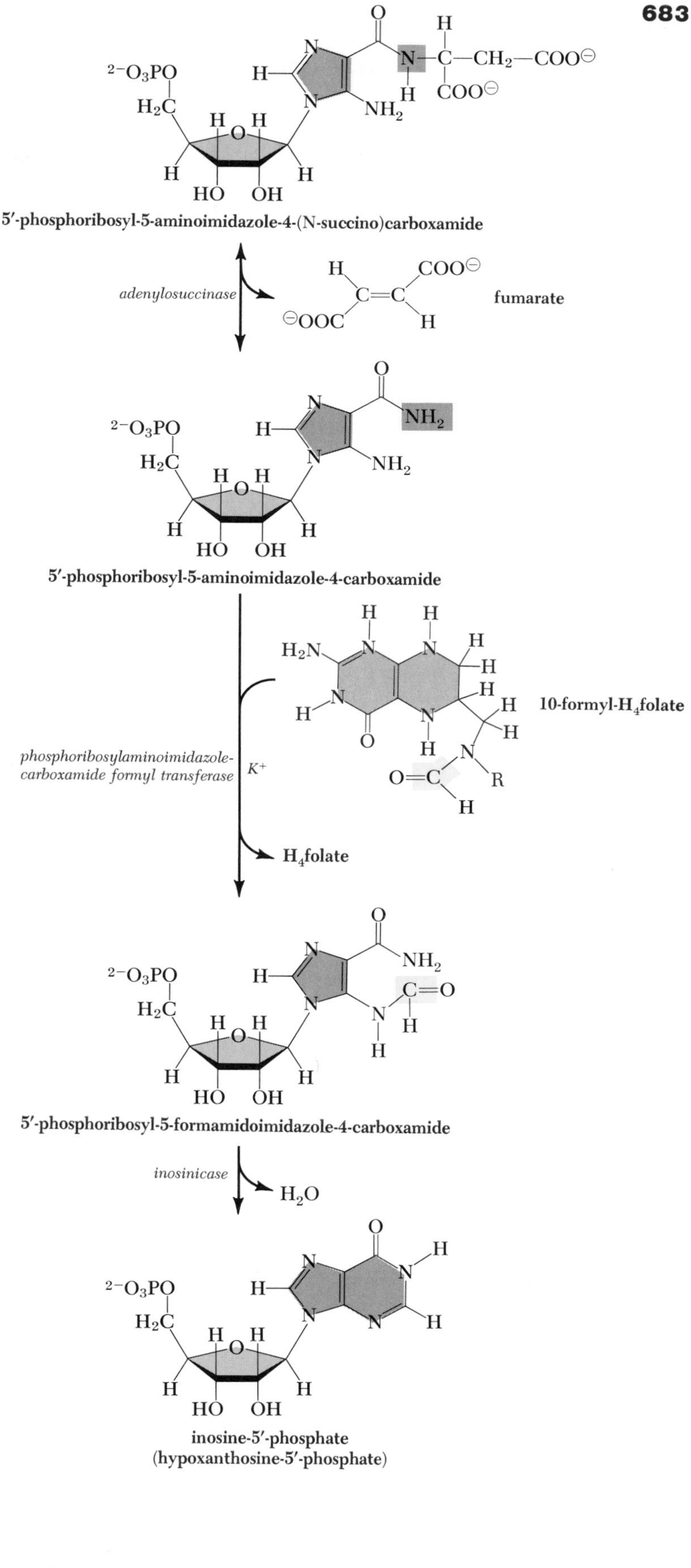

inosinicase

H_2O

inosine-5′-phosphate
(hypoxanthosine-5′-phosphate)

The route to **adenosine 5′-monophosphate** (Fig. 35–5, *right*) involves the transfer of the amino group of asparate *via* the intermediate N-succino compound, adenylosuccinate, in a way completely analogous to the transfer shown two steps earlier. The same enzyme cleaves fumarate in this and the previous set of reactions. However, the synthetase that forms adenylosuccinate differs in utilizing GTP rather than ATP as an energy source. The significance of this difference is not clear.

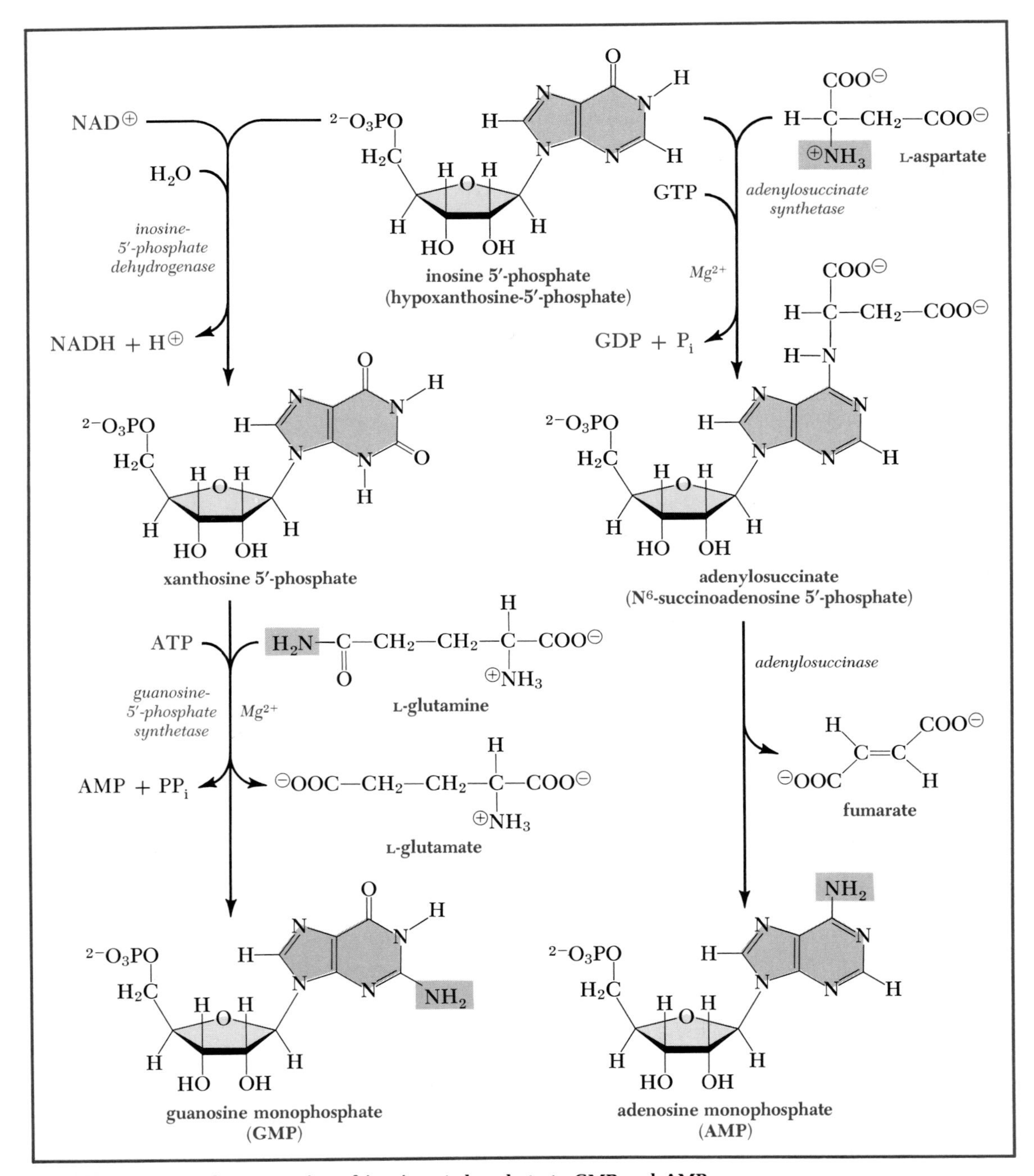

FIGURE 35–5 The conversion of inosine 5′-phosphate to GMP and AMP.

Guanine is a more oxidized purine than is adenine. The first step in the formation of **guanosine monophosphate** is therefore the oxidation of inosine 5′-phosphate by NAD to form **xanthosine 5′-phosphate,** the nucleotide of xanthine (2,6-dioxopurine). Guanosine monophosphate is then formed by a transfer of the amide group of glutamine, with the reaction driven by ATP. This reaction differs from the earlier ones in which glutamine is an amide donor in that ATP is cleaved to AMP and inorganic pyrophosphate. The reaction is irreversible, even without removal of inorganic pyrophosphate, so the functional value of the loss of the extra high-energy phosphate is not clear.

Hyperuricemia

Of each million otherwise normal adult males in the United States, approximately 5500 will have frank gout. This condition is caused by deposition of sodium urate crystals in the joints, with the subsequent ingestion of the crystals by leukocytes causing an exquisitely painful inflammation. The joint at the base of one great toe is frequently affected in initial attacks. If untreated, the accumulating deposits can cause permanent damage to the joints; nodules (tophi) of sodium urate crystals will also appear under the skin.

In the absence of other disease, gout is a condition of the male, with only 3 to 7 per cent of the primary cases occurring in women. It often first appears in the fourth decade of life. The normal range of blood serum urate concentrations in adult males runs up to 70 mg per liter—the limit of solubility, whereas it only ranges to 60 mg l^{-1} in adult premenopausal females. (The upper limit is defined here as 2 S.D. above the mean value.) Many males have even higher values without ever developing gout. Indeed, most individuals with hyperuricemia do not develop gout; the concentration must exceed 4 S.D. above the mean before the probability of becoming gouty exceeds 90 per cent.

The defects in primary gout are not well characterized. The known points of control of purine nucleotide formation are the generation of phosphoribosyl pyrophosphate and its conversion to phosphoribosylamine (Fig. 35–6). Overproduction of purines could result from a failure of control at either point. An excessively rapid breakdown of nucleotides could also cause excess synthesis. A defect in the mechanism by which the kidney excretes urate would also cause hyperuricemia.

Many cases of gout are secondary consequences of other conditions. Only recently has it been realized how many resulted from **chronic lead poisoning** that damaged the kidneys. Gout itself was formerly thought to result in life-threatening loss of kidney function, but it now appears that this is true only when it is exacerbating the effects of lead.

Any condition in which there is an increased rate of cell turnover can cause gout. These include **chronic hemolytic anemia, pernicious anemia, toxemia of pregnancy, starvation, glucose-6-phosphate dehydrogenase deficiency,** and so on. **Malignancies** are in themselves likely to contribute to hyperuricemia, and the use of **cytotoxic drugs** to destroy cancer cells can release enough purines to cause renal failure from uric acid deposition. (It is not known if prior exposure to lead is a factor in these cases.) It is common practice in the treatment of hematologic malignancies to start treatment to prevent hyperuricemia (see later discussion) before the cytotoxic drugs are given.

Perhaps the most common cause of hyperuricemia is the use of **thiazide drugs** to control hypertension. It can also result from **acidosis,** as in diabetic ketoacidosis or lactic acidosis.

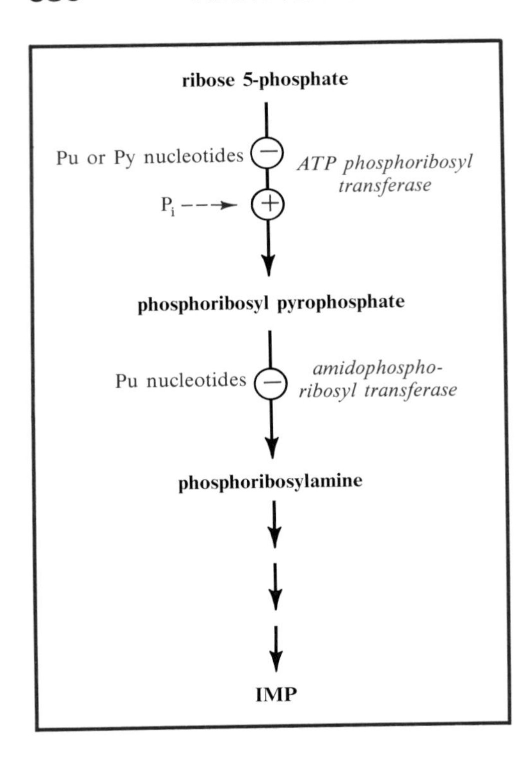

FIGURE 35–6

***De novo* synthesis of purine nucleotides is regulated at the formation of phosphoribosyl pyrophosphate and its conversion to phosphoribosylamine. Either purine or pyrimidine nucleotides inhibit phosphoribosyl pyrophosphate formation, while only purine nucleotides inhibit phosphoribosylamine formation.**

A dramatic, although rare, cause of hyperuricemia is a **hereditary deficiency of the salvage enzyme, hypoxanthine phosphoribosyltransferase.** This causes a loss of control over the synthesis of phosphoribosyl pyrophosphate. It is not clear why, but the result is a rapid *de novo* synthesis of IMP, with a consequent formation of large amounts of urate, more per unit of body mass than is seen in any other condition. Children with this condition, known as the **Lesch-Nyhan syndrome,** have grossly distorted behavior. While said to be very likeable and open, quick to laugh and capable of warm affection, they lapse into excessive aggression, directed at themselves as well as others. They bite; characteristically they bite off the tips of their own fingers, and their lips if unprotected. In later years, they curse, use obscene gestures, fling feces, and so on. Why does this simple enzyme defect have these bizarre effects? No one knows.

Although it may be unrelated, the Lesch-Nyhan syndrome returns us to a more fundamental question. Why have the primates lost their ability to make the more soluble allantoin? Folklore has persistently associated gout with high living. More recent understanding of the regulation of purine metabolism has discredited an intake of a diet rich in purines as an important cause of primary gout, but there is still a residuum of a belief that people with gout are likely to be superior folk. (Male chauvinists take comfort.)

This notion was given only a faint breath of life by comparison of the blood urate concentrations and intelligence test scores in a large sample of World War I draftees; there was a significant, but very small, correlation. However, a more recent study of male high school students supplied a new twist. The results indicated a correlation between urate concentration and drive and ambition. People with high urate tended to be overachievers who got ahead, not by greater abilities, but by using what they had to the utmost. (It is interesting to compare the emotional responses people have when they learn of this finding, and it will be even more interesting to see the results of more comprehensive studies if they can be done and interpreted in an unbiased way.)

Treatment of Hyperuricemia. Gout was treated in ancient times by eating corms of the autumn crocus, *Colchicum autumnale,* and more recently with the active principle of the corms, **colchicine.** This yellow compound combines with tubulin in cells, preventing motility. It arrests mitosis, but it also is very effective in diminishing gouty inflammation. (The patient is effectively titrated with 500 μg doses at hourly intervals until nausea or diarrhea demonstrate a dangerous cessation of cell division in the gastrointestinal tract.)

Colchicine still is a useful tool for differential diagnosis, and some physicians employ it yet for treatment of the acute attack or even for preventive maintenance in small doses. However, it has been replaced in most centers by anti-inflammatory drugs (indomethacin or phenylbutazone) for acute attacks and by drugs that promote the excretion of uric acid (sulfinpyrazone or probenecid), or by allopurinol:

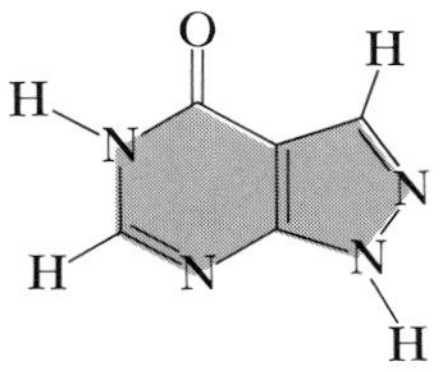

allopurinol

This compound is an analogue of hypoxanthine, differing in the distribution of ring nitrogens, and it was developed as an inhibitor of xanthine oxidase. The rationale for its use is that hypoxanthine will not and xanthine may not precipitate in the joints. Hypoxanthine is more soluble than sodium urate. Xanthine is not, and people with hereditary deficiencies of xanthine oxidase, as well as those treated with allopurinol, sometimes have trouble with xanthine stones and kidney damage. (The incidence of difficulty is much less than it is with untreated gout.)

Allopurinol has also been used in some cases of **adenine phosphoribosyltransferase deficiency.** The gene frequency of this condition may be as high as 0.5 per cent, but it is usually undetected because its consequences are mild. Even homozygotes may be normal except for precipitation of 2,8-dihydroxyadenine stones in the urinary tract. (Adenine and 8-hydroxyadenine are also excreted in abnormal amounts.) Allopurinol blocks the oxidation of adenine in this condition.

PYRIMIDINE METABOLISM

Degradation (Fig. 35–7)

The pyrimidine nucleotides, like the purine compounds, are converted to nucleosides by 5′-nucleotidases. Cytidine is converted to uridine by hydrolysis of the amino group:

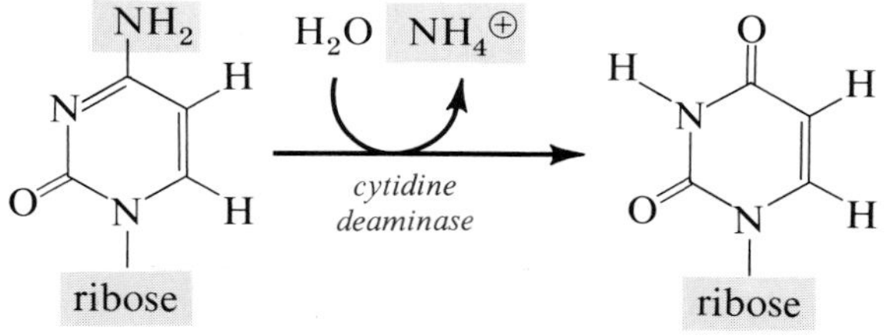

H_2O $NH_4^{\oplus}$

cytidine deaminase

cytidine **uridine**

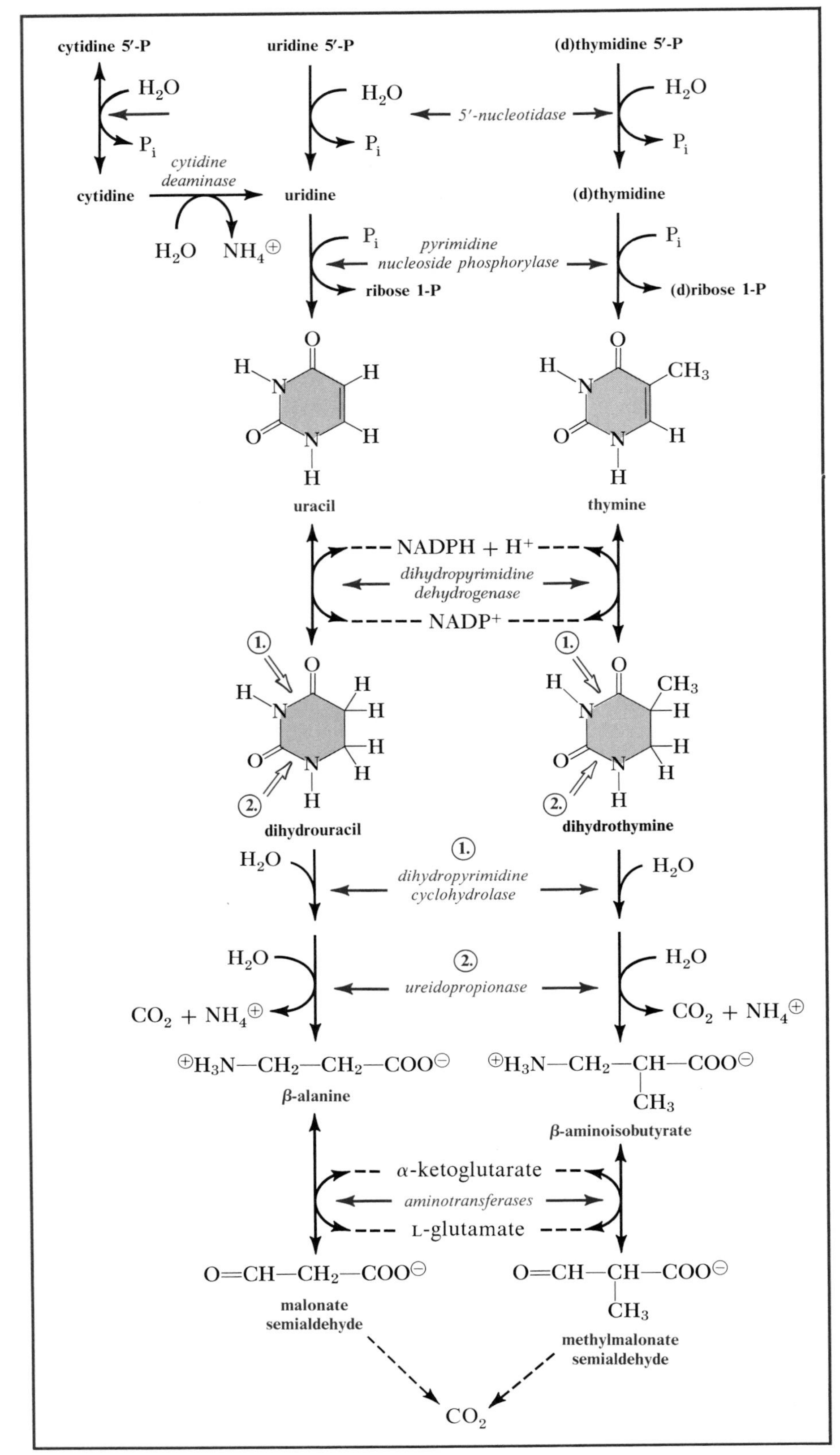

FIGURE 35–7 Degradation of pyrimidine nucleotides. Thymine occurs in both ribo- and deoxyribonucleotides.

Uridine and thymidine are cleaved by phosphorolysis, and the released uracil and thymine undergo similar reactions, in which the ring is reduced, and the amide groups are cleaved by successive hydrolysis. The 3-amino acids, β-alanine and β-aminoisobutyrate, along with NH_4^+ and CO_2, are the products. The amino acids are known to undergo transamination, and the carbon skeletons are oxidized to CO_2, but the exact route is not known.

β-Aminoisobutyrate is only slowly metabolized, and a surprisingly large proportion of the human race excrete significant quantities of the compound in the urine. Five to 10 per cent of Caucasians are hyperexcretors, and up to 90 per cent of Orientals. Hyperexcretion has no known deleterious effects.

The metabolic pool of pyrimidine nucleotides is much smaller than that of the purine nucleotides; although their turnover is rapid and of great qualitative significance, they make little dent in the total fuel or nitrogen economy. A pyrimidine nucleotidase in erythrocytes is very sensitive to lead. This may account for the hemolytic anemia caused by lead poisoning, since the maturation of erythrocytes involves extensive degradation of RNA. An accumulation of pyrimidine nucleotides may be deleterious.

Synthesis of Pyrimidine Nucleotides

Synthesis of the pyrimidines, unlike that of the purines, involves several steps before attachment of the ribose phosphate moiety. The elements of the ring are contributed completely by **glutamine, CO_2,** and **aspartate.**

The process begins by a reaction between carbamoyl phosphate and aspartate to form carbamoylaspartate (Fig. 35–8). The formation of carbamoyl phosphate is also the initial step in urea synthesis (p. 583). We saw that urea formation largely occurs in the liver, but many tissues are able to make the pyrimidines for use in nucleic acid synthesis. Carbamoyl phosphate is formed for this purpose by a different enzyme, which occurs in the cytosol*. The process differs from the reaction used for making urea. Acetyl glutamate is not a required cofactor, and glutamine rather than ammonia is the direct nitrogen donor. The content of the enzyme is quite low because the demand for pyrimidine synthesis is orders of magnitude lower than the demand for urea synthesis. Most importantly of all, **the enzyme is inhibited by UTP,** which therefore regulates the initial step in its own production.

The simple removal of water from carbamoylaspartate forms a reduced pyrimidine, dihydroorotate, which is then oxidized to **orotate.**

Orotate reacts with phosphoribosyl pyrophosphate to form the nucleotide **orotidine 5′-phosphate,** liberating pyrophosphate. As with other pyrophosphorylases, the removal of the inorganic pyrophosphate drives the reaction toward nucleotide formation.

Orotidine 5′-phosphate is converted to **uridine 5′-phosphate** by a simple decarboxylation.

Uridine monophosphate, like the other nucleoside phosphates, is converted to its diphosphate by a specific kinase utilizing ATP, and the diphosphate is converted to the triphosphate by a general nucleoside diphosphokinase, also at the expense of ATP. Part of the uridine triphosphate is consumed to make **cytidine triphosphate** by a

---

*We know that some carbamoyl phosphate can leak from the mitochondria to the cytosol in the liver, because increased pyrimidine synthesis and resultant elevated concentrations of uracil and orotate occur in children with a deficiency of ornithine transcarbamoylase (p. 591). This is a minor source of carbamoyl phosphate for pyrimidine synthesis in normal individuals.

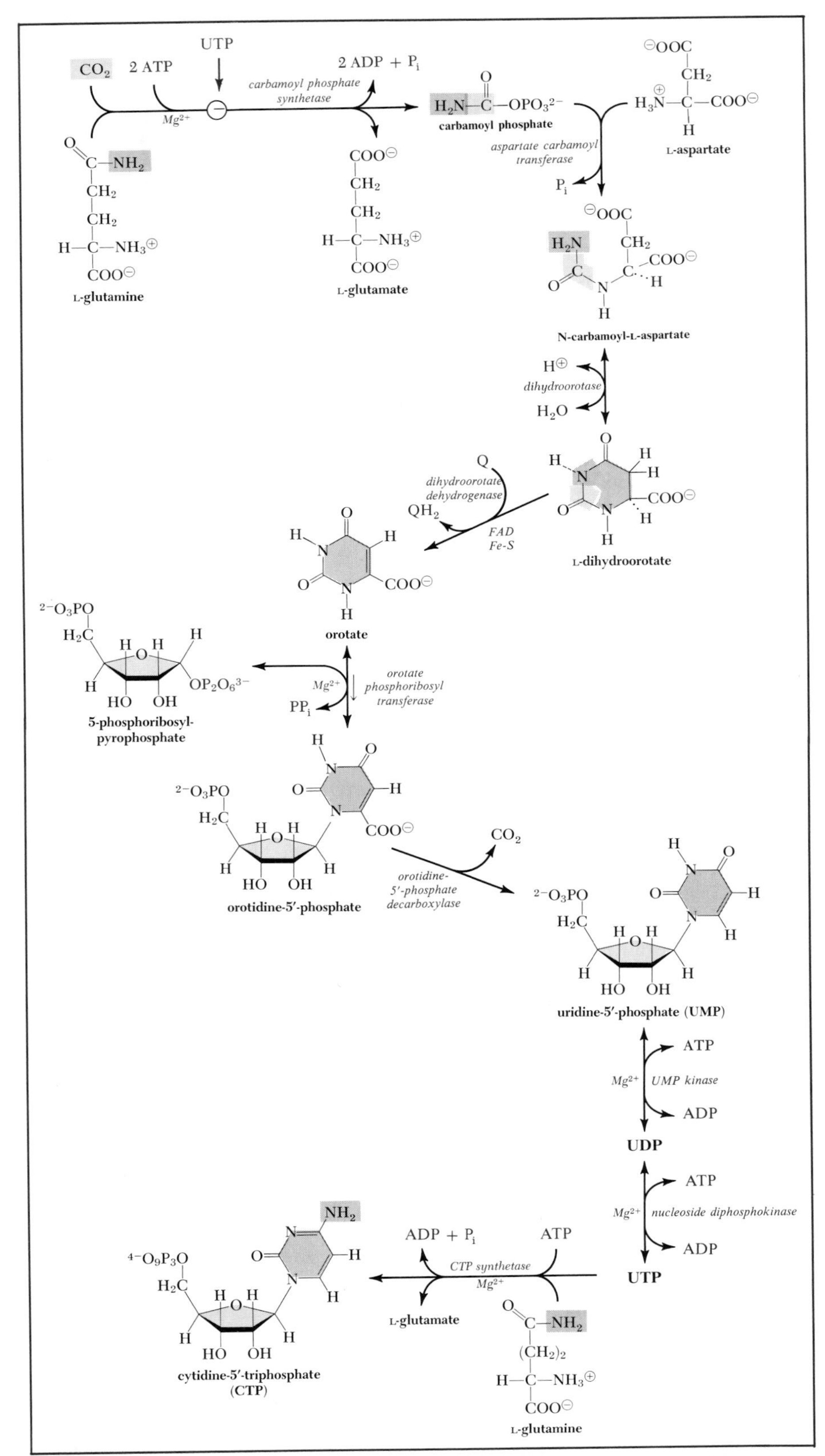

FIGURE 35–8 Synthesis of pyrimidine nucleotides. The first three reactions are catalyzed by separate domains on a single protein. The fifth and sixth reactions are catalyzed by separate domains on still another protein. See the text for details.

transfer of the amide group of glutamine, driven by the cleavage of ATP. This reaction is similar to those using glutamine in the formation of purine nucleotides.

Pyrimidine synthesis has a remarkable feature. Only three proteins are required to catalyze the first six reactions. We discussed the enzymatic reactions as separate entities, but the first three reactions are catalyzed by a single polypeptide chain folded into three separate catalytic domains. This protein occurs in the cytosol. Dihydroorotate dehydrogenase is a separate protein occurring on the external surface of the inner mitochondrial membrane. Another cytosol protein has in one polypeptide chain both the phosphoribosyltransferase and decarboxylase activities necessary for the next two reactions. Presumably there is a kinetic advantage from limiting the distance the intermediates must diffuse in reaching one enzymatic site from another when these sites are physically linked. (There is also evidence that the enzymes catalyzing tetrahydrofolate-dependent reactions in purine synthesis are part of a complex, although not covalently linked in a single polypeptide chain. Association of enzymes catalyzing related or sequential reactions may be more common than we realize.)

FORMATION OF DEOXYRIBONUCLEOTIDES

The deoxyribonucleoside triphosphates required for DNA synthesis and repair are generated by reduction of 2′-carbon atoms in the ribonucleoside diphosphates (Fig. 35–9). The electrons are contributed by NADPH through an uncharacterized route. The ribonucleotide reductase catalyzing the reaction contains two differing subunits. The catalytic subunit contains iron, and probably contains a stable free radical derived from a tyrosyl residue. (Such a free radical has been demonstrated in a prokaryotic enzyme.) The second kind of subunit is a nucleotide-binding subunit, involved in regulation of the enzyme. dADP, dGDP, and dCDP are made directly by ribonucleotide reductase and are converted to the triphosphates for use in DNA synthesis by the same nucleoside diphosphate kinase that equilibrates ribonucleoside diphosphates and triphosphates.

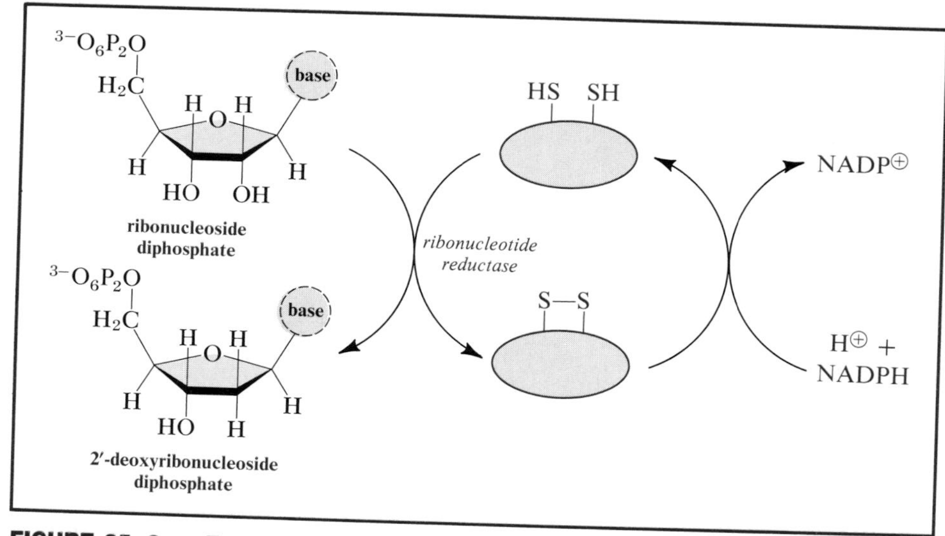

FIGURE 35–9 Formation of deoxyribonucleoside diphosphates by reduction of ribonucleoside diphosphates. Electrons are passed from NADPH through some uncharacterized thiol-containing protein.

The formation of dTTP as the fourth precursor of DNA involves a special route. The thymine moiety is created by an unusual reaction catalyzed by an enzyme named **thymidylate synthase.** In this reaction (Fig. 35–10), the methylene group of methylene-H_4folate is transferred to dUMP. If nothing else happened, the result would be the appearance of a hydroxymethyl group. In order to make a methyl group, there must be a simultaneous reduction, and the reducing agent in the reaction is the tetrahydrofolate that also serves as a one-carbon carrier. Tetrahydrofolate contributes two electrons and is converted to dihydrofolate. Cells producing DNA must therefore use dihydrofolate reductase (p. 621) to regenerate H_4folate.

The other reactions of the route for producing dTTP are designed to minimize formation of dUTP, which would otherwise be incorporated into DNA. The nucleoside monophosphate kinases (p. 366) that interconvert the mono- and diphosphates will equilibrate dCMP and dCDP, but not dUMP and dUDP. The dUMP necessary for dTMP formation is generated instead from dCMP by hydrolytic deamination (Fig. 35–11), and dUDP is made from it only slowly.

Regulation of Deoxynucleotide Synthesis. Ribonucleotide reductase is built with elaborate controls to ensure a balanced formation of the different deoxyribonucleotides. The nucleotide binding subunits appear to have two different sites. One is an activity site that regulates the overall rate. ATP stimulates and dATP inhibits

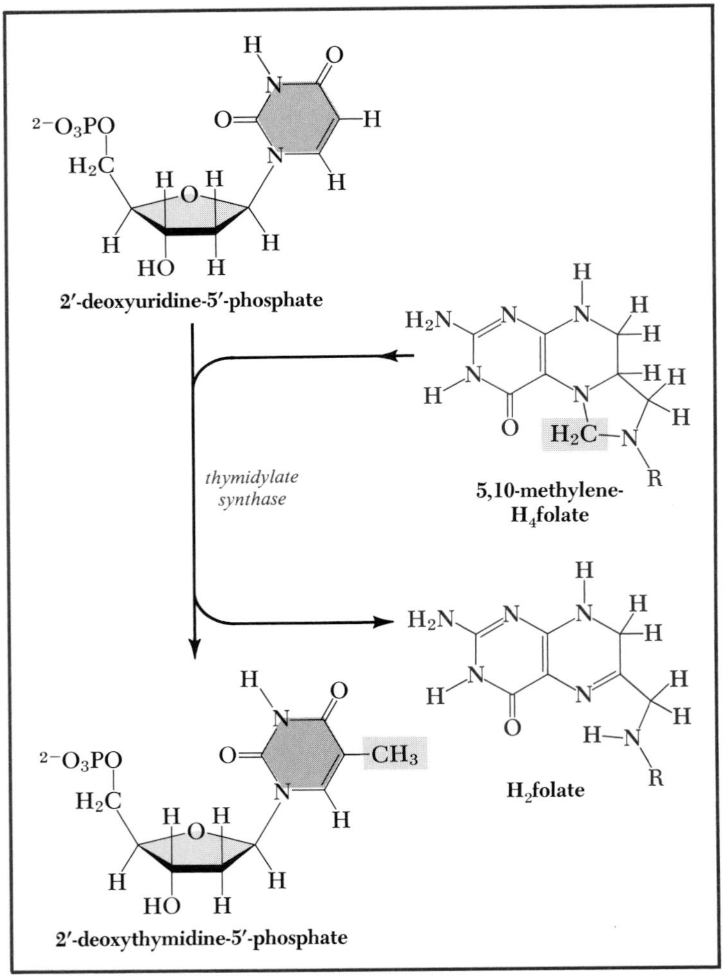

2′-deoxyuridine-5′-phosphate

thymidylate synthase

5,10-methylene-H_4folate

H_2folate

2′-deoxythymidine-5′-phosphate

FIGURE 35–10

dUMP is converted to dTMP by transfer of a methylene group and two electrons from 5,10-methylene-H_4folate.

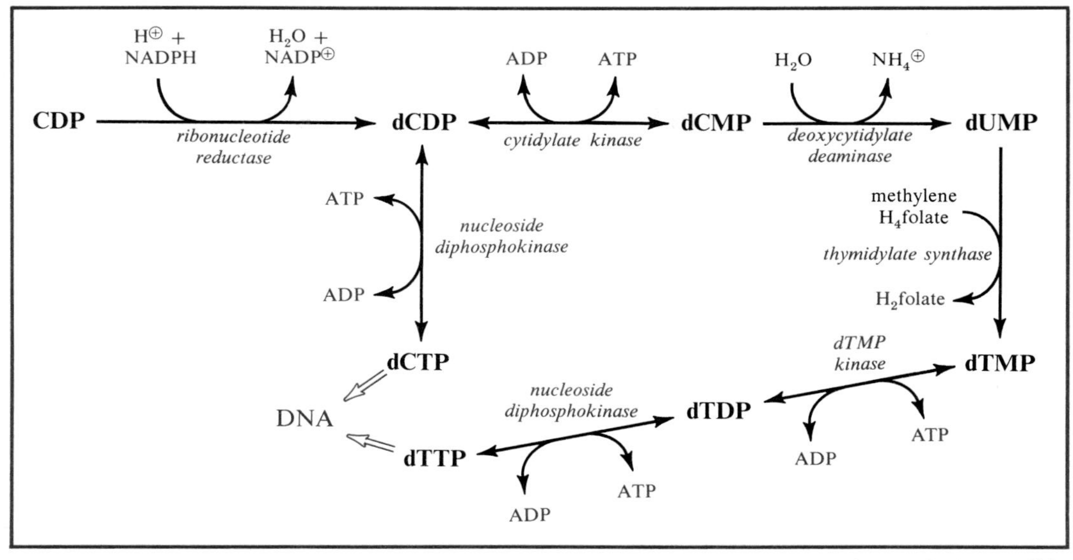

FIGURE 35–11 **The formation of deoxythymidine triphosphate (dTTP) for DNA synthesis. The irreversible conversion of dUMP to dTMP and the inability of dTMP kinase to interconvert dUMP and dUDP insure that there is little formation of dUTP for incorporation in place of dTTP.**

when bound to this site. Binding of nucleotides at the other site determines the specificity of the enzyme in the following manner:

Bound Effector	Activates Formation of	Inhibits Formation Of
ATP	dCDP	
dTTP	dGDP	dCDP
dGTP	dADP	dCDP, dGDP

Regulation of dTTP concentration is an important device that affects both the reductase and deoxycytidylate deaminase. The deaminase has typical sigmoid allosteric kinetics that become hyperbolic Michaelis-Menten kinetics in the presence of dCTP as a positive effector. This activation is not only overcome by dTTP as a negative effector, but micromolar concentrations of this nucleotide make the kinetics even more exaggeratedly sigmoid. The result of all this is to achieve a balanced mixture of the four ingredients of DNA—not too much or too little of any component. Figure 35–12 summarizes the various controls that are involved.

Some genetic defects in ribonucleotide metabolism cause immune deficiency diseases, apparently owing to disturbances in ribonucleotide reductase function. **A deficiency of the general purine nucleoside phosphorylase** causes an accumulation of inosine, deoxyinosine, guanosine, and deoxyguanosine, which are its normal substrates. Only deoxyguanosine can be phosphorylated by a kinase; the other nucleosides are excreted in the urine. The abnormal amount of deoxyguanosine triphosphate inhibits the reduction of CDP and GDP, thus preventing normal cell division in lymphocytes.

A deficiency of adenosine deaminase also prevents normal development of the immune system. Adenosine and deoxyadenosine accumulate and are phosphorylated to regenerate the nucleotides. DeoxyATP will then inhibit ribonucleotide reductase, preventing formation of any of the other DNA precursors. In addition, adenosine causes an accumulation of 5-adenosylhomocysteine, which is a potent inhibitor of the methyltransferases.

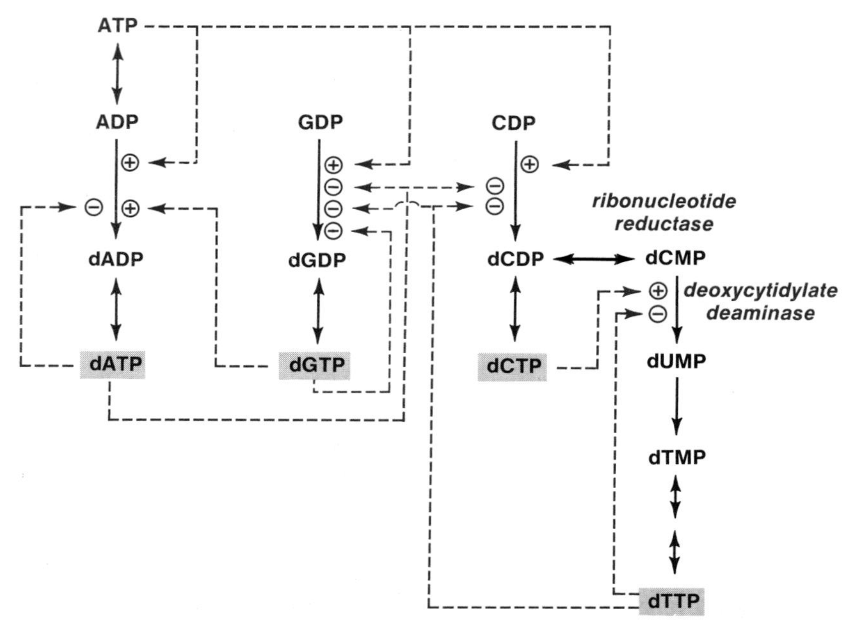

FIGURE 35–12

The precursors of DNA (brown boxes) are allosteric effectors of ribonucleotide reductase and deoxycytidylate deaminase.

CHEMOTHERAPY AND NUCLEOTIDE SYNTHESIS

Earlier in the book we examined drugs that are effective in cancers because they prevent the synthesis or function of nucleic acids in those rapidly dividing cells. Now we are in a position to examine drugs that are effective because they interfere with the formation of precursors of the nucleic acids or associated polyamines.

Methotrexate, a derivative of folate, is a useful drug:

methotrexate
(4-deoxy-4-amino-10-methylfolate)

It is a competitive inhibitor of the reduction of dihydrofolate to tetrahydrofolate, and of folate to dihydrofolate. It can create a general deficiency of H_4folate, which is required twice during the *de novo* synthesis of purine rings. H_4folate is also necessary for the formation of deoxythymidine monophosphate; however, recall that this is a special case, in which tetrahydrofolate is reduced to dihydrofolate in addition to carrying a one-carbon group. Methotrexate is therefore selectively toxic to dividing cells, since these are the only ones rapidly synthesizing deoxynucleotides with a concomitant rapid formation of dihydrofolate. Methotrexate prevents the reconversion of the dihydrofolate to tetrahydrofolate. It affects all rapidly dividing tissues, such as the intestinal mucosa, but it is valuable in the therapy of some malignancies. A strategy sometimes used is to give massive doses—many times the lethal dose—followed by rescue of the normal cells through administration of 5-formyl H_4folate ("folinic acid").

Other successful drugs are analogues of the purine and pyrimidine bases themselves. These include, for example, 6-mercaptopurine and 5-fluorouracil:

SH H N N N N H H

6-mercaptopurine

O H N F O N H H

5-fluorouracil

6-Mercaptopurine is sometimes used in acute leukemias, and it has its effect because it is incorporated into nucleic acids (as thioguanosine), thereby preventing normal protein synthesis, and also because its nucleotides inhibit the formation of phosphoribosylamine and the conversion of inosine monophosphate to adenosine monophosphate.

5-Fluorouracil is converted to deoxyfluorouridine monophosphate, which is a potent inhibitor of thymidylate synthase, and therefore prevents normal DNA synthesis. It is also converted to fluoroUMP and incorporated into RNA. The critical site of therapeutic action is still under study.

The following table lists a number of drugs that are in use, or are actively being investigated for possible use, in the treatment of cancer. The enzymes known to be affected are also listed, but it is necessary to recognize that there may be other effects of the drugs that are more significant.

Drug	Affected Enzyme	Illustration of Reaction
avicin	carbamoyl phosphate synthetase (glutamine hydrolyzing)	p. 690
alanosine	adenylosuccinate synthetase	p. 685
aminoguanosine	purine nucleoside phosphorylase	p. 678
azacytidine	tRNA (cytosine-5)-methyltransferase	p. 629
coformycin	adenosine deaminase	
deazauridine	CTP synthetase	p. 690
dichloroallyllawson	dihydroorotate dehydrogenase	p. 690
difluoromethyl-ornithine	ornithine decarboxylase	p. 657
5-fluorouracil*	thymidylate synthase	p. 692
guanazole	ribonucleotide reductase	p. 691
hydroxyurea*	" "	
methotrexate*	dihydrofolate reductase	p. 621
phosphonoacetyl-L-aspartate	aspartate carbamoyltransferase	p. 690
pyrazofurin	orotidine-5′-phosphate decarboxylase	p. 690
tetrahydrouridine	cytidine deaminase	p. 687
virazole	inosinate dehydrogenase	p. 684

*Drugs regularly used.

Many of these drugs are activated by addition of pentose phosphate or by other modifications. The effects on nucleotide pools are complex, and the expectation is that proper combinations of drugs will be more effective.

FURTHER READING

J.B. Stanbury, J.B. Wyngaarden, and D.S. Frederickson (1978). The Metabolic Basis of Inherited Disease, 4th ed. McGraw. Includes a section on disorders of purine and pyrimidine metabolism in which both normal and abnormal events are discussed.

M.E. Jones (1980). Pyrimidine Nucleotide Biosynthesis in Animals: Genes, Enzymes, and Regulation of UMP Biosynthesis. Annu. Rev. Biochem. 49:253.

J.N. Davidson, P.C. Rimsky, and J. Tamaren (1981). Organization of a Multifunctional Protein in Pyrimidine Biosynthesis. J. Biol. Chem. 256:2220.

L. Thelander and P. Reichard (1979). Reduction of Ribonucleotides. Annu. Rev. Biochem. 48:133.

A. Holmgren (1981). Regulation of Ribonucleotide Reductase. Curr. Top. Cell Regul. 19:47.

S.H. Snyder et al. (1981). Adenosine Receptors and Behavioral Actions of Methylxanthines. Proc. Natl. Acad. Sci. (U.S.) 78:3260.

V. Batuman et al. (1981). The Role of Lead in Gout Nephropathy. N. Engl. J. Med. 304:520.

D.W. Martin, Jr. and E.W. Gelfand (1981). Biochemistry of Diseases of Immunodevelopment. Annu. Rev. Biochem. 50:845. Discusses defects in purine metabolism.

G.R. Boss and J.E. Seegmiller (1979). Hyperuricemia and Gout. N. Engl. J. Med. 300:1459.

T.W. Kensler and D.A. Cooney (1981). Chemotherapeutic Inhibitors of the Enzymes of the *de novo* Pyrimidine Pathway. Adv. Pharm. Chem. 18:274.

CHAPTER 36

METABOLISM OF STRUCTURAL COMPONENTS

This chapter summarizes the synthesis and breakdown of proteoglycans, glycoproteins, phospholipids, and glycolipids without attempting to examine the many individual reactions in detail. Some generalities can be stated. All of the carbohydrate constituents in these compounds are supplied in the form of nucleotide sugars, similar to the UDP-glucose from which glycogen is made. The hydrophobic lipid chains are obtained from long-chain acyl coenzyme A compounds. Indeed, the starting materials for making these structural components are all found among the familiar intermediates of carbohydrate, fat, and amino acid metabolism.

PROTEOGLYCANS

The proteoglycans consist of long heteropolysaccharide chains laid down on a specific polypeptide core (p. 193).

Carbohydrate Precursors

All of the carbohydrate precursors of proteoglycans are UDP derivatives made from UDP-glucose. They include **hexosamines, uronic acids, galactose,** and **xylose.**

Use of Epimerization. Some of the proteoglycan precursors are made by inverting the configuration on one carbon atom of a glucose derivative, for example:

$$\text{UDP-glucose} \longrightarrow \text{UDP-galactose}$$
$$\text{UDP-N-acetylglucosamine} \longrightarrow \text{UDP-N-acetylgalactosamine}$$

These particular reactions involve epimerization at C4; each of the enzymes is really an oxidoreductase, using NAD as a cofactor. That is, it oxidizes the hydroxyl group on one carbon atom of the sugar residue to a carbonyl group and then reduces it back to an alcohol, but the reduction randomly forms either configuration:

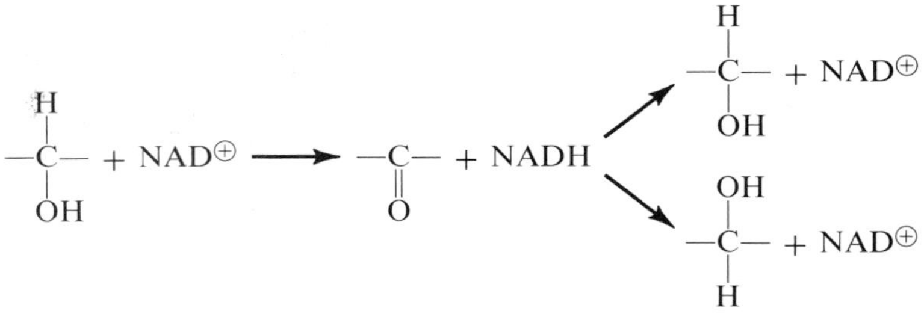

$$\mathrm{H{-}\overset{|}{C}(OH){-}} + \mathrm{NAD^{\oplus}} \longrightarrow \mathrm{{-}C({=}O){-}} + \mathrm{NADH} \longrightarrow \begin{cases} \mathrm{H{-}\overset{|}{C}{-}OH} + \mathrm{NAD^{\oplus}} \\ \mathrm{HO{-}\overset{|}{C}{-}H} + \mathrm{NAD^{\oplus}} \end{cases}$$

The intermediate carbonyl compound and NADH are not released. (Contrast ribulose 5-phosphate epimerase, p. 542, which does not require NAD.)

Glucosamine 6-phosphate is the progenitor of all of the hexosamines. The amino group is obtained by transfer from glutamine. Acetyl coenzyme A contributes an acetyl group:

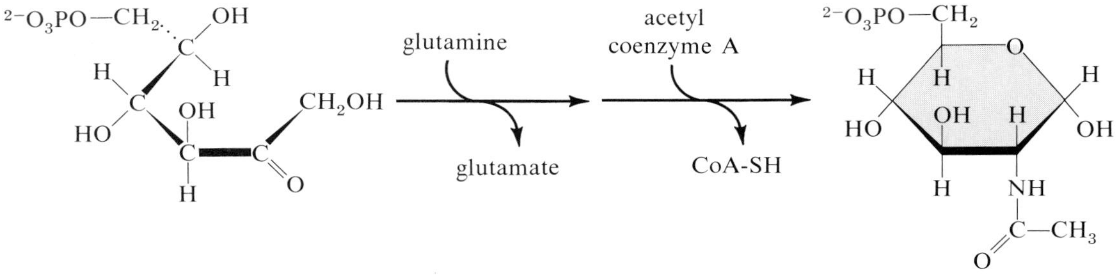

$$\text{D-fructose 6-phosphate} \xrightarrow[\text{glutamate}]{\text{glutamine}} \xrightarrow[\text{CoA-SH}]{\text{acetyl coenzyme A}} \text{N-acetyl-D-glucosamine 6-phosphate}$$

D-fructose 6-phosphate **N-acetyl-D-glucosamine 6-phosphate**

UDP-N-acetylglucosamine is then made by a series of reactions analogous to the steps in glycogen synthesis:

glycogen:

glucose 6-P ⟶ glucose 1-P ⟶ UDP-glucose ⟶ glycogen

heteropolysaccharide:

GlcNAc 6-P ⟶ GlcNAc 1-P ⟶ UDP-GlcNAc ⟶ polysacch.

The equilibration of UDP-N-acetylglucosamine and the corresponding galactosamine derivative by an epimerase supplies the hexosamines for proteoglycan synthesis:

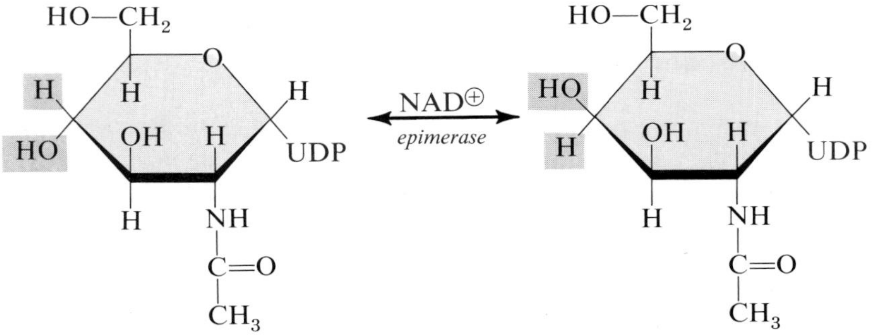

$$\text{UDP-N-acetyl-D-glucosamine} \underset{\textit{epimerase}}{\overset{\mathrm{NAD^{\oplus}}}{\rightleftarrows}} \text{UDP-N-acetyl-D-galactosamine}$$

UDP-N-acetyl-D-glucosamine ⇓ *used as GlcNAc donor*

UDP-N-acetyl-D-galactosamine ⇓ *used as GalNAc donor*

UDP-glucuronate is the parent of uronic acid residues. It is made by two successive oxidations of C6 on UDP-glucose:

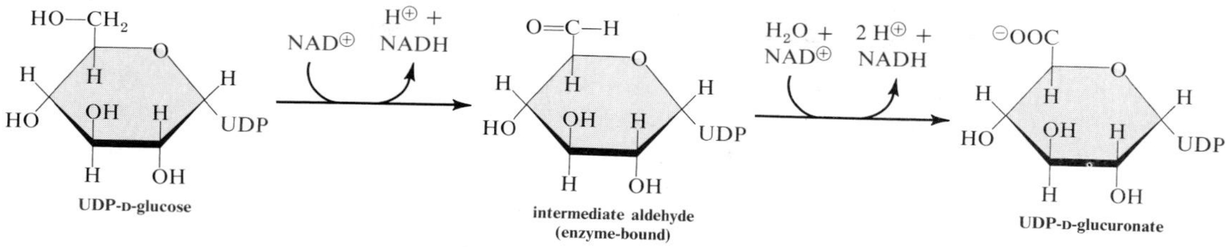

$NAD^{\oplus}$ → $H^{\oplus}$ + NADH; H_2O + $NAD^{\oplus}$ → $2\,H^{\oplus}$ + NADH

UDP-D-glucose → intermediate aldehyde (enzyme-bound) → UDP-D-glucuronate

(The intermediate aldehyde is not released.)

A small amount of UDP-glucuronate is decarboxylated to form **UDP-xylose:**

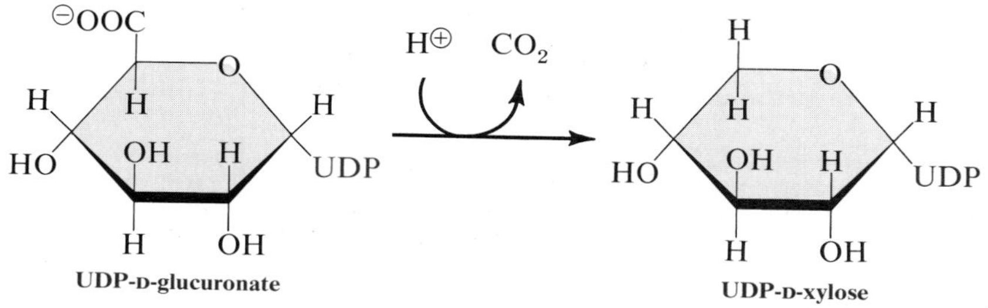

$H^{\oplus}$ → CO_2

UDP-D-glucuronate → UDP-D-xylose

Sulfate Groups

Some of the proteoglycans contain sulfate ester groups (pp. 195, 196). These groups are obtained by transfer from 3′-phosphoadenosine 5′-phosphosulfate:

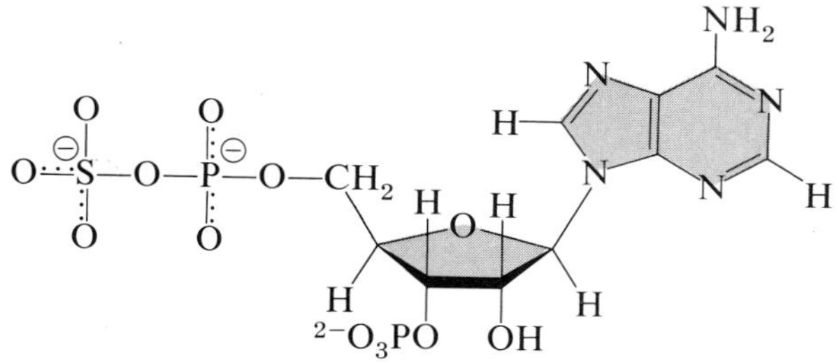

3′-phosphoadenosine-5′-phosphosulfate

This compound is a general source of sulfate groups; it is made from inorganic sulfate and ATP in two steps. First, the high-energy phosphosulfate bond is made by cleaving ATP with sulfate:

$$\text{sulfate} + \text{ATP} \longleftrightarrow \text{adenosine 5}'\text{-phosphosulfate} + PP_i$$

ATP is then used to phosphorylate the 3′ hydroxyl group of the product. (Compare coenzyme A, p. 423.) The free energy of hydrolysis of the phosphosulfate bond is greater than that of a pyrophosphate bond; the second phosphorylation ensures that the reaction can go even if the pyrophosphate concentration is significant.

When sulfate is transferred by enzymes to specific acceptor groups, adenosine 3′,5′-bisphosphate remains. A phosphatase removes the 3′ phosphate group, leaving 5′-AMP to be reused in the adenine nucleotide pool.

Synthesis of Proteoglycans

The heteropolysaccharide chains attached to the protein cores in proteoglycans (p. 193) are built by transferring one sugar at a time to a growing chain from the corresponding UDP derivatives, much in the way that amylose chains are built during glycogen synthesis (p. 504). Unlike glycogen, the proteoglycans are destined for export from the cell, so the transferases catalyzing chain formation are bound to the endoplasmic reticulum and the Golgi apparatus. Most of the chains are attached to serine or threonine side chains through a -Gal-Gal-Xyl-Ser(Thr) sequence. Chain formation begins by transfer of xylose from UDP-xylose to particular serine (or threonine) residues. Different transfers then add successive galactose residues obtained from UDP-galactose. These early steps may begin on the rough endoplasmic reticulum before translation of mRNA into a core polypeptide chain is completed. Additional transferases then build up the remainder of the heteropolysaccharide chain by alternate addition of the repeating residues. For example, chondroitin chains will be constructed from UDP-glucuronate and UDP-N-acetylgalactosamine, first one and then the other repeatedly being used as a donor of a modified sugar residue.

The transferases catalyzing these reactions all require Mn(II) for activity. They also invert the configuration at C1 upon transfer. The UDP-sugar precursors all retain the α-configuration, but the sugar residues in the heteropolysaccharides have the β-configuration. (This is not true of all transferases making the oligosaccharides of glycoproteins discussed below.)

Those cells making dermatan sulfate, heparan sulfate, or heparin modify some of the glucuronate residues in the growing polysaccharide chain to form iduronate residues. This is done by an epimerization at C5:

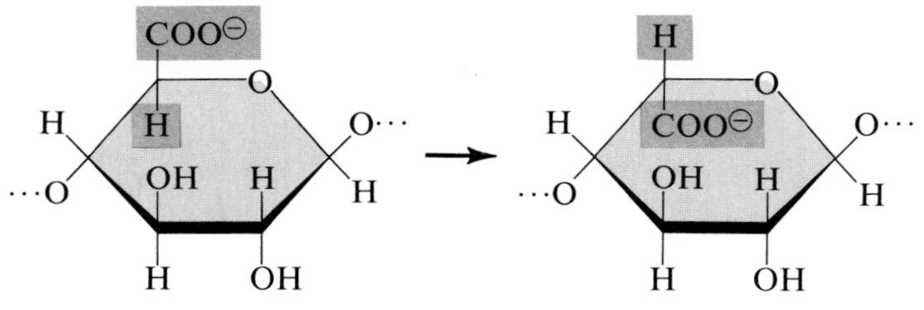

D-glucuronyl group → L-iduronyl group

(It is not known if this epimerase uses NAD.) Addition of sulfate and epimerization to form iduronate residues occur at the same time; it is believed that addition of sulfate makes the epimerization irreversible.

GLYCOPROTEINS

Additional Precursors

The oligosaccharide chains of glycoproteins contain mannose, fucose, and acetylneuraminate residues in addition to the carbohydrate residues found in proteoglycans. Some asparagine-linked oligosaccharides are mainly polymers of mannose. Fucose and acetylneuraminate are used as caps for more heterogeneous oligosaccharide chains.

Mannose residues are first made as mannose 6-phosphate from fructose 6-phosphate. During fuel metabolism, fructose 6-phosphate equilibrates with glucose 6-phosphate by an isomerase-catalyzed shift of the carbonyl group between C2 and C1. Cells synthesizing glycoproteins have another isomerase that catalyzes the same reaction, but forming mannose 6-phosphate:

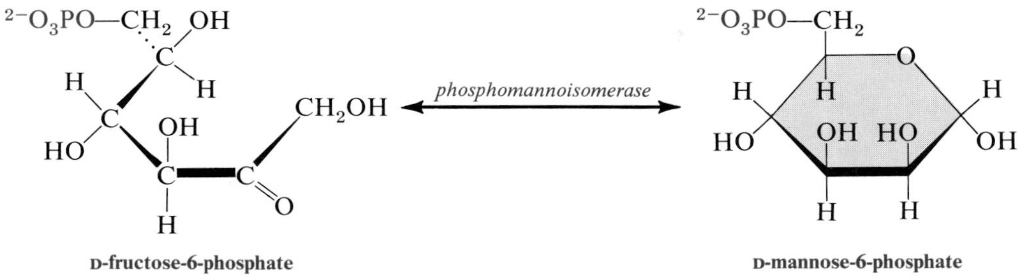

phosphomannoisomerase

D-fructose-6-phosphate ⟷ **D-mannose-6-phosphate**

GDP-mannose is made as a mannose donor, much in the way that UDP-glucose is made as a glucose donor:

$$\text{fructose 6-P} \longrightarrow \text{glucose 6-P} \longrightarrow \text{glucose 1-P} \longrightarrow \text{UDP-glucose}$$
$$\text{fructose 6-P} \longrightarrow \text{mannose 6-P} \longrightarrow \text{mannose 1-P} \longrightarrow \text{GDP-mannose.}$$

Why is it GDP-mannose rather than UDP-mannose? Probably to provide an additional element of specificity for the enzymes so that glucose or mannose is never mistakenly used in place of the other sugar.

Fucose is also transferred from the GDP derivative, which is made from GDP mannose by a series of reactions involving oxidation at C5 and reductions at C6 and C5:

3. reduce to OH and invert configuration on C6
1. oxidize to carbonyl
2. reduce to CH_3

GDP-D-mannose —NAD⊕→ —NADPH→ **GDP-L-fucose**

Man donor

Fuc donor

Acetylneuraminate is made from phospho-*enol*-pyruvate and UDP-N-acetyl-glucosamine (Fig. 36–1). An unusual enzyme strips the UDP group and simultaneously inverts the configuration on C2 to form N-acetylmannosamine:

$$\text{UDP-N-acetylglucosamine} \longrightarrow \text{UDP} + \text{N-acetylmannosamine}$$

A series of reactions follows, culminating in the formation of the cytidine monophos-

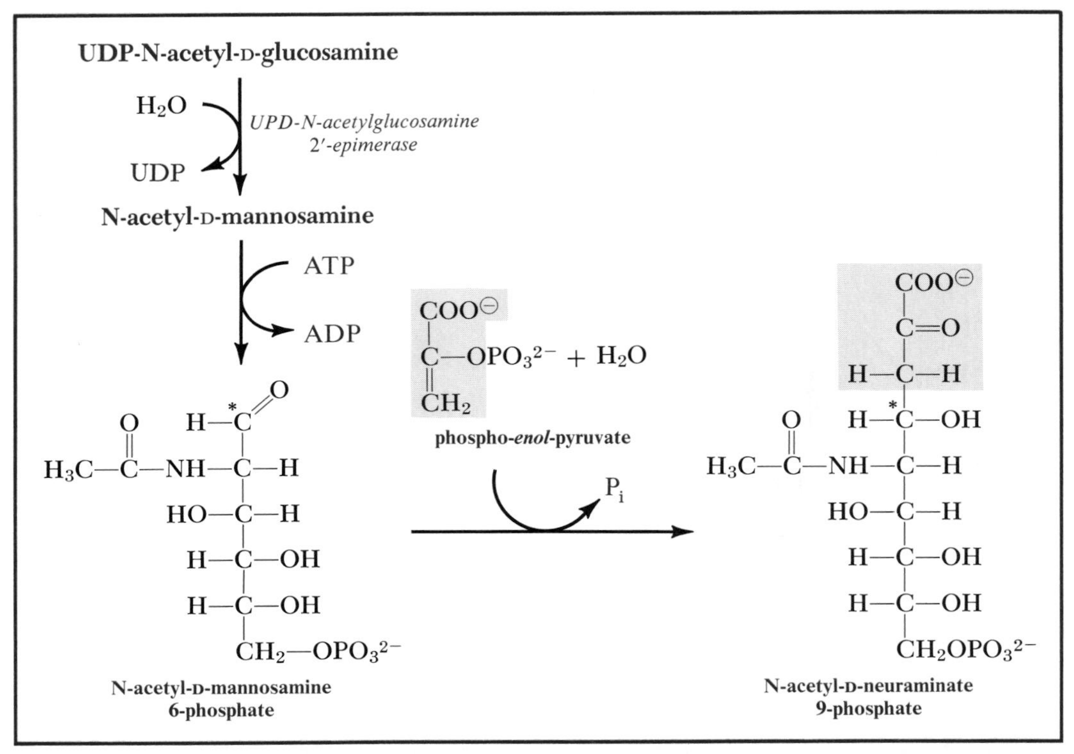

FIGURE 36–1 **Formation of neuraminic acid residues as the phosphate ester of the N-acetyl derivative.**

phate derivative of N-acetylneuraminate as the donor for N-acetylneuraminate groups:

N-acetyl-D-neuraminate 9-phosphate —(H_2O → P_i)→ N-acetyl-D-neuraminate —(CTP → PP_i)→ CMP-N-acetyl-D-neuraminate

cytosine

CMP-N-acetyl-D-neuraminate

The compound is shown in open chain form to illustrate the origin of the carbon atoms and in its more stable ring form. The use of CMP as the combined nucleotide certainly eliminates any chance of enzymatic confusion with the other oligosaccharide components.

The individual reactions are:

N-acetylmannosamine + ATP ⟶ N-acetylmannosamine 6-phosphate + ADP
N-acetylmannosamine 6-phosphate + phospho-*enol*-pyruvate + H_2O ⟶
N-acetylneuraminate 9-phosphate + P_i
N-acetylneuraminate 9-phosphate + H_2O ⟶ N-acetylneuraminate + P_i
N-acetylneuraminate + CTP ⟶ CMP-N-acetylneuraminate + PP_i

Note that the final reaction is a variation on the usual theme, with the nucleoside monophosphate group being transferred directly to a sugar hydroxyl group to create a CMP-sugar, rather than to a phosphate group on the sugar to create a CDP-sugar.

Synthesis of Glycoproteins

Getting highly polar carbohydrate polymers through a membrane is a problem. It is not certain exactly how this is done with proteoglycans or with glycoproteins in which complex oligosaccharides are attached to serine or threonine hydroxyl groups. However, the problem is solved for highly branched oligosaccharides destined for attachment to asparagine residues in a glycoprotein by constructing them on a very hydrophobic molecule as a base.

For example, dolichol is a long-chain polyprenyl alcohol, with 80 to 100 carbon atoms:

CH_2—OH

a dolichol (C_{80}-C_{100})

It is easy to visualize the long chain firmly anchored in the lipid core of the endoplasmic reticulum while polar carbohydrate residues are added one at a time to the hydroxyl group near the cytosol surface. According to this picture, the polypeptide chain of the glycoprotein is synthesized, as usual, with an amino terminal signal sequence that penetrates the endoplasmic reticulum. In the meantime, appropriate oligosaccharides are constructed on dolichol in the membrane. When asparagine residues at proper locations pass by, the oligosaccharide is transferred as a unit from dolichol to polypeptide. (The asparagine must be in the sequence Asn-X-Thr, where X is a residue of any amino acid except aspartate.) After transfer, and during passage of the nascent glycoprotein through the lumen of the endoplasmic reticulum and in vesicles to its destination (lysosomes for some hydrolases and the Golgi apparatus for secreted proteins), further processing of the oligosaccharide occurs.

As an example, Figure 36–2 shows how a mannose-rich oligosaccharide is made.

The exposed hydroxyl group of dolichol is first phosphorylated by CTP. (Why CTP rather than ATP? No rationalization is yet at hand.) An N-acetylglucosamine 1-phosphate group (P-GlcNAc) is transferred to the dolichol phosphate to form dolichol diphosphate N-acetylglucosamine (Dol-PP-GlcNAc). Another transferase then adds a second N-acetylglucosamine group without accompanying phosphate. A mannose residue is added from GDP-mannose. This completes the linking groups upon which oligosaccharide synthesis now begins in earnest.

Eight mannose residues are added to build a branched structure. The mannose donor in each case is dolichol phosphate mannose. The use of this intermediate presumably permits the polar mannose to be brought into the endoplasmic reticulum membrane for transfer. Finally, three glucose residues are added. These are supplied as dolichol phosphate glucose (made from UDP-glucose).

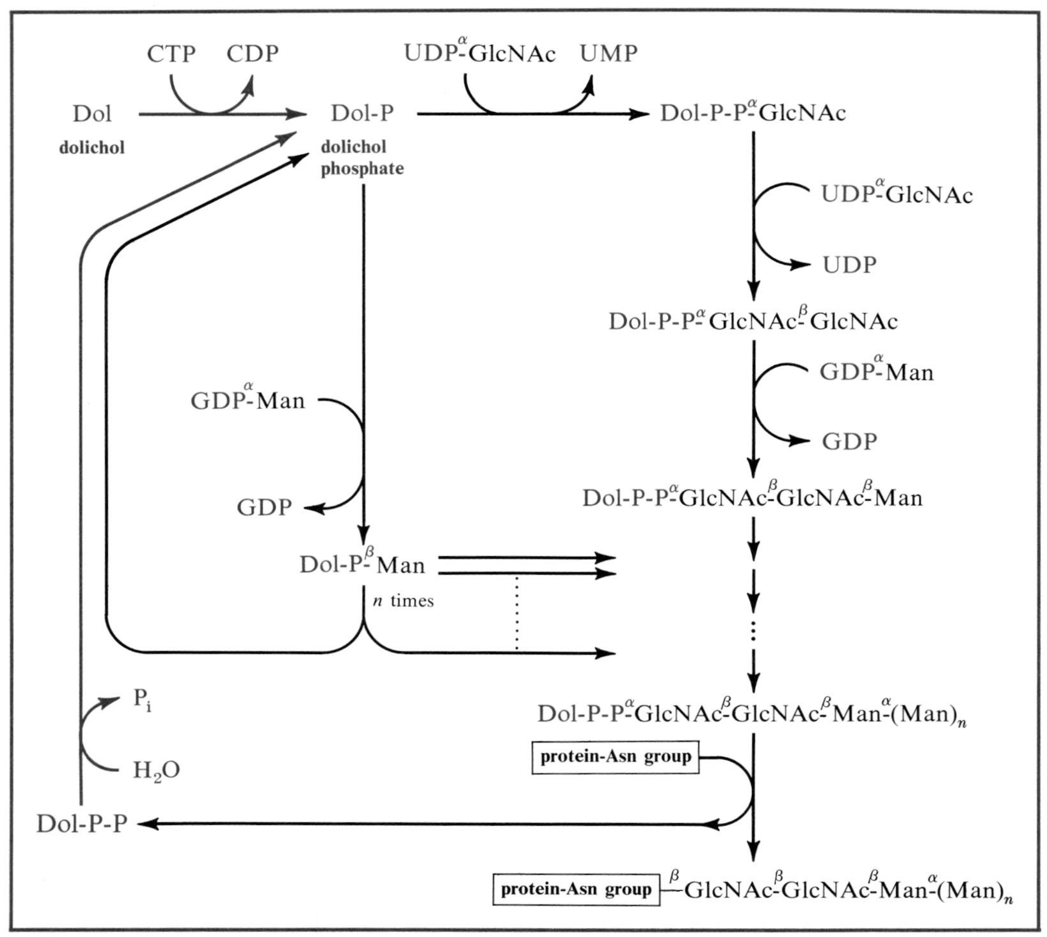

FIGURE 36–2 **Construction of heterooligosaccharide chains on dolichol for transfer to asparagine residues in glycoproteins. Note the differing initial transfers from UDP-GlcNAc. A phosphate group is transferred with N-acetylglucosamine in one (*top*), giving UMP as the product. Only the hexosamine residue is transferred in the second (*upper right*), giving UDP as a product.**

After completion of the oligosaccharide, it is transferred to the appropriate polypeptide chain, probably as it emerges into the lumen of the endoplasmic reticulum. Shortly thereafter, the three glucose residues are removed by hydrolysis. (Perhaps they served some protective or identifying purpose, or perhaps their removal simply serves to prevent reversal of oligosaccharide transfer.)

Phosphorylmannose groups are markers for proteins destined for incorporation into lysosomes. At some stage, mannose residues are phosphorylated in these proteins so that they will be recognized by receptors on lysosomes. Proteins destined for secretion may have some mannose groups removed and other carbohydrate residues added in the endoplasmic reticulum or the Golgi apparatus.

Less is known about the synthesis of the more complex oligosaccharides bound through serine or threonine hydroxyl groups. They appear to be made on the polypeptide chain by transfer from nucleotide sugars, in much the same way as are the heteropolysaccharides of proteoglycans.

PHOSPHOLIPIDS

Degradation

Phospholipids are attacked by hydrolases, both in the small intestine during digestion, and as part of the normal turnover within cells. Phospholipases are classified according to the bond attacked:

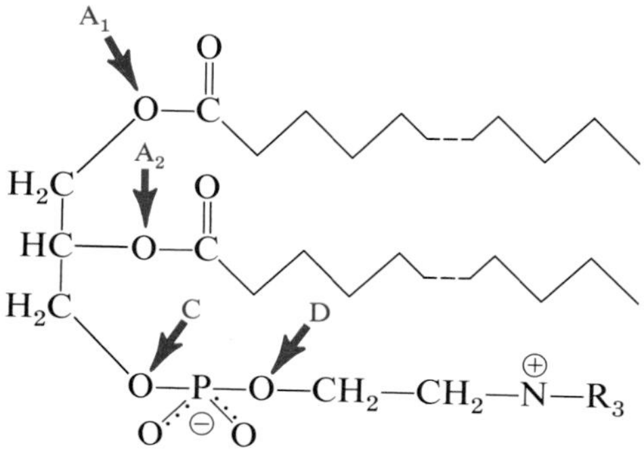

phosphatidyl base

In addition to those indicated, there are type B phospholipases, which are lysophospholipases that hydrolyze the single acyl group from lysophosphatidyl compounds.

The best known of these hydrolases in mammalian tissues are the type A and type B enzymes. The pancreas secretes an A_2 phospholipase as a proenzyme, activated by trypsin, and a lysophospholipase. When these enzymes act, for example, on a dietary phosphatidyl choline, they release the two molecules of fatty acid and a glycerylphosphocholine. The intestinal mucosa also contains a hydrolase that splits glycerylphosphocholine into glycerol-3-phosphate and free choline. The result is cleavage of the dietary phospholipids into their components. Similar hydrolases act on the phospholipids within cells, making their constituents available for reuse.

Synthesis of Phosphatidyl Bases

We already discussed the complete synthesis of the phosphatidyl bases, beginning with free choline or ethanolamine (p. 631). They are converted to the cytidine diphosphate derivatives, which act as phospho-base donors for phospholipid synthesis. That is, the phosphorylated base is transferred to a 1,2-diglyceride, creating a phosphatidyl base and leaving CMP as the other product.

However, this transfer does not complete the synthesis for all purposes. For example, membrane phosphatidylcholines are rich in arachidonoyl groups (20:4) on C2. We have already seen that the major diglycerides have mainly 18:1 on C2 (p. 208). The same diglycerides are incorporated into phospholipids, and something else must happen to change the phospholipid fatty acid composition before they are incorporated into membranes.

The intracellular phospholipase A_2 removes the 18:1 acyl group from newly synthesized phosphatidyl cholines. Another enzyme is present that selectively transfers a 20:4 group from arachidonoyl coenzyme A, creating a phosphatidyl choline of proper composition:

phosphatidylcholine (*newly synthesized*) [H_2C—O—palmitoyl; HC—O—oleoyl; H_2C—O—P-choline] + H_2O —*phospholipase A_2*→ oleate + lysophosphatidylcholine [H_2C—O—palmitoyl; HC—OH; H_2C—O—P-choline]; + arachidonoyl coenzyme A —*arachidonoyl transferase*→ CoA—SH + phosphatidylcholine (*modified for membranes*) [H_2C—O—palmitoyl; HC—O—arachidonoyl; H_2C—O—P-choline]

Cardiolipins, the diphosphatidylglycerols, are made in mitochondria by a sequence involving the formation of CDP-diglycerides (Fig. 36-3). This compound is also a phosphatidyl donor for the synthesis of **phosphatidyl inositol.**

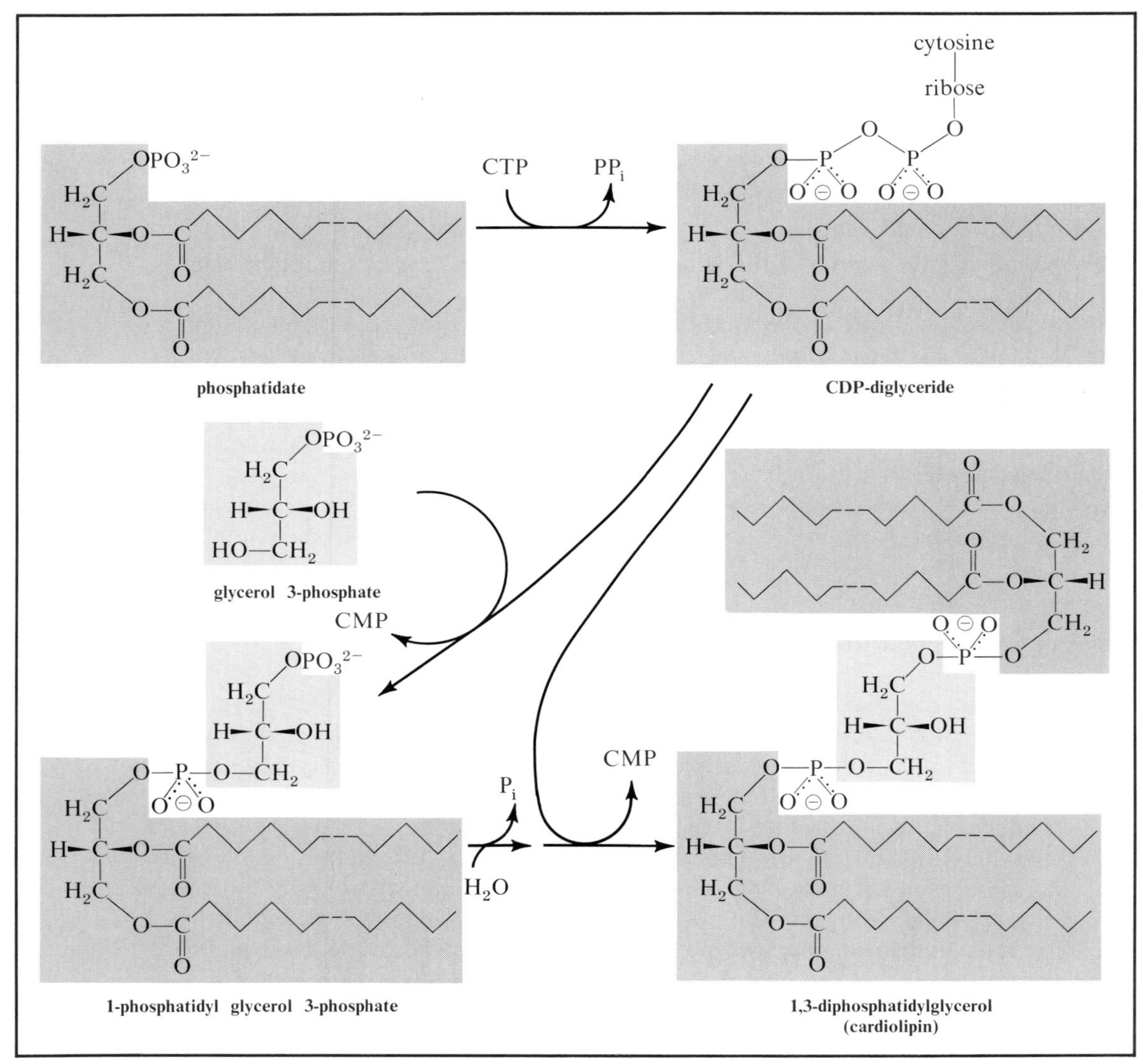

FIGURE 36–3 Conversion of phosphatidate groups to CDP-diglycerides for transfer to a glycerol backbone, generating mono- or diphosphatidylglycerols.

Phosphatidylinositols are made by a combination of CDP-diglycerides with inositol. This cyclic hexitol is made from glucose 6-phosphate by an intramolecular oxidation-reduction, followed by removal of the phosphate group:

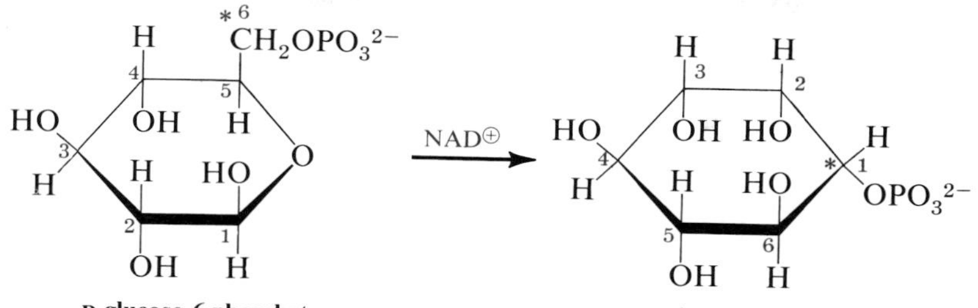

D-glucose 6-phosphate $\xrightarrow{NAD^{\oplus}}$ *myo*-inositol 1-phosphate

D-glucose 6-phosphate **_myo_-inositol 1-phosphate**

Phosphatidylinositols are also subject to exchange of the acyl groups. For example, the predominant compound in platelets has a 1-stearoyl-2-arachidonoyl composition, and is believed to be the source of arachidonate for formation of prostaglandins (next chapter). In general, phosphatidylinositols in membranes turn over especially rapidly upon external stimulation of cells, but the nature of their special role in membrane function is not clear.

Sphingomyelins

Sphingomyelins are ceramidylphosphocholines; we already noted their structural resemblance to phosphatidylcholines (p. 213). The **ceramide** group is a long-chain amine to which a fatty acyl group is attached. The long-chain amine is a **sphinganine**—either sphinganine itself, **sphingenine,** or some modified form of the compounds. **The sphinganine bases are made by condensing palmitoyl coenzyme A with serine,** with a simultaneous decarboxylation (Fig. 36–4). An acyl group, typically 18:0 or 24:0, is added to make a ceramide. (The 24:0 residue may be modified by a hydroxyl group at C2, or by a central double bond.) A sphingomyelin is then made by transferring phosphorylcholine from CDP choline to the terminal hydroxyl group of the ceramide.

Glycosphingolipids

If sugar residues rather than a phosphorylcholine residue are placed on a ceramide, the result is a glycosphingolipid. The same nucleotide sugars that are precursors of proteoglycans and glycoproteins are also the donors of carbohydrate residues in glycolipids. Indeed, some of the oligosaccharide groups are identical in cell surface glycoproteins and glycolipids. The oligosaccharides in glycolipids have cores made completely from UDP-glucose, UDP-galactose, UDP-N-acetylgalactosamine, and UDP-N-acetylglucosamine, with some cap and side groups contributed by GDP-L-fucose and CMP-N-acetylneuraminate. Sulfate groups are also added from phosphoadenosinephosphosulfate to make some glycosphingolipids. Not only the order in which the groups is arranged, but also the anomeric nature of the linking bonds (α or β) can be varied, so a wide variety of glycosyl transferases must be present in the body. Each cell creates its own assortment to make the glycolipids necessary for its function, and the enzymes apparently occur as an aggregate on the endoplasmic reticulum. The oligosaccharide is built one unit at a time directly on the ceramide; dolichol is not used in glycolipid synthesis.

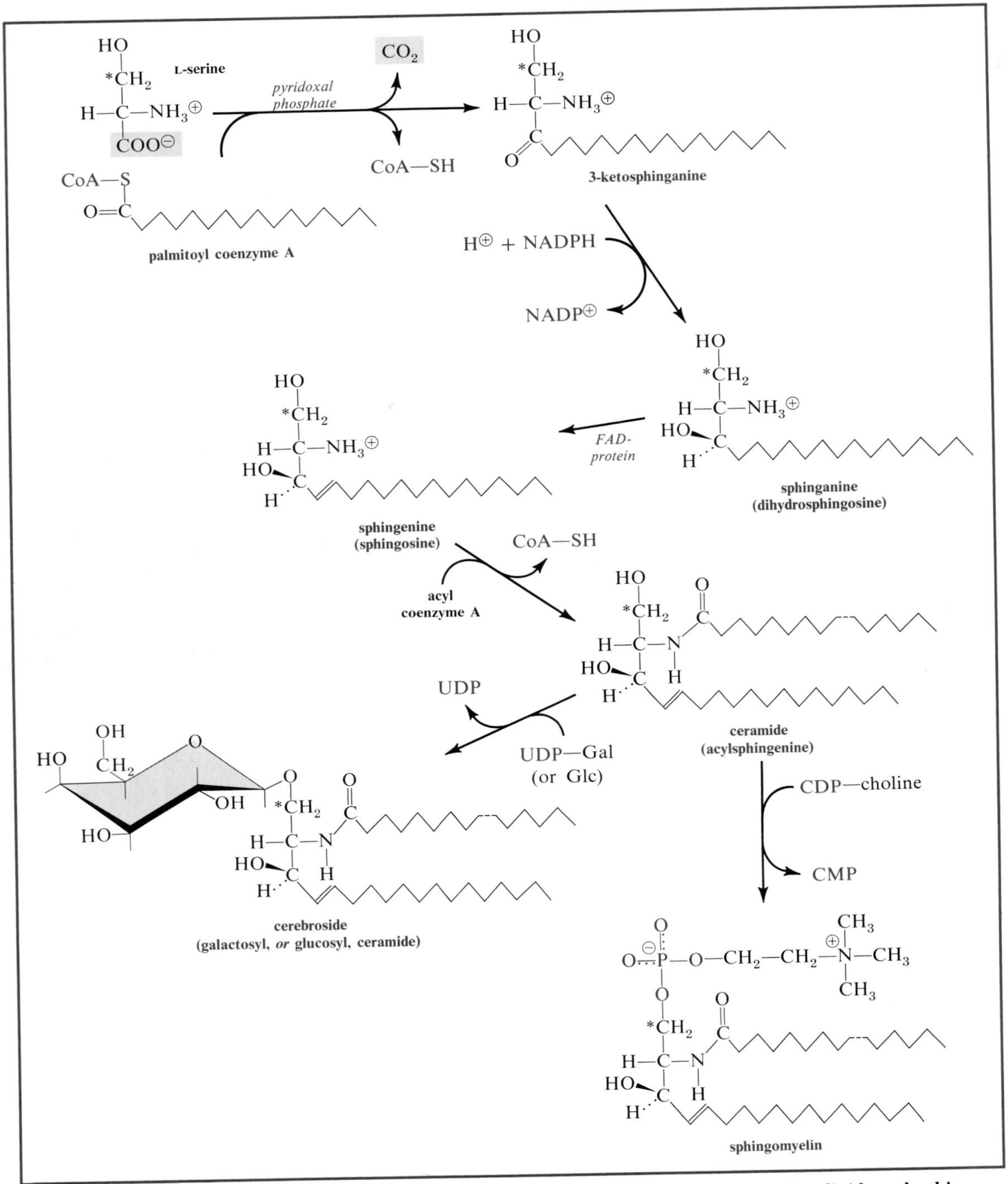

L-serine

palmitoyl coenzyme A

pyridoxal phosphate

CO_2

CoA—SH

3-ketosphinganine

$H^{\oplus}$ + NADPH

$NADP^{\oplus}$

sphinganine (dihydrosphingosine)

FAD-protein

sphingenine (sphingosine)

CoA—SH

acyl coenzyme A

ceramide (acylsphingenine)

UDP

UDP—Gal (or Glc)

cerebroside (galactosyl, *or* glucosyl, ceramide)

CDP—choline

CMP

sphingomyelin

FIGURE 36–4 **The formation of sphinganine and sphingenine for use in synthesizing glycolipids and sphingomyelins. Sphingenine is used more often in synthesizing ceramides, as shown, but sphinganine also is used.**

The completed oligosaccharides occur in five major types, with core structures as follows:

Gala	Galα1→4Galβ1→1Cer
Ganglio	Galβ1→3GalNAcβ1→4Galβ1→4Glcβ1→1Cer
Globo	GalNAcβ1→3Galα1→4Galβ1→4Glcβ1→1Cer
Lacto	Galβ1→4GlcNAcβ1→3Galβ1→4Glcβ1→1Cer
Muco	Galβ1→3Galβ1→4Glcβ1→1Cer

In addition, the ganglio and lacto series are decorated with N-acetylneuraminyl or fucosyl groups. For example, typical gangliosides have this structure:

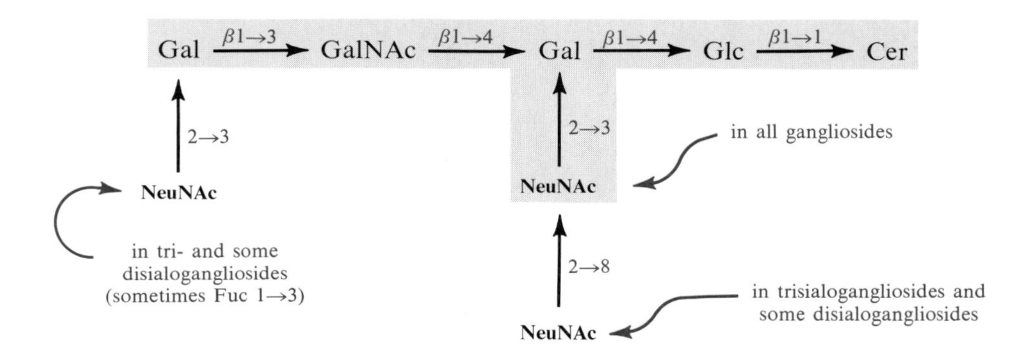

Note that either a β-glucosyl or β-galactosyl group is first placed on the ceramide. This forms a simple cerebroside (p. 214). Indeed, all but the Gala series have a lactosyl group (Galβ1→4Glc) as the first disaccharide unit, like a lactocerebroside. The series differ in the other groups; some residues are linked 1→3, others 1→4. Galactose may be combined in an α or β configuration.

LYSOSOMAL DEGRADATION AND GENETIC DEFECTS

Just as it requires a large battery of enzymes to synthesize the diverse saccharides in proteoglycans, glycoproteins, and glycolipids, it requires an almost equally large battery of hydrolases to degrade them in the lysosomes. Mutations may create a deficiency in any one of these hydrolases, and defects in more than 20 of the enzymes have been found. In each case, the lysosome becomes engorged with the substrate for the defective enzyme, usually a partially degraded polysaccharide or glycolipid. Cell function is compromised, and the accumulating material begins to appear in the urine.

As with other genetic defects, the symptoms depend upon the type of cell that is damaged, and the exact nature of the defect in the enzyme. Different mutations affecting the same gene will cause varying losses of activity and disturbances in specificity with corresponding variation in clinical severity. Some of the enzymes are affected frequently enough to justify alertness by pediatricians; most are quite rare, although the nature of the accumulating compounds has given us much insight into glycolipid structure. All of the conditions we shall discuss are autosomal recessive unless otherwise specified.

Mucopolysaccharidoses

The many defects that occur in the degradation of proteoglycans have some effects in common. They tend to cause coarse facial features. (The descriptive term gargoylism was abandoned in deference to the sensibilities of the affected families.) The liver and spleen become enlarged through engorgement with the accumulating debris. Skeletal changes and limited motion of the joints are common. Many of the conditions cause the cornea of the eye to become opaque. The mucopolysaccharides appear in the urine.

Hurler syndrome is caused by a **defect in α-L-iduronidase.** It therefore prevents the normal degradation of dermatan and heparan sulfates (p. 197). The gene frequency is estimated at 0.003 in the United States, with an estimated 30 new homozygotes born each year. It appears in infancy with a steady mental and physical deterioration, usually causing death by the tenth year. **Scheie syndrome** is an allelic variant with less serious impairment, making it compatible with normal intelligence and survival, although corneal opacities appear. Its gene frequency is only 0.0014, but this is enough to make the compound **Hurler-Scheie syndrome,** which is devastating, occur as often as homozygous Hurler's syndrome (0.003^2 vs. $2 \times 0.003 \times 0.0014$).

Much more rare mucopolysaccharidoses are Sanfilippo syndrome, Morquio syndrome, and Maroteaux-Lamy syndrome.*

Sanfilippo Syndrome. Deficiency of **heparan sulfatase** or **α-N-acetylglucosaminidase** will cause the same clinical picture. The patient is unable to degrade heparan sulfate and has progressive severe mental retardation with only mild somatic defects and no corneal clouding.

Morquio syndrome is an uncharacterized deficiency causing accumulation of **keratan sulfate.** Signs are severe knock-knees, short trunk and neck, loose joints, and broad mouth. A characteristic underdevelopment of the second cervical vertebra enables accidental partial dislocation of the spine (subluxation), causing quadriplegia. There is no demonstrated impairment of intelligence. Severely affected patients usually die in young adulthood from cardiopulmonary disease; some live much longer.

Maroteaux-Lamy syndrome is a deficiency of **N-acetylgalactosamine-4-sulfatase.** It resembles Hurler syndrome in somatic changes, but with retention of normal intelligence. Death comes before age 30 in the severe form. Dermatan sulfate, and perhaps chondroitin sulfate, accumulate.

A comparison of these conditions suggests some inferences concerning function of specific proteoglycans. Heparan sulfate is evidently more important than any of the others to the brain (Hurler and Sanfilippo syndromes). Dermatan sulfate is necessary for normal development of the skeletal system and the cornea; joint movement is restricted when it is not metabolized. Keratan sulfate, prominent in cartilage, must also be removed for normal skeletal development, but joints are loose, and there is only mild clouding of the cornea when it is not.

Sphingolipidoses

Genetic defects in the lysosomal degradation of sphingolipids also have varied effects, depending upon the affected compounds and the nature of the defect. The bonds normally split by the enzymes deficient in some of the conditions are summa-

---

*It is not always easy to find data for incidence of genetic diseases. Although not an infallible rule, if the discussion is long and the data absent, the condition is rare.

rized in Figure 36–5. The anionic sulfatides, gangliosides, and lactosides are mainly confined to the brain, so defects in the hydrolysis of these compounds impair mental development. The neutral glycolipids and sphingomyelin are more widely distributed, and genetic defects in their removal may impair development of the nervous system, other tissues, or both. Enlargement of the liver and spleen occurs in most of the conditions from the accumulated products.

Many of these conditions are much more prevalent among the Ashkenazic Jews than in other populations. Since their separation from Sephardic Jews occurred less than two millenia ago, it is not clear why this is so.

Gaucher's disease, for example, is the most common sphingolipidosis throughout the world, but even it is 30 times more prevalent among the Ashkenazic Jews, with one case in 2500 births. It is a defect in a **β-glucosidase acting on glucocerebrosides.** In the past, it has usually been first detected in the second decade, with episodes of pain in the extremities and back, sometimes with high fever. The lungs are involved, and bone destruction occurs. Mental impairment occurs only in a rare

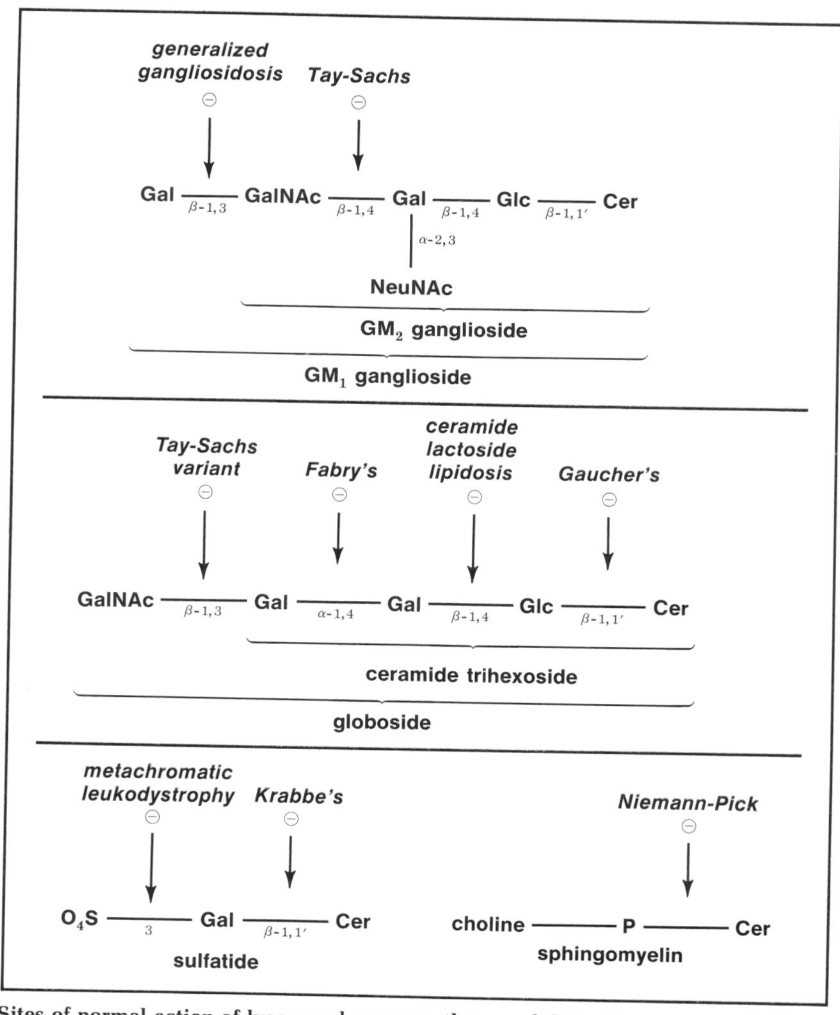

FIGURE 36–5

Sites of normal action of lysosomal enzymes that are deficient in some genetic diseases. The names of the compounds accumulating as a result of the deficiencies are shown below the structures.

infantile form, because most of the glucocerebroside probably arises from turnover of the platelets and leukocytes in the blood, with less from other cells. (As is true with many genetic enzyme deficiencies, some glucosidase activity remains, and it is sufficient to destroy all but a small fraction of the daily turnover of glucocerebroside. That is why it usually takes so long for symptoms to appear.)

Tay-Sachs disease is a deficiency in an **N-acetylgalactosaminidase isozyme.** This enzyme contains $\alpha\beta$ subunit dimers; it is known as the "A" isozyme. The "B" isozyme contains only β subunits, whereas an "S" isozyme contains only α subunits and is of less physiological importance. It is the α subunit that is defective in Tay-Sachs disease, leading to an accumulation of GM2 ganglioside (Fig. 36–5). Estimates of the gene frequency in Ashkenazic Jews range around 0.02. A screening program in the United States found over 100 mated heterozygotes out of 125,000 individuals examined. Homozygous infants have doll-like faces, enlarged heads, and seizures, and develop early blindness. Over 90 per cent have a characteristic cherry-red spot in the retina. Symptoms usually appear between three and six months of age, with death occurring between two and five years.

Sandhoff disease is a similar condition in which formation of the β subunit is defective, so neither the A nor the B isozyme of the acetylgalactosaminidase is active; it progresses more rapidly and presumably results in an accumulation of mucopolysaccharide as well as glycolipid products.

Niemann-Pick disease is a deficiency of **sphingomyelinase,** which hydrolyzes the ceramide-phosphorylcholine bond. Several types are known. Severe sphingomyelinase deficiency causes a general loss of motor and intellectual function, with death by the third year usual. Roughly half of the afflicted infants have the cherry-red retinal spot also seen in Tay-Sachs disease.

Fabry's disease is known in a few hundred adult males. This X-linked condition is caused by a deficiency in an **α-galactosidase.** The α-galactoside linkage occurs in globoside and galaside oligosaccharides, so there is little impairment of mental function in the condition. However, beginning in childhood or early adolescence, it is accompanied by episodes of severe burning pain referred to the palms and soles and continuing for minutes to days. Corneal opacities develop, and cardiac and renal dysfunction usually causes premature death (41 years is the average lifespan.) The earliest symptom is the development of clusters of telangiectasias in the skin of the lower trunk and thighs.

Even rarer genetic defects are summarized in the following paragraphs.

Farber's lipogranulomatosis is a deficiency in **ceramidase,** which hydrolyzes the fatty acyl-sphingenine bond. Symptoms appear between one and 16 weeks of age, with vomiting, edema, hoarseness, respiratory difficulty, arthritis, and painful movement. Those surviving into adolescence still have normal intelligence.

Krabbe's disease is a deficiency in **galactocerebroside β-galactosidase.** Incidence may be 1:50,000 in Sweden, but it is rare elsewhere. Galactocerebrosides are almost absent outside of the brain and the kidney, so the symptoms mainly involve progressive retardation of development after three to six months of age with death by the third year.

Metachromatic leukodystrophy is a deficiency of **cerebroside sulfatase,** hydrolyzing sulfate ester bonds. (Sulfate bonds in proteoglycans and steroids are hydrolyzed by distinct enzymes.) The name arises from the characteristic staining reaction in degenerating white matter of the brain. A late infantile form, in which the infant is first normal and then slowly fades away to an incoordinate vegetative state, occurs once in 40,000 births. A juvenile form, developing later, is only one fourth as common.

Generalized gangliosidosis is caused by a deficiency in a **β-galactosidase,** which removes galactosyl groups from both gangliosides and keratin. It therefore causes both mental retardation and skeletal deformities, as well as the cherry-red retinal spot.

Fucosidosis, mannosidosis, and **sialidosis** all have characteristics in common with the mucopolysaccharidoses. The respective deficient enzymes are **α-fucosidase, α-mannosidase,** and **N-acetylneuraminidase.** Fucosidosis also results in the same telangiectasias seen in Fabry's disease, and sialidosis causes the appearance of the cherry-red retinal spot.

FURTHER READING

E.C. Heath, S.A. Brinkley, and T.H. Haugen (1981). The Biosynthesis and Turnover of Glycoproteins. p. 331 in P.J. Randle, D.F. Steiner, and W.J. Whelan, eds. Carbohydrate Metabolism and Its Disorders. Academic Press.

W.J. Lennarz, ed. (1980). The Biochemistry of Glycoproteins and Proteoglycans. Plenum.

J.W. Callahan and J.A. Lowden, eds. (1981). Lysosomes and Lysosomal Storage Diseases. Raven.

E.F. Walborg, ed. (1978). Glycoproteins and Glycolipids in Disease Processes. Am. Chem. Soc.

T. Maeda and F. Eisenberg, Jr. (1980). Purification, Structure, and Catalytic Properties of L-*myo*-Inositol-1-phosphate Synthase from Rat Testis. J. Biol. Chem. 255:8458.

S.M. Prescott and P.W. Majerus (1981). The Fatty Acid Composition of Phosphatidylinositol from Thrombin-Stimulated Human Platelets. J. Biol. Chem. 256:579.

U. Lindahl and M. Hook (1978). Glycosaminoglycans and Their Binding to Biological Macromolecules. Annu. Rev. Biochem. 47:385.

S.C. Hubbard and R.J. Ivatt (1981). Synthesis and Processing of Asparagine-Linked Oligosaccharides. Annu. Rev. Biochem. 50:555.

S-i. Hakomori (1981). Glycosphingolipids in Cellular Interaction, Differentiation, and Oncogenesis. Annu. Rev. Biochem. 50:733.

R.M. Bell and R.A. Coleman (1980). Enzymes of Glycerolipid Synthesis in Eukaryotes. Annu. Rev. Biochem. 49:459.

R.O. Brady (1980). The Sphingolipidoses. p. 523 in P.K. Bondy and L.E. Rosenberg, eds. Metabolic Control and Disease, 8th ed. Saunders.

W.S. Sly (1980). The Mucopolysaccharidoses. p. 545 in Bondy and Rosenberg, ibid.

R.N. Rosenberg (1981). Biochemical Genetics of Neurologic Disease. N. Engl. J. Med. 305:1181. Includes short summaries of sphingolipidoses and mucopolysaccharidoses.

W. Stoffel (1973). Chemistry and Biochemistry of Sphingosine Bases. Chem. Phys. Lipids 11:318.

CHAPTER 37

THE HORMONES

At this point in our study, we have seen many circumstances requiring adjustment of the routes and rates of biochemical reactions. Survival of a multicellular, multiorgan animal hinges upon the organization and coordination of responses in its constituent parts. Rhythmic factors (diurnal, monthly, seasonal) and nonrhythmic factors (stress, work, age, diet) keep us from setting our biological chemostats at fixed points. The cells in diverse organs only communicate through the nervous system and through changes in concentration of circulating compounds. Among these compounds are the hormones.

Hormones are chemical signals released into the circulation with the specific purpose of coordinating and modulating the activities of target cells. The target cells may be present in only a few or in many different organs.

This type of control is called endocrine; it initially appeared distinct from control through the nervous system (physiologically separate but equal). As techniques for measuring the hormones improved, it became apparent that many are also neurotransmitters, that is, they have neurocrine function. Further studies showed some hormones moonlight as locally diffusible products acting differentially on neighboring cells; they have paracrine function.

Our focus will be on endocrine effects. Our purpose in this chapter is to develop the general principles of the action and regulation of endocrine hormones through a survey of specific examples. We want to lay a basis for the more detailed consideration of their physiology and clinical applications that comes later.

GENERAL MECHANISMS OF ACTION

Most, probably all, hormones bind with receptors specific for the particular hormone in target cells. The density of the receptors, as well as the concentration of the hormone in the interstitial fluid, is an important factor in determining the extent to which a particular cell will be influenced by a hormone. The combination of many hormones with their specific receptors causes an automatic **down-regulation,** in which pinocytosis of the receptor or receptor-hormone complex diminishes further response.

Hormones may be divided into two general classes according to the sites and nature of their action on cells:

1. **Some hormones initiate their effects at the plasma membrane itself.** These include the **catecholamines, prostaglandins,** and all of the **peptide hormones** such as glucagon, insulin, and the hormones of the pituitary gland.

2. **Some hormones initiate their action in the nucleus.** They may enter cells by combining with specific transport receptors in the plasma membrane or cytoplasm. These hormones influence the nature and rate of gene expression after reaching the nucleus. They include **triiodothyronine** and all of the **steroid hormones.**

Action at the Plasma Membrane

Activation of adenylate cyclase (p. 513) is a major mechanism for endocrine action; the primary effect is activation of specific protein kinases in response to increased concentrations of 3′,5′-cyclic AMP as a second messenger relaying the information brought by a hormone. Three components have been identified in complexes embedded in the plasma membrane: specific hormone receptors exposed on the outer surface, regulatory subunits in the membrane, and the adenylate cyclase catalytic subunits exposed on the inner surface. Combination of a receptor with its specific hormone apparently causes a conformational shift, enabling access of GTP to the regulatory subunit. It is the combination with GTP that triggers adenylate cyclase activity; the hormone permits the combination to occur. Different specificities to hormones acting in this way are created by varying the nature of the receptor proteins; the regulatory and adenylate cyclase proteins are the same.

Calcium ion is used as a second messenger in some cells. The mechanism is less well-defined, but the combination of hormone and some receptors in some way makes the plasma membrane leaky to calcium ions. They may act as such in affecting intracellular processes or by combination with calmodulin (p. 512). In either case the effects are at least in part due to activation of specific protein kinases.

Other potential second messengers include **3′,5′-cyclic GMP** and **polypeptides** derived from plasma membrane proteins. The role of guanylate cyclase in cell regulation is not clear.

Action on Gene Expression

The number of different kinds of mRNA being made at a given moment by a mammalian cell uses only a small fraction, sometimes less than 3 per cent, of the genetic information stored within the cell.

The fertilized ovum divides into identical or nearly identical cells, which change their character with successive divisions. The cells begin to differentiate, first into precursors of various classes of mature cells, and then through intermediate forms into the individual types present in the completed organism. Each stage in this process, from fertilized ovum to fully differentiated cell, involves the transcription of a different set of genes, all of which must be present in the original fertilized ovum. The presence of some hormones is a necessary part of this selective transcription, both during differentiation and the subsequent life of the differentiated cells.

It is not known how the expression of eukaryotic genes is regulated. In order for a gene to be transcribed, it must be accessible to RNA polymerase. Exposure of DNA involves unfolding of the chromatin and dissociation of histones, and perhaps other proteins. The histones cannot in themselves be the primary regulating proteins because they occur at regular intervals along the DNA fiber and are not associated with specific genes.

Current speculations revolve around an interaction of an effector molecule, that is, a signal, with a regulating protein in such a way as to cause the protein to be bound at a specific site on a chromatin fiber, akin to the mechanisms of operon regulation in prokaryotes (p. 73), but not identical to them. The binding may occur on DNA itself, or to still another protein that is one of the nonhistone components of chromatin. The critical feature is that the binding in some way affects the transcription of a segment of DNA by RNA polymerase. The effector that precipitates the sequence may be any molecule, such as a metabolite, but in many cases it is a hormone.

Those hormones directly affecting gene expression combine with cytoplasmic receptor proteins after entering the cell. (These are not to be confused with receptor sites on the cell surface.) The hormone-protein complex combines with the chromatin so as to change the transcription of selected DNA sequences into mRNA.

ADRENALINE AND NORADRENALINE

We previously discussed the synthesis and catabolism of these catecholamines (p. 649ff.) and their effects on glycogen (p. 513) and fat (p. 551) mobilization. They are both hormones and neurotransmitters. The adrenal medulla, likened to a group of nerve cells without axons, secretes more adrenaline than noradrenaline. (The adjacent adrenal cortex, discussed later, is responsible for stimulating the methylation of noradrenaline to adrenaline.)

Agonists, Antagonists, and Receptor Types. Agonists are compounds combining with a receptor to trigger the normal response. Antagonists are compounds preventing this response. Selective uses of agonists and antagonists of the catecholamines range from alleviation of minor nasal stuffiness to treatment of life-threatening shock or bronchoconstriction. Selectivity of drug action arises from differential reactivity of receptors to particular agonists or antagonists.

Catecholamine receptors have two general types of relative response to the synthetic agonist isoproterenol:

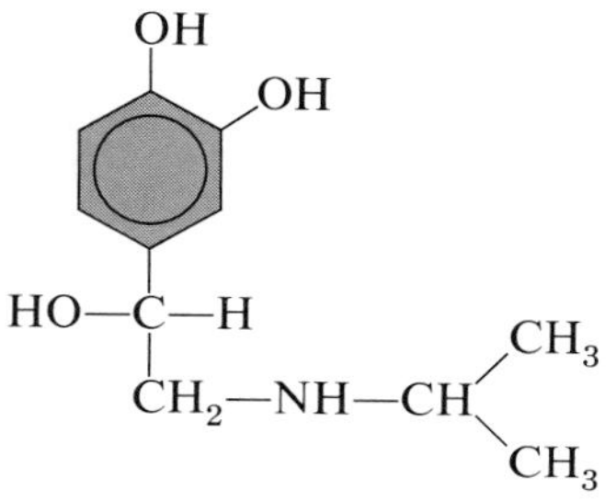

OH, OH (on ring)
HO—C—H
CH_2—NH—CH(CH_3)(CH_3)

isoproterenol

Alpha receptors are much more sensitive to adrenaline than they are to isoproterenol. Noradrenaline is equally or somewhat less active than adrenaline with these receptors. **Beta receptors** are more responsive to isoproterenol than they are to either adrenaline or noradrenaline. They occur in two subtypes, differentiated by their responses to various antagonists. (A second type of alpha receptor also occurs at presynaptic locations on adrenergic nerves, where it inhibits transmission as a negative feedback device. We shall not be concerned further with these receptors.) In general, combination with alpha adrenergic receptors triggers an influx of calcium or some other mechanism not involving cyclic AMP. Combination with beta adrenergic receptors activates adenylate cyclase to produce cAMP as a second messenger. Some of the effects mediated through the different types of receptors are:

Alpha-Adrenergic Responses
(*adrenaline* $\geqslant$ *noradrenaline* $\gg$ *isoproterenol*)
- mobilization of liver glycogen
- increased gluconeogenesis
- vasoconstriction in skin, mesentery, and legs
- mydriasis (dilation of eye pupil)
- relaxation of intestinal smooth muscle

Beta$_1$ Adrenergic Responses
(*isoproterenol* $>$ *noradrenaline* $\geqslant$ *adrenaline*)
- increased mobilization of cardiac glycogen
- increased mobilization of stored fat
- increased thermogenesis in brown adipose tissue
- increased heart rate (chronotropic effect)
- increased contractility of heart (inotropic effect)
- relaxation of coronary arteries

Beta$_2$ Adrenergic Responses
(*isoproterenol > adrenaline ≥ noradrenaline*)
increased mobilization of skeletal muscle glycogen (promotes lactic acidemia)
increased skeletal muscle contraction
increased potassium ion uptake
relaxation of vascular smooth muscle
relaxation of tracheal and bronchial smooth muscles
inhibition of release of anaphylactic mediator

Adrenaline is approximately ten times more potent than noradrenaline in causing metabolic effects. The blood concentration of noradrenaline must rise to 1500 to 2000 ng/l from the normal resting level of 200 (supine) to 500 (standing) ng/l before there are measurable hemodynamic or metabolic changes. This level is exceeded during heavy exercise and in patients with catecholamine-secreting pheochromocytomas; it is barely reached in some upon moderate exercise, after surgery, or with ketoacidosis or myocardial infarction.

The effective blood concentrations of adrenaline are much lower (normal resting levels are 30 to 50 ng/l):

50 to 100 ng/l
Causes increased blood [glycerol], heart rate, systolic blood pressure.
Exceeded during cigarette smoking, mild exercise.

150 to 200 ng/l
Causes increased blood [glucose], [lactate], [β-hydroxybutyrate], diastolic blood pressure.
Exceeded during moderate hypoglycemia (3.3 mM), during and after surgery, with pheochromocytoma.

>400 ng/l
Causes decrease in plasma [insulin].
Reached during severe hypoglycemia (<2.2 mM), ketoacidosis, heavy exercise, and after myocardial infarction.

In one amusing study, the differential effects of strenuous exercise and emotional stress upon blood catecholamines were demonstrated in nine young clinicians (house staff and junior faculty) by making measurements when they were running up and down stairs and when they were speaking to an audience (all concentrations in ng/l):

	Baseline	Exercise	Speaking
noradrenaline	583	1726	919
adrenaline	117	164	262

All of the comparisons between these numbers have a $p < 0.025$ except the baseline *vs.* speaking values for noradrenaline (*t* test for significance). Noradrenaline (from the sympathetic nervous system) was elevated by exercise, but it was the adrenal medulla that betrayed the tension in these outwardly blasé young professionals when they stood up before their peers.*

*Imagine what the adrenaline level must be in professors lecturing to hypercritical first-year medical students.

TRIIODOTHYRONINE AND THYROXINE (T3 and T4)

The action of these hormones and the regulation of their production by the thyroid gland provide examples of many general principles, in addition to being important in their own right. The thyroid regulates metabolic activity and promotes growth and development through the synthesis and release of thyroxine (tetraiodothyronine, T4) and triiodothyronine (T3):

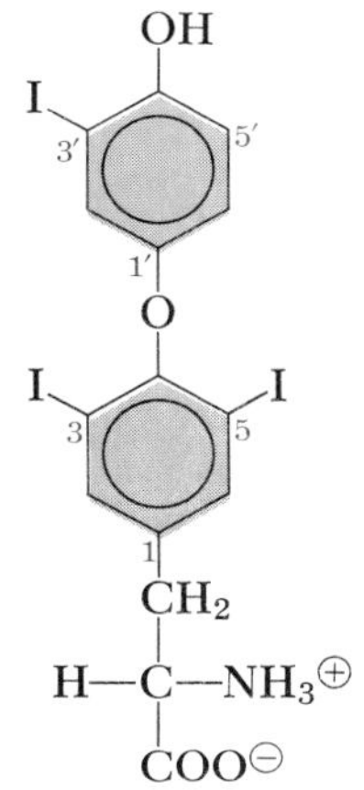

3,3′,5-triiodothyronine (T3)

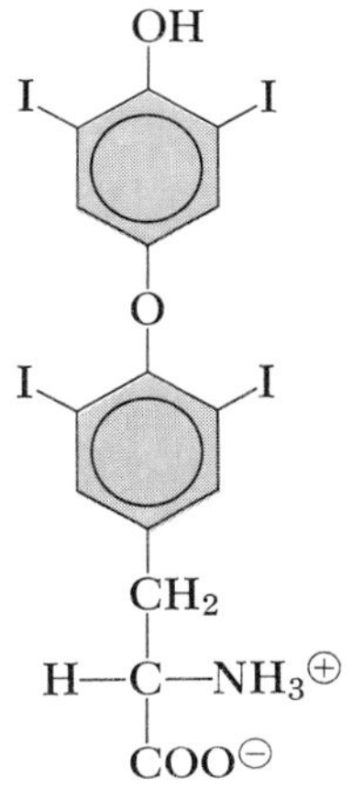

thyroxine (T4, 3,3′,5,5′-tetraiodothyronine)

Thyroxine is the major product of the gland, but triiodothyronine is more active and may be the only form bound to receptor proteins in the nucleus, thereby altering gene expression. We shall see that these hormones are made by iodinating specific tyrosine residues in a protein, **thyroglobulin.**

Specific effects of thyroid hormone include striking changes in gross appearance and activity. Lack of hormone can produce a listless, constipated, coarse-haired slow-pulsed individual who complains of being cold. Oversecretion of thyroxine or triiodothyronine leads to a garrulous, hyperkinetic individual with rapid heart rate, diarrhea, and huge appetite who complains of being hot. While these classic extremes do occur, milder effects are frequent. Indeed, thyroid malfunction is often diagnosed when it is absent and not diagnosed when it is present. A battery of tests is available to prevent these sins of commission and omission, but they sometimes cause their own problems, as we shall touch upon.

The thyroid gland is a bilobed organ in the anterior portion of the neck. It is really a collection of individual glands in the form of follicles, which are circular in cross section with a central lumen in which the newly synthesized thyroid hormone still present in thyroglobulin is stored. Apical portions of the cuboidal cells comprising the follicles contain numerous microvilli and secretory granules on the luminal side.

Synthesis of Thyroxine

Transport of Iodide. The thyroid consumes about 70 to 100 μg of iodide per day for hormone synthesis, which it obtains by re-utilizing the iodide released upon degradation of the hormones, making up any deficit from the dietary intake. The daily intake of iodine in the United States typically ranges from 200 to 500 μg a day. Dietary iodine is reduced to iodide and almost completely absorbed into the blood stream from the intestinal tract. Iodide is actively transported into thyroidal cells through linkage to a $(Na^+ + K^+)$-ATPase system. A thyroid in a normal adult contains *ca.* 6,000 μg of iodide, whereas all of the rest of the body has only about 75 μg of inorganic iodide and 500 μg of organic iodide. Follicular cells are avid collectors of iodide from the blood stream, much more so than any other cells in the body. Indeed, hyperactive thyroid glands can be selectively and therapeutically destroyed by drinking the radioactive isotope ^{131}I, which is concentrated in the thyroid gland and destroys it by emitting gamma rays and electrons.

Thyroglobulin is made by the thyroid follicular cells and serves as a matrix for iodination and condensation of tyrosine residues, and as a reservoir for the still bound triiodothyronine and thyroxine. About 75 per cent of the follicular cell protein is thyroglobulin. It is large (660,000 daltons) and heavily cross-linked by disulfide bonds involving most of the approximately 200 cysteine residues. It is a glycoprotein (p. 201), in which about 280 carbohydrate residues are in oligosaccharides linked to asparagine, serine, and threonine residues. There is some dispute over the number of different subunits translated to form the original molecule. According to one scenario, two identical 330,000-dalton polypeptides are partially degraded as iodination proceeds, with a part of the polypeptide chains having tyrosine residues especially oriented to favor a high yield of iodinated thyronine from iodotyrosine residues. In this view, the hormone-rich segment appears as a 26,000-dalton segment from a cleavage caused by the iodination process. The molecule as a whole is relatively poor in tyrosine residues, and most of the iodine in all but the 26,000-dalton segment is present as scattered iodotyrosine residues.

Maturation of thyroglobulin occurs near the apex of the cell, and the prohormone product is then secreted into the follicular lumen. (The term prohormone is technically correct, since the hormones are released by hydrolysis of polypeptide chains.)

Iodination of the tyrosyl residues in thyroglobulin is a complex process (Fig. 37–1), not yet completely elucidated, which occurs in the apical portion of the cell. The iodinating enzyme is a heme-containing peroxidase, which also travels through

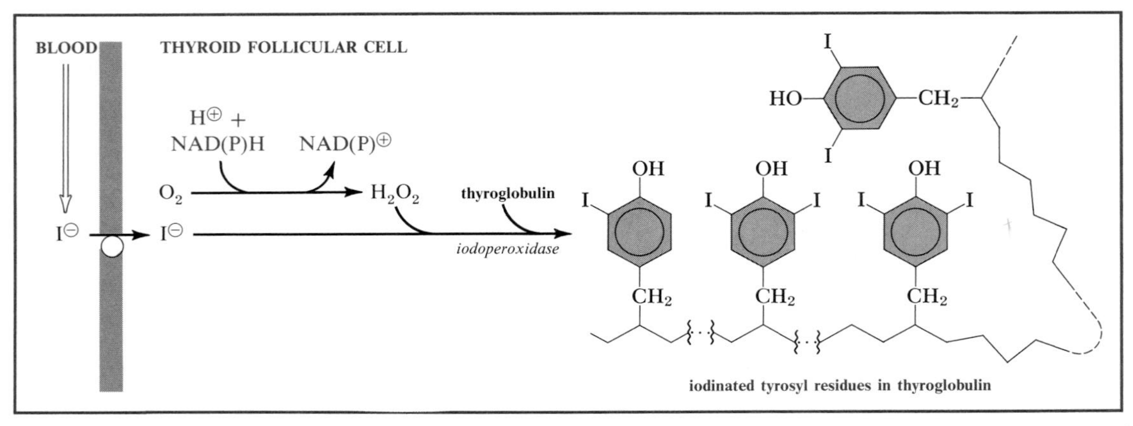

FIGURE 37–1 **Iodide is actively pumped into the thyroid gland. A peroxidase uses it to iodinate tyrosine residues in thyroglobulin. Both mono- and diiodotyrosine residues are formed.**

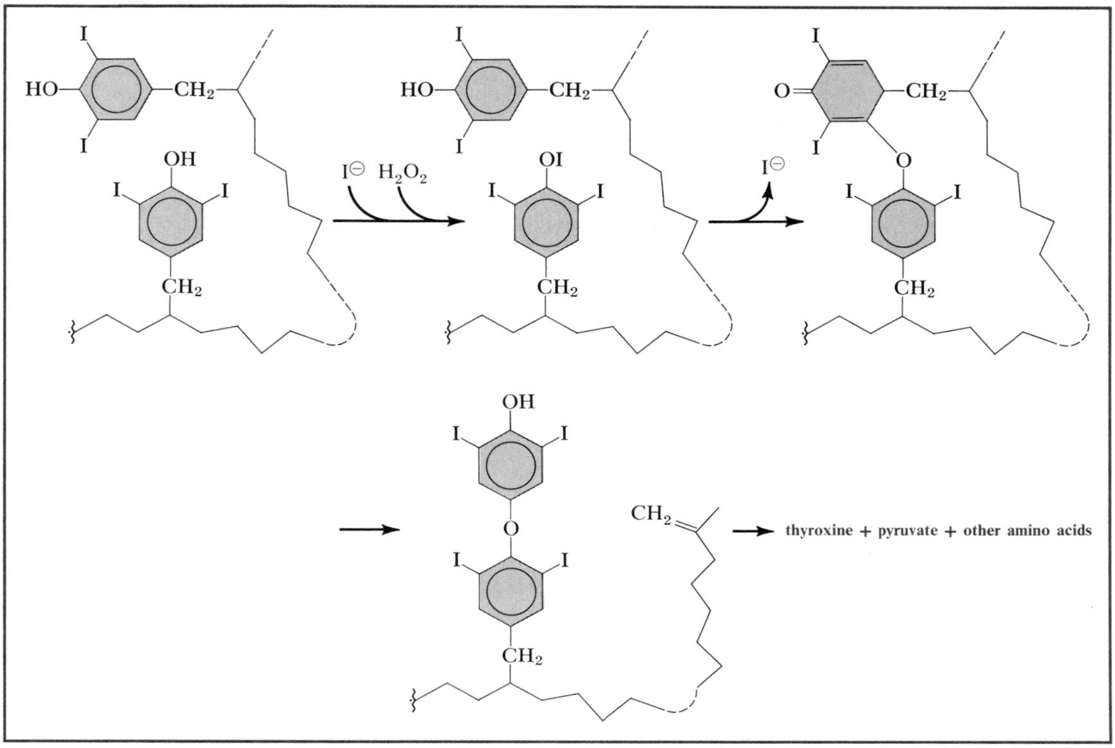

FIGURE 37–2 Possible mechanisms for coupling of two adjacent iodinated tyrosine residues in thyroglobulin. The example shows the formation of a thyroxine residue; triiodothyronine residues are created by a similar condensation of a monoiodotyrosine residue with a diiodotyrosine residue. According to this postulated mechanism, iodine or an iodonium ion also acts as the necessary oxidant for the condensation.

the cell as if it is to be secreted, although it is probably retained in the plasma membrane or other structures at the cell-lumen interface. (This interface has an intricately interdigitated morphology.) The sources of the necessary H_2O_2 also have not been defined; likely possibilities are the reduction of oxygen by NADPH or NADH (p. 418). The mechanism of iodination probably involves a two-electron conversion of iodide to the iodonium cation. A mixture of mono- and diiodotyrosine residues is formed.

The coupling of two molecules of diiodotyrosine to form thyroxine (tetraiodothyronine) may follow the scheme shown in Figure 37–2. Triiodothyronine is generated by coupling mono- and diiodotyrosine residues in the same way. When small amounts of iodine are available, nearly all of it appears in the 26,000-dalton subunit. The efficiency of formation of the iodothyronines increases as the iodine content increases *in vivo,* indicating that those residues are preferentially iodinated that are in appropriate positions for conjunction to form the iodothyronines, of which more than 80 per cent will be the tetraiodo compound (thyroxine) in individuals with an adequate iodine supply. Many proteins can be iodinated *in vitro* by the thyroid peroxidase system, but little thyroxine is formed owing to the lack of a conformation that brings pairs of tyrosine residues into favorable proximity.

Secretion of the Iodothyronines. Secretion is initiated by the return of iodinated thyroglobulin to the cell through fusion of droplets of the lumen contents with lysosomes to form phagosomes, in which the protein is hydrolyzed to its constituent amino acids. The released iodinated residues include both mono- and diiodotyrosine, as well as the coupled tri- and tetraiodothyronines. Iodine is removed from the iodotyrosines and becomes available for re-utilization. The iodothyronines pass through the plasma membrane and basement membrane to enter the blood stream where they circulate almost entirely bound to protein.

Circulating thyroxine and triiodothyronine are bound almost quantitatively to three proteins: **thyroxine-binding globulin,** which is the most important carrier, **thyroxine-binding prealbumin,** and **albumin.** Thyroxine-binding globulin (60,000 daltons, approximately 0.3 to 0.4 micromolar in blood serum) has about a tenfold greater affinity for thyroxine than for triiodothyronine. It is the concentration of the free (unbound) hormones, not the total, that is important. The values regarded as normal by the University of Virginia Hospital laboratories are:

Free		Total	
Thyroxine	**Triiodothyronine**	**Thyroxine**	**Triiodothyronine**
ng/dl 0.68–1.0	*ng/dl* 0.23–0.67	*μg/dl* 4.5–11.5	*μg/dl* 0.08–0.20
pM 9–13	*pM* 3.5–10	*nM* 60–150	*nM* 1.2–3.1

Changes in the concentration or structure of the proteins binding the hormones can change the total hormone concentrations in the blood without concurrent changes in the free hormone levels. For example, the concentration of thyroxine-binding globulin increases with pregnancy, liver disease, or the use of oral contraceptives; it decreases with nephrosis or the use of androgens (male sex hormones).

A new nondisease, **familial dysalbuminemic hyperthyroxinemia,** has been described in which an abnormal albumin appears with a greater affinity for thyroxine relative to triiodothyronine than are the relative affinities with thyroxine-binding globulin. Affected people have no signs of increased thyroid activity but have laboratory results suggestive of hyperthyroidism.

Peripheral Metabolism of Thyroxine. Only 20 per cent or so of the triiodothyronine in the periphery is secreted as such by the thyroid gland. The remainder arises from deiodination of thyroxine, primarily in the liver, kidney, and heart. At least 12 iodinated compounds derived from thyroglobulin can be detected in plasma. Only 30 to 40 per cent of the thyroxine is converted to 3,3′,5-triiodothyronine, the active isomer. About 40 per cent is deiodinated on the inner phenyl ring to produce 3,3′,5′-triodothyronine, so-called reverse T3.

Effects of Thyroid Hormones Upon Cells. Triiodothyronine appears to cross the plasma membrane by passive diffusion. Once in the cell, it is bound within the nucleus by a protein receptor associated with chromatin, where it initiates transcription of some genes and stops transcription of others. Physiological concentrations of triiodothyronine are absolutely necessary for fetal development and growth; in that sense the hormone is anabolic and promotes protein synthesis. Manifestations of increased levels of hormone in adults reflect stimulation of the adrenergic nervous system, especially β-receptors. Some of the adverse effects, such as irritability, hyperkinetic behavior, and restlessness are ameliorated by β-adrenergic antagonists. To the degree that adrenergic stimulation occurs, thyroid hormones are catabolic; they promote degradation of amino acids, loss of muscle mass, and mobilization of fat.

CONTROL OF THYROID ACTIVITY

Regulation by the Adenohypophysis

Some of the endocrine organs are arranged in a pecking order, with a chain of command running from the hypothalamus to the anterior pituitary gland (adenohypophysis) and from there to individual endocrine glands, which relay the orders to the working cells. The thyroid is an example of such a subaltern gland.

Some of the cells in the adenohypophysis secrete a polypeptide hormone, **thyrotropin (thyroid stimulating hormone, TSH),** which reaches the thyroid gland through the blood and stimulates it to release thyroxine and triiodothyronine. The thyrotropin-forming cells of the anterior pituitary gland are in turn stimulated by another hormone, the **thyrotropin releasing hormone,** which is an oligopeptide formed in the **hypothalamus** and is transported to the anterior pituitary gland through a portal circulation in the pituitary stalk. This sequence of cascade activations, hypothalamus to anterior pituitary gland to thyroid gland, is typical of a sequence affecting other endocrine glands, and therefore deserves more detailed attention. Like other cascade mechanisms, it greatly amplifies signals, with one nanogram of hypothalamic hormone causing the release of many times as much thyrotropin, which in turn stimulates the release of much more thyroxine from the thyroid gland.

The pituitary gland is a collection of differentiated cells that act as a message center. Signals reach it from the hypothalamus, cerebrospinal fluid, blood plasma, and nerve terminals. In response to these signals, the involved cells transmit their messages in the form of peptide hormones. Anatomically, the pituitary gland is enclosed in a bony box, the sella turcica, with a stalk connecting the gland to the hypothalamus. It is really two distinct glands. The **posterior pituitary gland,** or **neurohypophysis,** secretes the hormones **vasopressin** and **oxytocin,** which reach the gland for storage in secretory vesicles through the axons of specialized nerves arising in the hypothalamus, where these hormones are made in the cell bodies. We shall say more about the neurohypophyseal hormones in the next chapter.

The anterior pituitary gland synthesizes, as well as secretes, several polypeptide hormones. The controlling messages in the form of hypothalamic hormones reach it through a portal system of capillary vessels draining the median eminence of the hypothalamus and passing the blood through the anterior pituitary gland before it is returned to the heart. All of the hormones of the anterior pituitary gland are polypeptides (Table 37–1), and their secretion is under the control of other factors in addition to the hypothalamic hormones. Now let us consider the steps in this general scheme that directly affect the formation of thyroid hormones.

Thyrotropin releasing hormone is a tripeptide; it is almost certainly made by cleaving a larger precursor because it contains a pyroglutamyl group:

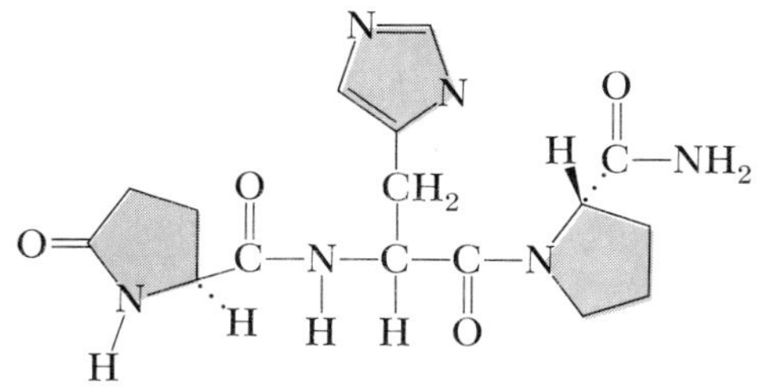

thyrotropin releasing factor
(pyroglutamyl histidyl proline amide)

TABLE 37–1 HORMONES OF THE ANTERIOR PITUITARY GLAND

Name	Abbreviation	Target Organ	Subunits
thyrotropin	TSH	thyroid gland	2
luteinizing hormone	LH	gonads	2
follicle-stimulating hormone	FSH	gonads	2
corticotropin (adrenocorticotropic hormone)	ACTH	adrenal cortex	1
melanocyte-stimulating hormone	MSH	melanocytes	
lipotropin	LPH	adipocytes	
endorphins		pain receptors	
prolactin		mammary glands	1
somatotropin (growth hormone)	GH	many	1

The cells synthesizing this hormone in the hypothalamus release it upon stimulation. The nature of the responsible neurotransmitters is not settled.

Over 70 per cent of the brain thyrotropin releasing factor occurs outside the hypothalamus; it may serve as a neurotransmitter. It is also found in the pancreas and gastrointestinal tract for unknown purposes.

Thyrotropin releasing hormone acts upon particular basophilic cells in the anterior pituitary gland known as thyrotropes. (There is a curious admixture in the literature of the stems *tropic,* meaning turning, and *trophic,* meaning feeding, in connection with these hormones.) These cells are stimulated to release thyrotropin upon the binding of thyrotropin releasing hormone to their plasma membranes. (There is a rise in the blood thyrotropin level within 15 to 30 minutes after injection of releasing hormone.) The mechanism by which the releasing hormone stimulates thyrotropes is not clear-cut, although some investigators emphasize stimulation of adenylate cyclase.

The major control of thyrotropin secretion is an inhibition by triiodothyronine or thyroxine. As the circulating iodothyronines increase in concentration, they shut off the release of thyrotropin, the signal for their own formation (Fig. 37–3). This very sensitive feedback loop is the device by which the blood hormone concentration is kept relatively constant; similar devices are used for other glands under the control of the anterior pituitary. Thyrotropin releasing hormone appears to regulate the set-point of response by the thyrotropes to the circulating hormone concentration.

There is also a diurnal variation in blood thyrotropin concentration that may be mediated by dopamine, and the secretion of thyrotropin is also inhibited by somatostatin from the hypothalamus.

Thyrotropin consists of an α and a β subunit. The same polypeptide chain is used to make the α subunit of other hormones from the anterior pituitary (luteinizing hormone and follicle stimulating hormone), which resemble thyrotropin in being glycopeptides. A variable number of residues are removed from the end of the α chain in these hormones. Thyrotropin and the other hormones get their distinctive characters from their β subunits.

Within minutes of administering thyrotropin to experimental animals, the cells of the thyroid gland begin synthesis of mRNA, active transport of iodide into the cells, and reabsorption of thyroglobulin from the lumen. Again, these responses may be mediated in part by activation of adenylate cyclase and in part by other effects on the plasma membrane of the thyroid cells.

Thyrotropin has less well-defined functions in other tissues. Perhaps the clearest demonstration came from the discovery that the hormone could be partially hydrolyzed by pepsin to produce a large fragment containing most of the β chain, but only part of the α chain. This fragment is devoid of activity on the thyroid gland, but it

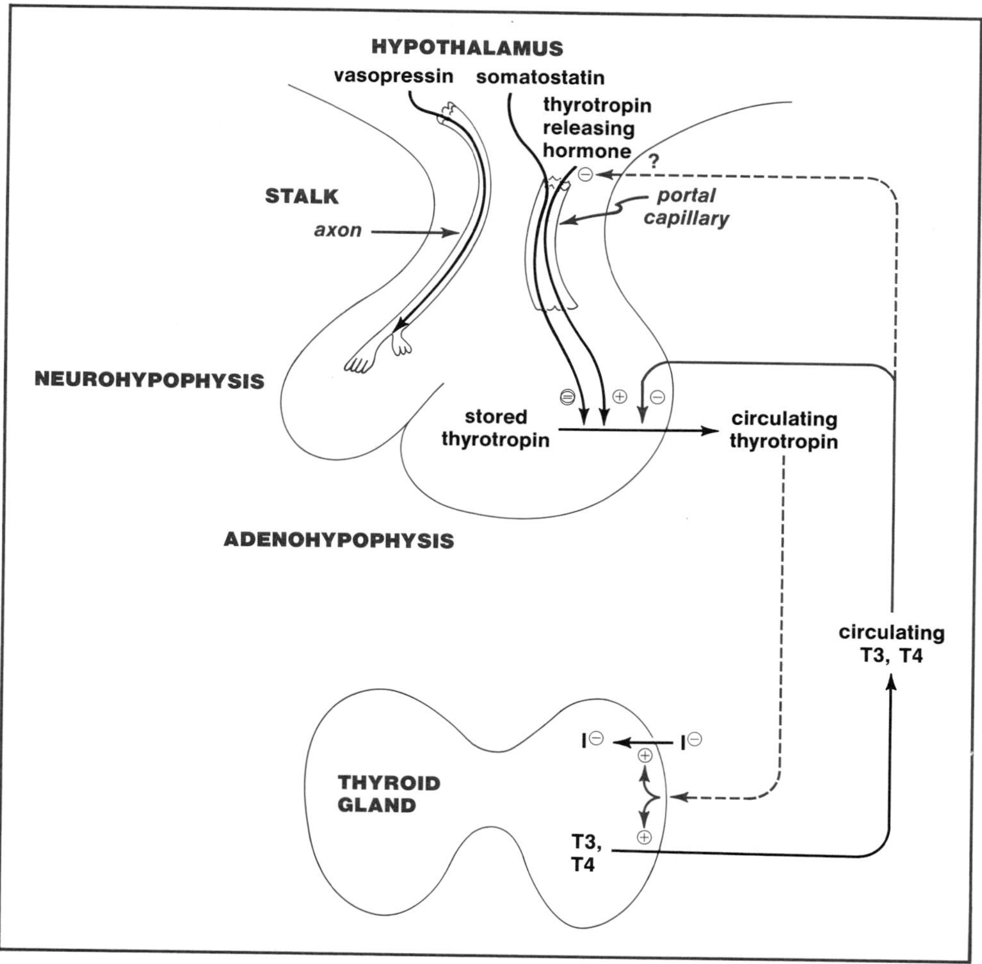

FIGURE 37–3 Regulation of the release of thyroid hormones through the hypothalamus and anterior pituitary gland.

stimulated development of the retro-retinal tissues in the guinea pig to produce the exophthalmos (protruding eyeball) sometimes associated with hyperthyroidism. These tissues were being stimulated by thyrotropin, not by the iodothyronines.

Regulation by Iodide Concentration

Changes in the concentration of circulating iodide cause inverse changes in the release of the iodothyronines. Part of the effect comes from a direct inhibition of the follicle cells by iodide; part can be indirect. Iodide has an inhibitory effect on the thyrotropes in the anterior pituitary; as its concentration rises, less thyrotropin is released.

Clinical Disruption of Thyroid Hormone Production. Administering radioactive ^{131}I or surgical removal are widely used and effective treatments for hyperactive thyroid glands. More sophisticated attacks upon the biochemical pathways involve blocking of specific sites with drugs. Monovalent anions (thiocyanates, perchlorates, and nitrates) inhibit active transport of iodide. Perchlorate can be used in humans.

Propylthiouracil and methimazole are clinically useful drugs that interfere with iodination of tyrosyl residues:

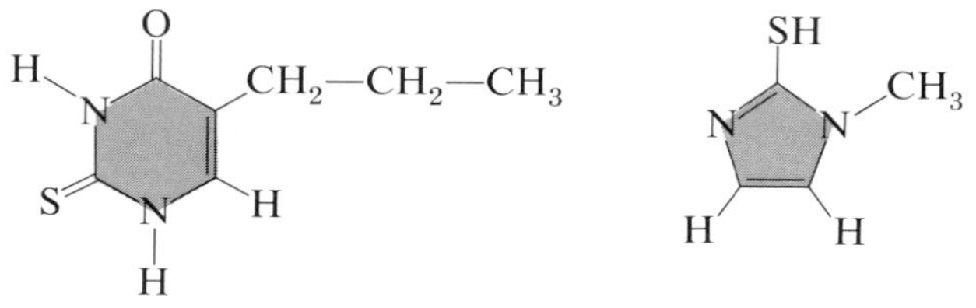

propylthiouracil **methimazole**

Propylthiouracil also interferes with the deiodination of thyroxine to triiodothyronine in target cells. Other drugs are useful in blocking the acute symptoms of excessive thyroid hormone. Propranolol, a beta adrenergic blocker, and reserpine, which depletes the supply of catecholamines, will relieve nervousness, fever, and hyperkinetic activity.

STEROID HORMONES

Steroids are characterized by a particular fused ring system. We have previously encountered cholesterol (p. 215) and the bile acids (p. 332). Just as cholesterol is the parent of the bile acids in the liver, it is also the parent of a variety of steroid hormones in some endocrine glands. They are classified on the basis of activity:

glucocorticoids made by the adrenal cortex modify certain metabolic reactions and have an anti-inflammatory effect;

mineralocorticoids made by different cells in the adrenal cortex modify excretion of salt and water by the kidney (next chapter);

androgens, made principally by the testis, affect male sexual characteristics;

estrogens, made principally by the ovary, affect female sexual characteristics;

progesterone, made by the ovary, the adrenal, and the placenta, affects uterine and ovarian development.

hydroxylated cholecalciferols, made from an intermediate of cholesterol synthesis by the combined action of several organs, affect calcium metabolism.

This is a somewhat simplistic list, but it will serve the purpose.

Those human steroid hormones made from cholesterol are synthesized by sequences involving reactions already encountered. Debranching enzymes, hydratases, dehydrogenases, monooxygenases (mixed-function oxidases), and isomerases are used in the steroid synthetic pathways to modify specific sites. Frequently, more than one pathway may be used to manufacture a given hormone, as the reactions can occur in different permutations of sequence.

There is a confusing array of natural steroids with endocrine activity, but we can sort out the principal ones into three main groups based only on the carbon skeleton:

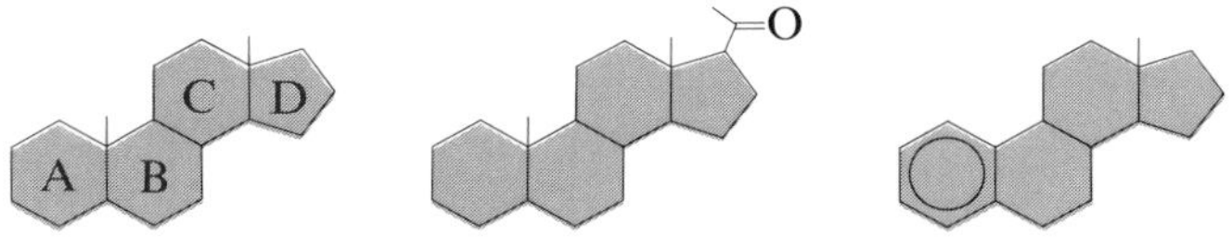

androgens **glucocorticoids mineralocorticoids progesterone** **estrogens**

It is important, however, to recognize that not all steroids with these unadorned skeletons have endocrine activity. The skeleton is a device for placing groups capable of polar interactions or H-bond formation in particular arrangements on a nonpolar background that has its own specific geometry. The nature of these groups is critical for activity—the black nonpolar sky is arranged so as to make these polar stars seem especially bright.

Transport. A glycoprotein, **transcortin,** in the blood tightly binds several steroid hormones, one molecule of hormone per molecule of transcortin. The bound molecules are in equilibrium with molecules free in solution or loosely attached to serum albumin. The free molecules can cross plasma membranes and become bound to receptor proteins within the cytosol of target cells. The hormone-receptor complex then moves to the nucleus, where it acts by affecting gene expression in most, perhaps all, instances. (It is possible that the complex sometimes affects post-transcriptional events.)

Adrenal Steroids

The adrenal cortex is made of different types of cells. Cells in a thin outer zone, the **zona glomerulosa,** make the mineralocorticoids; those in the wide middle zone, the **zona fasciculata,** and a narrower inner zone, the **zona reticularis,** make glucocorticoids.

The Biosynthetic Pathways. Synthesis of all steroid hormones made from cholesterol begins with the conversion of cholesterol to pregnenolone in mitochondria. This conversion is catalyzed by a complex containing cytochrome P-450, and it involves three successive monooxygenase reactions, utilizing NADPH as the second electron donor:

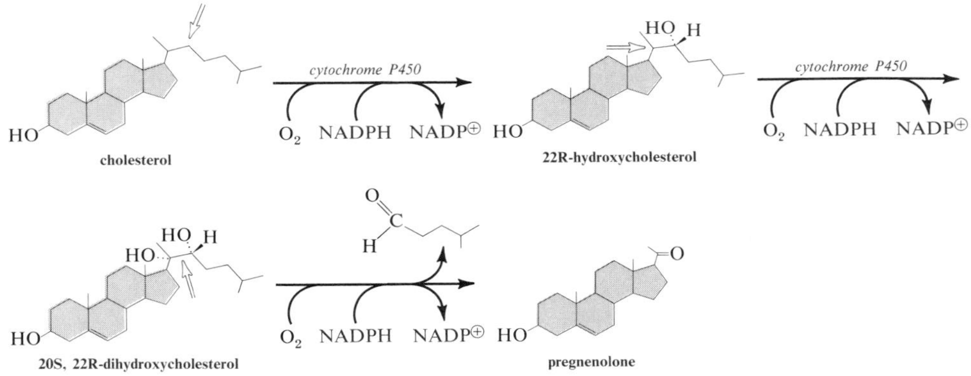

cholesterol —(*cytochrome P450*; O_2, NADPH → NADP⊕)→ 22R-hydroxycholesterol —(*cytochrome P450*; O_2, NADPH → NADP⊕)→ 20S, 22R-dihydroxycholesterol —(O_2, NADPH → NADP⊕; releases isocaproaldehyde)→ pregnenolone

This common pathway for endocrine steroidogenesis, cholesterol to pregnenolone, can be inhibited by aminoglutethimide:

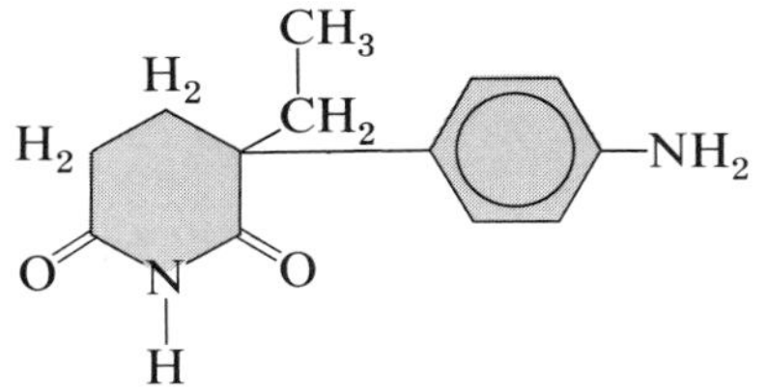

aminoglutethimide

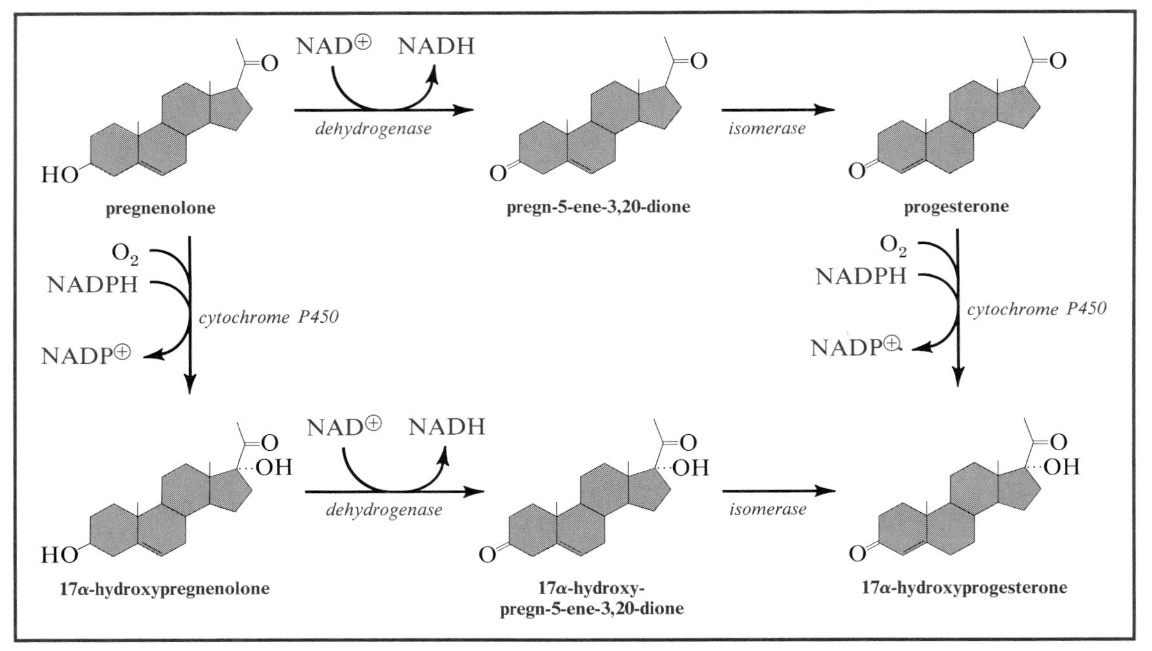

FIGURE 37–4 **Pregnenolone can be converted to 17α-hydroxyprogesterone by either of two routes in the adrenal cortex.**

The first, and an important, choice of alternative pathways occurs in the metabolic disposition of pregnenolone (Fig. 37–4). One pathway begins with an oxidation to a ketone at C3 and a shift of the double bond from the 5- to the 4-position, producing progesterone. These reactions occur on the endoplasmic reticulum of cells in all regions of the adrenal cortex and lead to both mineralo- and glucocorticoids. The other alternative is to hydroxylate C17, a reaction that leads to the glucocorticoids and therefore occurs primarily in the zona fasciculata and zona reticularis. This, like the other hydroxylases used in adrenal steroid synthesis, is a monooxygenase (mixed-function oxidase) that uses NADPH as a second electron donor and transfers electrons to oxygen through **adrenodoxin,** an iron-sulfide protein, and cytochrome P-450. The 17-hydroxylation may precede or follow the oxidation and isomerization by which progesterone is created. In either case, 17-hydroxyprogesterone is formed.

The route to the glucocorticoids in the inner zones continues with the successive action of a 21-hydroxylase and an 11β-hydroxylase to form **cortisol:**

O_2 NADPH NADP⊕ — O_2 NADPH NADP⊕

17α-hydroxyprogesterone → 11-deoxycortisol → cortisol

The route to the mineralocorticoids in the zona glomerulosa also involves hydroxylations at C21 and C11, but without the hydroxylation at C17.

The 18-methyl group side chain is then attacked by another mitochondrial complex containing cytochrome P-450 to produce an aldehyde group linked with the 11-hydroxyl group as a hemiacetal. The result is aldosterone:

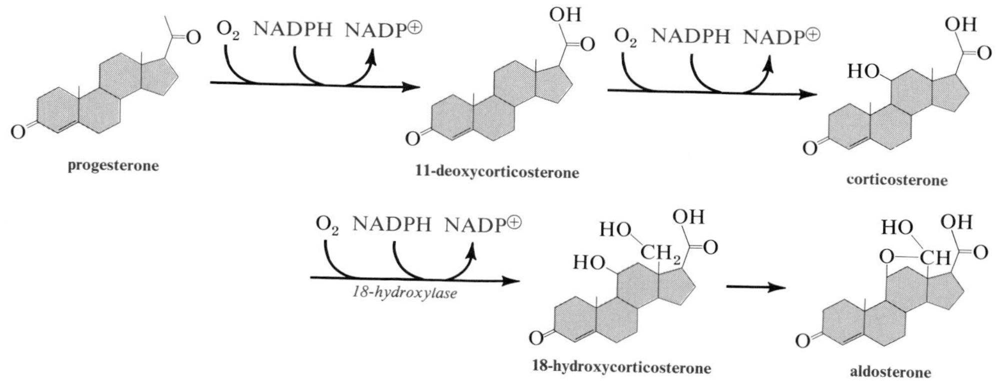

O_2 NADPH NADP⊕

progesterone

11-deoxycorticosterone

O_2 NADPH NADP⊕

corticosterone

O_2 NADPH NADP⊕

18-hydroxylase

18-hydroxycorticosterone

aldosterone

The 18-hydroxylase also occurs in the inner zones of the cortex, resulting in the formation of other 18-hydroxy derivatives. The estimated 24-hour production of the major compounds by the human adrenal gland is:

cortisol	8–24 mg
corticosterone	1.5–4 mg
11-deoxycortisol	0.5 mg
11-deoxycorticosterone	0.2 mg
aldosterone	0.04–0.20 mg
18-hydroxycorticosterone	0.15–0.45 mg
18-hydroxy, 11-deoxycorticosterone	*ca.* 0.1 mg

Glucocorticoids

Action. The gross effects of cortisol on metabolism are almost the opposite of those of insulin. It causes a mobilization of amino acids from the peripheral tissues and accelerated gluconeogenesis from amino acids in the liver. It suppresses peripheral glucose utilization and accelerates lipid mobilization, with a concomitant increase in the production of ketone bodies by the liver.

The metabolic effects of cortisol are summarized in Table 37–2. The table also shows the earliest times at which measurable effects appear (sometimes in the intact animal and sometimes with isolated tissues). All of these effects appear to be caused by changes in the rate of synthesis of particular proteins.

Regulation of Glucocorticoid Synthesis. There is little storage of glucocorticoids within the adrenal gland so that the blood level of glucocorticoids closely reflects their rate of synthesis. In the blood, glucocorticoids are mostly bound to plasma proteins, particularly **transcortin,** a corticosteroid-binding globulin. Less than 10 per cent is free. The blood half-life of cortisol, the major glucocorticoid in humans, is 80 minutes. (The blood half-life is considerably shorter than the tissue half-life.)

The blood cortisol level varies frequently during a 24-hour period, because it is discharged in pulses rather than in a continuous flow, but a pattern of diurnal varia-

TABLE 37–2 EFFECTS OF CORTISOL

Liver	Hours
Increased glycogen	4–6 (*in vivo*)
Increased glucose production	2–6 (*in vitro*)
	2–6 (*in vivo*)
Increased oxybutyrate production	3–24 (*in vivo*)
Increased urea production	4–8 (*in vivo*)
Increased amino acid uptake	1.5–2 (*in vitro*)
	2–4 (*in vivo*)
Increased RNA synthesis	2–4 (*in vivo*)
Increased protein synthesis	8–20 (*in vivo*)
Adipose Tissue	
Increased fatty acid release	1–2 (*in vitro*)
Decreased glucose utilization	2–4 (*in vitro*)
Muscle	
Decreased glucose utilization	2–4 (*in vitro*)
Lymphatic Tissue	
Decreased glucose utilization	2–4 (*in vitro*)
Decreased nucleic acid synthesis	2–4 (*in vitro*)

Claims have been made elsewhere for suppression of glucose uptake by thymus cells in 15 to 20 minutes.

Adapted from J. Ashmore (1967) *in* A. D. Eisenstein, ed.: *The Adrenal Cortex*. Little Brown. By permission.

tion is clearly seen. Around midnight cortisol levels are at their lowest (~0.15 μM, ~5 μg/dl) and begin to rise at about 0200, reaching their peak (~0.3 to 0.7 μM; 10 to 25 μg/dl) at 0600–0800 hours.

Stress or other unpleasant stimuli can rapidly alter the rate of cortisol synthesis. Within minutes after the start of a surgical procedure, the patient's cortisol blood level rises.

The Hypothalamic-Pituitary-Adrenal Axis. These changes in cortisol secretion are brought about through a system of control resembling the system described for the thyroid gland. The hypothalamus under appropriate stimulation releases an as yet uncharacterized **corticotropin releasing hormone,** which stimulates the anterior pituitary to synthesize and release the hormone **corticotropin (adrenocorticotropic hormone, ACTH).** Stress stimulates liberation of corticotropin releasing hormone, whereas release may be inhibited by increasing cortisol concentrations. (A neurohypophyseal hormone, vasopressin, also stimulates synthesis and secretion of corticotropin. This hormone primarily serves as a regulator of water balance, as is discussed in the next chapter.)

Corticotropin (ACTH) is a 39-residue polypeptide that stimulates the adrenal cortex to synthesize and secrete glucocorticoids. It acts on the plasma membrane of the affected cells to increase the rate of conversion of cholesterol to pregnenolone. The mode of transmission of the signal from plasma membrane to mitochondria is not known.

The half-life of corticotropin in the blood is less than 10 minutes. Normal adult concentrations range between 2 and 18 picomolar (10 and 80 ng/l) in the morning and are less than 2 picomolar (10 ng/l) by midnight. Corticotropin synthesis and secretion are directly controlled through negative feedback by free cortisol. High levels of cortisol inhibit and low levels stimulate corticotropin synthesis. External stimulation is absolutely necessary for adrenal function; in its absence, the fasciculata and reticularis zones of the cortex atrophy. This may occur upon prolonged therapeutic use of glucocorticoids in high doses because the medication suppresses synthesis of corticotropin in the pituitary. Abrupt discontinuation of the medication may then result in acute adrenal insufficiency.

This untoward event is avoided by using glucocorticoids that have a short duration of action and by giving them in the morning. The glucocorticoid level again becomes low at around midnight and remains so during the small hours, without interference in the normal diurnal appearance of early morning corticotropin secretion, and the resultant stimulation of the adrenal cortex.

Metyrapone is a compound used in testing pituitary function because it combines with a cytochrome P-450 to inhibit 11-hydroxylase activity in the adrenal gland:

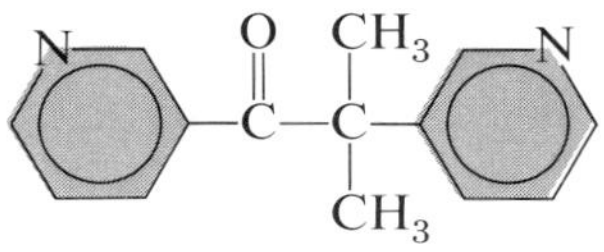

metyrapone

Individuals with normal pituitary function respond to metyrapone with a sharp increase in corticotropin secretion as a result of the drop in cortisol production. The stimulated adrenals accumulate the intermediate 17-hydroxy steroids, which cannot be converted to cortisol and are excreted in the urine. Failure to show a rise in urinary 17-hydroxysteroid excretion is an abnormal result. (Normal adrenal function is previously demonstrated by injecting corticotropin in the absence of metyrapone and measuring the excreted 17-hydroxysteroids.)

Pro-opiocortin is the parent molecule of corticotropin and several puissant siblings. A single polypeptide chain contains within it sequences, some overlapping, having a variety of potent biological effects in addition to stimulation of the adrenal cortex. Some sequences cause fat mobilization; they act as **lipotropins.** Others stimulate melanocytes to produce melanins; they act as **melanotropins.** Still others can relieve pain; they are the **endorphins** and **enkephalins.** The borders of these various sequences are flanked by pairs of basic amino acid residues, Lys-Arg or Arg-Arg, which are likely sites for cleavage by proteases. The sites in the sequence that are attacked are not necessarily the same in different organs, or even in different cells of the same organ.

The amino acid sequence of bovine pro-opiocortin has been deduced from the cDNA sequence. (See p. 136ff for the use of cDNA.) It may be seen from Table 37–3 how one biological activity may overlap or be nested within another in a single polypeptide.

LOCATION OF BIOLOGICAL ACTIVITIES IN PRO-OPIOCORTIN — TABLE 37–3

The first residue (Ser) of corticotropin is designated +1. A methionyl group 131 residues upstream (−131) is presumed to be the initial residue of the translated polypeptide chain.

Activity	From		To
γ-melanotropin	−55	through	−44
α-melanotropin	+1	through	13
corticotropin	+1	through	39
β-lipotropin	42	through	134
γ-lipotropin	42	through	101
β-melanotropin	84	through	101
Met-enkephalin	104	through	108
β-endorphin	104	through	134

Cushing's syndrome is a condition in which effects of excessive glucocorticoid secretion dominate the clinical picture. It may result from the growth of secreting tumors of the anterior pituitary or the adrenal cortex, from nodular hyperplasia of the adrenal cortex, or from carcinomas, frequently carcinoma of lung, that produce an ectopic corticotropin (ectopic = out of place). The most common cause of the symptoms of Cushing's syndrome is the prolonged administration of glucocorticoids for medical treatment. There is a general depletion of proteins to supply amino acids for gluconeogenesis with loss of muscle mass and thinning of the skin and bones. As a result, there is general weakness, easy bruisability, bone fractures, and a redistribution of fat, creating a characteristic moon-face and buffalo hump. Serious mental changes are frequently present. The excessive output of glucose causes hyperglycemia, and the picture of full-blown diabetes. Excessive secretion of aldosterone also occurs.

Failure of the adrenal cortex causes **adrenal insufficiency (Addison's disease** of the adrenal). Infection (tuberculosis, systemic fungal infections), metastatic cancer, atrophy, and therapeutic removal of the adrenal glands are the usual causes. The loss of glucocorticoids causes hypoglycemia and decreased muscle tone, whereas the loss of mineralocorticoids (next chapter) causes disturbances of salt balance (sodium loss and potassium retention). Weakness, fatigue, weight loss, increased skin pigmentation, nausea, vomiting, diarrhea, muscle pain, and salt craving are common symptoms. The patients are hypotensive, hypothermic, and usually have a low glucose and sodium level. Their grip on life is tenuous and any stress, infection, cold, or even noise can precipitate a crisis leading to death.

Inborn errors of steroid metabolism result from absence or impaired function of one of the many synthetic enzymes. The low production of cortisol leads to high secretion of corticotropin by the anterior pituitary gland, which in turn causes bilateral hyperplasia of the adrenal gland. The enzymes commonly involved are 21-hydroxylase, 11β-hydroxylase, 17α-hydroxylase, 18-hydroxylase, and 3β-hydroxysteroid dehydrogenase.

The treatment of advanced breast cancer can include the removal of the adrenal glands. Some patients experience a remarkable destruction of the cancer following adrenalectomy. Recently, a "medical" adrenalectomy has been used, which involves giving aminoglutethimide (p. 727) to block pregnenolone formation and cortisol to exert negative feedback upon the anterior pituitary, which otherwise would release ACTH and override the block. (Aminoglutethimide also blocks the conversion of androgens to estrogens in peripheral tissues; see next section.)

One of the most useful actions of cortisol from a therapeutic standpoint is its suppression of the inflammatory response. The maddening itch of poison ivy and the inflammation associated with some serious systemic diseases can be ameliorated by glucocorticoids. Continued use of glucocorticoids can also cause serious adverse reactions, including thinning of bones, failure to grow, increased susceptibility to infections, worsening of diabetes, and suppression of symptoms usually present with an acute abdominal process. A little stomach ulcer has been known to silently grow into a grossly hemorrhagic lesion with cortisol therapy. The availability of over-the-counter cortisol ointments has even been suggested as a possible cause of a current rash of previously obscure illnesses among male homosexuals. (There is no supporting evidence for either the inferred use or the presumed causal sequence.)

Gonadal Steroids

Both the androgens and the estrogens are made to some extent by the adrenal cortex, although in much lesser amounts than are the other steroid hormones. The major sources of these steroids are the gonads, which carry pregnenolone through the same initial steps involved in adrenal steroid synthesis, but control the rates of the reactions by varying enzyme concentrations so as to arrive at a different spectrum of products. The differences in function of the glands producing steroid hormones is, therefore, more of a matter of quantitative variation of gene expression than it is of presence or complete absence of particular enzymes.

Testosterone, the male androgen, is mainly made by the interstitial cells of the testis. A typical pathway of synthesis is:

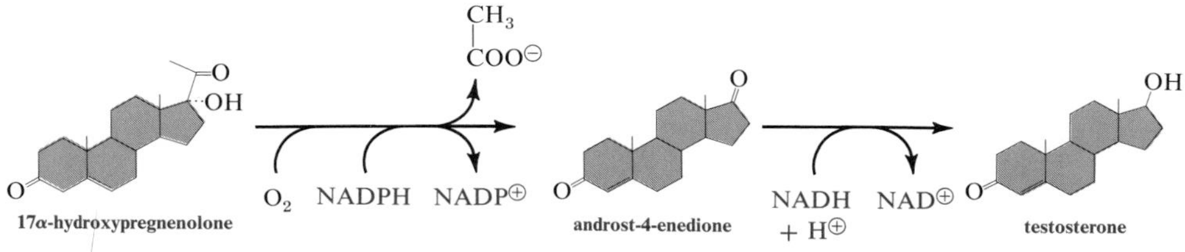

$CH_3COO^{\ominus}$

O_2 NADPH NADP$^{\oplus}$ NADH + H$^{\oplus}$ NAD$^{\oplus}$

17α-hydroxypregnenolone → androst-4-enedione → testosterone

Many cells responsive to testosterone contain a 5α-reductase to convert testosterone to a more potent androgen, **dihydrotestosterone,** which combines with a receptor protein in the target cells. As with other steroid hormones, the receptor-steroid complex is translocated to the nucleus, where it affects gene transcription. In this case, phosphorylation of nonhistone proteins in the nucleus has been observed within 30 minutes of the administration of testosterone to castrated animals.

Not only is testosterone necessary for the development of adult male sexual characteristics, its presence during fetal development is necessary for normal development of both male physique and psyche. Some evidence indicates that testosterone is required for the development of a greater spatial discrimination in males. If testosterone production is impaired after puberty, no loss of spatial ability occurs. (It is important to recognize that any comparison of average properties of groups is not a comparison of individuals.)

However, impairment of androgen function during later stages of fetal development is at least partially reversible. Consider pseudohermaphrodites with 5α-reductase deficiency. Infants with this autosomal recessive defect are born with seemingly feminine external genitalia and are often raised as females. However, the pubertal surge of testosterone production, which is normal, causes virilization. The previously undescended testes may enter the scrotum (previously regarded as the labia), the misdiagnosis of a clitoris becomes obvious, the voice deepens, and so on. Out of 38 subjects identified in a group of families carrying the defect in the Dominican Republic, 18 were raised as females, and 17 out of the 18 changed their identity to male after puberty.

Male Sex Pheromone? Earlier French romantic literature reports that a man can become irresistible to women if he wears a handkerchief with which he has previously rubbed his armpits while fully aroused. This belief may account for a rather odd feature of male attire—the exposed breast pocket handkerchief—and it now seems it may have some physiological basis. The testes form a particular steroid, 5α-androst-16-en-3α-ol, for which the trivial name **priapol** is suggested:

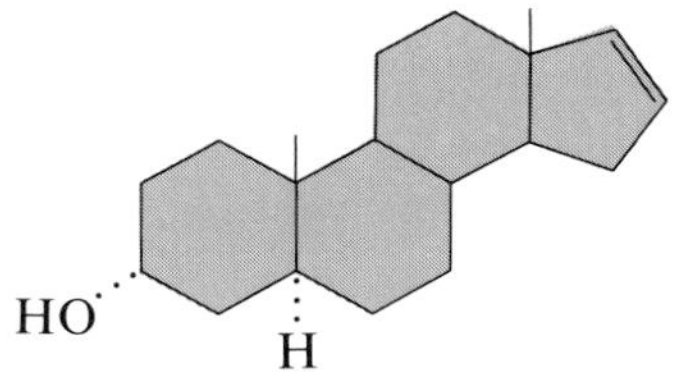

HO H

5α-androst-16-en-3α-ol
sex pheromone?

Priapol is transported through the blood, and is secreted by the axillary glands along with the corresponding ketone.* The alcohol has a musky odor, whereas the ketone is said to smell like urine. Only sketchy trials have tested humans for positive reaction to exposure to the scents; the best evidence for the potency of the alcohol comes from the behavior of swine. The concupiscent boar generates a salivary foam rich in priapol, and the odor induces the sow to stand for him. Curiously, truffles are rich in the compound, which explains why sows are dedicated truffle hunters.

Estrogens are mainly made by thecal cells in the ovarian follicles, through modification or extension of the routes by which testosterone is made:

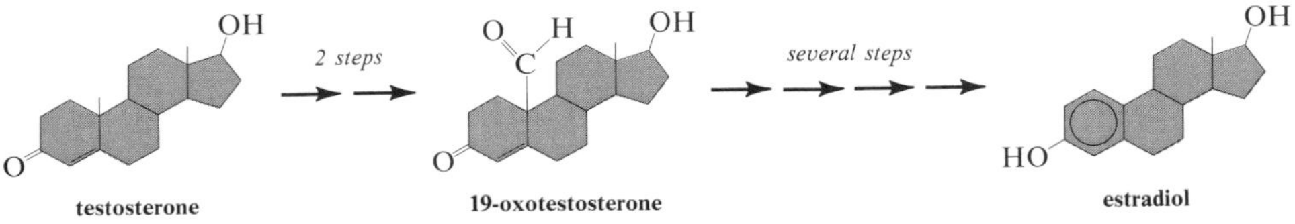

testosterone — *2 steps* → 19-oxotestosterone — *several steps* → estradiol

A branched methyl group (C_{19}) is removed and ring A is made aromatic.

Approximately 20 per cent of the total estrogen production by females occurs in other than ovarian tissues. (Even a normal adult man forms about 45 μg of estrogen per day, compared to the 500 μg produced on the day of ovulation by a woman.) Estrone, a weaker estrogen, is made principally by adipocytes from androstenedione. Hydroxylated estrogens may be important physiological effectors, but their exact role is yet to be defined:

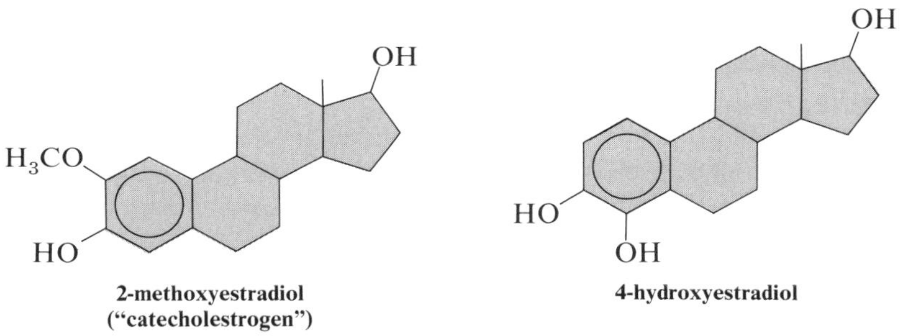

2-methoxyestradiol ("catecholestrogen")

4-hydroxyestradiol

Progesterone is a precursor of the other steroids we have mentioned, but it is also an important hormone in its own right, being supplied primarily by the corpus luteum that develops in ovarian follicles after release of the ova. Its synthesis and secretion is therefore cyclical, increasing about two days before ovulation, and surging after ovulation to a peak about eight days later, then returning to basal levels by 12 to 14 days if fertilization and implantation have not occurred.

Detection of cytosol receptors for estrogen and progesterone has become an important part of the management of patients with carcinoma of the breast. The presence of **estrophilin,** an estrogen receptor protein, in the malignant tissue is an indicator that the cancer is likely to be adversely affected by an alteration of the steroid hormone milieu. The correlation between presence of receptor and a good clinical response tc hormonal manipulation is far from perfect. The higher the level of the receptor, the more likely a favorable response will be elicited. There is also evidence to indicate that one action of the estrogen translocated to the nucleus by the

*It is not certain if there is a causal relationship between the parallel rise in the sale of underarm deodorants for men and the divorce rate.

receptor is activation of chromatin leading to the synthesis of progesterone receptors. High response rates occur when the breast cancer has both estrogen and progesterone receptors. Scientific mysteries lurk here. It is hard to explain why such diverse and even contradictory procedures as giving estrogens or antiestrogens, removing the ovaries, or removing the adrenals while giving glucocorticoids and mineralocorticoids can produce regression in breast cancer that contains estrophilin.

Hypothalamic-Pituitary-Gonadal Axis. The hypothalamus forms a single releasing hormone that stimulates the synthesis and secretion of two pituitary hormones, **luteinizing hormone (luteotropin, LH)** and **follicle-stimulating hormone (follitropin, FSH).** It is curious that one releasing hormone is used to stimulate the secretion of two hormones having quite different actions. It has this short sequence:

$$\text{Pgl-His-Trp-Ser-Tyr-Gly-Leu-Arg-Pro-Gly-NH}_2.$$

Both of the pituitary hormones are important in the male and female. The plasma levels vary independently, and the mechanism of control appears to be complex. They belong to the same family of glycoprotein hormones as does thyrotropin, with common α chains and differing β chains. (The fourth member of the family is chorionic gonadotropin, made by specific cells in the placenta. It is also formed by certain ovarian, placental, and testicular neoplasms.) Both of the pituitary gonadotropins appear to act by stimulating adenylate cyclase in the target cells, which in turn accelerates the conversion of cholesterol to pregnenolone.

Luteinizing hormone promotes the secretion of progesterone by cells of the corpus luteum in the ovary (the structure developing in a follicle after release of an ovum). It also acts upon Leydig cells in the testes to stimulate testosterone formation. Complete testicular femininization is thought to be caused by a failure of feedback control of testosterone concentration over luteinizing hormone release. The resulting high levels of the gonadotropin cause both estrogen and testosterone secretion.

Follicle-stimulating hormone promotes the secretion of estradiol by granulosa cells in developing ovarian follicles. It also is necessary for normal development of the Sertoli cells in seminiferous tubules in the male.

Variations in the concentrations of progesterone, estradiol, and the gonadotropins during a menstrual cycle are shown in Figure 37–5.

Somatotropin, Prolactin, and Chorionic Somatomammotropin

These hormones are a structurally related family of polypeptides. Each is a single chain of 191 to 198 amino acid residues, with considerable homology in sequence and in the location of two disulfide groups, although prolactin has an additional disulfide group. Somatotropin (growth hormone) and prolactin are synthesized by the anterior pituitary gland; chorionic somatomammotropin (placental lactogen) is made in the placenta.

Somatotropin stimulates normal growth of all tissues, but it is particularly active in stimulating the growth of connective tissues, especially bone and cartilage. Somatotropin make up as much as 10 per cent of the dry weight of the anterior pituitary gland. Its secretion is under complex control. Secretion is stimulated by insulin, arginine, dopa, and glucagon, which are used to diagnose a deficiency of the hormone. Secretion is inhibited by somatostatin, glucose, β-adrenergic agonists, and obesity. A releasing hormone has recently been demonstrated.

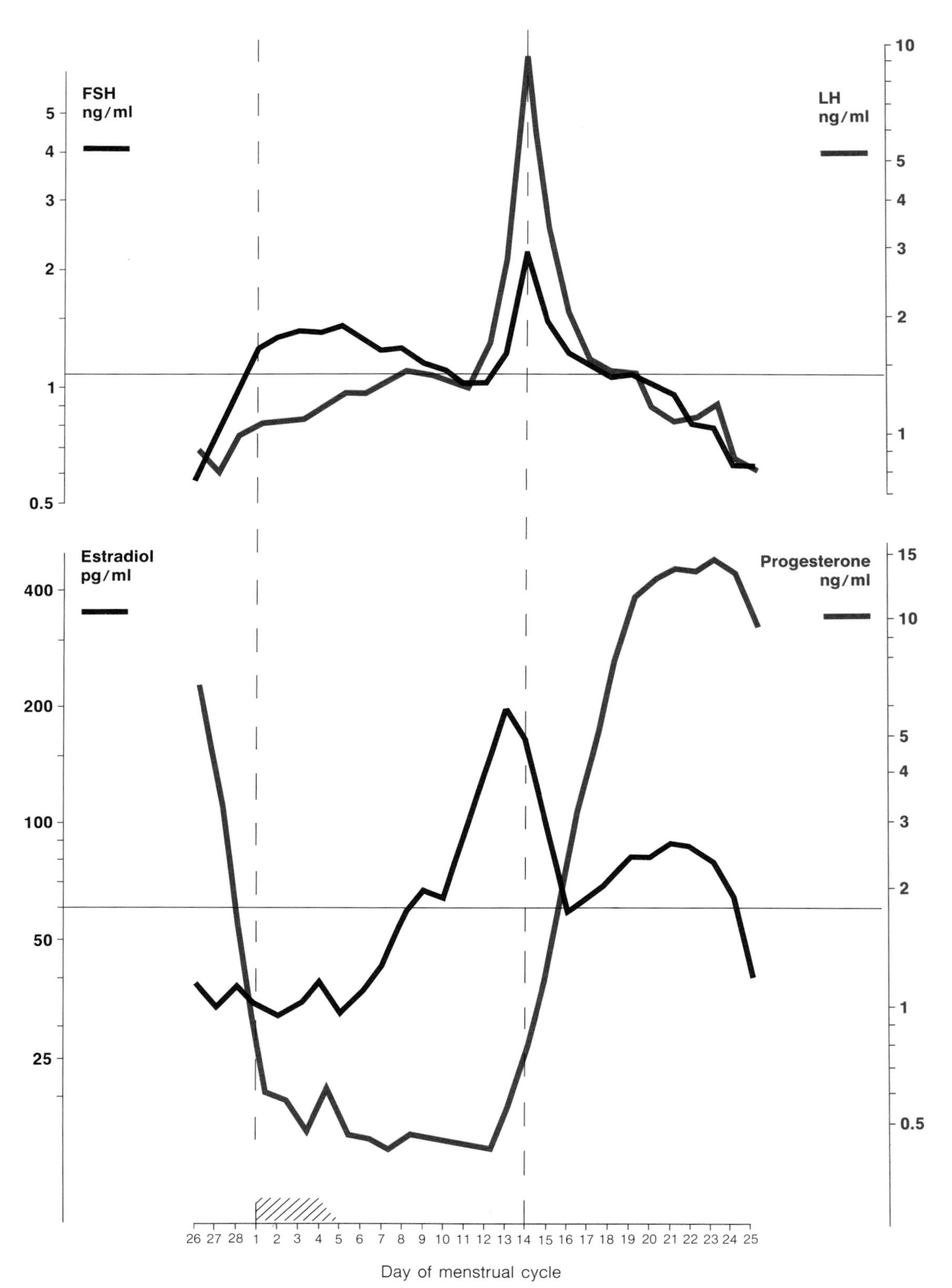

FIGURE 37–5 Concentrations of circulating hormones during the menstrual cycle. Reproduced by permission from L. Wide (1976), p. 87 in J.A. Loraine and E.T. Bell, eds.: Hormone Assays and Their Clinical Applications, 4th ed. Churchill.

Somatotropin may act in part through intermediary proteins, the **somatomedins** and **insulin-like growth factors.** These are relatively small proteins (7,000 to 10,000 daltons), which circulate bound to larger carrier proteins. They are formed by the liver in amounts varying with the rate of somatotropin release from the pituitary, and also with the insulin concentration.

Cloning for Growth—A Reality

While basketball coaches may only dream of cloning a Ralph Sampson, cloning the growth hormone gene has already arrived for clinicians. Using techniques described earlier (p. 131ff), *E. coli* has been made to synthesize human somatotropin, differing only from the *in vivo* material in having methionine as an extra residue at the amino terminal of the 192-residue hormone.

An estimated one in 4000 children fails in growth owing to a deficiency of somatotropin. The human hormone was formerly available only after a difficult purification from blood or from glands obtained at autopsy; in the near future, it will be generally available. This will increase the chance for misuse in children with normal hormone levels. Confirmation of somatotropin deficiency requires measuring the blood concentration after stimulation with insulin in combination with either dopa, arginine, or glucagon. Currently, treatment consists of three injections per week for a year. (The preparation is referred to as hGHr for human growth hormone recombinant.)

Prolactin is required for growth and function of glandular breast tissue. Hypothalamic control in this case involves inhibition, rather than stimulation, of hormone production. (Some drugs have the undesirable side effect of stimulating prolactin release, causing galactorrhea, in which the breasts synthesize and secrete a milk-like fluid. These drugs include tranquilizers and antidepressants inhibiting the hypothalamus and dopaminergic compounds.)

Epilogue on Pituitary Hormones

The preceding sections only touch on the complex actions of the many hormones involved in the hypothalamus–pituitary–end organ chains. Many of the active polypeptide sequences are made to some extent in other parts of the body, including the brain, liver, pancreas, and gastrointestinal tract. The brain is sufficiently isolated from the pituitary by the blood-brain barrier to permit the use of peptides as neurotransmitters that behave as hormones elsewhere. However, the hormones are not insulated from each other within the hypothalamus–portal blood–pituitary axis, and some of the difficulty in unraveling their action comes from the influence of other hormone concentrations on the action of one.

PANCREATIC ISLET HORMONES

Hormones of the islets of Langerhans play a vital role in the regulation of fuel metabolism. One of the commonest serious disorders of man, diabetes mellitus, represents a disturbance of the ability of the islet cells to properly regulate the glucose concentration. Diabetes mellitus affects 5 per cent of the United States population,

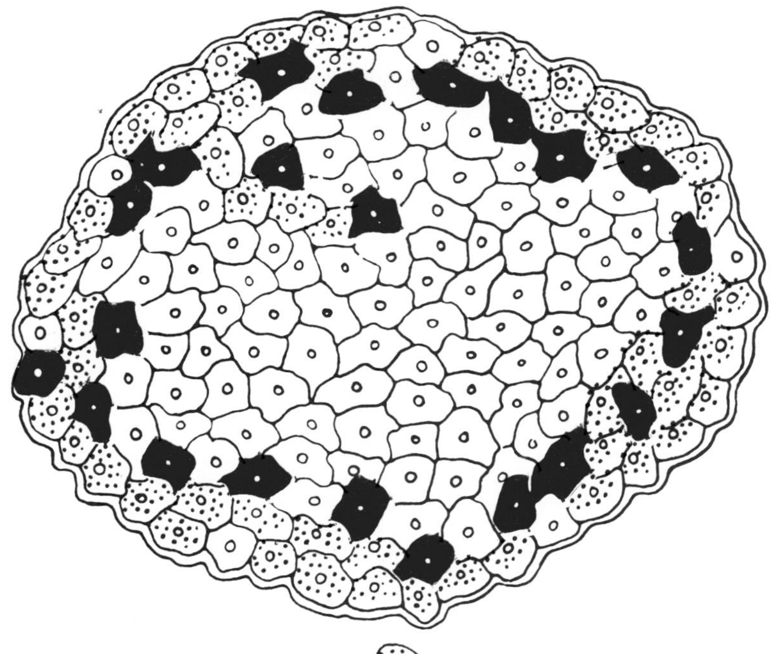

A CELLS Glucagon
D CELLS Somatostatin
B CELLS Insulin

FIGURE 37–6

Arrangement of cells in the pancreatic islets of Langerhans. Reproduced by permission from R.H. Unger and L. Orci (1977). Role of Glucagon in Diabetes. Arch. Intern. Med. 137:484. Copyright 1977, American Medical Association.

and 1.74 per cent of the deaths in 1981 were ascribed to it. The distribution of the various cell types in the islet is shown in Figure 37–6. Glucagon is synthesized by the A cells, insulin by the B cells, and somatostatin by the D cells. About 60 per cent of the cells, occupying the central zone of the gland, are B cells. Cell to cell contact is largely B to B cell. At the margin of the organ, a rim of A cells, one to two cells thick, makes up about 30 per cent of the total. Interspersed between A and B cells, or occasionally between A cells, are the somatostatin-secreting D cells. There are numerous gap and tight junctions between cells. The areas where the three cell types meet are invested with a rich blood and nerve supply. The islets act as a sensor of the glucose concentration and its rate of change, constantly adjusting the rate of secretion of glucagon and insulin to match conditions.

Glucagon

Glucagon (3485 daltons) is cleaved from a larger precursor, and intermediate products are known. One, **glicentin,** contains about 100 amino acid residues and is also found in secretory cells of the small bowel.

Release of glucagon is stimulated by decreasing glucose concentrations. The plasma glucagon concentration ranges from 30 to 200 ng/l after fasting and is lowered after a carbohydrate meal. Increases in amino acid concentration or stimulation of the sympathetic nervous system also stimulate glucagon secretion.

We have already seen that glucagon stimulates adenylate cyclase in the liver and adipose tissue, affecting the activity of enzymes of glycogen and fat metabolism, among others. Functionally, glucagon can be pictured as a hormone designed to raise blood glucose levels so as to protect the brain and other tissues.

Insulin

Active insulin is made by progressive modification of preproinsulin (p. 150). A rise in the concentration of blood glucose is a primary signal for secretion of insulin, and the release is prompt, beginning within one minute. Insulin is also secreted in response to a rise in blood amino acid concentration; arginine is the most effective signal, and arginine-loading tests are used to test beta cell function. The sympathetic nervous system inhibits insulin secretion through alpha adrenergic receptors.

Receptors for insulin are present in a variety of tissues, although the dependence of the tissues on insulin varies. Even the brain has some receptors, but it can use glucose very well in the absence of insulin. The number of receptors in cells of a given tissue varies between individuals, and an inverse relationship has been demonstrated between receptor concentration and the ambient insulin level. Where non-obese normal individuals had a basal insulin level of 35 to 145 pM,* obese nondiabetic individuals had the high basal insulin levels of 180 to 440 pM in their blood, but they had a decrease in the concentration of insulin receptors. This may explain a seeming resistance to insulin in some of these patients.

The classic result of insulin action is a dramatic increase in the rate of transport of glucose into skeletal muscle and adipose tissue. Insulin also promotes the uptake of amino acids by skeletal muscles and increases protein synthesis. It accelerates lipid synthesis and inhibits lipolysis and gluconeogenesis.

Although the effects of insulin occur rapidly, the concentration required to elicit different effects varies. Release of free fatty acids from adipose tissue is inhibited by 200 to 350 pM. Suppressing gluconeogenesis requires 700 to 1400 pM, levels approaching the maximum physiological insulin concentrations. Inhibition of hepatic glycogenolysis requires less insulin than does inhibition of gluconeogenesis. Glucose uptake by peripheral tissues increases with insulin concentration until a maximum is reached at about 1400 pM. Disposal of D-3-hydroxybutyrate requires concentrations of 350 to 700 pM. The mechanism of insulin action is not known in detail, although there is evidence that it causes release of a protein as a second messenger (p. 520).

Somatostatin

The pancreatic islets are a major source of somatostatin, a tetradecapeptide:

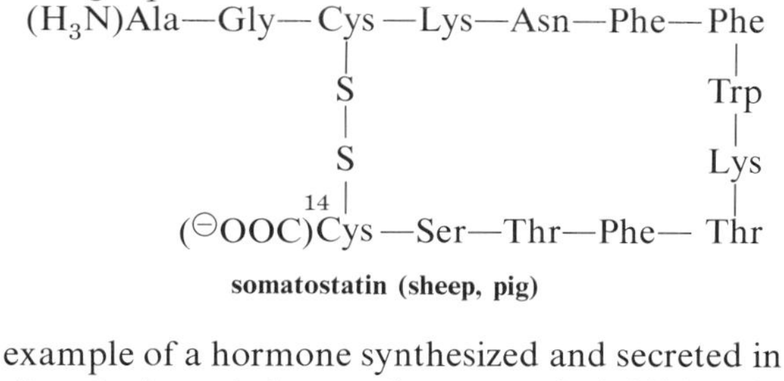

```
 ⊕   1
(H3N)Ala—Gly—Cys—Lys—Asn—Phe—Phe
              |                 |
              S                Trp
              |                 |
              S                Lys
           14 |                 |
     (⊖OOC)Cys—Ser—Thr—Phe—Thr
```

somatostatin (sheep, pig)

Here we have an example of a hormone synthesized and secreted in different regions of the body, including the hypothalamus. Somatostatin inhibits release of the following hormones, some of which we shall not discuss: thyrotropin, corticotropin, and somatotropin (growth hormone) by the adenohypophysis; insulin and glucagon by

*Insulin quantities are frequently expressed in units (U), with a unit defined as the amount giving the same biological activity as 42 micrograms of pure insulin (7.2 nanomoles). Blood concentrations are often given in microunits per ml; 1 μU ml^{-1} = 7.2 picomolar.

the pancreas; gastrin by the gastric mucosa; secretin by the intestinal mucosa; and renin by the kidney (next chapter). Somatostatin also inhibits the emptying of the stomach and secretion of both gastric acid and pancreatic enzymes.

Diabetes Mellitus

Diabetes* is a group of diseases having in common an insulin effect that is inadequate for the uptake of glucose from the blood. The resultant **hyperglycemia** frequently causes glucose to appear in the glomerular filtrate at rates exceeding the capacity of the kidney to reabsorb it (next chapter). This results in **glucosuria** and the voiding of large volumes of urine **(polyuria),** frequently at night **(nocturia).** Large volumes of water are drunk to replace the losses **(polydipsia).** Polyuria and polydipsia are also symptoms of diabetes insipidus, but they demand further attention in any event. Even in the absence of such overt symptoms, diagnosis is easily made when blood glucose concentrations are measured after an overnight fast. Single fasting values in excess of 7 mM (120 mg/dl), or in excess of 9 mM (160 mg/dl) one hour after a breakfast containing 100 grams of carbohydrate, are suggestive, and repeated excessive values are diagnostic. (The reported concentrations vary among clinical laboratories, owing to differing responses of the methodology to other compounds in the samples and to calibration errors. Each laboratory must establish its own normal upper limit of blood glucose concentrations by experience. Furthermore, concentrations of glucose are higher in serum or plasma samples than they are in the whole blood.)

Diabetes is also accompanied by abnormalities in fat and protein metabolism. Not only is there an increased mobilization of fatty acids, but the liver converts a larger fraction to the ketone bodies rather than to the triglycerides and phospholipids:

$$\text{palmitoyl group} + 7\ O_2 \longrightarrow 4\ \text{acetoacetate}^{\ominus} + 4\ H^{\oplus}.$$

Up to one mole of oxybutyrates may be excreted per day, and the concomitant production of H^+ results in acidosis, or **ketoacidosis** as it is commonly called in view of the simultaneous production of both ketone bodies and hydrogen ions.

Diabetes mellitus occurs in two forms. **Insulin-dependent diabetes** is also known as **juvenile-onset** because it tends to occur in younger people with an early requirement of insulin for its control, but it also may develop at a later age. Genetic predisposition to this disease is complex and uncertain. Hereditary factors are believed to play a part, mainly in determining susceptibility to environmental factors, infections, or autoimmunity that provoke irreversible damage to the B cells making insulin. There is an association between certain tissue antigens (histocompatibility antigens) and susceptibility to insulin-dependent diabetes.

Non-insulin-dependent diabetes develops more slowly. Perhaps some indications may be seen early by intensive study, but firm diagnosis is usually only possible in the middle and later years. The cause is usually a failure of the tissues, especially the skeletal muscles, to respond to normal concentrations of insulin, rather than a failure of insulin production.

While most emphasis is placed upon the role of insulin in diabetes, some investigators view it as a bihormonal disease, with the balance between insulin and glucagon activities being disturbed.

---

*The word *diabetes* means an excessive volume of urine. Diabetes mellitus is sweet urine. Diabetes insipidus is dilute urine, and so on. Used alone, diabetes is taken to mean diabetes mellitus, by far the more common of the two conditions.

A patient with diabetes may present with **diabetic ketoacidosis**—dehydrated with acidosis, glucosuria, and ketonuria, and perhaps comatose. This frequent medical emergency may be the first indication of the juvenile form of the disease, or it may have occurred in an insulin-dependent patient who failed to take his insulin or who developed an infection. In any event, it is a lethal condition, with a mortality remaining between 5 and 10 per cent in major medical centers.

A diabetic patient may also be comatose from hypoglycemia caused by taking too much insulin. This **insulin shock** represents starvation of the brain, and patients on insulin may carry supplies of readily absorbable sugar to counteract its first indications.

Complications. Leonard Thompson, the first person to have his life prolonged by insulin injections, still died later from complications of the disease, perhaps aggravated by following the carbohydrate-poor diet then used in treating diabetes. While the mortality from diabetic ketoacidosis has been dropping with better management and patient education, there has been a disappointing persistence of morbidity and mortality from other complications of the disease. Vascular disease involving any and all organs and regions of the body is likely to appear, causing blindness, renal failure, coronary artery disease, gangrene, and so on. We mentioned earlier (p. 175) the considerable thickening of the basement membrane in blood vessels likely to be seen upon microscopic examination in patients with diabetes.

It is not certain that tight regulation of the glucose level aids in avoiding these complications. Sensors permitting continuous measurement of glucose concentration are being developed, and kits are now available for use by patients in estimating their own blood glucose concentration; these may lead to better self-control of glucose concentration over longer periods. In the meantime, the measurement of concentrations of hemoglobins A1 (p. 260), is being used for assessing average glucose concentrations over long intervals. The level of Hb A1c was observed to decrease within a few weeks in patients kept under strict control in a hospital.

Two possible mechanisms are known for damage by excessive concentrations of glucose. One is the spontaneous glucosylation of proteins other than hemoglobin, and this could account for the damage to basement membranes, the lens of the eye, and so on. Another possibility, emphasized before protein glucosylation was recognized, is an accumulation of sorbitol. Many tissues, including peripheral neurons, the lens of the eye, and the liver, contain an aldose reductase that equilibrates aldohexoses and the corresponding polyols:

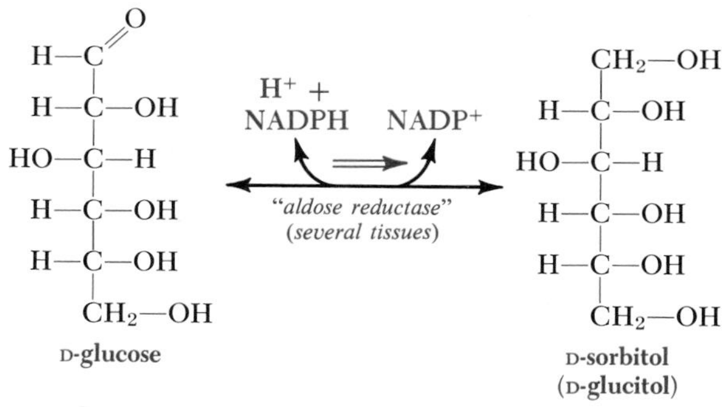

```
       O
      //
 H—C                                        CH2—OH
    |           H+ +                           |
 H—C—OH        NADPH      NADP+           H—C—OH
    |                                          |
HO—C—H      <——————————————————————>     HO—C—H
    |          "aldose reductase"              |
 H—C—OH        (several tissues)          H—C—OH
    |                                          |
 H—C—OH                                   H—C—OH
    |                                          |
   CH2—OH                                    CH2—OH
 D-glucose                                 D-sorbitol
                                          (D-glucitol)
```

D-Sorbitol is the polyol obtained by reducing D-glucose.*

*The name *sorbitol* came from its discovery in the mountain ash, *Sorbus* spp., along with L-sorbose (the 5-keto derivative of sorbitol as shown above). Sorbose is obscure compared to glucose, so glucitol is a more appropriate name, although less commonly used.

Therapeutic Measures. We must remember that the basic disturbance in diabetes is a failure to achieve adequate results of insulin action within the receptor tissues that are appropriate to the metabolic state. It is not a low insulin concentration *per se;* the circulating insulin concentration may be high in a definitely diabetic person who requires supplemental insulin for control. Many obese diabetic patients regain effective tissue responses to the level of insulin that they can produce simply by losing weight. Exercise also helps through adaptive increases in the capacity to transport glucose into the muscles effectively.

Patients with severe diabetes lose weight; they are losing fuel in the urine and have a diminished effectiveness for utilizing amino acids to make proteins. Not only is amino acid transport into the tissues impaired, but increased amounts of the amino acids are being degraded to make glucose, which is spilled into the urine. In contrast to the usual maturity-onset diabetic, these patients require insulin to avoid loss of weight.

Giving exogenous insulin one or more times during the day does not provide the adjustment in insulin level that normal pancreatic islet cells do as they constantly adjust the delivery of insulin (and glucagon) in response to the changing glucose level and other signals. Although insulin has been modified to provide different times of action (immediate, 6–8 hours, etc.) and although intake of food can be altered to try to match insulin effect and glucose level, proper control of diabetes with avoidance of both hyperglycemia and hypoglycemia can be extremely difficult in some people. An artificial pancreas is now in the stage of human experimental application. In this system, venous blood is continuously analyzed for glucose level. A computer is programmed to respond to the glucose concentration and rate of change of concentration by releasing insulin or glucagon into the blood stream. It is not certain if the procedure is implicated in the deaths of some subjects.

HORMONAL CONTROL OF CALCIUM METABOLISM

The concentration of calcium in blood is controlled closely: normal values lie between 2.1 and 2.6 millimoles per liter of serum. Part of the ion is chelated with proteins, so the concentration of the free ion is near 1.2 millimoles per liter. It is the free ion that determines the balance with the tissues.

Close control appears to be necessary because of the importance of calcium ion in the function of tissues generally. A role for calcium ion has been proposed for many of the hormone target cell effects described in this chapter. We have mentioned its importance in muscular contraction, and a fall in calcium ion concentration to 50 per cent of its normal value will cause tetany. Calcium appears to be required for normal function of most membranes.

Turnover of bone tissue enables adjustment of form to the kind of loads an individual routinely imposes upon his skeleton. Bone also serves as a reservoir of calcium ions; the bones of a 73 kg man have around 30 moles of calcium available for maintaining the 20 millimoles (0.8 gram) found in the body fluids.

Three hormones regulate the turnover of calcium. **1,25-Dihydroxycholecalciferol,** a steroid derivative, is made by the combined action of the skin, liver, and kidneys. **Parathormone,** a polypeptide secreted by the parathyroid gland, is another. The third is **calcitonin,** a polypeptide secreted by cells located in the connective tissue between follicles in the thyroid gland.

1,25-Dihydroxycholecalciferol

It was shown in 1919 that the childhood disease rickets, characterized by failure of proper development of the bones, could be prevented independently by a factor in

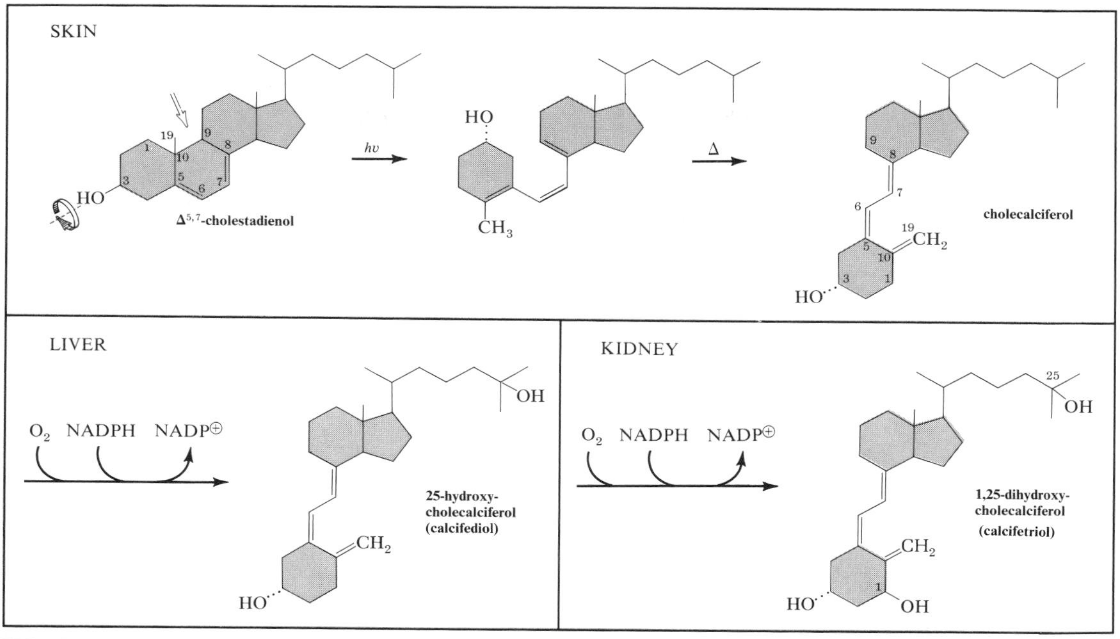

FIGURE 37–7 **The synthesis of the hormone 1,25-dihydroxycholecalciferol involves the successive action of three organs.**

the diet or by irradiation with ultraviolet light. The emphasis on the dietary factor led to the term **vitamin D** for any compound with curative action on rickets. It was later demonstrated that compounds with vitamin D activity were generated by the action of ultraviolet light on steroids containing conjugated double bonds, and that one of these, then designated vitamin D_3, was formed in the skin by the action of light on 5,7-cholestadienol, a normal intermediate in cholesterol synthesis (Fig. 37–7). The recognition that rickets was not a dietary deficiency disease but a sunlight deficiency disease led to the introduction of the name cholecalciferol to replace vitamin D_3.*

Cholecalciferol is transported in the blood in combination with a specific transport globulin, and it is taken up and stored in the liver. A 25-hydroxylase in the endoplasmic reticulum forms **25-hydroxycholecalciferol.** This enzyme is a typical monooxygenase, using NADPH as the second electron donor. It is regulated by feedback inhibition, so the product is maintained at a relatively low concentration.

The 25-hydroxy derivative is carried by the same transport globulin to the kidney. The kidney contains both a 1α- and a 24-hydroxylase that act on 25-hydroxycholecalciferol. The resultant **1,25-dihydroxycholecalciferol** is physiologically active, whereas the 24,25-dihydroxy compound is inert, and the 1,24,25-trihydroxy compound is less active.† The 1-hydroxylase is regulated by parathormone and by $[P_i]$ (next section); like the adrenal mitochondrial steroid hydroxylases, it acts through a cytochrome P-450, and contains an iron-sulfide protein (renal ferredoxin) for transport of electrons to oxygen from NADPH. The mechanism of the 24-hydroxylase is not as well known.

*There has been an unfortunate effort in recent years to resurrect the misleading vitamin D_3 terminology. It is sometimes useful to speak of vitamin D activity when talking about the several compounds that may be included in the diet to replace cholecalciferol, but there is no reason to use this name for a compound made by the body in adequate amounts under normal conditions.

†The trivial name calcifediol has been proposed for 25-hydroxycholecalciferol, which is a dihydroxy compound. By analogy, 1,25-dihydroxycholecalciferol would be calcifetriol.

TABLE 37–4 NORMAL PLASMA CONCENTRATIONS OF CHOLECALCIFEROL DERIVATIVES

	nM	μg/l
25-hydroxycholecalciferol	25 to 125	10 to 50
1,25-dihydroxycholecalciferol	0.07 to 0.12	0.03 to 0.05
24,25-dihydroxycholecalciferol	2 to 20	0.8 to 8

Action. 1,25-Dihydroxycholecalciferol is carried through the blood to target cells by the same cholecalciferol-transporting globulin; it is carried within the cells to the nucleus by attachment to an intracellular transport protein, and the complex is bound to chromatin. The apparent action of the compound is to promote the transcription of genes that facilitate transport of calcium and phosphate ions through the plasma membranes. In any event, this hormone has effects upon the intestine and bone that are clearly established. It increases absorption of calcium and phosphate ions from the digested food in the small intestine. The cytosol of intestinal mucosal cells contains a calcium-binding protein whose synthesis requires 1,25-dihydroxycholecalciferol, but this is evidently not the rate-controlling event in calcium absorption. 1,25-Dihydroxycholecalciferol also causes resorption of bone, with release of calcium and phosphate into the blood (along with citrate and other ions present in the bone). It appears that the prime role of this hormone is to maintain the calcium ion concentrations above hypocalcemic levels, thereby preventing tetany, with concomitant effects on phosphate metabolism. (The action on other tissues has been difficult to ascertain clearly.) Ability to increase the renal reabsorption of phosphate has been claimed but not yet proved.

Rickets

Infants and young children who are not exposed to ultraviolet light and who also lack cholecalciferol or related compounds in the diet develop rickets, a condition characterized by deficient mineralization of the bones. At the front where new bone is to be formed, an osteoid matrix is formed and then mineralized. In rickets this mineralization does not occur. Large areas of soft osteoid are formed—**osteomalacia.**

The relationship between lack of sunlight and incidence of rickets is quite clear. It is said that in the days before dietary supplementation all of the children in a New York hospital during the winter months had some signs of the disease, whereas few had the signs in the summer. The condition was mainly confined to the very young, who had the least exposure to daylight, especially in the tenement districts. Rickets is now rare in children reared in Western countries, although it has been noted in adolescent children and pregnant women among Asian immigrants in Great Britain. As the lifespan increases, more cases are being seen among the elderly, many of whom are housebound and on marginal diets.

Because of the prevalence of rickets, manufacturers were encouraged to fortify common foods such as milk, bread, and margarine. In addition, mothers were urged to give routine daily doses of vitamin D, principally as modified fish liver oils such as oleum percomorpheum (8,500 International Units, equivalent to 212 micrograms of cholecalciferol, per gram of oil). Infants require something in the neighborhood of 5 micrograms of cholecalciferol per day ***if they synthesize none,*** and 10 micrograms (400 I.U.) is usually specified to be safely above the minimum requirement.

The result of the program was a rapid decline in the incidence of rickets. This decline was greatly augmented by changes in the care of infants, who now receive more exposure to sunlight. However, cases of rickets still appeared and the pressure for more widespread use of fortified foods continued. The apparent consequences of this effort to supply vitamin D to those remaining deficient by increasing the intake of the whole population have diminished the joy of a nearly complete victory over rickets.

It has long been known and repeatedly stated in textbooks that large doses of vitamin D are toxic. The few who stopped to think saw that the intake of some children could be approaching the toxic level, since they had a high content of the compound in the diet, plus the amount synthesized internally during the summer, plus the large doses easily given from preparations commonly found in most households. Unfortunately, the few who saw were not heeded until frank cases of poisoning were finally recognized in a number of children. Most of the severe cases turned out to be children unusually prone to hypercalcemia, but further inquiry suggested the existence of marginal cases in children of normal response. The result has been a quiet de-emphasis on fortification of foods with cholecalciferol-like compounds, a drop in the amount of those compounds in preparations sold without prescription, and a plea to physicians to make some effort to find out how much an individual is already ingesting before prescribing more.

Humans are not alone in swallowing toxic doses of vitamin D. Cholecalciferol is formed in some plants, including the common European grass *Trisetum flavescens*. Cattle grazing the grass sometimes develop deposits of calcium salts in soft tissues (calcinosis).

Cholecalciferol-Related Bone Diseases in Chronic Renal Failure. Patients with inadequate kidney function who are maintained on hemodialysis develop a group of bone diseases called **renal osteodystrophy,** with osteomalacia and cystic bone changes (osteitis fibrosa cystica). Hemodialysis apparatus mimics kidney function in removing many toxins and natural waste products, but it does not restore the diminished ability to hydroxylate calcifediol in the 1 position to make calcifetriol. Calcifetriol, along with calcium supplements and phosphate binders, is used to treat the condition.

A variety of metabolites and analogues of cholecalciferol are available for treatment. Mindful of previous experiences with toxicity, current enthusiasm about the activity of individual compounds is coupled with information about the duration of toxic effects:

	Effective Daily Dose	Duration (days) of Toxic Effects
ergocalciferol	1 to 10 mg	17 to 60
calcifediol	0.05 to 0.5 mg	7 to 30
dihydrotachysterol	0.1 to 1 mg	3 to 14
calcifetriol	0.5 to 1 μg	2 to 10

Parathormone

The parathyroid glands respond to a fall in extracellular calcium ion concentration by secreting an 84-residue polypeptide hormone, parathormone, which acts upon bones and kidneys. Parathormone acts in concert with 1,25-dihydroxycholecalciferol to stimulate bone resorption. It also has important effects on the kidney: It causes increases in the activity of the 1α-hydroxylase that produces 1,25-dihydroxycholecalciferol, and it increases the reabsorption of calcium from the glomerular filtrate and decreases the reabsorption of phosphate. A low [P_i] is an independent stimulus for increasing the 1-hydroxylase activity, and it may be that the parathormone effect is an indirect result of its effect on phosphate absorption. Parathormone and 1,25-dihydroxycholecalciferol therefore independently act to raise the calcium level in the blood and extracellular fluid, but their effects are also interrelated. Parathormone stimulates synthesis of 1,25-dihydroxycholecalciferol, thereby indirectly assuring increased absorption of calcium by the small intestine. On the other hand, 1,25-dihydroxycholecalciferol evidently promotes the transport of calcium into the parathyroid cells, thereby enhancing the signal for cessation of parathormone formation.

Parathormone is synthesized as a preprohormone of some 115 residues. As with other preproproteins, the "pre" sequence is hydrophobic and serves to attach the protein to the endoplasmic reticulum. Following emergence, an approximately 25-residue sequence is removed, leaving the proprotein. Another six residues are removed, probably in the Golgi apparatus, to form the active hormone. Biological activity resides in the amino terminal 30 residues.

The 84-residue hormone is rapidly cleaved at the proximal end (residues 33 to 37) after release from the parathyroid glands. The active amino terminal fragment and the intact hormone have half-lives of about 5 minutes in the blood, whereas the carboxyl-terminal fragment has a half-life of 30 to 40 minutes. The latter fact is mainly of importance in interpreting immunoassays of blood parathormone concentrations. Some use antibodies specific for the amino terminal fragment, some for the carboxy terminal fragment, and some for determinants on both.

Parathormone acts at multiple sites to increase absorption of calcium in the gastrointestinal tract, to activate osteoclasts to demineralize bone, and to increase the reabsorption of calcium by kidney tubules. It is thought to activate adenylate cyclase in the target cells.

PSEUDOHYPOPARATHYROIDISM

A relatively uncommon disease with a prevalence of one in 100,000 to 1,000,000.

End organs are resistant to the hormone.

May be due to absence of the coupling portion of the adenylate cyclase complex.

Calcitonin

Clusters of calcitonin-secreting cells are found in the connective tissue between thyroid follicles; these cells, unlike the follicular cells, are derived from the embryonic neural crest and are related to cells in the adrenal medulla. Calcitonin is a 32-residue polypeptide that is secreted when the calcium ion concentration rises. Gastrin (p. 147), secreted by the stomach after ingestion of a meal, also promotes calcitonin secretion. Perhaps this aids in preventing hypercalcemia upon absorption of the ion from the food.

Calcitonin acts to lower extracellular calcium ion concentrations in at least three ways: It decreases the resorption of bone by osteoclastic cells, it increases the formation of osteoblasts that deposit bone minerals, and it increases the loss (decreases the reabsorption) of both calcium and phosphate ions in the urine. Calcitonin decreases the activity of the 1α-hydroxylase acting on 25-hydroxycholecalciferol. In general, calcitonin appears to be a device for protection against hypercalcemia, causing increased deposition of bone, decreased absorption from the diet, and decreased recovery from the glomerular filtrate. It also has a phosphate-lowering effect.

Hypercalcemia can cause almost every organ of the body to malfunction, sometimes creating a medical emergency. Unfortunately, the symptoms are not specific and may be ignored or misinterpreted. Bad dreams, bizarre behavior, hallucinations, depression, nausea, vomiting, and constipation are frequent. Progression to coma or presentation with coma may be required to force attention to the abnormal calcium concentration.

PROSTAGLANDINS, THROMBOXANES, AND LEUKOTRIENES

Swedish investigators exploring the effects of semen on the uterus discovered that it indeed contained compounds affecting uterine motility.* Believing them to be formed by the prostate gland, the investigators called them prostaglandins. (They are actually elaborated in quantity by the seminal vesicle.) This small beginning led to a still continuing explosion of effort that revealed whole families of compounds, having in common an origin from polyunsaturated fatty acids such as arachidonic acid and a limitation of effects to the cells that make them and the neighboring cells. Although they are effective at low concentrations, they are at best local hormones, paracrine rather than endocrine, but they fit the definition of hormone.

*The 1982 Nobel Prize in Physiology or Medicine was awarded to Sune Bergström and Bengt Samuelson from Sweden, along with John Vane from England, for their work on this class of hormones.

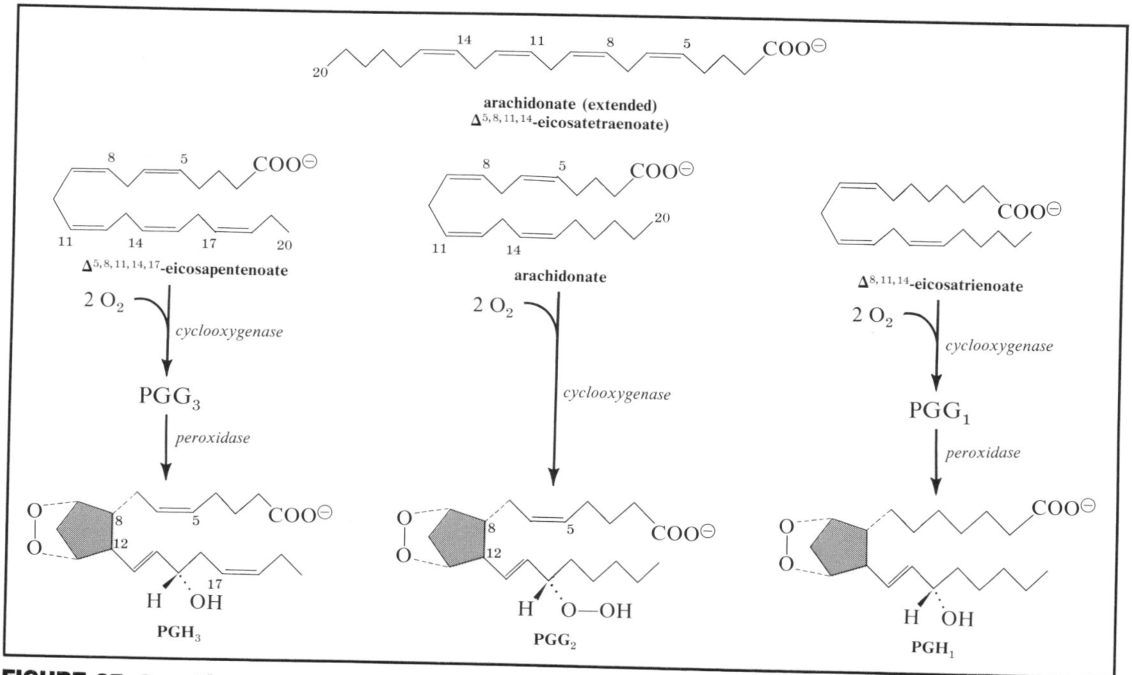

arachidonate (extended) ($\Delta^{5,8,11,14}$-eicosatetraenoate)

$\Delta^{5,8,11,14,17}$-eicosapentenoate → ($2\ O_2$, *cyclooxygenase*) → PGG_3 → (*peroxidase*) → PGH_3

arachidonate → ($2\ O_2$, *cyclooxygenase*) → PGG_2

$\Delta^{8,11,14}$-eicosatrienoate → ($2\ O_2$, *cyclooxygenase*) → PGG_1 → (*peroxidase*) → PGH_1

FIGURE 37–8 The prostaglandins are synthesized from 20-carbon polyunsaturated fatty acids with varying numbers of double bonds. The synthesis first forms a 15-hydroperoxy derivative of a cyclic endoperoxide, designated as PGG (*illustrated at bottom center*), which is converted to a 15-hydroxy derivative by a peroxidase (*illustrated at bottom left and right*).

These compounds modify plasma membrane responses, and in bewilderingly diverse ways. There are many different prostaglandins and relatives, often with quite different physiological actions. The net effect on a given tissue is the result of these actions rather than a definitive change we can describe in simple terms.

Our goal is to introduce these structures and point out the features in which they differ. First a note on nomenclature. The prostaglandins are designated by symbols such as PGE_1. The PG stands for prostaglandin, and the third capital letter indicates the type of substituents found on the hydrocarbon chain. The subscript number

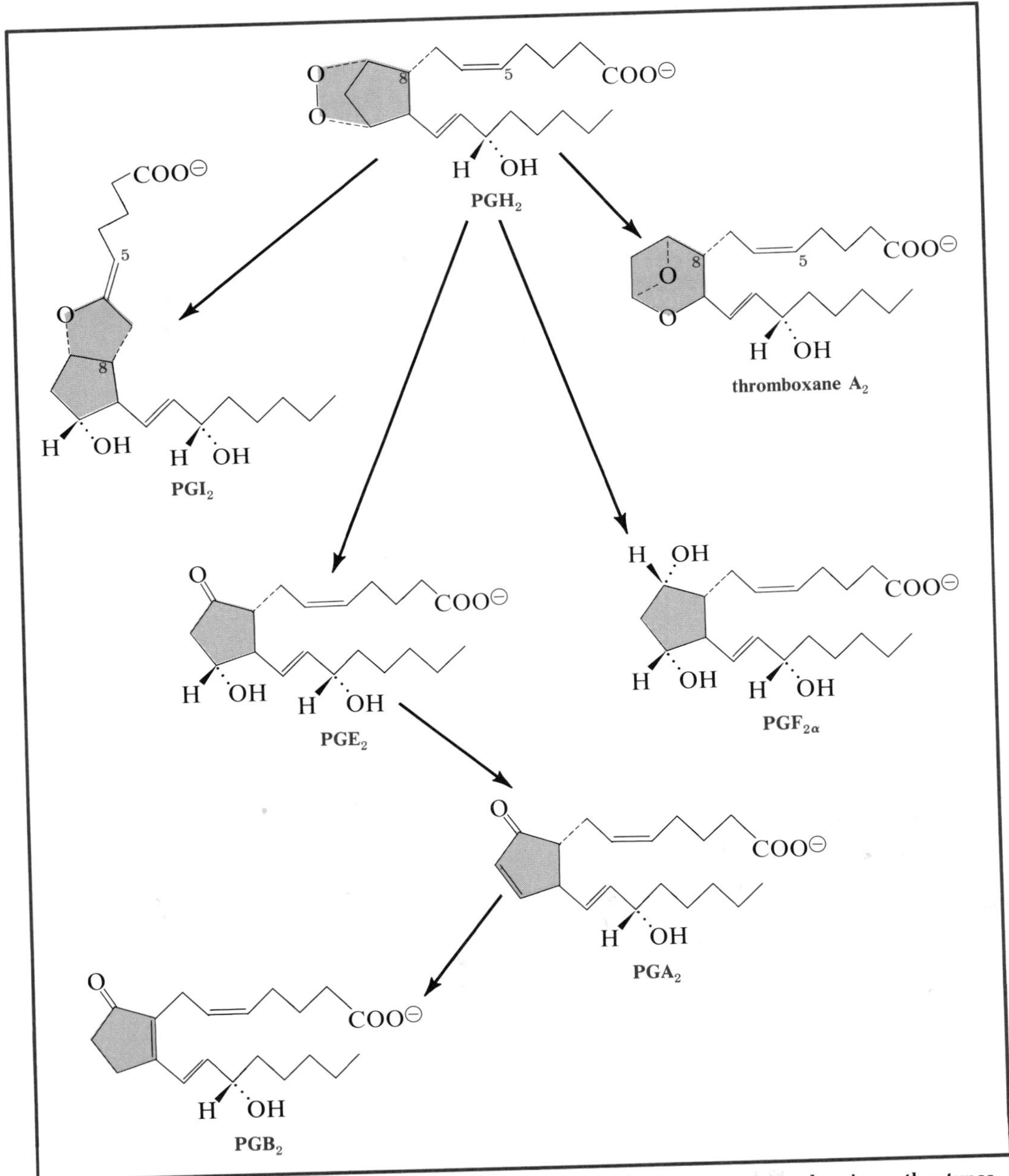

FIGURE 37–9 **Conversion of prostaglandins G to thromboxanes (*upper right*) and various other types of prostaglandins.**

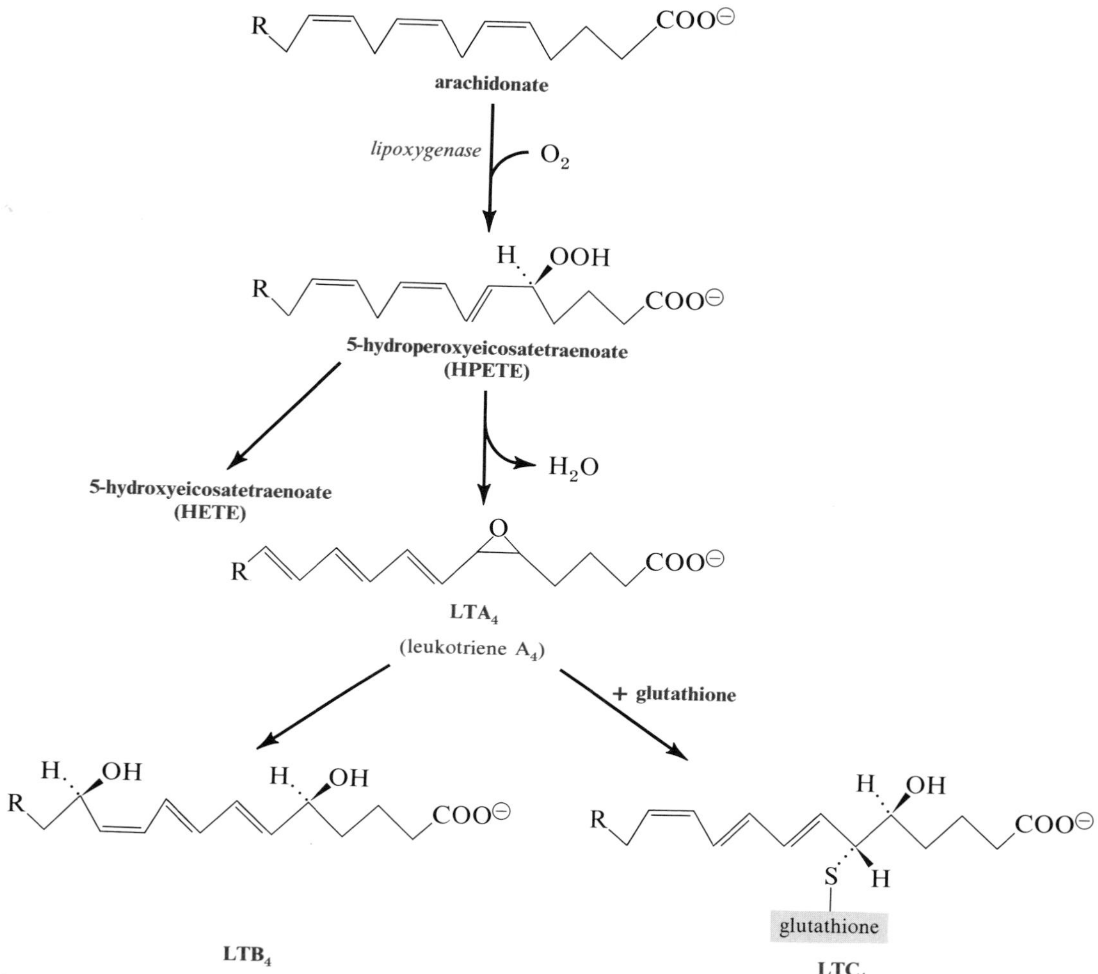

arachidonate

lipoxygenase O_2

5-hydroperoxyeicosatetraenoate (HPETE)

H_2O

5-hydroxyeicosatetraenoate (HETE)

LTA_4 (leukotriene A_4)

+ glutathione

LTB_4

glutathione

LTC_4

FIGURE 37–10 **Formation of leukotrienes involves an oxidation of arachidonate by a lipooxygenase. The leukotrienes include a glutathione conjugate.**

indicates the number of double bonds in the molecule. Thromboxanes are designated by TX and leukotrienes by LT.

The first step in the synthesis of prostaglandins and thromboxanes is the addition of oxygen catalyzed by **cyclooxygenase,** a cyclizing dioxygenase (Fig. 37–8). **The anti-inflammatory action of agents such as aspirin and indomethacin is due to their inhibition of this initial step in prostaglandin formation.** The product is a hydroperoxy prostaglandin (PGG) that is converted to a hydroxy prostaglandin (PGH) by a peroxidase. Figure 37–8 shows how the synthesis of 1, 2, and 3 series is initiated using different polyunsaturated fatty acids, but let us concentrate on the more abundant 2 series derived from arachidonate.

The endoperoxide produced by the cyclizing dioxygenase is the precursor of the thromboxanes and other prostaglandins (Fig. 37–9). The regulation of these diverse pathways is not understood. Study of the compounds is difficult owing to their low concentrations, their very short half-life (the thromboxanes have half-lives on the order of seconds), and to the presence of so many different compounds.

We can savor the complexity of the responses from the following observations: The endoperoxides, PGG_2 and PGH_2, as well as the thromboxanes derived from them, induce rapid and irreversible clumping of blood platelets. Contrariwise, the prostacyclin PGI_2 is a potent inhibitor of platelet aggregation. The opportunities for speculation on how it is that normal platelets are kept from aggregating and promptly caused to aggregate upon exposure to damaged vessels are obvious. Turning to another tissue, the smooth muscle of the bronchus or of the uterus in nonpregnant animals is relaxed by PGE_2, but it is contracted by PGF_2. Both compounds cause contraction of the uterus in pregnant animals.

Leukotrienes are made from arachidonate by another pathway, beginning with the addition of oxygen to C5 catalyzed by lipoxygenase (Fig. 37–10). This reaction is not affected by anti-inflammatory drugs. Neutrophils make one class of leukotrienes to alter motility and act as chemotactic agents. Mast cells make another class, formerly known as the **slow-reacting substance,** which is responsible for bronchial constriction and other anaphylactic allergic reactions. Contact with IgE (p. 236) causes prompt release of arachidonate, the precursor of the slow-reacting substance, from phosphatidylcholine and phosphatidylinositol.

FURTHER READING

Receptors

J.B. Baxter and J.W. Funder (1979). Hormone Receptors. N. Engl. J. Med. 301:1149. Review.

J.M. Stadel, A. DeLean, and R.J. Lefkowitz (1982). Molecular Mechanisms of Coupling in Hormone Receptor-Adenylate Cyclase Systems. Adv. Enzymol. 53:1.

Catecholamines

G. Kunos, ed. (1981). Adrenoreceptors and Catecholamine Action. Wiley. A comprehensive review.

J.E. Dimsdale and J. Moss (1980). Plasma Catecholamines in Stress and Exercise. J. Am. Med. Assoc. 243:340.

P.E. Cryer (1980). Physiology and Pathophysiology of the Human Sympathoadrenal Neuroendocrine System. N. Engl. J. Med. 303:436. Review.

W.H. Frishman (1981). β-Adrenoreceptor Antagonists: New Drugs and New Indications. N. Engl. J. Med. 303:500.

B.B. Hoffman and R.J. Lefkowitz (1980). Alpha-Adrenergic Receptor Subtypes. N. Engl. J. Med. 302:1390. Review.

Thyroid Hormones

P.R. Larsen et al. (1982). Thyroid Hypofunction after Exposure to Fallout from a Hydrogen Bomb Explosion. J. Am. Med. Assoc. 247:1571.

J.-M. Gavaret, J. Nunez, and H.C. Cahnmann (1981). Formation of Dehydroalanine Residues During Thyroid Hormone Synthesis in Thyroglobulin. J. Biol. Chem. 255:5281.

S. Ohtaki et al. (1981). Analyses of Catalytic Intermediates of Hog Thyroid Peroxidase during Iodinating Reactions. J. Biol. Chem. 256:805.

J.T. Dunn, P.S. Kim, and A.D. Dunn (1981). Favored Sites for Thyroid Hormone Formation on the Peptide Chains of Human Thyroglobulin. J. Biol. Chem. 257:88.

A.J. Van Herle, G. Vassart, and J.E. Dumont (1979). Control of Thyroglobulin Synthesis and Secretion. N. Engl. J. Med. 301:239,307. Review.

I.J. Chapra (1981). New Insights into Metabolism of Thyroid Hormones: Physiological and Clinical Implications. Prog. Clin. Biol. Res. 74:67.

D.J. Gruol and E.S. Kempner (1982). The Size of the Thyroid Hormone Receptor in Chromatin. J. Biol. Chem. 257:708.

A.J. Perlman, F. Stanley, and H.H. Samuels (1982). Thyroid Hormone Nuclear Receptor. Evidence for Multimeric Organization in Chromatin. J. Biol. Chem. 257:930.

J.H. Oppenheimer (1981). The Molecular Basis of Thyroid Hormone Action: Scattered Pieces of a Jigsaw Puzzle. Prog. Clin. Biol. Res. 74:45.

Anterior Pituitary Hormones

J.G. Pierce and T.F. Parsons (1981). Glycoprotein Hormones. Annu. Rev. Biochem. 50:465.
R.M. Bergland and R.B. Page (1979). Pituitary-Brain Vascular Relations: A New Paradigm. Science 204:18.
I.M.D. Jackson (1982). Thyrotropin-Releasing Hormone. N. Engl. J. Med. 306:145.
P.R. Larsen (1982). Thyroid-Pituitary Interaction. Feedback Regulation of Thyrotropin Secretion by Thyroid Hormones. N. Engl. J. Med. 306:23.
J.A. Phillips et al. (1981). Molecular Basis for Familial Isolated Growth Hormone Deficiency. Proc. Natl. Acad. Sci. (U.S.) 78:6372.
D.T. Krieger and J.B. Martin (1981). Brain Peptides. N. Engl. J. Med. 304:876,944.
U. Gubler et al. (1982). Molecular Cloning Establishes Proenkephalin as Precursor of Enkephalin-Containing Peptides. Nature 295:206.

Adrenocortical Hormones

A.M. Neville and M.J. O'Hare (1982). The Human Adrenal Cortex. Springer-Verlag.
R. Neher et al. (1982). Compartmentalization of Corticotropin-Dependent Steroidogenic Factors in Adrenal Cortex: Evidence for a Post-Translational Cascade in Stimulation of the Cholesterol Side-Chain Split. Proc. Natl. Acad. Sci. (U.S.) 79:1727.
R.J. Santen et al. (1982). Aminoglutethimide as Treatment of Postmenopausal Women with Advanced Breast Carcinoma. Ann. Intern. Med. 96:94.

Sex Hormones

J. Imperato-McGinley et al. (1979). Androgens and the Evolution of Male-Gender Identity Among Male Pseudohermaphrodites with 5α-Reductase Deficiency. N. Engl. J. Med. 300:1233.
D.B. Hier and W.F. Crowley, Jr. (1982). Spatial Ability in Androgen-Deficient Men. N. Engl. J. Med. 306:1202.
R. Claus, H.O. Hoppen, and H. Kang (1981). The Scent of Truffles: A Steroidal Phenomenon? Experientia 37:1178.

Pancreatic Hormones

L.S. Phillips and R. Vassilopoulou-Sellin (1980). Somatomedin. N. Engl. J. Med. 302:371,438.
R.H. Unger and L. Orci (1981). Glucagon and the A Cell. N. Engl. J. Med. 304:1518,1575. Review.
V. Schusdziarra (1980). Somatostatin—A Regulatory Modulator Connecting Nutrient Entry and Metabolism. Horm. Metab. Res. 12:563.
M. Kasuga, F.A. Karlsson, and C.R. Kahn (1982). Insulin Stimulates the Phosphorylation of the 95,000-Dalton Subunit of its Own Receptor. Science 215:185.
J.H. Kinoshita, P. Kador, and M. Catiles (1981). Aldose Reductase in Diabetic Cataracts. J. Am. Med. Assoc. 246:257.
G.F. Cahill and H. McDevitt (1981). Insulin-Dependent Diabetes Mellitus: The Initial Lesion. N. Engl. J. Med. 304:1454. Review.
D.S. Schadeet et al. (1982). A Remotely Programmable Insulin Delivery System. Successful Short-Term Implantation in Man. J. Am. Med. Assoc. 247:1848. Also see 247:1918 for report of possible connection of use of insulin pumps with deaths of diabetic patients.
D.C. Robbins et al. (1981). A Human Proinsulin Variant at Arginine 65. Nature 291:679.
G.N. Borrow, B.E. Hazlett, and M.J. Phillips (1982). A Case of Diabetes Mellitus. N. Engl. J. Med. 306:340. Discussion of first patient treated with insulin. Also see 307:127 (1982).

Cholecalciferols and Parathyroid Hormones

A. Veis, ed. (1981). The Chemistry and Biology of Mineralized Connective Tissue. Elsevier.
H. Zucker, H. Stark, and W.A. Ramsbeck (1980). Light-Dependent Synthesis of Cholecalciferol in a Green Plant. Nature 283:69.
D.R. Fraser (1981). Biochemical and Clinical Aspects of Vitamin D Function. Br. Med. Bull. 37:37. Review.
J.S. Adams et al. (1982). Vitamin D Synthesis and Metabolism after Ultraviolet Irradiation of Normal and Vitamin D-Deficient Subjects. N. Engl. J. Med. 306:722.
M.R. Haussler and P.E. Cordy (1982). Metabolites and Analogues of Vitamin D. Which for What? J. Am. Med. Assoc. 247:841.
E. Slatopolsky et al. (1982). Current Concepts of the Metabolism and Radioimmunoassay of Parathyroid Hormone. J. Lab. Clin. Med. 99:309. Review.
L.A. Austin and H. Heath, III (1981). Calcitonin. Physiology and Pathophysiology. N. Engl. J. Med. 304:269. Review.

Prostaglandins, Leukotrienes, and Thromboxanes

B. Samuelson et al. (1978). Prostaglandins and Thromboxanes. Annu. Rev. Biochem. 47:997.

R.A. Lewis and K.F. Austen (1981). Mediation of Local Homeostasis and Inflammation by Leukotrienes and Other Mast Cell-Dependent Compounds. Nature 293:103. Review.

A.J. Schafer (1982). Deficiency of Platelet Lipoxygenase Activity in Myeloproliferative Disorders. N. Engl. J. Med. 306:381.

R.P. Mason and C.F. Chignell (1981). Free Radicals in Pharmacology and Toxicology—Selected Topics. Physiol. Rev. 33:189. Includes discussion of prostaglandin synthesis.

S. Moncada and J.R. Vane (1979). Arachidonic Acid Metabolites and the Interactions Between Platelets and Blood-Vessel Walls. N. Engl. J. Med. 300:1142. Review.

J.B. Lee, ed. (1982). Prostaglandins. Elsevier. General review. (One of a continuing series on hormones.)

Addendum

R.A. Roth and D.J. Cassells (1983). Insulin Receptor: Evidence That It Is a Protein Kinase. Science 219:299.

CHAPTER 38

CONTROL OF WATER AND ION BALANCE

Maintenance of a constant concentration of most mineral ions hinges upon balancing intake and excretion, because they are not chemically modified during their sojourn in the body. (Sulfate and phosphate compounds are sometimes exceptions.) Control is exerted at both ends, but the physiological mechanisms concerned with thirst and salt hunger are beyond our scope. Let us concentrate on the biochemical devices for regulating water and ionic excretion by the kidney.

THE KIDNEYS

A kidney is essentially a device for filtering a large volume of blood and passing the filtrate through a long tubule, which is lined with cells that selectively transport substances into and out of the filtrate. Most of the selective transport involves uptake of water and solutes from the filtrate for re-use within the organism. Some of the transport is by active secretion from the cells into the filtrate. The product of all this processing is the urine, which, if all goes well, contains any surplus of ingested water and electrolytes, along with the daily production of urea, uric acid, creatinine, and other waste products not eliminated elsewhere. The result is shown in Figure 38–1.

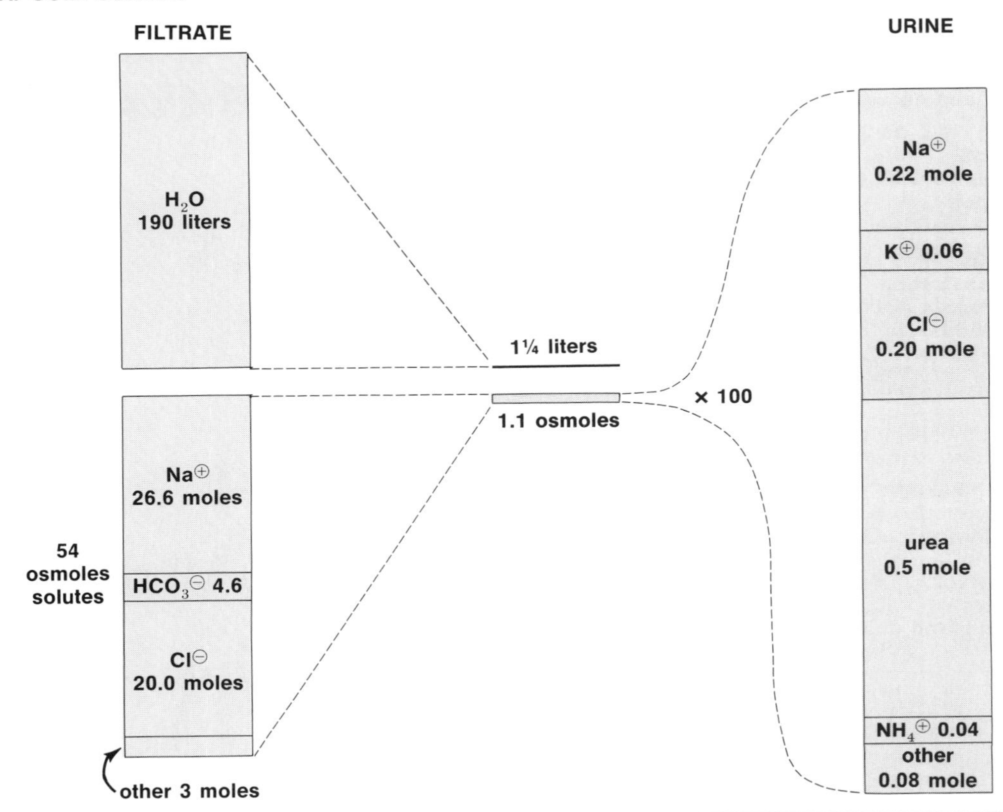

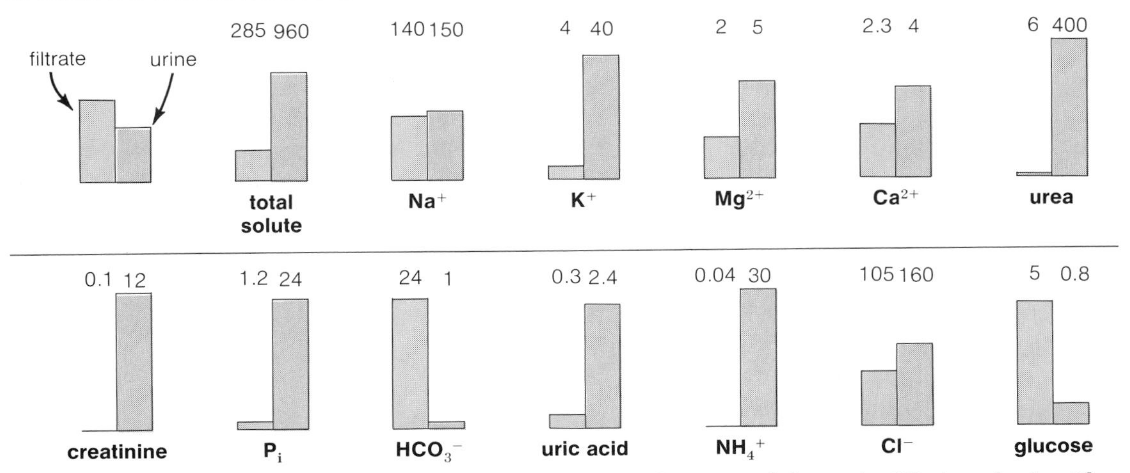

FIGURE 38–1 *A.* Changes in the total composition of the daily output of glomerular filtrate and urine. The scale of solute composition of the urine is expanded on the right. Typical values are given, but the quantity of water and solutes in the urine of normal individuals depends upon their dietary intake and varies widely from that shown.

B. Relative concentrations of individual components in the filtrate and urine are shown by bars. Actual millimolarities are also listed: filtrate in gray, urine in brown.

Anatomy

The anatomical features most important for our discussion are summarized in Figure 38–2. We are concerned with the flow of two fluids—the blood and the filtrate. A veritable torrent of blood, one fifth of the cardiac output at rest, enters the kidneys, where it passes into tufts of capillaries within glomeruli. A glomerulus is a filtration chamber; the space around the capillary tuft is at relatively low pressure, but the blood within the tuft is kept at relatively high pressure because the efferent arteriole carrying blood from the tuft is constricted more than the afferent arteriole. This high pressure gradient causes a rapid weeping of fluid through the capillary walls. Small solutes pass with little constraint; only a very minor portion of serum albumin and other proteins get through, and virtually no cells.

The volume of the resultant **glomerular filtrate** is roughly one fifth of the volume of the blood plasma passing through the kidney. The cardiac output of a median adult male at rest is about 6 liters per minute, of which one fifth, or 1.2 liters, passes through the kidneys. In that volume are 0.65 liter of plasma, of which one fifth, or 0.13 liter, appears as glomerular filtrate.

The filtrate is drained from each glomerulus by a **tubule,** which has a highly **convoluted proximal portion** located in the cortex (outer layer) of the kidney that then becomes relatively straight and dips toward the interior, where it forms the **loop of Henle.** Those tubules issuing from glomeruli deeper in the cortex have especially long loops, going nearly through the medulla before returning to the cortex. The descending limb and part of the ascending limb of the loop of Henle are very thin-walled for easy passage of water.

The **distal portion** of the tubule is again convoluted before turning once more toward the medulla as a **collecting tubule** of different structure. Collecting tubules join within the medulla as **collecting ducts** of still different properties, which deliver urine into the renal pelvis.

The tubule cells are surrounded throughout their length by capillary networks fed by the efferent arterioles of the glomeruli. The blood is at a relatively low pres-

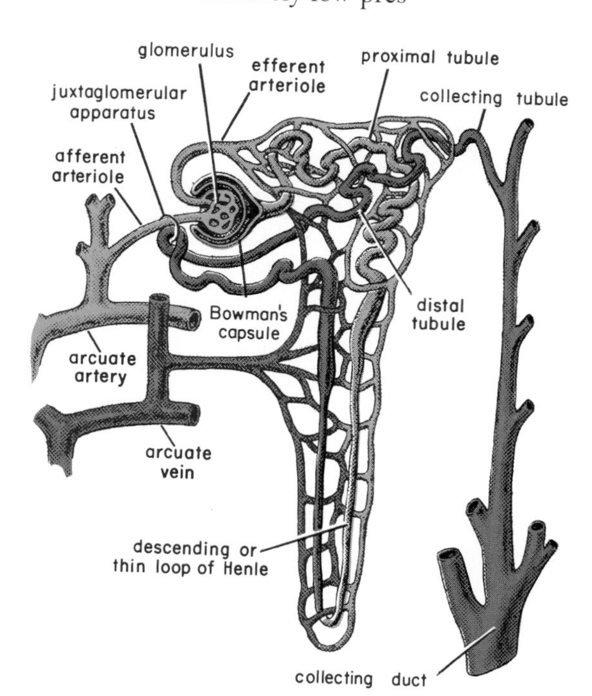

FIGURE 38–2

Schematic outline of the anatomy of a nephron in the kidney. Modified from A.C. Guyton (1971). Textbook of Medical Physiology, 4th ed. Saunders.

sure in these networks, which facilitates absorption. To summarize, there is a high pressure gradient from capillaries to lumen in the glomerulus to promote filtration, and a low adverse pressure gradient from lumen to capillaries in the tubules to promote absorption. Most of the blood passes through networks around the proximal and distal tubules; a smaller portion, a few per cent, passes through the **vasa recta** ("straight vessels"), which carry the blood deep into the medulla and back again, much like the flow of filtrate in the neighboring loops of Henle.

Absorption in the Proximal Tubule

The proximal tubule removes about 65 per cent of the water and salt from the filtrate passing through it, so that the 130 ml min^{-1} entering the tubules diminishes to approximately 46 ml min^{-1} at the beginning of the loops of Henle. The absorption requires energy. Most of the energy is supplied by (Na^+ + K^+)-ATPase; the Na^+ concentration gradient created by this enzyme is used to move other ions in and out of the tubule cells. A lesser amount of energy is provided by the contraction of the heart, which of course also involves the hydrolysis of ATP. It is the pressure of arterial blood that creates an almost protein-free filtrate in the kidney glomeruli. Owing to this lack of protein, the total concentration of solutes in the filtrate is less than it is in blood plasma. Put another way, the filtrate has a higher water concentration than does the blood plasma. The filtrate therefore has a lower osmotic pressure than does blood plasma. Let us summarize events in more or less causal sequence.

Water will tend to flow from the filtrate through the tight junction and the cells toward the blood:

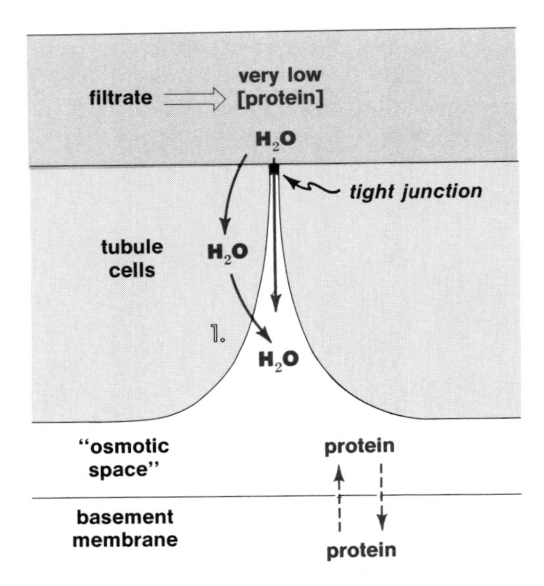

In doing so it will pass first into an **osmotic space** between the tubule cells themselves and also between them and the basement membrane. The diagram is simplistic; the tubule cells have intricate invaginations to magnify the surface exposed to the osmotic space, but the separation between lateral membranes of the tubule cells and between their basal membranes and the basement membrane is very small. **The osmotic space is tiny compared to the volume of the tubule cells.** As indicated, there is some movement of protein across the basement membrane in and out of the osmotic space, and this alone would make it more concentrated than the filtrate in the lumen.

Mineral ions are returned from the glomerular filtrate to the blood as a result of the action of ($Na^+ + K^+$)-ATPase coupled with differential permeabilities of the plasma membrane surfaces in the tubule cells:

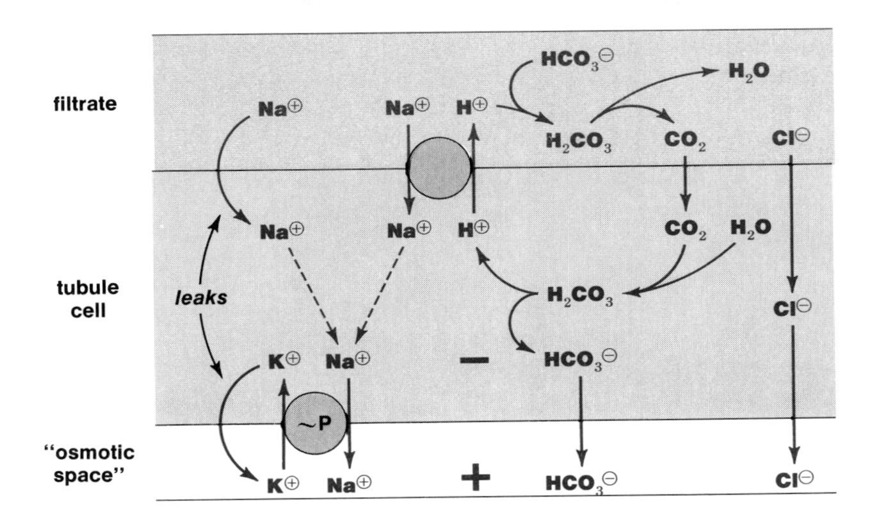

The ATPase is widely distributed in the **basolateral membrane** (the portion of the plasma membrane facing the basement membrane and adjacent tubule cells). The basolateral membrane is permeable to K^+, but not to Na^+, so the K^+ pumped into the cytosol can leak out of the cell, but the Na^+ cannot leak back. An electrochemical gradient is established; the $[Na^+]$ is lower in the cytosol than it is in either the osmotic space or the glomerular filtrate, and the cytosol is negative relative to the osmotic space.

Na^+ moves from the filtrate to the cytosol by several routes. It passively diffuses across the luminal membrane, which, unlike the basolateral membrane, is leaky to Na^+. The luminal membrane also contains transporters for Na^+. Among these is an Na^+/H^+ antiporter, which uses the Na^+ concentration gradient to pump H^+ into the filtrate from the cytosol. This generates an H^+ concentration gradient across the luminal membrane—lumen more acidic than cytosol. (The sequence of events described here pertains to the usual circumstances in which the urine is made more concentrated and more acidic than the blood plasma.)

Bicarbonate ions are actively transported from the glomerular filtrate to the cytosol by using the H^+ concentration gradient as an energy source. The luminal membrane is impermeable to bicarbonate ions, but like other membranes, it is freely permeable to CO_2. Dissolved CO_2, carbonic acid, and bicarbonate ions are in equilibrium both in the filtrate and the cytosol, but the respective positions of equilibrium are shifted in opposite directions by the differences in H^+ concentration:

lumen: $$H^+ + HCO_3^- \longrightarrow H_2CO_3 \longrightarrow H_2O + CO_2$$

$$CO_{2\,(lumen)} \longrightarrow CO_{2\,(cytosol)}$$

cytosol: $$CO_2 + H_2O \longrightarrow H_2CO_3 \longrightarrow HCO_3^- + H^+$$

Bicarbonate ions are taken up in the cytosol by dissipating the H^+ concentration gradient. The basolateral membrane is permeable to bicarbonate ions; both the increased intracellular bicarbonate concentration and the electrical potential between cytosol and osmotic space push bicarbonate from the cytosol into the osmotic space.

The transport of bicarbonate requires an active carbonic anhydrase in the tubule cells to accelerate the too-slow spontaneous equilibration of dissolved CO_2 and carbonic acid. **Acetazolamide,** an inhibitor of carbonic anhydrase, causes excretion of an increased volume of an alkaline urine. (It affects many tissues, and is used not only as a diuretic in congestive heart failure, but also in treating epilepsy and glaucoma, a condition causing increased pressure within the eye.)

Chloride ions may also be actively transported from lumen to cytosol by a Cl^-/H^+ symporter (not shown in the preceding diagram). The positive potential in the osmotic space would also attract them from the cytosol.

As the mineral ions are moved from the filtrate to the osmotic space, they create an osmotic gradient. The gradient never gets large because water rapidly flows through the cells into the osmotic spaces:

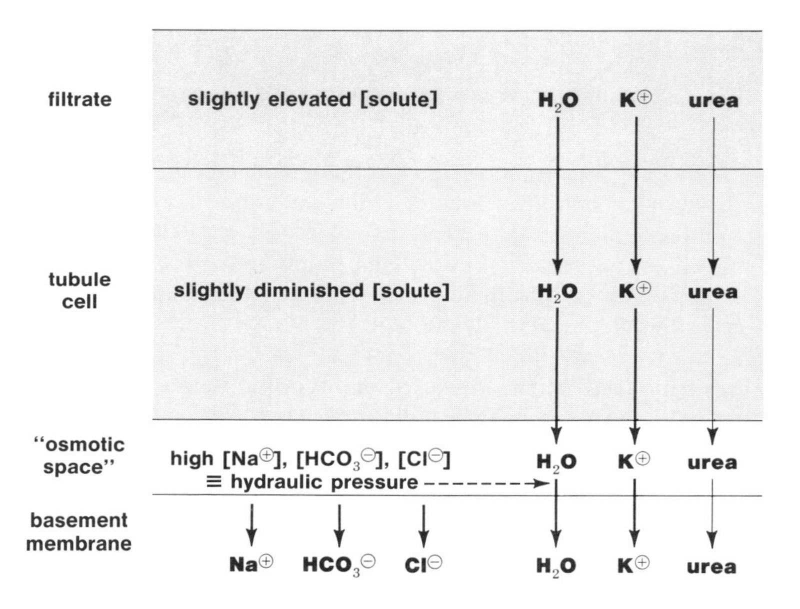

Since the osmotic space is small, this creates a hydraulic pressure in the space, which forces water across the basement membrane and into the surrounding capillary network. Water moves so readily that the concentration differences and hydraulic pressure gradient necessary for effective flow are slight, and the remaining filtrate after passage through the initial portion of the proximal convoluted tubule has nearly the same total solute concentration (osmolarity) as does the blood plasma. The loss of water from the filtrate elevates its concentrations of K^+ and urea, so they also travel with the water toward the blood plasma by passive diffusion through the intervening membranes.

Some organic solutes, especially glucose and the amino acids, are also actively transported from the lumen into tubule cells. Symporters simultaneously move Na^+ into the cytosol to provide the driving force:

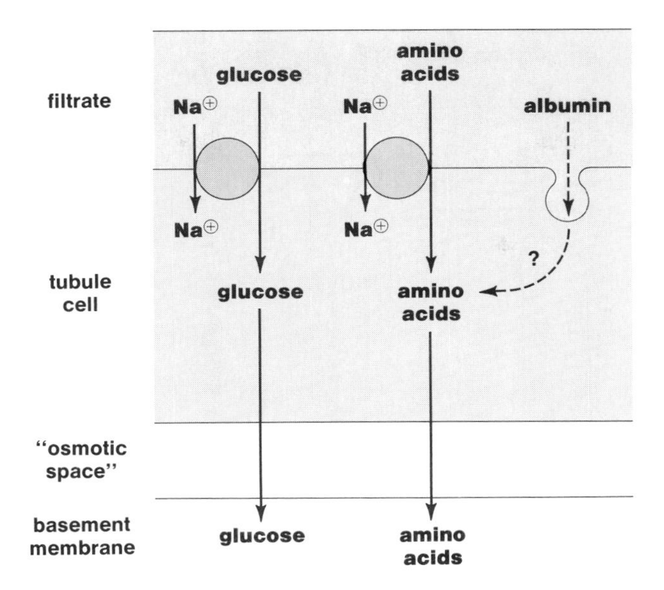

The small amounts of protein, mainly albumin, also present in the filtrate are taken up, presumably by pinocytosis, and degraded to the constituent amino acids within the tubule cells.

The Countercurrent Mechanism

The luminal fluid leaving the proximal tubule has the same total solute and Na^+ concentrations as does the blood plasma. The urine usually does not. Therefore, the further conversion of the tubular fluid to urine requires additional water and salt absorption, but with the relative amounts adjusted to fit circumstances. These deviations of urinary concentrations from blood concentrations hinge upon the triple passage of the luminal fluid through the medullary region of the kidney—once down the loop of Henle, back up, and then back down again through the collecting ducts. During each of these passages, the fluid encounters luminal membranes with different transport properties (Fig. 38–3). Let us list these properties and then discuss how they are used to create urine.

The thin descending limb of the loop is freely permeable to water, but only slightly permeable to Na^+, Cl^-, or urea. The thin ascending limb is almost the opposite; it is nearly impermeable to water, and very permeable to Na^+ and Cl^-, although much less permeable to urea. The thick-walled section of the ascending limb, nearer the cortex, is different. The cells are loaded with mitochondria to drive an active pump for Cl^-. (The mechanism is unknown.) They may also have some (Na^+ + K^+)-ATPase activity, but most of the Na^+ that leaves the cells does so because of the electrical potential created by the Cl^- pump. This section of the tubule is also nearly impermeable to water and urea.

Most of the distal tubule has little water permeability. However, the collecting tubules, and the adjacent segments of the distal tubules, have a regulated permeability to water, which depends upon the presence of an antidiuretic hormone, vasopressin, discussed below. The collecting tubule has almost no permeability to urea; however, the collecting ducts in the inner medulla do permit some urea to leave the lumen.

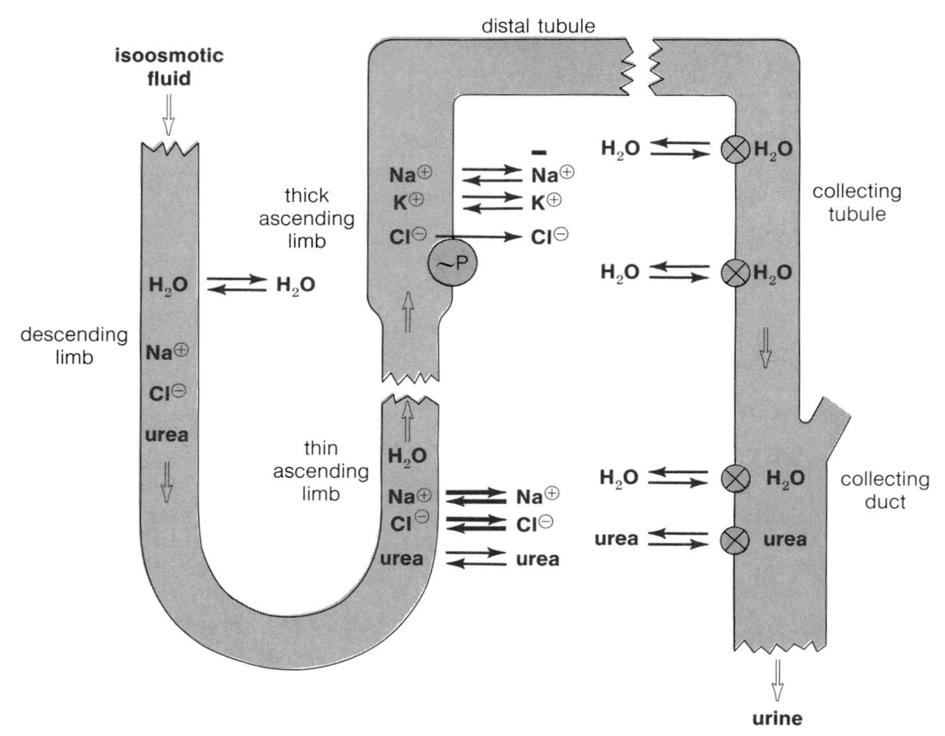

FIGURE 38–3

Diagram of permeabilities of different sections of the nephron after the proximal tubule, which delivers an isoosmotic fluid to the loop of Henle. This schematic diagram does not show the tubule cells and indicates only the luminal membrane. The greater diameter of the thick ascending limb is due to larger cells, not a thicker membrane. ⊗ indicates vasopressin-regulated channels for water passage.

The Concentration Gradient in the Medulla. There is an all-important result of the different transport properties of tubular segments. Let us simply state it as a fact for the moment:

> **Whenever the kidney is producing a concentrated urine, the interstitial fluid surrounding the loop of Henle and the collecting ducts becomes more and more concentrated with respect to NaCl and urea from the cortex through the medulla to the pelvis.**

The result is that the fluid within the tubules is also more and more concentrated as it moves down the descending limb of the loop, less and less concentrated as it moves up the ascending limb, and again more and more concentrated as it moves down the collecting tubules and collecting ducts. However, these fluctuations in total concentrations within the tubules involve *changing proportions of salt and urea,* and this is also an important part of the story that follows (Fig. 38–4).

1. As fluid moves through the lumen of the **descending limb,** it continuously loses water to the increasingly concentrated interstitial fluid. It becomes more concentrated to both NaCl and urea, but it is always a little more dilute than the interstitial fluid around the tubule until it reaches the turn of the loop and starts back toward the cortex. However, the NaCl concentration is higher than that in the interstitial fluid, because the interstitial fluid is rich in urea; the luminal fluid becomes correspondingly rich in NaCl as it loses water.

2. As the fluid passes up the **ascending limb,** it shortly reaches regions where the interstitial fluid has a lower osmolarity than the lumen contents. However, this limb has a low permeability to water and a high permeability to salt. The result is that Na^+ and Cl^- move out, along with a lesser amount of urea, which does not pass so readily through the luminal plasma membrane. The total solute concentration therefore progressively diminishes in the luminal fluid as it passes up the limb, being only a little greater than that of the interstitial fluid at any point, but most of the loss in solute is in NaCl, not urea.

3. This sequence culminates upon reaching the thick portion of the ascending limb, where Cl^- is actively pumped out and accompanied by Na^+. **This is the source of energy for creation of the concentration gradient in the medulla.** As a result of its action, the osmotic pressure in the luminal fluid falls below that of the interstitial fluid and blood plasma—it becomes hypoosmolar, although its urea concentration is high.

4. An important feature of the countercurrent mechanism is the small solute gradient between the ascending and descending limbs of the loop of Henle at any level, even though there is a large gradient along the length of each limb. The interstitial fluid at any level is a little more concentrated than the fluid within the descending limb and a little more dilute than the fluid within the ascending limb. The pumps in the ascending limb do not have to work against impossible concentration gradients, even though the total gradient they create is very large.

5. When the hypoosmolar fluid reaches the **region of controlled water permeability in the distal and collecting tubule,** water may be permitted to pass out of the lumen into the surrounding interstitial fluid, which has a higher solute concentration. Water loss will continue, if permitted, upon passage through the increasingly concentrated regions of the medulla, and the result will be recovery of most of the water that entered the loop of Henle and excretion of a concentrated urine. If water loss is not permitted, the urine will remain at the hypoosmolar concentration of the fluid leaving the loop of Henle, and a large volume of dilute urine will be excreted.

6. Free passage of water through the medullary collecting duct membrane is associated with passage of urea, and this is an obligatory step in the establishment of the high concentration gradient in the medulla necessary for water absorption. Although the total solute concentration of urine in the collecting duct is lower throughout its course than it is in the surrounding interstitial fluid, the urea concentration is higher. Therefore, urea will move from the urine into the region of lower concentration in the interstitial space, if its passage is permitted. NaCl cannot flow back into the collecting duct to restore osmotic balance, so the movement of urea creates a progressively higher osmotic pressure in the surrounding interstitial fluid.

The countercurrent concentration gradient that can be established is much lower if an adequate supply of urea is not available. Persons on a low-protein diet excrete a higher volume of dilute urine as a result.

7. The water and salts entering the interstitial fluid in the medulla are removed by blood flowing through the vasa recta loops. Blood flowing toward the region of high concentration accumulates solutes; on the return trip toward the cortex, it loses most, but not all, of the solutes and accumulates water. The concentrations in the blood parallel those in the loop of Henle as it flows through the vasa recta. Transport is efficient because the actual volume of the interstitial fluid between the tubule cells and the blood is small, and it need not contain high quantities of solute to establish large concentration gradients.

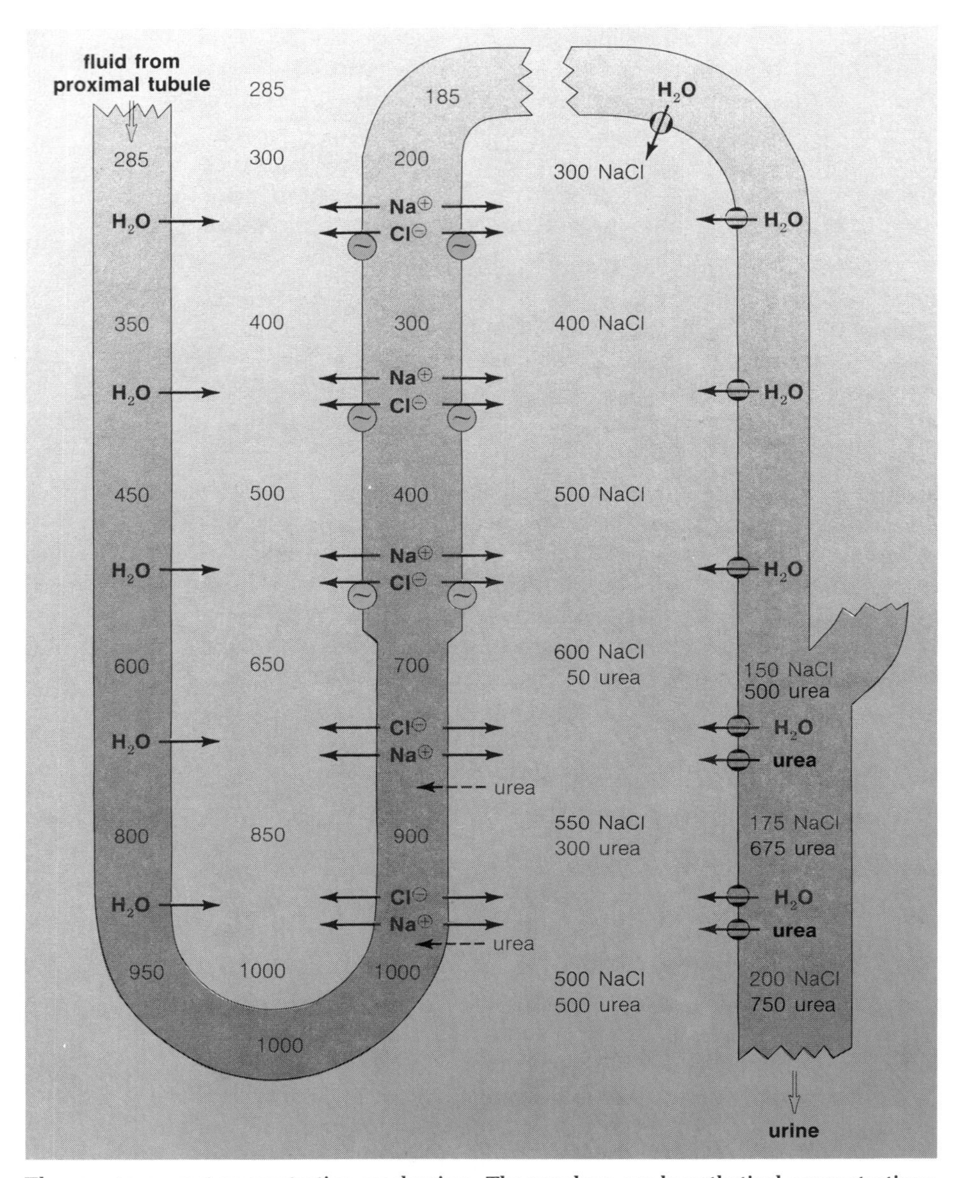

FIGURE 38–4

The countercurrent concentration mechanism. The numbers are hypothetical concentrations selected to illustrate the principles when urine is being concentrated; the values chosen for the fluid at the top and the bottom of the diagram are realistic under conditions of low water intake. The volume of the interstitial space is grossly exaggerated in diagrams of this type.

Concentrations are given in milliosmoles per liter (total millimolarity of all solutes, including individual ions). Concentrations within the tubular fluid are given in brown; the total milliosmolarity of the interstitial fluid is shown by the left center column in black. The contributions of NaCl and urea to that total are shown in brown and black, respectively, in the right center column in the interstitial space. The important features are: (*left*) water passes from the tubule lumen into the interstitial space, with the solute concentration constantly increasing to a maximum of 1,000 mOsm. (*Bottom center*) The fluid in the lumen of the thin ascending tubule is now more concentrated than the surrounding interstitial fluid, and it loses NaCl, but does not gain water. It does gain some urea. (*Top center*) Chloride ions are actively pumped out in the thick-walled ascending tubule, accompanied by sodium ions. This pump is the driving force by which the concentration gradient from cortex to pelvis is established. (*Right top*) Water passes out of the fluid in the distal and collecting tubules through channels that open in the presence of vasopressin. (*Right bottom*) Vasopressin also opens channels for passage of urea out of the collecting ducts. Urea leaves because the concentration in the urine is greater than that in the interstitial fluid, even though the solute concentration is lower in the urine.

FIGURE 38–5

Sodium is removed from the urine and potassium is added to it by the action of the $(Na^+ + K^+)$-ATPase in the basal membrane of cells in the distal tubules and collecting ducts.

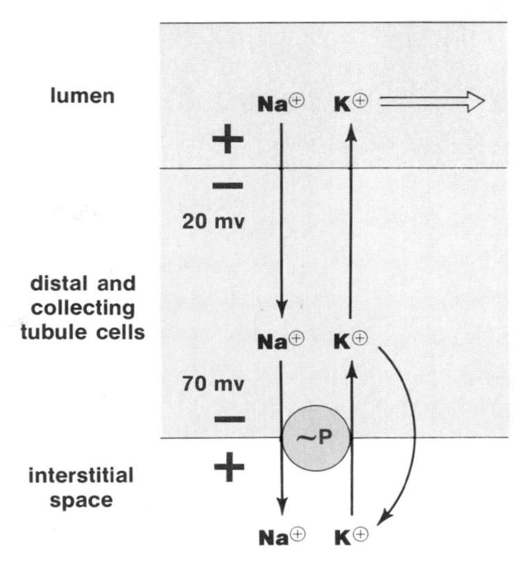

Sodium and Potassium Balance

So far, we have been dealing with the massive movements of Na^+, and the much lesser movement of K^+, associated with recovery of the glomerular filtrate and concentration of the urine. There are additional mechanisms in the distal tubule and collecting tubule for regulation of the Na^+ and K^+ concentrations of the blood. These mechanisms need have only the capacity for handling ions in the amounts likely to accumulate from dietary intake or dehydration.

The major mechanism appears to be a regulated $(Na^+ + K^+)$-ATPase on the interstitial side of the tubular cells. It raises $[K^+]$ and lowers $[Na^+]$ within the cells. K^+ leaks out into both the lumen and the interstitial fluid. The loss into the lumen is a major source of excreted K^+. The leakage also makes the interior of the cell more negative, causing more Na^+ to move from the lumen into the cell (Fig. 38–5). In addition to this passive loss, there appears to be secretion of K^+ into the lumen by an active pump (not shown). The result of these mechanisms is that there is always excretion of K^+ regardless of intake, whereas the excretion of Na^+ can be reduced to very low levels when there is little intake.

REGULATION OF SALT AND WATER BALANCE

Vasopressin, the Antidiuretic Hormone

A specialized group of neurons located above the pituitary gland and behind the optic chiasm (the supraopticoneurohypophyseal tract) synthesizes two small polypeptide hormones, vasopressin and oxytocin, which differ in only two amino acid residues:

$$\overset{\oplus}{(H_3N)}\text{—}\overset{1}{\text{Cys}}\text{—S—S—}\overset{6}{\text{Cys}}\text{—}\overset{7}{\text{Pro}}\text{—}\overset{8}{\boxed{\text{Arg}}}\text{—}\overset{9}{\text{Gly}}\text{—}\overset{\displaystyle O}{\overset{\|}{C}}\text{—}NH_2$$
$$\overset{1}{\text{Cys}}\text{—}\overset{2}{\text{Tyr}}\text{—}\overset{3}{\boxed{\text{Phe}}}\text{—}\overset{4}{\text{Gln}}\text{—}\overset{5}{\text{Asn}}\text{—}\overset{6}{\text{Cys}}$$

vasopressin

$$\overset{\oplus}{(H_3N)}\text{—Cys—S—S—Cys—Pro—}\boxed{\text{Leu}}\text{—Gly—}\overset{\displaystyle O}{\overset{\|}{C}}\text{—}NH_2$$
$$\text{Cys—Tyr—}\boxed{\text{Ile}}\text{—Gln—Asn—Cys}$$

oxytocin

This difference is sufficient to change the physiological activity. The physiological function of oxytocin is not known. Therapeutic doses cause contraction of the uterus and expulsion of milk from the mammary gland.

Vasopressin causes water to be absorbed from the distal and collecting tubules of the kidney, and this is its physiological function. Administration of higher doses causes constriction of arteries, resulting in higher blood pressures in anesthetized animals; this is the origin of the name. The hormone from humans and most mammals is sometimes named Arg-vasopressin to distinguish it from the pig hormone, which contains Lys in place of Arg.

Vasopressin (M.W. *ca.* 1 K) is synthesized in the cell bodies of the specialized neurons as part of a 17.3 K precursor. Other products from the precursor are a small glycopeptide and a larger polypeptide (M.W. *ca.* 10 K), a neurophysin, which binds vasopressin and the glycopeptide to form granules.

The granules are discharged from the neurohypophysis in reponse either to hypertonicity (increased solute concentration in the interstitial fluid) or to lowered blood volume. The released vasopressin travels through the blood, where its concentration is only 10^{-11} to 10^{-12} M, and binds to receptors on the distal portion of nephrons. (Nephron is used as a collective term for the entire glomerulus-tubule apparatus.) These specific receptors activate adenylate cyclase in the target cells, and the resultant increase in 3′,5′-cyclic AMP causes the luminal membranes to become water permeable. This is the permitted passage of water that we referred to earlier. The mechanism is not known, but it seems likely that a cyclic AMP-sensitive protein kinase causes a modification of protein conformation in the luminal membrane.

Aldosterone, Renin, and Angiotensin

Vasopressin provides a prompt response to transient alteration in salt and water balance. A slower regulation of basal conditions is obtained by the effect of aldosterone (p. 729) on cation exchange in the distal nephron. This steroid hormone affects gene expression in the distal nephron so as to increase the recovery of Na^+ and increase the loss of K^+. The protein(s) whose synthesis is being affected is not known. The result could be partially rationalized by effects on luminal permeability or on the $(Na^+ + K^+)$-ATPase, or on both.

Salt balance is therefore partially determined by the rate of secretion of aldosterone by the zona glomerulosa of the adrenal cortex. Corticotropin must be present for secretion to occur, but it is not the primary regulator. Aldosterone synthesis is accelerated by an increase in $[K^+]$ or a large decrease in $[Na^+]$; these responses are appropriate for maintenance of cation homeostasis.

One of the most effective signals for aldosterone secretion is the formation of an oligopeptide known as **angiotensin II,** which appears in response to changes within the kidney. Figure 38–2 sketched the location of the specialized structure known as the juxtaglomerular apparatus; this structure forms a proenzyme that upon stimulation is secreted into the blood as an active protease, **renin.** The signals causing release are (1) a loss of stretch of receptors in the afferent arteriole of the glomerulus indicating diminished blood volume, (2) a drop in Na^+ concentration in the adjacent distal tubular cells, (3) catecholamines in the circulation, and (4) stimulation of renal sympathetic nerves.

Renin attacks a protein, angiotensinogen, in the blood (Fig. 38–6) so as to split only the initial decapeptide segment from this approximately 100,000 M.W. glycoprotein. The decapeptide, angiotensin I, has only modest physiological activity, but

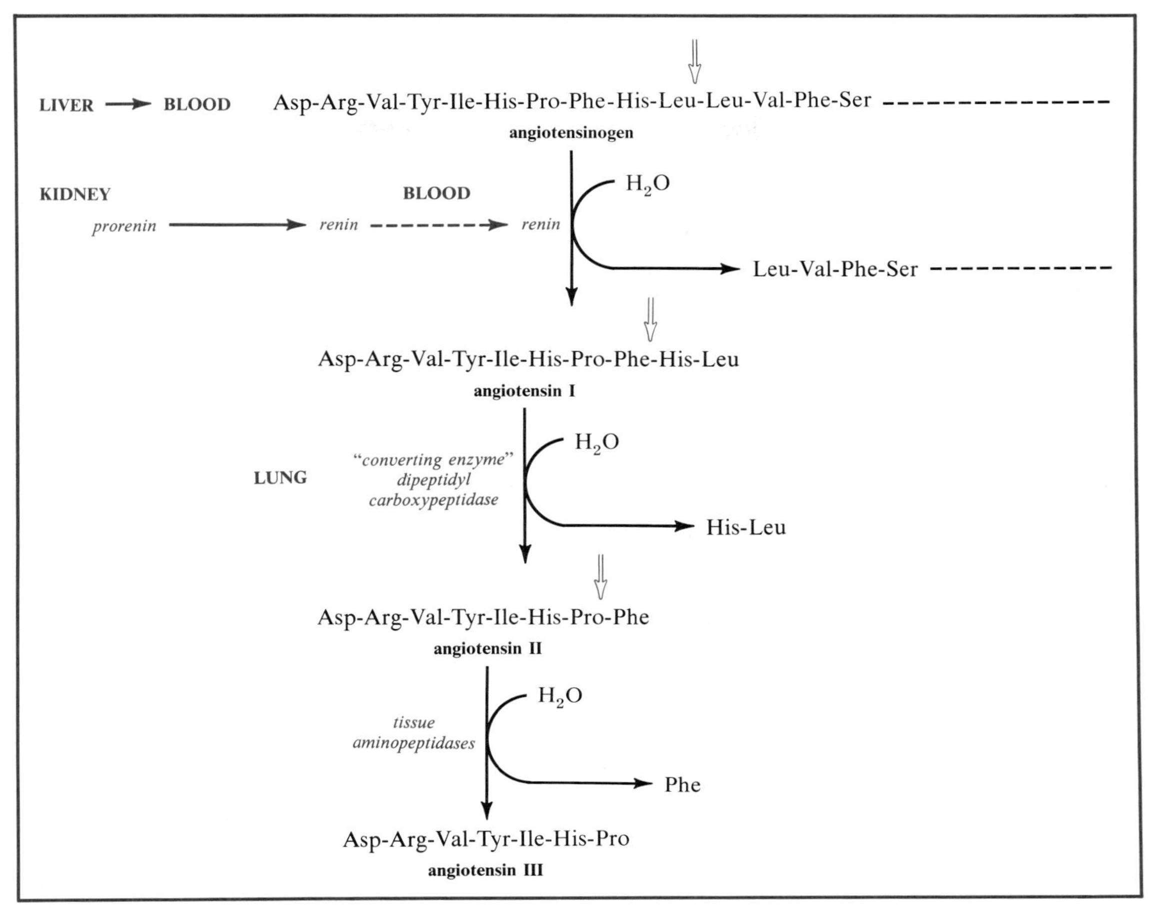

FIGURE 38–6 Formation of angiotensins.

when it passes through the lungs, it is attacked by a relatively nonspecific enzyme that removes the two amino acid residues at the carboxyl terminus as a dipeptide. (This is the same enzyme that inactivates bradykinin, p. 352.) The remaining octapeptide, angiotensin II, has profound physiological effects, including an increase in cardiac output and constriction of the vascular smooth muscles, thereby raising the blood pressure but diminishing fluid loss in the kidney by restricting flow. The effect of most interest for our purposes is a stimulation of the conversion of cholesterol to pregnenolone in the zona glomerulosa, thereby increasing aldosterone production and release.

Angiotensin II is also hydrolyzed by aminopeptidases in the blood and various tissues to create angiotensin III, a heptapeptide that is more active on the adrenal cortex but less active on the vascular smooth muscle. It is not clear if this is a physiologically important transition. Further hydrolysis of angiotensin III by tissue peptidases destroys the activity.

The renin-angiotensin system therefore conserves fluid by constricting flow through the kidney, increases blood pressure so as to counteract the effects of low blood volume, and conserves salt by increasing the recovery of Na^+ in the distal tubule. (The accompanying loss of K^+ is quantitatively smaller, especially when the urine flow is low.)

Licorice roots contain a compound, glycyrrhizinic acid, that has mineralocorticoid activity. Lovers of licorice were at risk to the effects of hyperaldosteronism (hypertension, abnormal cardiac function, muscular weakness, and other symptoms of low K^+ concentration) before artificial flavoring replaced extracts from the plant (*Glycyrrhiza glabra*). Even today an occasional case may appear. Some chewing tobacco is still made with the root extract.

Diuresis

Disturbances in almost any part of the concentrating mechanism can cause an increased loss of water and salts from the kidney. Diuresis is a consequence of uncontrolled diabetes mellitus because the amount of glucose in the glomerular filtrate exceeds the capacity of the proximal tubule to absorb it. This extra solute in the tubular fluid raises its osmotic pressure to the point that the distal nephron cannot reabsorb as much water.

Damage to the supraopticoneurohypophyseal stalk, especially above the median eminence, prevents the normal output of vasopressin. Without vasopressin, the distal nephron is nearly impermeable to water, so there is a massive flow of very dilute urine. The flow is so fast that glucose cannot be completely reabsorbed at even normal blood concentrations, and the condition is named diabetes insipidus. (A rare congenital lack of vasopressin receptors in the distal nephron has the same result.) An opposite effect is seen in patients with some neoplasms, which secrete peptides with antidiuretic activity. These people have a high urine osmolality and a low blood osmolality.

Drugs are frequently used to decrease the body salt content, and therefore tissue fluids, as a means of lowering the blood pressure. These diuretic agents, such as the thiazides, furosemide, and ethacrynic acid, inhibit the chloride pump in the thick ascending limb of the loop of Henle. This not only inhibits absorption of NaCl, it diminishes the medullary concentration gradient necessary for water reabsorption, so both salt and water excretion are enhanced.

CONTROL OF H^+ BALANCE

Hydrogen ion concentration affects the rates of enzymatic reactions, the conformation of proteins, and ion exchanges across membranes; its regulation is critical to survival. Excessive alkalinity causes tetany, and excessive acidity causes coma. The $[H^+]$ within tissues that results in these conditions cannot easily be measured directly, but the changes can be detected by accompanying shifts in the pH of arterial blood, which is normally near 7.40, and routinely determined in clinical laboratories. Tetany usually occurs if the pH rises to 7.8, and coma if it falls to 7.0 ($pH = -\log [H^+]$).

H^+ Concentration, pH, and Buffers

To begin, determinations of $[H^+]$ are made with electrodes or other devices that respond to the activity of the ion rather than its total concentration. Purists balk, but we shall use the activity as $[H^+]$ throughout.

Before going further, let us review some of the fundamental chemistry of acids and bases. Unfortunately, much of the older literature, and some of the current literature, makes heavy going out of what are quite elementary mass-action relationships.*

Consider lactic acid. It dissociates according to the equation:

$$\text{lactic acid} \longleftrightarrow \text{lactate}^- + H^+,$$

and by the classic definitions, lactic acid is an acid because it liberates H^+, and lactate is a base because it combines with H^+. Lactic acid and lactate are a conjugate pair of acids and bases. We can write the general equation for such conjugate acids and bases:

$$\text{acid} \longleftrightarrow \text{base} + H^+$$

and state a general mass-action equilibrium expression for the relationship between them:

(1)
$$K' = \frac{[\text{base}][H^+]}{[\text{acid}]}$$

or, by dividing by $[H^+]$:

(2)
$$\frac{K'}{[H^+]} = \frac{[\text{base}]}{[\text{acid}]}.$$

This simple equation is the basis for all of our discussion. If any three of the terms are known, the fourth can be determined. Some people use it as is; others prefer to take the negative logarithm so as to cast the expression in terms of pK and pH, which are $-\log K'$ and $-\log [H^+]$:

(3)
$$\log K' - \log [H^+] = \log \frac{[\text{base}]}{[\text{acid}]};\ pK' = -\log K';\ pH = -\log [H^+]$$

(4)
$$pH - pK' = \log \frac{[\text{base}]}{[\text{acid}]}$$

(5)
$$\text{or:}\quad pH = pK' + \log \frac{[\text{base}]}{[\text{acid}]}.$$

The equilibrium expression in this form is known as the Henderson-Hasselbalch equation. The advent of pocket calculators that readily handle negative fractional exponents has largely destroyed whatever advantage this equation possessed. Since pH and pK are still with us, some find it handy to use the logarithmic form.

Here are some examples of the value of the equilibrium expression, and the essential simplicity of its use:

Problem 1. The K' for ammonium ion at 38° C and 0.3 M salt is 3.98×10^{-10}. Estimate the proportion present as ammonia in capillaries at pH 7.40 and in a urine at pH 5.62.

---

*I still remember the feeling of total frustration as a young biochemist trying to make sense out of the textbook descriptions of acid-base balance, even though I had no trouble dealing with buffer relationships in the laboratory.

The equilibrium in question is: $NH_4^+ \longleftrightarrow NH_3 + H^+$.

Therefore,
$$\frac{K'}{[H^+]} = \frac{[\text{base}]}{[\text{acid}]} = \frac{[NH_3]}{[NH_4^+]}.$$

So, at pH 7.40
$$\frac{[NH_3]}{[NH_4^+]} = \frac{K'}{[H^+]} = \frac{3.98 \times 10^{-10}}{10^{-7.40}} = \frac{3.98 \times 10^{-10}}{3.98 \times 10^{-8}} = 0.010.$$

(Many pocket calculators will solve the ratio as first given without recasting $10^{-7.4}$ as 3.98×10^{-8}.) Therefore, there is 100 times as much ammonium ion as ammonia in the arterial blood, no matter what the total concentration of the two. At pH 5.62,

$$\frac{K'}{[H^+]} = \frac{3.98 \times 10^{-10}}{10^{-5.62}} = \frac{3.98 \times 10^{-10}}{2.40 \times 10^{-6}} = 1.66 \times 10^{-4} = \frac{1}{6{,}027}$$

and there is only one ammonia molecule for each 6,027 ammonium ions.

Here is the same problem solved by the logarithmic expression: 3.98×10^{-10} is the same as $10^{-9.40}$, so $pK' = 9.40$

$$\log\frac{[\text{base}]}{[\text{acid}]} = \text{pH} - pK' = 7.40 - 9.40 = -2;\ \frac{[\text{base}]}{[\text{acid}]} = 10^{-2} \quad \text{(blood)}$$

$$= 5.62 - 9.40 = -3.78;\ \frac{[\text{base}]}{[\text{acid}]} = 10^{-3.78} \quad \text{(urine).}$$

Problem 2. How much 1 M NaOH must one add to one liter of 0.1 M NH_4Cl to bring the pH to 8.9?

At pH 8.9
$$\frac{[\text{base}]}{[\text{acid}]} = \frac{K'}{[H^+]} = \frac{10^{-9.40}}{10^{-8.90}} = 10^{-0.50} = 0.316 = \frac{[NH_3]}{[NH_4^+]}.$$

The result says that for every 316 molecules of ammonia formed in the final solution, there must be 1,000 molecules of ammonium ion remaining. That is, of 1,316 original molecules of NH_4^+, only 316, or 24 per cent, are to be converted to ammonia. Therefore, using the relationship $V_1M_1 = V_2M_2$, we find one wants to add 0.24×0.1 or 0.024 liter of 1 M NaOH.

Buffers are solutions that minimize a change in $[H^+]$ upon addition of acid or base. They consist of roughly comparable and moderately high concentrations of both a conjugate acid and its conjugate base. That is, a solution must contain both lactic acid and lactate in concentrations within an order of magnitude of each other to be a good buffer. Another good buffer solution at a much higher pH would contain both ammonium ions and ammonia at concentrations within the same order of magnitude. It is necessary that both the acid and the base be present in order to resist a change in $[H^+]$ upon addition or removal of H^+ by some other source. The more nearly equal their concentrations, the better the buffer action.

Example: Consider two solutions of lactic acid ($K' = 10^{-3.74}$) and sodium lactate:

	A	B
[lactic acid]	0.05 M	0.015 M
[sodium lactate]	0.05 M	0.085 M
pH	3.74	4.49

(It is a useful exercise to verify the pH values.)

Now suppose that NaOH is added to each solution in an amount equivalent to 0.005 M. The hydroxide ion will remove H^+ and more lactic acid will dissociate to replace it. As a first approximation, the new situation will be:

	A	B
[lactic acid]	0.045 M	0.010 M
[sodium lactate]	0.055 M	0.090 M
pH	3.83	4.69

The $[H^+]$ in solution A has changed by 19 per cent, whereas it has changed by 37 per cent in solution B. (To do a better approximation, we would have to allow for part of the OH^- being used to react with H^+ in solution, but the error is not great at these H^+ concentrations.)

A truism emerges: Conjugate acids and bases are effective buffers only in solutions in which the $[H^+]$ is within an order of magnitude of the K' of the acid, if the total concentration of acid and base combined is held constant. That is, the pH must be within 1.0 of the pK'. The lactic acid–lactate pair buffers between pH 2.74 and 4.74. Ammonium ion–ammonia buffers between pH 8.40 and 10.40.

The internal pH of the organism is usually near 7. What then are compounds with pK' values near 7 that resist a change in $[H^+]$ within cells or in the extracellular fluids? By far the most important are groups in proteins, especially the imidazolium group in histidine residues. Its pK' value varies with location in a protein, but it frequently is not far from 7.0.

Inorganic phosphate and phosphate monoesters such as glucose 6-phosphate and glycerol phosphate are effective as buffers. The pK' values are near 6.8 for the dissociations:

$$H_2PO_4^- \longrightarrow HPO_4^{2-} + H^+$$
$$R—O—PO_3H^- \longrightarrow R—O—PO_3^{2-} + H^+.$$

Notice it is the second dissociations in each case that are in the appropriate range. Phosphodiesters with only one dissociation, such as phosphatidyl compounds and the phosphate groups in nucleic acids, are too strong acids for effectiveness as a buffer. Nucleoside polyphosphates, such as ATP, contribute little buffering power because of their high affinity for Mg^{2+}, which displaces H^+.

The Carbonic Acid–Bicarbonate Equilibrium

Another buffer system, the carbonic acid–bicarbonate pair, requires special discussion. The effective pK value for carbonic acid in blood plasma is only 6.10, but it still is very important for $[H^+]$ homeostasis because the concentration of the conjugate acid, H_2CO_3, is held relatively constant through regulation of respiration. Additions of acid and base therefore tend to affect only the bicarbonate concentration. The equilibria of interest are:

$$CO_2(gas) \longleftrightarrow CO_2(solution) \xleftrightarrow{\pm H_2O} H_2CO_3 \longleftrightarrow H^+ + HCO_3^-.$$

The usual equilibrium constant defines the relationship between the concentrations of carbonic acid, H^+, and HCO_3^-:

$$K' = \frac{[H^+][HCO_3^-]}{[H_2CO_3]}.$$

However, it is difficult to determine just how much carbonic acid is present in a solution compared to the amount of dissolved CO_2 that isn't hydrated. It is therefore customary to lump together the concentrations of dissolved gas and of carbonic acid and treat them as a single entity, and this sum is what we shall mean when we speak of [dissolved CO_2].

The second practical problem is that we are dealing with blood, which is a complex tissue, and it is not possible to sort out all of the possible equilibria involving CO_2, such as the combination with amino groups of proteins that occurs to a small extent spontaneously:

$$R\text{—}NH_2 + CO_2 \rightleftharpoons R\text{—}NH\text{—}COO^- + H^+.$$

These unknown reactions prevent us from applying equilibrium constants determined with more pure compounds to the relationship between dissolved CO_2, H^+, and HCO_3^- in blood. The solution is to be pragmatic and not worry about the side reactions, and this is what is done. An apparent equilibrium constant is determined by measuring the actual relationship between the three components of the carbonic acid equilibrium in the plasma compartment of whole blood, and it is designated K'' to indicate its empirical nature:

$$K'' = \frac{[H^+][HCO_3^-]}{[\text{dissolved } CO_2]} = 10^{-6.10} \text{ (blood at 38° C)}$$

in which all of the various forms of CO_2 are lumped together as an entity. The concentration of dissolved CO_2 can be calculated from the partial pressure of the gas, p_{CO_2}, in equilibrium with the blood:

$$[\text{dissolved } CO_2]/p_{CO_2} = 2.25 \times 10^{-7} \text{ M pascal}^{-1} = 3 \times 10^{-5} \text{ M torr}^{-1}.$$

We are primarily interested in the way in which the concentration of H^+ changes with changes in the concentration of dissolved CO_2, so let us rearrange the ionization equation by dividing by $[H^+]$ so as to obtain:

$$\frac{K''}{[H^+]} = \frac{[HCO_3^-]}{[\text{dissolved } CO_2]}.$$

Now, let us apply the equation by examining normal arterial blood. The pH is 7.40, so $[H^+] = 10^{-7.40}$. Since we know that $K'' = 10^{-6.10}$, we can now determine the ratio of concentrations of bicarbonate ions and of dissolved CO_2 in the plasma of that blood:

$$\frac{10^{-6.10}}{10^{-7.40}} = 10^{1.3} = 20 = \frac{[HCO_3^-]}{[\text{dissolved } CO_2]}.$$

We find that there is 20 times as much bicarbonate as dissolved CO_2 in arterial blood. (We round off $10^{1.3} = 19.953$ as 20.)

The arterial blood is equilibrated with CO_2 at the partial pressure existing in the alveoli of the lungs, which is normally near 40 torr (5.3 kPa). Therefore,

$$[\text{dissolved } CO_2] = 3 \times 10^{-5}\, p_{CO_2} = 3 \times 10^{-5} \times 40 = 1.2 \text{ mM}.$$

The concentration of bicarbonate ions is 20 times greater than this, or 24 mM.

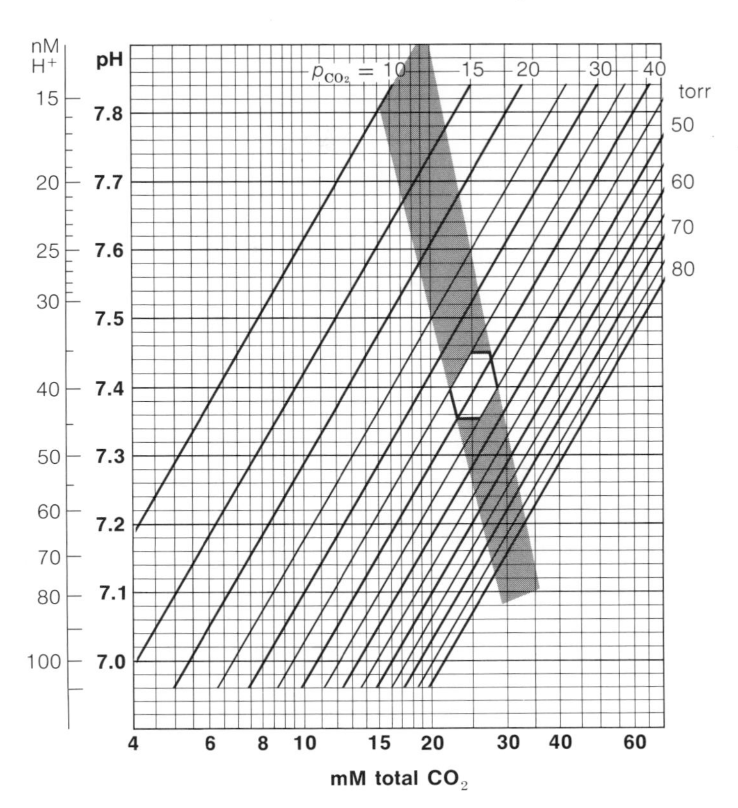

FIGURE 38–7

The carbonic acid–bicarbonate equilibrium in blood. The brown shading is the normal buffer band; rapid rises or falls in the carbon dioxide tension create pH values lying within this band. Concentrations in normal individuals usually lie within the clear area in the center of the buffer band.

The carbonic acid–bicarbonate pair is a useful indicator of acid-base status because of the accessibility of blood. All buffers equilibrate within the same cellular compartment according to the relationship:

$$[H^+] = \frac{K'[\text{acid}]}{[\text{base}]}.$$

The ratios of acidic and basic forms of each buffer will shift by the same factor when there is a change in $[H^+]$, so a shift in one is an index of a shift in all. Hydrogen ions do not fully equilibrate in all cellular compartments; even so, changes in the readily accessible blood plasma are a guide to changes within the tissues, and the blood is important in itself as a carrier of H^+. Two measurements of clinical interest are commonly made: the pH and p_{CO_2} of arterial blood. These parameters define the total CO_2 in a blood sample, and the relationship is shown in Figure 38–7.

ORIGIN OF CHANGES IN pH

Respiratory Loads

The high influx of CO_2 from the combustion of fuels and its discharge in the lungs inevitably causes fluctuation of the $[H^+]$ in the blood. Figure 38–8 shows the changes that occur in exercising men. The components of blood, especially hemoglobin, are designed to minimize these changes. In addition to these cyclical changes from arterial to venous circulation, changes also occur in the level of CO_2 that

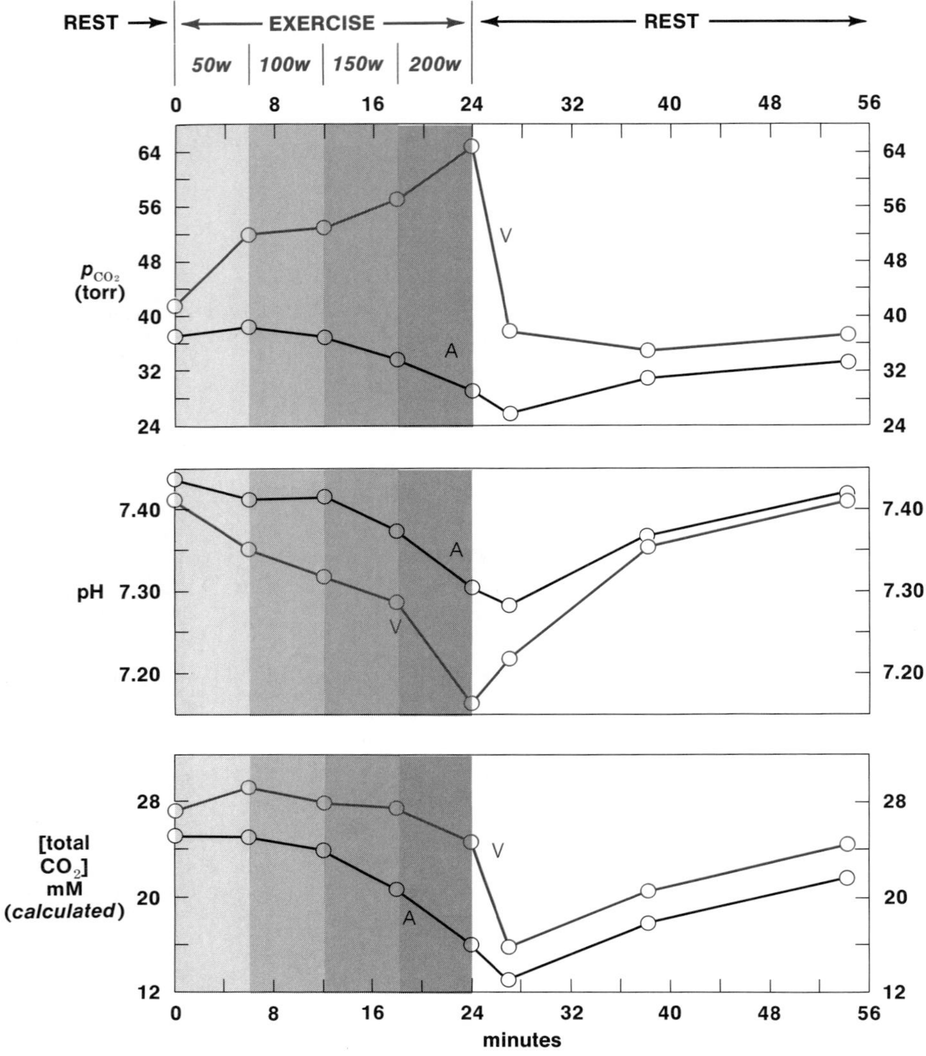

FIGURE 38–8 **Changes in the carbonic acid–bicarbonate equilibrium in blood circulating through exercising muscles. Individuals pedaled a bicycle ergometer for six-minute periods at increasing power outputs, ranging from 50 to 200 watts (*top left*). Analyses were made of arterial (A) and femoral vein (V) blood at the beginning and end of each exercise period, and during the resting period thereafter. (Data from J. Keul, E. Doll, and D. Keppler (1972). Energy Metabolism of Human Muscle. Med. Sport 7:72.)**

returns from the lungs, that is, in the basal level around which the arterial-venous fluctuations occur. These shifts in basal level occur where there is an alteration in the effectiveness of removal of CO_2 by the lungs. Increasing the rate of breathing when there is no corresponding increase in delivery of CO_2 to the lungs will diminish the CO_2 tension in the alveoli, and therefore in the blood returned to the peripheral tissues. The $[H^+]$ falls (pH rises), and a **respiratory alkalosis** has occurred. Holding the breath or impairment of gas exchange by partial obstruction of the bronchi or by drug overdose will diminish the discharge of CO_2 and elevate the CO_2 tension in the blood. The resultant rise in $[H^+]$ (fall in pH) is a **respiratory acidosis.**

The **Bohr effect** provides part of the compensation for changes in CO_2 concentration from arterial to venous blood. When oxygen is released, the affinity of hemoglobin for H^+ increases (p. 251). Approximately 0.4 H^+ is taken up for each O_2 released.* In a resting person liberating 0.8 CO_2 per O_2 consumed, over 50 per cent of the CO_2 can be transported without any change in other buffer systems:

$$\begin{array}{lrcl} \text{Tissue:} & O_2 + \text{fuel} & \longrightarrow & 0.8\ CO_2 \\ \text{Blood:} & HbO_2 + 0.4\ H^+ & \longrightarrow & Hb + O_2 \\ & 0.4\ CO_2\ (+H_2O) & \longrightarrow & 0.4\ H^+ + 0.4\ HCO_3^- \\ \text{SUM:} & HbO_2 + \text{fuel} & \longrightarrow & Hb + 0.4\ HCO_3^- + 0.4\ CO_2 \end{array}$$

Since the ratio of dissolved CO_2 to bicarbonate is 1:20 at pH 7.4, the 0.4 HCO_3^- will balance $0.4/20 = 0.02$ of the remaining CO_2 without changing pH. The total **isohydric carriage** (carriage without change in $[H^+]$) is therefore $0.4 + 0.02 = 0.42$ CO_2 out of the 0.8 formed.

The balance of the CO_2 reacts with the buffer systems until they are all equilibrated at constant pH:

$$CO_2 + H_2O + \text{buffer base} \longrightarrow HCO_3^- + \text{buffer acid.}$$

As we noted, hemoglobin itself is the most powerful buffer in the blood other than the bicarbonate system. (We are adding carbonic acid to the carbonic acid–bicarbonate system; it cannot buffer itself.) The shaded band in Figure 38–7 shows the changes in pH that occur in normally buffered individuals with changes in p_{CO_2}. That is, the pH of blood lies within this band with respiratory acidosis or alkalosis until compensatory mechanisms occur. There is little shift in the pH upon going from arterial to venous circulation and back again in resting individuals. Here are typical values for normal young adult males at rest:

	Arterial Blood	Femoral Venous Blood
pH	7.43	7.41
p_{CO_2} (torr)	37	42
[total CO_2] (mM)	25	27

The bicarbonate concentration increases in venous blood owing to the reaction of CO_2 with the buffer systems, mainly hemoglobin. This involves a series of events. CO_2 as such moves into the blood from the tissues and carbonic anhydrase within the red blood cells accelerates its hydration to carbonic acid. This is convenient, because the carbonic acid is generated in proximity to hemoglobin, which acts as the principal buffer system:

$$\begin{array}{rcl} \text{Erythrocyte: } CO_2 + H_2O & \longrightarrow & H_2CO_3 \\ H_2CO_3 & \longrightarrow & H^+ + HCO_3^- \\ H^+ + Hb & \longrightarrow & HHb^+ \\ \text{SUM: } CO_2 + H_2O + Hb & \longrightarrow & HCO_3^- + HHb^+ \end{array}$$

*This value is an approximation determined in the presence of CO_2 and bisphosphoglycerate. It therefore includes an adjustment for the change in carbamoylation of hemoglobin by CO_2, which releases H^+:

$$^+H_3N{-}Hb + CO_2 \longrightarrow {}^-OOC{-}HN{-}Hb + 2\ H^+$$

Approximately 10 per cent of the extra CO_2 put into the blood is transported to the lungs in this form in humans.

(We are distinguishing the buffer action of hemoglobin from the Bohr effect shift by indicating the acidic forms as HHb^+.) Additional events occur to restore osmotic balance. The formation of bicarbonate within the erythrocyte not only raises the total solute concentration, but it also increases the ratio $[HCO_3^-]_{in}/[HCO_3^-]_{out}$ over the corresponding ratio for chloride ions. Bicarbonate diffuses out of the erythrocyte and chloride in until these ratios match. Osmotic balance is restored by movement of water into the erythrocyte. This sequence of events, sometimes called the **chloride shift,** is detailed as a matter of general interest and to call attention once more to the importance of carbonic anhydrase in equilibrating CO_2 and bicarbonate in the presence of buffer systems.

Dietary and Metabolic Loads

Hydrogen ions may be added or removed as a result of eating particular foods, or by metabolic changes. An important point must be emphasized: **There is no substitute for writing a complete stoichiometry for any physiological event in analyzing the effects of the event on $[H^+]$.** This is neglected with surprising frequency, even by some workers with no small reputation in the field, with consequent mistakes in logic. It is critical in establishing the stoichiometry that the ionic state of the reactants and products be stated at least approximately, taking note of the fact that most meals are near neutrality, or slightly acidic. Some examples will make the idea clear, and show some important principles:

1. What is the effect on H^+ balance of ingesting starch and oxidizing it to CO_2 and H_2O? The overall reaction is:

$$(C_6H_{10}O_5)_n + 6n\ O_2 \longrightarrow 6n\ CO_2 + 5n\ H_2O.$$

Note that all of the reactants come from outside the body and all of the products are excreted from the body; note also that nothing is said about high-energy phosphate, the citric acid cycle, and so on. Including these reactions would be a mistake, because all of the intermediate compounds are recycled in relatively steady-state concentrations that do not affect the overall result. There is no net production or consumption of H^+ so long as the CO_2 passes out in the lungs as fast as it is formed in the tissues.

2. What is the effect of ingesting acetic acid (vinegar) and oxidizing it to CO_2 and H_2O? The ingestion of an acid increases the total body $[H^+]$ until it is removed. After it has been oxidized, the overall stoichiometry becomes:

$$C_2H_4O_2 + 2\ O_2 \longrightarrow 2\ CO_2 + 2\ H_2O.$$

Again, all of the components are outside of the body, and there is no net *long-term* change in H^+ concentration from the ingestion of a metabolizable acid. (This includes citric acid, malic acid, succinic acid, and so on.)

3. What is the effect of ingesting a solution of disodium hydrogen citrate ($Na_2C_6H_6O_7$) and oxidizing the citrate to CO_2 and H_2O? The solution is acidic (pH 5.0 at 0.1 M). If we write a stoichiometry for the combustion as such, we must formally balance the charges in some way:

$$2\ Na^+ + C_6H_6O_7^{2-} + 2\ H^+ + 4\tfrac{1}{2}\ O_2 \longrightarrow 2\ Na^+ + 6\ CO_2 + 4\ H_2O$$

or: $$2\ Na^+ + C_6H_6O_7^{2-} + 4\tfrac{1}{2}\ O_2 \longrightarrow 2\ Na^+ + 6\ CO_2 + 2\ H_2O + 2\ OH^-$$

or: $$2\ Na^+ + C_6H_6O_7^{2-} + 4\tfrac{1}{2}\ O_2 \longrightarrow 2\ Na^+ + 4\ CO_2 + 2\ H_2O + 2\ HCO_3^-$$

Any way we do it, we come to the conclusion that Na^+ is accumulating and the body will become more alkaline—H^+ is removed directly, or by combination with OH^-, or by a shift in the buffer pairs to match an increased bicarbonate concentration. The only difference between ingesting sodium bicarbonate and the sodium salts of metabolizable carboxylic acids is that the latter increase the bicarbonate concentration more slowly. Sodium lactate has been used therapeutically to diminish $[H^+]$ in this way. Citric acid cycle intermediates are common constituents of fruits and vegetables, where they are partially or completely ionized. Their solutions are acid, but complete combustion takes up H^+.

The unsuspected presence of salts of metabolizable organic acids can cause trouble in patients. For example, at least one patient with severe loss of kidney function was thrown into metabolic alkalosis by the administration of a commonly used plasma protein fraction, which was then found to contain substantial amounts of sodium acetate.

4. What is the effect of ingesting alanine and oxidizing it to CO_2, H_2O, and urea, which are excreted?

$$\underset{\text{alanine}}{C_3H_7O_2N} + 3\,O_2 \longrightarrow \tfrac{1}{2}\,\underset{\text{urea}}{CH_4ON_2} + 2\tfrac{1}{2}\,CO_2 + 2\tfrac{1}{2}\,H_2O$$

In general, metabolism of amino acids or polypeptides containing only C, H, O, and N that are ingested near their isoelectric points has no effect on H^+ balance if urea is excreted.

5. What is the effect of ingesting methionine and oxidizing it to CO_2, H_2O, urea, and sulfate, which are excreted?

$$C_5H_{11}O_2NS + 7\tfrac{1}{2}\,O_2 \longrightarrow \tfrac{1}{2}\,CH_4ON_2 + 4\tfrac{1}{2}\,CO_2 + 3\tfrac{1}{2}\,H_2O + SO_4^{2-} + 2\,H^+$$

Oxidation of cysteine and methionine-containing proteins causes a production of H^+. (This is the major source of H^+ in most normal people.)

6. What is the effect of ingesting phosphate esters and oxidizing the carbon skeletons to CO_2 and H_2O? The phosphate esters may be monoesters like glycerol 3-phosphate, or diesters as in the nucleic acids and phospholipids. (Remember that pK_2' for the monoesters and P_i is approximately 6.8, whereas the single pK' of the diesters is much lower.) The approximate changes in neutral foods (pH 6.8) and the resultant stoichiometries to excrete P_i in a urine at pH 6.8 are:

$$R\text{—}PO_4H_{0.5}^{1.5-} + x\,O_2 \longrightarrow y\,CO_2 + 2\,H_2O + H_{1.5}PO_4^{1.5-}$$
$$R,R'\text{—}PO_4^- + x\,O_2 \longrightarrow y\,CO_2 + 2\,H_2O + H_{1.5}PO_4^{1.5-} + 0.5\,H^+.$$

(These equations were obtained by calculating the ratios of acidic to basic forms of the compounds at pH 6.8.) If food and urine are at the same $[H^+]$, then the monoesters will cause no H^+ production. (This point is missed by many.) However, the nucleic acids and phospholipids are the most abundant phosphate compounds in foods, and their metabolism is another source of H^+ production. (Question: Compare the effects of the two compounds if the urine is at pH 5.8.)

In addition to the continuing loads of the sort just outlined, alterations in H^+ balance also occur from changes in the metabolic processes, either as normal transients or as pathological events. For example, the formation and accumulation of the anion of any carboxylic acid with a pK' value well below 7 is accompanied by a concomitant production of H^+. Here we must distinguish between acute and chronic loads.

The H^+ produced by the formation of lactate during strenuous exercise is an example of an acute load. (Part of the changes during heavy exercise shown in Figure 38–8 were due to lactate formation.) The constant formation of the ketone bodies in uncontrolled diabetes mellitus, or of propionate or methylmalonate in the aminoacidopathies, is an example of a chronic load in which there is constant spillage of the anion in the urine, and there must be an equally constant discharge of H^+ if it is not to accumulate to lethal levels.

The base forms of the protein and phosphate buffers bind H^+ liberated by metabolic events, just as they do H^+ formed by respiratory acidosis, but the metabolic pH changes are also minimized by alterations in the bicarbonate level. The increased H^+ is partially dissipated by conversion of bicarbonate to CO_2, which is blown off by the lungs:

$$H^+ + HCO_3^- \longrightarrow H_2CO_3 \longrightarrow H_2O + CO_2.$$

If nothing else happens, a **metabolic acidosis** will be characterized by a drop in bicarbonate, or total CO_2 concentration, whereas an uncompensated respiratory acidosis will be characterized by an increase in bicarbonate, or total CO_2 concentration.

Conditions that cause a loss of H^+, thereby creating a **metabolic alkalosis,** include direct losses through vomiting of the dilute HCl normally present in the stomach. Hereditary defects in urea formation that result in excessive excretion of ammonium ions also cause metabolic alkalosis. For example, the total stoichiometry of alanine combustion then becomes:

$$C_3H_7O_2N + 3\,O_2 + H^+ \longrightarrow NH_4^+ + 3\,CO_2 + 2\,H_2O.$$

There is a loss of one H^+ for each NH_4^+ excreted that does not occur when urea is synthesized and excreted.

COMPENSATORY MECHANISMS

Buffer action can only *minimize* the change in $[H^+]$ that occurs with a given gain or loss of H^+. Other compensatory mechanisms diminish the burden on the buffers by instituting processes that have the contrary effect on the turnover of H^+. In crude terms, acidosis is corrected by processes that would in themselves create an alkalosis and *vice versa*. The compensatory mechanisms involve changes in the rate of three processes: CO_2 exchange through the lungs, elimination of H^+ through the kidneys, and metabolic reactions that create or consume H^+.

Respiratory Compensation

The rate of pulmonary ventilation is controlled by receptors that separately respond to changes in concentrations of CO_2, H^+, and O_2, listed in the order of decreasing sensitivity to deviations from physiological levels. That is, the O_2 concentration must change relatively more than the CO_2 concentration to cause equal changes in the rate of breathing. The effect of increasing $[H^+]$ is to diminish the $[CO_2]$ necessary to achieve a given rate. This provides a compensatory device for changes in $[H^+]$; any increase makes the respiratory centers more sensitive to CO_2.

Therefore, the breathing rate will increase until the [CO_2] has fallen. The drop in carbonic acid concentration will cause a corresponding shift of all buffers from their acidic to basic forms at the expense of bicarbonate, thereby eliminating part of the H^+ load.

Picture the sequence this way: A metabolic acidosis creates high [HHb^+] and low [Hb]. Compensation by blowing off additional CO_2 creates low [CO_2] leaving [HCO_3^-] at its normal high level. This cannot persist because the two buffer systems must be at equilibrium; H^+ will exchange between them with the net effect being a reaction of the components in high concentrations:

$$HCO_3^- + HHb^+ \longrightarrow H_2O + CO_2 + Hb.$$

The extra HHb^+ created by the acidosis has now been partially removed. The contrary events occur, although to a lesser extent, in compensation for metabolic alkalosis. The rate of ventilation is decreased in order to elevate the [CO_2], which contributes H^+.

Renal Compensation

The normal course of metabolism in most people produces an acidic urine, depending upon their diet. The more protein, the more H^+ formed; the more fruits and vegetables, with their partially ionized carboxylic acids, the more H^+ removed. The kidney is constructed to eliminate H^+, or conserve it, in order to maintain constant [H^+] in the face of this metabolic load. Two mechanisms involve the proximal convoluted tubule and the collecting ducts.

The H^+ pump of the proximal tubule can also be used to eliminate excess H^+. However, it can only do so until the H^+ concentration in the lumen is four times that in the cells (pH 0.6 unit lower in lumen). **Its action is supplemented by another pump in the collecting ducts,** which is capable of generating almost a 1000-fold concentration gradient, with urinary pH as low as 4.5 (Fig. 38–9). (Increases in H^+ excretion interfere with K^+ excretion in the distal nephron.)

Even at the minimal pH of 4.5, the actual [H^+] in the urine is only 32 micromoles per liter. However, this does not mean that excretion of an acidic urine is removing only this much H^+ from the body. For one thing, there is no bicarbonate remaining in the urine; all of the bicarbonate present in a volume of glomerular filtrate equal to the volume of the urine has been added back to the remaining volume of body fluid. If the blood bicarbonate concentration is 10 millimolar in an acidotic patient, each liter of urine excreted at pH 5.2 or below (the pH of a carbonic acid solution) is in effect adding 10 millimoles of bicarbonate to the remaining body fluid, almost equivalent to the removal of 10 millimoles of H^+.

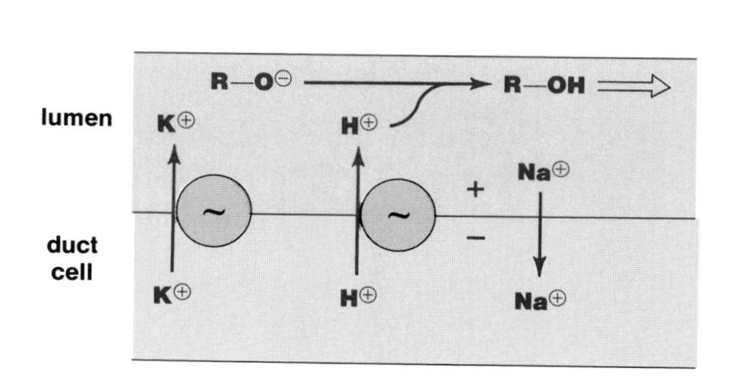

FIGURE 38–9

The collecting duct cells have active pumps for K^+ and H^+ in the luminal membrane. The amount of H^+ pumped is greater than its concentration in the urine indicates, because much of the H^+ combines with conjugate bases to form undissociated acids.

In addition, the urine contains other acids and bases, especially phosphate. The ratio $[HPO_4^{2-}]/[H_2PO_4^-]$ is 2.0 at pH 7.10 in the blood, but it is almost 0 at pH 4.5, and 0.67 mole of H^+ must be added for each mole of phosphate excreted before the pH can fall from 7.10 to 4.5 during metabolic acidosis. This can account for another 13 to 17 millimoles of H^+ excretion at the levels of phosphate found in normal urines, but severe acidosis also causes a mobilization of bone that leads to even more phosphate in the urine. Excretion of a urine at pH 5.5 to 6.0 is adequate under normal metabolic loads to maintain blood pH at 7.40.

Part of the H^+ generated in diabetic ketoacidosis can be eliminated in the urine as undissociated 3-hydroxybutyric acid. Since its dissociation constant is $10^{-4.39}$, 44 per cent of the compound is present as the undissociated acid at pH 4.5, which means that an additional 0.44 mole of H^+ must be added to the urine for each mole of 3-hydroxybutyrate excreted before the pH can fall to 4.5. (The dissociation constant of acetoacetic acid is too low for it to be a significant factor in this way. However, the spontaneous decarboxylation of acetoacetate may contribute a small loss of H^+:

$$\text{acetoacetate}^- + H^+ \longrightarrow CO_2 + \text{acetone.)}$$

Metabolic Compensation

Alterations in body composition and in metabolic processes can also be used to consume or produce additional H^+. **The dissolution of bone** during metabolic acidosis is an example. Much of the bone consists of hydroxyapatite, which has a lattice with a repeating unit of $Ca_{10}(PO_4)_6(OH)_2$, but it also contains substantial amounts of carbonate ions—of the order of one carbonate per seven phosphates in a typical long bone. This mass of crystalline salt represents a large reservoir of buffer capacity. Within a few hours of the appearance of metabolic acidosis, there is substantial loss of carbonate from the bones although with little loss of calcium or phosphate. If the acidosis persists over several days, more substantial alterations in composition occur. Part of the hydroxyapatite may then be converted to calcium phosphate, with loss of the hydroxyl ions, and part is dissolved. During metabolic alkalosis, there is increased deposition of carbonate in the bone, in both the short and long-term conditions.

Diminution in the formation of urea and increase in the excretion of ammonium ions are important adaptations to severe metabolic acidosis. Half, or more, of the nitrogen in the urine may be present as NH_4^+. In our sample stoichiometries, we saw that the conversion of nitrogen in amino acids to urea results in no change in H^+ balance, whereas the conversion to NH_4^+ results in a consumption of one H^+ per NH_4^+ formed:

$$C_3H_7O_2N + 3\,O_2 + H^+ \longrightarrow NH_4^+ + 3\,CO_2 + 2\,H_2O$$

There is a loss of one H^+ for each NH_4^+ excreted that does not occur when urea is synthesized from alanine and excreted (p. 775). There is one less H^+ to handle when nitrogen is excreted as NH_4^+, and the overall effect is the exchange of NH_4^+ in the urine for otherwise excreted Na^+ (Fig. 38–10).

Much of the nitrogen excreted as NH_4^+ during acidosis is brought to the kidney as glutamine. The kidney hydrolyzes glutamine to ammonium ion and glutamate and oxidizes the glutamate to ammonium ion and α-ketoglutarate. Two ammonium ions are available from the metabolism of one amino acid, when glutamine is used as a carrier.

UREA EXCRETION:

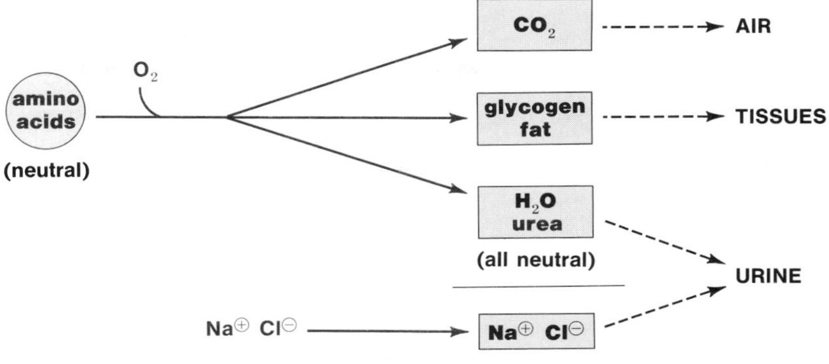

RESULT: NO CHANGE IN $H^{\oplus}$ BALANCE

AMMONIUM EXCRETION:

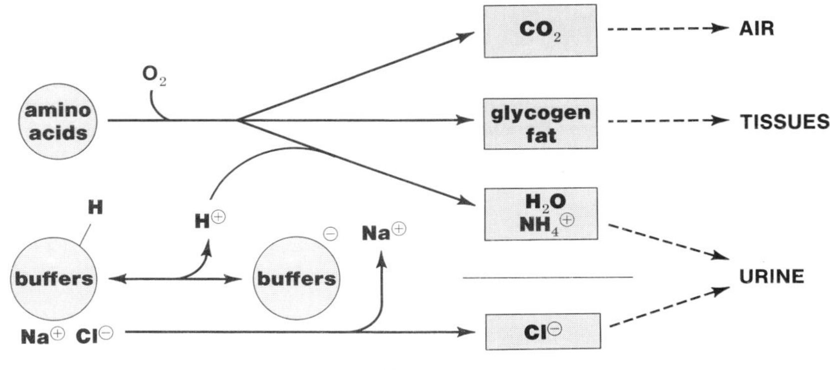

RESULT: LOSS OF THE $H^{\oplus}$ AS $NH_4^{\oplus}$, WITH COMPENSATORY RETENTION OF $Na^{\oplus}$

FIGURE 38–10 When the nitrogen from amino acids is excreted as urea (*top*), all of the major products containing C and N atoms are neutral. Since the mixture of amino acids used as a fuel is also nearly neutral, there is no net change in H^+ balance in the whole body, and neutral ion pairs, such as Na^+ and Cl^- are excreted.

(*Bottom*) When the nitrogen of amino acids is excreted as ammonium ions, H^+ is consumed. In terms of whole body balance, the H^+ is obtained from buffers, and their now greater negative charge is counterbalanced by retaining Na^+ and K^+ (not shown). Excreted excesses of anions such as Cl^- are now paired in part with ammonium ions as the cation in the urine.

There are two aspects to excretion of ammonium ion as a compensating mechanism in acidosis. One is the compensation and the other is the mechanism by which the excretion is caused to occur, and the two are often confused. The overall event is the excretion of ammonium ion formed by the combustion of the general pool of amino acids, and this is the source of the compensation, as illustrated above using alanine as an example. The cooperation of many tissues may be required to bring this event about, but it makes no difference how the steps are distributed among the organs if the overall result is the same.

The mechanism by which the compensation is caused to occur, however, involves the metabolism of glutamine by the kidneys in human beings (Fig. 38–11).

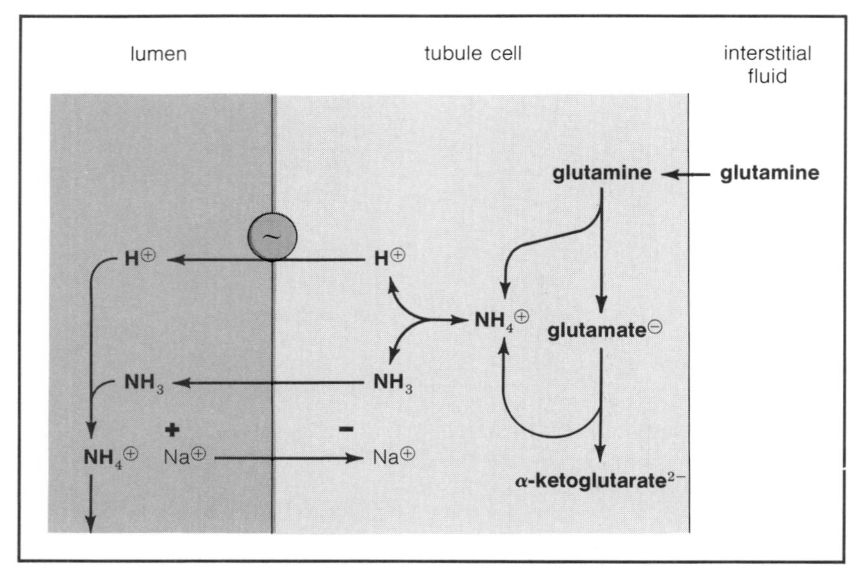

FIGURE 38–11

Diffusion of ammonia and the proton pumps in the luminal membrane of proximal tubule cells effectively add H^+ to the luminal fluid without making it strongly acidic. Ammonium ions are generated from both nitrogen atoms of glutamine in the kidneys of acidotic humans. When the α-ketoglutarate formed is converted to glucose for export (*not shown*), the net result will be the exchange of H^+ within the cell for Na^+ from the lumen.

Glutamine is formed from other amino acids, especially in the skeletal muscles.* Glutamine enters the mitochondria of tubule cells, where it is hydrolyzed to ammonium ion and glutamate, and the glutamate is oxidized to ammonium ion and α-ketoglutarate. **Ammonium ion cannot pass through the plasma membrane as such, but the ammonia in equilibrium with it can.** The cells move H^+ into the lumen in exchange for Na^+, and the diffusing ammonia recombines with H^+ in the lumen to form ammonium ions, which are excreted. The net movement of ammonia hinges upon a higher $[H^+]$ in the lumen than in the cytosol. The more acidic the lumen, the faster ammonia moves.†

The large load of α-ketoglutarate created during metabolic acidosis is more than can be oxidized in the kidney; the excess fuel is converted to glucose by the ordinary route of gluconeogenesis (p. 588) and exported to other tissues (not shown in Fig. 38–11).

The net effect of these complex events can be summarized by the following conversions, using alanine as an example of a general amino acid:

Other Organs

$$2\ \text{alanine} \longrightarrow NH_4^+ + \text{glutamate}^- + CO_2$$

$$NH_4^+ + \text{glutamate}^- \longrightarrow \text{glutamine}$$

*Reactions forming glutamine must be included in a proper stoichiometry for the compensation.

†Presumably this also applies to the mitochondrial inner membrane. The diffusion of NH_3 from matrix to cytosol discharges the proton concentration gradient, with the effective loss of 0.25 ~P per NH_3 moving.

Kidney

$$\text{glutamine} \longrightarrow NH_4^+ + \text{glutamate}^-$$

$$\text{glutamate}^- \longrightarrow NH_4^+ + \alpha\text{-ketoglutarate}^{2-}$$

$$2\ NH_4^+(\textit{internal}) + 2\ Na^+(\textit{lumen}) \longrightarrow 2\ NH_4^+(\textit{lumen}) + 2\ Na^+(\textit{internal})$$

$$2\ Na^+(\textit{internal}) \longrightarrow 2\ Na^+(\textit{lumen})\ (\text{glomerular filtration})$$

$$\alpha\text{-ketoglutarate}^{2-} + 2\ H^+(\textit{internal}) \longrightarrow CO_2,\ \text{glucose}$$

Sum for Whole Body

$$2\ \text{alanine} + 2\ H^+(\textit{internal}) \longrightarrow 2\ NH_4^+(\textit{lumen}) + CO_2,\ \text{glucose}$$

These conversions show only the charged components, and CO_2 and glucose. Note that the conversion of α-ketoglutarate to glucose is a key event in the mechanism. Writing the complete stoichiometries, including the neutral O_2 and H_2O components, is a useful exercise that will reach the same conclusion if correctly drawn. (Nucleotide components are omitted because they maintain steady state concentrations.)

A misconception of the role of ammonium ion excretion in compensation for acidosis was common in the renal physiology literature until very recently, and it still persists in many textbooks. It was believed that compensation itself requires the use of glutamine, and it is worth noting how this error came about, if for no other reason than avoiding similar mistakes. This incorrect version is outlined in Figure 38–12. According to it,

1. Glutaminase in the tubule cells hydrolyzes glutamine to glutamic acid and ammonia:

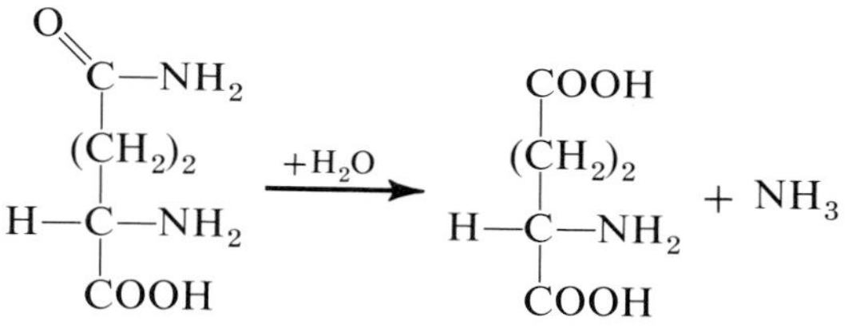

```
O
 \\
  C—NH2                         COOH
  |                              |
 (CH2)2          +H2O          (CH2)2
  |            ——————>           |          + NH3
H—C—NH2                       H—C—NH2
  |                              |
 COOH                           COOH
glutamine                    glutamic acid
```

2. The ammonia diffuses into the lumen of the tubules, where it reacts with H^+ to form NH_4^+. The neutralization of H^+ in the lumen diminishes the gradient for H^+ across the wall of the lumen and permits more H^+ to be pumped into the filtrate in exchange for Na^+. The overall stoichiometry becomes:

$$(\text{glutamine} + H_2O + CO_2)_{\text{blood}} + Na^+_{\text{filtrate}} \longrightarrow (\text{glutamic acid}\ (\textit{sic}) + Na^+ + HCO_3^-)_{\text{blood}} + NH_4^+{}_{\text{lumen}}.$$

Two things ought to have prevented the persistence of this explanation: writing the proper stoichiometry for the postulated events and considering the fact that some animals adapt to acidosis by excreting NH_4^+ but have little change in the glutaminase activity in their kidneys.

Accumulation of carboxylates provides partial compensation for alkalosis. For example, respiratory alkalosis is often accompanied by an increased accumulation of lactate, and metabolic alkalosis causes increased accumulation of the ketone bodies and citrate. It is not known what causes these accumulations.

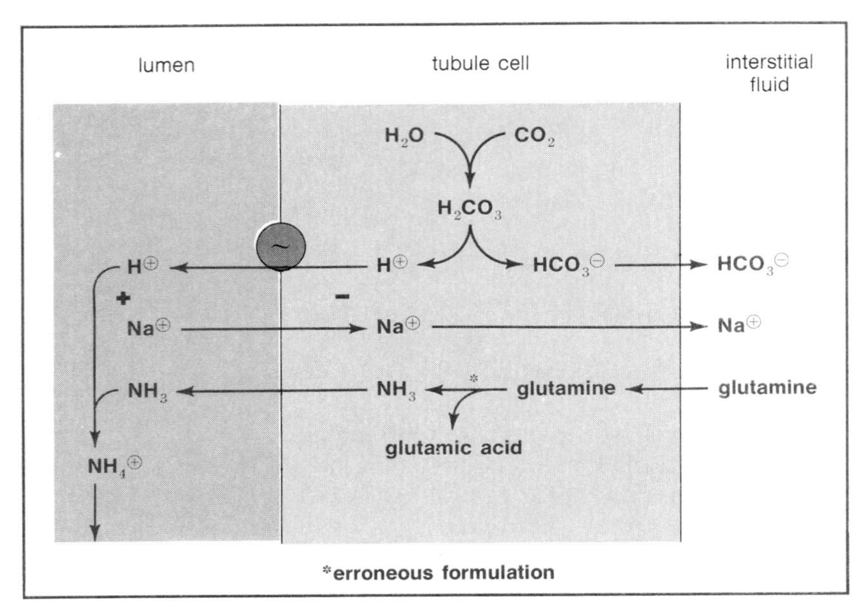

FIGURE 38–12

A common erroneous formulation of ammonium ion excretion in the kidney, purporting to show a net loss of H^+ and gain of bicarbonate ion as a result. The fallacy lies in the improper formulation of the hydrolysis of glutamine (compare p. 581).

Uncontrolled formation of lactate (lactic acidosis, p. 497) often accompanies a variety of conditions and is sometimes caused by a hereditary defect of unknown mechanism. A severe acidosis would tend to suppress lactate utilization through an adverse shift of the lactate dehydrogenase equilibrium:

$$\text{lactate}^- + \text{NAD}^+ \longleftrightarrow \text{pyruvate}^- + \text{NADH} + \text{H}^+$$

which forms H^+ when lactate is converted to pyruvate, and this may account in part for the lactic acidosis that sometimes aggravates severe diabetic ketoacidosis. (The concomitant shift of the equilibrium from acetoacetate to D-3-hydroxybutyrate sometimes causes trouble because measurement of acetoacetate is used as an index of the degree of ketosis.)

Assessment of Disturbances of H^+ Balance from Laboratory Data

The blood analyses of interest in interpreting acid-base changes are those correlated in Figure 38–7: pH, p_{CO_2}, and total CO_2, along with additional measurements of K^+, Na^+, and Cl^- concentrations. The pH value, obtained directly or by calculation, tells whether a significant acidosis or alkalosis exists. The effects of acute uncompensated changes are shown in Figure 38–13. Sudden changes in ventilation rate shift the pH and p_{CO_2} values within a characteristic band on the chart, as we noted earlier. The position of this band is determined by the total buffer capacity and the relative proportions of acidic and basic forms present at a given p_{CO_2}.

If a metabolic acidosis or alkalosis appears without respiratory compensation, the proportion of acidic and basic forms of the buffers is being changed without altering the p_{CO_2}. The effect is then a shift of the buffer band up or down the line of constant p_{CO_2}.

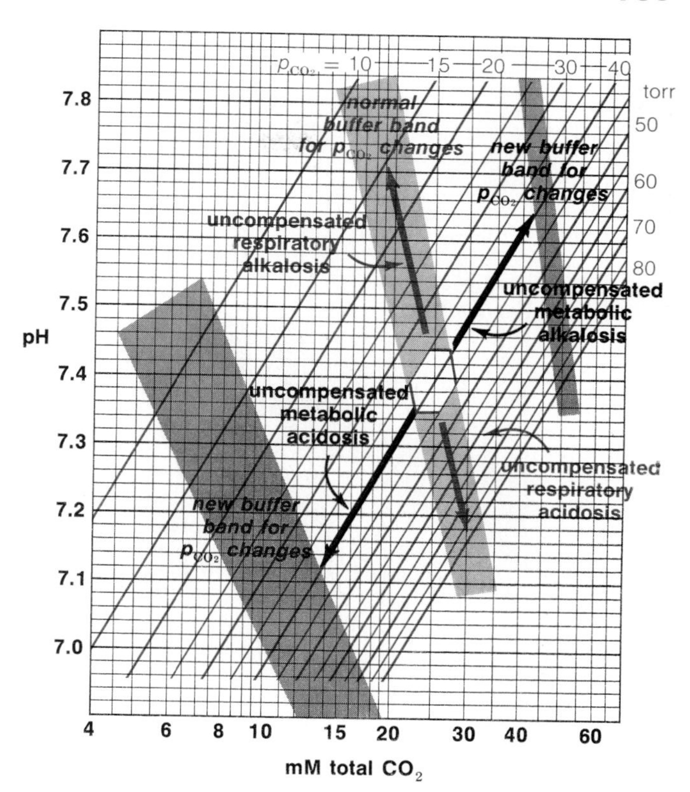

FIGURE 38–13

The effects of uncompensated changes in acid-base balance. Faster or slower exchange of CO_2 produces a respiratory alkalosis or acidosis within the normal buffer band, as shown by the heavy brown line. A rapid addition of acid or alkali produces a metabolic acidosis or alkalosis along the line of normal p_{CO_2}, as shown by the heavy black line. Compensatory mechanisms promptly begin to shift these values from the lines shown.

Compensated Changes. The compensation for alkalosis and acidosis in effect superimposes respiratory changes and the opposite renal metabolic changes. That is, the kidney creates a metabolic acidosis by conserving H^+ in order to compensate for a respiratory alkalosis caused by the hyperventilation in an overly anxious person. Contrariwise, the ventilation is sharply increased in the lungs to create a respiratory alkalosis to compensate for the increased production of H^+ associated with ketone body formation in an uncontrolled diabetic. If there is a clear-cut deviation from normal pH, the primary cause and the compensating event can be distinguished by making use of data on p_{CO_2} or [total CO_2] (Fig. 38–14):

$[H^+]$	pH	[total CO_2]*	p_{CO_2}	Primary Event	Compensation†
(1) high	low	low	low	metabolic acidosis	respiratory alkalosis
(2) high	low	high	high	respiratory acidosis	metabolic alkalosis
(3) low	high	high	high	metabolic alkalosis	respiratory acidosis
(4) low	high	low	low	respiratory alkalosis	metabolic acidosis

*The concentration of total CO_2 is regarded as high or low when it lies outside the normal buffer band for the measured pH or p_{CO_2}.

†Renal compensations are regarded as included in metabolic compensations.

Problems in interpretation arise when the $[H^+]$ is nearly normal, and the total CO_2 and p_{CO_2} are not. Which is the compensating acidosis or alkalosis, and which is the primary event? The concentrations of the other ions, in addition to giving important information on the water balance and the possible existence of deficits or surpluses of K^+ in particular, provide important clues to the existence of ions in the

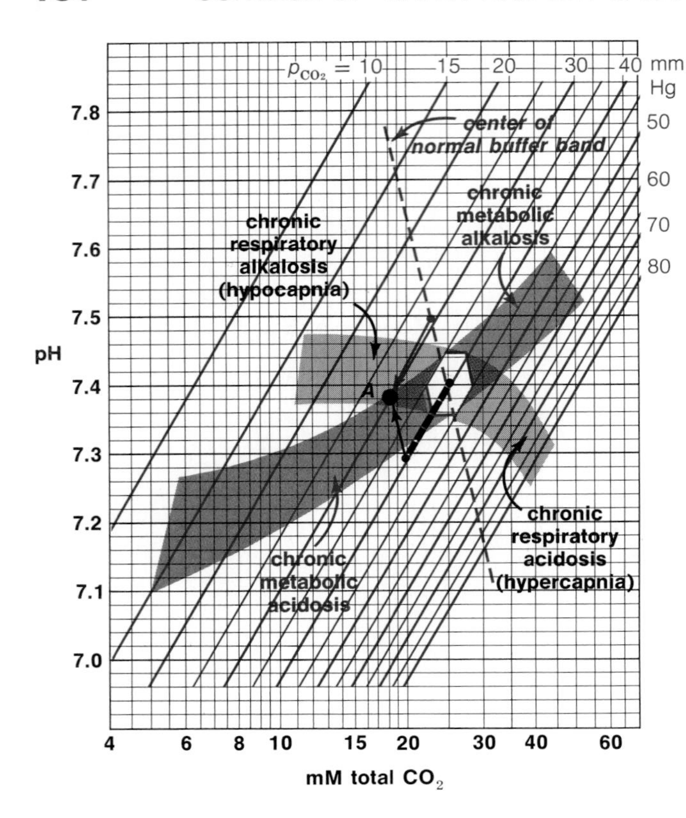

FIGURE 38–14

Typical ranges of values for chronic acidosis or alkalosis after compensation has proceeded. Decisions as to the primary events are frequently difficult. Even with values within the ranges shown, as at point A, similar concentrations could be produced by two routes—a respiratory alkalosis followed by compensatory renal acidosis, as shown by the brown band, or a metabolic acidosis, followed by compensatory respiratory alkalosis, as shown by the black band.

blood that are ordinarily not measured. That is, the sum of the measured anions (bicarbonate and chloride) is normally less than the sum of the measured cations (sodium and potassium), because there are more unmeasured anions, such as lactate, phosphate, and so on, than there are unmeasured cations like calcium and magnesium. This **anion gap** becomes much larger in metabolic acidosis caused by accumulations of 3-oxybutyrates, lactate, and so on, and this is an important diagnostic clue in some conditions. If there are no previous measurements to give a clue as to the direction things are going, measurement of the anion gap can be helpful. In practice, the normal range of the excess of $[Na^+] + [K^+]$ over $[Cl^-] + [\text{total } CO_2]$ is taken as 11 to 17 mM. (This range will vary from laboratory to laboratory.) A low total CO_2 concentration and a high anion gap is in itself an indication of primary metabolic acidosis.

FURTHER READING

E.J. Masoro (1982). An Overview of Hydrogen Ion Regulation. JAMA 142:1019.

J. Hood and E.J.M. Campbell (1982). Is pK OK? N. Engl. J. Med. 306:864.

R. Nuccitelli and D.W. Deamer (1982). Intracellular pH: Its Measurement, Regulation, and Utilization in Cellular Functions. Liss.

B.M. Brenner and F.C. Rector, Jr., eds. (1981). The Kidney, 2nd ed. Saunders. This two-volume treatise is the best detailed reference to kidney function.

The following small books give informative summaries; they differ in their coverage of various aspects:

E.J. Masoro and P.D. Siegel (1977). Acid-Base Regulation, 2nd ed. Saunders.

B.D. Rose (1977). Clinical Physiology of Acid-Base and Electrolyte Disorders. McGraw-Hill.

A.B. Schwartz and H. Lyons, eds. (1977). Acid-Base Balance and Electrolyte Balance. Grune & Stratton.

Other References

M.H. Maxwell and C.R. Kleeman (1979). Clinical Disorders of Fluid and Electrolyte Metabolism. McGraw-Hill.

R.L. Jamison and W. Kriz (1982). Urinary Concentrating Mechanisms: Structure and Function. Oxford.

W.N. Scott and D.B.P. Goodman, eds. (1981). Hormonal Regulation of Epithelial Transport of Ions and Water. Ann. N.Y. Acad. Sci. v. 342. Conference papers.

J. D. Veldhuis et al. (1980). Inborn Error in the Terminal Step of Aldosterone Biosynthesis. N. Engl. J. Med. 303:117.

L.F. Francisco and T.F. Ferris (1982). The Use and Abuse of Diuretics. Arch. Intern. Med. 142:28. Short review.

M.J. Brownstein, J.T. Russell, and H. Gainer (1980). Synthesis, Transport, and Release of Posterior Pituitary Hormones. Science 207:373.

K.J. Sweadner and S.M. Goldin (1980). Active Transport of Sodium and Potassium Ions. N. Engl. J. Med. 302:777. Review.

D.G. Vidt, E.L. Bravo, and F.M. Fouad (1982). Captopril. N. Engl. J. Med. 306:214. Review of an example of an "angiotensin-converting enzyme" inhibitor, a drug blocking the hydrolysis of angiotensin I to angiotensin II.

J.P. Knochel (1982). Neuromuscular Manifestations of Electrolyte Disorders. Am. J. Med. 72:521.

CHAPTER 39

NUTRITION: FUELS AND ENERGY BALANCE

Nutrition is the study of the effect on the organism of changes in the diet. It is fitting that we close our discussion of biochemistry with this aspect of the subject because the search for food is one of the things life is all about, and proper analysis of what happens when we consume the results of the search requires the broadest sort of background. We are all aware of the growing problems of feeding the human population, and we are also aware of the many admonitions to the affluent against overeating. The very importance of nutrition creates some ancillary problems that we may as well face head-on.

Until recently, many biological scientists and clinicians avoided discussing or even thinking about nutrition, despite its obvious importance as a practical matter and as an ultimate test of abstract rationalization. Nutrition acquired the taint of a pseudoscience in the eyes of many, who therefore denied themselves one of the more useful bodies of information in understanding the living organism. We now frequently encounter the converse problem. Nutrition has become chic to the public and an attractive source of funds for the research administrator. The desire to be a true believer in the new faith has led even professionals to occasional uncritical acceptance of dogmatic pronouncements on nutrition from an astonishing spectrum of people. To avoid falling into the same pits, old or new, let us take the time to discuss how these problems came about.

Part of the difficulty comes from confusing the science of nutrition with the art of dietetics, a confusion that is augmented by the little maxims about good food that are first delivered as the revealed word to nearly everyone in the primary schools and repeated at frequent intervals thereafter without any examination of their basis. To put it bluntly, most people talking about hygiene, health, or menu-planning know little about nutrition.

The next part of the problem paradoxically comes from the eminent practicality of much nutritional knowledge. It provides a basis for value judgments on the character of the food supply. The provision of the supply and its delivery is a major part of the economy of all nations: changes in any aspect can bring financial ruin to some people and great wealth to others, and can make or break a political career. Even a President of the United States publicly drank a glass of milk and proclaimed its necessity for good health when a decline in the market threatened the income of the producers. Spectacles in the form of White House conferences and Senate hearings have been convened with intensive press coverage announcing that a large fraction of the U.S. population is "suffering" from nutritional inadequacy. The result of all this is that the hard facts are sometimes misstated, ignored, or presented out of context to promote a particular product, or justify a particular action, and for some reason these misuses are rarely challenged by those who know better. (It doesn't help objective appraisal to have some of the best-staffed university centers of nutritional research totally dependent upon their own fund-raising activities.) Our task here is not to assess economic or political decisions, but to uncover some of the scientific bases upon which they can be made.

The task is made more difficult by our own tendency to confuse objectives with proven facts. We talk about malnutrition and frequently forget that it means bad nutrition, with its implication of failure to reach some desired goal. The science of nutrition does not define goals; it only gives information on effects. When we talk about a **minimum requirement** for a nutrient we are stating an estimate based on scientific information to achieve some purpose that is not defined by the science, and there will inevitably be discrepancies in the definition of the requirement because of discrepancies in the objectives. An adviser in a rich nation may strive to keep all of his people saturated with the nutrient as defined by laboratory measurements while one in a poor nation might be content with avoiding obvious functional impairment in 99 per cent of his people. The major defined objective for children frequently is maximum growth—somewhat as if people were Angus cattle. This is an easily measured criterion; however, there is also some evidence from experimental animals that maximum intelligence and maximum longevity are achieved at something less than the maximum possible growth rate. Here is a case in which the science of nutrition has yet to give us all the necessary information we need to make a considered judgment in light of our personal aims.

Perhaps the most difficult misconception for a layman to shake is the idea that there is some essential value in a given type of food—that without fruit, or leafy vegetables, or meat, or milk, and so on, a diet is inadequate, if not immoral. **There is no nutritional basis for an absolute requirement of any particular food.** Humans have been around for some time, and we have evidence from archeological explorations, historical records, and current observations to show that cultures can thrive and their adherents perpetuate their kind on a wide variety of dietary intakes—in some cases almost completely deficient in one or more of the "basic foods" touted in some quarters.

Having now shaken our preconceptions, let us go on to consider some of the truly usable information. Nearly all of the various components of foods have been encountered in the earlier chapters, and we have seen how they are used within the body. Our remaining problem is one of organizing the information and gaining a more quantitative concept. Let us begin by considering the dietary fuel supply.

THE FUEL SUPPLY IN FOODS

Most animals fit in ecological niches by confining themselves to foods of restricted types, and their survival hinges upon the maintenance of an appetite for tissues from particular classes of organisms, plant or animal. We don't think of cows eating meat or cats eating grass as a significant part of their dietary intake. This is in part due to biochemical differences. Cats don't have the intestinal flora necessary for degradation of cellulose in grass, and probably don't have the enzymatic capacity to handle propionate in the concentrations absorbed by the intestines of cows. Cows probably can't handle amino acids in the concentrations absorbed by cats. Much of the difference, however, is a result of adaptation and habits to fit the ecology, with the qualitative nature of the biochemistry not being so altered as might appear. Cows can utilize meat and cats can utilize grains if they are presented in forms that the animals will eat.

The problem of the relative value of energy sources in food comes into its own with the omnivores such as man, who may be gulping down the still-warm hindquarters of a wild pig (1.5 per cent carbohydrate, 25 per cent fat, and 16 per cent protein) brought down by a spear to provide the week's ration, while aspiring to a life of self-contemplation in the land of milk (4.9 per cent carbohydrate, 3.9 per cent fat, and 3.5 per cent protein) and honey (79.5 per cent carbohydrate, 0.3 per cent fat, 0 per cent protein).

Since man is capable, both by anatomy and by taste, of eating a wide variety of foods, we might expect the proportion of carbohydrate, fat, and protein available from the diet for utilization as fuels to vary quite widely.*

Units of Biological Energy. Studies of the energy economy of whole organisms began in earnest during the 19th century, when nothing was known about high-energy phosphate, proton concentration gradients, glycolysis, and so on. Instead, early students of biological energetics adopted the practice of thinking in terms of heat equivalents and so it was that **calories** entered the language of the nutritionist, now being slowly replaced by **joules.** We balance in calories or joules the energetic cost of biological activities (physical exertion, kidney function, and so on) and the energetic yield from fuels burned to sustain those activities.

An important point must be made. The standard unit of heat energy in scientific circles formerly was the calorie, then defined as the amount of heat necessary to raise

---

*A lead article in *Science* stated:

"The traditional precompetition meal of athletes continues to be a protein-rich one, and the favorite main dish is a large steak. The fact that protein is not used as an energy source by the body except in starvation or when the diet is grossly deficient in fat and carbohydrate seems to have little impact on the practice."

Whatever the merits of the traditional meal, the grounds given here for attacking it are wrong. An adult maintaining constant weight uses all of his dietary amino acids as fuel, except for the small fraction excreted in the form of uric acid, creatinine, and the like.

the temperature of one gram of water by 1° at 15° C and now defined in terms of the joule: 1 cal = 4.184 J. The heat of combustion of the common foodstuffs lies in the range of several thousand calories per gram. Because of a reluctance to write large numbers, it became the practice to specify the values in **kilocalories.** Someone made the mistake of designating this as a large calorie, spelled with a capital letter (Calories or Cal.). This trivial typographical distinction was soon neglected (what is a thousand-fold among friends?), and we must face the fact that laymen and dieticians mean kilocalories when they say calories. There is no excuse for such license in professional circles. When we say calories, we mean it, and 1000 calories is a kilocalorie, not a Calorie. It will take time to think in terms of megajoules (MJ) rather than kilocalories, but we shall make a beginning on this. 1 MJ = 239 kilocalories; 1000 kilocalories = 4.18 MJ.

Actual Consumption

The estimated production of major foodstuffs for the human race is plotted in Figure 39–1 in terms of the number of people that could be supported by the crop at an energy expenditure of 8 megajoules per day, a maintenance level only a little above basal. The figure is misleading in that it assumes total ingestion of the crops without waste or diversion for animal feed or other purposes. Even so, it is informative.

The majority of the world's population lives on diets in which cereals are the primary nutrient, and this has probably been true for a long time. The maintenance of the ever-normal granary was a desideratum in most early civilizations; cities and grain grew together. Even today, disastrous famine has been averted only through the development of high-yield varieties of wheat and rice, which have made India self-sufficient in most years, and several other overpopulated countries nearly self-sufficient.

Most cereals are rich in starch and low in fat, so oxidation of carbohydrate generates more than half of the total energy for most humans. Even children, for whom the dietary requirements are most demanding, will thrive on diets in which

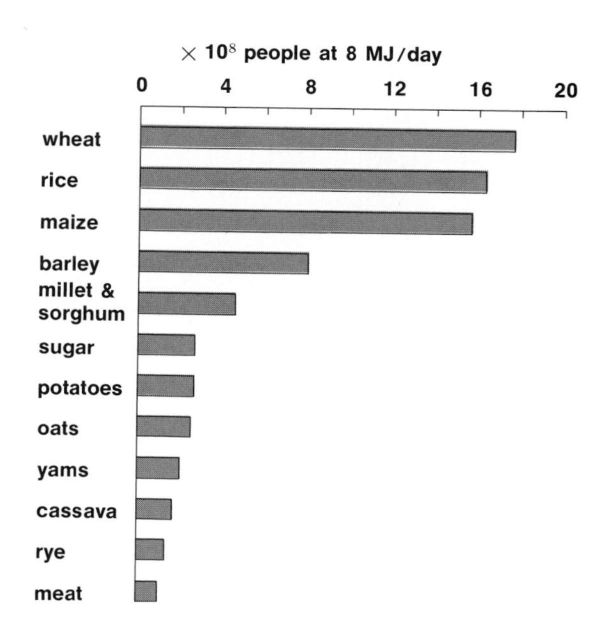

FIGURE 39–1

World production of major food energy sources, shown as the number of people who could be given 8 megajoules of energy per day from the crop if it were possible to use it with perfect efficiency. Calculated from rough estimates by the Food and Agriculture Organization of the United Nations.

the content of carbohydrate is 13 times greater than the content of fat. If we define a normal human diet as the diet of most adults alive today, then the normal partition of potential energy is approximately 65 per cent from carbohydrate, 25 per cent from fat, and 10 per cent from protein. For comparison, over 40 per cent of the energy yield from the average United States adult diet is derived from fats, and approximately 14 per cent from protein.

Regional preferences in foods differ widely. In parts of Africa, the small grains (millet, sorghum, and teff) and the tubers (yams and cassava) constitute the major part of the fuel supply. Much of the maize* supply is used as animal feed, causing large losses of the potential energy upon combustion before reaching the human mouth.

The Cost of Fuel. The economic factors making the cereals the major human foods can be appreciated from Figure 39–2, which plots the retail cost in Virginia supermarkets of 10 megajoules of potential fuel energy in individual foods as of June, 1982, considering economy rather than quality as the criterion of choice. (Ten

*The word corn is used for the principal grain in English-speaking countries. In the United States it means maize, in England wheat, and in Scotland oats.

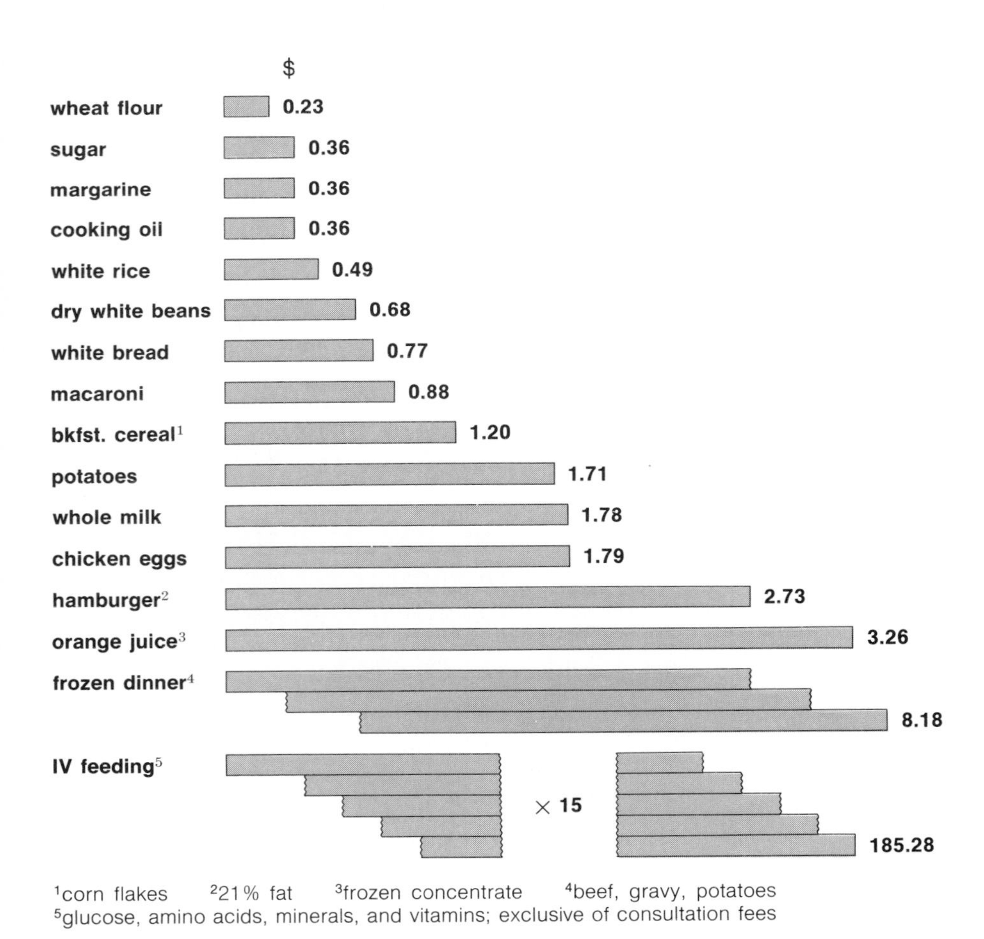

FIGURE 39–2 The cost of supplying 10 megajoules of energy in the form of various foods available at retail stores. Calculated from lowest price items available in Charlottesville, Virginia, in June, 1982.

megajoules is a day's supply for a lightly active adult woman of median size.) Even after the distortion caused by packaging and delivery in small lots, wheat flour is the cheapest fuel for humans, and rice is not too far behind, despite its lower demand.

Other points of interest in Figure 39–2 are the low cost of margarine and sugar and the very high cost of prepared meals. The cost of feeding patients a complete diet parenterally* (intravenously) is given for comparison. In considering the costs, it ought to be remembered that the human diet must contain other components than a fuel supply, and it is not possible to keep a human alive on $0.23 a day, at least not in Charlottesville, Virginia.

Sucrose is a major dietary fuel, especially in many of the wealthiest nations. It is hydrolyzed to glucose and fructose in the small bowel. High-fructose syrups are made by passing glucose solutions (prepared by hydrolyzing corn starch) over a bacterial isomerase covalently bound to insoluble supports in columns. (Glucose is less sweet than sucrose, fructose is sweeter.)

Fructose is absorbed by facilitated diffusion. It is metabolized mostly by the liver, but also by the small bowel and the kidneys. It is first phosphorylated to fructose 1-phosphate by a specific kinase, which is then cleaved to trioses by an aldolase (Fig. 39–3). The triose phosphates are then handled by the usual routes, with storage as fat or glycogen in times of carbohydrate surplus, or conversion to glucose at times of carbohydrate deficit. When fructose is rapidly absorbed, a significant fraction can be converted directly to fructose 6-phosphate in the muscles, because the muscle hexokinase will use fructose as well as glucose as a substrate.

The Tolerable Range of Composition. How much variation in the nature of the fuel supply can man tolerate? The usual diet is high in carbohydrates over much of the world and relatively high in fats in highly developed countries. We know that there is a minimum limit on the amount of protein in the diet, which is fixed by the necessity of precursors for proteins and other nitrogenous compounds, and not by the utilization of amino acids for energy production. Is there a maximum limit? Several lines of evidence suggest that there is. We should expect it, because we have seen that the processing of amino acids begins in the liver, and this involves the consumption of oxygen. The liver is like other organs in having a finite capacity for electron transport.

Good measurements of the oxygen consumption of human liver are not available, but by extrapolation from other animals and from *in vitro* measurements with human specimens, we find that the liver of a 73-kg man probably consumes between two and five moles of oxygen per day. Since the metabolism of amino acids according to the stoichiometry we previously cast involves the consumption of one mole of oxygen per mole of amino acid nitrogen handled in the liver, we can assume that the total capacity of the liver is the metabolism of two to five moles of amino acid nitrogen per day if *all* of its oxidative capability is devoted to amino acid metabolism. Total metabolism of amino acids containing this quantity of nitrogen is sufficient to supply from 40 per cent to all of the basal requirements.

Since the metabolism of the liver cannot be completely devoted to amino acids, we might expect that the actual ability of the body to handle amino acids as fuels is limited to somewhere near one half of the total energy requirements. (Of course, any degeneration of liver tissue would reduce this value even further.)

Human experience reinforces the idea of a limited capacity to handle protein. William Clark and Meriwether Lewis, two boys from Albemarle County, Virginia,

---

*Parenteral is any route outside the gastrointestinal tract, including subcutaneous, intramuscular, intraperitoneal, and intravenous injection or infusion. In practice, parenteral feeding is done intravenously.

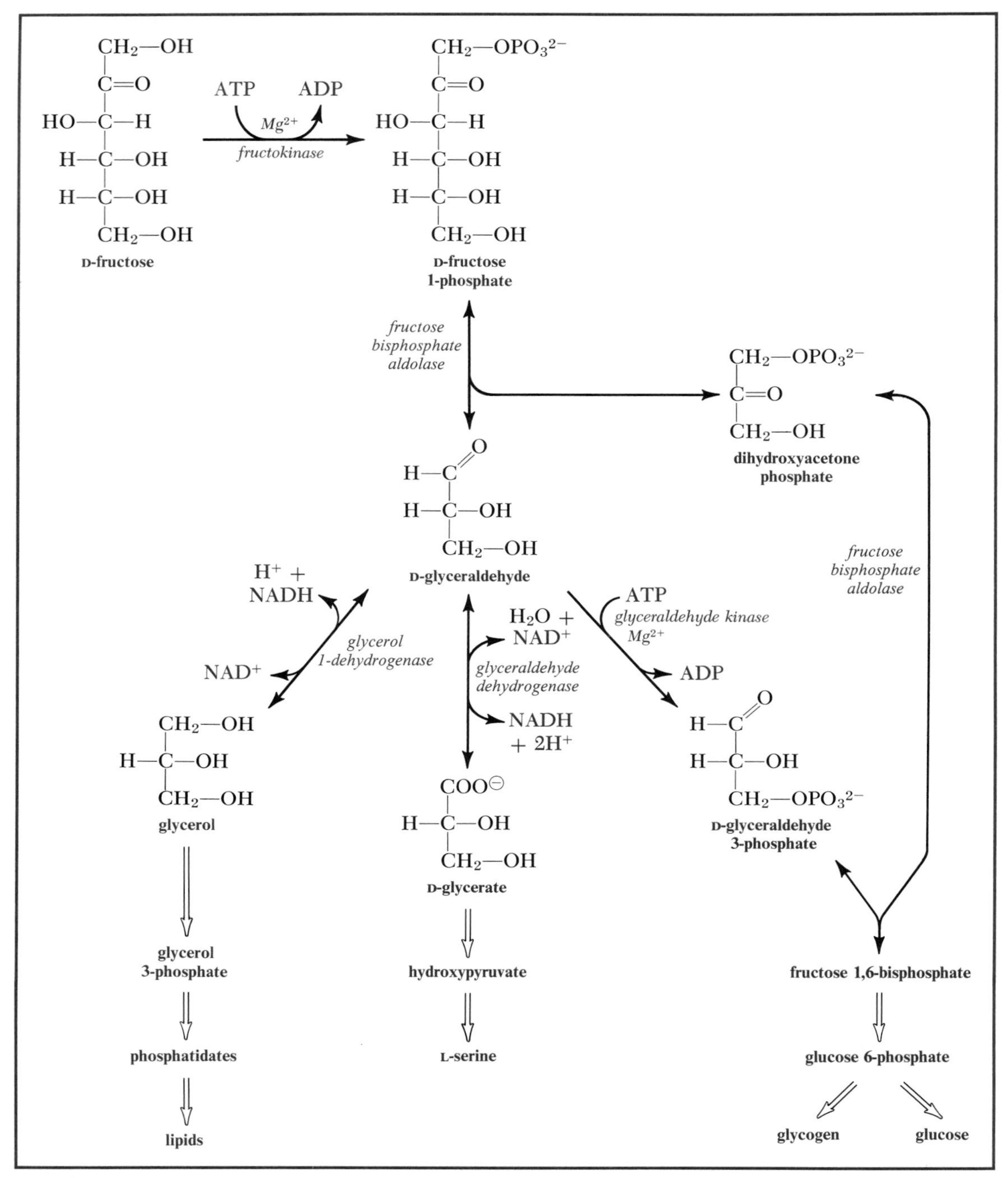

FIGURE 39–3 **The routes of fructose catabolism in humans.**

led a group from the Mississippi River to the Pacific Ocean and back in 1803–1805. They lacked scientific credentials, but proved themselves to be reasonably accurate and astute observers. They were dependent upon animals as the sole source of food during much of the journey, and fat dogs became not only a staple, but the preferred diet. However, they were able to obtain only lean deer at some stages during the early spring. They found that, although the meat was available in sufficient quantity, they lost weight and developed gastrointestinal distress along with other symptoms, justifying the conclusion they made: fat is a necessary component of a meat diet. (The word protein hadn't been invented.)

The conclusion is supported in a somewhat better study in more recent times in which an Arctic explorer, Vilhjamur Stefannson, and friends ate an animal tissue diet for more than a year to counter skepticism that such a diet could be consistent with health. The results of the study are published in the papers by McClellan *et al.* cited at the end of the chapter. In McClellan's words, "At our request, he began eating lean meat only, although he had previously noted, in the North, that very lean meat sometimes produced digestive disturbances. On the 3rd day nausea and diarrhea developed. When fat meat was added to the diet, a full recovery was made in 2 days. This disturbance was followed by a period of persistent constipation lasting 10 days. The subject had a craving for calf brain, of which he ate freely. (Calf brain has 9 grams of lipid and 11 grams of protein per 100 grams.) On March 12, poor appetite, nausea, and abdominal discomfort were present and a second but milder attack of diarrhea occurred which responded quickly to a proper proportionment of lean and fat meat."

As is always true of these kinds of observations, there is room for conjecture. Was the discomfort with lean meat a result of mental bias or of biochemical aberration caused by the diet? In light of other observations, and the general proclivity of those in pure hunting cultures to gulp fat, it seems likely that protein cannot be used as the primary source of energy by humans.

The discussion is somewhat academic as it pertains to some people in the world, for whom the principal problem is supplying enough, rather than too much, protein. Beyond this, the amount of protein consumed by choice is amazingly constant with humans from a variety of cultures, accounting for 10 to 15 per cent of the total energy yield. This is not always true, because many humans still live almost entirely on meat. The Eskimos are prime examples. However, the protein content in such cases rarely reaches one third of the potential energy yield, with the remainder being mainly fat.

The hunter's diet thus represents another of the potential extremes, in which there is little carbohydrate, and fat represents the bulk of the fuel. The "proper proportionment" of protein and fat referred to by McClellan in the preceding quotation was such that protein represented approximately one quarter of the energy yield, and fat represented three quarters. What happens in these circumstances? The evidence at hand suggests that adult humans can get along reasonably well for extended periods on such diets. Beyond the McClellan experiments, little has been done in careful observation, and there are no reliable data on relative morbidity and mortality of people living on meat compared to others in similar circumstances. Northern American Indians relied heavily on pemmican, which is ground dried meat in melted buffalo fat, as the primary food on long journeys, and the production of pemmican later became a quite substantial enterprise because the trappers of the Canadian trading companies relied on it exclusively during the winter. (There was a little-known "Pemmican War" between two of these companies for control of the supply during the first half of the nineteenth century. Interestingly enough, settlers on Hudson Bay supplied with conventional food from England developed serious signs of nutritional inadequacy, whereas no indication of this sort of difficulty was reported by the trappers living on pemmican.)

However, the meat-fat diet does cause the appearance of an asymptomatic ketosis, with the excretion of some 50 millimoles of ketone bodies per day in the urine. (There are great individual and daily variations in that amount.) This is akin to the ketosis seen in the starved obese, except that the obese frequently adapt so as to have little ketosis. The excretion of ketone bodies during ketosis represents the loss of potential fuel, but the quantity involved is only a few per cent of the total.

Adaptations to Fuel Changes

Whenever there is a shift in the quantity or the nature of the fuel supply, the enzymatic constitution changes so as to make most efficient use of the materials at hand. This is especially true of the liver, which plays a major role in balancing the flow of different metabolites. Most of the changes are a result of altered rates of gene transcription, sometimes caused by changes in hormone secretion, and sometimes by changes in metabolite concentration.

Adaptations to Starvation. The study of dietary adaptations is particularly satisfying because the observed changes are nearly always what we think they ought to be. Table 39–1 shows some examples from starved rats (everything we know about the omnivore, man, leads us to believe he has similar changes). When an animal doesn't eat, many of the enzymes responsible for handling excesses of foodstuffs become excess baggage, and we might expect them to disappear so that their constituent amino acids can be used for more pressing purposes. Starvation sharply diminishes the need for digestion, for conversion of glucose to fatty acids, for storage of fatty acids in adipose tissue, and for adjustments in fatty acid composition. At the same time, there is an increased mobilization of fatty acids. Even though the total utilization of amino acids as a fuel may fall with starvation, it is important that the liver divert the carbon skeletons as much as possible toward glucose formation when gluconeogenesis provides the only source of carbohydrate for emergency effort or for the construction of cellular components.

It is especially impressive to note the selective cut-off in synthesis of enzymes

TABLE 39–1 ENZYME ADAPTATIONS TO STARVATION*

Increased	Decreased
Enzymes Secreted By The Pancreas	
None	All hydrolytic enzymes
Enzymes of Fatty Acid Synthesis (Adipose Tissue and Liver)	
None	Acetyl-CoA carboxylase Fatty acid synthase Acyl-CoA desaturase (liver) Lipoprotein lipase (adipose tissue) NADP-malate dehydrogenase Citrate cleavage enzyme
Enzymes of Fatty Acid Utilization	
Carnitine palmitoyl transferase (liver)	None
Enzymes of Glucose Utilization	
None	Glucokinase (liver)
Enzymes of Gluconeogenesis From Amino Acids	
Serine dehydratase Alanine aminotransferase (liver cytosol) Pyruvate carboxylase Phospho-*enol*-pyruvate carboxykinase Glucose-6-phosphatase	None

*Only some of the more significant changes are listed. This is also true of later tabulations.

peculiar to fat storage during starvation, even including the liver desaturase that introduces more double bonds into fatty acids and the lipoprotein lipase that serves to clear transported fats from the blood for storage. The pyruvate carboxylase of liver does not decrease even though it is involved in fatty acid synthesis from glucose because it is also an essential component of the route of gluconeogenesis. Those enzymes peculiar to gluconeogenesis, such as phosphopyruvate carboxykinase, are made in increased amounts during starvation. Glucose-6-phosphatase, responsible for releasing glucose from the liver, increases during starvation, but the glucokinase that takes up glucose in the liver when the concentration is high declines. This is as it ought to be, because a starving animal does not have a high concentration of glucose.

When we think on it, we see that most animals go through a period of food deprivation every day while they sleep. The levels of many of the enzymes in the preceding tabulation have been shown to rise and fall every day in response to the diurnal variation in habits. It is a little unsettling to know that our metabolic machinery is being taken apart and rebuilt to new specifications while we sleep, only to have the alterations rescinded when we eat, but if it weren't so we should really be stuck in a rut.

Adaptations to Changes in Dietary Glucose and Fat. There are only two degrees of freedom for variation in the proportion of major fuels in a diet. A diet can't be simultaneously rich in carbohydrates, fats, and proteins. If the carbohydrate content is high, the fat content frequently will be low, and the enzymes necessary for handling carbohydrates will increase, while those peculiar to routes utilizing *dietary* fat and to routes of gluconeogenesis from amino acids decrease. The reverse effects occur if there are little glucose and abundant fat.

The responses shown in Table 39–2 are very much along the lines that would be predicted, with especially dramatic responses in the enzymes necessary for synthesizing fatty acids from glucose and for synthesizing glucose from amino acids. As a minor point, the selective response of the pancreas to the two types of diet, so that only the appropriate hydrolytic enzyme is elaborated in increased amounts, indicates that the decline in synthesis of all hydrolytic enzymes during total starvation mentioned earlier is a real adaptation and not merely the result of some general debilitation of the secretory cells.

Adaptations to Diets Rich in Protein. Eating diets in which proteins supply a major fraction of the oxidizable substrates mainly causes increases in the activity of enzymes involved in nitrogen metabolism. Many of the changes are exaggerated if the diet also is low in carbohydrates. All of the following changes occur in the liver unless otherwise noted. In some cases a specific amino acid is required for the change, rather than a general increase in nitrogen intake, and these are also noted parenthetically:

Increased

peptidases (pancreas)
ornithine carbamoyl transferase (Arg)
argininosuccinate synthetase
argininosuccinase
arginase
serine dehydratase (repressed by glucose)
tyrosine aminotransferase (separate inductions by Tyr and by mixed amino acids)
alanine aminotransferase (cytosol)
aspartate aminotransferase (cytosol)
cystathionase (Met)
glutaminase (kidney)
tryptophan dioxygenase (Trp)
fructose-bisphosphatase
ornithine aminotransferase

Decreased

methionine adenosyl transferase (Cys)

TABLE 39–2 ENZYME ADAPTATIONS TO GLUCOSE OR FAT DIETS

Increased with High Glucose, Decreased with High Fat	Increased with High Fat, Decreased with High Glucose
Amylase (pancreas)	Lipase (pancreas)
Glucokinase (liver)	Glucose-6-phosphatase (liver)
Glucose-6-phosphate dehydrogenase (liver and adipose tissue)	Fructose bisphosphatase (liver)
6-Phosphogluconate dehydrogenase (liver and adipose tissue)	Carnitine palmitoyl transferase (liver)
Acetyl CoA carboxylase (liver and adipose tissue)	Serine dehydratase (liver)
Fatty acid synthase (liver and adipose tissue)	Tyrosine aminotransferase (liver)
Citrate cleavage enzyme (liver and adipose tissue)	Ornithine aminotransferase (liver)
NADP-malate dehydrogenase (liver and adipose tissue)	

ENERGY BALANCE

The study of animal energetics began with Antoine Laurent Lavoisier, who put a guinea pig in a bell jar along with a piece of ice and found that CO_2 accumulated and the ice melted at proportional rates. There is a twist of fate here. Two phenomena, gas exchange and heat exchange, were seen to go nearly hand-in-hand, and it was only the great interest in heat engines during the subsequent years that led to the blessing of heat as the true measure of fuel balance. It is one of the ironies of history that Lazare Carnot, the military genius responsible for the successes of the Directory that executed Lavoisier, had a son Nicholas Carnot who essentially founded thermodynamics, because thermodynamics tells us that heat cannot do work at constant temperature, and yet heat evolution persists as a measure of fuel consumption or availability.

Measurement of Energy Demand and Supply

The early workers in biological energetics were not fools; they understood thermodynamics as it developed, so it was in many ways a bold step for them to use heat evolution as a measure of the energy expenditure of an individual, and the heat of combustion of the consumed foods as a measure of his energy supply. Heats of combustion would be valuable as a guide only if they had a constant relationship to free energy changes with all types of nutrients.

The validity of heat exchange as an approximation to total biological energy balance was established empirically. Calorimeters were made large enough to enable humans or other animals to live in them for extended periods. This enabled proof that the heat of combustion of a compound within the organism was the same as the heat of combustion within a bomb calorimeter. Humans, because of their cooperativeness, proved to be the most valuable subjects for this exacting type of experimentation, which required close control of activity, food consumption, collection of excreta, and so on.

The agreement between heat evolved during biological combustion and during bomb calorimetry proved to be excellent for readily absorbed fats and carbohydrates. However, there was an important and unsatisfactory correction necessary in the case of proteins because nitrogen is converted to nitrogen gas and oxides of nitrogen in the bomb, whereas it is converted to various other products mainly urea, in the mammal.

The next step was establishing that heats of combustion did bear some relationship to the relative yield of utilizable energy from the major fuels in the biological system. Perhaps the most reassuring finding was the result of comparing fats and carbohydrates for the maintenance of constant weight. If the heat of combustion of the dietary intake was held constant, it was found that there was little change in weight if fat was substituted for a substantial amount of carbohydrate in the diet. Variation of protein content gave less consistent results. We shall also see that heat of combustion, even after correction for urea output, is an inherently less satisfactory guide to the energy available from utilization of protein.

Is Calorimetry Valid? The real energy balance in large part is the result of the consumption and synthesis of high-energy phosphate bonds, which we cannot measure with present techniques in an entire living person. However, we do have estimates of the theoretical stoichiometry for the complete oxidation of the various fuels, and we have noted in the preceding chapters various lines of reasoning that suggest the actual yield of high-energy phosphate is not far from the theoretical value. If we compare the values for the common fuels, we see that the caloric values are nearly proportional to the theoretical high-energy phosphate yield for fats and carbohydrates, but differ significantly for proteins (Table 39–3). This is not disturbing, because proteins are a small part of the fuel supply in most diets, and any error in the yield of energy from their combustion is of correspondingly small significance.

The high energy-phosphate yields in this tabulation were computed with the following assumptions:

1. The bulk of the ingested glucose is directly stored as glycogen in skeletal muscles, and completely oxidized in that tissue, so the Cori cycle can be neglected.

2. Ingested triglycerides are hydrolyzed and stored in adipose tissue without modification by the liver.

3. Amino acids are metabolized according to the stoichiometry given in Chapter 32 (p. 640). All of the nitrogen is converted to urea. The carbon skeletons are converted to glucose and acetoacetate, which are metabolized in other tissues without intermediate storage.

4. The amount of high-energy phosphate necessary for transport across cell membranes is negligible compared to the total yield from oxidation. This is the most shaky premise, and we can hope only that the relative loss on transport is roughly the same with all types of compounds.

What is the practical result of our re-examination of the caloric values? The usefulness of caloric values for comparing fats and carbohydrates has long been established through experimentation without the aid of our theoretical justification.

ENERGY FROM COMMON FUELS **TABLE 39–3**

Fuel	RQ	kJ Liberated Heat *per gram of food*	Moles ~P *per gram of food*	Ratio kJ/mole ~P
Ingested starch	1.00	17.5	0.222	78.8
Ingested fat	0.71	39.6	0.502	79.9
*Ingested protein	0.82		0.202	94.1
†Ingested protein	0.80	18.1		
††Stored glycogen	1.00	17.5	0.235	74.7
Stored fat	0.71	39.6	0.510	78.7

*Calculated from amino acid composition of beef muscle proteins; grams protein = 5.65 × grams N.
†Accepted values for mixed proteins, corrected for fecal losses; grams protein = 6.25 × grams N.
††Muscle glycogen.

Indeed, the tables are turned, and this large body of observations generates confidence in the stoichiometric calculation of high-energy phosphate yields for assessing the potential energy of foods. This is important both for immediate use and for its future potentialities. The immediate applications are in adjusting the accepted values for the energy equivalent of protein as a fuel and in making a distinction between the ingested and stored fuels, which have identical heats of combustion but different capacities for supporting work. The future potentialities lie in assessing energy balance under conditions in which processes other than oxidation are proceeding at significant rates and in being able to take advantage of the more detailed analyses of foods now becoming available for computing their nutritional value.

We come, then, to the following conclusions at this point in the discussion:

1. The multitude of tables showing caloric equivalents of foods are useful and sound, provided that the protein content of the food is less than 0.2 or so of the calculated caloric equivalent. The major objection lies in the implication of using true thermochemical values and of being able to use heat for work.

2. The energy yield from stored glycogen is significantly higher than that of ingested starch, and the "caloric" value of oxygen consumed for the metabolism of glycogen ought to be raised by 5.5 per cent.

3. The accepted values for the energy yield from meat proteins are too high. If the caloric basis is to be used, and if the caloric yield from starch is taken as 17.5 kJ per gram, the corresponding yield from meat proteins would be near 1.18 MJ (282 kcal) per mole of N, or 84.5 kJ (20.2 kcal) per g of N. (The current accepted value is 111 kJ (26.5 kcal) per g of N. Meat has a relatively high content of glycine, owing to the presence of collagen, and the energy yield per mole of N from glycine is low. It so happens that the fudge factors applied to the original calculation for meat proteins make the value of 111 kJ per g of N about right for proteins such as casein and zein, with their larger content of high molecular weight amino acids. This occurred by accident, not design, and illustrates the need for re-evaluation based on amino acid content of proteins as fuels.)

The Respiratory Quotient and Indirect Calorimetry

Direct measurements of heat exchange are rarely used today. The heats of combustion for the constituents of foods are well established for everything except individual proteins, so the caloric yield of foods is calculated from analyses of their composition. Direct calorimetery of heat evolution by individuals is expensive, cumbersome, and time consuming. It has largely been supplanted by measurements of respiratory exchange—oxygen consumption and carbon dioxide evolution, from which the equivalent heat evolution is calculated. For precise work, the nitrogen excretion is also measured, but an approximate value is usually postulated because of the small differences between most people.

The respiratory quotient, which is the volume of CO_2 liberated divided by the volume of O_2 consumed, is the key to the calculation, because it enables an estimation of the proportion of the oxygen used to burn fats and the proportion used to burn carbohydrates:

$$\mathbf{RQ} = \frac{\Delta \mathbf{CO_2}}{-\Delta \mathbf{O_2}}$$

First, the measured values of respiratory exchange are corrected for protein metabolism. This can be done if the nitrogen excretion is known. For example, the stoichiometry for the metabolism of some proteins, if all of the nitrogen appears as urea, is as follows:

STOICHIOMETRY OF PROTEIN METABOLISM

	Moles per Kilogram of Food				
	CO_2	$-O_2$	N	CO_2/N	$-O_2/N$
Beef muscle proteins	37	45	12.6	3.0	3.6
Casein*	37	45	11.2	3.3	4.0
Zein*	39	47	11.4	3.4	4.1

*Casein is the principal protein of milk and zein is the principal protein of corn.

If we assume that an individual is metabolizing a mixture of amino acids approximating the composition of muscle proteins, we can multiply the measured nitrogen excretion by 3.0 and 3.6 to obtain the respective values for carbon dioxide output and oxygen uptake due to the combustion of amino acids. Of course, the values may be in error by 0.1 or more if the mixture of amino acids consumed more closely approximates the composition of casein or zein.

The second step in the calculation is to subtract the values for gas exchange due to protein metabolism, estimated as above, from the total gas exchange. This leaves the values of oxygen consumption and carbon dioxide production due to the metabolism of fats and carbohydrates. Now, the stoichiometry for the metabolism of typical examples of these fuels is as follows:

STOICHIOMETRY OF FAT AND CARBOHYDRATE METABOLISM

	Moles per Kilogram of Food		
	CO_2	$-O_2$	$CO_2/-O_2$
Starch	37	37	1.00
Sucrose	35	35	1.00
Lactose	35	35	1.00
Corn oil	65	90	0.72
Pig fat	64	91	0.71

Since the respiratory quotient (RQ) for the oxidation of carbohydrates is 1.00, and the value for the oxidation of most natural fats is near 0.71, we can estimate the fraction of the oxygen consumption used for carbohydrate metabolism by the following relationship (after correction for amino acid metabolism):

$$\frac{-O_{2\ (\text{carbohydrate})}}{-O_{2\ (\text{fat}+\text{carbohydrate})}} = \frac{RQ_{(\text{observed})} - RQ_{(\text{fats})}}{RQ_{(\text{carbohydrates})} - RQ_{(\text{fats})}} = \frac{RQ_{(\text{obs.})} - 0.71}{0.29}$$

Knowing the stoichiometry for the combustion of fats and carbohydrates, it is possible to use the oxygen consumed in burning each of the three types of fuels to calculate the total heat evolution. One can also calculate the theoretical high-energy phosphate yield from the appropriate stoichiometries.

This method of estimating energy consumption is termed **indirect calorimetry** because heat production is calculated from gas exchange rather than by direct observation in a calorimeter. This name gives the implication of lesser reliability, but the measure of oxygen consumption is in many ways a more useful guide to the metabolic economy than is heat evolution.

The factors commonly used in indirect calorimetry are shown in the following table. (The data are given in terms of liters of gas as well as moles, because these units are still in common use in many laboratories.)

ENERGY PRODUCED

	RQ	Per Gram of Food		Per Mole O_2		Per Liter O_2	
		kJ	*kcal*	*kJ*	*kcal*	*kJ*	*kcal*
Proteins	0.80	18.1	4.32	418	100	18.7	4.46
Fat	0.71	39.6	9.46	439	105	19.6	4.69
Starch	1.00	17.5	4.18	473	113	21.1	5.05

1 mole of urinary N = 1.56 MJ, 372 kcal; 3.00 moles CO_2 produced and 3.70 moles O_2 consumed.
1 gram of urinary N = 0.11 MJ, 26.5 kcal; 4.8 liters CO_2 produced and 5.9 liters O_2 consumed.

The Basal Metabolic Rate (BMR)

The basal metabolic rate (BMR) is the rate of oxygen consumption, or the calculated equivalent heat production, of an awake individual lying at rest, who has had no food for at least 12 hours. This measurement is an important physiological tool because it assesses the energy requirement for maintenance of tissues and any dissipation for heat production through oxidations not coupled to phosphorylations.

The usual determination of BMR is really only a determination of the rate of oxygen consumption. Much of the success of this approach results from the specified conditions, which minimize many potential sources of error. A previously well-fed individual will depend mainly on his fat stores for energy after an overnight fast, with a respiratory quotient near 0.82 if he has been eating the usual American diet. Only 15 per cent of the oxygen consumption will be used for the metabolism of amino acids, thereby minimizing error due to lack of knowledge of the exact composition of the amino acids being metabolized.

It is therefore reasonably safe to assume that most people in the basal condition will be metabolizing a proportion of fuels such that **1 mole of O_2 is equivalent to 452 kJ, 108 kcal; 1 liter of O_2 is equivalent to 20.2 kJ, 4.82 kcal.** (The latter value is frequently quoted as 4.8205 kcal per liter, which is a beautiful example of confusing accuracy in arithmetic with experimental precision. One part in 100 is an optimistic expectation for precision in this field. When measurements of BMR were commonly used in assessing thyroid function, any value within 10 per cent of the accepted mean was taken as normal.)

The Use of Surface Area

It is common practice to compare the metabolism of individuals on the basis of the area of the body surface. This practice is even extended to such things as the calculation of drug dosages. The advantages and defects of the custom deserve examination.

It ought to be apparent from what we have learned to this point that metabolic processes do not occur predominantly at the body surface; therefore, metabolism and body surface do not have a direct anatomical relationship. However, it was realized over a century ago that mammals have a proportionately slower basal metabolism as their size increases. A large mammal has a greater total metabolism than does the smaller mammal, but the metabolism per unit of body weight decreases with size. Many later studies have shown that the basal metabolism is more nearly proportional to the surface area of the adult animals in various species than it is to the weight of the animals.

We really ought to expect this to be so. As we mentioned in the earlier discussion of brown adipose tissue, there is no inherent reason that the cells of an elephant could not be conducting oxidations at the same rate as do the cells of a mouse. The limiting factor is the ability to dispose of the heat generated by metabolism. The primary mechanisms for heat loss for land animals are direct radiation to the environment and evaporation of water on the surface, both of which can be expected to be proportional to the surface area. (There are other mechanisms, such as evaporation in the lungs, so heat loss is not directly proportional to surface area among all animals.)

In short, we are dealing with an evolutionary development in which the metabolic processes of many kinds of animals have been adjusted to conform, at least roughly, to the surface area of the animal rather than to its tissue mass. This is the basis for using the surface area in comparing metabolic rates of species.

The justification for the use of surface area in comparing individuals of one species is considerably more shaky. One would have to argue that the metabolism of individuals somehow adapts to keep their heat output proportional to area, and there is no real evidence that this is so. The evidence at hand suggests that metabolic rate is more nearly proportional to the total amount of protein in the body, and if we could estimate the number of mitochondria, this would probably be the best guide of all. It is well known that, given two men of equal height and weight, the more muscled specimen has the higher metabolic rate. It is known that metabolic rate per unit surface area differs among various groups of people, and differs with age and sex in the same group of people. This is shown in Figure 39–4.

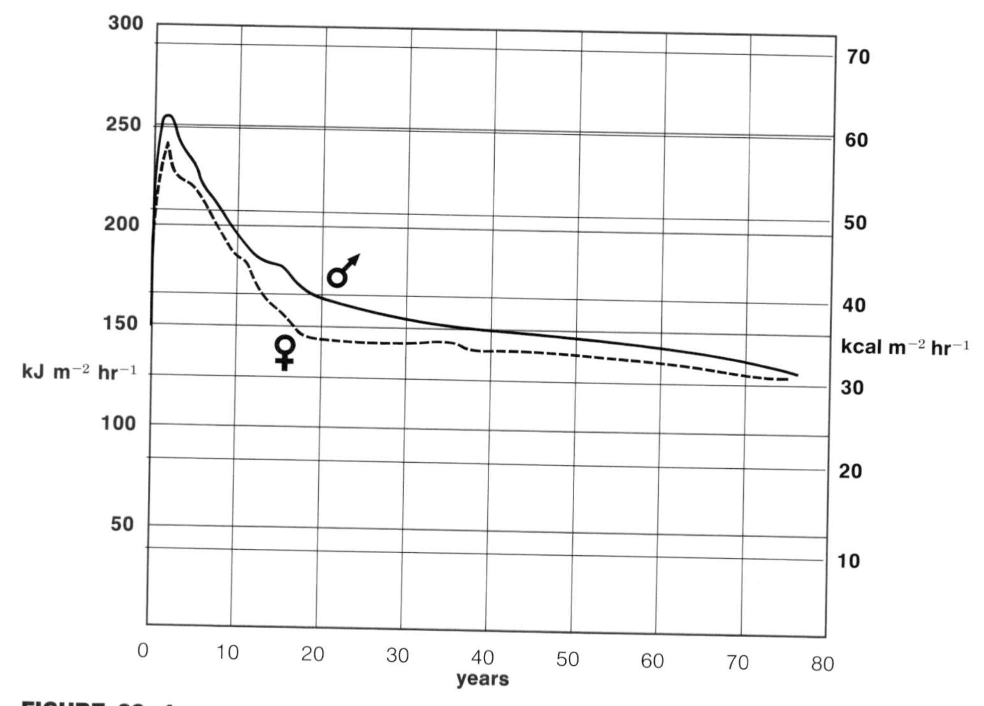

FIGURE 39–4 The basal metabolic rate of males and females of varying ages in the United States, given per square meter of calculated body surface area.

Why, then, aren't metabolic measurements made by simply measuring oxygen consumption per unit time and comparing the results with the average range of values obtained for people of similar age, sex, height, and weight? Perhaps the major reason is the striving for logical unity in science. Since we obviously can't compare the oxygen uptake of a lean person with that of an obese person on the basis of weight—the droplets of triglyceride have no oxygen uptake—we seize on the surface area as the best available means of minimizing individual variations. This evidently comforts many even though the practice has no biochemical basis, and the calculated values still must be compared with empirically determined standards for similar people.

It is exceedingly difficult to measure surface area accurately. Hence, there have been many attempts to estimate this value from more easily measured parameters, such as the overall length of the body and the body weight. Metabolic laboratories have charts for this purpose, which have been constructed from the relationship:

$$\log A = 0.425 \log W + 0.725 \log H - 2.144$$

in which A = area in square meters; W = weight in kilograms; H = height in centimeters.

One has to be alert for some curious examples of circular logic in the literature on surface area measurements. The object is to determine how closely metabolism is related to surface area. Some proceed to contend that their formulas for determining surface area are better because they result in more constant values of metabolic rate per unit of surface.

Specific Dynamic Action

The oxygen consumption and heat output of an individual rise upon eating, even if he remains at rest. This elevation in metabolic rate over the basal rate was named the specific dynamic action of foods. It is well established that the specific dynamic action is something above and beyond the result of the exertion of mastication, swallowing, and the increased motility of the gastrointestinal tract. The increased metabolism follows the absorption of the digested foods.

There have been many differing estimates of the magnitude of the specific dynamic action of various foods. These range from near zero for the fats, through 5 to 10 per cent of the total caloric equivalent of ingested carbohydrate, to 20 to 30 per cent of the total caloric equivalent of ingested proteins, when these foodstuffs are tested individually.

The phenomenon is real, and much effort has been expended in attempting to explain it. The problem has been complicated by the inconsistency of the results under varying conditions, and by a general increase in metabolic rate caused by eating any meal. For example, the specific dynamic action is greater for foods given to a starving animal than for foods given to an animal that is fed repeatedly. The specific dynamic action of a mixed meal is less than the sum of the effects seen with the constituents given individually.

It is self-evident that the handling of nutrients places variable demands on the oxidative metabolism. Digestive juices must be made, nutrients must be processed and stored, urea synthesized, and so on. The state of previous nutrition and the composition of the diet alter the routes that are used and therefore the obligatory increased energy expenditure of fuel processing. In addition, there is an uncertain contribution from thermogenesis, such as occurs in brown adipose tissue (p. 553).

A compromise in interpretation has been reached by which a specific dynamic action of 10 per cent of the total caloric value is assigned by many to the usual mixed meal. They then rationalize that the immediate output of heat represents something extra, not useful, and additional fuels must be supplied to compensate. Therefore, in computing the necessary food intake for a given individual, it is common to estimate his caloric requirements from his basal metabolic rate, and then add an extra 10 per cent to the intake to compensate for the specific dynamic action. Should that be done? No, because it has no rational basis. Fuel processing includes the first stages of fuel catabolism and is part of the normal energy balance; the magnitude of thermogenesis due to eating is unknown and probably trivial compared to other errors involved in calculating a desired dietary intake.

CHANGES IN ENERGY DEMAND

Measurements of oxygen consumption show in a general way the partition of energy demand among the various tissues and the changes in demand with work. The relative values in Table 39–4 were calculated from a tabulation by Dr. Richard Havel (quoted by permission):

These data reinforce a point already made—the oxygen consumption of working muscles far outstrips all other demands. The demands of the viscera account for about one third of the total oxygen consumption at rest. The brain utilizes one fifth. These demands remain relatively constant during work, but the utilization by the skeletal muscle and heart increases dramatically.

Essentially, we are seeing in these data the requirements for fuel represented by the basal metabolism and the additional fuel required for daily activities. The additional amounts required have been measured through calorimetry for a wide range of human activities, and such measurements provide the basis for estimates of the food consumption required for maintenance of body weight. As a first approximation, activities can be sorted into a few categories (energy expenditures given as fractions of basal):

Activity	Fraction of basal
lying at rest	1.0
sedentary work (reading, writing, talking, waiting, either sitting or standing)	1.6
light movement (dressing, sorting, strolling)	3
moderate work (gardening, shoveling, tennis, coitus (male))	5
heavy work (hand sawing, climbing, stacking 50-kg loads, swimming, skiing on level)	9

RELATIVE OXYGEN CONSUMPTION (whole body at rest 1.00: actual value near 0.17 mmoles min^{-1} kg^{-1}) — **TABLE 39–4**

	At Rest	Light Work	Heavy Work
Skeletal muscles	0.30	2.05	6.95
Abdominal organs	0.25	0.24	0.24
Kidneys	0.07	0.06	0.07
Brain	0.20	0.20	0.20
Skin	0.02	0.06	0.08
Heart	0.11	0.23	0.40
Other	0.05	0.06	0.06
SUM	1.00	3.00	8.00

The time actually spent on activities in each category can then be estimated. (It is important that this be done carefully. Eight hours at the manufacturing plant is not necessarily eight hours of continuous activity, and one hour at the club is certainly not one hour of continuous swimming.) The errors will be large, so large that finicky nitpicking over specific dynamic action and the like is a useless refinement. Even so, the results are a valuable guide.

To illustrate the principles, assume two 22-year-old males, with a median 177 cm height and 73 kg weight, and a basal metabolic rate of 310 kJ hr^{-1} (74 kcal hr^{-1}). The hours each devotes to particular grades of activity and corresponding hourly equivalents of basal energy expenditure are given in the following table*:

Activity	Fraction of Basal Energy	Student		Warehouse Laborer	
		hrs	*hrs × fraction*	*hrs*	*hrs × fraction*
Asleep	1.0	7	7.0	8	8.0
Sedentary	1.6	14.5	23.2	8	12.8
Light movement	3	2	6.0	2	6.0
Heavy work		0.5	4.5	6	54.0
SUM		24	40.7	24	80.8

The total daily expenditure of a student is 40.7 times his hourly basal expenditure, or 12.6 MJ (3,010 kcal), whereas the total for the laborer is 25.0 MJ (5,980 kcal), nearly twice as much. The lesson here is that a half hour of heavy workout per day, whatever its other advantages, will not require the fuel intake of a working man. (The recommended daily allowance for caloric intake of young men in the United States is 2,900 kcal per day. This reflects the sedentary character of our population. We shall discuss this, and other recommendations, in the last section of the book.)

Another factor that influences the energy requirement of a given person is the environmental temperature. This is really self-evident. Since the body temperature must be maintained within narrow limits, and exposure to cold increases the heat loss, there must be a corresponding increase in metabolism. Part of the increase comes from brown adipose tissue, part from shivering, and part from some increase in metabolism of muscle through mechanisms that have not been defined. Acute exposure to cold, such as immersion in icewater, will increase metabolic rate as much as two-fold.

Utilization During Work

Let us now translate the energy demand into biochemical events. The sequence of utilization of potential substrates for the generation of ATP during heavy activity is simply described: Glycogen is used preferentially, the fats are used next, and the proteins contribute little.

Let us get a clear picture of the changing fuel economy by studying what happens in a single individual working over an extended period of time. Table 39–5 shows what happened in a young man accustomed to physical work who ran intermittently over a period of six hours. His respiratory exchange and nitrogen excretion were measured during this period, and these data enable us to calculate the amount and kind of fuels being used.

*X-rated work is quantitatively insignificant.

THE MARATHON RUNNER **TABLE 39–5**

Rate of Running	Body Weight	Blood Sugar	$-O_2$	RQ	~P Production				Weight of Fuels	
					Total	*From Glycogen*	*From Fats*	*Fraction From Fat*	*Gly-cogen*	*Fats*
km hr^{-1}	*kg*	*mM*	*moles*		*moles*	*moles*	*moles*		*g*	*g*
0	59.61	5.6								
11.3	59.32	4.6	2.75	0.97	17.1	15.5	1.6	0.09	67	3
9.3	59.03	4.9	2.34	0.96	14.5	12.7	1.8	0.12	55	4
11.3	58.82	4.8	2.71	0.94	16.6	13.5	3.1	0.19	58	6
9.3	58.65	4.4	2.31	0.88	13.8	8.5	5.3	0.38	37	11
11.3	58.25	4.6	2.76	0.86	16.5	9.0	7.5	0.45	39	15
9.3	58.15	4.5	2.40	0.82	14.1	5.7	8.4	0.60	25	16
11.3	57.93	4.2	2.85	0.82	16.7	6.8	9.9	0.59	29	19
9.3	57.78	4.3	2.44	0.79	14.1	4.2	9.9	0.70	18	20
11.3	57.48	3.8	2.85	0.82	16.7	6.8	9.9	0.59	29	19
9.3	57.30	4.1	2.49	0.79	14.3	4.3	10.0	0.70	19	20
11.3	57.55	3.7	2.88	0.81	16.9	6.3	10.6	0.63	27	21
9.3	57.35	3.2	2.48	0.77	14.2	3.2	11.0	0.77	14	22
					185.5	96.5	89.0		415	175

1. Blood sugar concentrations were measured by methods now known to give high values owing to the detection of compounds other than glucose in the assay. The values are still useful as a relative guide.
2. The listed totals differ somewhat from the sum of the figures given, owing to the rounding off of the latter figures.
3. This is an experiment reported by Edwards, Margaria, and Dill, Am. J. Physiol. *108*:203 (1934), and their data have been recalculated to determine high-energy phosphate production. Protein metabolism accounted for only 0.02 of the total energy production and has been neglected.

First let us describe the circumstances. The man was previously well-fed on a mixed diet, which means that we can expect his glycogen reserves were reasonably typical. He was somewhat on the lean side, and therefore did not have gross rolls of fat. He worked quite hard, running at 3.1 meters per second for 25 minutes out of each hour and at 2.6 meters per second for another 25 minutes. However, this is not the maximum possible effort for six hours of work, because the comparable world record for continuous running is 5.4 meters per second.

Now, let us look at the data. They are simple, but they have a wealth of information concealed within them that is applicable to the general question of fuel economy.

1. Despite his leanness, the man used his stored triglycerides and glycogen for 98 per cent of his energy production, with only 2 per cent supplied by protein. This is the usual circumstance in the well-fed.

2. His body weight (column 2) fell by 2.3 kilograms, even though he was supplied with water. This is a guide to the magnitude of the effort.

3. His blood sugar concentration (column 3) fell by nearly one half, and the drop was precipitate during the last hour.

4. His oxygen consumption (column 4) went up and down with the varying rates of running, as might be expected, but increased during periods of equivalent effort as time went on, despite the fall in body weight. This would indicate that more oxygen was needed to maintain the same supply of ATP.

5. His respiratory quotient (column 5) declined throughout, indicating a shift from carbohydrate oxidation to fat oxidation. The RQ was generally higher during the periods of greater exertion, indicating increased mobilization of glycogen.

6. The total yield of high-energy phosphate (column 6) calculated from the gas exchange was remarkably constant throughout for periods of equivalent effort, even though the nature of the fuel was changing.

7. During the first period, 90 per cent of the ATP was supplied by oxidation of glycogen, but the fraction declined steadily, and the oxidation of fat increased until it was supplying nearly 80 per cent of the total ATP during the final period (columns 7, 8, and 9). This shift accounts for the increased oxygen consumption, because the ATP yield per mole of O_2 is less for fats than it is for carbohydrates. Using the values of 6.3 and 5.6 moles of ATP produced per mole of O_2 consumed in oxidizing glycogen and stored triglycerides, respectively, the calculated yield per mole of O_2 consumed for the first and final periods of running at 9.3 km hr^{-1} are 6.20 and 5.73 moles ~P/mole O_2. To maintain constant ATP supply, the O_2 consumption of 2.34 moles during the first period at that rate would have had to rise to 2.53 moles during the final period. The observed consumption during the final period was 2.48 moles. This is remarkable agreement. (There would be a slight drop in the requirement for ATP owing to the decline in body weight, but the variation with body weight during running is small—not at all proportional to the weight.)

8. The final two columns show the calculated weights of the glycogen and fat consumed. In discussing the storage of glycogen (Chapter 27), we noted that the total in the skeletal muscles and liver of a male adult of this size would be in the neighborhood of 450 grams. The total consumption shown in the table is over 90 per cent of this value. This agrees with the conclusion that might be drawn from the sharp fall in blood sugar concentration: The man had almost exhausted his carbohydrate reserves, and any further effort would have had to be sustained almost completely by the oxidation of triglycerides.

(The weight of the glycogen and fat consumed, together with the loss of the water associated with glycogen, accounts for approximately 60 per cent of the observed weight loss. However, it ought not to be assumed that loss in body weight can be predicted accurately from such data for a few hours of exertion. Changes in water balance have a large effect, and are not directly dependent on the loss of fuels.)

The conclusions we can come to with this experiment are reinforced by much more evidence than we have shown. Glycogen reserves are preferentially used for physical activity, but this activity is sustained with equal efficiency by either glycogen or triglycerides. That is, the metabolism of carbohydrate produces the same fraction of the predicted high-energy phosphate as does the metabolism of triglycerides. *There is no basis for believing that either fat or carbohydrate is an inherently inefficient fuel.* This in itself cuts the ground from under most dietary fads, which reverse the usual ambition and try to get nothing for something.

There is an important qualification that must be added. The availability of a glycogen reserve enables more intense short-term activity. We pointed this out in Chapter 26, showing that lactate production reflects a more rapid formation of pyruvate and associated production of ATP than can be accommodated by oxidative phosphorylation. In addition, the production of ATP is greater per mole of O_2 consumed during the oxidation of carbohydrates. It follows that mitochondria in which electron transport is proceeding as rapidly as possible will generate more ATP from the oxidation of pyruvate, which is derived from glycogen, than they will from the oxidation of fatty acids. The availability of glycogen therefore enables a greater maximum effort up to the time the glycogen is exhausted. This accounts in part, but not completely, for the inability of individuals with glycogen storage diseases of the muscle to do heavy work. Normal humans can get along reasonably well on a diet containing little or no carbohydrate. Their RQ falls to levels consistent with almost total dependence on fatty acids as fuels. They cannot be expected to win short races in competition with people on a mixed diet, but they can carry on normal physical activities without difficulty.

High Power Exertion

Even in these civilized times, many humans are concerned with the rate at which they can do work. Delivering maximum power for a set period of time may gain fame and fortune in the arena, earn a livelihood in more prosaic ways, or be life-saving, especially in the metropolitan jungles. The level of maximum power depends upon the time it is to be sustained. We can picture this by plotting the rate of running necessary to establish world records as a function of the logarithm of time required for different distances (Fig. 39-5). The rate is used as an index of power output. Everyone knows that a person must pace himself to complete a given distance. An ordinary person can go all out for perhaps 100 meters, but must begin at a slower pace if he is to complete a kilometer, and still slower if he is to travel 20 kilometers. Even so, with world-class athletes there are two sharp breaks in the decline of sustained power output with the time it is to be sustained, one near 20 seconds, and the other near 200 seconds.

Two factors are involved. One is the power output of single muscle fibers, and the other is the type of fiber recruited for the effort. We have noted before that the white fibers—fibers of low oxidative capacity and fast twitch—will produce more lactate than do the red fibers of high oxidative capacity, some of which are slow twitch for sustained effort and some fast twitch. The power output in each case sets

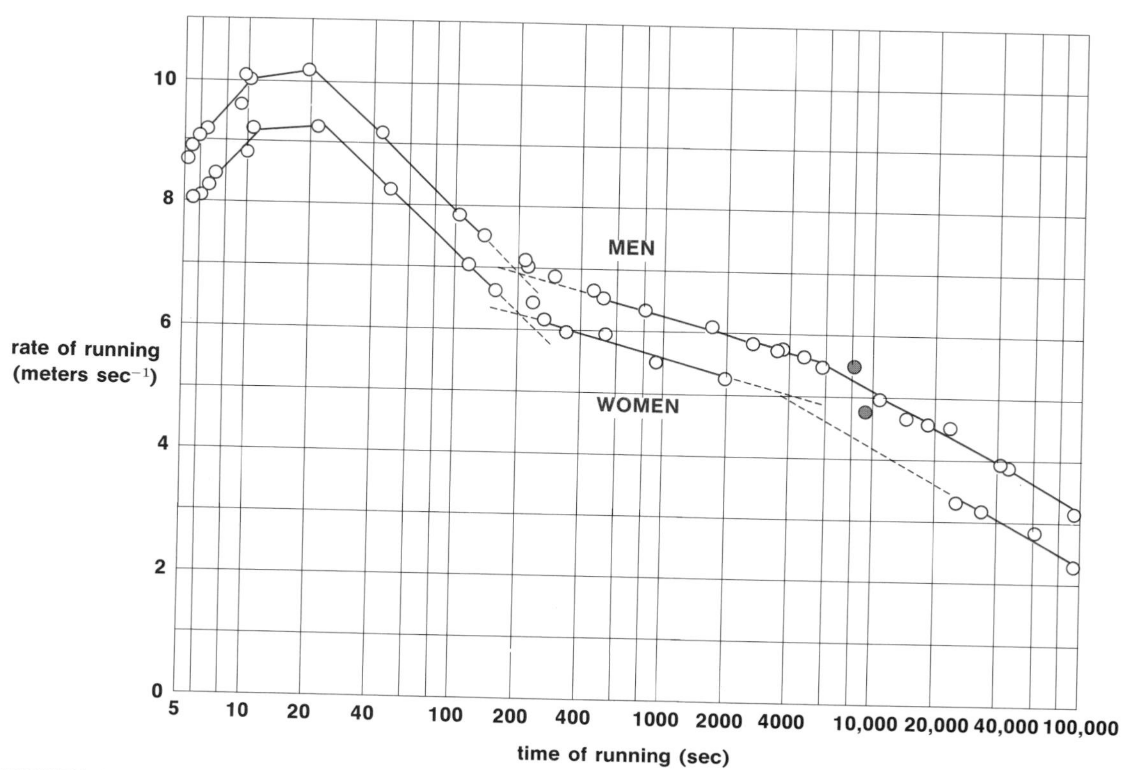

FIGURE 39–5 World record track performances as of January, 1982, plotted to show the relationship between the rate of running as an index of power output and the time for which a given power output can be sustained. The data suggest that humans exhaust some source of energy when running at the maximum rate possible for 20 seconds, and that there is some additional source that cannot be used if running is to be continued beyond 200 seconds. Another adjustment is necessary if running is to be continued beyond 70 minutes. (The solid points are values from less well-controlled races.)

the drop in phosphate potential (p. 409) necessary before metabolism is stimulated enough to make the production of high-energy phosphate balance its consumption.

For very fast efforts, the high-energy phosphate store suffices. Leaping from a crouch discharges high-energy phosphate at a rate of approximately 6 millimoles per kg of muscle per second, but the effort lasts for only 0.5 second, or so, and the resultant decline in phosphate potential is barely sufficient to cause a noticeable acceleration of oxygen consumption, with little lactate formation.

The maximum power output over several seconds is one that will bring the phosphate potential to the level of exhaustion, with essentially total discharge of phosphocreatine. It is not a simple matter of dividing the high-energy phosphate store by the rate of discharge, because the falling phosphate potential activates the generation of high-energy phosphate, finally reaching a level at which there is massive lactate formation. The break near 20 seconds represents the power output which will in that time exhaust the muscle despite maximum activation of lactate formation.

The next limitation on the power output is the availability of the glycogen supply for lactate formation. If a muscle were all white fibers, lactate formation would theoretically sustain contractions for 240 seconds with total dissipation of the glycogen. It is not all white fibers, and there is a transition from dependence on lactate formation to dependence on oxidative phosphorylation as power output is diminished. When the run is at a rate that can last longer than 200 seconds, steady state is being maintained completely by oxidative phosphorylation after the first few seconds of effort (the phosphate potential must fall before the rate of oxidation reaches its maximum). Beyond that point, the metabolic factor governing maximum sustainable power is the necessary shift from glycogen to fat oxidation as the supply of glycogen becomes exhausted.

Persistent Oxygen Consumption. Increased oxygen consumption will persist so long as the phosphate potential is low. The mitochondria do not "know" that muscles have ceased contracting. There is an immediate demand for continued high-energy phosphate production to restore the phosphocreatine concentration and ionic balances, but there are longer-lived demands for rebuilding of glycogen and triglycerides, and the conversion of lactate to glucose. It is unfortunate that this continued oxygen consumption was associated with lactate formation, which in turn was regarded as something occurring primarily in the absence of oxygen, because it led to the term "oxygen debt" for the state of the tissue at the end of exercise. One could equally well talk about an "oxygen credit," since the use of phosphocreatine and the formation of lactate diminish the necessity for supplying oxygen to meet metabolic demands. The phrase ought to be abandoned.

Utilization During Starvation

Starvation differs from heavy exertion primarily in the protracted demand on the stored fuels. The glycogen supply is largely depleted within a few days during total starvation, with nearly all of the glycogen disappearing from the liver, and the content in the muscle falling to the level that can be sustained by glucose derived from the metabolism of amino acids in the liver.

After the loss of the original glycogen stores, the starved person becomes totally dependent upon the metabolism of triglycerides from the adipose tissue and amino acids derived from his tissue proteins. The length of time he can survive is a function of the amount of triglycerides he carries, and the ability to exist on stored fat is demonstrated nicely by the use of starvation in the treatment of the obese. Patients have been deprived of all food for periods as long as eight months, taking only water and vitamin supplements. One such patient lost 33.7 kilograms of body weight. This proved to be a risky procedure with occasional sudden deaths, but in the main, most of the obese people who were subjected to such a regimen got along well. Within 30 days, the daily loss of nitrogen fell to less than 0.3 mole, so there was only rarely any difficulty from failure to maintain protein synthesis. (However, the adjustment of the kidney to the decreased demands on excretion did lead to an increase in the uric acid concentration, occasionally causing gout.)

Many of the starved have blood glucose concentrations as low as 2 mM without ill effect. This leads us to a consideration of a major change in the metabolic economy occurring with starvation. The production of acetoacetate and 3-hydroxybutyrate by the liver sharply increases, and these compounds become a major fuel for the brain.

Glucose is the major fuel for the brain in a well-fed individual on a mixed diet. Since the RQ of the brain remains near 1.0 under all circumstances, it was thought until very recently that the brain always depended upon an available glucose supply, believed to be derived by gluconeogenesis from amino acids during starvation. However, simple calculations with long-available data showed that the total supply of glycogen at the beginning of a fast and the amount of degradation of amino acids during a fast could account for only a small fraction of the total metabolism of the brain, which is responsible for 20 per cent of the total oxygen uptake at rest. It was then shown by placing catheters in the neck vessels of volunteers and measuring the arterial-venous differences in metabolite concentrations, that the brains of starved humans use the ketone bodies as fuels. The RQ values for the oxidation of acetoacetate and 3-hydroxybutyrate are 1.00 and 0.89 respectively, which partially explains why the utilization of these compounds was not observed earlier.

We saw earlier that acetoacetate may be formed from several compounds. It is a direct product of the metabolism of tyrosine and phenylalanine. It is also formed from 3-hydroxy-3-methylglutaryl coenzyme A, which in turn is created by condensation of acetyl coenzyme A and acetoacetyl coenzyme A, in addition to being a product of the metabolism of leucine. We previously emphasized the utilization of these routes in disposing of the carbons of amino acids. However, acetyl coenzyme A and acetoacetyl coenzyme A are also formed by the oxidation of fatty acids, and acetoacetate and 3-hydroxybutyrate are potential end products of this process. Elevation of the free fatty acid concentration causes increased production of ketone bodies by the liver.

Under ordinary circumstances, the output of the ketone bodies by the liver is relatively small, rising in the morning before breakfast, and subsiding after carbohydrates and proteins are eaten. The heart can effectively use these compounds at low concentrations, as will the skeletal muscles at somewhat higher concentrations. Together, these tissues will remove the compounds from the blood fast enough to keep the concentrations low. This explains why little is found in the blood during exercise strenuous enough to cause rapid mobilization of fatty acids from adipose tissue; the level sharply rises shortly after the contractions are stopped because synthesis continues while consumption diminishes. Current estimates indicate that this process accounts for only a few per cent of the total energy yield on mixed diets.

Starvation causes a sustained massive mobilization of fatty acids from the adipose tissue. At the same time, owing to the declining supply of glucose, an increased fraction of any citric acid cycle intermediates available in the liver is diverted to the formation of glucose. This combination of circumstances results in an increased oxidation of fatty acids by the liver with a resultant increased elaboration of the 3-oxybutyrates. Although the exact mechanism of these events is still in doubt, there are likely contributing factors. One is a decline in the concentration of oxaloacetate, owing to its removal for gluconeogenesis, which would slow the oxidation of acetyl coenzyme A. Another is increased oxidation of fatty acids to supply ATP within the liver for protein synthesis and other normal demands, thereby compensating for the decreased utilization of lactate and amino acids. Still another is an increase in the glucagon:insulin ratio.

Estimates indicate that the metabolism of the ketone bodies may account for as much as one quarter of the total energy production during starvation. The formation of these compounds provides a mechanism for feeding the brain and conserving the meager supply of glucose generated from amino acids.

Whether obese or not, total deprivation of food must eventually result in the exhaustion of the triglyceride supply. As this occurs, the only remaining source of energy is in the metabolism of the amino acids derived from tissue proteins. This is not adequate to sustain anything near normal activity, which would require the loss of something like 400 grams of protein (2 kg of tissue) per day, and death occurs rapidly after the triglyceride supply is gone.

In less severe starvation, with some food available, a person with ordinary amounts of adipose tissue gradually becomes more apathetic as the triglycerides disappear, thereby diminishing his energy requirements. He may lose so much tissue and still be alive that most of the external anatomical features of his skeleton are plainly molded by his skin.

PROBLEMS WITH THE FUEL SUPPLY

Famine

There is an inherent dichotomy of approach to nutritional problems. Shall we concentrate on feeding the multitude, or note each sparrow's fall? Let us first sketch the demographic problems and then examine individual problems of clinical interest.

Lack of sufficient food is so common that it may be considered a part of the normal human condition, now and at all past times for which we have evidence. More people die from starvation than are killed by war injuries, and even more people experience starvation and then recover. Demand for food and its supply run so close together that a wide variety of causes can create a mismatch. Too little rain, too much rain, a plague of locusts—we in the United States grasp that such disasters befall people in the sub-Sahara, or Bangladesh, or similarly remote regions. We forget that famine nearly always accompanies war—Norwegians, Dutch, Russians, Germans, all have experienced varying periods of deprivation in this century and as a consequence have provided some of the best studies of the physiological effects of starvation. Nearly one quarter of the eight million Irish died between 1845 and 1851 when a blight wiped out the potato crop and relief from the English was too little and too late. On a much lesser scale, but closer to home, the United States had a "year of no summer" in the 19th century that resulted in widespread hunger, but with poorly documented consequences.

The metabolic changes occurring during starvation were summarized above: An initial shift from carbohydrate to fat oxidation, followed by increased utilization of protein as the fat stores are depleted. There is a rapid loss of protein for gluconeogenesis up to a maximum of about 3 per cent of the total store of 80 moles of N during the initial days of acute starvation, before adaptation has occurred. The loss mainly comes from the liver in experimental animals, and the amount depends upon the previous protein intake, with those having ingested high protein diets losing more upon sudden deprivation. The proteins lost quickly without measurable impairment of function are designated "labile" proteins. After the initial high rate of loss, nitrogen excretion tapers to a level ranging around 0.14 mole per day at the close of a month of total starvation of trim subjects. The tissues losing protein most rapidly during this period are those with high rates of protein synthesis at times of normal nutrition: the liver, pancreas, and intestinal mucosa. The skeletal muscles lose a smaller fraction but represent a major supply during the initial weeks because of their large mass. There is significant loss of muscle proteins during the normal overnight fast, with the proteins being rebuilt after eating. The duration of starvation that can be withstood obviously depends upon the quantity of stored fat. An extremely lean person faces a quick death because he will begin to lose as much as 6 per cent of his protein per day when the fat is exhausted. Young children are also in a precarious position because of a more limited fat store and a greater metabolic rate.

People adjust to less than total starvation. Part of the adjustment comes from adaptations of the biochemical economy that we noted earlier. Part comes from a slowing of processes utilizing high-energy phosphate at rest, as is reflected by a drop in the basal metabolic rate. Another part comes from changes in habit and temperament. Starving people don't like to move and become self-centered and withdrawn. The often-invoked specter of starving hordes ravaging their well-fed neighbors probably never occurs; the starved aren't good candidates for group efforts. (A fear of starvation is something else again.) Charles Dickens was quite correct in suggesting that rambunctious boys might be more tractable if they were fed less meat and more gruel. The increased apathy explains some of the isolated cases of death from starvation in the United States, many of which occur among elderly recluses.

Perhaps the most surprising thing about starvation is the relative lack of specific deficiencies in many instances. Emaciation may be severe without symptoms ascribable to the lack of a particular coenzyme or amino acid, although this is not always true.

Consequences of Famine. We have statistics that are informative about the effect of natural famines on the very young. The Dutch experience has been especially valuable in this regard, because detailed data are available for those born or conceived during the famine of World War II at the time of their later compulsory military service. The conclusions, which are borne out by data on other starved populations, are that starvation of a pregnant woman leads to the birth of a shorter, lighter baby with a smaller head, who is somewhat more likely to die in the early years. However, given adequate food in the subsequent years, babies who survive grow to normal stature and weight without any impairment of intelligence.

Similarly, a study of Korean orphans adopted in the United States within three years after birth suggested that severe postnatal deprivation might result in slight residual impairment of physique or intelligence, but not of the magnitude implied by experiments with laboratory animals. The Dutch data show a drastic drop in fertility during famine; it is possible that humans have evolved so as to ride out famine by conceiving few children, but with preferential utilization of the limited nutrient supply for development of the brain in those who are conceived.

Problems can appear from dependence upon particular foods to stave off hunger. For example, cassava roots, a major source of fuel in Africa, are rich in nitrile-containing glycosides that yield thiocyanate upon degradation. The thiocyanate inhibits iodine uptake, so inhabitants of regions in which foods are low in iodine, such as parts of Zaire, have a high incidence of hypothyroidism with a consequent high incidence of mental retardation (cretinism).

Anorexia Nervosa

Some people starve themselves, even to the point of death, without obvious reasons. There may be some preceding hypothalamic imbalance, but the fundamental causes are probably psychogenic. A diagnosis of anorexia nervosa is justified if five criteria are fulfilled:

1. an age below 25 years at onset;
2. at least a 25 per cent loss in weight;
3. a warped and unrelenting aversion to food;
4. no discernible medical illness;
5. no other discernible psychiatric disorder.

A typical patient will be a female (fewer than 10 per cent are males) who is an excellent student with a disgust toward sex, and who was overweight as a child and began dieting after 12 years of age. A zinc deficiency must be excluded.

Obesity

An obese person by definition is one who has too large a fat store. What is too large depends upon who makes the definition. There is a continuum from what is a cultural ideal, itself subject to change, to an obviously disabling accumulation of masses of triglyceride. There is an association between being fat, diabetes mellitus, and risk of serious cardiovascular disease, which has increased the social pressure for maintaining a less spherical silhouette.

Why do people become fat and what can be done about it? Thinking rationally about these matters requires repeated emphasis on some fundamental conservation equations that cannot be evaded. No modification of diet, or behavior, or of metabolic efficiency can alter the truths stated by these equations. They are precepts at the grammar school level, but nearly every layman and a surprising number of professionals put them aside when they think about food.

1. The mass of nutrient intake (oxygen, water, food) equals the gain in body mass plus the mass of excreta (urine, feces, menses, and net losses through the lungs and skin).

2. The mass of nutrient intake equals the mass absorbed plus the mass not absorbed.

3. The mass of nutrients absorbed equals the gain in body mass plus the mass of nutrients catabolized and excreted.

4. The mass of nutrients catabolized and excreted equals the mass of fuels consumed for energy transduction and the mass of fuels consumed for futile cycles, cytosol oxidations, and the like.

5. The mass of fuels consumed for energy transduction equals the respective masses of fuels consumed for muscular work, for generation of ionic gradients, and for chemical syntheses.

What, therefore, causes a gain in body mass, and what can be done about it? The body mass increases because the mass of nutrient intake exceeds the mass of excreta. More narrowly, the mass of nutrients absorbed exceeds the mass of nutrients catabolized and excreted. These are the variables that must be altered if there is to be no further gain, and perhaps a loss, in body mass. Let us systematically examine what can be done to correct obesity, aside from surgical removal of fat deposits—a temporary respite to be deplored.

The food intake may be diminished. No matter how efficient the obese person's metabolism is alleged to be, there is some nutrient intake at which excretion will exceed intake. The problem is reaching that intake. Eating, like cigarette smoking and sexual intercourse, is a difficult habit to change, once a pattern is established. Gross obesity is most common among women from lower income families in the United States. The usual protective defenses erected against feelings of guilt are hardly alleviated by those physicians who approach even the moderately fat with all the zeal of a Puritan divine confronting a candidate for the scarlet letter. This zeal has led to such barbarisms as wiring the jaws closed to prevent surreptitious snacks. As might be expected, those with motivation so poor as to require such measures promptly begin to gain weight upon removal of the barriers.

The causes for some cases of adult obesity may lie in childhood eating habits. Overfeeding children promotes an increased cellularity in the adipose tissue, whereas overfeeding adults merely stuffs existing cells. Being overweight as a child greatly increases the prospects of being intractably overweight as an adult:

PERCENTAGE OF CHILDREN BECOMING OVERWEIGHT ADULTS

Weight as a Child	*Average*	*Overweight*
Males	42%	86%
Females	18%	80%

Experience seems to indicate that the prospects for becoming sufficiently motivated to diminish food intake permanently are inversely proportional to the degree of obesity. Assuming that no reason for heroic measure exists, the necessary diminution may be more acceptable if achieved in this fashion: A lean body weight is estimated from height and build, and the caloric requirement for a person of that weight with similar activities is estimated. (The estimates ought not be deliberately sloppy, but they cannot be expected to be precise.) A reasonable schedule of fat loss is then projected. In most instances, a loss of 0.5 kilogram of body weight per week is acceptable. This will require a deficit of approximately 2.4 megajoules of potential fuel energy per day (570 kcal). Food tables are then used to adjust the amount of the usual diet to that range. For many, it is helpful to take four small meals per day of approximately equal size, except that the customary main meal is made somewhat larger than the others. Unless the usual diet is bizarre, it is not necessary to pay attention to vitamin and mineral content at this point.

If the patient realizes that weight loss at this modest pace goes in irregular spurts, depending upon water intake, activity, bowel motility, and so on, he will not expect to see much happening in less than two weeks. If there isn't a clear loss of at least 1.5 kilograms after a month, re-examination of the eating habits is encouraged. The problem is nearly always one of social dining, or of between-meal snacks, especially in the case of those doing the cooking, but the estimates of energy needs also may be far from the mark. Unusually large losses may signal a potentially dangerous self-starvation.

A critical test of motivation is made upon reaching a desired weight. The additions to the food intake necessary for maintenance of the weight are small, and a self-congratulatory splurge can undo all of the control over appetite achieved with such difficulty. It is probably wise to have in mind well in advance exactly what those additions are to be, so they can constitute the reward.

The mass of nutrients absorbed may be diminished. The goal of not using your cake but eating it anyway is professed by quacks, gained by surgery, and sought by ethical drug houses. Given a life-threatening condition and a willing patient, surgical bypass of long segments of the small bowel can prevent effective absorption of food, but this is no procedure for casual application. Some diminution of food absorption occurs when fecal solids are increased by nonabsorbable but hygroscopic materials ("fiber"), but the impairment of absorption is slight compared with the diminution of intake required in even moderate obesity.

One increasingly popular device is eating an inhibitor of amylase action, a so-called "starch blocker." Inhibitory proteins are found in many seeds. Indeed, amylase inhibitors in wheat cause substantial passage of unhydrolyzed polysaccharide into the large bowel upon eating baked goods prepared from all-purpose flour. Bacteria in the large bowel attack the undigested carbohydrate and produce varying amounts of gas. The products include hydrogen gas, which readily diffuses into and out of the blood. The content of H_2 in expired air is used as an index of the amount of unabsorbed carbohydrate nutrients. (This is not a foolproof index. The bacterial flora in some large bowels produces little H_2, and the activity of those organisms that do is suppressed as the bowel contents become more acid from fermentation products.)

The amylase inhibitor sold by health food outlets without a prescription until its recent ban by the Food and Drug Administration is a preparation from beans (*Phaseolus vulgaris*). The consequences of its continued use include serious abdominal distress from fermentations in the large bowel. It is conceivable that the deleterious effects of a carbohydrate-free diet could also appear if the inhibition is indeed effective; a recent report claims it is not.*

Another approach at diminishing the absorbable nutrients is the substitution of a polyacylated sucrose for fats in the diet. **Sucrose polyesters** of long-chain fatty acids closely resemble cooking oils, and are not digestible. Foods prepared with them have satisfied the palate, or at least been eaten unnoticed, in preliminary trials, while apparently quenching the appetite.

The final variable determining the changes in body mass is the **proportion of the absorbed mass catabolized and excreted.** This is believed to be the site of individual differences in the metabolic responses to eating. It is generally accepted that some people gain less mass from excessive food intake than do others. It has also been convincingly demonstrated by studies with convict volunteers that even normally lean persons become blubbery when they stuff themselves with food, but the increase in solid tissue was substantially less than the amount predicted from their excess calorie intake. Their fat was stored by increasing the triglyceride volume in existing cells, rather than by an increase in the number of cells, as expected, and the fat store promptly shrank upon their return to *ad libitum* feeding. At least some obese people are believed to have a diminished thermogenic response to overeating.

How can the fraction of the absorbed nutrients catabolized and excreted be increased? The objective is to increase fuel combustion. A sure-fire way is to increase the expenditure for muscular work. Another way would be to use a drug that mimics thermogenesis by increasing the rate of futile cycles or of uncoupled metabolism. Thyroxine will do it, but the side effects make it unusable. Decades ago, dinitrophenol was eaten by the fat. The uncoupling of oxidation and phosphorylation did increase the metabolic rate, but it also destroyed the kidneys and eyes in some.

In the end, the only really desirable ways of correcting obesity are diminishing food intake and increasing muscular activity. For those without an immediate threat to life, any other measure is essentially a refusal to accept the personal cost of correcting their obesity.

---

*G.W. Linn et al. (1982). N. Engl. J. Med. 307:1413.

Food Intolerance

Some people cannot tolerate particular foods because of allergies. Others have partial or complete deficiencies of an enzyme necessary for utilization of a particular fuel. We have already mentioned the problems created by lack of a digestive hydrolase attacking disaccharides such as lactose or sucrose, with microbial fermentation of the undigested sugar causing flatulence, diarrhea, and cramps. A more serious problem is **celiac disease,** which appears to be caused by a defect in the hydrolysis of some polypeptide sequences. It is alleviated by elimination from the diet of gluten, a mixture of gliadin and glutenin, the principal endosperm proteins in wheat. The alcohol-soluble gliadin appears to be the offender. The condition causes atrophy of the villi of the intestinal mucosa, sometimes severe enough to be life-threatening. Even those with mild cases are plagued by passage of watery, foul-smelling stools.

Other defects affect the intracellular metabolism of sugars. The problems fall in two classes, those involving loss of a kinase, and those involving loss of an enzyme attacking a phosphorylated sugar derivative. Consider the metabolism of **galactose,** derived from the lactose in every nursing infant's diet. Galactose is utilized by first phosphorylating it and then exchanging the galactose phosphate for a glucose phosphate moiety in uridine diphosphate glucose:

$$\text{galactose} + \text{ATP} \longrightarrow \text{galactose 1-phosphate} + \text{ADP}$$
$$\text{galactose 1-phosphate} + \text{UDP-glucose} \longrightarrow \text{UDP-galactose} + \text{glucose 1-phosphate}$$

An epimerase then transforms the UDP-galactose into UDP-glucose:

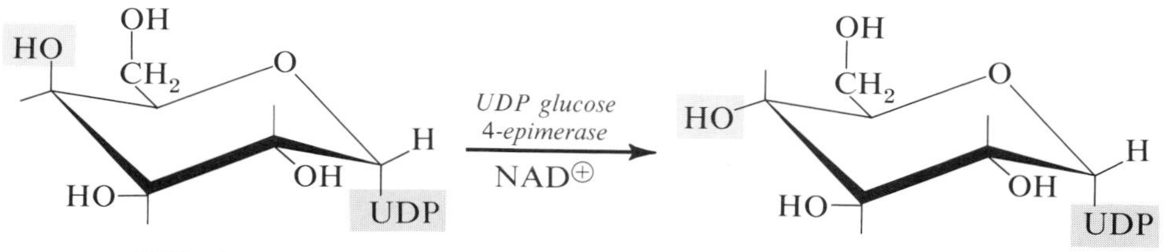

The sum of all of the reactions is a conversion of galactose to glucose 1-phosphate.

The effects of a deficiency of the specific galactokinase are akin to some of the effects of diabetes. The accumulating galactose is reduced to **galactitol** by a nonspecific **aldose reductase** using NADPH. There is no route for disposal of the galactitol other than loss in the urine; its accumulation in the lens of the eye, which has an active aldose reductase, results in the formation of cataracts.

Loss of the uridyl transferase activity that utilizes galactose 1-phosphate is even more serious. Not only is there an accumulation of galactose and galactitol, owing to feedback inhibition of the kinase, there is also a loss of inorganic phosphate, and perhaps some deleterious effects from galactose 1-phosphate itself. In any event, **galactosemia,** as the condition is called, results in liver failure and mental retardation, in addition to cataracts. The defective gene has an estimated incidence near 1 per cent in the United States; 1 in 35,000 infants in New York were found to have a full-blown galactosemia. There is no known benefit from the heterozygous condition to explain the high gene incidence. Treatment consists of scrupulous avoidance of all foods containing galactose, either free or in oligosaccharides.

Similar defects occur in the metabolism of fructose (Fig. 39–4). Deficiency of a specific kinase leads to an accumulation of the sugar after ingestion of sucrose or other fructose-containing foods. However, this ketose is not affected by aldose reductase. Indeed, one of the functions of the aldose reductase is to convert glucose to sorbitol from which fructose can then be generated by a specific dehydrogenase:

D-glucose		D-sorbitol (D-glucitol)		D-fructose
H—C=O		CH_2—OH		CH_2—OH
H—C—OH	H^+ + NADPH → NADP$^+$	H—C—OH	NAD$^+$ → H^+ + NADH	C=O
HO—C—H	⇌	HO—C—H	⇌	HO—C—H
H—C—OH	*"aldose reductase" (several tissues)*	H—C—OH	*NAD—sorbitol 2-dehydrogenase (seminal vesicle, placenta)*	H—C—OH
H—C—OH		H—C—OH		H—C—OH
CH_2—OH		CH_2—OH		CH_2OH

(Fructose is the main carbohydrate fuel in semen and is also important for fetal nourishment in many mammals.) A deficiency of fructokinase therefore leads only to fructose accumulation, which causes the compound to appear in the urine. This **essential fructosuria** is a benign condition.

In contrast, deficiency of the aldolase that removes fructose 1-phosphate is a highly deleterious condition. Inorganic phosphate is removed to make the accumulating phosphate ester, and again the ester itself may have deleterious effects. The resultant **fructose intolerance** causes liver failure.

FURTHER READING

General Texts

Nutrition Reviews' Present Knowledge in Nutrition, 4th ed. (1976). Nutrition Foundation. This authoritative paperback is skimpy on fuels and fuel metabolism.

S.S. Davidson, R. Passmore, and J.F. Brock (1975). Human Nutrition and Dietetics, 6th ed. Williams & Wilkins. A good summary.

R.S. Goodhart and M.E. Shils (1980). Modern Nutrition in Health and Disease. Lea & Febiger. Also skimpy on fuels and fuel metabolism.

Energy Balance

G. Lusk (1928). The Science of Nutrition, 4th ed. Saunders. The best source for the foundations of calorimetry.

R.W. Swift and K.H. Fisher (1946). Energy Metabolism. p. 181 in G.H. Beaton and E.W. McHenry, eds. Nutrition, A Comprehensive Treatise, vol. I. Academic Press.

M. Kleiber (1947). Body Size and Metabolic Rate. Physiol. Rev. 27:511.

N.J. Rothwell and M.J. Stock (1981). Regulation of Energy Balance. Annu. Rev. Nutr. 1:235.

A.L. Merrill and B.K. Watt (1973). Energy Value of Foods. USDA Handbook No. 74. Includes valuable discussion. Tabulations of food composition and calculated energy content are found in USDA Handbooks Nos. 8 and 456. A comprehensive looseleaf revision of No. 8 is slowly appearing.

W.S. McClellan et al. (1930–1931). Clinical Calorimetry: XLV, XLVI, and XLVII. Prolonged Meat Diets . . . . J. Biol. Chem. 87:651,669 and 93:419.

J.S. Garrow (1978). Energy Balance and Obesity in Man, 2nd ed. Elsevier. A classic, but like most European monographs, grossly overpriced.

Famine

A. Keys et al. (1950). Human Starvation. 2 vols. Univ. Minnesota Press. A classic study.

O.E Owen et al. (1967). Brain Metabolism During Fasting. J. Clin. Invest. 46:1589.

Z. Stein et al. (1975). Famine and Human Development. The Dutch Hunger Winter of 1944–1945. Oxford. A classic study.

W.R. Aykroyd (1974). The Conquest of Famine. Dutton. Includes an illuminating account of the Bengal famine of 1943 in which the loss of 1.5 million lives is believed to be an underestimate.

CHAPTER 40

NUTRITION: THE NITROGEN ECONOMY

Proteins in the food are the major precursors of most of the nitrogenous compounds in the body. While it is true that variable amounts of purines, pyrimidines, creatine, choline, nitrogen-containing vitamins, and so forth are obtained as such from the diet, they represent a relatively small portion of the total flow of nitrogen through the body. The error is not large if the food proteins are considered to be the only sources of nitrogen.

NITROGEN BALANCE

We are concerned with the requirements of the organism for nitrogen compared to its supply, that is, with its state of nitrogen balance. **Nitrogen equilibrium** is the characteristic condition in the adult, with the losses just balanced by the intake so that the body composition remains relatively constant, although there will be moment to moment and day to day variations. The need for dietary proteins to maintain nitrogen equilibrium is a result of metabolic losses. (The hibernating bear cuts its losses to nearly zero; the higher plants have no devices for excreting nitrogen.) Losses from bleeding and exhalation of ammonia are less important. Milk secretion is a major loss of nitrogen to lactating women. Contrary to adolescent folklore, protein loss in ejaculated semen requires no special compensatory diet.

Negative nitrogen balance, or losses exceeding the supply, is an obvious result of an inadequate intake. However, it is also characteristic of the ill and the injured, in whom cellular damage causes more nitrogen to be lost than is taken in. There is also a small negative nitrogen balance associated with aging, so small as to be imperceptible by direct comparisons of intake and excretion.

Positive nitrogen balance occurs when tissues are growing. It is the characteristic state of the pregnant woman and the convalescent adult, as well as the young. Maintenance of positive nitrogen balance requires sufficient dietary protein from which to construct the additional tissues being formed in addition to the amount required to replace metabolic losses.

USE OF DIETARY PROTEINS

The efficiency with which dietary proteins can be used for maintenance of tissues is dependent upon many factors. **Amino acid composition** is one of the most important; the amount of one amino acid can influence the metabolism of another amino acid. The **total caloric intake** affects the combustion of amino acids as fuels. The **previous dietary history** affects short-term events by determining the enzymatic adaptations that have occurred. **Differing heredities** cause variations in individual management of nitrogen metabolism. In addition, not all dietary protein is converted to absorbed amino acids. Digestion and absorption are affected by other components of the diet, by prior processing of the food, and by the state of the gastrointestinal tract.

Effect of Amino Acid Composition

Essential Amino Acids. The mixed population of proteins comprising tissues contain all of the 20 amino acids. When tissue proteins are being synthesized, either as replacements of existing molecules or as additional components, all of the 20 must be present. If the supply of any one runs low, then the concentration of the corresponding aminoacyl tRNA will fall below the level necessary for prompt incorporation into newly synthesized polypeptide chains. The rate of protein synthesis will decline, and the functions of the tissue will suffer. In that important sense, all of the 20 amino acids are essential. However, we have seen that most cells are capable of making some of the amino acids from glucose and almost any source of amino nitrogen. Some of those that cannot be made by a particular tissue can be supplied through the blood by the liver, which can make them. A total of eleven amino acids can be made in this way. All of the other nine must be obtained from the diet and are therefore essential dietary components. They are:

histidine	**phenylalanine**
isoleucine	**threonine**
leucine	**tryptophan**
lysine	**valine**
methionine	

The critical experiments on the amino acid requirements of human adults were performed from 1942 to 1955 by W. C. Rose at the University of Illinois with volunteer graduate students maintained on artificial diets. One amino acid at a time was omitted, and if negative nitrogen balance resulted, the compound was restored until the requirement, as indicated by a slightly positive balance, was satisfied. In most cases, deprivation of one or more of these essential amino acids has immediate deleterious effects, including negative nitrogen balance during the first day, and loss of appetite and other psychological symptoms within a few days. However, this is not necessarily true; Rose could find no effects from omission of histidine, and this amino acid was regarded as nonessential for humans. This was contrary to the observations with every other mammalian species tested, and it seemed highly unlikely that young Midwestern men had suddenly remade the genes that were gone so long. Later experiments with human infants (much criticized on ethical grounds) and with uremic adults established that histidine is indeed essential in the human diet. Rose's failure to discover the inability of adults to make histidine may be due to its relatively low abundance in proteins, a slow catabolism, and the large reservoir present in the form of carnosine in the muscles (p. 659).

Tissues contain different amounts of the various amino acids, and they are catabolized at different rates. There is a pattern of dietary amino acid composition that will enable the most efficient use of nitrogen for repair and growth, a pattern in which no one of the essential amino acids is in great excess or deficit of the demand for it when compared to the other amino acids. If the amino acid composition of food follows that pattern, a total nitrogen intake including enough of one of the essential amino acids to meet its requirement will also include enough of all of the others. The composition of the remainder of the necessary total nitrogen intake is not so critical. Indeed, it has been shown that the balance of the nitrogen required after providing sufficient essential amino acids can be supplied as ammonium acetate.

If one of the essential amino acids is in short supply compared to the others, protein synthesis will continue only until the supply is used, and other amino acids then remaining will be used as fuel. Remedying the deficit of that one will require the ingestion of more of the unbalanced mixture of amino acids with most of the additional intake except for the needed amino acid burned as fuel.

Contrariwise, increasing the total nitrogen intake diminishes the requirements for the essential amino acids. Less of these compounds will be used for making other nitrogenous compounds when the total supply of nitrogen is high. Similarly, the requirements for total nitrogen and for the essential amino acids depend upon the total supply of fuel ingested. If the fuel supply is low, more of the amino acids will be burned; it if is high, more amino acids will be left intact for protein synthesis. Indeed, the amino acid requirement is lowest in someone who is becoming fat.

Digestibility

The extent to which the hydrolases of the gastrointestinal tract can convert dietary proteins into absorbable amino acids and oligopeptides depends upon the source of the proteins and the preparation of the food. The amount of nitrogen escaping in the feces after eating eggs, milk, meat, and other animal proteins is usually only a few per cent of the amount ingested, and much of that may represent unhydrolyzed enzymes and other endogenous proteins, which have not been exposed to the denaturing acidity in the stomach. More nitrogen escapes on high fiber diets. (This is one of several important findings made in German laboratories during the First World War; it discouraged a proposal to extend the deficient food supplies by mixing sawdust with the meat.) Covalent modifications occur in foods exposed to high temperatures. These include destruction of amino acid side chains, formation of unhydrolyzable cross-links between segments of polypeptide chain, and combination with carbohydrates (the browning reaction). Determinations of the usefulness of various protein foods have not allowed for the backyard barbecue.

Contrariwise, seeds must be heated to have maximum value as sources of protein. This is especially true of the legumes, notably beans, peas, and soybeans. The proteins within the seeds occur in organelles known as protein bodies, which resist denaturation in the digestive tract. In addition, the seeds contain powerful inhibitors of the proteases. The result is that nearly half of the protein in the raw beans passes into the feces. Heating by extended boiling or by roasting disrupts the protein bodies and destroys the protease inhibitors.

Even so, the nitrogen absorbed from cooked plant products is as low as 40 per cent of the total (wheat bran) and only reaches 90 per cent for highly milled products, such as white wheat flour, whereas it ranges around 95 per cent for most animal protein foods.

Determination of Protein Quality

Quantitative Units. When work began on making quantitative comparisons of foods as sources of protein and of protein requirements, the qualitative description of the constituent amino acids had not even been made, let alone quantitative analyses of composition. Even today, amino acid composition is known for only a limited number of foods. Analyses for total nitrogen content of foodstuffs were developed early, and this was used as a basis for comparison. Since the proteins and free amino acids contain over 95 per cent of the total nitrogen in most foods, this is a useful guide to the content of amino acids. Unfortunately, these early workers decided to translate the nitrogen analysis into weights of protein, and after comparing the samples available to them, they decided that an "average" protein contains 16 per cent nitrogen. To this day, we speak about a content of protein when we really mean a content of nitrogen multiplied by 6.25 (100 ÷ 16). Modern tables of food composition include columns of protein contents that have been laboriously calculated by multiplying the analysis for total nitrogen by factors estimated for the particular type of food. Although these numbers are sometimes instructive in making assessments of the proportion of the mass occupied by protein, they created extra work for most purposes. **It is the nitrogen content itself that must be used for assessing quantitative metabolism.**

Direct comparisons of the relative supplies of different amino acids are best facilitated by speaking of moles of amino acids in the individual cases, and of moles of protein nitrogen when discussing total balance. (Conventional calculated "weights" of protein will sometimes be given parenthetically for comparison.) The relationships between the common units are:

Moles N	Grams N	Grams "Protein"
0.071	1	6.25
1	14	87.5
1.14	16	100

(moles N) × 87.5 = (grams "protein")
(grams "protein") × 11.4 = (millimoles N)
(grams "protein") × 0.16 = (grams N)

To get an appreciation of scale, one mole of animal protein N is a generous daily intake for most adults.

Net Protein Utilization. The value of a natural food as a protein source hinges upon its total protein content, its amino acid composition, and its digestibility. One method of gauging the effect of all of these factors involves feeding the food in question and measuring the excreted nitrogen. The test subjects are first fed a protein-free diet to deplete their labile proteins and to enable measurement of their basal nitrogen losses. They are then fed measured amounts of the food in question, and the nitrogen losses are again measured. The rationale is that all of the nitrogen from a perfect food will be retained to replace the depleted tissue components. In less than perfect foods, only part of the nitrogen will be retained and the remainder will be catabolized and cause an increase in nitrogen excretion above the basal level. The difference between the amount of nitrogen fed and the *increased* nitrogen loss caused by the feeding is taken as the amount of nitrogen retained. The retained nitrogen expressed as a percentage of the nitrogen fed is the net protein utilization.

For example, suppose a woman is fed a protein-free diet and her nitrogen excretion falls to 150 millimoles per day. Upon eating biscuits containing 200 millimoles of nitrogen, her nitrogen excretion becomes 230 millimoles per day. This is an increase of 80 millimoles per day over the basal level (230 to 150). She is assumed to have retained and used 120 millimoles of nitrogen (200 − 80) out of the 200 fed, so the measured net protein utilization is 60 ($100 \times 120/200$) for the biscuits.

Another way of assessing the value of a food protein is to compare its amino acid composition to the composition of some protein known to be used effectively. The proteins of hen's eggs were formerly used as a standard, but more recently a pattern based upon the composition of human milk has been developed as a reference because nitrogen deficiency is most likely to occur in the very young, and the infant utilizes human milk more efficiently than any other known protein source. The composition is adjusted so that the amount of protein satisfying the minimum requirement for total nitrogen will at the same time just satisfy the minimum requirement for the essential amino acids. The reference values are given in Table 40–1, together with the **estimated** minimum requirements of the essential amino acids in both infants and adults. Other foods can be given a **chemical score** based on a comparison of their amino acid composition with that of the reference pattern. The chemical score is the relative quantity of the limiting amino acid. Table 40–2 compares the chemical score and net protein utilization index for important foodstuffs.

An example will illustrate how the chemical score is determined. Whole wheat flour contains 16 millimoles of lysine residues per mole (1,000 millimoles) of total nitrogen. The reference pattern contains 31 millimoles. Therefore, wheat flour has a chemical score of $16/31 \times 100 = 53$, with respect to lysine. The score for no other

ESSENTIAL AMINO ACIDS **TABLE 40–1**

Amino Acid	Reference Pattern *millimoles/1,000 millimoles of total N*	Requirements (*millimoles kg⁻¹ d⁻¹*) *Infants*	*Adults*
His	10	0.21	0.02 (?)
Ile	28	0.63	0.073
Leu	47	1.03	0.095
Lys	31	0.68	0.064
Met*	9	0.20	0.010
Met + Cys	20	0.36	0.098
Phe†	16	0.36	0.019
Phe + Tyr	35	0.81	0.069
Thr	26	0.57	0.055
Trp	4.7	0.10	0.014
Val	36	0.79	0.091

*With cysteine also present.
†With tyrosine also present.

TABLE 40–2 VALUE OF SOME FOODS AS NITROGEN SOURCES*

	Chemical Score	Net Protein Utilization†
human milk	100	95
whole hen egg	100	87
cow's milk	95	81
soya (bean)	74	—
(flour)	—	54
peanuts	65	57
maize	49	36
polished rice	67	63
whole wheat	53	49

*Taken from p. 67 in WHO Technical Report No. 522.

†Not all measurements were made under comparable conditions. The last three were made with children aged 8 to 12 years, the others with children aged 3 to 7 years. The percent of the food energy derived from protein also varied, although no marked effect on the values was obtained when one food was studied at different levels of intake.

amino acid is this low; therefore, lysine is the limiting amino acid in wheat flour, and 53 is taken as the chemical score for the entire food. By this index, one ought to feed enough wheat flour so that $100/53 = 1.89$ times as much nitrogen is absorbed as is absorbed from human milk in order to meet the protein requirement. The measured net protein utilization of wheat flour in young children is 49 per cent, which is in close agreement with the chemical score.

However, chemical score and net protein utilization are not always close, even if the score is corrected for digestibility of the protein. A protein completely lacking an essential amino acid would have a chemical score of zero. This is fitting if it is the only nitrogen source, because it cannot sustain life. However, such a protein has a measured net protein utilization of as much as 30 per cent. The reason for this is that the test subjects adapt to a limited supply of any essential amino acid except threonine by diminishing the activity of the catabolic enzymes for that amino acid. (The content of threonine has never been shown to be limiting in natural foods, which probably explains why no protection against its deficiency has ever been evolved.) As a result, they conserve the amino acid from their own proteins and re-utilize it along with the dietary amino acids to make complete proteins. (Continued lack is eventually lethal, however.)

We see that neither index of protein quality is completely satisfactory. The chemical score is a more reliable indicator of defects in a single source of protein, whereas the net protein utilization index assigns some value to any source of nitrogen, which is appropriate when one is considering a food as an addition to other components in a diet.

THE DIETARY PROTEIN REQUIREMENT

Definition of a Requirement

We have emphasized earlier that definition of a requirement needs the statement of a goal. The goal has both qualitative and quantitative elements, and usually requires much more sophisticated definition than is generally recognized. Most would agree that a basal value of nitrogen intake must at least maintain adults in nitrogen equilibrium, or maintain a desired rate of tissue growth in the young or the convalescent. Which adults? Men, women, young, old, fat, thin? What is a desired rate of tissue growth? As great as can be forced? Is the objective to be reached for every individual that can be found, including those with serious genetic impairments in their metabolic capacity, or is the minimum requirement to be defined as the level adequate for half of the population?

Here we have the origin of much of the debate and misunderstandings over nutritional requirements. We are dealing, as is the case with so many questions of public health or proper therapeutics, with a probability. Most minimum requirements today are specified as being two standard deviations above the mean quantity necessary to attain the desired objective in a population. The minimum would therefore achieve the objective in 97.5 per cent of the population.

It is important to emphasize again that this does not remove the potential for disagreements over a proper objective and over the adequacy of the defined minimum. One group might look at it and say, "If you eat this much, you have only one chance in 40 of failing to meet the objective." Another might say, "If this is all you feed the American people, you will have 5,750,000 who are suffering from a deficiency."

This is dwelt on at some length, because it may shed some light on the shifts in attitude toward the minimum protein requirement that colored the estimates over the years.

Estimating the Requirement

Most estimates of the protein requirement begin with the postulate that the amount of nitrogen a person loses through normal wear-and-tear while eating an ordinary diet can be estimated by measuring his losses while eating a protein-free diet. That is, the urea, uric acid, and so forth in the urine and the nitrogen lost in the feces, along with losses from the skin when a person is not eating any source of nitrogen, are assumed to represent a minimum turnover of body constituents, a minimum that must be replaced by nitrogen absorbed from the diet. (The measurements are made a week or more after the rapid loss of labile protein.)

The first task, then, is to estimate the nitrogen losses. This varies, but values for a 73 kg man are sometimes like these, given in millimoles of N per day:

urine	175
feces	65
skin	20
other	10
SUM	270

The fraction of nitrogen lost from the skin depends upon circumstances. It is increased by frequent bathing and by heavy sweating. Approximately 8.5 millimoles more are lost for each megajoule of increased heat production during exercise. The value for miscellaneous losses is approximately 50 per cent higher in women, owing to menstruation.

The next task is to add to this minimum value any additional requirement created by special circumstances. It is especially difficult to define the required intake for growing children. Hegsted approached this question in a purely pragmatic way by defining the way things actually seem to be, rather than the way they ought to be, with a sample of American children presumably in reasonable health and growing

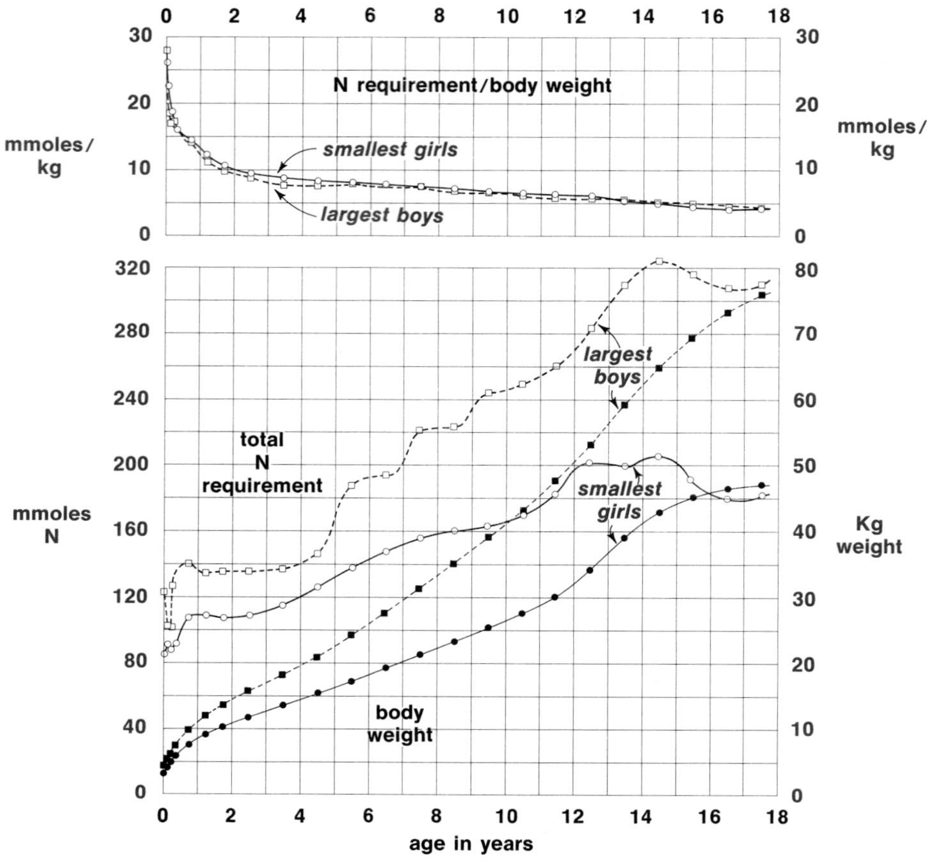

FIGURE 40–1 Theoretical estimates of the protein requirement of children as a function of age. The amount of protein nitrogen that must be absorbed in order to balance the utilization for tissue formation and the losses in the urine and feces is shown for the smallest girls (tenth percentile of weight) and for the largest boys (ninetieth percentile of weight) in terms both of the amount per kg of body weight and of the total amount. The body weights are plotted with solid points (*bottom graphs*). (Data from Hegsted: (1957). J. Am. Diet. Assn. 33:225.)

the way a random sample of the population in Boston does. He used the expected basal loss of nitrogen and the increment of the total nitrogen body content of the body because of growth to estimate the required dietary supply. A plot of the estimates is given in Figure 40–1; two curves are shown, one for girls in the tenth percentile of weight (only 10 per cent of the girls are smaller) and one for boys in the 90th percentile (only 10 per cent of the boys are larger), so these two curves encompass a range including the requirements for all but 10 per cent of the children. The calculated minimum amount of nitrogen absorbed is given per kilogram body weight and as a total. The body weights are also plotted. The relatively high requirement for protein during the early years is especially striking; small girls need half as much nitrogen absorbed when they are 3-kg infants as they do when they are 47-kg adults. It is also apparent that girls and boys have nearly the same protein requirement per unit body weight despite their different sizes, and the value declines throughout life.

The Estimated Requirement. Using the methods outlined, a committee of the World Health Organization estimated the nitrogen from high-quality proteins (eggs or milk) necessary to be absorbed to cover **mean** losses and growth at various ages:

Age	Estimated Mean Requirement for Absorbed Nitrogen			
	mmole N/kg		*g "protein"/kg*	
6 to 9 mo	11		0.96	
1 yr	8.6		0.75	
4	7.1		0.62	
7	6.3		0.55	
	M	F	M	F
10	5.6	5.5	0.49	0.48
13	5.2	4.6	0.46	0.40
16	4.4	3.9	0.38	0.34
Adult	3.9	3.5	0.34	0.31

Having estimated the minimum amount of nitrogen that must be absorbed, the next step is to estimate the amount of food that must be eaten in order to supply that minimum. Although small amounts of animal proteins are almost totally retained by a protein-starved individual, other experiments indicate that this is not true when the protein intake is increased to be near the minimum requirement. Consequently, many workers add 30 per cent to the tabulated values to compensate for this possible inefficiency. (All of the experiments are relatively short-term, and carried out with few subjects, most of whom are young and previously well-fed. More information is needed to assess the true minimum requirement.)

According to this estimate, half of the 73-kg men will obtain sufficient nitrogen if they eat a diet including 370 millimoles ($3.9 \times 73 \times 1.3$) of nitrogen (32 grams of "protein") from animal proteins. Since the coefficient of variation is roughly 15 per cent, raising the intake another 30 per cent to 480 millimoles of nitrogen (42 grams of "protein") would be sufficient for all but 2.5 per cent of the 73-kg men. Despite the upward corrections, some authorities believe these values are still too low for maintenance of healthy individuals, and much too low for those prone to infection or other debilitating diseases. On the other hand, there are many reports in the literature of individuals getting along quite nicely on much lower protein intakes. Some investigators assert that a nitrogen deficit is rare among adults who have an adequate fuel intake from natural diets; others believe that this is not so, particularly because it ignores the increased stress of episodes of illness that are more frequent with marginal nutrition. All agree that children frequently have an inadequate protein intake, as we shall discuss in the next section.

KWASHIORKOR: PROTEIN STARVATION

Simple starvation is caused by a generalized restriction of food supply. In some areas of the world, especially in or near the tropics, the total quantity of food may not be so obviously limiting, but the material available may consist mostly of starchy plant substances without an adequate content of protein. The diet is usually one of necessity rather than choice, and in an undernourished family it is frequently the children who are most affected, not only because lack of food tends to destroy any feeling of selflessness by the adults, but also because the very young have not built up a reserve of protein and fat upon which to live.

A long-standing inadequacy of protein intake in children leads to a condition known as **kwashiorkor,** from the Bantu word meaning displaced child; the symptoms appear in infants after they are no longer suckled by their mother (owing to the appearance of still another baby). Lack of protein is a central part of the condition—if the intake of carbohydrates and fats is also deficient, the child is simply being starved and wastes away. Such infant starvation has been termed **marasmus,** and there obviously is a spectrum of conditions ranging from total starvation with "pure" marasmus to a "pure" kwashiorkor in which the total supply of ingested fuels would be ample for maintenance of high-energy phosphate under ordinary conditions, but in which there is not sufficient protein to maintain cellular constituents. The latter results when the mother has access to quantities of starchy vegetables or sugar but not to foods containing enough protein, and the starchy part of the original food is frequently fed to the infant as a thin gruel. This circumstance is common in parts of Latin America and Africa.

Unlike children who are simply starved, those fed a protein-deficient diet may live for a considerable period, perhaps surviving into adulthood even though irreversibly impaired by the consequences of inadequate cellular development. Since kwashiorkor essentially represents a failure to synthesize normal amounts of protein, its consequences could theoretically appear in every metabolic process of the body, and in many ways do. However, some of the more striking phenomena can be given speculative interpretations in terms of isolated segments of the metabolic economy. The metabolic load in kwashiorkor is quite different from that in marasmus. Carbohydrates are still being supplied and the metabolic machinery for handling these compounds will tend to remain intact. At the same time, it is not possible to maintain all of the proteins in the carcass because of the lack of dietary amino acids. The result is an uneven depletion affecting some processes more than others, rather than the relatively smooth, general decline seen in marasmus.

A striking finding in kwashiorkor is the deceptively plump appearance of the youngsters; they are called "sugar babies" in the Caribbean area because they are fed on sugar and starches, but the name also evokes an image of round cheeks and bellies. The plumpness is not an expression of overfeeding and storage of fat, but is due to edema. The youngsters have a general accumulation of water to such a degree that their weight actually falls when they are put on restorative diets.

What can cause edema? One suspects congestive failure of the heart, but that is usually not the case in kwashiorkor; a mild failure is more likely to occur during initial recovery than it is during the active condition. Another possibility is a fall in the protein concentration of the blood, especially of serum albumin, causing a drop in the osmotic pressure. The synthesis of serum albumin is indeed impaired in kwashiorkor, sometimes falling from its normal level of 40 mg per ml to less than 10 mg per ml, and this kind of fall will invariably cause loss of fluid from the blood into the tissues. However, edema may appear without a precipitous fall in the albumin concentration (this is usually the case in adult starvation), and additional factors must be sought. Another possibility would be disturbances in the electrolyte and water balance, and these events also occur, both at the cellular level and in the kidney. There is an especially marked loss of potassium from the cells. It may well be that some of the proteins involved in ion transport are being destroyed and utilized for the formation of other proteins elsewhere, or to put it more accurately, there may be less of the transport proteins formed from the diminishing amino acid pool than there is of some other proteins.

The combination of high carbohydrate intake with deficient protein with which to build tissues leads to high blood insulin levels during the early stages of kwashiorkor. This aggravates the situation by increasing the uptake of already limiting essential amino acids into the skeletal muscles, while promoting fat synthesis from the excess carbohydrate. A lack of lipoprotein synthesis and increased fat synthesis may account for the fatty liver that forms in these children.

Diarrhea is almost invariably a result of kwashiorkor and contributes to the potassium loss. Runny bowels are so frequent an accompaniment of disease in general that we give little thought to cause. Many nutritional deficiencies are accompanied by diarrhea. This indicates a failure of intestinal function. The intestinal mucosa has the highest known rate of cell turnover in the whole body and, therefore, is especially vulnerable to any failure in the supply of nutrients needed for constructing cells or to an interference in protein synthesis.

The stools in kwashiorkor may also be fatty. Fatty stools may result from the loss of emulsifying agents secreted in the bile, which are steroid derivatives formed by the liver, or from a failure in secretion of lipase by the pancreas. Since the pancreas requires a high rate of protein synthesis for activity, one might expect a decrease in function when there is protein deficiency, and this is the case. The pancreas, along with the intestinal mucosa, atrophies in kwashiorkor.

The skin, which is constantly being shed and replaced in a normal individual, develops gray and scaly or ulcerating patches. This is probably due to a mixture of protein deficiency and nicotinamide deficiency (see the next chapter). Similarly, the growing hair becomes fine, dry, brittle, and abnormally light in color. Most of the people in areas where kwashiorkor occurs have naturally dark hair, so that the reddish or even blonde sparse hair of kwashiorkor is striking. In some cases the hair will be banded with light color along its length; the bands indicate times of inadequate protein supply in the way that tree rings indicate the passage of the seasons.

Failure of hemoglobin synthesis causes anemia, and the iron accumulating from continued erythrocyte destruction appears as deposits of hemosiderin. The accumulation of fat in the liver that often occurs is probably due to a failure in apolipoprotein synthesis necessary for fat transport. (A low supply of methyl groups accompanying a methionine deficiency may also contribute by failure of phosphatidylcholine synthesis.)

Finally, kwashiorkor is frequently accompanied by the symptoms of deficiency of one or more other nutrients, including many of the minerals and vitamins, as described in the next chapter. (Cholecalciferol is rarely deficient.) While the dietary supply of these nutrients may be low, a general failure of gastrointestinal function is probably the most important factor, preventing absorption of proper amounts.

SPECIAL REQUIREMENTS

Pregnancy. Women need additional protein with which to construct a placenta and fetus. From calculation of the average increment of tissue for which nitrogen must be supplied by foods with a net protein utilization index of perhaps only 70 per cent, followed by a further 30 per cent allowance for two standard deviations from the mean, one arrives at the following values:

Quarter of Pregnancy	Extra Nitrogen Requirement* total millimoles per day	g "protein" per day
1	10	1
2	50	4
3	90	8
4	100	9

*Values rounded off to nearest gram or 10 millimoles

Lactation. Lactating women need even more protein than pregnant women to sustain the outpouring of milk proteins. Provision of an extra 190 millimoles of N (17 grams of "protein") per day is suggested as a safe allowance.

Injury or Surgery. Massive intravenous feeding of nitrogen in the form of amino acids, up to 20 millimoles $kg^{-1}d^{-1}$, has been used to minimize negative nitrogen balance. We shall discuss the full composition of solutions for parenteral nutrition in Chapter 42.

Liquid Protein Diets. A dietary fad involves the use of protein hydrolysates as supplements to partial starvation in the correction of obesity. Many of these are hydrolysates of collagen, and therefore have grossly unbalanced amino acid compositions, including a very high concentration of glycine. The contribution of the imbalance to the sudden deaths that sometimes accompany this regimen is unknown.

Renal Disease. The accumulation of urea and other products of nitrogen metabolism is an especially serious consequence of renal failure. The accumulation can be relieved by dialysis. Considerable attention has been given to elimination of the need for dialysis in patients retaining a small fraction of normal kidney function, or to prolongation of the interval between dialyses in others, by restricting the protein intake to the minimal level consistent with survival.

Early regimens* involved restriction of amino acid intake to as little as 200 millimoles of N per day (18 grams of "protein"), sometimes as essential amino acids, sometimes in the form of two eggs, but cooperation of the patients was greatly improved by increasing the allowance to 450 millimoles of N (40 grams of "protein"), and cooperation is critical for any treatment that must be continued indefinitely.

A promising new approach to treatment has been the administration of the 2-keto analogues of the essential amino acids, which cause part of the nitrogen to be utilized through transamination to re-form the amino acids. This is both experimental and expensive at present and probably would not be useful for extended periods.

*These regimens are sometimes known as G-G diets, after C. Giordano and S. Giovannetti, who (together with Q. Maggiore) originally developed them.

FURTHER READING

J.R. Whitaker and S.R. Tannenbaum (1977). Food Proteins. AVI. A generally valuable source.

The following FAO Nutritional Studies prepared by the Nutrition Division of the Food and Agriculture Organization of the United Nations are useful: No. 1. Rice and Rice Diets (1948); No. 9. Maize and Maize Diets (1953); No. 19. Legumes in Human Nutrition (1964, by W.R. Aykroyd and J. Doughty) is especially good; No. 23. Wheat in Human Nutrition (1970) is another good Aykroyd-Doughty monograph; No. 27. Milk and Milk Products in Human Nutrition (1972).

M.L. Orr and B.K. Watt (1957). Amino Acid Content of Foods. Home Economics Research Report No. 4. U.S. Government Printing Office. The important source of older data. The slowly appearing revisions of Agricultural Handbook 8 contains more recent information.

W.C. Rose et al. (1955). The Amino Acid Requirements of Man. J. Biol. Chem. 217:987.

D.M. Hegsted (1957). Theoretical Estimates of the Protein Requirement of Children. J. Am. Diet. Assoc. 33:225.

R.E. Olsen, ed. (1975). Protein-calorie Malnutrition. Academic Press.

Energy and Protein Requirements (1975). WHO Technical Report No. 522.

J.C. Waterlow and A.A. Jackson (1981). Nutrition and Protein Turnover in Man. Br. Med. Bull. 37:5. One of a generally excellent series of articles on human nutrition.

W.A. Coward and P.G. Lunn (1981). The Biochemistry and Physiology of Kwashiorkor and Marasmus. Br. Med. Bull. 37:19.

R.M. May (1981). Useful Tropical Legumes. Nature 294:516. Discussion of potentially exploitable plants.

J.C. Waterlow and P.R. Payne (1975). The Protein Gap. Nature 258:113. An assertion that protein starvation is not a general problem.

N.S. Scrimshaw (1976). Strengths and Weaknesses of the Committee Approach. N. Engl. J. Med. 294:136,198. An important student of kwashiorkor defends the necessity for higher protein intakes.

Also see the general references cited for Chapter 39.

CHAPTER 41

NUTRITION: MINERALS AND VITAMINS

MINERAL ELEMENTS

Humans, like other animals, have evolved so as to retain a sufficient supply of the needed mineral components from their customary diets. Even so, circumstances and individual genetic variations sometimes prevent maintenance of adequate concentrations of one or more of the minerals, necessitating additional supplementation in order to restore normal function.

What are the essential mineral nutrients? We have already encountered most of them in connection with a discussion of their functions. There are others with no known function, but which appear to be necessary in that experimental animals do not thrive when fed a diet deficient in the element. The confidence with which these findings are extrapolated to humans varies with the degree of observed impairment and the ease of reproducibility of the experiments.

The essential mineral elements may be divided into classes on the basis of either function or the magnitude of the daily turnover. The **bulk minerals** include the major electrolytes and the constituents of bone and teeth. Moles are convenient units with which to describe their estimated content in a 73-kg man:

Ca—34 moles	**Na—4.8 moles**	**Cl—2.5 moles**
P—24 moles	**K—3.6 moles**	**Mg—1.2 moles**

The **major prosthetic minerals,** those most commonly occurring in proteins, are present in millimole quantities:

Fe—70 mmoles	**Cu—2 mmoles**
Zn—40 mmoles	**Mn—0.2 mmoles**

The **more specialized prosthetic minerals,** those occurring as constituents of only a few proteins, are probably present in micromole quantities. Good analyses are not available for most:

I—50 micromoles	**Mo—?**
Se—?	**Cr—?**

Other elements that may be essential include **tin** and **vanadium.** In addition, there is evidence that normal bone and tooth formation is aided by the presence of **silicon** and **fluoride.**

The minimal requirements of most mineral elements for normal human development have not been clearly defined. Table 41–1 gives typical amounts required for normal growth of young domestic animals in terms of the amount of element per kilogram of diet (dry weight). If human requirements are comparable, a four-year-old child requires roughly the amount in one half kilogram of diet per day. Adults would require less in most cases owing to previous accretion of the necessary total content, even though the dry weight of their total daily food intake approximates one kilogram.

Toxicity and Minimum Requirements

Continued ingestion of high levels of many of the mineral elements causes toxicity, even death. The average minimum intake for toxic effects has been shown to be 40 to 50 times the average minimum intake of several of the minerals required for normal growth in domestic animals. It is expected that evolution will have adapted mineral metabolism so as to cope with dietary variations. It is only when animals are living in areas where the soil content of an element is especially high or low that toxicity or deficiency becomes common. However, genetic variations make the question of toxicity of both minerals and vitamins pertinent to human welfare.

There is variation among people in both the amount of a nutrient required for normal growth and for maintenance as an adult. There is also variation in the amount of the nutrient that can be ingested and absorbed before deleterious effects appear. When the margin of safety is small, overenthusiastic emphasis on increasing the dietary intake of a nutrient in order to eliminate any possibility of dietary deficiencies in those less able to utilize the component results in a real risk to those with an unusual sensitivity to overdoses. We have seen how this happened with vitamin

TABLE 41–1 TYPICAL MINERAL REQUIREMENTS OF YOUNG ANIMALS*

Given in millimoles of element per kilogram dry weight of diet per day.

P – 160	Fe – 1	Se – 0.001
Ca – 150	Zn – 0.75	Cr – ?
K – 80	Mn – 0.6	Sn – ?
Na – 45	Cu – 0.08	Ni – ?
Cl – 30	Mo – 0.002	V – ?
Mg – 25	I – 0.0015	Si – ?

*Calculated from a compilation by W. G. Hoekstra, (1972) Ann. N.Y. Acad. Sci., *199*: 182.

D, but it is also a possibility with other nutrients, especially with the promotion of over-the-counter supplements of both minerals and vitamins.

Phosphorus as Phosphate

Over 80 per cent of the large store of phosphate is present in bones. The balance is mainly present in the tissues, bound in the many metabolites that have occupied so much of our attention. The large mass of the muscles, as well as their phosphocreatine content, makes them the reservoir of some 15 per cent of the phosphorus.

Since phosphate compounds are universal constituents of living cells, they are present in the natural diets of all animals. Phosphate is absorbed through the action of a sodium/phosphate symporter in the intestinal brush border and the renal brush border. Two Na^+ are transported with one HPO_3^{2-} ion, so the symport is electrically neutral. Intestinal absorption of phosphate is stimulated by 1,25-dihydroxycholecalciferol, and renal reabsorption of phosphate is inhibited by parathormone. The renal reabsorption is also inhibited by inorganic phosphate and by hydrogen ions. Acidosis therefore accelerates the loss of phosphate both by increasing mobilization of phosphate from the bone and by the lower pH. However, the close endocrine regulation of phosphate concentration and its universal dietary occurrence prevent isolated phosphate deficiencies, even though the total amount of phosphate required for development is quite large. (Isolated deficiency in this context refers to a deficiency of phosphate alone, without symptoms of other deficiencies.) Indeed, the minimum human requirement is not known.

Phosphate deficiency does occur in infants with genetic defects in the mechanism of absorption of phosphate in the renal tubule. They develop a **vitamin D-resistant rickets.** Repeated daily feedings of phosphate appear to be a more successful treatment for this condition than the massive doses of vitamin D previously used, the excess vitamin D being both toxic and without important effect on the renal loss.

Hypophosphatemia in general is a result of some pathological disturbance of metabolism rather than of grossly limited dietary intake. It may result from the excessive parathormone secretion caused by primary malfunction of the parathyroid gland, by malignancy or by rickets. It occurs when phosphate is transiently taken up by cells for phosphorylation of carbohydrates or deposition as bone and when there is chronic excessive loss in the urine, which will occur with continued metabolic acidosis. Correction of a diabetic ketoacidosis may therefore give a double whammy; the acidosis itself has caused loss of phosphate, and the action of insulin results in uptake of phosphate for phosphorylation of glucose. Attention to phosphate and also K^+ balance is therefore a necessary part of the correction. Refeeding after any dietary deprivation is likely to require similar attention. Hospitalized diabetics or alcoholics, and even some patients convalescing from surgery, may develop symptoms of phosphate deficiency for this reason. Parenteral feeding of phosphate-free solutions for extended periods, such as the commonly employed glucose-saline mixture, is to be avoided. Chronic use of magnesium or aluminum-containing antacids, which form unabsorbable complexes with dietary phosphate, also has caused deficiencies.

Calcium

Calcium is absorbed from the basolateral membranes of the intestinal mucosal cells and the renal proximal tubule cells by a sodium/calcium antiporter. The antiport is electrogenic because three Na^+ are moved for each Ca^{2+}. There is also a Ca-ATPase that pumps the ion by hydrolyzing ATP. Calcium enters the brush border by facilitated diffusion.

The necessity for calcium in the diet is so heavily emphasized to the layman that one would suspect a primary dietary deficiency to be common. It is not. This is rather surprising, since a human infant contains only some 0.7 mole of calcium at birth, and must have an average daily increment of 5 millimoles over a period of 18 years in order to achieve an adult content of 34 moles. A peak retention of 8 to 10 millimoles per day occurs in 13- to 14-year-old girls and 15- to 16-year-old boys (at least in Scotland and England, where the measurements were made). This increment is supplied by the diet. The nursing infant obtains its supply from its mother; human milk is 8 mM calcium. After weaning, the supply comes from other foods.

Hypocalcemia does occur within the first two weeks of life, when it is the most common cause of convulsions. Why it happens in the first day or two is not clear, but a contributing cause to a second flurry near the end of the first week is the use of cow milk in formulas. This milk, and some commercial formulas based on it, has a relatively high content of phosphate compared to calcium. The molar calcium:phosphate ratio is 1.7 in cow milk and 2.7 in human milk. The increase in phosphate lowers the solubility of calcium both in the bowel and the blood. Phosphate and calcium concentrations tend to vary inversely because a rise in either can exceed the solubility product of the calcium phosphates, causing increased formation of insoluble complexes, which pass into feces in the case of the bowel.* Less of the calcium from cow milk may be available even though its total concentration is greater than it is in human milk. This may well be a transient failure of normal regulatory mechanisms during the first days of life, since experimental changes in the calcium:phosphate ratio did not affect calcium balance in small experimental samples of older infants or adults.

Signs of calcium deficiency in young adults who have normal cholecalciferol content and parathyroid function are rare because the regulation of balance is so good. People adapt to the dietary intake of calcium. If it is small, little passes into the feces; if it is large, only a small fraction is absorbed. This adaptation has confused efforts to define the calcium requirement through balance studies. In short-term studies, an adult will "require" continuation of his previous intake to maintain calcium balance until he can adapt to a lower intake. As a result, there are sharply divergent opinions on the minimum calcium requirement. One school holds that the obligatory endogenous losses amount to 6 to 7 millimoles per day, and that it requires an estimated dietary intake of 15 millimoles per day for absorption of that amount on mixed diets. Phytates (inositol polyphosphates) and other poorly digested ligands in plants cause increased fecal loss of calcium and other readily chelatable metals.

Other studies, and an analysis of intakes in some populations, suggest that 5 millimoles per day over a long period is sufficient for an average adult, although some believe the average intake of this amount by the Japanese before World War II may account in part for their short stature. Intakes of 10 to 12 millimoles (0.4 to 0.5 gram) appear to offer sufficient margin for losses and incomplete absorption, except that pregnant or lactating women will require more. Approximately 6 millimoles per day will be incorporated into the fetus during the last trimester of pregnancy and secreted into the milk during lactation. The total drain on the calcium supply is approximately 1.5 moles due to pregnancy and subsequent lactation. This is of the order of 5 per cent of the supply in the skeletal reservoir, which is not an excessive loss in the absence of any additional source of calcium, but repeated pregnancies on a low calcium diet can lead to detectable losses of bone mineral (**osteomalacia**). The

---

*Adults can get into trouble from severe hypocalcemia caused by using phosphate-containing cathartics. At least one death has been ascribed to repeated use of such a preparation to remove stubborn feces prior to study of a patient.

loss of bone mineral that occurs in the aged, especially in postmenopausal women, is of a different nature (**osteoporosis**). The cause is not clear. It cannot be corrected in most patients with increased calcium alone; the proper preventive treatment in those whose bones have not become dangerously fragile is stimulation of bone formation through exercise, coupled with calcium supplementation.

Physicians ought to take note of the form in which patients are taking supplemental calcium. The ground dolomite popular in health food circles too often is rich in undesirable trace metals. Elevated concentrations of lead, mercury, arsenic, and aluminum have been found in the hair of patients on a neurological service who had been taking dolomite. Convenient and inexpensive sources of supplemental calcium include some commercially available antacid tablets. (Tums in the United States contains 200 mg of calcium per tablet.) Excessive calcium ingestion is mainly a problem for those prone to stone formation in the urinary tract.

Milk is a major source of calcium for the young, but its high caloric content and the presence of lactose make it less desirable for sedentary adults in affluent societies. Much of the calcium and the fuel, although little lactose, is retained in cheese. Broccoli, collards, and other greens (but not lettuce) are rich sources for vegetarians.

Sodium, Potassium, and Chloride

The average minimum requirement for sodium has been variously estimated at 1.7 to 8 millimoles per day for adults living in cool climates, with another 11 millimoles per day required during growth. These low requirements are possible because of tight control over the losses of sodium. The urinary losses can be diminished to trivial levels, leaving sweat as the major site of loss. Without adaptation, sweat contains some 68 mM sodium. Beginning work under hot and humid conditions may result in the loss of several liters of sweat, and therefore of hundreds of millimoles of sodium. However, a remarkable adaptation occurs within 10 to 14 days to diminish the secretion of Na^+ by the sweat glands. The concentration may go as low as 2 millimolar, although 6 to 7 millimolar is more typical. (Measured losses were 32 to 55 millimoles in 5 to 9 liters of sweat in one study.) The dietary requirement is therefore easily met by natural foods for nearly all except those sporadically doing heavy work in hot environments. Man's craving for salt is a matter of taste rather than a reflection of some desperate nutritional need. The use of salt as such is a relatively recent event in human history. It began as a luxury, with "sitting below the salt" equivalent to social inconsequence. While the demand for salt became an important political and economic force in post-Roman cultures, its major use was as a preservative. Perhaps it was an acquired taste for salt pork and the like that set the stage for the current habit of heavily salting food. The measured excretion of sodium by 71 asymptomatic males eating their customary diets was found to average 180 millimoles per day.

Much emphasis is now being given to diminishing the consumption of salt by the population at large with the expectation of forestalling the development of hypertension, or alleviating it, once developed. This will be difficult in a society so heavily dependent upon prepared foods, almost all of which are rich in sodium. Nearly all spicy or pickled foods must be avoided. (One large dill pickle may contain 80 millimoles of sodium. Each gram of soy sauce contains 3 millimoles.) Even frozen vegetables frequently contain added salt. Severe restriction of sodium intake for patients with heart failure—who are dangerously prone to retain extracellular fluid with increased salt intake—requires specially prepared, and relatively expensive, foods.

In normal individuals, it requires an intake of some 600 to 700 millimoles of NaCl to cause visible edema, so there is yet some margin of safety before we are acutely salted away.

Those who lose salt and water through excessive sweating can easily compensate by eating salted foods and drinking water. Young people in generally good health who exercise vigorously on hot days may lose enough salt to develop painful cramps, which are relieved by NaCl tablets. (Cramping of abdominal muscles has on occasion been so severe as to cause surgical exploration for a nonexistent acute abdomen.) Severe dehydration and salt loss causes coma, and infusion of isotonic sodium chloride is then indicated.

Potassium is continually lost in the urine of normal people and must be replaced. Those with potassium deficiency are found to require 5 to 10 millimoles per day. Approximately 300 millimoles are acutely toxic, but lesser amounts will cause diarrhea.

Potassium is a constituent of all cells and is therefore present in all normal diets. The amount is adequate except in the presence of abnormal losses, such as occur with continued vomiting, or diarrhea and excessive aldosterone secretion or treatment with some diuretics. Mild supplementation of potassium can be provided by feeding orange juice (5 mM K^+). More intensive supplementation with potassium salts requires care, not only because of local irritation of the gastrointestinal tract, but also because the amount of potassium necessary to restore intracellular balance can create toxic extracellular levels as it passes through the blood if it is absorbed too fast.

Chloride deficiency without a concomitant sodium or potassium deficiency was unheard of before the days of infant formulas and parenteral feeding. It was so taken for granted that a manufacturer forgot to put it into his infant formula. The infants fed the formula failed to thrive and had metabolic alkalosis with low concentrations of both sodium and potassium. Chloride deficiency causes alkalosis because bicarbonate becomes the principal anion available for reabsorption along with sodium and potassium. This also accounts for the failure to reabsorb adequate amounts of those cations.

Magnesium

The minimum requirement of magnesium has not been established. Less than 0.5 millimoles are lost in the urine per day during deprivation. Because of its wide distribution in all cells and its close regulation, primary dietary deficiencies are very rare. Deficient states are caused by extensive losses or failure in absorption. Conditions in which these may occur include kwashiorkor, phosphate deficiency, use of cisplatinum in chemotherapy of malignancies, alcoholism, congestive heart failure, use of diuretics, intoxication by digitalis, diabetic acidosis, malabsorption syndrome, and aldosteronism. The tissue content of magnesium is sometimes low without a corresponding fall in blood concentration. The effects of magnesium deficiency in humans are rather diffuse. The nervous system bears the brunt, as shown in weakness, nausea, tremor, stupor, coma, and cardiac arrhythmia. This is in sharp contrast to the effects of magnesium deficiency in some animals. Experimental deprivation in rats causes them to develop strikingly red ears and an excessive response to stimulus manifested first by biting the hand that feeds them and then by exhibiting running fits in response to loud noises. Veterinarians in general practice are well aware of the magnesium-deficient cow, who changes from placid Bossie to a horn-waving candidate for the corrida. Humans also are reported to show irritability, as well as confusion, a semicomatose state, and cardiac arrythmias during magnesium deficiency.

Magnesium intoxication has been reported to occur from excessive ingestion of antacids, but it more commonly occurs when parenteral magnesium is given to counteract the toxemia of pregnancy, with early nausea, cutaneous flush, followed by loss of neuromuscular function, and eventual heart block.

Iron

The body content of iron both as a prosthetic component and in storage forms increases from birth to maturity; indeed, the amount of stored iron increases in many males throughout life. As we noted in Chapter 34, the requirement for iron is fixed by the amount necessary to build new tissue from the time of conception to adulthood, and by replacement of losses, mostly due to bleeding. Typical values of iron content and the requirement for absorbed iron are given in Table 41–2. These data could be used as a guide for the dietary requirement of iron were it not for the fact that the efficiency of absorption and the magnitude of the losses are highly variable, not only from person to person, but also from day to day, depending upon the nature of the diet and the physiological state of the gastrointestinal tract, as well as the variations in the reproductive cycle in women.

If the absorption of iron fails to match losses in those without a large store of iron or fails to provide for growth in the young, the replacement of the total hemoproteins and iron-sulfide proteins will lag behind their destruction. This will lead to a diminished content of these proteins, usually hemoglobin in the first instance, followed by impairment of function.

Impairment does not happen overnight. A typical female adult with no iron stores upon which to draw and no iron intake will require 200 days to lose 10 per cent of her functioning iron pool. (However, 11 per cent of the women in a Swedish survey had such large menstrual blood losses that it would require them less than 130 days to reach the same level.) A typical male adult would require 400 days without iron stores and iron intake to lose 10 per cent of his larger iron pool. It is well to keep these slow responses of the iron pool in mind when considering short-term deprivation of iron. However, significant deficiencies do occur.

Some consequences of **negative iron balance** are clearly related to loss of function when the hemoglobin concentration falls: breathing is more labored (dyspnea), especially upon exertion, the heart rate increases (tachycardia), palpitations may occur, and there is general fatigue. The effects of deprivation of other iron-containing proteins are less clearly ascribed to their function, but they include the loss of papillae and inflammation of the tongue (glossitis), causing it to be smooth and bright red, appearance of fissures at the angles of the mouth, and changes in the growth of the nails.

BODY CONTENT OF IRON **TABLE 41–2**

Age Years	Body Content *millimoles*				Physiological Requirement *millimoles/year*	
	Storage		*Total*			
	male	female	male	female	male	female
birth	1.0	1.0	5.1	5.3	2.6	2.0
4	2.7	2.7	13.9	13.3	2.5	2.6
19*	13.0	8.3	62	40	5.0	8.3†
60–70	12.3	9.0	60	44	1.2	1.4

*Values for virgin females.
†World Health Organization estimate is 18 mmoles/yr for menstruating adults.

Iron deficiency can impair oxidative capacity before overt symptoms appear. Several studies indicate a direct correlation between daily work output, as well as exhaustive work capacity, with increasing hemoglobin concentrations up to levels approaching what are taken to be normal minimum values. Lactic acidosis appears at a lower level of acidosis in the iron deficient. This is ascribed to a selective decline in the activity of the mitochondrial oxidative enzymes, particularly glycerol phosphate dehydrogenase, which is necessary for removal of electrons from the cytosol in white skeletal muscle fibers (p. 470). One of the more striking symptoms of iron deficiency is **pica** (a bizarre appetite), and a craving for ice is especially common.

Direct assessments of iron balance are not practical for routine clinical use. The indicators that are routinely available include measurements of the blood **hemoglobin concentration,** the proportion of the blood volume contributed by the cells (the **hematocrit,** or **packed cell volume,** given in per cent), the **iron-binding capacity** of the serum (mainly due to transferrin), the **percentage saturation** of that capacity with iron (calculated from the measured iron concentration), and the **serum ferritin concentration.** Normal ranges have been established for these measurements. Table 41–3 lists the minimum normal values given by a committee of the American Medical Association and those published by Massachusetts General Hospital. Measurement of erythrocyte-free protoporphyrin concentration is not yet widely used, but it is an indicator of insufficient available iron for hemoglobin formation. Now come some all-important definitions. If the quoted values are taken as the true lower limits of the normal range, then a person with a lower packed cell volume or hemoglobin concentration has an anemia by definition. A person with a low saturation of iron-binding capacity has either an iron deficiency or a chronic inflammation. If the iron-binding capacity (transferrin concentration) is elevated, then the low saturation is caused by iron deficiency. A serum ferritin concentration between 15 and 500 μg/l is directly proportional to the concentration of stored iron; multiplying that ferritin

TABLE 41–3 MINIMUM NORMAL HEMATOLOGICAL VALUES

Committee on Iron Deficiency*

Age (years)	[Hemoglobin] g/dl	[Hemoglobin] mM	Per cent Packed Cell Volume
0.6–4	11	1.7	33
5–9	11.5	1.8	34.5
10–14	12	1.9	36
adult male	14	2.2	42
adult female			
not pregnant	12	1.9	36
pregnant	11	1.7	33

Massachusetts General Hospital†

	[Hemoglobin] g/dl	[Hemoglobin] mM	Per cent Packed Cell Volume	Iron-binding Capacity‡ μg/dl	Iron-binding Capacity‡ μM	[Iron] μg/dl	[Iron] μM	[Ferritin] μg/l
adult male	13	2.0	42	250	45	50	9	20§
adult female	12	1.9	40					

*J.A.M.A., *203*:407 (1968).

†N. Engl. J. Med., *298*:34 (1978).

‡<16% saturation of the iron-binding capacity indicates iron deficiency or inflammation (Finch, 1982).

§>200 μg/l of ferritin indicates iron overload.

concentration by 140 gives the stored iron in μg/kg body weight. A low ferritin concentration therefore indicates iron deficiency. (Inflammation elevates the ferritin concentration and may mask a concurrent iron deficiency.)

Iron deficiency is often stated to be the most common nutritional deficiency. How serious is the problem? Perhaps no aspect of nutrition currently represents a more serious challenge to objective appraisal than this question. The problem is essentially one of defining terms and eliminating some of the connotations. No one would disagree that a person in whom the signs and symptoms we have just described are relieved by iron supplementation according to objective appraisal has indeed been afflicted with an iron deficiency. Given a person in good health, but with low iron stores, so low that physiological impairment might develop with another pregnancy, continued hemorrhage, or chronic infection, is that person to be regarded as iron deficient? There is room for valid disagreement as to that point.

Cook and Finch have summarized the use of laboratory measurements as follows: A fall in the serum ferritin concentration below 12 μg/l indicates iron depletion ("pre-latent" iron deficiency). A drop in the saturation of the iron-binding capacity below 16 per cent, or a rise in the free erythrocyte protoporphyrin level over 1 mg/l, in addition to the fall in serum ferritin concentration, indicates iron-deficient erythropoiesis ("latent" iron deficiency). Finally, a fall in the hemoglobin concentration below the accepted levels for sex and menstrual state (11 g/dl in pregnant women, 12 g/dl in menstruating women, and 13 g/dl in men) in addition to the preceding flags indicates an iron deficiency anemia ("overt" iron deficiency).

The critical point that must not be missed is that a person need not be functionally impaired if he is anemic or iron deficient according to some defined laboratory measurement. When defined in this way, anemia or iron deficiency is a condition like pulse rate, or blood glucose concentration, and not in itself a disease. A Swedish survey illustrated the problem. Women were tested for true iron deficiency by giving them oral doses of iron and testing for an increase in hemoglobin concentration. Of those having normal hemoglobin concentrations according to World Health Organization standards, 17 per cent were found to be iron-deficient. Of those said to be anemic by those standards, 35 per cent did not respond to iron feeding. Among pregnant women, one-quarter were found to be misclassified if anemia was equated with iron deficiency, and more normal women were again misclassified as iron-deficient.

Iron and Infection. There would be little dispute over the value of aggressive administration of additional iron to nearly everyone if there was an indisputably large margin of safety. Unfortunately, there *is* dispute over this margin. Bacteria, like us, require iron for aerobic metabolism, and have developed means of concentrating the metal from the environment. Our own devices for trapping the metal include, in addition to transferrin in the blood and ferritin within the cells, an iron-binding protein in milk. These liganding proteins not only carry iron, they prevent iron-requiring bacteria from growing. There is strong evidence that increases in the blood iron concentration also increase the risk of infection. Haptoglobin, which tightly binds hemoglobin, also protects from infection by aerobes. (Tight binding of iron is not without some risk. The presence of iron is deleterious to anaerobes, including the Clostridia generating botulinum toxin. Cases of botulism have been reported among infants given honey-supplemented milk. The honey provides the Clostridia spores, which thrive in the iron-free environment. However, adding iron to the infant diet causes colic, because iron-requiring bacteria replace the normal lactophilic flora.)

Thus there are several possible rationalizations for our tendency to live at the brink of crippling iron deficiencies. Those who wish aggressive iron supplementation argue that we are trapped by an ancestral shift from being carnivorous hunters to being sedentary cereal farmers. Others argue we adapted to the iron age, which we have now left for the aluminum and Teflon-coated age. (The two groups obviously have different views on the speed of evolutionary adaptation.)

Those who resist general administration of supplementary iron point out the unassessed risk of increased infections, as well as the certain danger to those with hemochromatosis (p. 672). The number with idiopathic hemochromatosis is small, but the number with secondary iron overloads owing to thalassemia and hemolytic anemias is larger.

The availability of dietary iron varies widely. Ascorbic acid aids absorption, whereas tea inhibits absorption. Iron added to wheat flour, which is the usual route of supplementation in developed countries, is only fractionally absorbed. With typical lunch menus, the amount of iron absorbed by normal people in one study ran from a high of 45 per cent with a meal of sauerkraut and sausage to a low of 2 per cent with a vegetarian meal of navy beans, rice, cornbread, apple, nuts, and yogurt. Similarly, 23 per cent of the iron in an Italian dinner of antipasto misti, spaghetti, meat, bread, wine, and an orange was absorbed, whereas only 7 per cent was absorbed from an almost-Italian dinner of pizza, olives, tomatoes, anchovies, cheese, and beer.

Copper

The minimum daily requirement for copper is not well established; it evidently lies between 15 and 30 micromoles for adults. The minimum chronic toxic dose is also not established. Gram quantities are known to be fatal because copper sulfate is a favorite poison for suicides in India.

Cells, including those in the intestinal mucosa, contain proteins to bind copper. The first discovered was named **metallothionein** because it is rich in sulfhydryl groups, but this is not true of all of the binding proteins. Copper moves through the body as Cu(II), mainly in combination with serum albumin, which has one tight binding site per molecule. It is used to form several oxidative enzymes, including cytochromes $a + a_3$, lysyl oxidase, ferroxidase, and dopamine hydroxylase.

A dietary deficiency of copper is exceedingly rare except in infants and patients given parenteral feedings with deficient formulas. Anemia is an important early sign. The requirement for copper is increased by zinc and by fiber in the diet, which hinder transport and absorption, and decreased by protein, which presumably aids absorption through formation of amino acid chelates.

Two genetic disturbances of copper metabolism are known. **Wilson's disease,** or **hepatolenticular degeneration,** is an autosomal recessive condition with a gene frequency that may be as high as 0.02 (equivalent to a carrier frequency of 0.04 and a disease frequency of 0.0004). This condition involves some uncharacterized disturbance of the movement of copper within the liver, so that biliary excretion and the formation of ferroxidase (ceruloplasmin), which are the principal means of export from the liver, are below normal. The failure in excretion causes an accumulation of copper in various tissues. A pathognomonic sign is the occurrence of greenish-brown deposits in a ring around the outside edge of the cornea (Kayser-Fleischer ring), which occurs in no other condition. The disease usually is diagnosed in the first or second decade, and early diagnosis is important to minimize subsequent development of cirrhosis of the liver and neurological degeneration. Treatment involves the administration of a chelating agent, usually penicillamine, to remove the excess copper.

Menke's disease, the kinky hair or steely hair syndrome, previously mentioned as one or more defects in the utilization of dietary copper, affects the formation of normal connective tissue because of the resultant loss of lysyl oxidase activity (p. 169). The loss of other enzymatic activities has more widespread effects, and death within the first three years is expected. This X-linked recessive disease may have an incidence as high as 1 in 35,000 births. Diagnosis has usually been made too late to determine if intravenous copper supplements would be an effective treatment.

Zinc

The minimum daily requirement for zinc is greater than that for iron. Estimates range around 0.1 millimole for preadolescents and 0.2 millimole for adults. Again, the minimum chronic toxic dose is not known, but a 30 millimole acute dose causes gastrointestinal distress and vomiting. Absorption of zinc requires a low molecular weight ligand, evidently citrate, which is provided by the pancreas and the mammary gland. Copper competes with zinc to some extent for a place on binding proteins, and therefore increases the zinc requirement.

Zinc is a component of a number of enzymes, including RNA and DNA polymerases, porphobilinogen synthase, carbonic anhydrase, and carboxypeptidase. The blood serum zinc concentration is ordinarily 14 to 19 micromolar, but the concentration falls sharply without a continued dietary supply because there is little functional store of the metal. Any surplus becomes deposited in the bones where it is only slowly available. Several pathological conditions deplete the available zinc; they include malignancies, myocardial infarction, and infections.

Frank deficiencies of zinc are rare. Several dwarf males in Iran and Egypt were found to be prepubertal in sexual development, and to have anemia, enlargement of the liver and spleen, and mental lethargy. The anemia could be corrected by iron supplementation, but the most welcome improvement of the arrested development so as to give some hope of reaching a state of eupareunia required zinc. These people had been subsisting on a diet mainly composed of whole wheat bread and beans. Like many people on a deficient diet, they ate large amounts of earth, as much as 400 grams of clay per day, a practice also associated with iron deficiency.

Zinc deficiency also is caused by hereditary defects in its absorption, resulting in **acrodermatitis enteropathica,** a disease marked by severe chronic diarrhea, loss of skin around the anus and mouth, and a rash on the extremities. This condition is alleviated by human milk, but not by cow milk, and the apparent reason is that the zinc in human milk is bound to polypeptides of lower molecular weight, making the zinc more accessible for absorption. A similar condition has been observed upon parenteral feeding with formulas deficient in zinc.

The effects and incidence of more moderate zinc deficiencies are not known. The blood and hair concentrations in some American infants appear low compared to those observed in other countries, and some American formulas have been low in zinc concentration, as well as being based on cow milk with its higher molecular weight binding protein. Supplementation of the diet with zinc has in some cases aided growth.

Deficiencies in adults cause a loss of normal taste. The taste buds have no circulation and the saliva is evidently an important source of nutrients for them. The saliva includes **gustin,** a 27K molecular weight polypeptide that is high in histidine (8 per cent) and contains two zinc atoms. (Histidine supplementation can be used to create an experimental zinc deficiency.) Gustin is a close relative of, if not identical to, a nerve growth factor. The presence of this protein, perhaps as a source of zinc, appears to be necessary for normal development of taste buds. Some individuals have a hereditary defect in taste function that is correctable by large doses of zinc. (This would be consistent with a mutation causing a lowered affinity of some protein for zinc.)

Other Minerals

Selenium deficiency has long been known to cause a muscular dystrophy (white muscle disease) in domestic animals reared in areas where the selenium content of soils is low. It has only recently been recognized as a probable factor in **Keshan disease,** a sometimes fatal cardiomyopathy of children occurring in a Chinese province. A case has also been seen with parenteral feeding. Selenium is unusual among the essential minerals in that it commonly occurs in toxic levels in several kinds of plants growing in regions where the soil level is high. Animals grazing in these areas lost hoof and hair from chronic exposure and had severe damage to many organs in acute cases. Although people in those areas are alleged to be affected on occasion with the loss of hair, brittle nails, and a garlic odor on the breath characteristic of human selenium poisoning, clear confirmation is lacking.

The only known function of selenium is as a constituent of **glutathione peroxidase.** The enzyme is found in most tissues as an agent for removing hydrogen peroxide and organic peroxides:

$$2\,\text{GSH} + \text{ROOH} \longrightarrow \text{G-S-S-G} + \text{H}_2\text{O} + \text{ROH}$$

Deficiencies of **manganese, molybdenum, chromium,** or **vanadium** are not known in humans except in some patients fed parenterally for extended periods with deficient formulas. Manganese is a cofactor for several enzymes; molybdenum is a constituent of some oxidases, in which it occurs as a complex with a novel sulfur-containing pterin. The only known function of chromium is its occurrence in a complex that facilitates the effect of insulin on glucose transport in experimental animals. This **glucose tolerance factor** contains glutamate, glycine, and cysteine along with two moles of some nicotinate derivative, and it is evidently volatile enough to be removed from samples dried at 100° C or less.

The function of vanadium is not certain, except that it has been shown to be a potent inhibitor of $(Na^+ + K^+)$-ATPase.

WATER-SOLUBLE VITAMINS

The functions of the vitamins were discussed as we encountered them, and page references are noted in the following. As with other nutritional components, the vitamins are subject to bouts of overattention, sometimes followed by a period of rebound neglect. We first see assertions that a deficiency within the United States is a prime cause of a variety of frank illnesses and less well-defined general malaise. Then, after much discussion and an occasional experiment, the claims prove to be ill-founded. However, occasional deficiencies of some vitamins do occur even in our well-fed nation, and with sufficient frequency to warrant awareness on the part of the physician, but not so often as to warrant political interference.

Ascorbate (Vitamin C—pp. 174, 649)

The adult minimum daily requirement for ascorbate is approximately 50 micromoles (10 mg). Prokaryotes do not use the compound. A route for ascorbate synthesis (Fig. 41-1) was developed at some point during eukaryote evolution, but it was discarded by animals eating ascorbate-rich diets; these include the primates, bats, and many fishes and birds.

Unlike the other water-soluble vitamins, ascorbate is not an established component of any enzyme, and its exact function is not known, except that it is involved in some way with the action of prolyl and lysyl hydroxylases, and p-hydroxyphenylpyruvate oxidase, and in noradrenaline formation. It has an unusually varied distribution for a vitamin, both within an organism and among species. There are large concentrations in the adrenal gland and the aqueous humor of the eye. The high adrenal content suggests some relationship with the many hydroxylases involved in hormone production, a notion reinforced by the large discharge of ascorbate caused by corticotropin. However, cultured adrenal cells make the steroid hormones very

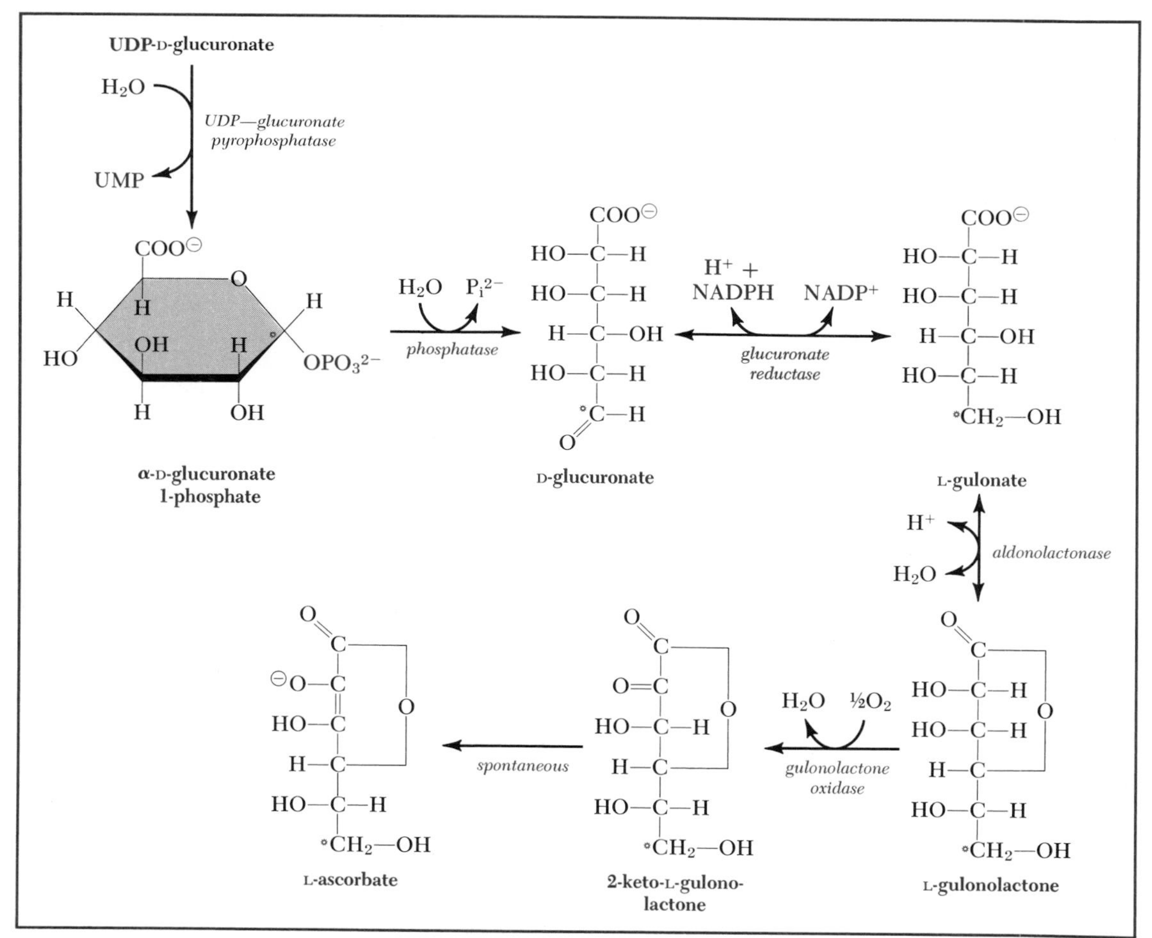

FIGURE 41-1 The synthesis of ascorbate uses UDP-glucuronate derived from UDP-glucose as a precursor. Primates lack the gulonolactone oxidase that catalyzes the final enzymatic reaction in the sequence. 2-Ketogulonolactone is a likely intermediate in that reaction. (Note that the gulonate configuration is the reverse of the glucose configuration; that is, C-1 of glucuronate becomes C-6 of gulonate, and vice versa.)

well in the apparent absence of ascorbate. This does not completely kill the idea of some role as an electron donor for the hydroxylases, since the other ascorbate-sensitive oxygenases also function *in vitro* without it, but it chills it.

The best known sources of ascorbate are fruits (especially citrus fruits, although berries and green peppers are better), but it also occurs in broccoli in high concentrations. The potato is a good source. Pure carnivores gain an ample supply by eating liver, brain, or kidney; the muscles provide only marginal amounts.

Either ascorbate or dehydroascorbate, the oxidized form, can be utilized within the body, but formation of dehydroascorbate in foods results in a subsequently slower, but irreversible, isomerization to 2,3-diketo-L-gulonate (Fig. 41–2). The formation of dehydroascorbate by reaction with oxygen, which is accelerated by trace metals, therefore results in loss of biological activity. (Moral: Don't store large opened containers of orange juice, and keep it cold.)

Laymen are frequently exhorted to have a daily supply of ascorbate because it is not stored. This is false. The total content in adults varies from a high of approximately 23 millimoles (4 grams) to as little as 1.7 millimoles (0.3 gram) without signs of deficiency. This level is determined by the intake, with the metabolic losses being approximately 3 per cent. The pool size adjusts until intake balances loss. According to this estimate, an intake of 0.7 millimole (140 mg) per day would maintain maximum content. Signs of scurvy do not appear in people saturated with the vitamin until approximately 90 days, and relatively small intakes can delay their appearance for months more. These observations explain why people can survive the winter without eating fresh fruits, vegetables, or meat in any quantity.

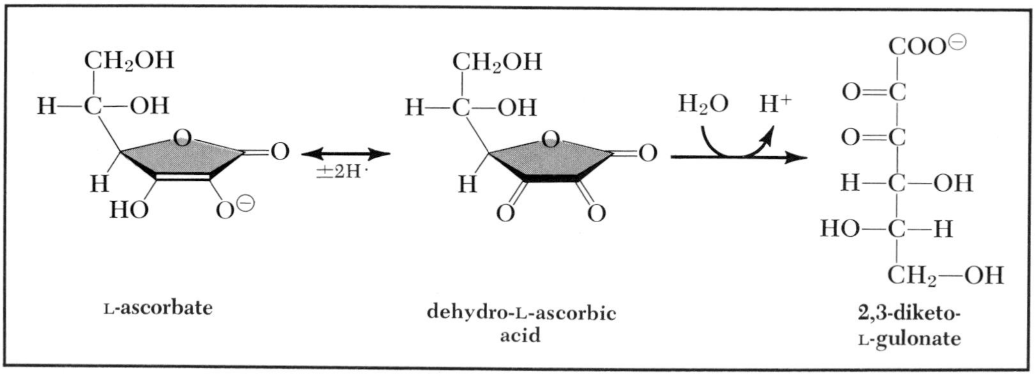

FIGURE 41–2 The dehydroascorbate formed by oxidation of ascorbate is spontaneously and irreversibly hydrolyzed to diketogulonate. Only a small fraction of the dehydroascorbate exists in the form shown; most is cyclized to an inner hemiacetal and the corresponding dimer:

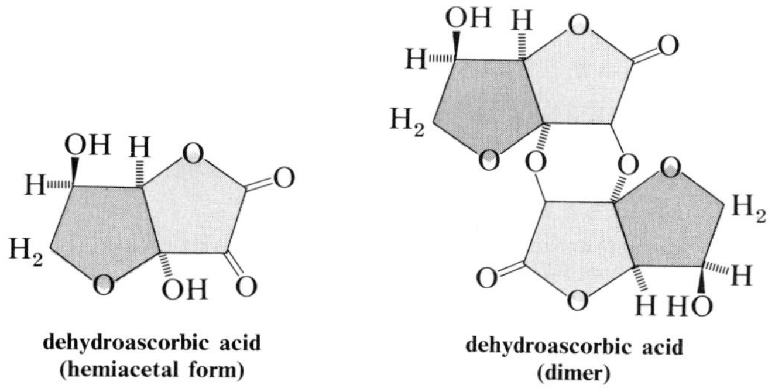

Excess ascorbate is excreted or metabolized. None of the postulated benefits or harmful effects of the very large doses that are currently the fad have been clearly proven, with one exception. Those prone to develop oxalate stones have their problems exacerbated by the increased formation of oxalate, which is one of the metabolic products of ascorbate. Both benefits and damage may be so marginal for most people that only huge trials with impeccable statistical analysis will detect them.

In experiments with convict volunteers fed an ascorbate-free diet, the extent of depletion of the body content necessary for development of particular symptoms varied widely among the individuals, sometimes appearing when over 2 millimoles was still present, and in other cases not until the content had dropped to nearly 0.6 millimole. In general, a whole blood concentration of 17 micromolar or less (3 mg l^{-1}) is a signal for attention. The early symptoms of scurvy include petechiae, hyperkeratosis, congested hair follicles with coiled hairs, joint effusions, swollen gums, and arthralgia.

Scurvy is now mainly seen in the elderly, especially in men living alone, and in infants. There is some indication that large intakes by the mother during gestation cause an adaptive high rate of ascorbate turnover in the infant, so that it must have more ascorbate after birth to avoid deficiency. It is not known to what extent the adult adapts.

Biotin (*p. 495*)

Only minute amounts of this cofactor of some carboxylase proteins are necessary to balance losses, and part of this small requirement may be met through synthesis by intestinal bacteria; deficiencies are rare. Some infants develop a dermatitis that is corrected by biotin, and a few men had a similar condition caused by eating large amounts of raw eggs, which contain **avidin,** a protein that binds biotin in a tight complex that escapes absorption.

Cobalamins (Vitamin B_{12}—*p. 599*)

In the earlier discussion, we noted that these large molecules require the presence of a glycoprotein, the **intrinsic factor,** to be absorbed and that the daily requirement is of the order of one nanomole or less. (Some estimates place it as low as 200 picomoles per day for adults.) Since plants do not use cobalt and do not make the cobalamins, the ultimate sources of the compound in the food chain are those microorganisms that have evolved routes that require it. The compound enters animals through ingestion of the microorganisms or of other animals. The liver can accumulate a milligram or so, a generous supply that will stave off appearance of symptoms for years in people who have lost their capacity to absorb the compound (for example, by removal of the stomach).

If intrinsic factor is not present to fix cobalamins for absorption, they remain bound to other proteins elaborated in the bile and are lost, although a small fraction of massive doses (700 micromoles, 1 mg, or more) will be absorbed. The intestinal receptors can handle only 400 to 1100 picomoles (0.5 to 1.5 μg) per day. Once absorbed, the cobalamin is somehow exposed to at least three binding proteins in the blood. Only one of these, transcobalamin II, acts as an immediate carrier of the cobalamins through the blood. The others only slowly turn over, so the measured blood cobalamin concentration may remain high for a time after the development of an overt deficiency.

Deficiencies of cobalamins cause a combined system disease with lack of acid secretion by the stomach (and an increased risk of developing carcinoma of the stomach), neurologic degeneration primarily in the posterior columns of the spinal cord, and anemia in which large red cells appear in the circulation and the bone marrow contains large megaloblasts instead of the normal precursor of erythrocytes, the erythroblast. The megaloblastic anemia is at least in part due to a concomitant deficiency of folate, owing to accumulation of 5-methyltetrahydrofolate (p. 634). The anemia, but not the neurological defects, is relieved in many cases by administration of folate. (It is for this reason that supplementation with folate is discouraged in this country; it may mask the presence of pernicious anemia or other causes of cobalamin deficiency until irreversible neural defects have appeared.) Methionine also diminishes the folate deficiency by inhibiting the reduction of methylene H_4folate to methyl H_4folate.

Cobalamin deficiency is a possible adjunct of any chronic disturbance of gastrointestinal function. Among normal people, those on vegetarian diets are at risk, especially if they are purists who shun even milk as an animal tissue. The many people in the world who subsist on a mainly cereal diet frequently have multiple dietary deficiencies rather than an isolated cobalamin deficiency. Indians migrating to England are said to be relatively free of deficiency symptoms when they arrive, but develop the condition later on the same diet. Perhaps dirtier food has some benefits. How many of those who are vegetarians in developed countries by choice rather than necessity escape deficiency without supplementation of their diet is not clear. One suspects backsliding from strict application of doctrine or insufficient elapsed time since leaving a carnivorous life, but a closer examination of such possible sources of cobalamins as the bacteria in the mouth or the flora in some pickles, and the like, is needed. However it is that most vegetarians escape, some are not so lucky, including infants suckled by marginally deficient mothers. Scandinavians have sometimes become deficient owing to infestations of fish tapeworms that compete successfully for the incoming supply.

Folate (p. 620)

The minimum daily requirement for folates is less than 0.1 micromoles, as determined in a small sample of American adults. Assays of American diets showed a total content of 0.9 to 2.5 micromoles in the daily intake. The availability for absorption varies. As little as 0 per cent or as much as 96 per cent may be absorbed. In usual mixed diets, roughly half is absorbed.

Folate deficiency, rather than cobalamin deficiency, is the usual cause of megaloblastic anemia. It is common in those areas of the world in which the diet is limited and mainly composed of cereals, and it also occurs in both the chips-and-cola set and the vegetarians in more affluent societies.

Megaloblastic change in the bone marrow is common among pregnant women. The problem here is similar to that of iron deficiency: It is difficult to draw the line between natural changes without deleterious consequences and sufficient deficiency to cause functional impairment. There is no doubt that frank impairment does occur in the United States and may result in death of the mother. Between 2.5 and 5.0 per cent of the pregnant women in developed countries develop megaloblastic anemia.

The plasma concentration of pteroylglutamates is a poor guide to development of a deficiency; it may fall abruptly without any functional changes. The erythrocyte concentration appears to be a better index, with anything lower than 100 μg/l (0.23 μm) of cells indicating deficiency, and a level lower than 140 μg/l (0.32 μM) indicating a high risk.

A deficiency may develop from too little food, too much dependence on refined cereals, or from prolonged cooking, which destroys the compound. (Dependence upon baked beans, stews, and the like as main dishes is risky.)

Nicotinamide (p. 290)

Nicotinamide, the active component in NAD and NADP, can be formed from tryptophan in the liver (Fig. 41–3). The reactions by which this is done diverge from the main route of catabolism at 2-amino-3-carboxymuconic semialdehyde (p. 614). In the main route, this compound is decarboxylated. However, it also spontaneously cyclizes to form quinolinate, which condenses with 5-phosphoribosyl pyrophosphate and loses CO_2 to form nicotinate mononucleotide. This nucleotide is converted to NAD by acquisition of an adenylyl group from ATP and an amide nitrogen from glutamine.

Dietary nicotinate and nicotinamide are also used as precursors of NAD and NADP through direct reaction with phosphoribosyl pyrophosphate. It is obvious that the diet need not include these compounds if it contains a supply of tryptophan sufficient to meet the needs for protein synthesis and for nicotinate formation, and if the rate of conversion of tryptophan to nicotinate matches the rate of loss of the nicotinate moiety. Man is an example of an animal that has no dietary requirement for nicotinate in the presence of an adequate supply of tryptophan. Cats, on the other hand, have a high level of the enzymes in the main route of tryptophan catabolism and are able to form little nicotinate.

In human adults, the molar conversion of extra tryptophan to NAD is roughly 3 per cent of the total degraded, so that 60 milligrams of tryptophan yield 1 milligram of nicotinate as NAD. The daily turnover of NAD is equivalent to about 1 mg of nicotinate per megajoule equivalent of dietary intake, so an adult with a 10 megajoule daily requirement would require roughly 600 milligrams of tryptophan to

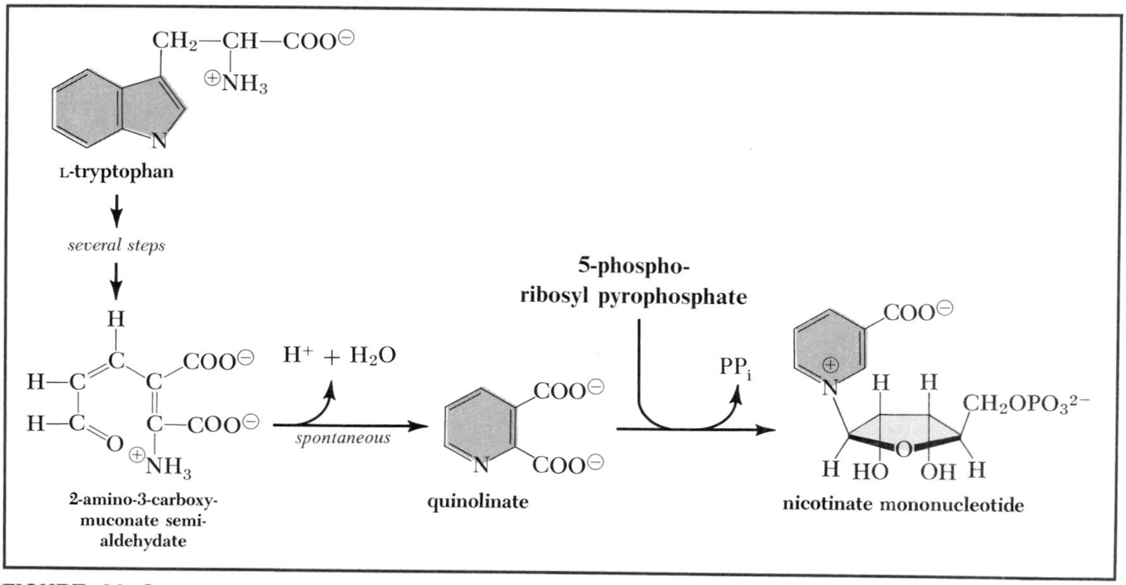

FIGURE 41–3 The nicotinate moiety of NAD and NADP is formed from tryptophan in animals.

supply his nicotinate requirement if that was the only source. This estimate is probably high, because there is some reason to believe that the efficiency of nicotinate formation becomes greater when a deficiency threatens. This amount of tryptophan would be supplied by approximately 250 grams of beefsteak, 450 grams of whole oats, or 1330 grams of whole maize.

These figures tell the tale. People subsisting mainly on maize (corn in the United States) are subject to tryptophan deficiency. Corn contains somewhat over 10 milligrams of pre-formed nicotinate per kilogram, but for some reason much of this is not available. However, treatment of the corn with mild alkali not only releases the nicotinate in a useable form, but it also may improve the amino acid composition of the product. (Corn has an excess of leucine compared to isoleucine, which further impairs tryptophan utilization. The alkali treatment appears to remove part of the leucine.) In any event, most of the Amerindians who depended upon maize as an important foodstuff had developed procedures for its preparation involving exposure to lime. Farmers in the southern United States, who also became dependent upon maize at times of deprivation, either ate it without alkali treatment or steeped it in sodium hydroxide with discard of the liquor. As a result, pellagra was widespread. This condition is characterized by the three D's: diarrhea, dermatitis, and dementia. However, the poor diet leading to pellagra is lacking in components other than nicotinate and tryptophan so that the condition is usually caused by a number of concurrent deficiencies.

The malignancy known as carcinoid can also provoke the appearance of a pellagric dermatitis because the tumor cells have a high capacity for converting tryptophan to 5-hydroxytryptamine (serotonin). The major end product, 5-hydroxyindolacetate, may be excreted in amounts up to 1.5 millimoles per day compared to the normal maximum of 0.05 millimole.

Large doses of nicotinate are toxic. The compound causes vasodilation by some unknown mechanism and blocks the mobilization of fatty acids from triglycerides. (There is a concomitant fall in cholesterol production, leading to tests of the compounds as a therapeutic agent in atherosclerosis. Nicotinamide does not have this effect. Increased doses of nicotinamide do cause increased excretion of N-methylnicotinamide, a normal product of NAD turnover, which will cause methyl group deficiencies in the absence of excess methionine and choline supplies. (This probably accounts for the liver damage seen in a schizophrenic who was ingesting 9 grams per day.)

Pantothenate (p. 423)

Deficiencies of this widespread constituent of coenzyme A and fatty acid synthase are not known in humans except through experimental administration of a synthetic antagonist.

Pyridoxal (p. 284)

Natural deficiencies of pyridoxal are also unknown in adults owing to the widespread occurrence of it and its precursors, pyridoxine and pyridoxamine. Isolated cases occurred in infants fed deficient formulas, which led to convulsions. Induced deficiencies are created in adults either by administration of an antagonist, or through the use of drugs reacting with aldehydes, such as isonicotinic acid hydrazide used in tuberculosis or hydralazine used for hypertension. Very high doses of pyridoxine are toxic and also cause convulsions.

Riboflavin (p. 398)

Isolated deficiencies of riboflavin are rare, although deficiencies do occur in conjunction with other deficiencies in impoverished areas or in patients with bizarre eating habits. The symptoms are nondescript, including inflamed tongue, lesions at the corners of the lips, dermatitis, and anemia. In general, suspicion of riboflavin deficiency ought to be equated with suspicion of general malnutrition and treated accordingly. The estimated requirement for adults is approximately 0.2 micromole per megajoule energy equivalent. It is lost only slowly from the body owing to the tight binding of the flavin coenzymes to their apoenzymes.

Thiamine (p. 429)

We discussed thiamine deficiency in connection with glucose catabolism. The deficiency is still common in parts of the world, especially those in which polished rice is the major food. It is seen in the developed countries mainly in association with alcoholism or a chronic disease of the gastrointestinal tract or a raw fish diet. It can occur rapidly upon parenteral feeding without supplemental thiamine. Normal individuals have been experimentally depleted within 12 to 14 days. An impending deficiency can be detected by assay of transketolase in red blood cells. A 25 per cent increase in the measured activity upon supplementation with thiamine pyrophosphate indicates a deficiency. The total content of an adult body is approximately 80 micromoles, and it is not increased by ingestion of large doses. (As Davidson and colleagues put it, the body is an efficient machine for dissolving thiamine pills and transferring them to the urinal.)

LIPID-SOLUBLE VITAMINS

The hydrophobic vitamins as a class will not be absorbed unless fat digestion and absorption are proceeding normally, and deficiencies ought to be sought in anyone with malabsorptive disease. Chronic use of mineral oil as a laxative, formerly more common than it is today, may also provoke deficiencies.

Retinol, Vitamin A, and Vision

The image that we "see" in our brain is constructed by the organization of impulses that are generated by the absorption of quanta of light in the rods and cones of the retina. In order to translate incoming light into impulses, the eye must contain compounds that will absorb quanta of particular wavelengths, in this case 400 to 700 nanometers. This is a useful, indeed a necessary, range of wavelengths for good vision—useful not only because the peak energy of sunlight at the earth's surface lies in the middle of the range, but also because it is a range enabling great distinction of objects. Light of much shorter or longer wavelength is absorbed by a variety of chemical structures, and we distinguish objects because of selective absorption by their less common constituents. In the biological realm, peptides, carbohydrates, and lipids are colorless. In the surroundings, water and the common minerals, except those containing iron, are colorless. We distinguish rubies and sapphires not by the colorless aluminum oxide crystal lattice, but by contaminants of the lattice, and we detect the jaundiced person through the selective absorption of quanta by the relatively small concentrations of bilirubin.

It is apparent from this that the light-sensitive compound in the eye must also have a distinctive structure; it must exist in a number of electronic states of relatively small energy difference so that quanta of light in the visible range can cause excitation, and at the same time we would expect this important compound to differ sufficiently in structure from other components of the retinal cells so that it is not easily removed by the usual metabolic reactions.

We have noted that animals frequently do not synthesize unusual structures when compounds containing the structure are available from the diet, and this is the case with the visual pigment. The necessary highly resonant structure is an isoprenoid hydrocarbon chain containing conjugated double bonds that is contributed by compounds in the diet having vitamin A activity. Let us consider the nature of these compounds and then go on to consider how they contribute to vision.

The Nature of Vitamin A. Compounds having vitamin A activity in the diet of mammals are precursors of a 20-carbon polyprenol, **retinol** (Fig. 41–4). Some of the precursors are fatty acid esters of the *all-trans* isomers of retinol, the most abundant being **retinyl palmitate.** These esters are hydrolyzed in the small bowel to form free retinol, which passes into the intestinal mucosa.

The other dietary components with vitamin A activity are **carotenes,** the original polyprenyl precursors of retinol. The carotenes are synthesized by plants as light-harvesting pigments for photosynthesis. (The synthesis resembles that of squalene in that two polyprenyl pyrophosphates are condensed head-to-head. However, they are 20-carbon compounds rather than the 15-carbon precursors of squalene.) The most abundant carotene, β-carotene, is a symmetrical compound that is oxidized in the intestinal mucosa to form two molecules of retinal, the aldehyde analogue of retinol. Retinal and retinol are equilibrated by an NAD-coupled alcohol dehydrogenase present in most tissues. Some other carotenes have the retinal configuration in only one half of the molecule, while others lack it completely.

Part of the carotenes in the diet is absorbed as such and dissolved in the lipid phase of lipoproteins. The normal plasma concentration is between 2 and 4 micromolar, but it becomes so high in some leporine people from their proclivity for eating carrots and the like in wholesale quantities that they turn yellow. This carotenemia is harmless, as is the lycopenemia rarely seen in some tomato enthusiasts. The latter become a bright red-orange from absorbed lycopene, a polyprenyl compound not useable by humans. In any case, the absorbed carotene can later be oxidized as a precursor of retinal.

The retinol generated in the intestinal mucosa from either retinyl esters or from carotenes is converted to the palmitoyl ester and passes into the circulation by solution in the triglyceride phase of chylomicrons. The liver removes it for storage in Kupffer cells, and there may be as much as 2 micromoles of retinyl esters per gram of tissue (0.25 micromole is more common). The ester is constantly broken down and rebuilt by the liver, and the parenchymal cells export free retinol in combination with a specific transport protein, which in turn combines with a protein that also transports thyroxine. There are only 40 mg of the retinol-binding protein per liter of blood plasma, and the normal total plasma retinol concentration is between 1 and 2.5 micromolar.

FIGURE 41–4 **(*right*) Dietary sources of retinol include β-carotene (*top*), which is oxidized in the intestinal mucosa to retinal. Retinal and retinol are equilibrated by an alcohol dehydrogenase. The retinyl esters in the diet are hydrolyzed in the intestinal lumen, forming retinol, which is absorbed. Retinol is converted to a palmitate ester for transport and storage.**

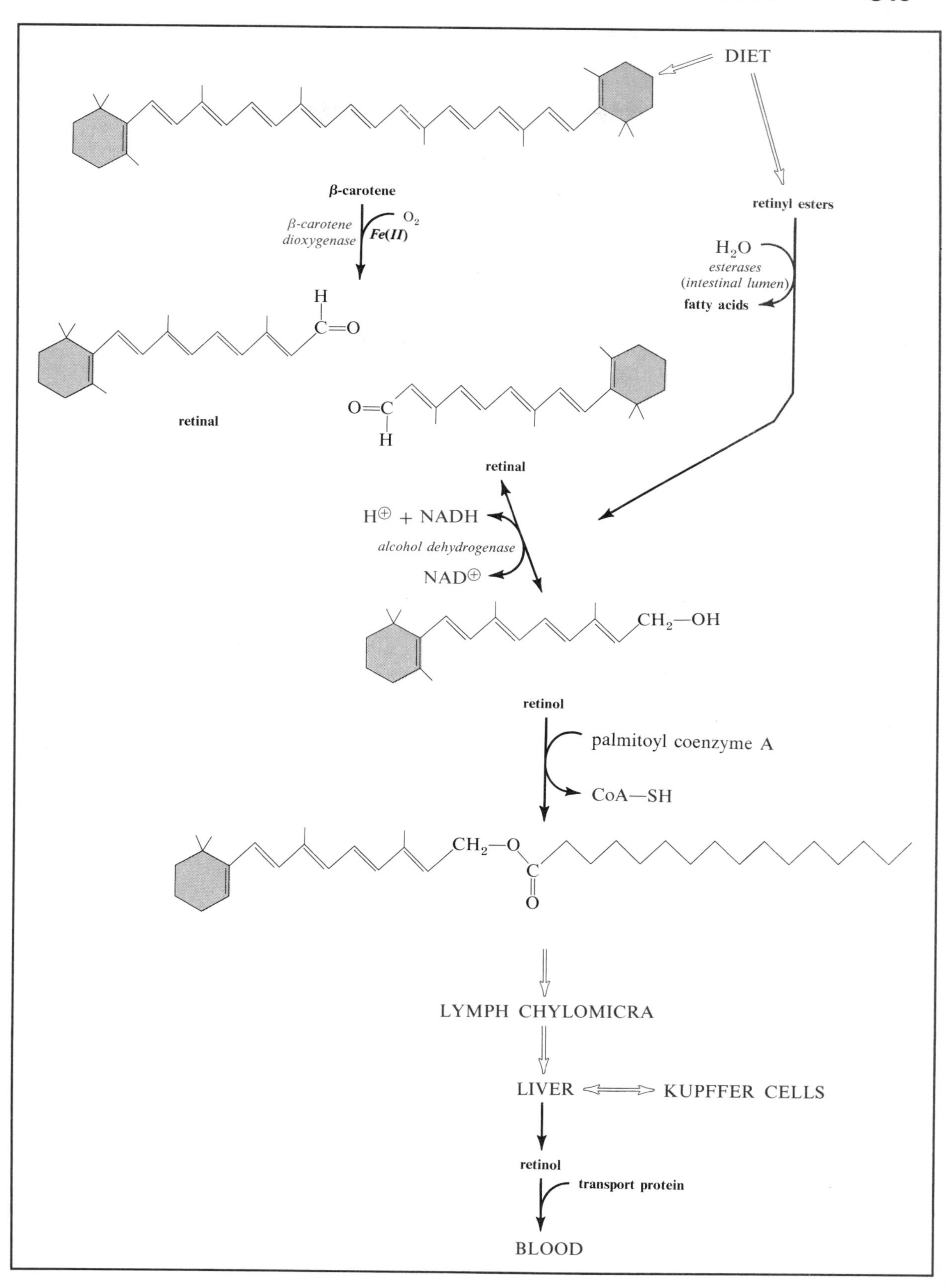
DIET
β-carotene
β-carotene dioxygenase
O_2
Fe(II)
retinyl esters
H_2O
esterases
(intestinal lumen)
fatty acids
retinal
retinal
$H^{\oplus}$ + NADH
alcohol dehydrogenase
$NAD^{\oplus}$
retinol
palmitoyl coenzyme A
CoA—SH
LYMPH CHYLOMICRA
LIVER
KUPFFER CELLS
retinol
transport protein
BLOOD

The Visual Pigment. The transport of retinol into the retina appears to involve the removal of the compound from its plasma protein carrier and the formation of a fatty acid ester within the retinal cells. The mechanism of concentration is quite effective, because the level of retinol must drop to half its normal range before any loss of the store in the eye becomes apparent. The retinyl esters of the eye are subject to hydrolysis, and the retinol liberated within the cells is oxidized by NAD in a reaction catalyzed by a specific dehydrogenase.

The equilibrium position of retinol dehydrogenase does not favor retinal production. However, the retinal that is formed is rapidly removed to form complexes with proteins, the *opsins,* of the rods and cones, and oxidation of retinol continues until no more of these complexes, which are the light-sensitive pigments, can be formed—until the opsins are saturated. The combination of retinal and opsin may involve Schiff's base formation with a lysine residue. The rod pigment, which has been studied most, is named **rhodopsin.**

One additional step is necessary before the visual pigment is made. The opsins are constructed so that they combine with the 11-*cis* isomer of retinal. There may be an isomerase in the retina catalyzing the equilibration of the *all-trans* isomer with this form, or the equilibration may be spontaneous. Although the position of equilibrium also favors the original *trans* isomer, the strong affinity of the 11-*cis* form for the opsins pulls the whole sequence toward pigment formation (Fig. 41–5).

The mechanism by which light absorption stimulates nerve transmission is another membrane-related phenomenon involving changing affinity for protons. Rho-

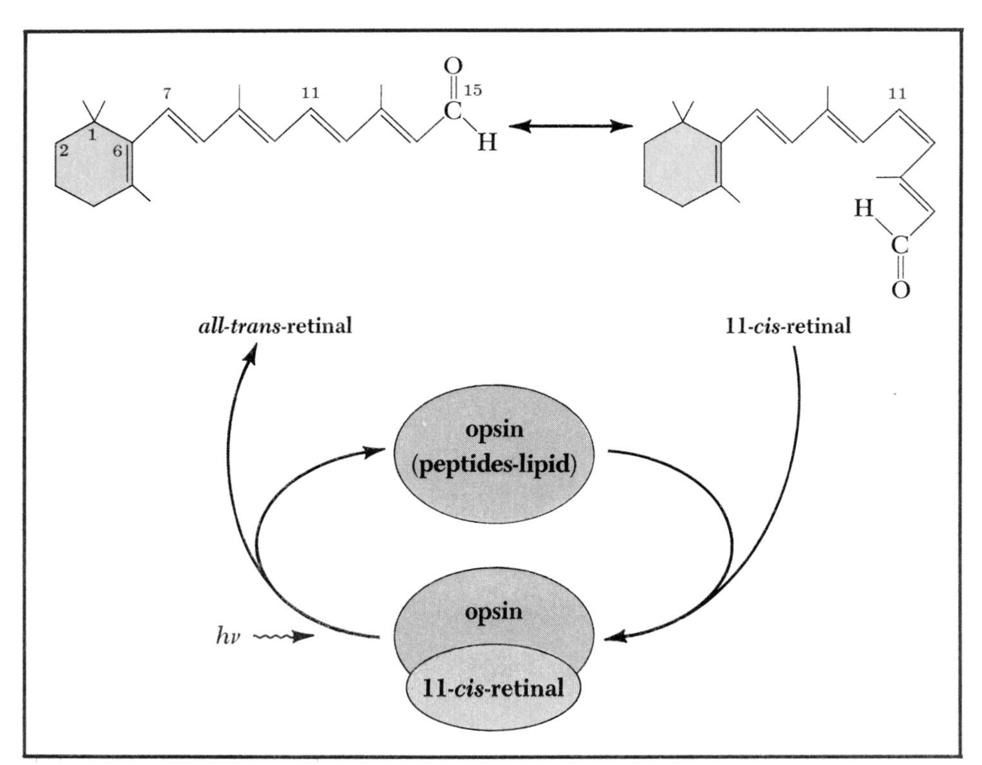

FIGURE 41–5 The visual pigments are complexes of 11-*cis*-retinal with opsins, which are lipophilic polypeptides in retina membranes. Excitation of the pigments causes a conformational shift that results in release of retinal in an *all-trans* configuration. Isomerization of 11-*cis* and *all-trans* retinal may be spontaneous or catalyzed by an enzyme.

dopsin occurs in the membranes of disc-like structures, which are stacked as cylinders at the interior end of rods. (The stacked array increases the likelihood that a photon will be absorbed as it moves the length of the rod.) The absorption of light triggers a series of conformational changes in the rhodopsin that cause proton uptake, owing to decreases in the ionization constants of some constituent groups. The process is driven by a simultaneous change in configuration of the attached retinal to the *all-trans* form, which has only a weak affinity for opsin.

The decreased negative charge on the opsin is believed by some to facilitate in some way a movement of calcium into the cytoplasmic space, which in turn decreases the conductance of sodium across the plasma membrane of the rods and generates an action potential. The calcium ion also promotes a slower phosphorylation of rhodopsin, as it regenerates from *cis*-retinal and opsin, making it more negative and causing proton release, thereby ending the stimulation. Later removal of the phosphate again sensitizes the system to a light-induced conformational change. Others believe the intermediate transfer of the signal involves changes in the formation of cyclic GMP, including light-activation of a phosphodiesterase.

Other Functions of Vitamin A. Animals deprived of vitamin A not only go blind; they die. Before they die, they develop abnormal deposition of keratin in the mucous membranes, a failure of bone remodelling leading to thick, solid long bones like the bones in the skull, lesions of the nerves, an increased pressure of the cerebrospinal fluid causing hydrocephalus, testicular degeneration in males, and abortion or malformed offspring in females. Obviously, vitamin A is involved in some critical function in most tissues. We don't know what it is, but we do know that it probably doesn't involve the retinol-retinal interconversion, because the corresponding acid, *retinoate,* will prevent malfunction in all of the tissues except the reproductive tract and the eye. Since it won't save vision, it is apparent that retinoate is not readily reduced to retinal. We may be dealing here with a considerably more fundamental and primitive function of the carotenes, with their use in the development of vision being a fortuitous result of the accompanying absorption spectrum.

Retinol itself has been shown to participate as a phosphate ester in the transfer of glycosyl groups, much like dolichol (p. 703), but this is not believed to be a major function of the compound.

Supply of Vitamin A. It is difficult for two reasons to make anyone with normal gastrointestinal function, except the very young, deficient in vitamin A. One is the common occurrence of the vitamin or its carotene precursors in plants, in most fish tissues, in eggs, and in mammalian and chicken livers. Most weaned humans eat one or more of these food classes. The other reason is the large capacity for storing the retinyl esters in the liver. The estimated requirement for humans past the weaning age is on the order of 6 micrograms per kilogram body weight per day, perhaps 1.5 micromoles per day for an adult. At this rate, the usual adult has a nine-month supply in his liver, and some may have a four-year supply.

Potential vitamin A deficiency is something to consider in any chronic disease of the pancreas, intestine, or liver. A failure of rod vision (night-blindness) is an early symptom, but it is also a result of other conditions, and vitamin A deficiency must be established by careful measurement, not by subjective report. The really characteristic sign is *xerophthalmia,* a dry keratinization of the cornea and conjunctiva; this is a late symptom and an indication for prompt therapy.

Retinol, like cholecalciferol, is toxic in excess. Acute poisoning has long been known from the fatal result of eating polar bear livers, which contain as much as 30 micromoles of retinyl esters per gram—a five-year supply for a human in each 100 grams. Severe headache, vomiting, and prostration result within a few hours. Doses of 300 micromoles in infants produced a transient hydrocephalus.

The more usual cases are those in which the intake has been relatively high for a long period of time, and chronic toxicity is frequently manifested by fatigue, loss of appetite, an enlarged liver, diffuse pains in the muscles, coarsening and loss of hair, scaly skin eruptions, and attenuation of the long bones that sometimes results in fractures. Diagnoses of pyschoneurosis or even schizophrenia seem to be made quite frequently. Most persons with severe cases have been ingesting on the order of 500 micromoles of retinol per day for a few years—sometimes from overzealous treatment for other conditions, and sometimes from self-dosage with easily available commercial preparations (or from maternal overdosage, in the case of infants). The minimum quantity necessary for the appearance of chronic toxicity has been suggested as 50 micromoles per day for 18 months (about 15 times the current recommended daily allowance).

Therapeutic Use of Retinoid Compounds. It was discovered that retinoate cured severe cases of acne. Compounds with fewer side effects were then developed, including **13-*cis*-retinoate (isotretinoin)** and an aromatic analog **(etretinate),** which are effective in a variety of skin diseases, including acne, some forms of psoriasis, and many others. The compounds have been shown to inhibit chemically induced malignant changes, and much effort is being expended to understand this effect.

Vitamin E—Tocopherols

Polyunsaturated fatty acids are subject to spontaneous attack by molecular oxygen through an autocatalytic mechanism involving free radicals (Fig. 41–6). The mechanism may be initiated through exposure to light or by complexes of the transition metal ions. The reason that the polyunsaturated compounds are vulnerable is a stabilization of the intermediate free radical through resonance. The tocopherols are compounds that have the ability to interrupt the free radical cycle by donating electrons, although the mechanism is not clear. They therefore supplement the action of glutathione peroxidase, which removes the peroxides formed by the cycle. (This explains why selenium and the tocopherols diminish, but do not eliminate, each other's dietary requirement.)

The tocopherols occur widely, and there is no evidence for a deficiency in adults. However, this may change, because the requirement for tocopherols depends upon the rate of oxidation of polyunsaturated fatty acids, which in turn depends upon the concentration of these acids and their exposure to initiators of free radical formation. Infants who are fed diets rich in polyunsaturated acids and supplemented with high levels of iron have been shown to develop frank tocopherol deficiencies, as manifested by hemolytic anemias. (The high peroxide levels make the erythrocytes and other membranes more fragile.) Similar circumstances are not hard to visualize in some adults.

Vitamin K (p. 349)

Dietary deficiencies of vitamin K have not been demonstrated in normal adults. The compound is widely distributed in plants and is also synthesized by intestinal bacteria. Although a major portion of the compound made by the bacteria is probably inaccessible for absorption, this source is probably imperative for people on a meat diet. In any event, experimental animals treated with antibiotics to suppress the intestinal flora become deficient. Newborn infants, especially the premature, are sometimes deficient. A deficiency is common in adults with common bile duct obstruction.

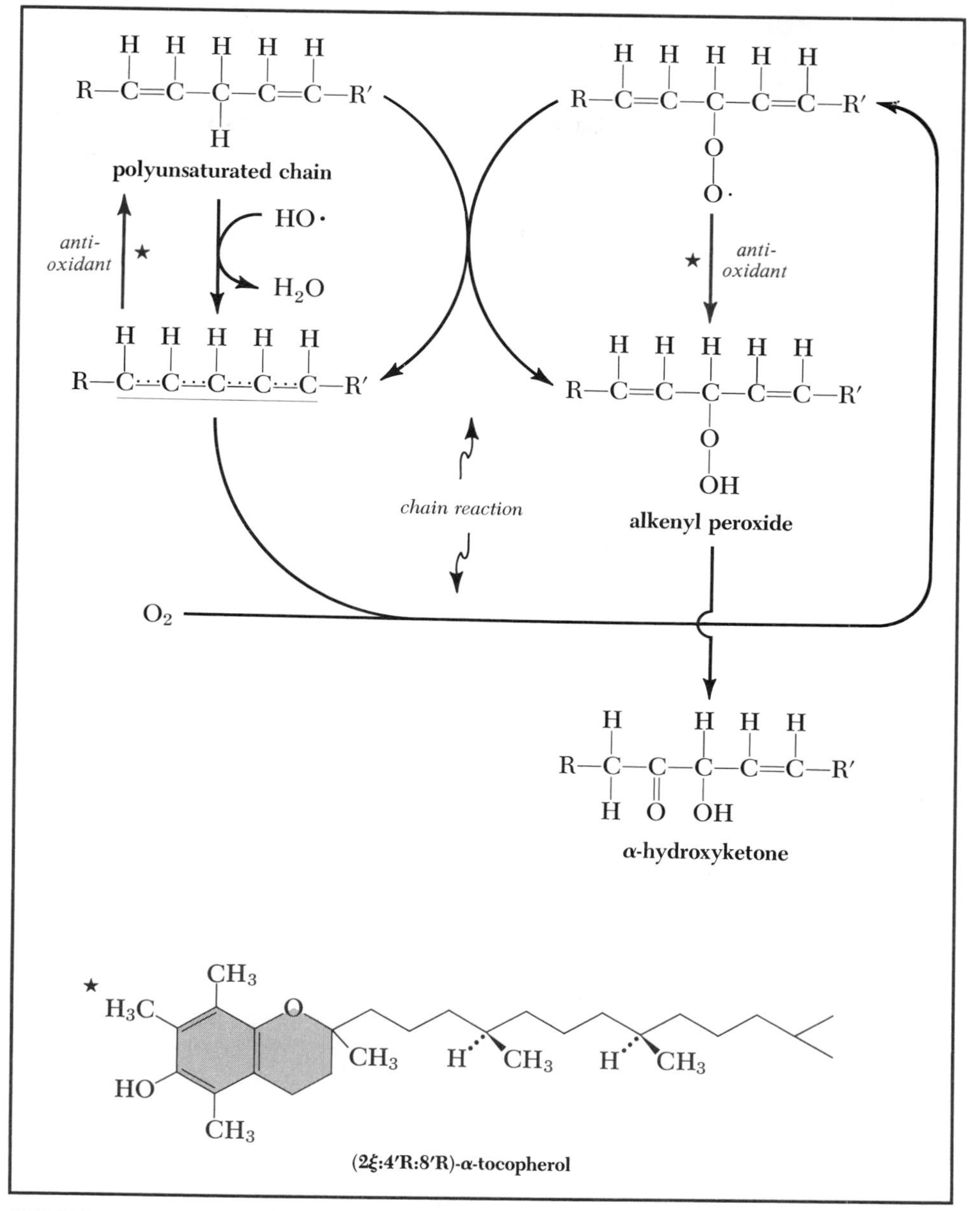

FIGURE 41–6 Polyunsaturated fatty acids are subject to oxidation by cyclical free-radical mechanisms, one of which is illustrated. The cyclical process can be interrupted by antioxidants that act as scavengers of the free radicals. The tocopherols (*bottom*) serve this function. Repeated oxidation at the 3-carbon spacing of polyunsaturated fatty acids releases malondialdehyde (not shown).

The natural compounds have long polyprenyl side chains and a correspondingly limited solubility in water. They are relatively nontoxic in excess, whereas the more water-soluble synthetic analogues are oxidized to reactive intermediates that cause hemolysis. Since these compounds are not available without prescription and are usually administered for brief intervals, this has not been a practical problem.

FURTHER READING

Also see the general references cited for Chapter 39 (p. 816).

Bone Minerals and Electrolytes

F. Bronner and M. Peterlik, eds. (1981). Calcium and Phosphate Transport Across Biomembranes. Academic Press.
L.H. Allen (1982). Calcium Bioavailability and Absorption: A review. Am. J. Clin. Nutr. 35:783.
B.E.C. Nordin et al. (1979). Calcium Requirement and Calcium Therapy. Clin. Orthop. Rel. Res. 140:216.
R.K. Rude and R.F. Singer (1981). Magnesium Deficiency and Excess. Annu. Rev. Med. 32:245.
Anon. (1979). Clinical Signs of Magnesium Deficiency. Nutr. Rev. 37:6.
S. Roy and B.S. Arant, Jr. (1979). Alkalosis from Chloride-Deficient Neo-Mull-Soy. N. Engl. J. Med. 301:615.
W.B. Schwartz, C.v. Ypersele, and J.P. Kassirer (1968). Role of Anions in Metabolic Alkalosis and Potassium Deficiency. N. Engl. J. Med. 279:630.

Iron

A. Jacobs and M. Worwood, eds. (1980). Iron in Biochemistry and Medicine, II. Academic Press.
C.A. Finch, ed. (1982). Clinical Aspects of Iron Deficiency and Excess. Sem. Hematol. 215:691.
L. Hallberg (1981). Bioavailability of Dietary Iron in Man. Annu. Rev. Nutr. 1:123.
P.C. Elwood (1977). The Enrichment Debate. Nutr. Today 12(4):18.
J.W. Eaton et al. (1982). Haptoglobin: A Natural Bacteriostat. Science 215:691.
R.B. Schifman et al. (1982) RBC Zinc Protoporphyrin to Screen Blood Donors for Iron Deficiency Anemia. JAMA 248:2012.

Zinc and Copper

M.H.N. Golden and B.E. Golden (1981). Trace Elements. Potential Importance in Human Nutrition with Particular Reference to Zinc and Vanadium. Br. Med. Bull. 37:31.
H.H. Sandstead (1982). Copper Bioavailability and Requirements. Am. J. Clin. Nutr. 35:809.
Z.A. Karcioglu and R.M. Sarper, eds. (1980). Zinc and Copper in Medicine. Thomas.
N.W. Solomons (1982). Biological Availability of Zinc in Humans. Am. J. Clin. Nutr. 35:1048.

Other Minerals

M. Mertz (1981). The Essential Trace Elements. Science 213:1332.
V.R. Young, A. Nahapetian, and M. Janghorbani (1982). Selenium Bioavailability with Reference to Human Nutrition. Am. J. Clin. Nutr. 35:1076.
V.R. Young (1981). Selenium: A Case for Its Essentiality in Man. N. Engl. J. Med. 304:1228.
K.V. Rajagopalan, J.L. Johnson, and B.E. Hainline (1982). The Pterin of the Molybdenum Cofactor. Fed. Proc. 41:2608.
H. Freund, S. Atamian, and J.E. Fischer (1979). Chromium Deficiency During Total Parenteral Nutrition. J. Am. Med. Assoc. 241:496.
R.L. Schilsky, A. Barlock, and R. F. Ozols (1982). Persistent Hypomagnesia Following Cisplatin Chemotherapy for Testicular Cancer. Cancer Treat. Rep. 66:1767.

Vitamins

National Research Council (1980). Recommended Dietary Allowances. Natl. Acad. Sci. (U.S.). Includes discussions of each vitamin.
C.G. King and J.J. Burns, eds. (1975). Second Conference on Vitamin C. Ann. N.Y. Acad. Sci. vol. 258. Many useful reviews.
R.E. Hodges et al. (1971). Clinical Manifestations of Ascorbic Acid Deficiency in Man. Am. J. Clin. Nutr. 24:432,444. Study of Iowa convicts.
C.J. Gubler, M. Fujiwara, and P.M. Dreyfus (1976). Thiamine. Wiley.
L.J. Machlin, ed. (1980). Vitamin E. A Comprehensive Treatise. M. Dekker.
C.E. Orfanos et al., eds. (1981). Retinoids. Advances in Basic Research and Therapy. Springer-Verlag.
D.S. Goodman, ed. (1979). Vitamin A and Retinoids: Recent Advances. Fed. Proc. 38:2501.

CHAPTER 42

THE COMPLETE DIET

Given an understanding of the requirements for the individual components of the diet, the formidable problem remains of translating this information into usable form. Even the most dedicated professional cannot calculate the contribution of every morsel a person might consume. Both the weight and the composition of foods often must be estimated. Every crop of wheat and every patty of hamburger is not identical, and even if they were, people forget or conceal what they do.

However, the person who desires to do so can achieve satisfactory control over his diet by the use of relatively simple guides. These prepared lists have food portions broken into categories such as meat, bread, or fat in a way that permits exchange of one food for another within the category. The Mayo Clinic Diet Manual* is a good example of a collection of exchange lists for various purposes that permits considerable latitude in satisfying individual tastes while avoiding detailed calculations that are beyond the reach of most people.

However, such lists can only be prepared by calculation, and detailed calculation is sometimes necessary for other purposes. The basic information is provided by tables of food composition, especially the Department of Agriculture Handbooks numbers 8 and 456 or the World Health Organization tables, supplemented by information provided by some manufacturers.

Recommended Dietary Allowances (Table 42–1)

Individuals vary in their requirements for each nutrient. If all consumed only the average minimum daily requirement for prevention of functional impairment, signs of deficiency might be expected in a large part of the population. When data permit, the requirement is usually estimated as the mean plus two standard deviations, a value that would be expected to avoid deficiency in 97.5 per cent of the population. Even this number of deficient individuals would be regarded as intolerable in a developed country. The Food and Nutrition Board of the National Academy

*5th edition, Saunders, 1981. Not the faddist and misnamed Mayo Clinic Diet.

TABLE 42–1 RECOMMENDED DIETARY ALLOWANCES

	Age	Weight *kg*	Height *cm*	Energy Yield *MJ*	Protein N *mmoles*	Retinol Equiv. *μmoles*	Cal-ciferol Equiv. *nmoles*	Toco-pherol Equiv. *μmoles*	Ascor-bate *μmoles*
Infants	0–6 mo	6	60	0.48/kg	25/kg	1.5	25	7	200
	6–12 mo	9	71	0.44/kg	23/kg	1.4	25	9	200
Children	1–3 yr	13	90	5.5	260	1.4	25	12	260
	4–6	20	112	7.1	340	1.8	25	14	260
	7–10	28	132	10.1	390	2.4	25	16	260
Men	11–14	45	157	11.3	510	3.5	25	19	285
	15–18	66	176	11.8	640	3.5	25	23	340
	19–22	70	177	12.2	640	3.5	19	23	340
	23–50	70	178	11.3	640	3.5	12	23	340
	51–75	70	178	10.1	640	3.5	12	23	340
	76+	70	178	10.1	640	3.5	12	23	340
Women	11–14	46	157	9.2	530	2.8	25	19	285
	15–18	55	163	8.8	530	2.8	25	19	340
	19–22	55	163	8.8	500	2.8	25	19	340
	23–50	55	163	8.4	500	2.8		19	340
	51–75	55	163	7.6	500	2.8		19	340
	76+	55	163	6.7	500			19	340
	Pregnant			+1.2	+340	3.5	25	23	460
	Lactating			+2.1	+230	4.2	25	26	570

Notes: In converting the Board allowances to a molar basis, the increments between ages used by the Board have been translated to the closest molar increment. That is, an increase of 10 mg of a nutrient might be converted to the nearest 0.05 millimole increase. The Board's estimate for the protein allowance is on the basis of proteins with a 75 per cent efficiency of utilization. The fat-soluble vitamin allowances have been translated to equivalent moles of retinol, calciferol, and α-tocopherol. The Board regards β-carotenes to be $\frac{1}{6}$ as effective as retinol in providing vitamin A activity on a weight basis. Those carotenes in which only one half of the molecule yield retinol are $\frac{1}{12}$ as effective per unit weight. The nicotinate allowance is calculated on the basis of 1 micromole of nicotinate being equivalent to 36 micromoles of tryptophan. The folate allowance is determined on the basis of a microbiological assay of the foods, and less than one quarter of the stated amounts given as pure folate will be equally effective.

of Sciences (U.S.) has, therefore, used expert judgment to gauge the amount of each nutrient that would diminish the occurrence of deficiency symptoms to an acceptable level. These are the Recommended Dietary Allowances, which are reviewed periodically, with the last assessment made in 1980. (Drug and food manufacturers are required to use the Recommended Daily Allowances of the Food and Drug Administration in their literature and advertising; these are similar to, and no doubt heavily influenced by, the Recommended Dietary Allowances cited here.)

It is apparent that these are not minimum daily requirements. Many individuals consume less than the recommended amount without developing a deficiency. The tables are intended as a guide for developing standards for large groups and for minimizing the risk of deficiency to an individual. In general, the expert committee has been generous in its margin of safety. However, it ought to be noted that the estimates for required fuel intake are adjusted to a sedentary life style, and were made before the current emphasis on physical exercise was so widespread. We already know that the fuel intake always ought to be appraised individually.

It also should be noted that some individuals are still likely to become deficient in a nutrient even with conscientious consumption of the recommended daily allowance. There are many possibilities for impairment of utilization of a nutrient—during absorption, transport, metabolism, excretion—and mutations thus can easily

Folate *µmoles*	**Nicotinate Equiv.** *µmoles*	**Riboflavin** *µmoles*	**Thiamine** *µmoles*	**Pyridoxine** *µmoles*	**Cobalamin** *nmoles*	**Calcium** *mmoles*	**Phosphate** *mmoles*	**Iodine** *µmoles*	**Iron** *mmoles*	**Magnesium** *mmoles*	**Zinc** *mmoles*
0.1	50	1.1	1.0	1.8	0.4	9	7.7	0.3	0.18	2.0	0.05
0.1	65	1.6	1.7	3.6	1.1	13.5	11.6	0.4	0.27	3.0	0.08
0.2	75	2.1	2.3	5.3	1.5	20	26	0.6	0.27	6.0	0.15
0.5	90	2.7	3.0	8	1.9	20	26	0.7	0.18	8.5	0.15
0.7	130	3.7	4.0	9.5	2.2	20	26	1.0	0.18	10.5	0.15
0.9	145	4.3	4.7	10.5	2.2	30	39	1.2	0.32	14.5	0.25
0.9	145	4.5	4.7	12	2.2	30	39	1.2	0.32	17.0	0.25
0.9	155	4.5	5.0	13	2.2	20	26	1.2	0.18	14.5	0.25
0.9	145	4.3	4.7	13	2.2	20	26	1.2	0.18	14.5	0.25
0.9	130	3.7	4.0	13	2.2	20	26	0.2	0.18	14.5	0.25
0.9	130	3.7	4.0	13	2.2	20					
0.9	130	3.5	3.7	10.5	2.2	30	39	1.2	0.32	12.5	0.25
0.9	115	3.5	3.7	12	2.2	30	39	1.2	0.32	12.5	0.25
0.9	115	3.5	3.7	12	2.2	20	26	1.2	0.32	12.5	0.25
0.9	105	3.2	3.3	12	2.2	20	26	1.2	0.32	12.5	0.25
0.9	105	3.2	3.3	12	2.2	20	26	1.2	0.18	12.5	0.25
0.9	105	3.2	3.3	12	2.2	30		1.2			
1.8	+20	+0.8	+1.0	15	3.0	30	39	1.4	*	19.0	0.30
1.4	+35	+1.3	+1.0	15	3.0	30	39	1.6	0.32	19.0	0.40

The stated protein nitrogen allowance may be converted to grams of "protein" by multiplying by 0.0875. The fat-soluble vitamin allowances may be converted to I.U. by multiplying by the following conversion factors:

I.U. vitamin A = µmoles retinol × 950
I.U. vitamin D = nmoles calciferol × 16
I.U. vitamin E = µmoles tocopherol × 1.6

The major allowances for the other vitamins and the minerals can be converted to a weight basis by multiplying by the molecular weights: ascorbate = 176, folate = 441, riboflavin = 376, thiamine = 301, pyridoxine = 170, cobalamin = 1,355, calcium = 40, phosphate = 31 (as P), iodine = 127, iron = 55.6, magnesium = 24, zinc = 65.4.

* Supplementation of 0.55 to 1.1 mmoles of iron per day recommended for pregnant women.

magnify the amount of the nutrient required to maintain function. It is probable that there is no level of supplementation that would completely eliminate deficiencies from the population. Indeed, extraordinary increases in the dietary content of one component are likely to cause trouble for another group with different mutations—those who are especially sensitive to the toxic effects of the component.

The Recommended Dietary Allowances are frequently misused. As an example, here is a quotation from an article in *Science,* the official organ of the American Association for the Advancement of Science, commenting on data from the Health and Nutrition Survey, an estimable project of the Public Health Service: "The preliminary report details the results from the first two tests; dietary intake and biochemical findings from a sample of 10,126 people. The most striking finding, which confirms those of earlier, smaller surveys, is that the population suffers from a widespread iron deficiency. According to the report, about 95 per cent of all preschool children and women of childbearing age have iron intakes below the standards set by the Food and Nutrition Board of the National Academy of Sciences." This quotation is misleading in two important respects. It equates iron deficiency defined by less than a prescribed level of intake with a disease state ("suffer"), and it equates the recommended dietary allowances with a requirement, an identification the Food and Nutrition Board specifically warns ought not to be made.

When to Calculate

Full-scale calculation of the composition of a diet is justified in at least three circumstances other than the preparation of guides. One is in the design of solutions for intravenous feeding, discussed later. Another is the assessment of a diet that is highly restricted in composition or quantity. The third is for assessment of population food supplies. Anyone who anticipates offering advice on nutrition ought to make such a full-scale calculation using commonly available foods in order to gain an appreciation of the actual circumstances.

Judgment is required in applying the calculations. For example, a patient at the University of Virginia Hospital allegedly was consuming massive quantities of fruit and juices as the sole diet. By calculation, his intake of essential amino acids and total nitrogen was ample, despite their low concentrations in fruits. However, he had clear indications of protein deficiency. Calculation did not reveal his actual consumption, or the unknown effect of that bizarre diet on the digestion and absorption of the dilute protein supply.

ADDITIONAL CONSIDERATIONS

It is a paradox that we who live in a time of unprecedented plenty use so much of our newfound freedom to worry over the nature of our diet. We have the luxury of attempting to fine-tune the national larder. This section addresses some of the factors that have attracted attention, but it ought to be recognized that many proposals are based upon alleged phenomena that are only recognized upon statistical examination of large samples of the population. The validity of the examination and the conclusions drawn are frequently subject to vehement challenge. There are philosophical questions, as well as biochemical questions, involved. When a person dies at age 70, the bereaved cannot say that he would have died at age 65 if he had not lowered his salt intake, or that he would have lived to age 75 if he had eaten less candy. For most people in developed countries, dietary risks are low probability risks according to the information we now have. The quality of life can be destroyed for many people by overemphasis on their diet, as well as by underemphasis. This is especially true for the obese. To take an extreme example, the zealous resident who places an elderly obese patient with limited life expectancy on a severely restricted diet is punishing a sinner, not treating a human. Milder examples are commonplace; flourishes and alarums about some speculative dietary danger, perhaps affecting one in 10,000 people at the most, abound to the point that one suspects a mass neurosis.

This is not to disparage a constant examination of the components of our foods and inquiry into their effects. It is only to suggest mildly that the attention of the well-fed American middle- and upper-class might be diverted to more demonstrably valid problems, and all efforts to change the diet of the entire population ought to be examined with great suspicion. On the other hand, most practicing physicians might well learn how to extract a detailed dietary history quickly and how to make a rough but quick appraisal of the quality of the diet.

Sedentary Life and Nutrient Volume

For some nutrients, such as thiamine or riboflavin, concentration in a diet appears to be more important than the total amount ingested per day. With other nutrients, absorption of a more fixed quantity suffices, regardless of the quantity of diet in which it is contained. The more one eats, the less chance there is of a marginal deficiency of those components. It follows that the more sedentary a person is, the greater is his risk of an at least marginal deficiency of some nutrient if he is to remain trim. Increased exercise, in addition to other virtues touted these days, permits one to avoid the Scylla of deficiencies without being sucked under by the Charybdis of obesity.

Dietary Fiber

Oligo- and polysaccharides that are relatively resistant to digestion are referred to as fiber, even though some are not fibrous in form. Most are plant cell wall materials or storage polysaccharides. They include:

cellulose—a β-1,4-glucose polymer
pectins—polymers of galacturonate + some other sugars
hemicelluloses—xylans, arabinogalactans, glucomannans, etc.
storage polysaccharides—levans, gums, mucilages.

In a typical whole wheat flour, only one quarter of the fiber is cellulose and lignin, the typical components of wood.

Not all fiber as measured in the laboratory and cited in food composition tables escapes digestion. The bacterial flora of the gut sometimes adapt so as to break down over three fourths of all of the major classes of dietary polysaccharides. The products include short-chain fatty acids that can be used by the host, along with unusable CO_2, methane, and hydrogen.

Emphasis on dietary fiber has a long history. Graham crackers are named for an early advocate (1839), and the American breakfast table was changed by Kellogg of corn flakes fame. The current interest in dietary fiber stems from comparison of the incidence of various diseases in populations on diets containing high and low amounts of fiber. Fiber promotes more rapid transit of the bowel contents, which may diminish the incidence of carcinoma and other bowel diseases that are important causes of death. High fiber diets also increase the fecal fat loss by roughly one gram per day and the fecal nitrogen loss by more than 30 millimoles per day. They also cause, at least temporarily, negative balances of calcium, zinc, iron, and cobalamins in many. The relative nutritional advantages and disadvantages are difficult to assess for a particular individual, but the young practitioner must realize the importance his older patients will place on avoiding constipation.

Polyunsaturated Fatty Acids

The earliest recognition of the essential nature of some fatty acids is attributed to St. Jerome (340 to 420 A.D.), who ate 180 grams per day of barley bread and vegetables between his 31st and 35th years, causing him to develop dim eyesight, scabby eruptions, and the mange. (So legend has it.) His ailments disappeared upon adding oil to his diet, and he continued his regimen until age 64, when he dropped the barley bread and ate only the vegetables and oil until expiring at age 80.

In more modern times, infants on parenteral nutrition without added lipid developed scaly skin and sparse hair and failed to thrive, with their levels of both 18:2 and 20:4 fatty acids becoming low. The condition was corrected by adding lipid to the feedings, providing safflower oil rubs, and feeding breast milk.

A guide to the state of fatty acid nutrition is the ratio of 20:3(5,8,11) fatty acids derived from oleate (p. 205), and 20:4(5,8,11,14) fatty acids derived from linoleate in the blood phospholipids. Dietary linoleate (18:2) and to lesser extents linolenate (18:3) and arachidonate (20:4) are used as precursors of prostaglandins, thromboxanes, and prostacyclins, as well as being incorporated into structural lipids. Estimates of the minimum requirement vary from a low of one gram per day to a high of 5 per cent of the total calories. The recommended dietary allowance of 3 per cent of the total calories is difficult to avoid on typical American diets.

Polyunsaturated Fatty Acids and Cardiovascular Diseases. A relationship between diet and cardiovascular disease has been sought for decades. Early students of nutrition blamed excessive protein in the diet. Not too long ago, carbohydrates were the culprit, and some who have not got the word still inveigh against starch diets, rich in "empty calories." Now it is the fats, especially saturated fats, that are the culprits, aided by sodium chloride. We have already reviewed the evidence for blaming the low density lipoprotein cholesterol and exculpating the serum triglyceride for increased cardiovascular disease (p. 570) in most people. Changes in total cholesterol concentration are directly related to changes in dietary fatty acid composition:

$$\Delta\text{cholesterol (mg/l)} = 13(2\,\Delta S - \Delta P) + 15(\Delta C)0.5$$

in which S = per cent calories in saturated fat; P = per cent calories in polyunsaturated fat; C = dietary cholesterol intake.

Demographic support comes from studies of Eskimos on their native diet. They are distinguished from Danes by a strikingly high proportion of 20:5 fatty acids in their blood lipids. These are obtained directly from their diet; marine fish oil fatty acids are about 10 per cent 20:5; scallops lipid about 21 per cent. The Eskimos have little coronary artery disease. However, they do have a tendency to bleed, which is ascribed to increased formation of PGI_3, which opposes platelet aggregation in contrast to the stimulation of aggregation by thromboxane A_2 formed from arachidonate (20:4).

There is persistent worry about adverse effects of loading the diet with excess polyunsaturated fatty acids. Unless there is a corresponding increase in tocopherol intake or in glutathione peroxidase activity, more free radicals will be present from photochemical and superoxide attack on the polyunsaturated compounds. These in turn would be expected to increase mutagenesis and carcinogenesis. However, the tocopherol intake usually does increase in parallel to the polyunsaturated fatty acid intake on natural diets.

Behavior and Diet

In theory, the behavior of some people may be affected by changing their dietary intake of precursors of neurotransmitters. Particular attention has been paid to tryptophan as a precursor of serotonin and 5-hydroxytryptophan. Clear proof of causal effects in normal individuals is yet to be obtained.

THE GOOD DIET

Demographic Considerations. Recommendations made to influence the eating habits of a population must be distinguished from recommendations made to a particular individual. The first are, often to the chagrin of their authors, political decisions; the second are, if properly designed, a summation based on both biochemical and personal considerations. The Food and Nutrition Board of the National

Research Council issued a report in 1980 essentially minimizing the necessity for changes in composition of the usual mixed American diet, while deploring obesity and overconsumption of salt, fats and oils, alcohol, and sugar.* This year, the National Research Council has reversed this stand, urging a shift from a meat and fat diet toward a more vegetarian diet.

Neglected in all of this are certain economic constraints. The percentage of the effective energy yield in foods now consumed in the United States is approximately as follows:

vegetables	2.7	fruits	3.6
potatoes	3.3	pulses and nuts	3.6
dairy products	11.6	sugar and sweets	14.6
fats and oils	19.0	flour and cereals	19.2
meats	20.3		

Casual inspection shows that all of the vegetables, fruits, potatoes, and pulses and nuts supply less than 25 per cent of the energy supplied by sugars and sweets, fats and oils, and meats. Put another way, it would take an explosive expansion of row crop farming, one of the more demanding forms of agriculture, to make vegetables, fruits, and so forth major energy sources in the American diet. A shift in eating habits toward breads and other flour products would be easier to manage.

Individual Diets

The management of the obese has already been touched upon. For the affluent average American, the Mayo Diet Manual and similar compendia offer excellent menus suitable for the obese, diabetics, and others requiring special self-administered diets.

The poor person is a special problem, able in theory to manage adequately on much less than is spent, but in fact lacking the training, the comprehension, and frequently the initiative to apply nutritional knowledge. The difficulty is frequently compounded by bad advice, emphasizing unnecessarily expensive, and sometimes frivolous, "basic" foods, such as lettuce.

The physician or the dietitian who can convey some simple principles to the impoverished will do much good. A diet of lowest possible cost will emphasize flour as the basic ingredient. Wheat flour is an imperfect source of protein, but it can be utilized effectively, even by the weaned infant, and is a cheap energy source. Baked goods can have sugar added for additional taste, and cooking oils as a source of essential fatty acids. No general prescription can be made for the balance of the diet, because of differing local supplies. Greens, when available, are the least expensive sources of most of the remaining nutrients.†

---

*The scientists responsible for the report were subjected to an extraordinary vilification in the mass media, including cartoons and editorials impugning their integrity on the grounds of their association with food manufacturers as consultants and grant recipients. (Few in nutrition research are in a position to escape such associations.) No mention was made of the deviation from objectivity caused by making a livelihood as a professional alarmist or by a thirst for power and publicity that might taint the views of their nonscientist opponents.

†Each member of a graduate class in nutrition was assigned the problem of designing a diet of minimum cost for 24-year-old women of average size. There was no difficulty in designing vegetarian diets meeting all known requirements (with cobalamin supplementation) costing less than $1.00 per day, and diets permitting some meat or fish at less than $1.50 per day (corrected to 1982 prices). I found to my surprise, when challenging palatability, that some students merely reported what they were already eating.

FAD DIETS

Most fad diets are designed for losing weight, and are relatively harmless because they are adopted as temporary titillations, rather than as serious continued efforts. The physician wishing to encourage reason rather than mysticism may wish to point out why the claims usually made for the diets are ridiculous.

Some diets are not so harmless, particularly when nursing infants and young children are involved. Adults are also endangered. We mentioned the Zen cult, but some people are imperiled by licensed physicians. One notorious example in Northern Virginia allegedly prescribed a diet containing 80 ounces of juice per day, linseed oil, enzymes, hydrochloric acid, insulin, cobalamins, calf liver, thyroid extract, adrenal extract, and ascorbate, along with coffee enemas.

DIETS IN CLINICAL SITUATIONS

Serious attempts to influence the progress of disease by nutritional measures are being made on many fronts. Space does not permit discussion in detail, but some of the governing principles and areas of ignorance may be noted.

Parenteral nutrition, frequently total, is becoming commonplace. The old dextrose-saline (glucose and NaCl) bottles for intravenous feeding were life-savers, but are now increasingly supplanted by more complete nutrient solutions. Some boners were made: collagen hydrolyzates of unbalanced composition, solutions lacking one or more essential vitamins or minerals created iatrogenic deficiencies sometimes more serious than the condition being treated. The emphasis today is on tight control of composition, particularly with the use of synthetic amino acid mixtures. However, the question still remains of the proper composition of a solution introduced directly in the circulation without being processed through the intestinal mucosa and liver.

Treatment of trauma, accidental or surgical, emphasizes attainment of positive nitrogen balance by heavy loading with amino acids or protein foods. The clinical benefit of this regimen has not been clearly demonstrated by carefully controlled experiments. Overloading the capacity to metabolize amino acids is dangerous in anyone, but there is also no clear-cut evidence that this is a factor in real situations. Resolution of wounds involves local negative nitrogen balance and nothing can be done to change this. It is not clear how storing protein or amino acids elsewhere is beneficial to previously well-nourished individuals.

Premature infants present special problems. Many enzymes required for independent metabolism only appear near or after normal term. The probabilities of difficulty with specific nutrients are not completely tabulated, and the pediatrician ought to be alert to an unexpected accumulation of metabolites.

IN CONCLUSION

The knowledge gained by your study of biochemistry enables you to improve many aspects of the human condition and to bolster the human spirit, if you will but use it. I hope the book has helped.

FURTHER READING

Also see the general references for Chapter 39 (p. 816).

National Research Council (1980). Recommended Dietary Allowances. National Academy of Sciences.

National Food Review. This quarterly magazine contains official estimates of U.S. food production, consumption, prices, etc.

J.H. Cummings (1981). Dietary Fibre. Br. Med. Bull. 37:65.

H. Trowell (1978). The Development of the Concept of Dietary Fiber in Human Nutrition. Am. J. Clin. Nutr. 31(Suppl.10):53. An introductory article in a generally valuable symposium report.

J.P.W. Rivers and T.L. Frankel (1981). Essential Fatty Acid Deficiency. Br. Med. Bull. 37:59.

S.H. Goodnight, Jr., W.S. Harris, and W.E. Connor (1981). The Effects of Dietary ω3 Fatty Acids on Platelet Composition and Function in Man: A Prospective Study, Blood 58:880. Examination of the fish oil diet.

Food and Nutrition Board (1980). Toward Healthful Diets. Nutr. Today 15(3):7

Also see the references concerning diet and cardiovascular disease after Chapter 29 (p. 571).

G.H. Anderson (1981). Diet, Neurotransmitters, and Brain Function. Br. Med. Bull. 37:95.

W.P. Steffee (1980). Malnutrition in Hospitalized Patients J. Am. Med. Assoc. 244:2630.

G.F. Sheldon (1979). Role of Parenteral Nutrition in Patients with Short Bowel Syndrome. Am. J. Med. 67:1621.

AMA Department of Foods and Nutrition (1979). Guidelines for Essential Trace Element Preparations for Parenteral Use. J. Am. Med. Assoc. 241:2051.

C.W. Woodruff (1978). The Science of Infant Nutrition and the Art of Infant Feeding. J. Am. Med. Assoc. 240:657.

S.J. Fomon and R.G. Strauss (1978). Nutrient Deficiencies in Breast-Fed Infants. N. Engl. J. Med. 299:355.

M.G.M. Rowland, A.A. Paul, and R.G. Whitehead (1981). Lactation and Infant Nutriture. Br. Med. Bull. 37:77.

ABBREVIATIONS

A adenosine
ACTH adrenocorticotropic hormone (corticotropin)
ADP adenosine diphosphate
Ala alanyl
AMP adenosine monophosphate
Arg argininyl
Asn asparaginyl
Asp aspartyl
ATP adenosine triphosphate
ATPase adenosine triphosphatase

C cytidine
cAMP 3′,5′-cyclic AMP
CDP cytidine diphosphate
Cer ceramide
CMP cytidine monophosphate
CoA coenzyme A (in names of compounds)
CoA-SH coenzyme A (in reactions)
CTP cytidine triphosphate
Cys cysteinyl
cyt cytochrome

dAMP 2′-deoxyadenosine monophosphate
dATP 2′-deoxyadenosine triphosphate
dCMP 2′-deoxycytidine monophosphate
dCTP 2′-deoxycytidine triphosphate
dGMP 2′-deoxyguanosine monophosphate
dGTP 2′-deoxyguanosine triphosphate
DNA deoxyribonucleic acid(s)
Dol dolichyl
DOPA dihydroxyphenylalanine
dTMP 2′-deoxythymidine monophosphate
dTTP 2′-deoxythymidine triphosphate
dUMP 2′-deoxyuridine monophosphate

E enzyme
EDTA ethylenedinitrilotetraacetate

FAD(H_2) flavin adenine dinucleotide (reduced)
Fru fructosyl
FMN(H_2) riboflavin 5′-phosphate (reduced)
Fuc fucosyl

G guanosine
Gal galactosyl
GalNAc N-acetylgalactosaminyl
GDP guanosine diphosphate
Glc glucosyl
Glc 6-P glucose 6-phosphate
GlcNAc N-acetylglucosaminyl
GlcUA glucuronyl
Gln glutaminyl
Glu glutamyl
Gly glycyl
GMP guanosine monophosphate
GSH, GSSG glutathione (reduced and oxidized)
GTP guanosine triphosphate

Hb Hemoglobin
HbO_2 oxyhemoglobin
HDL high-density lipoprotein
His histidyl
hnRNA heterogenous nuclear RNA
Hyl hydroxylysyl
Hyp hydroxyprolyl

I inosine
IdUA iduronyl
Ig immunoglobulin
Ile isoleucyl
IMP inosine monophosphate

K_{eq} equilibrium constant
K_M Michaelis constant

LDL low-density lipoprotein
Leu leucyl
Lys lysyl

Man mannosyl
Met methionyl
MgATP magnesium ATP
mRNA messenger RNA.
M.W. molecular weight

NAD (NADH) nicotinamide adenine dinucleotide (reduced)
NADP (NADPH) nicotinamide adenine dinucleotide phosphate (reduced)
NeuNAc N-acetylneuraminyl

~P high-energy phosphate
P_i inorganic phosphate
PAPS phosphoadenosine phosphosulfate
PCr phosphocreatine

Phe phenylalanyl
pI isoelectric pH
P : O high-energy phosphate/atom of oxygen consumed
PP_i inorganic pyrophosphate
$PP \cdots P_i$ inorganic polyphosphates
Pro prolyl

RNA ribonucleic acids
R.Q. respiratory quotient
rRNA ribosomal RNA

S Svedberg unit
S substrate
SDS sodium dodecyl sulfate
Ser seryl
SGOT aspartate aminotransferase (serum glutamic-oxaloacetic transaminase)
SGPT alanine aminotransferase (serum glutamic-pyruvic transaminase)

T thymidine (ribosylthymine)
ThPP thiamine pyrophosphate
Thr threonyl
tRNA transfer RNA
Trp tryptophanyl
TSH thyroid-stimulating hormone (thyrotropin)
Tyr tyrosyl

U uridine
UDP uridine diphosphate
UMP uridine monophosphate
UTP uridine triphosphate

V_{max} maximum velocity
Val valyl
VLDL very-low-density lipoprotein

XMP xanthosine monophosphate
Xyl xylosyl
Ψ pseudouridine

ABBREVIATIONS OF AMINO ACIDS

Three-letter abbreviations are used in text to designate amino acid residues in peptide sequences and the like, but not to designate free amino acids. One letter abbreviations were devised for computer storage and retrieval of sequence data, but are also being used for tabulation of long sequences.

Listed by Amino Acid

Amino Acid	Three-letter	Group	One-letter
Alanine	= Ala		= A
Arginine	= Arg		= R
Asparagine	= Asn	Asx = B	= N
Aspartate	= Asp	Asx = B	= D
Cysteine	= Cys		= C
Glutamate	= Glu	Glx = Z	= E
Glutamine	= Gln	Glx = Z	= Q
Glycine	= Gly		= G
Histidine	= His		= H
Hydroxylysine	= Hyl		
Hydroxyproline	= Hyp		
Isoleucine	= Ile		= I
Leucine	= Leu		= L
Methionine	= Met		= M
Phenylalanine	= Phe		= F
Proline	= Pro		= P
Pyroglutamate	= Pgl or <Glu		
Serine	= Ser		= S
Threonine	= Thr		= T
Tryptophan	= Trp		= W
Tyrosine	= Tyr		= Y
Valine	= Val		= V
Unknown	= AA		= X

Listed by Abbreviation

A = Alanine
B = Asparagine or Aspartate
C = Cysteine
D = Aspartate
E = Glutamate
F = Phenylalanine
G = Glycine
H = Histidine
I = Isoleucine
K = Lysine
L = Leucine
M = Met
N = Asparagine
P = Proline
Q = Glutamine
R = Arginine
S = Serine
T = Threonine
V = Valine
W = Tryptophan
X = Unknown
Y = Tyrosine
Z = Glutamate or Glutamine

INDEX OF CLINICAL CONDITIONS

The listed conditions are either described or mentioned in connection with some biochemical circumstance on the listed pages.

INDEX

Page numbers in **boldface** indicate figures illustrating structures or reactions of listed compounds, reactions catalyzed by listed enzymes, or other listed phenomena. Isomeric designations are neglected in the primary alphabetization. Thus, *cis*-Aconitate, S-Adenosyl-L-methionine, L-Alanine, and *p*-Aminobenzoate are listed as beginning with A.

B

F

H

I

J

K

M

N

O

P

Q

R

S

T

W

X

Z

RANGE OF CONCENTRATIONS IN NORMAL HUMAN BLOOD

Clinical analyses may be made on whole blood, separated plasma, or the serum remaining after clotting, as indicated by (B), (P), or (S). Venous blood is used unless otherwise specified. Standard values differ from one laboratory to another according to the conditions of analysis; those cited here are from the University of Virginia Hospital, except those designated by an asterisk are from an extensive tabulation in the N. Engl. J. Med. (1978) *298:* 34. The more useful SI units are calculated where appropriate, but the commonly used clinical units are also shown. The values are for adults. U = International Units of enzyme activity (one micromole of substrate reacting per minute).

Acetoacetate + acetone (S)*	0.05–0.35 mM	0.3–2.0 mg/dl
Alanine aminotransferase, SGPT (S)		4–30 mU/ml
Aldolase (S)		1.2–7.6 mU/ml
Ammonia (P)	11–35 μM	20–60 μg/dl
Amylase (S)		15–90 mU/ml
Aspartate aminotransferase, SGOT (S)		4–30 mU/ml
Bilirubin (S):		
direct	0–4 μM	0–0.25 mg/dl
total	3–20 μM	0.2–1.2 mg/dl
Blood volume (B)*	8.5–9.0 liters/100 kg body weight	
Calcium (S)	2.1–2.6 mM	8.5–10.5 mg/dl
CO_2, total (S)	24–30 mM	24–30 meq/liter
Chloride (S)	96–106 mM	96–106 meq/liter
Cholesterol (S):		
total, age 20–39 yrs	3.6–7.0 mM	140–270 mg/dl
esterified*	60–75% of total	
HDL, 0–19 yr (M)	0.8–1.7 mM	30–65 mg/dl
(F)	0.8–1.8 mM	30–70 mg/dl
HDL, 20–29 yr (M)	0.9–1.8 mM	35–70 mg/dl
(F)	0.9–1.9 mM	35–75 mg/dl
Cobalamins (S)*	0.15–0.65 nM	200–900 pg/ml
Copper (S)	11–26 μM	70–165 μg/dl
Corticotropin, ACTH (P)*	3–15 pM	15–70 pg/ml
Cortisol (P):		
8 A.M.	0.15–0.58 μM	5.4–21.2 μg/dl
4 P.M.	0.05–0.33 μM	1.7–12.0 μg/dl
Creatine kinase, CPK (S)		0–110 mU/ml

Table continues on following page

RANGE OF CONCENTRATIONS

Creatinine (S)	0.06–0.12 mM	0.7–1.3 mg/dl
Fat, triglycerides (S), age 20–39	0.1–1.8 mM	10–150 mg/dl
Fatty acids (S)*	7.0–15.5 mM	190–420 mg/dl
Ferritin (S)		10–300 ng/ml
Folate (S)*	14–34 nM	6–15 ng/ml
Glucose, fasting (P)	3.6–6.1 mM	65–110 mg/dl
Haptoglobin (S)	10–30 μM	100–300 mg/dl
Hematocrit (B):		
males	42–52%	
females	37–47%	
Hemoglobin (B):		
males	2.2–2.8 mM	14–18 g/dl
females	1.8–2.5 mM	12–16 g/dl
Hydrogen ion, arterial (B)	35–45 nM	
Immunoglobulins (S):		
IgA	6–20 μM	70–310 mg/dl
IgD	0–1.7 μM	0–30 mg/dl
IgG	45–85 μM	640–1350 mg/dl
IgM	0.7–3.0 μM	55–350 mg/dl
Insulin (S or P)*	40–190 pM	6–26 microunits/ml
Iron (S)	11–29 μM	60–160 μg/dl
Iron-binding capacity (S)	52–74 μM	290–410 μg/dl
Lactate (P)	0.5–2.2 mM	0.5–2.2 meq/liter
Lactate dehydrogenase, LDH (S)		100–350 mU/ml
Magnesium (S)	0.7–1.2 mM	1.8–2.8 mg/dl
Osmolality	295–315 mOsm/kg water	
Pco_2, arterial (B)	4.7–6.0 kPa	35–45 torr
pH, arterial (B)		7.35–7.45